CAMBRIDGE LIBRARY COLLECTION

Books of enduring scholarly value

Botany and Horticulture

Until the nineteenth century, the investigation of natural phenomena, plants and animals was considered either the preserve of elite scholars or a pastime for the leisured upper classes. As increasing academic rigour and systematisation was brought to the study of 'natural history', its subdisciplines were adopted into university curricula, and learned societies (such as the Royal Horticultural Society, founded in 1804) were established to support research in these areas. A related development was strong enthusiasm for exotic garden plants, which resulted in plant collecting expeditions to every corner of the globe, sometimes with tragic consequences. This series includes accounts of some of those expeditions, detailed reference works on the flora of different regions, and practical advice for amateur and professional gardeners.

A Dictionary of the Economic Products of India

A Scottish doctor and botanist, George Watt (1851–1930) had studied the flora of India for more than a decade before he took on the task of compiling this monumental work. Assisted by numerous contributors, he set about organising vast amounts of information on India's commercial plants and produce, including scientific and vernacular names, properties, domestic and medical uses, trade statistics, and published sources. Watt hoped that the dictionary, 'though not a strictly scientific publication', would be found 'sufficiently accurate in its scientific details for all practical and commercial purposes'. First published in six volumes between 1889 and 1893, with an index volume completed in 1896, the whole work is now reissued in nine separate parts. Volume 6, Part 4 (1893) contains entries from *Tectona grandis* (the common teak tree) to *Zygophillum simplex* (a flowering plant found in Sindh and the Punjab). The index to the entire work is included as an appendix.

Cambridge University Press has long been a pioneer in the reissuing of out-of-print titles from its own backlist, producing digital reprints of books that are still sought after by scholars and students but could not be reprinted economically using traditional technology. The Cambridge Library Collection extends this activity to a wider range of books which are still of importance to researchers and professionals, either for the source material they contain, or as landmarks in the history of their academic discipline.

Drawing from the world-renowned collections in the Cambridge University Library and other partner libraries, and guided by the advice of experts in each subject area, Cambridge University Press is using state-of-the-art scanning machines in its own Printing House to capture the content of each book selected for inclusion. The files are processed to give a consistently clear, crisp image, and the books finished to the high quality standard for which the Press is recognised around the world. The latest print-on-demand technology ensures that the books will remain available indefinitely, and that orders for single or multiple copies can quickly be supplied.

The Cambridge Library Collection brings back to life books of enduring scholarly value (including out-of-copyright works originally issued by other publishers) across a wide range of disciplines in the humanities and social sciences and in science and technology.

A Dictionary of the Economic Products of India

VOLUME 6 – PART 4:
TECTONA TO ZYGOPHILLUM AND INDEX

GEORGE WATT

CAMBRIDGE
UNIVERSITY PRESS

CAMBRIDGE
UNIVERSITY PRESS

University Printing House, Cambridge, CB2 8BS, United Kingdom

Published in the United States of America by Cambridge University Press, New York

Cambridge University Press is part of the University of Cambridge.
It furthers the University's mission by disseminating knowledge in the pursuit of
education, learning and research at the highest international levels of excellence.

www.cambridge.org
Information on this title: www.cambridge.org/9781108068819

© in this compilation Cambridge University Press 2014

This edition first published 1893–6
This digitally printed version 2014

ISBN 978-1-108-06881-9 Paperback

A

DICTIONARY

OF

THE ECONOMIC PRODUCTS OF INDIA.

BY

GEORGE WATT, M.B., C.M., C.I.E.,

REPORTER ON ECONOMIC PRODUCTS WITH THE GOVERNMENT OF INDIA.
OFFICIER D'ACADEMIE; FELLOW OF THE LINNEAN SOCIETY; CORRESPONDING MEMBER OF THE
ROYAL HORTICULTURAL SOCIETY, ETC., ETC.

(ASSISTED BY NUMEROUS CONTRIBUTORS.)
IN SIX VOLUMES.

VOLUME VI, PART IV.
[Tectona to Zygophillum.]

Published under the Authority of the Government of India,
Department of Revenue and Agriculture.

LONDON:
W. H. ALLEN & Co., 13, WATERLOO PLACE, S.W., PUBLISHERS TO
INDIA OFFICE.
CALCUTTA:
OFFICE OF THE SUPERINTENDENT OF GOVERNMENT PRINTING, INDIA,
8, HASTINGS STREET.

1893.

CALCUTTA
GOVERNMENT OF INDIA CENTRAL PRINTING OFFICE
8, HASTINGS STREET,

DICTIONARY

OF

THE ECONOMIC PRODUCTS OF INDIA.

The Teak Tree	(*J. Murray.*)	**TECTONA grandis.**

Teal, see **Ducks, Teals, etc.,** Vol. III., 196.

TECOMA, *Juss.; Gen. Pl., II., 1044.*

Tecoma undulata, *G. Don.; Fl. Br. Ind., IV., 378;* BIGNONIACEÆ.

227

Syn.—T.? GLAUCA, *DC.;* BIGNONIA UNDULATA, *Smith;* B. GLAUCA, *Dcne.;* TECOMELLA UNDULATA, *Seem.*

Vern.—*Rugtrora,* HIND.; *Rohira, roir, lahúra, lúár,* PB.; *Regdáwan, reodán, rebdún, rebdan, raidawan,* PUSHTU; *Roira,* MERWARA; *Lohíra, lohúri, lahéro, khen.* SIND; *Roira, lohuri, lohero, rakht-reora, rugtrora,* BOMB.; *Rakht reora,* MAR.

References.—*Roxb., Fl. Ind., Ed., C.B.C., 492; Brandis, For. Fl., 352; Gamble, Man. Timb., 275; Dals. & Gibs., Bomb. Fl., 161; Stewart, Pb. Pl., 149; Murray, Pl & Drugs, Sind, 177; Birdwood, Bomb. Prod., 334; Baden Powell, Pb. Pr., 599; Lisboa, Useful Pl. Bomb., 104; Stocks, Rep. on Sind; Lace, Notes on Quetta Pl. (MSS.); Settlement Reports, Panjáb, Kohát, 29, 30; Gazetteers:—Bombay, V, 27; Panjáb, Dera Ismáil Khán, 19; Bannu, 23; Peshawar, 27; Sháhpur, 69; Rohtak, 14; Muzaffargarh, 23; N.-W. P., I., 82; IV., lxxiv., Ind. Forester, IV., 232, 345; X., 61; XI., 388; XII., Apb. 18.*

Habitat.—A shrub or small tree, wild in Sind, the Panjáb, Gujrát, and Rájputána, distributed to Baluchistán and Arabia. It is frequently cultivated in gardens on account of its beautiful orange-coloured flowers, and readily adapts itself even to the steamy climate of Calcutta.

Gum.—Mr. E. A. Fraser, Assistant Agent to the Governor General, Rájputana, states, in a note to the Editor, that the plant yields a brown gum. The writer can find no account of this product in any book on Indian Economic subjects.

GUM.
228

Medicine.—The BARK of the young branches is often employed in Sind as a remedy for syphilis (*Murray*).

MEDICINE.
Bark.
229

Fodder.—The FOLIAGE is greedily browsed by cattle (*Stewart*).

FODDER.
Foliage.
230

Structure of the Wood.—Heartwood greyish or yellowish brown, close-grained, mottled with lighter streaks; weight 44℔ per cubic foot (*Brandis*), 64℔ (*Gamble*). It takes a fine polish, and is tough, strong, and durable. It is consequently highly prized for furniture, carved work, and agricultural implements; but Stewart remarks that it is rarely large or abundant enough to be much used except for native's ordinary work.

TIMBER.
231

TECTONA, *Linn.; Gen. Pl., II., 1152.*

Tectona grandis, *Linn.; Fl. Br. Ind., IV., 570;* VERBENACEÆ.
THE TEAK TREE.

232

Vern.—*Ságún, segun, sákhú,* HIND.; *Según,* BENG.; *Saguna,* SANTAL;

TECTONA grandis.	Area under Teak

Chingjagú, ASSAM ; *Singuru*, URIYA ; *Sag*, BHIL ; *Sipna*, MELGHAT ; *Sigwan*, BERAR ; *Sigwan, sagon, sag, khaka, teka, teak*, C. P.; *Teka*, GOND ; *Sagun*, N.-W. P.; *Sagún, sagwán*, PB.; *Loherú*, SIND ; *Saguán*, DECCAN ; *Ság, ságwán, sal, tégu, tékku*, BOMB. ; *Ság, ság a, ságwan, ságván*, MAR. ; *Ság a, sagach*, GUZ. ; *Tékku, tékkumaram, ték*, TAM. ; *Ték, téku, pedda téku, téku-mánu, adavitéku*, TEL. ; *Sagwani, téga, tegina, tyágada mara, jádi*, KAN. ; *Jati, tékka-maram*, MALAY. ; *Kyún, kywon*, BURM.; *Tekka*, SING. ; *Sáka*, SANS.I; *Sáj*, ARAB. ; *Sáj, sál*,* PERS.

* **Moodeen Sheriff** states that the name *sál* (properly a Persian synonym for teak) is in many Dictionaries incorrectly applied to **Shorea robusta**.

References.— *Roxb., Fl. Ind., Ed. C.B.C., 202 ; Brandis, For. Fl., 354, A. 44 ; Kurz, For. Fl. Burm., II., 259 ; Beddome, Fl. Sylv., t., 250 ; Gamble, Man. Timb., 283 ; Dals. & Gibs., Bomb. Fl., 199 ; Stewart, Pb. Pl., 166 ; Rev. A. Campbell, Rept. Econ. Pl., Chutia Nagpur, No. 8716 ; Graham, Cat. Bomb. Pl., 158 ; Mason, Burma & Its People 526, 793 ; Sir W. Elliot, Fl. Andhr., 150, 174 ; Rheede, Hort. Mal., IV., t. 27 ; Rumphius, Amb., III., t. 18 ; Pharm. Ind., 164 ; U. C. Dutt, Mat. Med. Hind., 316 ; Dymock, Mat. Med. W. Ind., 2nd Ed., 595 ; Cat. Baroda Darbar, Col. & Ind. Exhib., No. 170 ; Trans. Med. & Phys. Soc., Bombay (New Series), VI., 275 ; Hooper, The Mineral Concretion of the Teak (Nov. 1887) ; Baden Powell, Pb. Pr., 599 ; Drury, U. Pl. Ind., 413 ; Useful Pl. Bomb., XXV., Bomb. Gas., 107, 248, 393 ; Royle, Prod. Res., 56, 190, 191, 196, 198, 362 ; Liotard, Dyes, 36 ; Cooke, Oils & Oil-seeds, 77 ; Gums and Resins, 129 ; Darrah, Note on Cotton in Assam, 33 ; Forsyth, Highlands of Cent. Ind., 26, 27, 28, 105, 211-265 ; Hove, Tour in Bombay, 12, 97, 101 ; Aplin, Rept. on Shán States, 1887-88 ; Man. Madras Adm., I., 313 ; II., 52 ; Nicholson, Man. Coimbatore, 5, 41 ; Morris, Account Godavery, 67 ; Boswell, Man., Nellore, 95 ; Moore, Man. Trichinopoly, 80 ; Gribble, Man. Cuddapah, 262 ; Settlement Reports :—N.-W. P., Shájehánpur, ix.; Bundelkhand, I., 57 ; Central Provinces, Chanda, 107, App. vi.; Upper Godavery, 37 ; Seonee, 9 ; Hoshangabad, 280 ; Chhindwara, 110 ; Bilaspore, 77 ; Baitool, 125 ; Nimar, 305 ; Bhundara, 18 ; Port Blair, 33 ; Gazetteers :—Bombay, II., 42 ; IV., 24 ; V., 360 ; V., 12, 173 ; VII., 32, 34 ; VIII., 94 ; X., 37 ; XIII., 26 ; XV., 31 ; XVI., 17 ; XVII., 18 ; XVIII., 52 ; N.-W. P., I., 57 ; IV., lxxvi.; Central Provinces, 1870, 1, 6, 17, 46, 58, 123 ; Burma, II., 227 ; Mysore & Coorg, I., 48, 64 ; III., 21 ; Sel. from Bengal Govt. Records, No. IX., xxv.; Sel . from Rec. Govt. of Ind., Nos. IX., xxviii., xxxi.; For. Admin. Repts. for Lower Burma, Upper Burma, Central Provinces, Madras, Bombay, and Assam. ; Agri.-Horti. Soc. Ind. :— Trans., II., 50-55, App., 314 ; IV., Pro., 47 ; VI., 127, 240 ; Journ. (Old Series), III.. 218 ; Sel., 197 ; IV., 40-58 ; V., 185, Sel., 11, 12, 16 ; VI., 240-246, Sel., 154-173 ; VII., 73 ; VIII., Sel., 177, Pro., 29, 34 ; IX., 286-288 ; X., 24 ; XI., Pro. (1860), 37, 65 ; New Series, I., 180 ; V., Pro., (1876), 50 ; Quarterly Journ. Agri., X., 360 ; Ind. Agriculturist, June 16, 1888 ; Ind. Forester, I., 46-48, 50, 51, 109, 110, 191, 274, 363, 397, 406, 413 ; II., 19, 172, 182, 313, 318, 408 ; III., 22, 44, 63, 101, 204 ; IV., 215, 321, 424 ; V., 307, 328 ; VI., 76, 101, 299, 321 ; VII., 212, 213, 256, 260 ; VIII., 158, 240, 266, 301, 377, 387, 415 ; IX., 13, 94, 440, 475, 583 ; X., 60, 119, 190, 280, 403, 471, 545 ; XI., 48, 487, 562 ; XII., 72, 188, 313 ; XIII., 121, 512 ; XIV., 159, 198, 282 ; Smith, Econ. Dict., 408.*

Habitat.—A large, deciduous tree, indigenous to both peninsulas of India, in the eastern drier parts of Java, in Sumatra, and in some other islands of the Indian Archipelago. The distribution of the teak in its natural habitat in India is described by Sir D. Brandis as follows: "In Western India it does not extend far beyond the Mhye. In February 1870 I found it in the Sadri or Santola forests a few miles north of that river, about 20 miles south-west of Neemuch. In Central India it attains its northernmost point in the Jhánsi district at latitude 25°30', and from that point the line of its northern limit continues in a south-easterly direction to the Mahanadi river in Orissa. In Burma proper teak is known to extend to the 25th degee north latitude, and it is reported from Manipur at about the same latitude. There is no proof of its being indigenous in

Bengal, though there is a report of its having formerly been found wild in Assam, between Tezpur and Bishnath. It is, however, cultivated throughout Bengal, Assam, and Sikkim, and in North-West India without difficulty as far as Saharanpur. In the Panjáb it is difficult to raise, and has not been grown west of Lahore.''

No information can be given as to the area under teak in India, but in certain localities the supply may be considered practically unlimited. Brandis describes the forests of the various parts of the country as follows :—" On the Anamallys **Beddome** records trees with a girth of about 22 feet, and a straight trunk of 80 or 90 feet to the first bough. In the North Canara forests clear stems 70 to 80 feet long are not rare; in the Ahiri forests, latitude 19°30', **Col. Pearson** reports stems 60 to 70 feet high ; and even considerably farther north in the Khándesh Dangs, latitude 20°45', I have measured clear stems 60 to 70 feet long to the first branch. Teak attains a large diameter ; girths of 10 to 15 feet are not uncommon, and numerous instances of 20 to 25 feet are on record. The forest tracts, however, in India, which now contain teak of such dimensions, are neither numerous nor extensive. The forests richest in large timber on the west side of the Peninsula are the Travancore, Anamally, Wynaad, South-west Mysore, and North Kanara forests. The Dangs at the foot of the Khándesh gháts also have a considerable quantity of large timber. In the centre of the Peninsula are the Godavery forests, of which Ahiri, east of the Pránhita river, near the foot of the third barrier, is the most compact and valuable.

" In British Burma the sandstone hills of the Pegu Yomah, the outer valleys on both sides of the mountain range which separates the Sitang and Salween rivers, and the Thaungyeen valley, contain the best teak localities. Teak, however, is far more abundant beyond the frontier, in Burma proper, on the tributaries of the Irrawaddi and the head-waters of the Sitang river, in the Karenee country, the Shan States tributary to Burma, and in Siam on the feeders of the Salween, Thaungyeen, and Meinam rivers.

" It is estimated that the teak plantations of Burma, when mature, will contain, at the age of 80 years, about sixty trees per acre, measuring on an average 6 feet in girth, and yielding 3,000 cubic feet of marketable timber, which, with the thinnings, is expected to amount to a mean annual yield of 47 cubic feet per acre. The natural teak forests, not being pure or compact, do not distantly approach to this yield. As an instance of a particularly rich forest, I may quote **Colonel Pearson's** survey of a sample acre in Ahiri, stocked with eighteen large trees, containing an aggregate of 22 tons, or 1,100 cubic feet of timber. Most of these trees, however, were probably more than two centuries old.

" A great proportion of the teak on the Kymore and Satpura ranges consists only of coppice wood. The same may be said of most teak forests on the dry hills of the Dekkan, and of the Konkan teak forests a great portion consists of coppice word. Teak has great powers of reproduction ; it coppices vigorously, and the shoots grow up with great rapidity, much more rapidly at first than seedlings. This great power of reproduction is another point which favours teak in its struggle for existence against other trees, for most teak seedlings which come up naturally are cut down to the ground by the jungle fires of the hot season ; some are killed, but many sprout again during the rains ; and, though they are cut down repeatedly by the fires of successive seasons, yet, meanwhile, the root-stock increases In size every year by the action of the shoots which come up, and, at last, often after the lapse of many years, it produces a shoot strong enough to outlive the fire. Thus, in many cases, what appears

I A

TECTONA grandis.	Area under Teak

AREA under TEAK.

a seedling plant of teak is really a coppice shoot from a thick gnarled root-stock, bearing the scars of successive generations of shoots, which are burned down by the annual fires. The coppice shoots of teak attain a large size, and form good serviceable timber."

An idea of the importance of the various forests may perhaps be best gained by a short consideration of the outturn (as far as that can be ascertained from the Annual Reports of the Forest Department) during the past year — 1888-89.

Burma.
234

Burma.—The quantity of teak timber, worked out of British forests in Lower Burma, amounted to 53,236 tons. In the Tenasserim Circle 12,081 tons were extracted, or 2,245 tons more than in 1887-88. Besides this a large quantity was on its way to, but had not reached, the depôt when the year closed. The outturn in the Pegu Circle was 43,174 tons against 30,700 tons in 1887-88, and would have been larger had there not been much timber neaped *en route* to Rangoon, owing to the early closing of the rains. The total area of plantations, a large proportion of which was under teak, amounted in the Pegu Circle to about 2,000 acres; in the Tenasserim Circle 4,670 acres were under teak plantation on 31st March 1889. The exports from Rangoon amounted to 62,969 cubic tons (tons of 50 cubic feet), valued at R48,99,547; from Moulmein they were 93,465 cubic tons, valued at R77,21,819. A very large proportion of these exports were derived from Upper Burma.

The teak forests of Upper Burma are the most extensive and most important commercially of any under our possession. For many years previous to the last Burmese war they were leased by the Bombay Burma Trading Company from the Government of the late King. On the annexation of the kingdom, the Corporation claimed not only to hold leases stretching over a very large extent of country, but also to be entitled, under promises from the Government of the late King, to renewals of such leases after the expiration of the existing periods. The terms under which they held these leases involved the right of girdling and extracting as much timber as they could remove in return for the payment of a fixed rental for each forest tract. In August 1888 articles of agreement were drawn up by which licenses were granted to the Corporation to work, as contractors for the Government, seven forest tracts (of which they held leases from the King of Burma), for periods extending to 1904. This agreement provided for the extraction of a minimum quantity of timber during the continuance of the contracts and for the payment of a royalty at rates varying from R5 to R10 per ton. It was further provided that the forests should be subject to the rules and regulations of the Forest Department, and that all girdling should be carried out by, or under the direction of, officers of that Department. Similar conditions were made with the more important Burman lease-holders. The royalty thus charged represents at present the chief source of revenue to the Department from Upper Burma, but a minor income is obtained from local traders for trees felled for use in the vicinity, and from fees imposed on the extraction of general forest produce.

The amount of teak removed from the forests of Upper Burma in 1888-89 was:—by Government agency, 97,361 cubic feet; by purchasers 6,872,551 cubic feet, and by free grantees, 4,464 cubic feet, or a total of 6,974,366 cubic feet (about 139,500 tons). This represents a very large increase on the total of 1887-88, which was 78,379 tons. Mr. Hill, Conservator of the Circle, in his report for the year, notes on the great damage done by fire, and the difficulties met with in attempts at fire conservation. He states that owing to this cause, natural reproduction is everywhere in a most unsatisfactory state; "not only are the younger classes of tree ill

Cultivation in India.	(*J. Murray.*)	TECTONA grandis.

represented, but seedlings are scarcely to be found." But, as the work of the Forest Department progresses, when reserves have been extended and the limits of the permanent forest area have become definitely known, protection from fire on a large scale will doubtless be the first and most important measure to be instituted, and a great improvement in the existing state of matters may be looked for. (*Report on Forest Administration, Upper Burma, 1888-89.*)

AREA under TEAK.

Bombay.—In comparison with the vast teak forests of Burma, those of other parts of the Indian empire are very unimportant. But in Bombay much teak is produced, and timber of a quality that at least holds its own with that of other localities. No definite information can be given as to the annual outturn of the forests, owing to the fact that it is returned indiscriminately as cart-loads, head-loads, pieces, cubic feet, etc., in the reports on the Forest Administration of this Presidency. It is, in fact, impossible, without giving a complete table of the voluminous figures contained in the Forest Reports, to even indicate the extent of the outturn of teak in Bombay. This is prevented by want of space, so the reader desiring the information must be referred to the Reports of the Conservators of the Northern and Southern Circles. As a slight indication of the importance of teak in the forests of the Presidency it may be mentioned, however, that the selection cuttings alone, in the Northern Circle, amounted to 40,858 trees, and in the Southern to 6,328.

Bombay. 235

Madras.—In the Southern Circle the sales of teak in depôts and forests during 1888-89 amounted to 113,408 cubic feet, of which nearly the whole was in North and South Malabar and in South Coimbatore. In the Northern Circle the timber is unimportant, only 12,489 cubic feet having been collected during the year by Government agency, and 394 trees felled.

Madras. 236

Central Provinces.—In the Forest Department Reports for these provinces teak and *sál* (**Shorea robusta**) timber are considered under one heading. It is, therefore, impossible to give even an approximation to the amount annually collected of either. It may, however, be stated that the total yield of the two during the past year (1888-89) from Government forests was 198,808 cubic feet, or 3,976 tons.

Central Provinces. 237

Assam.—In this province teak is extremely unimportant. In 1888-89 the outturn amounted to only 180 cubic feet in the form of teak poles (*Forest Department Report*).

Assam. 238

In **Bengal** for the same year, 9 cubic feet is given as the only return of the timber. This was obtained from the Sunderbans. Many years ago attempts were made by **Lord Cornwallis**, on the advice of **Dr. Roxburgh**, to cultivate teak on a large scale and introduce it generally into the lower provinces of Bengal. In 1814 plantations were started at Sylhet, but after some years of trial they were given up in 1831, and in 1854 **Dr. Falconer** reported that only thirteen trees survived out of 1,800 said to have been standing in October 1891. Similar endeavours were made in the beginning of the century in Bankúra, Rajshahye, Rampur Bauleah and Kishnaghur, but **Dr. Falconer** reported on these very unfavourably in 1854, and all the attempts must be considered to have been failures (*Sel. Rec. Govt. of Bengal, No. XXV., 1857*).

Bengal. 239

Andamans.—An effort has recently been made, apparently with considerable success, to start plantations in the *Andaman Islands.*

Andamans. 240

CLIMATE.—Teak thrives best in regions with a mean temperature during winter of from 60° to 80° Fh., during the hot season 80° to 85°, during the rains, 77° to 87°, and during autumn 71° to 81°. The mean annual temperature which suits it best lies between 72° and 81° (*Brandis*). But it can withstand temperatures considerably lower. Thus, **Beddome**

Climate. 241

TECTONA grandis.	Area under Teak

<table>
<tr><td>AREA
under
TEAK.</td><td>states, that on the Anamally mountains, it grows to perfection as high as 2,500 feet, and in certain localities up to 4,000 feet, though of poor growth above 3,000 feet. In Burma it grows up to 3,000 feet (Kurz), an observation confirmed—in the case of the Shan States—by Mr. Aplin. As regards moisture, teak thrives best under a mean annual fall of 50 to 120 inches, and requires a rainfall of at least 30 inches to grow at all.</td></tr>
</table>

AREA under TEAK.

states, that on the Anamally mountains, it grows to perfection as high as 2,500 feet, and in certain localities up to 4,000 feet, though of poor growth above 3,000 feet. In Burma it grows up to 3,000 feet (*Kurz*), an observation confirmed—in the case of the Shan States—by Mr. Aplin. As regards moisture, teak thrives best under a mean annual fall of 50 to 120 inches, and requires a rainfall of at least 30 inches to grow at all.

Soil. 242

SOIL.—The tree grows well on a great variety of soils, but shows a decided preference for certain descriptions. It thrives on sandstone, limestone, and granite, and, in some of the valleys of the Khandesh Dangs, grows to great perfection on soil produced by the disintegration of basaltic rocks. The trees in the Nelambur plantation of Madras were found to succeed on hills of gneiss, while several laterite hills included in 1855-56 were found to be quite unsuitable—the attempt to plant on them failed signally. Brandis writes, " We find teak on light and sandy soils, as well as on those which are binding and heavy. But under all circumstances there is one indispensable condition—perfect drainage and a dry subsoil. To the absence of perfect drainage I ascribe the circumstance that teak does not seem to thrive on level ground with alluvial soil. Instances of natural teak forests in such localities are found on the head-waters of the Bieling and Domdamee rivers in Martaban, in the lower Bonce forests, and in some other places in the plains of Pegu. In such soil the teak grows freely and more rapidly than on the hills; but the trunks are irregular, fluted, and ill shapen; while on the adjoining hills the tree habitually forms tall, clean cylindrical stems."

Environment. 243

ENVIRONMENT.—Teak, like the oak in Europe, rarely forms natural pure forests. When a pure forest does exist it is generally met with on alluvial soils, in which the growth of the teak is unusually free and rapid, a fact which gives it an advantage over competing vegetation. The best teak forests are those in which bamboos or ordinary dry forest trees are found. Thus in the better localities of Burma, teak is estimated to constitute only one-tenth of the whole forest, but the proportion varies greatly; in certain instances it may form 50 per cent., in others scarcely 1 per cent. of the trees in the forest. It is hardly ever found in forests of **Shorea robusta** and but rarely in the *In* (**Dipterocarpus tuberculatus**) forests of Burma. Nor is it met with in the dense evergreen forests of Burma and the Western Ghâts. It is in fact a light-loving tree, and room overhead, and free circulation of air are necessary to its satisfactory growth. Sir D. Brandis, in a letter to the Bombay Government, on the subject of teak-planting, in 1879, strongly insisted on this point, and advocated that the whole area to be planted should not be covered with teak, but that cleared bands at certain distances should be planted between belts of jungle.

Cultivation. 244

CULTIVATION.—Under favourable circumstances as to climate, soil, and environment, teak forests require little attention save in supervising felling, and in strict fire conservancy. The latter is especially necessary owing to the time of seeding of the tree, which flowers during the rains, in July and August, and ripens its seed between November and January. " One of the greatest obstacles," writes Brandis, " to the spread of the teak is the circumstance, that the seed ripens and falls to the ground at the commencement of the hot season, before the annual fires pass through the forest. The tree produces seed at an early age and generally seeds freely and regularly every year; but a larger portion of the seeds are destroyed by the fires, and of those which escape numbers are washed away, in the hills at least, by the first torrents of the monsoon." It has been argued, from the difficulty of getting the seed to germinate in nurseries, that the hard outer covering is destroyed by the periodical fires

AREA
under
TEAK
CULTIVATION

Plantation.
245

in natural forests, and thus allows germination to take place; but experiments conducted in Madras, at the Conolly plantations, shew that even a slight application of fire destroys the vitality of the seed at once.

The oldest and largest teak plantation in India is that at Nelambur in Malabar, called after the late Mr. Conolly, Collector of the District, who commenced it in 1844. A short account of the history of this plantation taken from Mr. Atholl McGregor's memorandum on the subject, may be of interest, since it shows the difficulties met with and the methods which were found to combat these most successfully. The object of forming the plantation was, in Mr. Conolly's own words, " to replace those forests which have vanished from private carelessness and rapacity—a work too new, too expensive, and too barren of early return to be ever taken up by the Native proprietor." Land well suited as to climate and geological composition was secured by Government, and planting, commenced in 1844, was steadily carried on, till, in 1874, 2,730 acres were under the tree. Great difficulty was at first encountered in getting the seed to germinate, firing, soaking in water, removing the hard husk by hand, were all tried without success. Transplanting self-sown teak saplings had been simultaneously tried, but whether from injury to the long succulent tap-root or from some other cause this was also unsuccessful. The advice of Dr. Roxburgh having been requested (?) he recommended that the seed should be sown at the beginning of the rains in shaded beds lightly covered with earth and rotten straw. This was done with complete success, and in May, June, and July 1844, 50,000 seedlings were raised. In 1874, the date of the memorandum quoted the total outlay on the plantations had amounted to R2,29,000, of which R1,01,000 had been recouped by thinnings since 1863, leaving the cost at R1,28,000. After these thinnings the trees which remained were—at 10 years, 750; at 20 years, 500; and at 30, 150. Mr. McGregor enters into elaborate calculations, from which he deduces the conclusion that " eventually the result of the plantation must be to contribute to the wants of the country an immense stock of useful material, realising such a revenue as fully to reimburse the State for their outlay, even after compound interest for the unproductive period is allowed. Sir D. Brandis, as already stated, considers that a mistake has been made in the case of this plantation, in forming a pure teak forest, since teak, in its natural state, does not grow alone, but is associated with bamboos and a variety of other trees. Gamble appears to agree with this opinion, as is shown by the following passages: " No safe speculations can be formed regarding the future of a pure teak forest such as that of Nilambur; it is impossible to foresee the risk of damage by storms, insects, disease, or other causes to which pure teak forest may be exposed. It may be doubted whether, even on the best alluvial soil, the average mean girth of trees 85 years of age will be as much as 8 feet. " " The total area now (1881) stocked at Nilambur is 3,436 acres, of which 1,787 are stocked with a full crop on alluvial soil, the rest not being expected to yield a full crop. In his estimate of the future value of the plantations, Colonel Beddome only assumes 6,000 feet as the full crop expected on alluvial soil. " In the Forest Report of the Southern District, Madras, the Nelambur plantations are said to have covered 3,729 acres on the 31st March 1889, at a total cost up to that date of R2,33,927. Numerous valuable plantations also now exist in Burma; these have been already noticed.

The following note on the cultivation and planting of the tree by Mr. Ferguson, who for many years managed the Nelambur plantations and whose experience is therefore great, may be of value.—

"SEED.—Collect seed from trees with a clear stem, free from decay and of vigorous growth; February is the best month to collect in.

Seed.
246

**AREA
under
TEAK
CULTIVATION**

Nursery Beds.
247

" PREPARATION OF NURSERY BEDS,—Select good free soil, dig 12 inches deep, removing weeds, roots, and stones ; when caked, the soil should be reduced to a fine mould and the nursery levelled ; line off beds 3½ feet wide and one foot space betwixt each bed and its fellow, thenr aise an outer edging round each bed 3 to 4 inches high ; beds when thus finished will be about 2½ feet wide between the edgings, and 120 seers of seed will suffice for 150 feet in length of the above sized beds ; sow from 10th to 15th April. Before sowing steep the seed forty-eight hours in water, then sow and cover with a thin covering of fine soil nearly ¾ inch, then cover with straw to retain the moisture, placing betwixt the soil and straw a few very small twigs without leaves to prevent the straw from being washed into the earth by water ; if this be allowed, the young seedlings are apt to be destroyed on removing the straw. Water daily copiously, say, a common earthen-pot of water to each two running feet in length of bed less or more according to free soil, or otherwise : in this way the seed will germinate in from 10 to 15 or 20 days or more according to freeness of soil ; water less as the plant strengthens, but keep up sufficient moisture till the monsoon sets in from the first to the third week in June. The plants will then be from 4 to 8 inches high and ready for planting out permanently.

Site.
248

" SITE.—The site for planting should be selected and felled in December, allowed to dry till March, fired, then cross cut, piled, and burned off. After the soil is softened by the rains, line and mark off the pits the required distance apart ; from 6 to 7 feet answers well, the pits dug from 10 to 12 inches square and equal depth and filled in as dug with earth slightly raised around tops.

Planting.
249

" PLANTING.—The seedling should be put well down in the pit, taking care the tap-root is not twisted and turned up (to prevent which the tap-root is shortened to 6 inches as lifted from the bed). When planting the cooly inserts his hand the required depth perpendicularly, taking out the soil and putting the seedlings with the other hand (as above, without twisting or turning up the root), putting back the removed soil and pressing it firmly round, without damaging the plant, and this prevents its being wind-waved before taking root.

" Planting should take place after the soil is well saturated with rain ; from the 10th to 30th June and 8th July is the best season, since, if later, the seedling's tap-root rapidly swells like a carrot and does not throw out fibrous roots, nor establishes itself either so quickly or so well as before that state of growth. When the planting cannot be finished by the 8th of July, the small vigorous seedlings which continue to germinate up till August and will even germinate after twelve and fourteen months in the beds, should be selected in preference to larger, more robust ones, with carroty roots."

Felling.
250

FELLING.—As teak is, for the most part, removed from its native forest by water carriage, and since it does not float till thoroughly seasoned, a peculiar mode of seasoning is practised in many teak-growing regions. This practice, known as " girdling," consists in making a deep circular cut through the bark and sapwood so as to completely sever the communication in these layers above and below the incision. A tree thus treated dies after a few days if the operation has been effectual, but, if even the smallest band of sapwood remains to keep up communication, it frequently recovers completely. The girdled tree is allowed to stand one or two years, often longer, if large, and being fully exposed to the wind, rain, and sun, seasons more rapidly and more completely than a tree that has been felled green. Girdling has long been practiced in Burma and Travancore, but was also formerly common further north on the west coast. It is not

| Cultivation in India. | (*J. Murray.*) | **TECTONA grandis.** |

now practised in the Anamally, Wynaad, Mysore, and Canara forests, whence most of the " Malabar " teak is obtained, a circumstance that may account for the greater weight of West Coast when compared with Burma timber.

Opinions differ considerably as to the effect of this practice on the quality of the timber. **Brandis,** from whom the above description of the process has been condensed, evidently considers it beneficial to the timber, in addition to making it sufficiently light to float. Many writers in the *Indian Forester* have discussed the question, and certain hold the opinion that girdling decreases the durability of the wood by depriving it of a certain portion of natural oil. The verdict of **Mr. Thomas Laslett,** Timber Inspector to the Admiralty, is given as follows in his work on *Timber and Timber Trees* :—" The practice of girdling is, I think, objectionable, inasmuch as the timber dries too rapidly, is liable to become brittle and inelastic, and leads frequently to the loss of many fine trees by breakage in falling ; further, it must be regarded as so much time taken from the limit of its duration, which is of great importance." It may be remarked that in Malabar, where girdling has been long discontinued, it was supposed to cause or at least extend heart-shake ; whereas experiments made in Burma in felling green teak resulted in so many of the trunks being found with heart-shake that the trial was discontinued.

A remark by **Mr. Laslett** on the subject of felling is also of interest :— " I am of opinion," he writes, " that greater lengths of timber might be produced from teak than we generally receive, if only a little more care were taken to prevent waste in the forest. Ordinarily the practice is to cut off the bole or stem below the branches, whereas in many cases it would be easy to include in it the knots of some of the lower ones, and thus gain a foot or more of length in the log, which the ship-builders and many others would consider to greatly enhance its value."

GROWTH.—During the early portion of its life the growth of teak is very rapid, and in a comparatively few number of years it attains its full height. But the rate of lateral growth is slow and varies greatly in different localities. **Colonel Beddome** calculated that in Southern India the average contents of a tree was about 10·6 cubic feet at a mean age of 9 years, 23·8 at 19, 51·3 at 29, and that the annual increment increased steadily up to 30 years, and probably for a considerable time longer. In the Central Provinces and Berar the rate of growth is much slower, as might be expected from the dryness of the climate, and the fact of the locality being near the northern limit of the tree. From a survey made in the Nilambur plantation in 1868 by **Colonel Seaton,** the average girth, 6 feet from the ground, was 12 inches at 6 years, 16 at 12, and 29 at 24, while the heights were respectively 37, 45, and 65 feet. **Brandis** considers that the following may be accepted as a fairly accurate average estimate of the girth at 6 feet from the ground for trees in natural forests in Burma and South India :— at 19 years, 18 inches ; at 46, 36 inches ; at 88, 54 inches ; at 160 years, 72 inches.

DISEASES AND DEFECTS.—Besides the danger of damage by fire, winds, and the competition of surrounding vegetation, teak is infested by several insect pests which may attack the tree either when alive, when girdled, or when felled. For an account of these the reader is referred to the article **Pests** (Vol. VI., Pt. I., 148). One of the greatest defects in the timber is that the centre of the heart is rarely sound, but that a more or less irregular hollow, often surrounded by unsound wood, runs along the axis of the tree. This is probably caused mainly by the annual fires which scorch and frequently burn the bark of young trees. An additional cause may be the large mass of pith in the centre of young stems, which is often inhabited by

AREA under TEAK CULTIVATION

Felling.

Growth. 251

DISEASES. 252

T. 252

TECTONA **grandis.**	**Properties and Uses of Teak.**

DISEASES.

boring insects, and thus permanently injured. **Mr. Laslett** also comments on the frequency of "heart-shake," especially in logs taken from old trees. This defect is often found to extend to one-half, sometimes to two-thirds, the diameter of the tree, and may stretch along the whole length of it. If in one plane throughout, the conversion of the log involves no greater difficulty or loss than that occasioned by dropping out a piece large enough to include it. When, however, the cleft at the top is at right angles or nearly so to that at the base it obviously occasions a serious defect.

RESIN.
Tar.
253

Resin.—A rather liquid, black TAR may be prepared by destructive distillation of the wood, in the same way as that made from the various species of Pinus (*see* Pt. I., 243). It is made in small quantities in South India and Burma for medicinal purposes, but is neither prepared nor sold in large quantities. **Mr. Sterndale**, of Seoni, writing in 1860, describes the manufacture as follows :—" In the first place the wood,—that which has been cut about three months is the best—if too fresh the tar is thinner ; about 20 seers of the cut wood will yield one seer of tar, to extract which from one to two maunds of cowdung 'fuel (which is always used) is required ; this costs about two annas a maund. Allowing the full quantity of fuel, which would be four annas, and, say, one anna for the wood (which is over the price), and two annas daily hire to a man to attend the distilling, the maximum cost of one seer of tar is seven annas. If a larger quantity is made, of course it would be cheaper, as one man could attend to several distilleries. I should say 4 annas per seer is a fair average. Yesterday I measured off a cubic foot of good teak and had it distilled, the product was about one seer of tar, the fuel, one and a half maund of dried cowdung " (*Jour. Agri.-Horti. Soc. Ind.*, (*Old Series*) *XI.*, *Pro. lxv.*). The prices above given would now probably be increased owing to the enhanced value of labour. Teak tar was examined in the same year by **Mr. G. Evans**, who reported that it contained all the ingredients found in coal tar, but in different relative proportions. " I am of opinion," he added, " that if used in every way in which coal-tar is made available, its effects would be much less permanent, particularly if exposed to the action of the atmosphere, but this could only be proved by actual experiment, and by noticing its effects for a length of time. I am convinced that it might be rendered much more valuable by concentration, say, by exposure to the sun's rays for a short time in large evaporating pans, as it would then part with a large amount of watery vapour, which it contains in a free state " (*Jour. Agri. Horti. Soc. Ind.*, *XI.*, *Pro.*, *xlviii.*).

More recently teak wood has been examined by **Dymock**, who states that it yields on distillation an opalescent distillate impregnated with resinous matter, but no trace of essential oil could be obtained when operating with 126℔ of fresh sawdust. One pound of the sawdust exhausted with alcohol yielded a resinous extract, which, after having been well washed with hot water, weighed half an ounce ; the resin was black and had the characteristic odour of the wood.

DYE.
Leaves.
254

Dye.—The LEAVES are said to yield a red or yellow dye, of which very little is known. **Kurz** writes : " The leaves have been used and strongly recommended for dyeing silk yellow, olive, &c.," but he does not state whether they are so employed in Burma, the country to which most of his observations relate. **Drury** mentions that they yield a purple (?) colour, employed as a dye for silk and cotton. **Darrah** (*Note on Cotton in Assam*) does not describe them as in themselves tinctorial, but mentions that they are employed as a mordant with *thoiding*, a species of LABIATÆ, in dyeing black. [The Natives of many parts of India have a peculiar method of recognising the teak leaf. They scratch the surface, moisten the part with saliva, and rub, when if it be teak a red colour is produced.

| Teak—a valuable Timber. (*J. Murray*) | TECTONA grandis. |

This phenomenon may be connected with the tinctorial property of the plant. *Ed., Dict. Econ. Prod.*]

Oil.—Considerable confusion exists in Indian literature regarding an oil obtained from teak. An oil is said to be extracted from the wood, in Burma for medicinal use and as a substitute for linseed oil in painting. As already mentioned, Dymock failed to obtain any oil from the sawdust; and it appears to be not improbable that the oil frequently referred to may in reality be the tar already described. The seeds yield a bland, fatty oil, free from any peculiar odour, and said to be of medicinal value. This oil, however, is very difficult to extract, and probably will never be of much economic value owing to the expense which this fact would entail.

OIL.
Wood.
255
Seeds.
256

Medicine.—"Native physicians recommend a plaster of the powdered wood in bilious headaches, and for dispersion of inflammatory swellings; taken internally in doses of 90 to 200 grains it is said to be beneficial in dyspepsia with burning pain in the stomach arising from an overflow of bile, also as a vermifuge. The charred wood quenched in poppy-juice and reduced to a smooth paste is applied to swellings of the eyelids, and is thought to strengthen the sight. The bark is used as an astringent and the oil of the nuts, which is thick and has an agreeable odour, is used for making the hair grow, and removing itchiness of the skin" (*Dymock*, quoting *Makhzan-el-Adwiya*). Rumphius states that the bark was employed in Amboyna as a tonic and astringent and as a remedy for leucorrhœa. Waring remarks (*Pharm. Ind.*) that the value of the wood prepared as a wet paste, in allaying the pain and inflammation caused by handling the Burmese black varnish *thiet-tsi* (Melanorrhœa usitata) is worthy of note. Colonel Burney (*Jour. As. Soc. Beng., I., 170*) published some interesting remarks on its properties. It appears to be deserving of trial as a local application in inflammations arising from the action of the marking nut (Semecarpus Anacardium) and cashew-nut (Anacardium occidentale). Dr. Gibson states that he observed marked diuresis follow the application of an epithem of the bruised fruit to the pubes. He considers it worthy of the notice of future enquirers (*Pharmacopœia of India*). Dymock informs us that the tar is used in the Konkan as an application to prevent maggots breeding in sores on draught cattle.

MEDICINE.
Wood.
257

Bark.
258
Nuts.
259

Fruit.
260
Tar.
261

Structure of the Wood.—Sapwood white and small; the heartwood, when cut green, has a pleasant and strong aromatic fragrance and a beautiful dark golden yellow colour, which on seasoning soon darkens into brown, mottled with darker streaks. The timber retains its fragrance to a great age, the characteristic odour being apparent whenever a fresh cut is made. It is moderately hard, exceedingly durable and strong, does not split, crack, warp, shrink or alter its shape when once seasoned; it works easily and takes a good polish (*Gamble*). Teak owes its chief value to its great durability, which is ascribed, probably with justice, to the circumstance that it contains a large quantity of fluid resinous matter which fills up the pores and resists the action of water. [At the Kárlí caves near Poona the teak-wood-work, two thousand years old, seems perfectly good at the present day.—*Ed.*] Timber from different localities varies very greatly in appearance, weight, and strength. Thus Gamble gives a long table in which the weight observed by various authorities varies from 34 to 51℔ per cubic foot, and the value of P. from 467 to 953. He considers that for practical purposes the weight may be taken approximately at 40℔, and the value of P. at 600. Molesworth, however, in "*Graphic Diagrams for strength of Teak Beams*" gives,—Weight, 45℔; P. 800. Captain Dundas, in his report on experiments made at Lucknow in 1877 and 1878, gives the weight at much less, namely, 34℔, and the value of P. so low as 470. He remarks that the logs when received at Lucknow showed a weight of nearly

TIMBER.
262

Trade in Teak.

50℔, but that after being well dried and sawn into scantlings,the weight fell to 34 or 35℔. Gamble remarks that Captain Dundas' value for P. is a reliable one, for it was based on experiments made with beams of the large size of 10 feet×4 in.×6 in. After many experiments on timber from all the teak-growing localities of India Brandis remarks : " The comparative value of rapidly and slowly grown teak has not yet been determined in a satisfactory manner. It is well known that the rapidly-grown oak produced on alluvial soil in South and Central Europe is, for many purposes, considered equal, if not superior, in value to the slow-grown timber of Northern France and Germany or of England. It seems, however, to be a fact, established by experience at the Bombay dockyard, that the fast-growing saplings of the Malabar plantations are less valuable for oars than the slow-grown poles produced in the coppice woods of Severu-drúg and Colaba."

The many uses of Teak are well known. In India it is highly prized for construction, ship-building, bridge-making, and for making sleepers and furniture. In Europe it is chiefly employed for building railway carriages, for ship-building, for making decks and lower masts, and for the backing of armour plates in ironclads. It is peculiarly useful for the latter purpose, owing to the fact that the tarry matter which it contains acts as a preventive against rust, consequently the wood neither affects the iron, injuriously nor is affected by it. It is far superior to oak in this respect.

**TRADE.
263**

Trade.—The trade in Teak is very large and important, and, notwithstanding competition of other materials in ship-building, shews no signs of diminution. No statistics are available of internal trade by road, rail, and river, but the figures given by Mr. O'Conor in his *Statement of the Coasting Trade* shew the distribution of teak to various parts of India from Burma. Most of the timber shown in these tables, as well as that which goes to make up the large Foreign Trade, is floated down the Irrawaddy and Salween to Rangoon and Moulmein from the vast forests of Burma. Thence it is shipped to other Indian ports, or to foreign countries. The transactions by coasting vessels are very large. Thus, during the five years ending 1889-90, the registered imports averaged annually 132,788 cubic tons, valued at R83,86,253, while the exports averaged 149,534 cubic tons valued at R91,53,381. In the past year (1889-90) the imports amounted to 128,455 cubic tons, valued at R85,08,279. Of this amount 101,524 cubic tons were exported from Burma to other Presidencies ; while Madras exported 2,741 cubic tons and Bombay 1,919 cubic tons. The largest importer was Bombay with 70,392 cubic tons, followed by Bengal with 31,744, Madras with 17,934, and Sind with 2,403. The Bombay and Madras supplies came chiefly direct from Burma, while a great portion of those of Sind were shipped from ports in the Bombay Presidency. From this short analysis it will be seen that a large demand exists in India, a demand almost entirely supplied from Burma.

**Exports.
264**

The foreign exports have remained, on an average, fairly steady during the past twenty years, though fluctuations in the European ship-building industry have naturally influenced the trade considerably from year to year. The averages of the four quinquennial periods may be shewn in the following table :—

Quinquennial period.						Cubic tons.	R
1870-71 to 1874-75	.	.	.	.	.	43,905	31,72,499
1875-76 to 1879-80	.	.	.	.	.	47,738	34,61,905
1880-81 to 1884-85	.	.	.	.	.	55,043	53,51,350
1885-86 to 1889-90	.	.	.	.	.	47,683	49,58,410

T. 264

Trade in Teak.　　　　(*J. Murray.*)

The lowest export on record during these twenty years was in 1886-87. when it fell to only 23,946 cubic tons, with the low value of R22,31,543. With the revival of ship-building in 1887-88, the export again rose to 40,446 cubic tons; in 1888-89 it reached the height of 52,629 cubic tons, and the price revived to R57,52,634, or more than R100 per cubic ton, while in the past year (1889-90) the export was the highest on record, both in amount and value, namely, 71,342 cubic tons, costing R76,29,981.

The following table may be extracted from the *Statistics of Foreign Trade* to show the distribution of one year's exports, and the share taken by each Indian Presidency or Province in the trade :—

Analysis of Exports during 1889-90.

Countries to which exported.	Cubic tons.	R	SHARE OF EACH PRESIDENCY OR PROVINCE.		
			Presidency or Province.	Cubic tons.	R
United Kingdom . . .	60,213	64,78,888	Bengal .	792	74,734
Austria	8	1,300			
Belgium	50	6,025	Bombay .	1,373	1,42,490
France	36	4,393			
Germany	349	37,171	Sind .	...	8
Italy	1,244	1,34,978			
Malta	15	1,825	Madras .	1,144	74,729
Spain	663	1,28,707			
Cape Colony . .	156	15,780	Burma .	68,033	73,38,020
East Coast of Africa { Zanzibar .	355	40,156			
{ Other Ports	3	300			
Egypt	3,400	3,99,788			
Mauritius	1,292	1,09,723			
Natal	31	2,835			
Aden	221	24,778			
Arabia	845	55,282			
Ceylon	1,935	1,41,295			
Persia	39	4,157			
Straits Settlements . .	194	24,906			
Turkey in Asia . . .	199	12,329			
Australia	40	4,340			
Other Countries . . .	14	1,025			
TOTAL .	71,342	76,29,981	TOTAL .	71,342	76,29,981

It is an interesting fact that Egypt came next to the United Kingdom, though very far behind, as an importer of teak.

During the past fifteen years a small import trade has sprung up chiefly from Siam. In the years from 1875-76 to 1879-80 this averaged 1,237 cubic tons, value R73,396 annually; in the following quinquennial period it averaged 1,685 cubic tons, value R1,34,447, while in the five years ending 1889-90 it increased to 2,260 cubic tons, valued at R1,54,439. During the past year it amounted to 5,562 cubic tons, valued at R3,79,194. Of this amount 5,530 cubic tons were imported from Siam, 23 from the Straits, and 9 from other countries. Nearly the whole (5,534 tons) was imported into Bombay. A very unimportant quantity (average 45 cubic tons during the past five years, 60 cubic tons in 1889-90) is re-exported from Bombay. [For further information consult the brief article on Saw-mills—*Ed.*]

Prices and Freight.—During 1889-90 the market rate of Moulmein teak in Bombay varied from R65—R95 for squares in April 1889, to R65—R80

TEPHROSIA purpurea.	A Domestic Medicine.

PRICES.

in March 1890, while in Calcutta squares sold fairly regularly during the year between the prices of R70 and R90. Scantlings fell in price in Bombay from R90 to R95 in April 1889, to R85 to R90 in May, at which rate they continued during the year. In Calcutta they sold at from R85 to R95 from April to June, from R90 to R95 from June to October, and at R95 to R100 from October to March 1890. The home prices for Rangoon timber varied

Freight.
266

between £11 and £16 according to quality; for Moulmein teak from £11 to £12-10 in April 1889, to £9—£11 in February 1890. The freight per ton from the latter port varied from 47s. 6d. to 52s.-6d.

DOMESTIC.
Leaves.
267

Domestic and Sacred.—The LEAVES are used as plates all over teak-growing districts, and are also employed for packing and thatching. The timber frequently has a whitish mineral concretion in its cracks and crevices, which has been found to contain a large proportion of calcic carbonate. It is frequently used as a substitute for lime by the Natives of Southern India for eating with *pán*. For a full account of its chemical composition the reader is referred to **Mr. D. Hooper's** note on the subject (*Nilghiri Nat. Hist. Soc., Ootacamund*).

268

Tectona Hamiltoniana, *Wall.; Fl. Br. Ind., IV., 571.*

Syn.—THEKA TERNIFOLIA, *Ham.*

Vern.—*Ta-hat, ta-nap,* BURM.

References.—*Kurz, For. Fl. Burm., II., 259; Gamble, Man. Timb. 293; Mason, Burma & Its People, 526.*

Habitat.—A small, deciduous tree, met with in the Prome district and Upper Burma.

TIMBER.
269

Structure of the Wood.—Light brown, hard, close-grained; weight about 64℔ per cubic foot; a good wood, and likely to be useful (*Gamble*).

Telini Fly, see Mylabris cichorii, *Fabr.;* COLEOPTERA; Vol V., 309.

TEPHROSIA, *Pers.; Gen. Pl., I., 496.*

270

Tephrosia purpurea, *Pers.; Fl. Br. Ind., II, 112;* LEGUMINOSÆ.

Syn.—GALEGA PURPUREA, *Linn.;* G. LANCEÆFOLIA, *Roxb.;* G. COLONILA and SERICEA, *Ham.;* G. TINCTORIA, *Lamk.;* TEPHROSIA STRICTA, TAALORII, WALLICHII, LOBATA, TINCTORIA, GALEGOIDES, and LANCEOLATY, *Grah.;* INDIGOFERA FLEXUOSA, *Grah.*

Var. maxima,=T. MAXIMA, *Pers.;* GALEGA MAXIMA, *Linn.;* T. MITCHELLII, *Grah.*

Var. pumila,=T. PUMILA, *Pers.;* T. TIMORIENSIS, *DC.;* T. DIFFUSA, *W. & A.;* GALEGA DIFFUSA, *Roxb.;* G. PROCUMBENS, *Ham.;* T. PARVIFLORA, *Wight.*

Vern.—*Sarphónká,* HIND.; *Sarphónká, ban-nil gáchh,* BENG.; *Bánsabánsu, jhojhrú, sarphonka, sarpankh,* PB.; *Surpunka,* SIND; *Sarphúnkha, jangli kulthi, unhali,* BOMB.; *Sharapunkha,* MAR.; *Jhila,* GUZ.; *Hun, náli; jangli-kulthi,* DEC.; *Kolluk-káy-vélai,* TAM.; *Nempali, bonta vempali, tella vempali, mulu vempali, nela vempali, yampali, tella yampali,* TEL.; *Kozhinnila,* MALAY.; *Sarapunkhá,* SANS.

References.—*Roxb., Fl. Ind., Ed. C.B.C., 587, 588; Voigt, Hort. Sub-Cal., 215; Thwaites, En. Ceyl. Pl., 84; Dals. & Gibs., Bomb. Fl., 61; Stewart, Pb. Pl., 76; Mason, Burma & Its People, 479, 766; Sir W. Elliot, Fl. Andhr., 30, 180, 191; Rheede, Hort. Mal., I., t. 55; Ainslie, Mat. Ind., II., 49, 157; O'Shaughnessy, Beng. Dispens., 292; Irvine. Mat. Med. Patna, 120; Moodeen Sheriff, Supp. Pharm. Ind., 240; U. C. Dutt, Mat. Med. Hind, 317; Murray, Pl. & Drugs, Sind, 117; Dymock, Mat. Med. W. Ind., 2nd Ed., 217; Dymock, Warden & Hooper, Pharmacog. Ind., I., 415; Cat. Prod., Baroda Durbar, Col. & Ind. Exhib., No. 171; Atkinson, Him. Dist. (X., N.-W. P. Gaz.), 308, 751; Nicholson, Man. Coimbatore, 39, 192; Gribble, Man. Cuddapah Dist., 40, 227; Boswell, Man. Nellore, 128, 131, 139, 143; Gazetteers:—Panjáb, Gujrát, 12; Gurgáon, 17; N.-W. P., I., 80; IV., lxx.; Mysore, & Coorg, I., 59; Ind. Forester, IV., 233; VI., 240; XII., App. 2, 11.*

| A source of Indigo Dye. | (*J. Murray.*) | TEPHROSIA villosa. |

Habitat.—A copiously-branched, sub-erect perennial, found all over India from the Himálaya to Ceylon, Malacca, and Siam, ascending to 4,000 feet in the North-West; distributed throughout the tropics. *Var.* maxima is confined to the plains of the Western Peninsula and Ceylon; *var.* pumila has the distribution of the type.

Medicine.—Ainslie informs us that, in Southern India, a decoction of the bitter ROOT is prescribed by the *Vytians* in cases of dyspepsia, lientery, and tympanitis. O'Shaughnessy adds that it is given, in Bengal, as a cure for chronic diarrhœa, while in Ceylon, according to Thwaites, it is employed as an anthelmintic for children. Native works on Materia Medica describe the dried PLANT as deobstruent, diuretic, and useful in bronchitis, bilious febrile attacks, and obstructions of the liver, spleen, and kidneys. It is also recommended as a purifier of the blood, in the treatment of boils, pimples, &c. Muhammadan writers mention its use in combination with Cannabis sativa leaves as a remedy for bleeding piles, and with black pepper as a diuretic, which is said to be especially useful in gonorrhœa. When collected for medicinal purpose the whole plant is pulled up as soon as the flowers begin to appear and tied in bundles for sale. It appears to act as a tonic and laxative.

In certain localities of the Panjáb, an infusion of the SEEDS is believed to be " cooling " (*Stewart ; Dymock*).

CHEMICAL COMPOSITION.—The authors of the *Pharmacographia Indica* publish an analysis of the plant, from which it appears that it contains a resin, traces of wax, and a principle allied to quercitrin or quercetin. Cold water extracted gum, a trace of albumen, and colouring matter.

SPECIAL OPINION.—§" Fresh ROOT-BARK, ground and made into a pill with a little black pepper, is frequently given in cases of obstinate colic with marked success " (*Surgeon-Major E. Levinge, Rajahmundry, Madras*).

MEDICINE.
Root.
271

Plant.
272

Seeds.
273

Chemistry.
274

Root-Bark.
275

Tephrosia tenuis, *Wall. ; Fl. Br. Ind., II., 111.*   276

Syn.—MACRONYX STRIGOSUS, *Dalz.*
References.—*Dalz. & Gibs., Bomb. Fl., 61 ; Murray, Pl. & Drugs, Sind, 117 ; Gazetteers :—Mysore & Coorg, I., 59 ; N.-W.P., IV., lxx.; X , 308.*
Habitat.—Common in Sind, the Panjáb, and the Konkan.
Domestic.—The TWIGS are used by Natives to clean the teeth (*Murray*).

DOMESTIC.
Twigs.
277

T. tinctoria, *Pers. ; Fl. Br. Ind., II., 111 ; Wight, Ic.; t. 388.*   278

Syn.—T. HEYNEANA, *Wall.;* T. HYPARGYRÆA, *DC.;* T. NERVOSA, *Pers. ;* GALEGA HEYNEANA, *Roxb.*
Vern.—*Anil, alú-pilla,* SING.
References.—*Roxb., Fl. Ind., Ed. C.B.C., 587 ; Thwaites, En. Cey. Pl., 84; Gazetteers :—Mysore & Coorg, I., 56 ; Bombay, Kanara, XV , 431.*
Habitat.—An undershrub, met with in the Western Peninsula and Ceylon, ascending to 5,000 feet.
Dye.—A blue dye, similar to Indigo, is sometimes extracted from this PLANT in Mysore. (*Conf.* with Indigofera tinctoria, *Linn.* ; Vol. IV., 410, 412, 451.)

DYE.
Plant.
279

T. villosa, *Pers.; Fl. Br. Ind., II., 113.*   280

Syn.—T. ARGENTEA, *Pers.;* GALEGA VILLOSA, *Linn.* G. ARGENTEA, *Lamk. ;* G. BARBA-JOVIS, *Burm.*
Vern.—*Vaykkavalai,* TAM.; *Bú-pilla,* SING.
References.—*Roxb., Fl. Ind., 587 ; Burm , Fl. Ind., 172 ; Thwaites, En. Cey. Pl., 84; Dymock, Warden, & Hooper, Pharmacog. Ind., I., 416 ; Gazetteer, N.-W.P., I., 80.*
Habitat.—A much-branched perennial; native of the plains from the Himálaya to Ceylon.

T. 280

TERMINALIA Arjuna.	The Arjuna Myrobalan.

MEDICINE.
Leaves.
281

Medicine.—"In Pudukota, the juice of the LEAVES is given in dropsy (*Pharmacog. Ind*).

TERMINALIA, *Linn.; Gen. Pl., I., 685.*

282

Terminalia Arjuna, *Bedd.* ; *Fl. Br. Ind., II.,* 447 ; COMBRETACEÆ.

Syn.—T. BERRYI, *W. & A.* ; T. GLABRA, *W. & A.*; T. OVALIFOLIA, *Rottl.*; PENTAPTERA ARJUNA, P. GLABRA, & P. ANGUSTIFOLIA, *Roxb.*

Var. angustifolia,=PENTAPTERA ANGUSTIFOLIA. *Roxb.*; TERMINALIA ANGUSTIFOLIA, *Roxb.*, is an obscure form, apparently allied to T. CHEBUL, but distinct form P. ANGUSTIFOLIA, *Roxb.*

Vern.—*Arjan, kahú, árjún, khawa, ánjan, árjúna, ánjani, jamla, koha, kowa, kahúa,* HIND. ; *Arjún, kahu, árjúna,* BENG. ; *Gara hatana,* KOL ; *Kanha,* SANTAL ; *Orjun,* ASSAM ; *Arjun, hanjal,* URIYA ; *Arjun,* MELGHAT ; *Kowah, kow, kahua, saj, kowha,* C. P. ; *Kahu,* BAIGAS ; *Mangi, koha,* GOND ; *Kowa,* BANDA ; *Anjani, arjan,* N.-W. P. ; *Arjan, jumla,* PB. ; *Arjún, arjun, anjan, jamla, kowa, arjuna-sadra,* BOMB. ; *Sán madat, arjun, anjan, sadura, arjuna, arjun ladada, asun,* MAR. ; *Sádado, arjun sádada,* GUZ. ; *Vellai maruda, vella marda, vella matti, veila marúthú,* TAM. ; *Tandra,* CUDDAPAH; *Tella-maddi, tella madu, maddi, erra maddi, yer muddi,* TEL. ; *Maddi, tormatti, holematti, billi matti,* KAN. ; *Vella-maruta, pulla-maruta,* MALAY. ; *Toukkyan,* BURM. ; *Kumbuk,* SING.; *Kukubha, árjuna,* SANS.

References.—*DC., Prodr., III.,* 14 ; *Roxb., Fl. Ind., Ed. C.B.C.,* 382, 363 ; *Voigt, Hort. Sub. Cal.,* 37 ; *Brandis, For. Fl.,* 224 ; *Kurz, For. Fl. Burm., I.,* 458 ; *Beddome, Fl. Sylv., t.* 28 ; *Gamble, Man. Timb.,* 184 ; *Thwaites, En. Cey. Pl.,* 104 ; *Trimen, Sys. Cat. Cey. Pl.,* 32 ; *Dalz. & Gibs., Bomb. Fl.,* 91 92 ; *Stewart, Pb. Pl.,* 88 ; *Aitchison, Cat. Pb. & Sind. Pl.,* 59 ; *Rev. A. Campbell, Rept. Econ. Pl., Chutia Nagpur, No.* 7546, 9463 ; *Grah., Cat. Bomb. Pl.,* 69 ; *Mason, Burma & Its People,* 533, 743 ; *Sir W. Elliot, Fl. Andhr.,* 52 ; *Sir W. Jones, Treats. Pl. Ind.,* 147 ; *Ainslie, Mat. Ind., II.,* 193 (*under name of T. alata, Kœn.*) ; *Moodeen Sheriff, Supp. Pharm. Ind.,* 243 ; *U. C. Dutt, Mat. Med. Hind.,* 163, 291, 306 ; *Sakharam Arjun, Cat. Bomb. Drugs,* 209 ; *Dymock, Mat. Med. W. Ind., 2nd Ed.,* 323 ; *Birdwood, Bomb. Prod.,* 330 ; *Baden Powell, Pb Pr.* 350, 399 ; *Drury, U. Pl. Ind.,* 336 ; *Atkinson, Him. Dist. (X. N.-W. P. Gaz.),* 301 ; *Useful Pl. Bomb. (XXV., Bomb. Gaz.),* 74 *Forbes Watson, Ind., Survey,* 196, 277 ; *Econ. Prod. N.-W. Prov., Pt. I* (*Gums and Resins*), 16 ; *Liotard, Dyes,* 71, 90, *App. VI.* ; *Wardle, Rpt. Dyes* 15 ; *Cooke, Gums & Resins,* 26 ; *McCann, Dyes & Tans, Beng.,* 128, 133, 151, 161, 165, 166 ; *Watt, Selections Records Govt. India (R. & A. Dept.) 1888-89, pp.* 87-88 ; *Man. Madras Adm.,* 313 ; *Boswell, Man., Nellore,* 98, 127 ; *Gribble, Man. Cuddapah,* 263 ; *For Admin. Rep. Chota Nagpore, 1885,* 6, 31 ; *Settlement Reports:—Central Provinces, Nimar,* 305 ; *Beláspur,* 77 ; *Raepur,* 75 ; *Chhindwára,* 110 ; *Baitool,* 125, 135 ; *Seonee,* 10 ; *Upper Godávery,* 37 ; *Bhundara,* 18, 19, 20 ; *Gazetteers :—Bombay, V.,* 285 ; *VI.,* 12 ; *VII.,* 32, 37 ; *XV., Pt. I.,* 77, *Pt. II.,* 33 ; *N.-W. P., I.,* 81, *IV., lxxi.* ; *Orissa, II.,* 158., *App. IV.* ; *Agri.-Horti. Soc. Ind. :—Trans., VII.,* 57 ; *Jour. IX., Sel.* 44 ; *New Series, VII.,* 139, 140, 276 ; *Ind Forester, IV.,* 227, 322 ; *VI.,* 240, 303, 338 ; *VIII.,* 29, 116 ; *X.,* 31 ; *XI.,* 231 ; *XII., App.,* 13 ; *XIII.,* 121 ; *XIV.,* 288 ; *Balfour, Cyclop. Ind., III.,* 849 *etc. etc.*

Habitat.—A large, deciduous tree, common throughout the Sub-Himálayan tracts of the North-West Provinces, the Deccan, Southern Behar, Chutia Nagpur, Burma, and Ceylon.

GUM.
283

Gum.—A clear, golden brown, transparent gum, obtained from the tree, is met with in the bazárs of Northern India, as a drug (*Baden Powell*). A small sample was sent (from Madras?) by Dr. Shortt to the London Exhibition of 1862 (*Cooke*). Nothing is recorded in Indian economic literature as to its properties or uses, with the exception of Rev. A. Campbell's remark that it is edible (*Ec. Prod. Chutia Nagpur*).

DYE & TAN.
Bark.
284

Dye and Tan.—The astringent BARK in various localities is said to be sometimes used in dyeing. In Southern India the inner bark is broken into

T. 284

chips and the dye extracted by boiling in water. The tint produced is a dirty **DYE & TAN**
brown or *khaki* colour (*Liotard*). In Midnapur (Bengal) it is used to
dye cotton a light brown,—the method employed being as follows :—for
dyeing a yard of cloth a pound of the bark is cut or broken into very
small chips, and is boiled in about 5℔ of water until about 3℔ of water
remain. The solution is then allowed to cool. A *pice* weight of alum
(about ⅛ oz.) is then pounded and mixed with the solution. The cloth to
be dyed is washed in pure water, and the moisture well wrung out of
it. It is then steeped in the above solution, and afterwards put to dry
in the shade; this steeping and drying is repeated two or three times.
In Midnapur also it is employed in preparing a black dye, along with
the barks of *garán* (**Ceriops Roxburghiana**) and *babla* (**Acacia arabica**).
The price of the bark is given in Bengal as 3 annas per seer (*McCann*).
Samples examined by Mr. T. Wardle were found to contain a fair amount
of colouring matter readily soluble in boiling water. An infusion gave in
his hands by various processes and mordants colours ranging from light
yellowish or reddish drab to slate with *tasar* and *corah* silk and cotton.

The ASHES of the wood are used in the Central Provinces as an ad- **Ashes.**
junct (probably a fixing agent) in dyeing with *ál* (**Morinda citrifolia**) and **285**
with *kamala* (**Mallotus philippinensis**) (*Liotard*). The bark contains a
considerable amount of tamin (15˙1 per cent. according to Wardle) ; it
is used for tanning in several localities. In Midnapur it is generally
employed mixed with the bark, or bark, leaves, and fruit of **Acacia arabica**.
The FRUIT is not apparently used as a myrobalan ; indeed, it contains very
little tannin. A sample from the Colonial and Indian Exhibition examined
by Dr. B. H. Paul was found to yield only 1˙38 per cent. of gallo-tannic
acid (*Watt*).

Wax.—The TWIGS and BRANCHES of this species are frequently found **WAX.**
covered with vegetable WAX (*Conf.* Oils, Vol. V., 459). **Twigs.**
286
Medicine.—The BARK is considered by Sanskrit writers to be tonic, **Branches.**
astringent, and cooling, and is used in heart diseases, contusions, fractures, **287**
ulcers, etc. In diseases of the heart it is employed in a variety of ways. **MEDICINE**
Thus a decoction with milk is given as food, or is made with milk, treacle, **Bark.**
and water. A *ghrita* is prepared from the decoction and paste of the **288**
bark for internal administration. In fractures and contusions with excessive echymosis, the powdered bark is given with milk. A decoction is
employed as a wash for ulcers and chancres (*U. C. Dutt*). Ainslie,
describing what, from the vernacular names, is evidently this species,
states that the bark is considered by the *Vytians* to be febrifuge, and,
when powdered and mixed with *gingili* oil, to be a valuable application
for aphthæ. The juice of the LEAVES, he adds, is poured into the ears to **Leaves.**
allay the pain of earache. Baden Powell informs us that in Northern **289**
India the bark is considered "hot" and astringent, useful in bilious
affections, and as an antidote to poisons. The FRUIT is prescribed as **Fruit.**
a tonic and deobstruent. **290**

Structure of the Wood.—Sapwood reddish white; heartwood brown; **TIMBER.**
variegated with darker coloured streaks, very hard ; weight 48 and 54℔ **291**
per cubic foot (*Skinner*), 47℔ (*Cent. Prov. List.*), 57℔ (*Gamble*). It is apt
to split in seasoning and is not easy to work. It is chiefly used for
making wheels of country carts, for house-building, and for making agricultural implements, boats, and canoes. It appears to be generally regarded as an inferior timber, since it does not stand variations of temperature
and moisture, and is subject to the attacks of white ants.

Domestic.—Campbell states that the *tasar* silk-worm is often found on **DOMESTIC**
the tree in Chutia Nagpur. (*See* Silk, *Vol. VI., Pt. III.*) **292**

293 **Terminalia belerica,** *Roxb. ; Fl. Br. Ind., II., 445.*
BELLERIC MYROBALAN.

Var. 1, typica, =T. BELERICA, Bedd., *Fl. Sylv.,* t. *19 ;* T. EGLANDULOSA, *Roxb. ;* T. MOLUCCANA, *Roxb. ;* T. BELERICA, *W. & A. Prodr. (Excl. syn.) ;* T. GELLA, *Dals. ;* T. PUNCTATA, *Roth. ;* MYROBALANUS BELERICA, *Gærtn.*

Var. 2, belerica, *Roxb.,*—? T. MICROCARPA, *Dcne.*

Var. 3, laurinoides, *Miq.*

Vern.—*Bhairá, baherá, behara, behra, bhairah, sagoná, bharlá, bulla, buhura,* HIND. ; *Bohera, baheri, baherá, bhairah, bahira, buhuru, behera, bahura, bohorá, boyra,* BENG. ; *Lupúng, lihúng,* KOL ; *Lopong,* SANTAL ; *Hulluch, bauri,* ASSAM ; *Chirorœ,* GARO ; *Kanom,* LEPCHA ; *Thara, bhára, báhárá, bahadha,* URIYA ; *Sacheng,* MAGH. ; *Yehera,* BHIL ; *Behéra, bihara, bhaira, baherá, behara, toandi,* C. P. ; *Tahaka, taka, banjir,* GOND ; *Bahéra, buhéra, beharia,* N.-W. P. ; *Bahira, bahera, birha, balela, bayrah, behéra,* PB. ; *Bahera,* MERWARA ; *Ahera, jhera,* HYDERABAD ; *Bayrah,* SIND ; *Babra, baldá, balrá, batra, bairda, bálla, bherda, yehala,* DEC. ; *Behara, behada, beheda, behda, bherdha, behedo, balra, bhaira, bherda, bahudda, yella, hela, goting, yel,* BOMB. ; *Bherda, behedá, bahera, yelá, goting, beharda, behasá, sagwan, bedá, helá, berda, yehela béhadá,* MAR. ; *Behedan, beheda, behasá, ság,* GUZ. ; *Tani, thani, kattu elupay, tánrik-káy, tandi tonda, chattu-elupa, tamkai, táni-kái, tani-kaia, kattu-elupa, vallai-murdú, tanikoi, kattú-elupœ,* TAM. ; *Tani, tandi, toandi, thandra, thana, tádi, thani, tondi, katthu-olupœ, tándrakáya, thanddi, thandi, bahadrha, bahadha,* TEL. ; *Santi, táre, tanikayi, tárı-káyi, bherda, yehela, tarí,* KAN. ; *Thani, táni,* MALAY. ; *Thitsein, tissein, ban-kha, phánkhá-si, phángá-si, phangah, pan-gan, ruhira,* BURM. ; *Búlú, bulu-gaha,* SING. ; *Vibhitaki, vibhitaka, vipitakaho, bibhitaka, akasha, tusha, baheruka, baheruha, bahira,* SANS. ; *Batilj, béléyluj, balilaj,* ARAB. ; *Balela, belayleh, balilah,* PERS.

References.—*Roxb., Fl. Ind., Ed. C.B.C., 380 ; Voigt, Hort. Sub. Cal., 36 ; Brandis, For. Fl., 222 ; Kurz, For. Fl. Burm., II., 455 ; Pegu Rept., app. A., lix., B., 49 ; Beddome, Fl. Sylv., t. 19 ; Gamble, Man. Timb., 179 ; Cat., Trees, Shrubs, &c., Darjeeling, 39 ; Thwaites, En. Ceyl. Pl., 10 ; Trimen, Sys. Cat. Cey. Pl., 32 ; Dalz. & Gibs., Bomb. Fl., 91 ; Stewart, Pb. Pl., 89 ; Aitchison, Cat. Pb. & Sind Pl., 59 ; Rev. A. Campbell, Rept. Econ. Pl., Chutia Nagpur, No. 7526 ; Graham, Cat. Bomb. Pl., 69 ; Mason, Burma & Its People, 491, 743 ; Sir W. Elliot, Fl. Andhr., 172 ; Rheede, Hort. Mal., IV., t. 10 ; U. S. Dispens., 15th Ed., 1705 ; Fleming, Med. Pl. & Drugs (Asiatic Reser., XI.), 181 ; Ainslie, Mat. Ind., I., 236 ; O'Shaughnessy, Beng. Dispens., 341 ; Irvine, Mat. Med. Patna, 17 ; Med. Topog. Ajmere, 128 ; Moodeen Sheriff, Supp. Pharm. Ind., 241 ; U. C. Dutt, Mat. Med. Hind., 162, 323 ; S. Arjun, Cat. Bomb. Drugs, 54 ; K. L. De, Indig. Drugs, Ind., 117 ; Murray, Pl. & Drugs, Sind, 188 ; Cat. Baroda Durbar, C. & I. Exhbın., No. 172 ; Bidie, Cat. Raw Pr., Paris Exhıb., 29, 62, 112 ; Dymock, Mat. Med. W. Ind., 2nd Ed., 320 ; Dymock, Warden & Hooper, Pharmacog. Ind., I., 554 ; II., 5 ; Official Corresp. on Proposed, New Pharm. Ind., 238 ; Year-Book Pharm., 1879, 213 ; Trans., Med. & Phys. Soc. Bomb. (New Series), No. 4, 152 ; Birdwood, Bomb. Prod., 33, 152, 268 ; Baden Powell, Pb. Pr., 350, 453, 599 ; Drury, U. Pl. Ind., 417 ; Atkinson, Him. Dist. (X., N.-W. P. Gas.), 310, 777 ; Useful Pl. Bomb. (XXV., Bomb. Gas.), 73, 155, 218, 244, 396 ; Forbes Watson, Ind. Survey, 53, 101, 111, 125, 172, 277 ; Econ. Prod. N.-W. P., Pt. I. (Gums & Resins), 7 ; Pt. III. (Dyes and Tans), 85 ; Pt. V. (Vegetables, Spices, & Fruits), 73 ; Gums & Resinous Prod. (P. W. Dept. Rept.), 4, 16, 24, 26 ; Liotard, Dyes, 15-16, 21, 22, 132 ; Cooke, Oils & Oilseeds, 77 ; Gums and Resins, 26 ; McCann, Dyes & Tans, Beng., 48, 144, 146, 150-151, 161, 165 ; Darrah, Note on Cotton in Assam, 33 ; Christy, New Com. Pl., V., 40, 44 ; Ayeen Akbary, Gladwin's Trans., II., 124 ; Statistics Dinajpur, 150 ; Man. Madras Adm., I., 313 ; Nicholson, Man., Coimbatore, 42 ; Boswell, Man., Nellore, 100, 137 ; Gribble, Man. Cuddapah, 42 ; Butler, Top. Oudh & Sultanpur, 33 ; Settlement Reports:—Panjáb. Kangra, 22 ; N. W., P., Shahjehanpore, IX. ; C. P., Chánda, app. VI. ; Upper Godavery, 38 ; Seonee, 10 ; Bhundara, 19 ; Raepore, 75 ; Mundlah, 88 ; Chhindwára, 110, 111 ;*

Nimár, 305 *; Baitool,* 127 *; Gazetteers :—Bombay, V.,* 285 *; VI.,* 12 *; VII.,*
31, 36 *; VIII.,* 11 *; X.,* 402 *; XI.,* 24 *; XV.,* 33 *; XVI.,* 19 *; XVII.,* 19 *;*
XVIII., 41 *; XXII.,* 23 *; Panjáb, Siálkot,* 11 *; Hoshiárpur,* 10 *; N.-W.*
P., I., 81 *; IV., lxxi. ; Orissa, II.,* 158, *app. IV.,* 181, *app. VI. ; Mysore*
& Coorg, I., 60 *; Agri.-Horti. Soc. Ind. :—Trans., VII.,* 57 *; VIII., Pro.,*
380 *; Jour., IX.,* 422, *Sel.,* 41 *; X., Sel.,* 41 *; XI., Pro.,* 14 *; XIII.,* 318 *;*
New Series, V., Pro. (1876), 20 *; VII.,* 140 *; Ind. Forester, I.,* 78, 81, 363 *;*
II., 19 *; III.,* 202 *; IV.,* 321, 411 *; VI.,* 101, 104, 323, 331 *; VII.,* 250 *;*
VIII., 127, 279, 414, 438 *; X.,* 33, 325, 540, 550 *; XII.,* 311, 313, *app.*
13 *; XIII.,* 121 *; Spons, Encycl., II.,* 1396, 1694, 1987 *; Balfour, Cyclop.*
Ind., III., 849, *etc., etc.*

Habitat.—A large, deciduous tree, common in the plains and lower hills
throughout India, with the exception of the arid tracts to the west, and extending to Ceylon and Malacca. *Var.* **belerica** is met with in the Circars
also in Malaya if **T. microcarpa,** *Dcne.,* be the same species; while *var.*
laurinoides is found in Mergui, Ceylon, Java, and Malaya.

Gum.—Roxburgh was the first Indian writer to notice the gum of this
species. "From wounds in the BARK," he wrote, "large quantities of an
insipid gum issue; it much resembles Gum Arabic, is perfectly soluble in
water, and burns away in the flame of a candle, with little smell, into black
gritty ashes." In 1840, a Lieutenant Kittoe placed samples of the gum
before a meeting of the Agri.-Horticultural Society of India, stating that it
was largely partaken of by the Kóls and Chúars as food, and that it could be
collected in large quantities in the Midnapur forests. Subsequent writers
have done little more than repeat the above remarks of **Roxburgh** and
Lieutenant Kittoe, while Drury has apparently misread the statement of the
former, and writes that it is inflammable and burns like a candle. **Dymock,**
however, appears to have examined it afresh, and describes it as follows :—
"The gum is in vermicular pieces, about the thickness of a finger, of the
colour of inferior Gum arabic. Hardly at all soluble in water in which it
swells up and forms a bulky gelatinous mass; its taste is insipid. Roxburgh's statement that it is perfectly soluble in water, and Drury's, that
it burns like a candle, I am unable to confirm." The authors of the
Pharmacographia Indica describe the gum as "of the Bassora type," and
state that it is collected and mixed with soluble gums for sale as country
gum. They confirm **Dymock's** observation as to its insolubility and add
that it contains crystals of calcium oxalate in dumb-bell-like forms, sphæro-
crystals and groups of fine crystalline particles.

GUM.
Bark.
294

Dye and Tan.—The FRUIT is one of those exported from India under
the name of myrobalans, and is largely employed in India for dyeing and
tanning. Two kinds are said to be met with, one nearly globular $\frac{1}{2}$ to $\frac{3}{4}$
inch in diameter, the other ovate and much larger. Both narrow suddenly
into a short stalk, are fleshy and closely covered with a fulvous tomentum,
and when dried are obscurely five-angled. The stone is hard and penta-
gonal, and contains a sweet oily kernel, having three prominent ridges from
base to apex. In India it is largely employed for dyeing, as a mordant,
as a tan, and also medicinally. It may be used alone, in which case it
gives a yellowish or brownish-yellow colour to the cloth, or, with various
other dye-stuffs, to produce dark brown and black. The following detailed
description of the method employed in Hazáribágh for dyeing with the belleric myrobalan alone may be taken as typical :—"For each square yard of
cloth take $\frac{1}{4}$ seer of *bahera* nuts. Extract and throw away the stones, and
break the rind into as small pieces as possible. Put these into a seer of
water along with a tola-weight of pomegranate rind. Leave the whole to
stand for one night. Then boil the infusion, allowing it to boil over three
times. Then allow it to cool and strain through a coarse cloth. Wash
the cloth to be dyed well in water; when half dry wash again in water, in
which a tola of alum has been previously dissolved. Then dip the cloth

DYE & TAN.
Fruit.
295

2 A

TERMINALIA
belerica.   The Belleric Myrobalan.

DYE & TAN.

in the dye solution, working it about well so as to make the colour uniform. When the colour is deep enough, dry the cloth in the sun, and
afterwards wash frequently in clear water, so as to get rid of the smell of
the dye. The resulting colour is a snuffy yellow " (*McCann*). The drupe
is also used as a mordant, instead of *harra* (T. **Chebula**), in dyeing with

Leaves.
296

madder or *manjit*. In many localities it is employed as a tan in the same
way as *harra*, and McCann states that the LEAVES are similarly used in
Bírbhum. Buchanan mentions the BARK as also employed in tanning,
but it would appear to be less astringent, and consequently much less
valued than those of other species.

The fruit ripens during the cold season, from November to January,
and in Bengal costs about the same as inferior *harra*, *viz.*, from R1 in
Mánbhum to about R5 in Chittagong (*McCann*). Sir E. C. Buck, in his
Account of the Dyes and Tans of the North-West Provinces, gives the
export from Najibabad in 1874-75 as 36½ cwt., value R50, from Garhwál
as 135½ cwt., value R219. It is, however, impossible to separate the trade
accounts of this, from those of the other kinds of myrobalan (see T. **Chebula**).

Chemistry.
297

CHEMICAL COMPOSITION.—Analyses of the fruit of this, as of other
myrobalans, give very varying results and strongly indicate the necessity
of a thorough investigation into the effects of climate, soil, and age of the
fruit on its tanning value. Samples were submitted by Dr. Watt, from the
Colonial and Indian Exhibition, to Dr. Paul and Professor Hummel for
examination. Two samples, examined by the former chemist were found to
contain only 5·03 and 6·70 per cent., respectively, of gallo-tannic acid, while
that analysed by the latter contained 17·4 per cent. Professor Hummel
remarks, " The fruit consists of two distinct portions, an outer and an inner;
100 parts contain 75·4 parts of outer, and 24·6 parts inner. The inner portion only contains 1·25 per cent. of tannic acid." " This remarkable difference is worthy of note " Professor Hummel estimated the value of the
fruit at 5s. 8d. per cwt. compared with commercial ground myrobalans at 7s.
6d. per cwt. (*Watt, Selections from the Records of the Govt. of Ind., I., 88,
93*). The authors of the *Pharmacographia Indica* have recently examined
the pulp of the smaller kind of belleric myrobalan and the kernel separately with the following result :—

	Pulp.	Kernel.		Pulp.	Kernel.
Moisture	8·00	11·38	Ether extract	·41	·61
Ash	4·28	4·38	Alcoholic ,,	6·42	·61
Petroleum ether extract	·12	29·82	Aqueous ,,	38·56	25·26

The extracts obtained from the pulp were found to be as follow :—The
petroleum ether extract contained a greenish yellow oil. The ethereal extract contained colouring matter, resins, a trace of gallic acid and oil, but
no alkaloid. The alcoholic extract was yellow, brittle, highly astringent,
and partly soluble in warm water. The aqueous extract gave various
tannin reactions. Of the kernels, the petroleum ether extract consisted of
a pale yellow, thin, nutty-flavoured oil, non-drying and insoluble in alcohol;
the ethereal extract was also oily (see paragraph Oil); the alcoholic extract
was partly soluble in hot water, with acid reaction and tasteless, the aqueous
extract contained no sugar nor saponin. No alkaloid was detected.

OIL.
Seeds.
298

Oil.—The SEEDS yield a fatty oil to the extent of about 30·44 per cent.
which on standing separates into two portions, the one fluid, of a pale
green colour, and the other flocculent, white, semi-solid, with the consistence of *ghí*. It is used medicinally.

| The Beleric Myrobalan. | (*J. Murray.*) | TERMINALIA belerica. |

Medicine.—Belleric myrobalans (the DRUPE) are described by Sanskrit writers as astringent and laxative, and useful in coughs, hoarseness, eye diseases, etc. As a constituent of *triphala*, or the three myrobalans (T. **Catappa**, **T. belerica**, and **Phyllanthus Emblica**) they are prescribed for almost all diseases. The KERNEL is said to be narcotic and astringent, and is used as an external application to inflamed parts (*U. C. Dutt*). Dymock states that Muhammadan writers describe the fruit as astringent, tonic, digestive, attenuant and aperient, and useful as an application in inflammatory diseases of the eye. As long as the doctrines of the Arabian school prevailed, myrobalans were used medicinally in Europe, having been introduced by the Arabs from India. The μυροβαλανος of the early classical Greek and Latin writers was a fruit from which perfumers obtained oil for their unguents, and was probably the fruit of the African oil palm (**Elœis guineensis**); but later Greek physicians applied the terms μυβαλανος and μυρεψιχος to Indian myrobalans (*Pharmacographia Indica*). In modern Native practice the ancient Sanskrit and Muhammadan opinions are retained. Thus in the Panjáb it is chiefly employed in dropsy, piles, diarrhœa, and leprosy; also occasionally as a remedy for fever. When half ripe it is considered purgative; when fully ripe or dried, astringent. Mixed with honey it is employed as an application in cases of ophthalmia. The OIL is considered a good application for the hair; the GUM is believed to be demulcent and purgative. Dymock states that the kernel, with that of the marking nut, is sometimes eaten in the Konkan with betel-nut and leaf as a cure for dyspepsia.

Food and Fodder.—The FRUIT is eaten, when fresh, by goats, sheep, cattle, deer, and monkeys. The KERNEL is eaten by Natives; it tastes like a filbert, but is said to produce intoxication when eaten in excess. Many writers (*Roxburgh, Graham,*) notice this popular belief, but, according to recent researches by the authors of the *Pharmacographia Indica*, the narcotic property ascribed to the kernel is very doubtful. These writers state that Native evidence on the point is very conflicting; some people say that only the large-fruited variety is poisonous, others state that they have eaten both kinds freely without experiencing any narcotic effects, but that when water is taken after eating them giddiness and a sense of intoxication is experienced. The only authentic cases of poisoning by these seeds are said to have been recorded by Mr. Raddock, Sub-Assistant Surgeon in charge of the Malwa Bheel Corps. Three boys from five to nine years of age had eaten some of the nuts; two became drowsy, complained of headache and sickness, and vomited freely. The third, a weakly child, who had eaten the largest quantity of kernels, between twenty and thirty, presented no symptoms during the day, but on the following morning was found to be insensible, aud suffering from all the symptoms of collapse. Emetics and small quantities of strong tea produced an improvement in the symptoms. He gradually became sensible, but remained drowsy, complained of being giddy, and had a small and rapid pulse till next day, after which he recovered. Mr. Raddock infers that this was a case of mild narcotic poisoning, which, he is convinced, would have proved fatal had not the stomach pump been used. Dr. Burton Brown, in citing this case, remarks that the fruit is sometimes added to spirit in bazárs, in conjunction with the other myrobalans, so that it is possible that an accident might occur from the use of spirit so drugged. The experiments of Dymock, Warden, & Hooper lead to the conclusion that the kernels are perfectly harmless, when eaten in moderation; one of these authors has personally experimented on them without any ill effects. On one occasion the alcoholic extract from 9 grains of the kernels was injected into a cat's stomach, on another 13·2 grains (equal to about thirty-five to forty kernels) were administered to a

MEDICINE.
Drupe.
299
Kernel.
300

FOOD.
Fruit.
301
Kernel.
302

T. 302

TERMINALIA Catappa.	The Indian Almond.

FOOD.

starving cat, with, in both cases, negative results. These authors therefore regard their experiments as fairly conclusive that the kernels do not possess any toxic properties.

Leaves.
303

Stewart states that in Kangra the LEAVES are considered the best fodder for milch cows.

TIMBER.
304

Structure of the Wood.—Yellowish-grey, hard, no heartwood; similar in structure to that of **Ougeinia dalbergioides**; weight 43℔ per cubic foot (*Kyd, from Assam experiments*), 39℔ (*Cent. Prov. Lists*), 40℔ (*Brandis, Burma List*), 48℔ (*Gamble*). It is not durable, is readily attacked by insects, and is consequently of little value. It is, however, employed for planking, for making packing cases and canoes, and in the North-West Provinces for house-building after being steeped in water, which is said to have the effect of making it more durable. In the Central Provinces it is used for ploughshafts and carts when *bijasál* is not available. In South India it is employed for making packing cases, coffee boxes, catamarans, and grain measures.

DOMESTIC.
Tree.
305

Domestic and Sacred.—The TREE is an excellent one for avenue purposes, but has many superstitions connected with it which interfere with its utility. Thus the Hindus of Northern India consider it to be inhabited by demons, and consequently avoid it, never sitting under its shade. In Central and Southern India the people will not use the timber for building, under the impression that a dwelling-house which contains it is fated, and that no man can live in it long. The FRUIT is used for making country ink, and, by the Bhils, to poison fish (?) (*Liotard, Elliot*); the OIL as a dressing for the hair. The LEAVES have been used as an antiseptic to impregnate sleepers of *salei* (**Boswellia serrata**), which are said to have been rendered durable by soaking for five months in a tank filled with the leaves and water (*Indore Forest Rept., 1876-77; Conf.* Vol. I., 516). In the *Gazetteer of Sávantvádi* (Bombay) the WOOD-ASHES are said to be much used in the manufacture of molasses (*Conf.* with **Saccharum**—Sugar, *Vol. VI., Pt. II., 304*).

Fruit.
306
Oil.
307
Leaves.
308
Wood-ashes.
309

310

Terminalia bialata, *Kurz; Fl. Br. Ind., II., 449.*

Syn.—PENTAPTERA BIALATA, *Roxb.*
Var. cuneifolia, *Wall.*
Vern.—*Leinben,* BURM.
References.—*Roxb., Fl. Ind., Ed. C.B.C., 383; Voigt, Hort. Sub. Cal., 36; Kurz, For. Fl. Burm., I., 456; Gamble, Man. Timb., 182; Agri.-Horti. Soc. Ind.: Trans., VII., 57.*

Habitat.—A large, deciduous tree, which attains the height of from 80 to 100 feet; found in Burma and the Andaman Islands.

TIMBER.
311

Structure of the Wood.—Grey, beautifully mottled, similar to the timber of T. belerica; weight 39℔ per cubic foot (*Brandis, Burma List*), 48℔ (*Gamble*). Gamble remarks that Skinner's No. 124 gives weight 64℔, and P. 1042, but he considers that there may have been some mistake in these figures.

312

T. Catappa, *Linn.; Fl. Br. Ind., II., 444; Wight, Ic., t. 172.*
INDIAN ALMOND.

Syn.—T. CATAPPA & BADAMIA, *Tulasne;* T. MOLUCCANA, *Lamk.;* T. MYROBALANA, *Roth.;* T. SUBCORDATA, *Willd.;* T. INTERMEDIA, *Spreng.;* JUGLANS CATAPPA, *Lour.;* CATAPPA DOMESTICA, LITOREA, & SYLVESTRIS, *Rumph.;* BADAMIA COMMERSONI, *Gærtn.*

Vern.—*Jangli-bádám, hindi-bádám, bádámi,* HIND.; *Banglá-badám,* BENG.; *Bádám,* URIYA; *Desi-badám,* N.-W. P.; *Hindi-bádám, jangli-badám, bádáme-hindi,* DECCAN; *Bádám, jangli-bádám, bangáli-bádám, bádámi, desi-bádám,* BOMB.; *Bengali-bádám, jangli bádáma, nát-bádám,* MAR.; *Natvadom, nattuvadam-kottai, nathe-vadam-kottai, nattu-*

T. 312

| The Indian Almond. | (*J. Murray.*) | TERMINALIA Catappa. |

vadom, kottai, TAM.; *Vedam, nathe-badam-vittulu, vudam vittulu, vodamovettilla, badam-vittulu,* TEL.; *Tarí, taru, nát-bádámi,* KAN.; *Náttu-bádam, kotta-kuru, adamarram,* MALAY.; *Katappa,* MALAY; *Kotambá,* SING.; *Ingudi, hinghúdie,* fruit=*desha-vádá-mittee* SANS.; *Bádáme-hindí,* PERS.

References.—*Roxb., Fl. Ind., Ed. C.B.C., 380; Voigt, Hort. Sub. Cal., 36; Kurz, For. Fl. Burm., I., 454; Beddome, Fl. Sylv., t. 18; Gamble, Man. Timb., 182; Trimen, Sys. Cat. Cey. Pl., 32; Dals. & Gibs., Bomb. Fl., 33; Graham, Cat. Bomb. Pl., 69; Sir W. Elliot, Fl. Andhr., 19, 193; Rheede, Hort. Mal., IV., t. 3, 4; Rumphius, Amb., I., t. 68; Pharm. Ind., 89; Ainslie, Mat. Ind., II., 234; O'Shaughnessy, Beng. Dispens., 341; Moodeen Sheriff, Supp. Pharm. Ind., 241; S. Arjun, Cat. Bomb. Drugs, 214; Budie, Cat. Raw Pr., Paris Exhib., 29; Dymock, Mat. Med. W..Ind., 2nd Ed., 321; Birdwood, Bomb. Prod., 33, 152, 282; Useful Pl. Bomb. (XXV., Bomb. Gaz.), 76, 155, 218, 244; Gums & Resinous Prod. (P. W. Dept. Rept.), 16; Liotard, Dyes, 16; Cooke, Oils & Oilseeds, 78; Gums & Resins, 26; Wardle, Rpt. on Dyes & Tans, 12, 31, 43; Man. Madras Adm., I., 363; Boswell, Man. Nellore, 96; Gazetteers:—Bombay, V., 285; XV., Pt. 2, 22; N.-W. P., IV., lxxi.; Orissa, II., 181, app. IV.; Mysore & Coorg, I., 160; Agri.-Horti. Soc. Ind.:—Trans., III., 44; VII., 57; Jour., II., Sel., 539; III., Pro., 47; IV., 201; VI., 170, 173; IX., 421; Spons' Encycl., 1396, etc., etc.*

Habitat.—A large, deciduous tree, which attains 80 feet in height, and has branches in almost horizontal whorls; wild in the lowlands of Malaya, and perhaps of the trans-Gangetic Peninsula, largely planted all over India from the North-West Provinces to Ceylon and Burma, mostly from the sea-level to an altitude of 1,000 feet. It is raised easily from seed; and in a good light soil, well watered, will, in two years, grow to more than 10 feet in height, and blossom.

Gum.—The tree yields a GUM, known in the West Indies as "Indian Almond Gum;" it is dark-coloured, but soluble in great part, but contains fragments of the bark (*Cooke*). Nothing is known as to its properties or uses.

GUM.
313

Dye.—The BARK and LEAVES like those of most other species of Terminalia are astringent and contain tannin. In India they are mixed with iron salts to form a black pigment, with which Natives in certain localities colour their teeth and make Indian ink. Specimens of the bark examined by Mr. Wardle were found to contain 9 per cent. of tannin, and a small amount of colouring matter soluble in water, which, by various processes (not published), produced light brownish-yellow, light drab, golden fawn and slate colours in silk, light drab, olive and grey in cotton, and pale fawn in wool. The leaves were found to contain a moderate amount of colouring matter, which produced various shades of brownish-yellow on silk and wool.

DYE.
Bark.
314
Leaves.
315

Fibre.—The Rev. J. Long (*Journ. Agri. Horti. Soc. Ind., IX., Old Series, 422*) states that in Madras cloth is made from a FIBRE obtained from the LEAVES. This is probably a mistake; the writer can find no other mention of a fibre from any species of Terminalia.

FIBRE.
Leaves.
316

Oil.—The KERNELS yield a valuable oil, similar to almond oil in flavour, odour, and specific gravity, but a little more deeply coloured; it deposits stearine on keeping. It possesses the advantage of not becoming rancid so readily as true almond oil, and if it could be produced cheaply would doubtless compete successfully with it. As the tree is abundant everywhere and the fruit could be doubtless obtained very cheaply, "Indian almond oil" appears to merit the attention of dealers. It was first brought prominently to notice by a Mr. A. T. Smith of Jessor, who in 1843 wrote to the Agri.-Horticultural Society of India an account of its properties and method of preparation. Oil, made experimently by him, was expressed in the common native mill—a sort of pestle and mortar—from some fruit gathered during a few mornings from under the trees in the neighbourhood. After a sufficient

OIL.
Kernels.
317

T. 317

TERMINALIA
Chebula.   The Chebulic Myrobalan.

OIL.

quantity had been gathered and allowed to dry in the sun for a few days, which facilitates breaking the nut, four coolies were set to work with small hammers, to separate the kernels from their shells. In four days they broke a sufficient quantity for one mill, *viz.*, 6 seers. This quantity put into the mill produced in three hours about 3 *pucka* seers of oil. Mr. Smith remarks that the actual pressing of the oil is of no consideration, since the value of the oil-cake, to feed pigs, etc., is sufficient to cover the expense, but that the breaking of the nuts is a tedious and costly operation, and is a consideration requiring particular attention, with a view to its reduction, if manufacture of the oil on an extensive scale should be attempted. The product of the experiment, filtered through blotting paper, was of the colour of pale sherry, a circumstance which Mr. Smith explains is due to the rind being allowed to remain on the kernels. He concludes by remarking on the ornamental nature and utility of the tree for many other purposes, and recommends that it should be more extensively planted. A sample of the oil thus prepared was submitted for examination to Dr. Mouat, who reported as follows :—" I have compared the specimen with a good muster of the ordinary European almond oil in my possession, and find that in taste, smell, and specific gravity the former is very similar to the latter, but is deeper in colour, becomes turbid in keeping, and deposits a quantity of white stearic matter. For most ordinary purposes, medicinal and otherwise, the former, I think, might profitably be substituted for the latter in this country, and, if expressed with greater care and freed from every impurity, might become an article of commercial value and importance" (*Journ. Agri.-Horti. Soc. Ind., ii.*). Though easily made edible and pleasant in flavour, it appears to have been entirely neglected by the Natives, who are ignorant as to its existence.

MEDICINE.
Bark.
318
Kernel.
319
Leaves.
820

Medicine.—The astringent BARK may be used medicinally, though it does not appear to be much employed in Native practice. The KERNELS and OIL are mentioned in the secondary list of the Indian Pharmacopœia as substitutes for officinal almonds and almond oil. The JUICE of the young LEAVES is employed in Southern India to prepare an ointment for scabies, leprosy, and other cutaneous diseases, and is also believed to be useful internally for headache and colic (*Lisboa*).

FOOD.
Kernel.
321
Oil-Cake.
322
TIMBER.
323
DOMESTIC.
324

Food —The KERNEL resembles an almond or fresh filbert in flavour, and is largely eaten by Natives. It is very palatable, fairly wholesome and nutritious, and is a pleasant dessert fruit. The OIL-CAKE is said to be a good food for pigs.

Structure of the Wood.—Red, with lighter coloured sapwood, hard; weight, according to Skinner and Wallich, 32℔ per cubic foot, according to Gamble, 41℔. It is used for various purposes in Southern India, especially for making posts and well-levers.

Domestic.—This is one of the trees on which the *tasar* or *katkura* silk-worm is fed. The tree is highly ornamental, makes a good avenue, and is well worthy of extended cultivation.

325

Terminalia Chebula, *Retz. ; Fl. Br. Ind., II., 446.*

THE CHEBULIC or BLACK MYROBALAN.

Syn —T. RETICULATA, *Roth. ;* T. ARUTA, *Ham. ;* MYROBALANUS CHEBULA, *Gærtn. ;* EMBRYOGONIA ARBOREA, *Teys. & Binn.*
Var. 1, typica ; 2, (*the* T. citrina *of various authors*) ; 3, unnamed ; 4, TOMENTELLA, *Kurz* (SP.) ; 5, GANGETICA, *Roxb.*, (SP.) ; 6, PARVI-FLORA, *Thwaites,* (SP.).
Vern.—Tree=*Har, harrá, harara,* ripe fruit=*har, pilé-har, hár-pilé,* dried fruit=*bál-har, sanghi-har, kále-hár,* HIND. ; Tree=*Háritáki,* ripe fruit=*háritaki, hórá,* galls=*háritaki-phul,* BENG. ; Tree=*Rola, hadra,* KOL ; Tree=*Rol,* SANTAL ; Tree=*Hilikha,* ASSAM ; Tree & ripe fruit=*Herro,*

| The Chebulic Myrobalan. | (*J Murray.*) | TERMINALIA Chebula. |

NEPAL; Tree=*Silim*, ripe fruit=*silim-kung*, LEPCHA; Ripe fruit=*Hana*, PAHARI; Tree=*Karedha*, *haridra*, *haríra*, URIYA; Tree=*Kajo*, MAGH; Tree=*Harrá*, *hirdi*, C. P.; Tree=*Karka*, *harro*, *hir*, *horda*, *mahoka*, GOND; Tree=*Har*, *haraira*, *harara*, N.-W. P.; Tree=*Har*, *harrar*, *hurh*, *halela*, dried fruit=*har*, PB.; Tree=*?Bey-a-rah*, BERARS; Tree & ripe fruit=*Har*, SIND; Tree=*Halrá*, *harlá*, ripe fruit=*halrá*, *harlá*, *pilá-halra*, *haldá*, dried fruit=*bál-halré*, *zangi-halré*, DECCAN; Tree & ripe fruit=*Hírda*, *hardá*, BOMB.; Tree=*Hírda*, *hiradá*, ripe fruit=*hiradá*, *bála hirade*, galls=*hiradá-phúla*, MAR.; Ripe fruit=*Harle*, *pilo-harle*, *hardi*, *himagihira u*, galls=*harle-phúl*, *pilo-harle-phúl*, GUZ.; Tree=*Kadakái*, *kaduk-kay*, *pilá-marda*, ripe fruit=*kaduk-káy*, dried fruit=*kaduk-káy-pinji*, galls=*kaduk-káy-pu*, TAM.; Tree=*Karaka*, *kadukar*, *kurka*, *karaku*, ripe fruit=*karakkáya*, dried fruit=*pinda karakkáya*, galls=*karak-káya-puvvulu*, TEL.; Tree=*Hírda*, ripe fruit=*alale-káyi*, dried fruit=*alale-pinda*, galls=*alale-huvvu*, KAN.; Ripe fruit=*Katukká*, dri d fruit=*katukká-pinji*, galls=*katukká-pú*, MALAY.; Tree=*Pangah*, BURM.; Tree & ripe fruit=*Aalu*, *aralu*, galls=*aralu-mal*, SING.; Ripe fruit=*Haritaki*, *abhayá*, *pathyá*, galls=*haritaki-pushpam*, SANS.; Ripe fruit=*Halílaj*, *halilaje-asfar*, dried fruit=*halilaje-asvad*, ARAB.; Ripe fruit=*Halílah*, *halilahe-zard*, dried fruit=*halilahe-siyah*, PERS.

References.—*Roxb.*, *Fl. Ind.*, Ed. *C.B.C.*, *381*; *Voigt*, *Hort. Sub. Cal.*, *37*; *Brandis*, *For. Fl.*, *223*, *t. 29*; *Kurz*, *For. Fl. Burm.*, *I.*, *456*; *Pegu Rept.*, *app. A.*, *lviii.*, *B.*, *49*; *Beddome*, *Fl. Sylv.*, *t. 27*; *Gamble*, *Man. Timb.*, *180*; *Cat.*, *Trees*, *Shrubs*, *&c.*, *Darjeeling*, *39*; *Thwaites*, *En. Ceyl. Pl.*, *103*; *Trimen*, *Sys. Cat. Cey. Pl.*, *32*; *Dalz. & Gibs.*, *Bomb. Fl.*, *91*; *Stewart*, *Pb. Pl.*, *89*; *Aitchison*, *Cat. Pb. & Sind Pl.*, *59*; *Rev. A. Campbell*, *Rept. Econ. Pl.*, *Chutia Nagpur*, *8404*; *Graham*, *Cat. Bomb. Pl.*, *69*; *Mason*, *Burma & Its People*, *509*, *743*; *Sir W. Elliot*, *Fl. Andhr.*, *83*, *84*; *U. S. Dispens.*, *15th Ed.*, *1705*; *Fleming*, *Med. Pl. & Drugs* (*Asiatic Reser.*, *XI.*), *181*; *Ainslie*, *Mat. Ind.*, *I.*, *237*, *II.*, *128*; *O'Shaughnessy*, *Beng. Dispens.*, *340*; *Irvine*, *Mat. Med. Patna*, *36*; *Med. Topog. Ajmere*, *136*; *Taylor*, *Topog. Dacca*, *62*; *Moodeen Sheriff*, *Supp. Pharm.*, *Ind.*, *242*; *U. C. Dutt*, *Mat. Med. Hind.*, *160*, *299*, *313*; *S. Arjun*, *Cat. Bomb. Drugs*, *54*; *K. L. De*, *Indig. Drugs Ind.*, *117*; *Murray*, *Pl. & Drugs*, *Sind*, *189*; *Bidie*, *Cat. Raw Pr.*, *Paris Exhib.*, *29*, *III*; *Waring*, *Basar Med.*, *97*; *Dymock*, *Mat. Med. W. Ind.*, *2nd Ed.*, *317*; *Dymock*, *Warden*, *& Hooper*, *Pharmacog. Ind.*, *II.*, *1*; *Cat. Baroda Durbar*, *Col. & Ind. Exhib.*, *Nos. 173*, *174*; *Year-Book Pharm.*, *1879*, *214*; *Trans. Med. & Phys. Soc.*, *Bomb.* (*New Series*), *IV.*, *152*; *XII.*, *174*; *Birdwood*, *Bomb. Prod.*, *34*, *152*, *282*, *312*; *Baden Powell*, *Pb. Pr.*, *349*, *599*; *Atkinson*, *Him. Dist.* (*X.*, *N.-W. P.*, *Gaz.*), *310*, *751*, *777*, *779*, *816*; *Useful Pl. Bomb.* (*XXV.*, *Bomb. Gaz.*), *73*, *155*, *218*, *244*, *259*; *Econ. Prod. N.-W. P.*, *Pt. III.* (*Dyes & Tans*), *35*, *84*; *Croukes*, *Hand-book*, *Dyeing*, *&c.*, *500*; *Liotard*, *Dyes*, *15-23*, *25*, *69*, *116*, *122*, *131*; *Cooke*, *Oils & Oilseeds*, *79*; *McCann*, *Dyes & Tans*, *Beng.*, *16*, *31*, *33*, *48*, *74*, *86*, *88*, *125*, *134*, *135*, *138*, *144*, *145-150*, *152*, *161*, *165*, *166*, *167*, *168*, *169*; *Wardle*, *Dyes & Tans*, *42*, *53*; *Christy*, *New Com. Pl.*, *25*, *40*, *44*; *Sel. Rec. Govt. Ind.* (*R. & A. Dept.*), *1888-89*, *54*, *56*, *57*, *87*, *88*, *106*; *Linschoten*, *Voyage to East Indies* (*Ed. Burnell*, *Tiele*, *& Yule*), *II.*, *123-126*; *Statistics*, *Dinajpur*, *150*; *Man. Madras Adm.*, *I.*, *313*; *II.*, *98*, *123*; *Nicholson*, *Man Coimbatore*, *42*; *Boswell*, *Man. Nellore*, *98*, *116*, *125*; *Moore*, *Man. Trichinopoly*, *80*; *Gribble*, *Cuddapah Man.*, *262*; *Settlement Reports*:—*Panjáb*, *Kángra*, *21*; *C. P.*, *Bhundára*, *19*; *Chánda*, *app. vi.*; *Hoshungabád*, *179*; *Upper Godávery*, *38*; *Chhindwára*, *111*; *Seonee*, *10*; *Mandlah*, *88*; *Baitool*, *127*; *Jubbulpore*, *86*; *Gazetteers*:—*Bombay*, *XIII.*, *23*; *XV.*, *39*, *78*; *XVI.*, *18*; *XVIII.*, *46*; *Panjáb*, *Hoshiárpur*, *10*; *N.-W. P.*, *I.*, *81*; *IV.*, *lxxi.*; *C. P.*, *59*, *504*; *Orissa*, *II.*, *181*; *Mysore & Coorg*, *I.*, *48*, *437*; *III.*, *23*; *Agri-Horti. Soc.*, *Ind.*:—*Trans.*, *II.*, *app. 314*; *III.*, *68*; *VII.*, *57*; *VIII.*, *Pro. 443*; *Jour.*, *III.*, *Pro.*, *71*; *IV.*, *201*; *VI.*, *36*; *IX.*, *422*, *Sel.*, *42*; *XIII.*, *318*, *391*; *New Series*, *VII.*, *141*; *Trop. Agri.*, *January 1889*, *481*; *Ind. Forester*:—*I.*, *78*, *81*, *363*; *II.*, *19*; *III.*, *24*, *202*; *VI.*, *323*; *IX.*, *239*, *413*; *X. 33*, *325*, *545*; *XI.*, *326*; *XII.*, *313*; *XIII.*, *121*, *146*, *164*; *XIV.*, *302*; *Spons'*, *Encycl. II.*, *1087*; *Balfour*, *Cyclop*, *Ind.*, *III.*, *850*.

Habitat.—A large, deciduous tree, abundant in Northern India from

T. 325

Kumáon to Bengal and southward to the Deccan table-lands at 1,000 to
3,000 feet, also found in Burma, Ceylon, and the Malay Peninsula. In the
Madras Presidency it is common all over the forests; in Coimbatore it is
of large size; in Kanara and Sunda it abounds above the Gháts; in Gan-
jam and Gumsur it is tolerably plentiful; and it occurs in the Godáveri
tracts. In Bombay it is common on the higher forests on and near the
Gháts, and is very abundant in the forests of the highlands of the Satpurass
and above the Gháts in Belgaum and Kanara.

GUM.
Tree.
326

Gum.—The TREE yields a GUM which is said to be largely collected in
the Berars, mixed with those of **Acacia arabica, Anogeissus latifolia, Bassia
longifolia,** and **Melia Azadirachta.** The mixed gums of these trees are
taken to local markets by the Gonds who collect them, and sold either for
medicinal purposes or to dyers to mix with their colours (*P. W. Dept. Repts.
on Gums, &c., 69*). The vernacular name given, *viz., bey-a-rah,*seems to indicate that the above account may in reality refer to the gum of T. belerica.

DYE & TAN.
Fruit.
327
Galls.
328
Leaves.
329

Dye and Tan.—The dried FRUIT forms the "chebulic" or "black"
myrobalan of commerce, one of the most valuable of Indian tanning materials. In India it is occasionally used as a dye by itself, the rind of the
fruit being powdered and steeped in water. The cloth steeped in this
infusion acquires a dirty grey colour. With alum both the fruit, and the
GALLS produced in quantity on the LEAVES, are said to give with alum a
good permanent yellow. But the most extensive use to which *harra* is put
as a dye is in the production of various shades of black in combination
with some salt of iron, generally the protosulphate. In some cases *gúr* or
molasses is added, in others a little indigo is mixed with the dye to give
depth to the colour. In Dacca a deep black is obtained by using *gáb,* the
dried fruit of **Diospyros Embryopteris,** in combination with chebulic myrobalan and ferrous sulphate. In Chutia Nagpur a dark neutral tint called
kakraiza is obtained from *harra,* protosulphate of iron, and safflower. In
Chittagong the fruit is mixed with *tirí* pods (**Cæsalpinia Sappan**) to produce a black dye. A mixture of the fruit and ferrons sulphate in certain
proportions also produces a *khaki* or iron-grey colour (*McCann*). In
Madras it is used in the same way, also alone for dyeing cotton, wool, and
leather. In the North-West Provinces the chief shades, in producing
which the fruit plays a part, are black, as above described; green, in company with turmeric and indigo; dark blue, with indigo, and brown with
catechu. Excepting in the case of black it acts more as a concentrator of
colour than as contributing much colour of its own (*Sir E. C. Buck*). It
is commonly employed throughout the country as a mordant, or acces-
sory, to concentrate the colour in dyeing with safflower, *ál* (**Morinda citrifolia**), *manjít* (**Rubia cordifolia**), *huldi* (**Curcuma longa**), and *tesu* (**Butea
frondosa**). With iron-salts it is employed in making country ink, and
mixed with ferruginous mud it makes a black paste employed by harness
and shoe-makers as well as by dyers. The BARK is also occasionally used

Bark.
330

for dyeing *khaki* grey and black, and in Bengal and Manipur for dyeing
bamboos. A sample of the fruit examined by Mr. Wardle was found to
produce very dark shades of grey when mixed with salts of iron, on *tasar,
corah* and *eri* silk, and wool. A sample of the bark was ascertained to be
very astringent and to produce shades of colour much like those obtained
from *babul* pods, but of a somewhat yellower tone. The shades on un-
bleached Indian *tasar* varied from yellow-drab to slaty-drab, on bleached
Indian *tasar, eri* silk, *corah* silk, and wool they were a yellowish drab.
The galls were found to contain 13·1 per cent. of tannic acid and to produce a light yellow on wool.

The chief commercial value of chebulic myrobalan is, however, as a
tanning material; it forms the greater part of the ground myrobalans of

One of the most valuable of Tans. (*J. Murray.*) **TERMINALIA Chebula.**

DYE & TAN.

commerce, though belleric myrobalans are occasionally mixed with it. The liquor prepared from it is not only a powerful tan, but imparts a bright colour to the leather, and is hence highly esteemed to mix with other tanning agents (*Conf.* **Leather,** Vol. IV., 607). Thus Professor Hummel, in his report on Indian Tans at the Colonial and Indian Exhibition of 1886, writes : " Ground myrobalans are becoming more and more a favourite tannin matter, and practically combine every desirable excellence." At the Tanning Conference held at that Exhibition much interest was naturally evidenced in a product of so great value, and several facts of much moment to the success and future extension of the Indian trade were elicited. The following passage from the report, issued at the close of the Exhibition, is of interest : —

Commercial Forms.
331

" The gentlemen present were able to recognise and give the trade names for most of the forms exhibited. They pointed out that **Terminalia Chebula** must never be round or spongy in texture. The good qualities were known in the trade to be oval and pointed, and on section, of a pale greenish-yellow colour, and solid in structure. This oblong and pointed form was thought to be the product of a separate species, but Dr. Watt explained that, in his opinion, it was only the young or unripe fruit of **T. Chebula.**

" **Mr. Evans** kindly promised to furnish samples of the various commercial qualities, in order that these might be communicated to India, in the hope of an effort being made to disseminate a knowledge of what constitutes good and bad qualities. It seemed important, if the view was correct, that the oval and hard forms were but young fruits, that this fact should be published in India as widely as possible. The so-called Jubbulpur form of myrobalans was viewed as superior; and **Mr. Evans** picked out specimens of what he regarded as the best quality shown, in order to compare these with myrobalans procured in London." " Two samples of galls found upon the myrobalan trees were placed on the tables. That from **T. Chebula** was stated to be hopeful, but the very plentiful gall from **T. tomentosa,** obtained from the Reverend A. Campbell, in Chutia Nagpur, was, after it had been submitted to chemical examination, pronounced valueless. A number of other tanning materials were examined, but none seemed to afford sufficient interest to deserve special mention.

" At the close of the examination of tanning materials, the gentlemen adjourned to Dr. Watt's office, in order to discuss what action seemed desirable in the interests of the Indian tanning industry. It was urged that it was essentially necessary to have the better qualities of tanning materials carefully analysed ; and two gentlemen very kindly undertook to do this independently of each other, if they were supplied with samples, and they further agreed to communicate to the Government the results of their examinations.

" When asked what might be recommended to Government, it was stated that the only thing that could be done was to experiment with the production of extracts or half-stuffs, so as to overcome the heavy charges of transport and freight."

Results of Experiments.
332

In accordance with the promise mentioned above, samples were examined by Dr. R. H. Paul with very interesting results. Of three samples furnished to him, all of which were carefully identified and transmitted in such a way that no mistake could occur, one (No. 2) contained 32·82 per cent. of gallo-tannic acid, another (No. 6), 26·81, and the third (No. 3), only 6·11. Dr. Watt commenting on this curious and important result writes :—

" The writer gave the above samples to Dr. Paul personally ; and, together with the report, received back one of each fruit analysed, with its number written on it. The record of despatch of samples agreed with the return ; so that no room for doubt remained as to the botanical identification

TERMINALIA
Chebula.   The Chebulic Myrobalan—

DYE & TAN.
Results of
Experiments.

being correct. Moreover, those present at the Conference condemned the
sample No. 3 as a very inferior quality of the true myrobalan. Other experts
received corresponding samples, and their reports, had they been received,
would have been most valuable as placing this matter beyond the possibility
of doubt. Dr. Paul's analysis, however, of Nos. 2, 3, and 6 so completely
confirm the observations and valuations made by the experts present at
the Conference, that the theory then advanced, regarding the superior
quality of the oblong, pointed, and solid fruits, as compared with the
round inflated ones, will most probably be found correct, *viz.*, that myrobalans picked off the same individual tree during different stages of their
growth will be found to have a varying composition of from 6 to 30 per
cent. of gallo-tannic acid. And this theory is supported by a volume of
evidence in the history of all fruits, and particularly by **Professor Hummel's** observation, that in the case of **T. belerica** the tannic principle resides
chiefly in the outer pulp of the myrobalan. The transformation from the
bitter unripe apple to the ripe, sweetly-flavoured fruit, is so well known as
to scarcely require mention. This being so, it would seem desirable to
institute a thorough enquiry into the subject of these valuable tanning
materials which would have two objects in view—(*a*) to determine the exact
age in each locality when the maximum amount of tannic acid is present ;
and (*b*) the properties and value of the fruits of one district as compared with
those of another. In a country with so many different climatic features,
and such widely diversified peculiarities of soil as India, it neither follows
that the fruits will reach their perfection at the same time in various localities, nor even that different climates and soils will, when this point has
been determined, produce fruits of equal merit. Were these questions
determined, it would be possible for Government to encourage, with reasonable hope of success, the development of a large myrobalan trade ;
and for merchants to depend upon a good supply of superior fruit. As
matters stand, no dependence can be put on the supply or the quality of
Indian myrobalans. In the trade a form of the true myrobalan is known
as the Jubbulpur myrobalan ; and this may literally be grown at Jubbulpur, or may be but a form or condition of maturity first sent to Europe
from that district—that quality continuing to bear the name in spite of the
fact that it may be obtained from many other localities. So very much
superior are these oblong, pointed, solid, and pale green fruits to the large
rounded samples, that several of the experts seemed to feel hurt that their
belief should be questioned that these were not the fruits of a different
species" (*Sel. from Rec. Govt. of Ind., Rev. and Agri. Dept., I., 89*).

A sample from the Central Provinces (similar to that furnished as
No. 2 in Dr. Paul's set) was analysed by **Professor Hummel,** and was
found to contain 31 per cent. of tannic acid. The decoction it produced
was of a pale-yellow colour. The money value per cwt., as compared with
ground myrobalans at 7*s.* 6*d* per cwt., was 10*s.* 1*d.* The analysis of an
average sample of commercial ground myrobalans by the same author
revealed 23 per cent. of tannic acid, while the decoction differed from that of
the unadulterated fruit in being turbid. **Professor Hummel** remarks that
he did not observe such a distinction of parts as that found in **T. belerica,**
in the chebulic myrobalan, but it is worthy of notice that **Crookes,** in his
account of the tan quoted below, states that the kernel is inert.

The difference in quality of myrobalans at different seasons of the year
is, as **Dr. Watt** remarks, of the utmost importance. The following account
of the appearances and properties of the fruit when in the best condition
taken from **Crookes'** *Handbook of Dyeing*, may, therefore, be found a useful guide towards meeting the requirements of the market :—" In shape
and size myrobalans somewhat resemble shrivelled plums. They are of a

T. 332

| One of the most valuable of Tans. (*J. Murray*.) | TERMINALIA Chebula. |

DYE & TAN.
Results of Experiments.

pale buff colour, consisting of a dry pulp, varying in thickness, and en-
closing a stone-like kernel, which contains no tannin, and forms from 23 to
52 per cent. of the whole weight. The moisture present in the nuts, as
found in commerce, varies from 3 to 7 per cent., and the amount of ash
left on incineration is about 10 per cent. The tannic acid is mainly present in the pulp. Good unground myrobalans should be pale, plump,
free from dark blotches and from worm-holes. They should be hard and
firm, ringing like fragments of earthenware when shaken together, and if
crushed with a hammer yielding a dry, pale powder, mixed with hard
irregular fragments. If they can be crushed to a dark-coloured dust between the fingers, or if they work out into a paste under the pestle, they
are of poor quality. Ground myrobalans should form a pale buff powder,
dry, astringent to the taste, but free from a saline flavour or from intense
bitterness. If moistened and rubbed in the hand, they should form a
very tenacious paste.

"The nuts, when bought whole, are often found mixed with earth, sand,
mica, nux-vomica, betel nuts, etc. Ground myrobalans are sometimes
contaminated with divi-divi, spoiled sumac, and wild gall. To detect such
impurities a little of the powder is thinly sprinkled over a sheet of white
paper, and examined with a lens. If divi-divi has been added, fragments
of its bright brown flat seeds are sure to be found. The outer skin of a myrobalan may occasionally resemble a divi-seed in colour, but the minutest
portion of the former shows a wrinkled surface, whilst divi-seeds are smooth.
The leaf-stalks and midribs of sumac may also be distinguished, by the aid
of a lens, from the torn and jagged fibre of the myrobalan.

"Myrobalans, being cheaper than galls and stronger than sumac, have,
to a very great extent, superseded both. Along with salts of iron, they
dye cotton a fuller black than can be obtained with sumac. They are likewise preferred to sumac for fixing the coal-tar colours upon cotton, owing
possibly to the fact that they contain, along with tannin, certain oily and
glutinous matters. Gall-nuts and commercial tannin are, however, still
preferable."

SUPPLY.—As already stated, the tree is found all over the forests of
the Madras Presidency, and on the high table-lands and gháts of Bombay. It is also met with abundantly in the highlands of the Satpura
range, Central Provinces; in Palamow and Hazáribágh, Bengal (*Beng.
Govt. Rept., 1880*), "more or less common all over Bengal" (*McCann*);
in the Sub-Himálayan tracts of the North-West Provinces; and throughout the forests of Oudh. It is also met with more or less abundantly in
Assam and Burma. In the Panjáb it is occasionally cultivated in the Sub-
Himálayan tracts up to the Indus, in the Siwalik and Outer Himálaya up
to 5,000 feet and extending west to the Sutlej; according to Baden Powell
it is extensively grown in the Kangra district. The fruit is everywhere
an article of minor forest produce, which yields a greater or less revenue to
the Forest Department, but in Bombay only is this of any great account.
Up to 1887-88 the right of collecting the fruit was farmed out all over the
country by auction sale, but in that year a systematic collection and sale
of the myrobalans was instituted in the Southern Circle of the Bombay
Presidency by Lieut.-Colonel Peyton, the Conservator of Forests. During the ten years preceding, the revenue derived from the auction sales in
the division had varied from R18,000 to R40,000, the average from 1870-71
to 1876-77 having been R24,883. In 1877-78, the new system of departmental collection gave a net income of more than treble that amouut, *viz.*,
R76,966. In addition to the increase of revenue thus derived the system
had the advantages claimed for it :—(1) that it provided legitimate and
well paid employment for many people; (2) that by enlisting the interest

Indian Supply.
333

TERMINALIA
Chebula.   The Chebulic Myrobalan—

of the people it would tend greatly to reduce risk of forest-fires; and (3) that it did away with a great deal of oppression and bad treatment on the part of the former holders of the right to collect.

At the same time large quantities of fruit were sown in the reserves, resulting in a thick new growth of *hirda* all over the forests. The revenue derived from the fruit in other localities is comparatively small.

The Forest Administration Report for the Southern Circle during the past year shows the outturn and value of *hirda* collected by Government agency to have been as follows :—

DIVISION.	Outturn.		Receipts.		
	Khandis.	Maunds.	R	a.	p.
North Kánara	1,901	...	18,779	15	9
Central Kánara	366	...	3,893	4	9
South Kánara	2,131	...	21,846	1	9
Belgaum	6,608	...	56,723	8	0
Kolába	(Returned under "Minor forest produce.")		700	0	0
Ratnagiri	...	34	38	12	0
TOTAL .	11,006	34	1,10,981	10	3

In the Northern Circle the total yield was 1,386 khandis, 25 maunds 2 seers, which fetched R26,893-2-1. The cost of collection, &c., is estimated to have been R13,593-11-2. The revenue derived from sale of the fruit in other Presidencies and Provinces is very much smaller, and cannot be definitely arrived at owing to the practice of including it with other articles under the general heading of "Minor produce." In the Southern Circle of Madras 28 tons 1,678℔ are said to have been collected by departmental agency and sold for R901-2-9. In the Northern Circle myrobalans are not separated from other minor produce. In the case of the Central Provinces, the largest myrobalan-producing area in India, it is impossible, owing to myrobalans not being accurately separated from other minor forest produce in the returns, to give any idea of the amount collected from Government Forests. It will be observed below that the returns of internal trade show a very large export of myrobalans from these Provinces.

The Forest Administration Report of Bengal for 1888-89 shows a similar want of returns for myrobalans. Only one mention is made of the fruit, apart from other minor produce, from which it would appear that 20 maunds were removed by purchasers, from the Hazáribágh subdivision of Chutia Nagpur, on payment of R20, or R1 per maund. The system of collecting the fruit departmentally does not appear to have been adopted in the Lower Provinces. McOann states that the prices given, as a rule, for myrobalans in the various districts, are :—R2-8 per maund in Midnapur, R3 in Rájsháhi, R5 in Chittagong, R2-8 in Monghyr, about R2 in Cuttack, R5 for *jeonghi harra*, or large, picked, unripe fruit, and R1-4 for ordinary *harra* in Chutia Nagpur, and R3-8 to R5 in Palamow. In a report from the Bengal Forest Department (1880), however, cited by

T. 333

One of the most valuable of Tans.　(*J. Murray.*)　**TERMINALIA Chebula.**

Liotard, the prices are given at 10 annas 8 pie per maund in Palamow, annas 4 in Kurseong, and R2-8 in Chittagong, figures all very much lower than those given by McOann. Recent Forest Reports of the North-West Provinces make no return for myrobalans apart from other minor produce, but some idea of the probable outturn exported from the Province may be obtained from figures for 1874-75 given by Sir E. O. Buck. According to these the exports from the forests of Najibabád, Rehár, Dehra Dun, Garhwál, and Kumáon amounted to 451 cwt. of large fruit, valued at R974; and 205¾ cwt. of small, valued at R2,016. No approximate of the outturn in other localities can be hazarded. It must also be remembered that the figures of internal trade conclusively show that the amount collected by, or under the supervision of, the Forest Department, is very small in comparison with the large quantity which comes into the market. It would, in fact, appear that in Bombay only has the matter received sufficient attention to be productive of a noticeable revenue to that Department.	**DYE & Tan.** Indian Supply.
Trade.—It is not possible, in considering the trade, to separate the figures for chebulic, belleric, and emblic myrobalans, since all are returned under the general heading of myrobalans. But the article at present under consideration is vastly the most important and may be accepted as forming much the largest proportion of the material represented by the trade figures.	**TRADE.** 334
INTERNAL.—During the year 1888-89 (the year for which figures of the rail, road, and river traffic of all India are available) the total exports of myrobalans from one Province or Presidency to another amounted to 5,26,738 maunds, valued at R12,26,720. Of this quantity the Central Provinces is returned as having exported 3,03,696 maunds, nearly all of which went to Bombay port. Bombay comes next with 1,07,038 maunds. The only large importer is Bombay port with 4,36,515 maunds, followed by Madras seaports with 30,956, and Calcutta with 22,836 maunds.	**Internal.** 335

The average imports by coasting trade during the five years up to 1889-90 has been 228,207 cwt., valued at R7,88,508. In 1889-90 it amounted to 267,103 cwt., valued at R9,26,115. Of this quantity 261,442 cwt. represented the imports into Bombay ports, of which 6,384 cwt. came from Bengal, 7,108 from Madras, and 223,028 cwt. from other British ports within the Bombay Presidency, while 23,161 cwt. came from Goa. The transactions of other ports are unimportant.

FOREIGN.—The exports of Indian myrobalans to foreign countries show a marked tendency to increase, as will be seen by the following quinquennial averages :—

Foreign. 336 Exports 337

Five years ending	Quantity.	Value.
	Cwt.	R
1879-80　.　.　.　.　.　.　.　.　.	416,189	17,30,812
1884-85　.　.　.　.　.　.　.　.　.	435,962	17,52,036
1889-90　.　.　.　.　.　.　.　.　.	678,502	26,85,495

The year of maximum export was the first part, 1889-90, when a total of 781,741 cwt., valued at R31,75,330, was reached. The distribution of

**TERMINALIA
Chebula.**   The Chebulic Myrobalan —

DYE & TAN.

Foreign
Trade.

Exports.

the exports, and the share taken by each seaboard Presidency or Province, in the trade, during that year is shown by Mr. O'Conor as follows : —

COUNTRIES TO WHICH EXPORTED.	Quantities.	Value.	SHARE OF EACH PRESIDENCY OR PROVINCE.		
			Presidency or Province.	Quantities.	Value.
	Cwt.	R		Cwt.	R
United Kingdom .	633,235	25,82,944	Bengal .	25,009	73,447
Austria . . .	90,512	3,65,898	Bombay .	643,178	27,93,381
Belgium . . .	13,738	58,376	Madras .	113,554	3,08,502
France . . .	3,745	13,891			
Germany . . .	16,873	57,501			
Italy	12,418	48,574			
Russia . . .	4,796	20,883			
United States . .	525	1,607			
Australia . . .	5,709	24,815			
Other Countries . .	190	841			
TOTAL .	781,741	31,75,330		781,741	31,75,330

Imports.
338

OIL,
Kernels.
339
MEDICINE.

Fruit.
340

The trade will thus be seen to be almost entirely between Bombay and the United Kingdom.

One of the most curious facts in connection with myrobalans is the existence of a small import trade into India. It is difficult to understand why this should be so, in the case of a country in which the tree grows so plentifully and in which the fruit may be obtained so cheaply. During the three quinquennial periods ending 1879-80, 1884-85, and 1889-90, the average imports amounted to 1,330 cwt., valued at R5,498, 3,072 cwt., valued at R10,524, and 3,751 cwt., valued at R11,759. This trade is almost entirely between Ceylon and Madras.

Oil.—A clear, transparent, almost colourless, fluid OIL is obtained in small quantities from the KERNELS. It is used medicinally.

Medicine.—The chebulic myrobalan was highly extolled by the ancient Hindus as a powerful alterative and tonic. It has received the names of *Pránadá* or life-giver, *Sudhá* or nectar, *Bhishakpriya* or physician's favourite, and others of the same nature. Seven varieties of *haritáki* are described, of which only two are at present recognised, the large ripe FRUIT called *haritáki*, and the unripe fruit called *jangi haritáki*. A good *haritáki*, fit for medicinal use, should be fresh, smooth, dense, heavy, and round in shape. Thrown into water it should sink. Any fruit, however, which weighs over four tolas, is considered fit for use, although it may not possess some of the other properties. Fruit with small seeds and an abundant pulp is preferred ; the seed is always rejected. Chebulic myrobalans are described as laxative, stomachic, tonic, and alterative. They are used in fevers, cough, asthma, urinary diseases, piles, intestinal worms, chronic diarrhœa, costiveness, flatulence, vomiting, hiccup, heart diseases, enlarged spleen and liver, ascites, skin diseases, &c. In combination with emblic and beleric myrobalans, under the name of *triphalá*, or the three myrobalans, they are extensively used as adjuncts to other medicines in almost all diseases. As an alterative tonic for promoting strength, preventing the effects of age and prolonging life, chebulic myrobalan is used in a peculiar way. One fruit is taken every morning with salt in the rainy season, with sugar in autumn, with ginger in the first half of the cold season, with long pepper in the second half, with honey in spring, and with treacle in the two hot months (*Hindu Mat. Med.*). Numerous prepara-

MEDICINE.

tions prescribed by Chakradatta and other old Sanskrit writers are detailed at length by U. C. Dutt, to whose work the reader is referred for further information.

Myrobalans were known to the early Arabian writers and, through them, to the Greek writer, **Actuarius**, who describes five kinds. The author of the *Makhzan-el-Adwiya* distinguishes the following kinds, gathered at different stages of maturity :—" (1) *Halileh-i-zira*, gathered when the fruit is first set, being dried it is about the size of the *zira* (cummin-seed). (2) *Halileh-i-jawi*, more advanced ; it is the size of a barley-corn (*jao*). (3) *Halileh-i-zangi*, *halileh-i-hindi*, or *halileh-i-aswed*, a still further advanced stage of the fruit, which, when dried, is the size of a raisin, and black ; hence the names *zangi*=negro, and *aswed*=black. (4) *Halileh-chini*, is gathered when the fruit has attained some degree of hardness ; when it is of a greenish colour, inclining to yellow. (5) *Halileh-i-asfar*, or the very nearly mature nut, but still strongly astringent. (6) *Halileh-i-kabuli*, is quite mature. **Fleming** was the first European writer to iden-tify the fruit as that of **T. Chebula.** Commenting on the above de-scription from the *Makhzan* he writes : "The *zangi-har* is, as far as I can learn, more frequently used in medicine by the Hindus than any of the other myrobalans, being very generally employed by them as a purgative. It operates briskly, but without occasioning heat or irrita-tion. Persons liable to redundancy of bile, habitual costiveness, or any other complaint which requires the frequent use of gentle laxatives, will find this one of the most convenient which they can use." The authors of the *Pharmacographia Indica* translate the same passage from the *Makhzan* and remark, " Of these six varieties of chebulic myrobalans, the second, third, and sixth only are in general use for medicinal purposes, the fourth and fifth, also known as *rangari har* or *hirade*, are chiefly used by tanners." This observation is of considerable interest in connection with the remarks made above regarding the variability in the quantity of tannin matter in the fruit at different periods of its life, and indicates that the Persians, and probably the Arabs also, considered the fairly well formed, but still immature, fruit as most valuable for tanning. The authors of the *Pharmacographia Indica* continue, " Mahometans, like Hindus, attribute a great many fanciful properties to the drug; shortly, we may say, that the ripe fruit is chiefly used as a purgative, and is considered to remove bile, phlegm, and adust bile; it should be combined with aromatics, such as fennel seeds, carraways, &c. The Arabs say ' *Ihlilaj* is in the stomach like an intelligent housewife who is a good manager of the house.' The unripe fruit (*Halileh-i-hindi* or *Himaja*) is most valued on account of its astringent and aperient properties, and is a useful medicine in dysentery and diarrhœa ; it should also be given with aromatics. Locally it is ap-plied as an astringent. The first and second kind are supposed to have the same properties as the third in a less degree, and the fourth and fifth the same as the sixth in a less degree. The best way of administering my-robalans as a purgative is to make an infusion or decoction of from 2 to 4 drachms of fruit pulp with the addition of carraway seeds and a little honey or sugar."

Though myrobalans have long been known to European medicine, they have quite dropped out of use. During the early part of the Chris-tian era they were known to the Greeks. **Linschoten**, who visited India towards the end of the sixteenth century, describes five sorts of myrobalans as found in the country, endorsing the information on the subject pre-viously given by Garcia d'Orta. These, he states, were used for tanning leather as "tanners use sumach." The first and last of his five sorts " by physitions called *Citrinos* and *Quebulus*, by the Indians *arare* and

TERMINALIA
Chebula.   The Chebulic Myrobalans—

MEDICINE.   *aretean,*" appear to have been undoubtedly chebulic myrobalans. These,
he describes, as having been much used medicinally and as a food. His
commentator, **Dr. Paludanus,** states that all five kinds were regularly
imported from India in his time, either dried, pickled or preserved in
sugar. He describes *citrinos* as yellow, and valuable as purging the
stomach from bile, and " good against Tertians and other hot-burning
Feavers." **Linschoten's** *Quebulus* he calls *cepule* or *chebulæ*, writing,
" the greater they are the better. Blackish, and somewhat reddish, heavie
and sinking into the water, they purge fleame, they sharpen men's wits
and clear the sight. They are preserved in sugar and honnie, they doe
strengthen and purge the stomach, they heale the dropsie and are good
against olde agues, they likewise give a man an appetite and help
digestion."

Linschoten's other kinds he identifies as *Bellericos* (**T. belerica**), *Em-
blicos* (**Phyllanthus Emblica**), and *Inelus,* the last of which from its name
and description is probably the *halileh-i-hindi* or black chebulic myrobalan above mentioned. On the awakening of interest in Indian Materia
Medica towards the end of last century, myrobalans again and naturally
attracted attention. **Fleming,** with the help of **Roxburgh,** was able to
botanically identify the *phar* of the bazárs, and in an interesting and
exhaustive article on the subject, recommended it as a gentle purgative.
Ainslie noticed its value in the preparation of an application for aphthæ
in children and adults. **Buchanan-Hamilton** described it as a valuable
purgative and recognised its value as a tanning material ; he recommended
that it should be more generally planted, and its growth encouraged, near
villages. He also noticed an interesting medicinal use of the fruit by Natives.
" Men who have made a vow of chastity," he writes, " and who are inclined
to adhere to their resolution, endeavour to assist their virtue by eating this
preserve, which is supposed to diminish the desires of the flesh." **Hove,**
in his account of a visit to the myrobalan plantation at Bungar in the
Konkan in 1787, states that he found one fruit a sufficient purgative,
though the manager of the plantation told him that two were generally used.

On the compilation of the *Pharmacopœia of India* in 1868, the fruit
was admitted to a place on the secondary list, where much valuable
testimony as to its properties is detailed. **Waring,** the editor, found that
six fruits, bruised and given in decoction, acted efficiently and safely as
a purgative, producing four or five copious stools, unattended by griping,
nausea, or other ill effects. **Dr. Oswald** recommended a similar preparation as an application or injection for hæmorrhoids and vaginal discharges. **Rajah Kalikissen** extols their virtues and regards them as
combining mild purgative, with carminative and tonic properties. **Twin-
ing** (*Diseases of Bengal, I.,* 407) speaks very favourably of them in the
same character, and expresses surprise that a medicine with such useful
properties should be so little known in Europe. He gives a case of enlargement of the spleen, in which it was productive of good effect. The
Galls.   **Rev. J. Kearns** of Tinnevelly is quoted as testifying to the efficacy of
341   the GALLS in dysentery and diarrhœa, especially in infantile diarrhœa,
the dose for a child of one year old being one grain every three hours
(*Pharm. Ind.*). More recently the authors of the *Pharmacographia Indica*
add their evidence in favour of the value of the fruit, stating that they have
found it useful in dysentery and diarrhœa. It is therefore very remarkable
that a medicine of such reputed value should have been allowed to drop
out of European practice. Nothing appears to have been done in the way
of accurately determining its physiological action, at least since the days
of **Paludanus,** who wrote : " It purges in another kind of manner than
doth Cassia, manna or such like drug, but it does it by astriction or

binding." It is quite possible that its action may be that of a nervo-muscular stimulant like Nux-vomica, certainly the most valuable class of purgatives in the. treatment of Indian diseases. This quality, combined with its astringency, would render it valuable for diarrhœa and dysentery as stated by the authors of the *Pharmacographia Indica.* Recently M. P. Apery has brought to the notice of the profession in Europe the value of the drug in dysentery, choleraic diarrhœa, and chronic diarrhœa; he administers it in pills of 25 centigrammes each, the dose being from four to twelve pills or even more in the twenty-four hours (*Pharmacog. Ind., quoting Jour. de Pharm. et de Chim., Feby. 1st, 1888*). It is, therefore, possible that the therapeutic value of myrobalans may before long form the subject of systematic investigation.

CHEMICAL COMPOSITION.—The question of the percentage of gallo-tonnic acid, contained in the fruit, has already been dealt with (see paragraph **Dye and Tan,** p. (?). In 1884 Herr Fridolin reported the isolation from the fruit of a new organic acid which he called *chebulinic acid,* and considered to be probably the source of the gallo-tannic acid detected by previous observers. He suggests as a formula to represent its composition $C_{28} H_{24} O_{19} (C_7 H_6 O_5 ?)$. When decomposed by heating an aqueous solution in a closed tube it takes up the elements of water and the molecule splits up into two molecules of gallic acid and one of tannic acid. According to **M. P. Apery** black myrobalans contain an oleo-resin of a green colour, soluble in alcohol, ether, petroleum spirit, and oil of turpentine. He has called it *myrobalanin* (*Pharmacog. Ind., from Jour. de Pharm. et de Chim., Feby. 1st, 1888*).

SPECIAL OPINIONS.—§ "Preserve of *Harar, Murabha-Harar,* often used as aperient and taken at night by the Natives" (*Civil Surgeon J. C. Penny, M.D., Amritsar*). "The kernels commonly eaten raw" (*Brigade-Surgeon G. A. Watson, Allahabad*). "A conserve of the fruit is used as digestive. It can be used in diarrhœa and indigestion. It acts also as a mild laxative" (*Surgeon R. L. Dutt, M.D., Pubna*). A fruit, coarsely powdered and smoked in a pipe, affords relief in a fit of asthma. A decoction of the fruit is a good astringent wash. A fine paste, obtained by rubbing the fruit on a rough stone with little water, mixed with the carron oil of the Pharmacopœia and applied to burns and scalds, effects a more rapid cure than when carron oil alone is used" (*Surgeon-Major D. R. Thompson, M.D., C.I.E., Madras*). "The fruit with senna and confection of roses is an effective laxative" (*W. Forsyth, F.R.C. S., Edin., Civil Medical Officer, Dinajpore*). "It is very largely used in Native medicine, and forms an ingredient of most Native prescriptions. Water in which the fruits are kept for the night is considered a very cooling wash for the eyes. When cleaned the fruits form a common dentifrice. The ashes, mixed with butter, form a good ointment for soress" (*Surgeon-Major Robb, Civil Surgeon, Ahmedabad*). An effective purgative used in the form of decoction combined with cinnamon or cloves—myrobalans, six in number; cinnamon or cloves, one drachm; water four ounces. Boil for ten minutes and strain. The whole quantity for a dose in the early morning. The gall-like excrescence on the leaves (*Kadu-kai-pio*) is also used at the hospital in the following formulæ and found useful as an astringent in cases of diarrhœa, especially in children. Pulv terminal gall excrescence, one ounce; Pulv. cinnamon, one ounce; Pulv. catechu, half ounce; Pulv. nutmeg, half ounce—inft. Dose: from ten to twenty grains for an adult" (*J. G. Ashworth, Apothecary, Kumbakonam, Madras*). "Astringent. A decoction ℥iss-½ to ℥i is useful as a gargle in sorethroat and as a wash for piles" (*Surgeon-Major A. F. Dobson, M.B., Bangalore*). "Commonly used as an astringent, generally in combination with

FOOD.

Fruit.
343
Kernel.
344
FODDER.

Leaves.
345
TIMBER.
346

DOMESTIC.

Ink.
347
Dye.
348
349

gall-nuts, etc. It is also chewed by old people together with catechu to tighten their teeth " (*Civil Surgeon M. Robinson, Coorg*).

Food and Fodder.—The FRUIT when ripe is occasionally eaten. The oily KERNEL, like those of other species of the genus, tastes like a filbert, and is used as an article of food. The LEAVES are eaten as a fodder by cattle.

Structure of the Wood.—Brownish-grey, with a greenish or yellowish tinge, very hard, fairly smooth and close-grained, durable, and seasons well. It has no regular heartwood, but irregular masses of dark purple occur frequently near the centre. The weight has been variously given by different writers at from 42 to 66℔ per cubic foot, the value of P. at 825 to 1090. It takes a good polish and is fairly durable, though, according to Beddome, cross-grained and difficult to work. It is used for making furniture, carts, agricultural implements, and for house-building, and has been tried for sleepers in Bengal.

Domestic and Sacred.—So highly esteemed is the tree that a mythological origin was assigned to it by the ancient Hindus. It is said that when Indra was drinking nectar in heaven, a drop of the fluid fell on the earth and produced the *haritákí* plant. The fruit and galls are used for making country INK, and a black DYE for staining the teeth. The former is a constituent of an excellent preparation for preserving skins, commonly employed by sportsmen in India.

Terminalia citrina, *Roxb.; Fl. Br. Ind., II., 446.*

Syn.—MYROBALANUS CITRINA, *Gærtn.*
Var. **malayana,** *Kurz.*
Vern.—*Haritaki, harra,* BENG.; *Hilika, silikka, silika.* ASSAM; *Hortaki,* CACHAR; *Hariha, harira,* N.-W. P.; *Kyú,* BURM.
References.—*Roxb., Fl. Ind., 382; Voigt, Hort. Sub. Cal., 37; Brandis, For. Fl., 223; Kurz, For. Fl. Burm., I., 456; Gamble, Man. Timb., 181; O'Shaughnessy, Beng. Dispens., 340; Darrah, Note on Cotton in Assam, 30; Liotard, Dyes, 94, 116; McCann, Dyes & Tans, Bengal, 35, 152; Agri.-Horti. Soc.:—Ind., Trans. VII., 57, 58; Journ., IV., 124, 134; VI., 71.*

Habitat.—A large, deciduous tree of Assam, Eastern Bengal, Burma, and Tenasserim. Considerable confusion exists in Indian economic literature between this species and T. Chebula, of which the appearance and vernacular names are very similar. It differs from the latter in having a straight stem, brighter foliage, and narrower fruits. The fruit is described as nearly 2 inches long, oblong-lanceolar, and. while fresh, obscurely five-angled. Mr. C. B. Clarke regards it as doubtfully a distinct species from T. Chebula.

DYE & TAN
Fruit.
350
Bark.
351
MEDICINE.

Fruit.
352

TIMBER.
353

Dye and Tan.—The FRUIT is doubtless frequently used in the same way as that of T. Chebula. In Assam the BARK is said to be employed in producing a black dye; in Monghyr the fruit is used as a mordant in dyeing with *ál*.

Medicine.—The medicinal properties of this species are probably similar to those of the chebulic myrobalan. Mention is made by Fleming and various other writers of the FRUIT, as distinct from that of T Chebula, being used medicinally, but in all probability they refer to the old MYROBALANUS CITRINA of the shops, a form of chebulic myrobalan. Roxburgh is probably correct in stating that the fruit of this species is not so distinguished in Hirdu Materia Medica, and that both are employed indiscriminately.

Structure of the Wood.—Grey, darker towards the centre, hard, similar to that of T. Chebula; weight 60℔ per cubic foot (*Wallich*), 49℔ (*Gamble*). In Assam it is used for making planks, and for general purposes of construction.

T. glabra, *See* T. tomentosa, *Bedd.*

T. 353

Terminalia myriocarpa, *Heurck. & Muell.-Arg.; Fl. Br. Ind., II.,*
　Syn.—PENTAPTERA SAJA, *Wall.*　　　　　　　　　　　[448.
　Vern.—*Panisaj,* NEPAL; *Sungloch,* LEPCHA; *Hollock,* Ass.
　References.—*Kurz, For. Fl. Burm., I.,* 457; *Gamble, Man. Timb.,* 185;
　　Ind, Forester, VIII., 416; *XI.,* 355.
　Habitat.—A very large, evergreen tree, abundant in the subtropical
valleys of Sikkim and Bhutan, between 1,000 and 3,000 feet, also met with
in the Assam hills and Ava.

　Structure of the Wood.—Sapwood white, not broad; heartwood brown,　TIMBER.
beautifully mottled with dark streaks, similar to that of T. tomentosa;　354
weight 51 to 54℔ per cubic foot. Used for building and tea-boxes,
also for charcoal. Gamble writes, "A specimen cut from a log of wood
which had been lying for many years in the bed of the Chauwa Jhora, near
Sivoke, in the Darjiling Terai, and is now perfectly black, may be this
species."

T. paniculata, *Roth. ; Fl. Br. Ind., II.,* 448.　　　　　　　355
　Syn.—T. MONOPTERA, *Roth.;* T. TRIOPTERA, *Heyne;* PENTAPTERA
　　PANICULATA, *Roxb.;* HIPTAGE SP., *Wall.*
　Vern.—*Kinjal, kindal,* BOMB.; *Kinjal, kindal,* MAR.; *Pe-karakai,* TAM.;
　　Nimiri, TEL.; *Honal, huluvá, hulvé, hunáb,* KAN.; *Pú mardá, pillai
　　mardá,* ANAMALAIS; *Marwa,* S. KANARA; *Pilla-murda,* MALAY.
　References.—*Roxb., Fl. Ind., Ed. C.B.C.,* 384; *Voigt, Hort. Sub. Cal.,*
　　38; *Brandis, For. Fl.,* 226; *Beddome, Fl. Sylv., A. 20; Gamble, Man.
　　Timb.,* 182; *Dalz. & Gibs., Bomb. Fl.,* 92; *Elliot, Fl. Andhr.,* 134;
　　Dymock, Mat. Med. W. Ind., 2nd Ed., 323; *Lisboa, U. Pl. Bomb.,* 75,
　　244; *Birdwood, Bomb. Pr.,* 331; *Liotard, Dyes,* 22; *Madras, Man.
　　Adm., I.,* 313; *Gazetteers:—Mysore & Coorg, I.,* 60; *Bombay, XV.,*
　　33; *Ind. Forester, II.,* 19; *IV.,* 292; *XII.,* 313; *Agri.-Horti. Soc.,
　　Ind.:—Trans., I.,* 11, 12, 46, 47; *VII.,* 57; *Journ., VI.,* 25, 37.
　Habitat.—A large tree, common on the lower hills from Bombay to
Cochin, also met with in the Nilghiri and Coorg mountains.

　Dye and Tan.—The BARK contains tannin, and the FRUIT is a my-　DYE & TAN.
robalan (*Liotard*). Both are said to be used for dyeing and tanning,
but litte information is available regarding them.　　　　　　　Bark.
　Medicine.—"The country-people use the JUICE of the fresh FLOWERS,　356
rubbed with *parwel* root (Cocculus villosus) as a remedy in cholera, and　Fruit.
in poisoning with opium. Four tolas of the juice, with an equal quantity　357
of guava bark juice, is given frequently. In parotitis, the juice with *ghi*　MEDICINE.
and *saindhav* (rock salt) is applied. In cholera about 4 tolas of the juice　Juice.
with an equal quantity of *parwel* root is given every hour" (*Dymock,*　358
Mat. Med. W. Ind.).　　　　　　　　　　　　　　　　　Flowers.
　Structure of the Wood.—Weight 57 to 65℔ per cubic foot, valuable,　359
though not quite so good as that of T. tomentosa. It is said to be　TIMBER.
improved by immersion in water, after which it becomes more durable.　360
It makes good planking, and in Ratnaghiri is used for making the
handles of ploughs (*Gamble; Beddome; Brandis*).

　Domestic.—This tree and T. tomentosa are said to be the principal　DOMESTIC.
fuel of the *ráb* or ash manuring used in agriculture throughout the Ratna-　Manure.
ghiri district.　　　　　　　　　　　　　　　　　　　　361

T. tomentosa, *Bedd.; Fl. Br. Ind., II.,* 447.
　Var. 1, typica,—T. TOMENTOSA, *W & A., Prodr.; Wight, Ic., t.* 195;
　　T. GLABRA, *var.* TOMENTOSA, *Dalz. & Gibs.;* T. ALATA, *Roth.;* T.
　　OVATA, *Herb. Rottler;* T. CHEBULA, *Retz.;* β, MINOR, *Heurck. & Muell.
　　Arg.;* PENTAPTERA TOMENTOSA, *Roxb.*
　Var. 2, crenulata,—T. CRENULATA, *Roth.,* PENTAPTERA CRENULATA,
　　Roxb.; P. MACROCARPA, *Wall.*

T. 361

| TERMINALIA tomentosa. | The Asna or Saj Tree — |

Var. 3, coriacea,=T. CORIACEA, *W. & A.*; PENTAPTERA CORIACEA, *Roxb.*

Vern.—*Saj, sein, ásan, ássain, ásna, sadri, sain, ain,* HIND.; *Piásál, piáshál, usán, asan, áshán,* BENG.; *Hatana, matnak',* KOL.; *Atnak', SANTAL; Amari,* ASSAM; *Jhan,* RAJBANSHI; *Taksor,* LEPCHA; *Saháju, kala saháju, sáj, ansun,* URIYA; *Barsaj, sáj, sadur, sája, sijra,* C. P.; *Maru,* GOND; *Madge,* BHÍL; *Ain, saddra,* BERAR; *Athna,* MELGHAT; *Ság, hág, sáder, saddr, sádri, hadri,* NIMAR, GUZERAT, and adjoining parts of MEWAR; *Sain, ásin, asain, sáj,* N.-W. P.; *Sain, ásun, arjan, ásan, sein, aisan,* PB.; *Karkaya, sadora, holda, dudia maddi, jangli-karanj,* DECCAN; *Ain, sádada, ásna, sag, sadri, marthi, kenjal,* BOMB.; *Ain, madat, yén, sádada, sáj,* MAR.; *Ain,* GUZ.; *Karra marda, karú marúthú, anemúi, kurruppu-maruta, marutai, karu- maradu,* TAM.; *Maddi, halla naddi, nello-madu, nalla-maddi,* TEL.; *Murada, kali maruthai,* ARCOT; *Karakaya, sadora, holda, dudi maddi,* HYDERABAD; *Matti, kari-matti, banapu, tore matti madi. aini,* KAN.; *Karu-maruta,* MALAY.; *Toukkyan, taukkyan, hpan-kha,* BURM.; *Chou- chong,* TALEING; *Kúmúk, kúmbúk,* SING.; *Asana,* SANS.

References.—*Roxb., Fl. Ind., Ed. C.B.C., 383; Voigt, Hort. Sub. Cal., 38; Brandis, For. Fl., 225; Kurz, For. Fl. Burm., I., 458; Beddome, Fl. Sylv., t. 17; Gamble, Man. Timb., 182; xx.; Cat., Trees, Shrubs, &c., Darjeeling, 39; Thwaites, En. Ceyl. Pl., 104; Trimen, Sys. Cat. Cey. Pl., 32; Dalz. & Gibs., Bomb. Fl., 91; Stewart, Pb. Pl., 88; Rev. A. Campbell, Rept. Econ. Pl. Chutia Nagpúr, No. 7550; Graham, Cat. Bomb. Pl., 96; Sir W. Elliot, Fl. Andhr., 125; Pharm. Ind., 69; Ainslie. Mat. Ind., II., 193; O'Shaughnessy, Beng. Dispens., 340; Irvine, Mat. Med. Patna, 118; Moodeen Sheriff, Supp. Pharm. Ind., 243; U. C. Dutt, Mat. Med. Hind., 292; S. Arjun, Cat. Bomb. Drugs, 53; Bidie, Cat. Raw Pr., Paris Exh., 29, 112; Dymock, Mat. Med. W. Ind., 2nd Ed., 322; Baden Powell, Pb. Pr., 587; Drury, U. Pl. Ind., 420; Atkinson, Him. Dist. (X., N.-W. P. Gaz.), 310, 815; Useful Pl. Bomb. (XXV., Bomb. Gaz.), 75, 393; Gums & Resinous Prod. (P. W. Dept. Rept.), 2, 5, 7, 8; Liotard, Dyes, 22, 36; App. ii.; Wardle, Dyes, 8, 16; Cooke, Gums & Resins, 27; McCann, Dyes & Tans, Beng., 136, 151-52, 154, 160, 161, 162, 165, 167, 168. 169; Geoghigan, Silk in India, 139; Watt, Selections, Records Govt. India (R. & A. Dept.), 1888-89, 87, 88, 93, 98; Man. Madras Adm., I., 313; Nicholson, Man., Coimbatore, 401; Moore, Man., Trichinopoly, 80; Gribble, Man., Cud- dapah, 14, 262; Aplin, Rept. on Shan States, 1887-88; Settlement Re- port:—Central Provinces, Nimár, 305; Upper Godávery, 36, 37; Raepore, 75,77; Seonee, 9; Gazetteers:—Bombay, VII., 31, 36, 39; XV., 33, 78; XI., 23; XVI., 18; XVII., 19; XVIII., 41, 48; XXII., 23; Panjáb, Ho- shiarpur, 11; N.-W. P., IV., lxxi.; Orissa, II., 158; Burma, I., 137; Mysore & Coorg, I., 48, 52; III., 7, 20; Agri.-Horti. Soc. Ind.:—Trans., VII.,57, 58; Jour., VI., 170; XI., 489; XIII., 311; New Series, II., 229; VI.,272, 273, 276; VII., 142, 277; Ind. Forester, I., 86, 88, 275; II., 19; III., 202, 366; IV., 227, 292, 322, 366; V., 93, 212, 497; VI., 101, 303; VIII., 103, 105, 106, 117, 118, 126, 127, 128, 131, 132, 271, 378, 388, 391, 414, 415, 416, 417, 438, 439, 440; IX., 215, 216, 519; X., 92, 222, 325, 326, 543, 545, 550, 551; XI., 355, XII., 19, 188, 258, 259, 260, xxii., 311, 313, 419, App., 13; XIII., 121, 127, 139; XIV., 147, 151, 159, 199, 390.*

Habitat.—A large, deciduous tree, which attains a height of 80 to 100 feet, common throughout the moister regions of India. In the Siwalik tract and outer Himálayan valleys it extends west as far as the Rávi, and in places ascends to 4,000 feet. In Western India its limit appears to be in the forests south-west of Neemuch, where several places (Sadri, Bara Sadri, Chota Sadri) seem to have derived their names from it. It is also found on the western edge of the Malwa table-land, east of the Bunass river (Bassi forests). East and south of these points it extends throughout Central, Eastern, Southern India, and Burma. The tree thrives best in heavy binding soil, flowers in April, and the fruits ripen in February to April of the en- suing year. It coppices fairly well and bears long-continued pollarding. It does not generally lose its leaves until February, March or April, but is one of the latest trees, in dry forests, to come out in fresh leaf. In Burma it attains a larger size than that above given for trees of the Western Peninsula, 80 feet to the first branch, and a girth of 12 feet, being the average size of full-grown trees on good soil (*Brandis*).

Gum.—Cooke writes, " The GUM from this tree sent to the Panjáb Exhibition from Madras is described as a red gum, black outside the pieces. A specimen in this collection sent from Berar in 1873 is in rounded dull brown tears, soft, readily becoming agglutinated, and capable of being sliced with a knife, having a bitter disagreeable taste, and partly soluble in water. A specimen from **Mr. Broughton** (Madras) is much darker in colour, being pitchy brown and commercially valueless." In the reports on gums and resinous products published by the Government of India, Public Works Department, a letter occurs from the Officiating Conservator of Forests, Madras, dated 1868, forwarding 10℔ of this along with other gums. He states that it is used as an incense and cosmetic, that about 2 maunds are available annually, and that it would cost R27 to 30 per maund at the coast. The statement as to its utilisation for incense and as a cosmetic is repeated in the *Gazetteer of Mysore and Coorg.* Mr. Campbell, in his recent *Notes on the Economic Products of Chutia Nagpur*, gives quite a different description of the gum to that above quoted from Cooke. " It yields copiously a transparent gum," he writes, "which exudes in large globular tears, sometimes almost colourless, but oftener of a brownish tinge it is eaten by the Santals."

The LAC insect is sometimes found on the branches.

Dye and Tan.—The BARK is used occasionally, but very rarely, as a dye-stuff, being broken up and boiled in water to extract the dye. The resulting colour is brown or buff. In the Midnapur district it is employed, along with the bark of *bakul* (**Mimusops Elengi**) to produce a reddish dye, used in colouring gunny bags. A mixture of the barks of *asan* and *porashi* (a doubtful name, may be **Thespesia populnea**) produces a very good red dye, said to be a favourite with Native tanners, who employ it to produce the colour of the red leather shoes so much worn by the people (*McCann*). In Kolaba (Bombay), the bark is used for dyeing fishing nets. In many localities it is employed, with iron salts or ferruginous mud, to obtain a black dye. A sample examined by **Mr. Wardle** was found to contain 16·7 per cent. of tannin, and to have a moderate amount of brownish-red colouring matter, quite sufficient to bring it into use if it could be obtained at a cheap rate. With salts of iron it gave a brownish-black colour. The chief use of the bark is, however, as a tan, for which it is largely employed all over India. It is either used alone to form the tanning liquor, or in combination with the barks of Shorea robusta, Terminalia Arjuna, Mimusops Elengi, Ficus religiosa, Acacia arabica, Ceriops Roxburghiana, Cassia Fistula, or Mangifera indica. It is also sometimes mixed with chebulic, belleric, or emblic myrobalans, with the pods of Cæsalpinia digyna, and with the leaves of Phyllanthus Emblica and Terminalia belerica. In certain localities of Bengal skins are dyed black by steeping in a preparation of water and the bark, together with that of Lagerstrœmia parviflora (*McCann*). The FRUIT, like those of most other species of Terminalia, is a myrobalan, but is very much inferior in tanning power to the belleric or chebulic myrobalans. It is consequently much more rarely, though occasionally, employed as a tannin agent. Specimens from the Colonial and Indian Exhibition, examined by **Dr. Paul** and **Professor Hummel** were found to contain, respectively, 5·97 and 4 per cent. of gallo-tannic acid. The money value, compared with ground myrobalans at 7s. 6d. per cwt., was estimated by the latter chemist at 1s.

GUM.
362

DYE & TAN.
Bark.
363

Fruit.
364

TERMINALIA tomentosa.	The Asna Tree—on which Tasar-Silk-worm feeds.

DYE & TAN.

$3\frac{1}{2}d.$, a valuation which effectually precludes the possibility of its competing with more valuable tanning materials.

It is probable, however, that the bark has a more hopeful future in trade. As already stated, Wardle found it to contain 16·7 per cent. of tannin. Attempts have recently been made to obtain an extract from the bark and from that of *sál* (**Shorea robusta**) by the Forest Department of the North-West Provinces. The following is the account of the results, by **Mr. B. A. Rebsch**, Assistant Conservator of Forests, Gonda Division, Oudh :—

"I begun the *Asna* bark-boiling at Sungarha on the 11th January 1886; and the figures given below are the results of the process from that date to the 31st March 1886 (*viz.*, 79 days) : 184 maunds of *Asna* bark were boiled, and yielded 42 *ghurras* of the extract. The weight of a *ghurra*-full of extract was found to be $13\frac{1}{4}$ seers, so that the total outturn of extract was 13 maunds $36\frac{1}{2}$ seers. The expenditure amounted to R77-5-5.

" When comparing the results obtained from the two kinds of bark boiled, ample allowance must be made for the fact that the *sál* bark used at Ramgarh was taken from young and suppressed poles, while the *Asna* bark was taken from mature and decaying trees. This will at once account for the fact that in the latter case a smaller quantity of extract was obtained per maund of bark than in the case of *sál*. The most striking point, however, is that *sál* bark extract is heavier than an equal quantity of *Asna* bark extract; one *ghurra*-full of *sál* bark extract weighing 14 seers, while the same quantity of *Asna* bark extract weighed only $13\frac{1}{4}$ seers. The boiling has been continued both at Ramgarh and Sungarha; but I am not able at present to give the results, as the details are incomplete. But an extremely interesting fact has been observed, which is worth recording—namely, that after the 31st March the amount of *sál* extract obtained per maund of bark has increased, while in the case of *Asna* it has become less" (*Sel. from Rec. Govt. of Ind. (R. & A. Dept) l.c.*).

The *sál* bark extract obtained at the same time was analysed with very satisfactory results, but the *Asna* extract does not appear to have been chemically examined. It must, however, contain a large percentage of tannin matter, and from the extent to which it is utilised in India the tannin matter must, in all probabiliy, be suitable for tanning leather, and be of some commercial value. It is to be hoped that this question may speedily be settled. **Professor Hummel** has highly recommended the preparation of such extracts as the only means of bringing the valuable Indian tans obtained from barks and leaves into the market profitably. Further, **Captain Wood**, Conservator of Forests, Oudh Circle, has drawn the attention of the Government of India to the waste which is taking place and is likely to extend. He wrote :—" The utilisation of the bark of our *sál* and *asaina* coppice, which will now be cut over by the thousand acres annually owing to the entrance of railways into the forests, is, I consider, a point of vital importance to the interests of the Forest Department."

Galls.
365

The Rev. A. **Campbell**, Chutia Nagpur suggested that the GALLS often found in the calyx of the flower-buds, might prove useful as a tan, and he stated that they could be obtained in unlimited quantity. The examination to which they were subjected in connection with the Colonial and Indian Exhibition showed that they were valueless as a tan.

MEDICINE.
Bark.
366

Medicine.—The BARK is noticed in the secondary list of the *Pharmacopœia of India*, where it is said to have been favourably reported on by Dr. Æ. **Ross**, as an internal remedy, in the form of decoction, for atonic diarrhœa, &c., and locally as an application to callous ulcers. **Hunter** mentions it as used as a stimulant (? external) in Cuttack, and **Campbell** states that it is employed medicinally in Chutia Nagpur, but it does not

T. 366

| The Wall Germander. | (*J. Murray*) | TEUCRIUM Chamædrys. |

appear to be well known to, or much valued by, the Natives of India gene rally. Dymock states definitely that "it is not often used medicinally in Western India," and the authors of the *Pharmacographia Indica* have not alluded to it.

Food and Fodder.—The ASHES of the BARK are largely eaten by Natives as a substitute for lime with betel-leaf or *pán*. The LEAVES are lopped for cattle-fodder in the North-West Provinces and Oudh; the common *tasar* silkworm feeds on them. (*Brandis*). (**Conf.** with Silk Vol. II., Pt. III.)

Structure of the Wood.—Sapwood reddish-white, heartwood dark-brown, hard, beautifully variegated with streaks of darker colour, shewing on a radial section as dark streaks, generally wavy or undulating. It seasons well and takes a good polish. The weight has been given by various writers at from 50 to 71℔ per cubic foot, the value of P. from 675 to 1230 (the strongest being from Burma). The durability of the timber is uncertain; in Burma the heartwood decays rapidly, in Northern India beams are sometimes found to last well, at other times to perish from dry rot or the attacks of insects. It is largely used for house-building, furniture, carts, shafts and wheels, agricultural implements, ship and boat-building, and for making rice-pounders. It has also been tried for railway sleepers with fairly good results. Five sleepers laid down on the Oudh and Rohilkhand Railway in 1870 were reported in 1875 to be in a state of good preservation, but having been cut from small trees the sapwood had been eaten to a certain extent. Unless thoroughly seasoned it is very apt to split. It is an excellent fuel and makes good charcoal (*Gamble; Brandis*). When nicely polished it resembles walnut, and has been found one of the best woods for making stethoscopes at the Government Medical Store Depôt in Bombay (*Dymock*).

Domestic.—The BARK is said by Lisboa to be used by the *Bhils* for poisoning fish.

Terra japonica, see Uncaria Gambier *Roxb.*, below, p. 210.

TETRAMELES, *R. Br.; Gen. Pl., I., 845.*
[*Ic., t. 1956;* DATISCACEÆ.

Tetrameles nudiflora, *R. Br.; Fl. Br. Ind., II., 657; Wight,*

Syn.—T. GRAHAMIANA, *Wight;* T. RUFINERVIS, *Miq.;* ANICTOCLEA GRAHAMIANA, *Nimmo.*

Vern.—*Sandugasa,* BENG.; *Payomko,* LEPCHA; *Bolong,* GARO; *Mainakat,* NEPAL; *Jungli-bendi,* BOMB.; *Bolur, jermála,* KAN.; *Ugáda,* MAR.; *Thitpouk,* BURM.; *Tseikpoban,* MAGH.

References.—*Brandis, For. Fl.* 245; *Kurz, For. Fl. Burm., I.,* 535; *Beddome, Fl. Sylv., t.* 212; *Gamble, Man., Timb.,* 208; *List of Trees, etc., of Darjeeing,* 43; *Grah., Cat. Bomb. Pl.,* 252; *Lisboa, U. Pl. Bomb.,* 82; *Gazetteer, Bombay, XV.,* 78; *Ind. Forester, IX.,* 377.

Habitat.—A large, deciduous tree, which attains a height of 100 to 150 feet, found in Sikkim at 2,000 feet altitude, the western Gháts, from Bombay to Ceylon, Burma, Tenasserim, and the Andamans.

Structure of the Wood.—White, very light, soft. It may be found useful for tea-boxes (*Gamble*).

Tetranthera, *Jacq.;* see Litsea, *Lamk.,* LAURIUEÆ.; Vol. V., 81-85.

TEUCRIUM, *Linn.; Gen. Pl., II., 1221.*

Teucrium Chamædrys, *Linn.; DC., Prodr. XII.,* 587; LABIATÆ. THE WALL GERMANDER.

References.—*O'Shaughnessy, Beng. Dispens.,* 488; *Irvine, Mat. Med. Patna,* 112; *Birdwood, Bomb. Prod.,* 63; *S. Arjun, Bomb. Drugs,* 181.

MEDICINE.

FOOD. Ashs. 367 Bark. 368 FODDER. Leaves. 369 TIMBER. 370

DOMESTIC. Bark. 371

372

TIMBER. 373

374

T. 374

THALICTRUM
foliolosum.   Mamira—a powerful Febrifuge.

MEDICINE.
375

Habitat.—A native of Europe and certain parts of Asia, imported into India for medicinal purposes.

Medicine.—This is one of the ingredients of the celebrated *Triak jarúk* of the bazárs, which is the representative of the *Mithridatum, Theriaca Andromachi,* or *T. Damocratis* of the ancients Originally it consisted of but a few drugs, now it is said to contain as many as sixty-one, including opium. It is in fact, an aromatic opiate, a drachm of which is equal to one grain of opium (*Birdwood*). The little canisters found in the bazárs are said by Waring to be wrapped in paper on which is printed in Persian, "The *Theriakh of Andromachi,* an invention of Theron the Presbyter. It is prepared, measured, and made public by one John Baptist Sylvesticus in the Rialto, by authority of the excellent Government Physicians of ancient Righteousness, and of the Council of Apothecaries and learned Physicians, etc."

376

THALICTRUM, *Linn.; Gen. Pl., I., 4.*

Thalictrum foliolosum, *DC.; Fl. Br. Ind., I., 14 ;* RANUNCULACEÆ

Vern.—*Pinjari, shuprak* (root = *pili-jari,*) HIND. ; *Pila-jari, pengla, jari, barmat,* root = *mamira,* KUMAON ; *Gúrbiáni, pashmaran phalijari, chitra múl, keraita, chera, ? chireta, chitra,* (root = *pilijari mamira,* PB. ; *Chaitra,* KASHMIR ; *Mamiran,* BOMB.

References.—*Stewart, Pb. Pl., 5 ; Pharm., Ind., 5 ; O'Shaughnessy, Beng. Dispens., 160 ; Flück, & Hanb., Pharmacog., 5 ; Dymock, Mat. Med. W. Ind., 2nd Ed., 20 ; Dymock, Warden & Hooper, Pharmacog. Ind., I., 33 ; Baden Powell, Pb. Pr., 324 ; Atkinson, Him., Dist., 751 ; Drury, U. Pl , 421 ; Gazetteers:— Mysore & Coorg, I., 57 ; Simla, 12 ; Agri.-Horti. Soc., Ind., Trans., VII., Journ. (Old Series), XIII , 389..*

Habitat.—An erect, rigid perennial herb, found in the Temperate Himálaya from 5,000 to 8,000 feet, and in the Khasia Hills between 4,000 and 6,000 feet.

MEDICINE.
Root.
377

Medicine.—Two centuries ago Bernier mentioned "*mamiron,* a little ROOT good for the eyes, as being brought (along with rhubarb, musk, and the wood of China) from Cathay to Kashmír by a long journey" in which *jhúlas* are described as being crossed (*Stewart*). This *mamiron* is doubtless the root of the species under consideration, which is largely used as an *anjan* or application for ophthalmia in Afghánistán and throughout India to this day. (*Conf. with* Coptis Teeta *Vol. II., 521—526*). It is also considered a valuable antiperiodic and tonic in Native Materia Medica, and in the Panjáb is believed, in addition, to be purgative and diuretic (*Baden Powell*). Some fifty years ago specimens received from the Botanic Gardens, Saharunpur, were examined by Sir W. O'Shaughnessy, who describes his experiments with it as follows : "The bruised root having been given to large dogs in the quantity of 10 grs. to ʒii no particular effects were observed. It has been used in the Hospital of the Medical College in several cases of ague, and as a tonic in convalescence from acute diseases. Five grains of the powder or two grains of the watery extract, given thrice daily, have in some cases prevented, and in several moderated, the accession of fever, and at the same time acted gently on the bowels. The only sensation experienced was warmth at the epigastrium, and a general comfortable feeling." "It deserves extensive trial, and promises to succeed well as a febrifuge of some power, and a tonic aperient of peculiar value. Dose of the powder 5 to 10 grs. as a tonic and aperient, in the interval of intermittent fevers, and in convalescence from acute diseases."

On the compilation of the *Pharmacopœia of India* the plant was included in the secondary list, but since that time has attracted little atten-

T. 377

Cacao Butter.	(*J. Murray.*)	THEOBROMA Cacao.

tion. The authors of the *Pharmacographia Indica*, however, inform us that the root has recently been used with very satisfactory results in Bombay, as a remedy for atonic dyspepsia accompanied with slight fever. The remedy appears worthy of renewed investigation, and though supplies are generally obtainable in the shops under the above vernacular names, the fresh root would probably give more satisfactory results. This may be obtained if ordered from Mussorie, through the Superintendent of the Government Gardens. The ordinary bazár supply is chiefly exported from the Panjáb Himálaya and Kumáon.

CHEMICAL COMPOSITION.—" Thalectrum root contains a large quantity of *berberine*, so combined as to be readily soluble in water" (*Pharmacog Ind.*).

THAMNOCALAMUS, *Munro; Gen. Pl., III.*
[GRAMINEÆ.

Thamnocalamus spathiflorus, *Munro; Brandis, For. Fl., 563;*
Vern.—*Ringall,* JAUNSAR; *Purmiok,* LEPCHA; *Myoosay,* BHUTIA.
References.—*Brandis, For. Fl., 563; Gamble, Man. Ind. Timb., 427; List of Trees, etc., of Darjeeling, 87; Atkinson, Him. Dist., 320; Ind. Forester, III., 45; VII., 258; IX., 197, 198.*

Habitat.—The common small bamboo of Hattu and Deoban, found on the Himálaya generally from the Sutlej to Bhutan, above 8,000 feet.

Fibre.—It yields a fibre, of which little is known, but which might probably be useful for paper-making.

THEA.

Thea assamica, *Masters;* **T. sinensis,** *Linn.;* **T. bohea,** and **T. viridis,** see **Camellia theifera,** *Griff.,* TERNSTRŒMIACEÆ; Vol. II., 70; also the article TEA, Vol. VI., Pt. III.

THEOBROMA, *Linn.; Gen. Pl., I., 225.*

Theobroma Cacao, *Linn.;* STERCULIACEÆ.
References.—*DC., Orig. Cult. Pl., 313; Gamble, Man. Timb., 45; Drury, U. Pl., Ind., 424; Mason, Burma & Its People, 455, 754; Pharm Ind., 36; O'Shaughnessy, Beng. Dispens.,1227; Flück & Hanb., Pharmacog., 95; Ainslie, Mat. Ind., I., 47; Drury, U. Pl., 424; Christy, New Com. Pl., 11; Gazetteers:—Mysore & Coorg, I., 143; III., 48; Bomb., IV., 22; Agri.-Horti. Soc., Ind.:—Trans., III., 39; IV.,\Pro., 56; VI., 127; VII., 81, 83; Journ., (Old Series), II., 208, 397, 443, 591; IV., 140; VIII., Pro., 48; IX., 202; (New Series), I., Sel., 83; VI., Sel., 71, Pro., 25; (1879), 31; VII., 20; Sel., 29-40; Ind. Forester., I., 155; V., 393.*

Habitat.—A small tree, wild in the forests of the Amazon and Orinoco basins and of their tributaries up to 400 feet of elevation. It is also said to grow wild in Trinidad, to be naturalised by cultivation in many parts of South America and the West Indies. It has long been introduced into India, and is now cultivated in the Southern Presidency and Ceylon. The trees are raised from seed and come into full bearing when five or six years old.

Oil.—A light yellowish, opaque, solid oil, known as "Cacao butter," is prepared for use in pharmacy, by pressing the warmed SEEDS. These, when shelled, yield from 45 to 50 per cent. of oil. Cacao butter is dry at ordinary temperatures, and though unctuous to touch is brittle enough to break into fragments when struck, exhibiting a dull waxy fracture. It has the pleasant odour of chocolate, melts in the mouth with a bland agreeable taste, has a specific gravity of 0·961 and fuses at 20° to 30° C. (*Pharmacog*).

CHEMICAL COMPOSITION.—This fat consists of several substances which, by saponification, furnish glycerin and fatty acids. The chief in-

MEDICINE.

Chemistry.
378

379

FIBRE.
380

381

382

OIL.
Seeds.
383

CHEMISTRY.
384

T. 384

**THEOBROMA
Cacao.**   The Cocoa Nibs of Commerce.

CHEMISTRY.

gredients are *stearin, palmitin,* and another compound of glycerin containing probably an acid of the same series richer in carbon—perhaps *arachic acid,* $C_{20} H_{40} O_2$ or *theobromic acid,* $C_{64} H_{128} O_2$. A small quantity of *oleic acid* is also present.

MEDICINE.
Cacao-Butter.
385

Medicine.—CACAO-BUTTER possesses the valuable property of not becoming rancid from exposure, and it was introduced into European medicine, chiefly owing to this quality, for pharmaceutical purposes. It is official in the Pharmacopæias of India, and of the United Kingdom, and is now chiefly employed in the manufacture of suppositories, medicated pessaries, etc.

FOOD.
Seeds.
386

Food.—The tree bears a pod-like fruit, 6 to 10 inches long, and 3 to 5 in girth, which contains fifty or more SEEDS. These seeds dried and ground form the cocoa nibs of commerce, from which cocoa extracts and chocolate are prepared. These seeds were first brought to the notice of Europeans in 1513-1523, by the Spanish invaders of America, who found them current among the Tucatan, instead of money. Their value as a food-product was described by Benzoni about 1550 (*Pharmacographia*). The first notice of their having been brought to England occurs in 1659, from which date the popularity of chocolate as a beverage and confection has gradually increased, till, in 1880, the quantity entered for consumption in Great Britain amounted to over 10,000,000℔. As already stated, the tree has been introduced into India and is now cultivated to some extent. On the Malabar coast it is grown by the Roman Catholic missionaries, who make small quantities of cocoa regularly, for their own use and for local sale to Europeans. In Ceylon the cultivation has acquired considerable proportions, and the produce is said to be highly valued in the home-market.

The following account of the method of preparing the fruit as pursued in that Island may be quoted in full from the *Tropical Agriculturist :—*

"A coolie picks two bushels of cocoa beans per diem, and as five bushels wet are equal to 1 cwt. dry, the cost is only about 87 cents per cwt. for picking, as compared with R2 to R3 for Liberian-coffee. The pods are first cut from the tree, a small piece of stem being left on the tree; the coolie takes one in each hand and with a knock breaks them both in halves, and then with one draw of his fingers, dexterously strips all the beans off the centre pulp. The pods are then thrown down round the trees and act as manure, while the beans are removed to the fermenting cisterns. It takes from five to nine days to properly ferment the cocoa, and it is then ready for working. It is trampled first, as in coffee, with the feet, and then removed in baskets and carefully hand-washed, as washing with the *matapalagi* damages the beans. I have no doubt that ere long some means less expensive will be found for washing, and the "Clerikew" will be much improved on too.

"After washing" the cocoa is laid on mats to dry, as coffee is, if the weather be suitable; and at times it is advisable to give it a rub-over with small pieces of sack or cloth, which improves the appearance of the beans, and facilitates drying in showery weather. The difference in well-cured and badly-cured cocoa amounts to at least R20 per cwt, and the prices obtained for it, as in tea, will depend in a much greater measure on the careful attention of the curing than in the case of coffee."

A tree in full bearing is said to yield about 150℔ of seed annually. It favours hot, moist climates, and the young plants must be shaded and well watered. There would appear to be no reason why it should not be cultivated very successfully in the hotter and moister parts of Southern India. The dried seeds, roasted and ground, constitute the cocoa of commerce; chocolate is prepared by still finer grinding so as to

make a perfectly fine powder, after which it is flavoured (generally with vanilla) and moulded while hot. The nutritive and sustaining powers of cocoa and chocolate are too well known to require further detail.	**FOOD.**

THESPESIA, *Carr.; Gen. Pl., I., 208.*
[*Ic., t. 5*; MALVACEÆ.

Thespesia Lampas, *Dalz. & Gibs.; Fl. Br. Ind., I., 345; Wight,*

> **Syn.**—HIBISCUS LAMPAS, *Cav.*; H. TETRALOCULARIS & GANGETICUS, *Roxb.*; PARITIUM GANGETICUM, *Don.*
>
> **Vern.**—*Bankapas,* BENG.; *Bon kapsi,* SANTAL; *Bonkapash,* ASSAM; *Rán-bhendi,* MAHR.; *Adavi pratti, conda patti, rondapatti,* TEL.
>
> **References.**—*Kurz, For. Fl. Burm., I., 128; Roxb., Fl. Ind., Ed. C.B.C., 524; Thwaites, En. Cey. Pl., 26; Brandis, For. Fl., 28, 572; Rev. A. Campbell, Rept. Econ. Pl., Chutia Nagpur, No. 7563; Elliot, Fl. Andhr., 12; Atkinson, Him. Dist., 306; Gazetteers :—Mysore & Coorg, I., 57; Bomb., XI., 24; XV., 428; N.-W. P., IV., lxviii.; Ind. Forester, XIV., 297.*
>
> **Habitat.**—A small bush, common in the tropical jungles of India, Burma, and Ceylon, from Kumáon eastwards, ascending to 3,000 feet in Nepál.
>
> **Fibre.**—The young TWIGS yield a good fibre, used for binding loads of wood, etc.
>
> **Medicine.**—The ROOT and FRUIT are said by Mr. Campbell to be employed in Chutia Nagpur as a remedy in gonorrhœa and syphilis.
>
> **Structure of the Wood.**—Tough and pliant; weight 29℔ per cubic foot. It is said to be much used in certain parts of Bombay for making drum and other round frames, for which purpose it is planed, soaked in hot water, stained, and bent to the required shape (*Gamble*).

T. populnea, *Corr.; Fl. Br. Ind., I., 345; Wight, Ic., t. 8.*
> THE PORTIA TREE : the UMBRELLA TREE : or TULIP TREE of
> [Indian writers.

> **Syn.**—HIBISCUS POPULNEUS, *Linn.*; H. POPULNEOIDES, *Roxb.*; MALVAVISCUS POPULNEUS, *Gærtn.*
>
> **Vern.**—*Parsipu, pipal, páras-pipal, porush, bhendi, gajahanda,* HIND.; *Pares pipal, pálas pipal, prash, pórash,* BENG.; *Páras pipal* (corrupted into *pahári pipal*), PB; *Ranbhendi,* C. P.; *Bhendi, bhindi, pálas piplo, parsipú,* BOMB.; *Bendi, bhendi, rán-bhendi, parsacha-jháda,* MAR.; *Bendi, bhindi, párasa-piplo,* GUZ.; *Paris, páras-pippal,* DEC.; *Púrasha, puvarasam, púvarasu, purasa, púurusú, pursa, pursung, poris, portia,* IAM.; *Gangarénu, gangarávi, ganguraya, gangirana, muniganga rávi,* TEL.; *Asha, hurvashi, huvarase, kandarola, adavi bende,* KAN.; *Púvvarasha,* MALAY; *Sureya, súriya-gaha, gansuri-gahá,* SING.; *Gardha-bhánda, párisa, súparsha-vaka,* SANS.
>
> **References.**—*Roxb., Fl. Ind., Ed. C.B.C., 522; Voigt, Hort. Sub. Cal., 120; Brandis, For. Fl., 572; Kurz, For. Fl. Burm., I., 128; Beddome, Fl. Sylv., t. 63; Gamble, Man. Timb., 43; Thwaites, En. Ceylon Pl., 27; Dalz. & Gibs., Bomb. Fl., 18; Stewart, Pb. Pl., 24; Graham, Cat. Bomb. Pl., 15; Sir W. Elliot, Fl. Andhr., 57, 119; Rheede, Hort. Mal., I., 29; Pharm. Ind., 35; Ainslie, Mat. Ind., II., 333; O'Shaughnessy, Beng. Dispens., 218; Moodeen Sheriff, Supp. Pharm. Ind., 244; U. C. Dutt, Mat. Med. Hind., 298, 312; Dymock, Mat. Med. W. Ind., 2nd Ed., 105; Murray, Pl. & Drugs, Sind, 64; Irvine, Mat. Med., Patna, 85; Dymock, Warden & Hooper, Pharmacog. Ind., I., 213; Drury, U. Pl., 425; Lisboa, U. Pl. Bomb., 15, 226, 260, 400; Birdwood, Bomb. Pr., 276, 323; McCann, Dyes & Tans, Beng., 154; Liotard, Dyes, App. i.; Wardle, Dye Rep., 4, 22; Cooke, Oils & Oilseeds, 80; Boswell, Man. Nellore, 97, 123; Moore, Man., Trichinopoly, 80; Settle. Rep., Chánda, App. VI.; Gazetteers :—Mysore & Coorg, I., 53, 58; Bombay, V., 24, 285; III. 23; XV., 78; Burma, I., 139; Agri. Horti. Soc., Ind.; Journal (Old Series), IX., 406; Ind. Forester, III., 200; VI., 238, 321; X., 29; XI., 328.*

FIBRE.
Twigs.
388
MEDICINE.
Root.
389
Fruit.
390
TIMBER.
391
392

387

GUM.
393

DYE.
Capsules.
394
Flowers.
395
Bark.
396

FIBRE.
Bark.
397
OIL.
398

MEDICINE.
Heart-Wood.
399
Juice.
400
Fruit.
401
Bark.
402

Seeds.
403
Flowers.
404
Ro ots.
405
Leaves.
406
Chemistry.
407

Habitat.—A moderate-sized evergreen tree, found in the Coast forests of India, Burma, the Andaman Islands, and Ceylon; largely cultivated along roadsides, especially in Madras.

Gum.—It is said to yield a GUM, which was sent from Madras to the Panjáb Exhibition, but which may probably have been the yellow milk of the capsules, dried.

Dye.—The CAPSULES and the FLOWERS are said to give a yellow dye, which is apparently little used. Liotard states that the former are not articles of ordinary traffic, and that nothing is known of the process of dyeing with them. McCann states that the BARK of a tree called *páras* is employed in Mánbhúm with the bark of **Terminalia tomentosa** to produce a favourite red dye. He suggests that this may be the bark of **Thespesia populnea** since it cannot be that of the other *páras*, **Butea frondosa.**

The dried capsules and calyces were found by Mr. Wardle to contain a small amount of yellow colouring matter soluble in water, and capable of producing, by the aid of suitable processes, artistic though somewhat faint shades of brownish-yellow and light brown, on tasar and mulberry silk and wool. "This," he remarks, "would be a useful dye-stuff, but the fact of its containing so small an amount of colouring matter would be rather against it."

Fibre.—The BARK yields a strong fibre, rarely employed in India except in the rough state, for tying bundles of wood, etc. In Burma it is said to be used for cordage (*Gazetteer, I., 139*). It is said to be used in Demerara for making coffee bags.

Oil.—It yields a deep, red-coloured, and somewhat thick OIL—*huile amore*—the value of which is as yet unknown to the Natives, but which might be employed medicinally in cutaneous affections. Its expense precludes its use for other purposes.

Medicine.—Rumphius speaks highly of the value of the HEART-WOOD as a remedy for bilious attacks and colic, and in a kind of pleurodynia from which the Malays often suffer. Ainslie states that the yellow JUICE of the FRUIT is employed as an external application in various cutaneous affections, particularly in 'Malabar itch,' that a decoction of the BARK is used as a wash in the same complaints, and that the same preparation is given internally by the *Vytians* as an alterative, in doses of three or four ounces twice daily. Waring included the plant in the secondary list of the *Pharmacopœia of India*, where he states that he had made several trials with the juice of the fruit. In some cases it exercised a favourable influence, but in the majority it was productive of little or no benefit. Irvine remarks that in Patna the SEEDS are "used in horse-medicines and in purges;" in the *Report on the Settlement of the Chanda District* it is stated that the ROOT is taken as a tonic; the FLOWERS are said to be employed in the Konkan in the cure of itch, and **Dymock** informs us that the LEAVES are employed as a local application to inflamed and swollen joints.

CHEMICAL COMPOSITION.—The heart-wood, recommended by Rumphius, and apparently neglected by all writers since his time, has been examined by the authors of the *Pharmacographia Indica*. These chemists find that it contains a garnet-red resin which can be easily separated by digesting the wood in diluted alkali and using hydrochloric acid to precipitate it from the filtered solution. It is insoluble in water, but perfectly soluble in alcohol, chloroform, and the alkalies.

SPECIAL OPINIONS.—§ "The fresh leaves smeared with some bland oil, and applied hot over inflamed parts, form a soothing and valuable substitute for ordinary poultices" (*Surgeon-Major E. H. Levinge, Rajahmundry, Madras*). "The juice of the fruit is used as an application

| The Yellow Oleander. | (*J. Murray.*) | **THEVETIA**
neriifolia. |

for ring-worm; the leaves, heated and smeared with warm oil, make an excellent poultice. I used them largely at Bellary during the famine, with great success, as an application to the sores and abscesses caused by guinea-worm " (*Surgeon-Major Lionel Beech, Coconada*).

MEDICINE.

Structure of the Wood.—Sapwood soft, pale reddish to brown, with small dark-coloured, hard heart-wood; weight 50℔ per cubic foot. It is strong, even-grained, and durable, and is used in South India for making gun-stocks, carts, carriages, and furniture, in Burma for carts, wheel-spokes, furniture, and purposes of carpentery generally. It is said to have been much utilised at one time by the Ordnance Department for gun-carriages.

TIMBER.
408

Domestic and Sacred.—The Tulip Tree is largely planted in Southern India and Bengal in gardens and along road-sides to give shade. The LEAVES are employed by Hindus in the religious ceremonies attending death.

DOMESTIC &
SACRED.
Leaves.
409

THEVETIA, *Linn.; Gen. Pl., II., 699.*

Thevetia neriifolia, *Juss.; Kurz, For. Fl. Burm., II., 168;*

410

THE EXILE or YELLOW OLEANDER. [APOCYNACEÆ.

Syn.— CERBERA THEVETIA, *Linn.*

Vern.—*Zard kunél, pílá kanér,* HIND.; *Kolkaphul,* BENG.; *Berenjo,* SANTAL; *Pílá kanér, pile-phúl-ka-kanér,* DEC.; *Pila kaner, zard kunel, pivala kaner,* BOMB.; *Pivalakanhera,* MAR.; *Pilokanera,* GUZ.; *Pach-ch-ai-alari, tiruvách-chip-pú,* TAM.; *Pach-cha-gannéru,* TEL.; *Pach-cha-arali,* MALAY.; *Hpa-young-ban, molami-yái-pán,* BURM.

NOTE.—The vernacular names given to this plant, in most languages of India' mean "the yellow **Nerium odorum**" *Ed.*

References.—*Voigt, Hort. Sub. Cal., 531; Gamble, Man. Timb., 261, 265; Dalz. & Gibs., Bomb. Fl., Suppl., 53; Campbell, Rep. on Econ. Pl., Chutia Nagpur, No. 9458; Pharm. Ind., 138; Moodeen Sheriff, Supp. Pharm. Ind., 244; S. Arjun, Bomb. Drugs, 192; Dymock, Mat. Med. W. Ind., 2nd Ed., 503; Year-Book Pharm., 1878, 289; Cat. Baroda Durbar, Col. & Ind. Exhib., No. 87; Birdwood, Bomb. Pr., 54; Drury, U. Pl., 426; Lisboa, U. Pl. Bomb., 99, 266; Gazetteers :—Mysore & Coorg, I., 62; N.-W. P., I., 82; IV., lxxiv.; Ind. Agriculturist, Feb. 2, 1889.*

Habitat.—An introduced bush, native of America and the West Indies almost naturalised in Bengal, and common everywhere, scarcely a garden in the plains of India being without a few bushes, if not a hedge, of this plant.

Oil.—A bright yellow oil may be obtained from the SEEDS. It burns well without giving off much smoke, is of medicinal value, and from Dr. Warden's experiments would appear, if carefully prepared, to be not only inert, but wholesome. De Vry obtained 35·5 to 41 per cent. of this oil by expression, and 57 per cent. with benzol. The oil was found to be limpid, almost colourless, had an agreeable mild taste like that of almond oil; its density at 25°C. was 0·9148, at 15°C. it became pasty, and at 13°C. entirely solid. Oudemans found it to consist of 63 per cent. triolein, and 37 per cent. tripalmitin and tristearin. After expression of the oil De Vry obtained from the cake about 4 per cent. of a beautiful crystallised white glucoside, to which he gave the name of *thevetina*; he obtained the same substance in the bark also. Dr. Warden of Calcutta has described a blue colouring principle in the seeds, which he attributes to the action of hydrochloric acid upon pseudoindican (*Dymock*).

OIL.
Seeds.
411

Medicine.—The milky JUICE of the tree is highly poisonous. Its bitter and cathartic BARK is said to be a powerful febrifuge, the antiperiodic properties of which, first noticed by M. Descourtilz, have been confirmed

MEDICINE.
Juice.
412
Bark.
413

THYSANOLÆNA acarifera.	Thyme—mishk-i-taramashia

MEDICINE.
Kernels.
414

by Dr. G. Bidie and Dr. J. Shortt. It was tried in the form of a tincture in various kinds of intermittent fever, with highly satisfactory results. In large doses it acts as an acrid purgative and emetic, and in still larger doses as a powerful poison. The KERNELS are very bitter, and when chewed produce a slight feeling of numbness and heat in the tongue. The OIL extracted from them is said to be emetic and purgative, indeed, according to Dr. Shortt, it produces violent vomiting and hypercatharsis (*Pharm. Ind*). As already stated, however, Dr. Warden found the pure oil to be inert. The kernel is a powerful acro-narcotic poison, its property residing in a highly toxic principle (*thevetine*), which has been separated by Dr. Warden (*Conf.* para. on Oil, *above*). A case of poisoning by one of these kernels is recorded by Dr. J. Balfour (*Madras Jour. of Lit. and Science, 1857,* Vol. *III., N. S., 140*). "Recovery ensued ; but, from the symptoms detailed, they belong evidently to the class of acro-narcotic poisons. In all trials with this remedy, much caution is necessary." (*Pharm. Ind.*)

Thistle, see **Carduus nutans,** *Linn.* ; COMPOSITÆ ; Vol. II., 156.

Thítsí, see **Melanorrhœa usitata,** *Wall.* ; Vol. V., 208.

Thorn-Apple, see **Datura Stramonium,** *Linn. ;* SOLANACEÆ ; Vol. III., 40.

415

THYMUS, *Linn. ; Gen. Pl., II., 1186.*

A genus which contains about fifty species, natives of North Temperate regions. Of these only one is indigenous in India. A small dried Thyme of undetermined species is imported as a drug into Bombay from Persia. It is known as *mishk-i-taramashia, fahlin,* and *rang,* has a pleasant odour like peppermint but sweeter, and is stimulant and carminative in properties (*Dymock, Mat. Med. W. Ind., 2nd Ed., 613*). [This, I very much doubt being a species of **Thymus.** *Ed. Dict. Econ. Prod.*]. The medicinal oil of T. **vulgaris,** *Linn.* (*Bentley & Trimen, Med. Pl., t. 205*), is employed in European practice in India, but is not known to the Natives.

416

Thymus serpyllum, *Linn. ; Fl. Br. Ind., IV., 649 ;* LABIATÆ.

Syn.—T. LINEARIS, *Benth.*

Vern.—*Masho, rán gsbúr, marizha, shakei, kalandar satar,* PB. ; *Banajwáin,* N.-W. P.

References.—*Stewart, Pb. Pl., 173 ; O'Shaughnessy, Beng. Dispens., 491 ; Year-Book Pharm., 1874, 628 ; Agri.-Horti. Soc. Ind. :—Trans., III., 199 ; Journ. (Old Series), IV., Sel., 119.*

Habitat.—A small, aromatic shrub, common in the Western Temperate Himálaya from Kashmír to Kumáon, from 5,000 to 13,000 feet, and in Western Tibet, between 10,000 and 15,000 feet.

MEDICINE.
Seeds.
417
FOOD.
418
Leaves.
419
Twigs.
420

Medicine.—"On the Chenáb the SEEDS are given as a warm medicine, and Honigberger states that the plant is officinal in diseases of the eyes and stomach" (*Stewart*).

Food.—The LEAVES and TWIGS are employed as a flavouring agent in Kumáon (*Atkinson*).

THYSANOLÆNA, *Nees ; Gen. Pl., III., 1120.*

[*21 ;* GRAMINEÆ.

Thysanolæna acarifera, *Nees ; Duthie, Fodder Grasses, N. India,*

Syn.—AGROSTIS MAXIMA, *Roxb.*

Vern.—*Karsar,* SANTAL.

References.—*Roxb., Fl. Ind., Ed. C.B.C., 107 ; Rev. A. Campbell, Ec. P. Chutia Nagpur, No. 8178 ; Ind. Forester, XI., 233.*

T. 420

Products of India. **49**

The Tiger Grass; The Cat Tribe. (*J. Murray.*) | **TIGERS, CATS, Civets.**

Habitat.—A tall, handsome grass, with minute spikelets, native of Tropical Asia. It is not uncommon on the plains, and at low elevations on the hills, generally occurring in the vicinity of water.

Medicine.—A decoction of the ROOT is used in Chutia Nagpur as a mouthwash during fever (*Campbell*).

MEDICINE.
Root.
421

[GINEÆ; Vol. IV., 214·

Tiaridium indicum, *Lehm.*, see Heliotropium indicum, *Linn.*; BORA-

[317.

Tiger Grass, see Nannrorhops Ritchieana, *H. Wen il.*; PALMÆ; Vol. V.,

[In different parts of India various plants bear this name. By sportsmen, the one above all others that might be so designated would very probably be the spear-grass—Heteropogon contortus, see Vol. IV., 227.—*Ed., Dict. Econ. Prod.*]

TIGERS, CATS, AND CIVETS.

422

Tigers and Cats belong to the Family FELIDÆ, the most typical and highly specialised group of Carnivora; the Civets belong to the nearly related family VIVERRA. The Cat family comprises many species, and is largely represented in the fauna of India. All its members are closely allied and resemble each other in all details of structure. The whole organism is peculiarly adapted for capturing and killing other animals for food, the armature of teeth and claws, the power of speed for a short distance, the excessive muscular development and activity, all combine to enable the feline to seize and kill its prey, in many cases superior in size to itself (*Blanford*).

[(*Mammila*), 53-100.

423

Tigers, Leopards, Cats and Civets, *Blanfora, Fauna Br. Ind.,*
References.— *Jerdon, Mammals of India, 90-118, 120-123; Sterndale, Indian Mammalia, 156; Forsyth, Highlands of Central India, 266-326; Forbes Watson, Industrial Survey, 380-385; Balfour, Cyclop., III., 876; Abul Fazl, Ain-i-Akbari (Blochmann's Trans.), 288-290; Mason, Burma & Its People, 155-159; Pharm. Ind., 286; U. C. Dutt, Mat. Med. Hindus, 280; Ainslie, Mat. Ind., II., 479, 480, etc., etc.*

Although the members of the Cat tribe, annually killed in India, are too few in number to allow of their skins becoming an article of extensive trade, still those of the larger and more handsome are regularly exported, and good skins will always fetch a high price. Those of most importance have already been enumerated in the list of animals which yield FURS of economic value, but the Tigers, Cats, and Civets are also valuable for other economic qualities. A list of the species included under the FELIDÆ and VIVERRIDÆ may, therefore, be found useful for purposes of reference, while discussing their economic value. The species and habitats are detailed as given by Blanford in his recent work on the Mammalia of India, but arranged, for convenience of reference, as is customary in this work, in alphabetical order.

I.—FAMILY FELIDÆ.

1. **Cynælurus jubatus,** *Blyth; Blanford, Mam. Ind., 91.*
    THE HUNTING LEOPARD, or CHEETAH of Anglo-Indian writers.

SPECIES.
The Cheetah.
424

Vern.—*Chita, laggar,* HIND.; *Chitra,* GOND; *Chita puli,* TEL.; *Chircha, sivungi,* KAN.; *Yus, yus-palang,* PERS.

Habitat.—Found throughout Africa and South-Western Asia, extending from Persia to the countries east of the Caspian and into India. In this country it occurs throughout the greater portion of the Peninsula, from the Panjáb, through Rajputána and Central India, to the confines of Bengal and the Deccan.

TIGERS, CATS, Civets.	Species of the Cat Family.

SPECIES.

Leopard-Cat.
425

2. **Felis bengalensis,** *Kerr ; Blanford, Mam. Ind., 78.*
    THE LEOPARD-CAT.
    Vern.—*Chita billa,* HIND.; *Ban biral,* BENG.; *Wagati,* MAR.; *Rimau-akar,* MALAY.; *Kye-thit, thit-kyúk, kya-gyúk,* BURM.; *Kyoung,* ARAKAN; *Kla-hla,* TALAIN, KAREN.
    Habitat.—Common in the Himálaya as far west as Simla, in Lower Bengal, Assam, the Burmese and Malayan countries, Southern China, Sumatra, Java, Borneo, and the Phillipines. It is also found in the Syahádri Range or Western Ghâts, Coorg, Wynaad, Travancore, and in some of, perhaps all, the other forest regions of the peninsula, though not very abundantly.

Caracal.
426

3. **F. caracal,** *Güldenstädt ; Blanford, Mam. Ind., 88.*
    THE ·CARACAL.
    Vern.—*Siyah-gush* (black ears), PERS. & HIND.; *Tsogde,* LITTLE TIBET; *Ech,* LADAK.
    Habitat.—Found in the Panjáb, Sind, North-Western and Central India, and in the greater part of the Peninsula, except the Malabar coast, but rare everywhere. Ball met with it in Chutia Nagpur.

Jungle Cat.
427

4. **F. chaus,** *Güldenstädt ; Blanford, Mam. Ind., 86.*
    THE JUNGLE CAT.
    Vern—*Jangli-billi, khatás,* HIND.; *Khatás, banberál,* BENG.; *Berka,* HILL TRIBES OF RAJMAHAL; *Bául, bháoga,* MAR.; *Mant-bek,* KAN.; *Kada-bek, bella bek,* WADARI; *Katu-punai,* TAM.; *Jurka pilli,* TEL.; *Cherru puli,* MALAY.; *Kyoung tset-kun,* ARAKAN.
    Habitat.—The common wild cat of India from the Himálaya to Cape Comorin, and from the level of the sea to 7,000 or 8,000 feet or perhaps higher on the Himálaya. It is also found in Ceylon and extends east to Burma.

Lion.
428

5. **F. leo,** *Linn. ; Blanford, Mam. Ind., 56.*
    THE LION.
    Vern.—*Sher, babar-sher, singh,* HIND.; *Shingal,* BENG.; *Süh* or *suh,* ♂, *siming* ♀, KASHMIR; *Rastar,* BRAHUI; *Untia-bagh* (camel-tiger), GUZ.; *Sáwach,* KATHIAWAR.
    Habitat.—About twenty years ago the lion was common near Mount Abú ; several were shot near Gwalior, Goona, and Kota, and a few still existed near Lalitpur, between Saugor and Jhansi. In the early part of the century, it was common near Ahmedabad, and was found in Hurriana to the north-west, in Khándesh to the south, in many places in Rájputána, and eastward as far as Rewah and Palamow. Indeed, it was probably at one time generally distributed in North-Western and Central India. Now-a-days, however, it is verging on extinction, but there are probably a very few still living in the wild tract known as the Gir, in Kathiawar, and a few more in the wildest parts of Rájputána, especially Southern Jodhpur, Udaipur, and around Mount Abú.

Lynx.
429

6. **F. lynx,** *Linn. ; Blanford, Mam. Ind., 89.*
    THE LYNX.
    Vern.—*Gy,* TIBET ; *Patsalan,* KASHMIR.
    Habitat.—Found in the Upper Indus valley, Gilgit, Ladák, Tibet, etc.; also throughout Asia, north of the Himálaya, and Europe, north of the Alps.

Pallas's Cat.
430

7. **F. manul,** *Pallas ; Blanford, Mam. Ind., 83.*
    PALLAS'S CAT.
    Habitat.—Found in Tibet, extending into Ladák. It has not been observed on the south side of the main Himálayan range, but is found to the north as far as Siberia, and is common in Mongolia.

T. 430

Species of the Cat Family. (*J. Murray.*)	TIGERS, CATS, Civets.

8. Felis marmorata, *Martin : Blanford, Mam. Ind.,* 74.

    THE MARBLED CAT.

    Vern.—*Sikmar,* BHUTIA; *Dosal,* LEPCHA.

    Habitat.—Found in Sikkim and the Eastern Himálaya, and in the hilly regions of Assam, Burma, and the Malay countries, extending to Sumatra, Java, and, it is said, to Borneo.

9. F. nebulosa, *Griffith : Blanford, Mam. Ind.,* 72.

    THE CLOUDED LEOPARD, or CLOUDED TIGER of certain writers.

    Vern.—*Pungmar, satchuk,* LEPCHA; *Zik,* LIMBU; *Kung,* BHUTIA; *Lamchitia,* NEP.; *Thit-kyoung,* BURM.; *Arimau dahan* (tree-tiger), MALAY.

    Habitat.—Found in the South-Eastern Himálaya, Sikkim, Bhután, etc., at moderate elevations, probably not above 7,000 feet; also met with in the Assam hills, and throughout the hilly parts of Burma, Siam, the Malay Peninsula, Sumatra, Java, and Borneo.

Clouded Leopard.
432

10. F. ornata, *Gray : Blanford, Mam. Ind.,* 84.

    THE INDIAN DESERT CAT.

    Habitat. Found, on sandy plains and hills, throughout the drier regions of Western India, from the Panjáb and Sind to Ságar and Nágpúr, not extending to the Gangetic valley, and rare south of the Nerbudda. It is common in the deserts east of the Indus, in Sind, Western Rájputána, and Hurriana.

Desert Cat.
433

11. F. pardus, *Linn. ; Blanford, Mam. Ind.,* 67.

    THE LEOPARD or PANTHER.

    Vern.—*Tendwa, chita, sona-chita, chita-bágh, adnára,* HIND.; *Teon-kula,* KOL.; *Jerkos,* PAHARIA OF RAJMEHAL; *Burkál, gordág,* GOND.; *Sonora,* KURKU; *Syik, syiak, sejjiak,* LEPCHA; *Misi-batrai, kam-kei,* KUKI; *Hurrea kon, morrh, rusa, tekhu khuia, kekhi,* NAGA; *Kajengla,* MANIPURI; *Tidua, srighas,* BUNDELKHAND; *Bai-hira, tahir-hé, goral-hé, ghor-hé, lakhar-bagha* (the latter name used elsewhere for the hyæna), HILL-TRIBES NEAR SIMLA; *Sik,* TIBET; *Súh,* KASHMIR; *Diho,* BALUCH.; *Gorbacha, borbacha,* DECCAN; *Karda, asnea, singhal, bibia-bágh,* MAR.; *Tenduwa, bibla* BAURIS OF DECCAN; *Honiga, kerkal,* KAN.; *Chiru-thai,* TAM.; *Chinna, puli,* TEL.; *Puli,* MALAYL.; *Kutiya,* SINGAL.; *Kya-lak, kya-thit,* BURM.; *Klapreung,* TALAIN; *Kiché-phong,* KAREN; *Rimanbintang,* MALAY.; *Palang,* PERS.

    Habitat.—Found throughout Asia generally, with the exception of Siberia and the high Tibetan plateau, also throughout Africa. In India, Burma, and Ceylon it is generally distributed, except in parts of Sind and the Panjáb. Many Indian writers have separated the leopard, the panther, and the pard as distinct species, but Jerdon, Blyth, etc., agree in considering all to be merely varieties of one species.

Leopard or Panther.
434

12. F. rubiginosa, *I. G. Bélanger ; Blanford, Mam. Ind.,* 81.

    THE RUSTY-SPOTTED CAT.

    Vern.—*Namali pilli,* TAM. (MADRAS); *Verewa puni,* TAM. (CEYLON); *Kula diya,* SING.

    Habitat.—Found in Southern India, except on the Malabar coast, and in Ceylon. Sterndale also obtained one specimen at Seoni in the Central Provinces.

Rusty-spotted Cat.
435

13. F. temmincki, *Vigors & Horsf. ; Blanford, Mam. Ind.,* 75.

    THE GOLDEN CAT.

    Habitat.—Occurs in the South-Eastern Himálaya at a moderate elevation; rare in Nepál, more abundant in Sikkim, found also in Tenasserim, Sumatra, and Borneo, and probably throughout Burma and the Malay Peninsula.

Golden Cat.
436

14. F. tigris, *Linn. ; Blanford, Mam. Ind.,* 58.

    THE TIGER.

Tiger.
437

TIGERS, CATS, **Civets.**	**Species of the Cat Family.**

SPECIES.

Tiger.

Vern.—*Bágh, sher* (female=*bághni, sherni*), *náhar, sela vágh,* HIND.; *Go-vágh,* BENG.; *Tut, sad,* HILL-TRIBES OF RAJMEHAL; *Garúm kúla,* KOL.; *Kula,* SANTAL, HO, & KURKU; *Lákhra,* URAON; *Krodi,* KHOND; *Túkt, tuk,* BHOT.; *Sathong,* LEPCHA; *Keh-va,* LIMBU; *Sehi,* AKA; *Matsá,* GARO; *Kla,* KHASI; *Sa, ragdi, tekhu, khudi,* NAGA; *Humppi,* KUKI; *Sumyo,* ABOR; *Su,* KHAMTI; *Sirong,* SINGPHO; *Kei,* MANIPURI; *Misi,* CACHARI; *Tág,* TIBET; *Padar-suh,* KASHMIR; *Shinh,* SIND; *Masar,* BALUCH.; *Patayat-bágh, wahág,* MAR.; *Púli,* TAM., TEL., MALAYL., & GOND; *Púli redda-púli, peram-pilli,* TAM.; *Pedda-púli,* TEL.; *Perain-púli, kúdua,* MALAYL.; *Kuli,* KAN.; *Nari,* COORG; *Pirri, bursh,* TODA; *Kya,* BURM.; *Kla,* TALAIN.; *Khi, botha-o, tupuli,* KAREN; *Htso,* SHAN; *Rimau, harimau,* MALAY.

Habitat.—Found throughout India, Burma, and other parts of South-Eastern Asia, Java, and Sumatra, but not, it is said, Borneo. It occurs in suitable localities throughout a great part of Central Asia, and is found in the valley of the Amur, the Altai Mountains, around Lob Nor in Eastern Turkestan, about the Sea of Aral, on the Murgháb near Herat, on the southern coast of the Caspian (Hyrcania), and in the Caucasus, but not in Tibet, Afghánistán, Baluchistán, or Persia, south of the Elburz Mountains on the Caspian. In India it still occurs wherever large tracts of forest or grass jungle exist, but within the last twenty or thirty years the number of tigers has greatly diminished, and they are now becoming scarce, or have even, in some cases, disappeared entirely in localities where they were formerly common. This has been especially the case throughout a large area of the Central Provinces, in many parts of Bengal, and several districts of the Bombay Presidency. In the forests at the base of the Himálaya tigers are common, and they occasionally ascend the hills to an elevation of 6,000 to 7,000 feet, but none are found in the interior of the mountains.

Waved Cat.
438

15. Felis torquata, *F. Cuv.; Blanford, Mam. Ind.,* 85.
THE WAVED CAT.

Habitat.—This cat may be merely a descendant of the domestic cat which has run wild, but, according to **Blanford**, it is at least equally probable that it constitutes the original stock from which Indian domestic cats, and possibly those of other countries, are derived. It is probably widely dispersed through Northern India, since specimens have been obtained in Nepal, Kashmír, and Rájputána, but it does not appear to be common anywhere.

16. F. uncia, *Schreber; Blanford, Mam. Ind.,* 71.
THE OUNCE, or SNOW LEOPARD.

Snow
Leopard.
439

Vern.—*Ikar, zig, sachak, sáh,* TIBET; *Bharal he,* HILLS NORTH OF SIMLA; *Thurwágh,* KASHMIR.

Habitat.—High Central Asia, especially Tibet, extending north to the Altai, and west, it is said, into Persia. It is found throughout the Himálaya at high elevations, and is more abundant on the Tibetan side of the Snowy Range, where it is met with in the Upper Indus and Sutlej valleys. It is also fairly common in Gilgit.

17. F. viverrina, *Bennett; Blanford, Mam. Ind.,* 76.
THE FISHING CAT.

Fishing Cat.
440

Vern.—*Banbiral, báráun, khupya-bágh, bágh-dásha,* HIND.; *Mach-bágral,* BENG.; *Hándún-diva,* SING.

Habitat.—Found in marshy thickets, swamps, and tidal creeks, which it affects owing to its fish-eating habits, in Bengal, probably Orissa, and the Indo-Gangetic plain generally, extending as far as Sind, It is unknown in the Peninsula of India except on the Malabar coast, where it occurs from Mangalore to Cape Comorin, but not, so far as is known, to the northward near Bombay. It also occurs in Ceylon. Along the base of the Himálaya it is met with as far west as Nepál, and ranges through-out Burma, Southern China, and the Malay Peninsula.

II.—FAMILY VIVERRIDÆ.

18. Viverra zibetha, *Linn.; Blanford, Mam. Ind., 96.*
THE LARGE INDIAN CIVET.

Vern.—*Khatás,* HIND.; *Mach-bhondar, bágdos, pudo ganla,* BENG.; *Bhrán,* NEPAL TERAI; *Nit biralu,* NEPAL; *Kung,* BHOT.; *Saphiong,* LEPCHA; *Kyoung-myeng,* (horse-cat), BURM.; *Tangalong,* MALAY; *Gandha márjára,* SANS.

Habitat.—Bengal, Assam, Burma, the Malay Peninsula, Siam, and Southern China. It extends south and south-west from Bengal to Orissa and Chutia Nagpur, and probably some distance further south and west, and to the northward into Sikkim and Nepál, ascending the Himálaya to a considerable elevation.

Large Indian Civet. **441**

19. V. civettina, *Blyth; Blanford, Mam. Ind., 98.*
THE MALABAR CIVET-CAT.

Habitat.—This has been considered as one species with the above by several Indian writers, but Blanford remarks that as the area which it inhabits is separated from that of **V. zibetha** by a broad tract of country (there being no civet in the Central Provinces, Deccan, or Karnatik), it is probable that it is a distinct species. According to Jerdon, it is found throughout the Malabar coast, from Honawar to Cape Comorin, but may possibly extend further north.

Malabar Civet. **442**

20. V. megaspila, *Blyth; Blanford, Mam. Ind., 99.*
THE BURMESE CIVET.

Vern.—*Kyoung-myeng,* BURM.; *Músang-jebat,* MALAYS.

Habitat.—Burma, Malay Peninsula, Cochin China, and Sumatra; recorded from as far north as Prome.

Burmese Civet. **443**

21. Viverricula malaccencis. *Blyth; Blanford, Mam. Ind., 100.*
THE SMALL INDIAN CIVET.

Vern.—*Mashk-billa, katás, kasturi* (a name properly belonging to the musk-deer), HIND.; *Gandha gokal, gando gaula,* BENG.; *Sogot,* HO KOL; *Saiyar, bág-myúl,* NEPAL TARAI; *Jowádi manjúr,* MAR.; *Púnagin bek,* KAN.; *Púnagú pilli,* TEL.; *Uralawa,* SING.; *Koung-ka-do,* BURM.; *Wa-young-kyoung-byouk,* ARAKAN.

Habitat.—Found throughout India, except in Sind, the Panjáb, and the western parts of Rájputána; also in Assam, Burma, Ceylon, Southern China, the Malay Peninsula, Java, and some of the other Malay islands. It is frequently kept in confinement by Natives, for the purpose of yielding civet and becomes perfectly tame. Several other species of the family VIVERIDÆ exist, but the four above enumerated, especially the first and last, are the commonest sources of commercial civet; the others are not of sufficient economic interest to warrant giving an enumeration of them.

Small Indian Civet. **444**

Peculiarities and Properties of the Indian Tiger, Cats and Civets.

HABITS.—All the members of the family FELIDÆ are distinguished by their purely carnivorous habits, by their strength, activity, and, in certain cases, by their ferocity. The smaller species are, perhaps, the fiercest and least tameable, especially **F. bengalensis, F. chaus, F. lynx,** and **F. marmorata.** Of the larger species, **F. pardus,** the leopard, is, by universal consent, admitted to be the most courageous, and, when brought to bay, the most dangerous, but it, as well as the tiger, is rarely formidable unless when it has taken to man-eating. When it does so, it often becomes an even more fearful scourge than a man-eating tiger. Thus Sterndale, and also **Forsyth** (*Highlands of Central India*) relate the history of

Peculiarities Habits. **445**

**MAN-EATING
TIGERS.**

a leopard near Seoni which, in two years before it was shot, is said to have killed two hundred human beings. Leopards, when large, frequently kill cattle, poneys, donkeys, and large deer, such as *sambár*; but the smaller animals have to content themselves with inferior prey. Thus Blanford writes, "The leopard is absolutely without prejudice in the matter of food—all beasts, birds, and, I believe, reptiles, that are not too large to kill, or too small to catch, are the same to him : he will strike down an ox or bound upon a sparrow. If he has a predilection it is probably for dogs and jackals. He is a terrible foe to monkeys, and kills many of the *hanúmáns* or *langúrs* who inhabit the rocky hills in which he delights."

Great numbers of domestic animals are annually killed by tigers; indeed, many of the latter appear to live entirely upon cattle. Forsyth, in his interesting *Highlands of Central India*, states that tigers, as a rule, are entirely game-killers, during the more vigorous portion of their life; as they become older they grow more cunning, less afraid of man, and less able to find their prey amongst the swift big game, and naturally take to cattle-eating. From this stage many go on to that of the man-eater, "a tiger who has got very fat and heavy, or very old, or who has been disabled by a wound, or a tigress who has had to bring up young cubs where game is scarce,—all these take naturally to man, who is the easiest animal of all to kill, as soon as failure with other prey brings on the pangs of hunger" (*Forsyth*). Two classes are distinguished by the *shikáris* of the Central Provinces as *lodhia-bagh*, a game-eating tiger; *úntia-bagh*, a cattle-lifting tiger; a man-eater is generally an older and lighter coloured *úntia bágh*. A tiger that has once taken to man-eating, continues to live occasionally on the same prey, but, according to Blanford, it is the exception for even man-eaters to confine themselves to human food.

Tigresses with cubs are frequently very destructive to cattle, often quartering themselves in the neighbourhood of a village and feeding entirely on the herds within their reach. The tigress is said to be very destructive in such circumstances, partly in order to teach the young to kill their own prey; while the young tigers, according to Forsyth, appear to kill as many as they can among a herd, merely from wanton pleasure in the exercise of their developing strength. The destruction of such a dangerous animal has naturally been encouraged to the utmost by Government, which for many years has awarded sums varying from R5 to R50, according to the district, for every head killed. Notwithstanding these endeavours, coupled with the large extent to which both tigers and leopards are shot for sport by Englishmen throughout the country, the annual destruction of life is still very considerable.

The lion was at one time also very destructive, but for many years it has become so rare in India as to be hardly worthy of notice. It is said to feed chiefly on deer, antelopes, wild pigs, cattle, horses, donkeys, and camels, and used formerly to kill many of the latter (*Blanford*).

Almost all the other members of the tribe are very destructive to small game, and when in the neighbourhood of man, to sheep, goats, and poultry. In Tibet the lynx has the reputation of being extremely bloodthirsty and savage, a reputation which is more than confirmed by Scully's observation that a pair of them killed six sheep in one night near Gilgit. The leopard cat is said to be extremely destructive to poultry in South India; in Tibet and the inner Himálaya the ounce carries off sheep, goats, and dogs from villages, and even hill poneys, but, it is said, to have never been known to attack man. Two species, the hunting leopard or *chíta*, and the caracal, have long been employed in India to capture deer and other game. The former is always captured when mature, since, according to the native *shikári*, it never learns to kill properly when captured young. It is easily tamed, about six months being required to complete its training and render it quite obedient. When thus tamed it is said frequently to become as gentle and docile as a dog, delighting in being petted, and to become quite good-tempered even with strangers, purring and rubbing itself against its friends, as cats do (*Blanford*). For hunting purposes it is hooded and taken in a bullock cart to the neighbourhood of the antelope. When at the required distance (the game allows the cart to approach quite near, having no fear of an object it sees every day, and is accustomed to), it is unhooded and slipped. The leopard then either rushes directly on its prey, or, if at a greater distance, takes advantage of any inequalities or other advantages the ground may offer, to stalk the herd, running up till within distance for its rush. Its speed for a short distance is remarkable, far exceeding that of any other beast of prey, probably of any other mammal. It generally seizes on the buck, if there be one in the herd, and fells it, it is said, by stirking its legs from under it. It then seizes the quarry by the throat, and holds it until the keepers arrive, when it is rewarded with a bowlful of the antelope's blood. This sport is a very favourite one with Native princes in India, and, according to Abul Fazl, was much patronised by the Emperor Akbar, in whose time the system of training these animals must have been carried to great perfection. Thus it is stated that they were always allowed to remain loose, even towards evening, and yet made no attempt to escape. They were divided into eight classes according to their value, each of which got a certain stated allowance of food, while the best had "brocaded saddle cloths, chains studded with jewels, and coarse blankets or Gushkani carpets to sit on."

The caracal, a much smaller animal, is trained to catch birds and small deer, gazelles, hares, or foxes. According to Blyth, a favourite sport in certain parts of the country is to pit these cats against each other to kill pigeons out of a flock. While the birds are feeding on the ground the caracals are let loose together, and each is said often to strike down as many as ten or a dozen before the flock can take flight. Vigne states that their speed is, if possible, greater in proportion than that of the hunting leopard (*Blanford*). Like the *chíta*, these appear to have been favourites of Akbar's, for Abul Fazl writes, "His Majesty is very fond of using the plucky little animal for hunting purposes. In former times it would attack a hare or a fox; but now it kills black deer. It eats 1 seer of meat daily. Each has a separate keeper, who gets 100 *dam* every month."

Fur.—The skins of the larger species are much valued, and fetch a high price. Those of some of the smaller are remarkably beautiful and much sought after, that of **F. marmorata**, the marbled cat of Sikkim, being one of the finest.

Medicine.—The FLESH of the tiger and leopard are believed in many localities to be medicinal. That of the former is said by Ainslie to be boiled in mustard oil to form an unguent as a remedy for emaciation. In certain localities a medicinal oil is prepared from tiger FAT. Hamilton states that leopard's flesh is believed by the *Vytians* to be an efficacious remedy for epilepsy. The CLAWS, RUDIMENTARY CLAVICLES, and WHISKERS of the tiger have many fanciful properties attributed to them.

MEDICINE.
Flesh.
447
Fat.
448
Claws.
449
Rudimentary
Clavicles.
450
Whiskers.
451

Civet, the unctuous, highly odorous secretion from the anal glands of several of the VIVERRIDÆ, especially Viverra zibetha, and Viverricula malaccensis, is used to a considerable extent in India, under the name of *kustúri*, both for perfumery and for medicinal purposes. Valuable stimulant and aphrodisiac properties are ascribed to it, but probably it possesses no special powers in these respects. Jerdon states that Viverricula malac-

TIMBERS.	The Timbers of India.

MEDICINE.

censis is frequently kept by Natives for the purpose of yielding the secre-tion, and Waring, in the *Pharmacopœia of India*, mentions an establish-ment at one time kept up at the expense of Government in which civets were specially reared.

FOOD.
452

Food.—The Santals, Burman, and Malays, and several other aboriginal tribes, eat the flesh of the tiger and believe that it conveys with it the courage and sagacity of the animal. In most cases they refuse to allow their women to eat it, probably because they consider them better without any high development of these characters.

**DOMESTIC &
SACRED.**

**Clavicles.
453
Claws.
454
Whiskers.
455**

Domestic and Sacred.—Tigers or representations of them are objects of adoration or propitiation amongst the aboriginal tribes of Central India, and many of the less enlightened hill people of the Himálaya. They consider the CLAVICLES and CLAWS to be powerful charms; in certain lo-calities the WHISKERS are supposed to be a deadly poison, and are carefully burned off as soon as the animal is killed; in others they are believed to endow the possessor with unlimited powers over the opposite sex. Amongst the Santals the most solemn oath is on a tiger's skin, a circumstance which is, or at one time was, taken advantage of in the Courts of Justice. The claws are frequently mounted in silver or gold as bracelets, armlets, etc.

TILIACORA, *Coleb.; Gen. Pl., I., 36.*

456

Tiliacora racemosa, *Coleb.; Fl. Br. Ind., I., 99*; MENISPERMACEÆ.

Syn.—T. FRATERNARIA, CUSPIDIFORMIS, ABNORMALIS & ACUMINATA, *Miers.*; MENISPERMUM ACUMINATUM & RADIATUM, *Lamk.*; M. POLY-CARPUM. *Roxb.*; COCCULUS ACUMINATUS, *DC.*; C. RADIATUS, *DC.*; C. POLYCARPUS, *Wall.*

Vern.—*Karwanth, karrauth, rangoe, bága mushada,* HIND.; *Tiliakora, tiliakoru,* BENG.; *Tiga mushadi, tige mushidi, tige mushini, tige, tivva mushidi, naga mushini, pátéru, páta veru, kappa tige,* TEL.

References.—*Roxb., Fl. Ind., Ed. C.B.C.,* 733; *Kurz, For. Fl. Burm., I.,* 54; *Brandis, For. Fl.,* 10; *Gamble, Man. Timb.,* 11; *Elliot, Fl. Andhr.,* 83, 122, 146, 181, 182; *O'Shaughnessy, Beng. Dispens.,* 202; *Dymock, Warden, & Hooper, Pharmacog. Ind., I.* 64; *Gazetteers:—Mysore & Coorg, I.,* 56; *N.-W. P., IV., lxvii.; Agri.-Horti. Soc. Ind., Journ. (Old Series), V., Sel.,* 65; *VI.,* 5.

Habitat.—A large, climbing shrub, found throughout Tropical India, from Bengal and Oudh to Orissa, the Konkán, Ceylon, and Singapur.

**MEDICINE.
Root.
457**

Medicine.—Roxburgh informs us that the ROOT, rubbed between stones and mixed with water, is given as a drink for the cure of venomous snake-bites, though the Natives confess that they have little belief in its virtues. Three kinds of *mushadi* are believed to be antidotes of this kind, by the Telingás, *viz., mushade,* **Strychnos Nux-vomica**; *naga-mushadi,* **Strychnos colubrina,** and *tiga mushadi,* the plant now under consideration. The writers of the *Pharmacographia Indica* remark, "It is bitter like others of the genus" (?—Natural Order,—there is only one species of Tiliacora), "and it is hardly necessary to say, no antidote to snake-poison."

**DOMESTIC
Branches.
458**

Domestic.—The long flexible BRANCHES are used for thatching and basket-work (*Brandis*).

TIMBERS.

459

Timber Trees of India.

The quality of wood obtained from all the important timber trees of India, has, as the reader must have observed, been treated of in para-graphs under each species, headed **Structure of the Wood.** It was pro-posed to give here a collective article in which all the trees of any import-ance would have been enumerated, an article which would have served as a key to the descriptive accounts scattered throughout the Dictionary, and would also have proved of value in arranging specimens for Museums, but

Products of India.   57

Tin and Tin-ore.   (*J. Murray.*)   **TIN**
                                           and Tin ore.

want of space has precluded the carrying out of this proposal. The reader may, however, have noticed that advantage has been taken of such alphabetical headings as **Cabinet-work, Packing-cases,** etc., to give lists of timbers suitable for definite purposes. These lists, it is hoped, may be of value in referring any one, desiring the information, to the detailed accounts of Indian Timbers given under the scientific names of the trees from which the wood is derived—see **Agricultural Implements,** Vol. I., 145; **Boat & Ship-building,** Vol. I., 460; **Bows & Arrows,** Vol. I., 518; **Boxwood, Substitutes for —,** Vol. I., 518; **Bridges,** Vol. I., 535; **Cabinet-work, Furniture & General Carpentry,** Vol. II., 1; **Canoes, Dug-outs, Troughs, Water-pipes, Drinking-cups,** &c., Vol. II., 126; **Cart & Carriage-building,** Vol. II., 183; **Carving, Fancy-work, Images,** etc., Vol. II., 202; **Charcoal,** Vol. II., 264; **Combs, Fans, Brush-backs,** etc., Vol. II., 515; **Dandy, Banghi, & Palankin-poles,** Vol. III., 19; **Fuel & Firewood,** Vol. III., 452; **Gun-stocks & Gun-carriages,** Vol. IV., 189; **House-building,** Vol. IV., 300; **Packing-Cases,** Vol. VI., Pt. II., 1, **Pea-stakes, Pan-houses, Wattle,** etc., Vol. VI., Pt. II., 123; **Pounders & Presses,** Vol. VI., Pt. I., 333; **Saw-mills,** Vol. VI., Pt. II., 483; **Sleepers,** Vol. VI., Pt. III., 252; **Tinder & Gun-matches,** Vol. VI., Pt. IV., 62; **Tools & Tool handles,** Vol. IV., Pt. 70; **Walking Sticks,** Vol. VI., Pt. IV., 298; **Wattle,** Vol. VI., Pt. IV., 300; **Well-curbs,** Vol. VI., Pt. IV., 301; **Wheels,** Vol. VI., Pt., IV., 307.

TIN, *Ball, in Man. Geol. of Ind., III. (Economic), 313.*

460

Tin is said to be occasionally found native, or in the metallic, state; as the sulphide—stannite—it also rarely occurs. The only ore known to exist in abundance is the dentoxide or cassiterite, commonly known as "tinstone." This mineral, when pure, contains about 78 per cent. of the metal. Within the limits of peninsular India tin-stone has been rarely found, but in the southern portion of the Tenasserim provinces of Burma it forms extensive and valuable deposits, and constitutes the greatest mineral wealth of the region.

Tin, and Tin ore, *Mallet in Man. Geol. Ind., IV. (Mineralogy), 54.*

Vern.—*Kallai, ranga, ráng, kathel,* HIND.; *Kathir, ranga,* AJMÍR; *Taga-ram,* TAM.; *Tima, falagh,* MALAY.; *Khai-ma, khai-ma-phyu,* BURM.; *Vanga, ranga, trapu,* SANS.; *Kas-din, resás, abruz,* ARAB.; *Ursis,* PERS.

References.—*Ainslie, Mat. Ind., I., 568; Balfour, Cyclop. Ind., III., 889; Forbes Watson, Indust. Survey, II., 408; Baden Powell, Pb. Pr., 10, 103; Linschoten, Voyage to the East Indies, I., 104, 129; U. C. Dutt, Mat. Med. Hind., 68; Irvine, Mat. Med., Patna, 50; Mason, Burma & Its People, 568, 730; Bombay Adm. Rept., 375, 376; Gazetteers:— Bombay, IV., 128; VIII., 262; Delhi, 129, 131; Burma Govt. Proceedings, Revenue Branch, August 1881, Nos. 1 and 2B; also many passages in the publications of the Geological Survey, and the Asiatic Society, for an enumeration of which the reader is referred to the authorities quoted by Ball and Mallet, l.l.c.*

Occurrence.—Traces of tin ore have been observed in a few places in peninsular India, but generally merely in sufficient quantity to afford mineralogical specimens. In Bengal tin-stone has been found in two localities in the district of Hazaribagh—at Nurgo or Nurunga, and at Simratari, west of Pihra. Tin ore at the former locality occurred in three or four lenticular beds or nests in gneiss rock. The right to mine was obtained in 1867, from the Raja of Palgunj, on an annual payment of R2,000, a mine was opened and smelting was commenced. But at a depth of 20 yards the nests were found to be thinning out, the rock became harder and water troublesome, and the enterprise was abandoned. The deposit at Simratari is more of mineralogical than of practical interest. In Bombay, traces of tin-stone have been found here and there in the Dharwar District, and at Jambughora in Gujrat. Rumours also exist of tin being obtained in the Bustar State of the Central Provinces, but have not been confirmed by any discovery of the ore.

In certain parts of the Malay Peninsula and Burma, however, rich deposits of tin-stone occur along the base of the granitic ranged forming

OCCURRENCE.

Burma.

the former country. **Mr. Oldham** (*Select Records Govt. of Ind., X., 56*)
describes the occurrence of the ore in this region as follows :—" The greatest
mineral wealth of the southern portion of the Tenasserim provinces consists in the extensive and valuable deposits of tin-ore which they contain.
In the granite of the central dividing range, which separates these provinces from Siam, and more especially (so far as my opportunities for examination extended) towards the outer edge of this granite, or near its
junction with the highly metamorphosed slates with which it comes into
contact, tin-stone is an essential ingredient in the mass of rock, occurring,
disseminated through the granite, in small crystals, similarly arranged to the
quartz and felspar of the rock, and in some cases, as at Kahan hill near Mergui, veins of granite cut through and traverse the more recent (*sic.* older) ?*
—rocks, and contain large and abundant crystals of tin-stone" (*Quoted from
Mallet, Mineralogy, loc. cit.*). The deposits for which Tenasserim is famous
are situated from the neighbourhood of Yé in the Amherst District, southwards at intervals to Maleewoon, in the extreme south of Mergui. Throughout this tract ore, derived by degradation from the deposits in the rocks above
described, may be found in the sand of the beds of nearly every stream ; indeed, not only does it occur in this form, but through large tracts it actually
impregnates the soil. Owing to the density of the vegetation on the granitic
hills, they have never been thoroughly explored for lodes or veins, and therefore it is not known whether the ore may occur locally segregated from the
general mass of the rock. The Natives have always confined themselves
to working the stream tin, which is easily obtained and is generally of
great purity. In certain localities wolfram occurs in association with the
stream tin, thus decreasing the value of the ore, and that of the metal produced from it (*Ball*).

Ball mentions the following localities as specially noticed by various
writers :—

Amherst.
463

Amherst and Tavoy Districts. —Large quantities of tin are believed to
have been formerly manufactured in Tavoy, since the traveller **Ralph Fitch**
records that in 1586 the whole of India was supplied with tin from the
'island of Tair.' At present the collection of tin, if practised at all, is only
on a very small scale Tin-stone has been reported by English writers
from the river Kallee Ung, lat. 14° 48', long. 98° 10', to the south of Yé;
from the various streams which empty themselves into the Henzai basin
from the south, through the Oung-beng-quin, and from Myit-ta. In the latter
locality the streams containing the metal run into the basin of the Upper
Tenesserim river. The ore is stated to occur both in the alluvium,
and in the granitic detritus from the hills, and is apparently rather abundant in some places. It was worked in former times by the Burmans,
and some old pits are as much as 40 feet deep. **Captain Low**, in his
History of Tenasserim, mentions mines and smelting works at Bubein-
chaung near Ke-up-poch, which were being worked in 1825, also deserted
at Nayedaung, and Shenze near Kaleaung and Kamanula, one day's
journey to the north of Ke-up-poch. These mines were, at one time,
worked during four months of the year, gave employment to four hundred
men, and yielded a revenue probably not exceeding 1,500 *tickals*.

Mergui.
464

Mergui.—Stream-tin occurs in numerous localities in this district, and,
in a few instances, tin-stone has been found *in situ*. **Mr. Fryar**, who visited the principal works in 1871, states that they may be enumerated under
eleven heads as follows :—(1) Palouk.—This place was alluded to in a communication made to the Asiatic Society as early as 1829, in which it was

---

* An interrogation added by **Mallet** in quoting this passage.

stated, on the authority of some Chinese who had visited the locality, that **OCCURRENCE.**
the ore was of a superior description. In three tributaries of the Palouk **Burma.**
river, named Koosheelo, Walach, and Natheechoung, 18 miles from the
village of Palouk, **Mr. Fryar** found an abundant supply of stream-tin in
what he considered to be profitable quantity. There are no workings
there at present, owing, it is said, to the dearth of inhabitants. (2) **Mergui.**
Mergui.—A bed of quartzose tin in the island on which this town is built
contains stream-tin in small quantity, which used formerly to be worked.
(3) King's Island.—Stream-tin was found by **Mr. Fryar** in the bed of the
Kitan river near the sea; he believes a large quantity to be obtainable
from this locality. (4) Theandaw.—This river, a tributary of the Great
Tenasserim, was considered by **Captain Tremenhere** to be one of the
richest sources of tin in the whole district. Many old pits exist in its vicinity, which all appear to have been abandoned. (5) Thawbawleek river.
—In this river, a secondary tributary of the Little Tenasserim, there are,
and have been for many years, somehat extensive washings for stream-tin.
Mr. Fryar mentions two other localities in the neighbourhood, Belamo and
Seboopela, in which ore is or was formerly obtained. The tin-workers at
the last-mentioned place earned as much as R1 a day. **Captain Tre-**
menhere mentions another locality, Zahmon, in the Nunklai District,
where ore of a dark colour and inferior quality, owing to the presence of
wolfram, is obtained. (6) Yagnan.—This locality is 70 miles south of
Mergui. Tin washings exist some 2 or 3 miles up the river, and are
worked during the rain by twelve washers, who are said to earn an average
amount of R100 each, during the season. (7) Bopyng.—Is 30 miles
round the coast, south from Gagnay; the village is 10 miles up the river,
and the principal washings 5 miles further up. At the time of **Mr. Fryar's**
visit there were four Chinese and three Shan employers of labour, one of whom
owned a small furnace. The labourers received from R10 to 15 a month,
and a good workman could obtain 7·3℔ of ore by washing in a day. A
sample of ore, seen by **Mr. Fryar**, brought from a hill several miles further
inland, led him to conclude that an actual lode might exist in the hills.
(8) Kalathooree.—A locality 30 miles by the cart from Bopyng. There are
thirteen separate washings, but not more than fifteen men are employed.
When **Mr. Fryar** visited it the industry had fallen off probably owing to a
diminution in the supply of ore. (9) Choung Tanoung.—A place 4 miles
south from the Kala Thooree estuary; the washings are situated some 8
to 10 miles up the river. **Mr. Fryar** states that all the labourers employed,
averaging a dozen, were Chinese, and that the outturn in one year was
between 4 and 5 tons of tin pig which was sent to the Mergui market. He
estimated the cost of producing a ton at £60, £98-14 was received for that
amount at Mergui, and the profit was therefore £38-14 per ton. The
smelting furnace, which is probably similar to that in other localities, is
described as follows:—It is built of clay, strengthened and supported by
iron bars, which are bound to the clay by means of iron hoops. The blast
is produced by means of a long wooden cylinder, containing a piston
which is moved backwards and forwards by manual labour. The tap-
ping hole is permanently open, and the molten tin is accumulated in a
well, till there is sufficient to ladle out and cast into a pig. For each ton
of metal, 1·9 to 2 tons of tin-stone are required, since the percentage obtained varies between 52 and 68 per cent. One such furnace, worked by
six men, produces 400℔ of tin in the 24 hours. The slags are passed
through six times before being thrown away. (10) Pokchan—Is a village
situated about 40 miles up the river of the same name. The sand
is obtained at a place three days' journey from the village, sixty men
are employed, and all is sent to the furnace at Ma-lee-won (the next de-

OCCURRENCE.

Burma.

scribed locality) for smelting. Between the two places there are several other tin-washings. Tin-works also occur at Khenoung in Siamese territory on the opposite side of the river. The sand is brought on elephants from the stream a day's journey off, and 2,000 men are said to be employed. This locality was visited by **Dr. Oldham**, who was greatly impressed with the system adopted by the proprietor or Governor, an old Chinaman, everything connected with the works being conducted in the most orderly fashion. (11) Ma-lee-won.—A village situated on a tributary of the Pakchan river, the most southern locality in Tenasserim in which tin is worked. The principal washing is 8 miles from the village, where the river-bed is 300 feet wide. The tin-bearing stratum, from 15 to 27 feet beneath the surface, is only itself 2 feet thick, but is very valuable. Between the years 1860 and 1870 the annual rent for these works charged by Government was £272. About the year 1873 the township was leased to a firm of Rangoon merchants for a term of 30 years, originally at the rate of a ground-rent of £1,000 per annum, which, however, merged into a royalty of 7 per cent. Though it is stated that besides the abundant supply of stream tin veins of ore were discovered and opened in the hills, yet the lease was abandoned in 1877, as the expenditure had exceeded the outturn by a large amount. The Chinese again reinstalled themselves with from 80 to 100 workers.

The above information has been condensed from the compilation by Ball, published in 1880. Since then very little appears to have been done in furthering the industry, and the failure of the attempts by Messrs. Steel & Co., appears to have prevented any renewal of British enterprise in this direction.

In 1881 a correspondence took place between the Government of Burma and the Government of India, in which it was recommended by the former that the duty or royalty of 4 per cent. on all tin smelted or otherwise made marketable, which had been enforced since 1873, should be removed. It was urged that this condition had an adverse effect on the tin-mining industry, and that the royalty obtained had varied from R1,378 to R700, a sum too insignificant to be worthy of consideration. A report from the Deputy Commissioner of Mergui was forwarded, in which it was demonstrated that the main difficulties to be contended against in expanding the industry were;—(1) want of communication, (2) want of labour, (3) the difficulty and expense of obtaining supplies, and (4) the great fall in the market value of the metal from £150 per ton in 1872 to £73 in 1877. The advantage of opening free access to and encouraging Chinese immigration was also insisted on. In Upper Burma deposits of ore, or stream-tin, may doubtless be found in considerable quantity, as the mineral resources of the country are completely explored and opened out. Ore is said to have been found in the Shán country, south-east of Mandalay, in the Karenni hills between the Sitang and Salwin, and in the Toung-ngu district; and the Karens are said to work the ore at Kah-may-pew. The total production of tin ores in India during 1889 is said to have been 976 cwt., valued at R55,673, all of which was obtained in Burma (*Statement on Minerals, Rev. and Agri. Dept.*).

DYE.
465

Dye.—Tin is a highly important metal in dyeing as practised in Europe, but in this respect is apparently unknown to the Natives of India. It is employed as a mordant in three different states of chemical composition—namely, as the protoxide, the sesquioxide, and the peroxide or their corresponding chlorides. Generally speaking, the proto-salts are applied to wool, the persalts to cotton, and the sesqui-salts, the consumption of which is more limited, are used in certain cases to both fibres. The compounds in which tin plays the part of an acid, such as the stannites and stamrates, are almost exclusively restricted to cotton dyeing and printing (*Crookes, Handbook of Dyeing, 519*).

Medicine.—Tin has been known in India from a very remote period, and early held an important place in Sanskrit Materia Medica, being mentioned by Susruta. The supply was probably obtained from Burma, or from some of the tin-producing islands of the Malay Archipelago, between which and India a trade must have existed in very ancient times. Tin, like most of the other metals is used primarily in the metallic state, but is subjected to a complicated process of so-called purification which reduces it to the state of an impure oxide. To accomplish this it is melted, and the molten metal is poured into the milky juice of **Calotropis gigantea** (*arka*). It is then re-melted in an iron cup, one-fourth of its weight of *yavakshára* (impure carbonate of potash), and powdered tamarind husks are added, and the whole is agitated with an iron rod till the mass becomes reduced to a fine powder. It is then washed in cold water and dried over a gentle fire. The resulting product is a greyish white powder consisting chemically of oxide of tin with some impurities. It is chiefly valued as a remedy in urinary disorders, and may be given in 4-grain doses every morning with honey, or one of many elaborate compound preparations may be employed. The reader desiring further information regarding these is referred to U. C. Dutt's *Materia Medica of the Hindus*, from which the above information has been condensed.

Ainslie informs us that Muhammadan physicians prescribe "powder of tin" as an anthelmintic, in doses of ʒi or ʒii, mixed with honey, on two or three successive mornings. He quotes Dr. Good as mentioning a case in which a tape-worm was expelled by means of a dose of tin filings and jalap, but he remarks that he does not know of the metal ever being used in Native practice in the South of India. Irvine (*Materia Medica of Patna, 50*) states that the oxide is employed as a tonic and aphrodisiac. Surgeon-Major Robb, in a note to the Editor, mentions that the same preparation is administered as a remedy for chest affections.

Industrial Uses.—The metal is employed by the Natives of India for making bright toys and imitation jewellery, as well as for tinning copper vessels; it is also beaten out into leaf or thinfoil, and used for a silver paint. The industry of tin-plating is a large and fairly prosperous one, as all Muhammadans, Christians and Pársis, and certain classes of Hindus employ tin-plated vessels extensively. Copper vessels to be safely free from poisonous deposits must be tinned or *kalaied* once a month, thus affording constant employment to a large number of workmen, *qal'aigar* or *kalaigar*, who are all Muhammadans. In Northern India, vessels which are tinned for the first time are boiled in a solution of alum, verdigris, sulphate of copper and sal ammoniac. On all subsequent occasions they are coated with tin without any previous preparation save that of removing the old *kalai* by scrubbing the vessel with ground *kankar*. Tin reduced to powder is mixed with sal ammoniac and applied by means of a piece of cotton, the vessel being heated on a charcoal fire. It is then polished with sand and ashes.

Trade.—The trade in tin is comparatively unimportant, and consists for the greater part of imports from foreign countries. During the past fifteen years these have averaged; for the quinquennial period ending 1879-80, 35,591 cwts., valued at R16,45,547; for that ending 1884-85, 36,163 cwts., valued at R21,64,550; and for that ending 1889-90, 33,931 cwt., valued at R23,98,132. Notwithstanding this slight falling off in the average for the past five years, the imports in 1889-90 were high, *viz.*, 39,841 cwt., valued at R28,45,527. The countries from which these supplies were drawn, and the share taken by each importing Presidency

MEDICINE.
466

USES.
467

TRADE.
468

T. 468

TINDER
and Gun-matches.   Trade in Tin; Tinder.

TRADE.

or Province in the trade during the year are shewn as follows by Mr. O'Conor : —

Countries whence imported.	Quantities.	Value.	SHARE TAKEN BY EACH PRESIDENCY OR PROVINCE.		
			Presidency or Province.	Quantities.	Value.
	Cwt.	R		Cwt.	R
United Kingdom .	1,032	76,346	Bengal . .	22,592	16,46,883
China—Hong-Kong.	34	3,420	Bombay . .	14,010	9,65,315
Straits Settlements .	38,770	27,65,429	Sindh . .	13	1,218
Other Countries .	5	332	Madras . .	2,824	2,03,780
			Burma . .	402	28,326
TOTAL .	39,841	28,45,527	TOTAL .	39,841	28,45,527

In addition to the above a small quantity is annually imported for Government purposes. During the five years ending 1879-80, this amounted to 477 cwt., valued at R14,284; during a similar period ending 1884-85, the average fell to 234 cwt., valued at R26,555; while during the past five years it was 271 cwt., valued at R28,873. The largest imports were during the year of minimum price, *viz.*, 1878-79, when they amounted to 1,088 cwt., valued at R39,798. A small proportion of the general imports are re-exported, chiefly from Bombay to Persia and Turkey in Asia. The average of re-exports during the quinquennial period ending 1879-80 was 1,758 cwt., valued at R75,984; during that ending 1884-85, 2,130 cwt., valued at R1,23,134; and during the past five years, 2,814 cwt., valued at R2,01,987.

The exports of Indian tin are very unimportant, but show a slight increase. During the same three quinquennial periods for which figures are detailed above, the averages were, 1875-76 to 1879-80, 256 cwt., valued at R8,093; 1880-81 to 1884-85, 299 cwt., valued at R15,424; 1885-86 to 1889-90, 479 cwt., valued at R28,340. During the past year, 1889-90, the exports of Indian tin rose to 675 cwt., valued at R37,016, the largest recorded. The trade is entirely between Burma and the Straits Settlements. No figures can be given for the internal trade of India itself.

TINDER.

469

Tinder and Gun-matches. Substances used for—

Anaphalis contorta, *Hook. f.*; COMPOSITÆ; *Fl. Br. Ind., III.,* 284 (*Jhúla, bokla, gúfa,* PB.); used for tinder and moxas on the Sutlej (*Stewart*). It is not otherwise of economic value.

Bombax malabaricum, *DC.*; MALVACEÆ; Vol. I., 489, 492.

Borassus flabelliformis, *Linn.*; PALMÆ; Vol. I., 495. The light brown cotton-like substance from the base of the fronds is used for a tinder.

Butea frondosa, *Roxb.*; LEGUMINOSÆ; Vol. I., 51; the fibre from the root-bark.

Camel's dung, Vol. II., 50.

Caryota urens, *Linn.*; PALMÆ; Vol. II., 206; the cotton-like substance similar to that of **Borassus**.

Cocos nucifera, *Linn.*; PALMÆ; Vol. II., 415; the tomentum at the base of the leaf.

Cordia Myxa, *Linn.*; BORAGINEÆ; Vol II., 564.

T. 469

| Substance used for Tinder. | (*J. Murray.*) | TINOSPORA cordifolia. |

Cousinia sp.; *Stewart, Pb. Pl., 125;* COMPOSITÆ (*Kritz trutsa, húsh tsuk, tuse,* PB.) ; the leaves.

Echinops niveus, *Wall.;* COMPOSITÆ; *Fl. Br. Ind., III., 359* (*Brúgh laura brúsh, búsh, tso, púr-cho-bachá,* PB.); the leaves are used in the Panjáb (*Stewart*).

Gerbera lanuginosa, *Benth.;* COMPOSITÆ; Vol. III., 490.

Gnaphalium luteo-album, *Linn.;* COMPOSITÆ; Vol. III., 517.

Nannorhops Ritchieana, *H. Wendl.;* PALMÆ; Vol. V., 317.

Populus euphratica, *Oliv.;* SALICINEÆ; Vol. VI., 327.

Sambucus Ebulus, *Linn.;* CAPRIFOLIACEÆ; Vol. VI., Pt. II., 453.

Ulmus Wallichiana, *Planch.;* URTICACEÆ; Vol. VI., Pt. IV.

Tinkal, see Borax, Vol. I., 504-511.

TINOSPORA, *Miers; Gen. Pl., I., 34, 960.*

[*t. 385, 486;* MENISPERMACEÆ.

Tinospora cordifolia, *Miers; Fl. Br. Ind., I., 97; Wight, Ic.,*   470
   Syn.—T. PALMINERVIS, *Miers;* COCCULUS CORDIFOLIUS, DC.; C. VERRUCOSUS, *Wall.;* C. CONVOLVULACEUS, DC.; MENISPERMUM CORDIFOLIUM, *Willd.*

   Vern.—*Gurach, gúrcha, giloe, gulanchá, gul-bél,* extract=*palo, sut-gilo, satte-gilo,* root=*ghlanchá-ki-jar,* HIND.; *Gulanchá, gurach, giloe, nim-gilo, gadancha,* extract=*paló,* root=*gulanchá-ki-jar,* BENG.; *Gulancha,* URIYA; *Gurjo,* SIKKIM; *Garjo,* NEPAL; *Gulancha, gurcha,* extract=*gulancha, giloi,* KUMAON; Bark=*békh-gilló,* KASHMIR; *Gilo-gularich, sakhmi haiyat, gilo, garúm, garham, batindu,* extract=*palo, sat-gilo,* PB.; Extract=*palo, sut-gilo,* SIND; *Gúlwél,* C. P.; *Gulwail, guloe, gharol, giroli,* BOMB.; *Gula-véli, gulavela, gulwail, guloe, gharol,* MAR.; *Gado, gulvél, galo,* GUZ.; *Gúlwail, gul-bél, gulo,* extract=*palo, sat-gilo, gul-bél-ká-sat.* root=*gul-bél-ki-jar,* DEC.; *Shindil-kodi,* extract=*shindil-shakkarai,* root=*shindil, kodi-vér,* TAM.; *Tippa-tige, gulúchi, gudúchi, guricha, manapála, tippatingai,* extract =*tippa-tige-sattu,* root=*tippa-tége-véru,* TEL.; *Amrita-balli, amrúta balli,* KAN.; *Amrúta valli, citamerdú, amruta, chitrámruta,* MALAY.; *Sinza-manne, singo-moné,* BURM.; *Rassakinda, galúchi,* SING.; *Amrityel, amritwel,* GOA; *Gudúchi, amrita, amurta, sóma valli,* SANS.; *Giló,* extract=*satte-giló,* ARAB.; *Gul-bél,* extract= *stategiló,* PERS.

   References.—*Roxb., Fl. Ind.,* Ed. *C.B.C., 731, 734; Voigt, Hort. Sub. Cal., 330; Brandis, For. Fl., 8; Kurz, For. Fl. Burm., I., 52; Gamble, Man. Timb., 11; Thwaites, En. Ceylon. Pl., 11; Dalz. & Gibs., Bomb. Fl., 5; Stewart, Pb. Pl., 6; Elliot, Fl. Andhr., 57, 63, 111, 182; Rheede, Hort. Mal., VII., t. 21; Pharm. Ind., 9, 435; Ainslie, Mat. Ind., II., 377; O'Shaughnessy, Beng. Dispens., 198; Moodeen Sheriff, Supp. Pharm. Ind., 244; U. C. Dutt, Mat. Med. Hind., 105, 299; Dymock, Mat. Med. W. Ind., 2nd Ed., 30; Fleming, Med. Pl. & Drugs, as in As. Res., XI., 171; Flück. & Hanb., Pharmacog., 33; Bent. & Trim., Med. Pl., t. 12; Murray, Pl. & Drugs, Sind, 37; Moodeen Sheriff, Mat. Med. S. Ind.* (in *MSS.*), 9; *Dymock, Warden & Hooper, Pharmacog. Ind., I., 54; Med. Topog., Ajmir, 48; Irvine, Mat. Med. Patna, 74, 75; Trans., Med. & Phys. Soc., Bomb.* (New Series), *No. 4, 152; Cat., Baroda Durbar, Col. & Ind. Exhib., 175; Atkinson, Him. Dist., 304, 752; Drury, U. Pl., 427; Lisboa, U. Pl. Bomb., 261; Birdwood, Bomb. Pr., 4; Boswell, Man. Nellore, 139; Gribble, Man, Cuddapah, 199; Settle. Rep., Panjáb, Montgomery, 20; Central Provs., Chanda, App., vi.; Taylor, Topog. of Dacca, 53; Gazetteers :— Mysore & Coorg, I., 65; Bombay, V., 23; VI., 15; XVI., 323; N.-W. P., I., 78; IV., lxvii.; Orissa, II., 68; Montgomery, 20; Agri.-Horti. Soc. Ind.:—Journ.* (Old Series), *IV., 110, IX., 401,* (New Series), *V., Pro., 48; Agri-Horti. Soc., Panjáb, Select Papers to 1862; Ind. Forester XII., App., 5.*

T. 470

FIBRE.

Roots.
471
MEDICINE.

Stem.
472
Leaves.
473
Root.
474
Extract.
475
Plant.
476

Habitat.—A climbing shrub found throughout Tropical India from Kumáon to Assam and Burma, and from Behar and the Konkan to the Karnatic and Ceylon.

Fibre.—The ærial ROOTS are used for tying bundles (*Dymock*).

Medicine.—The STEM, LEAVES, ROOT, and a watery EXTRACT of this plant have long been valued in Hindu medicine. The entire PLANT is regarded as a valuable alterative and tonic, and is used in general debility, fever, jaundice, skin diseases, rheumatism, urinary diseases, dyspepsia, &c. It enters into the composition of many elaborate prescriptions by Chakradatta and other Sanskrit writers, most of which contain many other more or less powerful medicines. For a description of the most important of these the reader is referred to U. C. Dutt's *Mat. Med. Hindus.* The plant early attracted the notice of European physicians in India. Fleming remarked on its use as a febrifuge, and as a tonic in gout; Ainslie described the root as a powerful emetic, and as a popular remedy for snake-bite, and Taylor states that the watery extract was, in his time, administered for leprosy. A little later, a paper was published on the subject by Dewan Ramkamal Sen, in which he described the method of preparing the extract, also a decoction of the stem, root and leaves. O'Shaughnessy states the extract was found to be a very useful tonic in several trials made at the Calcutta College Hospital, though decided febrifuge effects could scarcely be attributed to it. The decoction or cold infusion he described as of very great utility in the treatment of several cases of chronic rheumatism, and of secondary syphilis, its action being decidedly diuretic and tonic in a very high degree.

Gulancha was accordingly admitted to the *Bengal Pharmacopœia* of 1844 and the Indian *Pharmacopœia* of 1868. In the latter a tincture, infusion, and watery extract (which correspond to the *palo* of the Natives) are officinal. Waring corroborates O'Shaughnessy's statements regarding its greater value as a tonic than as an antiperiodic. In Native practice it is valued as a tonic and alterative, antiperiodic and aphrodisiac. "The fresh plant is said to be more efficient than the dry; it is taken with milk in rheumatism, acidity of the urine and dyspepsia. The JUICE with *pakhanbed* and honey is given in gonorrhœa, and is an ingredient in *paushtiks*, given in phthisis. In Guzerát a necklace called *Kamalá-ni-málá* (jaundice necklace) made of small pieces of the stem, is supposed to cure that disease" (*Pharmacog. Ind.*). Moodeen Sheriff considers the drug alterative, tonic, and antipyretic, and states that the root and stem, but especially the extract, are useful in slight cases of fevers, in secondary syphilis and rheumatism, jaundice, general debility after long-standing sickness, and splenic affections. He suggests that one or other of the officinal preparations might be substituted for quinine, James' powder, potassic iodide and sarsaparilla. His remarks in the paragraph of Special Opinions below are interesting and worthy of attention. From all obtainable evidence there appears to be little doubt of the valuable tonic and alterative properties of the drug; whether it is in reality antiperiodic or antipyretic is on the other hand doubtful. It has never come into general use in European practice, nor does it even appear to have attracted much attention out of India. In 1884, however, it again formed the subject of investigation and was reintroduced to the notice of the profession in Europe, as a specific tonic, antiperiodic, and diuretic.

Chemistry.
477

CHEMICAL COMPOSTION.—"The extract called *palo* and *sat-i-galoe* is simply starch, which, though not having been washed, retains some bitterness; that sold in the bazárs is usually nothing but common starch. The stem has been examined by Fluckiger (1884) by boiling it with alcohol and a little hydrate of calcium, the alcohol was then evaporated and the

MEDICINE.

residue extracted by means of chloroform. The latter liquid was found to contain an alkaloid in very small quantity; on evaporating it and dissolving the residues by means of acidulated water, a solution was obtained, which proved to contain merely a trace of *berberin*. The alcoholic extract, after it had been exhausted by chloroform as above stated, was dissolved in boiling water and precipitated by tannic acid, avoiding an excess of the acid. The deposit thus obtained was mixed with carbonate of lead, dried and exhausted with alcohol, which, on evaporation, yielded the bitter principle. By boiling this bitter principle with dilute sulphuric acid, sugar was produced, and it lost its bitterness. Neither the original bitter principle nor the product derived from it could be crystallised" (*Pharmacog. Ind.*).

TRADE.—The stem and extract can be obtained in all bazárs. The former costs wholesale R1½ per maund (*Moodeen Sheriff*), R2½ per Surat maund of 37½℔ (*Dymock*); retail about 2 annas per ℔. The watery extract fetches R30 per maund wholesale, R1½ per seer retail (*Moodeen Sheriff*). The average annual export of the extract from the Kumáon forests is said to be about 2 maunds. The root is not generally obtainable in the bazár, but can be procured without difficulty at the cost of collection.

Trade
478

SPECIAL OPINIONS.—§ "There is a general belief amongst the Muhammadans of India that the *Gul-bél* growing on a Margosa tree is more efficacious as a medicine than that which may be found on other kinds of trees, hedges, etc, and they, therefore, cultivate the plant in their own houses and gardens, and make it run over **Melia Azadirachta**. The watery extract of the plant (*Satte-giló* or *palò*) is greatly used as a remedy in fevers, and is called 'Indian Quinine' by some *Hakims*. This is, of course, a great exaggeration of its value, but there is no doubt that it is a very useful drug, especially in some very obstinate, low, and long standing remittent and typhoid fevers. Its action is generally more satisfactory when employed in combination with other drugs of similar medicinal properties. Although it is not a costly drug, yet it is generally substituted by, or adulterated with, many cheaper substances in the bazárs of India, so much so, that out of the seven specimens I have received from different places, including Calcutta, Hyderabad, and Lucknow, none is found to be genuine. They are all very bitter and of various colour; whereas the real *Satte-giló* is either tasteless or slightly bitterish in taste, white in colour, if it is prepared from the roots of T. **cordifolia**, and greenish white or greenish brown, if from the stems. It occurs in powder or loose and flat cake-like pieces. It is prepared easily, by cutting the fresh stem into small pieces, which are bruised in a stone or wooden mortar and soaked in water from six to twelve hours. The mass is then rubbed, squeezed, and separated from the fluid by straining the latter through cloth. The fluid being evaporated to dryness in the sun, the residue is the *Satte-giló*. The last process is generally repeated several times with a view to make the drug whiter, but such repetitions, in my opinion, are not only unnecessary, but detrimental to its effects. The thicker the *gul-bél* or its root is, the better for the preparation of the *Satte-giló*, and the plant is known to attain sometimes the thickness of a man's arm. A transverse section of the stem and root shows the wood to consists of a very porous tissue, traversed by conspicuous medullary rays, with or without concentric zones. The specimens of the *Satte-giló* in my collection at Calcutta (Nos. 17 and 18) are genuine, being prepared by myself for the late International Exhibition at that place, the former from the stems, and the latter from the roots. The dose of the extract is from one to two drachms" (*Honorary Surgeon Moodeen Sheriff, Khan Bahadur, Triplicane, Madras*). "Have

5

MEDICINE.

used it a little in the shape of decoction of the root, as an antiperiodic, but not successfully" (*Surgeon D. Picachy, Purneah*). "*Gulancha*, which grows on *Nim* trees, is considered most efficacious for remedial purposes" (*Surgeon Anund Chunder Mukerji, Noakhally*). "Tonic and antiperiodic" (*Assistant Surgeon Nehal Sing, Saharunpore*). The bazár extract is starchy in its nature and does not contain the bitter principle of the plant" (*Surgeon-Major Robb, Civil Surgeon, Ahmedabad*). "Bitter tonic and diuretic in the form of decoction" (*Civil Surgeon C. M. Russel, Sarun*). "*Goluncho* is an excellent substitute for Calumba. The starch prepared from the plant is much used, but it is not so efficacious as the infusion" (*Surgeon R. L. Dutt, M.D., Pubna*). "Tonic,—useful in chronic rheumatism and fevers" (*Assistant Surgeon S. C. Bhattacharji, Chanda*). Useful in chronic rheumatism and secondary syphilis" (*Surgeon-Major H. J. Hazlitt, Ootacamund*). "Is a favourite dipensary remedy as a tonic and alterative in the Behar district" (*Surgeon R. D. Murray, M.B., Burdwan*). "In the form of infusion I have heard it recommended for chronic gonorrhœa, leucorrhœa, &c. It is, I believe, employed in leprosy, and the wandering devotees of this country use it as an anaphrodisiac" (*Surgeon J. Ffrench Muller, M.D., Saidpore*). "Used in cases of insanity" (*W. Forsyth, F.R.C.S. Edin., Civil Medical Officer, Dinagepore*). "An extract made from the stems by crushing and infusion, is much valued by the natives as an antiperiodic. It is given in doses of 5 grains, and commands a high price in the bazár. It is superior to cinchona febrifuge" (*Narain Misser, Kathe Bazar Dispensary, Hoshangabad, Central Provinces*). "An antiperiodic and bitter tonic, but inferior to other similar drugs" (*Civil Surgeon S. M. Shircore, Moorshedabad*). "An excellent tonic, preferably given as a decoction of root and stems" (*Civil Surgeon G. C. Ross, Delhi, Panjáb*). "It is much used here in the form of the infusion, either alone or combined with acetate of ammonia, and found useful in ordinary cases of intermittents and other mild forms of fevers" (*Apothecary J. G. Ashworth, Kumbakonam, Madras*). "Is a valuable tonic and febrifuge, used in intermittent fever with great benefit" (*Assistant Surgeon R. Gupta, Bankipur*).

FODDER.
Leaves.
479
Stems.
480
Ærial Roots.
481
DOMESTIC &
SACRED.
Plant.
482

Fodder.—The LEAVES are good fodder for cattle; the STEMS and ÆRIAL ROOTS are much liked by elephants.

Domestic and Sacred.—Ainslie remarks that the PLANT is often bruised and put into water, the liquid thus formed being "drunk by the Brahmins at some of their religious ceremonies. It is stated in the Nasik Gazetteer (*Bombay, XVI., 323*) that an intoxicating liquid called *bhoja* is prepared by boiling the seeds of old *joari* (**Sorghum vulgare**), *gulvel* (**Tinospora cordifolia**), *bháng*, and *kuchala* (**Strychnos Nux-vomica**) in water. [Could the Sanskrit name *Soma valli* be in any way connected with this remarkable property? *Ed.*]

483

Tinospora crispa, *Miers; Fl. Br. Ind., I., 96.*

Syn.—MENISPERMUM CRISPUM, *Linn.*; M. VERRUCOSUM, *Roxb.*; M. TUBERCULATUM, *Lamk.*; COCCULUS CRISPUS, *DC.*; C. VERRUCOSUS, *Wall.*

Vern.—Same as those of T. cordifolia.

References.—*Roxb., Fl. Ind., Ed. C.B.C., 730; Kurz., For. Fl., Burm., I., 52; O'Shaughnessy, Beng. Dispens., 202; Fleming, Med. Pl. & Drugs, as in As. Res., XI., 171; Pharm. Ind., II; Flück. & Hanb., Pharmacog., 34; Dymock, Warden & Hooper, Pharmacog. Ind., I., 64; Agri.-Horti.-Soc. Ind., Journ. (New Series), V., Pro., 48.*

Habitat.—Met with from Sylhet and Assam to Pegu and Malacca.

MEDICINE.
484

Medicine.—It possesses the bitterness, and probably the tonic properties, of *gulancha*, and is known by the same vernacular names. According

| Occurrence in India of Titaniferous Ores. (*J. Murray.*) | TODDALIA aculeata. |

to Captain Wright, quoted by Fleming, it is as powerful a febrifuge as Cinchona.

TITANIUM.

485

This metal does not occur native, but is generally found in combination with oxygen and iron as *titaniferous iron, mechachanite* or *ilmenite.* Combined with oxygen alone it forms *titanic-acid*, or *rutile.*

Titanium, *Ball, Man. Geol. Ind., III. (Economic), 323.*
References. -- *Mallet, Mineralogy ; Ball, Mem. Geol. Surv. Ind., XVIII., 43 ; Hackett, Rec. Geol. Surv. Ind., XIII., 249.*
Occurrence.—" The distribution of titaniferous iron ores in India is not very well known, but it is probable that with the black magnetic iron sands which are found in the beds of streams traversing the metamorphic rocks at intervals all over the peninsula, more or less titaniferous iron would generally be found to be associated.

OCCURRENCE.
486

Bengal.—" In the south-eastern portion of the district of Mánbhúm in Bengal, more especially in the neighbourhood of the village and thanna of Supur, large masses of ilmenite are sometimes to be seen weathered out from the quartz veins and lying strewn over the surface ; occasionally, too, lamellar plates or seams have been seen there *in situ* in quartz veins " (*Ball*).

Bengal.
487

Rajputana.—According to Mr. Hackett *rutile* occurs in certain quartz veins which traverse the Arvali rocks of the Motidongri range, a short distance south of Alwar.

Rajputana.
488

Tobacco, see Nicotiana Tabacum, *Linn.* ; SOLANACEÆ, Vol. V., 353-428.
Tobacco, Mountain—see Arnica montana, *Linn.* ; COMPOSITÆ, Vol. I.,
[318

TODDALIA, *Pers. ; Gen. Pl., I., 300.*

Toddalia aculeata, *Pers. ; Fl. Br. Ind., I., 497* ; RUTACEÆ.
Syn.—T. ASIATICA, NITIDA, ANGUSTIFOLIA, *Lamk.* ; T. ANGUSTIFOLIA, *Miq.* ; T. RUBRICAULIS, *Willd.* ; T. FLORIBUNDA, *Wall.* ; ZANTHOXYLON NITIDUM, & FLORIBUNDUM, *Wall.* ; SCOPOLIA ACULEATA, *Sm.* ; LIMONIA OLIGANDRA, *Dalz.* ; PANLLINIA ASIATICA, *Linn.*
Vern.—*Kanj, dahan, jangli-káli-mirchi, jangli-kali-mirch,* root-bark= *jangli-káli-mirch-ki-jar-ki-chhál,* HIND. ; *Kada-todali,* BENG. ; *Meinkara,* NEPAL ; *Saphijirik,* LEPCHA ; *Tundupará, tundapará,* URIYA ; *Kunj,* KUMAON ; *Dahan, lahan,* RAJ. ; *Jangli-kali-mirchi,* root-bark=*jangli-káli-mirchi-ki-jar-ki-chhál,* DEC. ; *Jún-li-káli-mirchi,* BOMB. ; *Limri,* MAR. ; *Mila-karanai, milkaranai, múlacarnai,* the root bark=*milakarnai-vér-pattai,* TAM. ; *Konda, kashinda, konda kahinda, konda-cahinda, mirapa-kándra, mirapagandra, varra kasimi, varragoki,* root-bark=*mirapa-kándra-véru-patta, kondakasin da-véru-patt ,* TEL. ; *Kyan-zah,* BURM. - *Kaka-toddali, kákka-totali, totuli, mulaka-táni,* root-bark=*totali-véra, tola, kakka-totali-véra-tóla,* MALAY. ; *Kudu-mirish, kúdúmiris-wel,* root-bark=*kudu-mirish múl-potta,* SING. ; *Kánchana, dahana,* SANS.
References.—*Roxb., Fl. Ind., Ed. C.B C., 207 ; Voigt, Hort. Sub. Cal., 186 ; Brandis, For. Fl., 46 ; Kurz, For. Fl. Burm., I. 183 ; Beddome, Anal. Gen , t , vi., f. 4 ; For. Man., xlii. ; Gamble, Man. Timb., 61 ; Thwaites, En. Ceyl. Pl., 69 ; Dalz & Gibs., Bomb. Fl., 46 ; Graham, Cat. Bomb. Pl., 37 ; Sir W. Elliot, Fl. Andhr., 96, 115, 189 ; Rheede, Hort. Mal., V., 81, t. 41 ; Thesaurus, Zey., 58, t. 24 ; Pharm. Ind., 47, 442 ; Flück. & Hanb., Pharmacog., 111 ; Ainslie,Mat. Ind., II., 200,A'O'Shaughnessy, Beng Dispens., 265 ; Irvine, Mat. Med. Patna, 55 ; Moodeen Sheriff, Supp. Pharm. Ind., 263 ; Mat. Med. S. Ind. (in MSS.), 75 ; Bent. & Trim., Med. Pl., t. 49 ; Dymock, Mat. Med. W. Ind., 2nd Ed., 129 ; Dymock, Warden, & Hooper, Pharmacog. Ind., I., 260 ; Official Corresp. on Proposed New Pharm. Ind., 230, 235, 293 ; Drury, U. Pl. Ind., 428 ; Atkinson, Him. Dist.(X., N.-W. P. Gaz.), 307, 752 ; Useful*

489

T. 489

TODDALIA
aculeata.

A valuable medicinal product.

Pl. Bomb. (XXV., Bomb. Gaz.), 148; Bidie, Prod. S. Ind., 5, 108; Boswell, Man. Nellore, 128; Gribble, Man. Cuddapah, 199; Gazetteers:—Bombay, XV., 429; Orissa, II., 181; Ind Forester, III., 238; IX., 451.

Habitat.—A rambling shrub, found in the Sub-tropical Himálaya, from Kumaon eastwards to Bhután, ascending to 5,000 feet; also in the Khásia mountains, ascending to 6,000 feet, and throughout the Western Peninsula and Ceylon.

DYE.
Root-Bark.
490

Dye.—Dr. Bidie remarks that the ROOT-BARK is used in Madras as a yellow dye-stuff. It yields its colouring matter to water, but nothing is known as to how the colour may be fixed.

OIL.
Leaves.
491

Oil.—The authors of the *Pharmacographia Indica* state that the LEAVES yield, on distillation, a pale yellowish-green limpid oil, having the odour of citron peel, and a bitter and aromatic taste. It has a specific gravity at 17°C of ·873; examined by polarized light in a tube of 200 mm., it rotates 15·30° to the left; it has no constant boiling point, but the greater part distils over between 190° and 210°.

MEDICINE.
Fruit.
492
Root.
493
Leaves.
494
Bark.
495

Medicine.—This plant early attracted European attention in India, as perhaps one of the most valuable of Indian medicinal products. Rheede mentions the unripe FRUIT and ROOT, stating that they are rubbed down with oil to make a stimulant liniment for rheumatism. Roxburgh writes, "Every part of this shrub has a strong pungent taste. The roots when fresh cut smell particularly so. The fresh LEAVES are eaten raw for pains in the bowels; the ripe berries are fully as pungent as black pepper, and with nearly the same kind of pungency; they are pickled by the Natives, and a most excellent one they make. The fresh BARK of the root is administered by the Telenga physicians for the cure of that sort of remittent, commonly called the hill, fever. I conceive every part of this plant to be possessed of strong stimulating powers, and have no doubt but under proper management it might prove a valuable medicine where stimulants are required." Ainslie repeats Roxburgh's remarks, and adds that, in Southern India, the root is considered stomachic and tonic, being given in weak infusion to the quantity of half a tea cupful in the course of the day. Later writers have done little to add to our knowledge of the virtues of this medicine. O'Shaughnessy simply reiterates the above, adding that the root-bark deserves most careful trial and will in all probability be found a very valuable medicine. Dr. G. Bidie, some years later, gave strong testimony in favour of its stimulant, tonic, and carminative properties, adding that he knew of no single remedy in which all these qualities were so happily combined. It was accordingly made officinal in the compilation of the *Indian Pharmacopœia*, where its administration is recommended in "constitutional debility and convalescence after febrile and other exhausting diseases." The officinal preparations are a tincture and an infusion of the root-bark, which is advised to be used fresh. Mooden Sheriff has lately written in the highest terms of the root-bark, which he considers to be decidedly diaphoretic, antiperiodic, and antipyretic, in addition to the qualities generally ascribed to it. His remarks will be found below in the paragraph of Special Opinions.

Flückiger & Hanbury inform us that the drug once enjoyed some celebrity in Europe under the name of "Lopez Root," or more precisely "Radix Indica Lopeziana," or "Root of Juan Lopez Pigneiro." It was first made known in 1671 from specimens obtained by Pigneiro at the mouth of the Zambesi. The drug was actually introduced into European medicine by Gaubins in 1771 as a remedy for diarrhœa, and acquired so much reputation that it was admitted to the *Edinburgh Pharmacopœia* of 1792. It appears to have been for sometime imported from Goa, but its

MEDICINE.

botanical origin and source were entirely unknown, and it was always extremely rare and costly. It has long been obsolete in all European countries except Holland, where, until recently, it was to be met with in the shops. The *Pharmacopœia Neerlandica* of 1851 says of it, "Origo botanica perquam dubia—Patria malacca?" (*Pharmacographia*). The disuse into which the drug has fallen in Europe contrasts strongly with the opinion expressed almost unanimously in its favour by Indian physicians. This may, however, most probably be due to the root-bark becoming inferior, if not, inert, when old and dry.

Chemistry.
496

CHEMICAL COMPOSITION.—"None of the constituents of the Toddalia root of India have yet been satisfactorily examined. The bark contains an essential oil, which would be better examined from fresh than from dry material. The tissue of the bark is but little coloured by salts of iron. In the aqueous infusion, tannic acid produces an abundant precipitate, probably of an indifferent bitter principle, rather than of an alkaloid. We have been unable to detect the presence in the bark of *berberine.*" (*Flückiger & Hanbury*). It is to be regretted that no Indian chemist has as yet worked up the composition of the fresh material, as suggested by the authors of the above passage. Such an analysis might go a long way towards confirming, or otherwise, the statements of those who hold in such strong terms that it is a valuable antipyretic, and might result in the separation of a valuable alkaloid. In any case it would be extremely satisfactory to have an authentic analysis of this, one of the most highly praised, and at the same time cheapest of Indian medicines.

SPECIAL OPINIONS. — § "I have been using the root-bark in my practice for the last sixteen or seventeen years, and do not hesitate to say that it is, as an antiperiodic and antiphyretic, equal, if not superior, to quinine and other alkaloids of Cinchona and to Warburg's tincture, respectively. As a diaphoretic, it is decidedly superior to Pulv. Jacobi Vera; and as a tonic to Gentian and Calumba. It is highly useful in effecting a cure in all idiopathic and uncomplicated fevers, whether periodical or continued. It is best used in tincture and decoction, and I make these preparations three or four times stronger than those generally in use. This is the chief reason, I think, which has rendered the drug so successful in my hands. The analogy between the medicinal properties of the root-bark of T. aculeata and those of the root of Berberis aristata and a few other species of Berberis is very great and complete, there being no difference whatever. Therefore, everything I have said about the preparations, doses, therapeutic uses and the manner of using the tincture and decoction of the latter, is quite applicable to those of the former" (*see* B. aristata, *D C.*; Vol. I., 443). "The drug under consideration, however, has one great advantage over the root of B. aristata and other species of Berberis— namely, that it is procurable in every large bazár of Southern India; whereas the root of the latter plants, must be procured from distant places, such as the Nilghiris, Shevaroy Hills, Central and Northern India, &c." (*Honorary Surgeon Moodeen Sheriff, Khan Bahadur, Triplicane, Madras*). "A very valuable tonic, &c." (*Apothecary T. Ward, Madanapalle*). "Used in Madras Hospital" (*Surgeon-General W. R. Cornish, F.R.C.S., C.I.E., Madras*). "This has been only lately used here in one case of debility after fever, and as a tonic proved useful" (*Apothecary J. G. Ashworth, Kumbakonam, Madras*).

Food.—The fresh LEAVES are eaten raw, and the ripe BERRIES are pickled by the Natives of the Coromandel coast; both have a strong pungent taste.

FOOD.
Leaves.
497
Berries
498

T. 498

Toddy, see **Borassus flabelliformis,** *Linn.,* Vol. I., 495 ; **Caryota urens,** *Linn.,* Vol. II., 208 ; **Cocos nucifera,** *Linn.,* Vol. II., 446, 448, 449, 451, 454, 457 ; **Melia Azadirachta,** Vol. V., 211 ; **Narcotics,** Vol. V., 334, **Phœnix dactylifera,** *Linn.,* and **P. sylvestris,** *Roxb.,* Vol. VI., 199, 215 ; **Sugar (Sacharum)** Vol. VI. Pt. II., 115, 118, 138, 226-227, 231, 266, 270, 310, 352, 361, 370 ; **Spirits,** Vol. VI., Pt. III.

Tomato, see **Lycopersicum esculentum,** *Mill. ;* Solanaceæ ; Vol. V., 100.

TOOLS & TOOL-HANDLES.

499

Tools & Tool-handles, Timbers used for—

Acacia arabica, *Willd.*
Anogeissus latifolia, *Wall.*
Bamboos, various species.
Carapa moluccensis, *Lam.*
Cassia marginata, *Roxb.*
Cratoxylon neriifolium, *Kurz.*
Dalbergia cultrata, *Grah.*
D. latifolia, *Roxb.*
Dodonæa viscosa, *Linn.*
Eugenia tetragona, *Wight.*
Grewia oppositifolia, *Roxb.*
G. tiliæfolia, *Vahl.*

Lagerstrœmia parviflora, *Roxb.*
Melanorrhœa usitata, *Wall.*
Mesua ferrea, *Linn.*
Murraya exotica, *Linn.*
Putranjiva Roxburghii, *Wall.*
Quercus Ilex, *Linn.*
Rhododendron arboreum, *Sm.*
Rhus Wallichii, *Hook.f.*
Shorea obtusa, *Wall.*
Terminalia paniculata, *W. & A.*
Xylia dolabriformis, *Benth.*

The above enumeration includes the timbers mentioned by such author-ities as Brandis, Stewart, Gamble, and others, as being specially em-ployed for making the handles and shafts of axes, hammers, chisels, and such tools. The reader is referred for information regarding each, to its alphabetical position in this work.

Toon Wood, see **Cedrela Toona,** *Roxb. ;* Meliaceæ ; Vol. II., 232.

TOPAZ, *Ball, Man. Geol. Ind., III.,* 530.

500

Topaz may be defined as a fluo-silicate of alumina, which may be represented by the formula $Al_2O_5Si.$, with part of its oxygen replaced with fluorine. In crystalline structure it belongs to the trimetric system, but the prisms are differently modified at either extremity. The Topaz is only found in metamorphic rocks or in the veins which traverse these. In colour it varies remarkably. The so-called "Brazilian ruby" is a yellow topaz, which becomes pale pink on exposure to heat. The stones obtained from Siberia have a bluish tinge. The Saxon topazes are of a pale wine-yellow but become limpid on exposure to heat. Those found on the Scotch Highlands are of a sky-blue colour. The present and most valued stones are obtained in Brazil, where they are termed *Goutte d'Eau* (drops of water), but in trade are often spoken of as "Brazilian Sapphires." They are of a deep celestial blue, and when cut in facets they closely resemble the dia-mond in lustre and brilliance. Brazilian rubies and sapphires may readily be distinguished from the true gems by their inferior hardness, though they are much superior in that respect to rock crystals. The "oriental topaz" is in reality a yellow sapphire or corundum. Of the occurrence of topaz in India, Ball says "there appears to be no authentic record, a reported dis-covery in the basalt of the Rajmahal hills being open to question. Ceylon, it is believed, yields a not inconsiderable proportion of the Topaz of Com-merce."

For further information consult the article on the Inferior Gems given under "Carnelian," Vol. II., 167.—*Ed., Dict. Econ. Prod.*

T. 500

TORENIA, *Linn.; Gen. Pl., II., 954.* 501

Torenia.—A genus of glabrous or pubescent herbs, which belongs to the Natural Order SCROPHULARINEÆ, and comprises some twenty species, indigenous in Tropical Asia and Africa. Of these about twelve are natives of India, but only one possesses any economic interest, and that of the smallest kind. This plant, **T. asiatica,** *Linn., Fl. Br. Ind., IV., 277,* is known on the Malabar coast as *kákápu.* Rheede informs us that the JUICE of its LEAVES was considered, during his time, to be a cure for gonorrhœa. Modern writers, however, do not mention it, with the exception of Dymock, who does so only to quote Rheede's remark, and to observe that in the Konkan, where it is pretty common, it possesses no reputation as a medicine.

MEDICINE.
Leaves.
502
Juice.
503

Tortoises and **Turtles,** see Reptiles, Vol. VI., Pt. I., 428-435.

TRADESCANTIA, *Linn.; Gen. Pl., III., 853.*

Tradescantia axillaris, *Willd.; Roxb., Fl. Ind., Ed. C.B.C., 280;* 504
[COMMELINACEÆ.

    Vern.—*Baga nella,* HIND.; *Gola gandi,* TEL.
    References.—*Mason, Burma & Its People,* 435; *Dymock, Mat. Med. W. Ind.,* 2nd Ed., 842; *Rheede, Hort. Mal., X., t. 13; Cor. Pl., 2, t. 107.*
    Habitat.— An annual, native of moist pasture ground, borders of rice-fields, etc., throughout India.
    Medicine.—According to Rheede a decoction of this PLANT is considered a useful remedy on the Malabar coast in cases of tympanitis.
    Food and Fodder.—In the last Deccan famine the SEEDS were largely used by the people as food-grains (*Dymock*). Roxburgh states that cattle are very fond of the PLANT.

MEDICINE.
Plant.,
505
FOOD &
FODDER.
Seeds.
506
Plant.
507
508

TRAGACANTH.

Tragacanth.—A gum obtained from several species of **Astragalus,** which mostly inhabit South Europe, Asia Minor, and Persia, but none of which are natives of India (*see* Vol. I., 348). It is of a dull white colour, translucent, inodorous, and tasteless. In India the gums of the following are employed as substitutes for Tragacanth :—

    Cochlospermum Gossypium, *DC.;* Vol. II., 413.
    Sterculia urens, *Roxb.,* Vol. VI., Pt. III.
    Hog Tragacanth, the produce of **Prunus Amygdalus,** imported into Bombay from Persia (*see* VI., Pt. I., 343). [For further information consult the article **Bassora.** I., 417.—*Ed.*]

TRAGIA, *Linn.; Gen. Pl., III., 329.*

Tragia involucrata, *Linn.; Fl. Br. Ind., V., 465;* EUPHORBIACEÆ. 509
    Var. α, **proper,**—*var.* α, RHEEDIANA, δ, GENUINA, & γ, HISPIDA, *Muell.;* T. HISPIDA, *Willd.*
      β, **cordata,**—*var.* CORDATA, *Muell.;* T. CORDATA, *Heyne; var.* MONTANA, *Thwaites;* T. MONTANA, *Muell.*
      γ, **angustifolia.**
      δ, **cannabina,** *var.* ε, INTERMEDIA & ζ, CANNABINA, *Muell, Arg.;* T. CANNABINA, *Linn.;* T. HISPIDA, *Herb. Russell;* CROTON HASTATUS & URENS, *Linn.*
    Vern.—*Barhantá,* HIND.; *Bichati,* BENG.; *Sengel sing,* SANTAL; *Kánchkúre,* DEC.; *Kánchkúri, kháj-kolti,* BOMB.; *Kán chúri-vayr,* var. cannabina=*sirru-kánchori-vayr,* TAM.; *China-dula gondi, révuli-dula gondi, triinadula gondi, duruda-gunti, tella dura dagondi,* TEL.; *Vrischikáli, dúst parisha, kásághini,* SANS.

References.—*Roxb., Fl. Ind., Ed. C.B.C., 651, 652; Kurz, For. Fl. Burm.. II., 398; Thwaites, En. Ceyl. Pl., 270; Dalz. & Gibs., Bomb. Fl., 228; Grah., Cat. Bomb. Pl., 186; Burm., Fl. Zeyl., 202, t. 92; Rheede, Hort. Mal., II., t. 39; Sir W. Elliot, Fl. Andr., 39, 47, 164; Rev. A. Campbell, Ec. Pl. Chutia Nagpur, No. 75, 79; Ainslie, Mat; Ind., II., 61, 62, 389; U. C. Dutt, Mat. Med. Hind., 324; Dymock, Mat. Med. W. Ind., 2nd Ed., 717, 718; Taylor, Topog. of Dacca, 54; Boswell, Man. Nellore, 122, 137; Bidie, Prod. S. Ind., 58; Gazetteers:— Bombay, XV., 443; N.-W. P., IV., lxxvii.*

Habitat.—A common stinging weed, found in dry places throughout India, from the Parjáb and Lower Himálaya of Kumáon eastwards to Assam, and southwards to Burma, Travancore, and Ceylon.

MEDICINE.

Root.
510

Leaves.
511

Medicine.—Rheede noticed the medicinal properties of the ROOT, stating that it was valued in febricula and in itching of the skin. Ainslie writes, "The *Vytians* reckon it amongst those medicines which they conceive to possess virtues in altering and correcting the habit, in cases of *mayghum* (cachexia) and in old venereal complaints attended with anomalous symptoms. By the Hindu doctors of the Coromandel coast it is given in quantity of half a tea-cupful of the decoction twice daily. Taylor states that the root forms the basis of an external application in leprosy, while the LEAVES dried, reduced to powder, and mixed with ginger and *kaiphul*, form an "errhine," which is prescribed in cases of headache. In the Konkan, according to Dymock, the root is used to aid the extraction of guinea-worm, a paste made from them being applied to the part. A paste with *tulsi* juice is also employed in the same locality as a cure for itchy skin eruptions. Campbell notes, that in Chutia Nagpur, the root is given when the extremities are cold during fever; also for pains in the legs and arms. Ainslie describes *var.* cannabina separately, writing that "The root, which is sometimes called *coorundootie vayr*, has, in its dried state, but little taste or smell, though in its more succulent condition it has a rather pleasant odour; it is considered as diaphoretic and alterative, and is prescribed in decoction, together with other articles of like virtues to correct the habit; an infusion of it is also given as a drink in ardent fever, in the quantity of half a tea-cupful twice daily."

Fruit.
512

SPECIAL OPINIONS.—§ "The FRUIT rubbed over the head with a little water is useful in cases of baldness" (*Civil Surgeon J. H. Thornton, B.A., M.B., Monghyr*). "The root, boiled in milk, is an occasional remedy for dry cough" (*T. Ruthnam Moodelliar, Chingleput, Madras Presidency*).

TRAGOPOGON, *Linn.; Gen. Pl., II., 530.*

513

Tragopogon porrifolium, *Linn.; Fl. Br. Ind., III., 418;*

THE SALSIFY.   [COMPOSITÆ.

References.—*Stewart, Pb. Pl., 131; DC., Orig. Cult. Pl., 44; Jour. Agri.-Horti. Soc., IV. (New Series), 37.*

Habitat.—A herb, with milky-juice, found wild in Greece, Dalmatia, Italy, and Algeria (*DeCandolle*); also in Western Tibet, and in cultivated places near Simla.

FOOD.
Root.
514
515

Food.—The young ROOT is eaten as a vegetable in Lahoul (*Stewart*). [Its uses are similar to those of **Scorzonera**, which see—*Ed.*]

TRAP.

Trap.—The Editor is indebted for the following note to Mr. H. B. Medlicott, late Director of the Geological Survey. For further economic information the reader is referred to the article **Stones**, *Vol. VI., Pt. III.*

| The Singhara Nut. | (*J. Murray.*) | TRAPA bispinosa. |

GREENSTONE, WHINSTONE (*including* Basalt, Dolerite, Diorite, etc.). TRAPP, *Fr.;* TRAPPE, BASALT, DOLERIT, ANAMESITE, *Germ.;* BASALTITE, *Ital.*

Rocks.

The greater part of Western India, particularly in the Deccan and the Central Provinces, is occupied by vast trappean or basaltic accumulations. From nor*h to south these rocks extend continuously from a point 100 miles south of Gwalior to the vicinity of Goa, and from west to east from Bombay to Amarkunkak, covering an area of above one-sixth of the Peninsula south of the Ganges. Great outliers of the same formation occur to the west in Káthiáwár and Cutch, and eastwards in Chutia Nagpur, and stretching southwards through Bustan to Rajamundry on the Godavari. A similar basaltic formation, but of different age, forms the whole of the Rajmahal hills, in Bengal, and reappears again over a small area at the base of the Khásia hills in Eastern Bengal.

In the form of dykes trap of the same ages occurs extensively in the coal fields and elsewhere. Dykes of much older dioritic trap occur freely, sometimes in great profusion, in the metamorphic rocks that cover such large areas in India. Trappean rocks have been, by preference, in many cases selected for temple-building and particular architectural or religious sculptures throughout India, not only when the rock is *in situ*, but often enough at long distances from its place of occurrence.

TRAPA, *Linn.; Gen. Pl., I.,* 793.

Trapa bispinosa, *Roxb.; Fl. Br. Ind., II.,* 590 ; ONAGRACEÆ.

516

THE SINGHÁRA NUT.

Syn.—T. QUADRISPINOSA, *Wall.,* nor of *Roxb.*

Vern.—*Singhárá,* HIND.; *Paniphal,* BENG.; *Gaúnri,* KASH.; *Gaúnri, singhárá,* PB.; *Shingodá, singodi,* GUZ.; *Shingádá,* MAR.; *Shingári, Dec.;* *Singhara,* TAM.; *Kubyakam,* TEL.; *Karim-polam,* MALAY.; *Sringátaka,* SANS.

References.—*Roxb., Fl. Ind., Ed. C.B.C.,* 143; *Kurz, in Jour. As. Soc.,* 1877, *Pt. ii,* 91; *Dalz. & Gibs., Bomb. Fl.,* 99; *Stewart, Pb. Pl.,* 89; *Rheede, Hort. Mal., XI., t.* 33; *Works of Sir. W. Jones, V.,* 83; *Ain-i-akbari, Gladwin's Trans., I.,* 85; *U. C. Dutt, Mat. Med. Hind.,* 319; *S. Arjun, Bomb. Drugs,* 57; *Murray, Pl. & Drugs, Sind,* 189; *Cat. Baroda Durbar, Col. & Ind. Exhib., No.* 176; *Baden Powell, Pb. Pr.,* 262, 263; *Drury, U. Pl.,* 437; *Lisboa, U. Pl. Bomb.,* 157; *Birdwood, Bomb. Pr.,* 155; *Royle, Ill. Him. Bot.,* 214; *Atkinson, Ec. Prod., N.-W. P., V. (Foods),* 13, 15; *Stewart, Foods of the Bijur Dist.,* 474; *Kew Off. Guide to the Mus. of Ec. Bot.,* 69; *Jour. As. Soc. Beng., Pt., II., No. II.,* 1867, 80; *Gazetteers:—Mysore & Coorg, I.,* 68; *N.-W. P., I.,* 57; *II.,* 375; *III.,* 225; *Panjáb, Gurgáon,* 86; *Settlement Reports:— Panjáb, Delhi,* 27; *C. P., Chánda,* 83; *N.-W. P., Allahabad,* 37; *Ind. Forester, III.,* 238; *XIV.,* 392; *Agri.-Horti. Soc. Ind., Journ. (New Series), VI., xi., xxiv.; Ind. Agri., Feb.* 1889, 91; *Trop. Agri., January* 1889, 502.

Habitat.—A floating herb, found on lakes, tanks, and pools throughout India and Ceylon.

Dye.—The ground FRUIT is employed, in certain parts of the country, for making the red *gulal* powder used during the *Hólí* festival. (*See* Abir, Vol. I, *pp* 6-7.)

DYE. Fruit. 517

Medicine.—The NUTS are considered by the Natives of the Panjáb and North-West to be cooling and useful in bilious affections and diarrhœa. They are also employed externally in the form of poultices.

MEDICINE. Nuts. 518

Food.—The KERNEL of the fruit is much used as an article of food in all localities where the plant is to be found in any quantity. It abounds in starch, resembles the chestnut in flavour, and is eaten either raw or

FOOD. Kernel. 519

TRAPA bispinosa.	The Singhara Nut.

FOOD.

cooked, especially by the Hindús, for it is *phalahár*, *i.e.*, it may be eaten during fasts. It may be boiled whole, after soaking a night in water, roughly broken up and made into a sort of porridge, or ground to a fine meal and made into *chapattis*.

It is cultivated extensively in Kashmír, and in the lakes, tanks, and freshwater reservoirs of the North-West and Central Provinces. In Kashmir it is said to furnish almost the only food of at least 30,000 people for five months of the year. Stewart, quoting from Moorcroft, states that, from the Wúlar lake in Kashmir, ninety-six to one hundred thousand ass-loads were, in his time, taken annually, the Government drawing R90,000 duty from this source. He further mentions on "good authority" that the Maharája annually obtained more than a lakh of rupees as duty from the Singhara nut. Many boat-loads filled with the fruit arrive daily at Srinagar.

Atkinson, quoting from Colonel Sleeman and other sources, gives the following account of *singhára* in the North-West Provinces :—" It is cultivated chiefly by *Dhímars* and *Kahárs*, who have spaces regularly marked out by bamboos, for which they pay rent to the landholder. The long stalks reach the surface of the water, on which float the green leaves, and amongst them the white flowers expand their petals towards evening. In the end of January the seed or fruit is scattered at the rate of a maund of 82℔ to a local *bígha*, over the water where it is sufficiently deep to preclude all chance of its drying up before the advent of the periodical rains. The seeds are then pressed into the mud with sticks or the feet, and in a month or two they begin to throw out shoots. In June, just before the rains set in, the excess is thinned out and transplanted, the produce of one *bígha* serving for three or four ; the roots are taken between the great and first toe of the planter's foot and thus fixed in the mud. The long stalks of the plant reach up to the surface of the water, on which the bright-green leaves float, supporting in the middle a pure white flower. In October the nut forms under the water; it is of triangular shape, covered with a rough brown integument, adhering strongly to the kernel and armed with spines ; hence its name. The kernel is white and esculent and of a fine cartilaginous texture. The fruit is gathered in November and December. In Hamírpur in the Mahoba Pargana, a local *bígha*, measured by bamboos 18 feet long, of which 12 by 2 make a *singhárá bígha*, yields 3 to 4 maunds of produce worth about R2 to R4 a maund. The rent is about R1 to R3 a *bígha*, *kankar* in the soil being esteemed unfavourable, and rich mud fetching the highest rent. The Dhímars of Hamírpur generally take the lease of a lake at a fixed rent, and divide it among themselves, their respective cultivations being marked by upright sticks, the removal of which, as of boundary marks on the shore, leads to many a quarrel.

Their great enemy is an insect called *bandu*, which in both stages of grub and fly feeds on the plant, eating through the husks, and thus destroying the fruit, which on being exposed to the water spoils. The labour of killing these and clearing away the weeds is great. In Cawnpore a kind of raft is made by joining two earthen vessels together by a bamboo, astride which, or resting his arms on which, the *kahár* paddles about to clear off the insects. For the cultivation flat-bottomed canoes are used, scooped out of the trunk of a single *mahúa* tree, costing about R5 each to make, and lasting fifteen to twenty years. They carry two men and are pushed on by bamboo, and when not used are sunk in the water till again wanted. The cultivation of this fruit forms one of the most important sources of the miscellaneous revenue in villages." In the Delhi District of the Panjáb the *singhára* nut is similarly cultivated, and is said to yield a considerable revenue. The following passage also occurs in the *Land*

| The Indian Nettle Tree. (*J. Murray.*) | TREMA orientalis |

FOOD.

Revenue Settlement Report of the Jubbulpore District of the Central Provinces: "In the highly cultivated central portion of the district the principal *sayer* income consists of rents from tanks devoted to the culture of the *singhára*, sometimes yielding a considerable item, of from R50 to R200. It constitutes an article of export." In Guzerát, also, the nut forms an important article of diet, and in Manipur, according to Dr. Watt, the immense lakes to the south of the valley afford food for a few months to a large community.

It has been suggested at various times that the food-supply of localities in which water abounds might be very greatly augmented by extending the cultivation of this plant. It affords a palatable food which, even in good seasons, would be freely used by the agricultural population, who would thus secure a considerable addition to their surplus outturn of rice and other food-grains for export. In bad seasons, on the other hand, the large adventitious source of food-supply would form a stand by, when other crops might altogether fail. It has been specially recommended that the plant should be experimentally grown in the great reservoirs and artificial lakes of the Madras Presidency.

But it must be remembered that the crop is not unattended with disadvantages. Thus Colonel Sleeman (*Rambles of an Indian Officer*) remarks that mud increases very rapidly from its cultivation, which should consequently be carefully prohibited where it is thought desirable to keep up the tank purely for the sake of the water. On the other hand, in stagnant and foul waters it may do good; the Chinese, indeed, believe that an allied species, **T. bicornis**, absorbs the putrid emanations which arise from such sources.

TREMA, *Lour.; Gen. Pl., III., 355.*

Trema amboinensis, *Blume; Fl. Br. Ind., V., 484;* URTICACEÆ.

520

Syn.—T. ORIENTALIS *var.* AMBOINENSIS, *Kurz;* T. VELUTINA & BURMANNI, *Blume;* SPONIA AMBOINENSIS, *Dcne.;* S. GRIFFITHII, *Planch.;* S. AMBOYNENSIS & VELUTINA, *Miq.;* S. VELUTINA, *Planch;* S. BURMANNI, *Planch;* CELTIS AMBOINENSIS, *Willd.;* C. TOMENTOSA, *Roxb.;* C. CAUDATA, *Wall.*

Vern.—*Jhávár,* SANTAL.

References.—*Roxb., Fl. Ind., Ed. C.B.C., 263; Kurz, For. Fl. Burm., II., 469; Beddome, For. Man., 219; Campbell, Ec. Pl., Chutia Nagpur, No. 8714.*

Habitat.—A small evergreen tree, found in hot valleys in the Sikkim Himálaya, Assam, Sylhet, and southwards to Singapore, also in the Andaman Islands.

DOMESTIC. Leaves. 521

Domestic.—Campbell states that the rough LEAVES are used in Chutia Nagpur for polishing wood in place of sand-paper.

T. orientalis, *Blume; Fl. Br. Ind., V., 484; Wight, Ic., t. 1971.*

THE INDIAN NETTLE TREE, or CHARCOAL TREE.

522

Syn.—SPONIA ORIENTALIS, *Planch.;* S. WIGHTII, *Planch.;* CELTIS ORIENTALIS, *Linn.* (*in part*).

Vern.—*Chikun,* BENG.; *Badu manu,* C. P.; *Kúail,* NEPAL; *Tugla,* LEPCHA; *Param,* MICHI; *Jupong, phakram, jigini, sapong, sempak, amphak, opang,* ASSAM; *Gol, khargul,* BOMB.; *Mini,* TAM.; *Gadanelli,* TEL.; *Gorklu,* KAN.; *Gol.,* MAR.; *Rukni,* BAIGAS; *Sap-sha-pen,* BURM.

References.—*Roxb., Fl. Ind., Ed. C.B.C., 262; Kurz, For. Fl. Burm., II., 468; Brandis, For. Fl., 430; Beddome, Fl. Sylv., t. 311; For. Man., 219; Gamble, Man. Ind. Timb., 344; Dals. & Gibs., Bomb. Fl., 238; Lisboa, U. Pl. Bomb., 132.*

T. 522

Habitat.—A small evergreen tree, met with along the foot of the Nepál and Sikkim Himálaya, in Bengal, Behar, and southwards to Travancore and Singapore, common in Ceylon. It comes up self-sown in forest clearings and waste places, often in great profusion, and may be utilised in plantations to keep down the grass jungle. According to **Van Someren**, it is often allowed to grow for shade in the Mysore and Coorg coffee plantations (*Gamble*).

FIBRE.
Bark.
523
TIMBER.
524

Fibre.—The inner BARK yields a FIBRE which is used for tying the rafters of native houses, and for binding loads; and in Assam for making the coarse *amphak* cloth (*Gamble*).

Structure of the Wood.—Light reddish-grey, soft, growth extremely fast; weight 28℔ per cubic foot. It is employed for making charcoal for the manufacture of gun-powder.

TREWIA, *Linn.; Gen. Pl., III., 318.*

(*1870-1 (excl. fem. fl.)*); EUPHORBIACEÆ.

525

Trewia nudiflora, *Linn.; Fl. Br. Ind., V., 423; Wight, Ic., t.*

Syn —T. MACROPHYLLA, *Roth.*; T. MACROSTACHYA, *Klotæsch*; ROTTLERA INDICA, *Willd.*; TETRAGAST<IS OSSEA, *Gærtn.*

Vern.—*Tumri, khamara, bhillaura, pindára,* HIND.; *Pitáli,* BENG.; *Morda,* URIYA; *Gara lohadaru,* KOL; *Gamhár,* MONGHYR; *Garum, gamari, kurong,* NEPAL; *Tungplam,* LEPCHA; *Bhillaur, bhillaura,* OUDH; *Tumri, khamara,* KUMAON; *Petári, tumri, bhillauri,* BOMB.; *Pitári,* MAR.; *Kat-kumbla,* KAN.; *Thitmyoke, ye-myot,* BURM.; *Hruprukban,* MAGH.; *Pindára,* SANS.

References. —*Roxb., Fl. Ind., Ed. C.B.C., 740; Brandis, For. Fl., 443; Kurz, For. Fl. Burm., II., 379; Beddome, Fl. Sylv., t. 281; Gamble, Man. Timb., 359; Graham, Cat. Bomb. Pl., 185; Dalz. & Gibs., Bomb. Fl., 231; Rheede, Hort. Mal., I., t. 42; U. C. Dutt, Mat. Med. Hind., 313; Atkinson, Him. Dist, 317; Lisboa, U. Pl. Bomb., 122, 171; For. Adm. Rept., Chutia Nagpore, 1885, 34; Gazetteers :—Bombay, XIII., 26; XV., 78, 442; N.-W. P., IV., lxvii.*

Habitat.—A large, deciduous tree, common in the hotter parts of India from Kumáon southward and eastward to Assam, Malacca, and Ceylon.

FOOD.
Pulp.
526
TIMBER.
527

Food.—The PULP under the rind of the fruit is said to be sweet and edible. It must, however, be small in quantity, for **Sir J. D. Hooker** describes the fruit as "almost woody."

Structure of the Wood.—White, soft, not durable; weight 28 to 29℔ per cubic foot. It is used for making Native drums and agricultural implements (*Gamble*).

TRIANTHEMA, *Linn.; Gen. Pl., I., 855.*

528

Trianthema crystallina, *Vahl.; Fl. Br. Ind., II., 660;* FICOIDEÆ.

Syn.—T. TRIQUETRA, *Rottl.*; T. SEDIFOLIA, *Visian.*; POPULARIA CRYSTALLINA, *Forsk.*

Vern.—*Alettié,* PB.; *Patar phor,* MERWARA; *Kukka* pála kúra,* TEL.

* This vernacular name is from **Roxburgh**; **Elliot** doubts its being in reality a synonym for any species of **Trianthema**.

References.—*Roxb., Fl. Ind., Ed. C.B.C., 384; Dalz. & Gibs., Bomb. Fl., 14; Kurz, in Jour. As. Soc., 1877, Pt. ii., 110; Stewart, Pb. Pl., 100; Elliot, Fl. Andhr., 102; Murray, Pl. & Drugs, Sind, 108; Gazetteers :— N.-W. P., IV., lxxii; X., 310; Bombay, V., 23; Rájputána, 36; Ind. Forester, IV., 234; XII., app., 14.*

Habitat.—A prostrate, branched herb, met with throughout India from the Panjáb to Ceylon, except in Bengal; very common in some of the desert tracts of the Panjáb and Rájputána.

T. 528

Food.—Near Múltán the SEEDS are swept up and eaten, in times of famine (*Edgeworth*).

| | FOOD. Seeds. 529 |

[*Ic., t. 296.*

Trianthema decandra, *Linn.; Fl. Br. Ind., II., 661; Wight,*

530

    Syn.—ZALEYA DECANDRA, *Burm.*

    Vern.—*Gada-bani,* HIND. & BENG.; *Bhis khupra,* DEC.; *Vallai-sharunnai,* TAM.; *Tella galijéru, galijéru,* TEL.; *Gaija soppu,* KAN.; *Punarnavi,* SANS.

    References.—*Roxb., Fl. Ind., Ed. C.B.C., 385; Dalz. & Gibs., Bomb. Fl., 15; Kurz, in As. Soc. Jour., 1877, Pt. ii., 110; Thwaites, En. Cey. Pl., 23; Burm., Fl. Ind., t. 31, f. 3; Elliot, Fl. Andhr., 57, 175; Ainslie, Mat. Ind., II., 371; O'Shaughnessy, Beng. Dispens., 353, 684; Bidie, Prod. S. Ind., 53; Boswell, Man. Nellore, 138; Gazetteers:—Mysore & Coorg, I., 55; N.-W. P., IV., lxxii.*

    Habitat.—A diffuse branched herb, native of the Deccan Peninsula and Ceylon.

    Medicine.—"The ROOT is aperient and is mentioned in some of the Tamool sastrums as useful in hepatitis, asthma, and suppression of the menses. Four pagodas weight of the BARK of the root made into a decoction, by boiling it in 1℔ of water till ½℔ remains, will open the bowels" (*Ainslie*).

| | MEDICINE. Root. 531 Bark. 532 |

    SPECIAL OPINION.—§ "The root ground up with milk and given internally is said to be a specific in orchitis. The JUICE of the LEAVES dropped into the nostrils relieves one-sided headache" (*Surgeon-Major D. R. Thomson, M.D., C.I.E., Madras*).

| | Juice. 533 Leaves. 534 |

T. hydaspica, *Edgw.; Fl. Br. Ind., II., 661.*

535

    Syn.—T. POLYSPERMA, *Hochst.*

    Reference.—*Murray, Pb. & Drugs, Sind, 108.*

    Habitat.—Found in Múltán, Sind, and certain localities of Bombay.

    Food.—It is said to be eaten as a pot-herb (*Murray*).

| | FOOD. 536 |

T. monogyna, *Linn.; Fl. Br. Ind., II., 660; Wight, Ic., t. 228.*

537

    Syn.—T. OBCORDATA, *Roxb.*; T. PENTANDRA, β OBCORDATA, *DC.*

    Vern.—*Swét, sabuni, lal-sabuni,* HIND.; *Bishkáprá,* PB.; *Nasurjinghi ke jurr, wurmah,* DEC.; *Kháprá, bishkáprá, sweta-punarnava,* BOMB.; *Sharunnai, shavalai, kirai,* TAM.; *Galijéru, yerra galijéru, erra galijéru, ambati mádu,* TEL.; *Muchchu góni,* KAN.; *Punaraví,* SANS.

    References.—*Roxb., Fl. Ind., Ed. C.B.C., 385; Dalz. & Gibs., Bomb. Fl., 14; Kurz, in Jour. As. Soc., 1877, Pt. ii, 110; Elliot, Fl. Andhr., 14, 52, 57; Ainslie, Mat. Ind., II., 370; Dymock, Mat. Med. W. Ind., 2nd Ed., 74; Murray, Pl. & Drugs, Sind, 108; Atkinson, Ec. Prod., N.-W. P., pt. V. (Foods), 91, 95; Boswell, Man. Nellore, 123, 144; Ind. Forester, III., 238; XII., app., 14; Gazetteers:—Mysore & Coorg, I, 55, 70; N.-W. P., IV., lxxii.*

    Habitat.—Common throughout India and Ceylon.

    Medicine.—Ainslie states that the *Vytians* consider the ROOT cathartic and give it in powder, in the quantity of about two tea-spoonfuls twice daily in combination with ginger.

| | MEDICINE. Root. 538 |

    Food.—The LEAVES and STEMS are eaten as a vegetable. Atkinson writes that they are said sometimes to have poisonous effects, producing paralysis and diarrhœa.

| | FOOD. Leaves. 539 Stems. 540 |

T. pentandra, *Linn.; Fl. Br. Ind., II., 660.*

541

    Syn.—T. OBCORDATA, *Wall., not of Roxb.*; T. GOVINDIA, *Wall.*

    Vern.—*Bishkáprá, itsit, narma,* PB.; *Bishkapra, narwa,* SIND.

    References.—*Stewart, Pb. Pl., 100; Murray, Pl. & Drugs, Sind, 108; Lisboa, U. Pl. Bomb., 200; Gazetteers:—N.-W. P., I., 81; IV., lxxii; X., 310; Ind. Forester, XII., app., 14.*

T. 541

| TRIBULUS terrestris. | The Tribulus or Gokhru Fruit—a useful medicine. |

Habitat.—A common weed in waste ground in the plains of the Panjáb, Sind, and North-West India.

MEDICINE.
Plant.
542

Medicine.—The PLANT is considered astringent in abdominal diseases, and is also stated to produce abortion (*Stewart*).

FOOD.
543

Food.—This, like the preceding species, is eaten as a pot-herb in times of scarcity, though stated to be apt to produce diarrhœa and paralysis (*Stewart*).

TRIBULUS, *Linn.; Gen. Pl., I., 264, 988.*

544

Tribulus alatus, *Delile; Fl. Br. Ind., I., 423;* ZYGOPHYLLEÆ.

Vern —*Gokhuri-kalan,* HIND.; *Lotak, bakhra, hasak, gokhrudesi,* PB.; *Kurkundai,* PUSHTU; *Nindo-trikund, trikundri, latak,* SIND.

References.—*Aitchinson, Bot. Afgh. Del. Com., 43; Murray, Pl. & Drugs, Sind, 91; Dymock, Mat. Med. W. Ind., 2nd Ed., 120; Dymock, Warden, & Hooper, Pharmacog. Ind., I., 245; Notes on the Ec. Pl., Baluchistan, No. 70; Stocks, Rept. on Sind.*

Habitat.—A prostrate herb, found in Sind and the Panjáb.

MEDICINE.
Fruit.
545

Medicine.—The FRUIT is used for the same purposes as those of T. terrestris, *Linn.*

FOOD.
Plant.
546
Seeds.
547
548

Food.—"The young PLANT is in some places eaten as a pot-herb; and the SEEDS are used as food, especially in times of scarcity."

T. terrestris, *Linn.; Fl. Br. Ind., I., 423; Wight, Ic., t. 98.*

Syn.—T. LANUGINOSUS, *Linn.*

Vern.—*Gókhrú, gokhuru, chota gokhrú, hussuk,* HIND.; *Gokhuru, gokshura, gókhru,* BENG.; *Gakhurá, gokshra,* URIYA; *Rásha, kokullak,* LAD.; *Lotak, bakhra, bhakhra, bhúkrí, gokhrú desi, bhakhra,* PB.; *Málkundai, kandalái,* PUSHTU; *Krúnda,* AFG.; *Trikundrí, gokhru,* SIND; *Gokhrú,* C. P.; *Lahana gokrú, gokhrú, saraté,* BOMB.; *Ghókaru, charátté, lahana gokharu,* MAR.; *Gókhru, gokharú, mithá gokhru, nhana gokhru,* GUZ.; *Ghókrú,* DEC.; *Nerunji, nerrenji kiray, nerunji-mullu,* TAM.; *Palléru-mullu, chiru-palléru, chiri palléru, palléru,* TEL.; *Negalu,* KAN.; *Neringil, nerinnil,* MALAY.; *Sule-anén, charatte,* BURM.; *Neranchi, neranji,* SING.; *Vanasrangátá, gókhurhá, trikantaka, sthala sringataka, gokshuri, ikshugandhá, súdúmstra, gokshura,* SANS.; *Bastitáj, khasak, busteyrúmi,* ARAB.; *Kháre-khasak, khussuck,* PERS.

References.—*Roxb., Fl. Ind., Ed. C.B.C., 371; Dalz. & Gibs., Bomb. Fl., 45; Stewart, Pb. Pl., 37; Rept. Pl. Coll. Afgh. Del. Com., 43; Sir W. Elliot, Fl. Andhr., 42, 143; Sir W. Jones, Treat. Pl. Ind., V., 134; Burmann, Fl. Zey., 265, t. 106, f. 1; Pharm. Ind., 39; Ainslie, Mat. Ind., II., 247, 248; O'Shaughnessy, Beng. Dispens., 259; Irvine, Mat. Med. Patna, 30; Med. Topog. Ajmir, 135; Moodeen Sheriff, Supp. Pharm. Ind., 247; Mat. Med. S. Ind. (in MSS.), 71; U. C. Dutt, Mat. Med. Hindus, 125, 298; S. Arjun, Cat. Bomb. Drugs, 28; Murray, Pl. & Drugs, Sind, 90; Dymock, Mat. Med. W. Ind., 2nd Ed., 119, 887, 889; Dymock, Warden, & Hooper, Pharmacog. Ind., I., 243; Cat. Baroda Durbar, Col. & Ind. Exhib., No. 177; Year-Book Pharm., 1878, 287; Birdwood, Bomb. Prod., 16; Drury, U. Pl. Ind., 432; Atkinson, Him. Dist. (X., N.-W. P. Gaz.), 307; Useful Pl. Bomb. (XXV., Bomb. Gaz.), 196; Econ. Prod. N.-W. P., Pt. V. (Vegetables, Spices, & Fruits), 91, 93; Bidie, Prod. S. Ind., 22; Stocks, Rept. on Sind; Boswell, Man. Nellore, 132; Settlement Reports :—Panjáb, Montgomery, 21; C. P., Chánda, app. vi.; Gazetteers:—Bombay, V., 24; XV., 428; Panjáb, Muzzafargar, 27; Montgomery, 21; Peshawár, 26; Gujrat, 11; N.-W. P., I., 80; IV., lxix.; Orissa, II., 159, 181; Rajputana, 3; Ind. Forester, III., 238; IV., 234; ^II., app. 2.*

Habitat.—This low trailing annual plant is common throughout India, ascending to 11,000 feet in Western Tibet, rarer in Lower Bengal, and absent from the vicinity of Calcutta; abundant in Behar and everywhere

| The Tribulus of the Greeks. | (*J. Murray.*) | **TRIBULUS terrestris.** |

throughout the Madras Presidency and the North-Western Provinces and Oudh.

Medicine.—The entire PLANT, but more particularly the FRUIT, is used medicinally throughout India. The latter is regarded by the Hindus as cooling, diuretic, tonic, and aphrodisiac, and is used in cases of painful micturition, calculous affections, urinary disorders, and impotence. It is one of the ten plants which go to form the *Dasamula kvatha*, a compound decoction often mentioned in Sanskrit works (*U. C. Dutt*). The plant is the τρίβολος of Dioscorides and the *tribulus* of Pliny, in modern Greek it is known as τριβόλια (*Ainslie*). It is used in Cochin China as an astringent. According to Bellew it is taken, in the Peshawar valley, by women, to ensure fecundity. Water rendered mucilaginous by the plant is drunk as a remedy for impotence, and an infusion of the STEM is administered for gonorrhœa. **Mr. Lace** informs the writer that the fruit of T. alatus is similarly employed in Bilúchistán, and is also a domestic remedy for uterine disorders after parturition. Several writers state that the fruit of T. terrestris is febrifuge, but probably this quality is secondary to its diuretic properties.

The plant is included in the secondary list of the *Pharmacopœia of India*, in which Waring writes, "In trials made with it by the Editor it was found, in some instances, materially to increase the urinary secretion, but in others it exercised no perceptible effect. The formula employed was as follows :—Tribulus fruit, bruised, two ounces ; coriander fruit, two drachms ; water, one pint ; boiled to one-half. This quantity was given in divided doses during the day." Another favourite mode of administration, adopted by the Natives, is to boil the fruit and ROOT with rice so as to form a medicated *congi* water, which is taken in large quantities. Here the amount of fluid may serve to act as a diuretic, irrespective of the presence of any medicinal agent. **Moodeen Sheriff** describes the fruit and LEAVES as demulcent, diuretic, and useful in cases of strangury, gleet, and chronic cystitis. He recommends a decoction similar to that above described, and the fresh JUICE of the leaves, in doses of one to three fluid ounces of the former, or one to two fluid ounces of the latter, four or five times a day.

CHEMICAL COMPOSITION.—A recent examination by the authors of the *Pharmacographia Indica* has proved the fruit to contain a body having the properties of an alkaloid, and associated with hydrochloric acid or alkaline chlorides. It also yields small quantities of a fat and a resin (the latter of which is fragrant when burned), and 14·7 per cent. of mineral matter.

TRADE.—The fruit may be collected in any of the more sandy districts of India, and is always procurable in the drug shops. **Dymock** states that in Bombay it costs R5 per Surat maund of 37⅓℔.

SPECIAL OPINIONS.—§"Diuretic, frequently given by Native *baids* in painful micturition" (*Assistant Surgeon S. C. Bhattacharji, Chanda*). "The dried fruit, powdered and given in doses of 30 grains with sugar and black pepper, is used in gleet, spermatorrhœa, and impotence" (*Hospital Assistant Lal Mahomed, Hoshangabad*). "An infusion made from the fruit has been found very useful as a diuretic in cases of gout, kidney disease, and gravel ; also used largely in this part of the country as an aphrodisiac" (*Civil Surgeon F. F. Perry, Jullunder City, Panjáb*).

Food.—The young LEAVES and STEMS are eaten as a pot-herb ; the prickly FRUIT is also gathered and used as food in times of scarcity, being ground to a powder and eaten in the form of bread. It is said to have constituted the chief food of many persons during the Madras famine and was also largely utilised in the Deccan famine of 1877-78.

MEDICINE.
Plant.
549
Fruit.
550
Stem.
551
Root.
552
Leaves.
553
Juice.
554
CHEMISTRY.
555
TRADE.
556
FOOD.
Leaves.
557
Stems.
558
Fruit.
559

T. 559

| TRICHOLEPIS glaberrima. | The Chota-kulpha. |

TRICHODESMA, *Br. ; Gen. Pl., II., 845.*

560

Trichodesma africanum, *Br. ; Fl. Br. Ind., IV., 154 ;* BORAGINEÆ.

Syn.—BORAGO AFRICANA, *Linn. ;* B. VERRUCOSA, *Forsk.*
Vern.—*Paburpani,* SIND.
References.—*Boiss., Fl. Orient., IV., 280 ; Murray, Pl. & Drugs Sind, 172.*
Habitat.—A coarse herb, met with in the Panjáb and Sind, distributed to Kábul, Balúchistán, Persia, and Mauritius.

MRDICINE.
Leaves.
561

Medicine.—The LEAVES are used as a diuretic (*Murray*).

562

T. indicum, *Br. ; Fl. Br. Ind., IV., 153 ; Wight, Ill., t. 172.*

Syn.—T. PERFOLIATUM, *Wall. ;* T. HIRSUTUM, *Edgew. ;* BORAGO INDICA, *Linn. ; ?* B. SPINULOSA, *Roxb.*
Var. subsessilis=T. SUBSESSILIS, *Wall.*
Vern.—*Chhota-kulpha,* HIND. ; *Choto-kulpa,* BENG. ; *Hetmudia,* SANTAL ; *Ratmandi,* KUMAON ; *Kallri-búti, ratmandú, nilakrái, andúsi,* leaves= *ratmandí,* PB. ; *Ratisurkh, nilakrái,* KASHMIR ; *Gaozaban,* SIND ; *Lahána kalpa,* MAR. ; *Kazuthai-tumbai,* TAM. ; *Guvva-gutti,* TEL.
References.— *Roxb., Fl. Ind., Ed. C.B.C., 154 ; Dalz. & Gibs , Bomb. Fl., 173 ; Stewart, Pb. Pl., 155 ; Rev. A. Campbell, Rept. Ec. Pl., Chutia Nagpur, No. 8483 ; Elliot, Fl. Andhr., 67 ; Pharm. Ind., 158 ; Murray, Pl. & Drugs Sind, 172 ; Baden Powell, Pb. Pr., 366 ; Drury, U. Pl. Ind., 432 ; Atkinson, Him. Dist., 314, 752 ; Bidie, Prod. S. Ind., 35 ; Off. Corresp. on proposed New Pharm. Ind., 239 ; Gazetteers :—Mysore & Coorg, I., 63 ; N.-W. P., I., 83 ; IV., lxxv. ; Ind. Forester, IV., 234 ; VI., 238 ; XII., App. 17.*
Habitat.—A coarse hispid herb, found throughout India except in the Bengal plains; also in British Burma.

MEDICINE.
Leaves.
563

Medicine.—In the Panjáb the LEAVES are considered cooling and depurative (*Stewart*). In Sind, the Deccan, and South India the drug has a great reputation in the cure of snake-bites, and a case of recovery, after its administration, in the practice of **Dr. Maxwell,** is on record. There is however, no evidence of its utility, and, like most other remedies of a similar nature, it has probably no virtue. In the Deccan the leaves are used to make an emollient poultice (*Pharm. Ind.*). The **Rev. A. Campbell** informs us that, in Chutia Nagpur, the ROOT, pounded and made into a paste, is applied to reduce swelling, particularly of the joints.

Root.
564

565

T. zeylanicum, *Br. ; Fl. Br. Ind , IV., 154.*

Syn.—BORAGO ZEYLANICA, *Linn.*
Vern.—*Hetenuria,* HIND. ; *Tirup-sing,* MANDARI.
References.—*Gazetteers :—Mysore & Coorg, I., 63 ; N.-W. P., I., 83.*
Habitat.—A common herb in the Deccan Peninsula and Ceylon.

MEDICINE.
Leaves.
566

Medicine.—Emollient poultices are made from the LEAVES.

TRICHOLEPIS, *DC.; Gen. Pl. II., 475.*

567

Tricholepis glaberrima, *DC. ; Fl. Br. Ind., III., 381 ;* COMPOSITÆ.

Syn.—? SERRATULA INDICA, *Willd.*
Vern.—*Bramhadandi,* MAR.
References. — *Dalz. & Gibs., Bomb. Fl., 131 ; Dymock, Mat. Med. W. Ind., 2nd Ed., 467.*
Habitat.—A stout annual, native of Central India, Merwara, the Konkan, and the Deccan.

MEDICINE.
Plant.
568

Medicine.—This PLANT is believed by the Natives of the region where it occurs to be a nervine tonic and aphrodisiac (*Dymock*).

T. 568

| The Snake Gourd. (*J. Murray.*) | **TRICHOSANTHES cordata.** |

TRICHOSANTHES, *Linn.; Gen. Pl., I., 821.*

Trichosanthes anguina, *Linn.; Fl. Br. Ind., II., 601;*   569
THE SNAKE GOURD.   [CUCURBITACEÆ.

Vern.—*Purwul, cháchenda, chachinga,* HIND.; *Chichingá,* BENG.; *Chha-chhindará,* URIYA; *Jhajhinda,* N.-W. P.; *Chachinga, chachinda,* OUDH; *Chachinda,* KUMAON; *Gálar tori, pandol, chichinda,* PB.; *Pandol, rebhri, kadotri,* SIND; *Pudola,* C. P.; *Pandolu, padval, porwar, pada, vala, parula,* BOMB.; *Padual,* MAR.; *Linga potla,, potla, potla káya, pollv káya, poallkáya,* TEL.; *Padavala káyi,* KAN.; *Pai-len-mwae,* BURM.; *Chichin-da,* SANS.

References.—*Roxb., Fl. Ind., Ed. C.B.C., 694; Voigt, Hort. Sub. Cal., 57; Kurz, in Jour. As. Soc., 1877, pt. ii., 98; Dals. & Gibs., Bomb. Fl. Suppl., 137; Stewart, Pb. Pl., 99; DC., Orig. Cult. Pl., 272; Elliot, Fl. Andhr., 107, 155, 156; Mason, Burma & Its People, 470, 747; Ainslie, Mat. Ind., II., 592; O'Shaughnessy, Beng. Dispens., 351; U. C. Dutt, Mat. Med. Hind., 295; Murray, Pl. & Drugs, Sind, 39; Atkinson, Him. Dist., 700; Ec. Prod., N.-W. P., Pt. v, 4; Duthie & Fuller, Field & Garden Crops, Pt. ii., 45, t. xlvi.; Lisboa, U. Pl. Bomb., 157; Birdwood, Bomb. Pr., 159; Royle, Ill. Him. Bot., 219; Stocks, Rep. on Sind; Madden, Note on Kumaon, 279; Smith, Dic., 381; Kew Off. Guide to the Mus. of Ec. Bot., 70; Settle. Reps., Kángra, 25, 28; Chánda, 82; Gazetteers :—Mysóre & Coorg, I., 61; II., 11; Bombay, V., 26; VIII., 184; XIII., 294; N-W. P., I., 81; IV., lxxii.; Orissa, II., 180l; Ind. Forester, IX., 201.*

Habitat.—An annual creeper, which probably was originally wild in India or the Indian Archipelago (*DeCandolle*). It has never been found truly wild, and was considered by **Mr. C. B. Clarke** to be a cultivated state of **T. cucumerina**, from which it differs only in the fruit.

CULTIVATION. — It is cultivated throughout India as a rainy season   **CULTIVATION.**
crop. Mr. Gollan of the Botanic Gardens, Sahâranpur, recommends   570
that two sowings should be made, the first in April, the second in May
(*Ind. Forester, IX., 201*). The general treatment and mode of cultivation
is the same as that of the cucumber. It is impossible to obtain information
as to the extent to which it is grown throughout the country.

Medicine.—The SEEDS are considered cooling.   **MEDICINE.**
Food.—The long, cucumber-like FRUIT is cooked and eaten as a   Seeds.
vegetable, either boiled or in curries. When ripe it varies in length from   571
1 to 3 feet, and is of a brilliant orange colour; when young it is prettily   **FOOD.**
striped with white and green. If gathered when very young, less than 4   Fruit.
inches in length, and cut into thin strips, it may be cooked in the same   5 72
way as French beans, and forms a very fair substitute for that vegetable.

T. cordata, *Roxb.; Fl. Br. Ind., II., 608.*   573

Syn.—T. TUBEROSA, *Roxb.;* T. PALMATA, *Wall, Cat 6688 F partly, & C.*
Vern.—*Bhoii kúmra, bhúmi kúmara, bha-khúmba, patol,* BENG.
References.—*Roxb., Fl. Ind., Ed. C.B.C., 695; Irvine, Mat. Med. Patna, 15, 87; O'Shaughneesy, Beng. Dispens., 350; Pharm. Ind., 96; Taylor, Topography of Dacca, 55; Drury, U. Pl. Ind., 433.*

Habitat.—An extensive climber, met with at the base of the Eastern Himálaya, from Sikkim to Assam and Pegu; frequent in the Khásia Tarai and Cachar.

Medicine.—The large, tuberous ROOT is considered a valuable tonic,   **MEDICINE.**
and is employed as a substitute for calumba (*Roxburgh*). Irvine remarks   Root.
that it is also deobstruent, and that in Patna, the dried FLOWERS are   574
believed to be stimulant in doses of 2 to 5 grains. Taylor states that in   Flowers.
Dacca the root, dried and reduced to powder, is given in doses of 10 grains   575
in enlargements of the spleen, liver, and abdominal viscera. The fresh
root, mixed with oil, forms a common application for leprous ulcers (*Topog. Dacca*).

6

TRICHOSANTHES cucumerina.   The Wild Snake Gourd.

576   **Trichosanthes cucumerina,** *Linn.; Fl. Br. Ind., II., 609.*

Syn.—T. LACINIOSA, *Klein;* T. PILOSA, BRYONIA UMBELLATA, CUCUMIS MISSIONIS, *Wall.*

Vern.—*Jangli-chi-chóndá,* HIND ; *Ban-chi-changá, ban-patol, ranacha-padavali,* BENG.; *Jangli chichinda, patol, ban-patol, kandori,* N.-W. P.; *Jangli-chachinda,* KUMAON; *Gwal kakri, mohakri,* PB.; *Rán parul, jangli-padavala, kadu-padavala, pudoli,* BOMB.; *Ránácha-padavali, kadú padavala, jangli-padavala, perula,* MAR.; *Patola, GUZ.; Káttup-pépudal, péy-pudal, pudel,* TAM.; *Adavi-potla, chédu-potla, patólamu, patólas, chéti-potla, chayud pottah chétipotla, chyaapotta,* TEL.; *Bettada-padavala, kiripodla káyi,* KAN.; *Kaippam-patólam, pata-valam, pépatolam,* MALAY.; *Tó-pelen-moye, tha-bwot-kha ,* BURM.; *Dúmmaala,* SING.; *Patola,* SANS.

References.—*Roxb., Fl. Ind., Ed. C.B.C., 694; Voigt, Hort. Sub. Cal., 57; Kurz, in Jour. As. Soc., 1877, Pt. ii., 98; Thwaites, En Ceylon Pl., 126; Dals. & Gibs., Bomb. Fl., 102; Elliot, Fl. Andhr., 12, 35, 37, 146; Mason, Burma & Its People, 470, 747; Ainslie, Mat. Ind., II., 296; O'Shaughnessy, Beng. Dispens., 350; Dymock, Mat. Med. W. Ind., 2nd Ed., 343; Cat. Baroda Durbar, Col. & Ind. Exhib., No. 178; Atkinson, Him. Dist., 310, 700; Ec. Prod., N.-W. P., Pt. v., 3, 4, 5; Drury, U. Pl., 433; Lisboa, U. Pl. Bomb., 158; Royle, Ill. Him. Bot., 219; Bedie, Prod. S. Ind., Paris Exhib., 53; Gazetteers :—Mysore & Coorg, I., 55; Bombay, XV., 435; N.-W. P., I., 81; IV., lxxvi; Boswell, Man. Nellore, 120; Ind Forester, III., 238; Agri.-Horti. Soc. Ind., Trans , VII., 64, 67; Journ. (Old Series), IV., 202.*

Habitat.—An extensive climbing annual, which grows on hedges and bushes; found throughout India and Ceylon.

GUM.
577

Gum.—A gum, said to have been obtained from this plant, was sent from Madras to the Panjáb Exhibition. It is, however, very doubtful if the product in question really was derived from a cucurbitaceous plant. It must in any case be unimportant, since no other reference to it can be found in works on Indian economic subjects.

MEDICINE.

Medicine.—The *patola* of Sanskrit writers, a plant which is mentioned by Chakradatta as febrifuge and laxative, is said by Dymock to be referred in Bombay to this species. In Bengal, on the other hand, T. dioica is believed to be the Sanskrit *patola*. However this may be, the species under consideration is supposed to possess several valuable properties. Thus Ainslie writes, "The tender SHOOTS and dried CAPSULES are very bitter and aperient, and are reckoned amongst the stomachic laxative medicines of the Tamools; they are used in infusion to the extent of two ounces twice daily." In South India, at the present day, the SEEDS are considered to be a remedy for disorders of the stomach, antifebrile and anthelmintic; the tender shoots and dried capsules are believed to have the qualities described by Ainslie, and are given in decoction with sugar to assist digestion; the JUICE of the LEAVES is thought to be emetic, that of the ROOT purgative, the STALK in decoction is reputed expectorant (*Drury*). In Bombay, Dymock informs us, the plant has a reputation as a febrifuge, and is given in decoction with ginger, *chiretta*, and honey. "Muhammadan writers describe it as cardiacal, tonic, alterative, antifebrile, and as a useful medicine for boils and intestinal worms." In the Konkan the LEAF-JUICE is rubbed over the liver, or the whole body, in remittent fevers (*Mat. Med. W. Ind.*).

Shoots.
578
Capsules.
579

Seeds.
580
Juice.
581
Leaves.
582
Stalk.
583
Leaf juice.
584

SPECIAL OPINION.—§ "The juice of the leaves and fruit is useful in cases of congestion of the liver and bilious headache; it also acts as a laxative. The roots act as a powerful cathartic" (*Civil Surgeon J. H. Thornton, B.A., M.B., Monghyr*).

FOOD.
Fruit.
585

Food.—"The ripe FRUIT is eaten in stews by the Natives; it is exceedingly bitter, for which it is reckoned the more wholesome" (*Roxb.*).

T. 585

The Patol Gourd.   (*J. Murray.*)   **TRICHOSANTHES lobata.**

Trichosanthes dioica, *Rox.b ; Fl. Br. Ind., II., 609.*   **586**

    Vern.—*Parvar, palwal, palwal,* HIND.; *Potól,* BENG.; *Patal,* URIYA; *Palwal,* PB.; *Potala,* GUZ.; *Kombu-pudalai,* TAM.; *Kommu-potla,* TEL.; *Patólam,* MALAY.; *Patola, putulika,* SANS.

    References.—*Roxb., Fl. Ind., Ed. C.B.C., 694; Stewart, Pb. Pl., 99; Elliot, Fl. Andhr., 94; Pharm. Ind., 96; Ainslie, Mat. Ind., II., 297; O'Shaughnessy, Beng. Dispens., 351; U. C. Dutt, Mat. Med. Hind., 169, 313; S. Arjun, Bomb. Drugs, 60; Official Corresp. on Proposed New Pharm. Ind., 228; Ec. Prod., N.-W. P., Pt. V., 4, 12; Drury, U. Pl., 433; Ind. Gard., 226; W. W. Hunter, Orissa, II., 160; Gazetteer, N.-W. P., IV., lxxii.*

    Habitat.—An extensive climber, common throughout the plains of Northern India, from the Panjáb to Assam and Eastern Bengal. It is extensively cultivated during the rains throughout the above-mentioned localities, in the same way as other gourds.

    Medicine.—The LEAVES, the fresh JUICE of the FRUIT, and the ROOT are all used medicinally. The leaves are described as a good, light, and agreeable, bitter tonic. The tender TOPS are regarded as tonic and febrifuge. The fresh juice of the unripe fruit is often used as a cooling and laxative adjunct to alterative medicines. The root is classified amongst purgatives by **Susruta.** In bilious fever a decoction of *patola* leaves and coriander in equal parts is given as a febrifuge and laxative. The fresh LEAF-JUICE is recommended by several writers as an application to bald patches (*Hindu Materia Medica*). An alcoholic extract of the unripe fruit is said to be a powerful and safe cathartic. According to Rai Kani Lal De Bahadur, "the bulbous part of the root is a hydragogue cathartic, operating in the same way as Elaterium, for which it can be substituted." He describes the PLANT itself as a wholesome, bitter, and useful tonic. Dr. Bowser, from personal trials, describes it as a febrifuge and tonic. The old Hindu physicians placed much confidence in it in the treatment of leprosy (*Pharm. Ind.*).

    SPECIAL OPINIONS.—§ "The leaves of *putwal* or *potol* are bitter and possess tonic properties. They are generally fried with flour paste in *ghi* and eaten. The fruit is an excellent vegetable, which agrees well with convalescents, even from bowel complaints. It is largely consumed as food. The conserve of the fruit is nice food for convalescents, and can easily be prepared" (*Surgeon R. L. Dutt, M.D., Pabna*). "The root is a drastic purgative, useful in dropsy" (*Assistant Surgeon S. C. Bhattacharji, Chanda*).

    Food.—The FRUIT is oblong, smooth, green when young, and yellow or orange when ripe. When unripe it is much used by Natives as a vegetable, being considered very wholesome, and specially suited for convalescents. The tender TOPS are also eaten as a pot-herb. By Europeans the tender fruit is valued as one of the most palatable of gourds. It is generally prepared in the following ways:—(*a*) Cut in half, boiled and served as a vegetable with butter, salt, and pepper;-(*b*) cut in half and fried; (*c*) cut in slices and stewed in sauce; (*d*) cut in half and preserved in syrup with cinnamon and vanilla.

T. lobata, *Roxb.; Fl. Br. Ind., II., 610.*   **596**

    Vern.—*Bun-chichinga,* BENG.; *Ban-chachinga, jangli-chichinda,* N.-W. P.

    References.—*Roxb., Fl. Ind., Ed. C.B.C., 694; Atkinson, Ec. Prod. N.-W. P., Pt. V., 3, 4, 5.*

    Habitat.—Found in hedges and among bushes in the Deccan Peninsula; probably a variety of T. cucumerina.

    Food.—It flowers during the rains and produces an oblong, acute FRUIT, which, however, is apparently not eaten. Atkinson remarks that the

Marginal entries:

MEDICINE.

Leaves. **587**
Juice. **588**
Fruit. **589**
Root. **590**
Leaf-juice. **591**
Tops **592**
Plant **593**

FOOD.

Fruit. **594**
Tops. **595**

FOOD. Fruit. **597**

---

reason of this is not evident, since it appears to be as edible as the other species.

598   **Trichosanthes nervifolia,** *Linn.; Fl. Br. Ind., II., 609.*

> **Syn.**—T. CUSPIDATA, *Lamk.*
>
> **Vern.**—*Parvar, palval,* HIND.; *Potól,* BENG.; *Kombu-pudalai,* TAM.; *Kommu-potta,* TEL.; *Podla káyi,* KAN.; *Patólam,* MALAY.
>
> **References.**—*Rheede, Hort. Mal., t. 16, 17; Pharm. Ind., 96; Moodeen Sheriff, Supp. Pharm. Ind., 248; Gazetteer, Mysore & Coorg, I., 55.*
>
> **Habitat.**—A native of the Deccan Peninsula and Ceylon.

MEDICINE.   **Medicine.**—Medicinal properties similar to **T. dioica,** *Roxb.*

599
600   **T. palmata,** *Roxb.; Fl. Br. Ind., II., 606; Wight, Ill., t. 104, 105.*

> **Syn.**—T. LACINIOSA, *Wall.;* T. ASPERA, *Heyne;* T. TRICUSPIS, *Miq.;* T. BRACTEATA, *Kurz;* CUCURBITA MELOPEPO, & BRYONIA PALMATA, *Wall.*
>
> **Vern.**—*Lál-indráyan, indráyan mʋkal,* HIND.; *Mákál,* BENG.; *Indrá-yan, parwar, palwal, makhál, lál-indráyan,* N.-W. P.; *Indráyan,* KUMAON; *Kaundal,* BOMB.; *Kavandala,* MAR.; *Lál-indrávan, gúdá-pandú, koundel,* DEC.; *Korattai, shavari-pasham, ancoruthai,* TAM.; *Avvagúda-pandu, ábuvva, káki donda, ábúba, donda, ágúba, avvagúda, avaduta,* TEL.; *Avvagude-hannu,* KAN.; *Titta-hondala,* SING.; *Mahákála,* SANS., *Anbaghól, hansale-ahmar,* ARAB.; *Hansale-surkh,* PERS.
>
> **References.**—*Roxb., Fl. Ind., Ed. C.B.C., 695; Kurz, in Jour. As Soc., 1877, Pt. ii., 99; Thwaites, En. Ceylon Pl., 126; Dals. & Gibs., Bomb. Fl., 103; Elliot, Fl. Andhr., 10, 18, 27, 77; Pharm. Ind., 96; Ainslie, Mat. Ind., II., Ind., 85; O'Shaughnessy, Beng. Dispens., 349; Moodeen Sheriff, Supp. Pharm. Ind., 248; U. C. Dutt, Mat. Med. Hind., 308; Dymock, Mat. Med. W. Ind., 2nd Ed., 345; S. Arjun, Bomb. Drugs, 60; Irvine, Mat. Med. Patna, 71; Official Corresp. on Proposed New Pharm. Ind , 239; Bidie, Prod. S. Ind., Paris Exh., 30; Atkinson, Him. Dist., 310, 699, 752; Ec. Prod. N.-W. P., Pt. v., 4; Drury, U. Pl., 433; Bird-wood, Bomb. Pr , 37; Gazetteers:—Mysore & Coorg, I., 61; N.-W. P., I., 81; IV., lxxii; Boswell, Man. Nellore, 118, 125.*

Habitat.—A very large climber, common in all moist thickets from the Himálaya to Ceylon and Singapore, ascending hills to the altitude of 5,000 feet.

MEDICINE.   **Medicine.**—Ainslie informs us that the FRUIT, pounded small, and in-
Fruit.   timately blended with warm cocoa-nut oil, is considered a valuable applica-
601   tion for cleaning and healing "those offensive sores which sometimes take
place inside the ears. The same preparation is supposed to be a useful
Root.   remedy, poured up the nostrils in cases of ozæna." The ROOT is described
602   by Wight as useful in inflammation of the lungs in cattle. O'Shaughnessy was induced by the singularly bitter taste of the rind to make experiments with a view to ascertaining whether it possessed purgative, tonic,
or aperient properties, but given in three-grain doses, thrice daily, it was
found to produce no sensible effect (*Beng. Dispens.*). Dymock states
that Natives in Bombay sometimes smoke the fruit as a remedy for
asthma. The root, with an equal portion of colocynth root, is rubbed into
a paste and applied to carbuncles; combined with equal portions of the
three myrobalans and turmeric, it affords an infusion which, when flavoured
with honey, is given in gonorrhœa (*Mat. Med. W. Ind.*).

SPECIAL OPINION.—§ "The juice of the fruit or the root-bark, boiled
with gingelly oil, is used with good effect as a bath oil for the relief of long
standing or recurrent attacks of headache" (*Surgeon-Major W. R. Thompson, C.I.E., Madras*).

T. 602

Food.—The bright-red FRUIT of the wild plant is not eatable, owing to its severely drastic properties, but, under cultivation, the fruit becomes a wholesome vegetable when well boiled. At the Cape of Good Hope its poisonous properties appeared to be removed by pickling (*Jour. Agri.-Horti. Soc., X. (Old Series), 3*).

Domestic and Sacred.—The poisonous FRUIT is said to be occasionally mixed with rice and thus employed to destroy crows (*Roxburgh*). It is used by the Hindus of Western India as an ear-ornament for their idol *Ganpatti*, who is dressed up and seated in state in every Hindu house once a year, to bring good luck to the inmates (*Dymock*).

| | FOOD.
Fruit.
603 |
| | DOMESTIC.
Fruit.
604 |

TRIFOLIUM, *Linn.; Gen. Pl., I., 487.*

Trifolium fragiferum, *Linn.; Fl. Br. Ind., II., 86;* LEGUMINOSÆ.

STRAWBERRY-HEADED CLOVER.

Vern.—*Chit-batto,* KASHMIR.

Habitat.—Confined to Temperate Kashmír, and much like **T. repens.**

Fodder.—Eaten by cattle. This plant receives its English name from fruit-like appearance of its calyces, which expand and take on a reddish colour after the flowers fade.

605

FODDER.
606

T. pratense, *Linn.; Fl. Br. Ind., II., 86.*

RED or BROAD-LEAVED CLOVER or COW-GRASS.

Vern.—*Trepatra, chit-batto,* PB.

References.—*DC., Crig. Cult. Pl., 105; Stewart, Pb. Pl., 76; Year-Book Pharm., 1873, 842; Atkinson, Him. Dist., 308.*

Habitat.—Extends from Kashmír to Garhwál at altitudes of 4,000 to 8,000 feet and is not uncommon.

Fodder.—This is one of the common forage clovers in the above-mentioned region, and is regarded as a good cropper where the commoner clover fails.

607

FODDER.
608

T. repens, *Linn.; Fl. Br. Ind., II., 86.*

WHITE or DUTCH CLOVER.

Vern.—*Shaftal, shotul,* PB.; *Ghurg,* PUSHTU.

References.—*Aitchison, Bot. Afgh. Del. Com., 48, Stewart, Pb. Pl., 76; Lace, Quetta Pl., in MS.; Atkinson, Him. Dist., 308; Gazetteer, Mysore & Coorg, I., 59; Agri.-Horti. Soc. Ind., Jour. (Old Series), XIV., 12.*

Habitat.—A slender, wide creeping herb, common in many parts or the Temperate and Alpine Himálaya, up to 16,000 feet; also found in the Nilghiris and Ceylon, where, perhaps, it has been introduced.

Fodder.—This is one of the most highly prized fodder plants of Europe. On the Himálaya unfortunately, however, it has the evil reputation of readily causing salivation. [The writer has seen several horses suffering very badly and one that died; in each case the attendants were confident that this was due to their having eaten the wild white clover.—*Ed., Dict. Econ. Prod.*]

609

FODDER
610

TRIGONELLA, *Linn.; Gen. Pl., I., 486.*

A genus of annual herbs which comprises some fifty species, of which eight are met with in India. The FENUGREEK is the only indigenous species of

611

T. 611

**TRIGONELLA
Fœnum-græcum.**   The Fenugreek.

any importance, though some of the other species might doubtless be utilised as fodder. The small crescent-shaped pods of an Asiatic, though non-
Indian, member of the genus, T. **uncata**, *Boiss.*, is imported into Bombay
from the Persian Gulf, under the name of *iklil-el-malik*, for medicinal purposes
(see *Pharmacog. Ind., I., 404*).

[MINOSÆ.

612   **Trigonella Fœnum-græcum,** *Linn.; Fl. Br. Ind., II., 87;* LEGU-

THE FENUGREEK OF FENUGRÆC.

Vern.—*Méthi, múthi,* HIND.; *Méthi, methi-shak, methika, hænugreeb,*
BENG.; *Methi, methun, methri,* PB.; *Shamli,* AFG.; *Mathi, mitha,*
SIND; *Méthi, methini, bhaji,* GUZ.; *Vendayam, ventayam,* TAM.; *Men-
tulu, menti kira,* TEL.; *Ménthyá, mente soffu, mente-palle, mente,*
KAN.; *Uluva, ventayam, venthiam,* MALAY.; *Pe-nán-ta-si,* BURM.;
Uluva, SING.; *Méthi, methiká,* SANS.; *Hulbah,* ARAB.; *Shanbalid,
shamlit, shamlis, shamlid,* PERS.

References.—*Roxb., Fl. Ind., Ed. C.B.C., 588; Stewart, Pb. Pl., 77; DC.,
Orig. Cult. Pl., 112; Sir W. Elliot, Fl. Andhr., 115; Flück. & Hanb.,
Pharmacog., 172; Fleming, Med. Pl. & Drugs (Asiatic Reser., XI), 183;
Ainslie, Mat. Ind., I., 130; O'Shaughnessy, Beng. Dispens., 291; Irvine,
Mat. Med. Patna, 66; Medical Topog., Ajm., 145; U. C. Dutt, Mat.
Med. Hind., 144, 309; S. Arjun, Cat. Bomb. Drugs, 44; Murray,
Pl. & Drugs, Sind., 113; Bent. & Trim., Med. Pl., 71; Dymock,
Mat. Med. W. Ind., 2nd Ed., 209; Cat. Baroda Durbar, Col. & Ind.
Exhb., No. 179; Year-Book Pharm., 1874, 624; Trans. Med. & Phys.
Soc., Bomb. (New Series), No. vi., 1860, 330; Birdwood, Bomb. Prod.,
31, 148, 220; Baden Powell, Pb. Pr., 151, 245; Atkinson, Him. Dist.
(X., N.-W. P. Gaz.), 308, 708, 752; Useful Pl. Bomb. (XXV., Bomb.
Gaz.), 151, 217; Econ. Prod. N.-W. Prov., Pt. V. (Vegetables, Spices,
and Fruits), 13, 15, 37, 40; Stock's Rep. on Sind; Nicholson, Man.
Coimbatore, 224; Morris, Descriptive & Historical Acct., Godavery, 68;
Bombay, Man. Rev. Accts., 102, 103; Madden, Note on Kumaon, 280;
Settlement Rept.:—Panjáb, Jhang, 90, 91; N.-W. P., Kumaon, App.,
33; Gazetteers:—Panjáb, Karnal, 172; N.-W. P., I., 80; IV., lxx.;
Orissa, II., 27 134, 180; Mysore & Coorg, II., 55, 59; II., 11.*

Habitat.—A robust, annual herb, wild in Kashmír, the Panjáb, and the
Upper Gangetic plain, widely cultivated in many parts of India, particularly in the higher inland provinces.

CULTIVATION
613

CULTIVATION.—No particulars can be given of the annual area under
the crop, except in the cases of Bombay and Madras, the former of which
had, in 1888-89, 1,358 acres, and the latter 257 acres, under Fenugreek. It
is, however, of considerable importance in other parts of India, especially in the Panjáb. The following may be accepted as typical of the
method of cultivation:—It is grown near wells and *sailáb* lands. On the
former it is generally sown after cotton, sometimes after *juár*, rarely
on uncropped ground. The seed, about 30℔ to the acre, is scattered
broadcast in the month of February, is trampled into the ground, and
watered. It seldom fails to germinate, and after the leaf appears requires no care beyond five or six further waterings. A top-dressing is
often given. The crop is ready to cut in April. On *sailáb* lands it is
sown in the end of October or beginning of November, good new alluvium or rich old clayey loam being generally selected. After one or two
ploughings the seed is scattered broadcast and ploughed lightly in. The
crop ripens about the same time as that on well lands (*Settlement Report,
Jhang District*).

DYE.
Seed.
614

Dye.—The SEED yields a yellow dye, and enters into the composition of
an imitation of carmine. The yellow decoction produces a fine permanent
green with sulphate of copper.

T. 614

The Fenugreek.	(*J. Murray*.)		TRIGONELLA Fœnum-græcum.

Medicine.—Fenugreek has been known and valued as a medicine from very remote antiquity. Sanskrit writers describe the SEEDS as carminative, tonic, and aphrodisiac. Several confections made with them are recommended for use in dyspepsia with loss of appetite, in the diarrhœa of puerperal women, and in rheumatism (*U. C. Dutt*). "Muhammadan writers describe the PLANT and seeds as hot and dry, suppurative, aperient, diuretic, emmenagogue, useful in dropsy, chronic cough, and enlargements of the spleen and liver. A poultice of the LEAVES is said to be of use in external and internal swellings and burns, and to prevent the hair falling off. The FLOUR of the seeds is used as a poultice, aud is applied to the skin as a cosmetic. The OIL of the seeds is also used for various purposes" (*Dymock*). Ainslie informs us that the seeds are much prescribed by Native practitioners in dysenteric complaints, being commonly toasted and given in infusion. In the Panjáb the seeds are used in fomentation, and are prescribed for colic, flatulence, and dysentery (*Stewart*). Made into a gruel they are given as a diet to nurses, to increase the flow of milk. In Western India the leaves are employed both externally and internally on account of their cooling properties. Dymock states that they have an aperient action in "bilious states of the system."

In European medicine, fenugreek at one time enjoyed as high a reputation as it now holds in Hindu and Muhammadan Materia Medica. It is the "*Fœnum Græcum*" of Latin writers, the τῆλις of Dioscorides and other Greek authors. Its mucilaginous seeds, "siliquæ" of the Roman peasants, were valued as a food and supposed to possess many medicinal virtues. Cultivation of the plant was encouraged by Charlemagne in Central Europe (A.D. 812) and fenugreek was grown in English gardens in the sixteenth century (*Flückiger & Hanbury*). Though officinal in most of the Pharmacopæias of the eighteenth century, fenugreek is now obsolete as a medicine in Europe. Flückiger & Hanbury, however, state that the seeds are still often sold by chemists for veterinary pharmacy.

CHEMICAL COMPOSITION.—The cells of the testa contain tannin, the cotyledons, a yellow colouring matter, but no sugar. The air-dried seeds give off 10 per cent. of water at 100°C., and on subsequent incineration leave 7 per cent of ash, of which nearly a fourth is phosphoric acid. From the pulverized seeds ether extracts 6 per cent. of a fœtid fatty oil having a bitter taste. Amylic alcohol removes a small quantity of resin ; alcohol, added to a concentrated aqueous extract, precipitates mucilage, which amounts when dried to 28 per cent. The percentage of nitrogen indicates an equivalent of 22 per cent. of albumen (*Pharmacographia*). Johns states that two alkaloids exist in the seeds, *choline*, a base found in animal secretions, and *trigonelline* ($C_7 H_7 NO_2 + H_2 O$), a substance which may be crystallised from alcoholic solution in colourless prisms ; it has a weak saline taste.

TRADE.—Large quantities of the SEED are annually exported from the higher northern plains in which the crop is grown, to other parts of India. The imports from Karáchí into Bombay alone amount to about 10,000 cwt. annually, and a considerable amount is received in the same town from the Gháts, the Deccan, and Guzerát. The value varies from R40 to R50 per candy.

SPECIAL OPINIONS.—§ "The seeds, made into a gruel, are used as a stimulant and tonic" (*Surgeon-Major A. S. G. Jayakar, Muskat*). "Boiled well with milk it is given internally in bleeding piles. The leaves fried with *ghí* are used in dysentery" (*Surgeon-Major D. R. Thomson, M.D., C.I.E., Madras*). "The leaves applied as a poultice are much used

MEDICINE.
Seeds.
615
Plant.
616
Leaves.
617
Flour.
618
Oil.
619

Chemistry.
620

Trade.
621
Seed.
622

T. 622

in cases of contusion" (*Hospital Assistant Lal Mahomed, Hoshangabad, Central Provinces*).

FOOD &
FODDER.
Leaves.
623
Seeds.
624
Pods.
625
Plant.
626

Food and Fodder.—The LEAVES, especially when young, are largely employed as a vegetable in India. They are boiled and afterwards fried in *ghí*; the taste is bitter and very disagreeable to Europeans. The SEEDS are chiefly used as a condiment to flavour curries made of rice, pulse, flour, and meat, or as a relish with unleavened bread. They have an unpleasant odour, with an unctuous, farinaceous taste, accompanied by considerable bitterness. The young PODS are eaten as a vegetable, being generally cooked by simply boiling in water. The PLANT is a valuable fodder, though believed to be heating and lactifuge. The seeds form an important constituent of many cattle foods, and are used to render musty hay and compressed fodder palatable. They are said to be also employed as an adulterant of, and substitute for, coffee.

627

Trigonella occulta, *Delile ; Fl. Br. Ind., II., 87.*

Syn.—T. ARGUTA, *Visiani.*

References.—*Boiss., Fl. Orient., II., 84; Murray, Pl. & Drugs, Sind, 113.*

Habitat.—A diffuse, densely cæspitose annual, found in the plains of Sind and the Upper Gangetic plain near Lucknow; distributed to Egypt and Nubia.

MEDICINE.
Seeds.
628
FOOD.
Plant.
629
Pods.
630

Medicine.—In Sind, the SEEDS are used in dysenteric affections (*Murray*).

Food.—"The fresh-gathered PLANT and PODS are eaten as a pot-herb" (*Murray*).

Trepe de Roche, see Lichens, Vol. IV., 638.

TRIPHASIA, *Lour.; Gen. Pl. I., 303.*

631

Triphasia trifoliata, *DC.; Fl. Br Ind., I., 507;* RUTACEÆ.

Syn.—T. AURANTIOLA, *Lour.;* LIMONIA TRIFOLIATA, *Linn.;* L. DIACANTHA, *DC.*
Vern.—*Chini naranghi,* HIND.
References.—*Dalz. & Gibs., Bomb. Fl. Supp., 12; Kurz, For. Fl. Burm., I., 192; Gamble, Man. Timb., 59; Burm., Fl. Ind., t. 35, f. 1; Mason, Burma & Its People, 453, 759; Kew Bulletin, 1889, 22; Lisboa, Useful Pl. Bomb., 149.*

Habitat.—Common as an escape in the Western Peninsula and in gardens throughout India. It is a native of China, but has been introduced into India for many years.

FOOD.
Fruit.
632

Food.—The FRUIT is eaten in Southern and Western India, and is frequently used in conserves and pickles. It is a very common ingredient of Chinese preserved fruits.

633

TRITICUM, *Linn.; Gen. Pl., III., 1204.*

A genus of annual or biennial grasses, erect, with flattened leaves, terminal, cylindrical, or elongated spikes, and a flexuous rachis, alternately hollowed for the reception of the spikelets, continuous or rarely jointed. According to Bentham & Hooker, the genus includes the two old genera, Crithodium, *Link*, and Ægilops, *Linn.*, and comprises in all some ten species, natives of the Mediterranean region and of Western Asia. Of these the only species of economic importance belong to the section of Triticum proper, or cultivated wheats. Hæckel, the latest monographer of the genus, whose arrangement will be followed in this article, considers that there are three species of that section, namely, Triticum monococcum, T. sativum, and T. polonicum. The first of these undoubtedly grows wild in Greece and Mesopotamia, and is cultivated in Spain and elsewhere. It was grown by the aboriginal Swiss lake dwellers, a fact demonstrated by the grain having been found near their dwellings. T. sativum, the ordinary cultivated wheat, is by Hæckel referred to three principal races, which will be considered below. The third species, T. polonicum, is a very distinct form, with long leafy glumes. It is of doubtful origin; like the first species it is not cultivated in India, and may, therefore, be excluded from consideration in this work. Though it is convenient to a dhere to this classification of the cultivated wheats into three distinct species, it must be remembered that there is every probability of the descent of all from one common stock; perhaps, as DeCandolle thinks, from a small-grained form of T. sativum, or from T. monococcum, formerly cultivated by the Egyptians, and by the lake-dwellers of Switzerland and Italy; or possibly from some of the wild species formerly included in the genus Ægilops.

Triticum sativum, *Lamk.;* GRAMINEÆ.   634

WHEAT, *Eng.;* FROMENT, *Fr.;* WEIZEN, *Germ.*

Hæckel recognises three principal races, namely, α, spelta, β, dicoccum, and γ, tenax.

Syn.—T. VULGARE, *Villars.;* T. HYBERNUM & T. ÆSTIVUM, *Linn.;* T. TURGIDUM, *Linn.;* T. COMPOSITUM, *Linn.;* T. COMPACTUM, *Heer;* T. DURUM, *Desf.;* T. DICOCEUM, *Schronk.;* T. AMYLEUM, *Seringe;* T. SPELTA, *Linn.*

Vern.—*Géhún, kunak, giún,* HIND.; *Giún, gom, gam,* BENG.; *Tro, shruk, tokár* (white), *tomár* (red), MICHI; *Ghúbot* (when the ear begins to form), *seonikar* (when the ear is out), C. P.; *Gehún,* N.-W. P. & OUDH; *Kanak, gehún, rozatt, dro, do, zud, nis, to,* PB.; *Gandam, ganam,* AFGHAN.; *Kank? gih.,* SIND; *Gehún,* DECCAN; *Gahu, gium, ghawn, mar-ghum, ghawut-ghum, kapale, gohum,* BOMB.; *Gahung,* MAR.; *Ghavum, gawn,* GUZ.; *Gódumai, godumbay-arisi,* TAM.; *Gódumulu,* TEL.; *Gódhi,* KAN.; *Kótanpam, gendúm,* MALAY.; *Giyonsabá, gyungsa-ba,* BURM.; *Tiringu,* SING; *Godhúma, saman,* (*yavá,* though sometimes applied to this grain more correctly denotes barley), (U. C. Dutt gives *Mahágodhuma,* a large-grained form; *Madhuli,* a small grained; and *Nihsuki,* a beardless wheat. The first was held, by Sanskrit writers, to have been introduced from the West and the second to have been indigenous to India. It was, therefore, very probably the most anciently cultivated form, but may have passed (as is customary with many other crops at the present day) as indigenous because its history was not known.—*Ed., Dict. Econ. Prod.*), SANS.; *Hintah, burr,* ARAB.; *Gandum,* PERS.

References.—*Dalz. & Gibs., Bomb. Fl. Suppl., 97; Stewart, Pb. Pl., 262, 263; Aitchison, Rept. Pl. Col. Afgh. Del. Com., 127; also Notes on Prod. W. Afgh. & Persia, 212; DC., Orig. Cult. Pl., 354, 359, 362, 365; Mason, Burma & Its People, 476, 818; Sir W. Elliot, Fl. Andhr., 61; Pharm. Ind., 254; O'Shaughnessy, Beng. Dispens., 632; Moodeen Sheriff, Supp. Pharm. Ind, 249; U. C. Dutt, Mat. Med. Hind., 267, 269, 298; S. Arjun, Cat. Bomb. Drugs, 154; Murray, Pl. & Drugs, Sind, 14; Birdwood, Bomb. Prod., 113, 243; Baden Powell, Pb. Pr., 225, 228, 383; Drury, U. Pl. Ind., 434; Duthie, & Fuller, Field & Garden Crops, 1-8; Useful Pl Bomb. (XXV., Bomb. Gaz.), 189, 208; Royle, Prod. Res., 426; McCann, Dyes & Tans, Beng. 36; Church, Food-Grains Ind., 34, 36, 90-98; Bidie, Prod. S. Ind., 70; Tropical Agriculture, 297; Rep. on the Col. & Ind. Exhbn., 125, 126; Ayeen Akbary, Gladwin's Trans., II., 44, 135; Hunter, Indian Empire, 3ʳᵈ 4, 385, 452; Man. Madras Adm., I., 288, 290; II., 85, 88, 109, 119; Nicholson, Man. Coimbatore, 222, 583, 585, 586; Man. of Kurnool, 172; Moore, Man. Trichinopoly, 366; Bombay, Admin. Rep.*

(1889), *76-79, 105* ; *(Statistical Returns), 5, 7, 9, 70, 72* ; *Bomb. Man. Rev. Accts., 101* ; *Bengal Admin. Rep.* (1889), *Pt. I., 27, 28* ; *Pt. II., 12, 106, 107, 115, 122, 136, 138, 146, 147, 163, 165* ; *British Burma, Rep. on Inland Trade (1884-85), App. ii., 4* ; *British Burma, Rep. on Trade & Navigation (1884-85), 9* ; *Settlement Reports* :—*N.-W. P., Kumaon, App. 32* ; *Central Provinces* :—*Baitool, 77* ; *Chanda, 81, 84, 96, 98* ; *Damoh, 87* ; *Jubbulpore, 86* ; *Hoshungabad, 277, 287* ; *Mundlah, 46* ; *Nagpore, 273* ; *Nimar, 196* ; *Nursingpur, 53* ; *Saugor, 98* ; *N. Godavery, 35* ; *Wurdah, 65-67* ; *Port Blair (1870-71), 27* ; *Gazetteers* :—*Panjáb, Karnal, 172* ; *N.-W. P., I., 86* ; *Central Provinces, 18, 114, 385, 471, 501, 516* ; *Sind, 306* ; *Burma, I., 465* ; *Mysore & Coorg, I., 68* ; *Rajputana, 96, 128, 254, 255, 279* ; *Ulwar, 87, 89, 127, 166* ; *Annual Reports of Dirs. Land Rev. and Agri. in many passages* ; *Reports of Chambers of Commerce, Bombay, Calcutta, Karachi* ; *Proceedings of the Govt. of India, Rev. & Agri. Depts., in many passages,* *Watt, on the conditions of wheat growing in India, Jour. Royal Agri. Soc. Eng., XXIV., 8* ; *Watt, The Trade of India and its Future Development, Pro. Royal Col. Inst. XVIII., 45, 60, 68* ; *Agri.-Horti. Soc. Ind.* :—*Trans., I., 19, 24, 27, 28, 166, 195, 197, 201, 207* ; *II., 157, 190* ; *III., 83, 90, 187* ; *IV., 82-85, 88, 91, 99, 102, 107, 117, 118, 124, 144, 145, 150* ; *V., 63, 68* ; *VI., 74, 75* ; *VIII., 94, 95, 171, 419* ; *Jour., I., 142-145, 158, 165, 335* ; *II. Pt. I., 257* ; *Pt. II., 176, 177, 256, 259, 267, 293, 409, 410, 442, 447, 448, 450, 480, 537-539, 547, 592* ; *III., 94, 98, Sel., 193, 194, 249* ; *IV., Pt. I., 120* ; *Pt. II., 29, 47, Pro. xxiii.-xxv., xlix.* ; *V., Pt. I., 135, 136* ; *VI., 148* ; *VII., Pt. II., 1, 2* ; *VIII., 57* ; *XIV* ; *133* ; *New Series, I., Pro. xxxii., xxxiii* ; *III., Sel. 34, 35* ; *VI., Pro. 26* ; *VIII., 79, 80, 82, 173, 178* ; *Bear The Indian Wheat Trade, Jour. Royal, Agri. Soc. Eng., XXIV., 50* ; *Indian Agriculturist, 1886, 1887, 1889, 1890* ; *Rev. & Agri. Dept., Notes on Wheat, 1885, 1889* ; *Basu, Rep. Agri., Lohardaga Dist., Pt. I., 50, 152* ; *Pt. II., 24, 30-32, 53, 54, 61, 72, 74* ; *Clifford Richardson, Investigation of the Composition of American Wheat and Corn* ; *Report on the Distribution and Consumption of Wheat, 1885, 1888* ; *Report, Impurities in Indian Wheats, 1888-1889* ; *Report Nagpore Experimental Farm, 1884-85* ; *Encyclop. Brit., XXIV., 531* ; *Balfour, Cyclop. Ind. 1069* ; *Morton, Cycl. Agri., II., 1004, 1127-1155* ; *Smith, Econ. Dict., 438.*

HISTORY.
635

Habitat & History of Wheat.—The question of the original habitat of wheat, and of the origin and history of its widespread cultivation, has been dealt with very elaborately by **M. A. deCandolle** in his valuable work on the *Origin of Cultivated Plants.* He adduces numerous arguments in support of the opinion that the cultivation is prehistoric in the Old World. " Very ancient Egyptian monuments, he says, older than the invasion of the shepherds and the Hebrew Scriptures, show otherwise this cultivation already established, and when the Egyptians or Greeks speak of its origin, they attribute it to mythical personages—Isis, Ceres, Triptolemus. " A small-grained wheat has been found at the earliest lakedwellings of Western Switzerland, the inhabitants of which were at least contemporary with the Trojan war, and perhaps earlier. The same form, the **T. vulgare antiquorum** of Heer, was found by Unger in a brick of the pyramid of Dashur in Egypt, to which he assigns a date of 3359 B.C. Another small-grained race, **T. vulgare compactum muticum**, *Heer*, was less common in Switzerland in the earliest stone age, while a third intermediate form was cultivated in Hungary at the same period. From philological data, combined with the absence of authentic records of wild wheat, DeCandolle believes that the culture of the plant in the temperate parts of Europe, Asia, and Africa is probably older than the most ancient known languages. The Chinese certainly grew it 2700 B.C., and considered it a gift direct from heaven. In the annual ceremony of sowing five kinds of seed, instituted by the Emperor Shen-nung or Chinnong, wheat was one of the species employed. After carefully considering all available information, DeCandolle concludes that the original home of the species in very early prehistoric times was in Mesopotamia, where it is

HISTORY.

said by **Berosus**, the earliest of all Western Historians, and a Chaldean priest who wrote some twenty-three centuries ago, to have occurred wild. "The area," writes **DeCandolle**, "may have extended towards Syria, as the climate is very similar; but to the east and west of Western Asia, wheat has probably never existed but as a cultivated plant; anterior, it is true, to all known civilization." Spelt is considered by the same author to be a distinct species, which is said to have probably had its origin in eastern temperate Europe and the neighbouring countries of Asia. This presumption is, however, based entirely on doubtful historical and philological data, the latter of which are certainly faulty. He remarks that spelt has no name in Sanskrit, nor in any modern Indian language, a statement that was shown by **Dr. Watt**, in dealing with the wheats of Bombay, to be erroneous. Spelt has been cultivated from an uncertain date in many localities of India, and bears distinct vernacular names, which are always applied to it and never to the commoner races of wheat. It is interesting to notice that support is given by **Olivier** to **Hæckel's** reduction of spelt to **T. sativum**, an authority, whose testimony, regarding the indigenous area of the latter species, is accepted by **DeCandolle**. **Olivier** writes that he several times found spelt in Mesopotamia, in particular upon the right bank of the Euphrates, north of Anah, in places unfit for cultivation. It is thus at least possible that spelt, if not distinct, may at very early times have become differentiated from ordinary wheat, in the original home of both, and that the seed of both forms may have spread together, giving rise to the irregular appearance of the former as a cultivated plant in many parts of the world. [But there is still another consideration. In addition to the testimony of **Berosus**, **DeCandolle** accepts that of **Strabo** as of some weight in any attempt to determine the origin of wheat. **Strabo** was born 50 B.C., and he affirmed, on the authority of **Aristobulus**, that a grain very similar to wheat grew wild upon the banks of the Indus. **U. C. Dutt** tells us that one of the wheats spoken of by Sanskrit writers was regarded as indigenous, while another was beardless. There is thus very nearly as strong presumptive evidence in favour of India being the home of some of the forms of wheat as can be shown for any other part of the globe. There is sufficient, at all events, to justify the reprehension of any arbitrary affirmation in favour of one country more than another. India possesses perhaps as comprehensive a series of timehonoured forms of wheat as can be shown for any other country. It has its hard wheats and soft wheats; its starch wheats and spelt wheats; its bearded and beardless wheats. It can also be abundantly demonstrated that most of these have been grown for countless ages on very nearly the same fields as they are to be found at the present day. Even were it demonstrated that the Indus Valley wild grain, mentioned by **Strabo**, was the wild rice which exists there now, the position here urged would not, in the least, be affected, namely, that wheat cultivation in India is as ancient as in Europe or any other part of the world. Its origin is, however, involved in an obscurity even more impenetrable than that which envelopes the historic records of the wheat cultivation in other parts of the world. It must, however, be remembered, in discussing the question of the origin of this anciently cultivated cereal, that it is very nearly impossible, whatever the original home of wheat may have been, to determine with accuracy the character of the first parent from which it was derived. But **DeCandolle's** concluding remark regarding the wheats found in association with the remains of the lake dwellers of Switzerland and Italy is therefore significant, *viz.*, that "None of these is identical with the wheat now cultivated, as more profitable varieties have taken their place." *Ed., Dict. Econ. Prod.*]

WHEAT
AREA.
636

REVIEW OF THE AREA UNDER WHEAT.

The wheat area of the Indian Empire has been described by Dr.
Forbes Watson, in his *Report on Indian Wheat (1879)*, as embracing the
whole of Northern India up to the Gangetic Delta, and, in Southern
India, the whole of the table-land above the Gháts. The crop is cultivated in all districts of Sind, the Panjáb, the North-West Provinces and
Oudh. The true wheat-growing region of Bengal is the valley of the
Ganges, though in several other parts of the province the cereal is cultivated to a small extent for local consumption. In Assam no wheat is grown
for exportation. In Lower Burma there is a trifling area under the crop, but
the soil and climate of Upper Burma are extremely well suited to wheat,
and this province may in the future attain an important position in this
respect. In Bombay the cultivation is general, except in Thana, Kolába,
Ratnagíri, and Kánara. In Madras it is grown in Cuddapah, Bellary, Kar-
nul, Coimbatore, and the Nilghíris, also to a small extent in the coast district
of Kistna. Wheat is also grown in nearly every part of Mysore, throughout Berar, and in all parts of the Central Provinces except Sambulpur; in
Coorg it is not cultivated, but in Ajmír a considerable acreage is under
the crop. Thus, excluding the coasts of the peninsula and of the north and
east of the Bay of Bengal, it may be said that all the territories of British
India, except Assam and Burma, contribute to the wheat supply of the
country.

Increase.
637

The increase of the area under wheat in India, and of the importance
of the country amongst the wheat-suppliers of the world, has been very
rapid during the past twenty years. This fact has naturally attracted
much attention, not always of a friendly nature, and has given rise to
much controversy as to the actual cause. Thus, at no more remote date than
1887, the Statistician of the Department of Agriculture, Washington, in
an official report, went the length of discrediting the statement of the Government of India that there had been such an increase; but in 1888 he
had to admit that an increase had occurred, though he made it out to be
but small in amount. It has also been urged from other quarters that
the increase in area shown by agricultural reports must necessarily entail
a diminution in the cultivation of other non-exported food-crops and
thus have led to a decrease in the food-supply of the people; but this has
been shown again and again not to be the case. Dr. Watt, in a lecture
on the *Trade of India and its Future Development*, read before the Royal
Colonial Institute in 1886, says:—"With reference to the remarkable
modern export trade in wheat it is customary to hear the most absurd
and misleading statements made in public. It is, for example, not un-
commonly urged that the trade will decline as rapidly as it has come into
existence. It has been pronounced forced and unnatural, the accumulated
surplus of food which used to be held by the people against a season of
scarcity being now sold. Such an opinion is opposed to all the facts
which have the least bearing on the case. In the first place, with the
single exception of the Panjáb, wheat has never been a staple food with
the people of India. In the second place, far from the area formerly
occupied by the food-stuffs of the poor (millets and pulses) having been
displaced by wheat cultivation, it has been greatly extended. Last year,
for example, there were 58,565,331 acres under rice, 48,000,000 under
pulses, 33,228,867 under millets, and 20,328,254 under wheat. Returns
have been called for over the length and breadth of India, and it has
been conclusively shown that had the wheat cultivation remained at what
it was twenty years ago, the increased cultivation of rice, pulses, and millets would alone have proved sufficient to feed the greatly enhanced popu-

Wheat Cultivation.   (*J. Murray.*)   **TRITICUM sativum.**

lation. Wheat has been grown on the lands suitable for it because it has proved remunerative, but were circumstances outside the limits of India to arise that would lessen the profits on wheat, other crops would be substituted for it. Nothing could be more clearly demonstrated than this fact, for an exceptionally good harvest in Europe and America is at once followed by a lessened cultivation of wheat in India. If wheat has displaced any crop more than another, it has been cotton, and few Natives would be so far lost as to cultivate the millets upon rich wheat soil. Their best lands have always been devoted to remunerative crops for the export trade. But, if further proof were needed that the fields formerly devoted to the supply of necessary food have not been taken by wheat, it can be had in the fact that, coincident with the great success of the wheat trade of India, the areas under oil-seeds and cotton have also greatly extended. But the necessity for such explanations is not difficult to find. The wheat trade has had a much more immediate effect upon the established industries of Europe and America, and has, therefore, attracted more attention; but the development of the oil-seed trade has been quite as rapid as that of wheat. During the past five years, for example, it has increased 78½ per cent. in quantity and 69½ per cent. in value.

"But still a third series of facts proves that the wheat trade of India is a perfectly good and natural one. Were it the case that the surplus wheat of the working classes was being removed from India, the prices of other food-stuffs would be expected to show a distinct rise. The most careful record has been kept of the prices of food in every district of India for every fortnight during the past twenty years. On a careful scrutiny those returns are found to indicate a constant adjustment which bears a most remarkable correspondence with railway extension." Dr. Watt then gives a series of tables demonstrating the fact that while during the years from 1864 to 1884 wheat and other food-stuffs had increased in price in certain localities, in others they had diminished. In the case of wheat, the grain was actually cheaper in 1884 than it had been in 1864, in the North-West Provinces, Oudh, the Panjáb, and in the Central Provinces; in other words, in the provinces in which wheat has always been an important article of food. "In 1864," writes Dr. Watt, "there was practically no wheat exported from India, and in spite, therefore, of the enormous exports which have taken place during the past ten or fifteen years, the local price has remained stationary, or, in some districts, has actually become cheaper. Surely this does not, by any manner of means, justify the statements one often hears made, that the surplus food of the people of India is being drawn out of it through the greed for money of certain members of the Indian community." The question of increase of area has been more recently taken up by the Revenue and Agricultural Department, which, in the end of 1889, published a Note on the subject, drawn up by Mr. F. M. W. Schofield. The whole question is most thoroughly dealt with in that note, and a table furnished which may be repeated in this work, with the addition of actual statistics for 1888-89, and estimates for 1889-90. From that table Bengal, Madras, Ajmere, Burma, Assam, and Coorg have been left out of account. In the case of Bengal about a million acres are estimated to be under the crop, but this area is approximate only, as there are no statistics. In Madras the area is small and fluctuates little, the normal is taken at 27,000 acres. Ajmere is supposed to have a normal area of 15,000 acres, and, as already stated, Burma, Assam, and Coorg are not wheat-growing countries to any note-worthy extent. For the purposes of comparison as to increase in area it is therefore advisable to con-

TRITICUM sativum.	Indian Area under

WHEAT AREA.

Increase.

sider only the remaining provinces, which are the chief wheat-producers :—

Table shewing the areas in the chief wheat-producing provinces during the quinquennial periods ending 1877-78, 1882-83, and 1887-88, also actuals in 1888-89, and the estimated forecast for 1889-90.

Provinces.	Quinquennial period ending			Actuals, 1888-89.	Estimates, 1889-90.§
	1877-78.	1882-83.	1887-88.		
	Acres.	Acres.	Acres.	Acres.	Acres.
North-West Provinces.	3,732,000*	3,773,000	3,814,000	3,479,279 }	4,490,600
Oudh . . .	991,000*	1,249,000	1,507,000	1,489,921 }	
Panjáb . . .	6,609,000	6,793,000	7,043,000	7,371,977†	6,222,900
Central Provinces .	3,536,000	3,466,000	4,002,000	3,531,941	4,085,900
Bombay { British Districts.	912,000	1,428,000	2,033,000	2,078,447	1,878,100
Bombay { Native States.	591,000*	591,000*	591,000	601,000	551,000‡
Sind { British Districts.	273,000	297,000	227,000	234,483	444,900
Sind { Native States.	35,000*	35,000*	35,000	32,000	...
Berar . . .	503,000	691,000	917,000	942,029	830,000
TOTAL .	17,182,000	18,323,000	20,169,000	19,761,077	17,952,400

Conf. with pp.

138.

123.

149.

132.

128.

155.
Central India.
p. 130.
Bengal.
p. 157.
Hyderabad,
Mysore,
Kashmir.
p. 166.
Assam,
Burma.
p. 166.

The above table exhibits the increase of area during the past sixteen years in the chief wheat-growing provinces of British India, for which alone statistics are available for such a lengthy period. But a nearer approximation of the actual area may be arrived at by adding the averages accepted for Bengal, 1,000,000 acres, Madras, 27,000, and Ajmere, 15,000 acres. These additions would increase the figures to 18,602,000 acres for the first, 19,743,000 for the second, and 21,589,000 for the three quinquennial periods. For 1888-89 accurate figures are forthcoming to supplement those given in the above table, with the exception of those for Bengal. They are, for Madras, 20,360 acres, Upper Burma, 9,185 acres, Assam, 12 acres, Ajmere-Merwara, 9,548 acres, and Pargana Manpur, Central India, 2,831 acres ;—a total of 41,936 acres. Assuming that the cultivation in Bengal remains at about 1,000,000 acres, the total during the year would thus be 20,803,013 acres. And from 1884-85 fairly reliable figures have been arrived at for Native States which formerly could not be obtained. The average of the four years, 1884-85 to 1887-88, have been accepted as follows :—

	Acres.
Rájputána	1,542,000
Central India	2,617,000
Hyderabad	1,111,000
Mysore	14,000
Kashmír	500,000
TOTAL .	5,784,000

* These figures are estimates, actuals not being available.
† Excluding proportion of the area under mixed wheat, a proportion included in the figures for the quinquennial period ending 1887-88. The average area so added is 341,000 acres. The above figures for the North-West Provinces do not include the areas under mixed wheat, but these are shown in the table p. 138.
‡ Includes Sind Native States.
§ Areas in 1890-91 see pp. 168, 198.

For 1888-89 the same figures are admitted, with the exception of that for Mysore, which is returned as 4,282 acres, shewing a marked falling off and reducing the total for these Native States to 5,774,282 acres. It is, of course impossible to shew to what extent the area in these States has increased, and it is therefore advisable to dismiss them from consideration in reviewing the question of expansion of area. But adding the figures to that above given for 1888-89, *viz.*, 20,803,013 acres, we get a total area under wheat, within the boundaries of the Indian Empire, of 26,577,295 acres.

From the above figures it will be seen that the area increased by 6·6 per cent. during the second, and by 10·7 per cent. during the third, quinquennial period. The returns for 1888-89 show a slight falling-off, but, as remarked elsewhere, the area under wheat, though increasing steadily and likely to continue to do so, exhibits marked fluctuations, dependent on various causes, from year to year. It will be observed that the greatest increase has been in Bombay and Berar, the two provinces closest to the chief port of shipment, and those in which a variety of wheat especially suited to the markets of Southern Europe, is grown. The forecasts for 1889-90 show a very marked falling off, especially in the Panjáb ; in the Central Provinces there is a considerable increase. There is some reason to believe that in these provinces and in Berar, however, a certain proportion of the crop has been replaced by cotton. In Sind wheat has probably taken the place of other crops. On the whole, however, the poor harvests and unfavourable climatic conditions of the past three years have caused a sensible, though probably only temporary, decline in wheat area.[*]

It may now be considered to what extent this increase in the area under wheat has affected other crops; but it is worthy of notice in passing that, as will be shown below, the increased production of wheat, implied by the increase of area, is of itself more than sufficient to meet the demand on the outturn made by the enhanced exports of recent years, without infringing on the surplus left for local consumption. In the note on wheat above quoted, returns of all the crops cultivated during the same periods have been compiled, with the result that a steady increase in other crops and in the total cropped area, coincidently with the development of that under wheat, has been clearly demonstrated. The total cropped area increased from an average of 102 million acres for the period ending 1882-83, to 109 million acres for that ending 1887-88, that is, by seven million acres or nearly 7 per cent. And, in 1888-89, notwithstanding a slight diminution in the area under wheat, the total cropped area increased very largely on the average of the preceding five years, *viz.*, to 148,811,480 acres, of which 14,158,424 acres was cropped more than once, or an *actual* area cropped of 134,653,056 acres.

The increase during the period ending 1887-88 is analysed by Mr. Schofield as follows, the figures representing millions of acres :—

	Wheat.	Other Food Crops.	Non-food Crops.	TOTAL.
Bombay, including Sind	+ ·53	+1·19	+ ·89	+2·61
N.-W. Provinces	+ ·08	+1·00	+ ·60	+1·68
Oudh	+ ·26	+1·06	+ ·18	+1·50
Panjáb	+ ·25	+ ·15	+ ·17	+ ·57
Central Provinces	+ ·54	+ ·03	+ ·10	+ ·67
Berar	+ ·22	— ·25	— ·07	— ·10
TOTAL.	+1·88	+3·18	+1·87	+6·93

[*] The Trade of 1891-92 necessitates a large increase in area.—*Ed., Dict. Econ. Prod.*

**WHEAT
AREA.**

Increase.

The total increase in wheat shewn in that statement is slightly different from that recorded in the table of area under wheat, owing to the fact that in the calculation above, the areas of all crops in the permanently-settled and hill districts of the North-West Provinces, are, for the sake of greater accuracy, excluded. It will be observed that the areas under each class of crop have increased in all the provinces with the exception of Berar, in which wheat has increased 32 per cent., while the other crops have decreased by 5·7 per cent. The decrease in food crops in this province has been explained as being due to the insufficiency of grazing land, in consequence of which cultivators allow parts of their fields to remain waste to admit of a supply of grass for their cattle (*Rev. Admin. Rept., Hyderabad Assigned Districts, 1887-88*); also to the fact that *juari* has been losing favour, and is being replaced by wheat. This has come about partly because of the better price obtained for the wheat, partly because it requires considerable manuring and improves the land by enabling a longer rotation. In the case of non-food crops, the decrease is largely due to the diminution of the area under cotton and its replacement by wheat (*Conf.* **Gossypium**, Vol. IV., 93).

In all the other provinces an increase of other food-crops and of non-food crops has gone on coincidently with that of wheat. Thus in Bombay wheat increased by 31 and other crops by 6·6 per cent., and in only two districts, *viz.*, the Karnatak and Khándesh, is it expressly stated that the cereal has ousted another crop to any marked extent,—" There is some probability that wheat is displacing cotton in the Karnatak and Khándesh " (*Rept. Dir. Land Rec. & Agri., 1887-88, 28*). In the North-West Provinces the percentage of increase has been small, partly owing to contraction of area, consequent on unfavourable seasons during the three years 1885—1887. In Oudh, on the other hand, a considerable augmentation of cultivation of all sorts is shown, probably chiefly due to recent railway extensions in the province. In the Panjáb cultivation of wheat seems to move *pari passu* with that of other crops (*Rept. Dir. Land Rec. & Agri., 1887-88*). In the Central Provinces " the expansion in wheat cultivation has been to a very large extent counterbalanced by an increase in cultivation generally, but the area has also expanded at the expense of the areas under linseed and cotton. It should be added that neither linseed nor cotton can be grown in Jabalpur or Hoshangabad so profitably as in other parts of the provinces, and that the substitution of wheat for them was to be expected, and is certainly not to be regretted " (*Rept. Dir. Land Rec. & Agri., 1887-88*).

These facts show that while wheat has been expanding in a greater ratio than other food-crops, it can by no means be said to have done so at their expense, since the area under them has also shown a considerable, though proportionately smaller, expansion.

**CULTIVATED
RACES.
639**

CULTIVATED RACES OF INDIAN WHEAT, AND THEIR COMMERCIAL VALUE.

As in the case of rice, very numerous races of wheat occur in India, distinguished by variation in the colour of the grains, by being " bearded " or " beardless," and by many other characters. It would be useless in such a work as the present to enumerate all the kinds distinguished by distinct names, but the more impoitant will be noticed in the account of cultivation under each province. From 1,000 samples sent to England by Government in 1879, it would appear that the various kinds may be classified, for commercial purposes, into four distinct classes, and the following report on them, taken from Dr. Forbes Watson's note, will be found to convey practically all the information available on this subject :—

CONNECTION WITH VERNACULAR NAMES.—Considerable confusion exists with regard to the Native names of the different varieties of wheat. Many samples from Bombay and the Panjáb had no names at all attached to them. Others are only distinguished as red or white wheat, or by similar generic designations in Native languages, such as *safed* (white), *lal* (red), *surkh* (brick colour), *safed gehoon* (white wheat), *lal gehoon* (red wheat). The other Hindi, Persian, or Sanskrit synonyms for wheat, *kunuk, gundum,* or *saman,* are also frequently employed, either alone or in connexion with the adjectives designating white or red. Whether by mistake or otherwise, the names appear sometimes to be misapplied. Thus, for instance, two samples called *lalia,* from Sitapur, consist of fine, soft, white wheat. In the same manner *lal desi,* from Muzuffarpur, *kathia lalia,* from Fyzabad, *lalia* from Lucknow, *lal pissia,* from Etawah, and some other samples similarly described, are all white, notwithstanding the use of the adjective *lal* or red. On the other hand, a "white *pissi*," from Bilaspur, is described by the valuer as a good, soft, dingy red.

"In addition to the generic designations, many specific names occur, which at first one would be inclined to consider as each applying to some well known variety of wheat with well defined characteristics. To a certain extent this seems to be the case. For instance, spread over all India from the Panjáb to Bengal, and south as far as the Deccan, appear names like the following,—*daudi, doodhia, dawoodi, daudkhani, dudhia,* which all seem to be derived from the same original name, *Daud Khani* (Prince David's wheat). Throughout the Gangetic valley the samples sent under these names are of a nearly uniform character, almost invariably white and soft, and frequently of a very superior quality; but occasionally, and especially in the Deccan, a very inferior hard white wheat will go under this name. Another wide-spread series of names which appear to apply to the same description of wheat are the names *mundi, mundia, mundwa, mandwa, mondha, mendha,* which appear to be equivalent with *muria, marua, marwa, muria, ratua,* and *raita, ratwa, ratua,* as also with *ujra.* These names occur throughout the North-West Provinces and Oudh, as also in the adjoining portion of the Central Provinces, such as *pissi mundi.* They apply almost invariably to a beautiful, soft, white wheat, as a rule of a very high quality. The same holds good of *setia, sitia, setwa,* and *satwa,* as also of *saman, bargehuna, sambhari, sambharia,* or *samaria,* names which are frequently met with in the same districts as the above. The name *pissi* frequently used in the Central Provinces, and occasionally in the North-West Provinces, seems invariably to imply a softness of grain, whether in white or red wheat.

"But in the case of almost every other name samples of the most different character may be found. Thus, for instance, the names *gajar, gajra, gaja,* designate, as a rule, in the Agra and Rohilkhand divisions and adjoining parts of Oudh, rather an inferior sort of white wheat. Very frequently the wheat is mixed, white and red, sometimes altogether red; but occasionally samples are to be found under the same name, consisting of the finest soft white wheat, equal in value to the best specimens produced in India. The same applies to the names *kathia, katteh, kuttya,* and other variations, which in most parts of India designate rather a poor sort of wheat, white or red, but mostly hard, and yet occasionally have been applied to the most beautiful samples of soft white. The names *jamali* and *gangajali* so frequently met are indiscriminately applied to wheat of the most different characteristics.

CULTIVATED
RACES.

"It may, indeed, frequently happen that, even in cases in which the grain is of a different hardness or colour, the plant, known under the same name, may be distinguished by a certain similarity in outward appearance, and that only the character of the seed has become changed or degenerated. It is impossible to decide this point without possessing specimens of the whole plant. But sufficient evidence has been shown of the uncertain meaning of the Native names as applied to the grain itself to make it unsafe to adopt them as a basis of a commercial grouping of the samples sent from India. This has been, therefore, effected entirely on the basis of the appearance of the samples themselves, without regard to the names under which they were sent.

Description.
641

DESCRIPTION.—"The whole of the 827 samples submitted for valuation may be arranged in four principal groups, embracing the white soft, the white hard, the red soft, and the red hard wheat. The differences between these varieties when pure are very striking.

Soft White.
642

"The *pure soft white wheat* has a grain usually of a bright straw colour, is opaque in appearance, and the fracture is white and floury, the inner portion of the grain being friable. This is the most valuable variety for the London market, as it yields the finest flour. The Indian wheat of this description is in special request on account of its dryness, which renders it useful for admixture with home-grown wheat containiug too much moisture when harvested in wet seasons. It is also liked by millers on account of the considerable increase in weight which it experiences in grinding in consequence of its power of absorbing moisture.

Hard White.
643

"The *pure hard white wheat* has a grain of a translucent, flinty, or 'ricey' appearance, varying in colour from a greyish or yellowish white to the lighter shades of brown, the fracture smooth and glass-like, and the grain hard and brittle. This kind of wheat is not much in favour in the London market, as the usual appliances of English millers do not seem to be so well adapted for dealing with it as with the soft white. It is, however, in considerable request in the Mediterranean, and especially in Italy, where it is used in the manufacture of macaroni. This is the reason why the quotation in Italy for wheat of this description is frequently as much as 5s. per quarter higher than that of the London market.

Soft Red.
644

"The *pure soft red wheat* is only distinguished from the soft white by the different colour of the skin, which varies in different varieties from an amber colour to a reddish brown. The fracture is as white and mealy as in the soft white wheat, and the grain as friable. It is eminently suitable for the English market. The Indian red varieties are, however, frequently rather smaller berried than the white varieties, and are usually much deteriorated by being mixed with barley, gram, and different oilseeds.

Hard Red.
645

"The *hard red wheat* is the darkest of any, being frequently of a dark brown colour. It is translucent in appearance, and the fracture is smooth and glass-like. It occupies the lowest position in the London market, as it is generally disliked by the millers.

"Only a certain number of the samples, however, present these characteristics with distinctness. A considerable number consist of mixtures of the primary varieties in all possible proportions, white and red, soft and hard, from which the grains of each kind may be picked out. A considerable number too, though consisting of grain of uniform quality, exhibit in the character of the individual grain a transition between some of the four varieties above mentioned. For instance, in certain samples of the Bansi wheat, from the Central Provinces, some portions of the same individual grain may be opaque and soft, others translucent and hard. In the same way, many of the samples of wheat from the Panjáb and from Bengal, although in general approaching in appearance the soft white wheat, are yet considerably harder than the pure soft wheat and yet not hard enough to be classed with the hard wheat. These were usually described as semi-hard, and are in general classed with the soft white wheats. The same may be remarked even more frequently of red wheat, in which, perhaps, the greater number of samples is neither completely soft nor yet decidedly hard. In colour, again, the hard white passes by imperceptible transitions into the hard red, as there are many specimens of a light brown translucent grain which might be arbitrarily classed with either the hard red or with the hard white variety.

"For the purpose of a simple classification, however, all the samples have been arranged in four groups, corresponding to the four distinctive varieties, the mixed or transitional samples being added to the group with the character of which they most nearly corresponded. Thus, all the semi-hard white wheats, or mixed wheats in which the soft white variety predominated, have been included in the same group with soft white. A peculiar description of short round-berried wheat, in appearance like pearl barley, with a hardness considerably in excess of the usual soft wheat, has been likewise included with it. In the same manner, most of the semi-hard red and brown wheats are included in the same group with soft red. The number of samples contained in each group appears from the following enumeration; the average price assigned to each group has been added in order to show in this way the relative estimation in which the different descriptions stand in the London market :—

	Number of Samples.	Average price per quarter of 496℔.
		s. *d.*
1. White, soft, and semi-hard, and pure or mixed, with a predominance of soft white.	357	41 9
2. Hard white	167	39 5
3. Red, soft, and semi-hard; pure or mixed, with a predominance of soft red.	161	38 5
4. Hard red	142	36 1
TOTAL	827	39 8

"The number of samples classed with soft white and soft red amounts to 518, against 309 of hard white and hard red samples, showing a considerable predominance of soft samples, even when taking into account that, among the inferior varieties of the samples grouped with the soft wheat, there are many semi-hard samples or samples of soft wheat largely mixed with hard grain. The great number of soft samples is an important fact, as it is the soft wheat which is most suitable for export to this country, as appears from the higher price realised by it.

GEOGRAPHICAL DISTRIBUTION.—"To a certain extent, the four distinctive varieties, the soft and hard white, and the soft and hard red, are cultivated side by side in the same districts, but, on the whole, a distinct geographical distribution of the several varieties may be perceived. It clearly appears that, while Northern India produces mainly soft wheats, the samples produced in Southern India and part of Bengal are chiefly hard. The cultivation of the soft white wheat appears to be comprised within the basins of the three great rivers, the Ganges, the Indus, the Nerbudda, and their tributaries. In fact, from the whole territory south of the Nerbudda basin only two samples of soft white wheat were sent, one from Khándesh on the Tapti, only a little south of the Nerbudda, and a sample of mixed soft red and white from the Belgaum district.

CULTIVATED RACES.

Description.

Hard Red.
646

Classification
647

Distribution.
648

7 A

| TRITICUM sativum. | Cultivated Races of Indian Wheat ; |

CULTIVATED RACES.

Distribution.

Spelt Wheat
649

Whilst the North-West Provinces and Oudh sent mainly soft white samples, the majority of the samples from the Panjáb are soft red. The soft red wheat also extends farther south than the soft white. Several fine samples were sent from Berar, from which not a single sample of soft white wheat arrived, and a sample (all but destroyed by weevil) came from as far south as Bellary. The hard white wheat occurs occasionally wherever the soft white is cultivated, but it is predominant in the Deccan, Berar, and parts of Bengal. The least valuable variety, the hard red wheat, extends the farthest to the south of all, and is, moreover, the only kind cultivated in the moist climate of the Gangetic delta, Orissa, and Burma. In the extreme south of the Madras Presidency, and in Mysore, the wheat appears to belong to a variety similar to the spelt,[*] in which the husk adheres so strongly to the grain that mere threshing is insufficient to separate them, so that it requires to be husked in the same way as paddy. The grain appears to be of an inferior hard red description.

"In the following table, the predominant character of the wheat culture in each province is shown by the number of samples of each of the four kinds of wheat sent from them, whilst the average price for each province illustrates the influence which the cultivation of the more valuable soft and white varieties exercises on the value of the whole provincial produce : —

| Name of Province. | Number of Samples of | | | | Total number of samples. | Average price per quarter of 496℔. |
	Soft White.	Soft Red.	Hard White.	Hard Red.		
						s. d.
Bengal . . .	27	21	12	8	68	39 10
North-West Provinces and Oudh . . .	251	71	37	15	374	40 9
Ajmere and Merwara . .	...	3	1	1	5	40 6
Panjáb	17	30	3	4	54	38 5
Sind	41	2	15	...	58	39 0
Bombay . . .	5	13	60	61	139	38 3
Central Provinces . .	16	13	8	12	49	40 1
Berar	...	7	31	16	54	38 10
Madras . . .	...	...	...	18	18	36 1
Mysore . . .	...	...	...	6	6	32 4
Burma . . .	...	...	...	2	2	34 3
All India .	357	160	167	143	827	39 8

" It appears from this table that the greatest preponderance of soft white samples occurs in the North-West Provinces and Oudh, 251 samples out of 374 belonging to this class. The average value of the samples from this province is therefore the highest of any in India. amounting to 40s. 9d. per quarter of 496℔. The two provinces in which the soft white likewise constitutes an important fraction of the samples, *viz.*, the Central Provinces and Bengal, realise also the next highest average values. *viz.*, about 40s. per quarter. The Panjáb would almost certainly have shown an equal, if not a higher, value, if the collection from it had been more complete, and the same would probably have applied to Sind, if the samples from there had not been weevilled throughout, in addition to being mixed frequently with barley and other grains. All the other provinces have a lower

[*] Compare with Dr. Watt's remarks on Spelt under " Bombay," pp. 129, 134.

CULTIVATED RACES.

Valuation.
650

average, the lowest occurring in the case of Madras, Mysore, and Burma, all the samples coming from which belong to the hard red description."

A later valuation than that above was made by **Messrs. McDougall Brothers** at the request of the Government of India in 1880. Five thousand pounds of each kind of wheat was operated on, half by the ordinary process of grinding under millstones, half by means of crushing between rollers. The amount of flour, middlings, pollard, and bran obtained was carefully measured, the percentage of impurity, of water, and of gluten was estimated and the value of each given. The tabular statement submitted is of great interest, since it exhibits the comparative qualities of Indian and other important commercial wheats, and may be reproduced in this place.

Synopsis and Comparison of Results obtained from Indian and other Wheats, by two Systems of Milling, by McDougall Brothers, at their Wheatsheaf Mills, London.

Figures in Roman.—System : Grinding under millstones. Figures in *Italics*.—System : Crushing between rollers.

No.	Wheat.	Value in London per 496lb net weight on day of valuation		Weight per bushel.	Weight of 100 separate grains of the wheat.	Quantity used.	Impurities removed.	Water absorbed to render mellow.	Yield. Flour.	Yield. Middlings	Yield. Pollard.	Yield. Bran.	Evaporation and loss.	Gluten by water test.
		s.	d.	lb	Grs. avdps.	lb	Per cent	Per cent.	Per cent.	Per cent.	Per cent.	Per cent.	Per cent.	Per cent.
1	Indian (fine soft white)	49	0	64	*55·4*	5,000	1·52	2·0	77·46	·82	8·8	12·0	*1·40*	6·4
	,, ,, ,,	*49*	*0*	*64*	55·4	*5,000*	*1·52*	*2·0*	*74·10*	*11·0*	*8·7*	*4·0*	*2·68*	*6·8*
2	,, (superior soft red)	45	0	62¾	*51·8*	5,000	·72	3·6	78·40	1·68	9·8	9·4	*3·6*	9·3
	,, ,, ,,	*45*	*0*	*62¾*	*51.8*	*5,000*	*·72*	*3·6*	*75·4*	*7·7*	*13·5*	*5·3*	*·98*	*10·5*
3	,, (average hard white)	44	0	60	*68·3*	5,000	3·7	8·4	80·52	·78	10·0	8·3	5·1	11·7
	,, ,, ,,	*44*	*0*	*60*	*68·3*	*5,000*	*3·7*	*8·4*	*73·2*	*10·3*	*14·3*	*3·1*	*3·8*	*12·6*
4	,, (average hard red)	43	0	61¼	*77·7*	5,000	1·2	7·6	79·88	·78	13·20	8·50	4·04	13·4
	,, ,, ,,	*43*	*0*	*61¼*	*77·7*	*5,000*	*1·2*	*7·6*	*74·2*	*10·3*	*13·8*	*3·0*	*5·1*	*13·1*
5	English	49	0	60½	*57·4*	5,000	1·5	None.	65·2	1·1	9·7	17·7	4·8	10·6
	,,	*49*	*0*	*60½*	*57·4*	*5,000*	*1·5*	,,	*70·3*	*7·6*	*7·2*	*9·2*	*4·2*	*11·4*
6	Australian	50	6	62½	80·5	5,000	1·0	,,	75·8	1·1	7·4	14·4	·3	11·6
	,,	*50*	*6*	*62½*	*80·5*	*5,000*	*1·0*	,,	*75·1*	*8·0*	*9·3*	*5·5*	*1·1*	*12·2*
7	New Zealand	48	0	62¾	67·6	5,000	·3	,,	70·1	·96	8·8	11·5	2·34	10·2
	,,	*48*	*0*	*62¾*	*67·6*	*5,000*	*·3*	,,	*76·1*	*7·8*	*6·6*	*5·6*	*3·6*	*9·0*
		(in Liverpool).												
8	Californian	48	0	59¼	*47·7*	5,000	1·7	,,	*71·1*	·72	9·2	15·3	1·98	10·5
	,,	*48*	*0*	*59¼*	*47·7*	*5,000*	*1·7*	,,	*70·1*	14·5	*6·3*	*3·9*	*3·5*	*8·7*
		(in London).												
9	American (winter)	49	6	61¾	49·6	5,000	·5	,,	73·8	·38	7·9	16·4	1·02	11·0
	,, ,,	*49*	*6*	*61¾*	*49·6*	*5,000*	*·5*	,,	*71·5*	*10·3*	*11·2*	*3·1*	*3·4*	*11·7*
10	,, (spring)	48	0	61	35·5	5,000	·9	,,	72·2	·24	7·2	14·7	4·76	15·3
	,, ,,	*48*	*0*	*61*	*35·5*	*5,000*	*·9*	,,	*69·5*	*12·1*	*10·4*	*3·8*	*3·3*	*14·6*
11	Russian (Saxonska)	52	0	60½	37·3	5,000	·9	,,	73·0	*1.2*	11·6	12·6	·7	22·1
	,, ,,	*52*	*0*	*60½*	*37·3*	*5,000*	*·9*	,,	*71·4*	*12·5*	*11·7*	*3·3*	*·2*	*23·2*
12	,, (hard Taganrog)	49	0	61½	54·7	5,000	·8	2·4	76·2	·2	12·7	8·1	3·4	17·6
	,, ,,	*49*	*0*	*61½*	*54·7*	*5,000*	*·8*	*2·4*	*72·0*	*9·6*	*12·1*	*5·0*	*2·9*	*15·6*
		(in Liverpool).												
13	Egyptian (Buhl)	47	0	58	50·1	5,000	2·7	3·1	72·9	1·0	11·0	10·0	5·5	4·4
	,, ,,	*47*	*0*	*58*	*50·1*	*5,000*	*2·7*	*3·1*	*72·6*	*10·4*	*8·5*	*3·5*	*5·4*	*7·9*
14	,, (Saida)	43	9	57½	61·4	5,000	13·1	2·7	66·9	·76	11·4	7·5	4·04	7·5
	,, ,,	*43*	*6*	*57½*	*61·4*	*5,000*	*12·1*	*2·7*	*67·8*	*7·2*	*6·5*	*4·9*	*4·2*	*6·6*

TRITICUM	
sativum.	Cultivated Races of Indian Wheat ;

**CULTIVATED
RACES.**

Valuation.

From this table it would appear that fine soft Indian white wheat is inferior to the finest descriptions of Russian and Australian grain only, and that it possesses several properties, especially its freedom from moisture, which ought to render it of great value to millers in England for mixing. **Messrs. McDougall,** in their detailed statement, also comment on the exceedingly high yield of flour obtained from it, on its fine white colour, and superior bloom. From the flour obtained, bread was made, which, when compared with that from other flours, was found to be too dense and close to be likely to find favour. **Messrs. McDougall** sum up their report with the following interesting paragraphs :—

**Admixture.
651**

"In addition to the particulars contained in the foregoing returns, we have to report that to any one experienced in the requirements of the wheat and flour markets of the United Kingdom, and indeed, of most other countries, it will be evident there is no probability of these Indian wheats coming into demand for manufacture into flour *without a liberal admixture* of other wheats. They all possess in a marked degree the same characteristics of great dryness, and a distinct beany and almost aromatic flavour, inseparable from wheats grown in the climates and soils of the tropics. Also, the flours are ricey, the texture of the breads is too close, and the crust is hard and brittle. But these characteristics do not detract from their usefulness in any important degree. As is well known a miller cannot show skill in his craft to greater advantage or profit than that with which he selects his wheats, and mixes his grists, so as to produce to best advantage a flour from which bread can be made of the colour, bloom, strength, and flavour desired, and withall a good yield.

" We pronounce them to be exceedingly useful wheats; in fact, hardly equalled for what is deficient and wanting in the English markets by any other wheats. Their chief characteristics are just those in which the wheats grown in our variable climate are most deficient. Their great dryness and soundness renders them invaluable for admixture with English wheats that are in any degree out of condition through moisture, and the great proportions of the wheats harvested here have been in that condition for some years past, a condition that must prevail in all other than that of wheats harvested and stored during fine and favourable weather; and this the English farmer knows, greatly to his cost, is a state of climate that is by a long way the exception rather than the rule. Added to their dryness, the thinness of the skins of these wheats and consequent greatness of the yield of flour, must always place them in the front rank as a ' miller's ' wheat, whenever they are handled with reasonable intelligence and skill.

"Such unprecedented yields of flour, as shown by these wheats, ranging (by ordinary grinding) from 77·46 to 80·52 per cent., against English 65·2, and American spring 72·2, speaks volumes in their favour, and their value is still further increased by another point of merit of almost equal importance, *viz.*, a larger percentage of bread may be obtained than from any other of the flours included in this review.

"That, for the best of these Indian wheats (the fine soft white), on the day they were valued on Mark Lane market, a price was offered as high as that for American winters, New Zealand or English (see list of values in synopsis), proves that the great value of the Indian wheats is becoming recognised here, a knowledge that will ere long extend to all our markets. The other lots of Indian (Nos. 2, 3, and 4) were lower in value to the extent of 4s. to 5s. per quarter, as might almost have been expected from the difference in colour and other characteristics; still, as these latter wheats become better known here, this difference in price will be somewhat lessened. Their beany flavour is not a serious obstacle, as fair average deliveries, when well cleaned and properly dealt with, can be employed in the proportion of 25 per cent. to 50 per cent. along with home-grown or other wheats, such as Americans, possessing a fine sweet, milky or nutty flavour. *Glancing at all the facts here elaborated, it is evident that these wheats afford a larger margin of profit, both to the miller and baker than any other.*

"We venture to record a conviction that we have long held, strongly emphasized by the results of these experimental workings, of the measureless importance of the great resources of the Indian Empire being developed to the utmost in producing wheat for this country. Farmers here are finding that to live they must produce beef and mutton rather than grain, hence the greater need of resources of supply under our own control.

"The character and general excellence of Indian wheats are improving with the deliveries of each successive season. The Indian wheats now specially under review were delivered to us in excellent condition, with freedom from dirt, barley, gram, and other impurities, also with a freedom from weevil rarely equalled by Indian wheats, and there is, no doubt, an outlet in this country and the Continent for unlimited quantities."

In addition to the above it may be remarked that the hard white wheat, which in England fetches from 4s. to 5s. a quarter less than soft white wheat, commands an extensive market and ready sale in Southern Europe, where it is largely employed for making maccaroni.

Dr. Forbes Watson discusses the commercial values of the various wheats as follows :—

SOFT WHITE WHEAT.—"The quality of the samples of white wheat, forwarded to this country, is, on the whole, surprisingly high, and Mr. Alexander Smith, to whom they were submitted, reports that a considerable number amongst them are far superior to any Indian wheat, ever seen in the London market. The better qualities of wheat coming from Calcutta are usually comprised under the classes No. 1 and No. 2 Club, the quotations for which at the time of the last valuations, in the beginning of February of this year, were 42s. to 43s. for No. 1, and 40s. to 41s. for No. 2. By adding, the samples, which depart only a little from these values, to the numbers contained within the above classes, and by distinguishing those either above or below them in value, the samples of soft white wheat may be arranged in the following five classes :—

CULTIVATED RACES.

Admixture.

COMMERCIAL VALUE.

Soft White.
652

		Price per Quarter of 496 ℔.
a.	Samples of superior quality	44s. to 48s.
b.	„ grade No. 1	43s. 6d. to 41s. 6d.
c.	„ „ „ 2	39s. 6d. to 41s.
d.	„ ordinary quality	37s. to 39s.
e.	Inferior samples	below 37s.

TRITICUM sativum.	Cultivated Races of Indian Wheat;

FORBES WATSON'S COMMERCIAL VALUATION. Soft White.	"The number of samples belonging to each class will appear from the following statement :—

NAME OF PROVINCE.	Number of Samples of Soft White Wheat (including mixed and semi-hard with predominance of soft white) of each of the Classes under-mentioned.					
	Superior quality, 44s. to 48s. per quarter of 496 ℔.	No. I. Club 43s. 6d. to 4·s. 6d. per quarter of 496℔.	No. II. Club 39s. 6d. to 41s. per quarter of 496℔.	Ordinary quality, 37s. to 39s per quarter of 496℔.	Inferior quality, below 37s. per quarter of 496℔.	Total.
Bengal . . .	6	9	5	5	2	27
North-West Provinces and Oudh .	81	96	44	26	4	251
Ajmere and Merwara . . .	...	...	...	...	...	...
Panjáb . . .	7	4	4	1	1	17
Sind . . .	...	8	14	17	2	41
Bombay . .	...	3	2	...	...	5
Central Provinces .	7	3	4	2	...	16
Berar . . .	...	...	...	...	...	...
Madras . . .	...	...	...	...	...	...
Mysore . . .	...	...	...	...	...	...
Burma . . .	...	...	...	...	...	...
All India . .	101	123	73	51	9	357

" Thus, out of a total of 357 samples, more than 100 samples are better than Grade No. 1, and more than 120 samples come up to that grade. The result would have been even more favourable if a certain proportion of the samples had not been to some extent damaged by weevil.

" Four different types of grain may be distinguished among the samples of soft white wheat.   These are :—

" *a.* Wheat similar to Australian or Californian, stout regular grains of a brilliant colour, mostly very soft, and usually very uniform in quality and in clean condition.

" *b.* Small berried wheat, more dull in colour than the foregoing, and less uniform in quality, hardly any sample being quite free from admixture with red or hard white wheat.   Frequently the admixture of red grains is so considerable that the wheat must be called mixed rather than white, and there may be such a preponderance of hard or almost hard grains, as to give to the sample the character of a semi-hard rather than of soft wheat.

" *c.* A large berried wheat, thick in the middle, pointed at the ends, rarely uniform in colour, some grains as bright as the first variety just mentioned, others more dull and grey, occasionally passing into red, but more frequently mixed with hard grain, or passing into it by imperceptible degrees, the individual grains being often partly soft and partly hard. In fact, the greater portion of the samples of this description must be described as semi-hard.

" *d.* Wheat with a peculiar, short, roundish grain, in appearance like pearl barley.   It is the wheat known under the names of *raighumbree, gilghit, giljit,* or *nikka* in the Panjab, and of *chunia, munia,* or *ræ muneer* in Oudh and the North-West Provinces.   It occurs locally in Oudh, the their Commercial Value. (*J. Murray*) **TRITICUM sativum.**

North-West Provinces, the Panjáb, and Sind, but it does not possess any great importance, as it does not appear to be very prolific. Although it is mostly semi-hard or even almost hard, the samples were very favourably reported on.

FORBES WATSON'S COMMERCIAL VALUATION.

Soft White.

"The sample sent from the Delhi district under the name *Gundun safed* may be accepted as the finest type of wheat grown in India. It is a soft white stout grain, remarkably uniform in quality, and in clean condition. The weight is very high, amounting to 65℔ per bushel. It is equal to the finest wheat grown in South Australia or California, the countries from which, of all others, the finest qualities of wheat are imported. It has been valued at 48s. per quarter of 496℔, this being the quotation for best Australian wheat at the time, and it has been declared equal in quality to the finest sample of wheat shown in the late Paris Exhibition. Nor is this an exceptional sample. Wheat of equal value and of equal excellence has been sent from Gya in Behar, *Dawoodie*, from Unao in Oudh, *samau*, from Bulandshahr and Meerut in the North-West Provinces, "*Safed*," from Dera Ismail Khan in the Panjab, as also from Hoshungabad, in the Central Provinces, white *pissi*. These places indicate the limits of the area within which the finest wheat is cultivated, and within which may be found a considerable number of samples ranging in value from 46s. to 48s., which are but little inferior to the eight samples selected as types of the best wheat, and which are mostly described by the valuer as being similar to Australian or Californian wheat. Not only most of the 45 samples ranging in value from 46s. to 48s., but also many of the 48 samples valued from 44s. to 46s., consist substantially of wheat of the same quality; all, more or less, like Australian or Californian, and only slightly depreciated in value by weevil, by slight admixture with red or hard grains, with barley, gram, or oilseeds, or impurities of other description. The local names vary from province to province, but no corresponding difference appears in the grain. The samples belonging to this class under the various names of *Pissi* in the Central Provinces; *shori* in Sind; *Zurd kanak* in the Panjab; *Doudia, Safeda, Setwa, Muria, Mundia*, or *Gajar* in the North-West Provinces and Oudh; *dawoodia* in Behar, are almost indistinguishable in appearance, and all of equal excellence, though under the same names many other samples may be found which are perfectly different in appearance, and often really inferior in quality.

"The varieties previously described as the small berried and the large berried respectively constitute the greater portion of the wheats of Grades Nos. 1 and 2, and of those of lower quality. But in the comparatively rare instances in which they are characterised by considerable uniformity and softness of grain they range as high in value as the samples resembling the Australian wheat, whilst the value of even this latter variety may be reduced by the admixture of impurities to the lowest level observed in any of the other varieties.

"It has been already explained that both the small-berried and the large-berried variety of soft white wheat are usually of less uniform appearance than the Australian-like wheat, and that they frequently contain a considerable admixture of red and of hard white wheat, so as to render it often difficult to decide whether a given sample should be classed as a red or a white, a hard or a soft wheat. This seems to afford an explanation why the same name so often serves to designate varieties of grain which appear to be very different from each other. Thus, under the name of *jamali*, a number of small-berried varieties were sent from, Bengal, some of which were soft and white, others perfectly hard, white or red, others again, a beautiful soft red, while most were rather mixed in character. In the same way the large-berried soft variety occurs frequent-

TRITICUM sativum.	Cultivated Races of Indian Wheat ;

FORBES
WATSON'S
COMMERCIAL
VALUATION.

Soft White.

ly under names like *burghona, anokha, kathia, jalalia,* and *bansi,* usually applied to hard white wheats, or even under the name of *desi* usually applied to red wheats, and, in fact, it is rarely free from a considerable admixture of hard grains, even if the individual grain does not, as is often the case with the *bansi* variety in the Central Provinces, possess a mixed character, partly soft, partly hard.

" Every defect in the way of colour, uniformity, and softness of the grain diminishes in a corresponding manner the value of the sample, but the quality is even more seriously deteriorated by the admixture of foreign substances, especially of barley, gram, linseed, and rapeseed or other grains. Other impurities, such as lumps of earth and clay, and chaff, likewise occur and often reduce the value of the wheat considerably In many of the samples mere screening would suffice to separate the fine grain from the impurities and to improve the value by as much as 5s. per quarter. For instance, a sample from the Etawah district, *mundia,* has been valued at only 37s. on account of the considerable admixture of barley and gram, and impure condition generally, but the valuer reported that by cleaning the value could be raised to 42s. It may be noticed that the Sind samples especially suffer from the presence of barley, and that the wheat, which is often of the finest quality, is thereby rendered often quite unsaleable, as being in its actual condition quite unfit for milling.

" The majority of the finest specimens, ranging from 45s. to 48s. per quarter, come from the North-West Provinces and Oudh. The Central Provinces come next, and there are likewise some specimens from Behar, and from the Panjáb. No samples equal to the foregoing appear from Bengal proper, although soft white wheat, of a tolerably good quality, mostly of the small-berried kind, is grown as low down as the Hooghly district, and a sample from Beerbhoom was fully equal to Grade No. 1. As regards the Panjáb again, the best samples all come either from the Delhi and Hissar divisions immediately adjoining the North-West Provinces, or else from the Trans-Indus districts, in which the wheat resembles the fine variety cultivated in Upper Sind. But from the Panjáb proper, between the Indus and the Sutlej, no perfect samples of pure soft white were available for report, even the best being rather semi-hard than soft, and the grain less uniform than in the samples included in the fore going list. Nevertheless, it would not be fair to conclude, on this account alone, that the cultivation of wheat is conducted less carefully and skilfully in the Panjáb than in the North-West Provinces, because the fragmentary nature of the Panjáb collection already commented on, and specially the highly weevilled condition in which most of the samples arrived, make it impossible to accept the Panjáb series of valuations as a fair representation of the cultivation of wheat in that province. It must be also remarked that the softest, that is, the finest, specimens, are those most exposed to destruction by weevil, and it is likely that but for their weevilled condition some of the samples from the parts of the Panjáb referred to would have been as favourably reported on as the finest samples from the North-West Provinces. The same applies to Sind, from which a number of samples were sent equal to the finest grown in India, but their weevilled condition prevented their being valued at anything like the prices assigned to the best samples. The districts under Bombay sent no samples which could be compared with the finest wheat of the Central Provinces, but a few samples were sent from Broach and from Khándesh which come up to Grade No. 1.

" The weights per bushel given in the above list of samples are very high, notwithstanding that some of the samples were touched by the weevil. None weigh less than 60℔ per bushel, whilst in some cases the weight goes up to as high as from 63 to 65℔."

T. 652

their Commercial Value. (*J. Murray.*)

FORBES
WATSON'S
COMMERCIAL
VALUATION.

Hard White.
653

HARD WHITE WHEAT.—"There is much less variety among the different samples of hard white wheat than among those of soft white wheat already noticed. A great number of samples resemble in appearance the large-berried soft or semi-hard wheat already described, only that they are completely hard and translucent. Others, with the same difference, resemble the small-berried soft or semi-hard variety. There is, in addition, a considerable number of samples with a long thin ricey grain, resembling some of the inferior varieties of hard red wheat. These latter samples, frequently known under the name of *Khattya, Kathe, Kuthya,* are really an inferior grain in every respect, and the samples are rarely in a good condition, most of them containing chaff, dirt, and foreign substances in considerable quantity. But among the large-berried samples there are many of a very high quality, with clear uniform grains in good condition and free from impurities. In fact, many of these samples are equal, if not superior, to the best hard St. Petersburgh or Kubanka, which is usually considered as the type of wheats of this kind. The highest prices realised by the hard white wheats are, however, much below those ranging in the London market for the soft wheats, and, as already remarked, the Mediterranean, and especially Italy, seems to afford the best market for wheat of this description, which is in demand there for the manufacture of macaroni, and frequently fetches prices as much as 5*s.* per quarter in excess of the quotations of the London market.

"No classification in any way resembling that of the soft white exists for the hard white wheat; but in order to obtain a view similar to that given for the samples of the former variety, the hard white wheats have been arranged in groups corresponding to the same ranges of price as those of the different classes of the soft white. The following table shows the result of this classification :—

NAME OF PROVINCE.	NUMBER OF SAMPLES OF HARD WHITE WHEAT OF A VALUE PER QUARTER EQUAL TO THAT OF THE UNDER-MENTIONED CLASSES OF SOFT WHITE WHEAT.				
	Class I., 43*s.* 6*d.* to 47*s.* 6*d.*	Class II., 39*s.* 6*d.* to 41*s.*	Ordinary quality, 37*s.* to 39*s.*	Inferior quality, below 37*s.*	Total.
Bengal	3	4	5	...	12
N.-W. Provinces and Oudh .	5	23	8	1	37
Ajmere and Merwara .	...	1	...	...	1
Panjáb	...	2	...	1	3
Sind	...	...	8	7	15
Bombay . . .	3	29	27	1	60
Central Provinces . .	...	5	3	...	8
Berar	2	19	10	...	31
Madras	...	...	...	...	...
Mysore	...	...	...	...	...
Burma . . .	...	...	...	...	..
TOTAL .	13	83	61	10	617

"It will be seen that a considerable proportion of the samples is supplied by Bombay and Berar, from which hardly any samples of the soft wheat were sent. These two provinces supply, by themselves, more than one half of the samples of hard white wheat. The number of samples sent from the North-West Provinces and Oudh is large in itself, but it appears small when compared with the number of soft white samples from the same parts, which was almost seven times larger.

"The finest samples of this wheat is *bansi,* from Khándesh, valued at 42*s.* 6*d.*, though even this sample would have been probably surpassed by a

**FORBES
WATSON'S
COMMERCIAL
VALUATION.**

Hard White.

sample of yellow *banshi* sent from Nassick, had not the latter arrived very much damaged by weevil. All parts of the Bombay Presidency south of the Nerbudda supply fine samples under the names of *banshi, bansi, bukshi, buxi, jalalia,* and others. But some of the samples from the North-West Provinces and Oudh sent under the names of *barhɔ, tamla, anokha,* and others are equally good. One of the best samples, *ganga-jali,* valued at 42s., comes from the Maldah district in Lower Bengal, and, altogether, the cultivation of hard white wheats, if restricted to a smaller area than that of the soft varieties, results in the production of many samples equal to the finest grown in any country, which, sent to a suitable market, would be likely to realise prices almost as high as those obtainable for the best descriptions of soft white wheat. It may likewise be noted that the hard wheat appears to be on the whole less liable to suffer from weevil than the soft variety. The weights per bushel are almost as high as in the soft white; all but a few are above 60℔ per bushel, some as high as 64½℔. Of the few samples below 60℔ most are so much weevilled that the loss of weight is accounted for."

**Soft Red.
654**

SOFT RED WHEAT.—" The condition of most of the samples of soft red wheat affords evidence that its cultivation is conducted with much less care than that of the fine white varieties. They are, as a rule, much mixed with barley, gram, or oilseeds, and foreign substances of all descriptions, and the grain is rarely so uniform in its quality as in the good white samples. The striking uniformity in the Native names for the red wheat is in itself a proof of the smaller estimation in which it is held by the Native agriculturist. In the greater number of districts, in which half a dozen or more different varieties of white wheat will be cultivated under as many different names, the red wheat will be known only under the generic deno-mination of *lal* or *surkh,* indicating that the process of selection of seed and cultivation of special varieties has been applied to red wheat in a much smaller degree than to the white wheat. The red wheat is, however, deserv-ing of more attention than it now receives, for the soft red varieties are ex-ceedingly suitable for the English market, and very readily saleable. The following table shows the number of samples and their range of prices for each province in India, the classes being those adopted for the soft white wheat, as there is no corresponding classification for the red variety:—

NAME OF PROVINCE.	NUMBER OF SAMPLES OF SOFT RED WHEAT OF A VALUE PER QUARTER EQUAL TO THAT OF THE UNDER-MENTIONED CLASSES OF SOFT WHITE WHEAT.				
	Class I., 41s. 6d. to 43s. 6d. per quarter of 496℔.	Class II., 39s. 6d. to 41s.	Ordinary quality, 37s. to 39s.	Inferior quality, below 37s.	Total.
Bengal	3	11	6	1	21
N.-W. Provinces and Oudh	2	23	38	8	71
Ajmere and Merwara	2	1	...	...	3
Panjáb	2	7	11	10	30
Sind	...	1	...	1	2
Bombay	...	8	5	...	13
Central Provinces	1	2	10	...	13
Berar	...	3	4	...	7
Madras	...	...	...	...	...
Mysore	...	...	...	...	...
Burma	...	...	...	...	...
TOTAL	10	56	74	21	161

their Commercial Value. (*J. Murray.*)

FORBES WATSON'S COMMERCIAL VALUATION.
Soft Red.

" As may be seen from the foregoing table, the provinces in which the soft red wheat is chiefly grown are the same as those in which the soft white is cultivated ; but he limit of cultivation of the former, as previously remarked, extends rather further to the south than that of the soft white variety.

" Many samples represent wheat which is very suitable for the English market. The finest sample is one from Ajmere, *Kharcha baja ;* but, there are numerous samples almost equal to it from Bengal, the North-West Provinces, the Panjáb, and the Central Provinces. The samples from the provinces just named are, as a rule, comparatively light in colour, and quite as soft as the softest white wheat. But from Bombay a good many samples of a large-berried wheat were sent which deserve rather the name of brown than of red wheat, on account of the darkness of their colour. They are also much harder than the usual soft wheat, in fact, almost hard. Still some of the samples grown in the State of Bhownagar are favourably reported on, with values up to 40s. 6d. per quarter. It may be likewise remarked that in the Panjáb, from which several of the best samples were obtained, there appears to be grown in many districts a very degenerate kind of red wheat, in which a large proportion of the grains are so thin and shrivelled that they resemble rather grass seeds than the berries of a cultivated grain.

" The prevalence of inferior varieties in the Panjáb may be gathered from the fact that out of 30 samples of soft red wheat 10 are classed as inferior, being in value below 37s. the quarter, whereas out of 131 samples sent from the other parts of India only 11 are inferior. The weights per bushel are as high as in the preceding group, rising to 64½ ℔, and only in three instances is the weight below 60℔."

HARD RED WHEAT.—" The following table shows the number of samples and their range of prices :—

Hard Red.
655

NAME OF PROVINCE.	NUMBER OF SAMPLES OF HARD RED WHEAT.		
	Ordinary quality, 37s. to 39s.	Inferior quality, below 37s.	Total.
Bengal	2	6	8
North-West Provinces and Oudh . .	4	11	15
Panjáb	...	4	4
Sind	...	...	...
Bombay	31	29	60
Central Provinces	11	1	12
Berar	13	3	16
Madras	5	13	18
Mysore	1	5	6
Burma	...	2	2
TOTAL .	68	74	142

Most of these samples come from the Deccan and Southern India, and a very large proportion are inferior in quality—very thin, hard, and very dark. Wheat of this description is not in good demand in the London market."

Though the general price of wheat has altered considerably since the above report was written ten years ago, the information regarding the comparative values of the various kinds, and their geographical distribution, remains thoroughly applicable to present conditions. A selected list

TRITICUM **sativum.**	Adulteration of Indian Wheat.

ADULTER-
ATION.
656
Refraction.
657
Conf. with
pp. 116, 168.

of the better forms of each commercial class will be found in the chapters on races under each provincial heading.

ADULTERATION. (*Conf. with pp. 110-116, 169-171, 186*).

Since the commencement of the Indian export trade in wheat, it has had to labour under a great disadvantage owing to the dirty condition of average consignments, and to the frequency of adulteration with linseed, gram, barley, and other seeds. These faults still persist to a large extent; thus Dr. Watt, in 1888, commented on this subject as follows : —

" There is a strong feeling that the severity with which the Calcutta merchants seem to wish to preserve the minimum refraction, at five per cent., is distinctly operating in the direction of lowering rather than enhancing the trade. This is too large a question to deal with in this place. Its issues extend to Europe. We were once walking through the Exchange where samples are exhibited on which the Mark Lane trade is transacted. One of the most influential corn-merchants in town thrust his hand into a sack of Indian wheat and exhibited the dirt it contained. Pointing to the gram in one sample, the barley in another, he remarked, ' Could you send us your wheats free of mud, and not adulterated with these other grains? It would command a much higher market.' The reply might fairly well have been given, ' When you use your influence to abolish a fixed rate of refraction, Indian wheat within a twelve-month will reach the market in a perfectly clean condition.'

" It would be absurd to expect a cultivator to sell clean wheat when he would be paid exactly at the same rate as if it contained five per cent. of dirt. That is precisely the position; and the Bombay Chamber of Commerce appears to be giving indications of a desire to lower the rate of refraction to two per cent. Why not abolish it entirely, and pay lower rates for all adulterated wheats? One Bombay firm has announced that it will pay at a higher rate for clean wheat than for wheat containing dirt and impurities.* " A most elaborate investigation has been instituted in every province of India into the question of this adulteration. With few exceptions, indeed, it has been found that the cultivator takes no part in the trade of adulteration. His methods of winnowing and storing are imperfect, but there is no inducement to modify this. He can clean now his grain, by the means at his disposal, considerably below the accepted rate of refraction. He makes no gain by producing clean; on the contrary, he is perfectly well aware that the middle-men employed by the exporting firms adulterate the grain before making it over to the firms that pay them at a minimum rate of five per cent. refraction. " An extensive correspondence has passed on this subject. The Indian Chambers of Commerce keep recommending to Government that the only action that can be taken is to urge upon the cultivators to grow only the better-class wheats, to avoid growing mixed crops, and to endeavour to produce as clean a grain as possible. For what purpose? That specially, prepared particles of mud, to the extent of the five per cent., may be added by the middle-man at the cost of the cultivator, who is paid as if it had been there originally." There is apparently little doubt that the question does, as stated by Dr. Watt, rest almost entirely with the merchant, and that a clearly expressed demand for cleaner wheat, and a reduction or withdrawal of the present standard of refraction might be productive of much good. Since 1886 the Bombay Chamber of Commerce have agitated

---

* It is understood that it has withdrawn from that offer; it may be added, however, that the Bombay rate of refraction has for some time been fixed at 2 per cent., while that of Calcutta remains at 5.—*Ed., Dict. Econ. Prod.*

ADULTER-ATION.

F. A. Q. Contracts.
658
Conf. with pp. *169-171.*

strongly in favour of the reduction of the standard of refraction to 2 per cent., and during the past two years the question has occupied the attention both of the Government of India and home authorities. A long series of *Reports and Papers* was published in 1888-89 containing the result of much careful enquiry into the subject. The following information on the forms of contract, etc., method of sales and general condition of Indian wheat in the London market, was supplied by Messrs. McDougall Brothers:—

"The form of contract used in the United Kingdom in the sale of Indian wheats stipulates that the quality shall be of fair average quality (f. a. q.) of that month's shipment, and does not make any mention of a 5 per cent. refraction. The Corn Trade Association arranges, with dock companies and others, to draw samples from each parcel received at the different ports, and from these samples the month's average is then mixed and prepared. Sales are generally made in lots of 100 tons each, and in shipping a bill of lading is made for each 100 tons, this having been found a convenient quantity. A very considerable trade is now done in buying and reselling Indian wheats on the f. a. q. basis.

"This helps to explain how it is that any parcel of special quality would receive little or no attention. Such a parcel would have to be sold by sample, and each buyer would have to inspect and pass on the sample, the first seller would have to seal it, and there would be much trouble and uncertainty if the parcel were sold several times. Thus the objection of the merchants and millers would be to selling or buying each parcel by its own special sample. If regular supplies of clean wheat could be ensured, there is not the slightest doubt but that they would much prefer, and would gladly pay for, clean wheats. We have personally inquired of many millers and dealers. and, without exception, they express a desire for clean wheat, some remarking, 'Do you not think we should give a less price for Australian wheats if they contained dirt?' Many of the largest millers have met the difficulty by erecting extensive washing and cleaning machinery, which gives them somewhat of an advantage, and so are willing that Indian wheats should continue as at present, but many millers are prevented using Indian wheats by the need of such machinery. This, we are informed, is still more the case on the Continent.

"During the last year or two special samples of 'selected Bombay' wheat have been offered upon the market, and command higher prices than No. 1, Bombay. They are guaranteed to contain—

Not more than 94 per cent. of white wheat
4 per cent. of red wheat
2 per cent. of dirt } 6*d.* to 9*d.* higher price than No. 1, Bombay.

Not more than 92 per cent. of white wheat
6 per cent. of red wheat
2 per cent. of dirt } 4½*d.* to 6*d.* higher price than No. 1, Bombay.

"It is important that different kinds of wheat should as much as possible be kept separate. The admixture of hard and soft, and white and red wheats prevents the miller using each to best advantage. Red wheat mixed with white prevents the white from being used for the finest flours. Hard wheats require damping to a much greater extent than the soft, so that if mixed together one is always either too much or the other too little damped.

"A remarkable point about the exportation of Indian wheats is that shipments of new wheat, *i.e.*, those shipped March, April, and May, are so very superior to those shipped later in the season, *i.e.*, August, September, and October. These latter are often so disappointing to millers who have expected to receive bulks equal to early shipments that it has

**ADULTER-
ATION.**

forced many to decline to deal in these latter shipments, for if the whole of a month's shipments were of low quality there would be no allowance on the f. a q. terms. The poor quality of the late shipments may arise from the storing of wheat in pits, some part of the wheat is almost certain to be damaged, and also gets a further admixture of dirt. A remedy for this would be to store the wheat in properly constructed public granaries until required for shipment.

"Large seeds, such as gram and peas, are easily removed by sifting, and also the small seeds, such as linseed and rape, but it is almost impossible to separate such seeds as barley, &c., they being about the same size as wheat. The presence of stones is the greatest difficulty the miller has to contend with, and these are found in the red Bombay and Atbara wheats.

"The information we have gathered is unanimous on the following points :—1, clean Indian wheats are much desired ; 2, extra price would be paid for clean wheats; 3, clean wheats would cause much increased use; 4, wheats carefully selected should be as near as possible of one sort, being then more valuable than when mixed together ; 5, that the practice of mixing dirt and seeds is most detrimental to the practical value of Indian wheats, and urgent steps should be taken against it."

A communication from the Director of the Agricultural Department in Bengal, in 1887, shews clearly ;—(1) that export houses have declined to pay any better prices for wheat with only one or two per cent. of dirt; (2) that it is, therefore, directly to the disadvantage of the Indian *rayat* or middle-man to deliver wheat with anything less than 5 per cent. of dirt; and (3) that these *rayats* and middlemen actually and systematically mix a certain proportion of dirt with their wheat before they deliver it to the export houses. In support of these facts many authentic statements might be brought forward, but the following quotation from a letter sent to Mr. Finucane, the Director of Agriculture, Bengal, may be accepted as forcibly representing the existing state of matters : —

**No
Encourage-
ment
given for
Production
of
Clean Wheat.
659**
*Conf. with
p. 170.*

"I have had a remarkable confirmation of these views from the Manager of the Dumraon Raj, the Honourable Jai Prokash Lall. The statements he made were so significant that I took a note of them at the time, and at the same time informed him that I should embody them in this report. The Manager said that, about two years ago, when the prospects of the wheat trade were apparently good, he seriously thought of cultivating wheat on a large scale. He estimated that on the Raja's estates there were 300,000 acres of land capable of growing wheat, and he proposed commencing operations with a capital of two lakhs. His idea was to induce the *rayats* to grow wheat alone by means of advances in coin and seed, and he intended purchasing machinery, such as he had seen at the Calcutta Exhibition, for cleaning the grain. All that he now required was a remunerative market. Last year, when in Calcutta on Council business, he called at the office of Ralli Brothers, and after telling them his plans asked what price they would give for clean grain. Ralli Brothers *informed him that, owing to the action of the merchants in England, they could not afford to pay more for a clean sample than they now did for grain with 5 per cent refraction. Upon hearing this the Manager abandoned the idea of growing and cleaning wheat on a large scale.* It is difficult to overrate the significance of this anecdote, which appears to prove conclusively that, so long as merchants will not pay a higher price for clean grain, it is useless for Government to think of inducing cultivators to change their present practice.

**Wilful.
660**

"I then asked the Manager about wilful adulteration. He said that he had a *gola* at Itari, near Buxar, from which he used to sell wheat on

| Adulteration of Wheat. | (*J. Murray.*) | TRITICUM sativum. |

rather a large scale to the agent of Ralli Brothers at Buxar. The wheat, as he got it, did not contain 5 per cent. of foreign matter. Accordingly, his servants were directed to mix two maunds of earth with every 100 maunds of grain, so as to bring the adulteration up to the required standard. This earth was treated with water and specially prepared for the purposes of adulteration. The suggestion for adulterating the grain in this way came, as the Manager says, from the employés of Ralli Brothers. This fully bears out what **Major Boileau** says, that grain dealers in Dinapore wilfully adulterate their grain, adding about two maunds and thirty seers of dry clay, *bhusí,* and other grains to every hundred maunds of wheat. **Mr. T. Gibbon, C.I.E.,** the Manager of the Bettiah Raj, told me that wilful adulteration was practised by the petty dealers in Chumparun, and **Mr. Carnduff,** writing from Hajipore, a large grain mart, says, ' In the hands of the middle-men, when the grain is lodged in their *golas,* such grains as *Akla pipra* are, I understand, intentionally added with a view to adulteration. **Mr. Jenkins,** from Buxar, who has clearly paid a good deal of attention to the subject, is of the same opinion.

" As regards the alleged imperfection of present arrangements for winnowing, it will thus be seen that the mixture of dust from the thrashing floor forms a very small portion of the impurities found in Indian grain, and that the present arrangements for winnowing are as good as can be hoped for under present conditions. It will, of course, be desirable to effect improvements in winnowing and thrashing, should any be found possible; but the root of the evil complained of can only be reached by the abolition of the system of allowing a minimum refraction of 5 per cent., a remedy which lies in the hands of the merchants themselves. The facts mentioned by the Manager of the Dumraon Raj show conclusively that clean grain will be forthcoming if the merchants pay for it, and that it will not be forthcoming, however perfect the winnowing and thrashing arrangements may be, so long as a minimum of 5 per cent. is allowed for impurities, be the samples ever so clean.

" These facts and arguments have been brought to the notice of the Calcutta Chamber of Commerce, who, while not denying their force, express regret that they are unable, in the present state of the trade, to alter the existing practice in this respect. As long as that practice continues, it would appear to me to be futile for Government officers to talk to cultivators of the advantages of producing entirely clean grain. On the contrary, if Government officials interfere at all in the matter, it should be by explaining to the cultivators that it is their interest to mix at least 5 per cent. of foreign matter with clean grain before offering it for sale."

It is obvious that this wilful admixture of dirt with fairly clean wheat must be a great disadvantage. Not only does it decrease the value of Indian wheat and renders its extensive employment by small millers who have no cleaning apparatus, out of the question, but it involves the wasteful expenditure of the freight of some 30,000 tons of impurity annually, and a comparatively large and similarly useless expenditure in conveying the grain from European ports by rail to the localities of consumption. For many years the Bombay Chamber of Commerce have made strenuous endeavours to do away with the f. a. q. system, and to introduce sale contracts on the scale of a refraction of not over 2 per cent. In 1888 they addressed representations on the subject to the London and Liverpool Commercial Trade Associations, urging them to assist in the matter. The former Association replied by stating that they did not see their way to making any alteration, but the latter agreed to alter the standards on which wheat sales were made, and advocated

ADULTER-ATION.

Wilful.

Directions given to Adulterate.
661

Winnowing.
662

5 per cent. refraction.
663

Steps to prevent.
664

8

TRITICUM sativum.

ADULTER-ATION.

Steps to prevent.

Conf. with p. 170.

that in future these standards should contain only 2 per cent. of dirt, seeds, and grain other than wheat.

In 1888-89 a series of questions were, at the request of the India Office, issued by **Messrs. McDougall Brothers** to a number of millers in the United Kingdom with the following results:—

Question 1.—Do you use Indian wheat in quantity? 249 millers state that they use Indian wheat in quantity; 259 millers only use Indian wheat in limited quantity; 2 do not reply to this question.

Question 2.—If not, are you prevented from so doing by its impurities? 348 millers state that they are partly prevented from using Indian wheats in consequence of its impurities; 41 millers having the necessary machinery to deal with the dirt, etc., are not prevented from using Indian wheat; 121 do not reply to this question.

Question 3.—Would you use larger quantities if free from admixture and impurity? 461 millers state that they would use a much larger quantity of Indian wheat if they could obtain it in a clean state; 27 millers state that even if clean they could not use a larger quantity of Indian wheat; 22 do not reply to this question.

Question 4.—Is the admixture of red wheat with white wheat of serious consideration to you? 229 millers state that the admixture of red (or hard) with white (or soft) wheat is of serious importance to them, as the red hard wheat can only be reduced by 'roller mills'; 256 millers, most of whom have roller mills, are indifferent as to the admixture; 25 do not reply to this question.

Question 5.—The shipments in the later months of the year show considerable increase of impurities. Do you in preference secure the earlier shipments; and, if so, do you pay a higher price for the same? 322 millers state that they prefer the early shipments and pay higher prices for them; 16 state that they are indifferent, it being merely a question of relative values; 172 do not reply; most of these millers dealing indirect with merchants are unable to give an opinion.

Question 6.—Would you approve of a form of contract limiting the admixture of dirt, seeds, and grain other than wheat to 2 per cent. in preference to the present 'f. a. q.' form? 429 millers express their warm approval of a form of contract limiting the admixture to 2 per cent.; 4 millers are against any alteration; 77 who do not reply are mostly millers unacquainted with the form of purchase; they buy locally from merchants.

Mr. McDougall, commenting on these results in his letter to the Under Secretary of State, dated March 1889, remarks: "The replies now received conclusively prove,—1, that the impurities in Indian wheats greatly restrict their use; 2, that clean Indian wheats are much desired, and would cause a largely increased demand and a higher price; 3, that millers earnestly desire a new contract form limiting admixture to under 2 per cent.

"And I have now to suggest several means which would ensure the object aimed at :—1, by the mutual consideration of the subject by the Indian Council and by representatives from the various Corn Trade Associations; 2, by the formation of a syndicate to purchase and export clean Indian wheats; 3, by the intervention, should it be found necessary, of the Government of India, to make it fraudulent to deal in, or export grain, to be used for human food, in any way adulterated.

"I am sanguine that the first of these suggestions may of itself prove successful in bringing about the desired reform, as I cannot but think that the selfish interest of a few large firms must give way to the unanimous desire of the millers of this country, and to the great ultimate benefit of all concerned. But, if not, I do not hesitate to strongly advise that the third of these suggestions should be promptly adopted.

T. 664

| Adulteration of Wheat. | (*J. Murray*.) | TRITICUM sativum. |

" In connection with this inquiry, I think the fact should not be overlooked that Italy, France, and Belgium are buyers of the better class Indian wheats, and if these wheats were clean there is no doubt this portion of the trade would receive a great impulse, as on the Continent they are less able to deal with the impurities than we are here, and consequently only the high class wheats are in demand."

" The first of these proposals was carried out by Government on the 8th May 1889, on which date a conference was held on the subject at the India Office under the presidency of Viscount Cross, G.C.B., Secretary of State for India. Besides members of the India Council and Departmental officials, many delegates and representatives of Chambers of Commerce, Corn Trade Associations, and large private firms were present. With the exception of the London Corn Trade Association, which maintained its former attitude of preference for the existing contract, nearly all the representatives recommended that an attempt should be made to raise the basis to 2 per cent. The general concensus of opinion, in opposition to that of the London Corn Trade Association, appears to leave little doubt that the latter body was in the wrong, a supposition confirmed by the following report of the Bombay Chamber of Commerce (1889) :—" At the London Conference a statement was read on behalf of the London Corn Trade Association which, in the opinion of your Committee, contained so many mis-statements both as to fact and theory, and was so misleading in purport, that they decided to address His Excellency the Governor on the subject, with the view of recording their protest against the arguments and figures used, and, if possible, leading to further action with the view of improving the cleanliness of Indian wheat. Owing, no doubt, to the promptitude with which the matter was taken up and discussed by your Committee, and the unanswerable character of their arguments, the London Corn Trade Association have so far modified their views that in a circular, dated 14th November, they have intimated that they had taken measures to get the various qualities of Indian wheat analysed, and, as the result of that analysis, have prepared a table showing the extent of impurities in the standard samples which the Association would recognise as allowable in shipments made before and after the monsoon. This table, however, so distinctly recognised a larger proportion of impurities than there was any necessity for doing, that your Committee addressed the London Corn Trade Association on the subject by the return mail, pointing out that in nearly every description of Bombay wheat, with the exception only of No. 1 Club and Red Club, the impurities allowed, even for ante-monsoon shipments, were in excess of the adulteration shown by their own analyses. This, the Committee showed, simply amounted to recognising a standard of impurity in excess of existing conditions, and so far, therefore, from assisting the movement towards greater cleanliness, would really be retrograde in effect. The Committee strongly urged the London Corn Trade Association to reconsider the subject and advocated a 2 per cent. refraction as one which would induce efforts being made to attain greater purity, and one at the same time which would be perfectly attainable without the necessity of extensive mechanical appliances for cleaning purposes. To this communication there has not as yet been time for receipt of a reply, but the Committee trust that this matter will not be allowed to rest, and that future Committees will continue to agitate for the introduction of a basis of contract which will ensure Indian wheats being exported in a cleaner, and therefore, more merchantable, condition than has hitherto been the case."

In addition to the accidental and introduced foreign matter in Indian wheat, a large amount of the impurity which exists is doubtless due to the action of weevils, especially in the later or post-monsoon consignments.

**ADULTER-
ATION.**

Remedy.

666

**Refraction.
667**
*Conf. with pp.
110, 169.*

Many proposals have been made regarding methods of remedying this evil, and here, certainly, the remedy rests almost entirely in the hands of the agriculturists. The subject has been already discussed and need not be again gone into (see **Pests, Insects,** Vol. VI., Pt. I., 145).

It is encouraging to observe that the endeavours made to improve the standard have met with a certain, though as yet very insufficient, amount of success. Thus the **Hon'ble Mr. Benett,** in the final crop report for the year 1889-90, writes:—"The Liverpool and London Corn Trade Associations have now reduced the refraction for Calcutta and Bombay shipments from a uniform amount of 5 per cent. to quantities varying for ante-monsoon shipments between 3 per cent. and 4 per cent., of which $1\frac{1}{2}$ per cent. may be dirt, and for post-monsoon shipments between $3\frac{1}{2}$ and 5 per cent., of which 2 per cent. may be dirt In the case of Karachi, however, it has been considered necessary to fix the high rate of 5 per cent. for ante-monsoon and 7 per cent. for post-monsoon shipments. The Bombay Chamber of Commerce has pointed out that there is really no difficulty in buying wheat with impurities not exceeding 2 per cent., and confirms the conclusion that the mixture with dirt is made between threshing and shipment. The concession of the Association therefore, it is to be feared, will but little affect the trade. But though slight, it is perhaps an indication of a tendency to give way under the pressure which has been brought to bear on the London Association by several of the other commercial bodies, and in particular by the valuable evidence brought to light at the conference held at the India Office. On that occasion it was clearly shown that although speculative buyers might all prefer the higher refraction, the millers of the United Kingdom were exceedingly anxious to obtain the cleaner article, and were largely prevented from using Indian wheat by its high percentage of impurities, due solely to the high refraction with which it is bought *

"The Government of India, at the instance of the Secretary of State, is in correspondence with Local Governments and Administrations regarding the expediency or otherwise of introducing grain-elevators into India with the view of cleaning, grading, and handling wheat. The introduction of these methods, it is to be feared, however, will be useless until the trade shows itself ready to buy clean, and nothing but clean. wheat The elevator system, however, has the merit of doing away with the necessity for arbitration and analysis of samples. It is being introduced by Russia and may serve to increase the advantages which that country already possesses over India in its competition for the wheat trade of Europe."

Prices.

**PRICES.
668**

The local prices of wheat and other food-stuffs have been recorded fortnightly in every district, and even in every large town in each district during the past thirty years. An examination of these elaborate returns reveals the fact that the price paid by the consumer to the local grain merchant has varied excessively from month to month and from year to year; indeed to such an extent has this been the case that it would be quite unsafe to attempt to express average prices for large areas India, unlike England, is entirely dependent on her own produce for her food-supply, hence scarcity or superabundance brings about an instantaneous change in local

**Purchased by
measure sold
by weight.
669**

---

* It is understood that an advantage of much moment to those who have desired to uphold the refraction standard exists in the fact that wheat, when purchased by measure and sold by weight, leaves a large margin of profit owing to the greater weight of the adulterants. If this be so in India, it seems desirable that this feature of the controversy should be more clearly brought out than has been done hitherto.— *Ed., Dict. Econ. Prod.*

prices. Both these conditions may prevail during the same season in two neighbouring provinces, or even in districts of one province, since inter-communication has as yet by no means reached a degree of perfection. But a tabular statement of the average annual prices at certain selected stations as published by **Mr. O'Conor** may afford a fair indication of the fluctuations and gradual tendency towards rise or fall of the prices : —

Prices of Wheat in certain selected stations, from 1861 to 1889, in seers to the rupee.

Year	Calcutta	Patna	Cawnpore	Fyzabad	Meerut	Delhi	Rawálpindi	Karáchi	Ahmednagar	Bombay	Jubbulpore	Raipore
	Seers	Seers	Seers	Seers	Seers	Seers	Seers	Seers	Seers	Seers	Seers	Seers
1861	16·76	23·24	17·25	25	16	14·7	29·04	19·08	18·5	12·6	25	48
1862	18·62	16·21	27·5	28·5	33·5	26·34	20·59	15·74	15·5	11·95	29	83
1863	20·48	31·17	28·5	27	33	25·32	20·74	15	13	9·05	23	61
1864	14·89	16·21	19·75	22	23·75	21·34	25·42	10·25	10	7·9	31	32
Average of 4 years	17·69	21·96	23·25	25·62	26·56	21·92	23·77	15·04	14·25	10·53	27	56
1865	14·89	16·89	16·75	16·5	20·75	14·77	27·12	11·56	10·5	8·47	20	22
1866	11·66	10·54	14	13·75	23·5	19·47	27·6	11·75	10·5	6·5	13	20
1867	21·42	15·10	16·75	23·5	21·5	21·08	21·37	13·33	11·75	8·95	21	36
1868	16·76	30·61	20	17·5	25·75	20·9	12·65	13·87	14·5	10·19	15	36
Average of 4 years	16·18	18·31	16·87	17·31	22·87	20·3	22·03	12·63	11·81	8·53	17·25	28·5
1869	14·43	13·04	11·5	13·75	13·25	11·66	13·29	13·62	10	8·47	11	15
1870	14·43	11·02	15·5	19	15·75	12·91	15·09	20·28	8·75	8·95	15	23
1871	15·91	23·94	24·11	24·65	24·35	21·87	17·94	11·67	14·3	10·21	22·87	48·15
1872	14·14	19·9	19·29	16·39	24·43	20·89	18·8	14·22	13·25	10·94	21·65	41·83
Average of 4 years	14·73	17·22	17·6	18·42	19·44	16·83	16·27	12·45	11·75	9·64	17·63	31·99
1873	11·53	15·95	10·58	15·25	20·44	19	18·47	15·41	17·99	11·04	10·75	42·27
1874	12·45	16·88	17·35	16·13	20·75	19·27	23·56	14·35	23·26	12·38	20·59	41·6
1875	16·	23·41	23·28	24·43	22·32	21·41	28·01	12·8	17·46	11·88	20·83	46·54
1876	17·1	25·39	25·1	25·95	25·83	24·75	31·48	12·81	16·95	11·64	26·53	53·74
Average of 4 years	14·27	20·41	20·58	20·44	22·28	21·11	25·85	13·84	18·91	11·88	23·42	46·04
1877	13·06	19·11	16·08	10·7	19·83	18·47	28·51	11·81	9·46	8·21	20·38	42·59
1878	11·19	14·56	13·8	12·74	14·99	14·1	18	9·37	7·42	6·58	11·54	19·48
1879	12·3	14·12	14·17	14·16	14·94	14·1	9·23	7·29	7·63	6·93	11·33	19·88
1880	13·06	19·94	18·34	18·08	18·53	17·48	10·9	10·31	11·09	8·93	16·8	25·54
Average of 4 years	12·58	16·93	15·6	15·39	17·07	16·04	16·66	9·84	9·05	7·66	15·01	26·87
1881	15·4	22·65	20·47	18·9	20·45	19·37	13·13	13·08	16·1	10·99	21	39·08
1882	13·2	19·44	18·68	17·18	18·55	18·59	20·27	13·30	14·01	10·37	18·52	31·53
1883	13·1	19·17	18·73	17·71	18·02	18·97	24·89	13·32	14·02	10·29	19·99	23·46
1884	16·02	20·9	22·01	19·76	20·42	20·02	30·8	14·01	13·89	11·22	23·59	20·25
Average of 4 years	14·43	20·54	10·97	18·30	19·36	19·24	22·27	13·4	15	10·72	20·77	30·83
1885	15·45	21·26	22·49	21·14	22·25	22·5	28·21	14·27	15·91	10·38	21·21	34·32
1886	15·14	20·95	19·65	18·9	18·36	19·08	21	11·85	14·19	9·93	16·44	26·58
1887	14·17	18·79	16·42	15·11	15·09	14·71	13·87	11·19	12·34	9·6	15·8	21·77
1888	12·98	17·72	16·19	14·33	15	14·75	14·58	11·61	12·83	10·32	14·83	21·53
Average of 4 years	14·43	19·68	18·50	17·19	17·67	17·76	19·41	12·23	13·82	10·32	17·07	26·05
1889	12·92	16·04	16·68	4·62	16·96	17·29	19·08	11·79	13·09	9·91	15·29	20·31

| TRITICUM sativum. | Production of Wheat in India. |

PRICES.

It is evident from these figures, that the price of wheat in India depends almost entirely on local conditions, and is practically quite unconnected with the question of foreign demand.

In Europe the prices of wheat from every source have undergone an almost continuous fall for many years. It is worthy of note that Indian wheat has relatively maintained a somewhat higher value than others. Thus, **Mr. McDougall** publishes the following comparative list of prices for No. 2 Club, Calcutta, and English wheat, from 1877 onwards :—

	No. 2 Club, Calcutta.	English.
	Per 492℔	Per 480℔
1877	50·6s.	56·9s.
1878	46·3	46·5
1879	50·0	43·10
1880	45·3	41·5
1881	45·0	45·4
1882	43·4	45·1
1883	39·1	41·7
1884	32·10	35·8
1885	33·2	32·10
1886	31·10	31·0
1887	31·1	32·6

In June 1888, the Mark Lane price was 32s., and, at the same date in 1889, it had fallen to 29s. per quarter.

In 1890 the prices ruled very closely with those of 1888, *vis.*, 29s. 10d. per quarter until the first half of May, when there was an upward tendency, the latest quotation up to June 12th being 31s. 7d. per quarter.

PRODUCTION.

PRODUCTION.
670

The question of the total wheat production in India is one which there is some difficulty in estimating, but is of much interest as bearing on a possibly increased demand for the Indian cereal. Dr. **Forbes Watson**, in his paper on Indian Wheat in 1879, made the following estimate of the wheat outturn of various countries :—

"All available facts point to the conclusion that, as regards wheat, India may shortly become one of the chief sources of supply for the United Kingdom. It must be borne in mind that India is one of the largest wheat-producing countries in the world. The production of the United Kingdom amounts to only about 10,000,000 to 13,000,000 quarters per annum. Austro-Hungary, Italy, and Spain each produce about the same quantity. Germany produces from 15,000,000 to 18,000,000 quarters, and the two countries which produce the largest amounts are France and Russia, each producing from 30,000,000 to 35,000,000 quarters per annum. Both are surpassed by the United States, which produced during each of the past two years upwards of 45,000,000 quarters. No complete stadtistics exist for India, but we know that the Panjáb alone produces about as much as the United Kingdom, Oudh about 3,500,000 quarters, the Central Provinces about 3,000,000, and Bombay not much less. The production in the North-West Provinces proper has never been estimated, but must be fully equal to that of the Panjáb, and that of Behar is also known to be considerable. Thus the yearly production of the provinces under direct British rule will amount to from 30,000,000 to 35,000,000 quarters, or to the same quantity as that produced by Russia

| Cultivation of Wheat in India. | (*J. Murray.*) | TRITICUM sativum. |

or France. But if the Native States in the Panjáb, Rajpútána, Málwa, Bundelkhand, and Guzerat be added, in all of which wheat is largely cultivated, it will be found that India must be considered as being, next to the United States, the largest wheat-producing country in the world."

The statistics from which the above estimate was founded were avowedly deficient, and more recently acquired data indicate that the 35 million quarters assumed by **Dr. Watson** are in excess of the actual. The Government of India, in its first large publication on wheat production, after the issue of **Dr. Watson's** report, calculated the outturn on a total area of 19,329,200 acres to be 26,548,000 quarters of 480℔ or 6 maunds; which, adopting the figures above given for other countries, would make India stand fourth in point of production, following the United States, France, and Russia. In the Central Provinces, the North-Western Provinces and Oudh, and the Panjáb the accepted average yields per acre were, in the same publication, stated to be 8 bushels, 11⅝ bushels, and 13½ bushels, or 6 maunds, 8¾ maunds, and 10 maunds, respectively. These figures must, however, be more or less approximate only, since the rates of yield necessarily vary greatly with the variety of seed cultivated, with the greater or less care bestowed on it, and with the degree of irrigation. These questions will be separately discussed in the account of each province.

The average outturn of the four years from 1884-85 to 1887-88, on an average total area of 26,508,000 acres, is estimated to have been 7,205,500 tons, or a little over 31 million quarters of 480℔ or 6 maunds. In 1888-89 an outturn of 6,510,979 tons was estimated on the area shown by the crop forecasts for the year, *viz.*, 26,381,765 acres, an estimate a little more than 200,000 acres under the actual. The former of these, which give an average acre-yield for all India of a little over 7 maunds, may perhaps be accepted as a fair average. The latter represents a crop which was on the whole under average, owing to an untimely and unevenly distributed rainfall, and the acreage yield is consequently less, a little under 7 maunds.

In addition to the 31 million quarters returned as the outturn for four average years, it must be remembered that a large quantity of wheat is grown as a mixed crop and, in agricultural statistics, comes under returns of unspecified food-grains. If the area and outturn for these crops be included, a due deduction being made for the proportion of the yield from barley, grain or other plants with which the wheat is mixed, it is not improbable that the total outturn of the grain in India might approximate very closely to, if it did not actually attain, **Dr. Forbes Watson's** estimate of 35,000 quarters.

WHEAT CULTIVATION.

In the foregoing chapters on area and production, the more important features connected with the development of Indian wheat cultivation have been dealt with, chiefly with the object of demonstrating the fact that the wheat trade up to its present stage, is a perfectly natural one. All the facts go to prove that this is so, and that the agricultural population are exporting only what they specially cultivate for that purpose. And there appears to be little doubt that so long as wheat proves a remunerative crop, its area will continue to increase, but that as soon as better profits can be realised on another crop, the *rayat* will turn from wheat and readily assume the cultivation of the more profitable crop with little or no inconvenience or pecuniary loss to himself. Wheat, however, is also grown, as already shewn, to a certain extent as a staple food-crop for home consumption, and will probably always continue to be so.

Some of the more important features having been thus dealt with as

**TRITICUM
sativum.**

Cultivation of Wheat

**CULTIVATION
in the
Panjab.**

a whole, we may now proceed to consider in detail the wheat cultivation of the several provinces. This has already formed the subject of many exhaustive and valuable official reports. After the issue of Dr. Forbes Watson's report, frequently alluded to above, Her Majesty's Secretary of State called for information to be furnished from all India, as to the nature of the soils on which the better wheats are cultivated, as well as details of the methods of cultivation. This stimulated detailed investigation, with the result that first one, and then another, volume on "*The Wheat Production and Trade in India*" was produced in 1879 and 1883. Since the appearance of the last of these, separate publications have been issued by several provinces on the subject. From these works and from the settlement, administration, and agricultural reports, the information obtained in the succeeding chapters has been mainly derived. The writer has also to acknowledge liberal quotation of many passages from a pamphlet published in London in 1888, by the editor, Dr. G. Watt, on "*The Conditions of Wheat-growing in India*," a paper which contains, in condensed form, a *résumé* of all available information up to the date at which it was written.

**PANJAB.
673**

PANJAB.

References.—*Gazetteers*:—*Ambala, 48, 49 ; Amritsar, 35-37, 47 ; Bannu, 132, 133, 135, 137-141, 144, 149-152 ; Delhi, 98, 100-103, 111, 113, 121, 139, 140 ; Dera Ghazi Khan, 31, 80, 81, 83, 84, 91, 92 ; Dera Ismail Khan, 122-125 ; Ferozepore, 65, 68-70, 74-77 ; Gujranwala, 53 ; Gujrat, 77-80 ; Gurdaspur, 51, Gurgaon, 81, 82, 90, 91 ; Hazara, 124, 125, 130, 131, 134, 135, 137, 143, 146-148, 150, 151 ; Hissar, 46, 48 ; Hoshiarpur, 86-89, 117 ; Jalandhar, 44 ; Jhang, 106-109 ; Jhelam, 98, 99, 104, 106; 108, 109 ; Kangra, I., 148, 149, 152, 154-156, 161 ; II., 57, 58, 64 ; Karnal, 157, 172, 180, 183-185, 196-200 ; Kulu, 57, 64, 96 ; Lahore, 85, 88, 102 ; Ludhiana, 138 ; Montgomery, 116, 117 ; Mooltan, 93 ; Musaffargarh, 93, 94 ; Peshawar, 144-146 ; Rawálpindi, 78, 80, 81 ; Rohtak, 90, 92-95 ; Shahpur, 64, 65 ; Sialkote, 63, 68, 69 ; Agri.-Horti. Soc. Panjáb (1853), 111 ; Indian Agriculturist, Jan. 25, 1890 ; Ind; Forester, X., 369 ; Settlement Reports :—Bannu, 40, 53, 76, 81-83, 85, 86 ; Delhi, 30-32, 38, 40, 43, 106, 107, 222, 224, App. xxxiv. ; Dera Ghazi Khan, 8, 9, 12, 44, 45, 63, 72 ; Dera Ismail Khan, 7-16, 24, 28, 342, 350, 351; App. xxiii ; Ferozepore (1855), 3, 7, 31-33, 83 ; (1875), 6, 8, 9, 12, 15, 22, 24, 25, 27 ; Gujarat (1860), 132, 136 ; Gujrat, 34, 77, 79, 84, 92, 96, 98 ; Gurdaspur, 13 ; Gujranwala, 11-13, 28, 32, 33, 37, 41, 44 ; Hazara, 82, 88-90, 102, 105, 171-174, 177 ; Hoshiarpur, 11, 22, 30, 31 ; Jhang, 57, 80-87 ; Jhelam, 37 ; Kangra, 24, 26, 76, 78 ; Kohat, 2, 3, 12, 120-122, 155, 160-165, App. xli ; Lahore (1858), 11, (1865-69), 7, 9, 32-37 ; Montgomery, 102-112, 127 ; Peshawar, 19, 135, 184, 188, 213, 215, 219, 225, 228, App. xxxiv., xlii., lii., liv., lxvi., lxxiv., lxxxiv., ciii.-cvii. ; Rawálpindi, 19, 20, 26, 59 ; Rohtak, 62, 86-93 ; Sialkote, 19, 31-33, 37, 54 ; Shahpur, 18, 19, 28, 89, 94 ; Simla, 13, 15, 31, 34, 42, 44, App. xi.-xv., xxiv., xxxv.-xxxix. ; Sirsa 57, 64 ; Forbes Watson, Rept. on Indian Wheats, 1879 ; Repts. on Wheat Production and Trades, Govt. of India, 1878, 1883, 1886 ; Wace, Panjáb wheat ; Crop Repts. and Forecasts.*

The cultivation of wheat in the Panjáb will be considered in some detail, for so much exists, common to all the wheat-producing areas of India, that the more characteristic features may be disposed of in one place, leaving only special modifications to be commented on afterwards. It may, perhaps, be advisable to commence with the consideration of the soils, which has been discussed by Dr. Watt as follows :—

**Soils.
674**

Soils.—"In the Panjáb, soils may be classified, first, according to the mode in which they are irrigated ; secondly, according to their composition. With slight local modifications the remarks which we here offer are applicable to the whole of the alluvial parts of India. One of these tracts of country or regions with a peculiar soil may pre-

CULTIVATION in the Panjab.

Soils.

dominate more in one province than in another; and in some instances the specific character of the soil may be modified or intensified. The main features are, on the whole, preserved. We shall establish, therefore, in this place, a standard from which, in our subsequent remarks under other provinces, we shall record departures and modifications.

"From the numerous mouths of the Ganges, and sweeping round the whole length of the Himálaya, at the same time isolating the great southern tableland, there extends a vast alluvial plain, which is only lost in the North-West Provinces and the Panjáb by blending into the drainage area of the Indus. From this point a similar alluvial region is continued to the mouths of the Indus, and may be said to widen until it embraces the northern division of Bombay. In the Bengal section of this vast expanse, the clay soil of the rice swamps can only be viewed as land, figuratively speaking, recently recovered from the sea; and immense portions of it are even now within tidal influence.

"The bulk of Bengal is rain-inundated. Passing higher up the alluvial basin, evidences of a more ancient soil, indeed, of a more ancient agriculture, are to be seen in the rich loam of Behar. This soil continues with varying degrees of fertility through the North-West Provinces to the Panjáb, and down the tributaries of the Indus to the basin of the combined stream, until it reaches the swamps of the western coast. Throughout this loam expanse there are two modifications. First, on the inundated tracts of the rivers and on depressed portions of the country (in most cases these are but the old beds of former streams, or the silted-up lakes which were thrown off as contortions of the river, isolated by the main stream taking the more direct course through a narrow isthmus), rich clayey loam occurs which merges in its character into the heavy mud soil of Bengal. Secondly, within the regions of climatic extremes, natural growth and cultivation alike have been checked, and loam is there found to be more and more intermixed with sand, until absolute sandy deserts are attained.

"Thus there exist four types of soil in the alluvial plains of India: a heavy loam, in which clay predominates (the muddy swamps of Bengal); a heavy loam, with a certain amount of sand, in which the clods remain firm (the lowlying and inundated tracts of Upper India); a light loam, in which the clods are pulverised on being let fall from the hand (the principal soil of Behar, the North-West Provinces, the Panjáb, and a certain portion of Bombay and Sind); and lastly, a poor loam with a large admixture of sand, passing into pure sand in which clods do not form at all (the soil of some parts of the North-West Provinces, of a large proportion of Central India and of Sind, with also certain parts of the Panjáb). The intimate relation of the two features of soil alluded to in the opening sentence of this paragraph has been thus exemplified. The absence of water, together with the extremes of heat and cold, have had much to say on the production of desert tracts, and annual inundations have greatly tended to preserve the heavy loams.

"There are certain agricultural terms used in the Panjáb, but fairly well understood throughout India.

"Land that is dependent on rain is known as *baráni*; if watered by canals it is *nahri*; *chahi* is watered by wells, and *abi* from tanks. The word *doáb* signifies a region between two rivers. The five great streams of the Indus break the Panjáb into vast interfluvial expanses or *doábs*, so that, to understand Panjáb agriculture, this feature must be fully appreciated. The tracts annually inundated by the rise of the rivers, or kept moist from being adjacent to flooded land, are known in the Panjáb as *bhet*, *banjar*, or *sailába*, and in other parts of India as *khadar*, but by the Hindustani-speaking population this name is even used in the Panjáb.

**TRITICUM
sativum.**

Cultivation of Wheat

CULTIVATION
in the
Panjab.

Soils.

The chief danger such regions are subjected to is the growth of the saline efflorescence known as *reh* (a crude sulphate or carbonate of soda).

"Land beyond the *bhet* influence is generally known as *desya* in the Panjáb, and to Hindustani-speaking people as *bángar*. This may be *chahi, abi, nahri,* or *baráni,* according to the source from which it derives its water. The interior or higher portions of the *doáb* are often spoken of as *des-utar,* (in contradistinction to *hetár*) or *máhjah.*

"The names given to denominate the physical character of the soil are:—1. *Nyái,* rich land around the homestead, on which vegetables, tobacco, poppies, etc., are grown.

2. *Dákar* or *chamb,* heavy clayey loam, too low for being drained. This is good for rice and grain. The term *rákar* in the Panjáb denominates bad *dákar,* on which rice only can be grown.

3. *Rausli* or *dosahi (dusháhi)* is the light, easily pulverised loam which we have spoken of as the most prevalent in Upper India. This yields all crops except rice. It is soft and easily worked, mixes readily with manure, and consists of clay and sand. It is probable that the term *dosahi* denotes a slightly inferior quality of *rausli* with more sand ; just as *rohi* would appear to be a rich soil approaching to *dákar,* only well drained. *Rohi* is admittedly the finest form of soil in the Panjáb.

4. *Bhúr* or *maira* is light sandy loam, suitable for the cultivation of millets. In this soil the sand predominates over the clay. *Tiba* is almost pure sand, *reti* being a soil with wind-blown hillocks of sand.

"Other terms are used in the hill tracts of the Panjáb, and nearly every province has special terms for local modifications of the soils we have indicated. As such names can be of little interest to persons not residing in India, we shall accept the above as conveying a general description of the characteristic soils of the alluvial basin of India. A separate account will be found under the Central Provinces of the soil, terrestrial character, and peculiarities of the southern tableland. From what has been said, a general idea, it is hoped, has been conveyed of the character and fruitfulness of the soils of the plains of India. The absence of a water-supply will, of course, make the best *rausli* land entirely dependent on the rains, and the inequality and insufficiency of the rains of the Panjáb leave neighbouring tracts either uncultivated or at most only occasionally thrown under crops. This is the field for the future operations of the canal engineer. A judicious control over the supply of canal water has made these arteries carry life and fertility where formerly rich undulations of fertile soil bore only a scanty herbage. Where artificial aid, in the form of canals, is not brought to the cultivator, it will be seen, from the account of soils, that there are narrow limits within which displacement of crops can be practised.

"The climate prescribes a limit to the *rabi* as to the *kharif* crop. The varied nature of the soils is such that a second check is given to the dangerous disturbance of established and natural conditions of agriculture through any greed the cultivator might manifest in desiring to reach a hand forward to the hard cash offered by an export trade like that of wheat. The extent to which the owner of a *desya* or *bángar* farm can supplant millets with wheat must depend on a chapter of accidents : the abundance of water in his wells (even should he possess such), the rainfall, the proximity of his fields to the irrigation canals, the character of the soil on which his labours from year to year have been expended. Should his fields fall under the class we have defined as *bhúr,* then, without manuring to an extent which would never pay, he must rest content with his millet and pulse crops, for in such soils, in the majority of cases, wheat-cultivation is a physical impossibility.

T. 674

| in the Panjáb. | (*J. Murray.*) | TRITICUM sativum. |

CULTIVATION in the Panjáb.

Irrigation.
675
Area.
676
Conf. with p. 94.

District Cultivation.
677

"While wheat-cultivation cannot, therefore, expand into the *bángar-bhúr* lands, there are immense tracts of *rausli* which wait only for means of export, or for a supply of water, to be at once thrown under the finest varieties of wheat."

AREA AND IRRIGATION.—The areas in each district of the Province during the year 1888-89 under pure wheat were as follows :—

District.						Irrigated.	Unirrigated.	Total.
Hissar	.	.	.	.	.	27,236	52,048	79,284
Rohtak	.	.	.	.	.	26,259	42,765	69,024
Gurgáon	.	.	.	.	.	22,968	39,587	62,555
Delhi	.	.	.	.	.	33,001	92,478	125,479
Karnál	.	.	.	.	.	59,890	93,436	153,326
Ambála	.	.	.	.	.	11,266	257,860	269,126
Simla	.	.	.	.	.	433	4,416	4,849
Kangra	.	.	.	.	.	44,725	149,167	193,892
Hoshiarpur	.	.	.	.	.	7,692	310,594	318,286
Jullundur	.	.	.	.	.	114,488	157,415	271,903
Ludhiana	.	.	.	.	.	73,141	159,065	232,206
Ferozepur	.	.	.	.	.	222,380	245,882	468,262
Múltan	.	.	.	.	.	242,714	34,963	277,677
Jhang	.	.	.	.	.	135,674	39,978	175,652
Montgomery	.	.	.	.	.	146,658	36,991	183,652
Lahore	.	.	.	.	.	332,589	65,507	398,096
Amritsar	.	.	.	.	.	217,556	107,978	325,534
Gurdaspur	.	.	.	.	.	70,585	237,971	308,556
Sealkot	.	.	.	.	.	209,374	159,408	368,782
Gujrat	.	.	.	.	.	136,170	169,345	305,515
Gujranwála	.	.	.	.	.	213,318	39,003	252,321
Shahpur	.	.	.	.	.	121,828	101,584	223,412
Jhelum	.	.	.	.	.	17,220	412,711	429,931
Rawálpindí	.	.	.	.	.	9,081	467,683	476,764
Hazára	.	.	.	.	.	11,330	112,532	128,862
Peshawar	.	.	.	.	.	94,312	175,146	269,458
Kohát	.	.	.	.	.	15,680	78,293	93,973
Bannu	.	.	.	.	.	83,326	221,599	304,925
Dera Ismail Khan	.	.	.	.		96,749	156,014	252,763
Dera Ghazi Khan	.	.	.	.		86,444	54,084	140,528
Muzaffargarh	.	.	.	.		164,810	47,574	212,384
	GRAND TOTAL			.		3,048,897	4,323,080	7,371,977

The expansion of this area has already been noticed in the chapter on AREA. Taking an average of all years the increase has been somewhat smaller than might at first sight have been expected in a province of which the staple has always been wheat, and which has still an immense area of cultivable land unoccupied. But this, owing to the extensive arid tracts, must necessarily depend entirely on the development of irrigation, therefore no rapid expansion can be looked for. Great efforts are, however, at present being made to extend the canal system through these tracts, and a slow but sure expansion may, therefore, be looked for. The Director of Land Records and Agriculture for the province states (*Report, 1887-88*) that the area under wheat of a given year depends largely on the rainfall, a bad *kharif* with a good rainfall in September resulting in extended wheat sowing and *vice versá*. The possible maximum variation must be very great, seeing that during the period from 1880 to 1886-87 the difference between the maximum and minimum annual area is as much

T. 677

TRITICUM sativum.	**Cultivation of Wheat**

CULTIVATION in the Panjab. Qualities. 678

as 27 per cent. This fact probably explains, at least to some extent, the large expansion during the past year.

RACES AND QUALITIES —The late Colonel Wace, in his elaborate report on Panjáb wheats, states that the " soft red " is grown on 5 million out of the total of 7 million acres under wheat cultivation. Notwithstanding the fact that this race is less suitable for export than the soft white, Colonel Wace deprecates the charge brought against the cultivators that they are careless as to the selection of seed, and states that they are by no means indifferent to securing seed of a good quality. It is, therefore, probable that if the advantages of growing the soft white form were pointed out to the agriculturalist and the small Native merchant, its cultivation might be taken up to a greater extent, and thus bring a larger quantity of more valuable wheat into the market.

Dr. Forbes Watson selects the following as among the best samples submitted to him :—

Commercial Class.	District.	Name.	Weight per bushel.	Price.	REMARKS.
				R a.	
SOFT WHITE .	Delhi . .	Gundun safed	65	48 0	Like Australian.
	Sarsa . .	. .	61	45 0	
	Dera Ismail Khan .	. .	63	48 0	Like best Californian.
HARD WHITE.	Sarsa . .	Brown wheat	60½	41 0	Small berried.
SOFT RED .	Gurgaon .	. .	64	41 6	
	Rohtak .	Red wheat .	62	41 0	
	Jhelum .	Lal . .	61½	41 0	
	Shahpur .	Rati . .	61	41 0	Large berried, mixed with barley.
	Dera Ismail Khan .	Rutti kanak .	60½	41 6	Long berried.

In the report from which the above is extracted, Dr. Forbes Watson made definite allusion to the " weevilled " condition of the samples received from the Panjáb and Sind. In consequence of this a second series was forwarded from the Panjáb during 1880 for supplementary valuation and report. Of 192 samples, 31 were soft white, 67 hard white, 31 soft red, 59 hard red, and 4 mixed. The average value of all was 46s. 10½d. per quarter, the prices ruling for all wheats at the time being about 4s. 6d. per quarter higher than at the period when the first report was written. The general quality of the samples was, however, very much superior, the hard reds especially having risen greatly in value.

The Gazetteers contain long lists of vernacular names of different forms of wheat, which it would probably serve no very useful purpose to reproduce. It may, however, be noticed that the red wheats are spoken of as being preferred, partly because its outturn is greater, partly because it can be grown in inferior soils and unirrigated tracts, since it requires less moisture. With the extension of the canal system, therefore, an increased proportion of the more valuable soft white wheats may be expected. Bearded, *kinjhari,* and beardless, *rodi,* wheats are also distinguished. A form known as *pamman* in the Muzaffargarh district is highly valued and cultivated as a luxury for the richer classes.

Methods. 679

METHOD OF CULTIVATION.—Dr. Watt continues :—" The wheat crop of the Panjáb is sown on *rausli* and *rohi* lands, and sometimes also on

CULTIVATION
in the
Panjáb.

Methods.

dákar. It occupies the soil for about six months—the first sowings commencing about the middle of October, and the harvest operations throughout the province being completed by the middle of May. The systems pursued vary, to some extent, in the various districts of the province, but mainly in consequence of the nature of soil and source of water-supply. We shall, therefore, comment specially on the systems adopted in Delhi, Umballa, Jullundur, Lahore, Jhang, Montgomery, Dera Ismail Khan, and Dera Ghazi Khan.

"The system followed in Montgomery for well-irrigated lands has been described thus :—During the rains in June or July the land is ploughed two or three times and smoothed. If rain has been plentiful and the ground remains moist, seed is sown broadcast in October, November, and December. The ground is then again ploughed and smoothed, and the beds formed. If there is subsequent rain, the fields are irrigated from wells or *jhálars* six or seven times : if there is no rain, nine or ten times. If there is little or no rain during the rainy season, or if the land does not remain moist up to October, it is irrigated before the seed is sown. If the seed is sown in October, a good crop is the result : if in November, about twenty-five per cent. less than if sown in October ; and if sown in December, about thirty per cent. less."

"On *bhet* or *sailába* lands.—At the last inundation during the rains (generally in August) the land is ploughed two or three times and smoothed. In October seed is sown through a drill ; no beds are formed, and no subsequent irrigation takes place, as the crop depends on rainfall."

"In the majority of the districts, sowing through a tube attached to the handle of the plough is followed in preference to broadcast sowings—the crop appearing in consequence in drills. In the Panjáb generally, drill-sowing is always practised where the character of the soil will permit of this system. In the sandy soils of Marwat the seed is drilled three or four inches into the ground without any preliminary ploughing.

"Manuring is practised if the cultivator can afford to do so, but chiefly only on well-watered lands. Canal-irrigated fields are nearly always cultivated without manure. *Dákar* or *dar* lands are considered rich enough to produce wheat without any manure. The degree of watering is indicated in the following paragraph regarding the Lahore district :—

" ' In October the field is irrigated and ploughed twice, the grain being dropped in at the last ploughing through a tube attached to the handle of the plough. The land is then smoothed by a rough roller called *sohaga.* After this the crop is irrigated once a month for three months, and periodically weeded, if the cultivator can afford this ; manure is rarely used, never at any distance from the villages. The people say there is something special in the soil, that when good seed is obtained it yields a good crop the first, and perhaps the second year, but afterwards deteriorates.' This same opinion, that without manure or a rotation of crops the soil deteriorates if wheat be continuously reared on it, prevails over the greater part of India. In Jullundur the ploughing is begun much earlier than we have indicated—the first ploughing taking place in January or February."

With reference to enquiry as to the period during which land has been under wheat cultivation in the Panjáb, instructive replies ha /e been received. "Wheat is considered the strongest crop, and to maintain the productive power of the land it is necessary to change this crop for some other, such as *jowar* (the larger millet), wheat being sown the second year." "Carefully manured land can remain for five or six years under wheat."

The report on Rohtak states that "the lands now growing wheat have been so used for a long time." Of Jullundur, the District Officer writes, "There is no reason to suppose that the land has deteriorated from

TRITICUM sativum.   Cultivation of Wheat

CULTIVATION
in the
Panjab.

Methods.

over-cropping. Except in highly-manured lands, wheat is grown year after year."

The opinions recorded are decidedly opposed to the view that the soil deteriorates under wheat. But one officer writes, " it is unquestionable that the finest crops are raised on lands newly brought under canal irrigation".

"About one-third of the whole cultivated area of the Panjáb is cropped with wheat. The acreage represented by this fraction is liable to considerable variation, due mainly to the character of the seasons, and the gradual increase of cultivation in general." Two-thirds of the annual cultivation consists of other than wheat crops, manuring is regularly resorted to when found necessary, and at least a seasonal if not an annual rotation is regularly observed, so that there is little reason to fear that the expansion of wheat cultivation in the province is in any way endangering the fruitfulness of the soil. A large proportion of the canal-irrigated area, and from $\frac{1}{3}$ to $\frac{1}{4}$ of the area irrigated by wells, were officially stated in 1883 to be double-cropped, giving a wheat, *rabí,* and some other *kharíf* crop every year. Other wheat lands are said to be generally cultivated on one or other of two plans. The first, which is generally followed by the best cultivators, is a two-year course, in which a wheat crop is first taken, immediately succeeded by an autumn pulse crop, after which the land is fallowed for a year. The other system consists in separating the lands for the spring crop from those for the autumn, and then maintaining the separation. The spring crop lands give a wheat crop every spring, and lie fallow during the autumn season. The autumn lands lie fallow during the spring, and give a pulse or other crop in the rainy season.

"The method of cultivation is essentially the same everywhere, but the skill and labour used in carrying it out are liable to indefinite variation, partly due to the differing character of cultivators, partly to local circumstances, and partly to the rotation of crops in common use. Reduced to its barest elements, the system is to plough and cross-plough as often as possible, then harrow, then sow the seed through a drill attached to the plough, and then plough over. The number of ploughings varies greatly once is enough for a Saiad, while a Ját thinks ten times hardly sufficient" (*Wheat Production and Trade, 1883,* 72).

Reaping.
680

REAPING, THRASHING, WINNOWING.—On these subjects Dr. Watt wrote :—" Reaping begins about the end of April, and the whole crop is in-gathered by the end of May or the beginning of June. The practice described in connection with the Montgomery district is fairly representative. The reapers are called *láwa,* and belong chiefly to the class of village servants. But they do not confine themselves to their own village,—they go wherever they can find work. The usual pay is one *pái* (seven seers) of grain, or four annas in cash per diem, with five sheaves. [This might be expressed as sixpence a day and the sheaves.] An ordinary reaper will cut down one *kanál* and a half in the day; and a strong and practised hand will do as much as two *kanáls* (*kanál* = half a rood). On an average five men will cut down an acre a day. Reaping is carried on during the moonlight nights in the last few hours before day if the straw is very dry, as the moisture of the night air is supposed to strengthen the stalk and prevent the ears falling off. If clouds gather, great efforts are made to get in the crops, as hail is much feared at this season ; but hail is very uncommon in this district. As soon as the grain is cut it is stacked. The reaper gets his share when the crop has been thrashed and divided."

Thrashing
681

"There are several ways of thrashing. The most common is to yoke a number of bullocks together, fasten the one at the left hand of the line to a post, round which the straw to be thrashed is piled, and drive them round and round from right to left. Wheat and barley are, however, first thrashed with the *phalha,* or thrashing frame.

CULTIVATION in the Panjab.

Thrashing.

"A pair of bullocks are yoked to the *phalha* and driven round the stake about which the straw is heaped; there may be several *phalhas* at work one after the other, but there are never more than four. One man is required with each, and a couple more to throw back the straw into the heap. One pair of bullocks with the *phalha* will thrash the produce of a quarter of an acre a day. They will work eight hours at a stretch in the sun. When wheat or barley has been thrashed with the *phalha,* the straw is shaken up with the pitchfork and thrown on one side, while the grain falls to the bottom.

"In the Bannu district, cows, and even donkeys, are used on the thrashing-floor. In Miánwáli thrashing is frequently done by bullocks drawing a weighted branch of some thorny tree over the outspread stalks. The floors are generally prepared by being well beaten, and on the hills are carefully paved, the circular thrashing-floor near each Himálayan homestead forming a striking feature of the scenery. In spite of every care, the dirt from the floor becomes mixed to a certain extent with the grain, and, moreover, the grains are often seriously injured. Thrashing is carried out as rapidly as possible, the owner generally sleeping beside his grain at night till it is all thrashed out."

The grain is separated from the chaff by being thrown up by long wooden shovels, the hot winds which prevail at the time readily blowing the dry chaff to a distance, while the grain falls on the thrashing floor. This is repeated till the desired degree of cleanliness is attained.

Winnowing.
682

STORING.—Wheat and other grains are stored in rooms of the cultivator's house, in large jar-like vessels made of mud, or wicker lined with mud, in large canvas bags called *théka,* which may hold as much as 50 to 100 maunds, or on prepared platforms in the open, carefully covered over and surrounded by a trench or hedge. Storing on the earthen floors of rooms or in mud vessels naturally tends to increase the amount of impurity.

Storing.
683

YIELD.—An average produce estimate is, as already stated, an impossibility, since the outturn must vary greatly with climatic and other conditions. It has been stated that "one year with another it is probably rash to expect more than 5½ maunds an acre from unmanured rain lands, 7½ from manured rain lands, and 10 to 14 maunds on lands manured and irrigated. Of course the greater certainty of the crop on the last class increases its comparative value over a long series of years. The yield on *sailab* lands varies very greatly. The average harvested is, very roughly speaking, 6½ maunds on a series of years. The crops are generally more secure than those on *baráni* lands.

Yield.
684

In the final wheat crop report for 1888-89 the outturn is estimated to have been 2,30,05,631 seers, or an average on the total estimated area (not actual, taken from the forecasts) of 332 seers=8·3 maunds to the acre. The average for lands irrigated by canals was 386 seers=9·65 maunds, for lands irrigated by wells, 435 seers=10·8 maunds, for flooded and alluvial land, 303 seers=7·5 maunds, and for dry land dependent on rain, 222 seers =5·55 maunds. The total outturn was somewhat over the average of the preceding year, *viz.,* 302 seers=7·55 maunds to the acre. The districts with the largest outturns were Shahpur, 11·27 maunds; Ludhiana, 11·2 maunds; Jhelum, 10·5 maunds; Jhang, 10·4 maunds; Amritsar, 10·2 maunds; and Dera Ismail Khan, 9·7 maunds to the acre. Hissar, Simla, Kangra, Kohat, Bannu, and Dera Ghasi Khan were much below the average, with yields varying from a little under 5. to 6 maunds. These figures, with the exception of those for Hazára and Simla, represent a crop

TRITICUM sativum.	Cultivation of Wheat

CUTLIVATION in Sind.

much above the average in two districts, above the average in sixteen districts, average in eleven districts, and below the average in no district.

SIND.

SIND.
685

References.—*Director, Land Rec. & Agri., Bombay, Reports ; Reports of Hyderabad Experimental Farm; .Govt. of Ind., Wheat Product & Trade in India, 1878, 1883, 1886 ; Forbes Watson, Rept. on Indian Wheat, 1879; Gazetteer of Sind.*

Very little of a special character can be said regarding the Sind wheats and wheat cultivation. In every feature Sind may be said to be intermediate between Bombay and the Panjáb. In certain parts of the country the methods of cultivation, the nature of the soil, and the character of the wheats are similar to those in the Panjáb, but in other parts of the province an approximation is seen to the wheats of Northern Bombay. The Sind wheats are generally pronounced superior to those of Bombay, and possess a larger proportion of soft white forms. The delta wheats are, however, specially liable to rust. Most of the Sind wheats are, as in the Panjáb, repeatedly watered or flooded during their growth. A dry crop (*see* the remarks under Bombay and Central Provinces) is, however, raised on lands that are inundated during the rains. On the water subsiding, these *band-baráni* soils are repeatedly ploughed, and the crop sown, no further watering being necessary.

Area.
686
Conf. with p. 94.

AREA.—The area under wheat in this province shows little alteration during the past sixteen years, and has, if anything, fallen off. Thus, during the five years ending 1877-78, it averaged in British districts 273,000 acres, in that ending 1882-83, 297,000 acres, and in that ending 1887-88, 227,000 acres, while in 1888-89 it amounted to only 234,483 acres. The average of four years from 1884-85 is 249,512 acres. The distribution of the area during the past year was as follows :—

District.	Irrigated.	Unirrigated.	Total.
Karáchi	28,554	...	28,554
Hyderabad	29,324	2,343	31,667
Shikárpur	138,811	...	138,811
Upper Sind Frontier	14,458	...	14,458
Thar and Párkar	Not available.		20,993
GRAND TOTAL .	...	...	234,483

In addition to the above, 32,438 acres are said, in the final crop report for 1888-89, to have been under wheat in the Native States of Khairpur, or a little under the average. The areas in Karáchi, Shikárpur, and the Upper Sind Frontier show a decided falling off, due to low inundation and scanty winter rain. In Hyderabad and Thar and Párkar the area was over the average of the preceding five years.

Races.
687

RACES.—The varieties of wheat grown on the Hyderabad farm, Sind, have included most of the commoner kinds grown in the province, and have been classified as follows by **Mr. Strachan,** the Superintendent

Class I.—Soft White.

Soft White.
688

Popri.—Flat, broad, short, club-like awnless heads, with roundish grain, short straw, and white husk.

Thori or Bhávalpuri.—Long, loosely packed, nearly square, dark-brown awnless heads, with round long grain, long straw, and rough brownish dark yellow husk.

T. 688

| in Sind. | (*J. Murray.*) | **TRITICUM sativum.** |

Jabalpuri.—Long heads, with loose spikelets, short awns, long and large grain, white husk, and long and strong straw.

Races.

Broach.—Loosely packed, slightly bearded heads, of medium height, with large Khano-like grain, white husk, and luxuriant straw.

Soft White.

Sind Soft White.—Rori-like square awnless heads, with white or cream-coloured husk, and long, thin, and weak straw..

Australian Purple Straw.—Thori-shaped, but more loosely-packed heads, with white straw, and sheath of peculiar brownish tint.

Tuscan.—Long, loose, roundish heads, with short awns from the upper spikelets, large grain, white husk, and 2¾ to 4 feet long straw.

Essex.—Slender loose heads with short awns on the upper spikelets, Thori-like grain, white husk, and weak long straw.

Class II.—Soft Red.

Soft Red.
689

Akola.—Medium-sized heads, with loose spikelets, short awns, few medium-sized grain, white husk, and ordinary-sized straw.

Ashby's Prolific.—Square heads, with loose straggling spikelets, only some heads having short awns from the upper spikelets. The husk is white, grain large, and straw short and strong.

Gerri.—Long red heads, with spikelets far apart and short awns, very much resembling the quills of a porcupine in bad humour. This variety has very dark, medium-sized grain, and very strong straw.

Gandio.—Loose heads with short awns, yellow or cream-coloured husk, and grain as big as that of *Thori.* A poor variety.

Class III.—Hard White.

Hard White.
690

Rodi.—Flat club-like awnless heads, with yellow brown husk, small grain, and strong straw. In this variety the two rows of grain on the two sides of the rachis widen or get broader towards the top of the spike.

Rari or Rari-Hidi.—Long, square awnless heads, with short thick grain, white husk, and long yellow straw.

Nágpuri.—Loosely packed and sparsely awned heads, with large grain, thick at the lower end, and tapering upwards to a point. It has white husk and medium-sized straw.

Káhno.—Closely packed, long, flat heads, with mostly black 6 to 8 inches long awns, strong white straw, white husk, and very large, long and thick grain.

Telhi or Maccain or Khudian.—Short, flat, 4-rowed, club-like awnless heads, with round grain as big as *juári* and straw occasionally purple, but oftener white. It has small white husk, and is a good dwarf variety.

Bakshi or Bombay Hard White.—Except perhaps in the colour of awns which is not a constant character, this variety is the same as the Sind variety called *Káhno.*

Class IV.—Hard Red.

Hard Red.
691

Bombay (no name).—Square, loosely-packed heads, with short, stubby white or light yellow awns, white or light yellow husk, and ordinary straw.

Pumban.—Flat 4-rowed heads, with long dark-coloured awns, and white or slightly yellow straw. The sheaths come nearly up to the head, and the grain is very large and difficult to be removed from the rachis or freed from the shell. It is a very strong variety.

Spelt Wheat.
692

The last variety *pumban*, is "spelt," and was probably grown from seed imported from Bombay (see pp. *100, 134, 135*).

Conf. with pp. 100, 134.

Outturn.
693

OUTTURN.—The estimated outturn for 1888-89 given in the Final Crop Report is 90,000 tons or 25,40,000 maunds, equivalent to an acre yield of 9·1 maunds. As the return was calculated, however, on the areas estimated

CULTIVATION
in
Sind.

Crop
Experiments.
694

Central India
and
Rajputana.
695

Area.
696
*Conf. with
p. 94.*

Outturn.
697

for the crop forecasts, which was about 50,000 acres too large, the actual acre outturn may also have been overestimated. The average acre outturn for the four years ending 1887-88 amounted to just a little under 8 maunds.

CROP EXPERIMENTS.—A long series of experiments has been carried out at the Hyderabad farm for the purpose of ascertaining the suitability of several foreign wheats and wheats from other parts of India for cultivation in Sind, also to ascertain the relative merits of different rotations and of the **Lois Weedon** system of alternate fallows. English and Australian wheats have been thoroughly tested, but with very poor results, and in 1887-88 the Director of Land Records and Agriculture remarks :—" It is questionable whether there is any use in attempting the growth of foreign wheats. The Indian varieties supply ample material for improvement."

Though the **Lois Weedon** system has shown its superiority over the rotation and continuous systems on the Bhadgaon Farm (see p. 133), it has not done so at Hyderabad, where wheat in rotation has all along occupied the first place. A large quantity of hand-picked selected seed has been issued to cultivators in the province with excellent results.

CENTRAL INDIA AND RAJPUTANA.

References.—*Agri. Statistics, Br. Ind., for several years; Govt. of Ind. Wheat Production & Trade in India, 1879, 1883, 1886; Rajputana Gazetteer, 96, 128, 254, 255, 279.*

There is little occasion to dwell upon this province. In climate and soil it closely approaches to the Panjáb, and its wheats are, accordingly, similar. The Commissioner of Ajmir-Merwara writes that the Natives invariably select the best lands for their wheat, generally that in the neighbourhood of a tank or well, from which it may be irrigated. The soil is of a light, sandy loam, unlike the stiff loams on which wheat is grown in England.

To obtain a full crop, the land is fallowed during the rainy season (June to September); during this period it is ploughed two or three times a month to a depth of 4 inches. At the close of the rains a heavy plank is drawn over the field, which serves the purpose of a roller in pulverising the surface, and also prevents the moisture escaping. The sowing season begins about October 25, and lasts till the end of November, the crop being reaped in April. The quantity sown is about 2 bushels to the acre, and, if manured and irrigated, the yield is about 34 bushels. When unmanured and unirrigated, the yield is perhaps not more than 7 bushels. If no winter rain falls the crop is irrigated three or four times.

AREA.—The average area under the crop during the four years ending 1887-88 is returned as 1,542,000 acres in Rájputána; 15,000 acres in Ajmir, and 2,617,000 in Central India, or a total of 4,174,000 acres. In 1888-89, the area in Rájputána is returned at 1,641,994 acres, an increase on the average, while those in Ajmir and Central India are estimated to be average.

OUTTURN.—The average outturn for the same four years is estimated at 1,08,92,000 maunds for Rájputána, 1,06,400 maunds for Ajmir, and 1,44,20,000 maunds for Central India, or a total of 2,54,18,400 maunds. These figures represent an average acre yield for the whole of Rájputána and Central India of a little over 6 maunds to the acre. In 1888-89, the figures estimated for Ajmir and Central India are the same as those of the average, but though the area in Rájputana is returned as higher the outturn is lower, *viz.*, 1,06,84,200 maunds.

BOMBAY.

References. — *Gazetteers :— II., 59-65, 269, 273, 277, 280, 284, 287, 291., 295, 390, 405 406, 536, 538, 541, 554; IV., 54; V., 105, 106, 294, 369-371; VI., 38, 39; VII., 77, 81, 89, 95; VIII., 175, 198; X., 144-153, XII., 145, 150, 222; XIII., Pt. I , 286-289; XVI., 91, 95-99; XVII. 241-258, 265-267; XVIII., Pt. II , 38, 39; XIX., 159-163 ; XX., 219, 237 ; XXI., 246, 252; XXII., 266-268 270-275; XXIII., XXIV, 316-321; 160 165; Govt. of Ind., Wheat Prod., & Trade in India, 1878, 1883, 1886; Forbes Watson, Report on Indian Wheats, 1879; Reports, Director of Land Records & Agriculture, annually in many passages; Experimental Farm Reports, Bombay, annually; Bombay, Man. Rev. Accts., 101.*

Though the figures of area given in the table at p. 94 show that wheat cultivation is rapidly expanding in this province, still, the crop is, in comparison with wheat in the Panjáb and North-West Provinces, but of secondary importance. The millets and pulses are infinitely more important. Thus the two principal species of the former class of food grain, *viz., jówári* and *bájri*, occupy more than six times the area of wheat, while the pulses collectively occupy about an equal amount of land. The cultivation of wheat is, therefore, naturally of less importance than in Northern India and receives a minor amount of care and attention.

Soils.—Dr. Watt writes :— " The soils of Bombay are much more diversified than in the Panjáb. Sind and certain parts of Bombay bordering on Sind and Central India possess almost identical soils to those we have described, light loams with a tendency to run into a superabundance of sand. But in many parts of Bombay a heavy red soil prevails, containing iron, and in other districts a heavy black soil which gradually approximates to the black cotton soil more immediately characteristic of the Central Provinces. Selecting a representative district for each of the divisions Gajárat, Deccan, Karnátik, and the Konkan, the following abstracts from the Gazetteers and other reports will give a general conception of the soils of Bombay :—

" In the *Broach* district the soil is said to consist of two kinds, a light soil and a black soil; but each of these types of soils is capable of sub-division. The light soil, *gorát, gorádu*, or *márwa*, varies from sand-drifts in the south to the richest alluvial loam, *bhálka*, found in the neighbourhood of the Narbada. So in a like manner the *káli*, or black soils, range from the rich alluvial deposits of the Narbada, the regular deep cotton mould, *kánam*, to the shallower and harsher soils, *bára*, near the sea-coast, on which little else but wheat can be grown. These black soils occupy more than three-fourths of the cultivable area.

" In *Nasik*, as representing the Deccan; land is primarily classed as hill land, *dángi*, and plains, *deshi*: The former are poor and wholly dependent on the rains for moisture, and, excepting the portions devoted to rice, the remainder cannot be cultivated for two years consecutively. Of the plains land there are said to be four kinds : black, *káli*; red, *mál*; red and black, *korál*; and light brown; *barad*. Except in the uplands, black soil is deep and very rich, and yields excellent cold-weather crops of wheat and gram. Red soil is found chiefly on hilly undulations, and yields good rainy season crops. The mixed red and black and the light brown soils are much inferior to the others, and often yield no crops at all when the rain is scanty.

" In the *Belgaum* district of the Karnátik, there are said to be two soils, red and black. The red soils are primary soils—that is, they are the direct result of the decomposition of the iron-bearing rocks. This soil is generally found all along the western border; but it occasionally occurs in the plains country. The black soils are secondary soils—that is, they are rock ruins changed by the addition of organic matter. The black soil

CULTIVATION
in
Bombay.

Bombay.
698

Soils.
699

Broach.
700

Nasik.
701

Belgaum.
702

TRITICUM sativum.	Cultivation of Wheat

CULTIVATION
in
Bombay.
Area.
703
*Conf. with
p. 94.*

covers most of the plains country, and is best suited for the growth of cotton, Indian millet, wheat, and gram.

AREA AND IRRIGATION.—The areas occupied by wheat, in the British districts of Bombay, during the year 1888-89 were as follows :—

District.	Irrigated.	Unirrigated.	Total.
I. *Gujarát.*	Acres.	Acres.	Acres.
Ahmedabad	40,288	143,369	183,657
Kaira	11,870	19,167	31,037
Panch Mahals	422	577	999
Broach	112	15,079	15,191
Surat	20	11,670	11,690
II. *Deccan.*			
Khándesh . . .	14,990	318,129	333,119
Násik	28,562	349,711	378,273
Ahmednagar	31,114	246,018	277,132
Poona	24,556	92,950	117,506
Sholápur	31,722	21,441	53,163
Sátára	24,571	44,732	69,303
III. *Karnátak.*			
Belgaum	2,797	121,802	124,599
Bijápur	1,894	204,005	205,899
Dhárwár	24	276,711	276,735
IV. *Konkan.*			
Thána	...	144	144
Kolába			
Ratnágiri	} No wheat grown.		
Kánara			
TOTAL .	212,942	1,865,505	2,078,447

It is an interesting fact in connection with the above that only 144 acres are grown in the Konkan, and that this should consist of 143 acres cultivated during the *kharíf* season, and 1 during the *rabí*. Throughout the rest of the presidency wheat is, as elsewhere, a *rabí* crop, with the exception of 2 acres in Broach returned as under wheat in the *kharíf*.

The area during the year was considerably lower than that of 1887-88, and about equal to the average for the four years ending with that year, *viz.*, 2,037,281 acres. The decrease occurred chiefly in Gujárat and the Deccan, while in the Karnátak a general increase occurred. The Officiating Director of Land Records and Agriculture, in his report for the year, remarks on this subject :—"The continuous increase in wheat in the Karnátak, followed by a corresponding decrease in cotton, give grounds for a belief that, as noticed in last report, wheat is here probably displacing cotton,—a result partly attributable to the increased demand for the staple for export, and partly to the facility afforded for export by the introduction of the Southern Mahratta Railway."

In addition to the area under wheat in British Districts, 600,975 acres (taken as 601,000 in the general table of area) are returned as having been devoted to the crop in Native States. The shares in each were,— Baroda, 97,129 acres ; Kathiawar, 223,047 ; Cutch, 44,550 ; other Gujárat States, 91,156 ; Satara jagirs, 21,248 acres; Akalkot, 6,003 ; Kolhapur, 14,483

acres; other Southern Mahratta States, 103,359 acres. This estimate is
considerably over that for 1887-88, and also exceeds the average adopted
for previous years, *vis.*, 591,000 acres. In Baroda nearly nine-tenths of
the crop, in Kathiawar nearly two-thirds, and in Cutch the larger proportion,
is irrigated.

RACES.—The wheats of Bombay may be said to be characterised by a
greater degree of hardness than those of the north of India, and are,
therefore, as a rule, less suited to the English market. They, however,
contain a large amount of gluten and are admirably suited to the Southern
European market, where they are employed in making maccaroni. The
greater numbers of the many kinds distinguished by separate vernacular
names, belong, in all probability, to the hard white commercial class.
Since the period when attention was first directed to India as a possible
wheat-producing country to meet the European demand, endeavours have
been made, with a certain amount of success, to introduce the soft white and
soft red wheats to a greater extent. The early reports of experiments with
English pedigree wheats are little more than records of failure, but the
results of experiments made during the past five years with seed from the
North-West Provinces are much more satisfactory. These trials were
conducted at the Bhadgaon Experimental Farm, where it was found that
the northern stock is much more prolific, yields more straw to an equal
amount of grain, and produces a heavier crop than *bansi*, the common
race cultivated in the district.

It has long been held by native cultivators all over India that the colour
and consistence of wheat are more dependent upon climate, soils, and
surrounding conditions in general, than on the original stock from which
the race is derived. Thus in many official reports from the Panjáb, North-
West Provinces, Central Provinces, and Bengal, as well as from Bombay, the
statement is commonly made that a soft white wheat, removed from óne
locality to another in which the grains grown are hard or red, tends to change
its physical characters, to become harder and to turn in colour. These
observations have been confirmed by experiments at the Bhadgaon Farm,
where it has been found that soft wheats, from whatever source introduced,
showed a sure tendency to harden, and the white wheats to become red, that is
to say, they assumed to a certain extent the characters of the crops prevalent
in the district. As a dry crop wheat, Jabalpur seed of the soft white class
was found to succeed best, but as an irrigated wheat, the hard red of the
district took the first place. Up to 1885-86, seed from other parts of India
was introduced, grown, carefully hand-picked, and distributed to cultivators, but the hand-picking was found to be costly, the outturn for at
least the first year or two was very small, the tendency to change in type
to that of the ordinary crop was marked. and, as a consequence, general
distribution was abandoned. From the reports of the last two years,
however, it would appear that the introduction of North-West Province
and Central Province white soft wheats has been more encouraging, and
that, even if the grain change in character, the mixture of acclimatized
seed of these kinds with that of the district is productive of benefit.

An extensive literature on wheat experiments exists in the Reports of the
Director of Land Records and Agriculture for the province, but space for-
bids more than a most cursory consideration of them. The result of introducing foreign grain has been briefly sketched above; in addition, it may be
noticed that, at the Bhadgaon Farm, the Lois Weedon system of interrupt-
ed fallow has been found to yield much better results than either the
continuous or the rotation systems, both in unirrigated and in irrigated
plots.

It is impossible to give a list of all the races distinguished in the

CULTIVATION
in
Bombay.

Races.
704

*Conf. with
p. 130.*

TRITICUM sativum.	Cultivation of Wheat

CULTIVATION in Bombay.

Races.

various districts of the presidency by distinct vernacular names, nor indeed would such an enumeration prove of much practical value. All four classes are represented, but, as already stated, hard wheats are predominant. The great majority of those submitted to Dr. Forbes Watson would appear to have been hard white and hard red, since, out of 139 samples, 60 belonged to the former, and 61 to the latter, while only 13 were soft red and 5 soft white. The accompanying list of those selected from the samples as of greatest commercial value may be of interest :—

CLASS.	District.	Name.	Weight per bushel.	Price per quarter.		REMARKS.
HARD WHITE	Bhownuggur	Hasia	60½	40	6	Long berried.
	Khándesh	Kali Kusal	61	42	6	Ditto.
	Ditto	Bansi	62½	42	6	very fine.
	Násick	Yellow Banshi	57	40	0	Finest of all, but weevilled.
	Punch Mahals	Daudkhani	62	42	0	Large berried.
	Ditto	Kathe Malvi	61	40	6	
	Dhárwár	......	58	40	6	
	Poona	Bakshi Gahu	62	41	0	Large berried, finest kind grown.
	Sátára	Buxi	60	40	6	Very fine.
SOFT RED	Bhownuggur	Vajia	58	40	6	Semi-hard.
		Patalia	58	40	6	Ditto.
		Ditto	59½	40	6	Ditto.

Spelt Wheat.
705
Conf. with pp. 100, 129, 135.

One of the most interesting, though perhaps least important commercially, of Bombay wheats, is the "spelt," which, there is now no doubt, is regularly cultivated, and probably constitutes the whole of the *kharíf* crop mentioned above. Dr. Watt drew attention to this form in his paper frequently quoted above. " In nearly every report," he writes, " a form of wheat known as *khaplé* is described as a wheat that requires much watering. There seems little doubt from the brief descriptions that have appeared of this wheat that it is a form of spelt-wheat. We have seen spelt-wheat sent from the mountains of South India, but have always suspected that it may have probably been a modern introduction. Here, however, there would appear to be no grounds for such an opinion. It is grown all over the Western Presidency, and it is quite possible its area of cultivation may extend to Southern India " After commenting on what has been shewn above in the chapter on **Habitat**, *viz.*, that this question has an important bearing on the theories generally held regarding the origin of wheat, Dr. Watt continues : " By way of showing that there is at least a strong probability that the *khaplé* wheat of Bombay is a form of spelt, we may reproduce one or two passages regarding it. In the *Poona Gasetteer* the following occurs :—'*Kaphlé* is the wheat usually grown in gardens. It is

Khaple.
706

very hardy. It owes its name to the fact that the grain cannot be separated from the husk without pounding. It is sown as a second or *dusota*
crop in January or February on irrigated land after *bájri*, maize, tobacco,
chillies, or wheat, with good results.'

"We have here in itself a fact of very considerable interest—namely,
that, as with rice, we *do* actually possess in India a wheat that may be
grown as an early *kharif* crop. Were there no other points of attraction
this alone is well worthy of being followed up and put to a final test. It
is much to be regretted that, while volumes have been written upon every
side issue of the wheat trade, no scientific investigation has been instituted into the subject of the varieties of wheat grown in India. Such an
inquiry would doubtless lead to decided advances towards establishing
the reasons for their peculiar adaptabilities. With such a knowledge, it
would not be necessary to grope so much in the dark in the matter of
efforts to introduce better varieties from one part of India to another.
We have not, however, at present the means at our disposal to verify the
suggestion contained in the above explanation of the *khaplé* form of
Bombay wheat, and as our readers may not have access to the numerous
records in which brief passages occur regarding it, we may extract one
or two more passages.

"In the *Ahmednagar Gazetteer* it is stated : '*Khaplé*, also called *jod*,
is very hardy; but requires pounding to separate the husk.' Of Kolhapur
it is said, '*Khaplé* is largely grown in watered lands as a crop alternately
with sugar-cane. The grain is coated with an adhering husk, which cannot be separated without pounding ' "

The above supposition that *khaplé* is **T. speltum** has been confirmed
by Mr. E. O. Ozanne, Director of Land Records and Agriculture in
Bombay, who, in a letter to the Collector of Hyderabad, dated August
1837, writes, " Considerable pains were taken to differentiate the varieties—local and imported – of wheat. The variety called *pamban* is
clearly the *khaplé* or *jod* of the Deccan, and is **T. speltum** " (*Conf.* Sind
p. 1229).

METHOD OF CULTIVATION.—The system pursued in growing the finer
wheats is briefly conveyed in the following extract :—*Bakshi* is the best
kind of wheat raised in the Deccan and Southern Máratha country. It
is either black-bearded or straw-colour-bearded. The grain is large and
hard and contains a large proportion of gluten. This wheat, not being
hardy, is not largely cultivated. The land is ploughed twice, once length-
and once cross-ways, with a six- or eight-bullock plough, according to the
nature of the soil. The land is then harrowed six times, thrice with a
four-bullock harrow and thrice with a two-bullock harrow, and then sown
with wheat. This is all that is considered necessary. It is not customary
to raise wheat on the same lands annually. The rotation generally adopted on dry crop land is as follows: first year *jowari*, second year *bájri*,
third year wheat. On garden lands two crops are annually raised as
follow :—

1st year.	2nd year.	3rd year.
1st crop *bájri*.	1st crop *bájri*.	1st crop *bájri*.
2nd crop wheat.	2nd crop gram.	2nd crop wheat.

Instead of wheat or gram for a second crop, onions, potatoes, etc., are
sometimes raised.

This system may well bear comparison with the careful methods pursued in the Panjáb and Northern India generally, but the method followed in Bombay is frequently of a much more careless nature. Thus it
is reported of Khandesh :—" Before sowing with wheat, the ground is

Marginal notes

CULTIVATION
in
Bombay.

Races.
Khaple.

Scientific
investigation
necessary.
707

Spelt wheat.
Conf. with
pp. 100, 129,
134.
708

Method.
709

TRITICUM sativum.	Cultivation of Wheat

CULTIVATION in Bombay.

Method.

never ploughed, only three or four times laid open with the hoe to the sun, rain, and wind. If the ground is so damp that the clay sticks in balls sowing begins in October or November, and in some of the Tapti Valley districts as early as September. The allowance of seed is from forty-five to seventy-five pounds an acre. A shower or two when the crop is shoot-ing is useful, though by no means necessary. With cool seasonable weather and heavy dews, wheat flourishes without rain."

One finds reports of similar different systems in each district through-out the Presidency, according to the kind of wheat grown. Thus in Ahmedabad it is reported that the finer kinds known as *chasia* are grown in light black soil, which is kept fallow and ploughed four times before the seed is sown. Sowing is commenced in the end of October and the harvest is in April. No crop precedes or succeeds it, but occasionally it is used as a substitute for cotton when that crop fails. The inferior kinds, *wadina* or *wajia*, are sown on irrigated light sandy soil, following rice, *jowári* or *bajrí*. One hundred and sixty pounds are required to sow an acre, while in the case of *chasia* 84 are deemed sufficient. The crop is sown in December, and fewer ploughings are given. In Kaira three sorts are said to be grown, *daudkhana* or *dudhia*, *dhola* or *kathia*, and *bhalia* or *wajia*. The first, a very superior soft white wheat (*Conf.* pp. 197-98), is grown on rich black soil only, the second is an inferior hard red or white grain, the last a mixture of the two. *Dudhia* we find again is cultivated with great care; the ground is allowed to lie fallow before and after the crop, it is manured if necessary, and ploughed from three to ten times. In the Panch Mahals wheat is generally sown as a second crop after rice or maize, and its rearing receives very little care at the hands of the culti-vator. In Broach, on the other hand, a system of alternate fallow is fol-lowed and manure is sometimes used. In Reports on Nasik it is stated that in certain parts of the district wheat follows *bajrí*, *kulthi*, or linseed, occa-sionally it is grown on dry crop land which is manured for it, in other localities it is grown on manured garden lands. In the latter case it fol-lows *konde*, and *tag*, hemp (*Wheat Prod. & Trade of Ind.*, 1879).

Diseases.
710

DISEASES.—Wheat in Bombay as in other localities in India, is sub-ject to the attacks of rust, known in this province as *geru*, *gerwar*, or *jeru*. The cultivators state that it attacks crops only when they are planted on irrigated land, and that it is favoured by showery or cold weather during the growth of the plant. *Chasia* wheat is said in the reports on Ahmed-abad to suffer from frost, *kapadi* (an insect pest), and other enemies (*Conf.* Fungi and Fungoid Pests, *III.*, *457*, also **Pests, Insects,** *VI.*, *Pt. I.*, *145*).

Yield.
711

YIELD AND PROFIT OF CULTIVATION.—The total production for 1888-89 is, in the Final Crop Report for the province, estimated on an assumed acreage of 2,654,342 (considerably above the actual) to have been 588,472 tons, or 1,64,77,216 maunds. Of this amount, 298,492 tons were estimated to be produced from dry, 289,980 tons from irrigated, crops. The total average outturn on these figures would be 6·2 maunds to the acre; the average for irrigated lands 13 maunds, for dry lands only 4·1 maunds to the acre. This outturn cannot, however, be accepted as typical, since it had decreased on that of the former year, in all districts and states ex-cept the Gujárat States. The diminution was especially marked in the Deccan, where the total yield amounted to only ⅓ of the former year's produce. It is, however, probable that the figures returned are consider-ably lower than the actuals, at least for dry land crops.

Profit.
712

The cost and profit of cultivation were worked out by several experi-ments in 1872. In six of the experiments made in good and over average soils, it was found that, without irrigation or manure, an acre yielded from 420 to 1,476 pounds. This outturn, calculated at prices about twenty

in the N.-W. Provinces and Oudh. (*J. Murray.*)

TRITICUM
sativum.

five per cent. below current market quotation at the time of the experiment, gave the following results :—

Statement showing the result of Wheat cultivation.

CULTIVATION
in
Bombay.

Profit.
713

COST OF CULTIVATION IN RUPEES.				OUTTURN PER ACRE IN POUNDS.		Value of crop per acre.				Net profit.			
Seed.	Ploughing to harvesting.	Rental.	Total.	Grain.	Straw.								
R	R	R a. p.	R a. p.			R a. p.	£ s. d.		R a. p.	£ s. d.			
2	6¼	4 0 6	12 4 6	432	368	20 11 4	2 1 5		8 6 10	0 16 10¼			
2	6¼	3 2 6	11 6 6	416	336	16 7 0	1 12 10½		5 0 6	0 10 0½			
2	6¼	3 11 0	11 15 0	1,476	1,846	59 15 4	5 19 11		48 0 4	4 16 0½			
2	6¼	5 12 6	14 0 6	620	1,104	30 2 0	3 0 3		16 1 6	1 12 2¼			
2	6¼	4 1 0	12 5 0	420	560	19 4 0	1 18 6		6 15 0	0 13 10¾			
2	6¼	5 14 6	14 2 6	684	880	31 2 8	3 2 4		17 0 2	1 14 0¼			

NORTH-WEST PROVINCES AND OUDH.

N.-W. P. &
OUDH.
714

References.—*N.-W. P. & Oudh, Duthie & Fuller, Field & Garden Crops, 1-8; Atkinson, Him. Dist., 321, 684; Settlement Reports :—Aligarh, 37, 47, 48 ; Allahabad, 31; Bulandshahr, 32, 33, 50, 55 ; III., 3, 6, 21 ; Bareilly, 26, 54, 59, 60, 61, 65, 66, 70, 71, 72, 78, 82, 100, 101, 102, 103, 162, 167, 170, 173; VI., 2, xiv., xv. ; App., A., 144, 145, 146, 147 ; VII., 2, 9, 10, 11 ; X., 49, 50, 70, 71 ; Lullutpur, 2, 3, 22, 23, 26, 140 ; XI , 5, 11, 17 ; Fatehpur, 5, 16, 17, 18, 19, 20, 21, 23 ; Basti, 1, 7, 26 ; XIII., 119, 120 ; XIV., 53, 54, 57, 58, 59 ; Agra, 6, 65 66, 67, 68, 69, 71, 72, 73, 74, 75, 76, 77 ; Hamirpur (Review), 1, 6, 8, 9, 27 (Rep.), 4, 75, 77, 79 ; Oudh, Lucknow, 77, 78, 80 ; Bara Banki, 39, 41, 44, 45 ; Sitapur, 19, 22 ; Gonda, 106, 107, 129 ; Bharaich, 150, 151 ; VI., 15, 20, 21, 22, 23 ; Gazetteers :—N.-W.P., I., 88, 90, 91, 94, 150, 151, 152, 251, 315, 316, 317 ; II., 167 ; III., 24, 26, 29, 30, 225, 226, 233, 234, 240, 243, 463 465, 483, 487, 709 ; IV., 27, 248, 249, 250, 251, 258, 259, 524 ; V., 26, 27, 542, 563, 564, 565 ; VI., 27, 29, 32, 66, 138-139, 148, 152-153, 201, 214, 231, 238, 245, 247, 256, 266, 303, 325-326, 329, 332, 334, 343, 411, 462, 477, 484, 503, 514, 533, 539, 548, 558-559, 587, 591, 593, 598-599, 603, 608, 646, 709, 704, 735, 745, 754, 762, 769, 775, 780 788, 792; F. N. Wright, Rep. on the Wheat Cultivation and Trade of the N.-W. P., 1878 ; Statement of Irrigation Operations of the N.-W. P., Aug. 1882; Revenue & Agricultural Dept., Memorandum on the Wheat Production of the N.-W P. & Oudh, Dec. 1884.*

Wheat cultivation in these provinces has attracted much attention and formed the subject of many useful reports and notes. In addition to the information contributed by various officers to the volumes of the Government of India on Wheat Production and Trade, the subject of cultivation may be found fully dealt with in the work of **Messrs. Duthie & Fuller** on *Field and Garden Crops*, while the question in all its bearings formed, in 1884, the subject of an elaborate report by the **Honourable Mr. W. C. Benett**, at that time Director of Agriculture and Commerce for the Provinces With such an extensive and complete literature already existing, it is perhaps unnecessary to go into the subject at any great length in this work. We shall, however, extract from the sources above enumerated some of the more noteworthy facts, and, when necessary, bring the chapters on area, production, and trade up to date from more recent publications.

SOILS.—Wheat is grown in almost every soil, except the very lightest sand; a rather heavy loam is considered best suited for it. In fact, what has already been said about the Panjáb wheats applies in its full force to those of these provinces The fields of loamy soil (*domat*), which cover a large portion of the Doáb, even when mere isolated patches in the midst of

Soils.
715

CULTIVATION
in the
N.-W. P. &
Oudh.

Solls.

ushr plains, are especially productive and suitable for wheat. Manure is applied to the better class of wheat-fields generally every second or third year, though in quantities which would sound ridiculously small to the English farmer, 4 tons (=100 maunds nearly) being about the average. Land is occasionally prepared by herding sheep in the fields. This same practice prevails in the Panjáb, and a case is recorded of a prosecution because a flock of sheep, which for years had herded on a particular farm, were by the owner taken to a neighbouring farm instead. A curious habit also prevails in Northern India of herding sheep, and even cattle, on the field while the crop is sprouting so as to top-manure the soil and cut down too rapid growth.

Area.
716
*Conf. with
p. 94.*

AREA.—In the general table of area for all India the figures given in the case of the North-West Provinces and Oudh are mere estimates, based on the degree of cultivation in later years and probably considerably over-estimate the actual area. From this cause cultivation shews very little development in the case of the North-West Provinces, though in Oudh it is shown to have undergone considerable expansion. Owing to the want of figures previous to 1879, it is impossible to make an actual comparison, but **Mr. Benett**, in the paper above referred to, believes the expansion even in the North-West Provinces to have been rapid and extensive after that year. "Since 1879," he writes, "the area under wheat has been steadily increasing, and there is nothing to show that its limit has even nearly been reached. The next table compares the areas under each class of cultivation in 1879 and 1883, in the North-West Provinces only; no comparison is possible for Oudh :—

	Pure wet.	Pure dry.	Mixed wet.	Mixed dry.	Total area under wheat in all districts except Kumaon Division.	Total area under all crops in 28 districts.
	Acres.	Acres.	Acres.	Acres.	Acres.	Acres.
1883 . .	2,223,634	1,541,297	756,278	1,700,011	6,221,220	21,334,803
1879 . .	1,756,876	1,517,691	640,954	1,449,995	5,365,516	20,402,718
Increase in five years . .	446,758	23,606	115,324	250,016	855,704	932,085
Percentage increase .	27	2	18	17	16	5

"The first fact brought out by this table is that in the North-West Provinces alone, excluding Oudh, nearly a million acres have been brought under wheat cultivation within the last five years. And this has not been to the detriment of other crops, for we find that the increase in the cultivated area in twenty-eight out of the thirty-four districts has been more than the increase under wheat only. Then we find that, although the area under crops has increased, the main increase has been under wheat, which is the crop which requires the most careful cultivation, the increase having been 16 per cent. in the case of wheat, while it has been only 2 per cent. in the case of all other crops, or nearly eight times in the first case what it has been in the second. Finally, more than half the increase under

wheat of all kinds, and nearly all the increase under wheat sown alone,   **CULTIVATION**
has been on irrigated lands, or with the best class of cultivation."   **in the N.-W. P. & Oudh.**

Mr. Benett then goes on to show that the figures for the two years   **Area.**
shown cannot be merely due to seasonal fluctuations, since a steady rise is
exhibited in the figures for the thirty temporarily settled districts during
the intervening three years. This being the case it is probable that the
figure in the table at p. 94, which has been accepted by the Revenue and
Agricultural Department as a maximum approximate (so that no error
could possibly be imputed in shewing expansion), may be considerably too
small, and that the increase has been greater than is there indicated.
However this may be, it is evident that the expansion during the five years
ending 1887-88 was very small, if Oudh be left out of account, and that
in the past year (1888-89) the cultivated area under pure wheat dropped by
some 330,000 acres, from the average of the previous five years. The
area in the United Provinces was 104,509 acres below the average for
the past ten years, which is 3,566,618 acres for the North-West Provinces and 1,507,091 for Oudh.

In addition to the above areas there is, as shown by **Mr. Benett's**
figures, an extent of some 2,500,000 acres under mixed wheat, wheat-gram,
wheat-barley, etc., in the North-West Provinces; an area which has been
excluded from consideration owing to the varying nature and uncertainty
of its outturn of wheat.

During the past year the area under pure wheat was distributed in the
proportions shown below : —

Division.		Irrigated.	Unirri-gated.	Total.
Meerut		542,353	514,933	1,057,280
Agra		348,543	87,217	435,767
Rohilkhand		176,019	753,031	929,050
Allahabad		173,685	36,372	210,056
Benares		356,927	166,350	523,279
Jhansi		23,145	56,135	79,280
Kumaon		21,679	222,890	244,567
	Total:N.-W. P.	1,642,351	1,836,928	3,479,279
Lucknow		242,612	43,078	285,690
Sitapur		251,985	178,344	430,329
Fyzabad		294,341	243,776	538,117
Rai Bareli		224,427	11,358	235,785
	Total Oudh	1,013,365	476,556	1,489,921
	GRAND TOTAL	2,655,716	2,313,484	4,969,200

From the above table it will be observed that the largest wheat growing divisions are Meerut and Rohilkhand, that irrigated crops form a
large proportion of the whole area, especially in Agra, Allahabad, Benares,
Lucknow, and Rai Bareli, and that, taking the United Provinces generally
more than half the area is irrigated. The above figures have been taken
from those published in the Agricultural Statistics of British India, and
differ somewhat from those enumerated in the Administration Report of
the Province for 1889.

RACES AND QUALITY.—The varieties and races of wheat grown in these   **Races.**
Provinces are, according to **Duthie & Fuller**, "countless," and testify   **717**

TRITICUM sativum.	**Cultivation of Wheat**

CULTIVATION
in the
N.-W. P. &
Oudh.

Races.

strongly to the importance of the cultivation and the lengthened period over which it must have extended. Here, as elsewhere, the forms may be conveniently classified into red and white, with the subordinate characters of hardness and softness. Hard wheats are said to be most highly valued by Natives, who consider them more wholesome for general consumption. A good deal has been written above in the chapter on RACES generally on the characters of the North-West Provinces wheats. It may, however, be repeated that *daudi* or *daudia* is, perhaps, the finest kind, and has been pronounced equal in value to the finest wheats in the English market. *Mundia, mundwa,* or *murilia* (*lit.* shaved) is a name generally applied to another class of white soft wheat of good quality, so designated from being beardless. In the western districts of the Provinces, hard white wheats are generally known as *badha* or *barha*; they are, however, much less frequently cultivated than the soft or mixed forms. *Pissi* is said generally to denote a soft red wheat, and *kathia* or *lallia*, a hard red wheat. *Gangajali* (a common term in the Bombay market) is, according to the authors of *Field & Garden Crops,* applied to many varieties, and its only general application appears to be to mixed red and white hard wheats. A curious round-berried form, which somewhat resembles peal barley, is called *paighambari,* and is said to have been an introduction from Arabia (*Field & Garden Crops,* 2).

With these preliminary remarks, a list of the samples valued in **Dr. Watson's** report may be given, as in the case of other Provinces:—

Class.	District.	Name.	Weight per Bushel.	Value per Quarter	REMARKS.
				R a.	
Soft White.	Azimgarh .	Daudi . .	61	46 0	
	Benares . .	Daudia . .	...	46 0	
	Basti . .	Gangajali .	...	45 0	
	Banda . .	Pisi gangajali .	...	47 0	
	Cawnpore .	Muria . .	...	48 0	
	Ditto .	Anokha . .	...	47 0	
	Ditto .	Desi . .	...	46 0	Large berry.
	Ditto .	Mudia . .	...	44 6	Fine drop wheat, probably 64 ℔.
	Futtehpore .	Pisi awwal .	...	46 0	
	Ditto .	Muria . .	...	45 0	
	Pertabghar .	Mundwa . .	62½	47 0	Like Californian.
	Ditto .	Setwa . .	60	46 6	
	Ditto .	Mundia . .	...	46 0	
	Bharaich . .	Daudi . .	61	46 6	Ditto.
	Ditto .	Sambodhwa .	...	46 0	
	Ditto .	Sandhua . .	...	46 0	
	Gonda . .	Daudi . .	...	45 6	
	Unao . .	Saman . .	...	48 0	
	Ditto .	Marua . .	...	46 0	Ditto.
	Ditto .	Ditto . .	...	46 0	Ditto.
	Ditto .	Safeda . .	...	45 0	
	Kheri . .	Sitia . .	60½	45 0	Like Danzig, but dirty.
	Sitapur .	Mundia . .	61	46 0	Like Californian.
	Ditto .	Muria . .	...	46 0	
	Lalitpur .	Pisi duem .	...	45 0	
	Etah . .	Mundia . .	62	47 0	Like Australian.
	Ditto .	Sambharia .	...	46 0	Like Californian.
	Ditto .	Ratta . .	...	46 0	Ditto.

	in the N.-W. Provinces and Oudh. (*J. Murray.*)					TRITICUM sativum.

Class.	District.	Name.	Weight per Bushel.	Value per Quarter.	REMARKS.	CULTIVATION in the N.-W. P. & Oudh. Races.
				R a.		
Soft White— (*contd.*).	Muttra	Safeda	...	46 6	Like Californian.	
	Mainpuri	Sambharia	...	47 0	Ditto.	
	Ditto	......	...	46 6	Ditto.	
	Bulandshahr	Safed	63	48 0	Like Australian.	
	Ditto	Gajar	63½	47 0	Like Californian.	
	Ditto	Safed	...	46 6	Ditto.	
	Ditto	Mendha	62½	46 0	Ditto.	
	Ditto	Gajar	...	46 0		
	Ditto	Rutta	62	45 6		
	Ditto	Ruta	63	45 0		
	Ditto	Munia	...	45 0	Like pearl barley.	
	Dehra Doon	Mihirta	...	45 0		
	Meerut	Safed	62	48 0	Like Californian.	
	Ditto	Ditto	...	48 0	Ditto.	
	Ditto	Monda	62	47 0	Ditto.	
	Ditto	Muria	...	45 6		
	Muzaffarnagar	Safeda	...	46 0		
	Saharanpur	Monda	62½	46 6		
	Ditto	Muria	...	45 0		
	Bareilly	Sambharia	...	46 0		
	Ditto	Pisia	60	46 0		
	Ditto	Khatia	...	45 0		
	Budaon	Bhambria	...	46 6		
	Ditto	Muria	61	46 6		
	Ditto	Ratua	61	46 0		
	Ditto	Rai munea	...	45 0	Like pearl barley.	
	Moradabad	Muria awwal	62	47 0	Ditto.	
	Ditto	Mundia	...	46 0	Small berry.	
	Ditto	Mandwa kada	...	46 0		
	Ditto	Ditto	...	45 6		
	Ditto	Muria safed	...	45 0		
Hard White	Gonda	Daudi	63	41 0	Mostly hard.	
	Barabanki	Murua	62	41 0	Mixed hard.	
	Lucknow	Tamla	64½	42 6		
	Unao	Samanbargehuna	...	41 0	Long berried.	
	Etah	Bhidia	61	40 6	Long berried, like Kubanca.	
	Mainpuri	Anokha	60½	42 0	Long berried.	
	Ditto	Ditto	...	40 6	Ditto.	
	Aligarh	Kathia	58	41 0	Ditto.	
	Bulundshahr	Barha	...	42 0	Ditto.	
	Ditto	Ditto	...	40 6	Large berried.	
	Meerut	Ditto	61	41 0	Long berried.	
	Ditto	Ditto	...	40 6	Ditto.	
	Budaon	Ratua	...	40 6	Ditto.	
	Shahjehanpur	Sambhari	60	40 6		
Soft Red	Azimgarh	Hurrah	63½	41 0	Small berried, semi-hard.	
	Allahabad	Raksa	...	41 0		
	Cawnpore	Pisia	...	41 0		
	Jalaun	Pisia, red	61½	41 6	Large berried.	
	Muttra	Lal	...	40 6		
	Bulandshahr	Gehun lal	62½	42 0		
	Meerut	Surkh	61½	41 0	Long berried.	

T. 717

CULTIVATION
in the
N.-W. P. &
Oudh.

Method
718
Seasons.
719
Rotation.
720

METHOD.—There is very little of any special character to record under this heading. The crop is entirely *rabi*, being sown in the end of October or beginning of November, and cut in March and April. As a rule, it is only sown in land that has lain fallow during the preceding *kharif* (known as *chaumás* or *púral*); but in highly manured lands near village sites it occasionally follows maize, that crop being cut only six or eight weeks before the wheat is sown.

No particular rotation is known to be followed, but in tracts where cotton is widely grown, wheat is generally said to follow it—probably, however, merely because cotton in the *kharif*, like wheat in the *rabi*, is the crop which is principally grown on the best land of the village (*Field & Garden Crops*). In the Meerut district, a very elaborate rotation is observed, in which wheat is grown only twice in five years.

Mixtures.
721

Wheat, as already indicated in the chapter on area, is not only grown alone, but is also cultivated to a large extent mixed with barley (when it is termed *gojai*) or with gram (*gochana*, or *bíra*). The latter mixture is but little grown north of the Jumna, but in Bundelkhand it forms one of the principal and most characteristic crops. A wheat field usually contains some rape or mustard, sown either in parallel lines across the field or as a border. These flower in the beginning of February, before the wheat has begun to ripen. Linseed and *duán* (**Eruca sativa**) are also occasionally, though less commonly, sown in wheat fields.

Tillage.
722

"The number of ploughings varies within very wide limits, depending not only on the character of the locality and soil, but on the energy and leisure of the cultivator. Thus timely ploughings are reported as not uncommon in Gorakhpur, while two or three are held sufficient in the black soil of Bundelkhand. Eight ploughings may be taken as the average. It is essential that the land should be ploughed at the very commencement of the rains, so as to lie in open furrow and drink in the whole of the rain which falls. Indeed, the ploughing of wheat land is often held to take precedence of preparations for the *kharif* crops. The clods are crushed and a fine tilth (which is absolutely essential in most soils) created by dragging a flat log of wood (*mai, pátha,* or *henga*) across the field, the bullock driver standing on it to increase the weight.

Sowing.
723

"If the ground is very damp the seed is sometimes sown broadcast and ploughed in, when it is not buried more than one inch below the surface, and is less likely to rot if buried deeply. But the two commonest methods of sowing are (1), by simply following the plough and dropping the seed into the furrow made by it, the seed being covered by the earth thrown up by the next furrow, and (2), by dropping the seed down a bamboo fastened to the plough stilt. It is said that the advantage of each practice varies with the condition of the soil, the former being best when the soil is very moist, and the latter when the soil has somewhat dried. But as a matter of fact the practices are strictly localized to tracts within which either one or the other is exclusively followed. The amount of seed used per acre varies from 100 to 140℔. After the sowing is completed the field is either left in furrow, or is smoothed with the clod-crusher, the latter practice being said to save irrigation by enabling the water to spread quicker over the surface. The field is then divided off into irrigation beds by scraping up little banks of earth with a wooden shovel."

The proportion of seed employed is very high, much higher than the average in most other localities. This fact has been frequently urged against the advisability of encouraging wheat cultivation in India. The poorer cultivators have to buy from the merchant (or rather get the grain on loan at high interest collected at harvest), and are at the same time compelled to accept whatever seed the trader of the district chances to have

in the N.-W. Provinces and Oudh. (*J. Murray.*) **TRITICUM sativum.**

CULTIVATION in the N. W. P. & Oudh.

Sowing.

in stock. A very extensive correspondence has passed on this subject between the various Governments, and attempts have been made to disseminate seed of good stock from the Government experimental farms. But these measures nave had very little permanent effect on the general character of wheat grown. As already shown, in the paragraphs on Bombay, the grain appears to tend very strongly to alter its colour and characters with change of soil and surroundings; the outturn, unless the new stock be thoroughly acclimatized, is generally very much lower than in the case of the wheats naturally cultivated in the locality, and the *rayat* returns to his old stock.

IRRIGATION.—"If the soil is sufficiently moist in October to allow of the seeds germinating properly, the necessity of irrigation depends in chief measure on the occurrence of winter rains. This is shown in the following table in which the normal winter rainfall of each division is contrasted with the percentage which irrigated wheat (grown alone) bears to the total:—

Irrigation. 724

	Meerut Division.	Rohilkhand Division.	Agra Division.	Allahabad Division, excluding Jaunpur District.	Benares Division, including Basti and Gorakhpur Districts only.	Jhansi Division.	Kumaun Division, including Tarai District only.
Normal rainfall between November 1st and May 31st*	5·56	4·73	2·55	2·26	3 55	2·06	6·53
Percentage of irrigated wheat to total . . .	53·1	20·1	74·3	63·7	71·0	27·4	32·7

The high percentage of the Meerut Division is due to unusual facilities for irrigation from canals. The percentage of the Allahabad division would have been far higher did it not include the two Bundelkhand districts of Banda and Hamirpur, where irrigation is rendered needless, as well as impossible, by the character of the soil.

Should the soil be too dry for germination, a watering (called *paleo*) must be given before sowing, and this—a comparatively easy matter in canal districts—occasions great labour and delay in districts which rely on wells for their water supply. The instance of Rae Bareli in the *rabi* season of 1879-80 shows, however, that nearly the whole of the usual crop area of a district can be sown entirely on well water, should the natural moisture be insufficient as it was in that year. The number of waterings given to wheat varies from one in Rohilkhand to seven or eight in the drier parts of the Doab; but, as a rule, three or four waterings are ample even in the driest localities, and when more water than this is used, it is probably merely a cover for bad cultivation, a state of things common enough in canal districts, where water is charged for by the crop and not by the amount used. Careful cultivators sometimes give their fields a weeding after the first watering, and benefit their crops almost as much by loosening the cake surface soil as by removing the weeds, but this is by no means a common practice, and if the land be in clear condition when sown, it is not as a rule weeded. The custom is reported from Bahraich District, and may prevail in other parts of the Provinces, of topping wheat which shows an undue

* Calculated from the normal rainfall at each district head-quarters in the Divisions.

TRITICUM
sativum.   Cultivation of Wheat

CULTIVATION
in the
N.-W. P. &
Oudh.

tendency to run to leaf and stalk, by cutting down the upper portion of the plants with a sickle. This is done when the crop is about 3 feet high, and care is taken not to cut down so low as to damage the ears which have formed in the leaf-covers, but not yet emerged. A similar custom obtains in parts of the Panjáb, where, however, the young plant is fed down by sheep.

Harvesting.
725

"The crop when ripe is cut down by sickles and carried to the thresh-ing floor, where, after having been allowed to dry for several days, it is trodden out by bullocks, and winnowed by the simple expedient of exposing the grain and chaff to the wind by pouring them out of a basket held some 5 feet from the ground. Should there be no wind, an artificial breeze is made by agitating a cloth, but this adds greatly to the expense and trouble, and is in no way an efficient substitute for the English win-nower" (*Field & Garden Crops*).

Cost.
726

COST.—Messrs. Duthie & Fuller have gone into the question of cost of production with much care. The results of their calculations may be stated briefly :—

						R	a.	p.
Expenditure for labour and seed	.	.	.	.	.	16	0	0
Irrigation, and labour in watering	.	.	.	.	.	5	7	0
Manure (100 maunds)	.	.	.	.	.	3	0	0
Rent of land (second class)	.	.	.	.	.	7	0	0
				TOTAL	.	31	7	0

It would be useless to attempt an estimation of the average profit derived, since this depends so largely on the price obtained, which has been shown elsewhere to be an extremely variable factor.

Outturn.
727

It was estimated by the Government of India in March 1884 in a cir-cular on the subject, that the area under wheat, mixed and unmixed, in the Provinces, was 6,200,000 acres, and that the average production per acre was 13 bushels or 9⅔ maunds per acre. **Mr. Benett**, criticising these figures in December of the same year, states that, in his opinion, they are under-estimates, since they are based on the supposition that more than half the area occupied by the crop is on third class lands, that is, lands that are inefficiently or carelessly cultivated. "But this," he writes, "is very far from being the case. Careless cultivation is less common with wheat than with any other kind of food crop, especially when it is sown unmixed. For the last three years careful experiments have been made on the Cawnpore Farm to ascertain the ordinary produce of wheat. The soil is poor, and the cultivation not superior than what is given by an ordinary skilled cultivator, either in point of manure or irrigation, while in the matters of ploughing and weeding it is probably inferior. The results have been as follows :—

					Area sown in acres.	Average produce of grain per acre in ℔.
1882 .	.	.	.	.	16·8	1,390
1882 .	.	.	.	.	16·1	1,309
1884 .	.	.	.	.	17·6	1,453
					Average of three years	1,384 or 23 bushels.

"(I have received the record of some very careful weighments done by Mr. Saunders, C.S., at Rae Bareli. He found the average of eight weighments of irrigated wheat to be 23·63, the lowest 11·81, and highest 36·2 bushels to the acre. Of fourteen weighments of barley and wheat

Products of India. 145

in the N.-W. Provinces and Oudh. (*J. Murray.*) TRITICUM sativum.

CULTIVATION in the N.-W. P. & Oudh.

Outturn.

the average was 23'7, the lowest 19'73, and the highest 26'18 bushels to the acre.)

"A careful review of all the evidence then in existence led Mr. Fuller to state in his work on *Field and Garden Crops* that the lowest average produce which could be assumed for irrigated land was 20 bushels (15 maunds) for wheat grown alone and wheat and barley, and 17 bushels (13 maunds) for wheat and gram. The harvests reaped in the canal-irrigated tracts of Meerut, and round the wells of Oudh, are hardly, if at all, inferior in quantity to good crops in England, where the average on all classes of soil is 28 bushels. In assuming 22 bushels ($16\frac{1}{2}$ maunds) as the average produce per acre on irrigated lands in these Provinces I feel that I am well within the mark.

"The greatest proportion of unirrigated wheat is grown in the *tarai* districts of Rohilkhand and the north of Oudh. The extremely careful investigations of Mr. Moens ascertained that the average produce in Bareilly was 12 maunds, or 16 bushels to the acre. As has been stated, wheat, even when unirrigated, is rarely sown on inferior soils, and to take the mean of the assumed produce on second and third class lands in the letter of the Government of India as the average production for all unirrigated wheat will be well within the mark. This is 12 bushels, or 9 maunds.

When wheat is sown as a mixture, it is, over the whole Provinces, about half the crop, and 10 bushels may be assumed for irrigated, and 6 for unirrigated land, as the average produce per acre.

"By applying the above produce rates to the area, we find the produce of the mean harvest of the last five years (1879—1883) for the North-Western Provinces and one year for Oudh to have been—

	Bushels.
Pure wet	62,096,584
„ dry	25,064,640
Mixed wet	9,074,190
„ dry	10,949,424
Total	107,184,838

or nearly three million tons. Before leaving the subject of produce it may be as well to remark that any comparison between English and Indian rates of produce with the view of discrediting Indian methods of agriculture is apt to be misleading. In England wheat has to compete with pasture, which is not the case here. Nearly every acre in this country is under some kind of cultivated crop, and the result is that large areas of inferior land are brought under the plough, which, in England, would be left to grass. On soils and under conditions really favourable to wheat culture it is doubtful whether the average outturn is much less than it is in England, and with the best classes of cultivators, such as *Kurmis* and *Káchhis*, it is probably at least as great."

Mr. Fuller agrees very closely with the estimates of Mr. Benett, accepting 15 maunds per acre for irrigated wheat, and 9 for unirrigated. More recent experiments than those quoted by Mr. Benett, at the Cawnpore Farm, also demonstrate the probability that the outturn accepted as normal in preparing the wheat forecasts in these Provinces is under rather than over estimated. Fuller estimates the proportion of wheat in mixed crops to be $\frac{2}{5}$ths of the outturn of wheat-barley and $\frac{2}{3}$rd of that of wheat-gram, except in the Allahabad and Jhansi Divisions where gram is the principal crop in the mixture, and the proportion of wheat is not much above $\frac{1}{3}$rd. Mr. Benett's figure of $\frac{1}{2}$ may, therefore, be practically assumed as a fair approximate for all mixed crops in the Provinces, and the average outturn of 20,000,000 bushels given above may

Cultivation of Wheat

CULTIVATION
in the
N.-W. P. &
Oudh.

probably be accepted as very near the average production of mixed crops. The outturn of straw varies in weight between half as much again and twice as much as that of grain. When crushed into small pieces, as it is in the process of treading out the grain, it forms an important cattle-fodder. During the past year, 1888-89, the area and outturn were both very much lower than in any previous year since 1884. Accepting the standard of full average outturn adopted in the wheat forecasts (an average much lower than that independently estimated by **Mr. Benett, Mr. Fuller,** and other authorities), the total outturn for the year is, in the final wheat report, estimated at 1,440,000 tons. This is based on the assumption that the condition of the crop varied from 50 to 25 per cent. below the average, and implies an average yield of only a little over 8 maunds per acre.

Experiments.
728

CROP EXPERIMENTS.—The following were selected for reproduction in the Government of India Report on Wheat Production and Trade in India, in 1883 :—Fairly accurate experiments were carried out in 1879-80 with the following main results. The average cost per acre was found to be—

	R	a.	p.
Seed	2	0	0
Ploughing, clod-crushing, sowing, weeding, and reaping .	6	8	3
Rent . . . . - . .	6	4	0
Irrigation from canal	3	12	0
TOTAL .	18	8	3

The labour was not under very careful supervision, and would cost a cultivator next to nothing, as it would be supplied almost entirely by himself and his family.

The price of manure varied with the kind and quantity given to each experimental plot. The net results were as follows :—

	Total cost of all kinds.			Outturn in bushels.	Value.		
	R	a.	p.		R	a.	p.
Bone superphosphate, 2 maunds . .	27	12	9	27·1	49	10	3
,, ,, 3½ ,, . .	31	4	5	31·5	58	0	11
Dissolved guano 1¾ ,, . .	34	6	5	23·7	43	11	2
,, ,, 2 ,, . .	39	9	7	29·6	55	0	8
Green soiling with indigo	30	8	3	23·8	47	0	3
No manure	19	12	3	19·5	35	4	7

Each one of these plots had been irrigated twice. On unirrigated land the produce of grain on four plots averaged little more than 10 bushels to the acre,—that is to say, it was considerably less than half what was obtained from land which had had only two waterings.

The result of experiments with thick and thin sowing on land dealt with similarly with regard to ploughing, watering, and manure, was as follows :—Sixteen pounds of the finest white country wheat to the acre produced a crop of 1,206℔ (or 20 bushels) at a cost of R25-6-9, while 116℔ of seed (the ordinary native allowance) produced 1,626℔ (or more than 27 bushels) at a cost of R27-0-9. The result showed that, though the rate of return to seed was far higher with thin sowing, thick sowing was the more profitable operation. In the succeeding year experiments were carried

out on similar lines, but with variations in the kind and quantity of manure CULTIVATION
employed. The results per acre were—

Name.	Cost per acre of all kinds.	Crop in bushels.	Value.
	R a. p.		*R a. p.*
Green soiling	28 13 0	30	58 4 3
Cattle-dung, 240 maunds . . .	33 7 0	21·7	42 12 8
„ „ *plus* 240℔ gypsum	38 11 0	20·7	41 0 5
Ashes of 240 maunds dung . .	33 7 0	18·6	36 11 1
Poudrette, 240 maunds . . .	33 7 0	31·4	62 1 11
Bone superphosphate, 225℔ . .	32 11 0	16·5	32 6 2
Bone dust, 320℔	25 7 0	13·3	26 11 7
No manure	21 7 0	11·8	24 13 6

In all these cases the plots were ploughed twice in the European fashion, and watered three times, besides, as the season was one of exceptional drought, having been watered, in order to admit of the sowing.

By far the most efficacious manures by these trials were green soiling with a leguminous plant and poudrette, and this goes some way towards demonstrating that it is in nitrates, and not in phosphates, that the soil chiefly requires to be re-inforced. The excellent results obtained from superphosphate in the preceding year were almost certainly due to a previous and unexhausted dressing with nitrogenous manure.

The value of early ploughing was shown in a crop of barley, which, having been ploughed for a month before the commencement of the rains, and exposed to the burning sun of May, yielded, with three waterings, but altogether without manure, an outturn of 30·7 bushels to the acre. A neighbouring plot which had been trenched and used as a latrine for 80 days gave the enormous outturn of 47·1 bushels to the acre.

As far as these experiments prove anything, they support the view that with the high and careful cultivation under which the best wheats are grown, the average produce is certainly not less than from 20 to 25 bushels to the acre.

These experiments do not seem to have included manuring with saltpetre, but the results of nitrogenous manures fully support the demonstration of the value of nitrogen given under the accounts of Bombay and Bengal.

CENTRAL PROVINCES.

References.—*C. P. Gazetteer, 18, 114, 385, 471, 501, 516 ; Settlement Reports :—Baitool, 42, 77 ; Chanda, 81, 84, 96, 98 ; Dumoh, 87 ; Jubbulpore, 86 ; Hoshungabad, 277, 287 ; Mundlah, 46 ; Nagpore, 273 ; Nimar, 196 ; Nursingpore, 53 ; Saugor, 98 ; Upper Godavery, 35 ; Wardha, 65-67 ; Government Report, Wheat Production and Trade in India, 1879, 1883, 1886 ; Forbes Watson, Report on Indian wheat, 1879 ; Reports, Dir Land Records and Agri. ; Crop Reports and Forecasts ; Report, Rev & Agri. Dept., 1885, 1886 ; Indian Agriculturist, 22nd February 1890.*

The subject of wheat production in these provinces has, as in the case of most other provinces in India, been fully dealt with in the Government reports. The communications, by Mr. J. W. Chisholm in 1877, and by Mr. Fuller in 1882, are especially valuable as giving a compact *résumé*, and will be freely quoted in this chapter.

Dr. Watt, in the *Journal of the Royal Agricultural Society*, commences his account of the Central Provinces as follows :—

"There are in these provinces 7,705,963 acres available for cultivation. In perhaps no other province can the literal meaning of this be more clearly demonstrated. The provinces are poorly inhabited, and within periods

**CULTIVATION
in the
Central
Provinces.**

recorded in our Settlement Reports large tracts of land have been taken up and brought into cultivation. The returns first obtained are well known, and important records have been kept of the deterioration of productiveness. The results have been identical with those obtained in America. The newly reclaimed land gave twenty and thirty fold for a few years, but rapidly deteriorated until it reached a fixed, or relatively fixed, position. The district officers repeat in their annual reports that there are still vast tracts of land on which this process might be repeated. In perhaps no other part of India is the principle of paucity of population, large holdings, and correspondingly low systems of agriculture more forcibly demonstrated. In the North-West Provinces small holdings and careful cultivation have produced results that, even with the present agricultural appliances, compare favourably with Europe. In the Central Provinces, on the other hand, the proprietor of a large estate is satisfied with the comparatively small results obtained by cheap and primitive means."

**Soils.
730**

SOILS.—" To understand the soils of these provinces, it is necessary to consider the geological formations from which it has been built up. The great basaltic formation which occupies nearly a third of the peninsula, crosses the north-western division of the Central Provinces, and slopes north-west in the drainage areas of the Narbada and Tapti rivers, and south-east in that of the Godaveri. To the south and south-east of this " Deccan trap " there extends the vast region of the archæan rocks of India. These two geological regions are broken here and there by isolated patches of the Gondwana and Vindhyan rocks, and the disintegration of all has contributed to the local peculiarities of the soils. The wheat fields of the northern section, bordering on the Narbada, owe their fertility to the Deccan trap, which is generally considered to have, by disintegration, produced the rich black " cotton soils " of the localities in which the formation occurs. There has been some controversy on this point, certain authorities maintaining that the black soil of the Central Provinces owes its origin to the action of lacustrine deposits. But, without entering into this discussion, it may be stated that there can be little doubt that, in certain cases at least, black soil does consist of disintegrated trap, since the process of disintegration can be traced *in situ*. Mr. Fuller's description of the local modifications of this soil is of interest, and may be quoted in entirety :—

" The black cotton soil is prominently characterised by its stubbornness under opposite extremes of damp and dryness. When soaked with rain, it becomes a sticky unworkable morass; and when dry, rapidly becomes of great hardness, splitting up into numerous dry fissures, which swallow up any water applied to the surface, and enormously increase the cost of irrigation.

" This black or ' cotton' soil is known by various names indicating the proportion in which it is mixed with lighter soil. It is of very variable depth, lying in much thicker deposits in flat valleys than on sloping ground. It is most suitable for wheat when at its greatest thickness, since, from the great capacity which it then enjoys of absorbing rain-water, it can, in a monsoon of average intensity, lay up a store of moisture sufficient to carry a wheat crop through a cold season in which the winter rains hold off entirely. When it is merely a shallow veneer of earth, covering rocks, it dries, of course, with far greater rapidity, and is in this case devoted to the production of rain (or *kharíf*) crops. This difference is brought out into strong relief by a comparison of the agricultural returns for the contiguous districts of Hoshangabad and Nimar. In Hoshangabad deep black soil predominates, and, in consequence, 63 per cent. of its cultivated area is returned as under wheat. The greater portion of the Nimar district is hilly or undulating ground, consisting of trap rock, overlaid with a shallow bed of black soil. Wheat only occupies 4 per cent. of its cultivated area."

AREA.—The area under wheat in these provinces exhibits an expansion from 3,536,000 acres in the five years ending 1877-78, to 4,002,000 acres in a similar period ending 1887-88 ; but in 1888-89, owing to an unfavourable season, it again fell to 3,531,941 acres, a figure which must be accepted as much below the normal. The expansion is doubtless largely dependent on the development of trade, but, according to **Mr. Fuller**, the apparent sudden advance after 1881 may to some extent simply indicate an improvement in the accuracy of the land returns to which the attention of District Officers had been specially directed in 1880. But even allowing this to be the case, the fact remains that, during the five years ending 1887-88, the area under the crop, notwithstanding more than one bad year, underwent an increase of 15·4 per cent. on that for the five years ending 1882-83. Concerning the substitution of wheat for other crops, the Director of Land Records and Agriculture states, that though it is impossible to deal with the question satisfactorily, the probability is, that the increase is fully accounted for by a falling off in the amount of land devoted to linseed, and also in certain localities in that under cotton. " It may be explained," he writes, " that such substitution, as would occur, would mostly affect the linseed and cotton areas. Less valuable crops are usually grown on land which would not carry wheat with advantage. It should be added that neither linseed nor cotton can be grown in Jabalpur or Hoshungabad so profitably as in other parts of the provinces, and that the substitution of wheat for them was to be expected, and is certainly not to be regretted " (*Rept., 1887-88*).

Regarding the distribution of the area through the various districts **Mr. Fuller** writes, " Wheat cultivation reaches its greatest importance in the northernmost districts and in the districts of the Nerbudda valley, after which it is most extensive in the Satpura hill region and the Wardha and Nagpur districts to the south-west of it. The insignificance of the area under wheat in Nimar is due (*Cf.* remarks on SOIL) to a peculiarity in its soil, which is not deep enough to grow cold-weather crops without irrigation. The cultivation of Bhandara, Balaghat, and Sambalpur is almost wholly engrossed by rice."

The areas in each district during the past year, and the proportion irrigated, may be most conveniently shewn in tabular form :—

CULTIVATION in the Central Provinces.

Area. 731 *Conf. with p. 94.*

District.				Irrigated.	Unirrigated.	Total.
				Acres.	Acres.	Acres.
Saugor	.	.	.	944	568,541	569,485
Damoh	.	.	.	211	183,374	183,585
Jabalpur	.	.	.	...	257,869	257,869
Mandla	.	.	.	...	38,977	38,977
Seoni	.	.	.	...	284,580	284,580
Narsinghpur	.	.	.	1,211	103,896	105,107
Hoshungabad	.	.	.	20	598,601	598,621
Nimar	.	.	.	4,784	35,166	39,950
Betul	.	.	.	35	138,088	138,123
Chhindwara	.	.	.	40	142,131	142,171
Wardha	.	.	.	16	290,984	291,000
Nagpur	.	.	.	10	465,889	465,899
Chanda	.	.	.	246	85,255	85,501
Bhandara	.	.	.	359	139,735	140,094
Balaghat	.	.	.	21	28,040	28,061
Raipur	.	.	.	...	121,123	121,123
Bilaspur	.	.	.	...	81,466	81,466
Sambalpur	.	.	.	...	329	329
GRAND TOTAL			.	7,897	3,524,044	3,531,941

TRITICUM
sativum.   Cultivation of Wheat

CULTIVATION
in the
Central
Provinces.

Races.
732

RACES.—The commonest kinds of wheat in these provinces are said to be *bansi, houra, daudkhani* or *pilalia,* a hard white wheat; *kathia, ghatka, hausia, bangasia* or *chawalkathi,* a large coarse red grain; *pisi,* the soft white wheat cultivated for export, and *botka,* white and red, a short, heavy, soft grain. The white soft wheats have increased in proportion very greatly during the past fifteen years. Thus Mr. Fuller writes :—

"Natives prefer, for ordinary consumption, the hard glutinous varieties to the soft white varieties, which are principally in demand for the English market. Before the commencement of the annual drain of wheat to Bombay, soft wheat of the kind known as *pissi* was held in very low estimation, and commanded a price which ruled from 8 to 10 per cent. lower than that of the hard *kathia* variety. Now its price is at least 12 per cent. higher than that of *kathia.* In old days it was no uncommon stipulation of a ploughman contracting for service that he should not have to eat *pissi* wheat more than twice a week. Now a ploughman who demanded it twice would certainly not receive it. *Pissi* wheat is grown on lighter land than *kathia,* and it is reported from both Saugor and Narsinghpur that the value of light land has risen in considerably larger proportion than that of heavy land, in consequence of the request in which *pissi* wheat now stands for the Bombay market."

The samples submitted to Dr. Forbes Watson were classified by him as follows :—

Class.	District.			Name.	Weight per bushel.	Price per quarter.		REMARKS.
	Baitul	.	...	Pissi . .	63	46	0	Like Californian.
	Hoshung-abad	.	...	Pissi white .	62	48	0	Like Australian.
SOFT WHITE	Ditto	.	...	Sohareea .	...	48	0	,, ,,
	Mandla	.	...	Pissi sookra-wali . .	60½	46	6	Like Californian.
	Saugor	.	...	Pissi . .	62½	46	0	Long berried.
	Seoni	.	...	Mundi . .	64	47	0	Like Californian.
	Ditto	.	...	Pissi . .	59½	45	6	
HARD WHITE	Saugor	.	164	Hansia .	62½	40	6	Long berried.
SOFT RED .	Nimar	.	194	Dhunya .	...	40	6	Fine specimen.
	Bilaspur	.	567	Khathia .	60	41	6	Small berried.
			566	Red pissi .	63	41	0	

Method.
733

METHOD OF CULTIVATION.—The system of cultivation which we have already described as practised in the heavy black soils of Bombay is practically that pursued in these provinces. It differs materially from the system followed in the North-West Provinces and the Panjáb, but it is probable that, while improvement is possible, the method followed in the latter localities could never be adopted in a country the natural conditions of which are so different. The system adopted has been fully described by the Director of Land Records and Agriculture, from whom we may again quote.

"Although these provinces are entitled to rank with the North-West Provinces or the Panjáb in respect to their wheat production, yet the conditions under which wheat is grown here are widely different to those of the two latter tracts. Manure and irrigation—all-important in Upper India—play but an insignificant part here. The thinness of population

T. 733

in the Central Provinces.　　(*J Murray.*)　TRITICUM sativum.

and consequent large extent of individual holdings, offers more inducement to low farming on a large, than to high farming on a small, scale. The outturn of the land is actually far smaller than in more crowded tracts, whilst the surplus produce per head is considerably larger, since the poorer the cultivation, the larger is that proportion of the produce which results from natural forces as opposed to the labour of the cultivator. It is doubtful, indeed, whether wheat cultivation in these provinces is susceptible of the elaboration which characterises it in Upper India."

"The alternation of *rabi* and *kharif* crops is not so common in these provinces as in Upper India, since the soils are of more marked diversity, and are, therefore, more strictly appropriated, some to autumn and others to cold-weather crops. The deep wheat land of the Nerbudda valley is reported to be almost unworkable in the rainy months, while shallow soils, with a substratum of rock, dry too quickly to be suitable for *rabi* crops. For ordinary wheat cultivation, operations commence in April or May, when the surface of the ground is scarified with a hoe-plough known as the *bakkar*. After the setting-in of the monsoon the surface is again scarified, once in July and once in August, if, as is hoped, there is long enough break in the rains to allow the ground to become sufficiently dry to bear the plough cattle. A fourth hoeing—the most necessary of all—is given in September towards the end of the rains, the importance of which arises from the fact that loss of moisture by evaporation is much checked if the surface of the ground be in loose condition. A final hoeing is given at the beginning of October, after which the field is ready for sowing. This represents the preparation which a careful cultivator will give his land under favourable circumstances, and, as a rule, land seldom receives more than two or three hoeings before it is sown. The seed is occasionally sown broadcast and ploughed in, but is more generally drilled in, the implement used for the purpose being an ordinary native plough, or more properly "grubber," fitted with a bamboo tube alongside of the stilt, down which the seed is dropped. From 80 to 120℔ of seed are sown to the acre, less being used when the soil is moist, and there is no fear of the seeds refusing to germinate, than when the soil is in dry condition. Weeding appears to be seldom, if ever, given, and the crop is left entirely to itself until harvest comes with the beginning of March.

"An entirely different system of cultivation is followed in a tract of country which includes a considerable portion of the Jubbulpore and a small portion of the Narsinghpur and Seoni districts. The fields are surrounded with banks so as to prevent all surface drainage, and the rain water is allowed to accumulate in them, converting them into tolerably deep ponds during the rainy months. The water is let off at the beginning of October, and wheat is then drilled into the ground without any preliminary ploughing or preparation whatever. Occasionally some rice is sown broadcast in the field when flooded, and is ready for cutting before the wheat is sown. But this rice crop, if taken at all, is generally a very small one, and the real object in embanking the land is to allow the soil to get saturated with water to as great a depth as possible, and increase the store of moisture which capillary attraction is to keep within the reach of the wheat crops during the dry cold-weather months. The process, therefore, corresponds, in some sort, to irrigation. It is reported to increase the outturn of wheat by at least 25 per cent., and to be of much service in clearing land infested with *kans* grass and other weeds. The principal reason why it has not spread over a wider area lies probably in the necessity for co-operation between one cultivator and another, since the system is believed to succeed only when practised on a block of contiguous fields which mutually assist one another to withstand the rush of surface drainage

T. 733

TRITICUM sativum.	Cultivation of Wheat

CULTIVATION in the Central Provinces.

Manuring.

734

Irrigation.

735

and the pressure of the water on the confining banks—causes which have hitherto ruined experiments tried with isolated fields.

MANURING.—" In regard to the use of manure. there appears considerable diversity of practice in different parts of the provinces. Thus, manure is reported to be hardly ever applied to land in the Nerbudda valley, while in Nimar and in the districts of the Nagpur Division its utility is fully recognised, each wheat-field receiving a manuring, fi possible, once in three years. The explanation may lie in the greater effectiveness of manure on shallow than on deep soils. On the former it makes all the difference between a fair crop and no crop at all, while on the latter it would merely add in some degree to a fertility which is as yet very far from being exhausted."

IRRIGATION.—" Irrigation is almost entirely confined to sugar-cane and garden crops, and is therefore rarely, if ever, applied to wheat, since if the rainfall is at all propitious, the harvest yields a sufficiently large surplus to satisfy the cultivator. The most favourable distribution of rainfall possible would be a heavy fall of 8 or 10 inches between the middle of June and the third week in July : then a break of a fortnight to allow of ploughing, followed by a second fall of 5 to 6 inches in August. September should open with a week or ten days of fair weather followed by a heavy downpour bursting on into the beginning of October. Under these circumstances, a fair crop would be reaped in the absence of all rain between sowing and harvest time, but there should be a fall of 2 or 3 inches during the cold weather, which would greatly add to the outturn if protracted damp weather does not develop the fungoid diseases of rust or smut. It is impossible, however, to say whether, with a greater press of population, irrigation may not be gradually extended to wheat. Experiments on the Nagpur Model Farm have shown conclusively that two waterings add very greatly to the outturn, but it is very doubtful whether the crops can ever stand the cost of irrigation if wells are the only source available."

In regard to the much-vexed question of the exhaustion of the soil by continuous cropping, it will be impossible to do better than cite the conclusions to which Messrs. C. A. Elliott and C. Grant arrived, after careful enquiry, when settling respectively the Hoshangabad and Narsinghpur districts. There appears to be no room for doubt that the black soil of the Nerbudda valley yielded a far larger return when first broken up than it does now. Tradition points to a rate of produce in the golden age—at the commencement of this century—which seems to have been as much as ten- or twelve- fold the seed sown; that is to say, 15 to 18 maunds to the acre. This is corroborated by recent experience of newly-broken land in the Tapti valley, an almost precisely similar tract, of which Mr. Elliott wrote that " the wheat stands breast-high in some of the new villages, and the ear is very large and full; the crop is nearly double the average of the Nerbudda valley." He estimated the average outturn in those villages at 12 maunds (=16 bushels) to the acre. The outturn appears to have fallen off very rapidly at first, the deterioration, indeed, was very apparent as early as 1844, and receives considerable attention in Colonel Sleeman's writings. It was in full progress during his time, but the decline seems now to have reached a point below which it does not continue. Mr. Elliott's enquiries led him to believe that the outturn fell very rapidly from 12 maunds to 8, and then rather slowly to 6 or 7. Mr. Grant expresses a similar opinion in the following paragraph :—

" Mr. Maloney and Captain Sleeman, who are the only authorities regarding the early condition of the valley, naturally attached great importance to the deterioration of the soil, for it was going on, and that rapidly, before their very eyes. All subsequent writers on the affairs of the district seem to have followed blindly in their footsteps, and it is almost a *reductio ad absurdum* of an undoubtedly true theory to find one of the district officers writing in or about 1830, that the returns had then sunk in places to two- or three-fold, and that ruin was hanging over the cultivating classes. The re-assuring feature in the otherwise disquieting decline of fertility in the soil is that the deterioration has not been gradually progressive, but that, commencing with a very considerable impetus, it has now become almost stationary. It will have been seen from the figures given above that while 20 years' cultivation reduced the returns from twenty-fold to six- or seven-fold, it has taken nearly double that time (from 1838 to 1866) to reduce them from five-fold to four-fold. And the present rate of diminution is so minute as to be imperceptible. Therefore, for all practical purposes, it may be assumed that the rates of produce will remain constant at the present point, even if improved modes of cultivation are not introduced in the course of the present settlement."

QUANTITY OF SEED, AND YIELD.—The quantity of seed sown per acre appears to vary within wide limits, but from about 80 to 120℔ may be taken as the average. Memoranda called for from all the districts in 1878 showed this difference in practice to a marked degree, the returns ranging from 50℔ to 150℔. About the same time a markedly high outturn was obtained at the Model Farm from only 40℔ an acre, but this was in the best fields and under irrigation.

The question of yield has been already discussed, to a certain extent, in the paragraph on the supposed deterioration of the Central Provinces soil. There appears to be little doubt that a virgin black cotton soil yields a wonderfully large outturn without the addition of manure, and that a certain amount of deterioration does occur. But the probability is that the average information given by Native cultivators regarding their profits and the yield from their fields is very much below the truth, and that many of the yields in the older reports are not even approximately correct. **Dr. Watt**, commenting on this fact, (1888) writes :—

"The question of yield has now in many provinces of India been put to a final test. From the supposed deterioration of the soil in the Central Provinces it was observed that, if this had actually occurred, the richest districts would long before this time have endured the utmost deprivation. The Deputy Commissioner of Raipur, for example, showed that if the rice crop of his district had been, in reality, what it was officially reported to have been, a large proportion of the population must have died of starvation— and this, too, in a district famous for plenty, and from which there has been, for many years past, a regular export of food-grain of an exceptionally large amount. This observation aroused the attention of the authorities, and instructions were, accordingly, issued that trial harvests should be held under the supervision of responsible European officers. Certain fields that had been cultivated by the owners were harvested in the presence of the officer appointed to supervise the experiment in each district. A large number of these trial harvests have been made, the result being that the normal yield per acre has been determined with the utmost degree of accuracy. This has shown a considerable increase in the yield of every crop experimented with.

"In the Raipur district the results of five harvests gave a mean of 1,048℔; of seventeen harvests in Nimar, 902℔; and of thirteen in Narsinghpur, 640℔. The lowest yield was that of Hoshangabad, where the mean of four harvests was 382℔. Without going into this matter in great detail, it may be added that the opinions held, both by Government and the public, as to the low yield in the Central Provinces, have been shown to have been founded on prejudiced returns. We have, in connection with

CULTIVATION in the Central Provinces.

Irrigation.

Yield.
736

TRITICUM sativum.	Cultivation of Wheat

the Panjáb, referred to the difficulty experienced in getting the Natives to furnish accurate information as to their profits. It may fully be anticipated that like results to the above will follow in Upper India when the Government feels called upon to direct test harvests to be made in the Panjáb as have now taken place in the Central Provinces. Reverting to the yield in the Central Provinces, it may, in conclusion, be said that in Raipur the yield in the older reports is put down at 368℔ (instead of what it has now been found to be, 1,048℔), and in Narsinghpur at 200℔ (instead of 647℔). These are test examples, and it may be inferred that, in the poor districts, the early records were found to be relatively more nearly correct than in the rich. Thus, for example, in Hoshangabad, instead of 382℔, the return was fixed at 328℔.

"The rents paid for wheat lands in these provinces vary considerably, according to the nature of the soil and the facilities of export. The average in Hoshangabad is 1*r*. 9*a*. 3*p*., in Saugor 1*r*. 14*a*., in Bilaspur 14*a*., in Jabulpur 2*r*. 4*a*."

The average yearly outturn during the four years ending 1887-88 is estimated, in the returns of the Agricultural Department, to have been 919,000 tons on an average acreage of 4,125,000 acres. These figures give an average acre yield all over, for these four years, of 6·2 maunds or 496℔ to the acre, an average very much lower than that derived from the experimental harvests above alluded to. The outturn for 1888-89 is reported to have been very much below even this small average, being estimated at 705,000 tons on a forecast area of 3,866,262 acres (about 330,000 acres in excess of the actual area). These figures give an average acre yield for the year of only 5·1 maunds, or 408℔ to the acre.

Mr. Fuller sums up his consideration of the question as follows :— " It is probable that the outturn at present gathered from land which has long been under cultivation may be considered as stationary for a very long period of years. The available food-substances which were found stored in the soil when it was first broken up have now been exhausted, and the crop depends for its support on the stock of nutriment which the soil acquires each year by absorption from the atmosphere, as well as by the regular process of disintegration which is at work within itself. Experience has yielded exactly similar results in America, where the produce of freshly-broken land has been found to decrease very rapidly during the first few years of cultivation, and then retrogress with shorter and shorter steps until a point is reached at which it becomes fixed, or below which the decrease is so small as to be almost imperceptible. And a further corroboration, if one be needed, is furnished by the results of certain experiments conducted by Messrs. Lewes and Gilbert at Rothemsted which point to an exactly similar conclusion. "

BERAR.

References.—*Wheat Production & Trade, 1878, 1883, 1886 ; Forbes Watson, Report on Indian Wheat, 1879 ; Crop Reports & Forecasts ; Agri. Statistics, Br. Ind.*

SOILS.—The richest wheat soil in this province, as in the Central Provinces, is the black cotton land. The crop is cultivated to a large extent in this class of soil in the Purna valley, also on the loam found above the Southern Gháts. It is never grown on the lighter soils, which are suited only for *kharif* millet crops, etc. Irrigation is rarely employed. Wheat lands are manured in their turn with other fields according to the means of the cultivator, who generally understands and fully appreciates its value. The only kinds available are, as a rule, farm-yard manure and the ashes of *jowari* roots.

in Berar.	(*J. Murray*)	TRITICUM sativum.

AREA.—The total area under wheat in Berar has undergone a very large expansion during the past fifteen years. This increase during the five years ending 1887-88 over a similar period ending 1882-83 amounted to the large proportion of 32 per cent., and was accompanied by a diminution in other crops of 5·7 per cent. In 1888-89 the area amounted to 942,029 acres, a considerable increase over the average for the preceding five years. Mr. Hobson, Settlement Officer, states that with the exception of the Buldana and Ellichpur districts, the area under *jowari* and cotton has decreased and been replaced by wheat and linseed.

The distribution of the area during 1888-89 is shown in the table below :—

District.	Irrigated.	Unirrigated.	Total.
Amraoti	1	193,120	193,121
Akola	476	126,773	127,249
Ellichpur	12	70,505	70,517
Melghat	...	11,269	11,269
Buldana	6,208	241.929	248,137
Wun	...	89,738	89,738
Bassim	1	201,997	201,998
TOTAL .	6,698	935,331	942,029

RACES.—Here, as in the Central Provinces, the hard red race was at one time the most extensively cultivated, and perhaps is still so. It is, however, certain that the recent great expansion of area has arisen to meet the demands of foreign trade, and it is therefore probable that the soft and hard whites are now cultivated much more extensively than formerly, to meet the requirements of the English and South European markets. The cultivators are said to be fully alive to their own advantages, and it has been only too well proved in the case of cotton how readily they take to a new race if they, rightly or wrongly, consider it preferable.

The samples submitted to Dr. Forbes Watson in 1878 were reported on as follows :—

Class.	District.	Name.	Weight per bushel.	Price per quarter.	REMARKS.
Hard white	Nagpore .	Botka . .	61	40 6	Long, thin.
	Amraoti .	Bansi . .	61	41 6	Long berried.
	Ellichpur .	Ditto . .	60½	41 0	Ditto.
	Ditto .	Ditto . .	60	40 6	Ditto.
	Akola .	Ditto . .	61	41 6	Long berried.
	Ditto .	Ditto . .	60	41 0	Ditto.
	Buldana .	Ditto . .	61	41 0	
	Ditto .	Ditto . .	60	41 0	
	Ditto .	Ditto . .	61	40 6	
Soft red	Bassim .	Casdee . .	61	41 0	

METHOD.—There is little difference between the method of cultivation pursued in the Berars to that already described under the Central Provinces. The following remarks by Mr. Dunlop (*Wheat Prod. & Trade, 1883, 94-99*) are, however, of interest :—

SOWING.—"Wheat is sown in October and reaped in February, and is cultivated in rotation with other crops. It is never sown in two consecutive

**TRITICUM
sativum.**

Cultivation of Wheat

CULTIVATION
in
Berar.
Rotation.
743

years on the same land. The system of rotation varies according to circumstances, or the ideas of particular cultivators. A not uncommon rotation has hitherto been, *1st*, wheat, *2nd*, lac (**Lathyrus sativus**), *3rd*, cotton, *4th*, jowari, *5th*, wheat. In other cases wheat is cultivated every second year, the intermediate crop being cotton or *jowari ;* and if the demand from Europe continues, this practice is likely to become more common, to the exclusion of gram and other pulses, for which there is only a local demand.

Ploughing.
744

PLOUGHING.—" Preparatory to sowing wheat, the land is carefully prepared with the native plough (*bukhar*), and manured, if possible. Beyond this no special attention is given to the crop. A careful cultivator will pass the *bukhar* (light plough) two or three times over his wheat land in the month of May, although the crop is not to be sown till October, the object being to soften the surface of the soil, so that it may absorb more of the rainfall. Being sown after the monsoon, weeding is unnecessary, and the cultivation is consequently less costly than that of cotton or other autumn crops.

Seed.
745

" I have come across cases in which the cultivators procured their seed from villages reputed to excel in the production of wheat, but the system of selecting seed does not generally prevail."

Outturn.
746

OUTTURN.—The average acre yield for the province has been estimated by the Agricultural Department, from actual measurements of selected fields during five years, ending 1887-88, to be 4⅓ maunds, a very low figure when compared with those of most other provinces. This outturn if applied to the acreage under wheat for the past year, 1888-89, yields an estimated gross outturn from the province of 42,39,130 maunds, or 151,400 tons, an increase of about 83,000 maunds or 2 per cent. on the average for the preceding five years, though a considerable decrease on that for 1887-88.

MADRAS.

MADRAS.
747

References.—*District Manuals :—North Arcot, 416 ; Coimbatore, 211, 222, 583, 585, 586 ; Cuddapah, 43 ; Kurnool, 172 ; Nilágiri, 134, 463, 466, 467, 469, 475 ; Salem, I., 148 ; II , 61, 67, 105, 225 ; Trichinopoly, 366 ; Settlement Reports : — Chingleput, 10 ; Kistna, 15 ; Kurnool, 18 ; App. O. paras. 127, 207 ; Nilágiri, 7 ; Salem, 13 ; Madras Manual of Administration :— I., 101, 288, 290, 296, 326 ; II., 60, 78, 83, 88, 109, 119, 545 ; Report Rev. & Agri. Dept., March, 1886 ; Govt. of Ind., Wheat Prod. & Trade in Ind., 1878, 1883, 1886 ; Forbes Watson, Report on Ind. Wheat, 1879 ; Agri. Statistics Br. Ind ; Crop. Reports & Forecasts.*

Wheat is a very unimportant crop in this Presidency, an average of 27,000 acres only being under the crop. The largest wheat-growing districts are Kurnool, Kistna, Bellary, Coimbatore, and Anantapur, but a small crop is cultivated in most of the others. There would appear to be but little chance of the area undergoing any material expansion, since the soil and climate of the Presidency are not generally favourable to wheat. Thus Mr. Glenny, then Acting Director of Settlement and Agriculture in the Presidency, wrote in 1886 :—

" So far as my observation goes, Karnool, is, as one would, from the geological conditions, expect, the district most likely to come to the front in wheat-growing. I took special pains to ascertain whether there were any substantial grounds for expecting the Karnool *rayats* to set about raising wheat for export. The merits of the question appeared to be fully understood both by the large farmers and by traders ; and there appeared to be a consensus of opinion, which I am afraid is but too well founded, that of all the more valuable crops grown on unirrigated land, wheat is by far the most delicate. It suffers, and suffers permanently, from spells of bad weather that hardly affect the millets, etc. It can be success-

T. 747

in Bengal.	(*J. Murray.*)	**TRITICUM sativum.**

fully raised on no land but the best *regada* overlying limestone, and this kind of land is so valuable for *cholum* and cotton that none but a very well-to-do man, of more than usually speculative turn, cares to devote more than a small patch of it to wheat."

RACES.—The wheats grown in Madras are of very poor quality, so much so that they have been stated to rarely meet a sale for making flour, the bakers being compelled to obtain the wheat required in their trade from other parts of India (*Wheat Prod. & Trade, 1883, 34*). A very large proportion of the wheat crop appears to consist of "spelt," since in the returns of outturn from the various districts the expression unhusked wheat is commonly used. Spelt wheat is very frequently met with in the bazárs and is usually sold in the husk.

YIELD.—Official returns on this subject, contained in the report to the Government of India (1881), are avowedly untrustworthy, and need not be repeated in this work. The total outturn estimated by the Agricultural Department is 6,800 tons, or 1,90,000 maunds, a figure which gives the average acreage yield to be a little over 7 maunds an acre. There is said to be ample evidence to shew that the yield has fallen off to a considerable extent in the Kaity valley and other wheat-producing districts of the Nilghiris.

BENGAL.

References.—*Govt. Reports on Wheat Production & Trade in India, 1879, 1883, 1886; Director of Agri. Dept., Bengal, Rept. on Cultivation & Trade in Wheat, 1886; Govt. of Ind., Rev. & Agri. Dept., Proceedings, February 1884, Nos. 1 to 4; December 1885, Nos. 10 & 11; September 1886, Nos. 4 to 7; Repts. of Agri. Department, many passages every year.*

AREA.—A long and instructive report has recently been issued by the Bengal Agricultural Department on the wheat grown in the Lower Provinces. From that and other official publications it would appear that the area under the crop varies a little on the one side or the other of 1,000,000 acres, but it must be remembered that the figures shewing cultivated areas are, in the case of this Province, merely estimates formed by the Collectors on the best data they can obtain.

The average estimate of 1,000,000 acres has been adopted as a round figure in the chapter on AREA, but it may be stated that the exact average for the five years ending 1888-89, estimated by the Agricultural Department, is 1,179,500 acres, while the area in 1889-90 is estimated at 1,041,300 acres (*Final Wheat Crop Rept., 1890*).

The distribution throughout the various districts was shewn in 1886 to be—

Side notes:
CULTIVATION in Madras.

Races. 748

Yield. 749

BENGAL. 750

Area. 751
Conf. with p. 94.

District.	Area.	District.	Area.
Patna Division.	Acres.	*Burdwan Division.*	Acres.
Patna	200,000	Burdwan	1,340
Gya	82,300	Bankoora	2,000
Shahabad	200,000	Beerbhoom	710
Durbhanga	12,200	Midnapore	600
Sarun	95,000	Hooghly	59
Champarun	89,000	*Chutia Nagpur.*	
Bhagulpore Division.		Hazaribagh	1,500
Monghyr	107,200	*Orissa Division.*	
Bhagulpore	162,000	Cuttack	500
Purneah	27,600	Balasore	600
Maldah	3,300		
Sonthal Pergunnahs	8,500		
TOTAL		TOTAL	989,409

TRITICUM sativum.	Cultivation of Wheat

CULTIVATION in Bengal.

Races.

752

RACES.—"Bengal, like other provinces, grows a variety of wheat-grains which cannot be reduced to a uniform or scientific classification. The four principal kinds exported from Calcutta are—

No. 1 club, containing 75 per cent. of white and 25 of red.

No. 2 club, containing 65 per cent. of white and 35 of red.

No. 3 hard red.

No. 4 soft red.

"Buxar sends No. 1 club—a very superior soft white grain. It is extensively grown in the southern part of that sub-division and in Sasseram, and is equal in quality to the best Delhi or Mozufferpore wheat. Bhagulpore grows hard and soft red, while Bengal principally sends soft red only—a grain of very inferior quality.

"A sample of Buxar wheat grown in the sub-division of that name and picked up in a local bazár, was sent to the Chamber of Commerce, and appraised by the Committee of the Calcutta Wheat and Trade Association. It was declared to be superior in quality to the ordinary No. 1 club of trade, as the sample contained no red grain, while the ordinary No. 1 club contains a mixture of 20 to 25 per cent. of red. It thus appears that there is grown within these provinces a species of wheat equal or superior to that of any other part of India, and, therefore, that there is no need to travel outside for the purpose of procuring superior seed. In ignorance of this fact attempts have, from time to time, been made to introduce seed-grain from Cawnpore, Delhi, and Mozuffernagar, even the Chamber of Commerce being itself, I believe, unaware that Buxar No. 1 club is largely produced in the Buxar and Sasseram sub-divisions and in other parts of Behar. There is reason to believe that the cultivation of this species might be profitably extended in the South Gangetic districts of Behar, should the Chamber of Commerce see its way to effecting an alteration in the present system of refraction, thus causing superior white grain free from dirt to command an adequate price. Samples of the red wheat produced in Bhagulpore were also appraised by the Committee of the Calcutta Wheat and Trade Association, and it was found that this grain was four annas per maund inferior in value to the Buxar grain. Since the imposition of a duty on wheat in France, the demand for this species has very much fallen off. Attempts have, accordingly, been made by **Mr. Hossein**, the Agricultural Officer at Bhagulpore, to introduce Buxar wheat in that division with very satisfactory results. Though the quality of Cawnpore wheat, which was supplied by the Chamber of Commerce for experimental cultivation in Bhagulpore, slightly deteriorated, yet it was found that Buxar seed, which was also tried in the same district, yielded a grain declared by the Committee to be 'a splendid description of wheat, soft, mellow, bold, regular, and of good colour.' Its value is two annas per maund over that of ordinary Buxar quality, which itself is four annas per maund superior in value to the grain now grown in Bhagulpore. **Mr. Hossein** has been instructed to endeavour to extend the cultivation of this grain in Bhagulpore. The soil and climate of the Shahabad, Patna, Bhagulpore, and Monghyr districts seem to be well suited to the growth of wheat, while these districts enjoy considerable advantages in point of cost of carriage of the grain to Calcutta amounting to four to five annas per maund, over cultivators of the districts of the North-Western Provinces."

It will be seen from the above remarks by **Mr. Finucane** (*Report, 1886*) that though an inferior soft red wheat is the general crop of Bengal, a very superior white soft wheat does occur in the province, and that its cultivation might be considerably extended. The samples enumerated

					TRITICUM

in Bengal.　　　　　(*J. Murray.*)　　**TRITICUM sativum.**

below were selected by **Mr. Forbes Watson** as the finest submitted to him from Bengal in 1878 :—

Class.	District.	Name.	Weight per bushel.	Price per quarter.	REMARKS.
				s. d	
Soft white	Chumparan	Dawoodia	63	47　0	Small berried.
	Gya	Ditto	62½	48　0	Like Californian.
	Patna	Maghie	64½	45　6	
	Shahabad	Doodhia	62	47　0	Like Californian.
Hard white	Moorshedabad	Jamali	62½	41　6	Short berried.
	Monghyr	Bogra	62	41　0	Long berried, flinty.
	Maldah	Gangajali	63½	42　0	Very fine, long berried.
Soft red	Tirhoot	Jamalkhani	64	41　6	Small berried.
	Beerbhoom	Jamali	61½	41　0	
	Burdwan	Desi	61	41　0	Small berried.
	Bhagulpore	Jamali	62½	41　0	Ditto.
	Hazareebagh	Gour	63	41　6	Small berried, yellowish.
	Lohardugga	Jogra	63½	42　0	Stout.
	Gya	Lall	64	41　6	Stout, short berried.
	Tirhoot	Sarhal	62	40　6	Fine, clean, semi-hard.

SOILS AND METHOD OF CULTIVATION.—The soil best adapted to the cultivation of the superior kinds of wheat is said to be *karail*, a black heavy loam.　Mr. Finucane writes : "In Bijhowra village, where the best white wheat of Buxar is grown, the soil is a mixture of *karail* and *banga* in proportion of ¾ to ¼; the land is first ploughed in *Assar* (second half of June and first half of July) and is reploughed eight times before *Asin* (second half of September and first half of October), ' when it is harrowed and the seed is then sown.　In fields where two or more different kinds of seed are to be sown, all the seeds are mixed and sown at the same time.'

Mr. Peppe of the Opium Department, who has made the wheat trade a subject of special enquiry, makes the following observations on this point :—

"There are two principal tracts in which wheat is largely grown in this district; the first is the alluvial soil called the *deara*, extending from Buxar to the Sone and reaching a maximum breadth of 10 to 15 miles with a length of, say, 40 miles, thus including less than 400 square miles. This is the tract in which the beautiful soft white wheat is grown; the land is a rich alluvium annually enriched by the overflow of the Ganges : it is needless to say that no manure is required in this soil for the finest wheats, and that it is in this tract possible to grow as fine wheat as anywhere in India; unfortunately, it is in this very tract that the cultivators sow three or four and even more crops in the same field, and this is a fertile source of dirty grain, as when reaped separately they are very often pulled out by the roots (not cut), and this carries an unusual amount of dirt with the grain on to the threshing floor, but it can be easily separated if the rayat chooses to take the trouble.　From what I can gather, about 40 per cent. of the best wheat comes from this tract.　The next in import-

Method.
753

TRITICUM sativum.	Cultivation of Wheat

ance is the black clay soil extending from the eastern part of the Sasseram sub-division well into the Bhabooah sub-division, and extending from the Grand Trunk Road on the one side, and on the other to within 15 miles of the Ganges, say 25 miles by 20 miles = 500 square miles.

" In this tract some of the finest wheat is grown, about 50 per cent. of the whole outturn of the district; the cultivators attempt to keep their seed pure white, but there is a tendency to revert to the red variety : the cultivators tell me that there is no market for red wheat, and that they endeavour to produce a pure white soft wheat without admixture of any kind, and sow it entirely by itself. Only a little linseed is sown round the margin of the field, and the whole *rabi* cultivation in some villages in this tract is confined to wheat and linseed, with small patches of barley and other crops for home consumption. The wheat from this tract, although very good, is not equal to the wheat grown on the alluvium lands, and sells for one to two seers less in the rupee. The third principal cultivation of wheat is the *koeries* and others, amounting to about 5 per cent., who sow on *dih* lands, (lands adjoining the village site), all over the district, one *bigha* or so at a time on rich *dih* land well manured and irrigated. This is probably as fine grain as could be produced in this district, and the only improvement that could be effected would be by sowing a better class of seed, which the *koeries* and others would be very glad to do; as it is, they carefully select their seed and endeavour to get rid of all bad grains before sowing. There is a fourth source of wheat, *viz.*, the small patches of wheat sown in good soil at the foot of the hills or in the valleys or local alluvium along the banks of the smaller streams; this does not amount to more than 5 per cent. of the wheat grown in the district, and is very variable in quality, a large proportion being red wheat in the south of the district. With regard to the question of increasing the cultivation, the whole of the *deara* lands could be sown year after year with wheat without any loss in quality, for the land is annually renewed by a rich deposit, and therefore the produce of this tract could be enormously increased, and no doubt the time will come when the value of a well-bred wheat crop, free from earthy particles or any other kind of grain, will be as well known to the wealthy cultivator of the *deara* as to the British farmer.'

" Mr. Allen correctly remarks that the conditions of soil and climate most favourable for the production of the best wheat cannot be determined off-hand. Experiments are being made under the supervision of agricultural officers at Dumraon and Arrah in Shahabad, Sripore in Sarun, in Wards' estates in Patna, in Bhagulpore, Monghyr, and elsewhere, in order to determine what species is best suited to the soil of these districts, what is the best method of cultivation, and by the application of what manures the largest outturn may be secured with the greatest economy. It will take some years before any of these questions can be positively answered. So far as our present knowledge extends, the soil and climate of Shahabad appear to be most favourable to the growth of the best grain, while Bhagulpore has the largest area under cultivation, but its grain is inferior in quality. Whether this inferiority is due to climatic and soil conditions it is not possible with our present limited knowledge to say for certain. Further experiment and observation will be made on this point. While the facts that red grain is so extensively grown, and that the white soft grain of Cawnpore has slightly deteriorated in Bhagulpore even in one year, appear *a priori* to indicate that the preference shown for red wheat by the cultivators of that division is not due to accident or ignorance, yet on the other hand the success which has attended the experimental cultivation of Buxar seed in Bhagulpore

T. 753

in Bengal.	(*J. Murray.*)	**TRITICUM sativum.**

and the results of the experiments made in the Patna Division afford good grounds for the hope that much may be done in both the Patna and Bhagulpore Divisions towards increasing the yield, and in the former division especially, towards improving the quality of the grain."

CULTIVATION in Bengal.

The method of cultivation followed appears to be a fairly careful one. The following description by Mr. Basu of the system pursued in Lohardaga may perhaps be taken as typical :—

SOIL.—"Wheat is found to do best on *kewál* or strong clayey soils, and sufficiently well on loams and alluvial deposits. On lighter soils the ear becomes weakly and shrivelled and the grains do not develop.

Soil.
754

ROTATION.—" The crop is, as a rule, grown on *bheetá* lands, and occasionally on *bári* lands which support a clayey soil. On *bheetá* it may be grown on the same land for two or three successive seasons, after which it is found advisable to crop it with some leguminous crop for a year or two. On *bári* land it usually follows maize or *máruá* in the same year. It is either grown as a single crop, or as a mixed crop with gram or lentils. The practice of sowing mixtures is not, however, so general as in Behar. Occasionally a wheat-field is fringed on the sides with linseed.

Rotation.
755

PLOUGHING, SOWING, ETC.—"The land is opened by two ploughings in *A'ssár*, if the *rayát* can spare any time for the work, and is thus kept free of weeds and exposed to rain and the sun. But most *rayats* are at this time too busy with paddy cultivation to bestow much thought and labour on their wheat-fields, which are consequently left fallow till the close of the rainy season. If the land has not been broken up in *A'ssár* (June-July), the first ploughings are given in the latter half of September. In either case they are continued at intervals of a week or so until the land is reduced to a sufficiently fine tilth to receive the seed. When the soil has become sufficiently dry on exposure, the plough is closely followed by the *hengá* or harrow. The total number of ploughings varies usually from 8 to 10. When the soil has been well pulverised and levelled by the last harrowing, the seed is drilled in furrows by the *táriá* or drilling plough. The *táriá* goes round and round the field in gradually narrowing circles, and finishes at the centre. A plough breadth of land is allowed between two contiguous lines of seed. The quantity of seed-wheat sown per *bigha* is a little variable, rich soil requiring less seed than comparatively poor ones. On the average, 36 seers of grain may suffice for sowing one acre of land. The season for sowing wheat extends over the whole of the month of *Kártic*, and the early part of *Aughrán*. *Doáb* lands or those forming the beds of *áhrás* take longer time in drying and are the latest to be sown.

Ploughing.
756

Seed per acre.
757

" After the seed has been drilled in, the crop requires no further treatment before it is harvested. Wheat-fields are neither manured nor weeded. Irrigation of wheat is also all but unknown in Palámau, and it is only here and there that a few *rayats* are fortunate enough to be able to command the use of a village well for this purpose. The people know fully well that by means of suitable manuring and irrigation, the outturn of wheat may be easily doubled, and the precarious nature of this as well as other *rabi* crops, depending as they do entirely on the chances of rainfall, safely guarded against. But unfortunately both these measures are, under their present circumstances, quite beyond their means.

Irrigation.
758

HARVESTING.—"The wheat harvest commences in the latter part of *Fálgun* and extends over the whole of *Cheyt*, and in rare cases to the early days of *Bysák*. The crop is cut when perfectly ripe and the straw has become quite dry and crisp. The crop is reaped by the sickle at about 6 inches above the ground. It is made up into loads which are carried on the same day to the threshing floor. It is spread out on the floor for a day or two to get fully dried up, when it is threshed in the

Harvesting.
759

T. 759

11

TRITICUM sativum.	Cultivation of Wheat

CULTIVATION in Bengal.

same way as paddy. The bullocks, usually 14 in number, and tied breast to breast to two rows, are made to tread the mass for two or more days in succession until the straw which has become crisp and friable by thorough exposure and dryage is reduced into small bits, rather soft to the touch, not unlike the condition of chopped straw. The straw thus reduced is known as *bhusá* and is deemed an excellent food for cattle.

" The grain is subsequently winnowed with the basket in the same way as paddy.

Storing. 760

STORING.—" There are four different ways of storing wheat, and in fact, all other valuable grains. These are—

" (1) In *morás* (or spherical baskets made of loose straw, bound round with straw, and *chop* or bark ropes), as in Chutiá Nágpur Proper. Storing grain in the *morá* is considered the best safeguard against weevils.

" (2) In *delis* or bamboo baskets. Although the top and the inside of the basket is plastered over with a mixture of cowdung and mud, grains are found to be liable to ravage by weevils and white-ants.

" (3) In pits or *kothis,* as is the common practice all over Behar and the North-Western Provinces."

Yield. 761

YIELD AND COST OF CULTIVATION.—The outturn estimated by the Agricultural Department on an average area of 1,179,500 acres for the five years ending 1888-89, is 346,000 tons, or 96,88,000 maunds. This amounts to an average acreage yield of a little over 8 maunds. This figure is slightly below the amount estimated by **Mr. Basu,** who writes :—" The yield of wheat averages, in Palámau, about 10 *kachhá* maunds per local *bigha,* or about 9 maunds per acre, valued in ordinary times at R2 to R2-4 per maund. On good clay soils the cultivation of wheat is found to be very remunerative, as the following calculations of its cost and outturn will clearly show. For the purposes of these calculations, the local standards of land and grain measure have been taken for the sake of convenience ; and the wages of a labourer reckoned at the rate of 6 pice per diem, and the hire of a pair of oxen at the same rate as the wages of one labourer.

Cost. 762

" *Cost of Cultivation of Wheat per local bigha.*

	R a. p.
" Eight ploughings with harrowings (three ploughs can plough one local *bigha* in 1 day : 24 ploughs =24 men and 24 pairs of bullocks, at 6 pice each)	4 8 0
" Seed-grain one local maund=27 seers (*pakka*) at R1-8 . .	1 8 0
" Reaping corn and carrying it to threshing floor (10 men at 6 pice each)	0 15 0
" Threshing and cleaning grain (four men and seven pairs of oxen for two days=11 men for two days at 6 pice per diem) .	1 0 6
" Rent of 1 *bigha*	2 0 0
TOTAL COST .	9 15 6
Produce 10 maunds (local) at R1-8	15 0 0

The above figures reduced to standard measures show :—

Cost of cultivation per acre	13 4 8
Produce 9 maunds per acre at R2-3-6 per maund . .	20 0 0

" On the *batwárá* system of dividing the produce between landlord and *rayat,* the outlay of the latter would be R9-15-6 *minus* the cost of reaping (15 annas), threshing (R1-0-6) and rent (R2), that is, equal to R6. His income of grain would be R7-8 *minus* $\frac{1}{12}$th of it, being the *patwári's* share, that is R6-14."

T. 762

CROP EXPERIMENTS.—Numerous careful experiments have been performed of late years with a view to exactly ascertaining the outturn, and estimating the effect of manure on yield and quality. The results of experiments made by Mr. Allen on the experimental farm of the Maharajah of Dumraon in the Patna Division are of special interest, and are described by that officer thus :—"The grain from each of the fourteen plots was weighed in my presence, so the results given below may be considered as absolutely correct.

CULTIVATION in Bengal.

Crop Experiments. 763

The first table shows the numbers and treatment of each plot :—

Number.	Kind of wheat.	Treatment.	REMARKS.
5A	Local . .	Deep ploughed . .	} With this wheat was sown a small amount of gram.
5B	Do. .	Shallow „ .	
10A	Dumraon prize.	Irrigated it from well .	The well was old and the water probably rich in nitrates.
10B	Ditto .	Ditto from canal .	Worst soil on the farm.
11A	Mozaffernagur .	Green soiled with hemp.	} These two plots were half a bigha each.
11B	Ditto .	Unmanured . . .	
12A	Ditto . .	Manured with ashes	
12B	Ditto . .	Ditto ashes and saltpetre.	
12C	Local . .	Ditto cowdung .	Good white wheat.
12D	Do. .	Unmanured . .	Ditto.
12E	Buxar . .	Manured with saltpetre.	
12F	Ditto . .	Unmanured.	
13A	Mozaffernagar	Ditto.	
13B	Local . .	Ditto . . .	Good red wheat.

The plots were arranged so as to make the comparison as fair as possible. Careful accounts were kept of the expenditure on each plot. Each plot (except 11A and 11B) measured exactly 1,600 square yards, or a standard *bigha*. The following table gives the results of each plot :—

1	2	3
Number of plot.	Grain per acre.	Straw per acre.
	Mds. s. c.	Mds. s. c.
5A . .	16 7 0	25 29 0
5B . .	13 31 0	22 8 0
10A . .	14 3 0	16 38 0
10B . .	8 38 0	11 13 0
11A . .	27 6 0	36 38 0
11B . .	15 22 0	22 2 0
12A . .	20 4 0	22 8 0
12B . .	23 11 0	25 15 0
12C . .	18 19 0	22 17 0
12D . .	13 28 0	16 35 9
12E . .	27 10 0	62 31 0
12F . .	14 5 0	17 12 0
13A . .	15 13 0	12 33 0
13B . .	12 27 0	18 4 0

The outturn in straw can only be taken as approximately correct owing to the native method of winnowing ; but every care was taken

II A

TRITICUM
sativum.   Cultivation of Wheat

CULTIVATION
in
Bengal.

Crop
Experiments.

to have the weight of straw ascertained with as much accuracy as possible. The following points are noteworthy :—

Plots 5A and 5B show the advantage of deep ploughing. Besides
the wheat, the former plot gave 39 and the latter 25 maunds of gram.
Plots 10A and 10B apparently show that well-water is far superior to
canal ; but the result of this experiment is misleading. In the first place
the well-water was unusually good, as it was full of organic matter, *e.g.*,
old leaves, etc., and then the soil on 10B plot is about the worst oh the
farm. It would be fairer to compare 10A with 12D, the plot sown with
good white local wheat which was also canal-irrigated. Here the advantage due to the well is reduced to 15 seers of grain per acre, while the outturn of straw is practically the same in both cases. The outturn from 11A
is remarkably good, nearly equalling that from 12E, the best plot on the
farm. The hemp was well ploughed in, and must have decomposed
thoroughly in order to give so large a yield. The advantage of green
soiling in this case amounts to nearly 12 maunds of grain and over 14
maunds of straw to the acre, which of course more than covers in the first
year the cost of the treatment whose benefits will continue for at least three
or four more croppings.

"Of the remaining plots the following points may be noted :—There
are four unmanured plots, *viz.*, 12D, 12F, 13A, and 13B. These four plots
were sown respectively with local white, Buxar, Mozaffernagar, and local
red wheat. Taking grain alone, the crops come in the following order :—
(1) Mozaffernagar, (2) Buxar, (3) white, (4) red ; but if we consider the
straw only, the order is (1) red, (2) Buxar, (3) white, and (4) Mozaffernagar. The average outturn in grain from the unmanured plots may be
taken as approximately equal to about 14 maunds to the acre. At Arrah
the average was about 10⅓ maunds, but the land here is in a good season
more suited for wheat than the lighter soil of the Arrah plots. Taking 14
maunds as the average, and disregarding for the present the difference
due to the kind of grain used, we may note that the advantage per acre
due to the use of ashes alone comes to 6 maunds, to the use of ashes
and saltpetre 9 maunds, to cowdung only 4 maunds, and to saltpetre alone over 13 maunds. In fact the outturn here is nearly doubled.
I consider the yield from ashes and saltpetre on 12B, which should have
given the best result, as abnormally low. *On the other hand, the outturn
from 12E is simply astounding ; 27¼ maunds of grain, which I found
equal to 34¼ bushels to the acre, would be considered a very good crop on
good soil in a good year in England, but here in India it is, I believe, unprecedented.* The outturn of straw, over 60 maunds to the acre, or say 2¼
tons, is also very high, for it more than doubles the weight of grain, which
is not usual with a good crop. As I saw and examined the grain weighed,
there can be no doubt of the accuracy of the figures given above.

I also took the weight of the different kinds of grains per bushel, which
reached in some cases over 66℔. In appearance the Mozaffernagar grain
looked best, and I have carefully set apart the grain from the beardless
ears in the hope of growing in time a beardless variety of wheat."

Saltpetre as
manure.
764

Mr. Finucane commenting on this passage writes :—" Mr. Hossein
also made similar experiments with saltpetre as a manure in Bhagulpore,
the results of which were not so satisfactory. They, however, also show
that saltpetre applied at the rate of one and a half to two maunds per
acre will give an increase of yield which more than pays the cost of the
manure. Applied in excess of that quantity, saltpetre as a manure did not
pay. Mr. Allen applied crude saltpetre at the rate of 1⅓ maunds per acre,
mixed with an equal weight of dry earth. The price of the crude saltpetre in the Patna bazár was R3 per maund, or R4-8 per acre. The increas-

in Hyderabad, Mysore, Kashmir.   (*J. Murray.*)   **TRITICUM sativum.**

ed yield, due to the application of the manure, is 13 maunds, valued at R30, thus showing a profit of R25-8 per acre. One year's experiments, it is true, *prove* nothing; but the results, so far as they go, are very encouraging, the outturn of Buxar wheat on a plot manured with crude saltpetre being, it will be observed, 34½ bushels per acre, which would be considered a very good crop on good soil in England. It may be here remarked that saltpetre is a forcing manure, and the continued application of it may lead to exhaustion of the soil; but, as Mr. Allen thinks, the exhausting effect will not be apparent for many years, and may be counteracted by green soiling once in five years. He accordingly advises the application of saltpetre as a top-dressing when the crop is a few inches high, followed by irrigation. Next to saltpetre, the best results were obtained from green soiling with hemp.

CULTIVATION in Bengal.
Crop Experiments.

" As to the effect of different soils in changing the colour of the grain, there appears to be no doubt that wheat brought from a distance will, under the influence of soil and climate, gradually change its character. The Committee of the Wheat and Trade Association report that the white Cawnpore wheat supplied to the Committee and introduced by Mr. Hossein into the Bhagulpore Division deteriorated in colour, and this result is in accordance with the general opinion of the cultivators in the wheat-producing tracts. In connection with this point the Sub-divisional Officer of Barh reported in 1884 that the cultivators of that sub-division understood the superior value of white soft wheat, but it will only grow on certain soils, and if sown on the soil known as *teliya kewal* will, they alleged, turn to red. Dr. King, Superintendent of the Botanical Gardens, explained the fact, supposing the cultivators' opinion to be well-founded, by the theory that *teliya kewal* is more suitable for red than for white wheat, that the few seeds of the red sort that were mixed with the seed grain or remained in the field probably throve better than the white, and thus produced a larger proportion of seed. This process probably went on for a few years, when the major part of the crop became red, and the belief thus arose that the soil caused the white wheat to change to red. Mr. Macpherson, in reporting on the point, after enquiry in Nasrigunge, expresses the opinion that the degeneration of white wheat into red on certain soils is so universal that Dr. King's explanation will not suffice. Whatever the explanation may be, there can be no doubt of the fact that the white *doodhya* wheat which grows well on *balsundar* degenerates when grown continuously in *kewal* (stiff clay) soil. Experiment will be made in order to ascertain under what conditions the degeneration occurs, and how far it can be obviated by importation from time to time of fresh seed. "

The results of crop experiments in other Provinces have been briefly noticed, but a very large literature on the subject exists, which it has been impossible, from want of space, to enter fully into. It may, however, be stated that the remarkable effect of nitre in increasing the outturn, above described, has been marked in nearly all cases in which it has been employed, in all sorts of soils.

HYDERABAD, MYSORE, KASHMIR.

In all these Native States wheat is an important crop, but it is impossible to deal with them in detail. Few accurate particulars are on record regarding them, and we can do little more than give an account of the areas and outturn estimated in the agricultural statistics of the Government of India. The averages during the four years ending 1887-88 and that for 1888-89 may be most conveniently represented in tabular form.

HYDERABAD, MYSORE, KASHMIR.
765

TRITICUM sativum.	Cultivation of Wheat.

CULTIVATION in Hyderabad, Mysore, Kashmir.
Conf. with p. 94.

The figures are taken from the final report of the Revenue and Agricultural Department on the wheat crop of 1888-89 : —

NATIVE STATE.	1884-85 to 1887-88.		1888-89.	
	Area.	Outturn.	Area.	Outturn.
	Acres.	Tons.	Acres.	Tons.
Hyderabad	1,111,000	87,000	1,111,000	87,000
Mysore	14,000	1,700	4,282	1,700
Kashmír	500,000	133,000	500,000	133,000

ASSAM & BURMA.
Conf. with p. 94.
766

ASSAM AND BURMA.

As already stated, the wheat crop in these provinces is at present of very little importance, though there is every probability that the area in Upper Burma, the climate and soil of many parts of which are admirably adapted to the growth of wheat, may undergo considerable expansion.

CHEMISTRY.
767

CHEMISTRY OF WHEAT.

The composition of wheat-grain varies considerably, but this variation is confined almost entirely to the relative proportions of starch and of nitrogenous matters, although the mineral and other minor constituents are not quite fixed in amount. The starch may vary from nearly 70 to less than 61 per cent., while the nitrogenous matter may be found to the extent of anything between 10 and 16 to 17 per cent. Of the minor constituents, the ash may be increased in proportion to a small degree by a wet season, a thin-skinned, well-developed sample contains less fibre, and a plump dark-coloured grain has a larger proportion of oil or fat.

All Indian wheats are characterised by dryness, containing at least 2 per cent. less moisture than average English wheat, and are further remarkable for a high percentage of albumenoids. Church states that he has never yet met with an Indian wheat which contained less than 10 per cent. of albumenoids; while a large number of samples of first-rate English, Canadian, and Australian samples give numbers between 8 and 9. According to the same authority the average percentage of albumenoids in Indian wheat is about 13·5, but some are as low as 10·3, and some as high as 16·7. Albumenoids are much more abundant in hard than in soft wheats, while soft opaque grains are richer in starch. This has been clearly shown by McDougall's analyses of the starch and gluten in the various Indian forms (see p. 101) which, though scientifically inaccurate as a guide to the exact amount of nitrogen, is of value as shewing the comparative variation. Differences in the chemical composition of wheat-grain are said to be found not only in various cultivated races, but in the same sort when cultivated under different conditions of climate or season. Even in a single grain such variations may exist. Forbes Watson and other writers have frequently noticed the fact that the grain of many varieties is partly horny and translucent, partly soft and opaque, in which case its composition corresponds with its mixed appearance. Church affirms that variations in the percentage of albumenoids may frequently be observed even in the grains from a single ear, analysis showing sometimes 3 or 4 per cent. more albumenoids in certain of such grains than in others.

According to the same authority, who has had the opportunity of carefully analysing many samples, the average composition of Indian wheat is, in 100 parts,—water, 12·5; albumenoids, 13·5; starch, 68·4; oil, 1·2;

T. 767

fibre, 2·7 ; ash, 1·7. The " starch " in reality comprises about 2 per cent. of the sugar or sugars found in many cereals, but this fact may be regarded as not appreciably affecting its food value. The ash, though not large in amount, is of great value as a source of mineral nutrients, when the grain is a staple article of dietary, containing as it does some 30 per cent. of potash and 45 per cent. of phosphoric acid.

CHEMISTRY.

" The various mill products obtained in grinding wheat differ much from each other and from the original grain, in several important particulars. For instance, the following figures were obtained in a series of analysis which I made of an entire series of such mill products :—

	Per cent. Nitrogen.	Per cent. Oil.
Whole wheat	1·692	2·02
Flour (white)	1·621	1·4
Flour (seconds)	1·967	1.82
Bran	2·143	2·75
Sharps (fine)	2·608	3·50

" These products represent but four out of a total of twelve, but they suffice to show how large a proportion of nitrogenous matter and of oil are rejected when fine flour is the sole product reserved for human food. It must not, however, be assumed that all the nitrogen, say, in fine sharps, is albumenoid; in fact, these fine sharps did not contain more than 13½ per cent. of albumenoids, though 2·608 of nitrogen corresponds to 16¼ per cent.; even the fine whiter or flour contained a little nitrogen in non-albumenoid forms " (*Church, Food-Grains of India,* 95).

USES OF WHEAT.

Food & Fodder.—The methods of employing wheat for human food in India vary somewhat, but the following may be accepted as the most important. From the GRAIN three chief kinds of flour are made, namely, *súji, maida,* and *atta.* The first is a granular meal obtained by moistening the grain overnight, then grinding it. The fine flour passes through a coarse sieve, leaving the *súji* and bran above. The latter is got rid of by winnowing, and the round, granular meal or *súji,* composed of the harder pieces of the grain, remains. This preparation is in reality a sort of *simolina,* and is most easily produced from the hard wheats which contain a larger percentage of gluten. It is highly appreciated by all classes, but is expensive, and consequently only within the reach of the well-to do. It is also employed in making confectionery, or in place of oatmeal in making a kind of porridge. The hard white wheats from which the best *súji* is made, are, as already stated, admirably suited for the manufacture of maccaroni, from their high percentage of gluten, and are exported to a considerable extent for this purpose to Southern Europe. *Maida* and *atta* may be prepared from the flour obtained in making *súji* by regrinding it and passing it through a finer sieve, the finer flour which passes through being known as *maida,* and the coarser as *atta.* They are, however, generally prepared without moistening and separating the *súji,* the grain being at once ground into *maida* and *atta. Maida,* or fine white flour, is, like *súji,* a luxury of the richer classes; *atta* is the ordinary form in which wheat is consumed by the people. It is generally cooked in the form of flat cakes of unleavened bread, resembling a girdle scone, and known as *chapatti,* which constitute one of the chief articles of diet in many parts of India, especially amongst the Muhammadan population. *Chapattis* are eaten with *dal, ghí,* or any other relish. They are very simply prepared, namely, by kneading the flour with water, passing the dough with the hands into a flat cake, and baking it over a fire on an iron plate, or on a hot

USES.
768

Grain.
769

Suji.
770

Maida.
771

Atta.
772

FOOD.

earthen platter. Fried with *ghi* and sugar, and seasoned with various condiments or spices, they form the chief substitute for bread amongst the well-to-do. In certain localities the *atta* employed is in reality produced not from pure wheat, but from the common mixed crops of wheat-barley or wheat-gram.

In the larger towns ordinary bread, prepared and leavened after the European manner, is said to be rapidly gaining favour with certain classes of the population; but, with this exception, fermented bread is little known in India. The qualities of the different Indian wheats as food-stuffs, and their adaptability to the requirements of bakers and millers in Europe, have been already discussed (see p. 101).

Wheaten straw, cut up as *bhusa*, is largely employed as a fodder for cattle, sheep, and horses, either alone or mixed with barley-straw and the haums of pulses. In the Panjáb and other parts of Northern India the young leaves are frequently cut for fodder, or sheep, and occasionally even cattle, are allowed to browse the young crop. Experiments have been made on Government Farms with wheat as a forage crop, but though much esteemed for this purpose in Japan, and later in Australia, the results in India have not been encouraging.

MEDICINE.
Flour.
773
Starch.
774
Bread-Crumb.
775
Bran.
776

Medicine.—Wheaten FLOUR, STARCH, and BREAD-CRUMB are officinal preparations largely employed for many purposes. The flour is esteemed as an external application in erysipelas, burns, scalds, and various itching or burning eruptions. A mixture of flour and water is used as an anti-dote in cases of poisoning by salts of mercury, copper, zinc, silver and tin, and by iodine. BRAN, though not officinal, is sometimes used in the form of a decoction or infusion, as an emmolient bath, and also internally as a demulcent. Bran poultices are useful for many purposes, and bran bread is slightly laxative and may be used with advantage in certain dyspeptic conditions, and, owing to its freedom from starch, in diabetes. Starch prepared from wheaten flour is employed for sprinkling over inflamed sur-faces, to absorb acid secretions and prevent excoriation. Mucilage of starch is valued in pharmacy for many purposes, and medicinally as a demulcent. In surgery it is occasionally employed for stiffening bandages, etc. BREAD-CRUMB is useful for giving bulk to pills, and for making poultices.

USES IN THE ARTS.—The well known properties of starch and muci-lage of starch, and the value they possess in the arts, require no comment in such a work as the present. Wheaten starch is too expensive to be much employed for similar purposes by the natives of India, and is gene-rally replaced by mucilage made from rice.

(*G. Watt.*)

TRADE IN INDIAN WHEAT.

INDIAN CON-SUMPTION
of
WHEAT.
777
Conf. with pp. 194, 197, 198.

It might almost be said that no branch of Indian Commerce has attracted so much attention in Europe as that of Wheat. In point of value, however, to the people of India, the foreign traffic in that cereal is of com-paratively little importance. To a very large extent Indian wheat is re-quired to supplement deficiencies in the European supply, so that the trade is liable to the most violent fluctuations. It has been estimated that the annual average exports up to 1890-91 would not have fed above $1\frac{1}{2}$ millions of people, and that that quantity was but about 10 per cent. of the total wheat crop. Worked out from a yield of 7 maunds an acre the surveyed acres of India would (in 1890-91) have yielded 5,046,990 tons (*Conf. with p. 94.*) The unsurveyed regions have been estimated to furnish $1\frac{1}{2}$ million tons. The foreign exports for the year named were 716,024 tons, so that the Indian consumption may be said to have been little under 6 million tons. These general statements exhibit, therefore, the position and re-lative value of the foreign and internal traffic in wheat.

T. 777

| Trade in Indian Wheat. | (*G. Watt.*) | **TRITICUM sativum.** |

TRADE.

Hotly contested opinions and theories, and even flagrant misstatements, have been thrust, however, on the public for serious consideration, as if the foreign export of wheat from India involved national and even international obligations. The European supply of wheat, the markets where the chief transactions are made, and the agencies employed in the trade, had been gradually matured. India, up to a certain year, played no part in the time-honoured calculations of the European merchants, and her appearance in the corn-markets seems by many to have been viewed as introducing not only new disturbing elements, but as bringing with her new competitors. The vast area of India rose as a dream of future trouble—a potentiality of danger. The delusion that rice was the staple food of India, and, therefore, her chief agricultural crop, seems never to have been more rudely dispelled than by the emphatically demonstrated fact that, she not only grew wheat but was prepared to contest the foreign supply of Europe in that grain. The natural restrictions imposed on the future expansion of the trade by the climate and facilities of transport, and by the prejudices and interests of the cultivators, appear to have escaped the consideration of certain sections of the wheat interests. The teeming population of India could grow and export wheat, and that, too, of a quality that must meet many demands. And what was even more perplexing, the grain could be actually landed at a price below that of the wheats then in the market. Why this trade should have had no existence prior to the opening of the Suez Canal, and why it had in little more than ten years from that date assumed commercial importance, were questions that for many years escaped consideration. Indeed, it might almost be said that a large proportion of those most interested in the wheat trade barred India from their careful consideration, until the traffic had accidentally or purposely drifted into a peculiar groove, from which it has since vainly struggled to escape.

Adulteration (Refraction) and Sale on Standard of Fair Average Quality (*f. a. q.*) —But the history of the past expansion of the traffic, such as it is, differs in no material respect from the records of the early transactions with America. About thirty years ago American wheat came to England with as much as 10 per cent. of prairie oats, rye, and other impurities. It was sold on a standard of "fair average quality," based on shipments, and it naturally therefore soon became of interest to certain shippers to establish as low a "fair average" as possible. The majority of the consignments fetched in consequence an unjustly low price. But the American producers, backed by a vigorous Board of Trade, soon saw the necessity of reform, and that they effected by producing a cleaner article and by establishing their own "fair averages" on the other side of the Atlantic in "the graded system" which has ever since prevailed. America thus became able to dictate her terms so to speak instead of being dependent on the purchasers for the valuation of her produce. At the present moment a similar reform is earnestly being initiated by the Russian Government. A series of questions, very similar to those issued to the millers of Great Britain by the India Office, has been addressed to the corn merchants of the world, the purport of which may be said to be that if a system of grading Russian wheat be desired, the Government of that country would be prepared to carry such into force. India alone, therefore, may be said to have her markets regulated by the buyers on an arbitrary standard of "fair average quality," established by themselves, and of so variable a character that it is of no value to the millers in their dealings with the importers. It would not be impossible, for example for a large shipper, by flooding the markets with inferior wheat (for a time at least), to establish the standard that

ADULTERA-TION.
778
Conf. with pp. 110-116, 186.

T. 778

TRADE.
Refraction.

suited his own purposes. Into this aspect of the Indian wheat trade must undoubtedly be placed the much talked of system of "refraction," by which a fixed percentage of adulteration has, it might be said, been legalised. The defects and imperfections of the Indian wheat were accordingly hailed as subjects of infinitely more pressing moment than the possibilities of future expansion and improvement. It was widely proclaimed that the crude appliances of the Natives of necessity resulted in a dirty article, and that their practice of growing mixed crops occasioned unavoidably the presence of other and injurious grains among the wheat. These defects, it was urged, necessitated protective regulations. But it was not contemplated (or, if it had been, the result has frustrated the good intention) to establish inducements that would ultimately remove these defects. An arbitrary rate of "refraction" was fixed above which deductions could be made from contracts, but below which should be the shippers' gain. That being so, was improvement possible? Certainly not, and therefore the cultivator or the local small buyer soon realized that it was to his interest to see that his wheat was directly adulterated to the prescribed rate. Without, for the moment, attempting to deal with the question as to whether any other system might have been designed that would have tended towards improvement, it may, without fear of contradiction, be affirmed that the dictates of the shippers, or whoever fixed the refraction, called into existence in India direct adulteration. That the Natives can and do produce, when they so desire, a good quality of fairly clean wheat, no one will deny who is familiar with India or who has perused the extensive official correspondence that exists on this subject. Nothing more is paid for clean wheat than for wheat adulterated to the established standard. Two new industries, therefore, took their birth, namely, (1) organized adulteration (for which certain muds and grains were found more serviceable than others), and (2) that of cleaning the wheat. A premium has, in other words, been paid to the larger and more prosperous millers who can afford to erect cleaning machinery against the smaller millers who have had (and at the present moment have still) to either purchase Indian wheat specially cleaned or be debarred from participation in the new traffic in Indian wheat. Mr. Samuel Smith, of the Victoria Corn Mills, Sheffield, states, for example, that not more than one in twenty millers are in a position to be able to clean Indian wheat.

CLEAN
WHEAT.
779
Conf. with
p. 112.

QUESTIONS
CIRCULATED
to
BRITISH
MILLERS.
780
Conf. with
p. 114.

That this state of affairs is not desired by the trade, as a whole, needs no further refutation than the replies furnished by the millers of Great Britain to the series of questions issued by the authorities of the India Office. One of those questions was—"Would you use larger quantities if free from admixture and impurity?" The answer sent by the vast majority was emphatically that they would. To whose interest is it, therefore, that the cultivators or small traders are compelled to adulterate, or rather that those of them who might find it to their interest to consign clean wheat, are practically prohibited from so doing? To whose interest is it, still further, that the millers of England are debarred from obtaining the full amount of Indian wheat that they desire? Mr. S. Smith says that among other reasons the English millers' objections to buy Indian wheat, on terms of *f. a. q.*, with a dirt clause of 5 per cent., are that the average sample which is to test the quality of delivery is a monthly preparation by the London Corn Trade Association over which the miller has no say. That the average sample varies considerably and thus affords no criterion to the millers what to expect in succeeding months. That in the case of arbitration the members of the London Corn Trade Association are the sole arbitrators. Further, that during arbitration the miller (however distant he may reside) has to attend the court at the port where the wheat was purchased. Also, that it is impossible to readily ascertain that not more than 5 per cent. dirt has been mixed with the wheat, etc., etc. As matters stand, the profits of the cultivator are, therefore, restricted, and a possible extensive wheat production is rendered impossible, very largely through the grain having to bear the enhanced railway charges and shipping freights on the percentage of dirt with which the wheat has to be adulterated before it is acceptable to the parties who have established and who maintain a fixed rate of refraction. All these additional charges the English consumer of Indian wheat has to pay, as also those occasioned through the process of cleaning. But it has been urged that in the present phase of the Indian wheat trade, there exist numerous difficulties that preclude the establishment of the graded system. Without admitting this argument as well founded, it may be repeated that a wholesome and much needed reform might in that case be effected by the establishment of the fair average quality (*f. a. q.*) on receipts not shipments. Arbitration is by no means saved by the existing system, and, indeed, it would not be materially increased were all sales to be made on individual samples instead of averages.

The Council of the National Association of British and Irish Millers point out, however, the difficulty of purchase on actual samples of arrivals as follows :—

It has been argued that millers need not buy on the form of contract issued by the Corn Trade Association, but can purchase on sample on arrival; but, as a rule, millers purchase Indian wheat for forward shipment in order to cover the sales of flour made for forward delivery, and are compelled to use the form of contract or not buy in this way at all. The Indian wheat coming by the Cape is very much purchased for its convenience as "cover," and if the millers so buying are not able to clean it, they have to sell on arrival. It is, therefore, of great importance that the contract form should be altered, and the proportion of dirt allowed very much reduced.

It would thus seem essential to the future prosperity of the trade, that, however effected, sales be made in such a way as to allow of a more direct reward for quality and purity. Such a position is the only natural incentive to reform and progression, for, by present arrangements, dishonesty is more likely to triumph than honesty. If, therefore, it be the case that any party of traders be benefited by the present system, they are so at the direct loss of the country. It would thus seem that the Government might almost step in and impose such restrictions as would enforce reform on the cultivator, if the fault lies at his door, but absolutely prohibit the traffic where it fosters individual gain at the expense of the community at large. This is, however, a different problem to the spasmodic prohibition of the trade when, through scarcity, the prices of food-stuffs have been raised. The one would be the adjustment of trade so as to remove injurious tendencies, the other, interference with free action within constitutional limits. Sir E. C. Buck has very properly pointed out that the existence of a large wheat trade is essentially a preventative measure against famine. Were non-food crops to be substituted for wheat, the country would be deprived of a store of food which restricts famine tendencies by becoming available when the price of the inferior grains rises to the level of wheat. It then becomes more profitable for the cultivator to sell his wheat for local use than to export it. A further rise in the price of food is thereby prevented.

Present Position and Character of the Trade —The wheat that is usually exported is essentially grown for the foreign market, and is, therefore, more than a surplus over local requirements. With the great majority of the wheat-growers that cereal is only to a very slight degree an article of food. It is produced purely and simply for a prescribed market and mainly as the rent-paying crop. The prosperity of the American wheat traffic, since the establishment of the graded system, affords abundant proof of the advantages of regulations that very largely fix the

TRADE.
Refraction.

Difficulties of purchase.
781

POSITION of Indian Trade.
782

TRITICUM sativum.	Present Position and Character

| TRADE, Character of. | standards of sale in the country of production. It has been contended by some writers, however, that the improved facilities of transport and the cheapening of the rates, sufficiently account for that prosperity and for the hold America has been able to take on the European markets. These improvements, it has been even argued, are the sole cause of the fall in the price of wheat that has occurred within the past ten years or so. But surely the carrying agencies work for their own gain, with as keen an eye to that object, as in any other branch of human enterprise, and it is contrary to all experience to suppose that reductions would have been effected and facilities extended if the necessity for these had not been felt. America had to contend inch by inch for her present supremacy, and not the least important advance was the security of transactions attained by the graded system. In India, reductions in transit and freight charges have been made on a scale quite as great comparatively as in America, but in spite of these, the Indian trade has not advanced relatively to the same extent. The cause of this, as would seem, must be sought for in the defective system of an *f. a q* standard and in the clause that provides for, instead of prohibiting, adulteration. But there are writers on this subject who think the American carrying agencies have reduced their charges to the point where the traffic possesses little attraction. It is in fact contended that the carrying power of the world is in excess of its requirements, and that, with the increasing population of America and a protective legislation that fosters home industries, the time is rapidly approaching when less wheat will be available for export to Europe. Indeed, it has even been predicted that within a very few years the American export is likely to be entirely stopped. Any such revolution would of necessity bring India into the field as a very necessity in the European supply of wheat. But without entertaining such extreme views, it may be said India is even now a factor of no mean importance, as may be seen by the fact that the exports of wheat in 1891-92 were more than double those of 1890-91. In spite of all disadvantages and adventitious charges, Indian wheat has for some years continued to undersell the produce of the old and established supplies, and is gradually assuming a recognised position in the grain markets of Europe. The outcry has, accordingly, in certain quarters, been raised against the objections to this new traffic. Philanthropy, that much abused ally of a weak cause, has been called to the rescue. The natural food and surplus stocks of the people, we have been told, were being drained away from them. For greed of the means to satisfy exotic desires of modern civilization, the people were being induced to part with their ordinary food, and were, in consequence, taking to the use of inferior and unwholesome grains. These and such like wildly exaggerated statements have been seriously advanced as ascertained facts—and that too by persons who might have been better informed. Such absurdities do not, therefore, call for a scrutiny of the village granaries, since they can be abundantly refuted without treading on the susceptibilities of the people, or awakening their fears of fresh taxation. The trade in wheat, though up to the present moment it might almost be said to have rested on an insecure and unsatisfactory basis, similar to the position Indian tea occupied a few years ago, does not stand alone as a branch of Indian commerce that manifests an expanding supply to foreign markets. The total exports from India were in 1866-67 valued at £44,291,497 (nominal sterling), in 1871-72 they stood at £64,607,020, and in 1889-90 at £105,238,800. The total trade of India (imports and exports) had, in fact, manifested an average expansion of very nearly 5 per cent. for each year during the ten years ending 1889-90. The writer chooses to deal for the present with those years, because within that period wheat rose steadily to a very high level and then slowly declined. It has within the past two |

T. 782

of the Indian Wheat Trade. *(G. Watt.)*

years manifested a most remarkable revival, and the traffic during the current year will very probably exceed even that of 1891-92. During 1886-87 the exports of wheat came to £8,625,863 (nominal sterling) and in 1890-91 they had declined by about £2,000,000, though in 1891-92 they had reached their highest recorded limit, *viz.*, £14,382,244 (nominal sterling). But this record of prosperity does not stand by itself. In 1850 the exports of cotton were valued at £3,474,789, and ten years later at £7,342,168, but in 1863 they suddenly rose to £35,864,795 and in the following year to £37,573,637. No obscure theory is necessary to account for this. The demand created by the American war was placed in touch with India's possibilities. With restoration of peace India's greatest agricultural competitor resumed a large portion of her former trade, and the exports of cotton rapidly sank to their present level of 13 to 15 millions nominal sterling value. These figures would show that a large percentage of the grain-producing area was possibly thrown under cotton, and that when the trade declined the new cultivators were in all probability ruined. But this was not the case; their energies were simply directed to the new channels of which the expanding foreign commerce of India gave them an extensive choice. But other and equally striking examples may be given. An export trade in oil-seeds had practically no existence thirty years ago. In 1867 it had run up to £1,853,000, and in 1889-90 it stood at £10,627,553. The export of rice was valued in 1867 at £3,647,007, in 1891-92 at £13,385,970. The export of raw jute was returned in 1828 at £62, in 1867 at £1,600,554, and in 1890-91 at £8,639,900. But it is perhaps needless to multiply examples of the corresponding expansion of the articles of India's foreign export trade with that of wheat. If, therefore, the people are being deprived of their wheat they do not seem to find it necessary to curtail their cultivation of cotton, oil-seeds, sugar-cane, millets, rice, pulses and the hundred and one crops that are exposed to them for selection. We have, in these facts, evidence, therefore, of a capability for expansion which, apart from the immediate cause, is a factor of vital importance. The limits of the future have not been even foreshadowed by past events, but one point has been clearly shown, *viz.*, that the growth of the export traffic in all its branches has not proceeded from any single cause, such as the depreciation in the value of silver. Striking fluctuations have occurred in one article of the trade and not in others, and these fluctuations have borne no fixed relation to the downward tendency in silver. Would it, accordingly, lead to any satisfactory conclusion to seek out for each separate item some obscure reason for its individual growth and prosperity? Is it not much more natural to see in India a vast agricultural country, which by civilization and facilities of transport is being year by year brought into the arena of European commerce. The fluctuations of the individual articles of that commerce are doubtless governed by specific causes, just as a stone thrown into the river produces a temporary disturbance in the onward current of the stream. The stream, so to speak, of the wheat trade flows from the fountain-head of all agricultural enterprise—the profits of the cultivator—and will be checked when these become less than can be obtained from other crops, and thus the traffic is primarily governed by the prices that prevail at the consuming markets. Hitherto, wheat has proved a profitable crop, and that, too, at the low prices which two or three years ago prevailed. It has been loudly contended that the wheat trade of India owes its prosperity, if not its very existence, to the depreciation in the value of silver. There are, however, certain broad principles that demonstrate the insufficiency of this argument. The growth of India's external commerce was year by year quite as striking before any such influence existed as since. The imports which should have been inversely affected have developed to a corresponding

TRADE,
Position of.

GREAT
POSSIBILI-
TIES
of
EXPANSION.
783

Depreciation
of
Silver.
784
*Conf. with
p. 182.*

TRADE.
Depreciation
of Silver.

degree with the exports. And, moreover, the fluctuations of the wheat trade manifest no synchronous relation to the value of silver. On the contrary, the exports, instead of continuing to expand from 1886-87 with the fall in the value of silver, seriously contracted, and in 1891-92 suddenly doubled those of 1890-91. Without wishing to deny absolutely any advantage from depreciation of silver, it may safely be affirmed that such advantage must be more than effaced by the adventitious charges incident to refraction. The export of wheat from India can, however, be shown to be governed by the demand for wheat in Europe.

Some few years ago the writer contributed a paper " On the Conditions of Wheat Growing in India " to the Journal of the Royal Agricultural Society of England. He there stated :—" It is of little consequence whether the depreciation in the value of silver acts favourably or unfavourably, unless it can be shown that the existence of the wheat trade is vitally dependent on the fluctuations of the silver market. Many causes have doubtless combined to assist in the establishment of the present remarkable trade. The question at issue may be stated briefly thus : *Is the trade a good and natural one? Has it reached its maximum development?* The former will have to be answered, among other considerations, by an enquiry in India as to whether it is profitable to the cultivator, and in Europe as to whether it is meeting a demand which another country in the future may not more successfully contest. The latter can alone be solved by a somewhat detailed analysis of the sources of food supply of the people of India taken in the light of the increasing population, the possible extension of agricultural operations, and the profitable establishment of new branches of industry or the growth of indigenous handicrafts. These are problems that represent the adjustment between productiveness of soil and man's inventive resources." " Far more will depend in the future on the growth of our cotton, jute, woollen, paper, oil, and other mills, than on the special demands of Europe for Indian wheat. Indeed, thoughtful men in India are beginning to speak in an undertone of India's agricultural prosperity as her greatest source of weakness. But it is an open question whether Europe would suffer most under the importation of a large surplus of cheap agricultural produce, or in having the Indian market closed to European goods through the growth of local industries." " Influences of a perfectly natural character have, during the past twenty or thirty years, been operating favourably to the wheat trade—have, in fact, been developing every branch of India's foreign commerce. Some of the more important of these may be here exhibited." The area of India is 1,382,624 square miles, and it possesses the climates and soils of the world. Its agricultural possibilities are, therefore, almost limitless. The population is rapidly increasing. The surveyed agricultural area of India is about 600 million acres, or a little less than half the geographical area of British India. The area of actual cultivation fluctuates from year to year, but on an average about 100 million acres are usually returned as available for future agricultural expansion. Among the modern facilities that have been effected, the opening of direct telegraphic communication (in 1865) between England and India may be mentioned. It was thus rendered possible to exchange a knowledge of the conditions of both markets at any given time. The Suez Canal, already alluded to, was opened in 1869, by which the time necessary to deliver goods from India was reduced from three or four months to as many weeks. The opening of the Prince's Docks, Bombay, enabled the shippers from that port to carry on work throughout the year, whereas formerly the monsoons practically stopped the export trade. The similar greatly improved facilities at Karachi have made that port one of vast importance and that, too, within the past few years. Railway communication has also been rapidly extended, thus not only increasing the

INFLUENCES
AFFECTING
INDIAN
WHEAT
TRADE.
785

Telegraphs.
786

Suez Canal.
787

Docks.
788

of the Indian Wheat Trade.	(*G. Watt.*)	**TRITICUM**
		sativum.

facilities of transport, but greatly cheapening the Indian (inland) charges.
The railway system of the Indus Valley placed an immense wheat area in
direct touch with Karachi, and the completion of the Midland Railway
system has closed many of the sources from which Calcutta formerly drew
its supplies of wheat, but has brought these into direct touch with Bombay
and Karachi, whence homeward charges are much lower than from Calcutta. This recent adjustment of the Indian trade doubtless accounts for
the facts exhibited in the table (page 189) in which the wheat exports from
Bengal are shown to have declined from 6,668,047 cwt. in 1881-82 to
1,340,355 cwt. in 1890-91, while during the same period the exports of
Karachi expanded from 1,852,334 cwt. to 6,767,300 cwt. The trade from
Bengal manifested a revival in 1891-92, but by no means recovered its lost
ground. Indeed, it is impossible to avoid the conviction that the rapidity
with which India is progressing may sometimes temporarily act almost
prejudicially in upsetting established agencies or channels of trade before
others have become fully able to take their places.

But significant though improvements in transport are, and of vital importance to the wheat trade, it must not be forgotten that the construction
of irrigation canals has thrown large tracts of fertile land under that crop,
which formerly, from the want of water, were almost sterile. The Ganges
Canal may be mentioned as an example of this nature, but only one out
of many such. It waters the Doáb (or interfluvial tract) between the Ganges
and the Jumna, and its main stream is 525 miles in length.

Within her own territory India can produce all the requirements of
modern trade. The facilities of interchange are even now sufficient (and
year by year these will be improved) to allow each cultivator to throw his
land under the crop best suited to it and the one most remunerative. Is
it to be wondered at, therefore, that he is gradually unlearning the lesson
of his ancestors and embracing the modern one that dependance for food
and clothing on the actual produce of his own small holding is impolitic and
wasteful? The inferences which, the facts briefly reviewed point to, may
thus be stated as (1) that India is a vast agricultural country the resources
of which are only beginning to be realised by the outer world : and (2)
that the extent to which the Natives are willing to adapt their agricultural methods and materials to the requirements of Europe is essentially
governed by personal gain. Nothing could more forcibly manifest this
fact than the response India gave in 1863 to England's demand for cotton.
Let a profitable market open out in Europe for Indian wheat, and the
supply would soon show the possibilities of this vast agricultural country.
It may now perhaps be admitted as quiet unnecessary to call in the aid
of the theory of the depreciation of the value of silver, or indeed of
any other obscure theory, when we pass in silence the thousand and one
benefits the country has derived from a peaceful administration, from the
practical liberation from customs duty, and from the independence and
equality conferred on the millions of the people, reforms that are daily bringing the actual producer into direct dealings with the consumer.

Influence of Indian Wheat on the European Markets.—In a paper by
W. E. Bear, Esq., which appeared in the Journal of the Royal Agricultural
Society of England (*XXIV., Second Series, 1888, pp 50-80*), the effort is made,
however, to show that the economies effected in India in transit and other
charges may be accepted as accounting for one-half the decline in the price
at which Indian wheat has been recently offered in England. The
other, half, he thinks, may, to some extent, be due to an advantage
gained by India through the depreciation in the value of silver. He holds
very emphatically, however, that the continuously decreasing rate at
which Indian wheat was being placed on the home markets had largely

**TRADE
INCREASED
FACILITIES.**

**Railways.
789**

**Irrigation
Canals.
790**

**Internal
Interchanges
791**

**FREEDOM
of
TRADE.
792**

**INFLUENCE
of
Indian Wheat.
793**

INFLUENCE
of
Indian Wheat.

assisted in the general fall in the price paid for wheat all over the world.
"It has been pointed out," he says, "that the exports of wheat from India
were not considerable until 1881-82, and, whether it be merely a coincidence,
or more than that, it is a fact that the average annual price of wheat in
England has been permanently below 45s. a quarter only since 1882. It
has further been remarked that we must consider the total supplies of Indian
wheat to Europe, and not those received in England only, in endeavouring
to form a fair estimate of their effect upon prices here." Mr. Bear then
furnishes the following table : —

Wheat and Flour imported into the United Kingdom.

	1881. Qrs.	1885. Qrs.	1886. Qrs.	1887. Qrs.
India	1,693,560	2,809,676	2,544,725	1,963,637
United States . . .	10,547,144	8,985,730	8,983,880	11,615,950
Russia	947,147	2,788,244	872,892	1,282,312
Other sources . . .	3,276,301	4,175,562	2,782,664	3,220,108
TOTAL . .	16,464,152	18,759,212	15,184,098	18,082,007

"These figures," he continues, " show that our receipts of wheat from
India, which in only one previous year had been as much as 5 per cent.
of our total foreign supplies, rose to 10·3 per cent. in 1881 to 15 per cent. in
1885, and to 16·7 per cent. in 1886. Surely such proportions are large enough
to account for a great fall in prices, considering that they represented
receipts from a new source of supply. It is true that the proportion fell to
10·9 per cent. in 1887, when American supplies were unusually large and
Russian contributions considerable ; but that was after prices had been
brought down to an extremely low level, and is to be explained by the
unusual deficiency of the crop of Indian wheat in 1887, following a crop
below the average in 1886." " Moreover, we received more wheat from
Russia on account of extensive Indian exports to Italy than would otherwise have come to us." It thus, as Mr. Bear forcibly demonstrates, becomes
imperative to take India's supplies of wheat to the Continent (as well as
to England) into account in any endeavour to fix the share which India
has played in lowering the value of wheat. And not the consignments to
the Continent alone, but those to Egypt as well, since the Egyptian imports
from India are largely re-shipped to Europe. The Table I. (p. 187) in
column I. shows the total exports of wheat from India during the twenty-one
years ending 1891-92, the returns for the last year being subject to revision
and correction. Columns V. and VI. analyse these exports into the two chief
sections, *viz.*, consignments to the United Kingdom and to Egypt, and the
Continent. The total of these two columns will be seen to be very nearly
the quantity shown in column I., and indeed all that has been omitted are
the supplies furnished to Malta, Arabia, etc. The Table II. (p. 189) will be
seen to analyse the wheat export trade of India more fully by displaying the
countries of the Continent to which the grain is consigned, as also the
shares taken in the traffic by the chief exporting provinces. For the
contention that it is at present desired to urge, however, attention need only
be directed to the former—Table I. It will there be seen that during the past
ten years at least the returns in column VI. (the Continent and Egypt) exceed those in column V.,—the United Kingdom. Any attempt, therefore,
to exhibit the effects of the exports to the United Kingdom on the price of

wheat in England must be fallacious since the supplies to the Continent must have released Russian and other wheats for the English markets, and hence the total supplies to Europe must be taken into consideration as well as those to Egypt. Last year (1891-92) these came to 29,472,137 cwt. or very nearly 1⅜ million tons. By way of passing, attention may be given to column vii. in Table I. which displays the total export of rice from India. It will there be seen that, until last year, the rice trade has for many years past been fully twice as valuable as the wheat, a fact which demonstrates the absurdity of the contention that in the wheat exports we see a manifestation of the greed of money, the natural food of the people being drawn away from them by the demands of Europe. Long before the wheat trade became of importance the rice exports were very large, and the traffic in that grain has during the past twenty-one years manifested a uniformity of expansion that conclusively demonstrates the increasing productiveness of India, in spite of the fact, which the new census has shown, of an increase of 3 millions per annum to the population.

INFLUENCE of Indian Wheat.

Prices : Economies and Reductions.—But if a difficulty exists in determining the effect of the Indian traffic with Europe in contributing to the significant fall of price in wheat in the world generally (up to 1890-91), an equally great difficulty exists in India itself in arriving at a knowledge of the influence of the exports on local prices. India is itself an empire of such magnitude that the problems of supply and demand within it are quite as complicated as in Europe collectively. If a limited area of India be dealt with, or if observation be confined to a prescribed number of years, errors creep in on every hand. Mr. O'Conor has, therefore, very justly condemned conclusions based on "figures as they stand, without regard to the conditions and circumstances which explain them." To illustrate that position he reviews the returns of prices of wheat in the North-West Provinces and Oudh. The price of wheat he points out, for the years 1881-84, was 18.94 seers the rupee, being a little higher than for the previous twenty years (19.12). If, however, the last eight instead of the last four years be taken into consideration, the price of wheat becomes 17.39 seers only, as compared with 19.94 seers for the previous sixteen years. Mr. O'Conor, therefore, concludes that " A more accurate method of ascertaining the present comparative level of prices will be to eliminate from the period of twenty years which preceded the last four years, those years in which the prices by reason of drought and scarcity ranged abnormally high, and to confine the comparison to years of fairly good harvests on each side." Mr. O'Conor thus rejects 1869, 1873, 1878, and 1879 as famine years, when he finds that "the average price of the sixteen years preceding 1881 was 20.47 seers, the average of 1881-84 was 18.94 seers or nearly 7½ per cent. higher." While admitting the value of this contention, the process of elimination to arrive at the average price of "fairly good harvests," might be carried a little further by the removal of the years of superabundance—years which fall as far on one side of the line of mean value as do those of famine on the other. Thus, for example, the price of wheat in the provinces dealt with was, in 1862, 27.89 seers; in 1863, 28.11 seers; in 1871, 23.55 seers ; and in 1872, 25.41 seers,—years which would show an average of 26.43 seers or 7.31 seers above the mean, while the years rejected by Mr. O'Conor were 5.37 seers below it. If, therefore, years of scarcity and years of superabundance be both removed, wheat is demonstrated to have become slightly cheaper in the North-West Provinces, during the period dealt with. The quotations of the Chambers of Commerce show, however, that the purchases have been made on slightly improved prices, though it would very probably be more correct to say that while fluctuations have taken place, the export of wheat from India has in no appreciable way

PRICES. 794

Mean averages 795

TRITICUM sativum.	Prices of Indian Wheat

PRICES.

affected the local market. But in the writer's opinion little good is derived by an analysis of the returns of prices in the light of *the possible influence on these of the export traffic.* The error of disregarding the balance between province and province, in food supply, which the extension of facilities of internal transport is effecting, is so great that, it might almost be said, the disturbances of famine and of superabundance are to some extent permanent in their effects. The liability to famine or scarcity, of necessity gives advantage to tractsnot so subjected. The nature of the agriculture changes in consequence. The crops grown in the regions exposed to the possibility of drought are those best calculated to withstand the danger. An interchange takes place in obedience to the controlling power of supply and demand. The extension of the irrigation measures is giving to regions once periodically liable to extremes of drought a security that is rapidly changing the nature of their agriculture, and in this change the wheat area is very largely expanded.

**Effect of increased Facilities.
796**

The effect of increased facilities of transport may be affirmed to be to lower rates where they were formerly abnormally high, and to raise them where once they were ridiculously low. Indeed, so keenly are the provinces of India entering into competition, that a famine may be seen to deprive a province or district of a once profitable industry or branch of agricultural enterprise—the temporary disturbance giving supremacy to its rival. This is the natural effect of improved facilities for interprovincial trade, a trade which year by year betokens the prosperity of the country at large, both in agriculture and in manufactures. The resources and enterprise of the people of India of to-day are not only effacing the deadening influences of the anarchy that formerly prevailed, but are acting and re-acting on the foreign imports and exports. That the production of food-stuffs in India has, during the British supremacy, immensely expanded, needs no further proof than the existence of a large foreign trade, which a quarter of a century ago did not exist. While the population has immensely increased, the advance in the price of the necessaries of life has not exceeded the enhancement of the value of labour. Indeed, it might with perfect safety be said that in no other country in the world has national prosperity and advancement, preserved to the labouring classes, a cheaper and more abundant supply of food than in India.

**Effect of Foreign Export Trade.
797**

But if it be admitted that the foreign exports of wheat have not raised the price of that grain to the Indian consumer, can it also be said that the keen competition witnessed within the past few years has preserved wheat cultivation as a remunerative industry? The conditions of one province are often so widely different from another; indeed, the conditions of one district or of one cultivator are often so dissimilar from another, that the statement of cultivation that might be framed by one investigator would be at variance with that of another. Little good can, therefore, be obtained by publishing estimates of cost of production and profit. The problem is better judged of by general than individual principles. In other words, the returns of the wheat area, of the foreign transactions, of the prices paid by the wholesale dealers and other such features of the trade, afford more trustworthy data than the returns of individual estimates. The subject of the area of wheat cultivation has already been dealt with ; the present chapter in the tables below furnishes a statement of the position of the trade ; and the subject of the prices paid for wheat by the exporters and of the price of the grain in India has repeatedly been shown, pp. 116-118. It may suffice, therefore, to deal here with this subject in its widest and most general aspects. It may be said that the fall in the price realized in Europe for Indian wheat down to 1891 has in no way affected the Indian cultivator. He is as willing to cultivate the grain to-day as he was ten

Economies and Reductions.	(*G. Watt.*)	TRITICUM sativum.

years ago. Indeed, as already shown, the exports in 1891-92 were more than double those of 1890-91, and, so far as the monthly returns of the current trade show, the year 1892-93 bids fair to exceed that of 1891-92. The future is, however, more influenced by the uncertainty of the trade to the merchant than to the cultivator. But, of course, as in Europe and America so in India, the amount available from year to year depends on two main considerations, *viz.*, the nature of the Indian harvest and the condition of the European markets. It has been affirmed by many writers that prices cannot fall much below what they have already touched, but in India these so-called "ruinously low rates" have been profitable to the cultivator, so that the possibilities of India hinge more on America than any other consideration. If it be the case that the point is rapidly being reached when America will find it more profitable to retain than to export her wheat, India may be expected to advance; but if America considers it possible and desirable to hold her supremacy in the wheat market, India is likely to fluctuate backwards and forwards, but to advance only slowly, until the wholesale reforms are effected that would place it in a position to contest the market on more rational grounds than at present. In a recent article in the *Economist*, however, it has been held that the consumption of wheat in the world has overtaken production, and that in consequence a rise in prices may be looked for. Such a rise has since taken place, and the Indian exports of what have accordingly manifested very great activity. That this is not due to the fall in the value of silver is at once established by the fact that wheat alone of the articles of Indian export has materially improved. Most articles of Indian export have in fact declined, during the recent term of rapid expansion in the wheat trade.

Conf. with pp. 182-3.

Some conception of the effects of past competition may be learned from the following table compiled from a series of returns issued by the Calcutta Chamber of Commerce:—

Charges.

798

		May 1871.	August 1871.	May 1886.	August 1886.
Exchange	D/Payment 6 M/S .	1s. 10½d.	1s. 11⁷⁄₁₆	1s. 5¹¹⁄₃₂	1s. 4⁵⁄₁₆d.
	D/Payment 3 M/S .	1s. 10¾d.	1s. 11⁵⁄₁₆	1s. 5¾⁄₃₂	1s. 4¼d.
Price .	England, per 492lb .	—	—	31s. 7½	31s.
	Calcutta, per Bengal maund .	R2-5-0	R2-3-0	R2-8-6	R2-11-0
Freight .	per 100 maunds, rail, Cawnpore to Calcutta . .	R83	R83	R53	R53
	per ton, steamer, Calcutta to London .	£3-10-0	£3-10-5	£1-10-0	£1-7-6
Shipping charges at Calcutta . .		R1 a ton		R0-12-0 a ton	
Export trade .	Bengal . . .	cwts. 205,138		cwts. 4,189,672	
	Bombay . . .	33,351		10,608,680	
	Sind . . .	7,323		6,241,017	
	Madras . . .	2,710		21,150	
	TOTAL .	248,522		21,060,519	

TRITICUM sativum.	Prices of Indian Wheat.

**RAILWAY
CHARGES
and
FREIGHTS.**

Reduction.
799

It will thus be seen that very nearly as great reductions on transit charges and shipping freights have been accomplished in India as can be shown to have been effected in America. The railway charges per 100 maunds have fallen since 1871 from R83 to R53. and more recent returns, due to increased facilities towards Bombay, have still further lowered the rates to and from that port. But to confine observation to the figures furnished by the Chamber of Commerce : the steamer freights from Calcutta to London have been lowered from £3-10-0 to £1-7-6. The shipping charges have also decreased from R1 to R0-12-0 a ton. But while all these reductions have been brought about, the price paid for the wheat was R2-5-0 per maund in 1871, with exchange at 1s. 10$\frac{13}{16}$d.; and in 1886 R2-11-0 for the same quantity with exchange at 1s. 4$\frac{19}{32}$d. The Chamber does not furnish the ruling price in England for Indian *f. a. q.* wheat during 1871, but in 1886 it had fallen to 31s. a quarter. An idea of the fluctuations of the London price paid for Indian wheat may be derived from the following quotations :—In 1879, 46s. 3d.; in 1880, 49—50s.; in 1881, 40—49s ; in 1882, 40—47s.; in 1883, 35—43s.; in 1884, 28—38s.; in 1885, 31—33s.; and in 1886, 30s. 9d. to 31s. 7$\frac{1}{2}$d. A vast improvement took place in 1891-92, with the result that the exports of that year were double those of 1890-91, while the exchange had not, on the whole, materially altered till towards the close of the year, when the depreciation effaced the temporary rise of 1890-91. On this subject Mr. O'Conor says, " It

PRICES.
800

may be said that all the world over, the months of March and April saw wild excitement and speculation in the wheat trade with a swift upward rush of prices. India was no exception. Between the end of January and the end of April (1891) prices rose in Karáchí by 17 per cent., in Bombay by 19 per cent., and in Calcutta by between 14 and 15 per cent. In all three places, however, prices have since fallen substantially." "On the whole, it seems clear that the excitement which temporarily prevailed in the wheat market in India, consequent on the sudden rise of prices in Europe, led to a very considerable rise in both wholesale and retail prices for a couple of months, but that as the excitement passed away, prices dropped again, and that they are not now, except in the Panjáb, very appreciably in excess of the prices prevailing about the end of December. Nor are they much, if anything, in excess of the prices which ordinarily ruled since first the wheat of India found an assured opening in Europe." While the English price thus steadily declined for several years the Indian fluctuated but if anything improved. In 1878 R3-9, R3-8, R3-8, and R3-6 in May, June, July, and August. respectively, and in 1886 for these months R2-8-6, R2-10-6, R2-10-0, and R2-11-0 per maund. It is thus probable that a large share of the advantage shown by the reduction of railway and other charges went to middlemen, not to the cultivator, since the rise in the price paid bears no relation to the reductions effected. In the Report of the Bombay Chamber of Commerce for the year 1891, the prices of wheat and freight charges are given. From the weekly returns the following abstract may be found useful :—

T. 800

EXCHANGE. FREIGHTS and PRICES.

Exchange; Freights to London; and Prices of Wheat (per cwt.) that ruled in Bombay during the first week of each month for the year 1891.

	January, 9th.	February, 6th.	March, 5th.	April, 3rd.	May, 1st.	June, 4th.	July, 2nd.	August, 6th.	September, 3rd.	October, 2nd.	November, 6th.	December 4th.
EXCHANGE—	£ s. d.	£ s. d	£ s. d	£ s. d.	£ s, d.	£ s. d.	£ s. d.	£ s. d.	£ s. d.	£ s. d.	£ s. d.	£ s. d.
Sight	0 1 $6\frac{1}{4}$	0 1 $5\frac{7}{8}$	0 1 $5\frac{1}{8}$	0 1 $5\frac{1}{8}$	0 1 $4\frac{13}{16}$	0 1 $4\frac{27}{32}$	0 1 $5\frac{9}{16}$	0 1 $5\frac{11}{32}$	0 1 $5\frac{7}{32}$	0 1 $5\frac{1}{32}$	0 1 $4\frac{11}{16}$	0 1 $4\frac{11}{16}$
3 months	0 1 $6\frac{7}{16}$	0 1 6	0 1 $5\frac{1}{4}$	0 1 $5\frac{1}{4}$	0 1 $4\frac{15}{16}$	0 1 $4\frac{31}{32}$	0 1 $5\frac{5}{8}$	0 1 $5\frac{7}{16}$	0 1 $5\frac{5}{16}$	0 1 $5\frac{1}{8}$	3 1 $4\frac{25}{32}$	0 1 $4\frac{25}{32}$
6 months	0 1 $6\frac{5}{8}$	0 1 $6\frac{3}{16}$	0 1 $5\frac{5}{16}$	0 1 $5\frac{3}{8}$	0 1 $5\frac{1}{16}$	0 1 $5\frac{3}{32}$	0 1 $5\frac{13}{16}$	0 1 $5\frac{9}{16}$	0 1 $5\frac{7}{16}$	0 1 $5\frac{7}{32}$	0 1 $4\frac{7}{8}$	0 1 $4\frac{7}{8}$
FREIGHT—												
A ton to—												
Liverpool	0 16 6	1 2 6	1 2 6	1 6 6	1 7 0	1 4 6	0 17 6	1 1 3	1 3 3	0 18 0	0 19 0	1 2 6
London	0 16 3	1 2 6	1 2 6	1 3 9	1 6 3	1 8 9	0 16 9	1 0 0	1 1 3	1 0 0	0 18 9	1 1 3
CLASSES OF WHEAT IN CWT.—	R a. p.	R a. p.	R a. p.	R a. p.	R a. p.	R a. p.	R a. p.	R a. p.	R a. p.	R a. p.	R a. p.	R a. p.
(1) White Pessy	4 9 9	4 9 3	4 12 6	5 1 0	5 1 0	5 2 0	5 3 0	5 4 0	5 8 6	5 9 0	5 11 0	5 14 0
(2) Ahmedabad (soft red)	4 5 6	4 5 0	4 8 6	4 10 6	4 14 0	4 14 6	4 13 6	4 12 6	5 0 0	4 14 0	5 0 6	...
(3) Laskari	3 8 0	...	...	...	...	...	...	...	...	...	...	...
(4) Yellow hard	3 11 0	3 12 0	4 2 0	4 3 0	4 7 0	4 7 0.	4 4 0 to 4 7 0	4 6 6	4 13 0 to 4 15 0	4 15 0 to 5 1 0	5 1 0 to 5 3 0	5 8 0 to 5 11 6
(5) Delhi No. 1 (White Pessy).	4 7 6	4 6 6	...	4 13 0	4 13 6	4 12 0	4 11 6	4 12 6	5 3 0	5 3 6	5 5 6	5 6 3
(6) Delhi No. 1 (Red)	4 4 0	...	4 4 6	4 6 6	4 9 0	4 8 0	4 6 0	4 8 0	4 14 0	4 15 0	5 0 0	5 1 6
(7) Hard Red Khatha	3 12 0	3 10 0	3 13 6	3 15 0	4 4 0	4 3 6	...	4 4 6	4 10 0	4 11 0	...	...
(8) Nagpore Khathi.	4 1 6	4 0 6	4 4 0	4 6 6	4 11 0	4 7 6	4 7 0	4 8 0	5 0 0	5 0 0	5 0 6	5 1 6

| TRITICUM sativum. | Prices of Indian Wheat. |

EXCHANGE FREIGHTS and PRICES.

In order to illustrate the relation of the trade to the value of silver, it may be said that the year 1889 showed, for the first week of each month, as follows:—Sight 3rd January, 1s. $4\frac{7}{16}d.$; 7th February, 1s. $4\frac{19}{32}d.$; 7th March, 1s. $4\frac{1}{2}d.$; 4th April 1s. $4\frac{11}{32}d.$; 2nd May 1s. $4\frac{3}{16}d$; 3rd June, 1s. $4\frac{1}{8}d$; 1st July, 1s. $4\frac{1}{8}d.$; 5th August, 1s. $4\frac{3}{16}d.$; 2nd September, 1s $4\frac{9}{32}d$; 3rd October, 1s. $4\frac{11}{32}d.$; 7th November, 1s. $4\frac{23}{32}d.$; and 5th December, 1s. $4\frac{7}{8}d.$ Now these figures fairly represent the rates that ruled throughout the year: exchange may be said to have been constant but to have preserved a very slightly lower level than during 1891. Prices of wheat were also fairly constant. The year opened with white Pessy at R5-2 and it fell slightly, the lowest record being R4-4 and the mean R4-7. To show the further history of exchange and prices, it may be useful to give the returns for 1890:—Exchange at sight 3rd January, 1s. 5d.; 6th February, 1s. $5\frac{1}{4}d.$; 7th March, 1s. $5\frac{7}{32}d.$; 3rd April, 1s. $5\frac{3}{32}d.$; 1st May 1s. $5\frac{23}{32}d.$; 5th June, 1s. $5\frac{7}{8}d.$; 3rd July, 1s. $6\frac{9}{32}d$; 1st August, 1s. $7\frac{7}{16}d.$; 4th September, 1s. $8\frac{3}{32}d.$; 3rd October, 1s. $7\frac{3}{8}d$; 7th November, 1s. $6\frac{1}{4}d.$; 5th December, 1s. $6\frac{13}{16}d.$ The American action which temporarily raised exchange is a subject too well known to call for any special remarks here, but the effect on prices of wheat is worthy of consideration. The year opened with white Pessy at R4-8-6 a cwt., and the following were the rates during the first week of each of the remaining months :— R4-6-0; R4-4-6; R4-6-0; R4-7-0; R4-6-0; R4-6-0; R4-6-6; R4-5-6; R4-6-6; R4-10-0, and R4-9-6. The remarkable rise in exchange did not, therefore, cause anything like a corresponding decline in the price of wheat at the port of Bombay. The selection of one week in each month and of one particular kind of wheat has been made to simplify the quotation of figures. This treatment exhibits nothing peculiar or different from what could be shown by

Incentive to speculation.
801

the analysis of the returns for any other term or kind of wheat. It will thus be seen that in comparison with the table for 1891, the behaviour of exchange cannot be regarded as affording indications of any direct gain to shippers, though, as pointed out elsewhere, a steady decline in exchange is an incentive to speculation. Exchange preserved a lower level in 1889 than in 1891, and yet the exports of the former year were less than half those of the latter. The financial year of India goes from the 1st April to the 31st March, and the Bombay Chamber of Commerce dates its year from 1st January to 31st December. This ambiguity is, however, in the wheat trade not of very serious moment, since the chief shipments are made in the months embraced by both systems. Thus while the foreign exports of 1890-91 (practically the year 1890 of the commercial returns) were only 14,320,496 cwt., those of

Conf. with pp. 179, 183.

1891-92 (*i.e.*, 1891) became 30,306,989 cwt. It may fairly be asked, therefore, what does the rate of exchange in these years manifest that could by itself account for so immense a development? Practically nothing, until the further consideration of the rise of prices in Europe is called into account; it then becomes apparent that the tendency to a fall in exchange became an inducement to traffic, seeing that prices in India had not manifested so great a rise as in Europe, and that freights were if anything easier.

DEPRECIATION OF SILVER.
802
Conf. also with p. 173.

Depreciation of Silver.—The subject of the depreciation of the value of silver has been incidentally alluded to in one or two passages above, but it may serve a useful purpose to attempt, very briefly, to exhibit here the direct bearings of this much-hackneyed controversy on the Indian wheat trade. It need scarcely, however, be said that what is true of the influences of a fluctuating currency on wheat must be true also of every article of Indian commerce. The exceptions (and these are only partial exceptions) would be in the traffic in articles, such as Indigo, Jute, Rice, etc., of which India may be accepted as holding a monopoly, sufficiently strong to enable her to control the market.

T. 802

| Depreciation in the Value of Silver. | (*G. Watt.*) | **TRITICUM sativum.** |

DEPRECIA-TION OF SILVER.

The rock upon which most persons have wrecked their theories and contentions on the silver question has been disregard of the constant adjustment that must take place in the markets of the world on any important article of trade becoming depreciated in value. Individual experience is too often used as an argument unanswerable and unerring, even although the inference drawn, be clearly at variance with the commonest principles of supply and demand. The other day, for example, a gentleman largely interested in Indian wheat assured the writer that " if a merchant can now purchase two shillings worth of wheat with one shilling and three pence, he is and must be a gainer by very nearly the difference between these figures." Now that statement ignores the fact that the one shilling and three pence, through the appreciation of gold, purchases the same or very nearly the same amount of return goods as was obtained by the two shillings formerly paid for the rupee's worth of wheat. These figures, are, moreover, inaccurate in more respects than one since they disregard the effect of reduction in transit charges, and put on one side the influence of the fall in the price of wheat itself. But we may accept these quotations of prices, for the sake of argument, without their absolute or even relative values being questioned. If then any such advantage has been secured, the exports of India relative to the imports should have immensely increased. So large a gain would certainly have augmented the demand and given a greater return to the cultivator, since the exporter would have received more and been thus in a position to pay more. Conversely the import trade should have declined. But, during the past 10 or 15 years, not only has the expansion of the imports kept pace with the exports but the Indian cultivator has not received the enhancement of prices that competition would of necessity have secured. Witness the actual returns of the exports and imports during the last two years : total exports of India in 1889-1890 £105,355,000 and imports £86,653,900 : exports in 1891-1892 £102,338,200 and imports £93,910,300 (nominal sterling). Now, while exchange reached its lowest ebb and wheat exports their highest level, the total exports of India declined by 3 millions and the imports improved by 7 millions. It is a fundamental principle in the commercial relations of all civilized nations that the exports pay for the imports. On this subject Mr. O'Conor writes :—" If exports increase while imports diminish the fact can only signify, either that the country is borrowing capital in foreign countries, or that it is giving away its produce for nothing. If the value of imports increases in greater proportion than the value of exports, the fact can only signify that the country has been lending money to other countries and is receiving back interest on its capital." But this line of argument may be extended. To allow of the profit assumed above, it must be accepted that the prices of the articles imported by India in exchange for her wheat, have remained stationary. Had they, meantime, risen in value, India would have got less for her wheat than formerly, and had they declined she might have got more. Now, as a matter of fact, the articles of European manufacture that India imports have fallen in value, so that the so-called one shilling and three pence worth sent to India for the rupee's worth of wheat is the same in amount and quality now as when the rupee had its par value. Were it otherwise the fall in the value of silver would have assumed the form of a direct bounty to local industries. An import duty was, some few years ago, levied by India on manufactured (piece) goods. The advocates of free trade regarded that duty as debarring the peasantry of this country from obtaining the full advantage of their dealings with England. Its removal was demanded and complied with, the result being the loss of a large revenue to the country without a concomitant reduction in the price paid by the people for their few yards of cotton goods.

Exports pay for Imports. 803

Conf. with p. 179.

| TRITICUM sativum. | Prices of Indian Wheat. |

**DEPRECIA-
TION OF
SILVER.**

The advantage went to the retail dealer, not to the consumer. But to be consistent, it is now necessary to ascertain clearly if any undue advantage be obtained by the exporters of Indian produce through the depreciation of silver, since such advantage would of necessity be a burden thrown on the rate-payers of India, far in excses of the loss annually sustained by the Government in the payment of its European liabilities. A gain on export, must, as already stated, amount to a bounty on local manufactures and against imported goods. The old duty on foreign piece goods was 5 per cent., the depreciation of silver, during the past 15 years, has averaged from about 20 to 37 per cent. If, therefore, a gain exists by depreciation, we have to assume that Manchester (by way of specific illustration) is now being permitted to swindle the people of India by sending literally only 1s. 3d. worth of the cloth formerly supplied in return for the 2s. worth of wheat. It goes without saying, therefore, that this illustration can be ac-

**England's
loss through
Fall in Silver.
804**

curate only were it possible to prove that the value of Manchester goods had remained stationary. If they have fallen in value till the purchasing power of 1s. 3d. is the same as the 2s. of former years, the gain must be regarded as a pure hallucination. Now, during the period when the mean fall in exchange may be accepted as approximately 25 per cent., Manchester goods became cheaper by fully 30 per cent., so that these articles can be (and have been) sold at very nearly 10 per cent. less than formerly. There would thus appear to be a gain to the consumer of foreign imported goods, but certainly no gain to the exporter; indeed, had railway charges and shipping freights not been materially reduced, the wheat trade of India must have ceased to exist. The fall in the price of wheat in India and the still greater fall that took place in Europe (during certain years) for that commodity, together with the reductions in transit and other charges, have been so adjusted on each other as to allow the trade a narrow margin of profit. But it would not be difficult to show that the gain on imports, illustrated above, must be more than effaced by the national losses on home remittances through exchange, so that the opinion would seem fully justified that Indian commerce is unfavourably influenced by the fluctuations of the silver market. These fluctuations when they show a downward tendency, afford, however, a distinct inducement to speculation. A temporary gain may thus be attainable by the buyer who is fortunate to find a fall in exchange since his purchase. Whenever the downward tendency assumes a degree of fixity, that is, when it manifests a likelihood to remain so, for an appreciable time or to fall still lower, prices are at once adjusted and the advantage becomes transitory or accidental, but is never inherent nor constant. The advantage, if it be called so, that is thus possible, partakes very largely of the nature of gambling rather than of legitimate commercial gains. It is more reprehensible than commendible and works evil rather than good, since it tends to keep India in the position which it largely holds at present, *viz.*, that of meeting occasional and speculative markets instead of assuming the status of a fixed and natural trade.

**Depreciation
encourages
speculation.
805**

Mr. O'Conor puts this feature of the trade forcibly thus :—" It is the constant tendency to a fall in exchange which has encouraged speculation, but if exchange instead of falling from 1s. 6½d. to 1s. 4½d. had fallen in the same interval from 1s. 9d. to 1s. 7d.,—that is, if the average rate had been largely higher than the rate which has prevailed—the effect would have been the same. While exchange is rising, importers hasten to bring in their goods, and when it is falling, exporters hasten to ship their goods, for, in each case, a transaction which has begun on a certain basis of exchange will bring in a much larger profit than was anticipated if, before the goods are delivered, exchange rises for imports or falls for exports. It is a purely temporary stimulus which dis-

| Depreciation in the Value of Silver. | (*G. Watt.*) | TRITICUM sativum. |

DEPRECIA-
TION
OF
SILVER.

appears with the readjustment of prices and freights which always takes place, but while it lasts it induces speculation, and if the course taken is steadily upwards or steadily downwards, it causes a profit to the importer or exporter. It is this temporary effect of fluctuations of exchange which has caused the 'practical man of business' to assert that a low exchange is beneficial to the country because it encourages the export trade." That such a gambling element should be a governing factor in India's trade is scarcely likely to be regarded as a recommendation for the present monetary system. Indeed, viewed from every aspect of the case, the direct and indirect advantages claimed are of so unsatisfactory a character that were they admitted as existing in every transaction, the wheat trade should be viewed as a national calamity, which it would be politic to check by an export duty, such as rice is made to bear at the present day. But it may emphatically be said that no direct gain can possibly be obtained through the fall in the value of silver and therefore, that the Indian wheat trade is a perfectly natural one which each year is likely to become more securely established than heretofore. It is perhaps needless to multiply further evidence against the theory of gain ; but it may be added that, if any advantage occurs through the depreciation of silver, would it not be natural to expect a synchronous relation in the fluctuations of the export traffic with the variations in the monetary standard ? That this has not been the case may at once be learned by an inspection of the statistical returns of the trade in relation to the quotations of exchange. In 1880-81 the exports first assumed commercial importance. Exchange was then from 1*s.* 7*d.* to 1*s.* 8*d.*, and freights from Calcutta were £2-12-6 to £3-2-6 a ton, with the English price 40—49*s.* a quarter. Since then the trade has manifested the most violent fluctuations, due mostly to influences of supply and demand outside India. Exchange was lower and freights were also lower in 1887-88 than in 1886-87, yet the exports fell from 22 to 13 million cwt., and the reason was that in the latter year prices were low in Europe, and relatively high in India. It became more profitable to retain than to export wheat. In 1891-92 exchange was, if anything, still lower than in 1887-88, but prices were rising in Europe and it became profitable to export. A rush was made, and India exceeded its previous record by 8 million cwt. A tendency to a fall in exchange with a rise of prices in Europe are, therefore, the conditions that encourage the Indian export trade as presently constituted, but the gain that results is a secondary, not primary, consequence of depreciation of silver. The late **Sir J. Caird** wrote on this subject that " The wheat trade of India is thus found to be a safety valve, for when the prospects of a material diminution in the food-grains leads to a serious rise in local prices relief is immediately afforded by the profitable retention in the country of wheat grown for export."

But enough may, perhaps, have been said on the various opinions that have been advanced on the subject of the effect of depreciation of silver on the Indian wheat trade. It may suffice, therefore, to furnish Mr. **Bear's** final conclusion (a writer whose utterances have, in one or two places, been quoted above) on the main issues of the wheat trade. Mr. **Bear** asks, " Then is the Indian wheat-grower benefited by the fall in the gold value of the rupee ? That is by no means certain. He is able to put rival growers in other countries at a disadvantage ; but he gets only about the same price for his wheat as he obtained when the rupee was at about what is conventionally considered its par value of 2*s.*, and any changes which would send it up to par again would almost certainly send the price of wheat up in Europe proportionately, so that he would still get the number of rupees he now receives for a quarter of wheat. By the unequal competition which existing circumstances enable him to carry

Gambling
Trade.
806

Conditions
favourable
to Indian
Exports.
807

TRITICUM sativum.	Prices of Indian Wheat.

DEPRECIA-TION OF SILVER.

on, he ruins wheat-growers elsewhere without, apparently, doing himself any good."

Mr. Bear then deals with the question whether there is not a limit to the quantity of the Indian wheats, such as now produced, which Europe will receive. " Indian wheat," he says, " certainly does not improve our bread, much as the bakers like flour made from it, because of the extra quantity of water which it will absorb, and if too much of it were used bread-eaters would rebel. This fact has been more patent than ever, since the finer qualities of Indian wheat have come in only very small quantities. Scarcely any No. 1 wheat of any kind has been imported of late, No. 2 Calcutta Club being about the standard quality of the great bulk of our supply. Moreover, the large buyers whom I have consulted, with only one exception, declare that the quality of the wheat sent here from India has, as

Adulteration.
808
Conf. with pp. 110-116, 169-171.

a whole, deteriorated." But the explanation of this decline may be seen in the sentence or two which Mr. Bear devotes to the subject of the adulteration of Indian wheat. " But the great difficulty, he says, has been— and there is nothing to show that it has been removed—that buyers of Indian wheat in this country prefer cheap ' dirty' wheat to comparatively dear ' clean' wheat." As the trade is presently constituted and controlled, it may therefore be safely affirmed that the evil and the good, the profit and the loss, is far more intimately related to " refraction " than to depreciation. The governing factor of the trade is, however, the low price at which the Indian cultivator can profitably produce wheat. Mr. Bear says that America has admitted that she cannot grow wheat profitably below from 21 to 24s. a quarter, and that the Indian cultivator is satisfied with 15s. 6d. The reductions that have and are still being effected in transit charges have and in the future will still further bring the vast wheat resources of India into the field of European commerce. The supplies of our granaries are poured into the European markets during the very months when prices are ruling high. Let India but effect therefore the greatly needed reforms and try to improve not only the purity but the quality of her wheats, and she need then fear no competitor for the peculiar class of wheats which she can produce.

FOREIGN TRADE.
809

FOREIGN TRADE.

Foreign Trade.—In order to convey some idea of the present position and character of the Foreign wheat trade of India, the following table may be furnished. It need only be remarked that the last year there exhibited has been derived from the monthly returns, and as these are found at times not absolutely correct the annual statement of the trade (which has not as yet appeared) occasionally modifies the monthly reports. It is believed, however, that the figures given for 1891-92 will be found very nearly correct; they demonstrate the very significant expansion that has undoubtedly taken place. This has been admitted on all hands as being, to a very large extent, accidental. To have been due to the anticipated scarcity of wheat in Europe, owing to the serious loss of the Russian supply. Prices rose in Europe, while at the same time exchange was falling. These are the conditions that make exportation from India advantageous, but the rush made proved disastrous to many since the Russian supply was by no means so deficient as had been anticipated at the beginning of the season. That the expansion of 1892 has given some foretest of being, however, more than an accidental fluctuation, it may be added that the currently accepted view, among those best qualified to judge, is that the Indian exports will very possibly never again fall to so low a position as they occupied in 1890-91. The reports of the traffic, so far as can be ascertained for the year 1892-93, give tokens of a higher level even than those for 1891-92.

T. 809

Foreign Wheat and Flour Trade of India.   (*G. Watt.*)   **TRITICUM sativum.**

TRADE, FOREIGN.

TABLE I.

Chief Items of the Foreign Wheat and Flour Trade of India.

| | WHEAT | | | | | | FLOUR | | | | | | | |
| | I. Exports, Indian produce. | | II. Exports, Foreign produce. | | III. Imports, Foreign produce. | | IV. Exports, Indian produce. | | V. Share of exports shown in column I taken by the United Kingdom. | | VI. Share of exports shown in column I taken by Egypt and the Continent. | | VII. Comparison of wheat with rice exports. Total exports from India of rice to foreign countries. | |
	Cwt.	R	Cwt.	R	Cwt.	R	lb	R	Cwt.	R	Cwt.	R	Cwt.	R
1871-72	637,099	23,56,445	...	...	84	242	No returns	No returns	243,993	7,28,678	594	3,193	17,311,285	4,49,91,611
1872-73††	394,010	16,76,900	...	...	3,304	11,858			198,571	8,88,611	1,187	5,073	23,93,956	5,76,10,301
1873-74	1,755,951	83,76,064	...	...	7,165	26,972			1,358,902	64,33,150	46,411	2,31,684	20,245,385	5,54,97,979
1874-75	1,069,070	49,04,352	4,579	10,154	46,799	1,65,596	No returns		399,492	17,95,174	414,677	19,65,556	17,392,938	4,76,53,337
1875-76	2,498,185	99,10,255	12,583	53,057	19,650	75,815	1,020,417	80,846	2,076,849	73,00,268	205,905	8,24,725	20,416,032	5,31,10,946
1876-77	5,583,336	1,95,63,335	3,268	13,074	60,392	2,03,465	1,809,982	1,59,063	4,337,208	1,47,69,240	807,446	32,07,120	19,014,234	5,81,52,309
1877-78	6,340,150	2,85,69,899	33,018	1,67,749	431,502	19,05,111	2,701,319	2,13,450	5,731,349	2,57,33,448	254,905	11,18,339	18,428,386	6,95,02,761
1878-79	1,044,709	51,37,785	12,011	63,590	411,282	19,29,258	3,430,632	2,53,652	855,682	40,90,848	11,012	56,039	21,437,233	8,97,37,126
1879-80	2,195,550	1,12,10,148	5,905	32,534	14,461		4,146,608†‡	2,93,417	1,626,976	80,46,261	130,062	7,12,317	22,165,765	8,40,25,017
1880-81	7,444,375	3,27,79,416			82	387	5,107,317‡	3,49,307	4,802,233	3,04,57,723	2,078,187	96,20,076	27,266,040	9,05,71,528
1881-82	8,604,815	8,60,40,815	37,485	1,59,803	206,639	8,09,257	6,816,048‡	4,11,971	9,379,230	3,96,36,431	10,060,123‡	4,45,12,270	28,888,421	8,30,81,669
1882-83	14,144,407	6,06,89,341	49,350	1,98,795	197,826	7,73,593	6,816,048§	4,59,418	6,575,160	2,74,31,833	7,109,046§	3,13,93,432	31,358,288	8,47,63,272
1883-84	20,956,495	8,87,75,610	44,917	1,82,502	188,310	7,35,477	7,800,884‡	5,93,349	10,508,210	4,33,59,172	10,121,485	4,40,35,640	27,039,859	8,36,10,798
1884-85	15,831,754	6,30,91,402	19,127	68,780	17,783	57,789	11,057,126‡	9,71,725	7,444,981	1,90,15,857	8,136,159	3,31,27,744	22,951,532	7,19,21,976
1885-86	21,060,519	8,00,23,504	8,405	29,807	28,473	1,30,871	18,879,201††	18,47,882	12,071,218	4,47,39,540	8,788,515	3,41,46,287	28,425,595	9,34,71,200
1886-87	22,263,320	8,62,58,638	304	1,218	28,671	1,11,156	35,744,009‖	—	9,667,591	3,70,74,464	13,180,389	4,74,66,936	26,879,272	8,83,68,266
1887-88	13,538,169	5,56,23,733			798	3,166	36,082,348¶	19,75,549	6,039,708	2,41,21,360	7,220,836	3,02,94,897	28,534,957	9,29,16,864
1888-89	17,610,681	7,53,26,759	1,227	6,035	106,680	4,39,035	36,390,371¶	20,79,254	9,037,830	3,90,24,062	8,434,919	3,55,68,831	23,144,641	9,01,54,487
1889-90	13,799,224	5,79,43,770	2,985	12,377	102,558	4,37,847	48,573,041‡‡	28,37,671	7,686,115	3,72,86,847	5,918,354	2,47,23,248	27,998,906	10,11,04,819
1890-91	14,320,496	6,04,24,260	...	...	124,135	5,31,816	47,035,831**	27,28,897	8,208,935	3,43,70,564	5,991,611	2,55,97,676	34,963,341	12,97,77,396
1891-92	30,306,989	14,38,42,442	...	...			61,028,494	36,15,874	12,345,453	5,95,73,272	17,126,684	8,00,94,386	33,166,929	13,38,59,706

* Years in which Egypt received Indian wheat to be chiefly re-shipped to Europe.

‡ This symbol alone or in combination with other marks, denotes major portion of Indian flour consigned to Aden and Ceylon ;

FLOUR.— { § In addition, smaller quantities to the United Kingdom ;

‖ Expansion due to exports to Arabia with a small quantity to Italy ;

¶ Consigned to United Kingdom, Italy, France and other European countries 1,154,668℔; the traffic with Europe in the following years was mainly with France and Italy ;

** Expansion due to recent large trade with Mauritius.

†† The export duty on wheat, removed on the 4th January 1873, was formerly 3 annas per maund.

| TRITICUM sativum. | Foreign Wheat and Flour Trade of India. |

TRADE, FOREIGN.

The above table will be seen to manifest the total foreign traffic: the exports to foreign countries of Indian wheat, columun I.; the re-exports of foreign wheat, column II ; the imports of wheat from foreign countries, column III ; the exports of flour from India, column IV.; the share of the exports of Indian wheat taken by the United Kingdom, column V.; the share taken by Egypt and the Continent of Europe collectively, column VI.; and to allow of comparison with column I., the exports of rice from India, column VII. It will be still further seen that during the past twenty-one years the exports of Indian wheat to foreign countries have expanded from 637,099 cwt. valued at £235,644 (nominal sterling) in 1871-72 to 30,306,989 cwt. valued at £14,382,244 (nominal sterling) in 1891-92. The imports of Foreign wheat and the re-exports of a portion of these again shown in columns III. and II. are unimportant and may, therefore, be set aside from all further consideration. The traffic in flour is, however, significant, and may almost be regarded as giving a foretaste of the possible future influence of this new and prosperous industry. For, were it possible for India to send to Europe in large quantities a good flour, the appearance of such a rival might rouse the millers to exercise their influence with the shippers against some of the pernicious practices of the present wheat trade. In 1871-72 the exports of flour from India were 243,093 ℔ valued at £72,867 (nominal sterling), and last year they were 61,028,494 ℔ valued at £361,587 (nominal sterling). But what is more significant, the traffic has never, during all these years, manifested the slightest fluctuation. Year by year it has steadily and surely advanced; the exports last year were 14 million pounds in excess of the previous year, and 25 million pounds greater than five years ago. The exports to Europe have not as yet assumed alarming proportions, the bulk of the flour being consigned to Aden, Ceylon, Arabia, etc., but the share taken by Europe has been recorded in one year as considerably over 1 million pounds and this traffic seems capable of immense expansion.

TRAFFIC in FLOUR. 810

Columns V. and VI. of the table manifest the shares taken by the United Kingdom and by the continent of Europe and Egypt. The average of the past ten years shows these two markets as of co-equal importance, the second being, if anything, slightly more valuable than the first. The Continental and Egyptian supply is, therefore, of great moment to India, since it very often represents the quantity, roughly speaking, of the Russian wheat liberated for England. The suitability of Indian wheat, for many purposes in the Continent of Europe, is as significant as is its want of favour with the bakers of Scotland. Its extreme dryness and ricey character commend it for many requirements, but these are the very features that render it unsuitable by itself in the Scotch baking system. To India, therefore, the Continental market has a stability about it that renders it in some respects more desirable than the English, where Indian wheat is used almost exclusively to correct the defects of other qualities and to meet deficiencies in supply.

Indian Flour appreciated on the Continent of Europe. 811 NOT LIKED in SCOTLAND 812

Before passing to consider an analysis of the figures given in Table I., it may be added that Indian wheat was freed from an export duty on the 4th January 1873, while rice bears to the present day a duty of 15 per cent. on the value. In spite of this fact it will be seen that the exports of rice have manifested a remarkable progression from £4,499,161 (nominal sterling) in 1871-72 to £13,385,970 in 1891-92.

REMOVAL of EXPORT DUTY. 813

But in order to manifest more clearly the shares taken by the various Continental countries, of India's wheat, the following analysis may be furnished of the returns for the past eleven years :—

T. 813

Shares taken by Continental Countries.   (*G. Watt.*)

TRITICUM sativum.

TRADE, FOREIGN.

TABLE II.

Analysis of the Exports of Wheat from India to Foreign Countries for the past eleven years; designed to show the share taken by the United Kingdom, Europe, Egypt and other Countries, as also the relative participation of the Indian Exporting Provinces.

	1881-82. Cwt.	1882-83. Cwt.	1883-84. Cwt.	1884-85. Cwt.	1885-86. Cwt.	1886-87. Cwt.	1887-88. Cwt.	1888-89. Cwt.	1889-90. Cwt.	1890-91. Cwt.	1891-92. Cwt.
United Kingdom	9,379,236	6,575,160	10,508,210	7,444,981	12,071,218	9,667,591	6,039,708	9,037,830	7,686,115	8,208,935	12,345,453
Austria	28,421	6,000	800	1,022	5,016	2,750	1,970	5,541	...	4	...
Belgium	2,625,227	1,458,898	2,393,577	1,738,684	2,661,583	2,403,785	596,088	2,477,730	2,329,510	1,920,138	4,654,949
Denmark	40,653	...	40,000	...	...	...	...	...	...	...	...
France	5,308,073	3,567,712	3,397,908	3,312,135	2,145,243	2,803,670	2,559,040	3,131,551	1,250,169	1,517,888	6,026,618
Germany	38,201	...	25,082	...	...	...	...	...	7,003	22,859	...
Greece	...	2,000	1,011	1,427	...	1,999	1,256	...	...	...	...
Holland	712,390	575,246	192,750	133,905	85,918	206,945	60,591	...	250,684	8,000	523,036
Italy	359,318	176,063	445,522	700,875	1,218,269	5,212,305	3,073,764	1,125,058	403,246	439,685	1,062,125
Portugal	...	24,979	12,000	3,998	6,000	46,798	10,000	...	...	52,000	...
Spain	...	1,500	17,600	...	202,086	139,146	127,400	...	600	...	...
„ (Gibraltar)	28,804	494,098	89,236	93,674	68,247	54,337	130,914	36,792	...	...	...
Malta	107,681	163,358	124,413	93,184	34,006	84,088	10,226	...	22,976	30,499	...
Total	9,248,768	6,472,854	6,939,399	6,078,904	6,526,468	11,946,823	6,571,259	6,776,672	4,264,188	3,991,073	12,266,728
Egypt	919,036	799,550	3,305,999	2,150,439	2,296,153	1,317,654	659,803	1,658,247	1,654,166	2,000,218	4,859,956
Other Countries	316,480	296,843	202,390	157,430	166,680	648,906	267,399	137,332	194,755	120,270	834,852
GRAND TOTAL	19,863,520	14,144,407	20,956,495	15,831,754	21,060,519	22,263,320	13,538,169	17,610,081	13,799,224	14,320,496	30,306,989
Bengal	6,668,047	4,439,405	7,611,535	2,563,204	4,189,672	7,037,957	4,334,768	2,959,985	1,550,839	1,340,355	4,746,938
Bombay	11,328,585	6,957,752	8,970,503	8,993,108	10,608,680	12,606,144	8,541,621	10,645,163	5,146,881	6,212,143	14,430,808
Sind	1,852,334	2,732,275	4,372,832	4,271,860	6,241,017	2,613,748	660,758	4,004,039	7,100,282	6,767,300	11,128,564
Madras	10,996	6,599	1,525	1,315	1,872	998	1,005	894	1,222	698	679
Burma	3,558	8,376	...	2,267	19,278	4,473	17	...	...	...	...
Total	19,863,520	14,144,407	20,956,495	15,831,754	21,060,519	22,263,320	13,538,169	17,610,081	13,799,224	14,320,496	30,306,989

(Rows Austria through Malta fall under the bracket **EUROPE**.)

TRADE,
FOREIGN.

The above table speaks so forcibly that comment seems scarcely necessary. The imports of Indian wheat obtained by the United Kingdom have fluctuated between 6 million and 12 million cwt. Up till recently that share of the trade represented fully one half the total, but last year the demands of the Continent of Europe were nearly as great as those of the United Kingdom, and, with the Egyptian supply, came to 17 million cwt. It has already been stated that it is necessary to view the Egyptian traffic in Indian wheat along with the European, since a large share of the Egyptian is destined ultimately for Europe. Next to England, France is the most important consuming country for Indian wheat, and last year's supply (6 million cwt.) was the highest recorded quantity taken by that country. Some five years ago, Italy took over 5 million cwt. of Indian wheat, but during the subsequent years the demands shrank considerably, though it seems possible a very large share of the Egyptian re-exports of Indian wheat go to Italy. By the returns of the past eleven years' trade, however, Belgium is seen to hold the third place in importance, the order being the United Kingdom, France, Belgium, Egypt, Italy.

Provincial
shares.
814

SHARE TAKEN BY INDIAN PORTS.—Turning now to the value of the Indian ports in the export traffic in wheat, the order of importance is Bombay, Karachi (Sind), Calcutta (Bengal), with Madras and Burma, taking very poor fourth and fifth places. It may be said that, with the prosperity of Karachi, the trade from Calcutta has declined, though it seems probable that the Bengal-Nagpur Railway may drain towards Calcutta large supplies that were either not exported from India at all or which formerly percolated towards Bombay. The competition of the Midland Railway system has undoubtedly diverted towards Bombay, and to some extent towards Karachi as well, a large quantity of the wheat that used to find its way to Calcutta. At the ports on the western side of India the charges borne by the grain are much lighter than at Calcutta. But it may be explained that the writer has been unable to see returns of the recently opened out Railway systems that bear on the wheat trade. He, therefore, only assumes that the improvement in the exports of 1891-92 from Bengal may, in some measure, be due to the Bengal-Nagpur Railway, but there would seem no doubt that the improved and cheapened railway communication with Bombay and Karachi largely accounts for the immense expansion of the trade that has taken place from these ports. It seems probable also that the lower rate of refraction that now prevails in Bombay, and the efforts that have been made to classify the wheat, have begun to tell materially in favour of that port. The chief cause, however, of the immense expansion of the total exports from India has primarily to be attributed to the rise in price that, for some time past, has ruled in Europe. But had the facilities of the trade effected in India not taken place and did India not in itself possess vast resources, the sudden expansion witnessed in 1891-92 would have been an impossibility.

Lower
Refraction
and
Classification
of wheat.
815

TRANS-FRON-
TIER.
816

Trans-frontier Trade.—Before turning from the subject of the Indian Foreign Trade to that of the Local Traffic and Consumption, it may be desirable to give here a few brief facts regarding the transfrontier land trade. For this purpose it does not seem necessary to do more than furnish a statistical statement of the transactions during the past three years. The imports shown in the table below from Khelat, Kandahar, Khorasan, and as carried by the Sind-Pishin Railway, are obtained by the province of Sind. The other imports are into the Panjáb and the North-West Provinces. The amounts furnished by Nepal will be seen to be under one-half, but considerably more than one-third the total supply. Of the exports by far the larger proportion goes from the Panjáb and mainly to Kashmír.

T. 816

Indian Trans-frontier Land Trade in Wheat. (*G. Watt.*) **TRITICUM sativum.**

TABLE No. III.

Indian Trans-frontier Land Trade in Wheat.

| | IMPORTED FROM | | | | | | EXPORTED TO | | | | | |
| | 1888-89 | | 1889-90 | | 1890-91 | | 1888-89 | | 1889-90 | | 1890-91 | |
	Cwt.	R	Cwt.	R	Cwt.	R	Cwt.	R	Cwt.	R	Cwt.	R
Khelat	7,006	27,241	2,780	10,423	3,043	10,285	53	213	119	431	22	75
Kandahar	37	137	...	...	...	...	...	...	...	...	...	...
Khorasan	44	195	2,441	7,728	2,020	6,558	...	...	...	...	...	...
Sewestan	1,748	7,230	281	748	998	2,848	13,003	46,977	14,443	41,839	8,316	24,911
Tirah	816	3,573	7,289	19,333	5,740	16,230	886	3,278	837	2,308	939	2,586
Kabul	8,270	33,258	6,961	18,577	7,472	20,344	1,028	4,060	2,217	5,966	504	1,374
Bajaur	1,842	6,742	4,280	10,195	8,517	24,074	52,524	1,48,594	38,054	98,193	36,297	1,06,240
Kashmir	7,844	25,552	15,550	50,389	28,655	1,01,291	1,919	6,519	2,092	7,411	720	2,572
Nepal	37,512	1,11,105	...	...	40	144	...	...	...	...	...	...
Bhutan	47	157	...	...	...	...	...	...	...	...	...	...
Lus Bela	...	...	...	...	...	...	51	209	1,438	5,675	458	1,879
Ladakh	...	...	...	...	...	...	306	916	463	1,258	279	570
Thibet	...	...	...	...	...	...	...	...	...	...	...	...
Manipur	...	...	...	...	...	...	...	...	...	...	...	...
Karennee	...	...	...	...	...	...	22	210	1	4	...	...
Sind-Pishin Railway	5,367	24,978	14,658	63,787	9,481	41,764	38,027	1,66,013	2,539	10,945	15,031	67,096
TOTAL	70,533	2,40,168	54,240	1,81,180	65,966	2,23,538	111,122	3,91,562	64,311	1,84,310	64,249	2,15,706

TRITICUM
sativum.   Internal Trade in Wheat.

TRADE:
INTERNAL.   An inspection of the above table will reveal the fact that the imports
are, as a rule, balanced by the exports, so that to the country at large the
traffic possesses little interest, as it neither adds to, nor removes from, the
annual supply of the grain.   The two chief items of the trade may be said
to be the market offered for a fairly considerable amount of Panjáb wheat
in Kashmír and in other countries across the North-Western Frontier of
India, and the very large amount of wheat annually furnished by Nepal to
the North-West Provinces.   The comparatively smaller traffic across the
Sind frontier usually shows a very small net export from that province.

817   INTERNAL TRADE.

Internal Trade in Wheat.—The necessary statistical information is not
forthcoming to allow of a satisfactory treatment of this branch of the
trade.   Of the vast population of India only a very small proportion ever
eat wheat in any form.   This feature of the subject has already been fully
dealt with in other chapters of this article, so that it may suffice to remind the reader of the fact without repeating statistical information in
support of it.   The returns of the internal trade most fully corroborate this
view however, and it becomes expedient to endeavour to bring before the
reader some of the leading indications of the internal consumption of
wheat as derivable from the returns of Coasting Trade, and of the traffic
recorded as conveyed along the Roads, Railways, and Rivers.   The movement of wheat is mainly, indeed almost entirely, towards the port towns
of Bombay, Karachi, and Calcutta.   The amounts recorded as imported
by these towns, very nearly corresponds, however, to the quantity shown as
exported from each, to foreign countries.   The small balance of imports
(by land routes and coastwise) over exports (by sea) to foreign countries
may therefore be accepted as roughly speaking representing the local consumption of wheat in the port towns.   The movement from province to
province, or from the provinces to the large inland towns, is remarkably
small indeed—a fact which conclusively demonstrates that, except in the
provinces of production, wheat may with safety be said to be scarcely if at
all consumed in India.

COASTWISE   **Coastwise Trade.**—The small consumption of wheat in the provinces
818   of India generally is significantly shown by the coastwise transactions.
The table No. IV. exhibits the total coastwise wheat trade during the
past fifteen years.   The one instructive feature is that by far the most important receiving province is Bombay ; the interchange between the other
provinces is quite unimportant.   The supply drawn coastwise by Bombay is,.
however, intended to in part meet the foreign exports, so that the wheat
shown in the returns can scarcely be regarded as consumed locally.   With
that exception the transactions (imports and exports) may be viewed as meeting local markets, but it is possible that the total trade for all India, shown
by these routes, rarely exceeds 200,000 cwt., after the necessary correction has
been made for the overlapping of returns and unavoidable duplication—due
to the exports of one province appearing again as the imports of another.
So far, then, the coastwise sea-borne trade by no means manifests a vigorous demand for wheat by the people of India.   On a further page (198)
particulars are given of the probable consumption of wheat.   For an
average year the total consumption of all India does not exceed six million
tons, and if to that figure be added the foreign exports the result may
be accepted as showing the total production.   But it is in Northern India
mainly that wheat is consumed, so that the provinces, such as Bengal,
Madras, and Bombay, that possess a sea-board, would natually be expected to manifest a very small interchange in this commodity.

T. 818

Indian Wheat Coasting Trade. (*G. Watt*) **TRITICUM sativum.**

TRADE: COASTING.

TABLE No. IV.

Coastwise Indian Wheat Trade.

	IMPORTS COASTWISE						EXPORTS COASTWISE					
	Into Bengal.	Into Bombay.	Into Sind.	Into Madras.	Into Burma.	Total.	From Bengal.	From Bombay.	From Sind.	From Madras.	From Burma.	Total.
	Cwt.	Cwt.	Cwt.	Cwt.	Cwt.	Cwt.	Cwt.	Cwt.	Cwt.	Cwt.	Cwt.	Cwt.
1876-77	4,496	10,801	4,535	123,244	9,986	153,063	92,733	106,413	40,556	26,755	11,440	271,997
1877-78	4,463	172,009	4,419	206,017	17,248	404,156	207,073	382,911	251,459	18,340	13,461	873,244
1878-79	748	166,367	1,840	112,030	12,517	293,502	132,073	310,267	21,112	5,380	9,672	478,504
1879-80	1,647	138,953	3,465	83,317	29,008	257,290	110,394	224,390	21,911	12,481	1,126	370,302
1880-81	3,320	170,976	54,960	61,798	16,366	307,420	47,552	240,258	926	11,117	793	300,646
1881-82	9,675	243,607	67,318	38,402	8,799	366,711	20,830	275,833	838	14,739	2,016	314,256
1882-83	22,165	360,114	1,008	42,888	3,745	429,920	13,850	430,202	61,207	14,810	17,754	537,823
1883-84	4,022	311,024	669	34,019	10,530	360,264	16,920	172,203	228,208	2,033	1,424	420,788
1884-85	10,841	95,077	7,037	54,605	5,296	172,856	12,979	160,473	41,405	22,925	12,318	250,100
1885-86	4,860	200,768	3,106	45,600	4,625	258,959	12,281	316,619	5,973	9,186	3,084	348,043
1886-87	8,684	366,665	24,658	32,538	8,045	440,590	54,456	334,008	7,033	1,956	3,652	401,115
1887-88	4,507	137,064	2,094	32,413	26,585	202,663	36,547	193,583	26,550	2,808	1,778	261,266
1888-89	8,315	364,252	168	37,546	58,632	468,993	66,435	199,399	59,844	3,464	555	339,697
1889-90	3,586	354,494	160	33,401	41,928	431,569	57,742	131,393	94,701	1,827	356	286,019
1890-91	1,677	801,712	869	38,694	50,596	893,548	48,213	114,587	382,552	1,646	22	547,020

T. 818

TRITICUM sativum.	Indian Wheat Coasting Trade

TRADE: COASTING.

The observation (deducible from the above table) may doubtless have been made by the reader, that the imports of the provinces of India, for some years past, have been in excess of the exports, whereas formerly the exports were in excess of the imports. The absence of a balance between imports and exports is customary in most returns of coasting trade and is due to many considerations. A large number of ships are often at sea, the cargoes of which have been recorded as exports, but not having been delivered when the year closed, they do not appear as imports and thus upset the relation of the record of imports and exports by being carried into another year. So, again (and this is particularly applicable to wheat), a province may have extensive transactions within its own ports, but little or none to external provinces. Thus, for example, the bulk of the Bombay imports coastwise are drawn from Sind and Goa—provinces which largely export, but practically import no wheat. Were Goa to receive a separate place in the coasting returns (similar to what is given to Sind, Bengal, etc., etc.), the balance sheet of the trade would come out more nearly correct; but, like the Native States, it is not so treated, and the result follows that its trade appears but on one side of the total account of the Indian transactions, *viz.*, as imports by Bombay, not as exports from Goa. The fact that the total coastwise exports of the provinces of India are not balanced by their imports, is a matter of less importance than the evidence which the figures afford of the comparative insignificance of the local trade in wheat, which the provincial exchanges demonstrate. Thus Bengal, in only one year (1882-83) during the past fifteen, has had a net import. Its average coastwise net exports for the five years, ending March 1891, were 47,326 cwt. The exports of Bengal go mainly to Burma, and the above average for the past five years provides for the corresponding average net import by Burma, which will be found to be 35,882 cwt. In only two years has Burma exported wheat in excess of its imports, *viz.*, in 1882-83 and in 1884-85. The production of wheat in Burma proper is very small indeed, and the Lower Province at least must be largely dependent on its coastwise supply of 35,882 cwt. Bengal, during the fifteen years dealt with in the above table, will be seen to have in only one year (1882-83) manifested a net import, so that from the coastwise trade returns we learn very little regarding local consumption. To obtain a knowledge of the Bengal consumption, reference must accordingly be made to the traffic by land routes. Madras, on the other hand, produces practically no wheat, and its supplies drawn by the railways amount on an average to about 100,000 cwt. from Bombay Presidency and to a smaller extent from the Nizam's Territory, so that if to that quantity be added the average net import (shown during the past five years) as carried coastwise, we learn that the consumption of wheat in Southern India does not on an average materially exceed 200,000 cwt. Its coastwise imports are obtained mainly from the northern ports of the Presidency, from Bombay port, and in a smaller degree from Bengal. Turning now to Bombay and Sind we learn that the former province has shown a net export as frequently as a net import (during the past fifteen years), but that for the latter half of that period, Bombay has manifested a distinct tendency to become an importing province, drawing its chief coastwise supplies from Sind and Goa. During the past three years (which consecutively have manifested an excess of imports over exports) the net import has averaged 357,659 cwt. Of that amount Goa in 1889-90 (for example) furnished 201,934 cwt. and Kattywar 14,651 cwt. The imports from Bengal and Madras may be said to be rendered of no moment, through the exports to these provinces balancing the record. The imports from Sind are, however, considerable; in 1889-90 they came

PROVINCIAL CONSUMPTION.
819
Conf. with pp. 168, 197.

in Bengal.
820

in Burma.
821

in South India.
822

in Bombay.
823

T. 823

By Rail, Road, and River.	(*G. Watt.*)	**TRITICUM** **sativum.**

to 59,347 cwt. But it need scarcely be said that these imports were to the port town of Bombay, and until evidence be obtained to the contrary it may safely be assumed the drain towards the Western capital has its existence in the foreign trade. During only three years of the fifteen here dealt with has Sind manifested a net import, *viz.*, 1880-81, 1881-82, and 1886-87. For the past four years the average net export has been 140,089 cwt., a quantity which will be found to have sufficed to meet the markets which Sind finds remunerative in Bombay port town, Cutch, and Kattywar.

Having thus demonstrated the lessons that may be learned from the study of the coastwise trade with regard to the Indian traffic and consumption of wheat, attention may be turned to the records (such as they are) of the internal transactions carried by

Rail, Road, and River.—It is often very difficult, in dealing with Indian trade, to obtain returns for a particular year and framed uniformly on the same plan for each route along which goods are carried and distributed. This difficulty the writer has been in the habit of combating (throughout the compilation of this work) by furnishing particulars of more than one year, so that the reader may be enabled to form a conception of the bearings of each section of the trade, even although it may be impossible to furnish particulars of the last year in one section of the trade though possible in another. The Government of India experimentally issued in 1888-89 an imperial review of all Rail and River returns. To construct a similar statement for 1891-92 would necessitate many weeks' labour, and the result when obtained might not even then be deemed very satisfactory. Even were the returns of rail and river accurately worked out, two important routes of transport would still remain untouched, namely, the Road and Canal traffic. The registration of the imports into the Provinces, Port towns, and Native States, of wheat in 1888-89 came to 2,98,67,722 maunds (or say 21,334,087 cwt.). Before attempting to deal with the provinces that furnished that amount, the receipts may be first exhibited. These were in the order of importance :—Bombay port town 1,53,64,191 maunds; Karachi 59,95,883 maunds; Calcutta 55,05,431 maunds; Sind province 13,11,281 maunds; Rajputana and Central India 4,35,781 maunds; the North-Western Provinces and Oudh 2,92,425 maunds; Bombay Presidency 2,87,560 maunds; Bengal province 1,68,614 maunds; Madras 1,52,632 maunds; the Panjab 1,06,391 maunds; the Central Provinces 1,10,277 maunds; Madras Ports 67,414 maunds; Mysore 48,294 maunds; Nizam's Territory 14,658 maunds; Assam 5,650 maunds; and Berar 1,240 maunds. So far then for the receipts of wheat, but to give full force to the interchange it becomes necessary to exhibit the producing provinces by a statement of the exports. By deducting from the figures about to be given those above (for the corresponding provinces) the net import or net export, as the case may be, will be manifested. The Central Provinces, during the year 1888-89, headed the list of exporting provinces with 1,00,16,387 maunds (or say 7,154,562 cwt.). Next came the Panjab 59,91,357 maunds; the North-West Provinces with 40,19,070 maunds; Bengal with 31,74,480 maunds; Sind with 25,25,888 maunds; Bombay with 24,07,680 maunds; Berar with 8,64,069 maunds; Rajputana and Central India with 6,94,844 maunds; Madras with 61,010 maunds; and the Nizam's Territory with 53,724 maunds. Smaller quantities were also returned to the provinces of India from the seaport towns, but these may be left out of consideration. We thus learn that the chief exporting provinces, during the year under consideration, were the above in the order mentioned.

| TRITICUM sativum. | Provincial Wheat Trade. |

Of the exports from the Central Provinces 98,72,724 maunds (or say 7,051,946 cwt.) were taken by the port town of Bombay. The presumption may be admissible, therefore, that the major portion of that import left India as part of the exports from Bombay to foreign countries. The balance of the Central Provinces' exports went to Calcutta (76,739 maunds); to Rajputana and Central India (23,447 maunds); to Bombay Presidency (25,092 maunds); and to the North-West Provinces and Oudh (17,108 maunds).

The Panjáb has been shown above to be, after the Central Provinces, the next most important exporting province. Of the total exports from the Panjáb (59,91,357 maunds) 34,70,428 maunds (or say 2,478,877 cwt.), figure as imported by Karáchi, the balance of the Panjáb exports being made up as follows:—13,08,222 maunds taken by Sind province; 8,20,673 maunds by Bombay; 2,04,186 maunds by the North-West Provinces and Oudh; 1,41,869 maunds by Rajputana and Central India; 35,846 maunds by Calcutta; 10,131 maunds by Bombay Presidency; and 2 maunds by Madras. *Thus, so far as the above returns show, the Central Provinces' wheat is mainly exported from Bombay and the Panjáb wheat from Karáchi.* The amount of the former that found its way to Calcutta in 1888-89 and of the latter to Bombay was almost unimportant. It has already been explained, however, that the improvements and extension of railway facilities have greatly altered the relative shares of the wheats that are now exported from Bombay, Karáchi, and Calcutta.

But in the above statement of the order of importance of the exporting provinces the North-West Provinces and Oudh stand next to the Panjáb. The total exports from these provinces in 1888-89 were 40,19,070 maunds, of which 22,25,226 maunds (or say 1,589,447 cwt.) were taken by Calcutta; 12,60,872 maunds (or say 900,623 cwt.) by Bombay town; 2,29,432 maunds by Rajputana and Central India; 1,56,203 maunds by Bengal province; 1,03,574 maunds by the Panjáb; and 43,686 maunds by Bombay province. *Thus the bulk of the North-West Provinces' wheat exports in 1888-89 went to Bengal,* but there is reason for believing that the Bombay share of these has (at the expense of Calcutta) been greatly increased through the facility effected by the Midland Railway System.

After the North-West Provinces and Oudh, Bengal, by the above returns is the next most valuable exporting province. Out of the total exports, 31,74,480 maunds, Calcutta, as would naturally be expected, took 31,66,895 maunds (or say 2,262,068 cwt.).

Sind in point of importance as an exporting province has now to be dealt with. The total exports in 1888-89 were 25,25,888 maunds, and of that amount very nearly the whole went to Karáchi, *viz.,* 25,25,455 maunds (or say 1,803,896 cwt.).

Lastly, of the large exporting provinces, comes Bombay with its exports in 1888-89, 24,07,680 maunds, of which Bombay port town took by far the major portion, *viz.,* 22,09,301 maunds (or say 1,578,072 cwt.). The production of wheat in the Bombay Presidency (it has already been remarked) shows signs of considerable expansion, but it may fairly be said of the present trade that by far the major portion of the foreign exports from the Port Town of Bombay is in Central Provinces' wheat.

Now, disregarding the remaining transactions, shown in the rail and river returns of India for 1888-89, we may bring together the chief items shown above under the headings of the exporting ports:—

Rail, Road, and River Traffic.   (*G. Watt.*)   **TRITICUM sativum.**

From	Imports by Bombay Port. Cwt.	Imports by Karáchi. Cwt.	Imports by Calcutta. Cwt.	Total of the Imports by the three Ports. Cwt.
Central Provinces . . .	7,051,946	*Nil.*	54,814	7,106,760
Panjáb	586,195	2,478,877	25,604	3,090,676
North-West Provinces and Oudh	900,623	*Nil.*	1,589,447	2,490,070
Bengal	*Nil.*	*Nil.*	2,262,068	2,262,068
Sind	*Nil.*	1,803,896	*Nil.*	1,803,896
Bombay	1,578,072	*Nil.*	*Nil.*	1,578,072
Total Imports .	10,116,836	4,282,773	3,931,933	18,331,542
Foreign Exports . . .	10,654,163	4,004,039	2,950,985	17,609,187
Balance available for local consumption and coastwise exports or where deficient the amount is indicated that has to be made up by coastwise and road traffic . .	—537,327	+278,734	+980,948	+722,355

It will thus be seen that the recorded transactions by rail and river tally so very nearly with the requirements to meet the exports to foreign countries that their accuracy is assured. Of the three exporting ports, Bombay alone requires the aid of the coastwise imports to bring its land route receipts up to the necessary standard. But it has already been shown that Bombay is the chief province of India that directly benefits by the coastwise trade. In the year for which the above review of the rail-borne trade was framed, Bombay had a net import coastwise of 164,853 cwt. Still that amount does not entirely remove the deficit so, that we have to presume that either the exports of that year were largely drawn from surplus stocks or that the road traffic is very considerable. One other feature of the above tabular statement need only be here alluded to, *viz.*, the very large amount shown as retained in Calcutta. The coastwise exports were, however, very considerably, namely, 58,120 cwt. net, and there is usually a fairly extensive return trade by rail to the province. What may also be admitted as worthy of consideration may here be mentioned, namely, the distance of the wheat-fields of Bengal, renders the error of defective road statistics of no moment in the case of Calcutta, while the nearness to Bombay necessitates a large margin being reserved for defective road trade in Western India. But even were it necessary to believe that Calcutta used up 900,000 cwt. of wheat per annum or retained large stocks in hand, either view would not be unreasonable. The City of Calcutta, with its approximately a million population, possesses a large European community of bread-eaters. But in this light it may be pointed out that the total imports by all provinces came to 21,334,087 cwt., and of that amount 18,331,542 cwt. have been accounted for as consigned to the great exporting ports. Deducting, therefore, the quantity drawn from the country by the ports, the balance would be the amount of wheat recorded as carried from province to province all over India to meet local demands. The remarkably small figure of that balance, *viz*, 3 002,545 cwt., or 150,127 tons, abundantly confirms the opinion advanced in more places than one of the comparative insignificance of wheat as an article of food in India generally. Of course in the great wheat-producing provinces, such as the

INDIAN WHEAT CONSUMPTION. 832 *Conf. with pp. 168, 194.*

| TRITICUM sativum. | Provincial Wheat Trade. |

LOCAL CONSUMPTION.

Central Provinces, the Panjab, Bombay, and the North-West Provinces, the local consumption of wheat is doubtless considerable, but the demand for the grain is exceptionally small in all non-producing provinces. It may, in fact, be said that wheat is the food of certain sections of the wealthy population, but except in the Panjáb it can hardly be classed as a staple article of human food in India. The remarkably low record of internal and coastwise traffic in the grain (when the transactions of the foreign trade are excluded from consideration) is, therefore, significant of the position wheat holds in India collectively. Contrast the facts furnished in this article, for example, with those of rice in another volume. It will (*Vol. V.,* 644), for example, be found that the Indian consumption of rice must be annually close on $25\frac{1}{2}$ million tons ; that of wheat does not materially exceed six million tons. The average outturn for the four years ending 1888, on the ascertained average area of 26,508,000 acres, under the crop, is found to have been 7,225,500 tons, or between 6 and 7 maunds an acre. This would be a little over 31 million quarters, of 480 ℔. But the foreign exports of these years showed an average of nine hundred thousand tons, so that the Indian consumption, for the years named, must have been annually close on 6 million tons, or say 27 million quarters. Expressed to the vast population of India this would represent but a nominal consumption—perhaps scarcely more than that of *one* of the numerous classes of grains known as millets (such as *juar*), but it has to be recollected that wheat consumption is very nearly confined to one or two regions. The Panjáb, for example, which has usually the largest area under wheat, exports comparatively the smallest amount, so that in the Panjáb wheat is an important article of diet. The Central Provinces, on the other hand, may with perfect safety be characterised as growing wheat almost exclusively for the export market (*Conf. with p. 197*). If, therefore, by a process of elimination the provinces that consume wheat be placed on one side, or if a figure be designed for each province to express the percentage which wheat plays to the total food of the people, there would remain by far the major portion of India for which no return could be furnished, and of which it will be correct to say that wheat is unknown as an article of diet (and scarcely as a luxury) to perhaps 95 per cent. of the population. These facts and statements explain, therefore, the remarkably small balance over foreign exports that, from the returns of rail, road, and river traffic, seems to be normally left in the non-wheat producing provinces and towns to meet local consumption. But to bring this brief review of the Indian wheat trade to a close, it does not seem necessary to do more than to furnish three tables of the internal trade : (1) one for Bombay in which the transactions with the port are isolated from those with the presidency ; (2) a similar statement for Karachi ; and (3) for Calcutta.

Outturn.
833
Conf. with p. 119.

Rail-borne Wheat Traffic of Bombay. (*G. Watt.*) **TRITICUM sativum.**

TRADE of BOMBAY.

TABLE No. V.

Analysis of the Rail-borne Wheat Traffic of the Bombay Port Town and of the Bombay Presidency during the past three years.

WHENCE IMPORTED AND WHITHER EXPORTED	IMPORTED INTO								EXPORTED FROM							
	1889-90				1890-91				1889-90				1890-91			
	Port Town		Presidency		Port Town		Presidency		Port Town		Presidency		Port Town		Presidency	
	Mds.	R	Mds.	R	Mds.	R	Mds.	R	Mds.	R	Mds.	R	Mds.	R	Mds.	R
	1	2	3	4	5	6	7	8	9	10	11	12	13	14	15	16
Provinces excluding Sea-ports—																
Madras	...	...	561	2,244	...	...	...	...	8	28	93,784	3,28,244	...	...	65,488	2,08,743
Bombay	1,007,905	35,27,663	...	...	1,703,607	54,30,247	...	...	41,634	1,45,719	...	...	5,720	18,264	...	...
Bengal	...	...	3	8	...	...	...	...	...	...	...	...	...	...	...	...
North-West Provinces and Oudh	648,725	17,03,903	7,612	19,082	216,558	5,95,562	805	2,214	8	...	2	7	8	25	15	48
Panjab	916,112	18,32,224	57,135	1,14,272	386,793	8,70,484	28,984	65,214	...	...	...	...	14	45	45	...
Central Provinces	4,430,927	1,95,23,452	16,766	39,819	5,897,021	1,32,68,302	25,253	56,819	2	7	372	1,302	5	16	1,251	3,988
Berar	342,178	8,55,445	93,022	2,32,555	291,108	7,45,904	34,883	89,388	...	...	33	115	...	...	49	156
Native States—																
Rajputana and Central India	991,939	34,71,786	148,511	5,19,789	760,165	24,23,026	94,624	3,01,614	8	28	34,800	1,21,800	191	609	14,841	47,306
Nizam's Territory	850	2,975	473	1,655	225	717	1,237	3,943	36	126	18,396	64,386	...	...	64,054	2,04,172
Mysore	...	...	1,084	3,794	...	...	326	1,039	38	133	87,667	3,05,834	3	9	89,697	2,85,909
Port Towns—																
Madras Ports	...	...	...	...	...	...	4	16	18	63	24,107	84,375	3	9	24,400	77,704
Bombay	...	...	41,634	1,45,719	...	...	5,730	18,264	...	...	1,007,905	35,27,668	...	...	1,703,607	54,30,247
Karachi	...	...	...	...	...	...	...	...	...	...	...	...	...	...	...	...
Calcutta	21	55	...	...	1	3	...	...	...	...	...	...	...	...	...	...
TOTAL { Mds.	8,333,557	2,19,15,508	366,802	10,79,837	9,255,490	2,33,34,105	191,846	5,38,511	41,744	1,46,104	1,267,066	44,34,731	5,954	18,977	1,663,406	63,58,363
{ Cwt.	5,956,183	...	262,001	...	6,611,064	...	137,033	...	29,817	...	905,047	...	4,253	...	1,402,434	...

If it be desired to arrive at the net imports (for example of Bombay port town in 1889-90) deduct the quantity at bottom of column 9 from that of column 1. If, on the other hand, it be desired to arrive at the net exports from the presidency in that year, deduct the total of column 3 from that in column 11. The table displays, at a glance, the sources from which Bombay draws its supplies.

TRITICUM sativum.

Rail-borne Wheat Trade of Sind.

RAIL-BORNE trade of SIND.

TABLE No. VI.

Analysis of the Rail- and River-borne Wheat Traffic of Sind, and of its Port Town Karachi, during the past three years.

Whence Imported and Whither Exported.	IMPORTS INTO								EXPORTS FROM							
	1889-90				1890-91				1889-90				1890-91			
	Karachi.		Sind.		Karachi.		Sind.		Karachi.		Sind.		Karachi.		Sind.	
	1	2	3	4	5	6	7	8	9	10	11	12	13	14	15	16
	Mds.	R	Mds.	R	Mds.	R	Mds.	R	Mds.	R	Mds.	R	Mds.	R	Mds.	R
British Provinces excluding chief Sea-ports—																
Sind.	25,37,122	72,94,226	...	...	24,42,967	74,81,587	...	...	3,535	10,163	...	...	4,418	13,530	...	...
North-West Provinces and Oudh.	83	218	...	...	...	...	...	...	...	...	...	...	...	...	...	...
Panjab	70,09,114	1,40,18,368	24,438	48,876	72,89,992	1,64,02,482	8,873	19,964	...	...	37	...	23	70	441	1,351
TOTAL	95,46,389	2,13,12,812	24,438	48,876	97,32,959	2,38,84,069	8,873	19,964	3,535	10,163	37	...	4,441	13,600	441	1,351
Chief Seaport—																
Karachi	...	...	3,535	10,163	...	...	4,418	13,530	...	...	25,37,122	72,94,226	...	...	24,42,967	74,81,587
TOTAL	95,46,389	2,13,12,812	27,973	59,039	97,32,959	2,38,84,069	13,291	33,494	3,535	10,163	25,37,159	72,94,226	4,441	13,600	24,43,408	74,82,938
Equivalent in cwt.	68,18,849	...	19,980	...	69,52,113	...	9,493	...	2,525	...	1,812,256	...	3,142	...	17,45,291	...

It will be seen from the above statement that the imports of wheat by Sind otherwise than those destined for Karachi, and that the exports from Sind otherwise than to Karachi, are unimportant in the extreme.

T. 833

Wheat Trade of Calcutta.	(*G. Watt.*)	TRITICUM sativum.

TABLE No. VII.

Analysis of the Rail- and River-borne Wheat Traffic of Calcutta during the past three years.

	Imports.			Exports.		
	1888-89.	1889-90.	1890-91.	1888-89.	1889-90.	1890-91.
By:—	*Mds.*	*Mds.*	*Mds.*	*Mds.*	*Mds.*	*Mds.*
East Indian Railway.	43,37,729	30,55,746	25,96,286	2,392	533	642
Eastern Bengal Railway.	1,48,867	60,405	1,05,716	931	1,501	1,633
Boat . .	9,54,341	6,71,732	6,65,774	7,822	6,642	5,828
Road . .	2,178	1,084	...	56,743	74,706	1,01,910
Inland steamer	64,494	67,592	1,00,702	2,203	3,142	2,202
Sea . . .	11,298	543	1,187	40,91,631	21,88,740	18,88,622
TOTAL .	55,18,907	38,57,102	34,69,665	41,61,722	22,75,264	20,00,837
Equivalent in cwt.	... 3,942,076	... 2,755,076	... 2,478,332	... 2,972,659	... 1,625,188	... 1,427,169

It has been thought unnecessary to prepare a detailed statement of the transactions with Bengal province alongside of that of Calcutta (similar to what has been furnished above for Bombay and Sind), but it may be said that the total imports during 1890-91 came to 65,222 maunds and the exports (excluding those to Calcutta) were 97,661 maunds. The province thus manifests a net export, but it is significant that its imports and exports are almost exclusively to and from the North-West Provinces, and from and to Behar. An inspection of the registration of the Bengal rail traffic reveals the fact that other than with Behar and with Calcutta no part of the province participates in the wheat trade to external blocks. Even the traffic with internal blocks is very limited, so that *Bengal may safely be said to practically not consume wheat.*

It need, therefore, be only necessary to furnish a statement of the Calcutta wheat supply in order to show the sources from which derived :—

	1888-89. Mds.	1889-90. Mds.	1890-91. Mds.
Behar	24,73,931	9,69,846	15,91,533
North-West Provinces and Oudh .	22,25,226	18,99,657	11,42,290
Bengal	6,94,095	4,95,291	4,84,798
Panjab	35,846	4,02,101	1,95,203
Central Provinces . . .	76,739	87,421	53,236
Other places	13,070	2,786	2,605
TOTAL .	55,18,907	38,57,102	34,69,665

The wheat from the North-West Provinces that drains to Calcutta, comes for the most part from Gonda. Bulandshahr, Barabanki, Allyghur, Hurdoi, Fyzabad, Goruckpore, Baraich, Sitapore, etc. The Behar wheat is, on the other hand, mainly derived from Monghyr, Sonthal Perganas, Bhagulpore, Shahabad, Patna, Saran, Maldah, Durbhanga, Gya, Chumparun, etc. The purely Bengal wheat, exported from Calcutta comes from the following

TRIUMFETTA rhomboidea.	The Paroquet Burr.

CALCUTTA WHEAT.

districts, for which the quantities furnished in the three last years may be given :—

	1888-89. Mds.	1889-90. Mds.	1890-91. Mds.
Nuddea	3,81,976	2,35,147	2,31,957
Hooghly	76,566	72,764	78,690
Moorshedabad	94,084	82,513	67,959
Other districts	1,41,469	1,04,867	1,06,195
TOTAL	6,94,095	4,95,291	4,84,798

(*J. Murray.*)

TRIUMFETTA, *Linn.; Gen. Pl., I., 234, 986.*

835

Triumfetta annua, *Linn.; Fl. Br. Ind., I., 396;* TILIACEÆ.

Syn.—T. POLYCARPA, *Wall.;* T. TRICHOCLADA, *Link.;* T. INDICA, *Lam.*
Vern.—*Aadai-otti*, TAM.; *Chikti*, HIND.
References.—*Atkinson, Him. Dist., 306; Gazetteer, Bombay, XV., 428.*
Habitat.—An herbaceous shrub, common in the Tropical Himálaya from Simla to Sikkim, the Khásia Mountains, Assam, the Konkan, Ava, and the Andaman Islands.

**FOOD.
Fruit.
836**

Food.—It produces orange-coloured flowers, and fruit of the size of a large pea. Green paroquets feed on the ripe FRUIT or burr, hence, in Jamaica, the plant is known as Paroquet Burr.

837

T. pilosa, *Roth.; Fl. Br. Ind., I., 394.*

Syn.—T. PILOSA; *var.* β, *Thwaites;* T. TOMENTOSA, *Wall.;* T. GLANDU-LOSA, *Heyne;* T. POLYCARPA, *Wall.;* T. OBLONGATA, *Link.*
References.—*Dalz. & Gibs., Bomb. Fl., 25; Thwaites, En. Ceyl. Pl., 31; Atkinson, Him. Dist., 306; Gazetteers :—Bombay, XV., 428; N.-W. P., IV., lxix.*
Habitat.—Found throughout the tropical parts of India from the Himálaya to Travancore and Ceylon.

**FOOD.
Fruit.
838**

Food.—It produces yellow flowers and small FRUIT of the size of a cherry. The remark made of the fruit of the above is equally applicable to that of this species.

839

T. rhomboidea, *Jacq.; Fl. Br. Ind., I., 395;* Wight, Ic., t. 320.

Syn.—T. BARTRAMIA, *Roxb.;* T. TRILOCULARIS, *Roxb.;* T. ANGULATA, *Lam.;* T. ANGULATA & ACUMINATA, *Wall.;* T. VESTITA, *Wall.*
Vern.—*Chikti*, HIND.; *Bun-okra*, BENG.; *Aadai-otti*, TAM.
References.—*Roxb., Fl. Ind., Ed. C.B.C., 390, 391; Dalz. & Gibs., Bomb. Fl., 25; Thwaites, En. Ceyl. Pl., 31; Stewart, Pb. Pl., 28; Gamble, Man. Timb., 52; Atkinson, Him. Dist., 306; Dymock, Warden, & Hooper, Pharmacog. Ind., I., 238; Gazetteers :—Mysore & Coorg, I., 58; Bombay, XV., 428; N.-W. P., I., 79; IV., lxix.*
Habitat.—An herbaceous plant, met with throughout Tropical and Sub-tropical India and Ceylon, ascending to 4,000 feet in the Himálaya.

**FIBRE.
Plant.
840**

Fibre.—The PLANT yields a soft glossy fibre, which is said to be considerably utilised in Madras.

**MEDICINE.
Fruit.
841**

Medicine.—All the species belonging to this genus are mucilaginous, and are used as demulcents, but this is the one generally employed. The mucilage is said to make a serviceable injection for inveterate gonorrhœa. The burr-like FRUIT is believed in India to promote parturition. The members of this genus are the Lappuliers of the French colonies, and bear the significant names of *Herbe à cousin, peu de moine,* and *tête à nègre (Pharmacog. Ind.).*

**FOOD.
Plant.
842**

Food.—In the Panjáb the PLANT is eaten as a pot-herb in times of scarcity (*Stewart*).

T. 842

TRUFFLES.

Truffles, *Baillon, Traite de Bot. Med. Cryptogam., 125.*
843

    References.—*Stewart, Pb. Pl., 268; Baden Powell, Pb. Prod., 258; Smith, Econ. Dict., 418.*

    Habitat.—Stewart describes truffles as being found in Kashmír, and Baden Powell mentions them as obtained from the *chír* forests of Kángra. Specimens sent by the latter to Kew were identified as **Melanogaster durissimus,** *Cooke* which see.

    The best truffles belong to the genus **Tuber,** which has given the name to the order TUBERACEÆ. Those most appreciated are T. **cibarium,** *Sibth.* ; T. **melanospermum,** *Vittad.* ; T. **æstivum,** *Vittad.* ; T. **magnatum,** *Pico* ; and T. **mesentericum,** *Vittad.* ; natives of France and Italy. The only European **Melanogaster** of economic value is M. **variegatus,** *Zul.* ; which though eatable is much less delicate in flavour than the French truffle. It is sold under the name of "Black Truffle."

    Food.—All the above-mentioned truffles are largely employed for culinary purposes, especially on the Continent. Baden Powell describes the Kángra truffle as occasionally eaten by the Natives. It is brown or black inside, "highly flavoured and of excellent quality." It grows to a large size, a diameter of 4 inches being by no means uncommon, and is said to resemble the Piedmontese truffle (T. **magnatum**) in shape and flavour. It is probable that, though used by Europeans in cookery, it, like the English species of the same genus, is of inferior quality. It is said that the Natives discover its presence in the soil by smell.
FOOD.
844

TULIPA, *Linn.; Gen. Pl., III., 818.*

Tulipa stellata, *Hook.; Linn. Soc. Jour., XIV., 288* ; LILIACEÆ.
845

    Vern.—*Bhúmphor, chamúni, padúna, jal kúkar, chamoti, piperi,* PB.; *Shandái gúl, ghentol,* PUSHTU.

    References.—*Stewart, Pb. Pl., 235; Baden Powell, Pb. Prod., 260; Atkinson, Him. Dist., 319; Gazetteer, Simla, 13; Agri.-Horti. Soc., Ind., Journal (Old Series), XIV., 14; (New Series), I., 106.*

    Habitat.—Common in the Western Panjáb, the Salt Range, the Siwaliks and the outer Himálaya to Kumáon.

    Food & Fodder.—The BULBS are frequently eaten by Natives, and are sold for that purpose in some of the bazárs, *e.g.,* in Peshawar. They are also eaten by animals.
FOOD & FODDER.
Bulbs.
846

Turmeric, see Curcuma longa, *Roxb.;* SCITAMINEÆ ; Vol. II., 659.

Turnip, see Brassica campestris, *Linn.;* sub-species Rapa, Vol. I., 523.

Turpentine, see Pinus, Vol. VI., Pt. I., 238, and Pistacia, Vol. VI., Pt. I., 271.

TURPINIA, *Vent.; Gen. Pl., I., 413, 999.*
[972 ; SAPINDACEÆ.
847

Turpinia pomifera, *DC.; Fl. Br. Ind., I., 698; Wight, Ic., t.*

    Syn.—DALRYMPELIA POMIFERA, *Roxb.;* T. NEPALENSIS, *Wall.* ; T. MICROCARPA, *W. & A.;* T. MARTABANICA & LATIFOLIA, *Wall.;* CANARIUM SAJIGA, *Ham.*

    Vern.—*Thali, nagpat,* NEPAL; *Janki-jám,* SYLHET; *Márgut, singnok,* LEPCHA; *Bundibru,* MECHI; *Nila,* NILGHIRIS; *Taukshama, daukyama,* BURM.

    References.—*Roxb., Fl. Ind., Ed. C.B.C., 213; Thwaites, En. Ceyl. Pl., 71; Kurz, For. Fl. Burm., I., 292; Gamble, Man. Timb., 102; Beddome, Fl. Sylv., t. 159; Gazetteer, Mysore & Coorg, I., 53; Ind. Forester, II., 22; IV., 241; Jour. Agri.-Horti. Soc., Ind. (New Series), V., Pro. (1875), 23.*

TYLOPHORA asthmatica.   A well-known medicinal Plant

FOOD & FODDER.
Fruit.
848
Leaves.
849
TIMBER.
850
Occurrence.
851

Habitat.—A moderate-sized deciduous tree, found in the Eastern Sub-tropical Himálaya from Nepál to Sikkim, at altitudes of 2,000 to 7,000 feet, also in the Khásia mountains, Assam, Sylhet, Cachar, Chittagong, Burma, and Penang, and in the Western Peninsula from the Konkan southwards.

Food & Fodder.—The FRUIT is edible, and the LEAVES given as fodder.

Structure of the Wood. —Grey or pale-brown, soft, fibrous but close-grained, soon attacked by insects; weight 30℔ per cubic foot. It is not used.

Turquoise, *Man. Geol. Ind., Vol. III*, *435.*

Occurrence.—The existence of the true turquoise in India is doubtful. From the presence of blue streaks in the copper ores of Ajmír, Mr. Prinsep suggested the possibility of the stone being found there. Subsequently Dr. Irvine reported its existence in these measures, but, according to Ball, the so-called turquoises of Ajmír are only blue copper ore. The principal turquoise mines in the world are at Ansar, near Nishapur, in Khorasan, Persia.

The turquoise is largely used by the Natives of India in jewelery but imitations are perhaps more generally employed than the true stone. *Conf. with* Carnelian, Vol. II.

852

TUSSILAGO, *Linn.; Gen. Pl., II., 438.*

Tussilago Farfara, *Linn.; Fl. Br. Ind., III., 330;* COMPOSITÆ.

Vern.—*Wátpan,* PB.

References.—*Stewart, Pb. Pl., 131; Year-Book Pharm., 1874, 626; Smith, Econ. Dict., 128; Agri.-Horti. Soc., Ind., Jour. XIV., 26, 55.*

Habitat.—A white, wooly herb, found in the Western Himálaya, from Kashmír to Kumáon at altitudes of 6,000 to 11,000 feet; distributed to North and West Asia, North Africa, and Europe.

MEDICINE.
Plant.
853
Leaves.
854

Medicine.—The PLANT is bitter and astringent, and contains a large quantity of mucilage. The LEAVES are sometimes used as a dressing for wounds in the Panjáb (*Stewart*); in Europe they are smoked like tobacco as a domestic remedy for asthma.

TYLOPHORA, *Br.; Gen. Pl., II., 770.*

[*Ic., t. 1277;* ASCLEPIADEÆ.

855

Tylophora asthmatica, *W. & A.; Fl. Br. Ind., IV., 44; Wight,*

Syn. —T. PUBESCENS, *Wall.;* T. VOMITORIA, *Voigt;* ASCLEPIAS ASTHMATICA, *Willd.;* A. TUNICATA, *Hort. Calc.;* A. VOMITORIA, *Kœn.;* CYNANCHUM VOMITORIUM, *Lam.;* C. VIRIDIFLORUM, *Sims.;* C. FLAVUM & BRACTEATUM, *Thunb.; Thwaites;* C. IPECACUANHA, *Willd.;* C., INDICUM, *Herb. Burm.*

Vern.—*Jangli-pikván, antamúl,* HIND.; *Anto-mul,* BENG.; *Mendi,* URIYA; *Pitmári, kharaki-rásna, anthamul,* BOMB.; *Pitakári,* MAR.; *Pit-kári,* DEC.; *Nach-churuppán, nanja-murich-chán, náy-pálai, peyp-pálai,* TAM.; *Verri-pála, kukka-pála, káka pála,* TEL.; *Valli-pála,* MALAY.; *Bin-nuga,* SING.

References.— *Roxb., Fl. Ind., Ed. C.B.C., 252; Voigt, Hort. Sub. Cal., 539; Thwaites, En. Ceyl. Pl., 197; (Excl. var. β); Dals. & Gibs., Bomb. Fl., 150; Mason, Burma & Its People, 479, 801; Sir W. Elliot, Fl. Andhr., 77, 102, 191; Pharm. Ind., 142, 458; Ainslie, Mat. Ind., II., 83; O'Shaughnessy, Beng. Dispens., 451, 455; Moodeen Sheriff, Supp. Pharm. Ind., 249; Dymock, Mat. Med. W. Ind., 2nd Ed., 519; Fleming, Med. Pl. & Drugs, as in As. Res., Vol. XI., 158; Fluck. & Hanb., Pharmacog., 427; Bent. & Trim., Med. Pl., t. 177; Official Corresp. on Proposed New Pharm. Ind., 284; Drury, U. Pl., 434; Lisboa, U. Pl. Bomb., 256; Bidie, Prod. S. Ind., 12; Gazetteers:—Mysore & Coorg, I., 62; Bombay, XV., 438; Hunter, Orissa, II., 181; Gribble, Man., Cuddapah, 200.*

Habitat.—Met with in N. and E. Bengal, Assam, Kachar, Chittagong, Deccan Peninsula, Burma to Malacca; common in Ceylon.

T. 855

A Substitute for Ipecacuanha. (*J. Murray.*) **TYLOPHORA asthmatica.**

Medicine.—The medicinal properties of this PLANT appear to have been long known to the Natives of those localities in which it occurs. It is, however, not mentioned in any of the ancient standard Sanskrit and Muhammadan works on Materia Medica, and was first brought to the notice of Europeans by Roxburgh, who writes as follows :—" On the coast of Coromandel, the ROOTS of this plant have often been used as a substitute for Ipecacuanha. I have often prescribed it myself, and always found it answer as well as I could expect Ipecacuanha to do. I have also often had very favourable reports of its effects from others. It was a very useful medicine with our Europeans who were unfortunately prisoners with Hydar Ally during the war of 1780, 1781, 1782, and 1783. In a pretty large dose, it answered as an emetic ; in smaller doses, often repeated, as a cathartic, and in both ways very effactually.

"I had made and noted down many observations of its uses when in large practice in the General Hospital at Madras in 1776, 1777, and 1778, but lost them, with all my other papers, by the storm and inundation at and near Coringa in May 1787. I cannot therefore be so full on the virtues of this valuable, though much neglected, root, as I could wish. I have no doubt but it would answer every purpose of Ipecacuanha.

"The Natives also employ it as an emetic; the bark, of about three or four inches, of the fresh root, they rub upon a stone, and mix with a little water for a dose; it generally purges at the same time."

" Dr. Russell was informed by the Physician General at Madras (Dr. J. Anderson) that he had many years before known it used, both by the European and Native Troops, with great success in the dysentery which happened at that time to be epidemic in the camp. The store of Ipecacuanha had, it seems, been wholly expended, and Dr. Anderson, finding the practice of the black doctors much more successful than his own, acknowledged, with his usual candour, that he was not ashamed to take instruction from them, which he pursued with good success ; and collecting a quantity of the plant which they pointed out to him, he sent a large package of the roots to Madras. It is certainly an article of the Hindu Materia Medica highly deserving attention."

Ainslie adds his testimony to the value of the drug, and remarks that the Vytians prize it for its expectorant and diaphoretic properties, and often prescribe it in an infusion to the quantity of half a tea-cupful, "for the purpose of vomiting children who suffer much from phlegm." Fleming repeats Dr. Russell's note on the opinion of the Physician General of Madras, and strongly recommends extended trial of the drug. Later, Sir W. O'Shaughnessy records his opinion that the emetic properties of the root are well established, but that it requires to be given in double the dose of the true Ipecacuanha, for which it affords an excellent substitute.

Owing to this concensus of opinion in its favour it was admitted as officinal in the *Bengal Pharmacopœia* of 1844. On the compilation of the *Pharmacopœia of India* in 1868, the opinions of the above writers were confirmed by numerous reports to the Committee who superintended the preparation of that work. The root was, however, superseded by the dried LEAVES which were found to be more uniform and certain in their action, and were therefore made the officinal part of the plant. They are described as one of the best indigenous substitutes for Ipecacuanha. Their actions are emetic, diaphoretic and expectorant ; they are recommended as useful in all cases indicating the necessity of emesis, and as a remedy for dysentery, catarrh and other affections in which Ipecacuanha is generally employed. The dose as an emetic is from 25 to 30 grains of the dried leaves powdered, as a diaphoretic and expectorant from 3 to 5 grains thrice daily or oftener. Moodeen Sheriff, in the long and interesting note given below, inclines to the opinion that the older

MEDICINE.
Plant,
856

Roots.
857

Leaves.
858

CHEMISTRY.
859

writers were right in preferring the root to the leaves, but that both are inferior to certain other indigenous substitutes for Ipecacuanha.

CHEMICAL COMPOSITION.—A concentrated infusion of the leaves has a slightly acrid taste. It is abundantly precipitated by tannic acid, by neutral acetate of lead or caustic potash, and is turned greenish-black by perchloride of iron. Broughton of Ootacamund obtained from a large quantity of leaves a small amount of crystals—insufficient for analysis. Dissolved and injected into a small dog they occasioned purging and vomiting (*Pharmacographia*).

SPECIAL OPINIONS.—§ "**Tylophora asthmatica** is one of the commonest plants in the fields and low and sandy jungles in Southern India. No part of this plant is sold in the bazár, but it can be very easily obtained in any quantity at the cost of collection. I have frequently used it during the last 16 or 17 years, and found every part of it, including the follicles, to possess the emetic property; but the roots and leaves are by far the best. Of the two latter, again, the roots are not only superior in their action to the leaves, but also much more easily reduced to a fine powder. The roots of **T. asthmatica** are involved in the greatest confusion as to their physical characters, and I shall, therefore, describe them here before speaking further of their medicinal properties. The root of this plant consists of many thin fibres or fibrils attached to a small woody portion, which is the axis or centre between them and the stem. The number of fibrils in each root is very variable, generally from 5 to 20, and sometimes upwards of 50. They are from 2 to 6 inches long, about a line in thickness, and of a pale or dirty white colour. They are seldom branched, but generally give attachment to very thin and hair-like fibres or rootlet. Their taste is slightly nauseous, but without the least bitterness. The roots of **T. asthmatica** in my collection of drugs at Calcutta have been gathered by myself with a view to avoid every doubt as to their genuineness. They correspond exactly with the above description. It is a matter of surprise that a doubt should exist in the description of the root of a plant which is found everywhere in this country and can be examined at any moment, if necessary. At one time I thought that **T. asthmatica** was the best substitute for Ipecacuanha in India, but from subsequent and more extensive experience of Native drugs I find that it is not the best, but one of the four or five best substitutes and ranks after **Randia dumetorum, Strychnos potatorum**, and **Luffa amara**. Dose:—Of the root and leaves, as an emetic, from 40 to 50 grains; and as a nauseant and expectorant, from 4 to 10 grains. Dose:—Of the root, as a remedy in dysentery, from 15 to 30 grains or more" (*Honorary Surgeon Moodeen Sheriff, Khan Bahadur, Triplicane, Madras*). "A liniment prepared with the root is applied to the head in cephalalgia and neuralgia" (*Native Surgeon T. Ruthman Moodelliar, Chingleput, Madras Presidency*). "Still used as one of our indigenous medicines" (*Surgeon-General W. R. Cornish, F.R.C.S., C.I.E., Madras*). "Diaphoretic and expectorant in doses of 10 and 15 grains, and emetic in 30-gr. doses; used in diarrhœa and dysentery and often employed as a substitute for Ipecac" (*Surgeon-Major A. F. Dobson, M.B., Bangalore*).

860

Tylophora fasciculata, *Ham.; Fl. Br. Ind., IV., 40; Wight, Ic.,*
Vern.—*Bhui-darí*, BOMB. [*t. 848.*
References.—*Dalz. & Gibs., Bomb. Fl., 151; Dymock, Mat. Med. W. Ind. 2nd Ed., 521; Lisboa, U. Pl. Bomb., 267.*
Habitat.—Found in South Nepál and the South Konkan.

MEDICINE.
Plant.
861

Medicine.—Dymock informs us that the PLANT is used in the Southern Konkan as a poison for rats and other vermin, and that Dr. Lyons records a case in which it proved fatal to man. He suggests that, since

| The Reed Mace or Elephant Grass. (*J. Murray.*) | TYPHA angustifolia. |

it possesses very active properties, its physiological effects should be investigated.

Tylophora mollissima, *Wight; Fl. Br. Ind., IV., 43; Wight, Ic.,*   **862**
                                                   [*t. 1275.*

    Vern.—*Moshiki,* KALADGI (BOMB.).
    Reference.—*Dymock, Mat. Med. W. Ind., 2nd Ed., 887.*
    Habitat.—A native of the Nilghiri and Pulney Mountains.
    Food.—This HERB is always more or less eaten in the Kaladgi District, and was especially utilised during the Deccan Famine of 1877-78 (*Dymock*).   **FOOD. Herb. 863**

        TYPHA, *Linn.; Gen. Pl., III., 955.* [TYPHACEÆ.   **864**
Typha angustifolia, *Linn.; Roxb., Fl. Ind., Ed. C.B.C., 648;*
    THE REED MACE, LESSER CAT'S TAIL, or ELEPHANT GRASS.
    Syn.—T. ELEPHANTINA, *Roxb.*
    Vern.—*Pater,* HIND.; *Hoglá,* BENG.; *Bora,* KUMAON; *Kúndar, dib, dab, pita, yira, boj, lúkh, patira, gond, pan, bori,* PB.; *Pita, yira,* KASHMIR; *Pun,* pollen=*búr, búri,* SIND; *Rámbána,* MAR; *Ghabajarin,* GUZ.; *Rámabána, ramban,* BOMB.; *Jammu gaddi, emiga-junum,* TEL.; *Eraka,* SANS.
    References.—*Grah., Cat. Bomb. Pl., 227; Stewart, Pb. Pl., 246; Elliot, Fl. Andhr., 72; U. C. Dutt, Mat. Med. Hind., 297; Dymock, Mat. Med. W. Ind., 2nd Ed., 843; S. Arjun, Bomb. Drugs, 149; Murray, Pl. & Drugs, Sind, 16; Cat., Baroda Durbar, Col. & Ind. Exhib., No. 180; Baden Powell, Pb. Pr., 262, 379, 514, 521; Atkinson, Him. Dist., 318, 752; Drury, U. Pl., 435; Lisboa, U. Pl. Bomb., 183; Birdwood, Bomb. Pr., 240; Christy, Com. Pl. & Drugs, VI., 48; Royle, Fib. Pl., 35; Stock's Rep. on Sind; Bidie, Prod. S. Ind., 76, 121; Settlement Report, Panjáb, Lahore, 13; Gazetteers:—Mysore & Coorg, I., 55, 66; N.-W. P., lxxvii; Panjab, Musuffargarh, 25; Hoshiarpur, 14; Ind. Forester, IV., 168; Agri.-Horti. Soc. Ind., Journal (Old Series), X., 354; (New Series), I., 105.*
    Habitat.—A rush found on margins of tanks and rivers throughout India.

    Fibre.—The fibrous STEMS and LEAVES are used for many purposes   **FIBRE.**
throughout the country. In Kashmír they are employed for making   **Stems.**
sieves, and for thatching huts and house-boats, in the Panjáb, Kulu,   **865**
and Kumáon for making soft matting, ropes, and baskets, and in Sind   **Leaves.**
for the same purposes and for building the rude wicker-work boats called   **866**
tirbo, used to cross the Indus during inundation. Graham states that, like
sedges in England, they are made up in bundles for buoys to support
swimmers. The fibre has been recently tried in the Bally Mills for paper-   **MEDICINE.**
making with success. It was described as easy of treatment. The fibre   **Down.**
has been examined in Europe, and is said to be of fine texture, tolerably   **867**
strong, and capable, with the aid of proper machinery, of being converted **Inflorescence.**
into textile fabrics.   **868**
    Medicine.—The DOWN of the ripe fruit, and the soft wooly INFLORES-   **FOOD FODDER.**
CENCE of the male spadix are applied like cotton to wounds and ulcers   **869**
    Food & Fodder.—The young SHOOTS are edible and taste like as-   **Shoots.**
paragus (*Baden Powell*). The ROOTS are eaten in Kashmír. In Sind   **Roots.**
the pollen is largely employed as flour and eaten when made into bread.   **870**
It was eaten in Bombay during the Deccan Famine. The plant is a   **DOMESTIC.**
favourite fodder for elephants.   **Roots.**
    Domestic, etc.—The long, tortuous, strong ROOTS penetrate the soil   **871**
to some depth, and are very valuable in binding the sandy banks of such   **Stem.**
a river as the Indus. The lower succulent part of the STEM is said to   **872**
have the property of speedily and effectually clearing muddy water. The   **Down.**
dried stalks are used for making Native pens. In Peshawar the DOWN   **873**
of the ripe fruit is mixed with mortar to bind it. The POLLEN, like the   **Pollen.**
                                                                      874

**TYPHONIUM
trilobatum.**   The Greater Cat's Tail.

---

spores of **Lycopodium**, is inflammable, and is employed in Europe as a substitute for that substance.

875   **Typha latifolia,** *Willd. ; Stewart, Pb. Pl., 246.*
        THE GREATER CAT'S TAIL.
            Vern.—*Patera,* BIJNOR; *Kanda-tella,* GARHWAL ; *Pis, yira,* KASH.
            Boj, lúkh, dile, kúndar, patíra, gond, PB.; *Mudo-pun,* SIND ; *Jungl,*
            bájrí, BOMB.
            References.—*Dymock, Mat. Med, W. Ind., 2nd Ed., 891 ; Murray, Pl.*
            & Drugs, Sind, 17 ; Atkinson, Him. Dist., 318.
            Habitat.—Common in similar situations as the preceding, in the Pan-
FIBRE. jáb, Sind, the Deccan, and probably also Kutch.
Leaves.   Fibre.—Stewart describes this as employed similarly in every way to
876   **T. angustifolia** in the Panjáb. Atkinson states that the LEAVES are largely
        used in Kumáon, in the manufacture of a coarse matting called *boriya,*
FOOD.   of which some 900 maunds are annually exported from the district.
Root.   Food.—The ROOT is eaten in Kashmír; in the Deccan the SEEDS are
877   used as an article of food during famine seasons.
Seeds.   Domestic.—The succulent lower portion of the STEM is said to have
878   the same property of clearing turbid water, as is ascribed to the preceding
DOMESTIC. species.
Stem.
879   **TYPHONIUM,** *Schott. ; Gen. Pl., III., 967.*   [*611,*
880
FOOD.   **Typhonium bulbiferum,** *Dalz. ; DC., Monograph. Phanerog., II.*
Bulbs.   Reference.—*Lisboa, U. Pl. Bomb., 183 207.*
881   Habitat.—Found in Malabar and the Konkan.
Leaves.   Food.—The BULBS and LEAVES are eaten boiled.
882
                                                       [*Wight, Ic., t. 801.*
883   **T. trilobatum,** *Schott. ; DC., Monograph. Phanerog., II., 614 ;*
            Syn.—ARUM TRILOBATUM, *Linn.;* A. ORIXENSE, *Roxb. ;* TYPHONIUM
            ORIXENSE, *Schott. ;* T. ROXBURGHII, *Saunders.* Brown in *Linn. Soc.*
            Jour., XVIII., 261, adds syn. T. TRISTE, *Griff.,* and reduces to this
            species T. SIAMENSE, *Zeigler.*
            Vern.—*Ghét-kochu,* BENG. ; *Karunaik-kizhangu, kár-karunaik-kizhangu,*
            TAM.; *Kanda-gadda, durada-kanda-gadda,* TEL. ; *Chéna,* MALAY.
            References.—*Roxb., Fl. Ind., Ed. C.B.C., 627 ; Pharm. Ind., 250 ;*
            Thwaites, En. Ceyl. Pl., 335 ; Moodeen Sheriff, Sup. Pharm. Ind., 249.

        Habitat.—Met with in the Indian Peninsula, Ceylon, and Cochin China.
MEDICINE.   Medicine.—Roxburgh writes, "The ROOTS (when fresh) are exceed-
Roots. ingly acrid. The natives apply them in cataplasms, to discuss or bring
884   forward schirrus tumours. They also apply them externally to the bite
        of venomous snakes, at the same time giving inwardly about the size of a
        field bean. It is certainly a most powerful stimulant, in proper hands it
        might no doubt be used to great advantage in the cure of several dis-
        orders." Roxburgh appears, however, to have overestimated its value,
        since the Editor of the *Pharmacopœia of India,* while including the plant in
        the secondary list, remarks that any good effect which could be expected
        from it might be more readily obtained from a mustard poultice. "The
        acrid principle," he writes, "is very volatile, and by the application of
        heat, or by simple drying, the roots become innocuous or even wholesome
        as articles of diet."
            SPECIAL OPINIONS.—§ " Is an article of food, it relaxes the bowels, and
        thereby relieves hæmorrhoids. The wild plant is used as a medicine for
        piles " (*Native Surgeon T. R. Moodelliar, Chingleput*). " The roots
        formed linto paste are used as an external application in the stings of
        bees, wasps, scorpions " (*Civil Surgeon J. H. Thornton, B.A., M.B.,*
        Monghyr).

        T. 884

ULMUS, *Linn. ; Gen. Pl., III., 351.*

1

A genus of deciduous trees, which comprises about sixteen species ; natives of North Temperate regions. **U. campestris**, *Linn.*, the European Elm, is said by Brandis to occur as a small shrub along river beds, and as a middle-sized tree, planted near villages, in the North-West Himálaya. Stewart describes **U. campestris** as probably the same species as **U. Wallichiana**, *Planch.*, and Sir J. D. Hooker, in the *Flora of British India*, while retaining the two as distinct species, states that in all probability the tree described by Brandis as **U. campestris** is only a form of **U. Wallichiana**. These two species are very closely allied : both vary greatly in foliage, and the leaves take similar forms, so that a mistake might easily occur. The vernacular names would be the same for both. They are, therefore, not recognised as distinct by the Natives, are used for the same purposes, and may, for the purposes of this work, be considered under one species—**U. Wallichiana**. The mucilaginous properties of elm bark, well known in England, though now little utilised in medicine, do not appear to be recognised by the natives of India.

Ulmus lancifolia, *Roxb. ; Fl. Br. Ind., V., 480 ;* URTICACEÆ.

2

Syn.—U. HOOKERIANA, *Planch.*
Vern.—*Lapi*, NEPAL ; *Thalai*, BURM.
References.—*Roxb., Fl. Ind., Ed. C.B.C., 263 ; Kurz, For. Fl. Burm., II., 474 ; Gamble, Man. Timb., 342.*
Habitat.—A large tree of the Sub-tropical Himálaya, from Kumáon to Sikkim, at altitudes from 1,000 to 5,000 feet, also met with in the Khásia Hills, Chittagong, Pegu, and Martaban.
Structure of the Wood.—Light-red, strong, hard ; adapted for house-building (*Kurz*).

TIMBER.
3

[261

U. integrifolia. *Roxb.;* see Holoptelea integrifolia, *Planch.;* Vol. IV.,

U. Wallichiana, *Planch ; Fl. Br. Ind., V., 480.*

4

[& T.

Syn. —U. EROSA, *Wall. ;* U. LÆVIGATA, *Royle ;* U. PEDUNCULATA, *H.f.*
Vern.—*Mored, pabuna, chambar máya,* HIND. ; *Yúmbok,* LADAK ; *Brári, breri, brankúl, bren, bran, amrái,* KASHMIR ; *Káin, bren, brera, bránkul, amrái, marári, marrún, marash, makshári, manderung, maldung, shko, kummar, hembra, mánú, mannú, bran, brahmi, kái, bhamji, bhamni, brannú, merú, chipal, marál, mandú, mányi, máurn mamji, mal dúng morún,* PB.
References. –*Brandis, For. Fl., 432, t. 52 ; Gamble, Man. Timb., 341 ; Stewart, Pb. Pl., 210 ; Aitchison, Flora of the Kuram Valley, 93 ; Baden Powell, Pb. Pr., 600 ; Atkinson, Him. Dist., 317 ; Royle, Ill. Him. Bot , 341 ; Gazetteers :—Ráwalpindi, 82 ; Gurdáspur, 55 ; Hazára, 14 ; Ind. Forester, VIII., 38 ; IX., 197 ; X., 318 ; XIII , 66, 67 ; Agri.-Horti. Soc. Ind., Journ. (Old Series), VIII., Sel., 154 ; XIV., 52.*
Habitat.—A large deciduous tree of the North-West Himálaya, from Kashmír to Nepál, between 3,500 and 10,000 feet.
Fibre.—The BARK contains a strong fibre, and is used for cordage and for making bed strings and sandals. According to Cameron an excellent FIBRE is made from the scape or FLOWER-STALK.
Fodder.—The LEAVES are a favourite cattle-fodder, on which account the trees are often very severely lopped.
Structure of the Wood.—Heartwood, greyish-brown, moderately hard ; weight about 35℔ per cubic foot. It is used locally in places where deodar is not available, and Pinus excelsa not very abundant, such as in Hazára, where it finds a ready sale at from R3 to R5 per tree (*Gamble*). Stewart states that it is not valued by Natives, except in Kanáwar, where it is employed for making ark-poles. He, however, remarks that it is

FIBRE.
Bark.
5
Fibre.
6
Flower-stalk.
7
FODDER.
Leaves.
8
TIMBER.
9

UNCARIA
Gambier.

TIMBER.

tough and has been found light, strong, and useful for the panels of dog-carts, etc. The timber of the smaller form, considered by Brandis to be **U. campestris,** is said by that author and by Gamble to be more valued than that of the larger typical **U. Wallichiana.** In Afghánistán the wood of the cultivated European Elm is much valued for making platters, small bowls, etc. (*Aitchison*).

DOMESTIC.
Bark.
10

Domestic.—The fibrous BARK is made into slow matches and gun-fuses in the Panjáb.

UNCARIA, *Schreb.; Gen. Pl., II., 31.*

11

Uncaria Gambier, *Roxb.; Fl. Br. Ind., III., 31;* RUBIACEÆ.

GAMBIER, PALE CATECHU or TERRA JAPONICA.

Syn.—NAUCLEA GAMBIER, *Hunter.*
Vern.—*Kath kutha,* HIND.; *Chinai katha,* BOMB.; *Ankudu kurra,* TEL.; *Gambír,* MALAY.
References—*Roxb., Fl. Ind., Ed. C.B.C., 173; Pharm. Ind., 117; Ainslie, Mat. Ind., II., 106; O'Shaughnessy, Beng. Dispens., 398; Dymock, Mat. Med. W. Ind., 2nd Ed., 413; Fleming, Med. Pl. and Drugs, in As. Res., XI., 187; Flück. & Hanb., Pharmacog., 335; Bent. & Trim., Med. Pl., 139; Murray, Pl. and Drugs, Sind, 195; Birdwood, Bomb. Pr., 45; Royle, Prod. Res., 397; McCann, Dyes and Tans., Beng., 128; Liotard, Dyes, 9; Smith, Dic., 189; Tropical Agricul-turist, Apr. 1889, 671, 675; Agri.-Horti. Soc. Ind., Trans., IV., 184; VI., 126; Journals (Old Series) III., 57; V., Sel., 112; VI., Sel., 141; 142; VII., Sel., 56, 58; VIII., 57.*

Habitat.—An extensive, scandent bush, met with, wild or cultivated, in Malacca, Penang, and Singapore; distributed to Java and Sumatra. It is largely cultivated at Singapore. In 1819 about 800 plantations were formed; in 1866 owing to scarceness of fuel and dearness of labour gambier-planting gave indications of disappearing from the island, but lately it is said to have again much increased.

TAN.
Leaves.
12
Gambier.
13

Tan.—From the LEAVES an extract is obtained, known as GAMBIER, or the officinal " pale catechu" of the Pharmacopœias of India and the United Kingdom. Much confusion existed in early literature regarding this substance which to a great extent was not separated from some of the forms of "true catechu" (see **Acacia Catechu,** Vol. I., 27-44). The first authentic account of gambier dates from 1780, and was written by a Dutch trader named Couperus. This author relates how the plant was introduced into Malacca from Poutjan in 1758, how gambier was made from its leaves, and names several sorts of drugs and their prices.

Dr. Campbell of Bencoolen described the drug and the process of making it to Dr. Roxburgh, who also mentions that he had seen it made in the "parts to the eastward of the Bay of Bengal." It is also described by Fleming. At the present day it is made as follows:—

Leaves.
14
Young shoots.
15

The LEAVES and YOUNG SHOOTS are boiled in water for about an hour, at the end of which time they are removed, thrown into a capacious sloping trough, the lower end of which slopes into the pan, and squeezed with the hand so that the absorbed liquor may run back into the boiler. The decoction is then evaporated to the consistence of a thin syrup, and baled out into buckets. When sufficiently cooled, the workman pushes a stick into the bucket in a sloping direction, works it up and down, and rubs off the mass which thickens round the stick. By the motion thus caused the whole mass is kept agitated, and gradually sets. It is then placed in shallow square boxes, and when somewhat hardened is cut into cubes and these are dried in the shade. The leaves are boiled a second time, and finally washed in water, which water is saved for another operation. Of late

| Gambier or Pale Catechu. | (*J. Murray.*) | UNONA. |

years the drug is made into cubical blocks by pressure instead of by cutting. A plantation with five labourers contains on an average 70,000 to 80,000 shrubs, and yields 40 to 50 catties (of 1⅓ ℔) of gambier daily (*Flückiger & Hanbury*).

Gambier is very highly valued for tanning purposes in Europe, since it imparts a softness to the leather, obtained from almost no other substance. It does not appear to be employed for that purpose in India (*Conf.* Leather, Vol. IV., 607).

MEDICINE.
Gambier.
16

Medicine.—GAMBIER appears to have been unknown to Hindu and Muhammadan medicine. Ainslie and also Fleming mention the drug. The former remarks that it is "employed by the Malays in all cases requiring astringent medicines, and is chewed by them with the betel-leaves." The latter simply states that it resembles catechu in its properties. It would indeed appear to have been altogether neglected by Native practitioners in favour of the indigenous catechu. It is official in the Pharmacopœias of Great Britain and of India, being, from its more ready solubility, preferred to catechu. The physiological actions and therapeutic properties of this useful astringent are too well known to call for remark in such a work as the present.

CHEMISTRY.
17

CHEMICAL COMPOSITION.—Gambier agrees in chemical composition with catechu, especially with the pale kind made in Northern India. Both substances consist mainly of *catechin*, and they contain the yellow colouring matter, *quercetin* (see Vol. I., 36-38).

FOOD.
18
TRADE.
Gambier.
19

Food.—Gambier is very largely employed in India for eating with *pán.*

Trade.—A considerable import of GAMBIER takes place into India, chiefly from the Straits Settlements. It arrives in large baskets, and, according to Dymock, fetches in Bombay from R4 to R6 per Surat maund of 37⅓ ℔. Up to the year 1884-85 "Cutch and Gambier" were returned under one head in the foreign trade reports, but since that date they have been separated, and it is now definitely shown that the imports consist almost entirely of gambier, the exports of cutch. During the six years in which these articles have been considered separately, the imports of gambier have averaged 16,287 cwt., valued at R3,16,690. The total during 1889-90 was 14,652 cwt., valued at R3,93,455, a large increase on the average price. Of that amount 14,585 cwt. came from the Straits Settlements, 67 cwt. from other countries. Bengal imported 13,778 cwt., Bombay 852, and Madras 22 cwt. Nearly the whole of the imports are consumed in India, the average re-export during the past five years having been only 275 cwt., valued at R7,005. During the past year it was 289 cwt., valued at R7,011. Of that quantity 154 cwt. went to Zanzibar, 49 cwt. to the Straits Settlements, and 86 cwt. to other countries. Bombay exported 209 and Bengal 80 cwt. During the past six years exports of Indian-made gambier are only twice mentioned, *viz.*, in 1886-87 and 1888-89, in each of which years 1 cwt. was exported from Bombay to Turkey in Asia.

UNONA, *Linn. f.; Gen. Pl., I., 24, 956.*

20

A genus of erect or climbing trees or shrubs, which belongs to the Natural Order ANONACEÆ. It comprises some twenty-five species; distributed throughout the tropics of Asia and Africa. Of these about eighteen are natives of India. None appear to be of economic interest except U. **pannosa**, *Dalz.* (*Fl. Br. Ind., I., 58*), a native of the Konkan and of the forests of Travancore. The inner BARK affords a strong FIBRE, said to be adapted for cordage and for paper-making (*Lisboa, U. Pl. Bomb., 226*).

Bark.
21
Fibre.
22

URENA lobata.	An ingredient of " Dasamula."

URARIA, *Desv.; Gen. Pl., I., 521.*

23

Uraria lagopoides, *DC.; Fl. Br. Ind., II., 156;* LEGUMINOSÆ.

Syn.—HEDYSARUM LAGOPOIDES, *Burm.;* U. RETUSA, *Wall.;* DOODIA LAGOPIOIDES, *Roxb.;* U. HAMOSA, *Wall.*

Vern. —*Pitvan,* HIND.; *Chákuliá,* BENG.; *Davala,* MAR.; *Dowla,* BOMB.; *Kóla ponna,* TEL.; *Prisniparni, aughriparnika, atiguha,* SANS.

References —*Roxb., Fl. Ind., Ed. C.B.C., 5*¿*1; Elliot, Fl. Andhr., 15, 93; U. C. Dutt, Mat. Med. Hind., 147, 314; Dymock, Mat. Med. W. Ind., 2nd Ed., 221; Dymock, Warden & Hooper, Pharmacog. Ind., I., 426; Agri.-Horti. Soc. Ind., Journ. (Old Series), VI., 43.*

MEDICINE.
Plant.
24

Habitat.—A native of the tropical zone from Nepál and Bengal to Burma; distributed to the Malay Islands, China, Polynesia, and Northern Australia.

Medicine.—"This PLANT is an ingredient of the *Dasamula,* and is thus much used in Native medicine. It is considered alterative, tonic and anticatarrhal, but is seldom used alone" (*Hindu Mat. Med.*). According to Susruta it was given with milk to women in the seventh month of their pregnancy to produce abortion. The properties attributed to it are probably entirely fanciful (*Pharmacog. Ind.*).

SACRED.
25

Sacred.—In Vedic times the plant was invoked as a goddess (*Pharmacog. Ind.*).

26

U. picta, *Desv.; Fl. Br. Ind., II., 155.*

Syn.—DOODIA PICTA, *Roxb.;* HEDYSARUM PICTUM, *Jacp.;* U. LINEARIS, *Hassk.*

Vern,—*Dábrá,* HIND.; *Sankar-jata,* BENG.; Seed=*deterdane,* PB.; *Prisniparni,* MAR.; *Pilavan, pitavan,* GUZ.; *Prisniparni,* BOMB.

References.—*Roxb., Fl. Ind., Ed. C.B.C., 582; Dals. & Gibs., Bomb. Fl., 65; Stewart, Pb. Pl., 77; Dymock, Mat. Med. W. Ind. 2nd Ed., 221; Dymock, Warden & Hooper, Pharmacog. Ind., 427; Atkinson, Him. Dist., 308; Gasetteer, Mysore & Coorg, I., 59; N.-W. P., I., 80; IV., lxx; Journals (Old Series), Agri.-Horti. Soc., N.S., VI., 43.*

Habitat.—An erect perennial, found from the Himálaya to Ceylon; it ascends to 6,000 feet in the North-West.

MEDICINE.
Fruit.
27
Plant.
28

Medicine.—In the Panjáb the FRUIT is used as an application to the sore-mouths of children (*Stewart*). In Southern India the PLANT is supposed by the Hindus to act as an antidote to the poison of the *phúrsa* snake, Echis carinata (*Dymock*).

Urceola elastica, *Roxb.,* and U esculenta, *Benth ;* APOCYNACEÆ; see Indian-rubber, Vol. IV., 361.

URENA, *Linn.; Gen. Pl., I., 205.*

29

Urena lobata, *Linn.; Fl. Br. Ind., I., 329;* MALVACEÆ.

Syn.—U. CANA, *Wall.;* U. PALMATA, *Roxb.*

Var. **scabriuscula**=U. SCABRIUSCULA, *DC.*

Vern.—*Bun-ochra,* BENG.; *Bhidi janetet',* SANTAL; *Bachita,* N.-W. P.; *Vana-bhenda,* MAR.; *Villiah,* KONKAN.; *Kat-sae-nai, wet-khyae-pa-nai,* BURM.; *Pattaappele,* SING.

References.—*Roxb., Fl. Ind., Ed. C.B.C., 519; Dals. & Gibs., Bom., Fl., 18; Thwaites, En. Ceyl. Pl., 25; Rev. A. Campbell, Rept. Ec. Pl., Chutia Nagpur, No. 7896; Mason, Burma & Its People, 520, 755; Murray, Pl. & Drugs, Sind, 61; Baden Powell, Pb. Pr., 228; Royle, Fib. Pl., 263; Cross, Bevan, & King, Rep. on Ind. Fibre, 9, 43; Liotard, Paper Mat., 31; Atkinson, Him. Dist., 306; Lisboa, U. Pl. Bomb, 228; Gazetteers:— Mysore & Coorg, I., 58; Bombay, XV., 427; N-W. P., IV., lxviii; Agri.-Horti. Soc. Ind., Journ. (Old Series), IX., 405, Sel., 47.*

U. 29

The Indian Squill.	(*J. Murray.*)	**URGINEA indica.**

Habitat.—A common herb, generally distributed throughout the hotter parts of India, very frequent in waste places, and in the bamboo and mango clumps of Bengal.

Fibre.—The BARK yields a good, easily extractable fibre, which is considered suitable for the manufacture of sacking and twine, and a fair substitute for flax. **Messrs. Cross, Bevan & King** found that it contained 77·7 per cent. of cellulose and lost by hydrolysis 11·9 per cent., when boiled for five minutes in 1 per cent. Na_2O; 18·5 when boiled for an hour. The length of the ultimate fibre is 1·5 to 2·0 mm.

> **FIBRE.**
> Bark.
> **30**

Medicine.—In Chutia Nagpur the ROOT is employed as an external remedy for rheumatism.

> **MEDICINE.**
> Root.
> **31**

Urena repanda, *Roxb.; Fl. Br. Ind., I., 330;* WIGHT, *Ill., I., 65.*

> **32**

　　Syn.—U. RIGIDA, *Wall. Cat. 1929 (in part)*; U. HAMILTONIANA, *Wall.;* U. SPECIOSA, *Wall.;* PAVONIA REPANDA, *Spreng.*

　　Vern.—*Sikuar*, SANTAL.

　　References.—*Roxb., Fl. Ind., Ed. C.B.C. 519; Campbell, Ec. Pl. Chutia Nagpur, No. 8740; Atkinson, Him. Dist., 306; Aplin, Rep. on Shan States; Gazetteer., N.-W. P., lxviii.*

Habitat.—An undershrub met with in North-West India, the Upper Gangetic plain, the Western Peninsula, and Burma.

Medicine.—The ROOT and BARK are believed by the Santals to be a cure for hydrophobia (*Campbell*).

> **MEDICINE.**
> Root.
> **33**
> Bark.
> **34**
> **35**

U. sinuata, *Linn.; Fl. Br. Ind., I., 329.*

　　Syn.—U. MURICATA, *DC.;* U. LAPPAGO, *DC.;* U. MORIFOLIA, *DC.;* U. HETEROPHYLLA, *Smith;* U. TOMENTOSA, *Wall.*

　　Vern.—*Lotloti, kunjáya,* HIND.; *Kunjia,* BENG; *Mota bhedi janetet',* SANTAL; *Tapkoté,* BOMB.; *Piliya mankena,* TEL.; *Hinappele,* SING.

　　References.—*Roxb., Fl. Ind., Ed. C.B.C., 519; Dalz. & Gibs., Bomb. Fl., 18; Thwaites, En. Ceyl. Pl., 25; Eurm., Fl. Zeyl., t. 69, f. 2; Rev. A. Campbell, Rept. Ec. Pl., Chutia Nagpur, No. 8492; Elliot, Fl. Andhr., 152; Lisboa, U. Pl. Bomb., 228; Gazetteers:—Bombay, XV., 428; N.-W. P., IV., lxv ii.*

Habitat.—A small bush, with deeply gashed leaves, found throughout the hotter parts of India.

Fibre.—The BARK yields a strong and tolerably fine FIBRE, which, like that from **U. lobata,** may be used as a substitute for flax. **Mr. Cameron** states that the plant attains its full size in wet land or by the margins of streams and tanks, and that, if necessary, it might be cultivated in the same way as jute.

> **FIBRE.**
> Bark.
> **36**
> Fibre.
> **37**

Medicine.—In Chutia Nagpur the ROOT is used as an external application for lumbago.

> **MEDICINE.**
> Root.
> **38**

URGINEA, *Steinh.; Gen. Pl., III., 810.*

[LILIACEÆ

Urginea indica, *Kunth.; Fl. Br. Ind. VI., 347; Wight, Ic., t. 2063;* INDIAN SQUILL.

> **39**

　　Syn.—SCILLA INDICA, *Roxb.;* U. SENEGALENSIS, *Kunth.;* S. CUNDRIA & S DENUDATA, *Ham.*

　　Vern.—*Kándá, janglí piyáz, kánde,* HIND; *Jongli piaáj, ban-piaáj, kánde,* BENG.; *Iskil, kúndri, kunda,* N.-W. P.; *Gheswwa,* KUMAON; *Phaphor, kachwassal,* PB.; *Jangli-piyáz, kandrá,* DECCAN; *Jangli-piás, kol-kánda, kochinda, janglí kanda, rána kándá,* BOMB.; *Ránácha-kándé,* MAR.; *Jangli-kánda, ran-kando,* GUZ.; *Nari-vengáyam,* TAM.; *Nakka vulli-gadda,* TEL.; *Adavi-irulli,* KAN.; *Káttulli,* MALAY.; *To-kesún, tankaet-tva, pa-daing-kyet-thwon,* BURM.; *Val-lúnu,* SING.; *Vunu-paldn dam,* SANS; *Isqíle-hindi aansale-hindi baslul-fáre-hindi, baslul-barre hindi,* ARAB.; *Piyáze-dashtié-hindi, piyáze-móshe-hindi,* PERS.

The Squill.

References.—*Roxb., Fl. Ind., Ed. C.B.C., 289 ; Stewart, Pb. Pl., 235 ;
Dals. & Gibs., Bomb. Fl., 250 ; Grah., Cat. Bomb. Pl., 220 ; Mason,
Burma & Its People, 814 ; Pharm. Ind., 241 ; Ainslie, Mat. Ind., I.,
402 ; O'Shaughnessy, Beng. Dispens., 662 ; Moodeen Sheriff, Supp.
Pharm. Ind., 250 ; Dymock, Mat. Med. W. Ind., 2nd Ed., 829, 887 ;
Flück. & Hanb., Pharmacog., 693 ; Irvine, Mat. Med. Patna, 41 ;
Official Corresp. on the Proposed New Pharm. Ind , 226, 235, 239, 295 ;
325 ; Atkinson, Him. Dist., 319, 752 ; Drury, U. Pl., 438 ; Birdwood,
Bomb. Pr., 91 ; Bidie, Prod. S. Ind., 45 ; Agri.-Horti. Soc. Ind., Trans.,
VI., 241.*

Habitat. -Found in sandy soil, especially near the sea, throughout
India, also in the drier hills of the lower Himálaya, and on the Salt Range
at altitudes of about 2,000 feet. The bulb is said by Atkinson to be
exported largely from the lower hills of the North-West Provinces.

MEDICINE.
Bulb.
40

Medicine.—The Hindus use the BULB in the preparation of *chandi-
bhasma* or "ashes of silver," which they employ medicinally. "Indian
Mahometan writers evidently consider the Indian squill as identical in
medicinal properties with the squill of the Greeks ; they prescribe it in
paralytic affections, also as an expectorant, digestive, diuretic, deobstruent
and emmenagogue, in many diseases, more especially in asthma, dropsy,
rheumatism, calculous affections, leprosy, and skin diseases" (*Dymock*).
European writers vary much in their opinions regarding the medicinal properties of the drug. Ainslie states that it "is chiefly employed by farriers
for horses in cases of strangury and fever." Roxburgh writes that the
bulb is quite as nauseous and bitter as that of the officinal squill ; while
O'Shaughnessy remarks that bulbs examined by him were inodorous,
nearly tasteless, and devoid of any medicinal property. Bidie, Atkinson,
U. O. Dutt, K. L. De, Dymock and others confirm the statement that
the drug is an efficient substitute for **Urginea Scilla**. Moodeen Sheriff
explains the discrepancy by stating that when young and small, not exceeding a lime in size, it acts as a diuretic, in doses of 10-20 grains, even
more powerfully than the officinal squill, but that as it grows larger it
becomes useless. The outer coats are always quite inert. It is also possible, as suggested by O'Shaughnessy, that the medicinal virtues may
vary with the season and locality of collection. The officinal squill is well
known to be thus affected. On the Spanish coast it has been found quite
inert in one locality, while as active as usual at the distance of a few miles.
A sufficient proof of its value, if collected and stored judiciously, is found in
the fact that, for many years, it has been used as a substitute for the officinal
squill at the Government Medical Store Depôt in Bombay. The dried bulb
met with in bazárs sells at from 1 to 2 annas per ℔ according to quality
(*Dymock*).

SPECIAL OPINIONS.—§ "The Indian squill is said to grow in abundance
in Pathankot, and to be as useful as the officinal squill" (*Assistant Surgeon Bhagwan Das, Rawalpindi*). "The bruised bulbs are applied as
a poultice to rheumatic pains or contusions and are much esteemed by
the people" (*Lal Mahomed, Hospital Assistant, Hoshangabad, Central
Provinces*). "Has been the only kind of squill used in the Bombay
Depôt for the last ten years ; it has proved quite satisfactory" (*W.
Dymock, Bombay*).

FOOD.
Leaves.
41
DOMESTIC.
Juice.
42
43

Food.—The LEAVES were eaten in the Khandesh District during the
famine of 1877-78 (*Dymock*).

Domestic.—The JUICE of the fresh bulb is said to be employed in the
North-West Provinces to give body to thread (*Stewart*).

Urginea Scilla, *Steinheil*.

THE SQUILL.

Syn —SCILLA MARITIMA, *Linn.* ; URGINEA MARITIMA, *Baker*.

U. 43

	URTICA diolca.
The Common Stringing Nettle. *(J. Murray.)*	

Habitat.—A perennial herb, found on the shores of the Mediterranean.

Medicine.—The BULBS are imported into India for use in European medicine. Their medicinal properties are too well known to require notice in this work.

MEDICINE.
Bulbs.
44
45

URINE.

Urine.

Vern.—*Pesháb,* HIND.; *Mutra,* SANS.

Reference.—*U. C. Dutt, Mat. Med., Hind., 84.*

Dye.—Stale urine is a common constituent of the indigo fermentation-vat (see **Indigo,** Vol. IV., 459). Indian Yellow, or *Peori,* is extracted from the urine of cows fed in a particular way (see **Peori,** VI., Pt. I., 132).

DYE.
46

Medicine.—The URINE of various animals has long been esteemed and much used in Sanskrit medicine. That of the cow is specially valued; it is employed in the purification of many metals for medicinal use, and is a common vehicle of iron prescribed for anæmia. U. C. Dutt (*Mat. Med. Hind.*) gives an interesting account, which may be quoted in entirety:—" The properties of the urine of various animals, such as the cow, buffalo, goat, sheep, horse, elephant, ass, and camel are minutely described. Of these cow's urine is much used both internally and externally in the purification of various metals and in the preparation of oils, decoctions, etc. It is described as laxative, diuretic, and useful in constipation, suppression of urine, colic, anasarca, jaundice, leprosy, and other skin diseases. Goat's urine is sometimes given internally. In congestive fever, with constipation, flushed face, and headache, an ounce of fresh and warm cow's urine is given as a domestic medicine. It is sometimes given as a vehicle for administering castor oil." Cow's urine is used in the preparation of various complicated medicines for the above enumerated diseases, of which Dutt gives two examples, from the *Bhávaprakása* and *Chakradatta.* The first is an extract of various drugs made with cow's urine, to which iron rust is added and administered internally; the second is an oily preparation said to be useful in leucoderma, chronic prurigo, and other obstinate skin diseases.

MEDICINE.
Urine.
47

Urostigma, see **Ficus,** *Linn;* URTICACEÆ; Vol. III., 342-362.

URTICA, *Linn. ; Gen. Pl., III., 381.*

48

This, the typical genus of the Nettle Family, comprises some thirty species, natives of Temperate and Sub-tropical regions, of which three are natives of India. In earlier works on Indian Botany this genus was made to include a large number of plants which by more careful study have been broken up into some twelve or thirteen genera—see, **Bœhmeria,** Vol. I , 465-484; **Debregeasia,** Vol. III., 52-54; **Girardinia,** Vol. III., 498-502; **Laportea,** Vol. IV., 587; **Maoutia,** Vol. V., 177-180; **Pilea,** Vol. VI., Pt I., 236; **Pouzolzia,** Vol. VI., Pt. I., 334; **Sarchoclamys,** Vol. VI., Pt. II., 476; and **Villebrunea,** p. 239.

Urtica dioica, *Linn.; Fl. Br. Ind., V., 548;* URTICACEÆ.

THE COMMON STINGING NETTLE.

49

Vern.—This, like other stinging nettles, is probably known in the Panjáb Himálaya as *bichu, bichúa,* or *chichrú*=the ' scorpion ' or ' stinger.'

Habitat.—Found in the North-West Himálaya from Kashmír and the Salt Range to Simla and Western Tibet, at altitudes from 8,000 to 12,000 feet. No information is available regarding the economic utilisation of this plant In India, but In Europe it has from the remotest times enjoyed the reputation of possessing many useful properties, which may be here briefly referred to.

URTICA
parviflora.

The Nettle Family.

DYE.
Root.
50
FIBRE.
Stems.
51
Fibre.
52
OIL.
Seeds.
53
MEDICINE.
Juice.
54
Root.
55
Plant.
56
FOOD &
FODDER.
Tops.
57
DOMESTIC.
Plant.
58
59

Dye.—A yellow colour, said to be extracted from the ROOT by boiling it with alum, may be employed as a dye.

Fibre.—The STEMS yield a well-known FIBRE which is said to rival in tenacity the best hemp. "Of late years it has become extensively cultivated in Germany and by dressing, the fibre is made to become as fine as silk" (*Smith*). In many parts of Europe it is employed for making fishing lines and even cloth.

Oil.—The SEEDS contain an OIL, which is itself edible, and renders the former a nutritious article of food.

Medicine.—The JUICE has been frequently used, and is even still employed as an external irritant. The ROOT is considered diuretic; the whole PLANT, in decoction, is believed to be diuretic, astringent, emmenagogue anthelmintic, and useful in nephutic disease, hæmorrhages, especially from the kidneys or uterus, consumption, and jaundice.

Food & Fodder.—The young TOPS are employed as a pot-herb and vegetable in soups, in certain parts of Europe. When dried they are given as fodder to cows, and cut up into small pieces they form a common food for fowls.

Domestic.—A salted decoction of the PLANT has the power of curdling milk.

Urtica hyperborea, *Jacquem.; Fl. Br. Ind., V., 548.*

Vern.—*Zatúd, dзatsutt, stokpo tsodma,* LAD.

Reference.—*Stewart, Pb. Pl., 15.*

Habitat.—A small, alpine species, found in Western Tibet, at 12,000 to 17,500 feet, and in Eastern Tibet, north of Sikkim, between 16,000 and 17,000 feet.

FOOD.
Leaves.
60

Food.—Stewart states that in Ladak the young LEAVES are eaten as a pot-herb.

61

U. parviflora, *Roxb.; Fl. Br. Ind., V., 548; Wight, Ic., t., 690.*

Syn.—U. ARDENS, *link.;* U. HIMALAYENSIS, *Kunth & Bouché;* U. VIRULENTA, *Wall.*

Vern.—*Berain, shishona, bichhu,* N.-W. P.

References.—*Roxb., Fl. Ind., Ed. C.B.C., 654; Baden Powell, Pb. Pr., 503; Royle, Fib. Pl., 371; Atkinson, Him. Dist., 317; Ec. Prod., N.-W. P., Pt. V., 91, 97; Gazetteer, N.-W. P., IV., lxxvii.*

Habitat.—Found in the Temperate Himálaya, from Kashmir to Mishmi, between 5,000 and 12,000 feet, also in the Nilgiris, at Ootacamund.

FIBRE.
62

Fibre.—This nettle yields a fibre of which little is known. As already stated under **Girardinia** (Vol. III., 499), considerable confusion exists in literature on the fibre obtained from Himalayan nettles. It is, however, not improbable that the following account published by Royle from the pen of Mr. C. Gubbins, C.S., may refer to this species:—
"The plant is cut in October, and dried in the sun; when brittle it is beaten, and the fibres separate easily. Seeing it stated that there was considerable labour required in cleaning the fibre, I made particular inquiries on this head; and as far as I can learn, there is no greater trouble in cleaning the fibre of the **Urtica** when merely dried, than is experienced with the hemp of the hills which is not retted in water." The fibre is probably employed in making ropes, etc., in the same way as that of the other fibre-yielding nettles, but information regarding it is very meagre.

FOOD.
Leaves.
63
Æcidial
hypertrophi-
cations.
64

Food.—In the North-West Provinces the LEAVES are cooked and eaten as a spinah (*Atkinson*). [The hypertrophied leaf-stalks, produced through the parasitic action of **Æcidium Urticæ**, *Schum.* var. **himalayense**, *Barclay*, are very generally eaten by the hill tribes (see *Barclay, Scientific Memoirs, etc., 1887, Pt. II., 30). Ed.*]

U. 64

	UVA ursi.
The Roman Nettle. *(J. Murray.)*	

Urtica pilulifera, *Linn.*
    THE ROMAN NETTLE,
    An introduced weed often seen in the vicinity of hill stations as, for example, at Simla. See **Utangan** below.

Usar, see **Reh,** Vol. VI., Pt. I., 400-427.

Ushnan, or **Soda-plants,** see **Barilla,** Vol. I, 396-399.

Usnea, see **Lichens,** Vol. IV., 635.

Utangan.—The *Utangan* or *Unjureh* of Muhammadan writers is the **Urtica prima** of the Latins (**Urtica pilulifera,** *Linn.*); but the drug which is found in the Bombay bazárs under the name of *Utingan* is the seed of **Acanthodium hirtum,** *Stocks.*

UTRICULARIA, *Linn.; Gen. Pl., II., 987.*

Utricularia bifida, *Linn.; Fl. Br. Ind., IV., 332;* LENTIBULARIEÆ.
    Syn.—U. BIFLORA, *Wall. (not of Roxb.);* U. DEANTHA, *A. DC.* (excl. most
    Vern.—*Arak jháwár,* SANTAL.   [syn.).
    References.—*Rev. A. Campbell, Ec. Pl., Chutia Nagpur, No. 7897;
    Gazetteer, Mysore and Coorg, I., 56.*
    Habitat.—Found throughout India from Nepál and Assam to Ceylon and Malacca, abundant during the cold weather in damp, moist situations.
    Medicine.—In Chutia Nagpur the PLANT is given medicinally when the urine is of a high colour, resembling that of the plant. This resemblance has probably suggested its use to the Santal *Ojhas (Campbell).*

UVARIA, *Linn.; Gen. Pl., I., 23, 955.*

Uvaria macrophylla, *Roxb.; Fl. Br. Ind., I, 49;* ANONACEÆ.
    Syn.—U. CORDATA, *Wall.;* GUATTERIA CORDATA, *Dunal.*
    Vern.—*Bagh-runga,* BENG.; *Thabwot-nway,* BURM.
    References.—*Roxb., Fl. Ind., Ed. C.B.C., 455; Thwaites, En. Cevl Pl, 6; Kurz, For. Fl. Burm., I., 28; Gamble, Man. Timb.. 8; Agri.-Horti. Soc. Ind., Trans., VII., 49; Journ. (Old Series), VI., 35.*
    Habitat —A large, sarmentose shrub, found in Eastern Bengal, Burma, and the South-Eastern districts of Ceylon.
    Food.—The FRUIT is eaten by the Singalese *(Thwaites).*

U. Narum, *Wall.; Fl. Br. Ind., I., 50; Wight, Ill., t. 6.*
    Syn. —UNONA NARUM, *Dunal.*
    Vern.—*Narum-panel,* MALAY.
    References.—*Dalz. & Gibs., Bomb. Fl., 3; Thwaites. En. Cevl. Pl., 6; Lisboa, U. Pl. Bomb., 222; Gazetteers:—Mysore & Coorg, I., 56; Bombay, XV., 426.*
    Habitat.—A large, woody climber met with in the forests of the Western Peninsula and the Central Provinces of Ceylon, where it ascends to 4,000 feet.
    Oil.—In Malabar a sweet-scented, greenish oil is obtained from the ROOTS by distillation.
    Medicine.—This OIL and the ROOT are used medicinally in various diseases. The latter is fragrant and aromatic, and the bruised leaves smell like cinnamon *(Rheede).*

Uva ursi, see **Arcostaphylos Uva Ursi,** *Spreng;* ERICACEÆ; Vol. I., 289.

Marginal notes (right column):
65
66
67
MEDICINE.
Plant.
68
69
FOOD.
Fruit.
70
71
OIL.
Roots.
72
MEDICINE.
Oil.
73
Root.
74

VACCINIUM, *Linn.; Gen. Pl., II., 573.*

[*Wight, Ic., t. 1188;* VACCINIACEÆ.

I

Vaccinium Leschenaultii, *Wight; Fl. Br. Ind., III., 455;*

Syn.—ANDROMEDA SYMPLOCIFOLIA, *Wall;* AGAPETES SYMPLOCIFOLIA, *G. Don;* A.? ARBOREA, *DC.*
Var. **arborea.**
Var. **rotundifolia**=V. ROTUNDIFOLIA, *Wight.*
Var. **zeylanica.**
Vern.—*Andúvan,* NILGIRIS.
References.—*Beddome, Fl. Sylv., t., 277; Madras, Man. Admin. I., 314; Ind. Forester, II., 23, 26.*
Habitat.—A tree of the mountains of Southern India and Ceylon, common at altitudes from 4,000 to 8,000 feet.

FOOD.
Fruit.
2

Food.—It produces an edible FRUIT, which is eaten by the Natives of the Nilgiris.

3

V. serratum, *Wight; Fl. Br. Ind., III., 452; Wight, Ic., t. 1184.*

Syn.—CERATOSTEMMA VACCINIACEUM, *Roxb.;* GAYLUSSACCIA SERRATA, *Lindl.;* AGAPETES SERRATA, *G. Don.*
Vern.—*Charu,* NEPAL; *Kesa prúm,* GARO.
References.—*Roxb., Fl. Ind., Ed. C.B.C., 374; Gamble, Man. Timb., 234.*
Habitat.—A shrub, often epiphytic, found in Sikkim, Bhután, and the Khásia Hills, from 4,000 to 8,000 feet.

FOOD.
Flowers.
4

Food.—Roxburgh states that the FLOWERS have an acid taste, and are eaten by the Natives of the Gáro Hills in their curries.

[LEGUMINOSÆ; Vol. I., 48.

Vachellia Farnesiana, *W. & A.*, see **Acacia Farnesiana**, *Willd;*

VALERIANA, *Linn.; Gen. Pl., II., 154.*

[*Ic., t. 1045-6;* VALERIANEÆ.

5

Valeriana Hardwickii, *Wall; Fl. Br. Ind., III., 213; Wight,*

Syn.—V. TENERA, *Wall;* V. ELATA, *Don;* V. JAVANICA, *Blume;* V. ACUMINATA, *Royle.*
Var. **Hoffmeisteri,** *Klotzsch* (Sp.).
Var. **Arnottiana,** *Wight* (Sp.).
Vern.—*Tágger, shumeo, asarún,* HIND.; *Tágger, balchur, úshur,* BENG.; *Nahâni, chár,* the root=*ásárun, bála, taggar,* PB.; *Shumeo, asárun,* KUMAON; *Tagger-ganthoda,* BOMB.; *Char,* C. P.
References.—*Stewart, Pb. Pl., 118; Pharm. Ind., 120; Irvine, Mat. Med. Patna, 17; Dymock, Mat. Med. W. Ind., 2nd Ed., 419; Year-Book Pharm., 1873, 78, 284; 1878, 289; Atkinson, Him. Dist., 311, 753; Royle, Ill. Him. Bot., 241; Settle. Rep., Belaspore, 77.*
Habitat.—A perennial herb of the Temperate Himálaya, from Kashmír to Bhután, at altitudes of 4,000 to 12,000 feet; also met with in the Khásia Mountains, between 4,000 to 6,000 feet.

PERFUME.
Root.
6

Perfume.—The ROOT is exported to the plains partly for medicinal use, but mainly as a perfume. It is chiefly employed to scent and clean the hair (*Irvine*), and is also, when dry, burned as incense (*Atkinson*).

MEDICINE.
Root.
7

Medicine.—The ROOT has probably long been used in India, but from the fact that it does not appear to be mentioned in any standard work on Hindu Materia Medica, it has possibly been always, as it is now, more valued as a perfume than as a drug. **Dymock** informs us that it is described by Muhammadan physicians as an Indian kind of *asárun* (Asarabaca). The author of the *Makhzan-el-Adwiya* mentions several kinds of *asárun,* but states that the Indian *tágger* is to be preferred. The medici-

V. 7

The Common Valerian.   (*J. Murray*.)   **VALLARIS Heynei.**

nal properties attributed to it by him resemble those of **Nardostachys Jatamansi**—see Vol. V., 338 (*Mat. Med. W. Ind.*). Stewart states that in the Panjáb it is employed medicinally, its properties being similar to those of the Valerians of Europe. In the North-West Provinces also, it is believed to possess anti-spasmodic properties (*Atkinson*). According to Dr. Adams the Syrian Nard of the ancients was probably the root of this plant (*Pharm. Ind.*). There is little doubt that it may prove an efficient substitute for the officinal Valerian; it is, at any rate, worthy of fair trial, and of chemical examination.

MEDICINE.

Valeriana officinalis, *Linn.; Fl. Br. Ind., III., 211.*

    Common Valerian.

    Syn.—V. dubia, *Bunge.*

    Vern.—*Kálávála*, Mahr.; *Jalalakan, billi-lotan*, Ajm.

    References.—*Bentley & Trimen, Med. Pl., t. 146; O'Shaughnessy, Beng. Dispens., 402; Pharm. Ind., 119; Flück. & Hanb., Pharmacog., 377; Smith, Econ. Dict., 425; Year-Book Pharm., 1874, 626; 1879, 468; Med. Top. Ajmere, 139.*

    Habitat.—Found in North Kashmír, at Sonamurg, altitude 8,000 to 9,000 feet.

    Medicine.—Valerian root is official in all modern Pharmacopœias, and its well known properties require no comment in this work. It is imported into India with other officinal drugs, in small quantities; but it is little known to Natives, by whom the common and more easily obtainable species are employed as substitutes.

8

MEDICINE. Root. 9

V. Wallichii, *D.C.; Fl. Br. Ind., III., 213.*

    Syn.—V. villosa, *Wall.;* V. jatamansi, *Jones;* V. spica, *Vahl.*

    Vern.—*Dálá, wálá, bálá, char, bala mushk, mushkwáli, char godar*, root= *bála, ásárun, taggar*, Pb.; *Mah kák, gúr-balchor-ák*, Afghan.

    References.—*Roxb., Fl. Ind., Ed. C.B.C., 55; Stewart, Pb. Pl., 118; Pharm. Ind., 120; O'Shaughnessy, Beng. Dispens., 403; Ainslie, Mat. Ind., II., 367; Fleming, Med. Pl. & Drugs in As. Res., XI., 181; Baden Powell, Pb. Pr., 354; Atkinson, Him. Dist., 311; Agri.-Horti. Soc. Ind., Journ. (Old Series), XIV., 15.*

    Habitat.—Grows in Temperate Himálaya from Kashmir to Bhután, at an altitude of 10,000 feet, also in the Khásia Mountains, altitude 4,000 to 6,000 feet.

    Perfume, Medicine, Domestic.—The root is used in every way similarly to that of **V. Hardwickii.** It has been confused by Ainslie, O'Shaughnessy, and others with **Nardostachys Jatamansi**, which see Vol. V., 338.

10

PERFUME, MEDICINE, DOMESTIC. Root. 11

    VALLARIS, *Burm.; Gen. Pl., II., 710.*

                       [438; Apocynaceæ.

Vallaris Heynei, *Spreng.; Fl. Br. Ind., III., 650; Wight, Ic., t.*

    Syn.—V. dichotoma, *Wall;* Echites dichotoma, *Roxb.;* Peltanthera solanacea, *Roth.*

    Vern.—*Rámsar, chamari-ki-vel*, Hind.; *Hápar máli, rámsar*, Beng.; *Dúdhi*, Kumaon; *Putta podara yárála, pála malle tívva*, Tel.; *Bhadravalli, bhadramunjá, visalyakrit*, Sans.

    References.—*Roxb., Fl. Ind., Ed. C.B.C., 247; Kurs, For. Fl., Burm., II., 181; Brandis, For. Fl., 327; Thwaites, En. Ceyl. Pl., 192; Dals. & Gibs., Bomb. Fl., 144; Gamble, Man. Timb., 262; Elliot, Fl. Andhr., 142, 161; U. C. Dutt, Mat. Med. Hind., 293, 324; Atkinson, Him. Dist., 313, 753; Gazetteers:—Mysore & Coorg, I., 62; N.-W. P., I., 82; IV., lxxiv.; Ind. Forester, X., 325; Agri.-Horti. Soc. Ind., Journ. (Old Series), X., 16.*

12

V. 12

| VANDA Roxburghii. | A Valonia Cups. |

Habitat.—A large climbing shrub, found in the Tropical Himálaya, ascending to 5,000 feet in Kumáon ; also met with in Sylhet, Burma, South India (from the Konkan southwards), and in Ceylon. It is commonly cultivated in gardens throughout the country.

Medicine.—The milky JUICE is employed as an application to wounds and old sores in the North-West (*Atkinson*).

MEDICINE. Juice. 13

SPECIAL OPINIONS.—§ "Useful in cases of fistula, the juice is corrosive" (*U. C. Mukerji, M.B., C.M., Civil Medical Officer, Dinagepore*). "The milky juice is very useful for chronic ulcers and sinuses, and in whitlow" (*Surgeon W. Wilson, Bogra*). "Especially useful in onychia (whitlow)" (*Surgeon A. C. Mukerji, Noakhally*). "The milky juice is a mild irritant. Applied to old sores and sinuses, it excites some degree of inflammation in them and thereby expedites the process of healing " (*Assistant Surgeon R. C. Gupta, Bankipore*). "The juice of the twigs is a useful application to old sores and sinuses" (*Civil Surgeon J. H. Thornton, B.A., M.B., Monghyr*).

14

VALLISNERIA, *Linn.; Gen. Pl., III., 451.*

Vallisneria spiralis, *Linn.; Fl. Br. Ind., V., 660.*

  Syn.—V. SPIRALOIDES, *Roxb.*

  Vern.—*Sáwala, syala,* HIND. ; *Punatsu, pancha-dub,* TEL.

  References.—*Roxb., Fl. Ind., Ed. C.B.C.,* 710 ; *Stewart, Pb. Pl.,* 241 ; *Baden Powell, Pb. Pr.,* 306 ; *Balfour, Cyclop., III.,* 988 ; *Gazetteer :— N.-W. P., I.,* 84 ; *IV., lxxvii ; Ind. Forester :—IV.,* 234 ; *XII., app.,* 21 ; *XIV.,* 393 ; *Agri.-Horti. Soc. Ind., Journ. (Old Series), XII., Pro.,* 12.

  Habitat.—A submerged herb, found in water throughout India and Ceylon.

DOMESTIC. Leaves. 15 16

  Domestic, etc.—The moist succulent LEAVES are employed to cover the surface of sugar in the Native process of refining, see **Phœnix dactylifera,** Vol. VI., Pt. I., p. 213 ; see Sugar, Pt. II , pp. 31, 267, 311, etc.

Valonia Cups.—The acorn cups of **Quercus ægilops,** the "prickly-cupped oak," are known in commerce under this name. The tree grows in the Morea, from which large quantities are shipped to Europe, where they are highly valued for tanning purposes.

VANDA, *Br.; Gen. Pl., III., 578.*

[*916;* ORCHIDEÆ.

Vanda Roxburghii, *R. Br.; Fl. Br. Ind., VI.,* 52 *; Wight, Ic., t.*

  Syn.—CYMBIDIUM TESSELLOIDES, *Roxb.* ; C. TESSELLATUM, *Swartz.* ; EPIDENDRUM TESSELLATUM, *Roxb.* ; ÆRIDES TESSELLATUM, *Wight.*

  Vern.—*Rásná, nái, vandá, bandá, persárá, perasárá,* HIND. ; *Rásná, nái,* BENG. ; *Dare banki,* SANTAL ; *Rásna,* MAR. ; *Rasno,* GUZ. ; *Rásná,* BOMB. ; *Kanapa chettu badanike, mardáru, chitteduru,* TEL. ; *Rásna, vandáka, nákuli, gandhanákuli,* SANS.

  References.—*Sir W. Elliot, Fl. Andhr.,* 44, 81, 112 ; *Sir W. Jones, Treat. Pl.Ind.,* 147 ; *Rev. A. Campbell, Rept. Ec. Pl., Chutia Nagpur, No.* 9292 ; *U. C. Dutt, Mat. Med. Hind.,* 259, 310, 315, 322 ; *Dymock, Mat. Med. W. Ind., 2nd Ed.,* 792 ; *Cat., Baroda Durbar, Col. & Ind. Exhb., No.* 181 ; *Year-Book Pharm.,* 1880, 251 ; *Gazetteer, Mysore & Coorg, I.,* 72.

MEDICINE. Root. 18

  Medicine.—Under the name of *rásna* the ROOTS of this orchid and of **Acampe papillosa,** *Lindl.* (see Vol. I., 64), are indiscriminately used by native physicians. They are believed to be fragrant, bitter, and useful in rheumatism and allied disorders for which they are prescribed in a variety of forms. They also enter into the composition of several medicated oils for external application in rheumatism and diseases of the nervous system (*U. C. Dutt*). It is also said to be considered a remedy for secondary syphilis. The Rev. A Campbell informs us that, in Chutia Nagpur, the LEAVES, pounded and made into a paste, are applied to the body during fever, and that the JUICE is introduced into the aural meatus as a remedy for otitis media.

Domestic.—"Santal girls split up the LEAVES and wear them as anklets, hence the name '*dare banki*,' or 'tree anklet'" (*Campbell*).

	MEDICINE. Leaves. **19** Juice. **20**
	DOMESTIC. Leaves. **21**

VANGUERIA, *Comm.; Gen. Pl., II., 111.*

Vangueria edulis, *Vahl.; Fl. Br. Ind., III., 136;* RUBIACEÆ.

THE VOA-VANGA OF VOA-VANGUER OF MADAGASCAR.

Vern.—*Moyen*, ASSAM; *Helu, alu*, MAR.; *Abú*, BOMB.; *Voa-vanga*, MADAGASCAR.

References.—*Kurz, For. Fl. Burm., II., 34; Dymock, Mat. Med. W. Ind., 2nd Ed., 890; Gazetteers:—Mysore & Coorg, I., 61; Cent. Prov., 223; Bombay, X., 402.*

Habitat.—A small tree, native of Madagascar, resembling **V. spinosa,** but unarmed; cultivated in India for the sake of its edible fruit.

Food.—Its FRUIT is eaten by the people of Madagascar, from whose vernacular name the botanical name of the genus has been derived. In India it is frequently used as an article of food.

Domestic.—The timber is occasionally employed for economic purposes in India (*Bomb. Gas., X., 402*).

	22
	FOOD. Fruit. **23**
	DOMESTIC. **24**

V. spinosa, *Roxb.; Fl. Br. Ind., III., 136.*

Syn.—V. MOLLIS, *Wal.*; V. SPINOSA & V. PUBESCENS, *Kurz*; V. EDULIS, *Miq.*; PYROSTRIA? SPINOSA, *Miq.*

Vern.—*Muyuana, muduna, moina*, HIND.; *Muyna, mainphal, muyuana, muduna, moina*, BENG.; *Gél, mainphal*, C. P.; *Alu, atu*, BOMB.; *Alu*, MAR.; *Chéga gadda*, TEL.; *Hsay-ma-kyi*, BURM.; *Pindituka*, SANS.

References.—*Roxb., Fl. Ind., Ed. C.B.C., 180; Kurz, For. Fl. Burm., II., 34; Dalz. & Gibs, Bomb. Fl., 114; Gamble, Man. Timb., 119; Elliot, Fl. Andhr., 35; Birdwood, Bomb. Prod., 163; Lisboa, U. Pl. Bomb., 87, 162; Buchanan, Statistics, Dinajpur, 152; Taylor, Top. Dacca, 49; Gazetteer:—Bombay, XIII., 23; XV., 436; XVII., 23; XVIII., 31; Cent. Prov., 223; Settlement Report, Central Prov., Chanda, App. vi.*

Habitat.—A large thorny shrub, found from Northern Bengal to Canara, also in Burma, Pegu, and Tenasserim. It flowers in the beginning of the hot season, after which the fruit ripens in three or four months.

Medicine.—The dry FRUIT is said to possess narcotic properties (*Cent. Prov. Gas.*), and to be a remedy for boils (*Chanda Settl. Rept.*).

Food & Fodder.—The FRUIT is round, of the size of a cherry, smooth, yellow when ripe, and succulent. Considerable differences of opinion are expressed regarding its quality. **Roxburgh** states that it is eaten by Natives. **Lisboa** writes, "Eaten cooked or roasted, but it is not palatable." **Taylor**, on the other hand, remarks that it is considered a fruit of great delicacy, and is common in the bazárs of Dacca during November and December; while **Buchanan** writes, "It possesses an intoxicating or rather deleterious quality, when fresh plucked, but after being kept a few days, may be eaten without danger and is said to be sweet and agreeable." In the Central Provinces it is said to be used as a vegetable when *green*, and to be narcotic when *dry* (*C. P. Gazetteer, 223*). From these conflicting statements the fruit would appear to vary in different localities, or with cultivation, or according to its age. The LEAVES are said to be a useful fodder.

	25
	MEDICINE. Fruit. **26** FOOD & FODDER. Fruit. **27**
	Leaves. **28**

V. 28

VANILLA.

29

Vanilla is the long pod-like fruits of the epiphytic Orchid named. **V. planifolia.** It is a native of the West Indies and of tropical America. It has been experimentally grown in India, but the degree of success hitherto attained has not justified the establishment of plantations. The grateful aromatic qualities of **Vanilla** are fully utilized in confectionery, perfumery, and medicine. It is relative to its bulk perhaps the most expensive article of the vegetable kingdom which can be classed as a commercial product. The major portion of the British supplies are drawn from Mexico. The active principle Vanilline is now imitated chemically in preparations from pine-wood and clove oil.

VARNISH.

30

Varnish.—A varnish is generally defined as a solution of a resin, or gum-resin in a liquid, which, when spread over a surface, evaporates and leaves the solid in the form of a transparent more or less coloured film. The chief resinous substances employed in the manufacture of varnishes are amber, benzoin, copal, colophony, dammer, elemi, lac, mastic, sanda-rach, **Vateria** resin, and the coniferous resins. The chief solvents are lin-seed oil, oil of turpentine, oil of rosemary, alcohol, and ether. The solutions thus formed are variously coloured by the addition of dyes of the desired tint. For an account of the above enumerated substances the reader is referred to the article on each, in its respective alphabetical position. In addition to these, however—the varnishes of commerce—India is rich in plants which yield natural varnishes, *i.e.*, resinous substances which naturally exist in a form suitable for immediate use as varnishes. The principal of these are derived from the following plants :—

> **Buchanania latifolia.**
> **Holigarna,** several species, the Black Varnish of Malabar.
> **Melanorrhœa usitata,** the Black Varnish of Burma.
> **Odina Wodier.**
> **Rhus,** several species—the Japan Varnish.
> **Semecarpus,** the Black Varnish of Sylhet, South India, and Ceylon.

Full accounts of these varnishes will be found in the article on each in its alphabetical position in this work.

VATERIA, *Linn.* ; *Gen. Pl., I.,* 193.

31

Vateria indica, *Linn.; Fl. Br. Ind., I.,* 313; DIPTEROCARPEÆ.

THE WHITE DAMMAR of South India, PINEY VARNISH, or INDIAN COPAL.

Syn.—V. MALABARICA, *Blume.*

Vern.—*Suféd-dámar, kahruba, sandras,* HIND.; *Chundrus,* BENG.; *Suféd-dámar,* DECCAN; *Rál,* BOMB.; *Vellai-kunrikam, vellai-dámar, velli-kúndricum, kúndricum, paini-pishin, kungiliyam, vellai-kungiliyam, piney maram, dhup maram,* TAM.; *Dúpa-dámaru, tella dámaru, dupada,* TEL., *Dupa maram, dhupa, paini, munda dhup, dhupadamara,* KAN.; *Payana, vella-kunturukkam, peinimarum. vella-kúdricum, painipasha, painimara, vella kondrikam,* MALAY.; *Hal, hal-dumula,* SING.

References.—*Roxb., Fl. Ind., Ed. C.B.C.,* 436; *Beddome, Fl. Sylv., t.* 84; *Gamble, Man. Timb.,* 41; *Graham, Cat. Bomb. Pl.,* 22; *Mason, Burma & Its People,* 486, 526, 757; *Rheede, Hort. Mal., IV., t.* 15; *Pharm. Ind.,* 33; *Ainslie, Mat. Ind., II.,* 482; *O'Shaughnessy, Beng. Dispens.,* 221; *Irvine, Mat. Med. Patna,* 25; *Moodeen Sheriff, Supp., Pharm. Ind.,* 253; *Mat. Med. S. Ind. (in MSS.),* 48; *Dymock, Mat. Med. W. Ind.,* 2nd *Ed.,* 93; *Dymock, Warden & Hooper, Pharma-cog. Ind., I.,* 196; *Official Corresp. on Proposed New Pharm. Ind.,*

| The White Dammar. | (*J. Murray.*) | **VATERIA indica.** |

238 ; *Birdwood, Bomb. Prod.*, 258 ; *Drury, U. Pl. Ind.*, 439 ; *Useful Pl. Bomb. XXV. (Bomb. Gaz.)*, 15 ; *Econ. Prod., N.-W. Prov., Pt. I. (Gums and Resins)*, 3 ; *Gums & Resinous Prod. (P. W. Dept. Report)*, 2, 3, 5, 6, 7-9, 10, 20, 30, 37, 57, 66, 67 ; *Cooke, Gums and Resins*, 87 ; *Bidie, Prod. S. Ind.*, 21 ; *Man. Madras Adm., I.*, 314 ; *II.*, 105 ; *Nicholson, Man. Coimbatore*, 41 ; *Gazetteers :—Bombay, XV., Pt. I.*, 31, 78 ; *Mysore & Coorg, I.*, 46, 53 ; *Agri-Horti. Soc., Ind., Journ. (Old Series), VIII., Sel*, 141 ; *IX.*, 293, 294 ; *Indian Agriculturist, March* 17, 1888 ; *Ind. Forester, II*, 21 ; *VI.*, 125.

Habitat.—A large evergreen tree of the Western Peninsula, from Kanara to Travancore, ascending to 4,000 feet.

Resin.—This tree yields a true resin of considerable value, known as RESIN. white dammar, or Piney resin. It occurs in three forms :—1st, COMPACT PINEY RESIN, in lumps of all shapes which varies in colour, on the outside, from bright orange to a dull yellow, has a bright vitreous fracture, and internally presents all shades of colour from a light green to a light yellow. It is very hard, and bears a general resemblance to amber, from which characteristics, added to its colour, it may be easily distinguished from all **Cellular Piney** other Indian resins. 2nd, CELLULAR PINEY RESIN, occurs either in **Resin.** small lumps or in large masses, generally of a shining appearance and **32** balsamic smell. It has a distinctly cellular structure, partly owing to the mode of collection, and partly to the age of the tree. Notches are cut in the trunk of the tree sloping downwards and inwards, the resin collects in the cavity, and is either permitted to dry *in situ*, or is collected and dried by heat. It varies in colour from light green to yellow or white, and is usually transparent, though occasionally from want of care in preparation, **Dark coloured** more opaque, of a dull green colour, and full of air bubbles. 3rd, DARK CO- **Piney Resin.** LOURED PINEY RESIN, is occasionally obtained on splitting open old **33** and decayed trees. It has the solid consistence of the first, but the inferior quality of the second variety (*Jury Rept. Madras Exhb.*, 1857).

This resin, from its valuable characters, has naturally given rise to considerable interest on the part of European writers, attended with not a little confusion as to its real nature. It has been confused, to begin with, with gum animi, a resin derived from Zanzibar. The following report by Mr. **Broughton**, late Quinologist to the Government of Madras, gives an interesting and fairly complete account of the preparation and uses of the substance, and may be quoted in its entirety :—

"This beautiful substance has long been known, and its properties and local uses have been repeatedly described. It is also not unknown in England, and I apprehend that its cost (and perhaps also ignorance of its peculiar properties) has prevented its becoming an article of more extended commerce. It should be remarked that the "East Indian Dammar," which is well known among varnish-makers, though frequently confounded with this, is the product of a very different tree, and is not produced in this Presidency. The finest specimens of Piney resin are obtained by making incisions in the tree, and are in pale green translucent pieces of considerable size. The resin that exudes naturally usually contains much impurity. In most of its properties it resembles copal, but it possesses qualities which give it some advantage over that resin. Like copal it is but slightly soluble in alcohol ; but as **Berzelius** pointed out in the case of copal, it can be brought into solution by the addition of camphor to the spirit. It is easily soluble in chloroform, and thus might find a small application as a substitute for amber in photographer's varnish. It differs most advantageously from copal in being at once soluble in turpentine and drying oils, without the necessity of the preliminary destructive fusion required by that resin,—a process which tends greatly to impair the colour of the varnish. The solution of Piney resin in

VATERI indica	The White Dammar
RESIN.	

turpentine is turbid and milky, but by the addition of powdered charcoal, and subsequently filtering, it yields a solution transparent and colourless as water, and a varnish which dries with a purity and whiteness not to be surpassed. The solution in turpentine readily mixes with the drying oils. It is on these properties of the resin that its chance of becoming an article of trade will depend. In price it cannot expel copal, the supply of which to the European market is regular and abundant. Major Beddome informs me that the cost of the Piney resin delivered on the sea-coast would be about R6 per maund of 25℔. The present price of the best copal in the English market is but £2-10 per cwt." In the list of resins of Mysore and Coorg contained in the Public Works Department report the same price is quoted for the inferior, and R8 per 25℔ for the finer, quality. The remark is appended, " of excellent quality and procurable in great abundance in the Nuggur, Mulnaad, Munjerabad, and Coorg :—the supply is almost unlimited. In Nuggur extensive avenues of this tree alone exist, besides numerous isolated trees in the jungles." Moodeen Sheriff quotes the present bazár price in Madras, as, of the superior variety, R4 per maund of 25℔, of the inferior, R2. Wight gives the following formula for the preparation of varnish from the resin :—" Into a new and perfectly clean earthen vessel put one part of Piney Dammar in coarse powder, cover closely, and apply a very gentle heat, until the whole is melted ; then add about two parts of linseed oil, nearly boiling hot, and mix well with a wooden spatula. Should the varnish prove too thick, it can, at any time, be reduced by the addition of more oil, or, if required, may be made thicker in the first instance. Close covering, complete liquidation of the resin, and boiling hot oil are absolutely necessary to the success of this process.

The varnish thus prepared would be valuable for coating carriages, furniture, and other work requiring a complete and finely finished protection to the coat of paint. A varnish made with camphorated alcohol, the camphor from which has evaporated, might be used with advantage for varnishing pictures. On the Malabar coast it is made into candles which diffuse an agreeable fragrance, and give a clear light and little smoke. For making these the fluid resin may be either run into moulds or rolled when soft into the required shape. It is said that these candles were at one time introduced into England, but, a very high duty having been imposed, the trade ceased.

In a recent communication to Government, however, the Collector of South Canara states, at the present time no candles are made from the resin. He conjectures that the above description of candles made from the resin may have occurred from confusion of sticks of resin used as incense, and candles made from the tallow. The best pieces of the resin are employed as ornaments, under the name of amber (*Kehroba*) (*Madras Jury Repts.*). In South Canara it is used for caulking boats, and for setting gold ornaments.

Chemistry.
34

CHEMICAL COMPOSITION.—The resin yields, on distillation, 82 per cent. of a volatile oil of agreeable odour, but which does not differ essentially from that obtained from cheaper resins.

OIL.
Seeds.
35

Oil.—The SEEDS contain a large quantity of a solid oil, which may be prepared as follows :—Clean the seeds, then roast and grind them into a mass. To 5 seers seed add 12 seers of water, and boil until the oil rises to the surface. Remove the oil, stir the contents of the vessel, and allow it to stand until the following day, when more oil will be observed on the surface, which may be collected and the process repeated (*Jury Rept., Madras Exhib., 1855*). The oil is solid even in hot climates, and appears to be well adapted for candle-making. In South Canara it is used for lamps,

	OIL.

for flavouring food, as a substitute and adulterant for *ghí*, and for medicinal purposes. In 1884 a paragraph appeared in the *Journal of the Chemical Society*, in which it was stated that for a few months previously, fatty seeds of remarkable size, the produce of this tree, had been brought into commerce by way of Marseilles and Trieste, and that large quantities of vegetable fat, "Malabar tallow," had also been recently imported into Europe. This extract was brought to the notice of the India Office by Sir Joseph Hooker, who suggested that the value of the seeds should be brought to the notice of the Forest Department, to which they might become a source of revenue. Enquiries were instituted which resulted in reports from various districts including that from South Canara above quoted, regarding the illuminating properties of the resin. The Collector of that district also stated that he had candles made of the oil, which burned well and gave a good light but were wanting in hardness, being softer and greasier than the ordinary dip. This fact is at variance with the results of chemical examination. Thus Mr. L. Hooper found the melting point to resemble that of ordinary tallow or wax, and consequently that Piney tallow should be specially adapted for the manufacture of candles in hot climates. No further information as to the extent of the trade with Europe in tallow or seed appears to have been forthcoming.

CHEMICAL COMPOSITION.—The authors of the *Pharmacographia Indica* write:—"The seeds have been examined by MM. Hoernel and Wolfbaner, who found that, when air-dried, they afforded 49·2 per cent. of a greenish-yellow solid fat, which bleaches rapidly on exposure to light and has a peculiar agreeable balsamic odour. This fat rapidly saponifies and consists of a mixture of fatty acids melting at 56·6°C. and solidifying at 54·8°C. The mixture contains oleic acid and 60 per cent. of a solid fatty acid melting at 63·8° (*Chem. Centr.; Jour. de Pharm. et de Chim; Jour. Chem. Soc., 1886*)."

Chemistry. 36

Medicine.—Fine shavings of the RESIN are said by Irvine to be administered internally to check diarrhœa. Dr. Bidie recommends it as an excellent substitute for officinal resin, stating that it combines with wax and oil, under the influence of a gentle heat, and forms a good resinous ointment. The OIL has obtained considerable repute as a local application in chronic rheumatism and some other painful affections, and might be employed as a basis for ointments, etc. It closely resembles the solid fats of Garcinia indica, Kokam butter (*Pharm. Ind.*).

MEDICINE. Resins. 37

Oil. 38

Structure of the Wood.—Sapwood white with a tinge of red; heartwood grey, rough, moderately hard, porous; weight 41℔ per cubic foot. It is not much in request, but is occasionally used for making canoes, the masts of native vessels, and coffins (*Gamble*).

TIMBER. 39

VATICA, *Linn.; Gen. Pl., I., 192, 981.* [p. 678.

Vatica laccifera, *W. & A.*, see Shorea Talura, *Roxb.* ; Vol. VI., Pt. II.

V. lanceæfolia, *Blume* ; *Fl. Br. Ind., I.,* 302. DIPTEROCARPEÆ.

40

Syn.—V. CANACA, *Ham.*; VATERIA LANCEOLARIA, *Roxb.*; V. LANCEÆFOLIA, *Roxb.* ; V. LANCEOLATA, *Roxb.*

Vern.—*Morhal*, ASSAM; *Moal*, SYLHET; *Panthitya*, BURM.

References.—*Roxb., Fl. Ind., Ed. C.B.C., 435 ; Kurz, For. Fl. Burm., I., 122 ; Gamble, Man. Timb., 33 ; Mason, Burma & Its People, 486 ; Gums & Resinous Prod. (P. W. Dept. Rep.), 16, 20 ; Cooke, Gums & Resins, 89.*

Habitat.—A large tree of the Eastern Himálaya, Assam, Eastern Bengal, Chittagong, and Burma.

Resin.—The tree yields a resin, of which very little is known. Roxburgh writes, "From wounds, etc., in the BARK, a clear liquid exudes,

RESIN. Bark. 41

VENTILAGO **calyculata.**	The Ventilago Fibre.

RESIN.

which soon hardens into a very pure, pale, amber-coloured resin, from which the Natives obtain, by distillation, a dark-coloured, thick, strong-smelling balsam, called *chooa*, or *chova*, by the people who prepare and sell it, and *ghoond* by the Brahmins who use it in their religious cere- monies and temples." Mason states that the resin is "precisely similar" to that of **Vateria indica**, a remark which would appear to be supported by Roxburgh's description of its colour and purity.

TIMBER.
42

Structure of the Wood.—Heartwood red, rough, hard; weight 35 to 52℔ per cubic foot; not very valuable and little used (*Gamble*).

Vatica robusta, *Steud.*, see **Shorea robusta,** *Gærtn. f.;* Vol. VI., Pt.
[II., p. 673,

43

V. Roxburghiana, *Blume; Fl. Br. Ind., I., 302; Wight, Ic., t. 26.*
Syn.—V. CHINENSIS, *Linn.;* VATERIA ROXBURGHIANA, *Wight, Ill., I.,*
88.
Vern.—*Let-touk, lettaub,* BURM.; *Mandora,* SING.
References.—*Gamble, Man. Timb., 33; Thwaites, En. Cey. Pl., 404; Beddome, Fl. Sylv., t. 95; Mason, Burma & Its People, 515, 757; P. W. Dept. Rept. on Gums & Resins, 32, 35; Cooke, Gums & Resins, 90.*
Habitat.—A tree of the Western Peninsula from Canara southwards, and of Ceylon.

RESIN.
44

Resin.—It produces a resin of which little is known, but which is said to resemble those of the preceding species and of **Vateria indica**.

45

V. scaphula, *Dyer; Fl. Br. Ind., I., 301.*
Syn.—HOPEA SCAPHULA, *Roxb.*
Vern.—*Boilshura,* BENG.
References.—*Roxb., Fl. Ind., Ed. C.B.C., 438; Kurz, For. Fl. Burm., I., 121; Gamble, Man. Timb., 33.*
Habitat.—A large tree found on the Chittagong coast at Mascal island.

TIMBER.
46
DOMESTIC.
Trunk.

Structure of the Wood.—Hard but not very durable.
Domestic.—The TRUNK is made into canoes by the Magh inhabitants of the island on which it is found.

47

V. Tumbuggaia, *W. & A.,* see **Shorea Tumbuggaia,** *Roxb.;* Vol.
[VI., Pt. II., p. 679.

VENTILAGO, *Gærtn.; Gen. Pl., I., 375.*

48

Ventilago calyculata, *Tulasne; Fl. Br. Ind., I., 631;* RHAMNEÆ.
Syn.—V. DENTICULATA, *Willd.;* V. MADRASPATANA, *Roxb.;* V. MAC-
RANTHA, SILHETIANA, SMITHIANA, & SULPHUREA, *Tulasne.*
Vern.—*Rai dhani,* HIND.; *Rúktupita,* BENG.; *Bonga-sarjun, doe-saraj, noduúr,* KOL.; *Bonga-sarjom,* SANTAL; *Raktapita, kala lag,* KUMAON; *Kyonti,* KHARWAR; *Papri,* C. P.; *Sakal yel,* MAR.; *Yerra chictali,* TEL.
References.—*Roxb., Fl. Ind., Ed. C.B.C., 211; Kurz, For. Fl. Burma, I., 263; Brandis, For. Fl., 96; Gamble, Man. Timb., 91; Rev. A. Campbell, Rept. Ec. Pl., Chutia Nagpur, No. 8462; Atkinson, Him. Dist., 307; Gazetteers :—Bombay, XV., 430; N.-W. P., IV., lxx.; For. Ad. Rep. Ch. Nagpur, 1885, 29.*
Habitat.—A large, climbing shrub, found throughout the hotter parts of India, from the Kumáon Himálaya and Nepál, to Bhután, Sylhet, and Burma; also met with throughout the western Peninsula.

FIBRE.
Bark.
49
OIL.
Seeds.
50

Fibre.—Mr. Campbell, who gives all the available economic informa- tion regarding this plant, states that the BARK yields a good cordage FIBRE.
Oil.—An OIL is obtained from the SEEDS, which resembles *ghí* in taste, and is used in Chutia Nagpur for cooking purposes (*Campbell*).

V 50

Medicine.—The JUICE of the BARK and young SHOOTS is, in the same locality, applied to the body as a remedy for the pains which accompany malarial fever. A ring made from the TENDRIL is worn as a charm against toothache (*Campbell*). [*t. 163.*

Ventilago madraspatana, *Gærtn.; Fl. Br. Ind., I., 631; Wight, Ic.,*

Syn.—V. BRACTEATA, *Wall.*

Vern.—*Pitti*, HIND.; *Raktapita*, BENG.; *Rakta pitta*, URIYA; *Chorgu,* HYDERABAD; *Keoti, pitti*, C. P.; *Lokandi, kanwail*, BOMB.; *Khand-vel, lokhandi*, MAR.; *Ragatarohado*, GUZ.; *Súri-chakka*, DEC.; *Pappili-chakka, suralpattai, surala chaki, súrute eheka, papli, vembádam*, TAM.; *Surati pette tige, surala tige, erra chiratali, surla tige, surugudu yerra chicatli, surala-tége-patta, yerra-chakatli-chakka, suriti-pette-chakka, súrúghúndu-putta*, TEL.; *Paipli-chakka, papli, popli-chukai*, KAN.; *Raktavalli*, SANS.

References.—*Rumphius, Fl. Amb., V., t. 2; Voigt, Hort. Sub. Cal., 146; Brandis, For. Fl., 96; Kurz, For. Fl. Burm., I., 262; Beddome, Fl. Sylv., t. 68; Gamble, Man. Timb., 91; Thwaites, En. Ceyl. Pl., 74; Dals. & Gibs., Bomb. Fl., 48; Elliot, Fl. Andhr., 51, 170, 171; Moodeen Sheriff, Mat. Med. S. Ind. (in MSS.), 106; Dymock, Warden & Hooper, Pharmacog. Ind., I., 355; Bidie, Prod. S. Ind., 109; Lisboa, U. Pl. Bomb., 231, 242, Liotard, Dyes, App. VI; Wardle, Dye Rep., 3, 21; Gazetteers :—N.-W. P., I., 80; Orissa, II., 181; Madras Man., Admin, II., 114; Gribble, Man. Cuddapah, 263; Ind. Forester, X., 547.*

Habitat.—An extensive climber, found in the Western Peninsula from the Konkan southwards, Burma, and Ceylon.

Gum.—It is said to yield a gum, of which no information is obtainable.

Dye.—The ROOT-BARK is a much valued dye-stuff in Southern India. Liotard writes, "At a certain period of the year immense numbers of coolies in Mysore proceed to the jungle to collect the root-bark which forms the dye-stuff. It is then conveyed to the nearest towns and sold to dealers, who export it to other districts of India, combined with *chay-root* (**Oldenlandia umbellata**); it yields a beautiful chocolate colour, and if galls be also added, a black dye. It appears to be exclusively used for dyeing cotton cloth." In another passage he states that it is a good deal utilised by carpet-makers in Bangalore. The dye is extracted by boiling the root-bark or chips of the WOOD (which also yields colour) in water; simple immersion in this liquid dyes a brownish-purple, which, however, is fleeting; with a mordant of alum and myrabolans a darker but fast colour is produced.

Samples of the root-bark were submitted to **Mr. Wardle** who reported as follows :—"It is exceedingly rich in a beautiful red colouring matter, and will, when proper methods are employed, produce many colours for which cochineal and madder are generally used. The fastness of these colours will be made a subject for further investigation by me, but still I should judge that they are fairly permanent. The dye is well adapted for *tasar* silk, excellent results being obtainable without the employment of extensive processes for bleaching the silk, before the application of the dye." "By some of the methods the colours obtained are inclined towards purple and chocolate, but generally reds, more or less pure, are produced. The bark contains little or no tannin, and when tannin matters, such as galls, are added, the only result in dyeing is the usual red colour, with much less brilliance and depth; but, when in addition to the tannin substance a salt of iron is used, a slaty black, reddened and deepened by the colouring matter, is produced. It is a substance which, no doubt, would be a valuable acquisition to the dye-house, and if quantities could be obtained in India, at the price stated in **Surgeon-Major Bidie's** report, *viz.*, annas 3-5 per ℔, it might be largely used."

MEDICINE.
Juice.
51
Bark.
52
Shoots.
52
Tendrils.
54
55

GUM.
56
DYE.
Root-Bark.
57

VERBASCUM Thapsus.	**The American Hellebore.**

DYE.

With reference to the last sentence it may be remarked that **Liotard** gives the price in one passage at 3-5 annas per ℔, in another at 2 annas per ℔. **Moodeen Sheriff** states that the wholesale price in Madras is R6 per maund of 25 ℔ or nearly 4 annas per ℔, while the authors of the *Pharmacographia Indica* state that during 1888-89 it was sold by Government agency at R2 for first class, R1-8 for second class, bark per maund (?Madras, of 25℔); or from 1 to 1½ annas per ℔. **Mr. Wardle's** remarks regarding the probability of a trade arising are thus likely to be confirmed. The bark is collected in large quantities in the Southern Presidency, especially on the northern slopes of the Nilgiris, and can easily be obtained in Madras. In the *Annual Report of the Madras Forest Department* for 1887-88, it is stated that 3 tons were collected, which realised a revenue of R62, the value of the permits. During 1888-89, 41 maunds of first class, and 66 maunds of second class, bark were collected and sold by Government agency at the prices quoted above (*Pharmacog. Ind.*).

Chemistry.
58

CHEMICAL COMPOSITION.—The colouring matter, extracted by water, is of a red colour and acid slight reaction. It gives violet-red precipitates with plumbic acetate, calcium hydrate, and barium hydrate, a rose-tinted cake with alum and potassic carbonate, and muddy mixtures with ferrous and ferric salts. It is probably one of the derivatives of *anthracene* (*Pharmacog. Ind.*).

FIBRE.
Bark.
59
Stems.
60
MEDICINE.
Root-Bark.
61

Fibre.—The BARK yields a FIBRE, which is said to be useful for cordage. According to **Rumphius** the Amboina fishermen use the long climbing STEMS instead of rope.

Medicine.— **Moodeen Sheriff** states that the powdered ROOT-BARK is carminative, stomachic, tonic, and stimulant; and useful in atonic dyspepsia, debility, and slight cases of fever. He recommends doses of from 30 to 90 grains three or four times in the twenty-four hours, and states that the drug may be employed as a substitute for cascarilla, pimenta, calumba, and cinchona. The powdered bark (mixed with gingelly oil) is also said to be sometimes used in South India as an external application for itch and other skin diseases.

VERATRUM, *Linn.; Gen. Pl., III., 834.*

[LILIACEÆ.

62

Veratrum viride, *Aiton.; Baker, in Linn. Soc. Jour., XVII., 471;* GREEN or AMERICAN HELLEBORE.

Syn.—HELONIAS VIRIDIS, *Ker;* MELANTHIUM VIRENS, *Thunb.;* M. BRACTEOLARE, *Desv.*

References.—*Pharm. Ind., 245; O'Shaughnessy, Beng. Dispens., 657; Flück. & Hanb., Pharmacog. 695; Year-Book Pharm.:—1874, 102; 1875, 219; 1876, 5, 213; 1879, 4, 128; 1880, 208.*

Habitat.—Found in the east coast of North America, from Canada to Carolina.

MEDICINE.
Rhizome.
63

Medicine.—The RHIZOME, which is officinal in the Indian and most other modern *Pharmacopœias*, is imported into India in small quantities with other drugs for European use. Its properties are too well known to require comment.

VERBASCUM, *Linn.; Gen. Pl., II., 928.*   [NEÆ.

64

Verbascum Thapsus, *Linn.; Fl. Br. Ind., IV., 250;* SCROPHULARI-

Syn.—V. INDICUM, *Wall.*

Vern.—*Gidar tamáku* (=jackal's tobacco), HIND.; *Vúlr, phúl, ban tamáku, phasrúk, bhún ke dúm, eklbír, kadanda, phúntar, kwispre, khar-*

gosh, kharkharnár, spín kharnár, gúrganna, karáthrí, ravand chíni, gídar tamáků, Pb.

References.—*Roxb., Fl. Ind., Ed. C.B.C., 188; Stewart, Pb. Pl., 163; O'Shaughnessy, Beng. Dispens., 477; Year-Book Pharm., 1874, 132. 627; Notes by Mr. Duthie's collector, Trans.-Indus; Atkinson, Him. Dist., 314; Gazetteers:—Mysore & Coorg, I., 64; N.-W. P., IV., lxxv.-*

Habitat.—A plant of the Temperate Himálaya, from Kashmír to Bhután; at altitudes of 6,000 to 11,000 feet; and of Western Tibet; dis tributed westwards to Britain.

Medicine.—In Bashahr the ROOT is given as a febrifuge; the name *rewand chíni* would seem to indicate that it is at times used as an adulterant of or substitute for rhubarb (*Stewart*). The SEEDS are supposed to be narcotic and are used for poisoning fish (*O'Shaughnessy*). Mr. Duthie's Trans-Indus collector states that the HERB is much employed by the Natives of that region for the treatment of asthma and other pulmonary complaints, that it possesses narcotic properties similar to those of tobacco, and that the seeds are considered aphrodisiac. The LEAVES warmed, and rubbed with oil, are employed as an application to inflamed parts. Dr. Watt states that he has known it imported from Europe for medicinal use by an officer resident in Simla. In Europe, and the United States of America, the thick woolly leaves were at one time much valued as demulcents and emollients, not only in domestic medicine, but by practitioners. They were used in the treatment of catarrh and diarrhœa, and as an external application for hæmorrhoids.

SPECIAL OPINIONS.—§ "Has long been used in Ireland as a remedy in phthisis and phthisical diarrhœa" (*Brigade-Surgeon G. A. Watson, Allahabad*). "It is a valuable remedy in phthisis; checks night sweats, relieves cough, and moderates looseness of the bowels. One ounce of the leaves, boiled in a pint of milk, given twice a day, relieves dyspnœa. Highly spoken of by Dr. Grinlan as a palatable and effective remedy" (*Surgeon-Major E. G. Russel, Asylums, Calcutta*).

Fodder.—It is eaten by camels and goats (*Stewart*).

MEDICINE.
Root.
65
Seeds.
66
Herb.
67
Leaves.
68

FODDER.
69

VERBENA, *Linn.; Gen. Pl., II., 1146.*

Verbena officinalis, *Linn.; Fl. Br. Ind., IV., 565;* VERBENACEÆ.

70

Syn.—V. SPURIA, *Linn.;* V. SORORIA, *DC.*

Vern.—*Pámúkh, kardíta,* Pb.; *Shamuki,* PUSHTU.

References. —*Stewart, Pb. Pl., 166; Year-Book Pharm., 1874, 628; Smith, Econ. Dict., 428; Atkinson, Him. Dist., 315; Gazetteers:—N.-W. P., IV., lxxvi.; Rájputána. 30; Peshâwar, 26.*

Habitat.—Common in the Himálaya from Kashmír to Bhután, at altitudes of 1,000 to 6,000 feet, and in the Bengal plain to the Sunderbands.

Medicine.—In the Panjáb the fresh LEAVES are considered febrifuge and tonic, and is said to be used as a rubefacient in rheumatism and diseases of the joints. The ROOT is believed to be a remedy for scrofula and snake-bite. At one time it was worn in Europe as a charm against evil, and for good luck. In Tuscany it is said to be still employed as a poultice for liver complaints, and taken internally for the same disease, and for dropsy.

MEDICINE.
Leaves.
71
Root.
72

VERNONIA, *Schreb.; Gen. Pl., II., 227.*

[COMPOSITÆ.

Vernonia anthelmintica, *Willd.; Fl. Br. Ind., III., 236;*

THE PURPLE FLEA-BANE.

73

Syn.—SERRATULA ANTHELMINTICA, *Roxb.;* CONYZA ANTHELMINTICA, *Linn.*

VERNONIA anthelmintica.

The Purple Flea-bane.

Vern.—*Sómraj, bákchi, káli-siri, buckshi, vapchi,* HIND.; *Somráj, buk-chie, babchi, bapchie, káli-siri, hákuch,* BENG.; *Somráj,* URIYA; *Kali-jiri,* KUMAON; *Káli siri, kálá sira, bukoki, kakshama, malwa bakchi,* PB.; *Kali-siri, kali-jiri, karvi-siri, kala-jira,* DECCAN; *Káli-jiri, kalen-jiri,* BOMB.; *Ránácha-jiré, kalenjiri, káralye,* MAR.; *Káli-jiri, kadvo-jiri, kalijiei,* GUZ.; *Káttu-shiragam, kát-siragam, nirnúchie,* TAM.; *Adavi-jilakara, visha-kanta-kálu,* TEL.; *Kádu-jirage,* KAN.-*Káttu-jirakam, kattasiragam,* MALAY.; *Sanni-náegam, sanni-násang; sanni-nayan,* SING.; *Atavi-jirakaha, kanana-jeraka, somaráji, aval, guja, rákuchi,* SANS.; *Atarilál, itrilál,* ARAB.; *Atarilál, itrilál,* PERS.

References.—*Roxb., Fl. Ind., Ed. C.B.C.,* 594; *Dalz. & Gibs., Bomb. Fl.,* 313; *Stewart, Pb. Pl.,* 131; *Sir W. Elliot, Fl. Andhr.,* 11; *Rheede, Hort. Mal., II., t.* 24; *Thesaurus, Zey.,* 210, t. 5; *Pharm. Ind.,* 126; *Ainslie, Mat. Ind., II.,* 54; *O'Shaughnessy, Beng. Dispens.,* 419; *Irvine, Mat. Med. Patna,* 128; *Medical Topog., Ajmere,* 128; *Moodeen Sheriff, Supp. Pharm. Ind.,* 254; *U.C. Dutt, Mat. Med. Hind.,* 183,318; *Murray, Pl. & Drugs, Sind.,* 181; *Dymock, Mat. Med. W. Ind.,* 2nd Ed., 421; *Baroda Durbar, Col. & Ind. Exhb., No.* 182; *Official Corresp. on Pro-posed New Pharm. Ind.,* 238; *Birdwood, Bomb. Prod.,* 50; *Baden Powell, Pb. Pr.,* 310, 358; *Drury, U. Pl. Ind.,* 441; *Atkinson, Him. Dist. (X., N.-W. P. Gaz.),* 311, 753; *Useful Pl. Bomb. (XXV., Bomb. Gaz.),* 257; *Cooke, Oils & Oilseeds,* 81; *Bidie, Prod. S. Ind.,* 32; *Gazetteers:—Bombay, VI.,* 15; *N.-W. P., IV., lxxiii.; Orissa, II.,* 160; *Mysore & Coorg, I.,* 56; *Agri-Horti. Soc. Ind., Journ. (Old Series), X.,* 11; *Smith, Econ. Dic.,* 235.

Habitat.—A tall, robust, leafy annual, met with throughout India, to Ceylon and Malacca, ascending to 5,500 feet in the Himálaya and Khásia Mountains.

OIL.
Seeds.
74
MEDICINE
Achenes.
75
Seeds.
76

Oil.—Lieutenant **Hawkes** states that the SEEDS yield an oil, which is never prepared for sale. It is probably used medicinally.

Medicine.—The ACHENES are highly reputed in Sanskrit medicine as a remedy for white leprosy (leucoderma) and other skin diseases. They are mentioned also as an anthelmintic, but are little used as such, except in combination with other drugs. In chronic skin diseases the SEEDS are taken alone or in combination with other medicines. In the severer forms, such as psoriasis and lepra, the remedy is recommended to be taken daily for one year, when a complete cure is said to be effected. Chakradatta describes several elaborate combinations for external and internal use. The former are oils and pastes; one of the latter, a decoction of the powdered seeds and sesamum in tepid water, is directed to be taken after perspiration has been induced by exercise or exposure to the sun. The diet prescribed is milk and rice (*U. C. Dutt*). The author of the *Makhzan-el-Adwiya* ascribes the drug as given internally to remove phlegm and worms from the intestine, and states that a poultice or plaster made of it is employed to disperse cold tumours. But, he remarks, it is not often prescribed as an internal medicine, as it is thought to have injurious effects, though it is frequently used as a drug for cattle (*Dymock*).

The medicine early attracted the attention of European writers on medicine in India. **Rheede** wrote that an infusion was given, on the Malabar coast, as a remedy for coughs and flatulency. **Ainslie** wrote that the seeds were reckoned a powerful anthelmintic, and also formed an ingredient of a compound powder prescribed in cases of snake-bite. **O'Shaughnessy, Taylor, Irvine,** and others notice the vermifuge properties of the seeds, but say little as to their opinion of their value. **Stewart** states that in the Panjáb they are given in anasarca, and used to make plasters for abscesses. **Baden Powell** writes that in the same province they are believed to possess most of the properties above noticed, and are also considered febrifuge, and a " valuable remedy for prolonging life, restoring youth, and preventing the hair turning grey." The plant is included in

| The Ash-coloured Flea-bane. | (*J. Murray.*) | VERNONIA cinerea. |

MEDICINE.

the secondary list of the *Indian Pharmacopœia*, where the seeds are said to enjoy a special reputation as an anthelmintic in cases of ascarides lumbricoides (round worms), which, under their use, are said to be expelled in a lifeless state, thus showing that the drug exercises a specific effect on the entozoa. The ordinary dose is about $1\frac{1}{2}$ drachm, given in two equal doses at the interval of a few hours, and followed by an aperient. Dr. Æ. Ross is quoted as recommending the drug as a vermifuge, given in doses of 10 to 30 grains, powdered; while Dr. Gibson, as the result of personal experience, is said to regard it as a valuable tonic and stomachic, in doses of 20 to 25 grains. Diuretic properties have also been assigned to it. In Travancore, the bruised seeds, ground up in a paste with lime-juice, are largely employed as a means of destroying pediculi. (*Pharm. Ind.*)

SPECIAL OPINIONS.—§ "In some cases I have used the seeds as an anthelmintic, but with unsatisfactory results. The administration was usually followed by a dose of castor-oil the following morning. Preparation—Powdered seeds and infusion. Dose : 30 grains twice a day, either in powder or infusion" (*Apothecary J. G. Ashworth, Kumbakonam, Madras*). "Seeds—tonic, stomachic, anthelmintic. Dose: 60 to 120 grains anthelmintic, 10 to 30 grains tonic" (*Apothecary Thomas Ward, Madanapalla, Allahabad*). "Four ounces fine powder seed, $\frac{1}{8}$ qt. rum, $\frac{1}{8}$ qt. water, put out in the sun for ten to twelve days. Dose, a wine-glassful at 5 A.M., after which patient is to lie on his side for half an hour. This prescription was given to me as a specific for spleen by a tea-planter, who professed to have never failed with it after trial in many cases. In my hands in hospital patients, it proved, like most specifics, useless (*Surgeon-Major, E. Sanders, Chittagong*).

Domestic.—The SEEDS, as already stated, are employed to destroy pediculi in the head and body, and may possess some antiseptic powers. The PLANT roasted in a room, or powdered and thrown about the floor, is believed to expel fleas—hence the popular English name.

DOMESTIC.
Seeds.
77
Plant.
78
79

Vernonia cinerea, *Less.; Fl. Br. Ind., III., 233; Wight, Ic., t. 1076.*

THE ASH-COLOURED FLEA-BANE.

Syn.—V. CONYZOIDES, *DC.*; V. RHOMBOIDEA & MONTANA, *Edgew.*; V. ALBICANS, *DC.*; V. ABBREVIATA & LEPTOPHYLLA, *DC.*; V. LAXIFLORA, *Less.*; V. PHYSALIFOLIA, *DC.*; CONYZA ABBREVIATA, BELLIDIFOLIA, CINERASCENS, INCANA, LINIFOLIA, ELEGANTULA, OVATA, & SUBSIMPLEX, *Wall.*; C. CINEREA, *Linn.*; C. MOLLIS, *Willd*; C. PROLIFERA & HETEROPHYLLA, *Lamk.*; SERRATULA CINEREA, *Roxb.*; CACALIA ROTUNDIFOLIA, *Willd.*

Vern.—*Kúksim, kúkshim, kala-jira,* BENG.; *Barangom, bahu tuturi, birlopong arak', jhurjhuri, durya arak',* SANTAL; *Sahadevi,* PB.; *Lalia, káli harr,* MERWARA; *Sira-shengalanir,* TAM.; *Gariti kamma,* TEL.; *Monera-kúdimbeya,* SING.; *Sahadevi,* SANS.

References.—*Thwaites, En. Ceyl. Pl., 160; Dalz. & Gibs., Bomb. Fl., 121; Rev. A. Campbell, Rept. Ec. Pl. Chutia Nagpur, Nos. 7874, 7877, 8452, 8744, 9862; Elliot, Fl. Andhr., 58; Rheede, Hort. Mal., X., t. 64; Burm., Thes. Zeyl., t. 96, f. 1; Ainslie, Mat. Ind., II., 363; Irvine, Mat. Med. Patna, 45; Dymock, Mat. Med. W. Ind., 2nd Ed., 423; Drury, U. Pl. Ind., 442; Atkinson, Him. Dist., 311; Gazetteers:— Mysore & Coorg, I., 62; Bombay, V., 26; N.-W. P., I., 81, 82; Agri.-Horti. Soc. Ind., Journ. (Old Series), X., 11; Ind. Forester, XII., App., 2, 15.*

Habitat.—One of the commonest Indian weeds, found throughout India, ascending to 8,000 feet in the Himálaya, Khasia mountains, and hills of the Peninsula.

Medicine.—Ainslie states that the whole PLANT, with its small, round, downy, tasteless flowers, is used in medicine by the Hindus, in decoction, to promote perspiration in febrile affections. Irvine writes that the SEED is

MEDICINE.
Plant.
80
Seed.
81

VIBURNUM **cotinifolium.**	The Guelder-rose.

MEDICINE.
Flowers.
82

employed in Patna as an alexipharmic and anthelmintic, and as a constituent of *masálas* for horses. In Chutia Nagpur the whole plant is given
as a remedy for spasm of the bladder and strangury; the FLOWERS are
administered for "blood-shot eyes"? conjunctivitis (*Campbell*). The
latter use is interesting, since, according to **Piso**, the leaves of another
species of the same genus are similarly employed in Jamaica.

This plant appears to be esteemed as a medicine by the Hindus of the
Eastern Provinces; in Bombay, **Dymock** informs us, it is not utilised, and
no writer on Northern India makes any mention of it.

FOOD.
Leaves.
83

Food.—The LEAVES are eaten as a pot-herb in Chutia Nagpur
(*Campbell*).

VERONICA, *Linn ; Gen. Pl., II., 964.*

[NRÆ.

84

Veronica beccabunga, *Linn.; Fl. Br. Ind., IV., 293;* SCROPHULARI-
  References.—*Stewart, Pb. Pl., 163 ; O'Shaughnessy, Beng. Dispens., 478 ;
    Year-Book Pharm., 1874, 134, 628 ; Atkinson, Him. Dist., 314; Agri.-
    Horti. Soc. Ind., Journ. (Old Series), XIV., 52.*
  Habitat.—A herb, met with in the Western Himálaya, from Kashmír

MEDICINE.
Plant.
85
Leaves.
86
Stems.
87

to Kanáwar, and in Western Tibet, at altitudes from 9,000 to 12,000 feet.
  Medicine.—Honigberger states that the PLANT is used medicinally in
Kashmír. In Europe and America, the LEAVES and young tender STEMS
were formerly much valued as an antiscorbutic; the former are still
occasionally employed as a styptic to wounds, and, when bruised, they are
applied to burns (*Stewart*).

VIBURNUM, *Linn. ; Gen. Pl., II., 3.*

[CEÆ.

88

Viburnum coriaceum, *Blume ; Fl. Br. Ind., III., 5 ;* CAPRIFOLIA-
  Syn.—V. CYLINDRICUM, *Ham.*
    Var. capitellata, *Wight, Ic., t. 1022, (sp.)*=V. HEBANTHUM, *Thw. (in
      part, not of W. & A.)*
    Var.—ZEYLANICA, *Thw. (in part)*=V. CORIACEUM, *var.* β, *H.f. & T.*
  Vern.—*Kala titmaliya,* KUMAON; *Bara gorakuri,* NEPAL.
  References.—*Brandis, For. Fl., 259 ; Gamble, Man. Timb., 214; Atkin-
    son, Him. Dist., 311 ; Ind. Forester, II., 23 ; V., 183, 184 ; VIII., 408,
    412.*
  Habitat.—A large shrub or small tree, common on the Himálaya from
the Panjáb to Bhutan, at altitudes of 4,000 to 8,000 feet; also found in the
Khásia Hills, the Nílgiris, and Ceylon.

OIL.
Seed.
89
TIMBER.
90

  Oil.—It is said that the Nepalese extract, from the SEED, an oil, which
they use for food and for burning (*Gamble*).
  Structure of the Wood.—Similar to that of **V. cotinifolium**; weight
50℔ per cubic foot.

91

V. cotinifolium, *Don ; Fl. Br. Ind., III., 3 ; Wight, Ill., t. 121.*
  Syn.—V. POLYCARPUM, *Wall.*
    Vern.—*Gwia. guya,* KUMAON ; *Richh úklú, bankúnᵣh, riᵣhhábi kilmich,
      gúch, bathor, pápat kalam, khimor, rájab, túmma, kátonda, jáwa, khatíp,
      tústús, sússú, marghwalawa,* PB. ; *Marghwalwa,* PUSHTU.
    References.—*Brandis, For. Fl., 258 ; Gamble, Man. Timb., 214 ; Stewart,
      Pb. Pl , 114; Atkinson, Him. Dist., 311 ; Gazetteers :—Bₐnnu, 23 ;
      Déra Ismail Khán, 19 ; Ind. Forester, XIII., 68 ; Agri.-Horti. Soc.
      Ind., Journ. (Old Series), XIV., 16.*
  Habitat.—A large deciduous shrub, met with in the Sulaiman Range,
common on the North-West Himálaya from Kashmír to Kumáon and
East Bhután between 4,000 and 11,000 feet.

V. 91

Food.—Produces a FRUIT which, when ripe, is sweetish, and is eaten in many places by the Natives.

Structure of the Wood.—White, hard to very hard, close-grained.

[*t. 1024.*

Vibernum erubescens, *Wall.; Fl. Br. Ind., III., 7; Wight, Ic.,*

Syn.—V. WIGHITIANUM, *Wall.;* V. PUBIGERUM, *W. & A.*

Vern.—*Ganné, asari,* NEPAL; *Kancha,* LEPCHA; *Nakouli, damshing,* BHUTIA.

References.—*Brandis, For. Fl., 258; Thwaites, En. Ceyl. Pl., 136; Beddome, Fl. Sylv., t. 124; Gamble, Man. Timb., 215; Ind. Forester, II., 23.*

Habitat.—A large shrub or small tree, common on the Himálaya from Kumáon to Bhután, between 5,000 and 11,000 feet; also found in the Nilgiris and Ceylon.

Structure of the Wood.—Very hard, reddish, close and even-grained; weight 59℔ per cubic foot. It might be used as a substitute for boxwood and for carving, and is employed for making house-posts in Sikkim.

V. foetens, *Dcne.; Fl. Br. Ind., III., 8.*

Vern.—*Gúch, úklú, kúnch, kílmích, kwillim, kulára, jamára, tilhanj, tianlandhá, púlmú, tiláts, túin, talhang, tandei, túndhe, tunánizenáni, talhang, thelain, tselain, thilkain,* PB.; *Guya,* KUMAON.

References.—*Kurz, For. Fl. Burm., II., 2; Brandis, For. Fl., 259; Gamble, Man. Timb., 215; Stewart, Pb. Pl., 114; Baden Powell, Pb. Pr., 601; Agri.-Horti. Soc. Ind., Journ. (Old Series), XIV., 13.*

Habitat.—A large shrub of the North-West Himálaya, from 5,000 to 11,000 feet. The flowers have a delicious scent, the name being derived from the foetid odour emitted when the branches are bruised.

Food.—The FRUIT is sweetish, when ripe, and is eaten by the Natives.

Structure of the Wood.—White, hard to very hard, close-grained; similar in appearance and structure to that of **V. cotinifolium**; weight 53℔ per cubic foot. It is chiefly utilised for firewood.

V. foetidum, *Wall.; Fl. Br. Ind., III., 4.*

Vern.—*Nárwel,* BOMB.

References.—*Kurz, For. Fl. Burm., II., 2; Dymock, Mat. Med. W. Ind., 2nd Ed., 603.*

Habitat.—Common at altitudes from 3,000 to 5,000 feet in Assam, the Khásia Mountains, and Burma; cultivated in gardens in other parts of India.

Medicine.—Dymock gives an interesting account of the medicinal uses of this shrub as follows:—"This shrub or small tree is not a native of Western India, but is very common in gardens; all the green parts emit a peculiarly unpleasant odour. It is customary for Hindu women who have been confined to hang a BRANCH over the room in which they lie, as a protection against evil spirits and post-partum hæmorrhage. Another superstition is, that if seven pieces of the STEM of this plant are knotted into a thread made from cotton picked by a virgin, the necklace thus formed will cure scrofulous glands. A cake made from the flour of eighteen kinds of grain with *narwél* JUICE, is scraped on one side while hot, well moistened with the juice and applied to the head in headache. A wineglassful of the juice of the LEAVES is administered internally in menorrhagia daily, also in post-partum hæmorrhage. It is remarkable that V. prunifolium, an American plant, has also been found useful in all uterine diseases characterised by loss of blood and in threatened abortion." (*Mat. Med. W. Ind.*).

FOOD.
Fruit.
92
TIMBER.
93
94

TIMBER.
95

96

FOOD.
Fruit.
97
Timber.
98

99

MEDICINE.

Branch.
100
Stem.
101
Juice.
102
Leaves.
103

V. 103

104   **Viburnum nervosum,** *Don; Fl. Br. Ind., III., 8.*

    Syn.—V. GRANDIFLORUM, *Wall.*

    Vern.—*Amrola, ámbre, árí, rís, dáb, thilkain, thalein,* PB.

    References.—*Brandis, For. Fl.,* 259; *Stewart, Pb. Pl.,* 115; *Atkinson, Him. Dist.,* 311.

    Habitat.—A large shrub or gnarled small tree, found on the Himálaya, from Kashmír to Sikkim, between altitudes of 10,000 and 13,000 feet; not common.

FOOD.
Fruit.
105
106
    Food.—It produces a pretty red FRUIT, which is eaten by the Natives of the locality in which it grows.

V. stellulatum, *Wall.; Fl. Br. Ind., III., 4.*

    Syn.—V. MULLAH, *Ham.*

    Vern.—*Lál tit-maliya,* KUMAON; *Amliácha, phulsel,* KASHMIR; *Jal bágú, eri, ira,* PB.

    References.—*Brandis, For. Fl.,* 258, 576; *Gamble, Man. Timb.,* 214; *Stewart, Pb. Pl.,* 115; *Atkinson, Him. Dist.,* 311; *Ec. Prod., N.-W. P., Pt. V.,* 44, 76.

    Habitat.—A shrub of the North-West Himálaya from 6,000 to 10,000 feet.

FOOD.
Fruit.
107
    Food.—The FRUIT, though sour, is eaten by Natives.

VICIA, *Linn.; Gen. Pl., I., 524.*

108   **Vicia Faba,** *Linn.; Fl. Br. Ind., II., 179;* LEGUMINOSÆ.

    THE GARDEN BEAN.

    Syn.—FABA VULGARIS, *Mœnch.*

    Vern.—*Bakla, anhuri,* HIND.; *Nákshan, nakhthan,* LAD.; *Bákla,* KUMAON; *Káiún,* KASHMIR; *Chastang raiún,* N.-W. HIMALAYA; *Chástang, bakla, kábli bakla, sein, mattz-rewari,* PB.; *Ful,* EGYPT.

    References.—*Roxb., Fl. Ind.,* Ed. C.B.C., 566; DC., *Orig. Cult. Pl.,* 316; *Stewart, Pb. Pl.,* 69; *Murray, Pl. & Drugs, Sind,* 121; *Birdwood, Bomb. Pr.,* 123; *Baden Powell, Pb. Pr.,* 242; *Atkinson, Him. Dist.,* 694; *Church, Food-Grains, Ind.,* 131; *Smith, Econ. Dict.,* 43; *Settle. Rept., Kángra,* 24; *Gazetteers:—Mysore & Coorg, I.,* 59; *N.-W. P., IV., lxxi.; Panjáb, Kángra,* 152; *Firminger, Man. Garden. Ind.,* 150; *Ind. Forester, IX.,* 452; *Rept. from Govt. of Burma.*

    Habitat.—According to DeCandolle the wild habitat of the Garden Bean was probably two-fold some thousands of years ago, one of the centres being to the south of the Caspian, the other in the north of Africa. "The nature of the plant," he writes, "is in favour of this hypothesis, for its seed has no means of dispersing itself, and the rodents or other animals can easily make prey of it. Its area in Western Asia was probably less limited at one time, and that in Africa, in Pliny's day, was more or less extensive. The struggle for existence which was going against this plant as against maize, would have gradually isolated it, and caused it to disappear if man had not saved it by cultivation." He brings many arguments to bear which prove that the bean was cultivated in Europe in prehistoric times. It was introduced into Europe probably by the Western Aryans at the time of their earliest migrations (*Pelasgians, Kelts, Slavs*); and was taken to China later, a century before the Christian era, still later to Japan, and quite recently into India (*Origin Cult. Pl.*). The last statement is probably true only of the plains. It is based on the want of evidence of ancient cultivation, together with the fact that there is " no Sanskrit name, nor even any modern Indian name." But it will be seen from the above list, that several names *do* exist, in Kashmír, Ladák, the Panjáb, and the North-West plains, and it is significant that Stewart wrote some twenty years ago, "Probably not grown in the plains except by Euro-

V. 108

peans, but *commonly* cultivated in Kashmír, 5,000, and Kanáwar, Spiti,
and Tibet, 8,000 to 12,000 feet." It would from these facts seem to be not
unlikely that the cultivation of the bean has survived from ancient times in
the parts of the higher Himálaya, though, as DeCandolle observes, of quite
modern introduction into India generally.

CULTIVATION.—This bean is, at the present day, cultivated in the **CULTIVATION**
plains here and there in European gardens, or for European consumption, **109**
but is only grown to any considerable extent in the North-West Provinces.
There are two distinct classes of the vegetable, the long-podded, and the
broad-podded—Broad or Windsor Bean. The former is the most prolific,
and succeeds best in India. The seed should be sown in succession from
the middle of September to the end of October ; two sowings are sufficient,
but it may be sown three times with advantage at intervals of a fortnight.
The seed should be soaked in warm water for six or eight hours before
sowing, and care should be taken that they are not dried again by being
put in hot and dry ground. The crop succeeds best in a deep, rich, and
somewhat heavy loam. Firminger recommends the broad bean for garden cultivation in India. Sowing should take place in the middle of October ; the treatment of the seed is similar to that above described. The seed
should be put in the ground 2 inches deep, in rows of double drills 4
inches apart, a space of 2 feet being left between each row. When the
plants come into full blossom about an inch should be nipped off the top
of each.

No information can be given as to the extent of cultivation in the
North-West Provinces. A recent communication from the Government
of Burma states that in the Pegu, Kyaukpyu, and Amherst Districts it is
grown by Chinese and Shan gardeners in moderate quantities, but that,
though it thrives well in any land, and finds a ready sale, it has not as
yet been taken up as a field crop.

Food.—The POD is tumid, leathery, spongy. At its base, on the lower **FOOD.**
side, there is a small hole, through which the internal water evaporates, so **Pod.**
that the seeds become dry before the dehiscing of the pod. In England **110**
the ripe SEEDS or beans are extensively used for feeding horses. In an **Seed**
unripe condition Europeans eat them at their tables as vegetables. Some- **111**
times the beans are ground into flour for food ; and are also sometimes
given to cattle. According to Church, Indian seed contains about 25 per
cent. of albumenoids, and 7·5 per cent. of fibre.

Vicia hirsuta, *Koch ; Fl. Br. Ind., II , 177.*   112
THE HAIRY TARE.

> **Syn.**—ERVUM HIRSUTUM, *Linn. ;* E. FILIFORME, *Roxb. ;* E. LENS, *Wall.*
> *Cat., 5954, C. (not of Linn.).*
> **Vern.**—*Jhunjhuni-ankari,* HIND ; *Musúr chuna,* BENG. ; *Tiririt°,* SAN-
> TAL ; *Masúri, masúr-chana, jhanjhaniya-kúri,* KUMAON.
> **References.**—*Roxb., Fl. Ind., Ed. C.B.C., 567 ; Dalz. & Gibs., Bomb. Fl.*
> *Suppl., 22 ; Rev. A Campbell, Rept. Ec. Pl., Chutia Nagpur, No. 7858 ;*
> *Atkinson, Him. Dist., 308, 694 ; Gazetteer, N.-W. P., IV., lxxi. ; Agri.-*
> *Horti. Soc. Ind., Journ. (Old Series), Trans., IV., 82 ; Journ., IV.,*
> *185-187, 189 ; IX., 416 ; XIII , 387 ; XIV., 28.*

Habitat.—An herb of the North-West Provinces, Panjáb, and Nepal
up to 6,000 feet, also of the Nilgiris Roxburgh says it is a native of
Bengal. Frequently met with in cultivated grounds during the cold
season.

Fodder.— In the inland provinces it is someti nes cultivated for fodder. **FODDER.**
113

V. sativa, *Linn. ; Fl. Br. Ind., 178.*   **114**
THE COMMON VETCH, or TARE.

VIGNA **Catiang.**	The Chowlee of India.

Var. **angostifolia**, *Roth.* (*sp.*)=V. BOBARTII, *Forst.*; V. PALLIDA, *Jacquem.*

Vern.—*Akra, ankra,* HIND.; *Ankari,* BENG.

References.—*Roxb., Fl. Ind., Ed. C.B.C., 566; DeCandolle, Origin Cult. Pl., 108; Stewart, Bot. Tour in Hazára, in Agri.-Horti. Soc. Journ. (Old Series), XIV., 19; Agri.-Horti. Soc. Ind., Journ. (Old Series), II., Sel., 174, 175; IX, 416; (New Series), II., Pro., 1870, 30, 31; Gazetteers, N.-W. P., IV., lxxi.; X., 308; Saidapet Exp. Farm Rept., 1877, 16.*

Habitat.—This annual leguminous herb exists wild throughout Europe, except in Lapland. **Roxburgh** pronounces it to be wild in Bengal and the North-West Provinces, but **Baker**, in the *Fl. Br. Ind.*, admits this only as far as the variety **angustifolia** is concerned. There is, however, no evidence of the plants being cultivated in India; if not truly wild, the seed must probably have been introduced with that of some European cereal which happened to contain it as an impurity. **Roxburgh** states that it is "seldom or never cultivated in Bengal," and no writer mentions it as a cultivated crop in any part of the country. In 1877 attempts were made at the Madras Experimental Farm to introduce it, but without success. The seed germinated well, but the young plants could not withstand the heat.

FODDER.
Plant.
115

Fodder.—Cattle browse the PLANT, wherever it occurs as a weed of, or near, cultivation.

VIGNA, *Savi; Gen. Pl., I., 539.*

116

Vigna Catiang, *Endl.; Fl. Br. Ind., II., 205;* LEGUMINOSÆ.

THE CHOWLEE OF INDIA and TOW COK OF CHINA.

Syn.—DOLICHOS CATIANG, *Linn.*; D. SINENSIS, *Linn.*; VIGNA SINENSIS, *Endl.*; DOLICHOS TRANQUEBARICUS, *Jacq.*; D. MONACHALIS, *Brot.*

Vern.—*Lobiá, chowli, rianish, rawás, rausa, souta, bora,* HIND.; *Barbati, ramhikolui, shim* or *chim,* BENG.; *Ghangra,* SANTAL; *Urohi mahorpat,* ASSAM; *Barbutiti,* C. P.; *Lobiá, rawas, rausa, souta,* N.-W. P. & OUDH; *Lobiya riánsh, ráish, riensh, souta,* KUMAON; *Lobiyá, rawán, souta, harwánh chota, rawangan, raongi, rawángi, ró-ín,* PB.; *Chaunro,* SIND; *Lobeh, chowli,* DEC.; *Chola, chowli, safed lobeh, hurrea, lobeh, gatval,* BOMB.; *Chaoli,* MAR.; *Chora, chola,* GUZ.; *Caramunny-pyre,* TAM.; *Boberlú, alusundi, duntú-pesalú, bobra,* TEL.; *Tadagunny, kursan-pyro, alasandi,* KAN.; *Alasendi,* MALAY; *Li-mæ,* SING.; *Rájamásha, nishpáva,* SANS.; *Lobiya,* PERS.

For a discussion on some of the above vernacular names, see the article on **Dolichos Lablab,** *Linn.;* Vol. III., 184-186.

References.—*Roxb., Fl. Ind., Ed. C.B.C., 559, 560; Stewart, Pb. Pl., 67; U. C. Dutt, Mat. Med. Hindu, 312, 315; Birdwood, Bomb. Prod., 118; Baden Powell, Pb. Pr., 241, 341; Drury, U. Pl. Ind., 186; Atkinson, Him. Dist. (X., N.-W. P. Gaz.), 309, 695; Duthie & Fuller, Field and Garden Crops, Pt. II., 12; Useful Pl. Bomb. XXV., (Bomb. Gaz.), 153, 341; Stewart, Food of the Bijnour Dist., 474; McCann, Dyes & Tans, Beng., 156; Church, Food-Grains, Ind., 156; Darrah, Note on Cotton, 35; Stock's Rep. on Sind,; Rept. on Kumaon by Madden, 279; Bombay, Man. Rev. Accts., 101; Settlement Report:—N.-W. P. Kumaon, app. 32; Central Provinces, Upper Godavery, 36; Rept., Dir. Agric. Dept. Beng., 1886, lxxiii.; Rep. on Land Rec. and Agriculture, Bombay, App. viii.; Gazetteers:—Bombay, VIII., 182, 189; N.-W. P., III., 225; IV., lxxi.; Eta, 27, 30; Mysore & Coorg, II., 11; Agri.-Horti. Soc. Ind., Journ. (Old Series), IV., 213, 214; IX., Sel., 58; Ind. Forester, IX., 203.*

Cultivation
117

Habitat.—Native, but also cultivated in the hotter parts of India.

CULTIVATION.—This crop is, as a rule, grown for its grain, and, like many others, forms an associate of the *kharíf* millets. Several races exist, which

The Chowlee of India.	(*J. Murray.*)	VIGNA Catiang.

differ in the colour of the flower and seeds ; one of these, with very long pods, is cultivated by market gardeners as a vegetable.

Panjáb.—No statistics can be given regarding the area under the pulse in this province. It is commonly cultivated as a mixed crop with millets in the *kharíf* season.

N.-W. Provinces.—" It is less frequently grown, in these provinces, as a sole crop than either *múng* or *úrd* " (see **Phaseolus**), " and the area which it occupies by itself is quite insignificant except in the Rohilkhand Division, where it amounts to 5,000 acres. On the other hand, it forms portions of the undergrowth in a large proportion of *kharíf* millet and cotton fields, with which it is sown at the commencement of the rains. It ripens in October or November, and yields a produce of about the same quantity as that of *úrd*. Its grain is less valued than that of *úrd* or *múng*, being difficult of digestion, and apt, according to Native ideas, to generate heat in the stomach. The leaves and stems are used as cattle fodder" (*Duthie & Fuller*).

Madras.—The crop is a fairly important one in this Presidency, and in 1888-89 occupied an area of 33,433 acres. During the last ten years the area has fluctuated between 30,000 and 47,000 acres.

Bombay and Sind.—*Lobíya* is one of the least important pulses in Bombay and Sind, occupying about the same area as the field-pea, lentil, and chickling vetch. In 1888-89 it is returned as having covered an area of 18,100 acres, of which 15,872 were in Bombay and 2,228 in Sind. Of the former 4,461 acres were in Bijápur, 3,973 in Thána, 2,370 acres in Kaira, and insignificant areas in most of the other districts of Gujarát, the Deccan, the Karnátak, and the Konkan. In Sind, 1,593 acres were in Hyderabad. The crop grows best in black soil, is sown soon after the first rain in June, and reaped in November. It is sometimes grown separately, but generally with *bajri* and *juár* (*Bomb. Gaz., VIII., 189*).

Bengal.—In the account of the crops of the Burdwan Division, two forms of the plant are said to be grown, of which one is known as *barbati* and the other as *rambha.* The crop is said to succeed in loam and sandy soils, but not to do well in clay soils. It comes after *áus* paddy, and is followed by the same crop, *begun* or *kachu.* The seed is sown in the end of September at the rate of five seers per *bigha*, either alone or mixed with *kajeli*, mustard. The crop is said to be subject to attacks of a small insect which appears in cloudy weather. The average yield is from $\frac{1}{2}$ to $2\frac{1}{2}$ maunds per *bigha*, the cost about R4 per maund.

No particulars can be given regarding the cultivation of this pulse in other parts of India.

Dye.—The LEAVES are said to be employed as a dye-stuff in Rájsháhí and Jalpáiguri in Bengal, but no particulars of the colour produced are obtainable (*McCann*). In Assam they are employed in the preparation of a green dye, as follows :—" Place a quantity of the leaves of *rúm* (**Strobilanthes flaccidifolius**) in an earthen vessel full of water, and having tied up the mouth, allow the vessel to stand for three or four days, or until the leaves rot. Then take out the rotten leaves, squeezing all juice out of them in so doing, and shake the liquid left behind well for some time. Then tie up the vessel once more, and let it stand for the night, next morning pour off any watery liquid that may be found and add to it one-fifth the quantity of ash-water (called *kharoni* in Assamese, and made by filtering water through wood-ashes), one-tenth the quantity of native liquor, and one-twentieth the quantity of the juice of *thekera* (**Ixora acuminata**, *Roxb.*). Then place the mixture in the sun for three consecutive days, after which the material to be dyed should be dipped into the liquid, squeezed out, and sun-dried, this process being repeated for three days.

Marginal notes:

CULTIVA-TION.
Panjab. 118
N.-W. Provinces. 119
Madras. 120
Bombay & Sind. 121
Bengal. 122
DYE. Leaves. 123

VILLEBRUNEA
frutescens.   The Villebrunea Fibre.

CULTIVA-
TION.

Then place in a mortar *urohi* leaves four parts, turmeric one part, and *thekera* leaves two parts; crush the whole well, and after rubbing the pulp so formed well into the cloth, dyed as above in *rúm*, leave the whole, cloth and pulp, to steep for the night. Next morning squeeze the juice out of the material, and dry in the sun. The process should be repeated till the desired shade of green has been obtained. The leaves of the plum-tree are said to answer as well as those of *urohi mahorpat*. There is another method of producing the dye, in which lime-water is used instead of ash-water, the rest of the process being identical with that described above" (*Darrah, Note on Cotton in Assam*). According to one account the latter part of the process alone is sufficient to dye previously uncoloured cloth a green colour. But this is probably incorrect; since from **Mr. Darrah's** careful account it may be presumed that the blue produced by the substitute for indigo, *rúm*, is converted by the second process, which of itself colours yellow to green. It is impossible to say how much of the yellow colour is due to the turmeric, and how much to the leaves of the *urohi*, but there is certainly no reason to believe that the latter will, by themselves, produce a green dye (*Cf.* Vol. IV., 451-455).

MEDICINE.
Seeds.
124

Medicine.—In the Panjáb the SEEDS are considered "hot and dry, diuretic and difficult of digestion. They are used in special diseases, and to strengthen the stomach" (*Baden Powell*).

FOOD &
FODDER.
Grain.
125
Pods.
126
Stalks.
127
Leaves.
128
129

Food & Fodder.—The GRAIN is eaten either as flour, or split, as *dál.* It is considered less wholesome than that of *úrd* or *múng* (see **Phaseolus**), and white seeds are reckoned the best. The green PODS, especially of the long-podded form, are plucked while young, boiled and eaten as a vegetable. The STALKS and LEAVES are used as fodder. One hundred parts of the husked bean contains, water, 12·5 parts; albumenoids, 24·1; starch, 56·8; oil, 1·3; fibre, 1·8; and ash, 3·5, of which 1·0 consists of phosphoric acid (*Church*).

Vigna pilosa, *Baker; Fl. Br. Ind., II., 207.*

    Syn.—DOILICHOS PILOSUS, *Roxb.;* PHASEOLUS DIFFORMIS, *Wall.*

    Vern.—*Jhikrái. kalúi, malkonia,* BENG.

    References.—*Roxb., Fl. Ind., Ed. C.B.C., 207; Agri.-Horti. Soc. Ind. Journ. (Old Series), IV., 213, 214.*

    Habitat.—A rare species, found in the Eastern Tropical Himálaya, Bengal, Western Peninsula, Orissa, and Prome.

FOOD &
FODDER.
Grain.
130

    Food & Fodder.—The GRAIN is eaten as *dál* by Natives; the STRAW is eaten by cattle.

Villarsia cristata, *Spreng.,* see **Limnanthemum cristatum,** *Griseb.;*
[Vol. IV., 641.

V. nymphæoides, *Vent.,* see **Limnanthemum nymphæoides,** *Link.;*
[Vol. IV., 641.

VILLEBRUNEA, *Gaud.; Gen. Pl., III., 390.*

131

Villebrunea frutescens, *Bl.; Fl. Br. Ind., V., 590;* URTICACEÆ.

    Syn.—MOROCARPUS MICROCEPHALUS, *Benth.;* URTICA FRUTESCENS, *Roxb.*

    Vern.—*Gar tashiára, poidhaula, kagshi, phúsar-patta,* KUMAON; *Kirma,* NEPAL; *Takbret,* LEPCHA. This seems also to be the *mesaki* fibre of many writers on the resources of the Panjáb.

    References.—*Roxb., Fl. Ind., Ed. C.B.C., 656; Brandis, For. Fl., 406; Gamble, Man. Timb., 325; Atkinson, Him. Dist., 317, 798; Royle, Fib. Pl., 365; Jour. Agri.-Horti. Soc. Ind., VII. (Old Series), 217; Watt, Sel. from Rec., Govt. of India, R. & A. Dept., I. (1889), 315.*

Habitat.—A small tree of the Tropical Himálaya from Kumáon eastwards, ascending to 5,000 feet in Sikkim; also found in the Khasia Hills at Shillong.

Fibre.—See **V. integrifolia,** *Gaud.*

FIBRE.
132
133

Villebrunea integrifolia, *Gaud. ; Fl. Br. Ind., V., 589.*

 Syn.—V. APPENDICULATA, *Wedd.;* OREOCNIDE ACUMINATA, *Kurz;* URTICA APPRNDICULATA, *Wall;* CELTIS ELONGATA & TETRANDRA, *Wall.*

 Var. sylvatica=V. SYLVATICA, *Blume;* BŒHMERIA SYLVATICA, *Hassk.;* OREOCNIDE SYLVATICA, *Miquel.*

134

 Vern.—*Lipic, lipiah,* NEPAL; *Ban rhea,* ASS.

 References.—*Roxb., Fl. Ind., Ed. C.B.C., 657 ; Gamble, Man. Timb., 325 ; Kurz, For. Fl., Burma, II., 427 ; Journ. Agri.-Horti. Soc., Ind. (Old Series), VI., 184; VII., 222 ; Cross, Bevan, & King, Report on Indian Fibres, 34 ; Watt, in Sel. from Rec., Govt. of Ind., R. & A. Dept., I., 315.*

Habitat.—A small tree or large bush, met with in the Eastern Himálaya, Assam, the Khásia Hills, Sylhet, Burma, Munipore, and Chittagong; according to **Stocks** it also occurs in the Deccan Peninsula from the Konkan southwards. The variety **sylvatica** occurs in Sikkim, Assam, Burma, the Andaman islands, the Western Ghats, and Ceylon.

Fibre.—Dr. **Watt** has recently written an exhaustive account of the fibre of this and the preceding species (*Sel. from Rec. Govt. of India, l. c.*), which, as containing all the information obtainable on the subject, may be given in full :—

FIBRE.
135

"Both **V. integrifolia** and **V. frutescens** are reputed to yield highly valuable fibres, and it is probably the case that they are of equal merit. At all events, the two plants are very nearly allied, and when both occur in the same locality are most probably not separately recognised by the Natives. The former is if anything a more tropical plant and prefers the damper eastern tracts of the Himálaya, whereas the latter takes the place of that species in the drier areas, and is distributed as far to the north-west as the basin of the Upper Sutlej. **Gamble** describes the fibre as 'brown in colour, strong and flexible, is made in Sikkim and Assam, into ropes, nets, and coarse cloth. The tree is of quick growth and coppices easily, and the fibre is likely to prove valuable.' **Kurz** makes a much more startling statement. 'This is the *ban-rhea* of the Assamese, which yields the fibre called China-grass cloth.' *Spons' Encyclopœdia* publishes a fact of the greatest importance (p. 932); speaking of **Villebrunea integrifolia,** the writer of that article says: 'The fibre is more easily separated than that of the preceding (**Maoutia Puya**), and is considered one of the strongest in India.'

"**Royle** throughout the whole of his admirable Chapter on Rhea (*Fibrous Plants of India*) alludes repeatedly to *Bon Rheea* or 'Wild Rhea.' In a special report on 'Rhea Fibres in Assam and Hemp in the Himálaya' submitted to the Board of Directors of the Honourable East India Company in 1853, **Royle** urges that every effort should be made to develop a trade in (*a*) the *Kunkhura* of Rungpore, Dinagepore, etc., which he says is also the *Pan* of the Shan States, and in (*b*) the *Bon Rhea* of Assam. The writer has read all the passages descriptive of the latter with the greatest care, so as to avoid, as far as possible, raising false hopes, through falling into mistakes regarding the remarkable fibre **Royle** designates as *Bon Rheea.* Compiling from the writings of others must necessarily be a less satisfactory procedure than reporting the results of original experiments, such as those performed by **Roxburgh** or by **Royle.** The possibility of falling into misinterpretations meets one on

FIBRE.

every hand, and the feeling of uncertainty is only equalled by the consciousness that we have no modern authorities from whom information can be culled that for a moment could be compared with those who wrote the greater part of a century ago. (Roxburgh's experiments with Rhea, for example, were performed in 1805.) The necessity for care, in deciding what **Royle** meant by *Bon Rheea*, will be at once apparent from the following table as given by the author of the *Fibrous Plants of India*, the more so since he remarks 'the specimens were very carefully prepared by **George Aston**, and their strength tried in the office of the Military Store.'

	Broke with a weight in ℔ of
Petersburgh Clean Hemp	160
Yercum (Calotropis gigantea)	190
China-grass, from China	250
Rhea fibre or China-grass from Assam	320
Wild Rhea, also from Assam	343
Hemp from Kote Kangra bore 400 without breaking.	

"Now, according to this result, the *Bon-rhea* of Assam is a stronger fibre than either Rhea or China-grass. Turning to **Royle's** work to discover the source of this fibre, there would appear to be little room for doubt that the *bon-rhea* fibre he experimented with was obtained from **Villebrunea integrifolia**. The following account of his *bon-rhea* may be reproduced entire (p. 363):—

"'In the preceding observations, the *Bon* or *Bun Rhea*,—that is, Jungle Rhea,—is so called as if it were the Dom Rhea or China Nettle in a wild state. Of this there is no proof, but considerable probability that it is a distinct species, possessed of many of the same properties as the Ramee or Rhea Nettle. Indeed, Major Hannay, who has chiefly brought it into notice, says of Bon or Jungle Rhea (*Bœhmeria* species) that it is a Jungle plant, common in the Assam forests, and thriving best in the vicinity of water or of running streams. When unmolested it grows to a tree, but, by proper management, any quantity of young shoots can be obtained; and as the divided roots afford numerous shoots, the plant can be propagated by slips as well as by the seed. Its cultivation for its fibre might be carried on as with the willow in Europe.

"'By the Chinese in Assam, it is said to be exported into Southern from Northern China. It is cultivated largely by the hill tribes north-west of Yunnan, and by the Singpoos and Dhonncas of our own north-eastern frontier, to a small extent only, for a coarse cloth, but chiefly for nets. The Nepaulese recognise it as the *Leepeeah* of Nepaul.'

"This fibre, in the state in which it has been sent, is well adapted for rope-making. It is about five feet in length, brown in colour, strong and flexible. **Capt. W. Thompson**, of the house of Messrs. Thompson, Rope-makers, of Calcutta, says of it : 'It is all that can be desired either for canvas or lines, and only requires to be known to be generally used for such purposes.' It was this fibre which was made into a 5-inch rope by Messrs. Huddart along with the Dom Rhea or China-grass, and broke with a weight of about nine tons, or precisely 21,025℔. Since then, it has been made up into ropes of various sizes, which have been carefully tested, and found in every case greatly to exceed in tenacity those made of Russian hemp of the same size. (Here a reference is given to the table of results which has been reproduced above.) 'It has been also made up into lines and cords, some of them almost fine enough for fishing-lines : in all which it displays its fitness for all such purposes from the union of strength and flexibility.' This is almost word for word Major Hannay's description

or the " Ban-rhea." (*J. Murray.*) VILLEBRUNEA
integrifolia.

[*Jour. Agri-Hort., Soc., VII. (Old Series), p. 222*], and further on that writer alludes to the *Mesakhee* fibre which he says is obtained from a plant very similar to the *Bon-rhea*. This would thus appear to be **Villebrunea frutescens**.

" At page 373 of his work **Royle** gives the following table :—

Fibre.	Size of rope.	Total No. of yarns.	Strength of rope in ℔.
Wild rhea, 1st experiment	4⅛	132	19,032
Ditto, 2nd experiment	4⅗	132	21,025
Rhea fibre	4⅝	132	20,488

" This then is the *Bon-rhea* which **Royle** extols so highly. It grows into a *tree* when not molested ; it is called *Ban-rhea* in Assam : *lĭpiah* in Nepál : and it yields a brownish fibre. That would seem a description of **Villebrunea integrifolia** and not of any of the **Bœhmerias** or of **Maoutia**, and **Royle's** information was not compiled in London from numerous sources, it was directly taken from the writings of **Major Hannay**, a gentleman whose name is so intimately associated with the development of the resources of Assam, that it is practically impossible to suppose him to have been mistaken. The samples of the fibre tested by **Royle** were obtained from **Major Hannay**, so that there would seem to be no possible mistake except the mistake made by all subsequent writers of ignoring the independence of *Ban-rhea* from Rhea itself. Here then the generic name Rhea has probably succeeded in diverting public attention from this most valuable fibre, and it is just possible **Kurz** may after all be correct. The China-grass from Northern China may be the fibre from this plant, and the China-grass from the south, the Rhea or Ramie, of India and of the Straits.

" Be that as it may, we have remained too long ignorant of the properties of the fibres allied to the Rhea of commerce. If all **Royle** has said, nay, even if half what he has said of this fibre be correct, the future may be expected to largely displace Rhea by the neglected *Ban-rhea* of Assam. That plant could be much easier cultivated than Rhea or China-grass, since it does not require the same damp sub-tropical climate. It is abundant throughout the Lower Himálaya and luxuriates on exposed hot valleys if only its roots have access to the damp soil of streams. It probably might not succeed well on the plains of India, since it may be found unable to withstand the extreme heat of the summer, but at all events its cultivation along canal courses should be experimentally tested. Even, however, were the plant found unsuited to the plains, its cultivation could be extended throughout the Himálayas and on all the mountain tracts of India, more especially in Assam, Burma, and on the Western Gháts. It is, moreover, much to be regretted that our experiments in testing the merits of Rhea-machinery were not extended to the allied fibres. It is just possible the difficulties offered by Rhea and China-grass do not exist with **Villebrunea**, but we should not at least be ignorant of this point, and, as cited above, we have the authority of one writer in thinking the fibre will be found to be more easily prepared than the *poi-rhea*." It may be mentioned that **Atkinson** describes the fibre of **V. frutescens** as follows :—
" The plant is cut down for use when the seed is formed. The bark or skin is then removed and dried in the sun for a few days; when quite dry it is boiled with wood-ashes for four or five hours and allowed to cool. When

16

FIBRE.

cold it is macerated [being beaten] with a mallet on a flat stone while cold water is applied. The woody matter gradually disappears, leaving a fine fibre which is admirably adapted for fishing lines and nets as well for its great strength as for its powers of resisting moisture."

Dr. Watt continues, " Modern experiments with **Villebrunea** fibre cannot, however, be discovered. No authentic samples of the fibre were shown at the Colonial and Indian Exhibition. The old delusion apparently prevailed so strongly, with the gentlemen who made the fibre collections, namely, that Rhea and *Ban-rhea* (=wild rhea) were essentially species of **Bœhmeria**, and hence no trustworthy collections of the fibres from the allied Rhea plants were furnished. And that not because they were unobtainable, but rather in consequence of undue attention having been given to the species of Bœhmeria.

" The writer cannot, therefore, commend in too strong terms the **Villebrunea** fibres to the attention of merchants and planters interested in Rhea. These plants could be grown as hedges throughout the whole tea and coffee districts of India and might afford annually two or three cuttings of fibre-yielding twigs practically at a nominal cost. Manufacturers who may think the grounds for the high expectations here held out sufficiently justified to warrant their embarking in experiments, would find little difficulty in having a ton or two of the twigs collected, dried, and baled to Europe, or even decorticated locally. The species most highly commended (**V. integrifolia**) is a plentiful jungle bush in Burma, Assam, Bengal, and the North-West Provinces. On the lower Himálaya and the mountain tracts of these provinces one or other of the species abounds. That they yield admirable fibres, we have the testimony of all modern observers (**Brandis, Kurz, Gamble, etc.**), but that they deserve to rank with the best Rhea, as Royle's *Ban-rhea* would appear to do, is a point which future re-investigation can alone establish. But this at least seems undeniable—they deserve to be rescued from the generic oblivion of ' Rhea.' Before leaving the subject of the **Villebrunea** fibres it may be as well to caution intending planters that there are several plants remarkably like the **Villebruneas** all of which yield good fibres though apparently inferior to those obtained from the *Ban-rhea*. In connection with the preparations for the Colonial and Indian Exhibition the Manager of the Glen Rock Fibre Company was good enough to forward to the writer botanical specimens and small samples of the fibres obtained from a few of the allied Rhea plants with which that Company was then experimenting. The following determination of these plants may be of interest, the more so since one or two of these belong to the class that might be mistaken as species of **Villebrunea** :—

1st—Two samples marked **Urtica tenacissima**.

" These appear to be large forms of **Bœhmeria nivea**, the China-grass, and not of **Bœhmeria tenacissima**, the Rhea as accepted by the writer.

2nd—A sample marked *Oreocknide*.

" This is **Villebrunea integrifolia**, *vár.* **sylvatica**, a plant regarding the fibre of which we have no information. Whether superior or inferior to **V. integrifolia** proper, is a point which can alone be solved by comparative tests. The present plant may be recognised by its glabrous leaves (except on the veins below), and sessile, small heads of flowers.

3rd.—A sample marked *Yaumúri nar*.

" This is **Trema amboinensis**, *Bl.*, a fibre-yielding plant very inferior to the Villebruneas.

4th.—**Debregeasia velutina (Conocephalus niveaus,** *Wight*)—the *Caþsi* of Bombay. The fibre sent along with this seems of good quality. The plant is common in the Concan and Ghát jungles and on the Nilgiri hills, etc.

FIBRE.

"The last three plants have long leaves with crowded heads of small flowers and to a non-botanist *might* be mistaken for **Villebrunea integrifolia.**

5th. — **Girardinia heterophylla,** *var.* **zeylanica.**
The Nilgiri Nettle

"The specimen No. 4 is more likely than any other plant to be mistaken for a **Villebrunea—Debregeasia velutina.** It is the **Conocephalus niveus,** *Wight,* and other writers, and of all the allied rhea fibres is the one best known in South India.

"The reader is referred to the Dictionary article on the species of **Debregeasia** (under which also will be found **Conocephalus**), but as a popular eye-mark it may be said that the **Debregeasias** are erect bushes having small densely tomentose leaves, and sessile flower-clusters, the females through the succulent growth of the perianth forming edible minute fruits that become agglomerated together like lac around the twigs. The species of **Villebrunea** and also of **Conocephalus** have the flowers borne on short flowerstalks, and in the former the leaves are large but not silvery tomentose below, and the margins almost entire, instead of being minutely and sharply serate as in the **Debregeasias,** while the species of **Conocephalus** are climbing shrubs with the leaves quite entire.

"Although the fibre obtained from the **Debregeasias** is most probably very inferior to that of the **Villebruneas,** the plants are more hardy and might with advantage be propagated all over India. They occur on the margins of neglected fields, especially along the foot of the Himálaya, and in South India; they ascend the hills to 7,000 feet. Dense undergrowths of these plants exist in the shady glades of the Himálaya, causing the hill-sides, with the ripplings of a gentle breeze, to appear as if sprinkled with snow. A perfectly inexhaustible supply, therefore, of **Debregeasia** fibre might be obtained.

"The record of a **Conocephalus** fibre, resting on a mistaken identification, all reference to that genus was omitted from the *Dictionary of the Economic Products of India.* No authentic information exists as to any of the species of **Conocephalus** being used by the people of India, though they doubtless possess strong fibres like most other Urticaceous plants. The fibre of **Debregeasia velutina** (the **Conocephalus niveus** of certain writers) is, however, of such high merit as to deserve special notice, and it should be critically examined along with **Villebrunea** fibre, since of the **Debregesias** that plant could be better grown on the plains of India than any of the **Villebruneas.** A sample of this so-called **Conocephalus** fibre was sent to England in 1883 by the Glen Rock Company and is said to have been valued at £70 a ton. If even half that sum could be realised it would pay handsomely to cultivate the plant, and as with **Villebrunea** it is probable that the separation of the fibre would be easier than that of Rhea or China-grass. Both these plants possess a property of great merit; they are small trees that stand coppicing freely, and might, as already urged, be grown by the coffee and tea planters as hedges and also in the deeper nullahs where tea and coffee cannot conveniently be grown. Once planted they would require little or no care and would yield a valuable crop of fibre and willow-like twigs for basket-making which might be profitably used on the estate, while the surplus would find a ready sale."

VINCA, *Linn.; Gen. Pl., II., 703.*

Vinca pusilla, *Murr.; Fl. Br. Ind., III., 640;* APOCYNACEÆ.

Syn.—V. PARVIFLORA, *Retz.;* CATHARANTHUS ROSEUS, *G. Don.*
Vern.—*Kapa-vila,* MALAY.; *Sangkhi, sangkhaphuli,* SANS.

134

VIOLA odorata.	The Sweet Violet.

References.—*Roxb., Fl. Ind., Ed. C.B.C., 242; Dals. & Gibs., Bomb. Fl., 144; Rev. A. Campbell, Rept. Ec. Pl., Chutia Nagpur, No. 8731; Rheede, Hort. Mal., IX., t. 33; Ainslie, Mat. Ind., II., 358; O'Shaughnessy, Beng. Dispens., 448; Dymock, Mat. Med. W. Ind., 2nd Ed., 509; Atkinson, Him. Dist., 313; Gazetteers:—Mysore & Coorg, I., 62; Bombay, XV., 438; N.-W. P., I., 82; IV., lxxiv.*

Habitat.—An erect, pale-green annual, found in the Western Himálaya at Garhwál, altitude 2,000 feet, in the Gangetic plain, and commonly in the Deccan.

MEDICINE.
Plant.
137

Medicine.—Ainslie writes, "Dr. F. Hamilton informs us (*MSS.*) that the *sangkhi* is a medicinal plant in Upper India, and that a decoction of the dried PLANT, boiled in oil, is rubbed on the loins in cases of lumbago." Royle, O'Shaughnessy, and other writers have repeated this statement without adding anything to our knowledge of the value of the remedy.

138

Vinca rosea, *Linn.; Fl. Br. Ind., III., 640.*

THE RED PERIWINKLE.

Vern.—*Ainskati,* URIVA; *Rattanjot,* PB.; *Sadaphúl,* MAR.; *Billa gannéru,* TEL.; *Them-ban-ma-hnyo-ban,* BURM.

References.—*Roxb., Fl. Ind., Ed. C.B.C., 242; Kurz, For. Fl. Burm., II., 178; Stewart, Pb. Pl., 143; Mason, Burma & Its People, 432, 799; Elliot, Fl. Andhr., 27; Gazetteers:—Mysore & Coorg, I., 62; N.-W. P., IV., lxxiv.; Orissa, II., 178.*

Habitat.—A West Indian plant, much cultivated in gardens and about pagodas, etc., in India; occasionally domesticated in waste places near villages.

MEDICINE.
Juice.
139
Leaves.
140

Medicine.—This species is mentioned as *rattanjot* under the name of V. minor by Honigberger, who attributes the properties of that drug to it (see Onosma echioides, *Linn.;* also Trichodesma). Surgeon-Major P. N. Mukerji, in a note to the Editor, states the JUICE of the LEAVES is employed in Orissa as an application to wasp stings.

VIOLA, *Linn.; Gen. Pl., I., 117, 970.*

141

Viola cinerea, *Boiss.; Fl. Br. Ind., I., 185;* VIOLACEÆ.

Vern.—*Banafsha,* SIND & PB.

Reference.—*Murray, Pl. & Drugs, Sind, 45.*

Habitat.—Common in the dry hilly region of the Panjáb and Sind.

MEDICINE.
Plant.
142
143

Medicine.—Murray states that this PLANT is used medicinally in Sind, in the same way as V. odorata.

V. odorata, *Linn.; Fl. Br. Ind., I., 184.*

THE SWEET VIOLET.

Vern.—*Banafshah,* HIND.; *Banosa,* BENG.; *Banafshah,* DEC.; *Banafshah,* BOMB.; *Baga banósa,* MAR.; *Banaphsa,* GUZ.; *Vayilettu,* TAM.; *Banafshaj, banafsaj, behussej,* ARAB.; *Banafshah,* PERS.

References.—*O'Shaughnessy, Beng. Dispens., 208; Irvine, Mat. Med., Patna, 12; Moodeen Sheriff, Supp. Pharm. Ind., 255; Mat. Med. S. Ind. (in MSS.), 30; Murray, Pl. & Drugs, Sind, 45; Bentl. & Trim., Med. Pl., t. 25; Dymock, Mat. Med. W. Ind., 2nd Ed., 65; Dymock, Warden & Hooper, Pharmacog. Ind., I., 140; Official Corresp. on Proposed New Pharm. Ind., 224; Year-Book Pharm., I., 8, 74, 622; Birdwood, Bomb. Pr., 8; Smith, Econ. Dict., 431; Gazetteer, Mysore & Coorg, I., 57; II., 13; Agri.-Horti. Soc. Ind., Trans., II., 121; VII., 71.*

Habitat.—Met with in Kashmír, at altitudes from 5,000 to 6,000 feet.

MEDICINE.
144

Medicine.—Dymock writes, "The Greeks made use of this herb as a medicine, and from them and their works the Muhammadans probably became acquainted with its properties; it does not appear to have been used by the early Hindu physicians. A long account of its properties will be found in most Arabic and Persian works on Materia Medica; it is gene-

rally considered cold and moist, and is especially valued as a diuretic and | MEDICINE.
expectorant, and as a purgative in bilious affections; it is seldom given alone
but is prescribed along with other drugs, which also have an aperient
action, such as tamarinds, myrobalans, etc. The diseases in which *banaf-shah* is recommended are too numerous to be mentioned here; suffice it
to say that they are generally those in which a cooling treatment is thought
to be indicated by the *hakims*."

O'Shaughnessy experimented with the dry plant as a substitute for
Ipecacuanha but without success. Moodeen Sheriff considers it antipyretic
and diaphoretic, and very useful in relieving febrile symptoms and excitement in all forms of fever, particularly in combination with other drugs of
the same class (*Mat. Med. Madras*).

CHEMICAL COMPOSITION.—"The flowers are said to contain, besides | CHEMISTRY.
colouring matter, slight traces of a volatile oil, three acids—one red, another | 145
colourless, and salicylic acid; an emetic principle called *violin*, probably
identical with *emetine; viola-quercitrin* in close relation to, but not identical with, *quercitrin* or *rutin (mandelin)*; and sugar, etc. The colouring
matter of the flowers is easily turned red by acids, and green by alkalis, and
hence the syrup of violets was formerly used as a reagent. The colourless
acid called violenic acid by Peretti is said to crystallise in silky needles, to
be soluble in water, alcohol, and ether, and to form yellow salts which stain
the skin. According to Boullay all parts of the plant contain violin"
(*Pharmacog. Ind.*).

SPECIAL OPINIONS.—§ "The FLOWERS are collected in large quantities | Flowers.
at and round Murree, Panjáb, and exported to the plains, to be employed as | 146
an emetic" (*Surgeon-Major J. E. T. Aitchison, Simla*). "An infusion
useful as a mild purgative in cases of fevers and hepatic disturbances"
(*Civil Surgeon F. F. Perry, Jallunder City, Panjáb*). "A *sherbet*
made of *bunafshah* has been found to be useful in fevers; cooling, diaphoretic" (*Assistant Surgeon S. C. Bhattacharji, Chanda*). "An infusion of
2 drams of the dried plant to 1 pint of water forms an excellent and certain
diaphoretic" (*Civil Surgeon C. M. Russell, Sarun*). "Diuretic, not purgative" (*Assistant Surgeon Nehal Singh, Saharunpore*). "The flowers
warmed in boiling water are used as poultice in inflammatory affections of the
throat and other parts of the body. It is extensively used by Natives as a
laxative, diaphoretic in fevers; in fact this drug forms one of the ingredients
of almost every prescription given by Native *hakims*" (*Assistant Surgeon
Bhagwan Das, Rawalpindi*).

Viola serpens, *Wall.; Fl. Br. Ind., I., 184.* 147

Syn.—V. WIGHTIANA *var.* PUBESCENS, *Thw.*; V. PILOSA, *Blume.*

Vern.—*Thungtu, banafsha*, HIND., KUMAON; *Banafsha*, PB.

References.—*Thwaites, En. Ceyl. Pl., 20; Stewart, Pb. Pl., 19;
O'Shaughnessy, Beng. Dispens., 209; Baden Powell, Pb. Pr., 331, 425;
Atkinson, Him. Dist., 305, 753; Ind. Forester, II., 24; Agri.-Horti.
Soc. Ind., Trans., VII., 71.*

Habitat.—Met with in the moist weeds throughout the Temperate
Himálaya, Khásia Hills, Pulney and Nilgiri Mountains, and Ceylon;
altitude 5,000 to 7,000 feet.

Medicine.—This species also yields part of the *banafsha* of the bazárs, | MEDICINE.
and is considered to have medicinal properties similar to those of V. odo- | 148
rata. Baden Powell states that a medicinal oil is prepared in the Panjáb
from it, called *raughan-i-banafsha*.

VISCUM, *Linn.; Gen. Pl., III., 213.*

Viscum album, *Linn.; Fl. Br. Ind., V., 223; LORANTHACEÆ.* 149

THE MISTLETOE.

VISCUM monoicum.	The Mistletoe.

Syn.—V. STELLATUM, *Don.*

Vern.—*Ban, banda,* HIND.; *Hurchu,* NEPAL; *Bhangra, banda, bamba, kahbang, ahalú, wahal, rini, reori, reng, ringi, jerra,* PB.; *Túrapanli,* AFGH.; *Dibki,* ARAB.

References.—*Kurz, For. Fl. Burm., II., 323; Brandis, For. Fl., 392; Gamble, Man. Timb., 319; Stewart, Pb. Pl., 112; Dymock, Mat. Med. W. Ind., 2nd Ed., 754; Agri.-Horti. Soc. Ind., Journ. (Old Series), IV., Sel., 262; XIV., 13.*

Habitat.—This parasite occurs commonly in the Temperate Himálaya, from Kashmír to Nepál, between 3,000 and 7,000 feet, where it chiefly grows on trees of the Natural Order ROSACEÆ, and in the walnut, elm, willow, alder, maple, poplar, olive, and mulberry.

MEDICINE. Plant. 150

Medicine.—The PLANT is used as a medicine in Lahoul. Honigberger states that, in the Panjáb, it is given in enlargement of the spleen, in cases of wound, tumours, diseases of the ear, etc. The dried berries imported into Bombay under the name of *kishmish-i-káwuliyán* (*vulg. kishmish-kawli*) are probably obtained from this species (*Dymock*).

DOMESTIC. 151 152

Domestic.—In Europe it is employed in making bird-lime.

Viscum articulatum, *Burm.; Fl. Br. Ind., V., 226.*

Syn.—V. ATTENUATUM, *DC.*; V. MONILIFORME, *Blume*; V. FRAGILE, *Wall.*; V. COMPRESSUM, *Poir.*; V. APHYLLUM, *Griff.*

Var. dichotoma, *Kurz*=V. DICHOTOMUM, *Don*; V. ELONGATUM, *Wall.*; V. NEPALENSE, *Spreng.*

Vern.—*Pan, púdú,* HIND.; *Kathom janga,* SANTAL; *Hurchu,* NEPAL; *Patha,* BANDA; *Banda,* C. P.

References.—*Kurz, For. Fl. Burm., II., 325; Brandis, For. Fl., 394; Rev. A. Campbell, Rep. Ec. Pl., Chutia Nagpur, No. 8431; Atkinson, Him. Dist., 316; Gazetteers:—Bombay, XV., 441; N.-W. P., I., 81.*

Habitat.—A native of the Sub-tropical Himálaya from Chamba eastward to Sikkim, ascending to 3,000 feet; also met with in Assam, Mishmi and the Khásia Mountains, where it ascends to 6,000 feet, and southwards to Travancore, Malacca, and Ceylon. *Var.* dichotoma occurs in the Himálaya, the Khásia Mountains, the higher hills of Pegu, and the Deccan Peninsula.

MEDICINE. Plant. 153

Medicine.—In Chutia Nagpur, a preparation from the PLANT is given in fever attended with aching limbs. The many joints in the plant have probably influenced the Santal *ojhas* in their application of it,—it is probably one of the many cases of the use of a remedy from a belief in the theory of signatures (*Campbell*).

154

V. monoicum, *Roxb.: Fl. Br. Ind., V., 224.*

Syn.—V. FALCATUM, *Wall.*; V. BENGHALENSIS & ? V. CONFERTUM, *Roxb.*

Vern.—*Kuchle-ka-malang,* HIND.; *Pet chamra banda,* SANTAL; *Kuchléki-sonkan,* DEC.; *Pulluri,* TAM.; *Pullurivi,* TEL.

References.—*Roxb., Fl. Ind., Ed. C.B C., 715; Kurz, For. Fl. Burm., II., 324; Gamble, Man. Timb., 319; Rev. A. Campbell, Rep. Ec. Pl., Chutia Nagpur, No. 8170; O'Shaughnessy, Beng. Dispens., 376; Moodeen Sheriff, Supp. Pharm. Ind., 255; Gazetteer, N.-W. P., IV., lxxvii.; Agri.-Horti. Soc. Ind., Trans., VII., 64; Journ. (Old Series), VI., 38.*

Habitat.—A large shrub met with in the Sikkim Himálaya, between 2,000 and 4,000 feet, the Khásia Mountains up to 3,000 feet, the Ganges Delta, Oudh, Martaban and Tenasserim, and the Nilgiri Hills.

MEDICINE. Leaves. 155

Medicine.—The LEAVES of a Viscum, doubtfully referred to this species, growing on Nux Vomica trees in the neighbourhood of Cuttack, have been found to possess poisonous properties, similar to those of the tree on which it grows. The subject was investigated by O'Shaughnessy, who detected strychnine and brucine in the powdered leaves. The powder of

V. 155

the dry leaf was used as a substitute for these drugs in the Hospital of the Medical College, Calcutta, with complete success, in doses of one to three grains thrice daily (*Beng. Dispens.*).

Viscum orientale, *Willd.; Fl. Br. Ind., V., 224.*   156

Syn.—V. VERTICILLATUM, *Roxb.*; V. HEYNEANUM, *DC.*; V. INDICUM, *Rottl.*

Vern.—*Banda*, HIND., SANTAL, KOL.; *Sundara badinika*, TEL.

References.—*Roxb., Fl. Ind., Ed. C.B.C., 715; Kurz, For. Fl. Burm., II., 324; Brandis, For. Fl., 393; Gamble, Man. Timb., 319; Rev. A. Campbell, Rep. Ec. Pl., Chutia Nagpur, No. 9224; Elliot, Fl. Andhr., 170.*

Habitat.—A rather large much-branched shrub, with purple fruit, found from Behar, Bengal, and Singapore, southwards to Singapore, Travancore, and the central provinces of Ceylon.

Medicine.—In Chutia Nagpore, "this PLANT is largely used medicinally, and is believed to derive some particular property from the tree on which it is found. It is employed in as many different diseases as the trees on which it is found" (*Campbell*). | MEDICINE. Plant. 157

VITEX, *Linn.; Gen. Pl., II., 1154.*

[VERBENACEÆ.

Vitex altissima, *Linn.; Fl. Br. Ind., IV., 584; Wight, Ic., t. 1466;*   158

Var. zeylanica—V. ZEYLANICA, *Turcz.*; V. ALTISSIMA (FORME SUB-GLABRA), *Thwaites.*

Vern.—*Ahay*, ASSAM; *Simyanga, gua*, KOL.; *Banalgay*, MAR. & BOMB.; *Maila*, TAM.; *Namili adogú*, TEL.; *Myrole, balgay, nauladi, sampaga-pala*, KAN.; *millilla, milla*, SING.

References.—*Roxb., Fl. Ind., Ed. C.B.C., 482; Beddome, Fl. Sylv., t. 252; Brandis, For. Fl., 370; Gamble, Man. Timb., 297; Dalz. & Gibs., Bomb. Fl., 201; Thwaites, En. Ceyl. Pl., 244; Drury, U. Pl., 442; Birdwood, Bomb. Pr., 335; Lisboa, U. Pl. Bomb., 108; For. Admn. Rep., Ch. Nagpur, 1885, 33; Gazetteer, Bombay, XV., 40, 87; Gribble, Man. Cuddapah, 262; Ind. Forester, III., 23, 178, 204; VI., 338; VIII., 29; X., 31, 33; XII., 551.*

Habitat.—A large tree of Bengal, South India, and Ceylon; especially common in Western India, up to 4,000 feet.

Structure of the Wood.—Grey with a tinge of olive brown, hard, close-grained; weight 50 to 53℔ per cubic foot (*Gamble*). Skinner gives 63℔ per cubic foot for Kanara specimens. Beddome describes the timber as one of the most valuable in South India; it does not split nor warp, polishes well, and is much used for building purposes, for cabinet work, and for making carts. It appears to be well worthy of attention. | TIMBER. 159

V. glabrata, *Br.; Fl. Br. Ind., IV., 588.*

Syn.—V. CUNNINGHAMI, *Schaner*; V. LEUCOXYLON, *Schaner* (not of *Linn* but described as V. LEUCOXYLON, *Linn. f.*, by *Kurz, Gamble*, and partly by *Brandis*); V. BOMBACIFOLIA, *Wall.*; V. PALLIDA, *Wall.*

Vern.—*Goda, horina, ashwal*, BENG.; *Bhodiya*, ASSAM; *Tokra*, MAGH; *Sheras, longarbis thiras*, BOMB.; *Longarbi thiras, sherasa, songarbi*, MAR.; *Luki, neva lédi*, TEL.; *Sengeni, karril, senkane*, KAN.; *Htoukshar*, BURM.

References.—*Brandis, For. Fl., 370; Kurz, For. Fl. Burm., II. 273; Dalz. & Gibs., Bomb. Fl., 201; Elliot, Fl. Andhr, 134; Rheede, Hort. Mal., IV., t. 36; Lisboa, U. Pl. Bomb., 108; Gazetteer, Bombay, XV., 78; Ind. Forester, X., 222.*

Habitat.—A small (very large, deciduous, *Gamble*) tree, common from South Assam and Cachar to Malacca. Considerable confusion exists in Indian literature between this species and V. leucoxylon, *Linn. f.*, a native of South India and Ceylon, which Kurz, and following him, Gamble,

V. 159

VITEX Negundo.	The Sambal.

MEDICINE.
Bark.
160
Root.
161
FOOD.
Fruit.
162
TIMBER.
163

appear to have united. It is probable, however, that most of the information here given refers to **V. glabrata.**

Medicine.—The BARK and ROOT are used as astringents in the Andaman islands (*Major Ford*).

Food.—The tree flowers in April, and produces a small, black FRUIT containing very soft pulp, which is eaten by Burmese in the Andaman islands.

Structure of the Wood.—Grey with a satiny lustre, hard, close-grained, durable; weight about 42℔ per cubic foot. It is used for cart-wheels, and deserves attention for furniture and other purposes (*Gamble*).

164

Vitex Negundo, *Linn.; Fl. Br. Ind., IV., 583; Wight, Ic., t. 519.*

Syn.—V. BICOLOR, *Willd.;* V. ARBOREA, *Desf.;* V. PANICULATA, *Lamk.*
Var. incisa=V. INCISA, *Lamk.*

Vern.—*Sanbhálú, nirgandi, sindhuca, nisindá, páni-ki-sambhálú, shi-wari, shawáli, nengar, mewri, sambhálú, sambhal, sinduari,* HIND.; *Nishindá, sámálú, nisinda, nirgúndi,* BENG.; *Ehúri, sindwor, hobaro, sinduari,* KOL.; *Sinduari,* SANTAL; *Beygúna, beguniyá,* URIYA; *Nir-gudi,* KURKU; *Semálu,* BERAR; *Nirgiri,* GOND; *Sindwar,* KHARWAR; *Shiwáli, simáli,* fruit=*filfil=bári,* KUMAON; *Marwan, moráun, máura, mora, wana, banna, torban, biuna, binna, tórbanna, morann, sanóke, swanján, shwári, bankahú, marwa, mawá,* root & leaves= *amalu,* fruit=*filfil=bari,* PB.; *Marwandai, mehrwán, warmande,* PUSHTU; *Shanbáli, shambáli, shamálú,* DECCAN; *Nirgundi, kátri, lingúr, nargunda, nirgur, shiwari, nisinda,* BOMB.; *Nirgunda, nirgúr, nirguda, lingúr,* MAR.; *Nirgari, nagoda,* GUZ.; *Vellai-noch-chi, noch-chi, nir-nochchi,* TAM.; *Tella-vávili, vávili, veyala, nalla vávili, vavali-padú,* TEL.; *Lakki-gidá, lakki, lakkle,* KAN.; *Vella-noch-chi, vel-noch chi, noch-chi,* MALAY.; *Kiyow-bhán-bin,* or *kiyubán-bin,* BURM.; *Nik-ka, súdú-nikka,* SING.; *Shvéta-surasa, vrikshaha, nirgundi, sindhuvára,* SANS.; *Aslag, fanjangasht, zúkhamsatilouráq, zúkhamsate-asábea,* ARAB.; *Sisbán, panj-angusht, banj-angasht,* PERS.

References.—*Roxb., Fl. Ind., Ed. C.B.C., 481; Brandis, For. Fl., 369; Kurz, For. Fl. Burm., II., 269; Beddome, For. Man., 171; Gamble, Man. Timb., 297; Dalz. & Gibs., Bomb. Fl., 201; Stewart, Pb. Pl., 166; Rev. A. Campbell, Rept. Econ. Pl., Chutia Nagpur, No. 8498; Sir W. Elliot, Fl. Andhr., 128; Sir W. Jones, Treat. Pl. Ind., V., 136; Rheede, Hort. Mal., II., t. 12; Rumphius, Amb., IV., t. 19; Lace, Quetta Pl., ; Pharm. Ind. 163; Ainslie, Mat. Ind., II., 252; O'Shaugh-nessy, Beng. Dispens., 485; Irvine, Mat. Med. Patna, 77; Taylor, To-pog. Dacca, 55; Moodeen Sheriff, Supp. Pharm. Ind., 256; U. C. Dutt, Mat. Med. Hind., 217, 311, 318; Murray, Pl. & Drugs, Sind, 175; Dymock, Mat. Med. W. Ind., 2nd Ed., 600; Cat. Baroda Dur-bar, Col. & Ind. Exhib., Nos. 183, 184; Official Corresp. on Proposed New Pharm. Ind., 238; Drury, U. Pl. Ind., 442; Atkinson, Him. Dist. (X., N.-W. P. Gaz.), 315, 753; Useful Pl. Bomb. (XXV., Bomb. Gaz.), 109; Moore, Man. Trichinopoly, 80; Gribble, Man. Cuddapah, 76; For. Admin. Rep. Chutia Nagpur, 1885, 33; Settlement Report:—Panjab, Peshawar, 13; Kohat, 30; Gazetteers:—Bombay, VI., 15; VII., 42; XV., 78; XVII., 25; Panjab, Dera Ismail Khan, 19; Karnal, 16; Bannu, 23; Hoshiarpur, 12; Peshawar, 27; Lahore, 95; N.-W. P., I., 83; IV., lxxvi.; Orissa, II., 181; Rajputana, 26; Agri.-Horti. Soc.:—Ind., Journals (Old Series), VI., 49, 225; X., 24; XIII., 319.*

DYE.
Ashes.
165
MEDICINE.
Root.
166
Leaves.
167
Juice.
168

Habitat.—A deciduous shrub, common throughout India and Ceylon, and ascending to 3,000 feet in the North-West Himálaya.

Dye.—The ASHES of this plant are largely used as an alkali in dyeing.

Medicine.—U. C. Dutt informs us that, according to Sanskrit writers, there are two forms of *nirgundi,*—that with pale blue flowers, *sindhuvára* (**V. trifolia**), and that with blue flowers, *nirgundi.* The properties of both are said to be identical, but the latter is generally used in medicine. The ROOT of **Vitex Negundo** is considered tonic, febrifuge, and expectorant, and the LEAVES aromatic, tonic, and vermifuge. The JUICE of the leaves is

| A valuable Medicine. | (*J. Murray.*) | VITEX Negundo. |

MEDICINE.

largely employed for soaking various metallic powders, before making the latter into pills. A decoction of the leaves is given, with the addition of long pepper, in catarrhal fever with heaviness of the head and dullness of hearing (*Bhavaprakasha*). A pillow stuffed with the leaves is placed under the head for relief of headache. The juice of the leaves is said to remove fœtid discharges and worms from ulcers. An OIL prepared with the juice of the leaves is applied to sinuses and scrofulous sores (*Chakra-datta*). Dymock states that Muhammadans consider *athlak* or *panjan-gusht* (which as sold in Bombay appears to be the fruit, not of **V. Negun-do** or of **V. trifolia**, but of **V. Agnus-castus** of Europeans) as astringent, resolvent, and attenuant.

Oil.
169

The Indian medicinal species of **Vitex** early attracted the notice of Europeans. "**V. trifolia** is highly extolled by Bontius, under its Malayan name; he speaks of it as anodyne, diuretic, and emmenagogue, and testifies to the value of fomentations and baths prepared with 'this noble herb,' as he terms it, in the treatment of *Beri-beri*, and in the obscene affection of 'Burning of the feet' in Natives. Of **V. Negundo**, Fleming remarks that its leaves have a better claim to the title of discutient than any other vegetable remedy with which he is acquainted, and he adds that their efficacy in dispelling inflammatory swellings of the joints from acute rheumatism, and of the testes from suppressed gonorrhœa, has often excited his surprise. The mode of application followed by Natives, and adopted, according to Dr. Fleming, by some European practitioners in India, is simple—the fresh leaves, put into an earthen pot, are heated over a fire till they are as hot as can be borne without pain; they are then applied to the affected part, and kept *in situ* by a bandage, the application is repeated three or four times daily until the swelling subsides" (*Pharm. Ind.*). Roxburgh describes both species as medicinal, and mentions that the leaves of **V. Negundo** are employed to form a warm-bath for women after delivery. Rumphius & Rheede both particularly notice **V. trifolia**, the first recommending it externally in swellings and diseases of the skin, while the latter asserts that the powdered leaves taken with water cure intermittent fevers. Ainslie writes that the fruit of the same species is supposed by the *Vytians* to be nervine, cephalic, and emmenagogue, and is prescribed in powder, electuary, and decoction. The medicinal qualities of **V. Negundo** he considered to be similar to but weaker than those of **V. trifolia**. He adds, however, that the ROOT of the former is a pleasant bitter and useful in fever, and that the Muhammadans smoke the dried leaves in cases of headache and catarrh. Irvine states that a decoction of the leaves is used in Patna as an internal remedy for fever. Taylor writes that in Dacca the leaves are given with garlic, rice, *gúr*, etc., as a remedy for rheumatism.

Root.
170

Both species are given a place in the Pharmacopœia of India, where, in addition to part of the above information, it is stated that Dr. W. Nigledew has described a very interesting method of treating febrile, catarrhal, and rheumatic affections in Mysore, by means of a rude vapour bath prepared with the plant. The dried FRUIT is considered vermifuge.

Fruit.
171

CHEMICAL COMPOSITION.—"Nothing is known of the chemistry of these plants, but the seed of **V. Agnus-castus** is said to contain a peculiar bitter principle called *castine*, a volatile acrid substance, a large quantity of free acid and fat oil. In Greece the fresh and rather unripe berries are said to be added to the merit of the grape to render the wine more intoxicating, and prevent it from turning sour" (*Dymock*).

Chemistry.
172

SPECIAL OPINIONS.—§ "The leaves, baked and applied to the head while warm or used as a pillow, relieve headache" (*Surgeon-Major Lionel Beech, Coconada*). "Given also in frontal head aches" (*Surgeon W. F.*

VITEX sp.	The Sambal.

MEDICINE.

Thomas, 33rd M. N. I., Mangalore). "The leaves (fresh) are credited with the power of destroying the smell of high or tainted meat or fish when boiled with it. The leaves, bruised and formed into cakes, may be applied to the temples to relieve headache" (*Civil Surgeon Banku Behary Gupta, Poori*). "I have often used a bath medicated with the leaves in cases of rheumatism and swelling of joints with excellent results" (*Honorary Surgeon E. A. Morris, Tranquibar*). "Leaves and root diuretic, diaphoretic and tonic. Tincture,—root bark 2 oz., Proof spirit 10 oz. Dose 1 to 2 drams three times a day is found useful in irritable bladder and rheumatism" (*Apothecary Thomas Ward, Madanapalla, Allahabad*).

TIMBER.
173

Structure of the Wood.—Wood greyish-white, hard; weight 42℔ per cubic foot. It is used for building purposes, and as a fuel and the branches for wattle-work.

[*IV., 587.*

174

Vitex peduncularis, *Wall., var.* Roxburghiana ; *Fl. Br. Ind.,*

Syn.—V. ALATA, *Roxb.*

Vern.—*Boruna, goda,* BENG.; *Osai,* ASS.; *Bhadu, marak',* SANTAL; *Krawru,* MAGH; *Hila auwal,* CACHAR; *Shelangri,* GARO; *Navaládi,* KAN.; *Kyetyo,* BURM.

References.—*Roxb., Fl. Ind., Ed. C.B.C., 482 ; Kurz, For. Fl. Burm., II., 272 ; Gamble, Man. Timb., 298 ; Dalz. & Gibs., Bomb. Fl., 201 ; Rev. A. Campbell, Rept. Ec. Pl., Chutia Nagpur, No. 9281 ; Lisboa, U. Pl. Bomb., 201 ; Aplin, Rep. on the Shan States ; Gazetteer, Mysore & Coorg, I., 48, 64.*

Habitat.—A tree met with in Behar, at Parisnath, in Eastern Bengal, the Khásia Tarai, and Pegu. According to Dalzell & Gibson it is also found in the Southern Mahratta country (Warri jungles) and the Konkan.

MEDICINE.
Bark.
175

Medicine.—In Chutia Nagpur the BARK is used for making an external application for pains in the chest (*Campbell*).

TIMBER.
176

Structure of the Wood.—Purplish or reddish-grey, heavy, hard, close-grained; weight 60℔ per cubic foot. It is a good timber, used in Cachar for posts and beams, in the Garo hills for sugarcane crushers, and in Chutia Nagpur for yokes.

177

V. pubescens, *Vahl.; Fl. Br. Ind., IV., 585; Wight, Ic., t. 1465.*

Syn.—V. ARBOREA, *Roxb.* ; PISTACIA VITEX, *Linn.*

Vern.—*Muria,* URIYA; *Nowli eragu, búsi, nevali adugu, nevalad ugu mánu,* TEL.; *Myladi,* TAM.; *Kyet-yob, htouk-sha,* BURM.

References.—*Roxb., Fl. Br. Ind., Ed. C.B.C., 482 ; Kurz, For. Fl. Burm., II., 271 ; Beddome, For. Man., 171 ; Gamble, Man. Timb., 297 ; Elliot, Fl. Andhr., 32, 124 ; Mason, Burma & Its People, 526, 792 ; Drury, U. Pl. Ind., 443 ; Ind. Forester, III., 204.*

Habitat.—A large tree of Eastern Bengal, Burma, the Andaman Islands, and South India.

TIMBER.
178

Structure of the Wood.—Wood smooth, grey, with an olive-brown tinge, when old, chocolate coloured ; very hard, close-grained ; weight about 55 ℔ per cubic foot (*Gamble*). It is durable, and is used for various purposes in South India ; the Burmans employ it to make wooden bells (*Mason*).

179

V. sp.

Vern.—·Seeds = *Hab-ul-fakad,* ARAB.; *Tukm-i-panjangusht,* PERS.; *Shambaloo kabij,* HIND.; *Renu kabij,* BOMB.

The above are given by **Moodeen Sheriff** as names for **Vitex Negundo,** but according to **Dymock,** though the seeds of this species are considered by Muhammadan physicians to be identical with the Indian *saubhálú,* they are not really so, but belong to another species.

References.—*Dymock, Mat. Med. W. Ind., 2nd Ed., 600; Year-Book Pharm., 1880, 250.*

V. 179

The Vine.	(*G. Watt.*)	**VITIS adnata.**

Medicine.—A small, dull grey, ovoid FRUIT, the size of a duck shot, half enclosed in the calyx, to which a portion of the peduncle remains attached. It is imported from Persia, and is considered to act as a resolvent and deobstruent in enlargements of the spleen, probably the fruit of **V. Agnus-castus.**

MEDICINE.
Fruit.
180

Vitex trifolia, *Linn.; Fl. Br. Ind, IV., 583.*

181

Syn.—V. INCISA, *Wall.;* V. AGNUS-CASTUS, var., *Kurz.*

Vern.—*Páni-ki-sanbhálú, suféd-sanbhálú,* HIND.; *Páni-samálú,* BENG.; *Páni-ki-shanbáli, ulji-shanbáli,* DEC.; *Nir-noch-chi, shiru-noch-chi,* TAM.; *Niru-vávili, shiru-vavili, tella-vavili,* TEL.; *Nir-noch-chi,* MA-LAY.; *Nira-lakki-gidá,* KAN.; *Kiyoubhán-bin, yé-kiyuban-bin,* BURM.; *Vaturu-nikka,* SINGH.; *Surasa-vrikshaha, jalu-nirgundi,* SANS.; *Asla, que-ábi,* ARAB.; *Panj-angushta-ábi, banj-angashte-abi,* PERS.

According to many writers the vernacular synonyms of **V. Negundo,** and of this species are the same. **Moodeen Sheriff** however, from whom the above has been quoted, states that the adjectives "white" and "water," which enter into the formation of so many of the terms, are properly applied to this species only.

References.—*Roxb., Fl. Ind., Ed. C.B.C., 481; Beddome, For. Man., 172; Brandis, For. Fl., 370; Mason, Burma & Its People, 413, 479, 792; Elliot, Fl. Andhr., 180, 190; Pharm. Ind., 163; O'Shaughnessy, Beng. Dispens., 484; Irvine, Mat. Med. Patna, 118; Fleming, Med. Pl. & Drugs (Asiatic Reser., XI.), 184; Dymock, Mat. Med. W. Ind., 2nd Ed., 600; Birdwood, Bomb. Pr., 66; Baden Powell, Pb. Pr., 364; Drury, U. Pl. Ind., 443; Cooke, Oils & Oil-seeds, 81; Aplin, Rep. on the Shan States, ; Settle. Rep., Chanda, App. VI.; Gazetteer, Mysore & Coorg, I., 64; Ind. Forester, XII., App. 19.*

Habitat.—A shrub or small tree, found scattered throughout India in the tropical and sub-tropical regions, from the foot of the Himálaya to Ceylon and Malacca; nowhere common.

Oil.—Drury says that a clear, sweet OIL of a greenish colour is extracted from the ROOT. It is supposed that the SEED also yields a fatty oil.

Medicine.—See V. Negundo.

OIL.
Root.
182
Seed.
183

(*G. Watt.*)

VITIS, *Linn.; Gen. Pl., I., 387, 999.*

Vitis acida, *Wall.;* see **V. setosa,** *Wall.;* p. 217; AMPELIDEÆ.

V. adnata, *Wall.; Fl. Br. Ind., I., 649; Wight, Ic., t. 144.*

184

Syn.—CISSUS ADNATA, *Roxb.;* C. CORDATA & KLEINII, *Wall.;* C. LATI-FOLIA, *Vahl.*

Vern.—*Bod-lar-nari,* SANTAL; *Pani-lara,* PAHARIA; *Kungchen-rik,* LEPCHA; *Kole-sán,* BOMB.; *Gudama tige, kokkitayárálu,* TEL.

References.—*DC. Prodr., I., 627; Roxb., Fl. Ind., Ed. C.B.C., 136; Brandis, For. Fl., 100; Gamble, List of Trees, Shrubs, etc. Darjeeling, 20; Thwaites, En. Ceylon Pl., 62; Graham, Cat. Bomb. Pl., 32; Dalz. & Gibs., Bomb. Fl., 39; Rev. A. Campbell, Econ. Prod. Chutia Nagpur, No. 8467; Sir W. Elliot, Fl. Andhr., 63, 92; Dymock, Mat. Med. W. Ind., 2nd Ed., 183; Gazetteers:—Bombay, XV., 430; N.-W. P., IV., lxx; Him. Dists., X., 307; Agri-Horti. Soc. Ind.:—Trans., VII., 53; Journ., VI., 36.*

Habitat.—A slender, far-climbing plant, met with in the hotter parts of India, from the Western Himálaya to Assam, Sylhet, Bengal, Tenasserim, the Western Peninsula, etc. Distributed to Ceylon, Java, Borneo, the Philippine Islands, etc.

Fibre.—The Revd. A. **Campbell** tells us that the Santals prepare a good cordage FIBRE from the STEMS.

Medicine.—Dymock says that in Western India "the dried TUBERS are used by the country-people as an alterative in the form of a decoction; they consider that it purifies the blood, acts as a diuretic, and

FIBRE.
Stems.
185
MEDICINE.
Tubers.
186

V. 186

VITIS carnosa.	The Horse-Vine.

MEDICINE.

Root.
187

renders the secretions healthy." **Mr. Campbell** remarks that in the Santal country the ROOT powdered and heated is applied to cuts and fractures.

FOOD.
Leaves.
188
189

Food.—Gamble says the LEAVES are eaten by the Lepchas of Sikkim.

Vitis araneosus, *Dalz. & Gibs.; Fl. Br. Ind., I., 657.*

Vern.—*Bendri, bender-wel, ghorwel* (or *ghorvel*=horse-vine), BOMB.

References.—*Dalz. & Gibs., Bomb. Fl., 41; Dymock, Mat. Med. W. Ind., 2nd Ed. 186; Pharmacog. Ind., I., 365.*

Habitat.—A slender, far-climbing plant, found in the Western Peninsula, the highest Gháts of the Concan, the Pulney mountains, etc.

MEDICINE.
Roots.
190

Medicine.—Dymock says this vine is often "given to horses, when it first springs up; it is said to be very beneficial once a year. The tuberous, starchy astringent ROOTS, sliced and dried, are sold by the Konkan herbalists under the name of "*Chamar-musli*.""

191

V. auriculata, *Roxb.; Fl. Br. Ind., I., 658.*

Syn.—CISSUS AURICULATA, *DC.*

Vern.—*Kúra palléru,* TEL.; *Wa-young-khyen, yinhnaung peinne,* BURM.

References.—*Kurz, For. Fl. Br. Burm., I., 274; Mason, Burma and Its People, 742; Sir W. Elliot, Fl. Andhr., 104; Dalz. & Gibs., Bomb. Fl. 40; Gazetteers:—Mysore and Coorg, I., 59; Bombay, XV., 430; Agri.-Horti. Soc. Ind., Trans., VII., 53.*

Habitat.—A large, woody climber, found in the Eastern Himálaya (3,000 to 5,000 feet), to Bengal (Chittagong) and Burma (tropical mixed forests of the Pegu Yomah).

TIMBER.
192

Structure of the Wood.—"Reddish, very coarsely fibrous" (*Kurz*).

193

V. barbata, *Wall.; Fl. Br. Ind., I., 651.*

Syn.—V. LATIFOLIA, *Hb. Ham.*; V. LANATA, *Hb. Roxb.*

Habitat.—An extensive climber, with very remarkable long spreading glandular hairs; found in Assam and the Khásia hills to Pegu and Tenasserim.

FOOD.
194

Food.—This is probably the edible species alluded to by **Mr. Darrah** under the name of **V. carnosa** (which see); but greater interest may, perhaps, be said to centre in this plant from its being identified as possibly the same as **V. Martini,** *Planch.,* which has been alluded to by many writers as of importance as a possible new source of grapes. The reader will find a full account of that plant in the *Kew Bulletin, 1888, pp. 134 and 135;* also in *Christy's New Commercial Plants and Drugs.*

195

V. carnosa, *Wall.; Fl. Br. Ind., I., 654; Wight, Ic., t. 171.*

Syn.—CISSUS CARNOSA, ACIDA, and AURICULATA, *Roxb.*; C. CRENATA, *Vahl.,* also *Wall.*; C. CINEREA, *Lamk.*

Vern.—*Amal-bel, gidar-drak, kassar,* HIND.; *Bundal, amal-lata, sone-kesur* (Patna), BENGAL; *Jarila-lara,* PAHARIA; *Takbli-rik,* LEPCHA; *Maimati, or marmarati,* ASSAM; *Kárik, ámal bel, gidardák, drúkri, vallúr,* PB.; *Odi, ambat-bel,* MAR.; *Khát, khatumáre, tamanya,* GUZ.; *Odi,* BOMB.; *Kuru dinne, kádépa tige, kanapatige, mandulamári tige, mékamettani chettu,* TEL.

References.—*DC., Prod , I., 630; Roxb., Fl. Ind., Ed. C.B.C., 137; Brandis, For. Fl.., 101; Gamble, Cat. Trees, Shrubs, etc., Darjeeling, 20; Stewart, Pb. Pl., 35; Graham, Cat. Bomb. Pl., 33; Sir W. Elliot, Fl. Andhr., 76, 81, 104, 111, 114; Rheede, Hort. Mal., VII., t. 9 (Kádi or "Yoke plant"); Rumphius, Amb., V., 450 f. t. 166 f. 2; Ainslie, Mat. Ind., I., 304; O'Saughnessy, Beng. Dispens., 254; Irvine, Mat. Med. Patna, 126; Sakharam Arjun, Cat. Bom., Drugs, 215; Dymock, Mat. Med. W. Ind., 2nd Ed.. 183; Dymock, Warden and Hooper, Pharmacog. Ind., Vol. I., 365; Darrah, Note on the Condition of the People of Assam, App., D; Gazetteers:—Bombay, III., 203; XV., 430; N. W. P., I., 79; IV., lxx; Rajputana, 25; Agri.-Horti. Soc. Ind.:—Trans., VII., 53; Jour. VI., 36; Indian Forester, IV., 227; XII., app., 2, 10.*

V. 195

The Himalayan "Virginian Creeper." (*G Watt.*)

Habitat.—A climber, found throughout the hotter parts of India, and ascending into the tropical Himálaya. Distributed to Burma, Ceylon, and Malacca.

Dye.—Ainslie suggests that a dye might be prepared from the acid FRUITS of this plant.

Medicine.—The names given to it in many parts of India denote one of its most general uses, namely, the treatment of yoke-sores on the necks of bullocks (*Elliot*). For that purpose a poultice of the LEAVES is most frequently employed. According to Irvine (*Mat. Med., Patna*) the SEEDS and also the ROOTS are employed as an embrocation. Stewart remarks that the root, ground with black pepper, is applied to boils. Dymock, speaking of this species conjointly with **V. setosa**, says "both occur in the Bombay Presidency; they are excessively acrid, and the roots and leaves are sometimes externally applied as domestic remedies to promote suppuration." In the *Pharmacographia Indica* V. **carnosa** is spoken of as "used as a domestic application to boils." It seems probable that most writers have confused **V. carnosa** with **V. setosa**; the latter species alone is the very acid plant used medicinally that has by some writers (*e.g*, Ainslie) been spoken of as **Cissus acida,** *Linn*. Roxburgh assigns no medicinal use to **V. carnosa**, and in this he is very probably correct, and subsequent writers incorrect. *Conf. with* **V. setosa**, p. 257.

Food.—Darrah (*Note on the people of Assam*) says that the YOUNG TOPS are boiled as a spinach and eaten either by itself or with other vegetables along with fish. The writer found a vine so eaten in Manipur, which he thinks may be the same species as that alluded to by Mr. Darrah, and if so, it is an undescribed species which is very different from **V. carnosa**. Stewart, while dealing with **V. carnosa** and **V. capriolata**, says that (it or) they are eaten by camels.

DYE.
Fruits.
196
MEDICINE.

Leaves.
197
Seeds.
198
Roots.
199

FOOD.
Tops,
Young.
200
Conf. with
p. 252.

Vitis discolor, *Dalz ; Fl. Br. Ind., 1., 647.*

Syn.—V. INŒQUALIS, & COSTATA, *Wall.*; CISSUS DISCOLOR, *Blume*; V. INŒQUALIS, *Wall.*; V. REPENS, *Dalz. & Gibs.*

References.—*Dalz. & Gibs., Bomb. Fl., 40; Gamble, Cat. Trees & Shrubs, etc., Darjeeling, 19; Gazetteer, Bombay, XV., 430.*

Habitat.—A very elegant, variegated-leaved species, found wild in the tropical East Himalaya, the Khásia hills, Sylhet, Manipur, Chittagong, Pegu, Tenasserim; also in the Western Peninsula. Distributed to Java. A favourite plant in gardens in India, but in some of its forms it is often quite colourless and seems to obtain depth of colour and breadth of leaf when grown in shade only, as, for example, in the wattle-orchid houses of Calcutta gardens.

201

V. himalayana, *Brandis ; Fl. Br. Ind. I., 655.*

Syn.—AMPELOPSIS NЃILGHERRENSIS, *Wight, Ic., t. 965*; A. HIMALAYANA, *Royle, Ill. 149*; CISSUS HIMALAYANA, *Walp.*

Vern.—*Phlankur,* SIMLA; *Zemardachan, semaro,* SUTLEJ; *Chappar tang,* KUMAON; *Bara churcheri,* PAHARIA; *Hlotagbret,* LEPCHA.

References.—*Brandis, For. Fl., 100; Gamble, Man. Timb., 93; Kurz, For. Fl. Burm., I., 73; Gamble, Cat. Trees, Shrubs, etc., Darjeeling, 20; Gazetteers, N.-W. Provinces (Himálayan Districts), X., 307; Ind. Forester, XIII, 68.*

Habitat.—An extensive, woody climber, which often covers with its foliage the tallest trees of the temperate Himalaya, at altitudes of 6,000 to 9,000 feet. In autumn these turn to a brilliant rosey tint, a fact which has given the plant the name of the "Virginian creeper" by the European residents at hill stations. Its area of distribution may be said to be from Kashmír to Sikkim, the Khásia hills, and Burma. It also occurs, however, in the Western Peninsula (Pulney hills).

202

| VITIS lanata. | The Indian Wild Vine. |

TIMBER.
203

Structure of the Wood.—Coarse and fibrous. The stems are more destructive to the trees on which it climbs than useful, though the young stems, like those of most other vines, are used for natural cords to tie bundles of grass.

DISEASES.
204

Diseases.—This is the only Indian vine, so far as the writer can discover, that has ever been described as bearing a Uredinous fungus. The late **Surgeon-Major Barclay** (Paper on Uredineæ occurring in the neighbourhood of Simla, *Journ. Asiatic Soc. Bengal, LIX, Pt. II., 98*) describes this parasitic plant under the name of **Uredo Cronartiformis**, from its appearance suggesting a **Cronartium**. He adds, however, that though he looked carefully and continuously for the teleutosporic form, he had never found any trace of such.

205

Vitis indica, *Linn.; Fl. Br. Ind., I., 653.*

THE INDIAN WILD VINE.

Syn.— V. RUGOSA, *Wall.*

Vern.—*Amdhauka, amoluka,* BENG.; *Jangli-angur,* HIND. & DEC.; *Sambara* or *shembara-valli,* TEL.; *Chempara valli,* MALYAL.; *Randraksha, kole-ján* MAR.; *Yen-doung,* BURM.

References.—*Roxb., Fl. Ind., Ed. C.B.C., 221; Brandis, For. Fl., 100; Graham, Cat. Bomb. Pl., 33; Dalz. & Gibs., Bomb. Fl., 41; Ainslie, Mat. Med., 334; Moodeen Sheriff, Supp. Pharm. Ind., 257; Rheede, Hort. Mal, VII., 11, t. 6; Mason, Burma and Its People, 460, 742; Indian Forester, XII., App. 10; Gazetteers, N.-W. P., I., 79; IV., lxx.; Bomb., XV. (Kanara), 430; Stewart, Bot. Journey Hazara in Agri-Horti. Soc., XIV., 9; Pharmacog. Ind. (Dymock, Warden and Hooper), I., 362; Kew Bulletin, 1889, 23.*

Habitat.—A slender, wooly species, with large, perennial, tuberous roots, found in the Central tableland of India, in the Western Peninsula, the Concan, and in Bengal. In point of foliage and shape of the bunches of small fruits, this much resembles the cultivated vine.

MEDICINE.
Juice.
206
Root.
207

Medicine.—Rheede was apparently the first European writer who assigned medicinal virtues to this plant. He says that the JUICE of the ROOT with the kernel of the cocoanut was, in his time, employed as a depurative and aperient. Dymock remarks that in the Concan, the country-folk also use it as an alterative in the form of a decoction, and they consider, he adds, that it purifies the blood, acts as a diuretic, and renders the secretions healthy.

FOOD.
Fruits.
208
Conf. with p. 263.
Tubers.
209

Food.—Few authors allude to the FRUITS being eaten, but this is probably an oversight, as they are regularly used, and the plant is often protected on hedges in a state of half cultivation. **Mr. Cameron** of Mysore speaks of the fruit as globose, the size of a large current, and as eaten only by the hill tribes. **Dymock** remarks that the TUBERS are rich in salts of potash and lime. When fresh they are acrid, owing to the mechanical irritation caused by the needles of oxalate of lime.

210

V. lanata, *Roxb.; Fl. Br. Ind., I., 651.*

Syn.— V. CORDIFOLIA, *Roth.;* V. HEYNEANA, *DC.;* V. RUGOSA, *Wall.;* V. LABRUSCA, *Linn. Var* γ; V. INDICA & PENTAGONA, *Hb. Ham.;* CISSUS VITIGINEA, *Roxb.* (*not Linn.*).

Vern.—*Kolo, kolo nari,* SANTAL.; *Jarila-lara,* PAHARIA; *Mikrum-rik* LEPCHA; *Asanjiya* or *asoja, pahár-phuta* (or *mountain splitter*) *purain,* KUMAON.

References.—*Roxb., Fl. Ind., Ed. C.B.C., 222; Brandis, For. Fl. 99; Kurz, For. Fl. Burm., I., 277; Gamble, Man. Timb., 93; Cat. Trees, Shrubs, etc., Darjeeling, 20; Atkinson, Ec. Prod. N.-W. Prov. Pt. V., 56; Him. Dist., X., 307; Gazetteer, N.-W. P., IV., lxx.*

Habitat.—A very variable plant in the size, shape, and vestiture of the leaves. Met with in the Himálaya, at altitudes from 1,000 to 7,000 feet; also in the hills of Eastern Bengal, the Circars, and Burma.

V. 210

Products of India. 255

The Himalayan Wild Vine. (*G. Watt.*) VITIS parvifolia.

Food.—This is one of the chief sources of the small WILD GRAPES to be met with here and there all over India. The other species are V. indica, and V. parvifolia. Atkinson says of the present plant that in Kumaon and Garhwal the names *asanjíya* or *pahar-phuta* denote the variety *rugosa* "the grapes of which are edible and ripen in September-October, hence the first vernacular name. The grapes of *purain* (the form of *lanata* proper) are small, about the size of a black current, and are unpalateable." (*Conf.* with remarks regarding Grapes below, pp. *259-60, 263-4, 270, 271-2, 274, 275, 278, 279, 282, 284, 285, 232, 235.*) **Mr. Camp**bell says of V. lanata that the ROOT is eaten.

FOOD.
Wild Grapes.
211
Conf. with pp. 256, 259-60, 263, 269, 275.

Root.
212

Vitis latifolia, *Roxb.; Fl. Br. Ind., I., 652.*

Syn.—V. KLEINII, *Wall.;* V. GLABRATA, *Heyne;* V. INDICA, *Wall.;* V. ZEYLANICA, *Russell.*

Vern.—*Govila,* BENG.; *I'c ewer, ic'er,* SANTAL, *Bédisa tiva,* TEL.; *Musal,* MERWARA.; *Chin-douk-nway-zouk,* BURM.

References.—*Roxb., Fl. Ind., Ed. C.B.C., 222; Brandis, For. Fl., 99; Gamble, Man. Timb., 93; Kurz, For. Fl., Burma, I., 277; Elliot, Fl. Andhr., 24; Indian Forester, IV., 227; XII., app. 10; N.-W. Prov. Gaz., IV., lxx; Bombay Gaz. (Kanara), XV., 430.*

Habitat.—A large, herbaceous climber, found in North-Western India, the Sub-Himálayan tract as far west as the Sutlej; frequent in Bengal and South India, also in Pegu, especially in the Sittang valley.

213

Food.—Though not specially mentioned as edible, the GRAPES are black and are largely eaten by birds if not by men.

FOOD.
Grapes.
214

V. Linnæi, *Wall.; Fl. Br. Ind. I., 649.*

Syn.—*Kurz,* in his *Forest Flora of Burma, Vol. I., 275,* endeavours to establish a form V. Linnæi, to which he refers V. repanda, *W. & A.,* of the *Flora of British India.* The writer is not in a position to be able to form an opinion on this subject, but as these forms (whether distinct or not) do not appear to be of any very great economic value (so far as is at present known), it is only necessary to provide a place for them provisionally. Elliot (*Fl. Andhr., 23,39,181*) gives, to what he calls V. Linnaei, the names *tige gummudu, china mandala mari* and *banka-baddu,* Telegu. Kurz tells us that his Burmese plant is known as *yin-noung-nway,* and that it is frequent all over Burma. The berries are said to be purplish black. V. repanda, *W. & A.,* Gamble tells us, climbs over the tallest trees. It is known as *pani-lara,* PAHARIA; *thym-rik,* LEPCHA. The wood, he adds, is very soft and fibrous; it holds a very large quantity of water. The writer found his attendants, while travelling in Manipur, cutting the long trailing stems of a vine to amuse themselves watching the stream of water flowing therefrom.

V. parvifolia, *Roxb.; Fl. Br. Ind., I., 652.*

HIMALAYAN WILD VINE. [DC.

Syn.—V. TRUNCATA, *Miq.;* V. VULPINA, *Linn,* var. γ; V. WALLICHII, **Vern.**—*Barain,* KUMAON.

References.—*Roxb., Fl. Ind., Ed. C.B.C., 222; Brandis, For. Fl., 99; Atkinson, Him. Dist., N.-W. P. Gaz., X., 307; Agri-Horti. Soc. Ind., XIX, 9.*

215

Conf. with pp. 259-60, 275, 291.

Habitat.—Roxburgh speaks of this as "a slender, perennial vine, of exactly the habit of the common grape vine," and that description, it may be said, is fully applicable. It is a grape-vine with small leaves and small black berries, found plentifully in the sub-temperate valleys of the North-Western Himálaya, from Kashmír to Nepal; also in Eastern Bengal.

Food.—The small black GRAPES are very sweet and delicately flavoured when ripe. They are regularly eaten by the Natives, and at the Simla Horticultural Society's shows were even exhibited as a small kind of grapes. The writer, as Secretary of that Society, had some opportunity of investigating this subject, and he believes that more careful study would

FOOD.
Grapes.

216

FOOD.

reveal as a fact that many of the small black grapes sold at hill stations in India are derived from the wild or semi-cultivated states of **V. parvifolia** (the very small grapes) and of **V. lanata** var. **rugosa** (the larger sizes). If this suggestion proves correct it would seem probable that some light might be expected to be thrown on certain obscure points of the origin of **V. vinifera**. The large black grapes of Simla and Bashrh, referable to **V. rugosa,** have a most peculiar favour, which might be described as something between a black currant and a grape. In these localities the true grape is also grown, but from what the writer was able to learn by the inspection of imperfectly dried specimens supplied to him, he is of opinion that the study of the living plants in Bashrh, for example, might reveal the existence of cultivated hybrids (*Conf. with pp. 259-60*) between **V. vinifera** and **V. lanata** (*Conf. with p. 255*). If not actual hybrids it seems likely that graftings of the true grape on to one or other of the wild forms may have been practised in the production of some of the Bashrh vines.

217

Vitis pedata, *Vahl.; Fl. Br. Ind., I., 661.*

Syn.—Cissus heptaphylla, *Retz.;* C. pedata, *Lank.;* C. serratifolia, *Hb. Rottl.;* Melothrix zeylanica, *Kœn.*

Vern.—*Goalilata,* Beng.; *Tungrútrikup,* Lepcha; *Edakula, mandula, kannem, puli máda, káni ápa tige, kádepatige,* Tel.; *Gorpadvel,* Mar.; *Mediya-wel,* Sinh; *Godhápadi,* Sans.

References.—*Roxb., Fl. Ind., Ed. C.B.C., 138; Dals. & Gibs., Bomb.Fl., 40; Grah., Cat. Bomb. Pl., 33; Elliot, Fl. Andhr, 49, 81, 82, 158; Gamble, List of Trees, Shrubs, etc., Darjeeling, 21; Kurz, For. Fl. Burm., I., 273; U. C. Dutt, Mat. Med., Hindus, 298; Rheede, Hort. Mal. VII., t. 10; Pharmacog, Indica, I., 365; Trimen, Cat. Ceyl. Pl., 19.*

Habitat.—A large, weak climber, met with in Bengal, Sylhet, Assam, the Khásia hills, Burma; also the Western Peninsula from the Concan southwards to Ceylon.

MEDICINE.
Plant.
218

Medicine.—This species appears to be sometimes used as a substitute for or adulterant of **V. setosa.** The Sanskrit name denotes a resemblance of the leaf to the foot of the Iguana. According to the authors of the *Pharmacographia Indica* this PLANT is used as a domestic medicine because of its astringency.

219

V. quadrangularis, *Wall.; Fl. Br. Ind., I., 645; Wight, Ic., t., 51.*
The Edible-Stemmed Vine.

Syn.—Cissus edulis, *Dals.;* C. quadrangularis, *Linn.;* Sœlanthes quadragonous, *Forsk.*

Vern.—*Hár-jorá. hadjora, nallar, harsankar, kándawel,* Hind.; *Hasj ora, hórjórá, hárbhángá, har,* Beng.; *Hárbhángá,* Uriya; *Nallér,* Deccan; *Harsankar, hárjorá, kándawel, nallar,* Bomb.; *Kándavela,* Mar.; *Harsankar, chódhári,* Guz.; *Perundei codie, pirandai,* Tam.; *Nalléru. nullerútigeh,* Tel.; *Mangarúli,* Kan.; *Tsgangelam-parenda, viranta, piranta,* Malay.; *Shazán-lese,* Burm.; *Hiressa,* Sinh.; *Vajra-valli, asthisanhara,* Sans.; *Har,* Pers.; *Di-xanh-voung,* Cochin-Chinese.

References.—*DC., Prod., I.; Roxb., Fl. Ind., Ed. C.B.C, 136; Brandis, For. Fl., 100; Trimen, Sys. Cat. Cey. Pl., 19; Dals. & Gibs., Bomb. Fl., 39, 40; Graham, Cat., Bomb. Pl., 33; Sir W. Elliot, Fl. Andhr., 129; Rheede, Hort. Mal., VII., t. 41; Ainslie, Mat. Ind., II., 303; Moodeen Sheriff, Supp. Pharm. Ind., 257, also Mat. Med. S. Ind. (in MSS.), 109; U. C. Dutt, Mat. Med. Hindus, 292; Sakharam Arjun, Cat. Bomb. Drugs, 215; Dymock, Mat. Med. W. Ind., 2nd Ed., 182; Dymock, Warden & Hooper, Pharmacog. Ind., Vol. I., 362; Drury, U. Pl. Ind., 443; Gazetteers:—N.-W. P., IV., lxx; Orissa, II., 181, App. VI.; Mysore and Coorg, I., 59; Agri.-Horti. Soc. Ind., Trans., VII., 53; Jour., IX., 411.*

MEDICINE.
Leaves.
220
Shoots.
221

Habitat.—A square-stemmed plant, found throughout the hotter parts of India, from the foot of the Western Himálaya in Kumaon, to Ceylon, and Malacca.

Medicine.—The LEAVES and young SHOOTS when dried are powdered and given in bowel complaints (*Ainslie*). Forskal states that the Arabs,

| The Grape-Vine. (G. Watt.) | VITIS setosa. |

when suffering from affections of the spine, make beds of the STEMS (*Graham*). This is the *asthisanhara* of Sanskrit writers (*Dutt*). Dymock says that the JUICE of the stem is dropped into the ear in otorrhœa, and into the nose in epistaxis, by the Marathas. It is also a remedy in scurvy and in irregular menstruation. In the latter disease, adds Dymock, "2 tolás of the juice, extracted by heating the plant, is mixed with 2 tolás of *ghi*, and 1 tolá each of *gopichandan* and sugar, and given daily." The late **Dr. Moodeen Sheriff**, in his work on the Materia Medica of South India (which he unfortunately did not live to finish) speaks of a preserve of the stem prepared by boiling it in lime-water, as a useful stomachic.

MEDICINE.
Stems.
222
Juice.
223

Food.—The STEMS of this vine are very generally eaten by the Natives of India in their curries. When young they are said to be very good, but as they get old they become very acid. The LEAVES and stems were greedily sought after during the Khandesh famine. The red BERRIES are said to be very acid.

FOOD.
Stems.
224
Leaves.
225
Berries.
226

SPECIAL OPINIONS.—§ "Tonic : Young shoots used in small quantity, cooked" (*Apothecary Thomas Ward, Madanapalle, Cuddapah*). "Used in dyspepsia in man, and in cattle disease" (*Native Surgeon T. Modoelliar, Chingleput*). "The young SHOOTS are lightly roasted and ground up into a chatney with other condiments and used as an appetiser and stomachic. The shoots are cut up into small pieces, put into a covered chatty or other earthen vessel, and placed over the fire until the contents are burnt to cinders. The ashes are then powdered and administered internally in cases of dyspepsia" (*Surgeon-Major D. R. Thomson, M.D., C.I.E, Madras*). "Used by the Madrasees as chatney" (*Surgeon-Major P. N. Mukherji, Cuttack, Orissa*).

Shoots.
227

Vitis setosa, *Wall.; Fl. Br. Ind., I., 654; Wight, Ic., t. 170.*

228

Syn.—CISSUS ACIDA, *Wall.*; C. SETOSA, *Roxb.*

Vern.—*Harmal*, HIND.; *Yek-gisam-ká-bachlá*, DEC.; *Kháj-goli-cha-vel*, MAR.; *Puli-pérandai, puli-naravi*, TAM.; *Barabutsali, barre bach-chali, warsi-pala, pulla bach-chali*, TEL.

The vernacular names and uses attributed to **V. carnosa** by many writers (and for the present retained above, page 212) should for the most part very probably be transferred to this species.

References.—*DC., Prod., I., 630; Roxb., Fl. Ind., Ed. C.B.C., 137; Dalz. & Gibs., Bomb. Fl., 41; Sir W. Elliot, Fl. Andhr., 24, 159; Ainslie, Mat. Med., II., 326; O'Shaughnessy, Beng. Dispens., 254; Dymock, Mat. Med. W. Ind., 2nd Ed., 183; Drury, Useful Plants of India, 443; Moodeen Sheriff, Mat. Med. S. Ind., 107; Pharmacog. Ind., I., 365.*

Habitat.—Found in the Western Peninsula, from the Circars and Mysore southward.

Dye.—Ainslie remarks that the BERRIES (**C. acida**) "might be turned to account in dyeing, staining or colouring, from the appearance of the dark coloured lamp-black looking substance, which can be squeezed out of it."

DYE.
Berries.
229

Medicine.—Every part of the plant is exceedingly acrid. The LEAVES, toasted and oiled, are applied to indolent tumours, to bring them to suppuration (*Roxb.*). The authors of the *Pharmacographia Indica* say that this plant "is sometimes applied as a domestic remedy to promote suppuration and assist in the extraction of Guinea-worms." **Moodeen Sheriff** remarks that it is a useful local stimulant in the form of a poultice. Is applied to sloughing and fœtid ulceration, and also in boils and small abscesses for the purpose of hastening suppuration. Is a good substitute for yeast poultice.

MEDICINE.
Leaves.
230

17

VITIS vinifera.	The Grape-Vine.

231

Vitis tomentosa, *Heyne ; Fl. Br. Ind., I., 650 ; Wight, Ill., I., t. 57.*

    Syn.—V. LANATA & CINNAMOMEA, *Wall.*; V. TRIFIDA, *Roth.*; V. TERNATA & TRILOBA, *Heyne* ; AMPELOPSIS? TERNATA, *DC.*

    Vern.—*Ghora lidi,* SANTAL ; *Atukula baddu,* TEL.

    References.—*DC., Prod., I., 634; Kurz, For. Fl. Br. Burm., I., 227; Sir W. Elliot, Fl. Andhr., 18; Gazetteer, Bomb., XV., 430; Rev. A. Campbell, Chutia Nagpur Econ. Prod., No. 9498; Trimen, Cat. Ceylon Pl., 19.*

MEDICINE.
Root.

    Habitat.—A densely wooly climber, found in the Western Peninsula, from Canara southwards ; also in Burma.

232

    Medicine.—With the Santals the ROOT is deemed useful to allay swellings.

233

V. vinifera, *Linn.; Fl. Br. Ind., I., 652.*

    THE VINE, ALSO GRAPES, *Eng.* ; VIGNE, RAISINS, *Fr.* ; WEINSTOCK, TRAUBEN, *Germ.* ; GRAPPI, *It.* ; UBAS, *Sp.* ; UVAS, *Port.* ; UVŒ, *Lat.* ; UZUM, *Turk.* ; RAISINS, *Eng.* ; RAISINS SECS, *Fr.* ; ROSINEN, *Germ.* ; UVE PASSE, *It.* ; PASAS, *Sp.* ; PASSA, *Port.* ; UVŒ PASSÆ, *Lat.* ; ISSUM, *Rus.*

The English word GRAPE, in its original sense, seems to have denoted a hook (*grapa,* Sp., a hold-fast; *grappare,* It., to seize ; *krapfe,* Middle High German, a hook) ; in its next meaning it became a cluster of grapes, and later on the sense altered and it became a single berry. The first conception was doubtless the outcome of the study of its clasping tendrils, and might have originated in every country and tongue where the plant grew, without in any way denoting a common origin either for the vine plant or for the word grape, in cognate languages.

    Vern.—*Angúr, dákh, drakh,* (raisins=) *kismis, manakká,* HIND. ; *Angúrphal, drakhyaluta,* (raisins=) *kismis, manakka,* BENG. ; *Angúr,* C. P. ; *Angúr, dákh,* (raisins=) *kismis, manakka,* N.-W. P. ; *Angúr, dákh, buri, tanaur, talor dach, newala, dakki, dehla, mámre, gandelí, láning,* (raisins=) *sirishk, mitha,* PB. ; *Kwar,* PUSHTU ; *Angúr* (grapes), *tak* (the vine plant), AFG. ; *Basho,* LADAKH ; *Uzúm* (grapes), *kurk uzum* (raisins), *sirishk* (currants), TURKI ; *Drákh,* SIND ; *Drákh, abai,* (raisins=) *kishmish,* BOMB. ; *Draksha,* MAR. ; *Drakasha, darákh,* GUZ. ; *Kodi-mun-dirrippasham, kodrimúndrie,* TAM. ; *Draksha-pondu, góstani dráksha, dracha,* (raisins=) *kisumisuchettu,* TEL. ; *Drakshi,* KAN. ; *Buaangúr,* (raisins=) *sabib,* MALAY. ; *Sabi-si, sa-pyih, sa-byit,* BURM. ; *Muddrap, wœlmidi, úwus,* SINH. ; *Drákshá, mridviká* (raisins=*laghu-drákshá*), SANS. ; *Ainab,* (raisins=) *zabib, mewiz,* ARAB. ; *Angúr,* PERS. ; *Pu-t'au,* (raisins=) *kan-pu-tau,* CHIN. (*Conf. with ver. names, p. 221.*)

    References.—*DC., Prod., I., 633 ; Voigt, Hort. Sub. Cal., 29 ; Brandis, For. Fl., 98, 574 ; Kurz, For. Fl. Burm., I., 277 ; Gamble, Man. Timb., 93 ; Cat., Trees, Shrubs, etc., Darjeeling, 20 ; Dalz. & Gibs., Bomb. Fl. Suppl., 15 ; Stewart, Pb. Pl., 35, 36 ; Aitchison, Cat. Pb. and Sind Pl., 34 ; Kuram Valley Rept., Pt., I., 41 ; Rept. Pl. Coll. Afgh. Del. Com., 46 ; Western Afghanistan and N.-E. Persia, 218 ; Graham, Cat. Bomb. Pl., 33 ; Mason, Burma and Its People, 460, 742 ; Sir W. Elliot, Fl. Andhr., 47, 62, 91 ; Darwin, Animals and Plants under Domestication, I., 332-33, 375, 382, 395 ; II., 228, 278, 308, 313 ; De Candolle, Orig. Cult. Pl., 191 ; Hehn and Stallybrass, Culti. Pl. and Domestic Animals in their Migration from Asia to Europe, 69-84 ; Stocks, Rept. on Sind ; Pharm. Ind., 57 ; British Pharm., 461-464 ; Fluck. & Hanb., Pharmacog., 159 ; U. S. Dispens., 15th Ed., 1506, 1522 ; Ainslie, Mat. Ind., I., 157, 333 ; O'Shaughnessy, Beng. Dispens., 251 ; Butler Medical Topog., Oudh and Sultanpur, 32 ; Moodeen Sheriff, Suppl. Pharm. Ind., 257 ; U. C. Dutt, Mat. Med. Hindus, 138, 297 ; Sakharam Arjun, Cat. Bomb. Drugs, 27 ; Murray, Pl. & Drugs, Sind., 78 ; Bidie, Cat. Raw Pr., Paris Exhb., 78 ; Bent. & Trim., Med. Pl., 66 ; Dymock, Mat. Med. W. Ind., 2nd Ed., 184-186 ; Dymock, Warden and Hooper, Pharmacog. Ind., Vol. I., 357 ; Trans. Med. & Phys. Soc. Bomb. (New Series) IV., (1857-58), 84 ; Birdwood, Bomb. Prod., 143, 144, 202, 242 ; Baden Powell, Pb. Pr., 271, 310-312, 334, 601 ; Atkinson, Him. Dist. (X, N.-W. P. Gaz.), 307,*

| The Grape-Vine. (G. Watt.) | VITIS vinifera. |

711; Useful Pl. Bomb. (XXV., Bomb.Gaz.), 150; Econ. Prod. N.-W. Prov., Pt. V. (Vegetables, Spices, and Fruits), 44, 55; Royle, Prod. Res., 361; Kew Bulletin, 1889, 23; Simmonds, Tropical Agriculture, 426; Official Correspondence, dated May and June, 1890; Proceedings, Rev. and Agri. Dept., April, 1890; Ayeen Akbary, Gladwin's Trans., I., 81, 83, 86; II., 30, 37, 39, 44, 47, 69, 135, 174, 175; Ain-i-Akbari, Blochmann's Trans., I., 65; Linschoten, Voyage to East Indies (Ed. Burnell, Tiele and Yule), I., 10, 103, 144; II., 266, 278; Davies, Trade and Resources, N.-W. Boundary, India, lxxxvii, cxxvi, clix; Man. Madras Adm., II., 52; Nicholson, Man. Coimbatore, 241; Bomb. Manual of Revenue Accounts, 102; Bomb. Admin. Rept. (1871-72), 374; Settlement Report:—Panjab, Gujrat, 135; Hazara, 94; Kangra, 22, 44; Kohat, 30; Simla, app. xlii; Central Provinces, Mandlah, 89; Nimar, 200; Gazetteers:—Bombay, XII., 177; XVI., 102; XVII., 274-276; XVIII., ii., 62, 63; XXI., 387; Panjab, Hazara, 131, 133; Jhelam, 108; Peshawar, 18; Simla, 11; N.-W. P., IV., lxx.; VI., 247; Mysore and Coorg, I., 53, 59; III., 48; Agri-Horti. Soc. Ind.:—Trans., I., 65, 96, 97, 105; II., 78, 202-205; app., 298; Pro., 340; III., 41, 67, 68, 102; Pro., 250; V., 69, Pro., 84; VI., 54, 55, 61-70; Pro., 52, 55; VII., 53; Pro., 39, 84, 101, 150, 151, 183, 184; VIII., Pro., 378, 398, 407; Jour., I., 294-296; II., Pro., 266; III., 89-94; Pro., 40, 185, 196; IV., Sel., 121; V., Pro., 29, 42; VI., Sel., 112; Pro., 41; VII., Sel., 7; VIII., Pro., 180, 181; IX., Pro., 147; 209; XI., 247, 248; Pro., 54, 57; XIII., 382, 386; New Series, VII., 264-267; Tropical Agriculturist (1881-82), 16, 26, 146, 772, 1024; (1882-83), 447, 491, 499, 502, 527, 536, 570-572, 617, 618; (1885-86), 69, 70, 444, 484; (1886-87), 176, 336, 444, 587, 629, 676, 758; (1887-88), 129, 232, 622, 736, 752, 702, 791; (1888-89), 431, 457, 472; Indian Forester, IX., 170; Quarterly Journal of Agriculture (1861-62), X., 547; Indian Agriculturist, 2nd March 1889; 20th July 1889; 30th November 1889; Encyclop. Brit., IX., 92; XII., 277; XX., 258, 520; XXIV., 237, 601, 611; Balfour, Cyclop. Ind., I., 1425; III., 343, 1018, 1028, 1077; Ure, Dict. Indus. Arts & Man., III., 1067-1087; Smith, Ec. Dict., 429, 430.

Habitat.—According to **Lawson** in the *Flora of British India*, this species is " perhaps wild in the North-West Himalaya ; cultivated extensively in North-West India, and rarely in the Peninsula and Ceylon." **Lawson**, however, adds that, according to **Regel**, the cultivated grape-vine is a hybrid between **V. vulpina**, *Linn.*, and **V. Labrusca**, *Linn.*, two American species, which he (Regel) identifies with the Indian **V. parvifolia**, *Roxb.*, and **V. lanata**, *Roxb.* Speaking of the American vines, **Darwin**, however, says that they " belong to a distinct species." The spirit of **Darwin's** observations is directed in support of the belief that, though extremely variable under domestication, the vines of Europe are not likely to be of hybrid origin. The grape sows itself freely in Southern Europe ; some of the forms reproduce their special properties when raised by seed ; **Knight's** attempts to cross-fertilize certain forms have not been successful in producing intermediary conditions nor new properties, although this has been accomplished by grafting. So, again, the extreme liability to variation has been proved by the multitude of sports found in the seedlings where crossing · had been prevented. The chief variations noticeable may be said to be mainly in conformity with acclimatisation. **Simon** classed grapes in two sections, according as the leaves are glabrous or tomentose. Other writers have founded their classification on the shape and colour of the berries. **Odart** had resort to a purely geographical system. Some grapes require a dry soil, others luxuriate under prolonged humidity. Certain forms have brittle, others tough stalks which resist wind. In liability to disease, remarkable differences have been noticed : the Chasselas group were seen to be all readily affected, while the American very largely escaped. In his chapter on the *Laws of Variation*, **Darwin** says " the line of practical culture has retreated a little southward since the middle ages ; but this seems due to commerce, including that of wine, being now freer or

Conf. with pp. (for wild and indigenous cultivated edible vines) 254, 255, 263, 270, 272, 273, 275, 278, 284, 285, 291.

Hybrids *versus* **Graftings.** 234

VITIS
vinifera.   The Grape-Vine.

more easy. Nevertheless the fact of the vine not having spread north-
wards shows that acclimatisation has made no progress during several
centuries. There is, however, a marked difference in the constitution of
the several varieties—some being hardy, whilst others, like the Muscat of
Alexandria, require a very high temperature to come to perfection. According to Labat, vines taken from France to the West Indies succeed

Conf. with   with extreme difficulty, whilst those imported from Madeira, or the Canary
p. 271.   Islands, thrive admirably." As with most other plants so with the vine,
acclimatisation must be undertaken by easy stages, not sudden translations.
To attempt to convey the vines of the northern sections of the European
area of cultivation to the tropical plains of India would be to court absolute failure, at least in the preservation of special characteristics. Any
such object can only be attained by a slow process of gradual acclimatisation from country to country in easy stages. That object might, how-

Hybridisation ever, be more readily accomplished than were the reverse course desired,
and   namely, to convey a tropical form to a temperate zone. According to most
Grafting.   authors the vine, whether it be viewed as a multiplicity of forms from one
common species, or the result of ancient hybridisation and grafting between
two or more species, was originally a native of Western Asia. Acclimatisation, therefore, into colder regions must of necessity be slower and more difficult than a return towards the conditions of its presumed original habitat.

While the writer is not prepared to agree with the statement made
above, that Vitis vinifera is "perhaps wild in the North-West Himálaya"
(*Conf. with p. 264*), he has seen it growing in the vicinity of gardens, in a

Indigenous   neglected condition. But what has already been stated is, he ventures to
Vines.   think, a prognostication of future interesting discoveries, namely, that the
235   fruits of V. rugosa and V. parvifolia constitute a large proportion of the small
Conf. with   dark-coloured grapes often sold at Himálayan stations (*Conf. with p. 256*).
pp. 255, 263-4, These plants may, in fact, be frequently seen in quite as great a degree of cul-
270-1, 274,   tivation as the true vine. Indeed, the writer would be prepared to venture
275, 278, 279,   even further, and suggest that future study may likely discover hybrid states
282, 284, 285.   between V. rugosa and V. parvifolia, extensively cultivated under the conviction that they are the true grape-vine. It must not, however, be forgotten
that many writers regard hybridisation as a much less powerful agent than
grafting (if even it be an agent at all) in the production of cultivated states
of the species of Vitis. The writer's attention was first directed to the possibility of the wild species yielding grapes, in connection with the Flower and
Fruit shows held by the Horticultural Society of Simla. Prizes were offered
for Native-grown fruits. Wild, semi-wild, and cultivated grapes were, accordingly, presented for competition. Among these, unmistakable forms
both of V. rugosa and of V. parvifolia were shown, as also certain that appeared transitory states between these two species. The peculiarities of
the seeds of these semi-cultivated grapes left little room for doubt but that
they were perfectly distinct botanically from the large grapes shown from the
green-houses of the European growers. As bearing on the possibility of
India, like America, possessing more than one species of grape-yielding vine,
the reader will find the passage quoted below, regarding the wild grapes of
Kashmír peculiarly instructive (*see p. 271-3*). The subject of the Himálayan
grapes would, therefore, seem well worthy of careful study, as it may not only
throw light on the nature of the disease alluded to (in the papers below) as
threatening the once prosperous industry of vine growing in Kanáwar, but
it may even extend its beneficial influence to the vine industry of the world.
It may, in fact, have been with some such benefit in view that French growers,
a few years ago, procured stock of the vines of Kashmír. There can be no
doubt, however, that the study of the Himálayan and Central Asiatic
grapes is more likely to be productive of good results than the investi-

The Grape-Vine.	(G. Watt.)	VITIS vinifera.

gation of the degraded forms of the true vine, presently to be found on the plains of India. The Himálaya may, indeed, be regarded as the extreme south-eastern boundary of the presumed wild source of this, one of the most anciently cultivated of plants. Many writers have dwelt on the injurious influence of the spread of the faith of Islam on vine-growing. In India, and more especially in Kashmír, the decline seems, however, to have dated from the close of the sixteenth century and after the death of the Emperor Akbar. That enlightened monarch is well known to have set the dictates of the Prophet on one side when these infringed or restricted the development of his vast empire. If we may judge from the record of his reign, furnished by his historian Abul Fazl, the Emperor fostered and encouraged grape cultivation in India, and by his direct aid grapes of high merit were successfully acclimatised in the Panjáb and throughout the greater part of Northern and Western India. And it is significant that to this day the better qualities of grapes in India bear the name of *Phakrí* as if brought originally by Muhammadan mendicants. The successors to the throne of Delhi restored the reign of bigotry that Akbar's personal character alone had stayed, and before it fell not only much that was beautiful but more still that was useful and beneficial. The order went forth for the destruction of the vineries of Kashmír, and grape culture in India shared in the neglect that followed. The seventeenth and eighteenth centuries, therefore, witnessed a decline of interest in vine culture sufficient to account for the low position that industry now occupies, and the very inferior nature of the grapes of this country. That there are large tracts of Upper India eminently suited for grape cultivation scarcely needs justification, and the wine produced in Kashmír at the present day shows that even the highest branches of the industry need not be despaired of. Sufficient may, therefore, be accepted as already denoted to justify the inclusion of India in the world's area of vine cultivation.

KASHMIR WINE. Conf. with pp. 269, 273, 296.

Medicine.—According to U. C. Dutt the dried FRUITS or RAISINS have for many centuries been employed medicinally by the Hindus. They are described as "demulcent, laxative, sweet, cooling, agreeable and useful in thirst, heat of body, cough, hoarseness, and consumption. Raisins also enter into the composition of numerous demulcent and expectorant medicines." Dutt describes the preparation of a medicinal wine known as *Drákshá arishta,* of which the chief ingredients are raisins, treacle, cinnamon, cardamoms, *tejpatra,* the flowers of Mesua ferrea, the fruit of Aglaia Roxburghii, black pepper, long pepper, etc., set aside until fermented. This liquor was deemed invigorating and nourishing, and was used in consumption, cough, difficult breathing, and hoarseness. In European pharmacy raisins are similarly employed in compound tincture of cardamoms, tincture of senna, on account of the saccharine matter they contain. Grapes are described by Dioscorides under the name of σταφυλη, raisins as σταφίς. Pliny speaks of *Uvæ,* grapes, and *Acini passi,* raisins. Argol (or the tartrate crust obtained from wine vessels) was the τρὐξȍινου of the Greeks, the *Fæx vini* of Latin, and the *Milh-el-tartir* of the Arabs. Noureddeen Mohammed Abdullah Shirazy, Physician to the Emperor Shahjehan, in his work *Ulfaz Udwieh* gives the uses of grapes, raisins, and wine among the Muhammadans in his time. A species of black grape known as *Asá ba ul Ghé-zá-ri,* PERS.; *Káli-dakh,* HIND.; and *Angur-zietúni,* ARAB., was regarded as hot and dry. He gives the grape the following generic names: *Dákh,* PERS. and SANS.; *Innh,* ARAB.; and *Angur,* HIND. The juice of the grapes, he tells us, is *U-má-sin* in Arabic; wine *khumr* in ARAB.; *Dakh-ká-mudh,* HIND.; and *Mey* or simply *Sherab,* PERS. Wine by the Muhammadan physicians was a hot and dry cordial. Raisins

MEDICINE. Fruits 236 Raisins. 237

NITIS vinifera.	History of the Vine

MEDICINE.

without seeds, were known as *Ze-bír* in ARAB. ; *Kishmish*, in HIND. ; and *Me-wíz* in PERS. ; they were regarded as emollient and suppurative. Abdullah Shirazy adds that the vine plant was in Arabic *Kerm*.

The Muhammadan writers regarded the fruit as one of the most digestible, purifying the blood, and increasing its quantity and quality. Dymock tells us that "The ASHES of the wood are recommended as a preventative of

Ashes 238

stone in the bladder, cold swellings of the testes, and piles ; in the two last named diseases they are to be applied externally as well as given internally. The JUICE of unripe grapes, *Husrum*, ARAB.; *Ghúreh*, PERS., is

Juice. 239

used as an astringent. It is the ὀμφάκιον of Dioscorides, and the *Agresto* of the modern Italians, who still use it in affections of the throat." "The *sharbat*, or syrup of grapes, says Moodeen Sheriff, is a very pleasant and cooling drink, and proves very useful in relieving thirst and other pyrexial symptoms in many forms of fever. I have also used it with advantage in ardor-urinæ, dysuria, strangury, and some cases of bilious dyspepsia. It is one of the best and most agreeable vehicles for other medicines, particularly those used in dyspepsia, dysentery, diarrhœa and dropsical affections. From their combined actions of demulcent, expectorant and laxative, raisins are a frequent ingredient in Muhammadan prescriptions for catarrhal and febrile complaints " (*MSS. of Mat. Med. South Ind.*).

Sap. 240

"The cut branches of the vine yield in spring an abundant SAP, which was formerly used as a remedy for skin diseases, and is still a popular remedy in Europe for ophthalmia" (*Pharmacog. Ind.*).

Vinegar. 241

VINEGAR or *Angúr-ki-sirkha* is the expressed juice exposed in the air till it ferments. It is used as an acid drink in indigestion and cholic, and sometimes even in cholera. Mixed with salt it acts as an emetic (*Baden Powell*). (*Conf. with Vol. I., 72—78.*)

SPECIAL OPINIONS.—§ "'*Manakka*' and '*Kismis*' are different species. The former are regarded as cooling and laxative" (*Assistant Surgeon Shib Chundra Bhattacharji, Chanda, Central Provinces*). "Grapes are largely imported from Afghanistan. The fresh fruit is grateful and useful in allaying thirst in febrile condition " (*Brigade Surgeon, G. A. Watson, Allahabad*). "*Zirishk-i-shirin* is the term in Leh for the English equivalent, (*Zanti*) currants that are produced in Iskardo, where they are called *Basho.* See Berberis vulgaris, Vol. I., 446" (*Surgeon-Major J. E. T. Aitchison, C.I.E.*).

FOOD. Fresh Fruit. 242

Food.—It is perhaps scarcely necessary to mention in this place that the FRESH FRUIT is largely eaten in all countries where the vine grows. So very highly is this fruit valued that in countries where the vine cannot be grown in the open air, special glass-houses are constructed for its cultivation. And to so great perfection has this (what might be called artificial) production been carried, that some of the most highly prized forms of the fruit have been produced under glass. Indeed it is mainly to the growers of this class that the world looks for the yearly increasing list of new and valued forms. The manufactures of the grape are briefly WINE, BRANDY, VINEGAR, RAISINS, AND CURRANTS. These products will be found dealt with below in the special paragraphs devoted to them, at least in so far as India is concerned. The present section is, therefore, intended to preserve the logical position only for the names of the chief edible products of the vine.

WINE. 243

WINE.

VIN, *Fr.* ; WEIN, *Ger.* ; OINOS, *Gr.* ; VINO, *It.*, *Sp.* ; VINUM, *Lat.* ; VINHO, *Port.* ; WINO, WINO-GRADNŒ, *Rus.*

Vern.—*Sharáb*, *angúri*, *kishmishi*, *arak*, HIND. ; *Madyá*, *madh*, BENG.; *Sheo* (wine), *rak*, *arrak* (brandy), KANAWAR; *Sharáb*, N.-W. P.; *Sharáb*, PB. ; *Angúri*, SIND.; *Sarayam*, TAM. ; *Sarayi*, TEL. ; *Bu-*

| and of Wine. | (G. Watt.) | **VITIS vinifera.** |

angúr, MALAY.; *Isa-pyit-ya,* BURM.; *Tsiu,* CHINESE; *Draksha-rasa, madira (maddhika,* brandy), SANS.; *Inub, khamr, wain,* ARAB.; *Sharáb, mei,* PERS.

HISTORY OF THE VINE & OF WINE.

HISTORY. 244

The late **Dr. U. O. Dutt**, in his valuable work on the *Materia Medica of the Hindus,* says : " Grapes have been known in India from a very remote period and are mentioned by **Susruta** and **Charaka.** " The best known names for the fruit in Sanskrit works are *Drákshá* and *Mridviká.* In Sanskrit works a spirit distilled from grapes, *Máddhika,* is distinguished from that from sugar-cane (*Sidhu*), from rice (*Surá*), from barley (*Kohala*), from wheat (*Madhuliká*), etc., *Váruní* (translated brandy) occurs in the Institutes of Manu (Lecture XI., 147), and three kinds of spirits are distinguished in Lecture XI., 95. **Atkinson** (*Himalayan Districts*) remarks that "the vines and apricots of Kanáwar are much praised in the Puranas." Kanáwar at the present day may be said to be one of the chief Indian localities where viticulture is a recognised industry, and one upon which the people are to a large extent dependent. The cultivation of the grape in Afghanistan, Baluchistan and Kashmír can also be carried into the most ancient records of these countries, and, at the present day, the trade in the produce of the vineyards of these regions is of no mean importance. The possession of an extensive vocabulary of names, both in the classic and modern languages of India, for the plant, its fresh fruit, its dried fruit, its wine and its spirit (brandy) when viewed in the light of the historic evidences that can be produced, and when taken in conjunction with the existence of several wild species which yield edible fruits, and, which, as already stated, are even partially cultivated along side of the true vine, are facts that leave little room for doubt that India manifests a strong claim to an ancient cultivation of the vine. The rise and fall of the industry in certain parts of the country have been governed by similar influences to those which have operated in Persia, Egypt, and Europe. It is unnecessary, therefore, to furnish any further illustration of this than has already been given in connection with Kashmír. But it may be said that in addition to political agencies the appearance of disease, due very largely to over-cultivation, has also modified or even ruined the industry in certain tracts of country, or has done so during certain periods. The importance of a careful study of the forms of the vine, met with in India, and the possible light that may be thrown thereby on the origin of the industry, has already been briefly touched upon. The late **Dr. Stewart**, than whom few more careful observers could be mentioned, tells us that he was unable to isolate the cultivated forms. Of **Vitis indica, V. lanata,** and **V. vinifera,** he furnishes conjointly the Panjab vernacular names given to them and then says : " I have not distinguished between these three species " but " *that* with velvety, white or red-backed leaves (**V. lanata**) which is only found wild, appears to run into the glabrous leaved one." He further furnishes the vernacular name for the first species (**V. indica**), *vis., Angúr,* and it may be pointed out that that is the name given throughout India for the cultivated grape. Speaking more especially of the cultivated forms, **Stewart** adds that "the fruit is of either colour and the crop is very precarious, especially at Kanáwar, on the Sutlej. Grapes of varying quality are raised at most places in the Panjab plains, those of Peshawar being the best ; currants are made from a small grape in Kanáwar, as well as in Western Tibet and Yarkand. In Kanáwar a spirit prepared from the juice is compared to grape brandy by **Hoffmeister.** This spirit is, according to **Longden,** called *rak* or *arrak,* and he mentions that a wine also (*Sheo*) is made there. The circumstance that the Hindu name is applied to this and the barley spirit of Lahoul would seem to imply, that the art of distillation

Conf. with pp. 254-5.

VITIS vinifera.	History of the Vine

HISTORY.

Conf. with p. 275.

had been introduced into these countries from below. In Afghanistan Bellew states that a grape wine is prepared, which is consumed by well-to-do Mussalmans, and a raisin wine for Hindus." Dr. Henderson remarks that the vine thrives in many parts of the Panjab quite as well as in Europe, and that, in his opinion, it is indigenous to Hazara, and, possibly also, to the Salt Range. Without more direct evidence in support of that opinion, it would perhaps be safer, however, to suppose that Henderson's wild Hazara vines were in reality one or other of the allied species to which reference has already been made. Henderson, in his *Lahore to Yarkhand,* says that "grapes of excellent quality are grown extensively in Yarkand, but not for making wine. In Ladak I got large bunches of black seedless grapes, the size of the Zante currant, but usually with one or two large grapes with seeds on each bunch. They were said to come from Skardo and to be exported in the dried state to Simla and the Panjab in large quantities."

As giving still further an idea of the importance of the vine cultivation in the countries bordering on the North-Western frontier of India, the following passage may be taken from Aitchison's *Notes on the Products of Western Afghanistan and North-Eastern Persia :* "The vine, *tak ;* the fruit grapes, *angúr ;* raisins, *kishmish ;* currants or corinths, *zirishk-shírín ;* wine, *sharáb ;* spirits made from raisnis, *arak ;* vinegar, *sirka ;* syrup of grape-juice, *shíra ;* sugar made from grape-juice, *kand-i-shira-ghi.* The vine is cultivated wherever there is a garden. At Herat and Meshad large gardens contain ground laid out in vines alone ; usually these are all trained as climbers, but at Bezel I saw some gardens in which were cultivated standard vines. The fruit is very variable in quality. The grapes of Herat are considered to be the finest. In Herat and its vicinity the largest amount of raisins is preserved, and much of both wine and spirits prepared. Throughout the country generally a syrup or very thin treacle is made from the juice of the grape ; this is much eaten by the people along with their food, and is a great improvement when added to their usual coarse bread. Grapes and raisins, more particularly the latter, form a great export trade to India." In another part of his most useful little book, Dr. Aitchison says that of raisins (*kishmish*) two well marked kinds are prepared—the red, *surkh,* and the green, *sabz.* In his still earlier work (*Hand-Book of the Trade Products of Leh*) he tells us that in Yarkand grapes are called *úzúm,* a word which recalls the Russian *issum.* He further remarks that there are four chief varieties, *viz., kúk úzúm* or green grapes : *yeshil úzúm,* also a green grape : *kassil úzúm,* a red grape : and *harah úzúm,* an almost black kind. Raisins are called in Turki *kurk úzúm,* and the currants of Iskardo are *zirishk.*

In order to exemplify the extensive and diversified knowledge that exists in India and its chief frontier countries regarding the vine, a whole volume of quotations might be furnished. But the influence of that contention on the history of the Indian vines may, perhaps, have been sufficiently shown by the passages already given. While, therefore, vine cultivation exists at the present day in India—and there is abundant evidence that the plant and its products have been known for perhaps 3,000 years— there is nothing to show that during any period of this country's history did viticulture attain the proportions it assumed in the Greek and Roman ages of Europe. Still less can the idea be entertained that vine culture emanated from the country of the early Sanskrit-speaking race, although there is a probability that grapes of a kind were grown in Southern Asia, from an indigenous stock, prior to the introduction of the superior qualities from the Semitic home of the vine of present European agriculture. Prior to the Muhammadan conquests of India we possess, however,

Standard Vines.

Conf. with p. 266.

Raisins.
Conf. with pp. 267, 269, 274, 283.

no very precise information as to the cultivation of the grape. The medi-cinal uses of the vine recorded in Hindu literature are mostly concerned with raisins, and spirit (*Máddhika*), and it is, therefore, probable that both the fresh fruit and the various forms of raisins were (as at the present day) imported across the northern frontier. The *Soma* of the Vedas was doubtless an alcoholic liquor, but it seems highly probable that it took its name from the bitter principle (or hops so to speak) much as the name "beer" was taken from the grain employed. There is little indeed that would go to show that the *Soma* was the juice either of the grape or of any other plant. The bulk of the evidence that has as yet been addu-ced points to a knowledge of fermentation, and possibly also of distillation which it was politic to withhold from the general public. The secrecy observed procured the sacred significance which it was impossible to retain when the knowledge was no longer confined to the priesthood. At one time the writer thought that certain passages descriptive of the *Soma* plant, which likened it to the joints of the human finger, might have reference to the long narrow green grapes of Afghanistan and Central Asia, but more direct evidence has convinced him that the *Soma* plant was in reality a flavouring or auxiliary ingredient but did not of itself afford the beverage. He has therefore, though somewhat reluctantly, rejected the very pleasing conception held by many writers that the *Soma* of the early Vedas was one of the most ancient allusions to wine. (*Conf. with Vol. III., pp. 246-251.*)

The more direct historic facts regarding Indian vines must, however, be traced from the sixteenth century only, when Akbar fostered the industry, and when his successors shortly after did all in their power to destroy the good he had achieved. During the past century or so, little real progress has been made, though the Journals of the Agri-Horticultural Society of India and other publications have recorded numerous experiments both in Kashmír and on the plains of India which have been undertaken with the object of trying to restore to India her lost position.

Before proceeding to furnish a few of the more important papers which have appeared on viticulture in India, the present chapter may fittingly be concluded by a few passages from **Victor Hehn's** great work on the migration of the vine from Asia to Europe. The quotations which follow will be found to not only tell the story of the various vicissitudes of the industry, but will be seen to mark the birth-places of the chief varieties of the vine, and of the systems of cultivation :—

"That wine reached the Greeks through the Semites, we learn from the identity of name (Hebr. *yain*, Ethiop. and Arab. *wain.* Gr. *voinos*, Lat. *vinum*). The course taken by civilization makes it extremely improbable that the Semites should have borrowed the word from the Aryans; that is, from the Græco-Italians, for the Iranians have it not. Attempts to show from Sanskrit that wine was an original possession of the unseparated Aryan races have fallen through, and in the eyes of the unprejudiced only prove the contrary. The true home of the vine, which is the luxuriant country south of the Caspian Sea, was also, as far as can be historically determined, close to the cradle of the Semitic race, or of one of its chief branches. There, in the woods the vine, thick as a man's arm, still climbs into the loftiest trees, hanging in wreaths from summit to summit, and temptingly displaying its heavy bunches of grapes. There, or in Colchis on the Phasis, in the countries lying between the Caucasus, Ararat, and Taurus, the primitive methods of cultivation, we read of in the works of the Greeks and Romans, are still practised; for instance, the dividing of vineyards by cross-paths running from north to south (*cardo*) and east to west (*limes decumanu*); the pitch-ing or chalking of the amphoral, the burying of them in the ground, etc. There grows the spicy, orange yellow wines of penetrating odour, and the precious Cachetian grape yields a juice so intensely dark-red that ladies write their letters with it. From those regions the vine accompanied the teeming race of Shem to the lower Euphrates in the south-east, and to the deserts and paradises of the south-west, where we afterwards find them settled, and developing the peculiar civilization which succeeded the Egyptian and long preceded the Aryan. To the Semites, then, who ever invented

VITIS vinifera.	History of the Vine.

HISTROY.

the distillation of alcohol, who accomplished the gigantic abstractions of monotheism, measurement, money, and alphabetic writing (a kind of mental distillation, on the threshold of which the Egyptians had halted), belongs also the dubious fame of having arrested the juice of the grape at the stage of fermentation, and so produced an exhilarating or stupefying beverage. From Syria the cultivation of the vine spread to the Lydians, Phrygians, Mysians, and other Iranian or half Iranian nations which had, in the meantime, moved up from the east. Thus it entered the Greek peninsula from the north, while at the same time Phœnician commerce, Carian colonies, and also old Greek communities that had crossed from Europe to Asia, brought the wonderful invention, and in time the plant itself, direct by sea. At the time of the Homeric Epos and Hesiod's poems, its introduction had long been accomplished and forgotten, the existence of the vine and of wine was taken for granted, and attributed, like all the blessings of life, to the instructions or the creating hand of a deity.

"The earliest voyages of the Greeks to the west must have introduced the intoxicating beverage to the Italian coast, for that wine came to Italy from Greece is proved by the word *vinum*, its neuter form being accounted for by imitation of the accusative *voinon*. The Greek sailors found a simple shepherd folk, on whom the foreign wine had the same stupefying effect as on the Centaurs mentioned by Pindar : 'When the Pheres became acquainted with the man-subduing power of sweet wine, they hastily pushed the white milk from the tables, drank out of silver horns, and wandered helplessly about.' That in Latium milk was older than wine is proved by ordinances attributed to Romulus, according to which white milk, and not wine, was to be poured out to the gods; and Numa decreed that wine should not be sprinkled on the bier, which shews that wine was not yet in use at the oldest funeral ceremonies. For there was a time when the Romans only practised agriculture, and the cultivation of the vine had not yet been introduced. It is remarkable that in that country, as in Greece, legends of battles between the nations are connected with the introduction of the vine. A much noted legend relates that Mezentius, King of Cære, demanded of the Latins the first-fruits of their vineyards, or the first wine from the press, but that they vowed these things to Jupiter, and so won the victory over the wicked tyrant. The rule of the Etruscans in Campania and Latium was probably broken by an alliance of the Greeks and Latins; and the faint remembrance of this victory got mixed up with that of the introduction of Greek viticulture and that of the institution of first fruits to Jupiter Liber and Venus Libera. The 19th of August, the day of the foundation of two temples to Murcia and Libitina, goddesses of the harvest, now also became the day of the *vinalia rustica*, which was preceded on the 23rd of April by the *vinalia priora*, both feasts connecting the younger cultivation of the vine with the older cultivation of the field. It was natural that Jupiter should be the patron of the new gift, and that his priest, the Flemen Dialis, should consecrate the harvest; for all fecundity and fruits of the earth were attributed to that god. The surname Liber, which makes *him* the god of wine, is a translation from Greek; the Greek genealogy, which makes Bacchus a *son* of Jupiter, had not taken root in Italy. The vine soon grew so luxuriantly on the mountains of South Italy, that already in the fifth century b. c. Sophocles calls Italy the favourite land of Bacchus, and Herodotus calls the southern point of Italy the land of vine-poles.

Œnotria.—Œnotria was the land where the vine was trained on poles, in contrast with Etruria and Campania, for instance, where it twined round trees; or Massilia and Spain, where it was cut short and left without support; with Brundusium, where it spread roof-like over trellis work or cords; or Asia Minor, where it crept upon the ground. These different methods of training resulted partly from the nature of the soil, which was either rocky and hot or damp and rich in *humus*, partly from the want of sufficient wood or cane, and partly from the habits of those by whom the cultivation had been introduced, and the kind of grape they brought with them. The abundance of timber in the country afterwards called Lucania and Bruttium—also called Italia from the cattle breeding connected with those woods—may have led to the general use of proper vine poles, and the name Œnotria may have been given by those Greeks who were accustomed to train the vine freely on the ground or on trees. In the districts at the mouth of the Po the vinestock must have been introduced very early by Greek maritime commerce, although the low and damp ground seemed little favourable to its cultivation. Even Strabo was surprised at the co-existence of marshes and flourishing vines. The vine grew well near Ravenna, bearing the heat and rains; nourishing itself on the mists, and yielding abundance of wine, and the same is remarked of other northern grapes. Wine in Ravenna was cheaper than water, so that the poet Martial says he would rather possess a tank of water than a vineyard, and complains that a cheat of a publican has sold him pure wine, instead of wine and water. Vicenum, where the geographical names and other things indicate that it was

Standard and climbing vines *Conf. with p. 264.*

| and of Wine. | (G. Watt.) | VITIS vinifera. |

anciently connected with the mouth of the Po, is very early described as rich in wine. We read in **Polybius** that **Hannibal** cured the horses of his army with the old wine of that country, of which there was an abundance; and long afterwards the wines of Picenum were still exported to Gaul and the East. There grew the celebrated Præ-tatian grapes, resembling the Istrian, and were identified by **Pliny** with the Pucinian grapes that grew on the river Timavus near Aquileia. The Picene vine had therefore been propagated from the old Greek times along the west coast of the Adriatic to the very head of that gulf. **Polybius**, who speaks as an eye witness, praises the wine of the extensive and fertile plains that stretched from the Po to the foot of the Alps. Most likely the vine grew there already when the Celts invaded Italy, tempted by its southern wines and fruits. **Martial** speaks of the vine-covered slopes of the volcanic Euganean hills near Padua. The Khœtian wines, that is, the wines of what is now Tyrol and Valtelin, were anciently celebrated. They really owe their immortality to **Virgil**, who considered them only second to Falernian wine; but perhaps he eulogiz-ed the Rhœtian wine because **Augustus** particularly liked it. **Strabo** joined in the song of praise, most likely echoing **Virgil**. The district of Verona, too, was celebrated long after for its wines."

"**Cato** was of opinion that of all kinds of culture that of the grape was the most profitable; and during the last years of the Roman Republic, Italy had become such a wine country that the relation between wine and corn was reversed; wine was exported and corn imported. But the cultivation of the vine had also long since begun to cross the borders of Italy and make itself at home in the North and West." **Hehn** then shows the progress into France with Cæsar's conquest over the greater part of Europe. "At that time, he says, Gaul stood in relation to Italy, as Italy had stood in primitive times to Greece, and Greece before that to Syria, Phrygia, and Lydia. Gallic wines pleased the Italian palate. Burgandy was drunk, though not under that name." "Gallic varieties of grape, which had been produced by transplantation to new soils, were transmitted to Italy and propagated there. The virtues ascribed to these Gallic vines entirely consist in greater resistance to an unfavourable climate, productiveness even on poor soil, and endurance of cold, rain, and wind; they all bear abundance of fruit, and yield a large quantity of *must;* they easily degenerate when removed to another soil, and have therefore no stable character; the grape called *helvennaca* does not thrive in Italy, but remains small and easily decays; the aroma of the Allobrogian wine is rapidly lost and so on." "It was in the natural course of things that during the Empire the culture of the vine should not only become permanent in Gaul, but be extended to the valleys of the Garonne, the Marne, and the Moselle, though it did not as yet cross the Rhine. But, if not the vine, yet wine itself soon became known to the neighbouring Germans, who by their acceptance of this product concluded the fateful compact with Gallic Roman civilization." "The cultivation of the vine in the Roman provinces threatened to choke the cultivation of grain to such a degree that the Emperor **Domitian** in an excess of anxiety, ordered that half and more than half of all the vineyards outside Italy should be destroyed—which order, naturally, could not be carried out. Prohibi-tion of the Oriental custom of castration being issued about the same time, **Apollo-nius** said that the Emperor spared men but eunuchized the earth."

"If we compare the present condition of viticulture with what it was in ancient times, we find that it has in some degree followed the general course of history; that is, it has declined in the countries of its origin, and stands at the highest point of de-velopment in those countries where it was introduced the latest. When Western Asia, the cradle of the vine, was overwhelmed by nations of the faith of Islam, it was natural that a product, the enjoyment of which was forbidden to the conquerors, should no longer flourish. In all countries that came under Arab government—in North Africa, Sicily, and Spain—the cultivation of the vine declined." "Modern Greece—after so many fatalities, after centuries of ethnologic and economic degrada-tion—produces, with few exceptions, very bad wine—the fame of her Chian, Lesbian, and Thasian has long since evaporated." "Perhaps *currants* are only a variety of grape produced by degeneration. They are said to have come from the Isle of Naxos and to have been unknown in the Morea before 1600. It is remarkable that they wander, as it were, from place to place. They have disappeared from Naxos, exist no longer at Cornith whence they received their name, and are now only found in Patras, Zante, and Cephalonia."

But after the fall of the Roman Empire Italian viticulture rapidly declined. The nobles and princes gradually came to care more for quantity than quality. It was the taste and wealth of the Roman nobles that had nurtured and developed the infinite

Currants.
Conf. with pp.
264, 269, 274,
283.

VITIS vinifera.	History of the Vine and of Wine.

HISTORY.

Vine Louse.
Conf. with
p. 290.

**Wines of
Palestine.**

**Grape
fungus.**
Conf. with
p. 291.

variety and quality of the wines. Thus Horace made known the qualities of the wines of Latium and Campania under the names of Falernian, Mussic and Cœcuban, but Pliny, two generations later, says that they were no longer valued. "Manufacturers of wine in Greece and Italy are now reproached with exactly the same thing." "It is universally admitted that in modern times the palm in the production of wine is due to Central and Southern France. Whilst Italy almost entirely consumes the thirty million hectolitres of her yearly produce, and has therefore little to spare for export, France, on the contrary, till the vine-louse began its ravages, produced double the quantity at a money value of about 2,000 to 3,000 million francs, became the chief exporting country, supplying all parts of the world with the first wines as well as common table wines." "It is a remarkable fact that vines now produce the best wines in places close to the northern limit of their extension, where the plant was only gradually and with difficulty, and last of all acclimatised; wines now famed over the world under the names of Burgandy, Johannisberg, etc. Here, of course, culture and technical skill have done their utmost; and who knows what they might not accomplish if adopted in the original homes of the vine? In this connection, a fact that meets our eyes in the first two or three centuries of the Middle Ages deserves serious attention. At that time, we find the western world thought of the wines of Palestine as the *strongest* and *finest*, just as we now quote the ports and sherries of the Pyrenean peninsula; and the wine of the Phœnician-Philistine coast was greatly valued at the Byzantine court. It was the Arab invasion that put an end to its production and the commerce founded upon it." After pointing to the results of the modern phases of the expansion of vine culture Hehn adds :—' Wine, we might say, loves not the west, and clings to the neighbourhood of its old home." "At two points only and quite at the end of the Middle Ages has the hand of man really extended the region of the vine, namely, in Madeira and the Canaries—which may in a sense be said to belong still to Europe and the Mediterranean. Prince Henry, the voyager, introduced shoots of the vine from the Peloponnesas and Crete into Madeira; and Alonza de Lungo transplanted vines from Madeira to Teneriffe about the year 1507. The wine yielded there by Grecian grapes became celebrated all over the world, but lately the grape-fungus has destroyed this culture, and it is being revived with difficulty. But the cultivation of the vine in those island is also interesting, because there it comes nearest to the climate of the tropics; the vineyards even of Southern Persia and of the Cape are further away from the equator than the Isle of Ferro at 27°48′ latitude."

The distribution of the vine throughout Europe is thus more or less a matter of historic record, although, as DeCandolle points out, the discovery of vine seeds in the lake-dwellings of Castione, Wangen and Switzerland carry a knowledge in the fruit into prehistoric times, a fact confirmed by the existence of vine leaves in the tufa of Montpellier and of Meyrargue,—deposits certainly older than the histories of man, though later than the tertiary epoch of geologists. "The dissemination by birds must have therefore begun very early, as soon as the fruit existed, before civilization, before the migration of the most ancient Asiatic peoples, perhaps before the existence of man in Europe or even in Asia. Nevertheless, the frequency of cultivation, and multitude of forms of the cultivated grape, may have extended naturalization and introduced among wild vines, varieties which originated in cultivation." "An absolute primitive habitation is more or less mythical, but habitations successively extended or restricted are in accordance with the nature of things. They constitute areas more or less ancient and real, provided that the species has maintained itself wild, without the constant addition of fresh seed" (*DeCandolle, Orig Cult. Pl.*). To some such influences must be attributed the existence of the vine on the temperate slopes of the Himálaya if indeed we are not forced, through future discoveries, to believe in Vitis vinifera having been a native of these mountainous tracts as well as of Western Asia, long anterior to its being cultivated. Indeed, there remains but one other possibility (already suggested), *viz.*, of India having possessed, in ancient times, grapes developed from indigenous species, prior to its having obtained the true vine. (*Conf. with p. 260.*) Such grapes might

| Vine Culture in India. | (*G. Watt.*) | **VITIS vinifera.** |

readily be supposed to have ceased to be cultivated on the introduction of the better qualities from the trans-Himálayan regions. That the earliest classic literature of India necessitates our acceptance of the vine or of a vine as having been known to the Sanskrit authors, is a matter upon which there can be little room for doubt. But we have to pass over a gap of many centuries before we obtain unmistakable evidence of its cultivation. There is, however, no ancient record of its introduction, and the classic and many of the vernacular names for it and for the preparations from it, seem purely spontaneous and bear little or no relation to the Semitic and European synonyms. **Victor Hehn** (in the passage quoted above) has very ably exemplified the effect of the conquests of Islam on the vine culture of the Mediterranean area, and it need scarcely be said that that power also dominated the arts and industries of the ancient peoples of this country for fully a thousand years. Is it matter for surprise therefore that the possibilities of India in the production of grapes and wine are still unsolved. So far as the experiments have gone, it may indeed be said they are hopeful in the highest degree. But that viticulture should have survived in Kanáwar (in the North-West Himálaya) in spite of the great disadvantages under which it has laboured, would seem to show that India need not confine her expectations of a future production of wine to the vale of Kashmír, nor regard increased facilities of trade with Afghánistán as essential to an expansion of her supplies of grapes, raisins and currants. The somewhat curious fact that the vineyards of Kanáwar (even to the present day) find an outlet for their raisins and currants in Tibet, illustrates no doubt the conservative nature of trade, but it confirms also the idea of the Muhammadan persecution of viticulture that for many centuries prevailed in the East. The simple cultivators of the higher fertile valleys of the Himálaya had either to abandon vine growing or had to seek seclusion rather than to court publicity for their industry. Is it to be wondered at, therefore, that few persons have thought of India as possessing vineyards or have heard even of the fact that for centuries she has produced within her own territory wine of so excellent a quality that it needs but the skill and experience of the West to place it among the commercial products of this vast empire.

HISTORY.

Conf. with pp. 255, 274.

Currants.
Conf. with p. 283 *et seq.*

VITICULTURE.

It is neither within the scope of this article to discuss the methods of vine culture practised in Europe, nor to deal with the forms of the plant so cultivated. Persons interested in these subjects will have no difficulty in procuring an extensive series of technical works. What more especially concerns the writer is to furnish such information as can be procured on the present vine industry of India, incomplete and fragmentary though the literature of that subject most undoubtedly is. The imperfections of the systems pursued and the unsatisfactory position of the industry, when fully made known, may, in the writer's opinion, very likely act as greater incentives to reform and progression than would an essay on Viticulture in general. This object cannot perhaps be better accomplished than by furnishing, province by province, a few passages from local works and official correspondence, illustrative of the degree of knowledge that prevails, and of the extent of the experiments that have been made towards improvement. It will be seen that well on to half a century ago, the subject of vine growing on the plains of India, to some extent occupied public attention, but that since then it has been allowed to drop into almost complete obscurity, and, but for the revival of interest (within the past few years) that has been taken in wine production in Kashmír and the raisin trade of Kanáwar, it might almost be said that we know little more at present

VITICULTURE.
245

Kashmír Wine.
Conf. with pp. 261, 296.

VITIS vinifera.	**Progress and development.**

VITICULTURE

than that some localities have the reputation of growing fairly good grapes, while in others that the humidity is said to prove too great for the successful rearing of the plant. The systematic study of the industry and the comparison of the results obtained, district by district by a qualified person, would doubtless remove much of the ignorance that prevails. At present, while we know that grapes of one quality are found here and of another quality there, we do not know for certain whether these differences proceed from climatic peculiarities, from the nature of the vines grown, or from the systems of cultivation pursued. We read of disease having once upon a time partially or totally ruined the industry in one locality and of devastation threatening another now; but in both cases we are alike unable to tell for certain whether it was or is due to fungoid or to insect *Conf. with* depredations. We are accordingly hopelessly unable to recommend cura- *pp. 274 289.* tive or protective measures. Indeed, all that can be written amounts to this, that in few branches of Indian agriculture is the need for investigation more imperatively demanded. It is not enough to know that vines are grown in India or can be grown. False hopes may have been even raised by what has been said in this article. We require to have it proved whether or not there is in the viticulture of India the inherent property necessary for commercial success.

PANJAB.
246

Panjab.

According to Mr. Baden Powell, "There are several varieties of grapes recognized in the Panjáb. The first is '*Kándaharí*,' being a purple grape. 2nd, the '*Kishmishi*,' small seedless grape (producing what are called in England 'Sultána Raisins'), these are of the varieties called '*Sahíbí surkh*', and '*Sahíbí ablak*'; the Khatan grapes produce the large common **Indigenous** raisins, called '*Munakka*'. 3rd '*Gholab dan*,' a white grape. 4th '*Husaini*,' **Grapes.** these are the grapes that come to Lahore from Kabul, in round boxes *Conf. with* packed in cotton wool. 5th, '*Sahíbí*,' a superior grape (white). 6th, '*Fakhri*,' *pp. 255, 259-* sometimes called '*askari*,' a black grape. 7th '*Munakka*' and '*abjosh* *60, 263-4, 271* *munakka*,' are grapes dried in the sun; to make *abjosh* the grapes are *274, 275, 278,* plunged into boiling water, and then dried in the shade. 8th, '*Rish bábá*.' *279, 282, 284,* 9th, '*Didah-i-gau*,' a white grape, with some spots on the skin, which are *285-6.* said to resemble a cow's eye, hence its name; pious Hindus refuse to eat this grape on this account. 10th, '*Karghani*' (white) called from the name of a place. 11th, '*Angúr Jálálábádi*,' called also '*Khátta angúr*,' grown at Chárbagh, a few miles from Jálálábád And 12th, '*Chárangur*,' grown also at Jálálábád." Mr. Baden Powell adds that the common grapes are known as *rocha-i-surkh* and *rocha-i-safed*, also *toran*. The white grapes sold in the cold season are the *hosaini* or *shaikh khalli*, and yet another the *akta* grape, which produces bloom raisins, called *dagh*, or more properly *kishímísh-i-daghi* or *abjosh*, which are prepared by dipping the ripe bunches of fruit into a boiling solution of quicklime and potash (hence called *abjosh* or infused in water) before drying in the shade.

Dr. Henderson, whose opinion has been in part already quoted, says "that in many parts of the Panjáb, the vine thrives quite as well as in Europe." He adds, however, that its only fault is its tendency to grow to wood and leaves. "This tendency might probably be counteracted by proper cultivation and by choosing a poor rocky soil, and selecting suitable varieties of vine. There seems to be no good reason why, if the best vines are obtained, good wine should not be produced in many parts of the Panjáb, particularly in the hills on stony ground, where little else will grow. I am not aware that any attempt that has been made, on a large scale, to grow grapes in the Panjáb for the purpose of making wine. In the plains, the grape ripens at a season when the heat is prob-

of Vine Culture in the Panjab. (*G. Watt.*)	**VITIS vinifera.**

ably too great to allow the juice to ferment properly without turning acid, but in the hills this does not hold good, the difficulty there will be to get either a climate where there is little or no rain, or to get the fruit to ripen before the rains set in. In an old number of the Agri.-Horticultural Society's Journal, to which I cannot at present refer, I recollect having seen a notice of a vine found in the south of India which ripens its fruit much earlier than the common vine of the country." The selection of forms, found to ripen their fruit more quickly or it may be later than others, has largely governed the production of some of the better known grapes of Europe, and doubtless some such process would result in much advantage to India. It is probable, however, that the translation of a plant from South India to the Panjáb would be attended with very nearly as serious departures from its recognised properties, as would occur in a vine taken from Italy or even France to the Panjáb, or of one sent from the Panjáb to Italy. Selection, to have permanent results, must be more restricted, at least in the first instance, than would be understood by Dr. Henderson's allusion to Madras grapes. There can be no doubt, however, that the Panjáb possesses excellent stock to allow of a start being made in improvement of quality by natural and scientific methods.

It is perhaps unnecessary to give the few scattered passages that occur in the Settlement Reports and Gazetteers on the subject of the Panjáb grapes. Those of Peshawar are said to be of excellent quality. The wild grapes of this district are spoken of as *kwar* in Pushtu. In Házara four or five kinds are grown ; the inferior are called *kali dakh* and *Jogan*, and the superior *bedana* (seedless), and *munnaka*. They are specially mentioned also in connection with Kohat, Guzerat, and Simla. But it may be said the main interest in the grapes of Upper India centres in the Native State of Kashmír.

KASHMIR.—Without calling in the aid of tradition or even that of ancient historic records, we possess information of two if not of three distinct efforts that have been made to place this Native State in the position of a wine-producing country. In the sixteenth century when the Emperor Akbar held his Court not infrequently in Kashmír, wine production would seem to have been a very general industry. Akbar took a personal interest in agricultural and horticultural improvements. Having learned from the foreign visitors to his Court of superior grains, fruits or flowers, he was in the habit of making efforts to procure these. As the result, many of the exotic plants now all but universally grown were originally brought to India by him. His historian speaks of the viticulture of Kashmír and Kábul in such terms as to lead us to suppose that he took as direct an interest in it as his successors are known to have persecuted the industry. " His Majesty," we read, looks upon fruits as one of the greatest gifts of the Creator, and he pays much attention to them." " Melons and grapes have become very plentiful and excellent." " Various kinds of grapes are to be had from *Khurdád* (May) to Amurdád (July), whilst the markets are stocked with Kashmír grapes during Shahriwar. Eight seers of grapes sell in Kashmír to one *dam*, and the cost of transport is R2 a maund. The Kahsmírians bring them on their backs in conical baskets which look very curious. From *Mihr* (September) till *Urdíbihisht* grapes come from Kábul." " Whenever His Majesty drinks wine, etc." " Grapes though in abundance in Kashmír are but of few kinds and those indifferent. In general, they let the vines twist round the trunks of the mulberry trees." " The inhabitants chiefly live upon rice, fish, and dried fruits and vegetables ; and they drink wine." These and other such like passages occur frequently, and in Blochmann's translation we read of the numerous " grandees " at the Court of Akbar who were addicted to the use of wine

VITICULTURE in the Panjab.

Conf. with p. 260.

Indigenous Grapes. *Conf. with p. 270.*

KASHMIR. 247

VITIS vinifera.	Progress and Development

VITICULTURE in the Panjab.

Indigenous Grapes of Kashmír.

(*See pp. 323, 339, 345, 363, 364, 378, 407, 412, 426, 447, 464, 468, 470 485, 492, 518, 545, etc.*). In Gladwin's Edition even more instructive particulars regarding the grapes of Kashmír are given, which indicate an almost universal cultivation. "The revenue," we are for example, told, "was collected in grapes either by measurement, or by agreeing to pay a certain sum. For measurement some experienced persons estimate the produce of the vineyards and the State exacts four *baberies* for every *kherwar*. In the reign of **Hemayun** the rate was two *baberies* and four *tunqhas*. A *babery* is one *miskal;* two and a half *baberies* are equal to one rupee. Besides these three (barley, wheat, and grapes) which are called *safeidtery*, there are taken upon the following articles, which are styled *subting* (or greens) seven and a half *baberies* for every *jereeb, viz.*, rice, melons, cucumbers, onions, turnips, carrots, poppies, and lettuce." It will thus be seen that the grape and the wine of Kashmír 300 years ago held a much more important place in Kashmír than at the present day.

Through the fostering care of the **Emperor Akbar** superior vines were also introduced into Lahore, Delhi, Agra, Allahabad, etc. During the reign of the **Emperor Jahangir** the grapes of Kashmír were improved, but very shortly after the decline set in which has been alluded to. Some idea may, however, be given of the position of the industry prior to the recent revival that has been undertaken. **Mr. Moorcroft** wrote in 1823 of the grapes he found in the State as follows :—Vines are of many varieties, both of exotic and indigenous origin ; of the former are the *Moskha, Sahibi Huseni* and *Kishmishi,* which last was introduced by the **Emperor Jahangir** from *Kábul*. The latter, or those indigenous and cultivated, are, *Pamuthil, Takrí, Upamahí, Bará kawar, Nika kawar, Kachibúr kanahepí, Harduch* and *Kathu husneni*. The wild grapes are *Deza, Kuwaduch,* and *Umburbarí*. The four first are good, but it is said that those of similar name in *Kábul* are still better. The skirts of the southern face of the northern hills were formerly largely clothed with vines, and under Hindu rule much wine was made. The practice was continued to, or revived in, the reign of **Jahangir**. A little brandy is occasionally distilled, even now, and under suitable management might vie with cogniac. Speaking of brandy, it may be added that according to **Mr. Baden Powell**, some was formerly made in Lahore of which **Ranjit Singh** drank, often too immoderately. There is thus no doubt that the art of wine-making was known in Kashmír (and even in India) three hundred years ago, and that it was even an article of every day use with the people of that State during the time when **Queen Elizabeth** ruled the destinies of England. Carrying the history of this subject, therefore, down to the present time we learn that whatever progress had been made in the sixteenth century was all but completely effaced by the reign of Muhammadan bigotry and degeneration, that ultimately culminated in the overthrow of the Emperors of Delhi. The subject next assumes interest in 1876, when **M. H. Dauvergne** made wine from the indigenous grapes and reported his success to the **Maharajah Ranbir Singh**. Another gentleman was then in Kashmír, **M. Ermiens** (a Belgian traveller), who had had some previous experience in vine growing, and to him was entrusted the institution of experiments on a large scale. Plant to the value of R80,000 was imported or constructed locally, and in 1880 the first vineyard came into bearing. It was now, however, ascertained that **M. Ermiens** had in reality no knowledge of wine-making. The Maharajah accordingly procured the services of two other foreigners, **M. Peychaud** of Chateau Margaux, as wine-maker, and of **M. Bonley** of Paris as distiller. The management of the enterprise was entirely left in the hands of these French experts until 1882, when the Native official in charge of the Agri-

Indigenous Grapes of Kashmír.
Conf. with pp. 255, 259-60, 263-4, 270-1, 274, 275, 278, 279, 282, 284, 285.

of Vine Culture in the N.-W. P. & Oudh.   (*G. Watt.*)

cultural Department of the State succeeded in obtaining the transference
of the vineyards to his supervision. But the Maharajah, for some time
after, continued to take an active personal interest in the enterprise.
Three new vineyards were planted, making a total of five, *viz.*, at Teed,
Mashri Bagh, and Neshat Bagh. A sixth, seventh, and eighth soon followed, until, as reported by **Dewan Lakhpat Ray**, there were in 1885 352,525
plants growing in the various vineyards (*Jour. Agri.-Hort. Soc. Ind.,
III., 264-267*). The heavy expenditure, with no return, began to tell, however, and apathy in time supervened. The industry in consequence soon
fell into the hands of Native agencies and was then threatened with complete extinction. At this stage serious changes, however, took place in the
State. The Maharajah sought and obtained the assistance of the Indian
Government in the administration of his country. The two best vines (introduced by the Maharajah)—that from Sauterne and from Medoc—had
continued to flourish, and, according to numerous reports, had yielded quite
as good fruit as in their original homes. The Indian Government, therefore, regarded the experiment, begun so admirably and which had attained
so great a degree of success, as worthy of being continued. Accordingly
a requisition was sent to Her Majesty's Secretary of State for the services
of two Italian experts, to strengthen the staff still available in the State.
This resulted in the appointment (in 1890) of **Messrs. Bassi and Benve-
nuti**, who in that year took over charge of the duties assigned to them.
It may thus be said that under the direct supervision of the Council of the
Kashmír State, a fresh effort has been made, and one in which the whole of
India may be viewed as deeply interested. The French experts, employed
in the early phase of this experiment, have spoken in such high terms of
the climate and soil of Kashmír, that it may safely be regarded that now
that the danger of starvation through Court intrigues has been removed,
the experiment may be looked upon as having obtained the opportunity
desired, namely, of ascertaining whether wine production in Kashmír can
or cannot be made a commercial success. If it can, the value of that State
may not only be incalculably increased but it is possible that in Kullu
and other parts of the North-West Himálaya, where a climate prevails
similar to that of Kashmír, vineyards may soon blossom forth and a new
industry take its birth, that may in the future astonish the world as completely as has that of tea. India can never very likely hope to take the
place of France nor even of Italy as a wine-producing country, but if it
succeeds to produce a wine that finds a ready sale, it will accomplish all
that is at present aimed at, and in so doing it will give another shock to the
European preconceived notion of this country as a land of tropical swamps,
tiger infested jungles, and rice fields.

It need scarcely be added that red and white wines, of a very superior
quality, have already been produced in Kashmír. Its white wine obtained a
gold medal at the Calcutta International Exhibition, and the samples of
both wines shown at the Colonial and Indian Exhibition were highly commended by many persons well qualified to give an opinion.

North-West Provinces and Oudh.

Although vines are cultivated here and there throughout these provinces, very little has been written of a definite nature regarding them. It
has already been said that **Abul Fuzl** speaks of Akbar having given to
Agra, Allahabad, etc., grapes of a superior quality So far as the writer
can discover, the grapes grown in these localities at the present day are
not much different, if different at all, from those to be found in other parts
of these provinces. Atkinson, in his *Economic Products*, says: "Throughout the plains the vine fruits well in every district, those raised at Agra,

18

**VITIS
vinifera.**

**VITICULTURE
in the
N. W. P. &
Oudh.**

Allahabad, Benares, Cawnpore, and Lucknow being well known for their excellence." But these grapes are quite unsuited (even were the climate favourable) for wine manufacture. It may at once, therefore, be admitted that the most sanguine enthusiast is not likely to be found ready to urge the claims of the Gangetic plains as a hopeful field for extended viticulture. Indeed, it might almost be said that, except in gardens where special care is bestowed upon them, the vines grown in these provinces cannot be made to fruit very freely. On the hills it is otherwise, and the local interest in this subject may, therefore, be said to centre mostly in that portion of the Himálaya placed under the Government of these provinces. The conditions that prevail in Kumaon, Kanáwar, etc., are, however, so similar to those of portions of the neighbouring district of Simla (in the Panjáb) that the information available may be treated of conjointly in this place. As remarked by Atkinson, vine culture has been pursued in Kanáwar since the early classic periods. **Mr. Coldstream** (*Deputy Commissioner, Simla*) in a letter, dated 24th January 1890, recommends that a scientist be deputed to Kanáwar to study the vine disease "which has for the last thirty years been very destructive to the vineyards of that district, so much so indeed that this industry once extensive is now almost a thing of the past." This subject has accordingly been under the consideration of the Government of India, and it may not be out of place to furnish here one of the earliest (of what may be called the modern) reports on the Kanáwar vine industry. **Dr. Cleghorn,** in a paper on *Notes on the Vegetation of the Sutlej Valley* (published *Journ. Agri.-Hort. Soc. Ind., XIII., p. 382*), wrote in 1863 as follows :—

**Kanawar.
249**

" In Upper Kunáwar the vine is extensively cultivated and ripens its crop at an elevation of from 6,000 to 9,000 feet. The first plants are seen at Nachar, but the climate there is not suitable; beyond the Miru ridge which intercepts the heavy clouds, the smaller amount of rain favours the ripening of the grapes. The vineyards occupy sheltered situations, generally on the steep slope facing the river. The vines are supported on poles three or four feet from the ground, connected by horizontal ones. The fruit hangs below the shade of the leaves, never exposed to the sun. A considerable portion of the crop is dried on the house top and stored as raisins for winter use, but without care, and many grapes are spoiled in the process. For several years the crop has been deficient, the grapes dropping off before they were ripe from unseasonable falls of rain and snow. This year, 1864, the rainfall was moderate, but the *Oidion* or vine disease appeared in the valley and destroyed many vineyards. The fresh fruit is exported to Simla for sale in *kiltas* or large hill baskets, and the small seedless grapes dried are also sold there as 'fine Zante currants.' At Akpa and Poari the price of fresh grapes is about one rupee for a *kilta*-full. ' Sangam is the highest point in the valley where the vine thrives.'"

**Manufacture
of
Raisins.**
*Conf. with pp.
283 et seq.*

**Ind'genous
Grapes.**
*Conf. with p.
270.*

In Kanáwar, says **Mr. Atkinson,** the fruit is called *dakhang* and the plant *lánang*, and there the vine is extensively cultivated as a field crop and ripens its fruits at an elevation of from 6,000 to 9,000 feet. He adds, a spirit called *rak* or *ark* is prepared from the juice and a wine called *sheo*. **Sir D Brandis** (*For. Fl , 98*) says : " In India, extensive vineyards were formerly found in Kanáwar, from Jani to Sangnam (between 5,500 and 9,000 feet), and in some of the other inner and drier valleys of the North-West Himálaya. But the vine disease broke out in these secluded valleys (between 1855 and 1860), and since then the cultivation has greatly diminished." It will be seen that **Dr. Cleghorn** (in the passage above) affirms that the disease in question was **Oidium,** and therefore not the incurable insect pest **Phylloxera.** This is a point that urgently calls for investigation, but it may be pointed out that whatever be the disease that is at present destroying the Kanáwar vineyards, it was not likely to have been **Phylloxera** in 1855, since the louse only appeared in Europe in 1866, and in America in 1854. Some few years ago, when it was proposed to introduce the better qualities of the vines of France into Kashmír,

Vine Culture in N.-W. P. & Oudh.	(*G. Watt.*)	VITIS vinifera.

Professor Maxime Cornu urged that rooted plants should certainly not be taken unless the insect-root-disease (**Phylloxera**) actually existed already in Kashmír. This point was accordingly referred to the Kashmír authorities and the reply came in the affirmative—the disease was already in the State. This definite statement must, therefore, be accepted, but the writer thinks it necessary that he should add that so far as the reports and correspondence which have been published are concerned, satisfactory proof has not as yet been given of the accuracy of the affirmation that **Phylloxera** is actually in Kashmír. It may be there now or may appear at any moment, for its progress throughout the area of viticulture has been almost unprecedented. Until we know for certain of its existence, however, by the publication of precise scientific reports, it may be affirmed that we possess at present only the vaguest generalizations regarding the vine diseases of India. That being so the extensive importation of the American vines, or indeed of any rooted vines, would seem highly impolitic and even unnecessary. To realize the full force of this remark, the reader has only to recall what has already been said under **Vitis lanata** and **V. parvifolia.** There are wild and semi-domesticated vines in India so nearly allied to the true **V. vinifera,** that a botanist is almost required to distinguish them. The *asaujiya* or *pahar-phuta*, for example, yields in Kumaon and Garhwal an edible grape. The question may, therefore, in all fairness, be asked—Have the indigenous species (wild or cultivated) shown symptoms of being attacked by **Phylloxera?** If they have not, the further question naturally suggests itself. Would it not be desirable (before attempting the acclimatisation of American mother-plants upon which to graft the true vine) to ascertain whether the indigenous wild stock would not serve the same purpose? The writer has already offered the suggestion that some of the wild forms of **Vitis,** found on the Himálaya, might easily have been the plants that yielded the grapes of the early writers prior to the introduction of the true vine. If there be any merit in that suggestion it would seem that India may not only be retarding her own natural development, as a vine-growing country, through neglecting these wild species, but that she may be depriving the world of the assistance she possibly could render, in the much needed renovation, so to speak, of the now exhausted and weakened stock of the wine-yielding grapes of Europe. In no other part of the globe where grapes and wine are produced has less interest at all events been taken in this great question than in India. We are alike ignorant as to whether we actually possess the diseases that have paralysed the wine industry of the world, or are in a position to assist in arresting the fatal tendencies of grape cultivation.

VITICULTURE in the N. W. P. & OUDH.

Conf. with p. 289.

Conf. with pp. 255, 256, 259-60, 263, 264.

Grafting.
Conf. with pp. 259, 260, 277.

Conf. with p. 291.

Central Provinces.

It is perhaps unnecessary to say more regarding these provinces than that the grapes of certain districts enjoy a high local reputation. They are never likely, however, to be grown for any other purpose than as a supply of fresh and refreshing fruit. The large sale for the boxes of imported fruits from Kábul would seem to show, however, that the Indian grapes are by no means, neither in quantity nor quality, able to satisfy the demands. The following passage from a district report furnish some particulars regarding the grapes of these provinces :—

"The vine is cultivated in the neighbourhood of Asirgurh to a small extent. It is a very remunerative, though laborious description of culture. The opening of the Railway has greatly extended the market, and the price obtained at the Chandni Railway station, where large quantities are disposed of, is now 5 annás a seer (3½*d.* per ℔). The grapes are of very excellent quality when thoroughly ripened ; but for facility of carriage are generally plucked unripe. The area under this crop may, I think, be expected to increase. The Asir vine, as well as the Kábul variety,

CENTRAL PROVINCES. 250

Indigenous Grapes.
Conf. with p, 270.

**VITIS
vinifera.**

**SIND &
BALUCHIS-
TAN.
251**

flourish well in the Khundwa public garden " (*Settlement Report, Nimar District, p. 200*).

Sind and Baluchistan.

It is well known that in many parts in these provinces, grapes of very superior qualities are grown, but the writer has failed to discover a detailed passage regarding either province worthy of citation. The late **Dr. Stocks**, in his report on Sind, says the name *Drakha* denotes the ordinary country grapes. Raisins (*kishmis*), he says, are not made in Sind, but two wines are prepared, *viz.*, *kishmishi* wine prepared from dry grapes, and *angúri* wine from fresh grapes; both these wines, he adds, are often strengthened by a spirit prepared from sugar. *Angúri*, he tells us, was in his day made at Hyderabad, Sehwan, and Shickarpore. The writer has not been able to discover whether these wines are still available. **Murray** (*Plants and Drugs of Sind*) says the crop is very precarious.

**BOMBAY.
252**

Bombay.

One of the earliest notices of the grapes of this province occurs in a review of the report of the Dapuri Botanic Gardens. On this subject **Royle** says : " The grape vine, **Dr. Gibson** mentions, as common in the eastern parts of the Deccan, where it afforded a cheap and delicious article of food, sold in some bazars, and yet that it was uncultivated in many places well suited to it. Into Khandesh it had been introduced by the Col'ector of Revenue, and was quite naturalized, though it had at one time been said that the air of this district was unfavourable to it." **Graham** (*Cat. Bomb. Pl., 33*) remarks that " the common grape vine, successfully cultivated in the Deccan, about Poona, Ahmednuggur, Aurangabad, etc. **Dr. Gibson** mentions that it is very susceptible of blight from fogs and heavy atmosphere; this may account for its failing below the Ghâts, where the atmosphere is comparatively moist."

The following passage from the transactions of the Agri.-Horticultural Society of India (*Vol. I., 96*) will, however, more fully convey an idea of the opinions that prevailed in Western India at the beginning of this century on the subject of vine culture :—

Mr. G. Ballard wrote in 1824 :—"In Bombay we find the soil best adapted to the culture of the vine, to be a light sandy earth free from moisture. When the soil has in the first instance been of a stiff clayish nature, I have seen it much improved by trenching it to the depth of thirty inches, or three feet, and filling up with sandy earth, from the cocoanut woods, and the small white sea-shell, in the proportion of two-thirds of the former and one-third of the latter in alternate layers each.

" We prune towards the end of October, taking care as soon as the rains are quite over to bare the roots of the vine completely in order to check vegetation, and give them a kind of artificial winter. Say, that the roo's are on an average opened about the 7th or 10th of October, they are allowed to remain exposed in the first instance for fifteen or sixteen days, when they ought to be pruned, leaving, according to the strength and age of the vine, two, three, or four shoots, with so much of the last year's wood as shall exhibit three or four healthy eyes, it being from these that the bearing wood of the season is produced. As soon as these eyes begin to bud, which they generally do in about a week (I think) after pruning, the earth should be filled up around the roots, with a considerable quantity of manure, and the vines be regularly watered morning and evening until the fruit attains nearly its full growth, and begins to swell and ripen, when it seems to be necessary to decrease the quantity of water allowing the vines only such a portion as will keep them healthy, perhaps I should say watering every third or fourth day.

" Vines pruned at the season I have mentioned, generally yield fruit fit for the table about the end of January, and as we cannot expect them from the same vine to continue in perfection above a month or five weeks, we have latterly, and with great success, allowed a portion of them to remain, unattended to, till the month of November, and some few till December, when the same process takes place in the treatment, that I have endeavoured to describe above.

| Vine Culture in Bombay. | (*G. Watt.*) | **VITIS vinifera.** |

VITICULTURE in BOMBAY.

" By these means we obtain a regular succession to the end of April, and sometimes till the middle of May. About a fortnight after the vines, which produce in January, have done bearing, say, the middle of March, the roots are again bared, as in October, and the system of pruning, manuring, and watering, once more resorted to, but as two crops in the year would weaken the plants, and that the second one could not attain perfection before the monsoon, we pick off all the promised fruit before it comes to any size, for once touched with heavy rain it loses all its flavour, and ceases to be of any worth.

"The vines which have been pruned in November, of course produce about the end of February, and are similarly treated with the above, at the proper season. Of the later ones, which, producing in March and April, would need pruning just at the commencement of the rains, I cannot with any confidence, though I think ——, who has been one of our most successful gardeners, told me he had not pruned these more than once a year. It is only by experience that we gain any knowledge of the treatment of European plants in this country, and it may perhaps require another year or two ere we can satisfy ourselves, as to the comparative strength and healthiness of vines managed in the latter mode.

"The change of seasons is so similar in Bengal and Bombay that I shall think the same kind of treatment adapted to both places, and it is more on this score than any other that I hope this memorandum may be useful, for, of the scientific part of gardening, I profess my utter ignorance. A great deal of most valuable information regarding the culture of the vine and other fruit trees is to be found in a work published by Forsyth, many years head gardner to the King, and which I have several times seen for sale in this place; I think it is entitled, " Forsyth on Fruit Trees."

In order to convey some idea of the position vine-growing occupies in Bombay at the present date, the following two passages from the Gazetteer may be furnished :—

Ahmadnagar. 253

AHMADNAGAR.—"Vines, *draksh*, are grown in the best garden lands near Ahmadnagar and to a limited extent in Parner, Shevgaon, Shrigonda and Jamkhed. The vine is grown from cuttings. In August or September the vine grower gets cuttings, each with three or four eyes, and puts them into a bed near the well, each cutting being buried till the lower eye is level with the ground and the top of the cutting is sealed with clay and cowdung to keep in the sap. These cuttings are watered daily, and in about ten days begin to shoot. The ground in which the vines are to be planted is ploughed several times till it is free from clods and weeds. At intervals of nine to twelve feet pits are dug a foot and a half square and as deep, and filled half with good soil and manure mixed in equal quantities. The sprouting cuttings are planted in pans in these pits, firmly set into their place with plain earth, and watered every six days. As the shoots grow four small stakes are placed round each cutting, and the other shoots are trained from one to the other, tying them in their places but keeping each vine separate. In five months they grow to the height of a man, when thick stakes of the coral tree *pángárá* (**Erythrina indica**) are planted near them as permanent supports and the topshoots of the vine are nipped off and they are trained on the coral trees. The coral tree is often a growing stump about five feet high and pollarded. For twelve months other garden produce, the egg plant, onion, and pumpkin are raised in the vineyard, care being taken to water the vines once a week unless the rainfall is heavy. In the following October all the branches are pruned to three eyes from the stem, the prunings being available as cuttings, and the flower soon appears. After the fruit has begun to form, water is not allowed to remain in the bunches, and every morning, for the first two months, the husbandman walks round and gently shakes each vine, holding a basket lid underneath, into which dead or diseased leaves, fruit, and insects fall and are carried away and burnt. A vineyard is calculated to yield a quarter crop at the end of the first year, a half crop at the end of the second year, and a full crop at the end of the third year, and, with a moderate amount of care, lasts for about fifty years giving a full crop each year. The vine is also trained in a small open trellis, which is set over the vineyard about six feet from the ground. The pollarded plant is said to give the best yield, but the rich prefer the trellis training, both for its look and its shade ; it is also said to keep the vine in strength to a greater age. The vines yield a crop of sweet grapes in January, February, and March, and a crop of sour grapes in August. The sour crop is large, but the husbandmen do not encourage it, as it is of little value; the sweet crop receives the greatest care, but is not easily brought to perfection. After each crop the vines are pruned, and after the sour crop they are manured with salt, sheep's dung, and salt fish which is particularly valued, as it is supposed to keep off white-ants. Once every five or six days the earth is loosened round the roots and the vines are

Grafting not practised.

Two Crops.

VITIS vinifera.	Progress and Development of

VITICULTURE in BOMBAY.

Blight.
Conf. with p. 292.

flooded. When the buds appear the vine is often attacked by a blight. To remove the blight the branches are shaken over a cloth into which the blight falls. It is then carried to a distance and destroyed. The diseased branches have to be shaken three times a day till the buds are an inch long. To grow vines requires an outlay much beyond the means of most market-gardeners. There is no profit for the first two or three years. Men from Bombay, Ahmadnagar, and Sirur buy the growing crop, the gardener agreeing to continue to water them and the buyers paying for the watchmen who are kept day and night, and in some cases for hoeing and manuring. The buyers, who pay only half the sum agreed, count the bunches and estimate their value at about 2*d*. the pound (6 *shers* the rupee). A vineyard, estimated to contain about thirty-five bullock loads of 120 pounds each, yields a crop worth about £35 (R350). No attempt is made to separate the ripe fruit from the unripe, the diseased from the sound. The bunches are wrapped more or less securely in grass, put into large baskets, and carried on bullocks to the nearest railway station, which sometimes takes two days to reach. From the Railway station the owner consigns them to a broker at the Bombay Crawford Market who puts them to auction, and, deducting his fee, remits the proceeds to the purchaser who pays the husbandman the remainder of the sum agreed. The grapes are sold at the Crawford Market at about 4 *d*. the pound (8½ shers the rupee) " (*Gas. Bomb., XVII., 275*).

Nasik.
254
Indigenous Vines.
Conf. with p. 270.

NASIK.—" Vines, *dráksh*, of three kinds, *abai, phakdi* or *phakiri*, and *káli*, have for long been grown by Kunbis and Malis in Nasik and Chandor. The vineyards are in rich garden lands carefully fenced. Cuttings are laid in September and get out in April and May. The land is ploughed and made ready as for sugarcane. Parallel lines eight feet apart are drawn along and across. At the crossings, which are marked with small sticks, holes, a foot and a half deep and a foot broad, are dug, and filled with half a basket of well seasoned manure. The cuttings are then planted in the holes and watered every fourth day, until they sprout strongly. Then they are regularly watered every ten or twelve days, and given poudrette and other rich manure. The shoots are at first trained on dry sticks, and after about four months, on forked *pángárá*, **(Erythrina indica)** stakes three or four feet high and three to four inches thick. These take root readily and are often trimmed so as not to grow too freely. In the rains most of them are allowed to grow, so that the upper shoots may supply the place of any stakes that die. When it reaches the fork, the top of the vine is lopped to force the stem to throw out side shoots. These side shoots resting on the *pángárá* branches keep the heavy weight of the top shoots and the fruit from dragging the plant to the ground. Vines bear fruit from the second year, and if properly cared for, go on yielding for more than a century. They are

Two Crops.
Disease.

trimmed twice a year in Chaitra (March-April) and Ashvin (September-October), and they bear fruit about four or five months after each trimming. The first are sour and are sometimes used for pickles and jams, but are generally allowed to decay on the tree. In Phalgun (February-March) four or five months after the second trimming, the vines yield good sweet grapes and the loppings then made are made for new vineyards. Vines were formerly largely grown in Nasik and at Satpur about four miles off, but about seven years ago they were attacked by a disease and most of the vineyard had to be destroyed. They also suffered considerably during the recent years—a scanty rainfall (1876-77), but their cultivation is still carried on " (*Gas. Bomb., XVI., 102*).

Bengal.
255

Bengal.

As vine culture in this Province from the high annual rainfall is never likely to be of much importance, it may suffice to furnish a few passages in illustration of the degree of success that has hitherto been obtained. In the upper divisions of the Province as, for example, in Behar, where the climate approximates to that of the North-West Provinces, greater success has been attained than in Bengal proper, and fairly good grapes are accordingly often produced as, for example, at Dinapore and in Tirhut.

Behar.
256

DINAPORE—BEHAR.—**Captain Sage**, Secretary to the Behar Branch of the Agri.-Horticultural Society, in a letter dated 9th August 1832 (*Trans. Agri.-Horti. Soc. Ind , II., 202*), wrote " In July, 1829, I took possession of a small estate at this place, the garden of which contained a vinery in its infancy running north and south, the sides consequently exposed to the east and west ; it is 110 feet long, 15 feet broad, and 7 feet high, having on either side eight vines. I am not sure that in 1829 a few bunches of grapes had not been produced, the number, however, was probably small,

Vine Culture in Bengal. (G. Watt.)	**VITIS vinifera.**

VITICULTURE in BENGAL.

Indigenous Grapes. *Conf. with p. 270.*

as, at this time, the vines were only two years old. In November of this year, I took down the whole of the vines, pruned them, stripped off such leaves as remained, and spread the branches, as far as they would extend, over the jaffry work, running some up the side and over the top, and carrying others horizontally along the sides. The roots were then laid bare, washed, and all filth and scab removed from them. In this manner they were exposed to the cold in the hope of creating an artificial winter, throwing down the sap, and destroying for the time all vegetation; thus giving the vines rest, to enable them in the spring to put forth all their vigour. They remained thus exposed all December and part of January, when they were treated with a composition of fish, *goor*, and black earth (from the native distilleries or toddy shops), mixed together and laid round the roots: the earth was then filled in, and the vines left to themselves. From absence, I am unable to state when the first buds burst, but on my return in the middle of February, 1830, vegetation was strong and appearances were very favourable. On my return from Sarun the end of March, I found the vines covered from the very ground with bunches; it was a display (ridiculous from excess) such as no one had seen at the station, or in the neighbourhood. I proceeded to thin the vines (not the bunches) and filled three large baskets. The *malli* said, he had irrigated them as soon as the young fruit was formed, and every third day, they were well flooded. I saw the vines again the end of May, when the fruit was ripe to some extent, the appearance of the vinery was most splendid; according to the *malli's* computation there were upwards of 3,000 bunches of fruit on it. The roof looked, from the entrance as if formed of one canopy of grapes. These vines are all of the white or common kind, and I had intended to give the Society an account of the success attending this mode of treatment; but was prevented by the late Dr. Charles Hunter, who was amazed at the quantity of fruit and declared, I had so forced the vines, that it was impossible they should bear the following year. I then determined to give a trial of three years, which including changes of weather, might give an average of success on which the Society could safely rely. In November, 1831, as last year, these vines were treated in the same way; the roots bared, washed, and exposed, manured with blood from the slaughterhouses, and such dead animals as I could procure, chiefly dogs and sheep; and they were exposed to the cold in the same manner and for the same length of time. The season was unfavourable for fruit, and a blight attacked the young grapes, and nearly half were destroyed. But in proportion to the success of these in the neighbourhood, I had a large crop, upwards of 1,700 bunches ripened, and were in size and flavour exceedingly fine. This year I added to the length of the vinery, and planted eight Constantia vines, four on each side. They were from Sarun, about 12 months old, from slips. In November, I again had the whole taken down, pruned, and spread out; the roots bared as deep as it was safe to go, washed, and a great quantity of fish given as manure, and the earth was filled in to obviate the horrible stench that was thrown out by the fish, nor could I have the roots bared again, although our winter was exceedingly cold and wet; indeed so great was the cold, that February passed without any show of vegetation, and it was near the 8th or 9th of April before they budded. About the 15th of April they were literally covered with blossoms, but at this time we had a heavy hailstorm, the stones of great size; I measured one 4 inches and 4-10 in circumference. The young grapes were just set, and the top and west side of the vinery from which the storm came, suffered very much; the bunches and young shoots were broken off, and the vines, I am afraid, considerably injured; as what would have been the bearing shoots of next year, are mostly broken off; they may still, however, as the plants are very healthy, be prolonged, so as to be available hereafter. In consequence of this storm, the eastern side has nearly all the fruit remaining, and it amounts to 865 bunches. Of the eight young Costantia vines only two have borne, and one has eighteen bunches on the principal stem, and the other seven, though they were only 3½ feet in height.

From the average 1,521 of these three crops, compared with the produce of other vineries in the immediate neighbourhood, I think it is satisfactorily proved, that allowing something for the vines being young, and the ground, comparatively speaking, new, opening the roots, washing them, and exposing them as much as possible to the cold so that the sap may be successfully thrown down and vegetation stopped, is the first thing to be done The next, a preparation of good rich manure to be laid over the roots and covered as the spring is about to commence. The branches must also be equally spread over the jaffry, so that light and heat may have access to the whole. It is also rather an advantage that in constructing a vinery, it should be partially protected from the violence of the easterly gales, for these gales bring the blight which consists of a great number of small brown iron coloured spots all over the grapes and leaves; the latter wither immediately. The grapes never increase in

Crops of 3,000 Bunches.

Blight. *Conf. with pp. 291-3.*

Hail storms. Destructive.

Blight. Brown spots over leaves and fruit. *Conf. with p. 293.*

**VITICULTURE
in
Bengal.**

size after the blight covers them; and while the bunches that have escaped its baneful influence, swell into large luscious fruit, the blighted ones retain the appearance of discoloured peas. We are in great want of good vines at Dinapore, and if I could procure some good ones from the Cape, I think of trying the hot house plan, which would bring them into use in March, April, and May, instead of having them destroyed by the rains, which the greater part of the grapes are annually "

By way of showing the success that has been attained in the vicinity of Calcutta through assiduous attention to every detail, the following account addressed by Capt. Milner to Dr. Wallich in 1837 may be furnished (*Trans. Agri-Hort. Soc. Ind. VI., 54*):—

"You were right in supposing the pale green grape to be the Muscatel. The purple cannot, I imagine, be the 'blue Hambro' as it is a cluster grape, and only attains its present perfection by being constantly thinned with a fine pair of sciss>rs when the fruit has formed. These vines, together with another fine sort, like the Portugal grape, the fruit of which, I am sorry to say, is not sufficiently advanced to enable me to send a specimen, were brought from the Cape about seven years ago, and were planted in a sheltered alley between two high godowns, where they have flourished surprisingly well. The purple vine is the more hardy, and produces much more abundantly than the Muscatel. Our largest purple vine could scarcely have had less than a hundred branches of grapes upon it this year. About the beginning of January the vine are removed from their trellices and extended on the earth, all the remaining leaves stripped off, and the branches *effectually* preserved; the roots laid open, superfluous fibres cut away, and a portion of the bark of the main stem of the vine removed, scraped off with a knife,—the whole length of the stem. After being suffered to remain a month in this state the roots are covered with fine fresh earth mixed with a little *surkee* and vegetable manure. We increase the richness of the soil by *burying fish* (*covered with oil-cake* to prevent vermin), at about half a yard distant from the root of each vine. Four pounds of fish for each vine will be sufficient.

"I must not omit to tell you that vines had been planted in various situations in this garden, and every possible means adopted to ensure their growth, but entirely without success until it occurred to **Mrs. Milner** to try the plan adopted at Pondicherry; and much to our amusement the alley before mentioned was cleared and the result you have now before you. Was it not a happy discovery?

"I am sure you will be delighted to see the specimens of grapes. They are so beautiful; and the plain and clear statement of **Mrs. Milner's** mode of treating her vines will likewise, I am sure, be interesting to you and the Society."

Assam.

**ASSAM.
257**

Very little has been said within recent years on the subject of vine culture in this Province, though it would seem that in the less rainy portions of the province fairly good grapes can be and are produced. The following passages will show the efforts that were made during the early decades of this century to establish an industry in vine growing:—

"*Successful cultivation of the Grape-vine in Lower Assam—beneficial effects of under-draining.*

"A long and interesting letter from **Major Jenkins** was next brought to the notice of the meeting. **Major Jenkins** states, he has succeeded in maturing very fine grapes at Gowhatti, and as the soil and climate are, he thinks, little propitious to vines, or any fruits of the colder countries, his success may be attributable to the management of the plants, and he therefore conceives that a note of his treatment of them may be useful to others. **Major Jenkins** adds, that he has lately reduced a piece of morass of the worst description to good garden ground, by under-draining, and, as he thinks, his mode of managing it may be deemed sufficiently useful to be placed on record, he has the pleasure of giving it to the Society.

"The best thanks of the Society were given to **Major Jenkins** for his useful communication, which was transferred to the Committee of Papers for the Journal" (*Jour. Agri.-Hort. Soc. Ind., III., 185*).

Major Jenkins fully dealt with his experiments in a paper designated *Hints for the management of the Grape-vine in an unpropitious*

Vine Culture in Assam.	(G. Watt.)	**VITIS vinifera.**

soil and climate, with an account of the beneficial effects resulting from a system of under-draining :—

"I have to thank you for a late note (whose date I cannot quote, for I have unfortunately mislaid it), forwarding opinions on the value of our hemp from the **Urtica nivea**

"I take this opportunity of mentioning that I have succeeded in maturing very fair grapes at this place, and as the soil and climate are both little propitious to vines, or any other fruit of the colder countries, my success may be attributable to the management of the plants, and a note of my treatment of them may, therefore, not be amiss. I may first premise, that the soil of Gowhatti, at least of my garden and the greater part of the station, is a deep diluvial red clay, which in the dry season is as hard and dry as a brick-bat, and in the rains, absorbs and holds water like a sponge. In the highest part of my garden, a spot relatively high with reference to all the adjacent ground, I sunk a well thirty or forty feet, with the hopes of meeting a spring for use in the cold weather, but I failed, and in the rains the waters rise and flow over the top of the well, swamping all my garden. The climate is as bad as the soil. After a dry season of three months, in November, December, and January, when the cold and drought is sufficient to stop vegetation almost entirely, a short vernal season in the beginning of February is followed by constant heavy rains and storms, with alternate bright and burning summer days, so that the fruit of this season that is slow in ripening is very liable to be rotted before it comes to maturity.

"The first vines I tried were planted round an open summer-house, built purposely to shelter them; the plants were set in a walk of rubbish round the house and under the leaves of it, from whence the branches were carried over a trelliced verandah. The plants bore abundantly, but very little of the fruit was sufficiently ripened to be at all eatable. It then occurred to me to plant vines round flat-roofed pucka buildings, with the view of maturing the fruit by the heat of the walls. I set the trees, as before, in broad walks of pucka bricks, stones, and bones, and trained up the vines against all the sides of the building by nailing them lists to the wall, and in this manner I have obtained this year, from trees three or four years old, a large quantity of fruit mostly well ripened. The plants are the common white sweet water of Hindustan. I have never seen branches of grapes more dense, and I should have had far better fruit had I taken the trouble to cut out with a scissors one-half the fruit.

"The building to which I have nailed the vines, have plain, upright walls, with little or no cornice and from the want of some protection on the southern walls, I have found that, after heavy rain, the bunches of fruit are liable to be burnt up by the fierce suns which we have at intervals; but this, I think, might be prevented, either by training the vines at a short distance from the southern walls on poles, or by bringing the upper branches out over a trellis, so as to shade the wall in some measure. The best fruit I got from a northern wall in the middle of July.

"My management of the vines otherwise has been left to my *mallee* and been careless enough; and the manure he has used has been a little rotten fish and sheep's blood.

"I should also mention that the branches of one of my vines strayed into a Guava tree, and with the shelter thus afforded, produced some well-ripened fruit.

"The foregoing hints may be of some service to those who have as yet a climate as this of Assam to deal with, and I have only further to note, that from my experience, I believe, it is essential in Bengal and the Eastern Frontier, that vines should be planted on an artificial substratum of rubbish of pucka buildings, or of stone, brick, and bones, and it in the vicinity of *pucka* buildings, the plants will fruit the better."

Madras.

On the plains of Madras it may fairly well be said the vine is never likely to rise above the position of being a triumph of laborious gardening. On the mountains and tableland it is quite otherwise. The public have been made familiar with the successful results that have been attained on the Nilghiri hills, in Mysore and Coorg. (*Conf. with Kew Bulletin, 1889, p. 23*) in the Wynaad, and other such localities through the appearance, in the local newspapers, of glowing accounts and high expectations. Sir Walter Elliot (*Flora Andhrica*) shows that in his time the subject of vine-

**VITICULTURE
in
Madras.**

growing must have been energetically prosecuted, as he furnishes numerous notes on the plant and its local and classic synonyms. So far it may be said that the grapes of the higher and drier tracts of South India resemble those of the Deccan generally, and possess very little of a purely local character. In the report of the Botanic Gardens of Bangalore for 1890-91 a highly encouraging statement is made of the vineyards under the charge of the Superintendent. The palace vineyard consists of 14 fruiting vines, covering an area of 600 square yards. The English vines introduced in 1888 are making fair progress. Large number of vines have been distributed to intending growers. A plant called the "Spanish Vine" has recently been added to the collection. Some new varieties of the Aurungabad vine have also been promised by the Director of the Department of Land Records and Agriculture, Bombay. The following account of vineculture in Bellary appeared in 1842, and may, therefore, be read with interest as manifesting the leading characteristics of the system of cultivation then pursued :—

**Bellary.
259**

**Indigenous
Grapes.**
*Conf. with p.
270.*

"Vines are sometimes propagated in this country by layers, but the more general method is by cuttings. These are procured at the time of pruning, which is usually in the months of October or November. The cuttings should be about 14 or 16 inches in length, and placed in a slanting position, about three-fourths of their length in the ground, so as to leave only three or four buds above the surface. They should be planted about 18 inches apart, so as to leave room for transplanting them, without injuring the roots. The soil in which the cuttings are planted should be light, and well manured, and they should be moderately watered every two or three days; when they have taken root they will probably throw out three or four shoots, the strongest and most healthy of which, as soon as it can be ascertained, should be selected for the stem of the future vine, and all the others carefully taken off. This should be trained and supported with much care by a bamboo or stick, and the lateral shoots taken off as they make their appearance, till it has attained the height of the *pandall* against which it is to be planted. In 10 or 12 months they will be ready for transplanting, or if necessary, might be removed earlier without injury.

"The *pandalls* used in this country are generally about 6 feet high, but I think one much lower is preferable. At Bellary I have one 6 feet in height, and three of 4 feet, and I have invariably found the latter to answer best; on this neither the blossom nor the fruit are so much exposed to injury from blights or strong winds, and I have generally found the fruit more abundant, much finer, and to ripen better on the low than on the high *pandalls*.

"At Bellary we commence pruning as soon as the rains cease, in October or November; and as we have several *pandalls*, leave an interval of about a fortnight between the pruning of each; by this means we obtain a constant succession of fruit from about the middle of March to the end of June.

"In pruning much care and judgment should be exercised, all dead and useless wood should be carefully taken away. In general, all the last year's shoots (except where the vines are very young and the *pandall* bare) should be cut away within a few inches of their commencement, leaving from two to four or five buds only. On those branches which are strong and healthy, four or five buds, or eyes, may be left, but if small and weak, not more than two or three; almost all the native gardeners err in pruning vines much too sparingly. By leaving too much wood the vine is weakened, and the fruit degenerates. There may be a few more clusters, but they will be small and inferior.

"I should have mentioned that previous to pruning vines, their roots should be opened and laid bare about 18 inches round the stem, and if they are five or six years old or more, may remain open 12 or 14 days: this will lessen the sap and cause the leaves to fall off. If they are very young, six or eight days may be sufficient.

"When the roots of the vines have been bared a sufficient time, then prune them, and at the same time, trim the roots also, by removing with the knife all the little fibres which surround the larger roots and have been thrown out during the last year, leaving only the larger roots. After pruning, let the branches be properly disposed in the *pandall*, and tied in their places, that they may not be afterwards displaced and broken by the winds.

"It is customary with many persons to use a very rich and expensive compost

	VITIS vinifera.

Vine Culture in Burma and Manipur. (G. Watt.)

VITICULTURE in Madras.

for their vines, composed of fish, toddy, etc., etc., etc., but this does not appear to be at all necessary. The manure used for the vines in the Mission garden composed of about two-fourths of red earth, one-fourth of sheep's dung, and the other fourth of about equal parts of common manure, and the contents of sheep's stomachs collected from the slaughter-house is usually collected about three months before pruning time, and thrown together to give it time to ferment and rot.

"In addition to this I sometimes put a small quantity (a quart or two) of sheep's blood to each root before the manure is thrown in, and I think it is an improvement. When all is ready, the hole is filled up with the manure, leaving a sufficient space to receive the water. When manured they should be watered as soon as possible, and to prevent too great a degree of heat or fermentation, some water should be given daily for two or three days: afterwards once in three days will be usually sufficient

"After the fruit is set, it will considerably improve it to examine the vines frequently and pinch off with the finger and thumb the numerous useless shoots on the fruit-bearing branches, which will continue for some time to rise, and if not removed, will not a little impoverish the fruit and entangle and shade it too much.

"Care also should be taken, while the fruit is growing, and especially when ripening, to keep the ground clean under the *pandall*, and clear from weeds, etc." *(Jour. Agri.-Horti. Soc. Ind., I., 294).*

More recently some attention has been given to the subject of vine-growing on the Nilghiri Hills. The reader will find a paper on this subject in the *Indian Agriculturist (November 30th, 1889)* which reviews a report by Mr. A. C. Lawford of the Madras Public Works Department. We are there told the grapes have been very successfully grown by Mr. Lascron of Longwood, by the late Mr. Misquith, by the late Mr. Breeks and at Wellington. Mr. Lawford's paper appears to review the leading methods pursued in California.

Burma and Manipur.

BURMA. 260

The late Dr. Kurz, in his *Forest Flora of Burma*, remarks that the vine is often seen cultivated by Europeans, and is said to bear good grapes in Ava. Mason (*Burma and Its People*) also briefly alludes to this subject, but it would hardly be expected that Burma should appear as a vine-producing country. It is probable, however, that in the drier and more interior sub-temperate tracts, bordering on, and in Manipur itself, a very different result might be obtained. Many portions of Manipur have a bright warm sun, a rich soil, and a rainfall that does not exceed 40 inches a year. By chosing the altitude and exposure for vineyards, it is possible large tracts of Manipur and of the hilly country of Burma would be found quite as suitable as Kashmír. Wild species of Vitis are plentiful and where the rhododendron, the berberry, the rose. and the mulberry luxuriate, forming a brush-wood on rounded hill-side, with, in the glades, a profusion of herbaceous plants of the warm temperate types, surely the vine might be grown. In the writer's opinion, large portions of Manipur and the neighbouring Burma hill tracts, afford, therefore, quite as hopeful a field for viti-culture as can be shown for any part of Southern Asia. While it might almost be said a marine influence exists that is wholly absent from Kashmír, there is at the same time a remarkable climatic depression all over the region indicated. Rhododendrons grow in rich profusion at 6,000 to 7,000 feet in altitude which nowhere occur on the Himálaya below 10,000 to 15,000 feet. Peaches are wild everywhere, so that when tapped by the proposed railways of Assam and Cachar, the little State of Manipur seems likely to become the orchard and vineyard of the neighbouring and populous province of Bengal, much as the Khasia Hills have for centuries supplied it with oranges.

MANIPUR. 261

MANUFACTURE OF RAISINS AND CURRANTS.

It would be beyond the scope and character of this work to deal fully with the subject of the manufacture of either wine or raisins, since neither of these branches of the trade of viticulture can be said to be at present

MANUFAC-TURE *Conf. with p 264, 267, 269, 274.* 262

VITIS vinifera.	Manufactures of

MANUFAC-TURES.

largely practised in India. Leaving, therefore, wine manufacture entirely out of consideration, a few brief passiges may be furnished on the subject of raisins and currants. Dr. Aitchison says that "In Herat and its vicinity the largest amount of raisins are preserved, and much of the vine and spirits prepared. Throughout the country generally a syrup or very thin treacle is made from the juice of the grape" Dr. Cleghorn's account of the manufacture of raisins in Kanáwar has already been quoted (see p. 234), and it does not appear that the system pursued anywhere in India differs materially from that in Afghánistán. Before, however, giving the recent information which has been collected officially, it may be of interest to furnish a passage regarding an e perimental manufacture of raisins in the Nizam's Dominions, Hyderabad, reported in 1840 : —

RAISINS.
263
Hyderabad Varieties of Grapes.
Conf. with p

Four boxes con aining speximens of various sorts of Raisins prepared under his inspection from Grapes, the produce of the country. Presented by Dr. v. Riddell.

"The Raisins here presented were prepared from three varieties of grape :—

The *Saiboe*, a white grape, which sells generally at two seers for the rupee, and is consequently too expensive to be converted into raisins, except as an experiment.

The *Fukerie*, a very luscious water-grape, sells from twelve to twenty seers the rupee.

The *Bokirie* is still more plentiful and is extensively cultivated, as from its cheapness the grape is consumed by the poorer classes. Sixty seers is the quantity that may be obtained for a rupee.

The plan pursued by Dr. Riddell, in the preparation of these grapes, was simply to get them in as ripe a state as possible, and expose them occasionally to the sun on mats, turning them as necessary; they were dry in ten or twelve days. One box contains raisins prepared from the Bokirie grape after the method laid down in Grey s *Pharmacopæia*. Dr. Riddell does not think he has been so successful with this latter as with the grapes treated in a more simple manner, but wishes to obtain information on this point from the Society, being anxious, now that he has quitted that part of the country (Runnur), to induce the natives to whom he has taken the trouble of pointing out the mode of preparing the grape, to carry on the experiments, as it is probable if properly attended to, that this staple may become a much greater article of commerce than it is at present."

Afghanistan
264

In April 1891 Mr. Dwyer of Mianwali, Bunnu, in the Panjáb, addressed the Government of India on the desirability of obtaining information on the subject of the methods pursued in Afghánistán in the manufacture of raisins. Mr. Dwyer asked some six questions. Mr. E. H. S. Clarke, Deputy Registrar of the Foreign Department, furnished the following reply, which will be seen to deal with the various questions raised by Mr. Dwyer :—

"I can personally answer nearly all Mr. Dwyer's questions. At different times I have seen a good bit of the grape-drying process in various parts of Afghanistan; and at one of the large villages of the southern slopes of the Hindu Kosh in 1886 I made some enquiries about the process, my attention being attracted to the quantities hanging up to dry in the shade of the houses. I will endeavour to answer Mr. Dwyer *seriatim*.

1. How is the Kishmis grape dried in Afghanistan?

The Kishmis* grape is a small green grape (not black), very sweet when ripe, and seedless. It is dried in two ways—by exposure to the sun; by hanging up in the shade. Those dried in the sun become of a red-brown colour; those dried in the shade of a dead green. The sun-dried grapes are considerably cheaper than those dried in the shade; and naturally so ; for the latter have to be hung up in the rooms or verandahs of the houses : and moreover, the bunches are picked while the Kismis is still rather tart, or little more than half ripe and consequently smaller. A man may have a very large vineyard, and yet very little space for drying in the shade, *i.e.*, in the verandah or rooms of his house. The local name for these grapes when dried is *Kishmis-i-sayagi*. The sun-dried grape is *Kishmis-i-surkh*. When quite ripe the bunches are picked and spread on the roofs of the houses, or on a patch of hard-

---

* The bunches contain an immense number of grapes, the skin of which is very thin.

Raisins and Currents.	(*G. Watt.*)	**VITIS vinifera.**

ened ground near the house, in the full blaze of the sun. Sometimes a child is left to keep off the wasps, but more often than not, they are just left alone.

2 and 3. What are the processes the grape is put through before it is exported? Why is the *Kishmis* not dried in bunches instead of loose?

The *Kishmis* is partly dried in bunches. Were it otherwise, the juice of these small grapes would be lost, and the roof or ground whereon the grapes are dried would become a mud-heap. When the *Kishmis* is half-dried, it is stripped from the bunch, and the drying completed, either in the sun or shade as the case may be. I believe they are subjected to no other process whatever.

4. To what degree of ripeness should the *Kishmis* be allowed to reach before removal from the vine for drying?

This I have answered above : half ripe or let us say while still rather tart for the green variety, and fully ripe for the *Kishmis-i-surkh*.

5. How many days does it take to dry the *Kishmis*? I am sorry I do not remember. I should think, however, not more than 20 days for either kind, judging from the fact that I saw many lots ready for export plucked from the same vines which still bore ripe grapes. Much would depend on the heat of the sun and the state of the atmosphere.

6. How should the *Kishmis* be packed for export—in bags or packing cases?

I have never seen them packed in cases. It is the invariable custom in Afghanistan to pack them in sacks, and it seems to answer well. It would certainly not be wise to attempt wooden cases in Afghanistan where the transport would be entirely camel."

The information desired by **Mr. Dwyer** being of considerable public interest, **Sardar Muhammad Áfzal Khan, British Agent at Kábul,** was invited to furnish any information he might be able to procure, and the reply obtained by the Government of India amplifying in many essential features the knowledge already possessed in India, may be here published :—

"Four vine lands of the Kábul district—Istalif, Gozar-i-mama-Khatun, Katakan, and Nanchi—are among the more notable vineyards of Afghanistan. The fresh fruit of the vine is termed *Angúr*, and the dried grape is generally known as *Kishmish*. The latter term is, however, specially applied to two of the varieties of the fresh fruit, probably because they were the only fruits that were generally dried for trade.

There were seventeen known varieties of grapes in Kábul, *viz.*, white *Angúr-i-kismish*, black *Kishmish, Khalili, Ashkari, Husaini, Kandari, Jozi, Goladani,* white *Sahibi,* red *Sahibi, Al-i-josh, Ayala, Rish-baba, Lal Monaka, Dida-i-gao,* and *Kaltá.*

Raisins are of three kinds in Kábul and are dried in two ways, *viz.*, first by hanging the bunches of the fruit in the shade, and second by exposing them to the heat of the sun ; the first is called *Kishmishi-i-sayagi* or the shade raisins, and the second is *Kishmish-i-aftabi* or the sun raisin. The fruits intended to be dried in the shade are severed from the vine in bunches or in clusters, and gathered generally in the beginning of September. They are then removed from the vineyards to the lofty roofed shades built of mud and straw, where they are hung upon bamboos or other long wooden poles, suspended by thick ropes. The time required for the drying of the fruits is generally three months, should the climate be dry; but in case of a wet or rainy climate, the time should be six months. At the end of the period of drying, the poles are shaken, and the dried fruit pour down with their tops. The shades are sometimes built so large and spacious as to contain 25 to 30 *Kharwars*—a *Kharwar* being equal to 16 maunds.

The grapes intended to be dried in the sun are also gathered in bunches at the end of September or the beginning of October and spread on the ground. The period of drying varies from 30 to 40 days according as the raisins are exported soon or late. Should they be exported early, they are exposed to the sun for 40 days, but if they are exported later they are dried for 30 days only. During this period the fruits are turned often up side and down. At the end of the period the grapes are severed from the bunches by the hand, the tops going always with the fruit.

In no case is the fruit left on the vine to dry, because the time required for this would be very long and the setting in of the winter puts it in the danger of being rotted by rain or snow. Besides, the fruit drops on the ground, and is eaten by birds and mice, and thus the cultivators lose produce.

The stated time of gathering the fruits is the only degree of ripeness[*] the Af-

---

[*] This means that the grapes to be dried in the shade are gathered about the beginning of September (before they are quite ripe), and that those to be dried in the sun are picked a month later, when ripe.—*E. H. S. Clarke.*

VITIS vinifera.	Method of preparation

MANUFAC-TURE.

ghans are said to know of. The wholesale rate of the sun dried raisins is 18℔ to 64℔ a Kábul rupee according to the fluctuations of the market and the forecast of the crop in Kábul; while the shade raisins, or *Sayagi*, are sold at 10 to 32℔ a Kábul rupee: The *Manakka* raisins, however, are dear; the wholesale rate never exceeds 24℔ a Kábul rupee.

Every kind of raisin is packed for export by the Afghans in gunny bags, along with dried tops, to prevent the fruits sticking together. If the fruit be packed in wooden boxes, they would reach foreign countries in a better condition, but the expense would be proportionately increased and consequently there would be risk of the demand being lessened and the market grow dull."

Since the above information on the subject of the manufacture of raisins, etc., passed to the press, the editor has had the pleasure to obtain two other communications on these subjects, dated June 1892. As these amplify very materially what has already been said, though going over the same ground, they may be published in full. The first communication, it will be seen, is a memorandum from the Political Agent and Deputy Commissioner, Quetta and Pishin. The second is a translation (also furnished by Major Gaisford) of a Report by Kazi Jelal-ud-din, Extra Assistant Commissioner, Quetta.

Indigenous. Grapes. *Conf. with p. 270.*

Memorandum regarding the method adopted by the Afghans for the preparation of the Kishmish grape for export.

The following descriptions of grape are used for preparing *kishmish* :—

(1) HAITA, a long white grape, called *Abjosh* when dry. This grape is collected when ripe and three mixtures are prepared.

(*a*) The first is made of 4 seers of *Barak*, a kind of grass which grows in Kandahar (**Ephedra pachyclada**) to 1 maund of water. The mixture is boiled. The bunches are dipped in this whilst boiling hot.

(*b*) The second mixture is prepared of lime 1, water 8 parts. This is boiled and then allowed to cool. The bunches, after being taken out from the *Barak* mixture, are dipped in this.

(*c*) The third mixture is prepared of potash crude (*khár*) 1 and water 6 parts. This is boiled and then cooled. The bunch is dipped in this after the lime bath. The grapes are then laid out on mats in the sun to dry, and are ready for export in about fifteen days.

The *Abjosh* is much prized and sells at Quetta from R12 to R15 per maund.

(2) The same grape, HAITA, is plucked when fully ripe and dried in the sun. This takes about thirty days. When dry it is called *Mánáká*, and sells for R5 per maund.

(3) The next in popular esteem to the *Abjosh* grape is the SAHIBI, a greenish-red grape, which when dry is known as the *Sahibi Kishmish*. It is picked when perfectly ripe, and the bunches laid out on mats in the sun. It dries in about fifteen days and is then ready for export. It sells at 3 seers per rupee.

(4) KHAYA GHULAMAN is another sort of grape. It is a long black grape, and when dry is known by the same name. This is taken off the plant when perfectly ripe, and dried in the sun on mats and is ready for export in fifteen days. It is sold at 3 seers per rupee.

(5) The "TAR ANGÚR, or black grape, is another sort that is dried, and when dry it is called in Pushtu *Tor Uchki*; in Persian *Kishmish-i-siab*, and known commercially as the *Drák*. This is dried and prepared in the same manner as the *Khaya Ghulaman*. It is sold at 4 and 5 seers per rupee.

(6) Another grape, which is white but nevertheless known as the *Lal Angúr*, is also dried and in this state goes by the name of *Lal Uchki*. This is the best sort of the *Kishmish*. It sells at the rate of 2 seers per rupee.

(7) The most extensively grown grape is the *Kishmisee* and which when dry is known commercially as *Kishmish*. It is a small round seedless grape. The bunches when perfectly ripe are hung up to dry in well-ventilated rooms – not in the sun—and are ready for export in a month. To be properly cured they must, however, be picked before the autumn and before it begins to get at all cold. When perfectly dry, the grapes are picked off the bunches and any that are bad are thrown away. This *Kishmish* is sold at 4 seers per rupee.

(8) Two other sorts of grape dried in the same way are the *Shundokhani*, an oval grape and the round red grape, the latter is known when dried as the *Lal Kish-*

of Kishmish.	(*G. Watt.*)	**VITIS vinifera.**

mish. The *Shundokhani* is sold at 3 seers and the *Lal Kishmish* at 5½ seers per rupee. Both these are seedless.

2. All the descriptions of grapes are dried in bunches, but when packed for exportation are freed from the stems, as these make holes in the fruit and spoil it.

3. The grapes should always be thoroughly ripe before they are gathered for drying.

4. The *Kishmish* for exportation are first packed in cotton cloth and then in thick woollen sacks called *jowals.* It might, however, be better to pack them in boxes of any dry wood. Care should be taken when packing that the cases are properly filled and that no empty spaces are left in the box.

G. GAISFORD, *Major,*
Deputy Commissioner, Quetta and Pishin.

Translation of a Report made by Kazi Jelal-ud-din regarding the method of preparation of Kishmish.

(1) For drying *Kishmish* special mud huts are built generally in gardens. The walls are made of mud and not of bricks. The dimensions of the building should be in proportion to the quantity of grapes in the garden, that is, if a garden contains a large quantity of grapes a large room should be erected, otherwise a small one, but in every case the height of the building should not be less than 20 feet. In all the four walls, holes of about 6 × 4 inches should be made for the purpose of ventilation. Generally there is one such hole in each square yard of the wall. While the walls are still wet, sticks of about an inch diameter are struck into the walls 9 inches deep. The whole surface of the walls on the inside is thus covered with sticks protruding from the walls. These sticks are 9 inches apart from each other. This hut is called the *Kishmish-khana.* When the *Kishmish* grape is well ripened and its colour becomes yellow, that is, when it gives no sour taste at all (which is at Kandahar about the end of August), the grapes are plucked in bunches from the vine. Sticks 4 feet long and 1 inch in diameter, in number equal to the holes in the *Kishmish-khana,* are provided. These sticks are put in the holes. As the holes are 9 inches deep, the sticks are firmly held there. As many bunches of grapes are hung on the sticks as the latter can easily bear (about 16 seers is the usual quantity). In the holes for ventilation thorns are placed so that, while the passage of air is not prevented, no birds can enter the room. Then the door is shut up for forty days. After that period the door of the *Kishmish-khana* is opened and all the sticks are pulled out and well shaken till the dry grapes fall off the stalks on the ground. Then the *Kishmish* is gathered and kept either in the same house or some other suitable place. The process of preparing *Kishmish* is then completed. Small portions of stalks still remain attached to the *Kishmish,* but as they are very thin and small they are a matter of no consequence.

(2) No other process is required for preparing the *Kishmish* than that described above.

(3) It will appear from the answer to question No. 1 that the grapes are dried in bunches *

(4) The test of the necessary ripeness of the grapes is that their colour becomes yellow and they give no sour taste in eating. (This degree of ripeness is attained in Kandahar about the end of August.) It is necessary that the grapes be plucked from the vine and hung in the *Kishmish-khana* before the summer has entirely passed away, otherwise the *Kishmish* will not dry properly.

(5) The *Kishmish* grape takes forty days to dry.

(6) There are two sorts of *Kishmish*† in Afghanistan, *viz.,* white (or green) and red, so called because the grapes of which the *Kishmish* are made are of these two colours. The red *Kishmish* is called *Aftábi* (vulgarly *Artawi*) because it is dried in the sun instead of in the *Kishmish-khana.*‡

---

* I understand the grapes are shaken off the stalks for convenience in packing and to reduce the weight.

H. S. B.

† *i.e.,* of the small Bedana, or stoneless grape, which is alone called *Kishmish* locally.

H. S. B.

‡ Of the two kinds of *Kishmish,* or Bedana grape, it is only the green or *Shana Kishmish* which is usually dried in the way described in paragraph 1. It is much more extensively grown than the red *Kishmish.*

H. S. B.

VITIS vinifera.	Varieties of Kishmish.

MANUFAC-TURE.

The people of Afghanistan put it in *palao* as a condiment. Its export is inconsiderable.

The *Haita* grape when dried in the sun is called *Manakka*, but if boiled in water it is called *Abjosh.* (The process of its preparation will be explained hereafter.) All sorts of grapes other than those mentioned above when dried are called *Woshki.**

Foreigners call all sorts of dry grapes *Kishmish,* and as the enquiry made appears to be about *Kishmish* in this wider sense, I will enumerate the different kinds. They are as follow :—

Lal, Sahibi, Shundu Khani,† Abjoshi, Shana Kishmish, Sara Kishmish, Manakka Torey Woshki and Sund Woshki.

The first three sorts are dried in the *Kishmish-khana.* They are only enough for presents to notable men and are seldom available for sale. *Abjoshi* is well fitted for trade and is exported to India in large quantities. *Shana Kishmish* is also suitable for trade and much exported.

Its method of preparation is described in the answer to the first question.

The process of preparing the *Sara Kishmish* has been dealt with in the answer No. 6 and the method of preparing the *Manakka* raisin is the same (*i.e.,* the *Haita* grape is dried in the sun). The *Tori Woshki* is made of black grapes which are dried in the sun. This species of grape is suitable for trade and can be exported to distant countries.

Tundi Woshki is made of *Tund* grapes, *i.e.,* of big green grapes. The *Tund* grape is the most inferior and least valuable kind of grape in Kandahar. It is white and round and rather sour and ripens later than other sorts of grapes. *Dushab,* or grape juice, is also made of *Tund* grapes. *Dushab* means that the juice of the grapes is boiled, and when it becomes thick it is taken off fire and eaten with *ghee,* etc.

As I have mentioned before, the *Lal Sahibi* and *Shund Khani* grapes are seldom abundant enough for sale. In Kandahar the price of raisins made of these grapes is R12 an English maund, that of *abjosh‡* is R7 per maund, that of *Shana Kishmish* R6 per maund, that of *Manakka, Sara Kishmish* and *Tori Woshki* R5 per maund, that of *Tund Woshki* R4 per maund. The method of preparing *Abjosh* is as follows :—

When the *Haita* grape is well ripened and its colour becomes yellow, it is picked in bunches from the vine. The bunches, about 6 seers in weight, are tied to a tender branch of the pomegranate which bends but does not break. All the bunches are thus tied. After this a large cauldron full of water is put on the fire. When the water begins to boil about 6 seers of Fuller's earth (Persian *Iskhir*) is put in it. After some minutes about 2 seers of dry lime is also added. Then the cauldron is taken off the fire and the water is allowed to remain unshaken, so that the sediment may settle down. When this is done the clear water is poured off and the sediment is thrown away. This clear water is called *Tezab.*

Another large cauldron of pure water is then put on the fire and when it begins to

---

* *Woshki* in Pushtu simply means dry grapes and corresponds to our word "raisin." The *Kishmish* proper is the small red and white Bedana grape, but as the Kazi says foreigners use the word in the sense of raisin to mean any dried grape.

H. S. B.

† *Lal, Sahibi,* and *Shundu Khani* are the names of the three better kinds of grapes in Kandahar. The raisins are called by the same names, but are only produced in small quantities for local consumption. There are also two other kinds called *Sheikh Ali* and *Khair Ghulamán.*

Abjoshi and *Manakka* are the dried *Haita* and *Hussaini* grapes.

The *Haita* and *Hussaini* grapes are the large green grapes exported so largely to India in small wooden boxes. *Shana Kishmish* is the white bedana (stoneless) grape and is prepared as in paragraph 1.

Sara Kishmish is the red bedana of less value and usually dried in the sun.

Tori Woshki are raisins of any other sort of black grape. "Tor" in Pushtu means "black."

H. S. B.

‡ The present rates in Kandahar for *Abjosh* is R9 per maund, for *Tori Woshki* R3-8, and for *Tund Woshki* R2-8.

H. S. B.

MANUFAC-
TURE.

boil about 3 seers of this *Tesab* are mixed with it. The branch of grapes which have been tied to sticks are then plunged thre times in the boiling water. At each plunge the grapes are allowed to remain in the boiling water for a minute. Then they are put in a basket and washed with fresh water. This is generally done near a stream in order that the grape may be washed profusely, so that the effect and taste of the Fuller's earth and lime may be totally removed. After the washing the grapes are dried in the sun. The sign of a good *Abjosh* is that some of the grapes dry up at once in one day. When all have been dried in the sun they are gathered and no other process is required (7 and 8). *Kishmish* is packed in bags for export, and no baskets or boxes are required.

COMMERCE.
265

Commerce in Grapes, Raisins, etc.—The grapes grown on the tableland of India are eaten fresh as a fruit, and are by the railways yearly being more thoroughly distributed from the centres of production of superior qualities. A very large trade is also done in the boxes of half-dried Afghan grapes, which reach the plains mostly during the winter months. But an even more extensive traffic exists in raisins and currants. These come for the most part from Afghánistán, very little of the Kanáwar dried fruit being taken to the plains. The chief qualities of raisins sold in India are the Sultanas from Kábul and Persia. These are often very large, are of a pale greenish yellow colour, and quite seedless. They are known as *Angúl Drákh*. Next may be mentioned the black bloom raisins—*Kála Drakh*—from the same countries. These are the raisins that are most frequently used medicinally in India. Lastly, an inferior kind of currants called *Munakha*. The prices, according to Dymock, in Bombay are: Indian grapes 2 to 4 annas a pound; Kábul grapes 4 annas a box containing about 100; raisins (Kábul and Persian) R5 to 7 per ℔; bloom, R5; *Munákha* R3; and *Angul Drakh* R6½ per Surat maund of 37½℔.

DISEASES TO WHICH THE VINE IS LIABLE.

DISEASES.
266

The reader will have come across, in the above remarks, isolated passages on the subject of the vine diseases met with in India as, for example, in Dr. Cleghorn's account of the Kanáwar vineyards; the closing sentence on the Nasik industry; and the remarks regarding Kashmír. There is no want of mention of great damage being done, but apparently no writer has scientifically examined the disease or diseases that have hitherto appeared. In consequence we cannot for certain say whether or not the greatest of all plagues to viticulture is or is not in India, *viz.*, **Phylloxera**, or Vine Root and Leaf Louse. Such information as we possess would, in fact, lead to the supposition that the Indian diseases, augmented as they are said to be by damp weather, are very possibly all of fungoid growth. The question was, however, raised the other day in connection with the importation by Kashmír of vines from France. **Professor M. Cornu** was consulted as to the best vines to be sent, and he strongly recommended that unless **Phylloxera** was already in Kashmír, rooted plants should on no account be taken from France. The matter was telegraphically referred to Kashmír, and the answer came that **Phylloxera** was already in the State. This assurance is the only evidence of the existence in India of that most destructive of all pests, that the writer can discover, in the fairly extensive correspondence and series of reports that he has had the opportunity of consulting, while preparing the present review of the literature of Indian viticulture. It may, perhaps, therefore be assumed that **Phylloxera** is or will shortly be in Kashmír, and, if so, it would seem highly desirable, in all efforts that may be made to resuscitate the industry in Kanáwar and elsewhere, that no rooted plants be taken from Kashmír, Europe, or America. That the diseases most prevalent on Indian vines are, however, of fungoid origin and, therefore (although in themselves these are sufficiently alarming), they are not beyond the means of remedial agencies, short of extermination and substi-

Phylloxera
in
Kashmir.
Conf. with pp.
268, 274, 275.

VITIS vinifera.	Diseases which infect
DISEASES.	

tution of new stock, seems unmistakably the case. The following passage will shew, for example, that many years ago the Natives of India had come to appreciate the value of sulphur in the treatment of one of these diseases.

In the Transactions of the Agri.-Horticultural Society of India (Vol. I., 104) **Mr. W. Leycester** furnished **Dr. Wallich** in (1824) with a translation of a book on gardening, written apparently by a Native of India. Neither the name of the book nor its date of publication has been furnished, but a remark occurs regarding the diseases of the vine (*dakh*) which would seem to prove that the disease then most serious was of fungoid nature. " To remove the diseases of the vines, says this author, you make a smoke under them with raisins, mustard seed, and sulphur." At the present day sulphur is considered the most effectual cure that is known for **Oidium,** but it is chiefly sprinkled in dry powder over the affected parts. It will be recollected that **Dr. Cleghorn** identified the disease seen by him in 1863 on the Kanáwar vines as the fungus known as **Oidium.** A specimen seen by the writer of a diseased leaf from Bashahr appeared to be a **Peronospora.** It may thus, perhaps, be admitted that the first and most rational course, by which aid could be rendered to viticulture in India, would be the institution of a thorough scientific investigation of the vine diseases at present existing in the vineyards of this country. It would be beyond the scope of this article to furnish a detailed account of all the vine diseases, indeed, of even the more common; but it may, perhaps, assist those interested in the subject to have the chief facts exhibited of the three great vine diseases, *viz.*, **Phylloxera, Peronospora,** and **Oidium.** It is quite possible, however, that Indian vines possess certain special and peculiar diseases of their own unknown to European students of this subject.

The Vine Louse. **267** *Conf. with p. 268.*	**1. Phylloxera vastatrix,** *Planch.* THE VINE LOUSE.

      Syn.—PEMPHIGUS VITIFOLIÆ, PHYLLOXERA VITIFOLIÆ, *Fitch.*

      References.— *Vines and Vine-culture by Barron, 156 ; Several volumes of Reports presented to the Institute de France (Acad. des Sciences) ; Observ. sur Le Phylloxera et sur Les Parasitaires de la Vigne, 1881 ; Nicholson, Dictionary of Gardening ; Kew Bulletin, 1889, 66, 230, 236, 255 ; 1890, 36 ; 1891, 44 ; Riley, State Entomologist, U. States—numerous Reports, etc., etc.*

      Habitat.—This insect first became known in 1854 as a destructive pest on the vines of America, when it was described by **Fitch.** On its appearance in Europe in 1863 to 1866 it was referred to its correct genus. In 1866 it assumed vast proportions in Tarascon, in the Department of Garo, whence infection spread to Avignon in the north, and to Arles in the south. The first Commission, convened to consider the measures that should be taken to check the ravages of this pest, met in 1868. Since then, Commission after Commission has been held, and a library of books and reports have appeared. Every feature in the life-history of the insect has been elaborately worked out, but as yet no material progress has been made towards staying the fatal calamity that has through this minute insect overtaken viticulture, and which is yearly extending its baneful influence. The rapidity with which it has spread over the greater part of the globe may be appreciated, when it is stated that the progeny of a single individual has been estimated to become in one year 5,904,000,000. The full-grown insect is almond-shaped and measures 1-30th of an inch in length by 1-50th in breadth.

Peculiarities. **268**	**Peculiarities and Characteristics.**—The **Phylloxeridæ,** says **Mr. A. Murray,** are intermediate between the **Coccidæ** and **Aphides.** In the early part of its cycle the Vine Louse appears under two distinct forms, both wingless, the one having tubercles on the back and the other destitute of these. The former is found exclusively on the roots, the latter exclusively on the leaves. **Mr. Riley** and other entomologists have clearly demonstrated that these are one and the same insect and, indeed, the roots have been infected with the leaf form. These insects are, however, so small that they can scarcely be seen by the naked eye, but under a lens appear of a fleshy texture and light yellowish brown colour. The condition found on the roots

| the Vine Plant. | (*G. Watt.*) | **VITIS vinifera.** |

PECULIARI-TIES of the Vine Louse.

inserts its sucker and thus remains fixed for the rest of its life. In this position the female lays her eggs in groups around her, and at first these look like fine sulphur powder, but afterwards they become smoky-grey or black. In about eight days the larvæ come out of the eggs. At first these are restless and crawl about, but in three days they become fixed like the mother. In about twenty days the females of this brood lay eggs generally thirty in number. But certain individuals are moulted five times instead of three times, and become winged. The anterior pair of wings are transparent except on the tips which are dark coloured. On escaping above ground the winged form lays its eggs in the down of the leaves and buds. These are of two sizes, the larger ones being females and the smaller males. But the insects that emerge from these eggs have certain peculiarities that may be here alluded to. They are again wingless, they are incapable of feeding, and the female only lays one egg, which is of a green colour, and is accordingly not easily seen. It passes the winter thus, and in spring a wingless insect is hatched, which exactly resembles those on the root and possesses from 20 to 24 ovaries full of eggs. Its descendants produce eggs without the intervention of males, some of them fix on the leaves and produce galls, the others find their way to the roots and renew the subterranean generations. How long this may continue without the intervention of the sexual males and females has not as yet been determined, but it has been noted that the continual renewal causes a reduction in the number of egg-bearing tubes until the process ends by the female producing but one egg which is sterile if not fecundated.

Professors **Balbiani & Girard**, from whom the above life-history has been taken, rest their theory of the destruction of **Phylloxera** on the possibility of being able to kill the winter egg by smearing the cane with coal tar or some other effective means, since it is that generation that renews the attack on the roots. Other writers hold that a better agent against this pest is bi-sulphide of carbon, while others hold that complete extirpation of vines for a number of years is the only really effectual method.

The course now being largely tried took its origin in the observation made by **Mr. Riley**, that certain American vines were less readily attacked by **Phylloxera** than others. These are therefore used as the root-stocks or mother-plants upon which to graft the superior qualities. The vines chiefly recommended for this purpose are **V. æstivalis**, *Michx.*—the Summer Grape; **V. riparia**, *Michx.*—the River Bank Grape; and **V. rupestris**, *Scheele*—the Bush or Sand Grape. The last mentioned is a purely wild species, never cultivated on account of its fruit, but the other two yield well-known American-fruits such as (**V æstivalis**) the Virginia seedling, the Cynthiana, Herbemont, etc., and (**V. riparia**) Taylor Bullet, the Delaware, the Clinton, etc. According to **Planchon**—(*DC., Monogr. Phaner., V., Pt. II.*) the most recent author on Vines—**V. riparia**, *Michx*, includes **V. vulpina**, *L.*, **V. incisa**, *Jacq.*, and **V. cordifolia** var. β riparia. While **V. palmata**, *Vahl.*, and **V. virginiana**, *Pair*, constitute but a variety under **V. riparia** and **V. Solonis**, *Planch*, is a hybrid form from that species. The interest in these American vines, at the present moment, is, however, chiefly as mother stocks on which to graft other grapes. It must be added, however, that a wholesale extermination of existing vines (unless these are hopelessly attacked by **Phylloxera**), and the substitution of American roots upon which to graft, would, by no means, be recommended by the majority of writers, since it is believed the advantages claimed are not at all likely to be permanent, and, indeed, many authors hold that the American stock does not give, in every climate and soil, immunity from this great scourge. The course known as the American system is, therefore, more palliative than curative. But, as already suggested, the advantage claimed by that system might, in India, be more easily attained by the study of the effects of **Phylloxera** on a fairly extensive series of indigenous species (some of the Indian species are closely allied to the American forms employed for grafting) in order to see if we do not already possess a serviceable stock, instead of the less hopeful course of acclimatising plants, which, in the process of acclimatisation, may lose their immunity to **Phylloxera**. The full force of this idea will be apparent when it is borne in mind that **V. riparia**, *Michx.* (one of the vines now recommended as a grafting stock), has by **Planchon** been regarded as identical with **V. vulpina**, *L.*, a form which is probably scarcely distinct from the Indian species, **V. parvifolia**, *Roxb.*

Hybrid. *Conf. with pp. 256, 260.*

Conf. with p. 275.

2. **Oidium Tuckeri.**

VINE MILDEW.

This disease is said to have been known in America long before it was observed in

Vine Mildew. 269 *Conf. with pp. 268, 269.*

VITIS vinifera.	Diseases which infect

DISEASES.

Vine Mildew.

Europe. It was first seen in England, at Ramsgate, in 1847 by a gardener of that town in honour of whom the Rev. M. J. Berkeley gave it the name it is now known by. It was detected in France in 1848. By 1851 it had spread over all the vineyards of Europe, and in the following year it had reached Madeira. From the remarks made above, regarding vine-growing in Kanáwar, in the North-West Provinces, it will be seen that Dr. Cleghorn identified the disease that had done so much harm there (from 1855 to 1860) as Oidium. There is no direct evidence in support of an importation of that disease by India from Europe or America, however, and it is therefore quite possible that Dr. Cleghorn may have been mistaken. It is, therefore, all the more to be regretted that this subject should, for so many years, have been left in a state of complete neglect, for, if the disease of Kanáwar be actually Oidium, it is by no means necessary to carry to India (as has been proposed) American vines in the hope of curing a malady which a few pounds of sulphur would accomplish more thoroughly and economically.

Conf. with so-called Blight, pp. 278-79.

The vine mildew appears to the naked eye like a coating of white powder, resting on the leaves, twigs, and fruit. The diseased parts are often seen to be pale-coloured and distorted; the coat grows in thickness and the patches become brown. On inspection through a microscope the surface of the affected part is seen to have the filaments of a fungus spread over the epiderm-cells, and further that from the sides of the filaments proceed little suckers which are pushed into the epiderm-cells to draw from these the nourishment needed by the fungus. From the upper side of the prostrate filaments arise erect branches, each formed of a row of cells, of which the terminal ones are conidia cells. These, on separating, furnish the spores, which, falling on the vine leaves, extend the malady. No other mode of reproduction has as yet been detected, though, as suggested by Mr. Berkeley, this curious fungus may belong to a more fully-developed state, such as the common Erisyphe communis.

Plants suffering from this disease have a mouldy disagreeable smell. The fungus spreads rapidly in moist weather, is checked by dry air or heavy rain, but seems to be most frequent when warm, moist weather succeeds on damp cold sunless days with a stagnant atmosphere. Where mildew has appeared, the only effectual cure is sulphur or certain preparations of sulphur. The modern usage regards the burning of sulphur, however, as dangerous, for although it will effectually kill the fungus, it may destroy the vine as well. The most successful method is to dust flowers of sulphur over the affected parts. This will, in the course of a few days, kill the fungus, when the plant should be thoroughly washed, otherwise the grapes may be injured.

American Mildew.

270

3.—Peronospora viticola.

AMERICAN MILDEW.

This parasitic fungus has been known on the American vines since 1834, but it was only introduced into Europe in 1878 on some vines intended to replace those destroyed by Phylloxera. It has since spread throughout France and Algeria. In its life-history it differs in no essential feature from Peronospora infestans — Potato Mildew or Peronospora arborescens — the Opium Poppy Blight (*see Vol. VI.*, 72-74). America thus not only gave to Europe "the most dreaded and dreadful of all the insects which attack the vine," but in her efforts, to try and eradicate the injury already done, she added a further scourge in Peronospora. This fungus attacks the under-surfaces of the leaves of the vine, about the time of the vintage in August; they become brown and shrivelled as if scorched by the sun or bitten by the frost. When the fungus first appears it looks like irregular patches of a whitish colour, which soon change to a leaden or brown colour and seem as if dry. "The tissues of the leaf are traversed by mycelium, furnished with *haustoria* or suckers, for taking food from the cells; and the white spots bear myriads of the erect, fruiting branches of the fungus, each repeatedly divided into three, less often into two branches. The terminal branches are short, and on the tip of each is an egg-shaped spore. In the spore there grow five or six smaller spores (zoospores), which escape by the bursting of the cell-wall of the spore, and can swim about in dew-drops and moisture of any kind; and at last the zoospores settle down on the leaves, push a slender tube through the epiderm, and give rise to a new plant. The diseased vines produce defective crops of fruit. The grapes also may be attacked. The resting, or sexual spores of the fungus have been found in Vitis æstivalis; they have a thick, smooth, yellow coat" (*Nicholson, Dict. of Gardening*). As is well known in the case of the potato mildew, sexual spores (*Oogonia* and their contained *Oospores*) are not essentially necessary for the perpetuation of a fungus of the class here dealt with (*see Vol. VI.*, 73), provided portions of the mycelium be sufficiently protected throughout the winter in perennial

| the Vine Plant. | (*G. Watt.*) | VITIS vinifera. |

stems or other portions of the host. Still it would seem worthy of consideration whether **V. æstivalis**, the host on which the Oogonia of this dangerous parasite live, should not be excluded from the vineyard.

DISEASES.
American Mildew.

It will be observed by the *Note* at the end of this chapter, that **Professor M. Cornu** regards copper a sure remedy for this disease. It would seem that the advantage claimed is more due, however, to the selection of cuttings likely to be free from portions of the mycelium than to any advantage gained by the copper treatment. The fungus must lie deeply imbedded in the tissue, and therefore beyond the reach of the copper or any other external application.

In addition to the three diseases described above, the vines of Europe and America are subject to the injurious visitations of many other pests, both of an insect and fungoid nature. Of the former may be mentioned by name the Red Spider (**Tetranychus telarius**); The Thrips (**Thrips minutissima**); The Mealy Bug (**Dactylopius adonidum**); The Vine Scale (**Pulvinaria** or **Coccus vitis**); The Vine Beetle (**Lethrus cephalotes**); The Vine Weevil (**Curculio Betuleti**); and the Vine Tortrix or Moth (**Tortrix vitisana**). Of fungoid diseases many more have been described. The roots have been found destroyed by four fungi, *viz*, the Rhizomorpha stage of **Agaricus melleus**; **Dermatophora necatrix**; **Roesleria hypogœa**; and by **Fibrillaria xylotricha**. **Glœosporium ampelophagum** often does great damage. It manifests itself by dark spots on the young shoots and leaves. These spots are at first round, but they soon become confluent. The centre becomes paler coloured or even pinkishgrey, owing to the pustles rupturing and discharging their conidia. This often does great injury, even killing the plant, but it is effectually cured by painting the stems in winter with a 10-15 per cent.solution of sulpha te of iron. Could this be the blight alluded to at page 279?

Blight spoken of in connection with Bombay Vines. *Conf. with p. 278.*

The present brief notice of some of the chief vine diseases may very fittingly be now concluded by the following translation (by **Mr. J. F. Royle**) of a report furnished by **Professor M. Cornu.** Certain cures and preventive measures, it will be seen, are discussed, as well as a few diseases purposely not dealt with above, since they are sufficiently indicated by **Dr. Cornu's** brief though highly instructive remarks.

Notes on the following Questions addressed to Dr. Cornu.

"The Government of the Cape of Good Hope having decided to import a large number of cuttings of American vines, it would be very advisable to have an opinion as to the best method of disinfecting them after their arrival, etc.

1st, PHYLLOXERA.—Is the introduction of **Phylloxera,** possible by means of cuttings which have been derived from infected districts?

The introduction does take place by means of the *winter eggs* which are deposited on the cuttings. These eggs are very rare on the cuttings, accidental, and difficult to find even by skilled entomologists; nevertheless they are met with.

Is there any known instance of contamination by cuttings alone?

Yes; such an instance occurred in the nurseries established by the Italian Government on the island of Monte Christo, although the precaution of disinfecting the cuttings was taken. The process employed was the use of a ten per cent. solution of *sulpho-carbonate of potassium;* this is an *excellent process in itself*, and I now know of none better, if the cuttings are left in the solution for a quarter of an hour. The two ends of the cuttings should previously be sealed with grafting mastic or wax. Theoretically this is very good, but in application it may happen that the solution *does not wet every portion*, especially the anfractuosities caused by the buds. At such points therefore the poison would have no effect, although it is precisely these points that the insect selects through its instinctive care for the safety of its progeny. The Italian Government wished this incident to be spoken of with the greatest discretion, and consequently the newspapers made very little mention of it.

Phylloxera then only existed in the extreme north-west of Italy, and the whole of the nursery on the island was destroyed (I think in 1886).

Another process of disinfection has been proposed by **MM. Balbiani** and **Couanon.** It is based on the fact that a temperature of 50° centigrade (about

Diseases which infect the Vine Plant

122° Fahrenheit) is fatal to organic bodies. The cuttings are plunged for from 10 or 15 minutes in water at the above temperature, and those eggs which are on the surface would be killed, the cutting being only partially heated, the vitality of the vine is not injured. Doubtless, theoretically, this process is much simpler than the other, and seems much easier of application, but if the temperature is only a little lower than it should be, the action may be absolutely *nil*, and this is what happens if bundles of cuttings are operated on. It is impossible to obtain a temperature of 50° in the middle of the bundles, and the cuttings should therefore be plunged in small lots and in a large quantity of water.

When the *sulpho-carbonate* is used, the cutting bears distinct evidence which proves the reality of the treatment, but if hot water be used there is no such evidence. This *proof* of the treatment is very useful, not in order to prevent fraud, but merely for fear of any forgetfulness through which a bundle of untreated cuttings might accidentally become mixed with the bundles of treated cuttings. The safest plan, however, seems to be to procure the cuttings of American vines from a region which is still free from **Phylloxera**. There are still such regions in France, and the appropriate methods of disinfection should still be applied. Of course the danger is not so great now that **Phylloxera** does exist at the Cape, and especially if the cuttings are to be taken into a district which is already contaminated.

2nd, CRYPTOGAMIC DISEASES (*Anthracnosis, Peronospora, Rot.*).—These different diseases, which do not as yet exist at the Cape, are very much to be dreaded, and every possible precaution against their introduction should be taken.

Against **Anthracnosis** a concentrated solution of sulphate of iron, acidulated with sulphuric acid, is used. The sulphate of iron should be freshly prepared in a 50 per cent. solution; the sulphuric acid should be added in the proportion of one per cent., and the cuttings should be painted with the solution by means of a brush or a rag. This method is preventive, and is said to be efficacious.

Against **Peronospora** copper is a sure and very efficacious remedy. Cuttings do not generally seem to be attacked by this disease, and I do not think they are ordinarily capable of transmitting its germs. It is necessary to insist on the absolute cutting away of all tendrils which have been herbaceous and may contain the dormant spores of the parasite. The thin, and often dry, extremities of the cuttings which have for a long time remained herbaceous should also, for the same reason, be cut away before they are packed. If these cuttings are plunged in a solution of sulphate of copper the **Peronospora** would probably be destroyed, or, if not, the presence of the salts of copper would destroy it in case the dormant spores were to germinate. A 10 per cent. solution ought to suffice.

Physalospora Bidwellii: **Coniothyrium Diplodiella**—These two new diseases, which have been recently observed in some of our vineyards, have been identified as the *Black Rot* and the *Grey Rot* of the Americans. They principally attack the grapes, but also the leaves. The salts of copper act effectually in their case also, but the doses should be strong. One application of the sulphate of copper bath would thus get rid of them. It would be quite possible to carry out the sulphate of copper and the sulphate of iron treatment in one and the same operation.

It may be asked, what influence would the climate of the Cape have on the germs of diseases which might be transported in the form of eggs or spores, on the cuttings? What influence is likely to be exerted by the voyage, and the high temperature which the cuttings will have to pass through on their journey?

As regards **Phylloxera**, this disease could only exist in the state of winter eggs. The temperature during the journey might doubtless cause the hatching of these eggs. But if the young insects do not meet with any tender and herbaceous parts they would probably perish. There is some reason to fear that the cuttings during this period might begin to vegetate, and might thus just supply the food required by the insects which greedily devour the aërial portions of the American vines. Therefore any parts of the cuttings which have begun to vegetate must be remorselessly removed and destroyed on the spot, after which it is prudent to apply the sulpho-carbonate treatment.

Are these different courses of treatment likely to have any influence on each other? Might they neutralize or destroy the action of one another?

The two salts of copper and iron, and the sulphuric acid which is added to the solutions may co-exist separately and act together without producing any inert compound by their reciprocal reaction. But it is not the same with the sulpho-carbonate and the salts in question. It would therefore be well to carry out in Europe before the

Trade in Wine and Brandy. (*G. Watt.*)

voyage the treatment with the sulphates. It would only be on the arrival of the vines at their destination that the treatment with sulpho-carbonate would take place. The mutual reaction would then produce two inert precipitates (the sulphides of copper and iron), which would have no effect on the fungi. It would therefore be well to leave a considerable interval, such as that of the voyage between the two treatments in order to allow the saline solutions to act for a longer period on the cuttings."

TRADE IN WINE AND BRANDY.

This may be said to be very unimportant when the immense population of India is taken into account. The consumption of foreign wines and spirits may, in fact, be said to be almost nominal.

Statement of the Imports from Foreign Countries of Wines, Brandies, etc.

	1886-87.	1887-88.	1888-89.	1889-90.	1890-91.	
	R	R	R	R	R	Gallons.
WINES—						
Champagne .	10,75,566	11,39,117	11,42,362	10,86,005	11,25,411	44,889
Claret . .	6,96,005	7,11,014	6,14,000	6,11,755	5,64,513	113,866
Port . .	4,96,137	5,59,091	6,04,858	5,27,283	5,90,157	77,998
Sherry .	4,17,705	3,92,651	3,53,252	3,23,807	3,31,281	42,672
Other Sorts	4,87,716	4,84,417	5,22,881	5,59,130	5,87,028	82,671
BRANDY .	29,05,067	26,55,781	25,64,080	22,63,543	22,63,448	323,237
LIQUEURS .	1,58,078	1,62,129	1,63,775	1,60,391	1,69,154	11,587
TOTAL .	62,36,274	61,04,100	59,65,208	55,31,914	56,30,992	696,920

It does not seem necessary to do much more than exemplify the countries of supply and the shares taken in the trade by the provinces of India during the last year of the above series.

Champagne.—Of the total supply shown for 1890-91, the United Kingdom furnished 39,383 gallons, France 4,120 gallons, and Belgium 1,201 gallons : the balance came from Italy, Germany, Austria, Hong Kong, etc. Of the receiving provinces, Bengal took 18,339 gallons, Bombay 13,184 gallons, Sind 5,747 gallons, Madras 4,587 gallons, and Burma 3,032 gallons.

Claret.—The United Kingdom furnished 56,791 gallons, France 42,444 gallons, Italy 4,684 gallons, Austria 3,338 gallons, the Straits Settlements 2,491 gallons, Germany 1,040 gallons, Ceylon 950 gallons, Australia, 743 gallons, and the balance in smaller quantities from Malta, Greece, Egypt, Aden, Turkey in Asia, etc. The receiving provinces were : Bengal 38,820 gallons, Bombay 33,973 gallons, Madras 20,870 gallons, Sind 10,464 gallons, and Burma 9,739 gallons.

Port.—The United Kingdom again headed the list, having supplied 67,706 gallons, France 5,390 gallons, Spain 3,711 gallons, etc. Bombay and the provinces it supplies appear to consume relatively very much more port than Bengal, *viz.*, 30,596 gallons to Bombay, 19,588 gallons to Bengal, 14,034 gallons to Madras, 8,797 gallons to Sind, and 4,983 gallons to Burma.

Sherry.—The United Kingdom furnished 34,851 gallons, Spain 3,907 gallons, Malta 1,544 gallons, Germany 1,290 gallons, and 281 gallons from the Straits Settlements. The receiving provinces were Bombay

VITIS : Wine.	Trade in Wine and Brandy.

TRADE.

276

277

Kashmír Wine.
278
Conf. with pp. 261, 273.
CONSUMP-TION of Wines and Spirits in India.

15,759 gallons, Bengal 14,110 gallons, Sind 5,171 gallons, Madras 5,028 gallons, and Burma 2,604 gallons.

Brandy.—The United Kingdom contributed 173,317 gallons, France 110,590 gallons, Germany 27,902 gallons, Belgium 10,293 gallons, Egypt 343 gallons, Italy 274 gallons, Ceylon 120. gallons, Straits Settlements 228 gallons, and very much smaller quantities were obtained from Spain, Malta, Aden, China, and Japan. The receiving provinces were Bombay 114,528 gallons, Bengal 81,856 gallons, Burma 57,404 gallons, Madras 36,143 gallons, and Sind 33,306 gallons.

It is perhaps undesirable to analyse the returns of liqueurs and other sorts of wines; sufficient has perhaps been indicated to show the position and extent of the Indian consumption of foreign wines and spirits. One of the most remarkable modern features, perhaps, has been the decline of the consumption of brandy and the substitution of whisky, especially so in Bengal.

Although Kashmír wine has been shown at several exhibitions, and has commanded high commendation, still it may be said that India practically produces no wine, so that the foreign imports represent her total consumption—a remarkably small consumption when compared with that of almost any country in Europe. Of course the imports here shown by no means represent the total consumption of alcoholic beverages, but they denote the extent to which it can be said that India is learning the "craving" for foreign wines and brandies. The amount of whisky imported last year was slightly in excess of the brandy, namely, 388,637 gallons, of which Bengal took 141,412 gallons; the gin imports were 70,267 gallons, Burma having taken the largest share, *viz.*, 29,847; and the rum imports were 27,402 gallons, of which Burma took 19,051 gallons. The total amount of brandy, whisky, gin and rum imported by India in 1890-91 was, therefore, 809,553 gallons. For some years past Government has practically imported no wines nor spirits, so that the above, less the re-exports, might be assumed to represent the consumption, since the stock in hand may be regarded as a fairly constant quantity. The exports of Indian spirits of all sorts were last year 34,975 gallons, and the re-exports of foreign spirits came to 11,577 gallons, so that the net import from foreign sources would have been close on 800,000 gallons. As in all other countries, India has two items of revenue from alcoholic beverages, *viz.*, a customs due on imports from foreign countries and an excise duty on local manufactures. The imports yielded a customs revenue last year of close on £600,000, but the corresponding receipts by Great Britain and Ireland on their imports came to over £6,000,000. The total customs and excise revenue of India in 1890-91 on wines, spirits and beers, etc., was a little over five millions, or if the receipts derived from the Indian consumption of opium and drugs be added, the total customs and excise income may be said to have been a little over six million pounds sterling, or one-fifth the corresponding revenue obtained in the United Kingdom. Thus the approximately 300 million inhabitants of India afforded a revenue to their country, from these sources, which was one-fifth only of that paid by the less than 30 millions of the mother country. From these facts some idea of the relative consumption of intoxicants in Great Britain and India may be obtained, but were the value of the articles consumed to be taken into consideration, the comparison would manifest still more seriously the luxury, and it might almost be called the alcoholic indulgence of the United Kingdom. (*Conf.* with article **Narcotics**, Vol. V., *332-338*, but in table on page 338 for " Rs. " read " Rx. ")

Vitriol, BLUE,—see **Sulphate of Copper**; Vol. II., 649.

Vitriol, GREEN,—see **Sulphate of Iron**; Vol. IV., 523.

Vitriol, WHITE,—see Sulphate of Zinc.

VOLUTARELLA, *Cass.; Gen. Pl., II., 476.*

[*t. 1139;* COMPOSITÆ·

Volutarella divaricata, *Benth.; Fl. Br. Ind., III., 383; Wight, Ic.,* | 279

Syn.—TRICHOLEPIS PROCUMBENS, *Wight;* T. CANDOLLEANA, *Wight;* MICROLONCHUS DIVARICATUS, *DC.;* CENTAUREA DIVARICATA, *Wall.;* CARDUUS RAMOSUS, *Roxb.*

Vern.—*Bádward,* PERS. & BOMB.

References.—*Roxb., Fl. Ind., Ed. C B.C., 595; Dalz. & Gibs., Bomb. Fl., 131; Dymock, Mat. Med. W. Ind., 2nd Ed., 466.*

Habitat.—An annual, straggling, stiff weed, found in Central, North-Western, Western, and Southern India, from Behar and the Upper Gangetic plain to Lahore, and from Sind to Mysore and the Deccan, ascending to 3,000 feet in the North-West Himálaya.

Medicine.—Dymock informs us that this PLANT is described by Muhammadan writers as the *shau kat-el-baida* of the Arabs, the *lufiniki* of the Turks, and the *sanakhurd* of the Syrians, and is also known in Persian as *kangar-i-sufed* and *asfar-i-bari.* It is considered by these writers to have tonic, aperient, and deobstruent properties, and to have the power of driving away noxious reptiles, when kept in a house (*Makhzan-el-Adwiya*). The indigenous plant does not appear to be known nor valued in India, but the dried drug is imported into Bombay from Persia (*Mat. Med. W. Ind.*).

MEDICINE. Plant. 280

(*J. Murray.*)

WAGATEA, *Dalz.; Gen. Pl., I., 568.*

[*1995;* LEGUMINOSÆ.

Wagatea spicata, *Dalz.; Fl. Br. Ind., II., 261; Wight, Ic., t.* | I

Syn.—CÆSALPINIA MIMOSOIDES, *Heyne;* C. FEROX, *Hohen.*

Vern.—*Wákerí,* MAR.; *Wakiry, wagati,* BOMB.; *Vágáti,* KAN.

References.—*Dalz. & Gibs., Bomb. Fl., 80; Gamble, Man. Timb., 135; Lisboa, U. Pl. Bomb., 217; Gazetteer, Bombay, XV., 79; Wardle, Letter on the Pods as a Tanning material.*

Habitat.—A robust, woody, prickly, climbing shrub, found on the Western Gháts.

Tan—The PODS contain a large proportion of tannic acid, and promise to become a commercial tan of some impertance. Samples were sent for examination to Mr. Wardle who, in a letter to Sir Louis Mallet, May 15th, 1879, wrote as follows:—"In these pods the relation by weight of the seeds to the husks is as 28 to 23, that is, 51 parts by weight of pods containing seeds consist of 28 parts seed and 23 parts husk or outer shell. If the seeds could be extracted from the pods when the latter are ripe, the husks would have the same value as sumach. I have used Lüventhal's permanganate of potash process which gives the permanganate value and is reliable Processes professing to give percentages of tannin are liable to error from each vegetable substance containing tannin of varying properties and constitution. which affects their accuracy. But, to prevent confusion, I have added the percentages of tannin calculated from the permanganate values.

"The figures showing the permanganate values represent the number of cubic centimetres of half decinormal solution of potassic permanganate

TAN. Pods. 2

W. 2

WALLICHIA densiflora.	Walking Sticks.

TAN.

equivalent to the tannin contained in 20 c. c. of an infusion of 5 grammes of the substance to be analysed, in one litre of water :—

Estimation of tannin—	Permanganate values of tannin, —accurate.	Percentages of tannin reliable only *inter se.*
—in seeds alone	0·7	2.13
—in pods alone . . . : . :	6.15	19·17
—in pods and seeds together	3.1	9·66"

OIL.
Seeds.
3

Oil.—Lisboa writes, " An oil used in lamps is obtained from this plant." The writer can find no further information as to this oil, but it is extremely probable that it is obtained from the SEEDS. If this be so, and if the oil be of any value, the seeds extracted from the pods in gathering the latter for tanning purposes, would be removed from the category of waste material to that of useful commercial products, and would thus enable the pods to be obtained more cheaply. Both the pods and the oil are worthy of attention.

WALKING STICKS.

4

Walking Sticks, Timber used for—

Balanites Roxburghii, *Planch.* ; SI-MARUBÆ.

Bamboos, various species ; GRAMI-NEÆ.

Cassia siamea, *Lamk.* ; LEGU-MINOSÆ.

Cocos nucifera, *Linn* ; PALMÆ.

Cotoneaster acuminata, *Lindl.* ; Ro-SACEÆ.

C. bacillaris, *Wall.*

Dichrostachys cinerea, *W. & A.* ; LEGUMINOSÆ.

Diospyros Kurzii, *Hiren.* ; EBEN-ACEÆ.

Dodonæa viscosa, *Linn.* ; SAPINDA-CEÆ.

Grewia populifolia, *Vahl.* ; TILIA-CEÆ.

Juniperus macropoda, *Boiss.* ; CO-NIFERÆ.

Parrotia Jacquemontiana, *Dcne.* ; HAMAMELIDEÆ.

Prinsepia utilis, *Royle* ; ROSACEÆ.

Prunus Puddum, *Roxb.* ; ROSACEÆ.

Pyrus Pashia, *Ham.* ; ROSACEÆ.

Sarcococca pruniformis, *Lindl.* ; EUPHORBIACEÆ.

Staphylea Emodi, *Wall.* ; SAPIN-DACEÆ.

Zanthoxylum alatum, *Roxb.* ; RU-TACEÆ.

For information as to these, the reader is referred to the article on each in its respective alphabetical position.

WALLICHIA, *Roxb.* ; *Gen. Pl., III.,* 916.

[*VI., 419* ; PALMÆ.

5

Wallichia caryotoides, *Roxb., Cor., Pl. III., t. 295* ; *Fl. Br. Ind.,*
Syn.—HARINA CARYOTOIDES, *Ham.* ; WRIGHTIA CARYOTOIDES, *Roxb.,* Hort. Beng.

6

W. densiflora, *Mart.* ; *Fl. Br. Ind., VI., 419.*
Syn.—HARINA OBLONGIFOLIA, *Griff* ; W. CARYOTOIDES, *Wall.*
Vern.—*Kala aunsa* (black reed), *gor aunsa,* KUMAON ; *Oh, úh,* LEPCHA ; *Takosu,* NEP. ; *Zanoung,* BURM.
References.—*Griffith, Palms of Br. E. Ind., t. 237, a. b. c.* ; *Brandis,* For. Fl., 549 ; *Gamble, Man. Timb.,* 419 ; *Kurz, For. Fl. Burm., II.,* 532 ; Ind. Forester, VIII., 407 ; XI., 6.

W. 6

| A violent emetic. | (*J. Murray.*) | **WALSURA piscidia.** |

Habitat.—The former species is a native of Chittagong and Burma. The latter is a small, stemless palm, common in the outer Himálaya, from Kumáon eastwards, up to 4,000 feet; also found in Eastern Bengal and Chittagong. The two species are closely allied and have the same properties.

Fodder.—In Darjiling the LEAVES are used as fodder for poneys.

Domestic.—In Kumáon the LEAVES are employed for thatch.

Wallichia disticha, *T. And., in Linn. Soc. Jour., xi., 49, p. 6; Fl.*
    Syn.—W. YVONÆ, *Kurz.*   [*Br. Ind., VI., 419.*
    Vern.—*Katong*, LEPCHA.
    References.—*Gamble, Man. Timb., 419; Cat. Pl. Darjiling, 84.*
    Habitat.—A handsome palm of the outer hills of Sikkim, and probably eastward to Pegu.

    Food.—The Lepchas fell the tree to eat the PITH of the stem near its summit. Anderson remarks that the BERRIES and perhaps the LEAVES irritate the skin (*Gamble*).

Walnut, see **Juglans regia,** *Linn.;* Vol. IV., 549; JUGLANDEAE.

WALSURA, *Roxb.; Gen. Pl., I., 336.*

Walsura piscidia, *Roxb.; Fl. Br. Ind., I., 564;* MELIACEÆ.
    Syn.—TRICHILIA CORIACEA, *Rottl.;* T. TRIFOLIATA, *Wall.;* HEYNEA TRIFOLIATA, *A. Juss.*
    Vern.—*Walasura, wallursi,* BOMB.; *Walsura,* TAM.; *Chadda-vakku, walsurai, kanná-kampu,* TAM. in Ceylon; *Válarasi, walurasi,* TEL. ? *Joe-boe,* BURM.; *Kiri-kon, kirrekóng,* SING.
    References.—*Roxb., Fl. Ind., Ed. C.B.C., 366; Dals. & Gibs., Bomb. Fl., 37; Beddome, For. Man., lvi.; Anal. Gen., t. 8, f. 6; Mason, Burma & Its People, 539, 759; Thwaites, En. Ceyl. Pl., 61; Trimen, Cat. Ceylon Pl.; Elliot, Fl. Andhr., 188; O'Shaughnessy, Beng. Dispens., 247; Dymock, Warden, & Hooper, Pharmacog. Ind., I., 340; Lisboa, U. Pl. Bomb., 44, 272; Gazetteer, Bombay, XV., 429; Ind. Forester, X., 33; Agri.-Horti. Soc. Ind., Journ. (Old Series), IX., Sel., 41.*
    Habitat.—A small tree of South and Western India and Ceylon, said to be also plentiful in the Pegu, Tounghoo, and Tharawaddy forests of Burma (*Mc'Lelland; Mason*). Kurz, however, does not describe it as found in Burma, so probably the above refers to some other species.

    Medicine.—Corre & Lejanne state that in the Antilles the tree is known as *Herbe à mauvaise gens* or *Herbe à méchants,* and that the BARK acts as a dangerous emmenagogue and violent emetic. Mr. Hœlingsworth of Madras has experimented with it, and finds it to be stimulant and expectorant. The FRUIT of another species of the same genus is said by Forskhal to be the *jauz-el-kai* or the emetic nut of the Arabs, with whom it is also used as hair wash to kill vermin, and as an ointment to cure itch (*Pharmacog. Ind.*).

    Chemical Composition.—The authors of the *Pharmacographia Indica* state that the bark contains a *resin* anhydride in the alcoholic solution. An aqueous extract contains *saponin,* and an alcoholic extract a large quantity of tannin.

    Structure of the Wood.—Heavy and strong, said to be good and used by the Natives of South India for various purposes.

    Domestic.—Roxburgh, and following him many other writers, state that the BARK is largely employed to intoxicate fish, and that fish so caught are not less wholesome than ordinarily. Mr. Hœlingsworth, according to the authors of the *Pharmacographia Indica,* finds that it

Marginal notes (right column):
FODDER. Leaves. 7
DOMESTIC. Leaves. 8
9
FOOD. Pith. 10 Berries. 11 Leaves. 12
13
MEDICINE. Bark. 14 Fruit. 15
CHEMISTRY. 16
TIMBER. 17
DOMESTIC. Bark. 18

WEBERA corymbosa.	Wattle Barks.

DOMESTIC.

acts effectually for this purpose, and corroborates the statement that the fish killed with it are quite wholesome.

19

Walsura robusta, *Roxb.; Fl. Br. Ind., I., 565.*

Syn.—MONOCYCLIS ROBUSTA, *Wall.* ; S:YTALIA GLABRA, *Ham.*
Vern.—*Upphing,* SYLHET ; *Gyopho, tsoukmayba,* BURM.
References.—*Roxb., Fl. Ind., Ed. C.B.C., 366 ; Kurz, For. Fl. Burma, I., 223 ; O'Shaughnessy, Beng. Dispens., 247 ; Agri.-Horti. Soc. Ind., Journ. (Old Series), VI., 41.*

TIMBER.
20

Structure of the Wood.—Light red, very hard ; weight 63℔ per cubic foot.

WATTLE BARK.

21

Wattle Bark—The bark of various species of Australian Acacias, used for tanning, but chiefly **A. decurrens,** *Willd.*—the Black Wattle—(Vol. I., 46), a species now being experimentally cultivated in several districts of India, chiefly on the Nilgiris. The "Golden" or "Broad leaf" Wattle - **A. pycnantha,** *Bth.*—is perhaps, next to the Black Wattle, the most valuable species for tanners' bark and gum. **A. melanoxylon** (Vol. I., 53) and **A. dealbata**—the Silver Wattle—(Vol. I., 46) are also much used. But **A. floribunda, A. affinis,** and others are amongst those that are now so largely exported to Europe as Tanners' Wattle ; in fact, vast tracts of Acacia forest are fast disappearing in Australia. The destruction of these forests reached such an extent in 1878 that the Australian Government ordered the matter to be enquired into by a special Board. This resulted in the following recommendations :—that regulations should be framed so as to conserve the trees on crown lands ; that wattle cultivation should be adopted by the State ; and that certain areas of poor land should be leased on the stipulation that the whole of the land should be devoted to wattle cultivation. Many Indian species of **Acacia** possess barks of great value in tanning and are used in place of wattle (see Vol. I., 17-61).

Wax, see **Honey & Wax,** Vol. IV., 263-271 ; also **Oils,** Vol. V., 448,
[457, 458.

WEBERA, *Schreb.; Gen. Pl., II., 86.*
[*t. 309, 584, 1064 ;* RUBIACEÆ.

22

Webera corymbosa, *Willd.; Fl. Br. Ind., III., 102; Wight, Ic.,*

Syn.—W. ASIATICA, *Bedd.* ; W. GLOMERIFLORA, *Kurz ;* W. CERIFERA, *Moon.* ; RONDELETIA ASIATICA, *Linn.* ; CUPEA CORYMBOSA, *DC.* ; STYLOCORYNE SP., *Wall.* ; S. WEBERA, *A. Rich.* ; S. RIGIDA, *Wight.* ; POLYOZUS ? MADRASPATANA, *DC.* ; PAVETTA WIGHTIANA, *Wall.*
Vern.—*Kankra,* BENG. ; *Jhanjhauka,* URIYA ; *Kachuria cháll,* CUTTACK ; *Komi,* TEL. ; *Tarana, karani,* SING.
References.—*Roxb., Fl. Ind., Ed. C.B.C., 234 ; Kurz, For. Fl. Burm., II., 47 ; Beddome, For. Man., 133 ; Anal. Gen., XVI., f. 2 ; Thwaites, En. Ceyl. Pl., 148 ; Dalz. & Gibs., Bomb. Fl., 119 ; Gamble, Man. Timb., 226 ; Rheede, Hort. Mal., II., t. 23 ; Agri-Horti. Soc. Ind., Journ. (Old Series), VI., 48 ; Ind. Forester, X., 31.*

Habitat.—A large shrub or small tree, found in the Western Peninsula from the Konkan southwards, and in Ceylon.

RESIN.
Shoots.
23

Resin.—The extremities of the young SHOOTS are often found covered with a white resinous matter (*Roxb.*).

TIMBER.
24

Structure of the Wood.—Yellowish-white, hard, close-grained ; weight 57℔ per cubic foot ; used in Ceylon for making fishing-boats.

Products of India. 301

Timbers used for Well-Curbs.   (*J. Murray.*)   WELL-CURBS.

WEDELIA, *Jacq.; Gen. Pl., II., 370.*
[*Ic., t. 1107;* COMPOSITÆ.

Wedelia calendulacea, *Less.; Fl. Br. Ind., III., 306; Wight,*

> **Syn.**—VERBESINA CALENDULACEA, *Linn.;* JOEGERIA CALENDULACEA, *Spreng.*
>
> **Vern.**—*Bhánrá, bhángrá,* HIND.; *Kesraj, kesarája, kesuriá, bhimráj, bangra,* BENG.; *Piwalá máká,* MAR.; *Pilo bhangro, bhángaro,* GUZ.; *Pila-bungra,* DEC.; *Pivala bhangra,* BOMB.; *Postaley-kaiantagerai,* TAM.; *Bhringarája, kesaraja, pita-bhringi,* SANS.; *Ranwan-kikirindi,* SING.
>
> **References.**—*Roxb., Fl. Ind., Ed. C.B.C., 606; Dalz. & Gibs., Bomb. Fl., 129; Thwaites, En. Ceyl. Pl., 165; Burm., Fl. Zeyl., 52, t. 22, f. 1; U. C. Dutt, Mat. Med. Hind., 181, 294; McCann, Dyes & Tans, Bengal, 152; Bidie, Prod. S. Ind., 55; Drury, U. Pl. Ind., 445.*

25

Habitat.—Met with in wet places in Bengal, Assam, Sylhet, the Eastern and Western Peninsulas, and Ceylon. The plant has a slight camphoraceous odour.

Dye.—The LEAVES of this plant are said by **U. C.** Dutt to be used in dyeing grey hair and for promoting its growth. McCann writes that, in Lohardagga, the ROOT is pounded and used as a black dye with salts of iron.

DYE.
Leaves.
26
Root.
27

Medicine.—According to Dutt some confusion exists in the Sanskrit and vernacular names of this species and **Eclipta alba,** *Hassk* (*cf.* Vol. III., 201). Both are called *kesaraj* or *kesuriá* in Bengali, and the two Sanskrit names above given are employed as synonyms for both. The Hindustani term, *bhángrá,* derived from the Sanskrit *bhringarája,* is, however, generally applied to the species now under consideration. The LEAVES are regarded as tonic, alterative, and as useful in cough, cephalalgia, and skin diseases, especially alopecia. The JUICE of the leaves is much used as a snuff in cephalalgia, and also in soaking various powders for the preparation of pills. Several elaborate preparations of the drug prescribed in the *Bhávaprakásha* are recommended for phthisis, cough, catarrh, and affections of the head (*U. C. Dutt, Mat. Med. Hind.*).

MEDICINE.
Leaves.
28
Juice.
29

Domestic.—The LEAVES and their JUICE are employed, as above stated, to dye and promote the growth of the hair. The juice is said to be used to tattoo the body, the colour produced being a deep, indelible, bluish-black (*U. C. Dutt*).

DOMESTIC.
Leaves.
30
Juice.
31

WELL-CURBS.

Well-Curbs, Timbers used for—

Many timbers are employed for this purpose, but the following are the more important :—

32

Acacia arabica, *Willd.;* LEGUMINOSÆ.
Albizzia Lebbek, *Benth.;* LEGUMINOSÆ.
Barringtonia acutangula, *Gærtn.;* MYRTACEÆ.
Bombax malabaricum, *D.C.;* MALVACEÆ.
Butea frondosa, *Roxb.;* LEGUMINOSÆ.
Cordia Myxa, *Linn.;* BORAGINEÆ.
C. vestita, *Hook. f. & T.*
Eugenia Jambolana, *Lam;* MYRTACEÆ.
Ficus bengalensis, *Linn.;* URTICACEÆ.
F. glomerata, *Roxb.*
Gmelina arborea, *Linn.;* VERBENACEÆ.
Phyllanthus Emblica, *Linn.;* EUPHORBIACEÆ.
Populus euphratica, *Oliv.;* SALICINEÆ.
Prosopis spicigera, *Linn.;* LEGUMINOSÆ.

WENDLANDIA
tinctoria.   **A good mordant.**

Soymida febrifuga, *Adr. Juss.*; MELIACEÆ.
Zizyphus Jujuba, *Lamk.*; RHAMNEÆ.

WENDLANDIA, *Bartl.*; *Gen. Pl., II., 50.*

33   **Wendlandia exserta**, *DC.*; *Fl. Br. Ind., III., 37*; RUBIACEÆ.
    Syn.—W. CINEREA, *DC.*; RONDELETIA EXSERTA, *Roxb.*; R. CINEREA,
      Wall.
    Vern.—*Chaulai, chila, chilkiya, tíla, birsa, tilki, tilai,* HIND.; *Hundro,*
      pichari baha, SANTAL; *Kangi, tilki, mimri,* NEPAL; *Kúrsi,* SEONI;
      Marria, GOND; *Tilliah,* BAIGAS in MANDLA; *Pansíra, pudhárá, chil-*
      kiyá, PB.
    References.—*Roxb., Fl. Ind., Ed. C.B.C., 176; Beddome, For. Man.,*
      130; Brandis, For. Fl., 268; Gamble, Man. Timb., 225; Stewart, Pb.
      Pl., 117; Rev. A. Campbell, Rept. Econ. Pl., Chutia Nagpur, Nos. 7542,
      9283; Atkinson, Him. Dist., 311; Gazetteer, N.-W. P., IV., lxxiii.;
      Ind. Forester, III., 203; IV., 241; VIII., 412; XIV., 343; Agri.-Horti.
      Soc. Ind., Journ. (Old Series), XIII., 319.
    Habitat.—A small, deciduous tree of the Sub-Himálayan tract, from
the Chenab eastwards to Nepál and Sikkim; also found in Oudh, Bengal,
Central and Southern India.

FODDER.
Leaves.
34
    Fodder.—In certain localities the LEAVES are given as fodder to
cattle.

TIMBER.
35
    Structure of the Wood.—Reddish-brown, extremely hard, close grain-
ed, fibrous and tough; weight 47℔ per cubic foot. Though small it is
used for building and for making agricultural implements, and in the
Sikkim Tarai, for house-posts.

36   **W. Notoniana**, *Wall.*; *Fl. Br. Ind., III., 40*; *Wight, Ic., t. 1033.*
    Vern.—*Rameneidelle* or *rawan-idala,* SING.
    References.—*Beddome, Fl. Sylv., t. 224; Dals. & Gibs., Bomb. Fl.,*
      117; Thwaites, En. Ceyl. Pl., 159; Trimen, Cat. Ceylon Pl. 41; Gamble,
      Man. Timb., 225; Gazetteers:—Mysore & Coorg, I., 70; Bombay, XV.,
      435.
    Habitat.—A small tree, common on the hills of South India and Ceylon,
ascending to 7,000 feet.

TIMBER.
37
    Structure of the Wood.—Red, and similar in structure to that of
W. exserta; it is strong and used for various purposes by the Natives of
Southern India.

38   **W. tinctoria**, *DC.*; *Fl. Br. Ind., III., 38.*
    Syn.—RONDELETIA TINCTORIA, *Roxb.*
    Var. normalis=W. PROXIMA, *DC.*; RONDELETIA PROXIMA, *Don.*
    Var. grandis=W. BUDLEOIDES, *Wall.*
    Vern.—*Túla-lodh,* BENG.; *Tilai,* SANTAL & KOL.; *Kangi,* NEPAL;
      Singnok, LEPCHA; *Telli,* URIYA; *Tamayoke,* BURM.
    References.—*Roxb., Fl. Ind., Ed. C.B.C., 175; Kurz, For. Fl. Burm.,*
      II., 74; Beddome, For. Man., 130; Brandis, For. Fl., 269; Gamble,
      Man. Timb., 225; Rev. A. Campbell, Rept. Econ. Pl., Chutia Nag-
      pur, No. 8439; Darrah, Note on Cotton in Assam, 32; Ind. Forester,
      III., 203; VIII., 416; Agri.-Horti. Soc. Ind., Journ. (Old Series), IX.,
      Sel., 45.
    Habitat.—A small, elegant tree, with large crowded panicles of small
white, sweet-scented flowers, terminating the boughs; common in the
forests of the Tropical Himálaya from Garhwál eastwards, and from
Oudh and Bengal to the Khásia Mountains, Assam, Chittagong, Pegu,
and Tenasserim.

DYE.
Bark.
39
    Dye.—The BARK is largely employed in Bengal and Assam as a
mordant in dyeing. The specific name, which would naturally imply that

W. 39

the plant itself yields a dye-stuff, has probably been derived from this fact, since no record exists of any part yielding a colouring matter.

Medicine.—In Chutia Nagpur the BARK is employed as an external application to the body to relieve the cramps of cholera (*Campbell*).

MEDICINE.
Bark.
40
41

(*J. Watt.*)
WHALES, PORPOISES, DOLPHINS, & DU-GONGS ; *Blanford, Fauna Br. Ind., Vol. I., 564-594.*

The aquatic MAMMALIA which, as a matter of convenience, have been grouped together in this place, belong to two very distinct Natural Orders—the CETACEA and the SIRENIA. The former embraces many genera, which, to a large extent, approach each other so closely as to, in some cases, make their separation a matter of considerable doubt, while the latter constitutes a well marked series perfectly distinct from the Cetacea and which possesses but two living genera with only a few species in all under these. In external form the members of these Orders resemble fish and in that respect differ from all the other Mammalia. They are adapted to an aquatic life. They have no external hind limbs ; the tail is expanded into lobes or "flukes" resembling that of the fish, though flattened horizontally instead of vertically ; the anterior limbs are formed into paddles ("flappers") or pectoral fins, in which the digits are completely incased by skin and destitute of nails ; the dorsal fin, present in many species, is composed of integument ; the skin (with the exception of the Dugongs) is for the most part hairless, although a few bristles often occur at the mouth, especially in the young ; the heat of the body is preserved by a layer of fat or "blubber" placed immediately beneath the skin ; the eyes are small and the ear-orifice minute and not protected by an external ear.

These are the chief peculiarities of the two Orders, and it need only be added that the Dugongs, or Sirenia, differ chiefly from the Whales, Porpoises, and Dolphins by being herbivorous animals that feed on the aquatic vegetation of shallow seas, estuaries and mouths of rivers. They possess special modifications, fitted for that life in place of the carnivorous adaptations of the Cetacea. The nostrils are separate and anteriorly situated ; the mouth, small ; the teeth, incisors and molars ; the muzzle truncated and having horny plates which appear to be used in mastication.

The CETACEA have been referred to two sub-orders :—

I.—**Mystacocœti.** *These have no teeth after birth ; baleen present ; breathing-orifice double.*
The following are the genera of this sub-order, BALÆNOPTERA, and MEGOPTERA.

II.—**Odontocœti.** *Teeth present throughout life ; no baleen ; breathing-orifice single.*
This sub-order has been referred to three families with the genera named below under each :—

  1st—Physeteridæ—functional teeth in lower jaw only—PHYSETER and COGIA.
  2nd—Delphinidœ—functional teeth in both jaws (upper deciduous in Grampus); ribs abnormally articulated—PHOCÆNA, ORCA, GLOBICEPHALUS, ORCELLA, LAGENORHYNCHUS, TURSIOPS, STENO, and DELPHINUS.
  3rd—Platanistidœ—functional teeth in both jaws, ribs normally articulated—PLATANISTA.

The SIRENIA, so far as India is concerned, are represented by one genus, the HALICORE.

The more important species of the above genera may now be dealt with very briefly in alphabetical order, as is customary in this work.

1st, Balænoptera edeni, *Anderson : Blanford, Fauna of British India, I., 568.*
  THE SMALLER INDIAN FIN-WHALE.
  Habitat.—Bay of Bengal. This is probably identical with **B. rostrata.** The adults are about 40 feet long.

42

2nd, B. indica, *Blyth ; Blanford, Fauna of British India, I., 567.*
  THE GREAT INDIAN FIN-WHALE.
  Habitat.—Bay of Bengal and the Arabian Sea. It is the largest of all known animals, living or extinct, and is presumed to be very probably identical with the northern fin-whale (**B. sibbaldi**). Total length about 80 to 90 feet.

43

WHALES Porpoises, etc.	Whales, Porpoises,

44 3rd, **Delphinus delphis,** *L.*; *Blanford, Fauna of British India, I., 587.*
   THE COMMON DOLPHIN.
      Vern.—*Pomigra,* TAM.
      Habitat.—Tropical and Temperate seas. In India recorded only
   from the Madras coast.

45 4th, **D. dussumieri,** *Blanford, Fauna of British India, I., 588.*
   THE INDIAN LONG-NOSED DOLPHIN.
   Habitat.—Malabar Coast.

46 5th, **D. malayanus,** *Lesson; Blanford, Fauna of British India, I., 588.*
   THE MALAY DOLPHIN.
   Habitat.—Indian Ocean; has been captured in the Sunderbans.

47 6th, **Globicephalus indicus,** *Blyth; Blanford, Fauna of British India, I.,*
   THE INDIAN PILOT WHALE.   [577.
      Habitat.—This species, which is nearly allied to **G. melas** of the
   European seas, has been captured in the brackish water of the Gangetic
   delta.

48 7th, **Gogia breviceps,** *Gray; Blanford, Fauna of British India, I., 572.*
   THE SMALL SPERM-WHALE.
      Vern.—*Wongu,* TEL.
      Habitat.—Found in the Indian and Australian seas. The type of the
   species was captured at Vizagapatam.

49 8th, **Halicore dugong,** *Illiger.; Blanford, Fauna of British India, I., 594.*
   THE DUGONG OR DUYONG.
         Vern.—*Talla mala; muda ura,* SING.; *Duyong, parampuan laut,*
      MALAY.
      Habitat.—The shores of the Indian Ocean from East Africa to Australia
   for about 15 degrees on each side of the Equator. They have been seen
   on the coast of Malabar, the Andaman Islands, Mergui Archipelago, and
   Ceylon. They feed on marine algæ and haunt shallow bays, but do
   not ascend rivers.
      Oil.—The part of this animal yields a clear limpid oil of great value.
OIL.
50 Food.—The FLESH is regarded as excellent.
FOOD.
Flesh.   9th, **Lagenorhynchus electra,** *Gray; Blanford, Fauna of British India,*
51   [*I., 580*
52   THE INDIAN BROAD-BEAKED DOLPHIN.
         Habitat.—Indian and Tropical Pacific Ocean. Collected at Vizaga-
      patam by Sir W. Elliot.

53 10th, **L. obscurum.** *True; Blanford, Fauna of British India, I., 580.*
   THE BEAKLESS DOLPHIN.
   Habitat.—Indian and Pacific Oceans.

54 11th, **Phocæna phocænoides,** *Blanford, Fauna of British India, I., 574.*
   THE LITTLE INDIAN PORPOISE.
      Vern.—*Molagan,* TAM.; *Bhulga,* MAR.
      Habitat.—The shores of the Indian Ocean from the Cape of Good
   Hope to Japan. The tidal rivers of Bengal, Madras, Malabar, Bombay,
   and Karachi.
55   According to Mr. W. F. Sinclair of Bombay, it feeds chiefly on prawns,
   cephalopods, and fish, and is generally seen singly or not more than
   four or five together. "The roll of this porpoise is like that of **Phocæna
   communis.** It does not jump or turn summersaults like **Platanista** and
   the **Dolphini,** and is, on the whole, a sluggish little porpoise."
      The genus NEOMERIS into which this animal has been placed, by some
   zoologists, differs only from **Phocæna** by having no dorsal fin.

W. 55

12th, Physeter macrocephalus, *L.; Blanford, Fauna of British India, I.,* 571. **56**
THE SPERM-WHALE or CACHALOT.
Habitat.—Found in nearly all tropical and sub-tropical seas, and was formerly much hunted in the Bay of Bengal and off the coast of Ceylon. Blanford says that the only recorded case of one having been stranded on the Indian coast was at Madras in 1890.
It is found in the open sea, generally in herds of from 10 to 15 or sometimes in very much larger numbers. The old males are said to live apart. Sperm-whales have been killed in the Atlantic with harpoons, that had been left in them in the Pacific Ocean.
Oil.—SPERMACETI AND AMBERGRIS—Spermaceti is obtained from the   OIL.
head, sperm-oil from the blubber, and ambergris from the intestines, though   **57**
it is sometimes found floating on the sea. See the general remarks in the concluding paragraphs.

13th, Platanista gangetica, *Gray; Blanford, Fauna of British India, I.,* 590. **58**
THE GANGETIC and INDUS DOLPHIN.
Syn.—P. INDI, *Blyth.*
Vern.—*Sús, súsú, sous susá,* HIND.; *Súsúk, sishúk,* BENG.; *Hiho, seho,* ASSAM; *Huh,* SYLHET; *Bhulan, súnsar,* SIND; *Sisúmar,* SANS.
Habitat.—The Indus, Ganges, and Brahmaputra, and all their larger tributaries, from the sea to the foot of the mountains. This Dolphin is common in the tidal waters but never enters the sea. It is believed not to be gregarious, although several individuals may often be seen together. It is perhaps also migratory to some extent, since it is not seen in the Hugli near Calcutta, from March to June, though frequent from October to March. It is said to be quite blind, and it is argued that sight would be useless in the thick muddy waters of the rivers in which it is found. It seems, therefore, to capture fish and prawns by feeling for them.
Oil.—The oil of this species finds a ready sale; it is used for burning   OIL.
and other purposes.   **59**
Food.—The FLESH is eaten by certain castes. It is captured in nets   FOOD.
or by harpooning.   Flesh.
60

14th, Orca gladiator, *Gray; Blanford, Fauna of British India, I.,* 576. **61**
THE GRAMPUS or KILLER.
Habitat.—Doubtfully found in Indian seas.

15th, Orcella brevirostris, *Anderson ; Blanford, Fauna of British India,* **62**
[*I.,* 578.
THE LARGE INDIAN PORPOISE.
Vern.—*Lomba-lomba,* MALAY.
Habitat.—Bay of Bengal, ascending the rivers as far as the tide extends; also found in Singapore and North Borneo.

16th, O. fluminalis, *Anderson ; Blanford, Fauna of British India, I.,* 579. **63**
THE IRRAWADDY PORPOISE.
Habitat.—The Irrawaddy river. A gregarious species very similar to O. brevirostris but with the dorsal fin placed further back, also smaller, lower and less falcate and with the pectoral fins rather shorter and broader.

17th, Steno frontatus, *Blyth; Blanford, Fauna of British India, I.,* 582. **64**
THE ROUGH TOOTHED DOLPHIN.
Habitat.—Indian and Atlantic Oceans. Captured in the Bay of Bengal.

18th, S. lentiginosus, *Owen ; Blanford, Fauna of British India, I.,* 584. **65**
THE SPECKLED DOLPHIN.
Vern.—*Bolla gadimi,* TEL.

WHALES, **Porpoises, etc.**	Whales, Porpoises, etc.

Habitat.—Indian seas. It has been captured at Vizagapatam and at Alibag, Bombay.   [*585.*

66   19th, **Steno? maculiventer,** *Owen ; Blanford, Fauna of British India, I.,*
Habitat.—A doubtful species recorded from Vizagapatam.

67   20th, **S. perniger,** *Blanford ; Fauna of British India, I., 583.*
ELLIOT'S DOLPHIN.
Vern.—*Gadamu,* TEL.   [tralia, etc.
Habitat.—Indian Ocean. Captured at Vizagapatam, Karachi, Aus-

68   21st, **S. plumbeus,** *Blanford ; Fauna of British India, I., 583.*
THE PLUMBEOUS DOLPHIN.
Vern.—*La-maing,* BURM.
Habitat.—Indian Ocean. Recorded from Madras, Malabar Coast, Karachi, Burma, and Ceylon.

69   22nd, **Tursiops tursio,** *Flower ; Blanford, Fauna of British India, I., 581.*
THE COMMON BOTTLE-NOSE PORPOISE.
Habitat. —Probably throughout temperate and tropical seas. Blyth records the skull of one captured in the Bay of Bengal.

OCCURRENCE
70

The economic facts regarding *1st,* the Dolphins and Porpoises may be here briefly reviewed and on the next page those of *2nd,* the Whales.
OCCURRENCE.—*1st,* " The Dolphins or Porpoises, as they are popularly called (the word dolphin is often restricted to the fish **Coryphæna,** celebrated for its changeable tints when dying), are found all over the world inhabiting seas, and many ascending large rivers. They generally associate in flocks or shoals, are very active, swimming and playing near the surface of the sea, and feeding on fishes, crustacea, cuttle fish, etc. They frequently accompany ships for miles" (*Jerdon*). The Indian species of the genus **Delphinus** are recorded from the Bay of Bengal and the coast of Malabar, that of **Globicephalus** from the same sea near the Hooghly, and those of **Platanista,** from the Ganges and Indus, respectively.

OIL.
Porpoise.
71
Black-fish.
72

Melon.
73

Oil.—From mammals belonging to this family is obtained the oil known in commerce as " porpoise oil," a term which includes " black-fish oil." It may be made from species belonging to any of the genera, but is principally obtained from the black porpoise, the white whale, and the grampus-all non-Indian or doubtfully Indian species. It may be noticed, however, that a particularly fine quality of oil is obtained from species of **Globicephalus** of which we have an Indian representative. Besides yielding ordinary " black-fish oil," these animals afford from the head a very limpid oil, commonly termed " melon oil," which has a very low solidifying point, has no corrosive effect on metal, and is a very superior lubricator for delicate machinery. Ordinary porpoise-oil is inodorous, burns with a brilliant light, congeals only in intense cold, and from its softness is valuable for lubricating and leather-dressing (*Spons' Encyclop.*). It might probably be prepared from any of the Indian species. The oil obtained from **Platanista gangetica** is esteemed in India as a valuable embrocation in rheumatism, for strengthening the loins, and for pains in the lumbar region generally. According to **Dr. Anderson** it pos

Sperm.
74
Spermaceti.
75

sesses high illuminating powers, and **Murray** mentions that it is used for burning by the fishermen of Sind. "Spermaceti" is the solid wax-like portion of Sperm-oil, or so-called " head-matter," found in the head of the Sperm-whale, **Physeter macrocephalus.** As already stated this. or a nearly allied species, occurs occasionally in the Indian seas. The "head-matter" is contained in a large cavity situated on the right side of the nose and upper portion of the head. By a process of purification this substance is split up into sperm-oil and spermaceti, both of which are of well known value in the arts, and the latter also in pharmacy.

W. 75

Timbers used for Wheels.	(*J. Watt.*)	WHEELS.

Medicine.—See above description of " Oil."

Food.—Dr. Anderson states that the Garhwals and certain other castes eat the flesh of the porpoise, **Platanista gangetica,** found in the Ganges and its tributaries, and **Murray** makes the same statement regarding the Indus form (**P. indi**), in Sind.

2nd, of whales there are three genera :—**Balæna** (the Right whales), **Negaptera** (the Humpbacks), and **Balænoptera** (the Fin-whales). Only the last-named genus has for certain been found in Indian waters.

Occurrence.—In India there are two species of whales belonging to the group which possess a dorsal fin, and hence called Funner, Finback, Finwhale, etc., also Pike-whale and Rorqual. **Balænoptera indica** was founded on a specimen cast up dead at Amherst Island, which measured 84 feet in length. Other large whales, supposed to belong to the same species, have been recorded at different times as thrown ashore on the Chittagong, Karachi, Malabar, and Ceylon Coasts. **Jerdon** states that they are hunted by whalers who make the Maldives and Seychelles their headquarters though they are not so much sought after as the " Right-whales " (**Balæna**), which yield much more blubber.

Oil.—WHALE-OIL, TRAIN-OIL, or BLUBBER, is too well known to require any description in this work. It is obtained much more abundantly from the true **Balæna,** all of which are inhabitants of Arctic or Antarctic seas— than from the other genera of the Family **Balænidæ.** Owing to the competition of mineral oils for illuminating and of other animal and vegetable oils for industrial purposes, and the substitution of various articles for the once almost indispensable WHALE-BONE, the industry of whale-fishing is undergoing a general decline (*Spons' Encyclop.*). For this reason, combined with the fact that the Indian whale is of an inferior kind, neither whale-oil nor whale-bone are ever likely to become important economic articles in this country.

(*J. Murray.*)

Wheat, see **Triticum sativum,** *Lamk.* ; pp. 88—202.

WHEELS.

Wheels, Timbers used in making—See under the following in their respective places in this work.

Acacia arabica, wheels.
A. Catechu, wheelwrights' work.
A. modesta, cart-wheels, persian water-wheels.
Ægle Marmelos, naves and other parts of carts.
Albizzia Lebbek, wheel-work.
A. odoratissima, wheels.
A. procera, wheels.
A. stipulata, naves of wheels and cart-wheels.
Anogeissus latifolia, axles.
Bassia latifolia, naves of wheels.
Carapa moluccensis, wheel-spokes.
Cassia marginata, naves of wheels.
Cordia vestita, wheels.
Dalbergia cultrata, wheels.
D. latifolia, cart wheels.
D. Sissoo, felloes and naves of wheels.

Feronia elephantum, naves of wheels.
Hopea odorata, cart-wheels.
Lagerstrœmia parviflora, cart-wheels
Odina Wodier, wheel-spokes.
Ougeinia dalbergioides, carriage-poles and wheels.
Pongamia glabra, cart-wheels.
Salvadora oleoides, persian wheels.
Streblus asper, wheels.
Tamarindus indica, wheels.
Tamarix articulata, persian wheels.
Terminalia Arjuna, cart-wheels.
T. Chebula, cart-wheels.
T. tomentosa, cart-wheels.
Vitex leucoxylon, cart-wheels.
Xylia dolabriformis, *Benth.* ; axles.
Zizyphus Jujuba, *Lamk.;* persian wheels.

Side margin notes:

MEDICINE.
76
FOOD.
Flesh.
77

OCCURRENCE.
78

OIL.
Whale.
79
Train.
80
Blubber.
81
Whale-bone
82

83

| WISSADULA rostrata. | Malaya or Borneo Rubber. |

White Lead, see **Lead,** Vol. IV., 603; also **Pigments,** Vol. VI., Pt. I., 231

WIGHTIA, *Wall.; Gen. Pl., II., 938.*

84   **Wightia gigantea,** *Wall.; Fl. Br. Ind., IV., 257;* SCROPHULARINEÆ.

Syn.—GURELINA SPECIOSISSIMA, *Don.*

Vern.—*Lakoú,* NEP.; *Bop,* LEPCHA.

References.—*Gamble, Man. Timb., 273; Cat. Pl., Darjíling, 58; Aplin, Rept. on Shan States, 1887-88; Ind. Forester, XIV., 343.*

Habitat.—A large tree, often half epiphytic, appressed to, and grasping the boles of forest trees by roots given off from the trunk, which are sometimes a foot or more in diameter. It is found in the Central and Western Himálaya, from Sikkim to Bhután, between the altitudes of 3.000 and 7,000 feet. Mr. Aplin recently found it also in the Southern Shan States of Burma.

TIMBER.   Structure of the Wood.—White, porous, light, and soft; it does not
85   warp, and is employed in the Himálaya to make Buddhist idols (*Gamble*).

WIKSTRŒMIA, *Endl.; Gen. Pl., III., 193.*
[THYMELÆACEÆ.

86   **Wikstrœmia canescens,** *Meissn.; Fl. Br. Ind., V., 195;*

Syn.—W. SALICIFOLIA, *Dcne.;* W. CHAMÆDAPHNE & INAMŒNA, *Meissn.;* W. VIRGATA, *Meissn.;* DAPHNE CANESCENS & VIRGATA, *Wall.;* D. SERICEA, *Don;* D. INAMŒNA, *Gardn.;* D. OPPOSITIFOLIA, *Ham.*

Vern.—*Chamlia,* KUMAON; *Bhat niggí, thilúk,* PB.

References.—*Beddome, For. Mun., 178; Brandis, For. Fl., 386; Gamble, Man. Timb., 314; Stewart, Pb. Pl., 189; Atkinson, Him. Dist., 316.*

Habitat.—A small shrub, found in the Temperate Himálaya from Kumáon to Central Nepál and the Khásia Hills, between 5,000 and 6,000 feet; also in Upper Assam and the Central Province of Ceylon.

FIBRE.   Fibre.—An inferior sort of Nepal-paper is made from the BARK in
Bark.   Kumáon; but it affords a strong cordage material, and ropes made of it
87   are used in Nainí Tál (*Madden*). (*Conf.* with the article on **Daphne,** *III.,* 20-24).

WILLUGHBEIA, *Roxb.; Gen. Pl., II., 691.*

88   **Willughbeia edulis,** *Roxb.; Fl. Br. Ind., III., 623;* APOCYNACEÆ.

MALAYA or BORNEO RUBBER.

Syn.—W. MARTABANICA, *Wall.;* PACOUREA GUDARA, *Ham.*

Vern.—*Luti-am,* BENG.; *Thit kyouk nway,* BURM.

References.—*Roxb., Fl. Ind., Ed. C B.C., 260; Kurz, For. Fl. Burm., II., 165; Drury, U. Pl., 445; Ind. Fore-ter, VII., 242; Agri.-Horti. Soc. Ind., Trans., VI., 29; Journ (New Series), VII., Pt. iii., 92.*

Habitat.—A large climber, found in Assam at Goalpara, Sylhet, Cachar, Chittagong, Pegu, Martaban, and Malacca, distributed to Borneo.

GUM.   Gum.—It yields a form of Caoutchouc, see **India-rubber,** Vol. IV., 363.
89   Food.—The FRUIT is eaten by Natives, and is considered good
FOOD.   (*Roxb.*).
Fruit.
90   **Wines,** see **Narcotics,** Vol. V., 319, 338; also **Vitis** above, pp. 251-296.

WISSADULA, *Medik.; Gen. Pl., I., 204.*

91   **Wissadula rostrata,** *Planch; Fl. Br. Ind., I., 325;* MALVACEÆ.

Syn.—W. PERIPLOCIFOLIA, *Thwaites;* W. ZEYLANICA, *Medik.;* ABUTILON PERIPLOCIFOLIUM, *G. Don;* SIDA PERIPLOCIFOLIA, *Linn.*

W. 91

References.— *Roxb., Fl. Ind , Ed., C.B.C., 516 ; Thwaites, En. Ceyl. Pl., 27 ; Royle, Fib. Pl., 263.*

Habitat.— Cultivated in India, naturalised in Ceylon, and very common in the south of the island ; a native of the Malay Peninsula, Java, Tropical Africa, and America.

Fibre.— "The BARK of this abounds in serviceable flaxen fibres, and as it shoots quickly into long simple twigs, particularly if cut near the earth, it answers well for procuring the fibre of a good length for most purposes" (*Roxb., Fl. Ind.*). Royle commenting on that passage writes : "Some of Dr. Roxburgh's original specimens, marked July 1804, are still in the India House ; the fibres are from 4 to 5 feet in length, and display a fine soft and silky fibre, as well adapted for spinning as jute, but are apparently superior." From this description the fibre would seem to resemble that of the nearly related **Sida rhombifolia** (*Cf. Vol. VI., Pt. II., p. 681*), and, like that fibre, to be well worthy of further examination.

FIBRE.

Bark.
92

WITHANIA, *Pauq.; Gen. Pl., II., 893.*

[*t. 1616 ;* SOLANACEÆ.

Withania coagulans, *Dunal.; Fl. Br. Ind., IV., 240; Wight, Ic.,* THE CHEESE-MAKER or INDIAN RENNET.

93

Syn.— PUNEERIA COAGULANS, *Stocks.*

Vern.— *Akri, punir,* HIND.; *Ashvagandá,* BENG.; *Spin bajja, shápiang, khúmazare, makhazura, panir, khamjira, kútílúna, khamjaria,* fruit= *akri, panir,* PB.; *Khamazora, shápránga, spin-bajja,* PUSHTU ; *Punir-band, punir-ja-fota,* SIND.; *Káknaj,* BOMB.; *Hindi-káknaj, nát-ki-asgand,* DEC.; *Amukkura,* TAM.; *Pennéru-gadda,* TEL.; *Asvagandhi,* KAN.; *Káknaje-hindi,* ARAB.; *Káknaje-hindi, panir-bad,* PERS.

References.— *Gamble, Man. Timb., 161 ; Pharm. Ind., 181 ; Moodeen Sheriff, Supp. Pharm. Ind., 258 ; Dymock, Mat. Med. W. Ind., 2nd Ed., 645 ; Baden Powell, Pb. Pr., 273, 362 ; Drury, U. Pl., 445 ; Birdwood, Bomb. Pr., 59, 345 ; Kew Reports, 1881, 36 ; Gazetteers :—Panjab, Dera Ismail Khan, 1883-84; Peshawar, 27 ; Agri.-Horti. Soc. Ind., Journ. (Old Series), XII., 384; XIII., 176 ; (New Series), I., 75 ; Rep. Exp. Farms, Madras, 1882-83, 1883-84, 1884-85 ; S. Lea, in the Proc. of Royal Society, Eng., 1883, No. 228 ; Chem. News., Dec. 7th. 1883.*

Habitat.— A small shrub, common in the Panjáb, Sind, Afghánistán, and Balúchistán.

Medicine.— The FRUIT, when fresh, is used as an emetic ; when dried as a remedy for dyspepsia, flatulent colic, and other intestinal affections, for which they enjoy a high reputation in Sind, Balúchistán, and Afghánistán. It is also prescribed in infusion with the leaves and twigs of **Rhazya stricta,** *Dcne.;* an excellent bitter tonic, known in Sind by the name of *sihar* or *sewar* (*Conf. Pharm. Ind.; Dict., Econ. Prod., VI., 443*). Dymock adds that it is alterative, diuretic, and believed to be useful in chronic liver complaints. Stewart states that it has anodyne or sedative properties. According to Honigberger the LEAVES are bitter, and are given as a febrifuge by the Lohanís. *Thirty-five years in the East by Honigberger, Vol. II., 325.*

MEDICINE.
Fruit.
94

Leaves.
95

Domestic, etc.— The FRUIT is commonly employed in Sind, North-West India, Afghánistán, and Balúchistán, as a substitute for Rennet to coagulate milk (*Conf. with Rennet, Vol. VI., Pt. I., 427*). For this purpose they are rubbed up with a small quantity of milk, and then added to the rest. This valuable property was first noticed and made known by Dr. Stocks in the *Journal of the Asiatic Soc., Bomb., 1849, 55.* In 1880-81, the question of obtaining an efficient vegetable Rennet, which was represented as necessary to the expansion of the consumption of cheese by Natives, and consequently to the development of cheese-making as an industry, attracted the attention of the authorities at Kew. Sir J.

DOMESTIC.
Fruit.
96

W. 96

WITHANIA
coagulans.   The Cheese-maker or Indian Rennet.

DOMESTIC.   D. Hooker, on the suggestion of **Surgeon-Major Aitchison**, recommended the trial of the fruit of this plant. A quantity of the dried
capsules were obtained at the Government Farm, Khandesh, and were
found to answer wery well, the Superintendent reporting as follows :—
" It has been ascertained that an ounce of the powdered capsules in a
quart of water is a very suitable strength for use ; a table-spoonful of this
decoction coagulates a gallon of warm milk in about half an hour." Ex-
Experiments   periments made in 1883 and 1884 at the Saidapet Farm, Madras, were
in cheese-   much less satisfactory, probably owing to the fruit having been old and
making.   inactive, or to the defective method employed. In this case the powdered
97   fruit was added to the milk without any previous soaking ; and as a
natural consequence very little coagulation occurred till the milk became
unfit for use (see **Streblus asper,** Vol. VI., Pt. III., 373). In November
1883, **Mr. S. Lea** published an account of certain experiments he had made
with the seeds, with the object of ascertaining whether they contained a
definite ferment with the properties of ordinary Rennet, and the applica-
bility of such a ferment to cheese-making purposes. The seeds, carefully separated from the capsule and all other foreign matter, were subjected for 24 hours to the action of various solvents which were then
added to milk. A five-per cent. solution of common salt in water was
found to be most efficient in the extraction of the ferment, the extract
rapidly curdling milk. The results of **Mr. Lea's** experiments may be sum-
marised as follows :—

(1) The sodic chloride solution extract loses its activity if boiled.
(2) The ferment is soluble in glycerine, and can be extracted from the
        seeds by means of it ; this extract possesses strong coagulat-
        ing powers, even in small amounts.
(3) Alcohol precipitates the ferment body, but it may be re-dissolved
        in an appropriate fluid without loss of its coagulating power.
(4) The active principle will cause coagulation when present in very
        small amount ; the addition of a larger quantity simply increases
        the rapidity of the coagulation.
(5) The coagulation is not due to the formation of acid by the fer-
        ment.
(6) The clot is a true clot, resembling in appearance and properties
        that formed by animal rennet, and is not a mere precipitate.
(7) Lastly, there is no doubt that the substance, which possesses the
        coagulating power, is a ferment, closely resembling animal
        rennet.

Having determined these points the analyst went on to endeavour to
prepare an active extract which should be applicable for cheese-making
purposes. The only difficulty encountered was in obtaining a colourless
solution, since all the extracts proved to be deep-brown, and all methods
of decolourizing them destroyed the activity of the ferment. In order to
obviate this disadvantage he found it best to prepare very concentrated,
active extracts of carefully cleaned seeds. Such an extract he prepared by grinding the dry seeds very fine in a mill, extracting them for
24 hours with a volume of five per cent. salt solution sufficient to render
the mass still fluid after the absorption of water by the fragments of the
seeds as they swelled up, and separating the fluid part of the mass by a
centrifugal machine, and subsequent filtering. Forty grammes of the seeds
thus treated with 150 cubic centimetres of five per cent. salt solution gave
an extract of which 0˙25 cubic centimetres, clotted 20 cubic centimetres of
milk in 25 minutes, and 0˙1 cubic centimetres clotted a similar volume of
milk in one hour, in both cases producing a perfectly white curd. But the
presence of a little colouring matter may be after all unimportant and scarce-

ly worthy of the trouble which this somewhat tedious process involves, and Mr. **Lea** found that in any case the greater portion of the colouring matter was dissolved out by the whey. He prepared an extract capable of being kept for some time by adding sufficient salt to the five-per cent. solution to raise the percentage to 15 per cent.; and also alcohol up to 4 per cent. The activity of the extract was not appreciably altered by this, and such a preparation ought to retain its activity, since it agrees in composition with ordinary commercial animal rennet extracts. It was also found to correspond very closely in activity with such a commercial extract.

Mr. **Lea**, in concluding his interesting paper, a report of experiments which have indisputably proved the value of this vegetable Rennet, writes:— "I may add that I have coagulated a considerable volume of milk with an extract such as I have described, and prepared a cheese from the curds. I have also given a portion of the extract to a professional cheese-maker who has used it as a substitute for animal Rennet in the preparation of a cheese. The product thus obtained, and the statements of the person who has made the experiment for me, lead me to suppose that the seeds of **Withania** can be used as an adequate and successful substitute for animal Rennet."

It may be added in conclusion that the fruit is readily obtainable in the bazárs of Northern India, where, according to **Baden Powell**, it sells for from 12 to 14 seers (24 to 28℔) per rupee.

DOMESTIC. Experiments in Cheese-making.

Withania somnifera, *Dunal; Fl. Br. Ind., IV., 239; Wight, Ic., t.853.* **98**

> **Syn.**—PHYSALIS SOMNIFERA, *Link.*; P. FLEXUOSA, *Linn.*; P. ARBORESCENS & TOMENTOSA, *Thunb.*
>
> **Vern.**—*Punir, asgand, asgandh,* HIND.; *Ashvaganda, asvagandhá, seed=kak-nuj,* BENG.; *Asud-gandhá,* URIYA; *Asgand nágori, isgand, ak, aksan,* root=*asgand nagauri, vaman, agsend,* PB.; *Kútilal, sin,* PUSHTU; *Asgund, asvagandha,* BOMB.; *Askandha, tilli,* MAR.; *Ghodá, asoda, asan,* GUZ.; *Hindi-kaknaj, nát-ki-asgand,* DEC.; *Amukkurá, amkúlang,* root=*amúlang kalung, aswagandhi,* TAM.; *Pennéru-gadda, pennéru, pilli véndram,* TEL.; *Yiremaddinagadde,* KAN.; *Amúkkará,* SINH.; *Ashwa gandha, asvagandhá,* SANS.; *Káknaje-hindi,* ARAB. & PERS.
>
> **References.**—*Roxb., Fl. Ind., Ed. C.B.C., 189; Thwaites, En. Ceyl. Pl., 217; Trimen, Cat. Ceylon Pl., 61; Dals. & Gibs., Bomb. Fl., 175; Stewart, Pb. Pl., 161; Rheede, Hort. Mal., IV., t. 55; Elliot, Fl. Andhr., 17, 151, 152; Pharm. Ind., 182; Ainslie, Mat. Ind., II., 14; O'Shaughnessy, Beng. Dispens., 466; Honigberger 35 years in the East, II., 324; Moodeen Sheriff, Supp. Pharm. Ind., 258; U. C. Dutt, Mat. Med. Hind., 210, 292; Dymock, Mat. Med. W. Ind., 2nd Ed., 643; Irvine, Mat. Med. Patna, 4, 50; Trans., Med. & Phys. Soc., Bomb. (New Series), No. 4, 154; Cat. Baroda Durbar, Col. & Ind. Exhib., No. 185; Med. Topog., Ajmere, 123; Official Corresp. on Proposed New Pharm. Ind., 238; Baden Powell, Pb. Pr. 363; Atkinson, Him. Dist., 314, 753; Drury, U. Pl., 446; Lisboa, U. Pl. Bomb., 268; Birdwood, Bomb. Pr., 59; Hunter, Orissa, II., 158; Boswell, Man. Nellore, 134; Settlement Rep., C. P., Chanda, App. vi.; Gazetteers:—Bombay, V., 27; N.-W. P., I, 83; IV, lxxv.; Ind. Forester, XII., App. 18; Agri.-Horti. Soc. Ind., Journ. (New Series), I., 48, 97.*

Habitat.—An erect shrub, found throughout the drier parts of India, frequent in the West and in Hindústán, but rare in Bengal.

Medicine.—According to **Dymock**, Indian literature on Materia Medica is quite untrustworthy as regards the medicinal properties of the ROOT of this shrub. He writes, "It has universally been confounded with a root met with under the same names in the bazárs, but which bears no resemblance to the root of **W. somnifera.** The *asgund* of the shops is the tuber of a CONVOLVULUS, which, though much smaller and different in

MEDICINE. Root. **99**

WOODFORDIA
floribunda.   A good tonic and diuretic.

MEDICINE.

habit, does not appear to differ botanically from **Ipomæa digitata**" (*Cf.*
Vol. IV., 484). Honigberger was, however, the first author who pointed
out the mistake in the two roots commonly sold as *asgund*. The Panjábís,
he says, call the plant *agsend*, not as stated by some authors *asgend*, the
latter is also officinal in the Panjáb, but is imported from Hindústán.
Roxburgh states that the "Telinga physicians reckon the roots alexephar-
mic." Ainslie describes the bazár *asgund*, which is probably not the article
Leaves.
100   at present under consideration, and states that the LEAVES (very likely
Seeds.   those of **Withania**), moistened with a little warm castor-oil, are a useful
101   external application in cases of carbuncle. Irvine describes the SEEDS as
Root.   diuretic and hypnotic, and the ROOT as narcotic and diuretic, a remark
102   confirmed by Dalzell & Gibson. In the *Pharmacopœia of India*, the
root is said to be used externally similarly to the leaves, to be regarded
by Rájpúts as useful in rheumatism and dyspepsia, and to be feebly
diuretic. Most of the above information probably refers in reality to this
species.

The *asgund* of the shops is quite different in appearance from the root
of **Withania** (see Vol. IV., 484); it has a mucilaginous and slightly bitter
taste, and is evidently the *asvagandha* of Sanskrit writers. According to
Dutt it is regarded in Hindu medicine as tonic, alterative, and aphrodisiac,
and is employed in consumption, emaciation, debility from old age, and
rheumatism. It enters into the composition of many tonic preparations
prescribed by **Chakradatta** and others, and is a favourite constituent of
aphrodisiac medicines (*Mat. Med. Hindus*). Dymock informs us that in
the *Makhzan-el-Adwiya* it is described as tonic and alterative, and is said
to have much the same properties as white Behen.

SPECIAL OPINION.—§ "Root, tonic and diuretic, juice of whole plant
a useful remedy for rheumatism, in doses of one to two ounces : no narcotic
effect observed" (*Apothecary T. Ward, Madnapalle, Cuddapah*).

FODDER.
Leaves.   Fodder.—In the Panjáb, the LEAVES are browsed by goats.
103   Domestic, etc.—In Bombay the SEEDS are employed similarly to those
DOMESTIC.   of **W. coagulans** in Sind, *viz.*, to coagulate milk (*Dals. & Gibs.*). Stewart
Seeds.   states that the ROOT is occasionally employed in the Panjáb to effect cri-
104   minal abortion, and that the same practice is believed to be common in
Root.   Sind.
105

Wood-apple Tree, see Feronia elephantum, *Corr.;* Vol. III., 324.

Wood-oil, see Dipterocarpus alatus, *Roxb.;* D. incanus, *Roxb.;* D. lævis,
Ham.; D. pilosus, *Roxb.;* D. tuberculatus, *Roxb.,* & D. turbinatus,
Gærtn. f.; Vol. III., 157-171.

WOODFORDIA, *Salisb.; Gen. Pl., I., 778.*

[RACEÆ.

106   **Woodfordia floribunda,** *Salisb.; Fl. Br. Ind., II., 572;* LYTH-
Syn.—W. TOMENTOSA, *Bedd ;* W. FRUTICOSA, *Kurz ;* GRISLEA TOMEN-
TOSA, *Roxb.;* G. PUNCTATA, *Ham.;* LYTHRUM FRUTICOSUM, *Linn.*
Vern.—*Dáwi, thawi, santha, dhaula, dhaura, dhái, dha,* HIND.; *Dhái,
dawai, dhawayi, dawa, dhowa, dhao, dhadki, dhan, dhainti, dhaura-*
BENG.; *Icha, dhawe,* KOL; *Ichak,* SANTAL; *Dahiri, laldairo, dhager-
ako,* NEPAL; *Chungkyek-dúm,* LEPCHA; *Jatiko, harwari,* URIYA;
Dadki, BHUMIJ.; *Khinni, dhi,* KURKU; *Dhuvi, surtari, dhaiti, dhowra,
dhowai,* C. P.; *Pit:a, petisurali, surteyli,* GOND.; *Datti,* BHIL; *Dhai,
N.-W. P. ; Dhewti,* OUDH; *Dhái, dhaula, dhaura, thawa, dhárla,*
KUMAON; *Gul daur, dhai,* KANGRA; *Tháwi, thái,* KASHMIR; *Táwi, thái,
tuu, dahái, dháwi, khúrd, dhá, dáwi, dhaur, dhas,* flowers=*gul-dháwi,
gul bahar,* PB.; *Dátki,* PUSHTU; *Dhái,* SIND; *Dhauri, dhayati, dhávri,
dhavshi,* BOMB.; *Phulsatti, dhasatichi,* MAR.; *Dhavadina,* GUZ.; *Jar-*

*gi, serinji, gaddaisinka, gáji, godári, dhataki, kusumamu, reyyi pap-
pu, jagi,* TEL. ; *Dhátaki, agnijvála,* SANS.

References.—*Roxb., Fl. Ind., Ed. C.B.C.,* 317 ; *Brandis, For. Fl.,* 238 ;
Kurz, For. Fl. Burm., I., 518 ; *Beddome, Fl. Sylv., Anal. Gen., t. XIV.,
fig. 4 ; For. Man.,* 117 ; *Gamble, Man. Timb.,* 200 ; *Dalz. & Gibs.,
Bomb. Fl.,* 97 ; *Stewart, Pb. Pl.,* 90 ; *Rev. A. Campbell, Rept. Econ. Pl.,
Chutia Nagpur, No.* 7536 ; *Mason, Burma & Its Peole,* 512, 759 ; *Sir
W. Elliot, Fl. Andhr.,* 47, 56, 60, 164 ; *Irvine, Mat. Med. Patna,* 27 ; *U. C.
Dutt, M t. Med. Hind.,* 165, 296 ; *Murray, Pl. & Drugs, Sind,* 144 ;
Dymock, Mat. Med. W. Ind., 2nd Ed., 306 ; *Dymock, Warden & Hooper,
Pharmacog. Ind., II.,* 40 ; *Cat. Baroda Durbar, Col. & Ind. Exhib., No.*
112 ; *Birdwood, Bomb. Prod.,* 298 ; *Baden Powell, Pb. Pr.,* 348 ; *Drury, U.
Pl. Ind.,* 235 ; *Atkinson, Him. Dist. (X., N.-W. P. Gaz.),* 310, 753, 778 ;
Useful Pl. Bomb. (XXV., Bomb. Gaz.), 245, 396 ; *Econ. Prod. N.-W.
Prov., Pt. III. (Dyes & Tans),* 37, 53, 58 ; *Gums & Resinous Prod. (P. W.
Dept. Rept.),* 16 ; *Liotard, Dyes,* 46, 69, 136, 137 ; *Cooke, Gums & Resins,*
18 ; *McCann, Dyes & Tans, Beng.,* 32, 33, 34, 152-153, 161 ; *Wardle, Dye
Rep.,* 8, 21, 23, 43, 45 ; *Cat. Col. Ind. Exhb., Raw Prods., No.* 146 ; *Selec-
tions, Records Govt. India (R. & A. Dept.),* 1888-89, 91 ; *Man. Madras
Adm., I.,* 313 ; *For. Adm. Rep., Ch. Nagpur,* 1885, 6, 31 ; *Settlement
Reports:—Central Provinces, Chanda. App. vi. ; Raepore,* 76, 77 ; *Ho-
shungabad,* 180 ; *Gazetteers:—Bombay, XIII.,* 24 ; *XV.,* 434 ; *Panjáb,
Ráwalpindi,* 15 ; *Peshawar,* 27 ; *N.-W. P., I.,* 81 ; *IV., lxxii ; Burma,
I.,* 138 ; *Agri.-Horti. Soc. Ind. :—Journ.* (*Old Series*) *I.,* 290-292 ; *IX.,
Sel.,* 54 ; *XIII.,* 307, 390 ; *XIV.,* 15 ; (*New Series*), *VI., Sel.,* 19 ; *Trans.
of Med. & Phys. Soc., Bomb.* (*New Series*), *IV.,* 156 ; *Ind. Forester, II.,*
175, 176 ; *III.,* 202 ; *IV.,* 228 ; *X.,* 222, 325 ; *XII., A.,* 14 ; *XIII.,* 121 ;
XIV., 296, 390.

Habitat.—A small, much-branched shrub, brilliantly purple in the hot
season owing to the numerous flowers all along its branches ; common
throughout India, ascending on the Himálaya to an altitude of 5,000 feet,
and in the mixed dry forests of Prome.

Gum.—According to Balfour the gum of this plant known as *dhaura*
or *dhau-ka-gond* is collected largely in Harauti and Mewar. It is said
to resemble gum tragacanth and to swell in water. It is employed in
dyeing to coat the parts of a fabric which are required to remain uncoloured ; one maund is said to cost R10.

**GUM.
107**

Little is known about the properties or value of this gum. **Cooke**
writes : " The Museum samples do not at all agree in character. One
sample from Allahabad is a good strong gum in tears ; a sample from
Bengal is in smaller fragments ; one from Ahmedabad appears to be mixed, and one from Indore is very much like the gum of **Conocarpus.** The
Dhokra gum from Gúnah is in rounded tears about the size of a filbert,
and may belong to this species." [It seems probable that there is some
mistake here, a confusion having been made with *dhává* or *dhau,* etc.—
the gum of **Anogeissus latifolia**—(*Conf. with Vol. I.,* 256). The writer has
no recollection of ever having seen a gummy exudation on **Woodfordia.**—
Ed., Dict. Econ. Prod.)]

Dye.—The FLOWERS are employed throughout India in dyeing either
to produce a colour of themselves, or as an adjunct or mordant, principally with *ál,* **Morinda citrifolia** (*Cf.* Vol. V., 272). The plant flowers
from February to April, during which period the blossoms are gathered and
dried, and in districts where the LEAVES are used as a tan, these are
gathered and dried in the autumn. The plant is everywhere a jungle product, so the cost of production is merely that of the labour of collecting
the flowers. McCann states that the flowers are rarely used as a dye by
themselves in Bengal, but nearly invariably as an adjunct to the process
of *ál* dyeing. When used by themselves, the flowers are either boiled in
water, or else steeped for a considerable time in cold (? Manbhum) or hot
water. To the solution thus prepared, alum, or lime and alum, is added

**DYE.
Flowers.
108
Leaves.
109**

DYE.

as a mordant, and the material to be dyed is immersed in this solution several times until a pink colour of the required depth is obtained (*Dyes & Tans of Bengal*). In the Central Provinces and Rájputána and the Panjáb, the practice of dyeing with the flowers alone appears to be more common than elsewhere; a pink or red colour is said to be obtained. Sir E. C. Buck (*Dyes of N -W. Prov.*, p. 37) makes no mention of their being used except in *ál* dyeing or with silk.

Samples of the flowers submitted for examination to **Mr. Wardle** elicited the following report:—"These flowers, a sample of which, in the dried state, I have examined with regard to their tanning and dyeing qualities, I find are principally valuable from the fact that they contain tannin equivalent to about 25 per cent. of oak-bark tannin, thus being almost equal to sumach. They contain also a small amount of yellowish-brown colouring matter soluble in water, which gives, by the use of various processes, faint though artistic shades of colour to *tussur* silk, mulberry silk, and woollen fabrics, and when the infusion is simply applied without the intervention of any other mordant, no doubt the tannin acts as a fixing agent. When the infusion is used as a dye in the presence of a salt of iron, peculiar slate and brownish shades may be obtained, owing to the dark colour produced by the action of the iron on the tannin being modified by the yellow colouring matter contained in the flowers." Sir E. C. Buck (p. 53) alludes to this fact being known to the calico-printers of India, as giving lustre to the black dyes of sulphate of iron. In another passage **Mr. Wardle** describes the tints produced on silk as "beautiful brown-yellow colours."

Twigs.
110
TAN.
Flowers.
111
Leaves.
112

The leaves and TWIGS yield a yellow dye called *nauti*, which is occasionally used in Northern India in calico printing (*Sir E. C. Buck*).

Tan.—Though the flowers would appear from **Mr. Wardle's** report to be of considerable value as a dye-stuff, still, as he remarks, their importance depends much more on the large quantity of tannin they contain. They have been long used to a small extent for tanning in certain parts of India, but appear to have been neglected for this purpose, for the LEAVES, which are one of the most commonly employed of Indian tanning materials. Perhaps the first person to notice the large amount of tannin contained in the flowers was **Dr. Balfour**, who thus explained their value in *ál* dyeing. As stated above **Mr. Wardle** found them to contain a large percentage, and predicted the probable value which they might thus obtain. At the Colonial and Indian Exhibition samples were submitted for analysis to **Professor J. J. Hummel**. He reported that they contained 20·6 per cent. of tannic acid, and yielded a deep red turbid solution. He classed them with the best Indian tanning materials, and remarked, "On examining the list (of selected tanning materials) it becomes evident that the best Indian tannin matters are already in the market, with the exception of **Woodfordia floribunda**; and even this has the disadvantage of giving deeply-coloured decoctions." "I was somewhat surprised to find the flowers contain such a large percentage of tannic acid; but it explains why the Hindus use these flowers in connection with alum as a mordant and with other dye-stuffs, *e.g.*, **Morinda**, as mentioned in the Exhibition Catalogue. It would be interesting to examine the leaves of the plant."

It is to be hoped that the last suggestion may be carried out. The leaves may prove of equal value and could probably be obtained in larger quantities and more cheaply. **Professor Hummel** estimated the value of the flowers at 5*s.* 7*d.* as compared with Divi-divi, 10*s.* 5*d.* as compared with Valonia cups, 14*s.* as compared with Ground Sumach, and 6*s.* 8½*d.* as compared with Ground Myrabolans. Regarding the available supply and cost in India, accounts vary greatly, and reliable information is in

W. 112

many cases wanting. In Bengal McCann states that no particulars are available except from Palamau in Lohardaga, where the annual produce is estimated at 200 maunds. In Manbhúm "any quantity" is said to be obtainable in January and February. The prices were reported to vary from annas 4 to R1 a maund in Manbhúm, to R5 per maund in Húghlí. According to Atkinson the average annual export from the tract between the Jumna and the Sárda is about 27 tons, of which about 200 maunds comes from the Kumáon forest division. Sir E. C. Buck states that in 1874-75, 539 cwt., valued at R980, were imported into the North-West Provinces from Garhwál, Kumáon, Bijnor, and Behar. From these figures it may be assumed that a considerable supply could be obtained should a demand arise, and at a price probably about R2 per maund.	TAN.
Medicine.—The dried FLOWERS are regarded in Hindu medicine as stimulant and astringent, and are much used in bowel complaints and hæmorrhages. Two drachms of the dried flowers are given with curdled milk in dysentery, and with honey in menorrhagia. The powdered flowers are sprinkled over ulcers to diminish discharge and promote granulation (*Hindu Mat. Med.*). Dymock states that the Natives of the Konkan, in cases of bilious sickness, fill the patient's mouth with sesamum oil, and apply the JUICE of the LEAVES to the crown of the head; this is said to cause the oil in the mouth to become yellow from absorption of bile; fresh oil is then given repeatedly until it ceases to turn yellow. In Northern India the flowers are considered cooling, astringent, and stimulant, and are prescribed in pregnancy, bilious and mucous disorders, and hæmorrhoids. The leaves are also employed medicinally in Native practice. In Chutia Nagpur, "a decoction of the flowers is given for the female complaint known as *pordhol*" (*Campbell*).	MEDICINE. Flowers. 113 Juice. 114 Leaves. 115
Food.—In Bengal the FLOWERS are largely employed in the preparation of a cooling drink (*McCann*). The flowers are said to be eaten in the Central Provinces (*Chanda Settl. Rept.*). "In Kangra part of the plant is stated to be used in the preparation of spirits?" (*Stewart*).	FOOD. Flowers. 116
Structure of the Wood.—Reddish-white, hard, close-grained; weight about 46℔ per cubic foot (*Gamble*). It is used only for fuel.	TIMBER. 117

Wool, see the article **Sheep and Goats,** Vol. VI., Pt. II., 549—672.

WORMIA, *Rottb.; Gen. Pl., I., 13 & 954.*

Wormia triquetra, *Rottb.; Fl. Br. Ind., I., 35;* DILLENIACEÆ.	118

    Syn.—W. DENTATA, *DC.;* DILLENIA DENTATA, *Thunb.*

    Vern.—*Diyapara,* SING.

    References.—*Thwaites, En. Ceyl. Pl., 4; Gamble, Man. Timb., 4.*
    Habitat.—A tree found in the moist, warm parts of Ceylon up to 2,000 feet.

Oil.—The NUT yields an oil (*Gamble*).	OIL. Nut. 119
Structure of the Wood.—Reddish, resembling that of Dillenia; weight 44℔ per cubic foot. It is used in building.	TIMBER. 120

Worm-seed, see Artemisia maritima, *Linn.;* Vol. I., 324.

Worm-wood, see Artemesia Absinthium, *Linn.;* Vol. I., 323.

WRIGHTIA, *Br.; Gen. Pl., II., 712.*

Wrightia antidysenterica, *Grah.;* APCCYNACEÆ; see Holarrhena antidysenterica, *Wall.;* Vol. IV., 255.	121

122   Wrightia tinctoria, *R. Br.; Fl. Br. Ind., III., 653; Wight, Ic., t.*
                                                              [*444.*

Syn.—NERIUM TINCTORIUM, *Roxb.*
Var. Rothii=W. ROTHII, *G. Don; Wight, Ic., t. 1319;* W. TINC-
TORIA, *Roth.*

Vern.—*Indarjou, mithá indarjou,* HIND.; *Indrajau, indarjou,* BENG.;
Khirni, MEYWAR; *Dudhi,* BANDA; *Kálakado, kála-kud ι, kuda, khirni,
bhúrkúri, kála-kúra,* BOMB.; *Kála kúdú, indrajou,* MAR.; *Indarjou,*
GUZ.; *Indarjou, mithá indarjou,* DEC.; *Pálá. veypalé, pilá, palak,
palavay-ravnú, vetpá-larishi, vetpála,* TAM.; *Tedluvál, tella pal, amkudu, tedlapála, tshil-ankalú, chit-ankalú, kodisha, kalinga,* TEL.; *Kod-
murki, beppalli, hale,* KAN.; *Kotakappála,* MALAY.; *Venál-arsi, vepál-
pál,* SING.; *Hyamoraka,* SANS.; *Lasánul aasáfir, lasánul-aasáfirul-
haló,* ARAB.; *Indarjou, indarjouve-shirin, ahar, ah ≀re-shirin, zabáne-
kunj ι∙hk, zabáne-kunjashke-shirin,* PERS.
Two kinds of *indarjau* are found in the bazárs and have been much confused,
namely, the seeds of this species—sweet *indarjau*—and those of **Holarrhena
antidysenterica**,—bitter *indarjau* (see Vol. IV., 255).

References.—*Roxb., Fl. Ind., Ed. C.B.C., 243; Brandis, For. Fl., 324;
Kurz, For. Fl. Burm., II., 193; Beddome, Fl. Sylv., t. 241; Gamble,
Man. Timb., 264; Grah., Cat. Bomb. Pl., 114; Dalz. & Gibs. Bomb.
Fl., 145; Elliot, Fl. Andhr., 14, 44, 174; O'Shaughnessy, Beng. Dispens.,
446; Moodeen Sheriff, Supp. Pharm. Ind., 259; Dymock, Mat. Med. W.
Ind., 2nd Ed., 500; Baden Powell, Pb. Pr., 601; Drury, U. Pl., 447;
Lisboa, U. Pl., Bomb., 100, 166, 247, 291, 391; Birdwood, Bomb. Pr., 55,
301; Buck, Dyes & Tans, N.-W.P., 30; Liotard, Dyes, 96; Man. Adm.
Madras, II., 98; Moore, Man. Trichinopoly, 81; Nicholson, Man.
Coimbatore, 41; Gazetteers:—Mysore & Coorg, I., 62; Bombay, VI., 14;
XIII., 25; XV., 79, 438; Agri.-Horti. Soc. Ind., Journ. (Old Series),
III., 232; IV., Sel., 86-88, 129, 130; Pro., 27, 36, 91; V., 7. Sel., 28, 30,
77; IX., 296, Sel., 55; XI., Pro., 52; Ind. Forester, III., 203; VI., 240;
XII., App., 16.*

Habitat.—A small deciduous tree met with in Central India, the Western Peninsula generally, and Burma.

DYE.   **Dye.**—The SEEDS are said to be used as an adjunct to other materials
Seeds.   in dyeing. From time immemorial the Natives of Southern India have
123   employed the LEAVES as a source of a blue dye or indigo. This fact ap-
Leaves. pears to have first attracted the notice of Roxburgh, who wrote a treatise
124   on the subject, recommended certain apparatus, boilers, etc., to be used
in the preparation of the dye, and in 1792 made and transmitted a sample
to England. He found that the colouring matter was contained in the
leaves alone; that the best time for gathering was in March and April,
but that the picking might be profitably continued till the end of August;
that the colouring matter might be completely extracted by boiling for
three hours, and that from two to three hundred pounds of the leaves
yielded one pound of indigo. After the date of Roxburgh's experiments
and treatise interest in the matter appears to have dropped till 1844. In
that year the subject was again raised in the publications of the Agri.-
Horticultural Society of India, and much interesting information was elicited. From papers by a **Mr. Fishcher** of Salem in particular, it appears
that the manufacture of **Wrightia** indigo had been carried on for some years
in that place, with an apparatus fundamentally similar to that proposed
by Roxburgh. The indigo obtained was good, and fetched from 4 to 5-6*d.*
per ℔. The leaves picked by coolies in the jungle cost from 3 to 4 annas
per 150 to 200℔, from which quantity one pound of dye was obtained.
The leaf was mixed with water, boiled in large boilers and the dye separated in a straining vat.

The objections to the further utilisation of the leaves of this common
tree in competition with ordinary indigo appear to have been, *1st,* the great

expense of the boilers and the fuel required for them ; *2nd*, the limited quantity of leaves obtainable at any one place and the large area over which they have to be collected

These facts must necessarily render the preparation of **Wrightia** indigo more expensive than that obtained from **Indigofera** (*Cf.* Indigo, Vol. IV., 451).

Medicine.—The ROOT-BARK of this plant, along with the SEEDS, have been the cause of much confusion in the literature of Indian Materia Medica. This has already been fully dealt with under the article **Holarrhena antidysenterica** (see Vol. IV., 255-258), and need not be again entered into here. The plant now under consideration is practically inert from a medicinal point of view.

Food.—According to **Lisboa** the tender LEAVES and PODS are eaten in Bombay.

Structure of the Wood.—White like ivory, hard, close-grained ; weight 49℔ per cubic foot. It is used for carving, turnery, and building.

Domestic & Sacred.—The tree bears handsome clusters of white, jasmine-scented FLOWERS, which are much esteemed by Hindus as fit offerings at temples.

[*Wight, Ic., t. 443, 1296.*

Wrightia tomentosa, *Ræm. & Schultes ; Fl. Br. Ind., III., 653 ;*

    Syn.—W. MOLLISSIMA, *Wall.* ; W. WALLICHII, *A. DC.* ; W. CORAIA, *Wall.* ; W. HAMILTONIANA, *Wall.* ; HUNTERIA EUGENIFOLIA, *Wall.* ; NERIUM TOMENTOSUM, *Roxb.* ; N. CORAIA, *Ham.* ; N. ? TINCTORIUM, *Ham.*

    Vern.—*Dudhi, dharauli, daira,* HIND. ; *Dudh-koraiya,* BENG. ; *Sandi-kuya,* KOL. ; *Atkura, buru machkunda,* SANTAL ; *Atkuri,* ASSAM ; *Karingi, kirra,* NEPAL ; *Selemnyok,* LEPCHA ; *Pal kurwan,* URIYA ; *Harido,* CUTTACK ; *Dúdhi, kildwa, keor,* PB. ; *Dudhi, kadu-inderjao, daira,* BOMB. ; *Kala inderjan,* MAR. ; *Teua pal, koila-mukri, koyila mokiri, pútta jillédu, pedda pála,* TEL. ; *Lettouk thein, lettop-thein,* BURM.

    References.—*Roxb., Fl. Ind., Ed. C.B.C., 243 ; Brandis, For. Fl., 323 ; Kurz, For. Fl. Burm., II., 192, 193 ; Beddome, Fl. Sylv., Anal. Gen., 159 ; Gamble, Man. Timb., 264 ; Grah., Cat. Bomb. Pl., 114 ; Dalz. & Gibs., Bomb. Fl., 145 ; Stewart, Pb. Pl., 143 ; Mason, Burma & Its People, 398, 799 ; Elliot, Fl. Andhr., 100, 149, 160, 161 ; Campbell, Econ. Prod., Chutia Nagpur, Nos. 9221, 9284 ; Atkinson, Him. Dist., 313 ; Lisboa, U. Pl. Bomb., 100, 248, 274, 391 ; Darrah, Note on Cotton in Assam, 34 ; For. Adm. Rep., Chutia Nagpur, 32 ; Agri.-Horti. Soc. Ind., Journ. (Old Series), XIII., 319 ; Gazetteers :—Bom bay, XIII., 24 ; N. W. P., IV., lxxiv.*

Habitat.—A small deciduous tree with corky bark, found throughout Tropical India from the Indus eastwards and southwards to Ceylon, Burma, and Penang, ascending to 2,000 feet in the Himálaya and to 4,000 in the Nilghiris.

Dye.—Every part of the tree discharges a yellow, milky JUICE on being wounded. **Roxburgh** states that this yields a fairly good yellow dye when diluted with water, and that pieces of cotton so coloured retained their colour unimpaired for two years. Its value for this purpose is apparently not known to Natives, nor has it been commented on by writers subsequent to Roxburgh.

Medicine.—A thick red-coloured medicinal OIL is said to be obtained from the SEEDS, but this is probably a mistake. The seeds of **Holarrhena antidysenterica**, with which this might easily be confused, certainly yield a medicinal oil. In Chutia Nagpur a preparation from the BARK is given in menstrual and renal complaints (*Campbell*). The BARK and ROOT-BARK are said by Lisboa, Gamble, and others to be believed useful in snake-bite and scorpion stings, probably again a mistake which has

Right margin notes:

DYE.

MEDICINE.
Root-bark.
125
Seeds.
126
FOOD.
Leaves.
127
Pods.
128
TIMBER.
129
DOMESTIC.
& SACRED
Flowers.
130

131

DYE.
Juice.
132

MEDICINE.
Oil.
133
Seeds.
134
Bark.
135
Root-Bark.
136

XANTHIUM
strumarium.   The Bur-weed.

arisen from the confusion between the species of this genus and **Holarr-hena**.

FOOD.
Leaves.
137

Food.—The LEAVES are eaten as a pot-herb by the Santals (*Camp-bell*).

TIMBER.
138
139

Structure of the Wood.—Yellowish-white, moderately hard, close-grained; weight 41·5℔ per cubic foot (*Gamble*). It is even-grained, easy to work, and used for making combs, and for carving and turnery.

Wrightia zeylanica, *Br.; Fl. Br. Ind., IV., 654.*

Syn.—W. ANTIDYSENTERICA, *Br.;* NERIUM ZEYLANICUM & N. ANTI-DYSENTERICUM, *Linn.;* N. DIVARICATUM, *Herb. Madr.*

Habitat.—A common tree in the south of Ceylon, which, like the other species of this genus, has been much confused with **Holarrhena antidy-senterica** [see Vol. IV., 255-258], but which is itself inert and valueless.

XANTHIUM, *Linn.; Gen. Pl., II., 355.*

I

Xanthium strumarium, |*Linn.; Fl. Br. Ind., III., 303;* COMPOSITÆ.

BUR-WEED; LAMPOURD, *Fr.;* SPITZKLETTE, *Ger.*

Syn.—X. INDICUM, *DC.;* X. ROXBURGHII, DISCOLOR, & BREVIROSTRE, *Wallroth;* X. ORIENTALE, *Blume.*

Vern.—*Ban-okra, chhota-gokhrú,* HIND.; *Ban-okra,* BENG.; *Agara,* ASSAM; *Tsúr, láne tsúrú,* KASHMIR; *Wangan tsúrú. chirrú, kúrí, jojre, súngtú, gúdal, gokrú?, khagarwal ?,* fruit=*g•khrú kalán,* PB.; *Bag-giárí,* PUSHTU; *Gókhrú kallán.* SIND; *Shankeshvara,* BOMB.; *Dutundi,* MAR.; *Marlumutta,* TAM.; *Verri tala noppi, párswapu, tala noppi, talnopi, marula mátangi, marulu jada, marulu tíge,* TEL.; *Cho-sa, kouk-pin,* BURM.; |*Arishta,* SANS.

References.—*Roxb., Fl. Ind., Ed. C.B.C., 660; Dalz. & Gibs.,Bomb. Fl., 127; Stewart, Pb. Pl., 132; Elliot, Fl. Andhr., 113, 145, 172, 191; Dymock, Mat. Med. W. Ind., 2nd Ed., 458; Murray, Pl. & Drugs, Sind, 182; Note on the Condition of the People of Assam, App. D; Gaz-etteers:—Mysore & Coorg, I., 62; N.-W. P., I.,81; IV., lxxiii.; Journ. (Old Series), Agri.-Horti. Soc. Ind., X., 11.*

Habitat.—A coarse annual herb, found throughout the hotter parts of India and Ceylon, usually near houses; it ascends the Western Himá-laya to 5,000 feet.

DYE.
Leaves.
2

Dye.—According to **Balfour** the LEAVES are used as a yellow dye.

OIL.
Seeds.
3

Oil.—The SEEDS are said to yield an oil used in medicine, also for illumination.

MEDICINE.
Fruit.
4

Medicine.—The FRUIT is employed medicinally in the Panjáb and Sind, being considered cooling, and efficacious in small-pox. The latter belief is due to the appearance of the fruit, from which it is used on the doctrine of signatures. In certain parts of the same provinces it is burnt and applied to sores on the lips and mucous membrane of the mouth. In Southern India the prickly INVOLUCRE is applied to the ear, or tied in a bunch to the ear-ring, to cure hemicrania (*Elliot*).

Chemistry.
5

CHEMICAL COMPOSITION.—" **Zander** (1881) obtained from 100 parts of the fruit, 52 ash, 38.6 fat, 36.6 albumenoids, 1.3 *Xanthostrumarin* and organic acids, besides sugar, resin, etc. *Xanthostrumarin* seems to be a glucoside, is yellow, amorphous, soluble in water, alcohol, ether, benzol, and chloroform, and yields precipitates with group reagents for alkaloids, and with ferric chloride, lead acetate, and salts of other metals, but is not pre-cipitated by tannin or gelatin. **M. V. Cheatham** (1884) obtained only 14.5 per cent. of fixed oil, and a principle which was precipitated by tannin" (*Dymock, quoting Amer. Journ. Pharm., 1881, 271, & 1884, 134*).

SPECIAL OPINIONS.—§ " Known here as *Bhukhra* Has proved very useful in urinary diseases, a good diuretic, diminishes the irritability of the

MEDICINE.

bladder. Very useful also in gleet and leucorrhœa, given as infusion (Ʒii to Ʒv of water) or in one-drachm doses in powder. Extensively used in Panjáb. Has also been given in menorrhagia" (*Civil Surgeon J. C. Penny, M.B., Amritsar*). "Hospital Assistant Gopal Chunder Ganguli states that the fruits are slightly narcotic" (*Surgeon A. C. Mukerji, Noakhally*).

FOOD.
Flowering-top.
6
Leaves.
7

Food.—The " young FLOWERING-TOP and the two LEAVES immediately below," boiled in *khár* water, are eaten by the people of Assam (*Note on the Condition of the People of Assam*). This statement is curious, and requires explanation, from the fact that the plant has been found in America and Australia to be poisonous to cattle and pigs. It is said to paralyse the heart causing coma and death without pain or struggle.

Xanthochymus pictorius, *Roxb.;* see **Garcinia Xanthochymus,** *Hook. f.;* Vol. III., 478.

XANTHOPHYLLUM, *Roxb.; Gen. Pl., I., 139, 974.*
[POLYGALEÆ.

Xanthophyllum flavescens, *Roxb.; Fl. Br. Ind., I., 209;*

8

Syn.—X. PANICULATUM, *Miquel;* X. ARNOTTIANUM, UNDULATUM, & ROXBURGHIANUM, *Wight, Ill., I., 50.*
Vern.—*Ajensak, gandi,* BENG.; *Thitpyú,* BURM.
References.—*Roxb., Fl. Ind., Ed. C.B.C., 313; Kurz, For. Fl. Burm., I., 81; Gamble, Man. Timb., 19; Beddome, Fl. Sylv., Anal. Gen., t. 3; Rheede, Hort. Mal., IV., t. 23.*
Habitat.—A large tree, found in Eastern Bengal, Burma, South India, and Ceylon.

TIMBER.
9

Structure of the Wood.—Heavy, close-grained, "useful to the Natives for many purposes" (*Roxb.*).

X. glaucum, *Wall.; Fl. Br. Ind., I., 209.*

10

Vern.—*Thetpyu,* BURM.
References.—*Kurz, For. Fl. Burm., I., 81; Gamble, Man. Timb., 19.*
Habitat.—Common in the swamp forests and around inundated jungle-swamps of the alluvial plains and base of the hills of Pegu, Martaban and Tenasserim " (*Kurz*).

TIMBER.
11

Structure of the Wood.—Light but comparatively strong, white and pinkish, soft, probably valuable for furniture (*Kurz*).

Xanthoxylon, see **Zanthoxylum,** *Linn.,* below, p. 323.

XIMENIA, *Linn.; Gen. Pl., I., 346.*

Ximenia americana, *Willd.; Fl. Br. Ind., I., 574;* OLACINEÆ.
FALSE SANDAL WOOD.

12

Syn.—X. RUSSELLIANA, *Wall.*
Vern.—*Konda nakkera, úranechra,* TEL.; *Pinlaytsi, penlay-hsí, pinlési, pin-lai-kú-yin,* BURM.
References.—*Roxb., Fl. Ind., Ed. C.B.C., 323; Kurz, For. Fl Burm., I., 233; Gamble, Man. Timb., 80; Mason, Burma & Its People, 751; Elliot, Fl. Andhr., 96; Drury, U. Pl. Ind., 448; Lisboa, U. Pl. Bomb., 149; Smith, Econ. Dict., 366; Ind. Forester, III., 238; Agri.-Horti. Soc. Ind., Trans., VII., 54, 55.*
Habitat.—A large straggling shrub, or low tree, found in the Eastern and Western Peninsulas, the Andaman Islands, Burma, Malacca, and Ceylon.

FOOD.
Fruit.
13

Food.—Produces, about the beginning of the hot weather, small, dull white, fragrant flowers, with an odour of cloves. These are followed by small oval red or yellow pulpy FRUITS, an inch long, of an acid-sweet, aro-

X. 13

XYLIA **dolabriformis.**	**The Burma Iron-wood Tree.**

FOOD.
Kernels.
14
TIMBER.
15

matic, slightly bitter taste. When ripe the fruit is eaten by the Natives. The KERNELS are also eaten and taste much like filberts (*Roxb.*).

Structure of the Wood.—Yellow, like sandal-wood, often powdered and substituted for true sandal-wood by the Brahmans of the Coromandel Coast in their religious ceremonies (*Roxb.*).

XYLIA, *Benth.; Gen. Pl., I., 594.*

16

Xylia dolabriformis, *Benth.; Fl. Br. Ind., II., 286;* LEGUMINOSÆ.

THE IRON-WOOD TREE OF PEGU and ARRACAN.

Syn.—MIMOSA XYLOCARPA, *Roxb.;* INGA XYLOCARPA, *DC.;* I. LIGNOSA, & DOLABRIFORMIS, *Grah.*

Vern.—*Jambu,* HIND.; *Boja, kongora,* URIYA; *Jamba, yerrul, suria,* BOMB.; *Jámba, jámb'ra, suria,* MAR.; *Irúl,* TAM.; *Konda tangédu, tanjédu, tanjedu mánu, eruvalu, bójeh, bója,* TEL.; *Jabmé, tirawa, shi've,* KAN; *Pyinkado,* BURM.

References.—*Roxb., Fl. Ind., Ed. C.B.C., 417; Brandis, For. Fl., 171; Kurz, For. Fl. Burm., I., 419; Beddome, Fl. Sylv., t., 186; Gamble, Man. Timb., 148; Dalz. & Gibs., Bomb. Fl., 85; Mason, Burma & Its People, 530, 772; Elliot, Fl. Andhr., 29, 97; Balfour, Trees of S. India, (Madras, 1862), 133; Lisboa, U. Pl. Bomb., 65; Cooke, Oils and Oil-seeds, 82; Aplin. Rep. on Shan States, 1887-88; Gazetteers:—Mysore & Coorg, I., 48; II., 64; Bombay, XV., 33, 79; Burma, I., 125, 436; Agri.-Horti. Soc. Ind., Journ. (Old Series), IX., Sel., 46; XI., 446; Ind. Forester:—I., 115; II., 19; III., 23, 189; IV., 249,292, 366; VI., 125; VII., 196; VIII., 403, 414, 415; IX., 14, 216; X., 38, 543, 544, 545; XI., 231, 321, 322, 374; XII., 72, xxii., 311, 313; XIII., 127, 133, 553.*

Habitat.—A large, deciduous tree, met with in the Central Provinces, South India, Arrakan, and Burma.

RESIN.
17

Resin.—It yields a red resin, which is said to be more abundant in Burmese than in South Indian wood. It is reported that the lac insect in Southern India is found on this tree (*Bomb. Gaz., XV., i., 79*).

OIL.
Seeds.
18
TIMBER.
19

Oil.—The SEEDS yield an oil of which nothing is known.

Structure of the Wood.—Sapwood small; heartwood dark-brown or reddish-brown, extremely hard, beautifully mottled, cross-grained; weight from 60 to 80℔ per cubic foot. The heartwood is very durable, and resists the attacks of white ants—a property it doubtless owes in great measure to the resinous substance which it contains.

DOMESTIC.
20

Domestic.—It is used for boat-building and for agricultural implements in Burma; also for carts and tool handles. In South India, it is employed for making railway sleepers, posts, boats, and carts. In Burma and Bengal it has been largely employed for telegraph posts, for which it has answered well. The large forests in Arrakan, of which **Dr. Schlich**, in his report on the iron-wood forests of Arrakan, dated 1st September 1869, says that "a third of the forest vegetation consists of *Pynkado*" produce large numbers of telegraph poles and railway sleepers. **Major Seaton**, in his report for 1876-77. stated that 10,000 such sleepers from Arrakan had then lately been sold at Calcutta at R5 each, and **Mr. Ribbentrop** states that large numbers of *Pynkado* pieces and sleepers are brought out from the forests in Pegu. Between 1865 and 1868 inclusive, 70,377 sleepers were obtained by the East Indian Railway Company from Arrakan (*Burm. Gaz.*). In the *British Burma For. Adm. Rept.* for 1884-85, 17,631 are said to have been sold in Calcutta at R2-1 each. The cost of cutting and freight are said to have amounted to about 12 annas per sleeper. In 1885-86, 81,569 sleepers were removed from Burma by Government agency, of which 75,000 went to Madras and the remainder to Calcutta. The profit which accrued from the sleeper works in the Pegu circle amounted to 36 per cent. (*For. Ad. Rep., 1885-86*). The wood is hard, heavy, and difficult to cut, but is valuable for

X. 20

Adam's Needle. (*J. Murray.*)	**YUCCA gloriosa.**

all ordinary purposes, and is a useful wood for piles and beams of bridges (*Gamble*). **Balfour**, writing in 1862, states that the wood was then largely used in the Madras gun-carriage factory for poles, axle-cases, and braces for transport limbers, poles and yokes for water-carts, and cheeks and axle cases for lighter mortar carts. **Captain Puckle**, writing at the same time from Mysore, says that it is largely employed for furniture, shafts, plough-heads, knees, and crooked timber for ship-building, and for railway sleepers. The hardness of this timber has recently been utilised by Mr. **Oliver**, Conservator of Forests, Burma, as a new form of boundary board. These he makes of this wood, pointed at one end, so that from their hardness they can be driven on to trees of softer wood.

Xylocarpus Granatum, *Kœn.;* see **Carapa moluccensis,** *Lam.;* Vol. II., 141.

XYLOSMA, *Forester; Gen. Pl., I., 128.*

Xylosma longifolium, *Clos.; Fl. Br. Ind., I., 194;* BIXINEÆ.   **21**
    Syn.—FLACOURTIA FEROX, *Wall.*
    Vern.—*Dandál, katári, kandhára,* HIND.; *Kattáwa,* OUDH; *Chopra, chíúndi, chirúnda, chírndi, chiraunda, drendú, thakola, kathágli,* PB.
    References.—*Voigt, Hort. Sub. Cal., 84; Brandis, For. Fl., 19; Gamble, Man. Timb., 18; Stewart, Pb. Pl., 191; Atkinson, Him. Dist., 305, Gazetteer:—N.-W.P., IV., lxviii.; Agri.-Horti. Soc. Ind., Trans., VII.; 73.*
    Habitat.—A small evergreen tree of the North-West Himálaya, from Kumáon to Marí, ascending to 5,000 feet; also found in Assam.
    Structure of the Wood.—Pinkish, moderately hard, even-grained; weight 55℔ per cubic foot. It is used for fuel and charcoal.   **TIMBER. 22**

XYRIS, *Linn.; Gen. Pl., III., 842.*

Xyris indica, *Linn.; Fl. Br. Ind., VI., 364;* XYRIDEÆ.   **23**
    Vern.—*Dábi-dúba,* HIND.; *China ghausa, chine ghás,* \dabídúbi, BENG.; *Dadumari,* SANS.
    References.—*Roxb., Fl. Ind., Ed., C.B.C., 60; Rheede, Hort. Mal., IX., 139, t. 7; Mason, Burma & Its People, 435, 820; Dals. & Gibs., Bomb. Fl., 259; Dymock, Mat. Med. W. Ind., 2nd Ed., 818.*
    Habitat.—Found in Bengal, the Southern Konkan, and Coromandel, generally on sandy soils or salt marshes.
    Medicine.—"The Natives of Bengal esteem it a plant of great value, because they think it an easy, speedy, and certain cure for the troublesome eruption called ringworm" (*Hon'ble John Hyde in a letter to Roxburgh*). This remark accords with the description of the plant as given by **Rheede**.   **MEDICINE. 24**

Yak, see **Oxen,** Vol. V., 664.

Yamamai, see **Silk.**

Yamani or **Moka Aloes,** see **Aloe succotrina,** *Lam.;* and A. Perryi, [*Baker;* LILIACEÆ; Vol. I., 184.

Yeast, see **Cerevisiæ Fermentum,** or **Torula cerevisiæ,** Vol. II , 257; also [**Malt Liquors,** Vol. V., 131.

Yellow, Indian—, see **Peori,** Vol. VI., Pt I., 132.

Yew, see **Taxus baccata,** *Linn.;* Vol. VI., Pt. III.

YUCCA, *Linn.; Gen. Pl., III., 778.*   [LILIACEÆ.

Yucca gloriosa, *Linn.; Baker, in Linn. Soc. Jour., XVIII., 225;*   **I**
    ADAM'S NEEDLE.
    Mr. **Baker** describes ten varieties, but these are not of sufficient economic importance to necessitate their enumeration in this work.

ZANONIA
indica.

Adam's Needle & the Bear Grass.

References.—*Lisboa, U. Pl. Bomb., 236 ; Baden Powell, Pb. Pr., 518 ; Royle, Fib. Pl., 57 ; Liotard, Mem. Paper-making Mat., 5, 15, 18 ; Tropical Agriculturist, 1st Feb., 1883 ; Smith, Econ. Dict., 5 ; Watt, Sel. from Rec., Govt. of Ind., R & A. Dept., 1889, 179 ; Spons, Cyclop., 999 ; Gazetteers :—Mysore & Coorg, I., 67 ; N.-W. P., I., 85 ; Ind. Forester, IX., 274 ; Agri.-Horti. Soc. Ind., Journ. (Old Series), II., Sel., 458 ; IX., 114, 120.*

Habitat.—A native of America from Northern Carolina to Florida (*Baker*) ; introduced into India as a cultivated plant of gardens, naturalised here and there in the Madras Presidency, and on the lower slopes of the Himálay, often seen in hedges.

FIBRE.

Leaves.
2

Fibre.—This species, together with the less important **Y. angustifolia,** *Pursh.,* and **Y. filamentosa,** *Linn.* (the Silk or Bear Grass), contains a large quantity of fibre in the LEAVES, which is in many respects similar to that of **Agave** and is applicable to like purposes. It is from 2 to 4 feet in length, rather wiry, fine, round, even, strong, and easily dyed. It is moderately flexible, but has naturally a certain amount of harshness. The fibre has apparently not attracted the attention it deserves, even in America where it occurs plentifully. Thus the writer of the article on the subject in *Spons' Encyclopœdia* remarks, " The whole genus has been utterly neglected from an industrial point of view, no real attempt having ever been made to grow the plants on a commercial scale, though their hardiness, their preference for arid, barren sands, and the quality of their fibre would seem to be special recommendations." The fibre, if obtainable at a sufficiently low price, would doubtless be specially suited to many of the requirements of the paper-maker.

MEDICINE.
Fruit.
3
Root.
4
Soap-substitute.
5

Medicine.—The FRUIT is purgative, the ROOT detergent (*La Maout & Decaisne*). Dr. Bidie writes that the latter statement is not correct as regards the root of **Y. gloriosa** in India. [The Indian public papers were recently greatly concerned in an announcement of supposed considerable importance regarding a plant the leaves of which yield fibre and the roots a useful soap-substitute. It seems likely that the plant referred to may have been a **Yucca.**—*Ed., Dict. Econ. Prod.*]

Zahr-i-mohra, see **Magnesia,** Vol. V., 108.

ZANONIA, *Linn.; Gen. Pl., I., 839.*

[CUCURBITACEÆ.

I

Zanonia indica, *Linn.; Fl. Br. Ind., II., 633; Wight, Ill., t. 103;*

Vern.—*Chirpoti,* HIND.; *Chirabuti,* MAR.; *Penar-valli,* MAL.; *Walrasakinda,* SING.; *Chirpota, dirghapatra, kuntali, tiktaka,* SANS.

References.—*Dals. & Gibs., Bomb. Fl., 99 ; Rheede, Hort. Mal., VIII., t. 47, 48, 49 ; Thwaites, En. Ceyl. Pl., 124 ; S. Arjun, Cat. Bomb. Drugs, 260 ; Pharmacog. Indica, II., 94 ; Drury, U. Pl. Ind., 450 ; Gazetteer, Mysore & Coorg, I., 61 ; Agri.-Horti. Soc. Ind., Trans., VII., 64.*

Habitat.—A climbing herb of Assam and East Bengal, the Deccan Peninsula, the Malabar Gháts, and Ceylon.

MEDICINE.
Leaves.
2
Plant.
3
Fruit.
4
Juice.
5

Medicine.—According to Rheede the LEAVES, beaten up with buttermilk, are used in South India as an anodyne application. The Sinhalese value the PLANT as a febrifuge (*Thwaites*). The FRUIT is said to possess acrid cathartic properties. The *Hakims* in Bombay assert that the fresh JUICE is very efficacious as an antidote to the venomous bites of the Gecko, known in the Deccan as *shal-i-alam* or " king of the world " (*S. Arjun*). " In Malabar a bath made by boiling the leaves in water is used to remove the nervous irritation caused by boils, and an antispasmodic liniment is made by pounding the leaves with milk and butter" (*Pharmacog. Indica*).

Z. 5

ZANTHOXYLUM, *Linn.; Gen. Pl., I.,* 297, 991.

[RUTACEÆ.

Zanthoxylum acanthopodium, *DC.; Fl. Br. Ind., I.,* 493;　　　　6

Syn.—Z. HOSTILE, *Wall.*; Z. ALATUM, *Wall., Cat. 1209, in part.*

Vern.—*Nipáli-dhanya, tumra, tejphal, darmar,* HIND.; *Tambul,* BENG.; *Bogay timur,* NEPAL.

References.—*Gamble, Man. Timb., viii; List Darjeeling Trees, Shrubs, etc., 14; Kurz, For. Fl. Burm., I., 181; Dymock, Warden & Hooper, Pharmacog. Ind., I., 255.*

Habitat.—A small tree of the hot valleys of the Sub-tropical Himá-laya, from Kumáon to Sikkim, ascending to 7,000 feet, and the Khásia Hills, from 4,000 to 6,000 feet. Flowers in short dense cymes ½-1 inch long.

Medicine.—See Z. alatum.　　　　　　　　　　　　　　　MEDICINE.

Structure of the Wood.—Yellowish-white, soft.　　　　　　　7

　　　　　　　　　　　　　　　　　　　　　　　　　TIMBER.

Z. alatum, *Roxb.; Fl. Br. Ind., I.,* 493.　　　　　　　　8

Syn.—Z HOSTILE, *Wall., Cat. 1210, in part.*　　　　　　　9

Vern.—*Tumru, tun, timbúr timúr, tesmal, nipáli-dhanya, darmar, tej-phal,* HIND.; *Gaira, nepáli dhaniá, tun,* BENG.; *Sungrú-kúng,* LEPCHA; *Tejbal, timúr, jwarán-tika,* N.-W. P.; *Tejbal, kabába, tesbal, timmal, timrú,* PB.; *Tumburu,* SANS.

References.—*Roxb., Fl. Ind., Ed. C.B.C., 717; Brandis, For. Fl., 47; Gamble, Man. Timb., 60, also List Darjeeling Trees, Shrubs, etc., 14; Pharm. Ind., 48; O'Shaughnessy, Beng. Dispens., 264; U. C. Dutt, Mat. Med. Hind., 321; U. S. Dispens., 15th Ed., 1539; Irvine, Mat. Med. Patna, 115; Dymock, Warden & Hooper, Pharmacog. Ind., I., 255; Baden Powell, Pb. Pr., 329, 601; Atkinson, Him. Dist., 753; Birdwood, Bomb. Pr., 17.*

Habitat.—A shrub or small tree, with dense foliage which possesses a pungent aromatic taste and odour, panicles 2-6 inches; found in the hot valleys of the Subtropical Himálaya from Jammu to Bhután, ascending to 6,000 feet, and in the Khásia Mountains between 2,000 and 3,000 feet. This and the previous species are much alike and often confused with each other. They have both the peculiar smell so characteristic of them. The leaves have a winged rachis in both forms, but in **Z. acanthopodium** the leaves are smaller and more crowded, the fruits form dense cymes and the flowers appear in March, those of **Z. alatum** not till a month or six weeks later.

Oil.—An essential oil has been separated from the CARPELS by Dr. 　OIL.
Stenhouse and later by Pedler and Warden. It is isomeric with oil of
turpentine, possesses a similar odour to Eucalyptus oil, and might therefore　Carpels.
prove of value as an antiseptic and disinfectant (see para. CHEMICAL 　10
COMPOSITION).

Medicine.—The authors of the *Pharmacographia Indica* state that 　MEDICINE.
Sanskrit writers call the CARPELS of this species and of **Z. acanthopodium**　Carpels.
by the name of *tumburú,* which signifies coriander. The FRUITS of these 　11
are so similar as to be very difficult of distinction, and both possess the 　Fruits.
peculiar flavour of coriander, which fruit they also resemble in size. " In 　12
Hindu medicines they are considered to be hot and dry. The Chinese
also use the carpels under the name of *Hwa-tseaou* or ' Pepper-flower"
and in Japan the carpels of Z. piperitum are used. The Arabians appear
to have obtained the carpels of **Z. alatum** or **Z. acanthopodium** first from
Northern India. Ibn Sina, under the name of *Fághireh* (open-mouthed),
describes them as ' a berry, the size of a chick-pea, containing a black seed
as large as a hemp-seed, brought from Sakála in Hindustan.' Sakála
or Sangla was an ancient town in the Panjáb, near the modern Sangla-
wala Tiba or Sangla Hill. It is the Sangala of Alexander, and was

ZANTHOXYLUM   Several kinds of
      alatum.

___

MEDICINE.

visited by the Chinese pilgrim Hwen Thsang in A.D. 630; it had then a
large Buddhist monastery and a stupe 200 feet high. Haji Zein el Attar,
who wrote A.D. 1368, gives a similar account of *Fághireh*, and says that
the Persians call it *Kabábeh-i-kushádeh* (open-mouthed cubebs)." "The
Mahometan physicians consider *Fághireh* to be hot and dry, and to have
astringent, stimulant, and digestive properties. They prescribe it in
dyspepsia arising from atrabilis, and in some forms of diarrhœa." The
Bark.
13
Root-bark.
14
BARK of these trees is tonic and aromatic, and may be used with advantage in rheumatism and in atonic dyspepsia; the ROOT-BARK is to be
preferred. Heckel & Schlagdenhauffen (*Académic des Sciences, Ap. 21st,
1884*) reported that a crystalline principle, obtained from the bark of a West
Indian **Zanthoxylum**, produced in frogs, rabbits, etc., general paralysis and
Branches.
15
Thorns.
16
Chemistry.
17
abolition of the functions of respiration and circulation" (*Pharmacog.
Ind.*). Baden Powell states that the small BRANCHES and THORNS are
employed in Northern India as an application for toothache, and that the
seeds and bark are prescribed in fever, dyspepsia, diarrhœa, and cholera.

CHEMICAL COMPOSITION.—"The bitter crystalline principle present in
the barks of the **Zanthoyleæ**, and formerly called *Zanthopicrite*, has been
recognised as identical with *berberine* by Dyson Perrins (*Trans. Chem. Soc.,
1862*). The bark also contains a volatile oil and resins. Dr. Stenhouse
has obtained from the carpels of **Z. alatum**, by distillation, an essential oil
to which the aromatic properties are chiefly due. This oil, which when pure
is called by Dr. Stenhouse *Zanthoxylene*, is a hydrocarbon, isomeric with
oil of turpentine. It is colourless, refracts light strongly, and has an agreeable aromatic odour similar to that of **Eucalyptus** oil; its composition is C_{10}
H_8. He also obtained a stearopten, *Zanthoxylin*, floating on the water
distilled from the carpels and separable from the crude essential oil. After
repeated crystallisations from alcohol, *Zanthoxylin* may be obtained in a
state of purity, and then presents the form of large crystals of a fine silky
lustre, insoluble in water, but readily soluble in alcohol or ether. It has a
very slight odour of stearine, and a slightly aromatic taste. It distils
unchanged, its fusing point before and after distillation remaining the
same, namely, 80°C., and its solidifying point 78°C. Its composition is
$C_{40} H_6 O_4$. The essential oil was obtained by Pedler & Warden (1888)
by distilling the crushed carpels with seeds in a current of steam. The oil
was dehydrated by fused Ca Cl_2. It commenced to boil at 175° to 176°C.,
the greater part passing over between 176° to 179°C., the temperature then
rose to 181°C., and rapidly to 183°C., when the distillation was stopped.
The rectified oil had a specific gravity of ·873 at 15.5°C. Its vapour density
determined by Meyer's method was 5.43. They were unable to obtain
the crystallisable stearopten isolated by Stenhouse. The freshly distilled
oil exposed to O°C failed to deposit any crystals. In addition to the
essential oil, they also detected the presence of a pale yellow, viscid, non-
drying oil, an acid resin, and a yellow acid principle, forming deep yellow
solutions with alkalis, and reprecipitated from its alkaline solution by
acids" (*Pharmacog. Ind.*).

FOOD.
Carpels.
18
Food.—The CARPELS are occasionally employed as a condiment.

TIMBER.
19
Structure of the Wood.—Close-grained, yellow; weight from 34 to 46℔
per cubic foot; used for making walking sticks, clubs, pestles, etc.
the hemp-plant.

DOMESTIC.
Fruits.
20
Bark.
21
Domestic.—*Fághireh* (FRUITS) is employed as an ingredient of *guráku*
(tobacco for the *hukka*) in certain localities, and in the preparation of a
ground-bait for fishing (*Pharmacog. Ind.*). Brandis says the BARK is
used for intoxicating fish. In Northern India the fruit is said to be
used for poisoning fish? (*Atkinson*); and to purify water (*Gamble*). The

small BRANCHES are frequently employed as tooth-sticks, to clean the teeth and are regarded relieving toothache.

DOMESTIC.
Branches.
22

Zanthoxylum Budrunga, *Wall.; Fl. Br. Ind., I., 495.*

23

Syn.—*Z.* CRENATUM, *Wall.;* ? FAGARA BUDRUNGA, *Roxb.*

Vern.—*Badrang,* HIND.; *Brojonali,* ASSAM; *Mayanin,* BURM.

References.—*Roxb., Fl. Ind., Ed. C.B.C., 140; Kurz, For. Fl. Burm., I., 182; Royle, Ill. Him. Bot., 157; O'Shaughnessy, Beng. Dispens., 264; Irvine, Mat. Med. Patna, 99; Dymock, Warden & Hooper, Pharmacog. Ind., I., 256; Agri-Horti. Soc. Ind., Trans., VII., 163; Journ. (Old Series), IX., Sel., 49, 53.*

Habitat.—A tree of the Tropical Himálaya, the Khásia Hills, Eastern Bengal, and Burma.

Medicine.—The CARPELS can hardly be distinguished from those of Z. Rhetsa, and are used similarly in medicine (see below).

MEDICINE.
Carpels.
24
TIMBER.

Structure of the Wood.—Rather heavy, soft, yellowish-white, close-grained (*Gamble*).

25

Z. Hamiltonianum, *Wall.; Fl. Br. Ind., I., 494.*

26

Vern.—*Purpuray timur,* NEPAL.

References.—*Kurz, For. Fl. Burm., I., 181; Dymock, Warden & Hooper, Pharmacog. Ind., I., 256.*

Habitat.—A climbing thorny shrub of Sikkim, Assam, and Burma.

Medicine.—The FRUIT of this species and that of Z. oxyphyllum, *Edgew.,* are employed medicinally as a sort of *Fághireh,* with which they have probably similar properties (see Z. alatum).

MEDICINE.
Fruit.
27

Z. ovalifolium, *Wight; Fl. Br. Ind., I., 492.*

28

Syn.—*Z.* LUCIDUM, *Wall.;* TODDALIA MITIS, *Miq.;* LIMONIA LEPTOSTACHYA, *Jack.*

References.—*Beddome, Fl. Sylv., Anal. Gen., xlii., t. vi., f. 3; Bombay Gazetteer, XV., Pt. I., 429.*

Habitat.—A large shrub found in Kanara, Coorg, Nilgiri Hills, and near Madras; also in the Khásia Hills, Assam, the Mishmi Hills and Singapore.

MEDICINE.
Fruit.
29
Bark.

Medicine.—The FRUIT and BARK are not reported to be used in medicine, but probably possess similar properties to those of other members of the genus.

30

Structure of the Wood.—Light yellowish-white, very hard, close-grained.

TIMBER.
31

Z. oxyphyllum, *Edgew.; Fl. Br. Ind., I., 494.*

32

Syn.—ZANTHOXYLON VIOLACEUM, *Wall.*

Vern.— *Timur,* NEPAL.

Reference.—*Dymock, Warden & Hooper, Pharmacog. Ind., I., 256.*

Habitat.—A climbing prickly shrub of the Himálaya from Garhwál to Bhután, between 6,000 and 9,000 feet; also met with in the Khásia Hills, from 4,000 to 6,000 feet.

MEDICINE.
33

Medicine.—See Z. Hamiltonianum.

TIMBER.

Structure of the Wood.—Yellowish-white, soft, porous.

34

Z. Rhetsa, *DC.; Fl. Br. Ind., I., 495.*

35

Syn.—*Z.* OBLONGUM, *Wall.;* FAGARA RHETSA, ? F. BUDRUNGA, *Roxb.,* not of *Wall.*

Vern.—*Sessal, tirphal, tisal, chirphal,* MAR.; *Tessul, kokli, chirphal, triphal, sessal, tijabal, tephal,* BOMB. & GOA; *Rhetsa-maram, rhetsa, rhetsa maum,* TEL.; *Jummina, jimmi-mara,* KAN.; *Kattú-kina-gass,* SING.

References.—*Roxb., Fl. Ind., Ed. C.B.C., 140; Beddome, Fl. Sylv., Anal. Gen., xli.; Thwaites, En. Ceyl. Pl., 69; Dalz. & Gibs., Bomb. Fl., 45; Graham, Cat. Bomb. Pl., 36; Gamble, Man. Ind. Timb., 60; O'Shaugh-*

nessy, Beng. Dispens., 264; S. Arjun, Cat. Bomb.Drugs, 29; Dymock, Mat. Med. W. Ind., 2nd Ed., 127; Dymock, Warden & Hooper, Pharmacog. Ind., I., 255; Lisboa, U. Pl. Bomb., 31, 222; Drury, U. Pl. Ind., 450; Gasetteer, Bombay, X., 404; XV., 79; Ind. Forester, III., 200; Rheede, Hort. Mal., V., t. 34.

OIL.

**Carpels.
36
MEDICINE.
Carpels.
37
Oil.
38
Bark.
39
Root-bark.
40
FOOD.
Carpels.
41
Seeds.
42
Bark.
43
DOMESTIC.
44**

Habitat.—A tree with corky bark and spreading leafy branches, found in the Western Peninsula, from Coromandel and the Konkan southwards, also in Tavoy.

Oil.—The essential oil, obtained from its CARPELS, probably similar to that of **Z. alatum**, is used medicinally.

Medicine.—The CARPELS constitute the Fagara major of the old pharmacologists. They are similar to those of **Z. Budrunga**, and much larger than the Fagara minor, or *Fághireh* described by Muhammadan writers (see **Z. alatum.**) In Southern and Western India they are used as a remedy for rheumatism, and the essential OIL for cholera. They probably possess precisely similar properties to those of **Z. alatum, acanthopodium,** and other species, and the BARK and ROOT-BARK are also probably equally valuable.

Food.—"The unripe CARPELS are like small berries; they are gratefully aromatic, and taste like the skin of a fresh orange. The ripe SEEDS taste exactly like black pepper, but weaker" (*Roxb.*). Both are largely employed in Southern India as condiments, especially with fish curries. The BARK is also aromatic, and is used as a substitute for limes and pepper. It is cooked with sugar or honey; and when mixed with onions, mustard-seed, and ginger, makes a good pickle (*Drury*).

Domestic.—"*Rhetsa* means a committee, or select number of men assembled to settle disputes, etc., and *maun*, means tree of the largest size. Under the shade of this tree the Hill people '(of South India)' assemble to examine, agitate, and determine their matters of public concern, deliver discourses, etc." (*Roxb.*).

[*ghiana, Benth.*; see Vol. III., 305.

45

Zanthoxylon triphyllum, *Wight, Ic., t. 204;* see **Evodia Roxbur-**
The following supplementary facts may be given to those furnished in Vol. III.

**RESIN.
46**

Resin.—Specimens of a resin obtained from this tree were sent to the Madras Exhibition, but the quantity produced did not warrant its being of importance from a commercial point of view (*Drury*).

**FOOD.
Capsules.
47
DOMESTIC.
Bark.
48**

Food.—The CAPSULES, which are of about the same size and shape as those of **Zanthoxyon alatum,** possess similar aromatic properties.

Domestic.—In Amboyna the women prepare a cosmetic from the BARK, which they employ to improve their complexions (*Rumphius, Amb, II., 188, t. 62*).

[see ARISTOLOCHIÆ, Vol. I., 316.

Zarawand-i-gird, the imported root of **Aristolochia rotunda,** *Linn.;*
Zarwand-i-tawil, the imported root of **Aristolochia longa,** *Linn.;*

[see Vol. I., 316.

(*G. Watt.*)

49

ZEA, *Linn.; Gen. Pl., III., 1114.*

This very striking genus possesses, so far as is known, only one species and it stands moreover almost by itself in the MAYDEÆ. Its nearest affinities are with **Euchlæna,** *Schrader,* but there are no transitionary forms between these very distinct genera, and as **Bentham** says: "With most of the general characters of the tribe to which Maize gives its name, it is exceptional not only in that tribe, but in the whole order, by the manner in which its numerous female spikelets are densely packed in several vertical rows round a central spongy or corky axis. How far this arrangement may have gradually arisen after so many centuries of cultivation can only be a matter of conjecture. Its gradual progress

Maize or Indian-corn.	(*G. Watt.*)	ZEA Mays.

cannot be traced through the numerous cultivated varieties, many of them described as species in Bonafous's splendidly illustrated monograph; and the idea that some of them are wild indigenous forms must be traced to the insufficiency of the observations recorded by travellers." Of Euchlæna, Bentham says that "like Tripsacum and Zea it has a terminal male panicle with the female spikes in the lower axils wrapped up in broad bracts, but the female spikelets are within each bract superposed in a single row on the articulate rachis of the single spike. The affinity to Zea appears to be recognised; for specimens of Euchlæna have been received from Schaffner purporting to be known as 'Wild Maize.'"

Zea Mays, *Linn.;* GRAMINEÆ.

50

MAIZE, INDIAN-CORN, *Eng ;* BLED DE TURQUIE, *Fr.;* TURKISCH-KORN, *Germ.;* GRANO TURCO O SICILIANO, *It.;* TRIGO DE INDIAS, TRIGO DE TURQUIA, *Sp.*

Vern.—*Bhutta, makka, makai, junri, bara-juár, kukri,* HIND.; *Janar, bhutta, jonár* (in Chutia Nagpur), BENG.; *Jondra,* SANTAL; *Butá, maká,* URIYA; *Makká* or *maká, makai, bhutta, junri, bara-juár,* N.-W. P.; *Mungari, júnala,* GARHWAL; *Bhútta, mukni, junala,* KUMAON; *Makki, makkei, mak, kúkri, bará-juár, chhale, kuthi, juár,* PB.; *Conác,* KASHGAR; *Jaori, jaodri, jaori-khurdáni,* W. AFG.; *Mukka,* RAJ.; *Bara-juár, makkái,* SIND; *Makka-jári, makka-jowári,* DECCAN; *Makai, buta,* BOMB.; *Maka,* MAR.; *Makkai,* GUZ.; *Makká-shólam,* TAM.; *Mokka jonna makká zonnalu,* TEL.; *Mekkejola, musuku jola, goin jol,* KAN.; *Chólam,* MALAY.; *Pyaungbú* (= *flowering-juar*), BURM.; *Yavanala* (according to Birdwood), SANS.; *Khandarús, khálávan, surratul-makkah, hintahe-rúnu, durah-kizan, durah-shámi,* ARAB.; *Gaudume-makkah, khoshahe-makki, bájri,* PERS.

NOTE.—The word *Makkai* or *Mekkai,* which is given so very frequently in India as the name of Maize means "of Mecca," and may, therefore, be rendered "Mecca corn." The origin of the equally prevalent name *Bhutta* or *Bhuta* is more obscure. It might be traced from *Bhukta,* Sanskrit, which, in Palí, is *Bhutta,* to eat. Or it is probable that it came from *Bhú* to be borne, to exist; *Bhúta* occurs as having in one of its meanings "a seed" or "rosary bean." It is somewhat significant that this same word *Bhutta* or *Bhuta* is, in the various languages of India, often given to widely different things, but mostly to introduced plants. Thus, for example, in Kumaon *Bhúta* means the Egg-apple (Solanum Melongena). Sir Walter Elliot assigns the name *Yavanála* to Sorghum vulgare, and he suggests that its derivation is from *Yavana,* a general term for a Greek, Muhammadan, etc. This is, however, very possibly not correct; the word may be *Yava* = barley, and *nala* (or in the adjective form *nála*) reed-like. But the name "The Reed-like barley" would be equally applicable to Sorghum vulgare and to Zea Mays. Sir Monier Williams does not mention *Yavanála* as a Sanskrit name for either of these grains, but he furnishes three words as denoting "Maize." These are *Sasyam, Stamba-kari,* and *Sasyavisesha.* The derivation of these words is probably as follows: *Sasyam,* grain, in fact any grain,—a word which may be seen specialized in the Burmese *Sabá,* rice in husk, and which occurs in the Palí *Sassá.* It is synonymous for "wealth." *Stamba-kari* denotes a grass which grows in clusters. And *Sasyavisesha* might be rendered "remarkable grain." They are thus descriptive words and probably very modern, and more probably denote Sorghum, or Pennisetum rather than Zea.

Conf. with pp. 333, 351.

References.—*Roxb., Fl. Ind., Ed. C.B.C., 649; Stewart, Pb. Pl., 263, also in report on Food of Bijnour District, 1862; Aitchison, Kuram Valley Rept., Pt., I., 105; Hooker, Him. Jour., I., 148; DC., Orig. Cult. Pl., 387; V. Hehn, Culti. Plants and Domestic Animals in their Migration, 384, 497; Crawfurd, Migration of Cultivated Plants in Reference to Ethnology (see Jour. Agri.-Hort. Soc. Ind. N.S., I., Sel., 6; Rev. A. Campbell, Rept. Econ. Pl., Chutia Nagpur, No. 8221 A.; Graham, Cat. Bomb. Pl., 240; Mason, Burma and Its People, 476, 817; Sir W. Elliot, Fl. Andhr., 116, also paper on Farinaceous Grains, etc., of South India, in Trans. Edinb. Bot. Soc., Vol. VII., 200; Forsyth, Report of the Mission to Yarkand, in 1873, 79; Stocks, Report on Sind; Moodeen Sheriff, Supp. Pharm. Ind., 261; U. C. Dutt, Mat. Med. Hindus, 270; Murray, Pl. & Drugs, Sind., 9; Bent. & Trim., Med. Pl., 296; Dymock, Mat.*

ZEA Mays.	Localities in which Maize is found.
	Med. W. Ind., 2nd Ed., 855; *Birdwood, Bomb. Prod.*, 113; *Baden Powell, Pb. Pr.*, 204, 212, 213-225, 230, 231, 251, 516; *Atkinson, Him. Dist. (X., N.-W. P. Gas.)*, 320, 687; *Forbes Watson, Indian Prod.*, 10, 43; *Royle, Prod. Res.*, 40, 214, 230, 381; *Liotard, Mem. Paper-making Mat.*, 34; *Church, Food-Grains, Ind.*, 65; *Wallace, India in 1887*, 201; *Mueller, Select Extra-tropical plants*, (*Ed. 8th*), 525; *Simmonds, Waste Products*, 292; *Duthie, Ind. Fod Grasses of North India*, 12; *Official Corresp., R. and A. Dept.*, 1876 to 1879; *Ain-i-Akbari, Bloch-mann's Trans.*, I., 83; *Buchanan-Hamilton, Kingdom Nepal*, 284, 312; *Port Blair, Settl. Rep.* (1870-71), 26; *Andaman Islands, Admin. Rep.* (1885-86), 54; *Kumaon, Official Rep.*, 279; *Gazetteers :— Bikanir*, 229; *Rajputana*, 128, 150; *Ulwar*, 87; *Agri.-Horti. Soc., Ind. :—Trans*, I., 165; II., 96, 212-215, 236, 311, *Proc.*, 358; III., 8, 9, 59, 69, *Proc.*, 236, 242, 252; IV., 78, 84, 102, 104, 107, 125, 146, 150, 236; V., 60-64, 80-82, *Proc.*, 48, 85,88; VI., 240, 243, 245, 247; *Proc.*, 7-9, 24, 35, 60, 104; VII., *Proc.* 37, 95, 138, 153, 193; VIII., 22, 96-97, 179-181, 225, 233, 235-237, *Proc.*, 336; *Journals :—*II., *Sel.*, 140, 294, 367, 541, 544; III., *Proc.*, 59, *Sel.*, 196; IV., *Sel.*, 31, 33, 152; IX., *Sel.*, 59; X., 359, *Sel.*, 24; XI., *Proc.*, 82; XIII., *Sel.*, 51; XIV., 44; *New Series :—*I., *Sel.*, 7, 14; II., *Sel.*, 57, 70-79; IV., 25; V., 80-94; *Proc.*, (1875), 12, 27, 43, (1876) 6-8, 16-18, 37, 38; VI., *Sel.*, 52-58, 82-85, *Proc.* (1880), 36; VII., 92-95, 169-203, 356, *Sel.* 37, 38; *Proc.*, 35, (1883), 104, 105, 107-108, 111, 174, 175; VIII., *Proc.*, 48, 68; *Indian Agriculturist*, 18th Sep. 1886; *Indian Forester*, IX., 203; *Quarterly Jour. Agri.*, I., (1828-29), 484; III., (1847-48), 81; IV., 1849-51) 74; VIII., (1857-59), 115; *Smith, Econ. Dict.*, 257, 258; *Treasury of Botany*, II., 1248; *Balfour, Cycl. Ind.* *Spons' Encyclop.*; *Encycl. Brit.*, XV., 309; *Morton, Cycl. Agri.*, II. 1171; *Ure, Dict. Inds. Arts & Man*, III., 20.

DISTRIBU-
TION.
51

Habitat, Distribution, and Forms.—This most useful grass is now cultivated, it might almost be said, throughout the world. DeCandolle presumes that it was originally a native of New Granada from the circumstance that, since it can be shown to have been cultivated from a remote antiquity by both the Peruvians and the Mexicans, it most likely was diffused from an intermediate region. Although unknown to Europe prior to the discovery of America, it has been cultivated for so many centuries in the New World as to have not only lost all trace of its wild habitat but to have become so obedient to man's necessities as to have yielded an extensive range of forms. In consequence there are conditions of this protean species suited to moist tropical regions, to burning arid tracts, to temperate zones, and it might almost be added to arctic climes. The modifications in stature, foliage, and grain are no less significant; but, as pointed out by Darwin (and subsequently fully confirmed), most if not all the forms of this plant are but climatic states and are hardly, therefore, entitled to be called races. They are, in fact, forms that tend to disappear when conveyed from one region to another. Thus, for example, some few years ago the effort was strenuously made to acclimatise the Cuzco maize in India. Large supplies of seed were freely distributed by Government. The verdict pronounced on the crop was, however, singularly uniform, *viz.*, that where it was found possible to secure fertile seed, the properties of the introduced plant were in a few years entirely lost. Degeneration at once took place and was only greater and more rapid in the localities most dissimilar to the Andes than in those that approached the climatic conditions of the valley of Vilcamaya. It was on this account that Markham wrote in terms of strong disapproval of the Indian experiments. "The Cuzco maize," he said, "should not have been sown in the plains of India;" nor does it, he added, grow on a high and rainless tableland. The Cuzco maize sown in Bengal grew so tall and rampant that it failed to mature seed at all. But it seems likely that, had the experiment been persisted in, till seed had accidentally been obtained, less and less trouble would have been experienced in seeding until a stock had

Cuzco Maize.
52

| Maize or Indian-corn. | (*G. Watt.*) | **ZEA Mays.** |

been produced that would have differed in no material respect from that of the country. And this result was arrived at in many stations throughout India, for the Cuzco maize gave origin, under each condition of climate and soil, to forms that differed only slightly from the recognised standards of the district in question and in the vast majority of cases lost all trace of its original characteristics. Not only Cuzco maize but many other famed American forms have been, for years past, systematically cultivated at Government Experimental Farms and Botanic Gardens, etc., until it might almost be said that it would save much unnecessary trouble and expense to have a paragraph prepared and stereotyped ready for reproduction as the special report of each fresh experiment. Such a paragraph might be briefly summed up in these words 'degeneration towards the existing or Native forms of maize.' It may in fact be said that with few crops do climatic influences exercise a stronger power than with Indian-corn. Improvement must, therefore, generally speaking, come from within, not from without. For one accidental acclimatisation of merit, thousands are worthless. But under careful treatment and by selection from existing stock, vast improvements might be effected. Fixity of merit however accomplished, must be secured for each district, if not for even areas of smaller size than districts. There can be no doubt but that maize came to India from America, and that it was cultivated for at least 100 years before forms were obtained for each tract of country of sufficient merit to justify extensive cultivation. But when India had in time evolved its own forms, maize moved rapidly over the length and breadth of this vast continent. It assumed tropical conditions in one part, and temperate and even arctic in others. We now find it grown on the swamps of Eastern Bengal, on the sandy desert tracts of Central India and Rájputána, in the humid temperate portions of the Eastern Himálaya, and in the drier and colder tracts of the western ranges even to Yarkand, Herat, and Tibet. The area of its cultivation, as a ripe grain, may, however, be said to be on the central tableland, the northern extremity of the plains, and the Himálayan slopes and inner valleys up to an altitude of 9,000 feet above the sea. On the lower or Gangetic plains it is grown chiefly as a green vegetable, as it was in Roxburgh's time when he spoke of it as "a delicacy." In the vicinity of all large towns the sale of the unripe cob is so remunerative that by peculiar systems of cultivation and selection, special forms have been matured that could scarcely be eaten in the condition of ripe grain. On the other hand, it might be said, that within the regions where maize is grown for its ripe grain, it is hardly possible to procure green cobs as a vegetable. The special forms with sweetness in the green state yield for the most part a much smaller proportion of glucose in the ripe condition. In most parts of India accordingly a classification into ripe grains rich in starch, and those rich in glucose, would very nearly correspond to the reputation the Natives give to certain crops that yield a sweet flour and those that do not. And this sweetness does not appear to be associated invariably with recognisable structural peculiarities that could be applicable outside the narrow limits of the districts in which they are found. For example, a short crop with white cobs is in one district regarded as best suited for flour, the yellow kind being viewed as inferior; while in another district a yellow grain has the reputation of greatest sweetness, the white being rejected. But adaptation to local conditions is perhaps more strikingly seen in the fact that in many parts of India there are forms of maize that require six months to mature, while in others maize is at most a three-months' crop. In some parts of the country, indeed, both kinds may be seen grown separately or as mixed crops. As a further manifestation it may be added that where the transition of the seasons into the

**DISTRIBU-
TION.**

Cuzco Maize.

*Conf. with
pp. 336, 337,
347, 348.*

**Area of Ripe
Grain
cultivation.
53
Area of Green
cultivation.
54**

**A
16 months'
Crop
or 3 months'
Crop.
55**

ZEA Mays.	Distribution of Forms
Kharif & Rabi crops. 56 *Conf. with pp. 337, 342.*	*kharif* and *rabi* crops allows of tropical cultivations during the former and temperate during the latter, two widely different classes of maize may be found. Maize in the greater part of the plains of India may be called a *kharif* crop, but *rabi* maize is by no means unusual, that is to say, maize sown in autumn and reaped in spring and thus produced along-side of wheat or barley.
FIBRE. Paper Material. 57 Spathe. 58	**Fibre.**—It yields a fibrous material which is said to be capable of being spun, but the chief use of the fibre hitherto has been as a PAPER MATERIAL. In Germany, Austria and Hungary the SPATHE seems to be largely utilized as a paper material, the article produced being regarded as superior to the paper made from any other grass, as it is remarkably tough and devoid of the silicious matter so much objected to in grass papers. The silicious property is said to destroy the type when employed by printers. The toughness of maize paper has, moreover,'made it serviceable for bank notes. Apparently the maize of India is not pitilized by the paper makers, probably from the expense that would be occasioned on its being carried by rail from the regions of chief production to the paper mills. In the numerous special reports, (reviewed by Mr. L. Liotord in his *Materials of Indian growth suitable for Paper-making*) there is only one allusion to maize fibre, *viz.*, in the report from Mysore. An article in the *Times*, September 13th 1865, describes maize paper as "so strong and durable that if ground short, it is even said it can be used as an excellent substitute for glass, so great is its natural transparency and firmness."
OIL. **Soap.** 59	**Oil.**—Dr. C. O. Curtman (*Chemist and Druggist, 1886*) says the OIL is made largely by the Woodchande Milling Company, ST. Louis, United States. The embryo yields from 13 to 15 per cent. by pressure. The oil at first is turbid, pale brownish yellow in colour, easily cleared by filtration or by letting it stand for some weeks. The specific gravity 0·916. Its taste is bland and of agreeable sweetness. It does not readily become rancid, and its odour being but slight is easily overcome by essential oils. It is of the non-drying class of oil. In some of its general properties it stands intermediate between olive oil and oil of sweet almonds, but approaches nearest to olive oil. It contains a large proportion of oleine; with Pontet's reagent it forms an abundance of elaidine, becoming nearly as solid as olive oil. SOAP—with alkaline lyes it saponifies promptly, forming a beautiful white soap, fully equal in appearance to the best Castile soap from olive oil. One of the most remarkable properties is its low congealing point, it remains entirely clear down to—8° C. (+ 17·6° F.); below that it begins to get slightly turbid and is congealed at—20° C. (+ 14° F.). As to the commercial value of the oil it is at present difficult to say. In its crude state it could not command a high price on account of the objectionable odour; but this might doubtless be removed by proper treatment, so as to bring the product up to the grade of refined cotton seed oil (*E. B. Shuttleworth, Canadian Pharma. Journ., August 1881*).
MEDICINE. Grain 60 Polenta. 61 Maizena. 62 Stigmas. 63 Meal. 64	**Medicine.**—Dr. Dymock (*Mat. Med. W. Ind.*) says: "It is considered by Mahometan physicians to have properties similar to those of S. vulgare, *viz.*, resolvent, astringent, and very nourishing; they consider it to be a suitable diet in consumption and a relaxed condition of the bowels. In Europe it is much used as a valuable article of diet for invalids and children under the names of *Polenta* (maize meal) and *Maizena* (maize flour). In Greece the silky STIGMATA are used in decoction in diseases of the bladder, and have lately attracted attention in America under the name of *Corn-silk*, of which a liquid extract is sold in the shops as a remedy in irritable conditions of the bladder with turbid and irritating urine; it has a marked diuretic action. The MEAL has been

| of Maize or Indian-corn. | (*G. Watt.*) | ZEA Mays. |

long in use in America as a poultice, and gruel is also made of it. In the Concan an alkaline solution is prepared from the BURNT COBS and is given in lithiasis." In the United States and elsewhere the meal is much used in the hospitals, and makes an excellent emollient poultice. Gruel prepared from it is also stated to be sometimes more grateful to the sick than that made from oat-meal (*Bentley and Trimen*).

SPECIAL OPINION.—§ "The centre of the COB (core) deprived of the seeds and reduced to an ash by burning, is given in combination with common salt in bronchial catarrh and hooping cough. The dose is 10 grains, repeated 2 or 3 times a day" (*Lal Mahomed. 1st Class Hosp. Asstt., Main Dispensary, Hoshangabad, Central Provinces*).

CHEMICAL OBSERVATIONS.—It has been suggested that the cobs of Indian-corn may yet form an important source for the supply of *Potash* salts. The average yield of 1,000 parts of cobs is 7·62 parts of carbonate of potash, or nearly twice as much as the best specimens of wood. The North American corn crop for 1871 was calculated to have yielded 7,700,000 tons of cobs, which are supposed to have contained 115½ million pounds carbonate of potash (*Journ. Soc. Arts, Dec. 1882*). Most writers agree in the opinion that the azotised matter is less in maize than in wheat, but that this grain contains a larger quantity of oil which accounts for its fattening property. In those unaccustomed to maize it is considered to excite and keep up a tendency to diarrhœa. The green cobs to some persons act almost as a drastic purgative, more especially in certain states of the system. The writer saw a case of acute dysenteric diarrhœa which, there could be no mistake, was caused through the patient having eaten roasted maize.

Food and Fodder.—In the Panjáb, the North-West Provinces and Oudh maize assumes the position of a staple article of food. It is, however, grown throughout the length and breadth of India, but in Upper India mainly is the ripe grain reduced to a FLOUR and made into bread. In some parts of the country it is ground into a MEAL and eaten as porridge. In others the GREEN COBS are eaten after being roasted or boiled. The ripe grain is also very often PARCHED and in that state is eaten as a midday meal. In Upper India there are generally two crops—the one a little earlier than the other and eaten as a green crop, the later crop only being allowed to ripen. The straw (STEMS and LEAVES) of the ripe crop is not of great value as a fodder (except for elephants), but the crop reaped in the green state affords a much valued fodder. In many parts of the country the stems of the ripe crop are not even cut, so little value is put upon them; indeed their chief use may be said to be as fuel.

It is perhaps scarcely necessary to have to repeat that in Europe and America maize is not only largely used as an article of horse-food, but the better qualities are extensively employed as human food. It is ground into several preparations, such as Hominy Maizena, Polenta or Indian-corn flour. Maize, owing to the large amount of oil which it contains, is specially valued for its fattening properties. A kind of beer is made from the grain in South America which is known as *chica*, and in Western Africa a similar beverage is made from it which is there known as *pitto* or *peto*.

In Bengal (Chutia Nagpur) a kind of porridge is made of maize meal which is known as *lapsi* or *gathá*, and the preparation, there called *satu*, which is eaten with sugar is the parched grain reduced to flour. Bread made of maize flour is largely eaten in the Panjáb. Professor Church says that maize is not considered so wholesome as wheat, since it is thought rather heating. Its nutrient-ratio is 1 : 8·3, and the nutrient-value 88½. The Indian forms of maize, Professor Church says, manifest a lower

MEDICINE.
Cobs.
65

Core.
66

CHEMISTRY.
67

FOOD & FODDER.
Ripe Grain.
68
Meal.
69
Flour.
70
Green cobs.
71
Parched Grain.
72
Stem.
73
Leaves.
74

Hominy.
75
Maizena.
76
Indian-corn. Flour.
77

Beer.
78

Satu.
79
Bread.
80

ZEA Mays.	An article of Food.

FOOD.

proportion of water and of oil than the average of European and American samples. Generally, he says, the American grown maize contains about 1 per cent. more fat or oil than the East Indian.

Starch.
81
Gluten.
82
Cattle-food.
83

In the European industry of starch-making from maize it is generally stated that flat-yellow American affords $53\frac{1}{2}$ per cent. of starch and the flat white and round yellow about $54\frac{3}{4}$ per cent. In its occurrence and association maize starch closely resembles that of wheat but it differs in the respect that the gluten forms a tough mass which may be separated without having recourse to fermentation. This by-product is of great value in cattle feeding, being for that purpose mixed with the hulls. The finer qualities of maize starch are largely used as a substitute for arrowroot, while the lower qualities find a sell for laundry purposes.

Sugar.
84

SUGAR.—At a meeting of the Linnæan Society held on the 21st February 1843, **Professor Croft** read a paper on the manufacture of sugar from the stalk of maize as experimented with in the State of Indiana. It was affirmed that the juice of the stalk contained three times the sweetening principle of that of the beet, five times that of the maple, equalling, if not surpassing that of sugar-cane grown in the United States. From maize, sugar was easily obtained, the cultivation of it easier, while the shorter period it occupied the ground, were the advantages it possessed over sugar-cane. Although by no means a new discovery, the process having been described previous to that date, those statements, however, having been reported in various scientific papers, attracted considerable attention, and various experiments were made to test the truth of them. Amongst the residents in India who took an interest in the subject, **Mr. C. B. Taylor**, Palamow, may be mentioned. On 11th November 1843, he wrote an account of his experiment to the *Agri.-Hort. Soc., Ind.* (*vide Journ. II., 541*). He there says that he had forwarded for inspection of the members "a box containing six earthen vessels of sugar, or more properly speaking, the CONDENSED JUICE expressed from the stalks of Indian-corn—it cannot, I apprehend, be called sugar, as it will not *grain ;* likewise two bottles of SPIRITS distilled from the same substance". Writing at a later date he expresses his disappointment in being unable to obtain any *grain* from the maize *gúr.* About the same date experiments were tried by **Mr. Frederick Nicol** of Jessore who states that after prolonged boiling of the juice, and afterwards allowing it to stand over several days, it presented no signs of granulation. It was submitted to a second boiling, and while upon the fire, some manufactured sugar was thrown into the pan to give a grain to the material, and allowed as before to cool. When the molasses were run off, nothing remained but dirt. Having failed in his attempt to produce crystallizable sugar, he next tried to manufacture rum, but was discouraged by the flavour of the spirit produced being unpleasant.

Spirits.
85
Condensed juice.
86
Failure to granulate.
87

Rum.
88

The subject of maize sugar or rather glucose is in European works generally treated of conjointly with that of **Sorghum** sugar, and as that subject has been fully dealt with already, it need not be gone into in this place. The above are the only special Indian reports on maize sugar which the writer has been able to discover. So far as the development of this subject has as yet gone, it may be said that maize glucose is chiefly used for adulteration of cane-sugar. It is held to impart to the sugar prepared a whiter and finer appearance, but to reduce the sweetening property. The production in America of maize glucose may be said to be on a very large scale.

DOMESTIC.
Thatch.
89

Domestic.—When the crop is allowed to mature the stalks become so hard as to be unsuitable for fodder and are then used mainly for *thatching* and as *fuel.* "The stalks with the leaves attached are used for

| History of Maize. | (G. Watt.) | ZEA Mays. |

fuel" (*A. Campbell, Nepal, Trans. Agri.-Hort. Soc., IV., 127*). "The STEMS are left on the ground and are gathered and burnt as fuel" (*Agri. Report, Lohardaga District*). Recently cobs compressed into a hard structure have in Europe been employed in the preparation of tobacco pipes.

HISTORY.

It is now universally admitted that Maize or Indian-corn is a native of America. Formerly, but upon the most unsatisfactory evidence, it was believed by some writers to have been known in Europe prior to the discovery of America. DeCandolle on this subject writes : "No one denies that maize was unknown in Europe at the time of the Roman Empire, but it has been said that it was brought from the East in ·the Middle Ages. The principal argument is based upon a charter of the thirteenth century, published by Molinari, according to which two crusaders, companions in arms of Boniface III., Marquis of Monferrat, gave in 1204 to the town of Incisa a piece of the true cross . . . . and a purse containing a kind of seed of a golden colour and partly white, unknown in the country and brought from Anatolia, where it was called *meliga*, etc." The seed referred to was by some thought to be Sorghum, by others maize : but adds DeCandolle "These old discussions have been rendered absurd by the Comte de Riaut's discovery that the charter of Incisa is the fabrication of a modern imposter." "I quote this instance, continues DeCandolle, to show how scholars, who are not naturalists, may make mistakes in the interpretation of the names of plants, and also how dangerous it is to rely upon an isolated proof in historical questions." The various names which it bears in Europe, Egypt and Asia only show that in each country it was supposed to come from some not very distant region. Thus it is Turkish-Wheat, Indian Corn, Roman Corn, Sicilian Corn, Spanish Corn, Barbary and Guinea Corn, etc. The Turks call it Egyptian corn and the Egyptians speak of it as Syrian grain. Its most general vernacular name in India may be rendered Mecca-corn (*Makkai*), but here again it would appear that this is but an appropriation of an older name *Durah-i-Makka* or *Gandum-i-Makka*, the synonym for which, by Muhammadan writers on Materia Medica, is given as *Khanderús*, the χóνδρος of the Greeks, or Sorghum vulgare. And it is remarkable that a very large percentage of its names in India denote the larger *Juár* or *Sholam* (Sorghum vulgare). By the Arabs maize is *Durah kizán* or *Durah shami*. *Durah* by itself is Sorghum vulgare. There is no authentic Sanskrit name for the plant, nor is the grain in any way associated with the religious nor even with the domestic observances of the Hindus. The Sanskrit word *Yavanála* sometimes given to maize, at other times to Sorghum, has been discussed in the concluding note on the vernacular names in the paragraph above, and need not be further dealt with. The one singular feature regarding the names given in the Old World for this, perhaps the most important gift of the New World—is the fact that it nowhere apparently carried with it any trace of its aboriginal names. DeCandolle appears to have misread Crawfurd's remarks regarding it when he attributes to that author the belief "that the species was a native" of the Malay Archipelago. What Crawfurd did say was that the names given for it " in some Oriental languages " " seem entirely native " and have a specific signification. He cites " *Bhutta** in Hindi, *Jagny* in most of the languages of the Indian Archipelago, *Katsalva* in Madagascar." But adds Crawfurd " This would lead to the belief that

DOMESTIC.

Fuel.
90
Tobacco Pipes.
91

HISTORY.
92

---

* See the note which concludes the paragraph above on vernacular names p. 327. Conf. also with remark regarding Behar names, p. 351.—*Ed., Dict. Econ. Prod.*

ZEA Mays.	Historical Sketch of

HISTORY.

the plant was indigenous where such names were given to it, but the probability is that they were taken from some native plant bearing a resemblance to maize." In another sentence he removes any possible misconception of his meaning, for he says "Maize is, beyond all question, a native of America, and before the discovery of the New World was wholly unknown to the old" (*Migration of Plants in reference to Ethnology*). DeCandolle in refutation of Crawford's *supposed* opinion, that it had or may have had an Asiatic origin, proceeds to show that the fact of Rumphius being silent regarding the plant, points to a later introduction than the seventeenth century. Here, again, it is possible an error may be inculcated, for although Rumphius was undoubtedly one of the greatest of the early Asiatic botanists he might easily have regarded an American plant of recent introduction as deserving of no special consideration by him. Royle (*Prod. Resources, p. 40.*) says that the Portuguese very probably introduced the richest products of America into India, such as Maize, Capsicum, Guava, Custard-apple, and Pine-apple. And there is much to be said in

Probable Introduction in the 16th Century.

93

favour of this suggestion. The *Ain-i-Akbari*, which may be designated the Administration Report of the Emperor Akbar for the year A.D. 1590, contains what the writer regards as an undoubted reference to Pine-apple and an accidental allusion to what has been translated maize. If there can be no mistake regarding the former, then it might perhaps be admitted there was at least a plausibility for the accuracy of the latter, since both plants are American. The pine-apple is described as having "leaves like a saw. The fruit forms at the end of the stalk, and has a few leaves on its top. When the fruit is plucked, they cut out these leaves, separate them, and put them into the ground; they are the seedlings. Each plant bears only once, and one fruit only." In the *Toozuk-i Jahángírí* (edited by Sayyid Ahmad, *p. 3*) it is stated that in the time of Jahangir pine-apples came from the harbour towns held by the Portuguese. While Abal Fuzl, in his long list of grains and pulses grown in India during the sixteenth century, does not include maize, under his chapter on the beautiful flowers to be seen at the Court of Akbar, he says of *kewrah*, that its leaves are like those of "maize" (*Blochmann's Transl. Ain-i-Akbari, p 83*). Linschoten, one of the most painstaking Indian explorers of the sixteenth century, makes no mention of maize as seen by him in the East. In the eighteenth century Burmann published his *Thesaurus Zeylanicus* and his *Flora India*, but in neither of these works does he allude to Zea. Hove, who at the close of the last century, visited Bombay in order to study its cotton, and whose report is full of information on all the crops seen by him, does not apparently mention Indian-corn. Now it would, as it seems to the

Not cultivated in the 18th Century.

94

writer, be unsafe to assume that, although the pine-apple was fully known in the sixteenth century, so extremely valuable a plant as the maize did not reach India until after the date of Rumphius' works. It is quite clear, however, that it was little more than experimentally grown for, perhaps, two centuries after its introduction, and that when once acclimatised and on its properties having been made known it was thereafter rapidly distributed over the length and breadth of India. Its name *makkai* may be regarded as manifesting the association of the distribution of the grain with the Muhammadan rulers of India, and its displacement or appropriation of the names formerly given to the introduced forms of Sorghum may be viewed as denoting the innate propensity of Asiatics to contrast all new ideas with previous conceptions. So very little progress had, however, been made with maize cultivation that Roxburgh wrote, about the beginning of this century, that Indian-corn was "cultivated in various parts of India in gardens, and only as a delicacy; but not anywhere on the continent of India, so far as I can learn, as an extensive crop." In 1819 Buchanan-

| Maize or Indian-corn. | (*G. Watt.*) | ZEA **Mays.** |

Hamilton published his account of the kingdom of Nepal ; while dealing with the ancient state of Yumila (its capital Chhina-chin) says that they had maize. In a further page he remarks of Kangra : " The poor people live much on maize." Very shortly after the appearance of **Roxburgh's** *Flora Indica,* however, **Graham,** in his Catalogue of the Plants of Bombay (published 1839), wrote of Western India that maize was "commonly cultivated." **Dalzell & Gibson,** some thirty years later (1861), said that it was "extensively grown in the early part of the rains, especially near large towns." And these authors add : "The grain is seldom used in India as a flour." But as illustrative of the extremely local character of the information often furnished by Indian writers it may be added that **Stewart,** in 1862, wrote of Bijnour that "much of the maize was ground into flour and made into bread, although very much less is here used in this way than in the Panjáb." It is thus very probable that in Upper India (a region, comparatively speaking, unknown to **Roxburgh**) maize was much more extensively grown at the beginning of the century than might be inferred from **Roxburgh's** words. At the present day it would be more nearly correct at any rate to speak of maize as of equal value to the people of India collectively with wheat, instead of its being grown purely as a garden " delicacy." It is a field crop upon which at least the bulk of the aboriginal tribes of the hilly tracts of India are very largely dependent for subsistence. Thus its diffusion over India, during the present century, might almost be said to be one of the most powerful arguments against the statement often made that the Natives of India are so very conservative that they can scarcely be induced to change their time-honoured customs, even when these can be shown as inimical to their best interests. So completely has India now appropriated the *Makkai* that few of the village fathers would be found willing to admit that it had not always been with them as it is now, a staple article of diet. They may even cite its supposed ancient names and quote wise sayings regarding it, oblivious all the while that a very few years ago these were universally accepted as denoting an altogether different plant. Thus, in many parts of the Panjáb, *juár* means maize, not **Sorghum vulgare,** the latter crop being known as *chari*—fodder. In some parts of South India *Chôlam* or *Sholam* means maize ; it has thereby appropriated the name for **Sorghum.** In the same way the Persians often give the name *bájri* to maize instead of to the millet **Pennisetum typhoideum.** Alluding to the manner in which maize may revive after a spell of drought the proverb runs—

" When maize* droops the farmer laughs
When wheat is laid he laments."

It has been pointed out that mention is made of maize in a Chinese work which dates from 1578 (according to **Bonafous**) and 1597 (according to **Mayers**). "If this be true, says **DeCandolle,** and especially if the second of these dates is the true one, it may be admitted that maize was brought to China after the discovery of America. The Portuguese came to Java in 1496, that is to say four years after the discovery of America, and to China in 1516. **Magellan's** voyage from South America to the Philippine Islands took place in 1520. During the fifty-eight or seventy-seven years between 1516 and the dates assigned to the Chinese work, seeds of maize may have been taken to China by navigators from America or from Europe."

Maize had reached Europe a short time before the dates mentioned for India and China. "The first botanist who uses the name, Turkish

---

* *Juári*—originally **Sorghum.**

ZEA **Mays.**	Cultivation of Maize

HISTORY.

wheat, is Ruellius, in 1536. Bock or Tragus in 1552, after giving a drawing of the species which he calls **Frumentum turcicum,** *Welschkorn,* in
Germany, having learnt by merchants that it came from India, conceived
the unfortunate idea that it was a certain *typha* of Bactriana, to which
ancient authors alluded in vague terms. Dodoens in 1583, Camerarius
in 1588, and Matthiole rectified these errors, and positively asserted the
American origin. They adopted the name *Mays,* which they knew to be
American" (*DeCandolle*). It is perhaps unnecessary to continue the quotation from DeCandolle's very instructive historic sketch of the literature
of this subject in order to show that so little was known of maize in Europe
that for some time the early travellers in the New World expressed their
astonishment regarding it. Nor is it necessary to review the records of
America that point to a great antiquity for the crop in that country.
Suffice it to recapitulate that philology and history alike confirm the opinion that maize was originally a native of America. Nor is botanical evidence wanting in support of that view. Mays is not only the sole representative to the genus **Zea** but it stands almost alone in the family. There
are no Asiatic wild plants in any way closely related to it. Darwin found
heads of maize imbedded on the shore in Peru along with several marine
shells, but at a height of 85 feet above the present level of the sea.
This necessitates a vast antiquity, a fact also indicated by an ancient cultivation denoted through the discovery of two forms (now extinct) in tombs
apparently prior to the dynasty of the Incas. The aboriginal form has
not as yet been found in a wild state. The effects of long cultivation are,
however, seen in the extreme variability of the plant. Its rapid adaptations both in size of plant and shape and colour of grain, to certain conditions of climate and soil, are very remarkable. A supposed aboriginal
form, discribed by **Saint Hilaire as Zea Mays tunicata** and which **B** onafous
figured under the name of **Zea cryptosperma,** is by no means the most remarkable form nor is the presence of a sheath to each grain (the character
that gave origin to these names) more stable under altered conditions of cultivation than many of the less striking characteristics of other races. Indeed,
it may be doubted whether any of the forms of maize are entitled to be
regarded as races. The peculiarities are for the most part climatic adaptations that rapidly disappear under altered environment. Thus **Darwin**
tells us "that tall kinds grown in southern latitudes, and therefore exposed
to great heat, require from six to seven months to ripen their seed ; whereas dwarf kinds, grown in northern and colder climates require, only from
three to four months. **Peter Kalm,** who practically attended to this plant,
says, that in the United States, in proceeding from south to north, the
plants steadily diminish in bulk." Adaptation to climate, **Darwin** adds, is
very nearly as striking as in the summer and winter wheats, and the change
from the one into the other may be effected only gradually. The influence
of the climate of Europe on American maize, according to **Metzger,** for
example, was a loss in height and a complete change in the shape and
colour of the grain. "In the third generation nearly all resemblance to the
original and very distinct American parent form was lost. In the sixth
generation this maize perfectly resembled a European variety." **Darwin**
adds : "These facts afford the most remarkable instance known to me of
the direct and prompt action of climate on a plant. It might have been
expected that the tallness of the stem, the period of vegetation, and the
ripening of the seed, would have been thus affected ; but it is a much more
surprising fact that the seeds should have undergone so rapid and great a
change" (*Animals and Plants under Domestication, I., 320-323*). These
observations, it will be seen, by the remarks in another chapter of this
article, have a peculiar significance in India. We possess not only tall

Climatic
variations,
97
Conf. with p.
329.

Character of
grain altered,
98
Conf. with pp.
340, 347, 348.

		ZEA
or **Indian-corn.**	(*G. Watt.*)	**Mays.**

HISTORY

tropical forms and dwarf alpine states, but kinds only serviceable as green vegetables aud others that yield grain rich in starch, others in glucose. The experiments performed hitherto in India (with a view to improving the local stocks) might almost be said to have failed mainly from a disregard of the accepted principles which should have governed their procedure. States of the plant suited to temperate regions have been experimented with in the tropics with the not unnatural result of failure. Tropical conditions that require six months to mature their grain have been experimented with in regions that either did not possess more than four months suitable weather or where the cultivator did not care to have his soil for such a long period under the crop. The result, disappointment or neglect. The whole subject of the maize cultivation of India is so very little understood, that comparatively few persons seem to be aware that in many parts of the country there are two crops a year which differ from each other very nearly as greatly as do the rampant conditions of the tropics from the dwarf states of the temperate zones. An expenditure of much time and money has been entailed by experiments at acclimatising certain American forms, the result of which might have been foreseen, *viz.*, the gradual loss of all the characteristics of the American stock and the production in a very few years of a form that differed in no material respect from those already in the country. Acclimatisation is, therefore, of comparatively little avail with a plant, which, like maize, is subject to rapidly change almost every one of its characteristic features under dissimilar conditions to those by which it was nurtured and developed. So, again, all writers are agreed that to preserve pure the varieties of maize, they must be grown at sufficient distances apart to prevent the crossing that will otherwise freely take place. With maize, it may in conclusion be said that fixity of characteristics, whether these have been attained by selection, crossing, or acclimatisation, is only possible within the very narrowest climatic changes. Each province, indeed almost each district, must, therefore, develop its own forms. Acclimatisation, while it may accidentally give a useful new kind of maize, will rarely if ever reproduce the conditions desired by the importation of a supply of foreign seed of reputed properties. These and such like features of Indian-corn are of the most potent character in the study of the agricultural aspects of this subject, but in the present historic chapter they have their bearing in the manifestation of ancient cultivation as given by multiplicity of form and variability or adaptability to man's requirements.

Indian Experiments. 99 *Conf. with p. 3. 9.*

Two Crops in India. 100 *Conf. with pp. 330, 342.*

Selection to local conditions essential. 101

CULTIVATION.

CULTIVATION 102

The material at the writer's disposal is too meagre and unscientific to justify an attempt at describing or even classifying the forms of Indian corn met with in this country. He must, therefore, rest satisfied with having briefly indicated, in general terms, under the paragraphs of **Habitat** and HISTORY, the wide range that must exist, from the diversified conditions under which the crop is raised and the influences and necessities that have controlled natural selection. In the paragraphs that follow, the reader will find that the customary proceedure of this work has been followed, *viz.*, to furnish under provincial sections a selection of passages, calculated to convey an idea of the position of the industry and the systems of cultivation adopted by the Indian *rayats.*

Area of Cultivation.—The area under maize in India can scarcely be determined, since nearly every peasant grows a few plants near his homesteads and these must of necessity escape estimation. Roughly speaking, it may be said that the bulk of the crop, which is eaten as a green vegetable,

Area. 103

CULTIVATION
Area.

Total
probably
5 million
acres.
104

is excluded from the calculated area. In the annual returns of agricultural statistics several provinces, such as Bengal, furnish no estimates, in others maize is grouped with millets. The seriousness of this latter fact may be here shown. The Panjáb does not give separately (in the report for 1890-91) its maize crop, but in a special report issued in 1885-86 it was ascertained that the average area under maize for the three previous years had been 1,215,206 acres. The surveyed maize area for the rest of India (in 1890-91) came to 1,789,057. These two returns show, therefore, a total of 3,000,000 acres, and were a provision made for Bengal, Central India, Rájputana, Burma, Assam, Hyderabad, etc. (provinces for which no returns of maize cultivation have been furnished), it seems likely that the total might exceed rather than fall far short of 5 million acres per annum as under this crop. The Panjáb appears, however, to be by far the largest Indian-corn producing province of India and is followed by the North-West Provinces with (in 1890-91) 978,653 acres, Oudh with 476,036 acres, Bombay with 137,457 acres, and the Central Provinces with 106,659 acres. So far as these figures go, therefore, it might be said the maize area of India closely corresponds with that of wheat and that the grain is least produced in rice-growing provinces.

PANJAB.
105

I.—PANJAB.

References.—*Gazetteers :—Rawalpindi, 52, 78, 80, 81 ; Ludhiana, 133-138, 140, 143, 159 ; Hazara, 52, 129, 130, 134-137, 147, 150, 151 ; Sialkot, 34, 65, 68 ; Gujrat, 77, 79, 81 ; Jhelam, 55, 98, 100, 108 ; Kangra, I., 60, 61, 153, 155, 157, 158, 161 ; Kangra, II., 24, 58 ; Shahpur, 37 ; Peshawar, 84, 144, 146, 157, 159 ; Gujranwala, 27, 47, 48, 51, 55 ; Jalandhar, 18, 44 ; Gurgaon, 73 ; Delhi, 44-46, 101, 113, 114, 139, 140 ; Montgomery, 54, 88, 103, 104, 106, 111 ; Amritsar, 19, 20, 35, 36, 47 ; Lahore, 48, 86, 89 ; Bannu, 53, 138, 145, 150 ; Kohat, 60, 101, 104, 105, 121, 122 ; Jhang, 48, 107, 115 ; Dera Ghazi Khan, 84 ; Ferozepore, 65-69, 74 ; Hisar, 46 ; Mooltan, 111 ; Dera Ismail Khan, 144 ; Hoshiarpur, 35, 86, 87, 91, 92 ; Ambala, 31, 32, 44 ; Karnal, 157, 172, 176, 185, 197 ; Simla, 37, 39, 53, 57, 78 ; Settlement Reports :—I. (Hoshiarpur), 11, 23, 31 ; (Kangra) (1850), 4, 24, 25, 27, 30, 31, 44 ; (1865), 78, 144 ;(Sialkot), 31, 33 ; II., Lahore, (1860), 11 ; (1865-69), 9, 33, 54 ; IV. (Dera Ghazi Khan), 130, app., cxvi ; (Peshawar), 184, 188, 215, 219, 223, 225, App. xxxiv, xlii, lxvii, lxxv, lxxxv, ciii-cviii, cxvii (ix) ; V. (Ferozepore), 3, 7, 31, 33, (Muktsar), 22, 25 ; (Rawalpindi), 59 ; (Shahpur), app., iii ; VI., (Gujrat), (1860), 136 ; (1870), 33, 34, 79, 84 ; (Montgomery), 102, 107-109, 115, 116, 126, 128 ; VIII. (Bannu), 53, 80, 85, app. xxi ; ix.; (Hazara), 81, 88-90, 102, 103, 173, 174, 178, 192, 194, 196, 202, 204, 207, app., lxxxviii, xc, civ, cvi, cviii, cx, cxii, cxiv, cxvi, cxviii, cxx, cxxii, cxiv, cxvi ; X. (Simla), 10-15, 42, 44, 45, xxxix ; XI. (Delhi), 43, 106; 224, app. xxxv, cclxvii ; (Kohat), 2, 12, 73, 74, 120, 122-124, 155, 162-164; Kohat, 122-124; Jhang, 85-94; Selections from Records Fin. Commissioner (1887), pp. 780-836.*

Area.
106

Area.—Some few years ago the Commissioner of Settlements and Agriculture (the late Colonel Wace) issued a circular letter calling for information on the subject of maize cultivation in the Panjab. The replies which were obtained brought together a detailed statement such as exists for no other province on the extent, position, and nature of the maize cultivation. These replies were published in the form of Selections from the Records of the Financial Commissioner's office in 1887. Reviewing these district reports it was pointed out " that little or no maize is grown in the western and south-western districts of Hissár, Rohták, Gurgáon, Jhelum, Shahpur, Jhang, Montgomery, Mooltan, Muzaffurgarh, Dera Ghazi Khan, and Dera Ismail Khan." The crop is " mostly grown where the summer rainfall is highest, that is, in the montane and sub-montane districts·

or Indian-corn in the Panjab. (*G. Watt.*)

The proportions of irrigated and unirrigated maize in the last two years were—

	1884-85	1885-86.
Irrigated .	477,713	464,468
Unirrigated	770,547	717,684
TOTAL .	1,248,260	1,182,152

and about five-sevenths of the unirrigated maize is grown in the hill districts of Hazára and Kángra, and the sub-montane districts of Hoshiarpur and Umballa, where the rainfall is good and practically certain. The crop will not thrive without abundant moisture, and where the rainfall is insufficient, there must be irrigation. But too heavy rain or long continued cloudy weather are nearly as bad as drought. Maize likes moderate and constant showers, with alternating sunshine. It also requires a rich soil, and is usually grown in the highly manured belts of land around the villages or detached farms. It is scarcely ever grown on poor soil. If the land growing maize is manured, it will usually be found that the soil is intrinsically rich, or has received a good alluvial deposit from stream. The Panjab cultivator generally recognises the importance of not growing maize plants very close together unless sown merely for fodder. It is nowhere the custom, as in America, to sow three or four seeds in separate little mounds of earth."

"The common maize in the provinces is yellow. The white kinds are little cultivated, though they are prized near towns for roasting in the cob. The plant has many uses. The cob is often roasted before quite ripe, or the ripe grains are parched, or the flour is made into porridge or bread. The grain is also sometimes given to horses or cattle; and the stalks and the leaves are used as fodder. As the farmer cultivates maize in his best land, he does not care for any variety that occupies the ground for more than 80 or 90 days, for he looks to cut his maize crop in September or October, so as to be able to clear the ground for wheat or some other winter crop. It would be useless to attempt to introduce any variety of maize that occupies the ground too long, or that is not hardy and will not stand a slight drought, or that is too easily blown down by high winds. Some attempt has been made to introduce good kinds of American maize, but no variety has as yet become popular."

The above brief review of the leading facts brought out by the district reports was furnished by the Panjáb Government, and space can hardly be afforded in this work for giving much more. It may be said, however, that in the appendix to the report a table is furnished of the area under the crop from which the following may be given in the order of importance as the chief districts :—

DISTRICTS.	Average of three Returns previous to 1885-86.	Percentage of Area under Maize to total cultivation.	Average rainfall for the months from 1st June to 30th September.
Hazára .	193,588	47	24
Kangra .	137,188	29	55
Hoshiarpur .	135,787	19	27
Amballa .	121,243	12	28
Peshawar .	91,114	10	5
Jullundar .	76,383	11	22
Siálkot .	67,231	7	27

| ZEA Mays. | Cultivation of Maize | | | |

DISTRICTS.	Average of three Returns previous to 1885-86.	Percentage of Area under Maize to total cultivation.	Average rainfall for the months from 1st June to 30th September.
Gurdáspur	53,992	6	24
Ludhiána	51,985	7	23
Ráwalpindi	49,609	4	20
Amritsár	36,109	5	20
Lahore	36,085	3	16
Bannu	31,443	4	7
Ferozepore	28,714	2	17
Simla	1,649	16	52

There are one or two considerations brought out by the above statement that are of no small importance. In Házára maize cultivation occupies 47 per cent. of the total cultivated area, in Kángra 29 per cent., in Hoshiarpur 19 per cent., in Simla 16 per cent., in Jullundar 11 per cent., and in Peshawar 10 per cent. Leaving out of consideration the wide range in altitude, embraced by the districts named, there is another fact exemplified by the table. The rainfall in two of the districts (and these might be called temperate regions), namely, Kángra and Simla, is normally 55 and 52 inches, respectively, during the months when the crop is grown, while in a third, Peshawar—one of the hottest tracts of India during those months—the rainfall is only 5 inches. Such facts as these forcibly display the extensive series of forms of maize that must exist even in this one province of India, and they demonstrate also the adaptations that have been accomplished before the crop could be one of such importance as it undoubtedly is to the people of the Panjáb.

Conf. with pp. 336, 347.

The circular letter issued by the Commissioner of Settlements and Agriculture (March 1884) to which the local reports constituted the replies, asked that attention should be given to certain points. These were:—

1st.—The varieties grown and the circumstances under which each is preferred.

2nd.—The system of cultivation and rotation pursued.

3rd.—The date of sowing, reaping, etc.

4th.—The estimated yield per acre.

5th.—The consumption of the grain and stalks.

6th.—The diseases to which the crop is liable.

The replies to these and such like questions are so highly instructive that it is unfortunate space cannot be afforded to republish the report in its entirety. It may, however, be useful to give here a few facts under each of these questions.

Varieties. 109 Amritsar. 110

Varieties cultivated.—In AMRITSAR there are said to be four qualities: the 1st quality produces cobs of about 9 inches in length, the grains are yellow, sweet in taste and more durable than any of the others. The 2nd quality has cobs not more than 6 inches in length. The produce is much lower than the first; it is sweet in taste, hard in grinding but durable. The 3rd quality grows commonly near the villages on the banks of the Ravi. Its cobs are 6 inches long, but the yield is much less than with the yellow maize. The 4th quality called *Lahori* maize has cobs only four inches long and the grains are only half the weight of those of the first quality. The stalks are, however, soft and are used for cattle food. It is also useful in another respect. If the times for sowing yellow and white maize have past this

	ZEA Mays.
or Indian-corn in the Panjab. (*G. Watt.*)	

form, even if sown in the beginning of *Bhádon* (15th August) grows till *Assu* (September-October) and ripens its crop in two to two-and-a-half months. In HAZÁRA, Mr. Kennedy says, there are four kinds of maize—two of which are "practically indigenous and two lately imported." The former are distinguished by the colour of their grains, the one having cobs of black with white grains intermingled, and the other white or yellowish white. The Natives prefer the so-called indigenous maize with yellow grains. It is sweeter than the others when made into bread. In GURDASPUR the people recognise three varieties by the difference in the grain, *viz.*, yellow, white, and red. The yellow is preferred to the white in Gurdaspur and Batala tahsils as it has larger cobs, bigger grain, and is sweeter. In PATHANKOT, on the other hand, a white variety is preferred. The yellow variety flourishes best on richly manured lands, and it requires greater attention than the white. It takes a longer time to mature. The white may accordingly be sown much later, say after the rainy season sets in, but it will then ripen at the same time as the yellow. But it may be added that while these differences exist, it is currently believed in Batala tahsil that if the white be grown for four years consecutively, it will become yellow. Of RÁWALPINDI the Settlement Officer reports that there are two forms all but universally cultivated. These are the white (*Suféd* or *chitti*) and yellow (*pili*), but in Murree tahsil varieties are grown which are known as *Sattri*, *Saithi*, and *Kari*. The last mentioned (*Kari*) appears to be grown chiefly as fodder. It is not very good to eat and thrives best in a cold climate. *Saithi* prefers a cold climate and may be grown on poor soil. *Kari* when taken to Ráwalpindi tahsil grows better than in its more common habitat of Murree. In HOSHIÁRPUR practically only one kind is grown—the yellow called *chhalli* sometimes *kathi*. Red cobs are occasionally to be seen in a yellow field, but they are never separately cultivated. The yellow form is, however, often seen with a lighter shade on the outside of the grain This is called *dhusár*. White maize is also very common and where found is preferred to the yellow for eating roasted. In LAHORE maize is almost invariably called *chhalian*, but the name *makki* is thoroughly understood. When the people speak of *Jowár* they mean maize and not the greater millet, which is grown as fodder, *chari*. Of PESHAWAR Colonel Waterfield reports that "generally speaking only one kind of maize is cultivated, *viz.*, the white variety." In the *bárání* tracts the yellow is sometimes found, but the area under yellow maize is comparatively small. In the Peshawar tahsil and sporadically elsewhere also, the red variety is found. The white variety is preferred as it is softer and sweeter than the others. But it requires more careful cultivation and irrigation. The white variety is preferred for roasting. The unripe cobs are also given to fatten horses. During the winter months, bread made of maize grain (cooked in ovens) forms the ordinary food staple of the poorer classes. In KARNAL there are said to be three forms, (1) *Pili makki* (yellow), (2) *Dhauli makki* (white), and (3) *makka*. The last named is larger and finer than the two former and may be either white or yellow, but it ripens late and requires too much moisture to be a popular crop. The white *makki* requires more water and ripens a fortnight later than the yellow, so that it is less popular. It is an important consideration to have maize off the soil in time to allow the preparation of the land for the *toria*, barley or wheat that has to follow. White flour sells at a dearer rate than yellow, but the people prefer the taste of the latter and only chose the former from the desire to have white and clean-looking bread. Red maize is not grown as a special form, though the yellow may often be seen shading off to the red. The Deputy Commissioner of KOHAT writes that maize in the Kohát and Hungu tahsils is nearly as important as wheat. There are two varieties grown, (1) *Sarda*,

CULTIVATION in the PANJAB. Hazara. **111**
Gurdaspur. **112**
Pathankot. **113**
Rawalpindi. **114**
Hoshiarpur. **115**
Lahore. **116**
Peshawar. **117**
Karnal. **118**
Kohat. **119**

ZEA Mays.	Cultivation of Maize

CULTIVATION
in the
PANJAB.

Ludhiana.
120

Kangra.
121

Amballa.
122

Seasons of
Sowing and
Reaping.
123
Two Crops.
Conf. with pp.
330, 337.

usually of a white colour—the early sowings; (2) *Garma*, a yellowish kind —the late sowings. The first of these is in most favour for making bread, while for parching the second is preferred. In Kohát maize has completely appropriated the name *jowar*. In LUDHIANA three forms of Indian corn are grown, *viz.*, yellow, white, and red. The yellow is the most abundant, but the white ripens sooner and is in some cases preferred on that account. The red is not grown by itself but appears among either of the other two kinds. In KANGRA (Kullu) Mr. Dane wrote there are three local forms grown, (1) *Sathu*, so called because it ripens in sixty days; (2) *Tandara*, named because of the length of the stalk and of the cob, and (3) *Rohru*, an inferior form so named because of its small size. The first is most generally preferred as it can be grown as a second crop, but the second gives the finest result, though it occupies the ground for six months and thus prevents a spring crop, while the third is only grown on inferior lands or by indolent cultivators. Of AMBALLA it has been said there are two varieties—a white with large, and a yellow with smaller cobs. "The yellow variety gives bread of better flavour; but the white, when parched, swells out to a larger size, and is said to be better suited to old, toothless people. There is said to be more gluten in the yellow variety, which renders it palatable, and being richer in nitrogenous flesh-forming compounds, it is probably more nutritious than the white and is certainly more largely grown."

The above briefly reviews the chief facts brought out in the special series of reports on the varieties of Panjáb maize. It is to be regretted that space cannot be afforded to bring together in a similar manner the answers to all Colonel Wace's questions. A *précis* of a few other facts may, however, be briefly attempted, namely, on the seasons of sowing and reaping; the rotation pursued; and the yield.

Seasons of Sowing and Reaping.—As might be expected the range of the periods of sowing and reaping is very great. Speaking generally it may be said that in the plains the period of sowing is dependent on the rains to moisten the soil and allow of its cultivation. In such cases it is therefore a *kharif* crop, sowings taking place from June to August (according to local peculiarities), and the crop comes into bearing of green cobs (from the earliest sowings) in August and the ripe grain (from later sowings) in September, October, or perhaps not even until November.

On the hills the sowings are generally much earlier, and the higher reaches are earlier than the lower. Thus, for example, in Hazára, Kullu, and Simla the early sowings are in April and May, but in these cases the crop occupies the field throughout the summer, the land being cleared in autumn to allow of the wheat sowings that lie in the ground throughout winter.

While these are the general principles of the crop, the most remarkable variations occur due to local climatic peculiarities, systems of cultivation, or the nature of the maize crop grown. In some districts, for example, there are two widely different crops that correspond very nearly to the *kharif* and *rabi* seasons. Thus, for example, in Ráwalpindi (Pindigheb tahsíl) the usual *kharif* crop comes into bearing in the beginning of August and the *rabi* crop yields its grain in December and January. In Jullunder the grain crop of maize is sown in the beginning of August and reaped in November and December, while the green cob crop is sown in June-July and reaped in September and October. In Kullu the *Sathu* crop is sown in July and the *tandara* in April and May. In Hoshiarpur the best crop is sown in June-July and reaped in September and October, but the green cob crop is sown in March-May and reaped in June-July. In Ludhiána maize appears to be a two-months or at most a two-and-a-half-months crop, being sown in August and reaped in September and October.

**CULTIVATION
in the
PANJAB.**

Rotation.—According to many of the reports maize is regarded as an exhausting crop which, although often followed by wheat or barley, gives an inferior return with these crops unless the maize soil has been previously highly manured. But this is often obviated by a selection from the extensive series of winter crops so as to avoid following by wheat. Carrots, for example, are frequently sown between the lines of maize, while the crop is standing. This is specially resorted to in threatened drought. The leaves of the carrots are given to cattle and the roots eaten by the people. On the other hand, *toria, kasumba* or gram follow maize in years of heavy rainfall. So in a like manner it is a frequent custom to follow maize with the poppy, but such non-graminaceous crops are only employed in the rotation by the more intelligent cultivators. In the province as a whole, the custom appears to be wheat, maize, barley, and it is often said that barley is distinctly a better crop to follow after maize than wheat, although in some districts wheat may be seen. In Kohat it is the custom to follow wheat or barley with melons, which are off the ground by the middle of July and thus allow one month's rest before preparing for the maize. In that district the usual rotation is first year maize, second rice, third cotton; the double crops of these years are filled up thus :— Maize followed by barley which is off the soil by April. In May rice is sown, and harvested in September; the land is then prepared for wheat which is sown in November, and is followed by cotton which is off the soil by the 15th December. The land remains then in fallow for the *rabi* but is manured so as to be ready to renew the rotation with maize.

**Rotation.
124**

Yield.—In the Appendix to the Panjáb special paper on Maize, the reports of the local officers are tabulated under certain headings, such as area irrigated and area not irrigated : colour of grain : description of soil : whether mixed with other crops or not : maximum, minimum and average yield per acre : price per seer ; uses of the grain and of the stalks : extent of exports, etc., etc. The figures shown under these columns are highly instructive, but those of yield only may be here reviewed. There were in all 82 returns, and if the average shown between the maximum and minimum yields be accepted as fairly accurate, it would appear that the average of all these averages would be 443 seers (886℔) to the acre or, say, 10¾ maunds. There are 59 averages out of the 82 that show 400 seers or over to the acre; 40, with 500 seers or over; 24, with 600 seers or over; 20, with 700 seers or over, and 3 with 1,000 seers or over; while only 5 show less than 200 seers. It would thus seem that if any value can be put in a figure to express a probable provincial average yield, 450 seers or 900℔ per acre, might be accepted. But it has been repeatedly pointed out in this work that no such provincial average can with safety be employed in any effort to arrive at a knowledge of the total production from surveyed acreage, until the relative extent of the lands that give the acreage has been ascertained. A few acres of high class cultivation might seriously raise the average, while providing only a few maunds to the provincial total production. But it may be pointed out that the lowest returns are those of certain forms of maize grown in Kangra and in Amritsar. The former showed 125 seers and the latter 130 seers. Such low returns very probably exercise as serious a disturbing influence on a provincial average as do the abnormally high figures of *chitti* maize in Attock (1,332 seers an acre) and the *chitti* and *pili* maizes of Fatteh Jang in Rawalpindi which had each 1,350 seers. There were in 1885-86 under these crops with high yield, however, only a little over 13,000 acres, and under those with an abnor-

**Yield.
125**

CULTIVATION
in the
PANJAB.
Yield
say half
million tons.
126

Practically
not exported.
127

mally low return about 52,000 acres. Deducting these areas of extremes
in yield there would have remained over 1,150,000 acres as under a normal return of from 450 to 500 seers an acre. Thus the Kangra district,
which has usually a very large area under maize, showed, in 1885-86, an
average of 450 seers an acre (if the crops that gave 125 seers and 168
seers be left out of consideration). But Hazara, on the other hand,
which has by far the largest tract under this crop of any of the Panjáb
districts, manifested an average of 827 seers, or 1,754℔ an acre. If,
therefore, it be assumed that the Panjáb normally possesses 1,000,000 acres
that yield 500 seers (or 1,000℔) an acre, this would be equivalent to a
production of 8,928,871 cwt of grain or, say, very nearly half a million
tons. The returns show little or none exported from the Province, so that
some conception from these facts may be obtained of the value of this
crop to the Panjáb population of 20,807,020. It would, however, be misleading to express this article of food to head of population because, except as a luxury in the condition of green cobs, it is an article of diet with
certain communities only. The consumption in the green state does not,
however, in the writer's opinion detract from this calculation, since if it did
not pay to sell in that stage it would be allowed to mature, and, moreover,
it seems likely that a large garden cultivation has not been provided for
since a few plants here and there could scarcely have been included in
the surveyed maize cultivation. It is, therefore, as it would appear perfectly safe to assume that Indian-corn furnishes annually half a million
tons of the food to this Province.

N.-W.
PROVINCES &
OUDH.
128

II.—N.-W. PROVINCES AND OUDH.

References —*Gazetteers :—I., 90, 115, 317; II., 28, 159, 160, 375, 479;
III., 24, 29, 225, 229, 305, 306, 463-467; IV., 19-22, 248, 251, 252, 254,
504, 521, 522; V., 26, 267, 541, 555; VI , 27, 28, 138, 539, 587; VII.,
34, 448, 449, 456; VIII., (Muttra), 41; (Allahabad), 29, 30; IX.
(Shahjehanpur), 45; (Moradabad), 40, 42; (Rampur), 24, X., 320;
XIII., 46, 97; XIV. (Benares), 25; (Mirzapur), 38, 40 ; Settlement Reports :—I. (Mozuffernuggur), 6, 14; (1866), 16, 89, 134; II., (Meerut),
34, 39, 56; (Bulandshar), 32, 33; (Aligarh), 37, 44; IV., (Bijnor),
16, 18, 88, 294; (Budan), 43, 46 ; V. (Bareilly), 71, 82, 86, 167, 173,
177, 214-217, 225, 288-293, 378 ; (Pilibhit), 102, 111 ; VI. (Shahjehanpur), XIV. (Jalalabad), 145; .(Muttra), 125, 179, 180, 208, 228, 244,
245 ; VII. (Faruckhabad), 11 ; (Kanauj), 21, 47 ; (Mainpuri), 13, 14,
115, 148, 163, 218, 237, 268 ; VIII. (Etawah), 10, 24, 106 ; (Etah), 18,
85, 98, 111, 127, 128 ; X. (Jhansi), 75 ; (Lalatpur), 25 ; XI. (Cawn-
pur), 4, 8, 17, 57, 119 ; (Allahabad), 15, 24, 28 ; XII. (Gorukhpur), 9,
146, 216 ; XV. (Agra), 65-69, 73-75 ; Azamghur, 115 ; Kumaon, App.
32d.; Agri. Dept. Reports :—(1880), 58, 59, 61, 71 ; (1881), 22 ; (1882),
27, 28, 31, 34 ; (1885), 25-28 ; Exp. Farm Reports (1880), 10 ; (1886),
7 ; (1887), 1 ; Duthie and Fuller, Field and Garden Crops, I., 21-24.*

Area.
129

Area.—The total area under crops in these Provinces in 1890-91
has been shown in the Agricultural Returns as 30,572,629 acres in the
North-West and 11,843,631 acres in Oudh, or a total of 42,416,260 acres.
Of that large productive region, 1,454,689 acres were maize. The cultivated acreage of the Panjáb (by way of comparison) was 23,536,126, of
which it has been calculated about 1,200,000 are normally under this
crop. It will thus be seen that in relation to the extent of these Provinces,
the Panjáb may be regarded as more especially the maize-growing
country, but the North-West Provinces, Oudh, and the Panjáb conjointly
afford India's maize fields. The whole of the other Provinces do not
very possibly possess very much more than 1½ to 2 million acres as under
this crop. In the North-West Provinces it may be said to be diffused
every where except in Bundelkhund in which it is very little grown. It

India's Maize
area,
viz.,
Panjab &
N.-W. P.
130

or Indian-corn in N.-W. P. and Oudh.	(*G. Watt.*)	ZEA Mays.

reaches its maximum in Gorakhpur and Basti. The area fluctuates greatly from year to year, and without any apparent reason it is a favourite crop in one district, and little grown in a precisely similar and perhaps neighbouring district.

CULTIVATION in the N.-W. P. & Oudh.

Although the writer has before him the very extensive series of papers quoted in the above paragraph on references, it does not appear that much information has been brought to light since the publication of Messrs. Duthie & Fuller's *Field and Garden Crops*. It may, therefore, suffice to give some of the leading paragraphs from that work in illustration of the methods of cultivation, varieties of maize grown, and yield, etc., etc.

Varieties cultivated. —"So far as the colour of the grain is concerned there are endless varieties, and the cobs may be of any tint from dark purplish red, through yellow and orange, to a pure white. But the most important variety is that grown in Jaunpur and Azamgarh, in which the cobs are of double the usual length, and the plants of taller growth than the ordinary. The grain of this variety is, however, nearly a month longer in maturing."

Varieties.
131

Seasons of sowing and Reaping. —"Maize is a *kharif* crop and ranks next after broadcasted rice in the rapidity with which it comes to maturity. It is sown, as a rule, when the rains break, but in localities where the green cobs are likely to command a sale as a vegetable, sowing often takes place in May, after the ground has been irrigated, since in this case it is of great importance to be early in the market. In the beginning of July a single cob will fetch a pice, while at the end of August a maund of them can be purchased for eight annas. If sown when the rains commence, the ordinary small cobbed varieties are ready for cutting at the end of August, and leave therefore ample time for preparation of the ground for a *rabi* crop. Hence maize is almost invariably followed by either wheat or barley, and nearly the whole of the area under maize may be presumed to bear two crops in the year." "As a rule, it is grown alone, since few other crops would keep pace with it in maturing; occasionally cucumbers are grown between the lines. It is not uncommon too to mix a certain proportion of the lesser millets (*kakuni* and *mandwa*) and a little pulse (*urd*), since these require but little more time to ripen and secure some measure of return in the not uncommon case of the maize completely failing."

Seasons of Sowing & Reaping.
132

One Crop.
133

Mixed Crops.
134

" If the cobs are to be sold as vegetables they are pulled while green and the stalks in that case are of some use as cattle fodder. Otherwise the cob is not harvested until the leafy envelopes surrounding the cobs are dry and shrivelled, when the stalks are so hard and desiccated as to be almost useless for any purpose but thatching. The cobs may either be pulled by themselves and the stalks left standing in the field until there is leisure to cut them, or the stalk may be cut with the cobs on them and heaped in stacks to dry before threshing. If the grain is to be separated from the cob before it is perfectly dry, the task is a slow and troublesome one, it being necessary to deal with each cob separately, forcing the grain from it by the fingers or the point of a trowel. When the cobs are perfectly dry, threshing can be easily and speedily performed by beating a heap of them with a rough flail or stick, or treading the grain out by cattle. The weight of grain varies from one-half to two-thirds of that of the cob. If the outturn of grain does not promise well, the stalks are sometimes cut while green and given to cattle, since the maize stalk when young and succulent contains a very large amount of saccharine matter and is a valuable fodder."

Thatch.
135

Green Fodder.
136

Z. 136

Cultivation of Maize

CULTIVATION
in the
N.-W. P. &
Oudh.

Yield.
137

Yield.—" The general average outturn for the Provinces may be taken as 10 maunds for unirrigated and 14 maunds for irrigated maize. The Settlement Officers of Bijnor and Aligurh arrived at averages of $7\frac{1}{2}$ and $10\frac{1}{2}$ maunds, respectively, while the careful experiments of **Mr. Moens** in Bareilly give $15\frac{1}{4}$ maunds for manured land, 12 maunds for unmanured land, and $12\frac{1}{4}$ maunds as the general average for the district. In both Etawah and Cawnpore the average outturn is returned as 12 maunds "

Messrs. Duthie & Fuller furnish no particulars regarding Oudh, but there is no reason for thinking that the system of cultivation and the results obtained differ in that Province in any material respect from the North-West. It will also be noted that the outturn in these Provinces (ascertained by **Messrs. Duthie & Fuller**) correspond very closely with that which the writer has worked out above from the numerous recent returns of the Panjáb. If, therefore, a yield of 500 seers an acre be accepted

Outturn
649,414 tons.
138
No Exports to
speak of.
139

as a fair average, these Provinces, from the acreage of 1,454,689, may be assumed to have given an outturn of 12,988,295 cwt., or 649,414 tons. There is no indication of more than a district to district exchange. The exports from these Provinces must be very small indeed, so that it may safely be concluded that this plant which, little more than a century ago was a garden crop, affords to-day over 1,000,000 tons of food annually to the people of the North-West Provinces, Oudh, and the Panjáb.

BOMBAY.
140

III.—BOMBAY.

References —*Gazetteers:—III., 45,232; IV., 53; VI., 39; VIII., 182, 188; XII., 137, 149, 151; XV., Pt. II., 18; XVI., 91; XVIII., ii., 35, 38; iii., 76, 96, 99; XIX., 160, 163; XX., 229, 395, 399, 403, 405; XXI., 247; XXII., 273; XXIII., 319; XXIV.. 156, 164, 167; Settlement Reports:— (South Division), I., (1874-75), 54; (1875-76), 89, 109. 133, 153, 176; (1876-77) 35, 41, 53, 57, 69, 85, 97, 113, 124, 126; (1877-78), 45, 56, 57, 59, 70, 71, 73, 85, 86, 89, 101-104; II. (North Division) (1876-77), 87, 99, 113, 119, 127, 151, 170, 171, 173, 186 (1877-78), 53, 65, 79, 87, 95, 107, 119, 127, 154, 155; (Central Division) (1877-78), 79, 91, 103, 117, 131, 145, 157, 158; V., (Kalol Taluk, Panch Mehals), 8; (Anklesar), 20, 37; VI., (Ahmednagar), 171, 174, 204; (Sholapur), 413; VII. (Haveli Taluk, Poona), 402; VIII. (Khandesh), 127; IX. (Omercote and Narra, Sind), 51; Agricultural Reports:—(1884-85), 18; (1885-86), 6; 1886-87), vi; (1887-88), vi; (1888-89), vi, xviii; Bomb. Man. Revenue Accounts, 101; Useful Plants Bombay, (vi, XXV., Bomb. Gaz.), 186, 208, 375; Statistical Atlas, Bombay Presidency, 7.*

Area.
141

Area, Outturn, etc.—It has already been stated that, according to the Agricultural Returns for 1890-91, the acreage of this crop in Bombay and Sind came to 137,457. It is, therefore, a very much less important crop in these Provinces than in the North-West and the Panjáb. The systems of cultivation do not appear to differ very materially from what has already been detailed, so that it may suffice to convey an idea of the crop to furnish the brief note given by **Mr. Ozanne** in the Statistical Atlas :—

Fodder.
142
Two Crops,
143

" *Makkai* (Guj.), *Maka* (Mar.), *Goinjol*, *Mekkejol* (Kan.) is chiefly cultivated in the Panch Mahals, where it ranks as a staple. Sátára and Sholápur grow a good deal, and in the Deccan it is mostly grown for early fodder, though the grain is allowed to ripen and the ears are readily sold in towns for roasting. Some success has been secured in Ahmadnagar with the American seed. It is a four-month crop and in the Panch Mahals is followed by wheat or gram. In this district maize is the early crop in light sandy lands, but in heavier land rice and maize are equally important, and both are followed by wheat or gram. The late crop seldom covers as large an area as the early one. It is confined principally to the portions of the field most retentive of moisture from position or depth of soil. The maize stock is a good fodder, especially when eaten green as soon as

		ZEA
or Indian-corn in Madras.	(*G. Watt.*)	**Mays.**

the ears have been plucked. In the Panch Mahals it is much wasted, being left to rot after the monsoon begins. It can be preserved by ensilage."

Seasons of Sowing and Reaping.—The two seasons of sowing and reaping pursued in many of the districts of Bombay are for the early crop sown in March or April and reaped in May and June: the late crop sown in June and reaped in November.

CULTIVATION in Bombay.
Seasons of Sowing & Reaping.
144

IV.—CENTRAL PROVINCES.

References.—*Settlement Reports :—I., (Chanda), 80-83 ; (Upper Godavery), 27, 30, 35 ; II. (Saugor), 21, 42, 50, 56, 85 ; (Seoni), app. iv ; (Mundlah), 58 ; III. (Hoshungabad), 99 ; (Nursingpur), app. iii., (Baitool), 77 ; (Chindwara), 153, 158 ; IV. (Nimar), 192 ; Exper. Farm (Nagpur) Reports (1883-84), 6 ; (1884-85), 4 ; (1885-86), 4.*

CENTRAL PROVINCES.
145

Area, etc.—It has already been stated that according to the last Annual Statement of the Agricultural Statistics of India these provinces (in 1890-91) had 106,659 acres under the crop. Although this cultivation is briefly alluded to in many of the Settlement Reports and other such local publications, no detailed special report has as yet been published. The reports of the Nagpur Experimental Farm, so far as maize is concerned, appear to deal solely with the efforts that have been made to acclimatise several of the better known American races, such as Golden dent, Maryland, etc. The attempt does not appear to have been made to improve the forms already in the country, nor, indeed, can the writer discover any description of the local or long acclimatised forms. It does not, therefore, seem necessary to attempt to build up a general statement of this branch of the agriculture of these provinces from the scattered and imperfect literature that presently exists. It may, however, be said that two crops— early and late—corresponding to those already mentioned in connection with other provinces, seem to exist and that the early is eaten in the green state, and the late to some extent employed in the preparation of flour used by the poorer classes, or the ripe grain is simply parched and eaten in that state. Thus, for example, **Mr. Morris** (*Descript. and Hist. Acc. Godavery, p. 90*) says that the Kois make a kind of porridge called *java* out of this grain. In another passage he remarks that there is both a *kharif* and *rabi* crop, chiefly the former.

Area.
146

Experiments.
Conf. with pp.
329, 336-7,
340.

Two Crops.
147

V.—MADRAS.

References.—*Man. Madras Adm., I., 288 ; Man. of Kurnool, 167, 269, 274 ; Boswell, Man. Nellore, 403 ; Gazetteer Nilghiri, 475, 479 ; North Arcot, 331 ; Salem, I., 148 ; Account Godavery District by Morris, 90 ; Settlement Report Kistna, 15 ; Exper. Farm Reports (1871) 11, 18 ; (1875), 27, 28 ; (1876), 50; (1877), 13, 97 ; (1879), 40, 109 ; Agri. Dept. Reports (1882-83), 22-27, 34, 35, 48, 51, 52, 95 ; (1883-84), 20, 28, 63, 64, 69, 70, 72 ; (1884-85), 18, 19, 27, 28, 48 ; (1885-86), 30, 31, 62, 68 ; Guide Saidapet Farm, 37-39 ; Shortt, Man. of Indian Agriculture.*

MADRAS.
148

Area, Outturn, etc.- Maize does not appear to be a crop of much importance in South India. It would, at all events, be safer to describe it as a plant of garden rather than of field cultivation. According to the Agricultural Returns of 1890-91 there would appear to have been only 42,040 acres under Indian-corn in South India. It is, therefore, a little difficult to understand the statement that it is one of the chief crops of the Presidency (*Madras Man. Admin., I., 288*). Shortt, in his *Manual of Indian Agriculture*, published a very long and somewhat disjointed statement which purports, apparently, to be applicable to the whole of India. He says : "It is not much cultivated in India as a field but as a garden cultivation, small quantities have been grown from time immemorial in most parts of India." It is perhaps scarcely necessary to remind the reader that there are perhaps five million acres annually under maize *as*

Area.
149

ZEA Mays.	Cultivation of Maize

CULTIVATION in Madras.

Yield. *Conf. with p. 349.*

Varieties said to be met with. 150

Acclimatisation and the results obtained. 151 *Conf. with pp. 329, 336-7, 340, 347.*

a regular field crop and that far from its having been grown "from time immemorial" in this country it would be safer to say that it has only recently been introduced. As a field crop it has probably not existed more than a century and very probably was originally brought to this country about the middle of the sixteenth century. But to revert to **Shortt's** essay on maize "The best varieties of maize when well cultivated is very productive and its value as human food is well known, being considered superior to rice and other dry grains of India. An acre in a good soil will produce 1,500 to 2,000 pounds of corn without any particular care being given to the plants and furnish 2 to 3 tons of fodder rich in saccharine matter and proves a valuable fodder for cattle and horses." "Four varieties are commonly met with in various parts of India; the large eared, small seeded pinkish, red eared and small grained black corn. The large white eared variety is the best and most largely produced, the others appear more the result of accident and are only occasionally found among the others." It does not seem necessary to continue quotations of this character as the article so far appears to have no bearing on India. Further on, however, when **Shortt** left the field of general compilation, and took to actual facts with which he was familiar, he furnished a few local particulars. These are distinctly of interest and are therefore worthy of a place in a notice of the maize cultivation of South India. "I have seen, he says, Indian-corn as a garden culture growing in most out-stations in South India and have also grown it myself. About a mile from Vellore on the great western trunk roadside, is situated a village called Totta Pállium, where Indian-corn is grown as a field culture on the same grounds year after year on rather a large scale. These fields can be seen distinctly from the roadside as they are located in a valley immediately below the road; some portions of the cultivation are irrigated and others not. The corn stalks attain from 8 to 10 feet in height and the produce of each varies from 3 to 5 cobs. It grows remarkably well at Palmanair and produces largely; two of the best cobs, I have ever seen, were grown at Palmanair, one was 12 and the other 10 inches long and covered from end to end with good sound seed; on another I counted 600 corn grains, and that was by no means a picked cob. Indian-corn grows well in most districts and produces 3 to 4 and sometimes 5 cobs on each stalk, the average is 2 cobs per plant." **Shortt** then continues with a long and detailed report by **Mr. W. Robertson**, Superintendent of the Government Experimental Farm of Sydapet, on the experiments with Queensland maize under dry cultivation. These and such like experiments the writer considers it desirable to exclude from consideration in this work. The object aimed at here is to try and bring together as much purely local information as possible, within the available and limited space. While not entirely disapproving of experiments at acclimatisation on a large scale (if funds can be provided for these) the writer does not think the results have as yet assumed the position of definite value to India. That being so, there does not appear any pressing necessity to publish either the failures or the successes. Speaking broadly, however, the conviction arrived at by the writer, while working up the material for the various volumes of this work, may be said to be that the successes attained in India in the acclimatisation of foreign animals and plants have been very much less than has been accomplished in other countries by natural selection alone. It would accordingly appear more urgently necessary to turn attention to the existing forms of maize and to try and improve these, rather than to waste the entire energies and available funds of the agricultural reformers in the more or less fruitless effort at acclimatisation of the triumphs of the agriculture of other countries.

Z. 151

or Indian-corn in Bengal.	(*G. Watt.*)	**ZEA Mays.**

With few subjects does this opinion carry greater weight than with maize All the known forms of this plant are regarded by those best qualified to judge as climatic conditions of a protean species, and as such they are liable to change if not to a complete transformation during acclimatisation. The result may be worse than what exists already in the country, or persistent acclimatisation may furnish a new form of great value, but which may possess few if any of the characteristics of the prized ancestor.

CULTIVATION in Madras.

In the Madras Farm Manual and Guide these views are abundantly confirmed. " *Mokka-jonora, mokka, buta* has become a regular crop of the farm, occupying an important place as one of the best food crops we possess ; it, however, requires better cultivation than that generally practised in Southern India, and so when good seed falls into the hands of the Native cultivators, its fate is inevitable and it rapidly deteriorates." The statement of a yield of 2,000℔ grain and 2½ to 3 tons of dry straw per acre is then given in the Manual, which, doubtless true of the high class methods pursued at the experimental farm, assumes a very different position when advanced as Shortt appears to desire, as the usual Indian return on a " good soil." The writer believes 1,000℔ an acre of grain the highest average that can be accepted for all India, though in some of the returns he has consulted an yield of 2,500℔ is mentioned.

Outturn. 152

The reader would do well to consult the brief passages in the Madras District Manuals for local details. These passages, so far as they go, do not differ to any extent from those already given regarding other provinces, and it has, therefore, been thought undesirable to publish them here.

VI.—BENGAL.

BENGAL. 153

References —*Statistical Account by Sir W. W. Hunter (numerous small passages in each of the volumes) ; Orissa (W. W. Hunter), II., 133 app. iv. ; Administration Report (1882-83), 12 ; Agri. Dept. Reports (1886-87), 16 ; (1887-88), 14, 15 ; (1888-89), 15, 18 app. A ; Report on Agri. Lohardaga, I., 13, 50, 51, 65-67, 152, 153 ; II., 24, 25, 28, 54, 61, 73, 74.*

Area, etc.—Indian-corn is not cultivated to the same extent, nor indeed has it obtained the same amount of popular favour in Lower Bengal as in the northern parts of India. Several reasons may be assigned for this. Maize can be grown to profit only on rich lands, such as are to be found near the homestead, and these (*bari*) lands are of necessity limited. The climate also, taken as a whole, is unfavourable to an extensive cultivation, while more trustworthy, and hence more profitable crops, especially that of rice, is likely to prevent an extensive cultivation of Indian-corn. However profitable a crop maize may be in one part of India, where during certain months the temperature does not exceed that of the southern parts of Europe, its cultivation, under widely different conditions, is of necessity an independent problem. During a century or more of maize cultivation in Lower Bengal, special forms have been evolved suited to the climate, the nature of cultivation, and the requirements of the people. Were it, therefore, contemplated to attempt the improvement of these by the importation of superior stock from the Panjab, nearly as much good might be looked for from an effort to acclimatise the maize of Norway on the inundated plains of Bengal. It has to be accepted, therefore, that the success in one province is no proof of greater returns being possible in another. This should not by any means, however, be regarded as an affirmation that there is no room for improvement. It is a caution against reckless experimenting. Bengal, so far as maize cultivation is concerned, may be said to manifest at least three widely different phases : (1) the homestead cultivation in Lower Bengal, to produce

Area 154

Adaptation to conditions. 155

Reckless experimenting. 156
Conf. with pp. 329, 336-7, 340, 347.

green cobs; (2) the cultivation as a staple food grain on the hilly tracts, such as in Chutia Nagpur; and (3) the cultivation in Behar which differs in no essential from that in the greater part of the North-West Provinces. All experiments towards improvement should bear these three widely different cultivations in view. An exchange between Madras and Lower Bengal would be more hopeful than between Behar and the Lower Provinces. So, again, interchange between the hilly tracts of the Central Provinces, or even the Himalaya might result in more good to the Chutia Nagpur maize than any other experiments that could be tried. Bengal, it may be said, is infinitely less suited for experiments at acclimatising foreign maize than any other part of India. If it be desired to obtain a foreign strain this should alone be looked for from the successes attained in other parts of the country.

In Bengal, taken as a whole, maize can hardly be regarded as of great moment. It is, in fact, among the hill tribes alone that it can be said to rank as a staple article of diet. Its importance in India may safely be gauged by the extent to which the crop is grown for its ripe grain and by the use of maize flour. In Lower Bengal as a whole, the ripe grain and flour might almost be said to be unknown. The crop is produced almost entirely for its unripe cobs, and the forms of the plant suitable for such cultivation differ in many respects from those grown for ripe grain.

The following account of maize cultivation, in a section of Chutia Nagpur, may be accepted as fully expressive of the system pursued by the hill tribes of the Lower Provinces :—

LOHARDAGA.—"Two varieties are usually distinguished, *viz.*, a dull yellow coloured variety, and a red or *lálká* variety. The first variety is in common cultivation. When the grains are not properly filled or matured, they have a whitish appearance. The grains of both the varieties are thin and small in size. In respect of time of sowing, cultivation, etc., the two varieties named above do not differ from each other.

"Maize is cultivated by almost every *rayat* in a small plot of *bari* or homestead lands. It requires a rich soil and will not grow in outlying upland." "Although cultivated in every village, it occupies a small area as compared with the other cereals of Chutia Nágpur. In jungly tracts, however, maize is looked upon as only inferior in point of importance to paddy." "Maize is usually grown for two or three successive seasons on the same *bari* plot and followed by some cold weather crop like mustard; but as it refuses to grow well on the same land for successive seasons, it is the practice with *ráyats* to take a crop of some other *Bhádoi* crop like *cárai* or *máruá* every third or fourth year, as the case may be.

"In *Asár* (June-July) after the soil has been well moistened by a shower of rain, the land is ploughed three times over, and the clods, if any, broken by passing the harrow over it. The seed is then sown broadcast at the rate of 4 or 5 seers per acre. Grains are rarely dibbled in, which is a better but tedious practice, and the small saving of seed is not enough to make up for the extra labour of dibbling. A poor *rayat* having a small quantity of seed is found here and there to economise it by dibbling the grains in at intervals of a cubit from one another. After the seed has been broadcasted, it is buried in by a light ploughing and the soil then levelled by the harrow. When the plants have come up about 4 inches high, all grasses and weeds are picked up by the hoe, which also works the soil about 3 inches deep; at the same time where the plants have come up too close to each other, the superfluous among them are spudded out and thrown away. The land is thus weeded and hoed two or three times in all during *Asar* (June-July) and *Sravan* (July-August).

"The flowers come up in the second half of *Sravan* 1st, to 15th August. Two cobs are usually formed on each plant; of these one is properly filled and the other generally empty. The cobs ripen in early *Bhadra* (15th August—15th September), but when eaten raw or after being roasted (as is usually the practice with the people), they are available a fortnight before the time of ripening. They are simply picked off the plants, and after being dried the corn is beaten out with a stick. It requires to be further dried in the sun, otherwise it is liable to grow musty, when kept in close damp places. The stems are left on the ground or are gathered and burnt as fuel.

"The outturn averages about five maunds per acre; eight maunds will be considered a very heavy yield.

Maize appears to suffer more than any other *Bhadoi* crop from adverse weather. For its successful growth it requires plenty of rain, which should not, however, be continuous, but be intervened by frequent stretches of fine weather. Too much rain, at once, during the first stages of its growth, is extremely injurious; the plants become stunted, the stems of a reddish colour, and the leaves blanched. The soil is beaten down, becomes pressed and close-grained, and thus interferes with the due spreading out of roots, on which the growth and vigour of the plants depend. Besides hoeing is not possible when the soil is too wet. Hoeing can be useful only when the loosened soil can be exposed for a day or two to the action of the sun. The partial failure of maize and other *bhadoi* crops in 1888 was neither general nor so heavy in Chutia Nágpur Proper as it was in Palámau, owing to the naturally light and loose character of the soils of the former, which do not succumb so readily to heavy rains as the comparatively closed grained soils of its sister sub-division." "In the neighbourhood of towns and large villages, maize is sown as early as the beginning of Jeyt (15th May), on land which is irrigated at suitable interval still the setting in of the rains. In this way the growers are able to offer the unripe cobs for sale in *Sravan, i.e.,* several weeks before they are available under ordinary cultivation.

"It appears that the cultivation of maize is restricted, owing to the limited extent of *bari* land. The outlying *taur* lands are naturally poor, and under the present condition of agriculture, do not admit of maize cultivation. For such lands the *rayat* can spare little or no manure from his scanty supply, which he scrupulously reserves for his paddy and *marua ;* consequently they are devoted to such poor crops as *gondli, kurthi, sawan,* etc. I am inclined to believe that if the value of nitrogenous and other artificial manures be well demonstrated to the people, and they are popularised, many lands which do not now admit of the maize cultivation can be rendered rich enough for it. As a cereal, maize is inferior only to paddy, wheat and barley, and is considered far superior to the millets. The extension of its cultivation is therefore very desirable " (*Agri. Report of the Lohardaga Dist.*).

PALÁMAU.—" Maize is the most important among the *bhadoi* crops grown in Palámau. It may be said to be grown more or less by every *rayat,* and is the most prominent crop in the hilly parts of the sub-division, where the cultivation of rice and of the *rabi* crops cannot be profitably carried on.

"The cultivation of maize in Palámau differs in no respect from that in the Chutia Nágpur sub-division. There are two varieties grown in Palámau—one with white grains called the *ch·rka,* and the other with yellow grains called the *piár.* These differ only in respect of colour and are equally valued. The average produce of grain per acre is about 6 maunds."

HAZARIBAGH.—Indian-corn (*makai*) is said to be sown in May and June and reaped in August and September. Two varieties are grown—the red and the white. Indian-corn may be regarded as the staple food of the lower classes during the year. It is eaten parched or is ground into flour (*satu*) and is eaten with sugar. When simply ground it forms a kind of meal used in the preparation of a dish of porridge known as *lupei* or *gatha.*

BEHAR.—Grierson (*Bihár Peasant Life,* 223) gives a very extensive glossary of the names in use in this province for Indian-corn (the field crop, the grain, the unripe cobs, the spathes, the fodder, etc., etc.). The most general name for the plant is, he says, *makai* or *makaiya,* but it is also called *janera, jinora*—names that more properly belong to the large millet. *Bhutta,* or *bál,* is the name for the ripe cob, and *pakthail* for the ripe grain. The roasted green cob is *horha* or *orha,* and the empty spike after the grain has been removed is *lenrha, nerha* or *lenruri.* The sheath of the cob is *khoiya balkhoiya* or *bokla.* The panicle of male flowers is called *bhanbal* or *bhanahra.*

These and many other names are in use throughout the Province—a series so extensive that but for the immense value of the crop it would be impossible to believe had been coined or appropriated to specific signification in little more than a century. But many of these names are not only adaptations to modern necessities, but some of them, such as *makai* and

Yield
5 maunds
an acre.
158

Palámau.
159

Yield
6 maunds
an acre,
160

Hazaribagh.
161

Behar.
162

Conf. with pp.
327, 333, etc.

ZEA Mays.	Dieases of Maize.

CULTIVATION in Bengal.

Gaya and Shahabad.
163

bhutta occur in every dialect and tongue throughout the length and breadth of India.

Of GAYA and SHAHABAD districts it may be said maize is sown at the commencement of the rainy season and cut up at the end. In an estimate framed some few years ago it was announced that the two crops of maize represented an area of 80,000 acres and yielded 2,40,000 maunds of grain. The early crop is generally transplanted and it yields a very considerable portion of the food of the poor classes. The grain is often made into *satú*. The stems and leaves are employed as fodder. On these being gathered the land is irrigated and prepared for its winter crop.

Sandarbans.
164

SANDARBANS & 24-PARGANAS.—In the Statistical Account of Bengal (*Vol I., 139*) a brief notice occurs of the Indian-corn of these districts, which may be said to fairly represent the conditions that prevail in the Lower Provinces generally. *Bhuttá* or *janár* is grown to a small extent only, though nearly every well-to-do peasant has a small patch. It is sown in May and reaped in September, the cobs being eaten in the green state.

DISEASES OF MAIZE.

DISEASES.
165

The replies obtained from the District Officers of the Panjáb, to a circular letter on Maize (already freely drawn upon), afford by far the most extensive and useful information hitherto published on the PESTS & DISEASES of this crop. Unfortunately, while these replies speak of the injury done, the diseases are alluded to under native names, and it is often impossible to discover whether a certain disease is fungoid or insect, or whether that spoken of in one district, is the same or different from a disease dealt with in connection with another. While it is thus impossible to furnish in this place more than a general statement, it may confidently be affirmed that the injury done is frequently of such serious consequence as to justify a scientific investigation in the future.

Too much rain and too much sun injurious.
166

Maize does not appear to flourish under too much rain nor too continuous sun-shine. Thus of Gurdáspur it is said: "If there are deficient rains, the stalk does not bear any cobs; and if the rains fail and the land cannot be irrigated at the time that the plant is about to produce cobs, the cob is found to have grain only in a portion of it. If, on the other hand, there are excessive rains at that time, then the stalk grows to a great height, and it does not bear cobs, or if any, the cobs are of an inferior kind, the grain is small, and the produce little." With late sowings

High winds destructive,
167

the plant stands a chance to be unable to ripen its grain from want of sun, or to be destroyed by frost or hail. High winds also do great damage and in consequence districts liable to such visitations have by natural selection developed a form of the crop, which, while inferior in some other respects, has the advantage of being able to withstand sudden gales. Other forms possess the property of enduring a fairly protracted drought and of reviving again and yielding a fairly good crop on being irrigated

Selection of properties to withstand above dangers essential.
168

artificially or naturally. So, again, natural selection has produced forms of the plant suitable in other respects to district requirements such as those that can mature grain within a certain limited period of favourable weather. Such then are the climatic considerations that must be borne in mind in the study of the maize crops of India or when it is contemplated to attempt the substitution of new qualities. But there are enemies and

Enemies.
169

pests to the crop that are often of very serious moment. The maize fields have to be carefully fenced when young, to protect them from domestic or wild herbivorous animals, and later on when the cobs appear watchmen have often to be stationed in the fields night and day to ward off the bears, jackals, dogs, pigs, monkeys, porcupines, rats, squirrels, and birds that would otherwise effect frightful depredations.

Z. 169

Diseases of Maize.	(G. *Watt.*)	ZEA **Mays.**

The following jottings from the letters alluded to above seem to denote INSECT PESTS:—

AMRITSAR—" Maize crop is liable to *tela*, a small sized worm or insect; *sundhi*, a large worm (insect) which eats up the cob."

HAZÁRA.—" Maize is subject to the attacks of grubs, especially in dry weather."

GURDÁSPUR.—" The white-ant (*dimnk*) attacks the plant sometimes."

HOSIÁRPUR.—" Soon after the young shoots come up, a small hairy red caterpillar, called *kutra*, appears, and frequently does much harm in eating the tops, but it generally disappears after 10 or 12 days. The *bhúndi*, a grey winged insect, also damages the young shoots. There is an earth worm, called *garuna* or *gut*, which destroys the roots; and in dry weather white-ants also attack the roots"

JULLUNDAR.—" In its early days maize may suffer from caterpillars (*sundhi*)."

PESHAWUR.—" White caterpillar (*Pishakai*) eats the young stalks. Red caterpillar (*changu hangu*) eats the young leaves. White caterpillar (*spin chinjai*) feeds on the cobs."

KARNÁL.—" If there is not seasonable rain soon after the plant has germinated, the young sprouts are eaten by an insect called *phirka*."

KOHÁT.—" *Chinji* or the "worm" is a white insect from half to three quarters of an inch in length; when there is a deficiency of rainfall it attacks the early sown maize at the root, on which the plant withers away soon after."

Fungoid Diseases.—Although scientific papers have not as yet appeared on the fungoid diseases of this crop, with a special reference to the extent of the injury done, it is well known that Smut and Rust, which do so much damage in other parts of the world, also occur in India. The former, being more striking than the latter, has been observed by many writers; but we are indebted to the late Surgeon-Major Barclay for particulars regarding rust.

1st.—SMUT (**Ustilago Maydis**) is a remarkable disease It attacks all parts of the plant above ground and forms larger or smaller irregular tubercles, which finally break and emit a black sanious matter. When the cob is so attacked no fruit is formed, and the spathe then seems like a large puff-ball of black powder. Few fields of maize can be seen in India without a large percentage of the plants so affected and the crop largely destroyed thereby. The late Surgeon-Major Barclay and the writer once attempted to count the number of affected plants in a field a little below Simla. We arrived at the opinion that in that particular case well on to 30 per cent. of the crop was utterly ruined. The cultivator admitted that his field was unusually badly attacked, but seemed to regard the circumstance as one over which he had very little control. He was wholly ignorant of the possible advantages of eliminating the diseased plants from the healthy, or of the disease being communicable to and perpetuated by the grain sown by him.

Ustilago carbo, the smut, so prevalent on Sorghum, has been recorded as seen on the male inflorescence of maize. The reader might consult the remarks on smut that will be found under **Sorghum vulgare**, *Vol. VI., Pt. III.*

2nd.—MILDEW or RUST (**Puccinia Sorghi**, *Schw.*). In an interesting paper in the *Journal of the Asiatic Society of Bengal* (July 1891), p. 214, the late Surgeon-Major Barclay gave an account of the Rust seen by him on this plant. The remarkable features of that discovery have been already noticed by the author under the article **Sorghum vulgare** (*l. c.*), but the chief point may be here recapitulated, *viz.*, the rust on Sorghum in

ZEA Mays.	Trade in Maize.

DISEASES.

India is not **Puccinia Sorghi,** while that disease appears in this country on a new host, *viz.*, **Zea Mays**.

Having thus briefly indicated the two chief fungoid diseases which are known to attack the maize crop of India, a few passages may be furnished from the Panjáb reports that would seem to relate to these or other fungoid diseases.

175 HAZÁRA.—" The most fatal disease to maize is *jhal* or *channi*, which causes the stalks to dry up. Excessive moisture brings this on."

176 JULLUNDAR.—" Is subject to a disease called *ukherá*, in which the roots dry up."

177 KOHÁT.—" *Channi*, Dew-fall — Towards the end of autumn the dew-fall in places shut in by the hills is excessive, and the crop when attacked by it is completely destroyed. In such places, therefore, even at the risk of obtaining a poor crop, maize is sown earlier than elsewhere, so as to be harvested before the dew-fall comes on." " *Spinki*—This disease is common to both the tahsils (Kohát and Hangu) and generally attacks the crop when the cobs are being formed. The stalks thereupon shrivel up and become white or yellowish, and little or no grain is formed on the cob." " *Tilli*—This disease is peculiar to the Hangu tahsil. It attacks the crop during the rains between the 30th June and the 15th July, when the plants are young, and is popularly supposed to be due to the heat engendered in the rain water, which has passed over waste and barren lands."

178 LUDHÍANA.—" When the crop is ripening, damage is sometimes done by a sort of ' Smut ' called *sundi* which destroys the stalks."

179 KULLU.—" The crop is liable to Mildew known as *buka*, if rain does not fall soon after the sowing, and to another disease called *arni*."

TRADE.
180

TRADE IN MAIZE.

Neither the Annual Statements of the Foreign Trade of India nor the Returns of Internal Rail, Road, and River Traffic show Maize. It is believed that India practically exports no maize to foreign countries, and that the internal transactions are shown along with those of millets. The statement made by several writers that India exports Indian-corn and has again to import the corn-flour she requires, instead of making corn-flour locally, is not, strictly speaking, correct. India, in fact, imports only the small amount of corn-flour required by the European residents, and the great bulk of the maize grown in this country is of too poor a quality to be utilized in the manufacture of that article. That India might, however, do a considerable traffic in the supply of maize to be used as cattle food seems likely; but whether it would pay to carry so low priced a grain such distances by land and sea as would be necessary seems highly problematic. India could easily increase her area of maize production were it possible to compete in the European supply, but so far as is presently known maize is purely grown for local consumption, and each district appears very nearly to produce its own supplies. There is at all events a much smaller provincial exchange in this grain than with any other article of food. If we assume an average production of 500 seers per acre, and accept the estimate that the normal total area under this crop is 5,000,000 acres, then it would appear that there would be produced 44,642,857 cwt. or 2,232,142 tons of this grain. This estimate must not, however, be accepted as including the cultivation necessary for the production of the cobs that are eaten in the green state. Maize might almost be said to be grown in every garden in India, and the sale of green cobs to city communities is one of the most profitable branches of market gardening. It will thus be seen that maize, although possibly the most recently introduced food crop, is by no means the least important article of food to the people of India.

Total Production.
181

Z. 181

The Black and Yellow Zedoary. (*J. Murray.*)	ZEUXINE sulcata.

(*J. Murray.*)

ZEHNERIA, *Endl.; Gen. Pl., I., 830.*

Zehneria umbellata, *Thw.; Fl. Br. Ind., II., 625;* CUCURBITACEÆ.

Syn.—Z. HASTATA & CONNIVENS, *Miq.;* KARIVIA UMBELLATA, *Arn.;* K· RHEEDII, *Ræm.;* MOMORDICA UMBELLATA, *Roxb.;* BRYONIA UMBELLA-TA, *Klein.;* B. SINUOSA, *Wall.;* B. AMPLEXICAULIS, *Lamk.;* B. SAGITTATA & RHEEDII, *Blume;* HARLANDIA BRYONIOIDES, *Hance.*
Var. nepalensis=BRYONIA NEPALENSIS, *Seringe.*

Vern.—*Amaut-múl, tarali,* HIND.; *Kudari,* BENG.; *At,* SANTAL; *Gulkuk-ru, gulále-kukri gulákri,* in Kullu; *Bankakra, bankakra* in Chamba, PB. (*conf. with Vol. VI., Pt. I., 306-307*); *Gametta, gometti,* BOMB.; *Tid-danda,* TEL.

References.—*Roxb., Fl. Ind., Ed. C.B.C., 697; Thwaites, En. Ceyl. Pl., 125; Dalz. & Gibs., Bomb. Fl., 101; Rev. A. Campbell, Rept. Ec. Pl., Chutia Nagpur, No. 9470; Rheede, Hort. Mal., VIII., t. 26; Dymock, Mat. Med. W. Ind., 2nd Ed., 346; Atkinson, Him. Dist., 310; Lisboa, U. Pl. Bomb., 160; Gazetteer, Bombay, XV., 435; Agri.-Horti. Soc. Ind., Journ. (Old Series), IV., 202.*

Habitat.—A climbing herb, very common throughout India and Ceylon; the variety **nepalensis** is peculiar to a limited area in Garhwál, Kumáon, and Kunawar.

Medicine.—Dymock writes, "Its medicinal properties do not appear to be generally known to European writers on Indian Materia Medica, nor does it appear to have had a place in the Sanskrit Materia Medica. In the Konkan the JUICE of the ROOT with cummin and sugar is given in cold milk as a remedy for spermatorrhœa, and the juice of the LEAVES is applied to parts which have become inflamed from the application of the marking-nut juice. As a *paushtik,* or restorative and fattening medicine, roasted onions, *gometta* root, cummin, sugar, and *ghí* are given, or *gometta* only with milk and sugar" (*Mat. Med. W. Ind.*). [The reader might consult the remarks that have been made regarding this plant under **Podophyllum emodi,** Vol. VI., 305.—*Ed., Dict. Econ. Prod.*]

Food.—"The ripe and unripe FRUIT are eaten by the Natives, as are also the ROOTS when boiled" (*Roxb. Campbell*). In Bombay the fruit is eaten together with that of **Capparis zeylanica,** *Linn.,* on *duadashis* which occur in the month of *Ashad.* The two fruits are invariably associated in the *bháji* or dish made for those days (*Lisboa*).

Zedoary, Black, see **Curcuma cæsia,** *Roxb.;* SCILAMINÆ, Vol. II., 658. [II., 65.

Zedoary, Yellow, see **Curcuma aromatica,** *Salisb.;* SCILAMINÆ, Vol.

ZEUXINE, *Lindl.; Gen. Pl., III., 599.*

Zeuxine sulcata, *Lindl.; Fl. Br. Ind, VI., 107;* ORCHIDEÆ.

Syn.—Z. BRACTEATA, BREVIFOLIA, & ROBUSTA, *Wight, Ic., t. 1724 bis, 1725, 1726;* Z. MEMBRANACEA, *Lindl.;* Z. TRIPLEURA, *Lindl.;* Z. INTEGERRIMA, *Lindl.;* Z. EMARGINATA, *Lindl.;* Z. PROCUMBENS, *Blume;* TRIPLEURA, PALLIDA, *Lindl.;* ADENOSTYLIS EMARGINATA & INTEGERRIMA, *Blume;* PTERYGODIUM SULCATUM, *Roxb.*
Vern.—*Shwet-húli,* BENG.

References.—*Roxb., Fl. Ind., Ed. C.B.C., 610; Griffith, Notul., III., 396; Ic. Pl. As., t. 349; Stewart, Pb. Pl., 238; Gazetteer, N.-W. P., I., 84; IV., lxxviii.; Ind. Forester, XII., App. 421; Agri.-Horti. Soc. Ind., Journ. (Old Series), X., 340; XIV., 7.*

Habitat.—This, the commonest of Indian orchids (*Hooker*), is found throughout India in the plains and lower hills, from the Panjáb and Sind, to Assam, Chittagong, and southward to Ceylon; distributed to Afghánistán, Java, China, and the Philippines.

Margin notes:

182

MEDICINE.
Juice.
183
Root.
184
Leaves.
185

FOOD.
Fruit.
186
Roots.
187

188

ZINC.	Localities in which Zinc occurs.

MEDICINE.
Tubers.
189

Medicine.—Stewart writes, "I have once been told that its TUBERS are locally used as *salep* by Natives."

ZINC, *Ball, Man. Geol. Ind., III., 312.*

190

Zinc, *Mallet, Mineralogy (Man. Geol. Ind., IV.), 18.*

ZINC, *Fr.;* ZINK, *Ger.;* SPELTER, *Dut.;* CHINCK, zinco, *It., Sp.;* SCHPAATER, *Rus.*

Vern.—*Dastá*, impure calamine=*dusta*, HIND.; *Dasta*, NEP.; *Jast, jasd*, oxide=*missi safed*, PB.; *Sung busri*, sulphate=*safed túta*, DEC.; *Tú-tánagam*, impure calamine=*madal tútum*, sulphate=*vulley tútam*, TAM.; Sulphate=*tútum*, TEL.; *Tambaga-putih*, MALAY.; *Thwot*, BURM.; *Ya-sada*, sulphate=*kaburni*, impure calamine=*kharpara*, SANS.; Impure calamine=*Kal-khúbri*, PERS.

References.—*Mason, Burma & Its People, 732; Balfour, Cyclop., 1129; Ainslie, Mat. Ind., I., 573; Rajputana Gazetteer, I., 15; Mallet & Medlicott, in the publications of the Geol. Survey Ind. (see Ball, l.c., & Mallet, l.c.); U. C. Dutt, Mat. Med. Hind., 71; Baden Powell, Pb. Pr., 101, 113.*

Occurrence.—Zinc occurs in nature, in combination with sulphur, forming the sulphide or *zinc blende*, with oxygen, forming *zincite;* and more rarely as a silicate, carbonate, sulphate or arsenate. All these minerals are by no means common in Peninsular India, but occasional crystals of blende do occur in association with the ores of other metals in several localities (*Ball*).

LOCALITIES

Madras.
191

1. MADRAS.—Blende containing a small admixture of gold and silver is said to have been found in the Madura District. Specimens of ferruginous carbonate of zinc, with barite, hornstone, and a little green blende have been obtained from Karnúl, possibly from the Baswapur, Gazúpally mines (*Mallet*).

Bengal.
192

2. BENGAL.—Blende has been found, associated with lead and copper ores, in the mines at Mahabank and at Baragund in the Hazáribagh District (*Ball*), also at Bairuki in the Sonthál Pergunnahs (*Mallet*).

Rajputana.
193

3. RAJPUTANA.—The only considerable deposit of zinc ores, which has been extensively worked in the peninsula, occurs in this region, at Jawar or Zawar in the Udepur State. Though these mines were at one time worked, information regarding them is by no means distinct. Thus in Colonel Tod's *Rajasthan*, the mines are alluded to incidentally as having yielded an annual revenue of R2,22,000, but the metal is said to have been tin. This, according to the *Rájputána Gazetteer*, is probably a mistake, as there is no tradition of tin having ever been found there. Captain Brooke states that the ore occurs in veins, 3 to 4 inches thick, and sometimes in bunches, in quartz rock. The Natives at one time collected the ore, pounded it, freed it from quartz, and obtained the metal by sublimation. This was carried out in crucibles from 8 to 9 inches in height, and 3 inches in diameter, with necks 6 inches in length and half an inch in diameter. The mouths of these were closed up, the crucibles were inverted, placed in rows on a charcoal furnace, and in two or three hours the ore completely melted, the metal subliming into the necks. The mines were closed during the famine of 1812-13, and have not since been re-opened. The ore appears to consist chiefly of zinc carbonate (*Ball*).

Himalaya
Panjab.
194

4. HIMALAYA—PANJAB.—"At Shigri in Lahoul, zinc blende was found by Mr. Mallet in no great abundance, disseminated through the gangue of the antimony ore which occurs there" (*Ball*). "It has also been obtained from the Belar copper-mine in Garhwál, in the Sabáthu lead-mine near Simla, and from some uncertain locality in Kashmír" (*Mallet*). "In the

Z. 194

Products of India. 357

Localities in which Zinc occurs.   (*J. Murray.*)   **ZINGIBER Cassumuna**

Sirmur-Jaunsar mines a distinct string of ore occurs, which consists of zinc blende, with some galena, iron pyrites, and quartz " (*Medlicott*).

    5. AFGHANISTAN.—"According to Dr. Lord, in certain volcanic regions near the Ghorband valley, and elsewhere in Northern Afghánistán, an efflorescence of zinc sulphate, locally called *zak*, was common; whether it was employed for any purpose, medicinal or otherwise, is not stated " (*Ball*).

    6. BURMA.—According to Dr. Mason zinc ore occurs in Tavoy; the same authority records that Dr. Helfer had obtained an ore of the metal in one of the Mergui islands. Nothing is known as to the occurrence of zinc in Upper Burma.

    Medicine.—" Zinc is not mentioned by the older Sanskrit writers such as Susruta, nor does it enter into the composition of many prescriptions. The *Bhávaprakása* mentions it in the chapter on metallic preparations, and directs it to be purified and reduced to powder in the same way as tin." " It is said to be useful in eye diseases, urinary disorders, anæmia, and asthma" (*U. C. Dutt*). A crude oxide and silicate of the metal called *kharpara* is also mentioned in most Sanskrit works, and enters into the composition of a number of prescriptions both for internal and external use. It is considered tonic, alterative, and useful in skin-diseases, fevers, etc. U. C. Dutt describes several compounds containing this substance, one of which, composed of *kharpara*, goldleaf, pearls, cinnabar, black pepper, *ghí*, and lemon-juice, is said to be " much used by up-country physicians in chronic fever, secondary syphilis, chronic gonorrhœa, leucorrhœa, etc." Ainslie states that Muhammadan physicians in India use impure calamine (an oxide and silicate, or carbonate of zinc) for nearly the same purposes for which it used to be employed in England, namely, as a dry application to excoriations, ulcers, and superficial inflammation. On the other hand, he states that Muhammadan physicians did not, in his time, appear to employ metallic zinc, while the Tamils used it freely. They prepared it by fusing the metal in an earthen crucible, adding some green leaves of Euphorbia neriifolia, *Linn.*, and constantly stirring the molten mass. This in time caught fire in the usual way, and the resulting ashes, after still further refining by fire, were preserved for medicinal use. The oxide thus obtained was employed by the Native practitioners " with the greatest confidence," in gonorrhœa, spermatorrhœa, " fluor albus," and hæmorrhoids. The sulphate and other salts, the value of which is well known in European medicine, are now employed to some extent by the better informed class of Native practitioners.

    Arts.—In British India zinc is chiefly used in making alloys. Of these the alloys with lead are chiefly employed on account of the facility with which they can be turned or filed, those with copper for their ductile properties, or for making pot-metal, a combination which is brittler when warmed. The many uses of the metal in the arts in Europe are too well known to require description.

ZINGIBER, *Adans.; Gen. Pl., III., 646.*

Zingiber Cassumunar, *Roxb.; Fl. Br. Ind., VI., 248;* SCITAMINEÆ.

    Syn —ZINGIBER PURPUREUM, *Roscœ; Z.* CLIFFORDII, *Andhr.*

    Vern.—*Ban-ádá,* HIND. & BENG.; *Nisan, nisana, penlékosht,* MAR.; *Káru allamu, kúra pasúpu, karpushpú,* TEL.; *Van árdraka,* SANS.

    References.—*Roxb., Fl. Ind., Ed., C.B.C., 17; also Monandrous Pl., 347, t. 5; Voigt, Hort. Sub. Cal., 562; Thwaites, En. Cey. Pl., 315; Grah., Cat. Bomb. Pl., 207; Elliot, Fl. Andhr., 85, 104; Rheede, Hort. Mal., II., t. 13; Rumph., Amb., V., t. 64; U. C. Dutt, Mat. Med. Hind.,*

Margin notes:

LOCALITIES.

Afghanistan.
195

Burma.
196

MEDICINE.
197

ARTS.
198

199

**ZINGIBER
officinale.**

255, 322; *Dymock, Mat. Med. W. Ind., 2nd Ed., 764; Year-Book Pharm.,
1880, 251; Atkinson, Him. Dist., 318; Birdwood, Bomb. Pr., 88; Agri-
Horti. Soc. Ind., Journ. VII. (New Series), Pro. (1885), cxii.*

Habitat.—A native of various parts of India (*Roxb.*), Coromandel,
the Konkans, Behar, Bengal, and Sylhet (*Voigt*), rare in Ceylon
(*Thwaites*). It flowers in July and August and fruits in November and
December.

**MEDICINE.
Root.
200**

Medicine.—"The ROOT of this plant Sir Joseph Banks and Dr. Combe
think the true *Cassumunar* of the shops. When fresh it possesses a strong
camphoraceous odour, and warm, spicy, bitterish taste; when dried considerably weaker" (*Roxb., Monand Pl.*). Though used medicinally by the
Natives, it appears never to have been an article of commerce in India; but
is said by Mr. Colebrooke, President of the Asiatic Society, in a note on
Roxburgh's account of the plant, to have been first introduced into European practice by Marloc, "as a medicine of uncommon efficacy in hys-
terick, epileptick, and paralytick disorders; but is gone out of repute."
At the present time it has a similar reputation in Indian practice to the
officinal ginger, and in the Konkans is considerably used as a carminative stimulant in diarrhœa and cholera (*Dymock*).

201

Zingiber officinale, *Roscœ; Fl. Br. Ind., VI., 246.*

Syn.—AMOMUM ZINGIBER, *Linn.*

Vern.—Plant=*adrak*, dried root=*sónth, sindhi,* fresh root=*adrak, adhruka,* HIND.; Plant=*ádá,* dried root=*sónt,* fresh root=*ádrok, adá,*
BENG.; Plant=*ádá,* ASSAM; Dried root=*Súnt,* NEPAL; Plant=*adá,*
URIYA; Plant=*ada,* dried root=*adrak,* fresh root=*sonth,* N.-W. P.;
Plant=*ada, adrak,* dried root=*zangzabil, sonth,* fresh root=*zunjbel,
adrak,* PB.; Dried root=*sónt,* fresh root=*adrak, adhruka,* DECCAN;
Plant=*ádu, ále,* dried root=*sont, sunt, sunta,* fresh root=*alen, alem,
alch, adrack, adu,* BOMB.; Plant=*álé,* MAR.; Dried root=*súnt,* fresh
root=*ádú, adhú,* GUZ.; Dried root=*shukku,* fresh root=*inji,* TAM.;
Plant=*allam,* dried root=*sonti, sonthi, allam,* fresh root=*allam,* TEL.;
Dried root=*vana-sunthi,* fresh root=*hasisunthi,* KAN.; Dried root=
chukka, fresh root=*inchi,* MALAY.; Plant=*khyen-seing,* dried root=
ginsi-khiáv, fresh root=*gin sin,* BURM.; Dried root=*velicha-nguru,
inguru,* fresh root=*amu-inguru,* SING.; Plant=*árdraka, sringavéra,*
dried root=*vishva-bhishagam, nágara, sunti, mahaushadha,* fresh root=
árdrakam, SANS.; Dried root=*zanjabil, zanjabile-yábis,* fresh root=
zanjabile-ratab, ARAB.; Dried root=*zanjabile-khushk,* fresh root=*zanja-
bile-tar,* PERS.

References.—*Roxb., Fl. Ind., Ed. C.B.C., 16; also Monandrous Pl. in
Asiatic Researches, XI., 345; Voigt, Hort. Sub. Cal., 561; Stewart, Pb.
Pl., 239; Graham, Cat. Bomb. Pl., 207; Mason, Burma & Its People,
803; Sir W. Elliot, Fl. Andhr., 13, 169; Rheede, Hort. Mal., II., t. 12;
Rumphius, Amb., V., t. 66, f. 1; Pharm. Ind., 228; Flück. & Hanb.,
Pharmacog., 635; U. S. Dispens., 15th Ed., 1156; Ainslie, Mat. Ind.,
I., 603; O'Shaughnessy, Beng. Dispens., 647; Irvine, Mat. Med. Patna,
93; Moodeen Sheriff, Supp. Pharm. Ind., 262; U. C. Dutt, Mat. Med.
Hind., 253, 291; S. Arjun, Cat. Bomb. Drugs, 142; K. L. De, Indig.
Drugs Ind., 124; Murray, Pl. & Drugs, Sind., 21; Bent. & Trim.,
Med. Pl., t. 270; Dymock, Mat. Med. W. Ind., 2nd Ed., 762; Cat.
Baroda Durbar, Col. & Ind. Exhib., No. 186; Year-Book Pharm., 1873,
112; 1879, 426; 1881, 393; 1882, 173; 1886, 156; Macleod, Med. Top.,
Bisnath, 16; Birdwood, Bomb. Prod., 231; Baden Powell, Pb. Pr., 298,
379; Drury, U. Pl. Ind., 455; Atkinson, Him. Dist. (X., N.-W. P.
Gaz.), 706, 734; Useful Pl. Bomb. (XXV., Bomb. Gaz.), 174; Econ.
Prod. N.-W. Prov., Pt. V. (Vegetables, Spices, and Fruits), 25, 34;
Bidie, Prod. S. Ind., 17, 88; Cat. Col. & Ind. Exhib., Raw Products,
No. 121; Tropical Agriculture, 481; Linschoten, Voyage to East Indies
(Ed. Burnell, Tiele, & Yule), II., 7, 79, 80; Milburn, Oriental Com-
merce (1825), 288; Buchanan, Journey, through Mysore & Canara, etc.,
II., 209, 507; Gribble, Man. Cuddapah, 200; Note on the Condition
of the People of Assam, App. D; Morris, Descriptive & Historical*

Account of the Godavery, 10 ; Madden, Note on Kumaon, 280 ; Bombay Man. Rev. Accts., 103 ; Settlement Reports :—Panjab, Kangra, 25, 28 ; N.-W. P., Kumaon, App., 34 ; Central Provinces, Baitool, 77 ; Agricultural Dept. Reports :—Madras, 1883-84, 72 ; Bombay, App. x. ; Bengal, 1886, App., xxvi., liv., lxxxiii. ; 1886-87, 12 ; Gazetteers :—Bombay, VIII., 183. XII., 171 ; XIII., 292 ; N.-W. P., I., 84 ; Oudh, III., 419 ; Orissa, II., 27, 179 ; Mysore & Coorg, I., 67 ; Agri.-Horti. Soc. Ind., Trans., I., 165 ; II., 196, 208, App., 314, Pro., 340 ; III., 12, 13, 67, 199, Pro., 228, 236 ; IV., 104, 132, 149, Pro., 32 ; V., Pro., 79, 105 ; VI., 126, 227, Pro., 12 ; VII., 87 ; VIII., 192, 193 ; Journ. (Old Series), I., 102 ; II., Sel., 323 ; IV., 229 ; IX., 395 ; X., 341 ; (New Series), II., Pro., 1870, 41 ; Tropical Agriculturist, 481.

Habitat.—The ginger is not known in a truly wild state, but is doubtless a native of Tropical Asia, in which it has been cultivated and exported from very remote times. From Asia it was introduced into the West Indies, where it is now abundant. From the East and West Indies it has now spread throughout the warmer parts of both worlds, a small portion of the ginger of commerce coming from Africa.

History.—According to **Fluckiger & Hanbury** the Sanskrit name *Sringavera* is probably derived from the Greek Ζιγγίβερι, but much more probably the Greek was derived from the Sanskrit, which is a very old name, through the Arabic *Zanzabil.* The drug was known to the Greeks and Romans as a spice, who appear to have received it by way of the Red Sea, and considered it to be a product of Southern Arabia. It is probable, therefore, that they may have adopted the Arabic name which they received along with the plant, and which in its turn was derived from the Sanskrit. The learned authors of the *Pharmacographia* give an interesting account of the history of ginger, from which it would appear that as early as the second century A.D., it was one of the spices liable to the Roman fiscal duty at Alexandria. During the middle ages it is frequently mentioned in similar lists and evidently constituted an important item in European commerce with the East. " In England it must have been tolerably well known even prior to the Norman conquest, for it is frequently named in the Anglo-Saxon leech-books of the eleventh century, as well as in the Welsh *"Physicians of Myddvai."* During the thirteenth and fourteenth centuries it was, next to pepper, the commonest of spices, costing on an average nearly 1-7d. per ℔, or about the price of a sheep. The merchants of Italy, about the middle of the fourteenth century, knew three kinds of ginger, called, respectively, *belledi, colombino,* and *micchino.* These three terms may be explained thus : *belledi* or *baladi* is an Arabic word, which, as applied to ginger, would signify " country " or " wild," *i.e.,* " common ginger." *Colombino* refers to Columbum, Kolam or Quilon, a port in Travancore frequently mentioned in the middle ages. Ginger termed *micchino* denotes that the spice had been brought from, or by way of, Mecca. Ginger preserved in syrup, and sometimes called " Green Ginger," was also imported during the middle ages and regarded as a delicacy of the choicest kind.

" The plant affording ginger must have been known to **Marco Polo** (*circa* 1280-90), who speaks of observing it both in China and India. **John of Monticorvino,** who visited India about 1292, describes ginger as a plant like a flag, the root of which could be dug up and transported. **Nicolo Conti** also gave some description of the plant, and of the collection of the root, as witnessed by him in India " (*Flückiger & Hanbury*).

John Huyghen van Linschoten in 1596 gives a most interesting account of the spice. He states that it then grew in many parts of India, but that the best, and that most exported, grew on the coast of Malabar (this kind

HISTORY. 202

ZINGIBER officinale.	Cultivation of the

HISTORY

was probably identical with the *colombino* mentioned above). He describes the methods of cultivation and preparation, which appear to have then been very similar to those now pursued. Regarding the trade he writes: "There is much shipped as well to the Red Sea as to Ormus, Arabia, and Asia, but little for Portingale, because it will not save ye freight and custome, onlie the gunner of the Indian shippes may lade and bring certain quintals without paying any customes which by the King of Portingale was of long tyme granted unto them, and is yet observed; and this they may sell to merchants, and so by this meanes there is some broughte, otherwise but very little, for that the most part of Ginger broughte into Spain, cometh from Cabo Verde, the island of St. Thomas, Brasilia, and the island of S. Domingo in ye Spanish Indies, which is much trafficked withall in Spaine; wherefore that of the Portingale Indies is little brought out of the country, because of the long way and great charges, and yet it is better than other ginger; as also all other spices, mettals and stones, that are brought out of the Orientale Indies, that is out of the Portingales Indies, are for goodnesse and virtue better than any other which the continuall traffiques hath sufficiently made known. There is likewise much ginger conserved in sugar, which cometh out of the countrie of Bengala, but the best cometh from China, it is verie good to eate, and much used in India and brought out of Portingale into these countries." The remark on the effect of West Indian competition in diminishing East Indian trade at that early date is of interest, and testifies to the latge extent to which the cultivation had been carried in the former country in a comparatively short period. It would also appear to indicate an earlier date for the commencement of the West Indian ginger trade than that assigned to it by **Fluckiger & Hanbury,** who write: "It was shipped for commercial purposes from the Island of St. Domingo as early at least as 1585, and from Barbadoes in 1654. On the other hand, it corroborates the statement made by **Renny** (*Hist. of Jamaica, Lond., 1807, 154*), and quoted with doubt by those authors, to the effect that in 1547, 22,053 cwt. were exported from the West Indies to Spain.

CULTIVATION 203

Cultivation.—The plant is cultivated in all the warmer and moister parts of India, up to an elevation of 4,000 to 5,000 feet in the Himálaya. It will be seen from the accounts of the method pursued in the various Provinces, that the cultivation is one on which much care and labour are exercised. The soil must be rich, but neither too heavy, nor too light and coarse. The amount of moisture allowed to the crop requires much care; contrary to the usual custom in India except in the case of betel, and a few other expensive crops, manure is freely used; weeding is carefully and frequently carried out, and in every way ginger cultivation is much more elaborate than that of most other agricultural products.

Madras. 204

1. MADRAS.—The best Malabar ginger, spoken of so highly by **Linschoten** some three centuries ago, is now said to be the produce of the district of Shernaad situated to the south of Calicut. The soil of this district is peculiarly suited for the purpose, being a good rich red earth. The cultivation generally commences about the middle of May, after the ground has undergone a thorough process of ploughing and harrowing. At the commencement of the monsoon, beds of 10 to 12 feet long by 3 or 4 wide are formed, and in these small holes are dug at ¾ to 1 foot apart, which are filled with manure. The rhizomes, hitherto carefully buried under sheds, are dug out, the good ones picked from those which are affected by the moisture, etc., and cut into pieces of 1½ to 2 inches long, suitable for planting. These are then buried in the holes, and the whole of the beds are then covered with a good thick layer of green leaves, which, whilst they serve as manure, also contribute to keep the beds from unnecessary damp-

ness, which might be otherwise occasioned by the heavy rains of the monsoon. Inundation entirely ruins the crop, but as a fair supply of rain is absolutely necessary, great care is taken in draining. Strict care is observed in choosing the leaves to cover the beds, only certain kinds are chosen, since others are supposed to breed worms and insects injurious to the future prospects of the crop (*Drury*). No particulars can be given as to the area, yield, or profit.

CULTIVATION in Madras.

2. BOMBAY.—Ginger is a crop of considerable importance in this Presidency. In 1888-89 it occupied 918 acres, of which 640 were in Gujarát, 99 in the Deccan, 6 in the Karnatak, and 173 in the Konkan. In Gujárat the chief ginger-growing districts were Ahmadabad with 265 acres, Surat with 186, and Kaira with 155 acres, and in the Konkan, Thána with 168 acres, while Sátára in the Deccan had 87 acres under the crop. The following account describes the method followed in Thána, but is probably applicable to the whole Presidency :—The ginger which is to be used for " seed " is dug up in March and April. When the plant withers, the best roots are washed, dried in the shade, and placed in a heap on dry sugarcane and ginger leaves. More of these leaves are laid above the roots, and the whole is covered with an air-tight covering of clay. They are thus preserved till the planting season, by which time they have begun to sprout. The crop requires much the same soil as sugar-cane, *viz.*, a loose, light, stoneless soil with at least one quarter of sand. The ground is used for a rice-nursery and for *náchni*, and when the *náchni* has been reaped, it is cleaned, watered, ploughed, and turned into furrows 13¾ feet long, half a foot broad, 3 inches deep and about 9 inches apart. The pieces of ginger are then laid in the furrows at intervals of about 9 inches, the earth between the furrows is thrown into them, and the whole is levelled. The planting season is from April to July. If April is chosen, the ginger must be watered every fifth day, and to keep the ground moist and cool, hemp or *vál* (Dolichos Lablab) is sown along with it, and the young plants are covered with grass and plantain leaves. If, on the other hand, it is planted after the rains set in, there is no need to sow hemp or *vál*, or to cover the plants with grass. The ginger garden is divided into beds, *vápha*, with a waterway between each; and in each waterway, red-pepper and turmeric are grown. When the young ginger plants are about a foot high, oil-cake manure is applied at the rate of about 5℔ to each bed, and this process is repeated in August and September. The first and second layers of manure are not covered with earth, but the third layer is. In about nine months the rhizomes are ready for gathering; they are dug up, the rind rubbed off with tiles, and, when baked and dried in the sun, the ginger is ready for use (*Bomb. Gaz., XIII., Pt. I., 292*). In the Khándesh District the manure applied is said to be equal parts of horse, cow and sheep dung. In curing, the rhizomes are first partly boiled in a wide-mouthed vessel, then after drying for a few days in the shade they are steeped in weak lime water, sundried, steeped in stronger lime-water, and buried for fermentation. When the fermenting is over, the ginger, now called *sunth*, is ready for the market (*Bomb. Gaz., XII., 171*). The produce, according to Dymock, is from 50 to 150 maunds (of 25℔) per *bígha*. In a green state it sells at from ¾ to 1¼ maund (of 25℔) per rupee; when dried at from 5 to 10 seers per rupee.

Bombay. 205

3. BENGAL —Ginger is largely grown in many parts of this Province, but no returns of the approximate area under the crop can be given. The cultivation extends, as in other localities bordering on the Himálaya, for some distance on the hills; indeed, Mr. Campbell (*Agri. & Rural Economy of the Valley of Nepal*) states that ginger is carefully grown in Nepál, and

Bengal. 206

ZINGIBER officinale.	Cultivation of the

CULTIVATION in Bengal.

that the produce "is reckoned by the people of the neighbouring plains of Tirhoot and Sarun of very high flavour and superior to the produce of their own country." The following account of ginger-cultivation in Burdwán, taken from the *Report of the Dir., Agri. Dept., Bengal, 1886,* may be accepted as typical of the method pursued throughout the Province generally :—

Varieties.
207
Soil.
208

There is only one kind under cultivation, which, being grown under nearly the same conditions everywhere, and being propagated by buds and not by seeds, has not undergone much variation. The only SOIL on which ginger can be profitably grown is a fine sandy loam, both light coarse sand and stiff clay being quite unsuited to this crop. It is necessary that the soil for ginger should be loose. Sandy soil is loose when dry, but during the rains and after irrigation it sinks and becomes compact. Ginger does best after potatoes and *kachu* (**Colocasia antiquorum,** *Schott.*), but can also be grown after any of the pulses. The general principle on which the ROTATION of ginger is regulated is this :—It can be grown either after a crop which requires no irrigation, or after an irrigated crop if in the cultivation of this latter the ground had to be hoed constantly. In both cases the land is kept loose and mellow.

Rotation.
209

Tillage.
210

Ploughing begins in the end of March or beginning of April. After each fall of rain the land should be ploughed once. The soil should be thoroughly pulverised, stirred to as great a depth as possible, and get well weathered. It altogether receives from 12 to 14 ploughings. The PLANTING season is in the second and third weeks of May. When the field is ready for planting, it is levelled with the ladder, after which a number of main-water channels are drawn up and down the field from 60 to 80 feet apart. Then a number of smaller water channels are drawn at right angles to the preceding ones and about 8 feet from one another. Pieces of ginger about three inches long are now planted in parallel lines which extend from one of the smaller water channels to the next one. Earth is then raised by a *kodáli* from two sides of these rows and put over the pieces of ginger to a depth of nearly 9 inches. The field now appears to be laid in ridges, the furrows between which are closed at the upper end and opened into the smaller water channels at the lower end. At certain places are planted chillies, *beguns* (**Solanum Melongena,** *Linn.*), and *kachus.* The ginger pieces are placed at intervals of 9 inches in the rows, which latter are 18 inches apart.

Planting.
211

After-cultivation.
212

The plants may come out in 10 to 15 days, but sometimes take as much as two months. Throughout the rainy season every possible care is taken not to let water accumulate in the field, stagnant water being most injurious to the crop. As soon as weeds make their appearance, they should be pulled out with the *phor.* If on account of frequent heavy showers the earth sinks and the soil ceases to be friable, the field should be hoed with the *pashuni* or hoe. In the second week of September the plants are top-dressed with four maunds of oil-cake, consisting of two maunds of mustard and two of castor cakes, and then earthed up. If the field be dry, irrigation is needed in the end of October and beginning of November. If there be no rain in the cold weather, irrigation is required twice a month till the end of February or beginning of March, when the ginger is to be lifted. Four maunds of ginger are planted in one *bígha,* and the yield is from 40 to 60 maunds. At the time of ploughing, about 30 maunds of well-rotted dung is applied per *bígha* and then, as stated before, the plants are top-dressed with four maunds of oil-cake. Sometimes in the month of *Assin* (September-October) the ginger cuttings which were planted are carefully removed by the *phor,* without disturbing the rest of the plant, and sold at a high price. For this it is necessary that the cul-

Yield.
213

Manures.
214

Z. 214

tivators should know the exact spots where the cuttings were placed at the time of planting. Out of the four maunds of seeds nearly three maunds of ginger may be recovered in this way.

The COST of cultivation is said to be about R46 per *bigha*, of which R16 represents the price of the selected ginger used as "seeds." The manure costs about R7, and the rest is made up by several small items representing the cost of the many operations which attend this very carefully cultivated crop. At about R2 per maund the outturn of 40 to 60 maunds would represent a total money value for the yield of R80 to R120, or a profit of R32 to R74 per *bigha*.

4. NORTH-WEST PROVINCES.—Ginger is extensively grown in all hot valleys in Kumáon. The method is very similar to that already described in Bombay and Madras. A piece of ground not liable to be flooded is selected, and protected from excessive rainfall by trenching round the upper side. The soil is then well hoed and richly manured, and in April the ginger is planted in deep furrows. The earth is then heaped over the trenches, and the whole is covered with small leafy branches, preferably of oak, which are kept in their place by bamboo or wooden poles. The poles are not removed before the rains, but the leaves are not disturbed until the crop is dug up; all the weeding is done by hand. The rhizomes are gathered in February. Kumáon ginger is much esteemed, and its superior quality is unanimously believed by the hill-cultivators to be due to the leafy covering which they apply (*Atkinson*). No details can be given of area, yield, cost of cultivation, or profit.

5. PANJÁB.—Here, as in the North-West Provinces, ginger cultivation is chiefly carried on in the lower hot valleys of the Himálaya. The selected rhizomes for planting are preserved in heaps covered with a coating of cow-dung. In the end of June or beginning of July the land is ploughed, divided into beds, and saturated with water, but preserved from stagnant water by drainage. The rhizomes are planted and leaves applied as in Kumáon, but a layer of manure to the depth of $\frac{3}{4}$ an inch is applied over the leaves in addition. The rain water thus filters through this covering impregnated with manure, and carries much nourishment to the plants. After the cessation of the rains, artificial irrigation is necessary from October to January. In the latter month the rhizomes are dug out and removed to another place for a month, after which they are taken up, exposed to the sun for a day, and are then fit for use. The crop is weeded three times, in August, September, and October, respectively. A *bigha* requires 8 maunds of ginger to plant it, and yields 32 maunds in a good crop. Selected rhizomes for planting sell at 8 to 10 seers per rupee, the ordinary crop at 24 to 32 seers per rupee (*Baden Powell*). The late Captain Pogson states that the best ginger in the neighbourhood of Simla is grown in the Sabáthú District. Throughout the Province ginger is dried into *sonth* by placing it in a basket suspended by a rope, and shaking it for two hours daily for three days. They are then dried in the sun for eight days, and again shaken in the basket. The object of this shaking the roots together is to remove the outer skin and scales. Two days further drying completes the process. *Sonth* is very much more expensive than the green root, and well repays the labour spent in its preparation, as it sells for 3 to 4 seers per rupee (*Baden Powell*).

Medicine.—Ginger has long been known both to Sanskrit and Muhammadan medicine. By writers on the former it is described as acrid, heating, carminative, rubefacient, and useful in dyspepsia, affections of the throat, head and chest, hœmorrhoids, rheumatism, urticaria (nettle-rash), dropsy, and many other diseases. A favourite carminative remedy frequently prescribed by the older Sanskrit writers is *trikatu*, or the three

Margin notes:
CULTIVATION in Bengal.

Cost. 215

North-West Provinces. 216

Panjab. 217

MEDICINE. 218

MEDICINE.
Rhizome.
219

Juice.
220

acrids (see **Piper longum**, and **P. nigrum**, *Vol. VI., pp. 259, 263*). The dried RHIZOME is believed to possess all the properties of the green and to be laxative in addition. Ginger with salt, taken before meals is highly praised as a carminative, is said to purify the tongue and throat, increase the appetite and produce an "agreeable sensation." In cephalalgia and other affections of the head, ginger JUICE mixed with milk is used as a snuff, the fresh juice taken with honey is supposed to relieve catarrh, cough, and loss of appetite (*U. C. Dutt*). Many prescriptions of **Chakradatta** and from the *Bhávaprakásha* are translated in the *Hindu Materia Medica*, to which the reader is referred for further information. The properties ascribed to the drug by Muhammadan writers are similar. Fresh ginger is much employed as a domestic medicine, the juice with sugar or honey being prescribed for colds, coughs, and with the addition of lime-juice, in bilious dyspepsia. The juice with an equal portion of *tulsi* juice and a little honey and burnt pea-cocks' feathers is a popular remedy for vomiting in Bombay (*Dymock*).

The uses of ginger in European medicine, in which it is one of the most highly valued of all mild carminatives and enters into many officinal preparations, are too well known to require mention in this work.

"The gingers at present found in the London market are distinguished as Jamaica, Cochin, Bengal, and African. Jamaica ginger is the sort most esteemed; and next to it the Cochin. Scraped or decorticated ginger is often bleached, either by being subjected to the fumes of burning sulphur, or by immersion, for a short time in a solution of chlorinated lime. Much of that seen in grocers' shops looks as if it had been white-washed, and in fact is slightly coated with calcareous matter" (*Pharmacographia*).

CHEMISTRY.
221

CHEMICAL COMPOSITION.—Mr. **J. O. Thresh** has very completely analysed the different gingers of commerce. He found a sample of Cochin ginger to contain:—volatile oil, 1.350; fat, wax (?) and resin, 1.205; neutral resin, .950; α and β resins, .865; *gingerol* .600; substance precipitated by acids, 5.350; mucilage, 1.450; indifferent substance precipitated by tannin, 6.800; extraction soluble in spirits of wine, not in ether or water, .280; alkaloid, a trace; metarabin 8.120; starch, 15.790; pararabin, 14.400; oxalic acid, .427, cellulose, 3.750; albumenoids, 5.570; vasculose, etc., 14.763; moisture 13.530, and ash, 4.800 per cent. *Gingerol*, the pungent or active principle of ginger, is a viscid fluid of the consistency of treacle, of a pale straw colour, devoid of odour and with an extremely pungent and slightly bitter taste. The essential oil is of a pale straw colour, has a somewhat camphoraceous odour, and aromatic but not pungent taste, a sp. gr. of about .883 at 63° F., and is lævo-gyrate. An interesting result of Mr. **Thresh's** analysis was the fact that a fine selected sample of Jamaica ginger contains only about half the quantity of essential oil found in the Cochin and African samples, and less of the active principle than the African, though about as much as the Cochin gingers. Though less in quantity, however, the volatile oil of the Jamaica ginger possessed a much finer bouquet than the others (*Year-Book of Pharmacy, 1879, 1881, and 1882*).

FOOD.
222

Food.—Ginger is sold in every bazár throughout India, and is very largely employed as a condiment, especially in the preparation of curries. It is also pickled, and an excellent preserve, similar to the well known Chinese preserved ginger, is made by cooking the fresh younger rhizomes in syrup. The quality of the ginger produced in different localities varies much. Thus in Bombay three kinds of dried ginger are met with in the market, namely, Ahmadabad, which costs about R12 per cwt. Calcutta, valued at about the same, and Malabar or Cochin, which fetches more than double the price, namely, from R24 to R40 according to quality

| Trade in Ginger Spice. | (*J. Murray*.) | ZINGIBER officinale. |

FOOD.

(*Dymock*). Besides these chief commercial classes, other smaller and more unimportant kinds are distinguished in bazárs. In the North-West Provinces, Kumáon ginger is said to be most highly esteemed ; in the Panjáb, that grown in the Sabathu District, and other Himálayan tracts ; in Bengal that obtained from Nepál, etc.

TRADE.
223

Trade.—The internal trade in ginger is fairly large and important. In 1888-89 the total quantity registered as transmitted by road, rail, and river amounted to 1,03,168 maunds, valued at R6,27,421. The chief exporting province was Bengal with 52,035 maunds, Bombay port followed with 12,703, then the Panjáb with 12,314, Calcutta with 8,334, Bombay with 5,750, the North-West Provinces and Oudh with 4,693, and Madras with 4,042. It is noteworthy that the exports from Bombay port and from Madras were of much higher comparative value than those from other localities. Much the largest importing centre was Calcutta with 50,953 maunds, followed by the North-West Provinces and Oudh with 13,142, Bombay with 8,342, and Rájputana and Central India with 8,078 maunds. The transactions between other localities were insignificant. The coasting trade is important, chiefly as regards Bombay,—the principal exporting Presidency, since it shews the source from which it derives its material for foreign export. The total imports into the various Presidencies and Provinces by this channel amounted in 1889-90 to 5,915,489℔, valued at R5,49,652. Of that amount Bombay received 4,705,811℔; Burma, 685,415; Madras, 317,783; Bengal, 171,929; and Sind, 34,451℔. Of the supply to Bombay 2,886,004℔ was received from Madras, 1,288,751℔ from Travancore, and smaller quantities from other sources.

It will be observed from the above that neither in the returns of external trade by rail, river, etc., nor in those of coasting trade, is there any explanation of the source from which the large foreign exports from the Madras seaports are derived. This must, therefore, as in the case of cotton, be due to a considerable unregistered trade by road and canal from the ginger-growing districts of the Presidency to its seaports.

The external trade is fairly important. Milburn informs us that in 1808 the total quantity imported by the East India Company was 2,245 cwt., valued at £5,629, or an average per cwt. of £2-10-2. The quinquennial average exports for the past fifteen years have been— 6,691,867℔, value R9,72,853, for the period ending 1879-80; 5,421,397℔, value R8,89,016, for the period ending 1884-85 ; and 10,377,710℔, value R13,94,213, for that ending 1889-90. The trade suffered a large diminution during the years from 1880-81 to 1883-84, but in 1884-85 it again revived, and in 1886-87 reached a maximum of 14,927,926℔. In the following year it again fell to 9,510,564℔, in 1888-89 it rose to 10,212,971℔, while in 1889-90 it again fell to 6,918,681℔—the lowest export recorded since 1883-84. The price also shews a considerable diminution, and, during the past year, was as nearly as possible R1 for 10℔, or a little over R11 per cwt. The following table, taken from Mr. **O'Conor's** statistics,

**ZINGIBER
officinale.**

Trade in Ginger.

TRADE.

Exports.

224

shows the distribution of the exports during last year and the share taken by each Indian Presidency or Province in the trade :—

Countries to which exported.	℔	R	SHARE OF EACH PRESIDENCY OR PROVINCE.		
			Presidency or Province.	℔	R
United Kingdom . .	3,827,990	4,21,323	Bengal .	913,352	57,351
Austria . . .	230,434	24,655	Bombay .	3,120,555	3,17,245
France . . .	57,042	3,520	Sind .	3,164	280
Germany . .	81,116	9,885	Madras .	2,881,710	3,29,105
East Coast of Africa. { Mozambique .	154	14			
Zanzibar	77,385	6,264			
Other Ports .	448	40			
United States . .	546,025	32,443			
Aden . . .	811,405	81,702			
Arabia . . .	708,682	70,844			
Ceylon . . .	106,609	10,591			
Persia . . .	328,198	29,591			
Turkey in Asia . .	121,569	11,286			
Other Countries .	21,624	1,895			
TOTAL .	6,918,681	7,03,981		6,918,691	7,03,981

[*t. 2003.*

225

Zingiber Zerumbet, *Roscœ; Fl. Br. Ind., VI., 247; Wight, Ic.,*

Syn.—Z. SPURIUM, *Kón.*; AMOMUM ZERUMBET, *Willd.*; A. SPURIUM, *Gmel.*

Vern.—*Mahá bari bach, nar kachúr,* HIND. & BENG.; *Kachúr, nar-kachúr,* PB.; *Kathu-inshi-kua,* MALAY.; *Wal-ingúrú,* SING.; *Sthula granthi,* SANS.

References.—*Roxb., Fl. Ind., Ed., C.B.C., 17 ; also Monandrous Pl. in As. Researches, XI., 346 ; Voigt, Hort. Sub. Cal., 562 ; Thwaites, En. Cey. Pl., 315 ; Rheede, Hort. Mal., II., t. 13 ; Rumph., Amb. V., t. 64, f. 1 ; Pharm. Ind., 229 ; Ainslie, Mat. Ind., I., 492 ; Irvine, Mat. Med. Patna, 71 ; U. S. Dispens., 15th Ed., 1783 ; U. C. Dutt, Mat. Med. Hind., 255 ; Birdwood, Bomb. Prod., 88 ; Baden Powell, Pb. Pr., 380.*

Habitat.—Found throughout both peninsulas and Ceylon.

DYE.
Rhizome.
226

Dye.—The RHIZOME is used as a dye (*Baden Powell*). This remark may very possibly be a mistake which has arisen out of the confusion between this species and **Curcuma Zedoaria,** *Roscœ,* the latter of which is employed in making *abír* (see Vol. II., 670).

MEDICINE.
Rhizome.
227

Medicine.—The RHIZOME has a slightly aromatic odour and possesses similar properties to those of officinal ginger, but in a minor degree. It is employed by Natives as a "hot" remedy for coughs, asthma, "special diseases," worms, leprosy and other skin diseases (*Baden Powell*). Much of the information regarding *serumbad,* **Zerumbet, Zedoary,** etc., is very confusing, since it is doubtful how much refers to this plant, and how much to **Curcuma Zedoaria,** *Roscoe* (see Vol. II., 670).

228

ZIZYPHUS, *Juss.; Gen. Pl., I., 3798.*

A genus of trees or shrubs which contains about fifty species, found in Tropical Asia and America and the temperate regions of both hemispheres. Of these some eighteen to twenty are natives of India.

The Indian Jujube, or Chinese Date. (*J. Murray.*) **ZIZYPHUS Jujuba.**

[*t. 282 ;* RHAMNEÆ.

Zizyphus glabrata, *Heyne ; Fl. Br. Ind., I., 633 ; Wight, Ic.,* **229**
    Syn.—Z. TRINERVIA, *Roxb., not of Poir.*
    Vern.—*Karukatá, karkattam, carúkúva,* TAM.; *Kakú-pala,* TEL.; *Vata-dalla,* SANS.
    References.—*Roxb., Fl. Ind., Ed. C.B.C.,* 204; *Beddome, Fl. Sylv. Anal., Gen., lxviii.; Ainslie, Mat. Ind., II.,* 69; *Drury, U. Pl. Ind.,* 457; *Moore, Man. Trichinopoly,* 81; *Nicholson, Man. Coimbatore,* 401.
    Habitat.—A tree of Eastern Bengal, Bhutan, and the Western Peninsula.
    Medicine.—The LEAVES are employed in Southern India in decoction as a remedy to purify the blood in cases of cachexia, and as an alterative in old venereal affections (*Ainslie*).   **MEDICINE. Leaves. 230**

Z. Jujuba, *Lamk.; Fl. Br. Ind., I.,632 ; Wight, Ic., t. 99.*   **231 ·**
    THE INDIAN JUJUBE, or CHINESE DATE.
    Syn.—Z. MAURITIANA, *Ham.;* Z. SORORIA, *Schult.;* RHAMNUS JUJUBA. *Linn.*
    Vern.—*Bér, baer, beri,* HIND.; *Kúl, bér, bór,* BENG.; *Janumjan, jom janum, janumjarom,* KOL.; *Dedhaori janum, jom janum,* SANTAL; *Bar koli,* URIYA; *Jibang,* MAGH.; *Bhér, bori,* C.P.; *Renga,* BHIL; *Bor,* BAIGAS; *Bogri,* RAJHANSHI; *Ringa,* GOND.; *Ber, bera,* N.-W. P.; *Ber, khalis, guter,* KUMAON; *Bér, 'unab, beri,* fruit=*kuchra, kurkunda, kokanber,*stone=*kmárkábij, hal-ká-bij,* PB.; *Berrá,* PUSHTU; *Ber,* RAJ.; *Ber jangri,* SIND; *Bér,* DECCAN; *Bor, bhor, bordi, búr, bhurmi,* BOMB.; *Bóra, bhor, bera,* MAR.; *Bór, bordi, boyedi,* GUZ.; *Elandap, yellande, elládu, elanda,* TAM.; *Yellantha,* MADURA; *Régu, ganga régu, karkan-dhavu, rengha, regi,* TEL.; *Yalachi, yelchi, ilanji, yagachi,* KAN.; *Elan-tap, elentha, elanta,* MALAY.; *Zi,* BURM.; *Ilanda, másánká,* SING.; *Badari, kola, badara,* SANS.; *Sidr, nabiq, aunnábe-hindi,* ARAB.; *Kunár,* PERS.
    References.—*Roxb., Fl. Ind., Ed. C.B.C.,* 204; *Brandis, For. Fl.,* 86, *t.* 17; *Kurz, For. Fl. Burm., I.,* 266; *Beddome, Fl. Sylv., t. cxlix.; Gamble, Man. Timb.,* 88; *Thwaites, En. Ceyl. Pl.,* 74; *Dalz. & Gibs., Bomb. Fl.,* 49; *Stewart, Pb. Pl.,* 43; *DC., Orig. Cult. Pl.,* 197; *Rev. A. Campbell, Rept. Econ. Pl., Chutia Nagpur, Nos.,* 8799, 9445; *Mason, Burma & Its People,* 458; *Elliot, Fl. Andhr.,* 57, 83, 165; *Cleghorn, Forests & Gardens, S. Ind.,* 244, 281; *Rheede, Hort. Mal., IV., t.* 40; *O'Shaughnessy, Beng. Dispens.,* 273; *Irvine, Mat. Med. Patna,* 82, 127; *Moodeen Sheriff, Supp. Pharm. Ind.,* 262; *Mat. Med. S. Ind.* (in MSS.), 106; *U. C. Dutt, Mat. Med. Hind.,* 293, 305; *S. Arjun, Cat. Bomb. Drugs,* 31; *Murray, Pl. & Drugs Sind,* 146; *Dymock, Mat. Med. W. Ind., 2nd Ed.,* 180, 888; *Dymock, Warden & Hoober, Pharmacog. Ind., I.,* 350; *Birdwood Bomb Prod.,* 145, 192, 260, 327; *Baden Powell, Pb. Pr.,* 266, 269, 337, 601; *Atkinson, Him. Dist.* (X., N.-W. P. *Gaz.*), 307, 779; *Useful Pl. Bomb.* (XXV., *Bomb. Gaz.*), 49, 149, 242, 250, 279, 388; *Econ. Prod. N.-W. Prov., Pt. III.* (Dyes and Tans), 83; *Gums and Resinous Prod.* (P. W. *Dept. Rept.*), 18, 21, 34; *Liotard, Dyes,* 33, 36, 65, 105, *App. viii.; Cooke, Gums and Resins,* 29; *McCann, Dyes and Tans, Beng.,* 50; *Kew Bulletin,* 1889, 23; *Stock's Rep. on Sind; Buchanan, Statistics Dinajpur,* 162; *Man. Madras Adm., I.,* 363; *Moore, Man., Trichinopoly,* 81; *Gribble, Man. Cuddapah,* 263; *For. Adm. Rep., Chota Nagpore,* 1885, 6, 29; *Settlement Report:*—*Panjáb, Lahore,* 15; *Guzrát,* 134; *Hazára,* 12, 94; *Jhang,* 20; *Déra Ghási Khán,* 4; *Peshawar,* 13; *Delhi, App. XXX., ccliii.; Kángra,* 22; *Central Provinces, Seonee,* 10; *Mundlah,* 88, 89; *Chundwara,* 110; *Nimar,* 306; *Chanda, App. vi.; Agri. Dept. Rep., Madras,* 1883-84, 57; *Gazetteers:*—*Bombay, II.,* 42, 355, 359; *IV.,* 24; *V.,* 24, 285; *VI.,* 13; *VII.,* 39, 40, 42; *VIII.,* 100; *XIII.,* 24; *XV.,* 79; *XVI.,* 18; *X. II.,* 18; *XVIII.,* 44; *Panjáb, Déra Ismail Khan,* 19; *Jalandhar,* 4; *Muzaffargarh,* 22; *Ludhiana,* 10; *Shahpur,* 70; *Bannu,* 23; *Hazara,* 133; *Sialkot,* 11; *Rohtak,* 14; *Delhi,* 18; *Jhang,* 15; *N.-W. P., I.,* 80; *III.,* 33; *IV., lxx.; Orissa, II.,* 153, 179; *Burma, I.,* 137; *Mysore & Coorg, I.,* 50, 60; *II.,* 7; *Agri.-Horti. Soc. Ind:*—

Z, **231**

*Trans., II., 1-5, 168, App., 306 ; VI., 48 ; VIII., Pro., 406 ; Ind. Forest-
er, I., 273, 274 ; II., 175 ; III., 201, 238 ; IV., 230, 322 ; V., 80, 93, 212 ;
VI., 108, 218 ; VII., 259, 277 ; VIII., 30, 82, 102, 119, 333, 373, 388, 410,
416, 438 ; IX., 401 ; X., 309 ; XII., 139, App., 4, 27.*

Habitat.—A small tree, wild and extensively cultivated throughout
India, from the North-West Frontier, Sind, and the base of the Himálaya
to Ceylon, Malacca, and Burma; distributed to Afghánistán, Tropical
Africa, the Malay Archipelago, China, and Australia. According to
DeCandolle the great number of known cultivated races indicates an
ancient domestication. Its abundance in a wild state in India and Burma,
together with the number of Sanskrit and vernacular names, and the fact
that botanists at an early date received it from Bengal, all point to an
Indian origin. Rumphius states that it had only been recently introduced
into the eastern islands of the Amboyna group, while he was living there,
and ancient Chinese authors do not mention it. Its extension and natur-
alisation to the east of the Indian continent seems, therefore, to have been
recent. It appears to have been introduced into Arabia and Egypt at a
still later date, and it must have spread to Zanzibar from Asia, and by
degrees across Africa, at a quite recent date (*DeCandolle, Cult. Pl., 197*).

In support of the theory of the indigenous nature of the tree in India,
the writer may quote an instructive passage, which appears to have escaped
the notice of later writers. In one of the earliest publications of the Agri.-
Horticultural Society of India an interesting paper on the *ber* is given, written
by Babu Radakant Deb, and read in April 1829. In that article we read :
"According to the *Purana*, there was, in former times, a celebrated place
of pilgrimage called Badarica Srama (the Badarináth of modern travellers, a town and temple on the west bank of the Alacananda river in the
province of Srinagar) [in Garhwál, North-West Provinces], which abound-
ed with the *badari* or jujube trees, and the devotees or sages of those
times lived upon its fruits ; whence the tree is supposed to have been introduced more generally into other parts of India." This tradition testifies at
least to a very ancient knowledge of the tree and of its fruit, and points to the
probability that the tree, or the knowledge of its cultivation, may, as stated,
have originally spread from Northern India. Cultivation by selection
and grafting has very much improved the wild jujube fruit in India, and as
a natural consequence many kinds exist, which differ markedly from each
other in size, shape, and flavour. The plant itself varies in size, from a
shrub or very small tree to a large tree. One in the Central Provinces,
carefully measured by Mr. Hooper, was found to have a girth of 16 feet
9 inches at 5 feet from the ground, and 23 feet at the base, with a height of
80 feet. [In this connection it may be added that, according to some
writers, a species of Zizyphus is supposed to have been the Lotus fruit,
but by others it is believed that the oblivion fruit as obtained from a
species of Diospyros (*Conf. with Vol. III. pp. 136, 147, 149*).—*Ed.*]

GUM.

Bark.
232

Lac.
233
DYE.
Bark.
234

Gum.—Frequent reference is made in works on the products of
India to a gum derived from the *bér*, but satisfactory evidence even as to
its existence is wanting. In the *Bombay Gazetteer*, Vol. XV., it is stated
that the BARK yields a kind of kino gum, employed in tanning and for
medicinal purposes. Sir George Birdwood states that a portion of gum-
gattie is derived from the tree, but on the other hand, the Catalogue of the
Madras Exhibition of 1855 contains the remark that the produce is not
a true gum. It appears to be most probable that the LAC which is frequently produced on the tree (see Coccus lacca, Vol. II., 411) may in certain cases have been inadvertently classed with gums.

Dye & Tan.—The BARK is said to be used for tanning purposes in
Northern India (*Stewart, Baden Powell, Brandis, Atkinson, Buck, etc.*),

| Medicinal properties of Jujube. | (*J. Murray.*) | ZIZYPHUS Jujuba. |

Bombay (*Lisboa, & several Gazetteers*), Madras (*Beddome, Drury*), and Burma (*Kurz*). It is apparently not employed for this purpose in Bengal proper, since McCann makes no mention of it, but Campbell states that it, along with the FRUIT, is used for tanning in Chutia Nagpur. The bark, LEAVES, and fruit, all contain tannin, but no information can be given as to the percentage, nor as to the respective tanning value of the different parts. The bark is occasionally thrown into indigo fermenting vats to aid in precipitating the fecula. In the Henzada district of Burma the fruit is employed as a mordant in dyeing silk a reddish pink colour with safflower (*Liotard*).

DYE & TAN.
Fruit.
235
Leaves.
236

Oil.—The KERNELS are said to yield an oil, of which nothing is known. It is stated in the *Ahmadabad Gazetteer* (Bomb., IV., 24) that the tree "yields a WAX much used by goldsmiths for staining ivory red;" this remark probably refers to lac.

OIL.
Kernels.
237
Wax.
238
MEDICINE.

Medicine.—In the *Bhávaprakásha* three different kinds of *bér* are said to be described,—*suvira, kola,* and *karkandhú*. The first is considered, in Sanskrit medicine, to be cooling, aperient, astringent, aphrodisiac and nourishing, and to be indicated in bilious affections, fever, hæmorrhages, consumption, and thirst. The second is described as sweet, yet slightly acrid, and full of flavour; it is considered a hot remedy and is recommended for flatulence, bilious affections, and constipation The third form is inferior and similar in properties to the second. In the *Rájavallabha,* old or preserved *bér* fruit is described as capable of removing "dryness and weariness," to act as a stimulant and to be easy of digestion. The KERNEL is said to be antibilious, and useful in cases of nausea and thirst from fever (*Babu Radakant Deb, in Trans. Agri.-Horti. Soc., Ind., II., 1*). Ainslie states that the ROOT is prescribed in decoction by the *Vytians* in conjunction with sundry warm seeds, as a drink in certain cases of fever, "but," he adds, "I am inclined to think that it has little virtue." In Northern India, the FRUIT is believed to purify the blood, and to assist digestion; the BARK is said to be a remedy for diarrhœa; the root is used as a decoction in fever and delirium, also, when powdered, as a dressing to ulcers and old wounds; the LEAVES are made into a plaster which is applied in strangury and other diseases, and the SEEDS are employed as an astringent in diarrhœa. The small wild fruit, *Kokanber,* is believed to have specific virtues in " special diseases " (*Baden Powell*). In Bombay the young leaves, pounded with those of **Ficus glomerata,** are applied to scorpion stings (*Dymock*).

Kernel.
239
Root.
240
Fruit.
241
Bark.
242
Leaves.
243
Seeds.
244

CHEMICAL COMPOSITION.—The fruit of Z. vulgaris, which probably is similar in composition to that of Z. Jujuba, contains mucilage and sugar; the bark and leaves contain tannin; the watery extract of the wood contains a crystallisable principle (*Ziziphic acid*), a tannin (*Ziziphotanic acid*), and a little sugar (*Lotom*) (*Pharmacog. Ind.*).

Chemistry.
245

SPECIAL OPINIONS.—§ "The tender leaves and TWIGS are used in the form of paste as an application to boils, abscesses, and carbuncles; they promote suppuration" (*Civil Surgeon J. H. Thornton, B.A., M.B., Monghyr*). " The root and bark are astringent and are used with *babul*-bark in preparing gargles" (*Surgeon-Major Robb, Ahmedabad*).

Twigs.
246

Food & Fodder.—The FRUIT of the wild *ber,* which ripens in the cold weather, resembles the crab-apple in flavour and appearance, is never larger than a gooseberry, but, notwithstanding its acidity, is much eaten by the poorer classes. In times of scarcity it is especially prized. In the Deccan Famine of 1877-78 it was powdered and made into a sort of meal (called *Berchúnt* in Hindí), which was largely consumed. By cultivation it is very greatly improved both in size and flavour. Certain kinds are long, others oval, or round; all are sweet, mealy, and palatable. The unripe

FOOD & FODDER.
Fruit.
247

24

**FOOD &
FODDER.**

Pulp.
248
Kernels.
249
Leaves.
250
TIMBER.
251
DOMESTIC
& SACRED.
Branches.
252
Leaves.
253

fruit is pickled ; the ripe PULP is dried, mixed with salt and tamarinds, to form a condiment ; the KERNELS are also eaten. The LEAVES are a good fodder for cattle and goats.

Structure of the Wood.—Hard, reddish, no heartwood, fine and close-grained, strong ; weight from 43 to 58℔ per cubic foot. It is largely used in ordinary constructive work, for making well-curbs, well wheels, ploughs and other agricultural implements, oil-mills, tent-pegs, *charpoy* legs, saddle-trees, camel saddles, clogs, combs, and other articles for which a hard, durable, close-grained timber is necessary. It has been recommended for furniture, and is said to make excellent charcoal.

Domestic & Sacred.—The BRANCHES are employed for making hedges in many localities (*Cleghorn*). The LEAVES are largely used as food for *tasar* silk-worms, in certain parts of the North-West Himálaya; the cocoons of this tree are said to be superior to those on any other. Stewart states that the silk obtained in Kangra from a wild silk-worm's cocoon, found on this tree, was at one time generally employed for tying the barrel on to the stock of the matchlock, " being found better for the purpose than sincus or leathern thongs" (see Silk, Vol. VI., Pt. III.). According to Bellew women near Peshawar make a lather (?) with the leaves in water for washing the head.

254

Zizyphus nummularia, *W. & A.; Fl. Br. Ind., I., 633.*

Syn.—Z. LOTUS, *Lamk.* ; Z. MICROPHYLLA, *Roxb.* ; Z. ROTUNDIFOLIA, *Lamk.* ; RHAMNUS NUMMULARIA, *Burm.*

Vern.—*Jar-beri*, HIND. ; *Jánd, kánta-ber*, BUNDEL. ; *Malla, bér, birár, jhari, kanta, jhar-ber*, N.-W. P. ; *Birota, jar-beri, jhar-beri, malla-bér, mallán, jand, ber, birár, mallá, kokni-ber, mara ber, zari, pála, kokan ber, jhár-pálá*, PB. ; *Karkanrá, karkana, karkan*, PUSHTU ; *Ber, bhor, jhalbhor*, RAJ. ; *Gangr, jangra, jangri, nando-jangro, bér*, SIND ; *Pali*, C. P. ; *Parpalli gidda*, KAN.

References.—*Roxb., Fl. Ind., Ed. C.B.C., 206 ; Brandis, For. Fl., 88 ; Beddome, Fl. Sylv., Anal. Gen., lxix. ; Gamble, Man. Timb., 89 ; Dals. & Gibs., Bomb. Fl., 49 ; Stewart, Pb. Pl., 43 ; DC., Orig. Cult. Pl., 196 ; Lace, Quetta Pl. ; Journ. Linn. Soc., xxviii., 294, 314 ; Murray, Pl. & Drugs, Sind, 146 ; Baden Powell, Pb. Pr., 337, 602 ; Atkinson, Him. Dist., 307 ; Ec. Prod., N.-W. P., Pt. V., 44, 54 ; Settlement Reports :— Delhi, 28 ; Rohták, 78 ; Jhang, App. xxv., cclii ; Hoshungábád, 284 ; Gazetteers :—Bombay, XV., 430 ; N.-W. P., I., 80 ; IV., lxx. ; Jalandhar, 5 ; Karnal, 16 ; Múltán, 102 ; Ludhiána, 11 ; Bannu, 23 ; Delhi, 20 ; Agri.-Horti. Soc. Ind., Journ. (Old Series), XIII., 320 ; (New Series), I., 85 ; V., 73 ; VI., Sel., 18 ; Ind. Forester :— III., 201 ; IV., 228, 233 ; V., 13, 31, 471 ; X., 168, 325 ; XI., 467 ; XII. App., 2, 9 ; XIII., 542.*

Habitat.—A prickly shrub, found in the Panjáb up to 3,000 feet, in the North-West Provinces, Sind, and Balúchistán ; also in Gujarát and the Western Peninsula from the Deccan and Konkan southwards.

TAN.
Bark.
255
MEDICINE.
Fruit.
256
Bark.
257
FOOD &
FODDER.
Fruit.
258

Tan.—The BARK is used for tanning in the Panjáb (*Baden Powell*).

Medicine.—In Northern India the FRUIT is considered cool, astringent, and of value in bilious affections (*Stewart, Baden Powell*). Mr. Lace informs the writer that in Balúchistán the BARK is employed to make a poultice for foul sores, and that a decoction of the same part is used as a gargle in sore-throat and ulcerated gums.

Food & Fodder.—The FRUIT is small, round, acid, and much inferior to the preceding, but is appreciated by the poorer classes, especially in times of scarcity. It is sweet, acidulous, has a not unpleasant flavour, and when boiled in milk is said to make a fairly good tart. Brandis states that during the famine of 1869, which drove large numbers of the inhabitants of Marwar and other parts of Western Rájputána from their homes, it served as food for thousands. "In the winter, 1869-70," he writes, "the crop of these berries had been plentiful; and when I marched through Rájputána, from Agra to Guzerát, in December 1869 and January 1870, I found the shrubs completely stripped of their fruit wherever the flocks of hungry emigrants from Marwar had passed through." Like the fruit of the former species this has been supposed to be the *lotos* of the ancients, but, as DeCandolle remarks, they must have been very poor or very temperate to be satisfied with such a fruit, and widely different opinions prevail as to what the fruit described by the ancients really was.

The LEAVES form a most valuable fodder for camels, goats, buffaloes, and cows, and are highly esteemed in the sandy districts of Sind, the Panjáb, and Balúchistán. They are stored for winter use by allowing the cut branches to dry, beating the leaves off, and gathering them into heaps. They may be grown either alone or with some form of chaff, straw or *bhúsa*, and are supposed to be heating, and to promote the secretion of milk. In the *Settlement Report of the Delhi District* it is stated that camels and goats prefer this fodder to almost any other. It is said to be cut in that district twice a year in April and November, and that it sells at from 3 to 5 maunds per rupee. **Mr. Coldstream** states in a note to the Editor that miles of it exist in the Hissar district, and that it is so valuable as a camel and cattle fodder (under the name of *palé*) that villagers often pay their *nonuce* off the produce.

Structure of the Wood. —Yellow, hard, compact, weight 43℔ per cubic foot; too small to be of value except for fuel.

Domestic.—The dried BRANCHES, from which the leaves have been shaken for fodder, are much used in making heaped-up fences. These are made either by burying *jhári* stumps in the ground, and using the barrier thus formed as a foundation on which to pile fresh bushes, or by simply laying the branches lengthwise on the ground, and weighting them with mud and stones to keep them in position. In certain localities they are more elaborately and permanently raised by sticking the branches upright and binding them with straw ropes. The BUSH is also not unfrequently planted as a hedge.

FOOD &
FODDER.

Leaves.
259

TIMBER.
260
DOMESTIC.
Branches.
261

Bush.
262

Zizyphus Œnoplia, *Mill.*; *Fl. Br. Ind., I., 634.* 263

Syn. —Z. ALBENS, NAPECA, & SCANDENS, *Roxb.*; Z. CELTIDIFOLIA, *DC.*; Z. FERRUGINEA, *Heyne*; Z. PALLENS, *Wall.*; Z. PEDICELLATA, *Wall.*; Z. RUFULA, *Miq.*; RHAMNUS ŒNOPLIA, *Linn.*

Vern.—*Makai*, HIND.; *Siákul, shyakúl, mahkoa*, BENG.; *Barokoli*, URIYA; *Siyáhkúl, mako, bamolan*, N.-W. P.; *Irún*, C. P.; *Paragi, paringi, parimi, paranu, porki*, TEL.; *Tawzinmé, tau-hsi*, BURM.; *Erraminya-wel*, SING.; *Srigálakoli*, SANS.

References.—*Roxb., Fl. Ind., Ed. C.B.C.*, 204, 205, 206; *Brandis, For. Fl.*, 86; *Kurz, For. Fl. Burm.*, *I.*, 266; *Beddome, Fl. Sylv., Anal. Gen., lxix.*; *Gamble, Man. Timb.*, 89; *Thwaites, En. Ceyl. Pl.*, 74; *Dals. & Gibs., Bomb. Fl.*, 49; *Elliot, Fl. Andhr.*, 144, 145; *Mason, Burma & Its People*, 760; *O'Shaughnessy, Beng. Dispens.*, 273; *U. C. Dutt, Mat. Med. Hind.*, 318; *Atkinson, Him. Dist.*, 307; *Ec. Prod., N.-W. P., Pt. V.*, 44, 54; *Balfour, Cyclop., III.*, 1130; *Gazetteers:—Bombay, XV.*, 430; *N.-W. P., I.*, 80; *IV., lxx.*; *Ind. Forester, III.*, 201; *IX.*, 451; *Agri.-Horti. Soc. Ind., Journ. (Old Series), XIII.*, 320.

Habitat.—A straggling or climbing shrub, very common throughout the hotter parts of India, from the Panjáb and North-West Himálaya to Assam, Malacca, and Ceylon.

Gum.—"The BARK affords a good deal of KINO, and dyes leather red" (*Balfour*). This remark probably, as in the case of Z. Jujuba, applies to the LAC occasionally found in it.

GUM.
Bark.
264
Kino.
265
Lac.
266

24 A

Z. 266

ZIZYPHUS rugosa.	Different kinds of Jujube.

MEDICINE.
Bark.
267
FOOD.
Fruit.
268
DOMESTIC.
Shrub.
269
270

Medicine.—A decoction of the BARK is said to promote the healing of fresh wounds (*Roxburgh*).

Food.—The FRUIT, which ripens in the rains, is eaten.

Domestic.—The prickly SHRUB is commonly used for hedges and heaped-up fences.

Zizyphus oxyphylla, *Edgw.; Fl. Br. Ind., I., 634.*

Syn.—Z. ACUMINATA, *Royle.*

Vern.—*Kurit rama*, SANTAL ; *Giggar*, N.-W. P. ; *Pitni, kokán bér, amlái, amnia, beri, shamor*, PB. ; *Ghar-guru*, PUSHTU ; *Kúrkun bér*, AFGH.

References.—*Brandis, For. Fl., 86 ; Gamble, Man. Timb., 89 ; Rev. A. Campbell, Rep., Ec. Pl., Chutia Nagpur, No. 8706 ; Lace, Quetta, Pl., Journ. Linn. Soc., xxviii., 309; Atkinson, Him. Dist., 307 ; Royle, Ill. Him. Bot., 168 ; Settle. Rep., Lahore, 14.*

Habitat.—A thorny shrub of the Temperate Himálaya, from the Indus to the Ganges, between 2,000 to 6,000 feet ; also found in Hazára. Mr. Campbell includes it in his list of Economic Products of Chutia Nagpúr.

FOOD.
Fruit.
271
DOMESTIC.
272
273

Food.—The FRUIT ripens in the cold season, and is eaten.

Domestic.—It is largely used in dry-fencing in Chutia Nagpúr, where it is believed not to rot so quickly as the other species of **Zizyphus.**

Z. rugosa, *Lamk.; Fl. Br. Ind., I., 636 ; Wight, Ic., t. 339.*

Syn.—Z. BURRÆA, *Ham. ;* Z. GLABRA, LAT'FOLIA, & TOMENTOSA, *Roxb. ;* Z. OBLIQUA, *Heyne ;* Z. PANICULATA, *Roth. ;* RHAMNUS GLABRATUS, *Heyne.*

Vern.—*Tshirka*, KOL. ; *Sekra*, SANTAL ; *Bogri*, RAJBANSHI ; *Rukh baer, harray baer*, NEPAL ; *Dhaura, dhauri*, OUDH ; *Kataila*, KHARWAR ; *Churni*, MELGHAT ; *Suran, churna*, C.P. ; *Turan, toran, torne*, BOMB. ; *Suran*, MAR. ; *Swarm*, NILGHIRIS ; *Myauksi*, BURM. ; *Maha-erraminsa*, SING.

References.—*Roxb., Fl. Ind., Ed. C.B.C., 204, 205, 206 ; Brandis, For. Fl., 89 ; Kurz, For. Fl. Burm., I., 265, 267 ; Gamble, Man. Timb., 90 ; Thwaites, En. Ceyl. Pl., 73 ; Dals. & Gibs., Bomb. Fl., 49; Rev. A. Campbell, Rep. Ec. Pl., Chutia Nagpur, 8433 ; Dymock, Mat. Med. W. Ind., 2nd Ed., 181, 886 ; Dymock, Warden, & Hooper, Pharmacog. Ind., I., 351 ; Atkinson, Him. Dist., 541 ; Lisboa, U. Pl. Bomb., 149 ; Aplin, Rep. on the Shan States, 1887-88 ; Gazetteer, Bombay, XIII., 27 ; XV., 430 ; XVI., 18 ; Ind. Forester, VIII., 133 ; XIII., 120 ; XIV., 159.*

Habitat.—A large, evergreen, scrambling shrub or small tree, of the Sub-Himálayan tract from Kumáon eastwards, Behar, Assam, Sylhet, Burma, Central and South India, and Ceylon.

GUM.
274
MEDICINE.
Bark.
275
Flowers.
276
FOOD.
Fruit.
277
TIMBER.
278
DOMESTIC.
Branches.
279

Gum.—Said to yield a gum, of which no information is obtainable.

Medicine.—The BARK, powdered and mixed with *ghí*, is applied to the swelling in the cheek caused by toothache, and is also given for ulcers in the mouth (*Rev. A. Campbell, Ec. Prod., Chutia Nagpur*). In Bombay it is used as an astringent in diarrhœa, and the FLOWERS with an equal quantity of the petioles of betel leaf, and half as much lime, are given in 4-grain pills twice a day for menorrhagia (*Pharmacog. Ind.*).

Food.—The FRUIT is eaten by Natives, in all localities in which the plant grows. In Bombay it is said to be a "great support to the people of the *Gháts* from March to the middle of May" (*Lisboa*), and was much eaten in the Poona district during the famine of 1877-78. It has a peculiar mawkish flavour.

Structure of the Wood.—Reddish, moderately hard, warps, readily attacked by insects ; weight 45℔ per cubic foot. It is only valuable for fuel.

Domestic.—The BRANCHES, like those of the other species, are used to make heaped-up fences.

Z. 279

The Common Jujube.	(*J. Murray.*)	**ZIZYPHUS vulgaris.**

Zizyphus vulgaris, *Lamk.; Fl. Br. Ind., I., 633.*   **280**

THE COMMON JUJUBE ; JUJUBIER, *Fr.*

Syn.—Z. FLEXUOSA, *Wall.* ; Z. NITIDA, *Roxb.* ; Z. SATIVA, *Gærtn.* ; *? Z.* SINENSIS, *Lamk.*

Vern.—*Titni* or *pitni-bér, bér, handika, kandiári, singli, simli, ban,* HIND.; *Sinjili, kandiari, bér, kúl, khalis, ghuter, bheri,* N.-W. P. ; *Phitni, konkan ber, sinjli, simli, bárj, bán,* KASHMIR ; *Phitni, kokan bér, ganyeri, kándika, kandi ari, barari, shamor, sinjli, amlai, amnia, amni, amrá, imlá, berí, pitni, ber, relnú,* PB. ; *Karkan ber,* PUSHTU ; *Ber, anab,* SIND ; *Unnáb, rán-bor,* BOMB. ; *Unnáh,* ARAB ; *Sinjid-i-jiláni, kunár,* PERS.

References.—*Roxb., Fl. Ind., Ed. C.B.C., 204, 205 ; Brandis, For. Fl., 85 ; Gamble, Man. Timb., 88 ; Stewart, Pb. Pl., 42, 44 ; Aitchison, Rept. Pl. Coll. Afgh. Del. Com., 46 ; DC., Orig. Cult. Pl., 194 ; O'Shaughnessy, Beng. Dispens., 273 ; S. Arjun, Cat. Bomb. Drugs., 31 ; Murray, Pl. & Drugs, Sind, 147 ; Dymock, Mat. Med. W. Ind., 2nd Ed., 180 ; Dymock, Warden, & Hooper, Pharmacog. Ind., I., 350 ; Year-book Pharm., 1874, 624 ; Baden Powell, Pb. Pr., 601, 602 ; Atkinson, Him. Dist. (X., N.-W. P. Gaz.), 307 ; Ec. Prod., N.-W. P., Pt. V., 44, 54 ; Stocks, Rep. on Sind ; P. W. Dept., Rept. on Gums & Resins, 36, 50 ; Cooke, Gums & Resins, 28 ; Settlement Reports:—Panjáb, Montgomery, 17 ; Hasára, 94; N.-W. P., Shahjehanpur, ix.; Gazetteers:—Bannu, 23 ; Déra Ismail Khán, 19 ; Montgomery, 17 ; Hasára, 133 ; Ind. Forester, II., 175, 407, 408 ; XIV., 390 ; Smith, Ec. Dict., 115, 229.*

Habitat.—A shrub or small tree with rigid spreading boughs and stiff branches, found, wild and cultivated, in the Panjáb up to 6,500 feet, and extending to the North-Western Frontier, occasionally cultivated as far south-east as Bengal (*Fl. Br. Ind.*). DeCandolle expresses the belief that it is not truly wild in India, but that it is simply an escape from cultivation. "It appears to me probable," he writes, "that the species is a native of the north of China ; that it was introduced and became naturalised in the west of Asia after the epoch of the Sanskrit language, perhaps two thousand five hundred or three thousand years ago ; that the Greeks and Romans became acquainted with it at the beginning of our era, and that the latter carried it into Barbary and Spain, where it became partly naturalised by the effect of cultivation." [But it may fairly be asked, is the plant, or was it ever, of such value as to justify an ancient cultivation and introduction into India, where several equally good if not superior species were indigenous ?—*Ed., Dict. Econ. Prod.*]

Gum.—A gum is said to be obtained from this species, and to be used for dyeing, and as a drug. A sample from the Central Provinces, examined by Cooke, was in irregular masses and broken pieces of a dull dark brown colour, with a lustrous fracture, soluble in water, but forming a dark coloured mucilage very like that of coarse dark *Babúl* gum. It was, in fact, very similar in every way to an inferior sample of the latter. The lac insect occasionally lives on this species ; frequently in Sind (*Stocks*).   **GUM. 281**

Medicine.—The dried FRUIT of this species, the Jujube of Arabic and Persian works on Materia Medica, takes the place of the Indian Jujube to a large extent in Northern and Western India. Dymock informs us that it is largely imported into Bombay from the Persian Gulf and China. Muhammadan writers regard it as suppurative, expectorant, and a purifier of the blood. The BARK is employed to clean wounds and sores, the GUM in certain affections of the eyes, and the LEAVES when chewed are said to destroy the power of the tongue to appreciate the taste of disagreeable medicines (*Dymock*). The fruit is used in Europe in the preparation of syrups, confections, and lozenges, which are taken to allay cough.   **MEDICINE. Fruit. 282 Bark. 283 Gum. 284 Leaves. 285**

CHEMICAL COMPOSITION.—See Z. Jujuba.   **CHEMISTRY. 286**

TRADE.—"The Indian market is supplied from China and the Persian Gulf. The Chinese fruit is preferred, as it is larger and sweeter. Value,   **TRADE. 287**

Z. 287

| ZORNIA diphylla. | The Tandi Jhapni. |

Chinese, R8 per Surat maund of 27½ ℔; Arabian, R4 to R5 " (*Pharmacog. Ind.*).

FOOD.
Fruit.
288

Food.—The FRUIT is very similar in every way to that of **Z. Jujuba**, being an oval pulpy drupe about the size of a plum. It varies much, and can be greatly improved by judicious cultivation and grafting. When fresh it is rather acid, but when dried is much sweeter. The small sour fruit of the spontaneous form is also eaten by the poorer classes. The LEAVES are used for fodder.

TIMBER.
289

Structure of the Wood.—Very similar in structure to that of **Z. Jujuba**, and used for the same purposes. In France it is employed for cabinet-work, under the name of *acayou d' Afrique.*

290

Zizyphus xylopyrus, *Willd.; Fl. Br. Ind., I., 634.*

Syn.—Z. CARACUTTA, *Roxb.;* Z. CUNEATA, *Wall.;* Z. ELLIPTICA, *Roxb.;* Z. ORBICULARIS, *Schult;* Z. RUMINATA, *Ham.;* Z. ROTUNDIFOLIA, *Roth.;* RHAMNUS XYLOPYRUS, *Retz.*

Vern.—*Kat-ber, béri, goti, gotáha, kakor, chittania, sitabér, ghóut,* HIND.; *Karkatta,* KOL; *Karkat,* SANTAL; *Got, gotoboro, kanta bohul,* URIYA; *Goit,* BHUMIJ; *Kankor,* KHARWAR; *Katber,* BERAR; *Ghota,* MELGHAT; *Ghato, ghouti,* C. P.; *Ghattól ghotia,* GOND; *Ghunt,* N.-W. P.; *Ghot, súti,* BOMB.; *Goti, bhorgoti, kánte gotti, guti,* MAR.; *Goti, gotte,* TEL.; *Challe, mullu káre,* KAN.

References.—*Roxb., Fl. Ind., Ed. C.B.C., 205, 206; Brandis, For. Fl., 90; Beddome, Fl. Sylv., Anal. Gen., lxviii.; Gamble, Man. Timb., 90; Thwaites, En. Ceyl. Pl., 74; Dalz. & Gibs., Bomb. Fl., 49; Rev. A. Campbell, Rep. Ec. Pl., Chutia Nagpur, No. 7881; Elliot, Fl. Adhr., 63; Atkinson, Him. Dist., 307; Drury, U. Pl., 459; Lisboa, U. Pl. Bomb., 50, 242, 278; Birdwood, Bomb. Pr., 342; Buck, Dyes & Tans, N.-W. P., 85; Liotard, Dyes, 33, 36; Settle. Rep., Seonee, 10; Gazetteers:— Mysore & Coorg, I., 50; Bombay, XIII., 24; XV., 79; N.-W. P., I., 80; IV., lxx.; Agri.-Horti. Soc. Ind., Trans., VI., 48; Jour. (New Series), VI., Sel., 18; Ind. Forester, I., 274; III., 201; IV., 228, 233, 318; VIII., 417; IX., 401; X., 222; XII., App., 10; XIII., 120.*

Habitat.—A large, straggling shrub or small tree, found in North-West India, Nepál, Banda, Rájputána, and Oudh, ascending the Himálaya to 2,000 feet; also in Behar, the Western Peninsula from the Konkan southwards, and in Ceylon.

TAN.
Berry.
291
Bark.
292

Tan.—"The BERRY contains a considerable amount of tannin; the BARK is also used in Bundelkhand as a tanning agent in company with the leaves of the *dha* shrub," Woodfordia floribunda, *Salisb.* (*Sir E. C. Buck*). They are similarly employed in Chutia Nagpur, Bombay, and other parts of India.

FOOD & FODDER.
Kernels.
293
Shoots.
294
Leaves.
295
Fruit.
296

Food & Fodder.—The pulp of the fruit is not eatable, but the KERNELS, which taste like filberts, are eaten by Natives (*Roxburgh*). The young SHOOTS, LEAVES, and FRUIT are eaten by cattle and goats.

TIMBER.
297

Structure of the Wood.—Yellowish-brown, hard, tough, heart and sap-wood not distinct; weight 60℔ per cubic foot (*Skinner*), 49℔ (*Gamble*). It is durable and easily worked; used for cart-building and making agricultural implements, and for torches.

DOMESTIC.
Bark.
298
Charred Fruit.
299

Domestic.—The BARK and CHARRED FRUIT, especially the latter, are largely employed in making a blacking, or black-dye for leather.

ZORNIA, *Gmel.; Gen. Pl., I., 518.*

300

Zornia diphylla, *Pers.; Fl. Br. Ind., II., 147.*

Syn.—Z. ANGUSTIFOLIUM, *Smith;* Z. DICTYOCARPA, *DC.;* Z. GIBBOSA & GRAMINEA, *Spanoghe;* HEDYSARUM DIPHYLLUM, *Linn.*
Var. zeylonensis=Z. ZEYLONENSIS, *Pers.;* Z. CONJUGATA, *Smith;* HEDYSARUM CONJUGATUM, *Willd.*
Var. Walkeri, *Arn. Pug. (sp).*

Z. 300

The Alethi.	(*J. Murray*.)	**ZYGOPHILLUM** simplex.

Vern.—*Tandi jhapni, bir môch,* SANTAL ; *Nelam mari,* MALAY.

References.—*Roxb., Fl. Ind., Ed. C.B.C., 576 ; Dals. & Gibs., Bomb. Fl., 62 ; Thwaites, En. Ceyl. Pl., 85 ; Campbell, Ec. Prod., Chutia Nagpur, Nos. 7844, 8228, 8738 ; Rheede, Hort. Mal., IX., t. 82 ; Gazetteers :— N.-W. P., IV., lxx ; X., 308 ; Bombay, XV., 432.*

Habitat.—A very common annual throughout the plains of India from the Himálaya to Ceylon and Burma, ascending to 4,000 feet in Kumáon.

Medicine.—" The ROOT is given, along with that of *bhadar jhapni,* to induce sleep in children. These plants shutting up their leaves at night have probably suggested the idea to the *Ojhas*" (*Campbell, Ec. Prod., Chutia Nagpur*).

MEDICINE.
Root.
301

ZYGOPHILLUM, *Linn. ; Gen. Pl., I., 266, 988.*

Zygophillum simplex, *Linn. ; Fl. Br. Ind., I., 424 ;* ZYGOPHYLLEÆ.

302

Vern.—*Alethi,* PB. ; *Aletthi, putlani,* SIND.

References.—*Stewart, Pb. Pl., 38 ; Murray, Pl. & Drugs, Sind, 92.*

Habitat.—A prostrate, much-branched herb of the arid, sandy tracts in Sind and the Panjáb.

Medicine.—The Arabs beat up the LEAVES in water and apply the infusion to the eyes in ophthalmia, etc.

Food & Fodder.—The SEEDS are swept up from the ground by the nomad tribes of the Panjáb and Sind deserts and used as food under the above name. **Stocks** states that camels are very fond of the PLANT, and eat it greedily ; but it is said to have such an offensive odour that no other animal will touch it.

MEDICINE.
Leaves.
303
FOOD & FODDER.
Seeds.
304
Plant.
305

A

DICTIONARY

OF

THE ECONOMIC PRODUCTS OF INDIA.

BY

GEORGE WATT, M.B., C.M., C.I.E.,

REPORTER ON ECONOMIS PRODUCTS WITH THE GOVERNMENT OF INDIA; OFFICIER
D'ACADEMIE; FELLOW OF THE LINNEAN SOCIETY; CORRESPONDING MEMBER
OF THE ROYAL HORTICULTURAL SOCIETY, ETC., ETC.

(ASSISTED BY NUMEROUS CONTRIBUTORS.)

INDEX.

PREPARED BY EDGAR THURSTON, SUPERINTENDENT, GOVERNMENT MUSEUM, MADRAS;
ASSISTED BY T. N. MUKERJI, F.L.S., ASSISTANT CURATOR, INDIAN MUSEUM,
CALCUTTA.

Published under the Authority of the Government of India,
Department of Revenue and Agriculture.

CALCUTTA:
PRINTED BY THE SUPERINTENDENT OF GOVERNMENT PRINTING, INDIA.
1896.

CALCUTTA
GOVERNMENT OF INDIA CENTRAL PRINTING OFFICE
8, HASTINGS STREET.

PREFACE.

— ◆ —

THE numerical references are to the series of numbers entered in the
margin of the Dictionary and commence afresh with each letter of the
alphabet.

Attention is drawn to the following erratum slip, which was issued to
possessors of early issues of Volumes III. to V. of the Dictionary: "Attention has been drawn to the fact that an error runs through Volumes III.
IV., and V. of the Dictionary, wherein the consecutive numbers of letters
G. and L., in passing from Volumes III.-IV. and from Volumes IV.-V., respectively, have been partially duplicated. The numbers in those volumes,
which had not been issued from the Press when the error was pointed out,
have been corrected, and will, as corrected, form the future reference numbers. It is suggested, that those who possess uncorrected volumes should
adopt a similar course, making the first reference number in Volume IV.,
G. 381, and in Volume V., L. 379." References to the letters G. and L. in
this index must be taken subject to the above erratum.

ENGLISH

INCLUDING

CLASSICAL OR FOREIGN NAMES

IN FAMILIAR USE.

Abor Vitæ . . . . . B. 518
Acacia, N. American Locust
    (Robinia) . . R. 492
———, Soap (Acacia) . . A. 200
Acalypha, Birch-leaved . . A. 304
Aconite, Atees . . . A. 401
———, Indian . . . A. 397
———, Monks'-hood . . A. 413
———, Wolves'-bane . . A. 413
Adam's Needle . . . Y. 1
Agar Agar . . . G. 653, I. 493
Agate . . . . . C. 617
Agrimony . . . . E. 492
Ajava seeds . . . . C. 691
Alabaster . . . . G. 769
Albumen . . . . A. 727
Alcohol . . . . A. 729
Alder, English . . . A. 791
———, Nepal . . . A. 797
Algaroba . . . . A. 743
Algarobilla . . . A. 743, P. 1258
Alkalis . . . . A. 759
Allspice . . . . A. 789
Almond (Prunus) . . . P. 1274
——— Indian (Terminalia) . T. 312
———, Java (Canarium) . C. 279
Aloe (Aloe) . . . . A. 815
———, American (Agave) . A. 603
———, Bastard ,, . A. 636
Aloe-wood (Aquilaria) . . A. 1251
Aloes, Barbados . . . A. 829
———, Cape . . . A. 822, 823
———, Indian . . . A. 829
———, Jafferabad . . A. 818
———, Moka . . . A. 824
———, Socortrine . . A. 824
———, Yamani . . . A. 824
Alum . . . . A. 897
Amarant, Globe . . . A. 914
———, Prickly . . . A. 943
Amber . . . . A. 955
Ambergris . . . . A. 596
Ammania, Blistering . . A. 958

Ammoniacum . . . . D. 810
Anime . . . . C. 1786
Anise (Pimpinella) . . P. 727
———, Japanese Sacred (Illicium) . I. 30
———, Star, of China ,, . I. 29
Antelopes . . . . S. 1225
Antimony, Black . . . A. 1224
Antlers . . . . H. 408
Apple (Pyrus) . . . P. 1463
———, Adam's (Citrus) . C. 1270
———, Custard (Anona) . A. 1158, 1166
———, Elephant (Feronia) . F. 53
———, Love (Lycopersicum) . L. 596
———, Malay (Eugenia) . F. 444
———, May (Podophyllum) . P. 1009
———, Otaheite (Spondias) . S. 2644
———, Paradise (Citrus) . C. 1263
———, Pine (Ananas) . . A. 1045
———, Rose (Eugenia) . E. 432
———, Star (Chrysophyllum) . C. 1050
———, Sugar (Anona) . . A. 1166
———, Thorn (Datura) . D. 166
———, Wood (Feronia) . F. 53
Apricot . . . . P. 1285
Arar . . . . C. 142
Areca . . . . A. 1294
Arnatto or Arnotto dye . B. 523
Arnica . . . . A. 1414
Arracacha . . . . A. 1418
Arrow-head . . . . S. 510
Arrowroot, West Indian (Maranta) M. 267
———, Wild or East Indian
    (Curcuma) . . C. 2385
Arsenic, Disulphide, Red, or Native
    Sulphuret . . R. 61
———, White . . . A. 1425
———, Yellow Sulphide . O. 242
Artichoke (Cynara) . . C. 2556
———, Jerusalem (Helianthus) H. 88
Arum, Egyptian . . . C. 1732
Asarabacca . . . . A. 1545
Asarabica . . . . A. 1545
Ash, Common (Fraxinus) . F. 685

B

Iron Wood (Memecylon)	M. 439	Leather	L. 156
——— (Xylia)	X. 16	Leeches	L. 245
Isinglass	A. 393, I. 490	Leek	A. 775
Ivory	E. 148	———, Stone	A. 773
Ivy	H. 52	Lemon	C. 1286
Jackals	D. 734	———, Sweet	C. 1304
Jack fruit Tree	A. 1489	Lentil	L. 252
——————— (India rubber)	I. 58	Leopards	T. 423
Jade	J. 1	Lettuce	L. 21
Jalap	I. 401	———, Strong-scented	L. 26
———, Indian	I. 415	Lichens	L. 332
Jarosse	L. 100	Lignite	L. 345
Jasmine, Arabian (Jasminum)	J. 35	Lilac, Persian	M. 393
———, Cape (Gardenia)	G. 111	Lily, Snake (Arisæma)	A. 1378
———, Red (Ipomæa)	I. 405	———, Water, White (Nymphæa)	N. 192
———, Spanish (Jasminum)	J. 18	Lime, Sour (Citrus)	C. 1296
Jasper	C. 620	———, Sweet „	C. 1301
Jesuit's Bark (Cinchona)	C. 1115	———, Wild (Atalantia)	A. 1601
Job's Tears	C. 1686	Lime, Carbonate	C. 489
Jujube, Common	Z. 280	Limestone	C. 489
———, Indian	Z. 231	Linseed	L. 412
Jungle-Fowl	P. 536	Liquidambar	L. 455
Juniper	J. 78	Liquorice (Glycyrrhiza)	G. 278
———, Weeping Blue	J. 104	———, Indian (Abrus)	A. 51
Jute (Corchorus)	C. 1847, J. 123	———, Wild „	A. 51
———, American (Abutilon)	A. 82	Litchi	N. 68
Kaolin	K. 25	Lithographic Stones	L. 459
Kapok Fibre (Bombax)	B. 641	Liverworts	L. 496
———, Floss (Eriodendron)	E. 289	Locust Tree	C. 933
Kavika Tree	E. 444	Lode Tree	S. 3062
Keersal	A. 196	Logwood	H. 1
Kerosine	P. 451	Longan	N. 72
Khaki	K. 26	Looking-glass Plant	H. 137
Khersal	A. 196	Loquat	E. 285
Khus-khus	A. 1097	Lotos (Zizyphus)	Z. 254
Kino, Bengal (Butea)	B. 944	Lotus, Sacred (Nymphæa)	N. 39
———, Indian (Pterocarpus)	P. 1370	Lovage	C. 691
Knol-kohl	K. 29	Love Lies Bleeding	A. 921
Kohl-rabi	K. 29	Lucerne, Purple	M. 334
Kokam	G. 36	———, Yellow	M. 331
Koosa	A. 1097	Lucrubau Seeds	G. 761
Kopeh	C. 1732	Lukrabo Seeds	G. 761
Krameria	K. 38	Lupin	L. 578
Laburnum, Himálayan (Sophora)	S. 2376	Mace (Myristica)	M. 885
———, Indian (Cassia)	C. 756	———, Reed (Typha)	T. 864
Lac	L. 1	Madder, European (Rubia)	R. 580
——— Insect (Coccus)	C. 1491	———, Indian (Oldenlandia)	O. 137
——— Tree of Kosumba, (Schleichera)	S. 950	———, „ (Rubia)	R. 564
Lacquer Industry, Japan	R. 334	Magnesia	M. 53
Lapis Lazuli	L. 75	Magnolia, Red (Magnolia)	M. 49
Larch, Himálayan	L. 82	———, White (Michelia)	M. 535
Laterite	L. 94	Mahogany (Swietenia)	S. 3029
Laurel, Alexandrian	C. 146	———, Indian (Cedrela)	C. 838
Lavender	L. 121	Mahuá Tree	B. 220
Laver, Purple	P. 1176	———, of Southern India	B. 265
Lead	L. 143	Maiden-hair	A. 498
		Maize	Z. 50

Malloes, Rose	. . .	L. 455
Mallow, Common (Malva)	.	M. 115
———, Country (Abutilon)	.	A. 80
———, Indian ,,	.	A. 82
———, Jew's (Corchorus)	.	C. 1861
———, Marsh (Althæa)	.	A. 880
———, Musk (Hibiscus)	.	H. 168
Malt Liquors	. . .	M. 89
Mandrake, American (Podophyl-		
lum)	. .	P. 1009
———— Officinal (Mandragora)	M. 128
Manganese .	. .	M. 131
Mango (Mangifera)	. .	M. 147
——— Ginger (Curcuma)	.	C. 2381
Mangosteen	. .	G. 55
Mangrove (Bruguiera)	. .	B. 898
——— (Ceriops)	.	C. 964
——— (Rhizophora)	.	R. 242
———, White (Avicennia)	.	A. 1655
Mangrove Barks .	.	M. 214
Manioc	. .	M. 216
Manna	. . .	M. 235
——— (Fraxinus)	. .	F. 697
———, Bamboo	. .	B. 84
———, Madar (Calotropis)	.	C. 187
———, Persian (Alhagi)	.	A. 745
Manures .	. .	M. 237
Maple, Indian	. .	A. 328
——— Sugar	. .	A. 326
——— Timber	. .	A. 329
Marble	. C. 489, M. 277
——— Wood, Andamanese	.	D. 603
Margosa Tree	. .	M. 363
Marigold (Calendula) .	.	C. 117
———, African (Tagetes)	.	T. 17
———, French ,,	.	T. 17
———, Marsh (Caltha)	.	C. 198
Marjoram .	. .	O. 220
———, Sweet	. .	O. 214
Marking-nut Tree	. .	S. 1041
Marmots	. .	R. 47
Marvel of Peru	. .	M. 606
Mast Tree .	. .	P. 1052
Mastic Tree .	. .	P. 841
Mats and Matting .	.	M. 319
——————, Sedges used	C. 2583
Matt, Cobaltiferous	.	M. 318
Medlar, Japan	. .	E. 285
Meerschaum	. .	M. 338
Melon, Musk (Cucumis)	.	C. 2263
——— (Cucurbita)	.	C. 2325
———, Sweet (Cucumis)	.	C. 2263
———, Water (Citrullus)	.	C. 1221
———, White Gourd (Benincasa)	B. 430
Mercury	. .	M. 472
Mezereon	. .	D. 124
Mica	. . .	M. 509

Mice	. . .	R. 47
Midnapur Clove-scented Creeper	R. 487
Mignonette	. .	R. 140
Milfoil	. .	A. 367
Milk-bush	. .	E. 553
Milk-hedge .	.	E. 553
Milk Thistle .	.	S. 2357
Milkwort, Common Indian	.	P. 1062
Millet, Bulrush (Pennisetum)	.	P. 384
———, Common (Panicum)	.	P. 63
———, Cumboo (Pennisetum)	.	P. 384
———, Great (Sorghum)	.	S. 2424
———, Indian ,,	.	S. 2424
———, Italian (Setaria)	.	S. 1212
———, Juar (Sorghum)	.	S. 2424
———, Kodo (Paspalum)	.	P. 332
———, Little (Panicum)	.	P. 67
———, Quick-Growing (Panicum)	P. 53
———, Sorghum (Sorghum)	.	S. 2380
———, Spiked (Pennisetum)	.	P. 384
Millets	. .	M. 548
Millstones	. .	M. 553
Mimusops, Obtuse-Leaved	.	M. 590
Mint, Marsh	. .	M. 447
——, Spear	.	M. 465
——, Pepper	.	M. 453
Mishmi Teeta	. .	C. 1789
Mistletoe	. .	V. 149
Mochi Wood	. .	E. 342
Moles	. .	R. 47
Molybdenum	. .	M. 620
Mongoose Plant	. .	O. 180
Monkey Bread Tree (Adansonia)	A. 455
——— Face Tree (Mallotus)	.	M. 71
Monkey's Horn	. .	C. 480
Monkeys	. .	M. 646
Monk's-hood	. .	A. 413
Moon-flower	. .	I. 368
Moon of the Faithful	.	P. 1285
Moon Plant	. .	S. 882
Morell	. .	M. 647
Moss, Ceylon (Gracilaria)	.	G. 653
——, Club (Lycopodium)	.	L. 598
——, Iceland (Cetraria)	.	C. 985
——, Jaffna (Gracilaria)	.	G. 653
Mowa Tree	. .	B. 265
Mucilage, Plants yielding	.	M. 779
Mulberry, Indian (Morus)	.	M. 756
———, ,, (Morinda)	.	M. 656
———, Paper (Broussonetia)	B. 883
———, White (Morus)	.	M. 747
Mules	. .	H. 414
Muntjac	. .	D. 219
Mushroom	. A. 590, M. 808
Musk and Musk Substitutes	.	M. 852
—— Mallow	. .	H. 168
Mustard, Black (Brassica)	.	B. 841
———, Indian ,,	.	B. 833

Vomit-weed	E. 542	Wool	S. 1379, 1618
Wall-flower	C. 997	Worm-seed . . .	A. 1452
Walnut (Juglans) . . .	J. 61	———, Levant . .	A. 1452
———, Belgaum (Aleurites)	A. 737	Wormwood . . .	A. 1441
———, Indian „ .	A. 737	———, Indian . .	A. 1469
Waras	F. 638	Wrack, Bladder . .	F. 709
Water Cress . . .	N. 28	———, Knobbed Sea .	F. 708
Wattle, Black . . .	A. 212	Yaka	P. 1
———, Green . . .	A. 271	Yam, Bulb-bearing .	D. 494
———, Golden . . .	A. 271	———, Chinese . .	D. 510
———, Silver . . .	A. 209	———, Common .	D. 513, 534
Wattle Bark . . .	W. 21	———, Dark Purple .	D. 490
Wax, Bees . . .	H. 341	———, Guinea . .	D. 481
———, Japan (Rhus) . .	R. 325	———, Kawan of Fiji .	D. 522
Wayaka	P. 1	———, Kidney-shaped .	D. 507
Weld	R. 138	———, Malacca . .	D. 490
Whales	W. 41	———, Prickly stemmed .	D. 481
Wheat	T. 634	———, Purple . .	D. 526
Whinstone . . .	T. 515	———, Rangoon . .	D. 490
Wicker-work, Plants used .	B. 210	———, Tivoli . .	D. 515
Willow, Crack (Salix) .	S. 574	———, Wing-stalked .	D. 484
———, Huntingdon (Salix)	S. 537	Yarrow . . .	A. 367
———, Jerusalem (Elæagnus)	E. 40	Yeast . . .	C. 954
———, Red Wood (Salix) .	S. 574	Yellow Dye (Peori) .	P. 401
———, Weeping „ .	S. 544	Yew	T. 93
———, „ , Indian (Salix)	S. 570	Zedoary, Black . .	C. 2422
———, White (Salix) . .	S. 537	———, Long . .	C. 2499
Wine	V. 243	———, Round . .	C. 2499
Wolves	D. 734	———, Yellow . .	C. 2406
		Zinc	Z. 190

SANSKRIT, ARABIC, PERSIAN.

Aa Aainu-ddík, *Arab.*, A. 51; aaimunnás, *Arab.*, *Pers.*, A. 1045; aaknak, *Pers.*, C. 1703; aalaq, *Arab.*, L. 245; aanabahehindi, *Arab.*, *Pers.*, C. 581; aansale-hindí, *Arab.*, U. 39; aáqarqarhá, *Arab.*, A. 1026; aáqúl, *Arab.*, A. 745; aasl, *Arab.*, H. 342; aaslum-nahal, *Arab.*, H. 342; aazbah, *Arab.*, T. 51, 70.

Ab Abab, *Arab.*, J. 52; abhal, *Arab.*, J. 78; abhayá, *Sans.*, T. 325; abhra, *Sans.*, M. 509; abnús, *Arab.*, D. 615; abnúse-hindí, *Arab.*, *Pers.*, D. 582; abqar, *Arab.*, S. 682; abraka, *Sans.*, S. 2712; abr-i-amber, *Pers.*, A. 956; abruz, *Arab.*, T. 460; abunom, *Arab.*, P. 87.

Ac Achchhuka, *Sans.*, M. 704.

Ad Adas (Lens), *Arab.*, *Pers.*, L. 252; adas (Phaseolus), *Pers.*, P. 468; adhakí tubariká, *Sans.*, C. 49; adhopushpí, *Sans.*, C. 2569; aditya bhakta, *Sans.*, C. 1367.

Af Afiún, *Arab.*, *Pers.*, P. 87; afkar, *Arab.*, C. 913; afsantin (Artemisia), *Arab.*, *Pers.*, A. 1441, 1467; afsantín (Grangea), *Arab.*, G. 660; afsantíne-hindí, *Arab.*, *Pers.*, A. 1469; afsantín-ul-bahr, *Arab.*, A. 1452, A. 1460; áftábi, *Pers.*, H. 74.

Ag Agar, *Pers.*, A. 1251; agare-hindí, *Arab.*, A. 1251; agarikún, *Arab.*, A. 597; agaru, *Sans.*, A. 1251; agaru-gandha-káshtaha, *Sans.*, P. 1381; agasti vaka vranári, *Sans.*, S. 1186; agati, *Sans.*, S. 1186; agha-lúkhí, *Arab.*, A. 1252; ágháta, *Sans.*, A. 382; agnijvála, *Sans.*, W. 106; agnimantha, *Sans.*, P. 1233; agnimatsya, *Sans.*, C. 2237; agnishikha, *Sans.*, P. 986; agnisikhá, *Sans.*, G. 243; agre-hindí, *Pers.*, A. 1251; agre-turkí, *Pers.*, A. 430; agri-turki, *Pers.*, C. 225.

Ah Áhak, *Pers.*, C. 489; ahan, *Pers.*, I. 440; ahar, *Pers.*, W. 122; ahareshírín, *Pers.*, W. 122; ahiphena, *Sans.*, P. 87; ahmur, *Arab.*, C. 448; ahu, *Pers.*, S. 1255; ahu, *Arab.*, C. 489.

Ai Ainab, *Arab.*, V. 233.

Aj Ajájí, *Sans.*, C. 2339; ajaka, *Sans.*, O. 31; ajmódum, *Sans.*, C. 691.

Ak Akarákarabba, *Sans.*, A. 1026; akara-karava, *Sans.*, A. 1026; akasha, *Sans.*, T. 293; akásvallí, *Sans.*, C. 805; ákhóda, *Sans.*, J. 61; ákhóta, *Sans.*, J. 61; akik, *Arab.*, C. 617; akitmakit, *Arab.*, C. 6; akrí, *Pers.*, C. 225; akshota, *Sans.*, A. 737; akshota áks-chóda, *Sans.*, J. 61.

Al Alábu, *Sans.*, L. 30; alarka (Calotropis), *Sans.*, C. 170, C. 191; alarka (Solanum), *Sans.*, S. 2315; alfafa, *Arab.*, M. 334; alfalfa, *Arab.*, M. 334; altasafat, *Arab.*, M. 334; alhaju, *Arab.*, A. 745, C. 224; almás, *Arab.*, *Pers.*, D. 364; alte, *Arab.*, E. 232; alu, *Pers.*, P. 1304; alu-ba-lu, *Pers.*, P. 1297; alu-bu-ali, *Pers.*, P. 1297.

Am Ámala, *Sans.*, P. 632; ámalaki, *Sans.*, P. 632; amaravela, *Sans.*, C. 2508; amba, *Pers.*, M. 147; amba-hindí, *Arab.*, *Pers.*, C. 581; ambar-báris, *Arab.*, B. 443, B. 458, B. 465; ambashta, *Sans.*, O. 547; ambashthá, *Sans.*, S. 2794; ambashthái páthá, *Sans.*, C. 1205; ambhoruna, *Sans.*, N. 39; ambia, *Sans.*, T. 28; ambu-prasáda, *Sans.*, S. 2960; ambuvetasa, *Sans.*, C. 77; ámelah, *Pers.*, P. 632; amlaj, *Arab.*, P. 632; amlaloniká, *Sans.*, O. 547; amlavetasa, *Sans.*, R. 650; amlíka (Oxalis), *Sans.*, O. 547; amliká (Tamarindus), *Sans.*, T. 28; amra, *Sans.*, M. 147; amrátaka, *Sans.*, S. 2649; amrita, *Sans.*, T. 470; amrita, *Sans.*, T. 483; amritaphala, *Sans.*, P. 1452; amrúd (Psidium), *Arab.*, *Pers.*, P. 1343; amrúd (Pyrus), *Arab.*, P. 1452; amruta-phalam, *Sans.*, P. 1343; amuleh, *Pers.*, P. 632; amulki, *Sans.*, P. 632; amurta, *Sans.*, T. 470; amurta, *Sans.*, T. 483.

An Anab-us-salab, *Pers.*, S. 2270; anantá, *Sans.*, H. 119; anar, *Pers.*, P. 1426; anbaghól, *Arab.*, T. 600; anbalah, *Pers.*, T. 28; anbar-báris, *Arab.*, B. 443; anbé-haldi, *Pers.*, C. 2412; an-bus-sa'lap, *Arab.*, S. 2299; anesun, *Arab.*, P. 727; angabín, *Pers.*, H. 342; angaraha (carbon), *Sans.*, C. 487; angaraha (coal), *Sans.*, C. 1414; angarœ-hindi, *Pers.*, H. 227; angharœ-hindi, *Arab.*, H. 227; angúr, *Pers.*, V. 233; anguza, *Pers.*, F. 76; anguzeh, *Pers.*, F. 76; anísún, *Arab.*, P. 727; anjana, *Sans.*, L. 143; anjanam, *Sans.*, A. 1224; anjir, *Pers.*, F. 149; anjíra, *Sans.*, F. 149; anjir-dashte, *Pers.*, F. 202; anjubár-i-rumi, *Pers.*, P. 1112; ankota, *Sans.*, A. 681; anu, *Sans.*, P. 63; anúk, *Arab.*, L. 143; anzarút, *Arab.*, A. 1592; anzrút, *Arab.*, A. 1591.

Ap Apámárga, *Sans.*, A. 382; apángaka, *Sans.*, A. 382; aparájitá, *Sans.*, C. 1403.

Aq Áquarqarhá, *Arab.*, A. 1026.

Ar Aragbadha, *Sans.*, C. 756; arah, *Arab.*, P. 841; arák, *Arab.*, S. 705; arák, *Arab.*, S. 717; aralu, *Sans.*, A. 658; aravin, *Sans.*, N. 39; árdraka, *Sans.*, Z. 201; árdrakam, *Sans.*, Z. 201; arena, *Pers.*, S. 771; arishta (Melia), *Sans.*, M. 363; arishta (Xanthium), *Sans.*, X. 1; arishta phalam (Sapindus), *Sans.*, S. 818; arit-tamunjayrie, *Sans.*, A. 306; arjuna, *Sans.*, L. 42; árjuna, *Sans.* T. 282; árka, *Sans.*, C. 170; arkamulá, *Sans.*, A. 1398; arkapushpiká, *Sans.*, G. 753; arka-vallabha, *Sans.*, P. 393; arminá, *Arab.*, A. 962; arruz, *Arab.*, O. 258; arsaghna, *Sans.*, A. 996; arsániyún, *Arab.*, O. 242; artániyáe-hindi, *Arab.*, H. 486; arú, *Pers.*, P. 1322; aruda, *Arab.*, R. 663; arukanla, *Sans.*, C. 2381; arúna-chitraca, *Sans.*, P. 979; arus, *Sans.*, A. 484;

C

arushkara, *Sans.*, S. 1041; arzan, *Pers.*,
P. 63; arzun, *Pers.*, S. 1212.

As As, *Pers.*, M. 921; asaba-ul-feteyat, *Arab.*,
O. 18; asáeaul-malik, *Arab.*, A. 1585;
asana, *Sans.*, T. 361; asárún, *Arab.*, A.
1545; asbáràghús, *Arab.*, A. 1567; ásbirí,
Pers., M. 921; asfidáj, *Arab.*, L. 143;
asfrak, *Pers.*, D. 271; ashur, *Arab.*, C.
170; ashwa gandha, *Sans.*, W. 98; asl,
Arab., T. 70; asla, *Arab.*, V. 181; aslag,
Arab., V. 164; aslak-asvad, *Arab.*, J.
116; asle-lubní (Altingia), *Pers.*, A. 892;
asle-lubní (liquidambar), *Pers.*, L. 455;
aslul armar, *Arab.*, T. 51; aslussús,
Arab., G. 278; asma-righna, *Sans.*, C.
2039; asoka, *Sans.*, S. 861; asp, *Pers.*,
H. 414; asperag, *Pers.*, D. 271; asphota
(Clitoria), *Sans.*, C. 1403; asphota (Jasminum), *Sans.*, J. 10, J. 35; as-sukkar,
Arab., S. 375; asthisanhara, *Sans.*, V.
219; astmabayda, *Sans.*, A. 554; asu,
Sans., H. 414; asus, *Pers.*, C. 278;
asvagandhá, *Sans.*, W. 98; asvakarna,
Sans., S. 1656; asvamáraca, *Sans.*, N.
80; asvattha, *Sans.*, F. 236; aswa, *Sans.*,
H. 414; aswaththamu, *Sans.*, F. 236.

At Át, *Arab.*, A. 1166; ataicha, *Sans.*, A. 401;
atarílál, *Arab.*, *Pers.*, V. 73; atasí, *Sans.*,
L. 385; átá sítáphal, *Arab.*, A. 1166;
atavi-jambíra, *Sans.*, A. 1601; atavі-
jírakaha, *Sans.*, V. 73; atavi madhuka
vriksha, *Sans.*, B. 220; atibalá, *Sans.*,
S. 1703; atiguha, *Sans.*, U. 23; lativishá,
Sans., A. 401; atkumah, *Arab.*, A. 382;
átmaguptá, *Sans.*, M. 786.

Au Aúd, *Arab.*, A. 1251; aúde-hindí, *Arab.*,
A. 1251; audul, *Arab.*, M. 869; aúdul-
qarhá, *Arab.*, A. 1026; aughriparnika,
Sans., U. 23; aulqum, *Arab.*, C. 1211;
auluk-bag-dadí, *Arab.*, P. 841; auma,
Sans., L. 385; aunnábe-hindí, *Arab.*, Z.
231; aurukesáfur, *Arab.*, C. 2433; ausá-
rahe-mahak, *Pers.*, G. 278; aushâre-
révand, *Arab.*, *Pers.*, G. 66; aushbahe-
hindí, *Pers.*, H. 119; aush-batunnár,
Arab., H. 119.

Av Aval, *Sans.*, V. 73; avalguja, *Sans.*, P.
1352; avi, *Sans.*, S. 1320; avinga, *Sans.*,
C. 596.

Aw Awartani, *Sans.*, H. 92.

Ay Ayam, *Sans.*, I. 440.

Az Ázád-darakhte-hindí, *Pers.*, M. 363; azanul-
fil, *Arab.*, C. 1834; azfur zukkum, *Arab.*,
E. 553; azúrí, *Arab.*, M. 869.

Ba Bábunah (Matricaria), *Pers.*, M. 314;
bábúnah (Anthemis), *Pers.*, A. 1184;
bábunaj, *Arab.*, M. 314; bábúnaj,
Arab., A. 1184; bádám, *Pers.*, P. 1274;
bádáme-hindí, *Pers.*, T. 312; bádámitte
Sans., P. 1274; badangan, *Arab.*, S.
2284; bádanján, *Pers.*, S. 2284; badara,
Sans., Z. 231; badarí, *Sans.*, Z. 231;
bádávard, *Pers.*, F. 2; bádáward, *Pers.*,
V. 279; badawurd, *Pers.*, F. 6; badiáne
huttáic, *Arab.*, I. 30; bádingán, *Pers.*,
S. 2284; badinj, *Pers.*, C. 1520; badin-
jan, *Arab.*, S. 2284; bádiyán (Illicium),
Pers., I. 30; bádíyán (Pimpinella),
Pers., P. 727; badlan, *Arab.*, O. 616;
badrúj, *Arab.*, O. 18; bagalá, *Sans.*,
C. 2316; baglatul-mulk, F. 721, F. 723;
baheruha, *Sans.*, T. 293; baheruka,

Sans., T. 293; bahira, *Sans.*, T. 293;
bahman abiad, *Arab.*, C. 912; bahman-
i-suffaid, *Pers.*, C. 912; bahoola, *Sans.*,
E. 151; bahu-bíja-phalam, *Sans.*, P.
1343; bahuváraca, *Sans.*, C. 1931;
baíbarang, *Arab.*, M. 910; bajri (Panicum), *Pers.*, P. 53; bájrí (Zea), *Pers.*,
Z. 50; bakam, *Pers.*, C. 35; bakam-i-
kirmyz, *Pers.*, C. 35; bakara, *Arab.*,
O. 574; bakila-i-misri, *Pers.*, L. 578;
bakul, *Sans.*, M. 570; balá (Sida),
Sans., S. 1688; balá, *Sans.*, S. 1694;
bálá (Pavonia), *Sans.*, P. 344; balaksh,
Arab., R. 619; balasán, *Arab.*, B. 54;
balela, *Pers.*, T. 293; balílah, *Pers.*, T.
293; balilaj, *Arab.*, T. 293; balva,
Sans., A. 534; bámiyá, *Arab.*, H. 196;
bámiyah, *Pers.*, H. 196; banafsaj, *Arab.*,
V. 143; banafshah, *Pers.*, V. 143; banafshaj, *Arab.*, V. 143; bandhujiva, *Sans.*,
P. 393; bandhújívaka, *Sans.*, I. 513;
bandhuka (Pentapetes), *Sans.*, P. 393;
bandhúka (Ixora), *Sans.*, I. 513; bándinj,
Pers., C. 1520; bandúkamu, *Sans.*, P.
393; bang, *Pers.*, C. 331; banj-angasht,
Pers., V. 164; banj-angashte-abí, *Pers.*,
V. 181; banj-angashte-siyáh, *Pers.*, J.
116; baqam, *Pers.*, C. 35; baqlatul-
aarabbíyah, *Arab.*, P. 1187; baqlatul-
humquá, *Arab.*, P. 1179; baqlatul-
yamániyah, *Arab.*, P. 1187; baqqam,
Arab., C. 35; baran-jásif kowhí, *Pers.*,
G. 660; barbara (Acacia), *Sans.*, A. 101;
barbará (Clerodendron), *Sans.*, C. 1388;
barg-a-banà, *N.-E. Persia*, P. 847; barge-
tanból, *Pers.*, P. 775; bárhang, *Pers.*, P.
926; barhissu, *Sans.*, I. 51; baring, *Arab.*,
M. 910; bártang, *Pers.*, P. 926; barzad,
Arab., F. 95; basbás, *Arab.*, M. 885;
basbásah, *Arab.*, M. 885; basl, *Arab.*,
A. 769; baslul-barre-hindí, *Arab.*, U. 39;
baslul-fáre-hindí, *Arab.*, U. 39; bassunta,
Sans., P. 468; bastaj, *Arab.*, B. 771;
bastítáj, *Arab.*, T. 548; batíkh, *Arab.*,
C. 2263; batilj, *Arab.*, T. 293; batú,
Arab., C. 2192; bátyálaka, *Sans.*, S.
1694; bazbáz, *Pers.*, M. 885; bazre-
katíma, *Arab.*, P. 932; bazr-el-katíf,
Arab., C. 1005; bazre-quatúná, *Arab.*,
P. 932; bazrul-bang, *Pers.*, H. 525;
bazrul-banj, *Arab.*, H. 525; bazrul-baq-
latul-humqá, *Arab.*, P. 1179; bazruli-
karafs, *Arab.*, C. 701; bazrul-mázarí
yúne-hindí, *Arab.*, C. 1403; bazrul-quissá,
Pers., C. 2278; bazrut-kattán, *Arab.*, L.
385.

Be Bed, *Pers.*, C. 104; bedana, *Pers.*, B. 465;
bedánjir, *Pers.*, R. 369; bedanjíre-khataі
(Croton), *Pers.*, C. 2192; bédanjíre-
khatái (Baliospermum), *Pers.*, B. 28;
bede-mushk, *Pers.*, S. 550; begapúra,
Sans., C. 1270; begbu-nuphsha, *Pers.*, I.
423; beh-dánah, *Pers.*, C. 2546; behen
abiad, *Arab.*, C. 912; behussej, *Arab.*,
V. 143; bekh-i-banfsa, *Pers.*, I. 425;
beladin, *Arab.*, S. 1041; belayleh, *Pers.*,
T. 293; béléy-luj, *Arab.*, T. 293; beng,
Pers., C. 335; benú-mash, *Pers.*, P.
513; béséd, *Arab.*, C. 18u8; bey-
khneelufir, *Pers.*, N. 39.

Bh Bhadramunjá, *Sans.*, V. 12; bhadra musta,
Sans., B. 2605; bhadra muste, *Sans.*, C.
2612; bhadrasri, *Sans.*, S. 790; bhadra-

Sanskrit, Arabic, Persian. 19

valli, *Sans.*, V. 12; bhalavaanga, *Sans.*,
J. 114; bhallátaká, *Sans.*, S. 1041;
bhallátamu, *Sans.*, S. 1041; bhan-dákí,
Sans., S. 2284; bhándíra, *Sans.*, C. 1380;
bhánga, *Sans.*, C. 331; bhantaka, *Sans.*,
C. 1380; bhantaki, *Sans.*, S. 2280; bhan-
tákí, *Sans.*, S. 2284; bhanti, *Sans.*, C.
1380; bháradváji, *Sans.*, H. 263; bhárgi,
Sans., C. 1394; bhavya, *Sans.*, D. 428;
bheka-parní, *Sans.*, H. 486; bhringarája,
Sans., W. 25; bhúchampaca, *Sans.*, K.
8; bhúdra, *Sans.*, P. 625; bhúdrá-búla,
Sans., P. 625; bhúdroudúní, *Sans.*, P.
625; bhúmi-champa, *Sans.*, K. 8; bhumi
darimba, *Sans.*, C. 580; bhumijambu,
Sans., P. 1231; bhúmi-jambúka, *Sans.*,
P. 1231; bhumikashmánda, *Sans.*, I.379;
bhumyá-malaki, *Sans.*, P. 657; bhunimba (Andrographis), *Sans.*, A. 1064;
bhunimba (Swertia), *Sans.*, S. 3018;
bhurjaputra, *Sans.*, B. 501; bhúrúndí,
Sans., H. 102; bhústrina, *Sans.*, A. 1079;
bhutakesi, *Sans.*, C. 1980.

Bi Bib-hitaka, *Sans.*, T. 293; biddari, *Sans.*,
G. 298; bihí-dánah, *Pers.*, C. 2546; bihi-
tursh, *Arab.*, C. 2546; bíja, *Sans.*, C.
2192; bíkhe-hayát, *Pers.*, C. 1403; bikhe-
mahak, *Pers.*, G. 278; bíkh-i-banafshah,
Pers., I. 428; biládur, *Pers.*, S. 1041;
bilin, *Sans.*, F. 53; bilva, *Sans.*, A. 534;
bilvaphalam, *Sans.*, A. 534; bimba,
Sans., C. 919; bimbika, *Sans.*, C. 919;
bindak, *Pers.*, C. 1985; bindake hindí,
Pers., S. 818; bin-kúk, *Arab.*, P. 1285;
biranj, *Pers.*, O. 258; bireez, *Pers.*, F.
95; birozeh, *Pers.*, P. 760; bish, *Arab.*,
A. 397; bíshnág, *Pers.*, A. 397; bizrul-
khashkhash, *Arab.*, P. 87.

Bo Bóé-jahúdán, *Pers.*, B. 43; bokel, *Arab.*,
R. 31; ból, *Pers.*, B. 48; bóla, *Sans.*,
B. 48; bóle-síyáh, *Pers.*, A. 824; bóraq,
Arab., B. 731; boz, *Pers.*, S. 1233; bóz-
ghánj, *Pers.*, P. 858; bozpásang, *Pers.*,
S. 1233.

Br Brahmadandi, *Sans.*, A. 1351; brahma-
darbha, *Sans.*, C. 691; brahma yashtiká,
Sans., C. 1394; brahmi, *Sans.*, H. 149;
brahmuní, *Sans.*, C. 1394; brahmunu
yushtika, *Sans.*, C. 1394; brahu maricha,
Sans., C. 455; bran maricha, *Sans.*, C.
455; brihat-upakun-chiká, *Sans.*, A. 976;
brinj-mógrá, *Pers.*, G. 761.

Bu Buchanaka, *Sans.*, A. 1261; buckum, *Pers.*,
P. 1381; budelut-ul-mobarik, *Arab.*, P.
1187; buka, *Sans.*, S. 1186; búkampa-
dáruka, *Sans.*, C. 1931; buklut-ul-gezal,
Arab., O. 220; buklut-ul-kukema, *Arab.*,
P. 1179; buklut-ul-zub, *Arab.*, O. 18;
buks-lat-ul-mulik, *Arab.*, F. 723; bun,
Arab., *Pers.*, C. 1641; bunduk, *Arab.*,
C. 6; búrakes-sághah, *Arab.*, B. 731;
buréh, *Pers.*, B. 731; burr, *Arab.*, T.
634; búrum, *Sans.*, S. 579; buruq-es-
sághah, *Arab.*, B. 731; bustán-afróz,
Pers., A. 939; busteyrúmí, *Arab.*, T.
548; bútankúshum, *Sans.*, A. 1132.

Ca Cadhi, *Arab.*, P. 26; cahwa, *Pers.*, C. 1641;
camala, *Sans.*, N. 39; carnicára, *Sans.*,
P. 338.

Ch Cháe-kashmíri, *Pers.*, A. 1079; chahár-
maghz, *Pers.*, J. 61; chahar-maghze-

hindí, *Pers.*, A. 737; chakramarda,
Sans., C. 799; chalgoza, *Pers.*, C. 1985;
champaka, *Sans.*, M. 517; chanaka,
Sans., C. 1061; chana-kámla, *Sans.*, C.
1061; chandana, *Sans.*, S. 790; chandáta,
Sans., N. 80; chandrá-bha, *Sans.*, C.
257; chandra-mallika, *Sans.*, C. 1043;
chandra-mulika, *Sans.*, K. 3; chandrasura,
Sans., L. 283; chandrika, *Sans.*, L. 283;
changeri, *Sans.*, O. 547; chapola, *Sans.*,
C. 331; chára, *Sans.*, B. 913; cháratí,
Sans., I. 335; charm, *Pers.*, L. 156;
charma, *Sans.*, H. 265; chár-maghz,
Pers., J. 61; charua, *Arab.*, R. 369;
chashmízaj, *Arab.*, *Pers.*, C. 728;
chashmkhurós, *Pers.*, A. 51; chashúm,
Pers., C. 728; chatr-i-mar, *Pers.*, F.
725; chatr-i-már, *Pers.*, A. 590; chat-
trak, *Sans.*, A. 590, F. 725; chaviká,
Sans., P. 797; chayapula, *Sans.*, C.
1221; chennuka, *Sans.*, C. 1061; chesh-
mak, *Pers.*, C. 728; chhá, *Pers.*, *Arab.*,
C. 244; chichin-da, *Sans.*, T. 569; chil-
gozá, *Pers.*, P. 746; chilpásah, *Pers.*,
R. 110; china, *Sans.*, P. 63; chinár,
Pers., P. 935; chini, *Sans.*, P. 63;
chirika, *Sans.*, B. 913; chirke-áhan,
Pers., I. 472; chirpota, *Sans.*, Z. 1;
chitra, *Pers.*, B. 443; chitraka, *Sans.*, P.
979, P. 986; chitraka-vrikshaha, *Sans.*,
P. 986; chittermúl, *Arab.*, P. 979; choba-
chini, *Sans.*, S. 2240; cholza, *Pers.*, P.
1179; chorapushpi, *Sans.*, C. 1053;
c'hosa, *Sans.*, P. 87; chukra, *Sans.*, R.
650; chúkrika, *Sans.*, O. 547; chundrika,
Sans., R. 57; chundruvala, *Sans.*, E. 151;
chupula, *Sans.*, P. 805; chúrna, *Sans.*,
C. 489; chutu, *Sans.*, M. 147; chuve,
Sans., P. 797.

Co Cotter mija, *Arab.*, L. 455.

Cu Cuseshaya, *Sans.*, N. 39; cusso, *Arab.*,
B. 856; cutch, *Pers.*, B. 465.

Da Da, *Sans.*, N. 39; daban-shah, *Pers.*, O.
18; dábk, *Arab.*, C. 1931; dadamardan,
Sans., C. 797; dádamari, *Sans.*, C. 797;
dadhi, *Sans.*, D. 15; dadhiphala, *Sans.*,
F. 53; dádima-phalam, *Sans.*, P. 1426;
dadima-vrikshaha, *Sans.*, P. 1426; dad-
rughna, *Sans.*, C. 732; dadumari, *Sans.*,
X. 23; dahana, *Sans.*, T. 489; dam-el
akwain, *Pers.*, C. 68; dam-eth-thuaban,
Arab., C. 68; dam-et-tinnin, *Arab.*, C.
68; dammul-akhvaine-hindí, *Arab.*, P.
1370; dam-ul-akhwain, *Arab.*, C. 68.
dand, *Arab.*, C. 2192; dandálu, *Sans.*,
D. 485; dande-barrí, *Arab.*, *Pers.*, J. 41;
dande-nahrí, *Arab.*, *Pers.*, J. 41; dan-
dotpala, *Sans.*, C. 382; danie-másha,
Sans., P. 513; dánti, *Sans.*, B. 28;
darákh-i-misvák, *Pers.*, S. 705; darakhte-
moryam, *Pers.*, S. 2649; darakht-i-mis-
vák, *Pers.*, S. 717; dár-al-sída, *Arab.*,
T. 28; darbha balbajamu, *Sans.*, I. 51;
dár-chíní, *Pers.*, C. 1196; dár-chób, *Pers.*,
B. 443; dardár, *Pers.*, M. 473; dár-filfil,
Arab., P. 805; dár-hald, *Arab.*, *Pers.*,
B. 443; dárivká, *Sans.*, H. 1196; dár-
shíshaán, *Pers.*, M. 869; dársíní, *Arab.*,
C. 1196; dárú-haridrakam, *Sans.*, C.
2007; dárulavanam, *Sans.*, C. 527;
dárumuch, *Sans.*, A. 1425; darúnaj-i-

C 2

akrabí, *Pers.*, D. 813 ; darvi, *Sans.*, C. 2007 ; dár-zard, *Pers.*, C. 2433 ; dátún, *Arab.*, C. 2192 ; dauna, *Arab.*, *Pers.*, A. 1467 ; davanamu, *Sans.*, A. 1469.

De Deíshar, *Arab.*, A. 89 ; désha-panjasaram, *Sans.*, S. 373 ; désha-sharkara, *Sans.*, S. 373 ; desha-vádá-mittee, *Sans.*, T. 312 ; devadáru, *Sans.*, C. 846 ; dévdár, *Pers.*, C. 846.

Dh Dhamni, *Sans.*, G. 714 ; dhanvana, *Sans.*, G. 673 ; dhánya, *Sans.*, O. 258 ; dhanyáka, *Sans.*, C. 1954 ; dhányaka, *Sans.*, C. 1954 ; dhárákadamba, *Sans.*, A. 514 ; dharmana, *Sans.*, G. 673, G. 714 ; dhátaki, *Sans.*, W. 106 ; dhátri, *Sans.*, P. 632 ; dhattúra, *Sans.*, D. 151 ; dhátukásis, *Sans.*, I. 481 ; dholasamudrika, *Sans.*, L. 232 ; dhonul-hal, *Arab.*, S. 1078 ; dhonul-jouze-hindí, *Arab.*, C. 1520 ; dhonun-narjíl, *Arab.*, C. 1520 ; dhonu simsim, *Arab.*, S. 1078 ; dhura, *Sans.*, S. 2424 ; dhúra, *Arab.*, S. 2424 ; dhurat, *Arab.*, S. 2424 ; dhustura, *Sans.*, D. 151.

Di Dibki, *Arab.*, V. 149 ; difli, *Arab.*, N. 80 ; difíí, *Arab.*, C. 225 ; dilpasand, *Pers.*, C. 1221 ; dirghapatolika, *Sans.*, L. 569 ; dirghapatra, *Sans.*, Z. 1.

Dj Djyl-djylan, *Arab.*, S. 1078.

Do Dokhu, *Arab.*, P. 63 ; dolá, *Sans.*, I. 145 ; dowálah, *Pers.*, L. 332.

Dr Dráksha, *Sans.*, V. 233 ; draksha-rasa, *Sans.*, V. 243 ; dronapushpi, *Sans.*, L. 323.

Du Dughdika, *Sans.*, O. 600 ; duhn, *Arab.*, S. 1078 ; duhnul-balasán, *Arab.*, B. 54 ; dukhn, *Arab.*, S. 1212 ; dúkú, *Pers.*, P. 456 ; dul-surkh, *Pers.*, P. 1381 ; dumparástma, *Sans.*, A. 853 ; dund, *Pers.*, C. 2192 ; dunti, *Sans.*, C. 2192 ; durah-kizan, *Arab.*, Z. 50 ; durah-shámí, *Arab.*, Z. 50 ; duralabha, *Sans.*, A. 745 ; duralabha, *Sans.*, C. 224 ; durbha, *Sans.*, E. 252 ; duri-haskhak arísa, *Pers.*, S. 2968 ; durvá, *Sans.*, C. 2558 ; dusparsha, *Sans.*, F. 2 ; dúst parisha, *Sans.*, T. 509.

Ea E-ahmar, *Arab.*, A. 838.

El Ela (Amomum), *Sans.*, A. 976 ; ela (Elettaria), *Sans.*, E. 151 ; elá, *Sans.*, E. 151.

En Ena, *Sans.*, S. 1226.

Er Eraka, *Sans.*, T. 864 ; eranda, *Sans.*, R. 369 ; erváru, *Sans.*, C. 2274.

Fa Faddah, *Arab.*, S. 2205 ; faham, *Arab.*, C. 487 ; fahm (carbon), *Arab.*, C. 487 ; fahm (coal), *Arab.*, C. 1414 ; fálseh, *Pers.*, G. 663 ; fanjangasht, *Arab.*, V. 164 ; farfí, *Yarkand*, A. 428 ; farfirán, *Arab.*, *Pers.*, G. 66 ; fazzeh, *Arab.*, S. 2205.

Fe Feel, *Pers.*, E. 83 ; féjan, *Arab.*, R. 663.

Fi Fifil-i-surkh, *Pers.*, C. 466 ; filfildray, *Pers.*, P. 805 ; filfile, *Arab.*, C. 448 ; filfile-ahmar, *Arab.*, C. 455 ; filfile-asvad, *Pers.*, P. 811 ; filfile-gird, *Pers.*, P. 811 ; filfile-siyáh, *Pers.*, P. 811 ; filfile-surkh, *Pers.*, C. 448 ; filfil-i-ahmar, *Arab.*, C. 466 ; filfil-i-daráz, *Pers.*, P. 805 ; filfiluswud, *Arab.*, P. 811 ; filzahrah, *Pers.*, B. 443 ; fil-zahraj, *Arab.*, B. 443 ; findak, *Pers.*, C. 1985 ; finduk-i-hindí, *Arab.*, S. 818 ; fioyl, *Arab.*, R. 31 ; firanj-mushk. *Pers.*, O. 18 ; fisfisat, *Arab.*, M. 334 ; fizah, *Arab.*, A. 1359.

Fo Fódanaje-hindí, *Arab.*, M. 447 ; fófal, *Arab.* A. 1294 ; fótanaje-hindí, *Arab.*, M. 447 ; foufal, *Arab.*, A. 1294 ; fóvvah, *Arab.*, R. 564.

Fu Fugil, *Arab.*, R. 31 ; fúlfil-i-súrkh, *Pers.*, C. 455 ; fulúz-máhi, *Pers.*, S. 2943 ; furanjmishk, *Arab.*, O. 28.

Ga Gaja, *Sans.*, E. 83 ; gaja-pippalí, *Sans.*, S. 970 ; gal, *Pers.*, S. 1212 ; gamal, *Arab.*, C. 205 ; gamán, *Arab.*, P. 355 ; gandaka (rhinoceros), *Sans.*, R. 237 ; gandhaka (sulphur), *Sans.*, S. 2999 ; gandha márjara, *Sans.*, T. 441 ; gandhamula, *Sans.*, H. 196 ; gandhanákuli (Acampe), *Sans.*, A. 317 ; gandhanákuli (Vanda), *Sans.*, V. 17 ; gandha-rasaha, *Sans.*, B. 48 ; gandhasra, *Sans.*, S. 790 ; gandhera, *Sans.*, R. 116 ; gandum, *Pers.*, T. 634 ; gangird, *Pers.*, S. 2999 ; ganikáriká, *Sans.*, P. 1233 ; gánjá, *Sans.*, C. 331 ; ganjika, *Sans.*, C. 331 ; ganna, *Sans.*, A. 950 ; gaoshir, *Pers.*, F. 95 ; gao-zubán (Bomb), *Pers.*, A. 1132 ; garákshi vrikshamu, *Sans.*, C. 2306 ; garanthiparni, *Sans.*, A. 1469 ; gardha-bhánda, *Sans.*, T. 392 ; garjara, *Sans.*, D. 173 ; garnikura, *Sans.*, H. 177 ; gaudi, *Sans.*, S. 65 ; gaudume-makkah, *Pers.*, Z. 50 ; gaushur, *Pers.*, O. 189 ; gawshír, *Pers.*, O. 189 ; gaz, *Pers.*, T. 70 ; gazangabin, *Pers.*, T. 70 ; gazánjabín, *Arab.*, T. 70 ; gazar, *Pers.*, D. 173 ; gaze-surkh, *Pers.*, T. 51 ; gazmájú, *Pers.*, T. 70 ; gazmázaj, *Pers.*, T. 70 ; gazmázaje-khurd, *Pers.*, T. 51 ; gazmázak, *Pers.*, T. 70 ; gazshakar, *Pers.*, T. 70.

Gh Ghalichah, *Pers.*, C. 626 ; ghantápátali, *Sans.*, S. 959 ; ghantáravá, *Sans.*, C. 2158 ; ghantáravamu, *Sans.*, C. 2105 ; ghatá, *Sans.*, P. 625 ; ghirriyus-samak, *Arab.*, A. 393 ; ghirta-kumàri. *Sans.*, A. 829 ; ghoda-sala, *Sans.*, R. 116 ; ghotak, *Sans.*, H. 414 ; ghour, *Pers.*, H. 414 ; ghrita, *Sans.*, G. 189 ; ghruttham, *Sans.*, G. 189.

Gi Giló, *Arab.*, T. 470 ; giló, *Arab.*, T. 483 ; girdagán, *Pers.*, J. 61 ; girdagáne-hindí, *Pers.*, A. 737 ; gird-chób, *Pers.*, A. 1294 ; girikarnika, *Sans.*, C. 224 ; girikarnikayavása, *Sans.*, A. 745 ; giyáhe-qaisar, *Pers.*, A. 1585.

Go Godhápadi, *Sans.*, V. 217 ; godhúma, *Sans.*, T. 634 ; godumbá, *Sans.*, C. 2306 ; godúnika, *Sans.*, P. 625 ; gojihbá, *Sans.*, E. 80 ; gojihwá, *Sans.*, E. 80 ; gokarna múl, *Sans.*, C. 1403 ; gókhurhá, *Sans.*, T. 548 ; gokshura, *Sans.*, T. 548 ; gokshuri, *Sans.*, T. 548 ; golgolan, *Arab.*, S. 1078 ; gópi-múlam, *Sans.*, H. 119 ; gosfand, *Pers.*, S. 1320 ; gosh-wára, *N.-E. Persia*, P. 847 ; gouzun, *Pers.*, D. 212.

Gr Gráminá, *Sans.*, I. 145 ; granthiparni, *Sans.*, A. 1469.

Gu Gubak, *Sans.*, A. 1294 ; guch, *Pers.*, S. 1278 ; guda, *Sans.*, S. 65, S. 316 ; gudatvak, *Sans.*, C. 1196 ; gudúchi, *Sans.*, T. 470, T. 483 ; guggilam, *Sans.*, S. 1656 ; guggulu, *Sans.*, B. 43 ; guja, *Sans.*, V. 73 ; gul, *Pers.*, R. 504 ; gula, *Sans.*, S. 371 ; gúla, *Sans.*, S. 317 ; gul-bél, *Pers.*, T. 470 ; gul-bél, *Pers.*, T. 483 ; gulchakán, *Pers.*, B. 265 ; gulchakáneahrári, *Pers.*, B. 220 ; gule-bábúnah, *Pers.*, A. 1184; gule–daudi, *Pers.*, C. 1043 ; gule-supéd, *Pers.*, J. 35 ; guli-aabbás, *Pers.*, M.606 ; gúli-aftab, *Pers.*, H. 74 ; guligafas, *Pers.*, S. 850 ; guli-surkh, *Pers.*, R. 504 ; gulkhairo, *Pers.*, A. 880 ; gulnár, *Pers.*, P. 1426 ; gum-bharí, *Sans.*, C. 287 ; gundrá, *Sans.*; C. 2612; gúndra, *Sans.*, S. 6 ; gunja, *Sans.*, A. 51 ; gurs-dusti, *Pers.*, E. 335 ; guzgiah, *Pers.*, D. 151.

Ha Habaghadi, *Arab.*, B. 41 ; habak-hádí, *Arab.*, B. 41 ; habaqulhind, *Arab.*, M. 447; habbatoussouda, *Arab.*, N. 158; habbe-khatáí, *Pers.*, C. 2192 ; habbul-aaraar, *Arab.*, J. 78 ; habbul-asl, *Arab.*, T. 70 ; habbul asle, *Arab.*, T. 51 ; habbul-balasán, *Arab.*, B. 54 ; habbul-fahm, *Arab.*, S. 1041 ; habb-ul-kalkal, *Arab.*, C. 551 ; habbul-mishk, *Arab.*, H. 168 ; habbul-mushk, *Arab.*, H. 168 ; habbus-safarjal, *Arab.*, C. 2546 ; hab-bussalátín, *Arab.*, C. 2192 ; habbussa-látíne-barrí, *Arab.*, B. 28 ; habbussala-tíne-sahráí, *Arab.*, B. 28 ; hab-but-tarfa, *Arab.*, T. 70 ; hab-el-kalb, *Arab.* S. 1041 ; hab-es-soudán, *Arab.*, C. 728 ; hab-úl-ás, *Pers.*, M. 921 ; hábul bálasán, *Arab.*, B. 54 ; hab-ul-bán. *Arab.*, M. 393 ; hab-ul-fakad, *Arab.*, V. 179 ; hab-un-níl, *Arab.*, I. 384 ; habzal, *Arab.*, C. 1211, hadíd *Arab.*, I. 440 ; haivah, *Pers.*, C., 2546 ; háj, *Arab.*, A. 745 ; hajrulmighnátís, *Arab.*, I. 472 ; háleh, *Pers.*, L. 283 ; half, *Arab.*, L. 283 ; halgún, *Arab.*, *Pers.*, A. 1567 ; halílah, *Pers.*, T. 325 ; halilahe-siyah, *Pers.*, T. 325 ; halílahe-zard, *Pers.*, T. 325 ; halílaj, *Arab.*, T. 325 ; halílaje-asfar, *Arab.*, T. 325 ; halílaje-asvad, *Arab.*, T. 325 ; haliyún, *Pers.*, A. 1567 ; hallaka, *Sans.*, N. 200 ; halwa, *Pers.*, C. 232 ; hamáz, *Arab.*, R. 650 ; hamzal, *Arab.*, C. 1211 ; hangá, *Sans.*, C. 331 ; hansráj, *Pers.* ; A. 510 ; hanzale-ahmar, *Arab.* T. 600 ; hanzale-surkh, *Arab.*, T. 600 ; hapushá, *Sans.*, P. 811 ; har, *Pers.*, V. 219 ; harenso, *Sans.*, P. 885 ; hari chandana, *Sans.*, S. 790 ; haridrá, *Sans.*, C. 2433 ; harikusa, *Sans.*, A. 324 ; harítaki, *Sans.*, T. 325 ; harítakí-pushpam, *Sans.*, T. 325 ; haritála, *Sans.*, O. 242 ; harmal, *Arab.*, C. 225, P. 372 ; harnia, *Sans.*, S. 1226; hasalban-achsir, *Arab.*, R. 547 ; hasti, *Sans.*, E. 83 ; hastid carnid, *Sans.*, C. 1737 ; hatisunadá, *Sans.*, H. 102 ; hayamáraca, *Sans.*, N. 80 ; hazár-dánah, *Pers.*, E. 549.

He Hebak, *Arab.*, O. 18 ; hel, *Arab.*, E. 151 ; hel-bava, *Arab.*, E. 151 ; hél-zakar, *Arab.*, A. 976 ; hemapushpiká, *Sans.*, J. 24 ; hémaságara, *Sans.*, K. 14 ; hemda, *Arab.*, O. 547; hememdab, *Arab.*, O. 547 ; herar, *Pers.*, A. 590.

Hi Hijjala, *Sans.*, B. 180 ; hil, *Arab.*, E. 152 ; hilamóchaka, *Sans.*, A. 923 ; hílamochiká, *Sans.*, E. 213 ; hiltut, *Arab.*, F. 76 ; himadruma, *Sans.*, M. 393 ; himma, *Sans.*, G. 733; hiná, *Pers.*, L. 126 ; hináb, *Arab.*, C. 331 ; hinda-vanahe-talkh, *Pers.*, C. 1211 ; hindi, *Pers.*, E. 553 ; hindíra, *Sans.*, S. 2284 ; hindyba, *Arab.*, C. 1104, C. 1108 ; hinghúdie, *Sans.*, T.

312 ; hingól, *Pers.*, M. 473 ; hingolí *Sans.*, S. 2284 ; hingu, *Sans.*, F. 76, F. 84 ; hinná, *Arab.*, L. 126 ; hintah, *Arab.* T. 634 ; hintaherúnu, *Arab.*, Z 50 ; hin-tála, *Sans.*, P. 582 ; híráka, *Sans.*, D. 364 ; hisán, *Arab.*. H. 414.

Ho Hodthai, *Arab.*, B. 57 ; homadmad, *Arab.*, O. 547 ; hotai, *Arab.*, B. 57.

Hr Hrivera, *Sans.*, P. 344.

Hu Hujed, *Arab.*, A. 455 ; hulbah, *Arab.*, T. 612 ; humar, *Arab.*, T. 28 ; bumarbostaní, *Arab.*, R. 650 ; humbíjít, *Arab.*, R. 650 ; humez, *Arab.*, C. 1061 ; hummus, *Arab.*, P. 885 ; hunkarú, *Sans.*, C. 416 ; hurf, *Arab.*, L. 283 ; hurita, *Sans.*, P. 513 ; hurmul, *Arab.*, P. 372 ; hursíní, *Sans.*, C. 331 ; hussilüban, *Pers.*, S. 2968; hussí-úl-jawí, *Pers.*, S. 2968 ; hút, *Arab.*, R. 287 ; huzízehindi, *Arab.*, B. 443.

Hy Hya, *Sans.*, H. 414 ; hyamaraka, *Sans.*, W. 122.

Ij Ijás, *Arab.*, P. 1304.

Ik Iklíl-ul-malik, *Arab.*, A. 1585 ; ikshu, *Sans.*, S. 25 ; ikshu, *Sans.*, S. 30 ; ikshugandhá (Hygrophila), *Sans.*, H. 508; ikshugandhá (Tribulus), *Sans.*, T. 548 ; ikshurasa, *Sans.*, S. 65 ; ikshwálika, *Sans.*, S. 25.

In Indar-jave-talkh, *Pers.*, H. 294 ; indarjou, *Pers.*, W. 122 ; indarjouve-shírín, *Pers.*, W. 122 ; indivara, *Sans.*, N. 209 ; indral-varuni, *Sans.* C. 1211 ; indrásana, *Sans.*, C. 331 ; indra-váruni, *Sans.*, C. 1211 ; indrayava, *Sans.*, H. 294 ; inghúlam, *Sans.*, M. 473 ; ingudam, *Sans.*, B. 13 ; ingudi (Terminalia), *Sans.*, T. 312 ; ingudí (Balanites), *Sans.*, B. 13 ; ingudí-vrikshaka, *Sans.*, B. 13 ; inqitriyún, *Arab.*, A. 955 ; intyba, *Arab.*, C. 1104 ; inub, *Arab.*, V. 243.

Ir Irak, *Arab.*, S. 717 ; irsá, *Arab.*, I. 428.

Is Isbaghól, *Pers.*, P. 932 ; isband, *Pers.*, C. 225, P. 372 ; isferem, *Pers.*, M. 921 ; is-finaj, *Pers.*, S. 2574 ; isfist, *Pers.*, M. 334 ; iskabínah, *Pers.*, S. 493 ; ismad, *Arab.*, A. 1224 ; ismar, *Pers.*, M. 921 ; ispaghol, *Pers.*, P. 932 ; ispanaj, *Arab.*, S. 2574 ; ispand, *Pers.*, P. 372 ; isparzah, *Pers.*, P. 932; ispoghul, *Pers.*, P. 932 ; isqíle-hindí, *Arab.*, U. 39 ; isrenj, *Arab.*, L. 143 ; iswarg, *Pers.*, R. 166.

It Itrilál, *Arab.*, *Pers.*, V. 73.

Iz Izaragi, *Arab.*, S. 2943 ; izaraki, *Pers.*, S. 2943 ; izkhir, *Arab.*, A. 1093.

Ja Jadhírdàh, *Arab.*, C. 1520 ; jadwár, *Arab.*, D. 253 ; jaintjaintar, *Sans.*, S. 1174 ; jajipatri, *Sans.*, M. 885 ; jáji-pha lam, *Sans.*, M. 885 ; jala-nirgundí, *Sans.*, V. 181 ; jalapippali langulí, *Sans.*, C. 1759 ; jalúkaha, *Sans.*, L. 245 ; jamal, *Arab.*, C. 205 ; jamba, *Sans.*, E. 432 ; jambeh, *Pers.*, E. 327 ; jambíra, *Sans.*, C. 1258, C. 1296 ; jambu, *Sans.*, E. 419, E. 432 ; jambula, *Sans.*, E. 419 ; jaó, *Pers.*, H. 382 ; jauz, *Pers.*, J. 61 ; japa-pushpam, *Sans.*, H. 227 ; jargadi, *Sans.*, C. 1686 ; jarjir. *Arab.*, L. 283 ; jata, *Sans.*, C. 1879 ; jatámánsi, *Sans.*, N. 17 ; játi, *Sans.*, J. 18 ; játiphala, *Sans.*, M. 885 ; jauz-ut-tríb. *Arab.*, M. 885 ; jawars, *Arab.*, S. 2424 ; jawashir, *Pers.*,

F. 95; jaya (Sesbania), *Sans.*, S. 1174;
jáyá (Cannabis), *Sans.*, C. 331; jayanti,
Sans., S. 1166; jayapála, *Sans.*, C.
2192; jazar, *Arab*, D. 173; jazmázaj,
Arab., T. 70.

Je Jerasayna, *Arab.*, P. 1297.

Jh Jhávuka, *Sans.*, T. 70; jhingáka, *Sans.*,
L. 556; jhinti, *Sans.*, B. 165.

Jí Jingini, *Sans.*, O. 38; jintiyáná, *Arab.*, G.
167; jiotishmati, *Sans.*, C. 854; jíraka,
Sans., C. 2339; jírana, *Sans.*, C. 2339;
ívah, *Pers.*, M. 473.

Jo Joba, *Sans.*, H. 227; jouz, *Arab.*, J. 61;
jouzbóyah, *Pers.*, M. 885; jouzbuvá,
Arab., M. 885; jouzebarri, *Arab.*, ι A.
737; jouze-hindí, *Arab.*, C. 1520; jouz-
masal, *Arab*, D. 160; jouz-másameasvad,
Arab., D. 151; jouz-másleabyaz *Arab.*,
D. 160; jouz-másle, asvad, *Arab.*, D. 151;
jouz-massel, *Arab.*, D. 151; jouzul-kosul,
Arab., K. 1; jouzuttíb, *Arab.*, M. 885;
jovari, *Sans.*, A. 1398; jowr, *Pers.*, C.
225; jowz-hind, *Arab.*, C. 1520.

Ju Judwar, *Yarkand*, C. 2504; judwar, *Arab.*,
C. 2406; júhar, *Arab.*, P. 355; jukutam,
Sans., S. 2284; júta, *Sans.*, C. 1879;
juvashur, *Pers.*, O. 189; juwashur,
Arab., O. 189; júz-ul-kueh, *Pers.*, R. 1.

Jy Jyantika, *Sans.*, S. 1174; jyautishmati,
Sans., C. 551.

Ka Kababah, *Arab.*, P. 801; kabáb-chíní,
Pers., P. 801; kabar, *Arab.*, *Pers.*, C.
431; kabare-hindí, *Arab. & Pers.*, C.
919; kabbar, *Arab.*, S. 717; kabikaj,
Pers., R. 28; kabiste-talkh, *Pers.*, C.
1211; kabit, *Pers.*, F. 53, *Arab.*, F. 53;
kábrit, *Arab.*, S. 2999; kabsún, *Arab.*,
B. 856; kabur, *Arab.*, C. 431; kaburni,
Sans., Z. 190; kachchí, *Sans.*, C. 1732;
kachla, *Sans.*, C. 838; kach rehn, *Pers.*,
C. 1221; kachwæ, *Sans.*, C. 1732;
kachwí, *Sans.*, C. 1732; kadali, *Sans.*,
M. 811; kadamba, *Sans.*, A. 1192;
kadar, *Arab.*, P. 26; kaddú, *Pers.*, L.
30; kadgin, *Sans.*, R. 237; kadi, *Pers.*,
P. 26; kadim-el-bint, *Arab.*, E. 7; kaf-es-
saba, *Arab.*, R. 28; kafral yahúd, *Pers.*,
P. 436; kafsún, *Arab.*, B. 856ι; kafter,
Arab., C. 833; káfúr, *Arab.*, *Pers.*, C.
257; kahkab, *Arab.*, S. 2284; kahkam,
Arab., S. 2284; kahrubá, *Pers.*, A. 955;
káhú, *Pers.*, L. 21; kahwa, *Arab.*, *Pers.*,
C. 1641; kaidary-ama, *Sans.*, M. 869;
kaikahr, *Arab.*, S. 1656; kaiz-másale-
siyáh, *Pers.*, D. 151; káj, *Pers.*, A 1166;
kakachinchi, *Sans.*, A. 51; kákádani,
Sans., C. 427; kákadumbar, *Sans.*, F.
202; kákajanghá, *Sans.*, L. 229; káka-
máchi, *Sans.*, S. 2299; káka mári, *Sans.*,
A. 1038; kákángah, *Sans.*, L. 229;
kakilahe-khurd, *Pers.*, E. 151; kakílahe-
saghir, *Arab.*, E. 151; kakinduka, *Sans.*,
D. 656; káknaje hindí, *Arab.*, W. 93;
káknaje-hindí, *Pers.*, W. 93; káknaje-
hindí, *Pers.*, W. 98; káknaje-hindí,
Arab., W. 98; kákulah, *Arab.*, E. 152;
kálá, *Sans.*, I. 145; kálabarbúra-niryá-
sam, *Sans.*, A. 102; kála-hémiká, *Sans.*,
D. 151; kálaklítaka, *Sans.*, I. 109; ka-
lambi, *Sans*, I. 343; kála-méshiká, *Sans.*,
R. 564; kalangúra, *Pers.*, S. 2968; ká-
lanusárivá, *Sans.*, L. 355; kalasáka,

Sans., C. 1846; kála-sarshapa, *Sans.*, B.
812; kaláya, *Sans.*, P. 885; kalháramu,
Sans., N. 200; kalikari, *Sans.*, G. 243;
kalínbak, *Pers.*, C. 1286; kálinga, *Sans.*,
H. 294; kalkas, *Arab.*, C. 1732; kal
khúbrí, *Pers.*, Z. 190; kamáduriyús,
Pers., S. 2518; kamáfítús, *Arab.*, B. 546;
kamala, *Sans.*, N. 200; kámalata, *Sans.*,
I. 405; kamalottara, *Sans.*, C. 637;
kamáaríyús, *Arab.*, S. 2518; kaméne-
kirmání, *Arab.*, C. 697; kamkam, *Pers.*,
S. 2968; kampilla, *Sans.*, M. 71; kamu,
Sans., P. 632; kamue mulúki, *Arab.*, C.
691; kamún, *Arab.*, C. 2339; kamúne-
asvad, *Arab.*, N. 158; kaná, *Sans.*, P.
805; kanab, *Arab.*, C. 331; kanakaphála,
Sans., C. 2192; kananakanda, *Sans.*,
D. 821; kanana eranda, *Sans*, J. 41;
kanana-jeraka, *Sans.*, V. 73; kanana-
kerundum, *Sans.*, J. 41; kananamullika,
Sans., J. 10; kanbélá, *Pers.*, M. 71;
kánchan, *Sans.*, B. 308; kánchana, *Sans.*,
T. 489; kanchata, *Sans.*, C. 1748; kand;
Pers., S. 370; kand, *Arab.*, S. 376;
kanda, *Sans.*, A. 996; kandar-i-rumí,
Pers., P. 841; kande-suped, *Pers*, S.
376; kandur, *N.-E. Persia*, P. 847; kan,
gai, *Pers.*, M. 115; kangoi, *Pers.*, M.,
115; kangu, *Sans.*, S. 1212; kanguni,
Sans., S. 1212; kanguruku, *Sans.*, S.
30; kanjak, *N.-E. Persia*, P. 847; kankéli, *Sans.*, S. 861; kantakari, *Sans.*, S.
2345; kantala, *Sans.*, A. 603, A. 636;
kanyá, *Sans.*, A. 829; kapardaka-
bhasma, *Sans.*, C. 489; kapi-ballí, *Sans.*,
S. 970; kapikachchhu, *Sans.*, M. 786;
kapíla, *Sans.*, M. 71; kapipriya, *Sans.*,
F. 53; kapittha, *Sans.*, F. 53; kapura-
káchalí, *Sans.*, H. 59; karafs, *Arab.*, A.
1227; karamadika, *Sans.*, C. 606; karamarda, *Sans.*, C. 596; karamardaka,
Sans., C. 596; karanja, *Sans.*, P. 1121;
karasb, *Pers.*, A. 1227; kára-valli-latá,
Sans., M. 626; kárave, *Sans.*, N. 158;
káravella, *Sans.*, M. 626; káraví, *Sans.*,
C. 551; kara-vira, *Sans.*, N. 80; karawya,
Arab., C. 681; karchura, *Sans.*, C. 2499;
karelah, *Pers.*, M. 626; kari-pippalí,
Sans., S. 970; karíra, *Sans.*, C. 402;
karka, *Sans.*, C. 2316; karkadan, *Pers.*,
R. 237; karkataka, *Sans.*, M. 634; kar-
katasringi, *Sans.*, R. 323; karkata-sringí,
Sans., P. 833; karkati, *Sans.*, C. 2278;
karmaranga, *Sans.*, A. 1646; karnikára,
Sans., P. 1389; karoyá, *Arab.*, C. 681;
karóya, *Pers.*, C. 681; karpas, *Sans.*, G.
404; kárpásamu, ι *Sans.*, G. 385; kar-
pasi, *Sans.*, G. 404; karpúra, *Sans.*,
C. 257; karpura-haridrá, *Sans.*, C. 2381;
kartik, *Arab.*, H. 111; karuntaka, *Sans.*,
B. 171; kasá, *Sans.*, S. 49; kasabi-sha-
kar, *Arab.*, S. 30; kásághini, *Sans.*, T.
509; kasakdánah, *Pers.*, C. 637; kása-
mara, *Sans.*, C. 780; kásamarda, *Sans.*,
C. 787; kas-din, *Arab.*, T. 460; kaseruka,
Sans., S. 977; kásha, *Sans.*, S. 49;
kashfa, *Arab.*, B. 768; kashmirja, *Sans.*,
S. 910; kasib shakar, *Arab.*, S. 30;
kasisa, *Sans.*, I. 481; kásmari, *Sans.*, G.
287; kásmíra, *Sans.*, C. 2013; kásmíra-
janmá, *Sans.*, C. 2083; kasni, *Pers.*, C.
1108; kasní, *Pers.*, C. 1108; kasrat-el-
azlaa, P. 930; kasta-ze-rambet, *Pers.*,

A. 860 ; kasturi, *Sans.*, D. 228; kat, *Arab.*, C. 833 ; kátaka, *Sans.*, S. 2960; katchú, *Sans.*. C. 1732 ; katha-ul, *Arab.*, C. 756 ; kathíra, *Arab.*, C. 1513; kati-mukki, *Arab.*, P. 1370; katíra, *Pers.*, C. 1513 ; katíra-í-hindí, *Pers.*, C. 1512; katphala, *Sans.*, M. 869; katrabungá, *Sans.*, A. 1395 ; kattán, *Arab.*, L. 385 ; katuka, *Sans.*, C. 1791; katuká, *Sans.*, P. 700 ; katurohini (Picrorhiza), *Sans.*, P. 700 ; katuróhini (Helleborus), *Sans.*, H. 111 ; katu-roum, *Sans.*, H. 111 ; ka-tutumbi, *Sans.*, L. 30 ; kaunti, *Sans.*, C. 1875 ; kazhirah, *Pers.*, C. 637 ; kazhúr, *Pers.*, C. 2499 ; kazi, *Arab.*, P. 26.

Ke Kebír, *Pers.*; C. 431 ; kechu-búh, *Arab.*, D. 151; kemúka, *Sans.*, C. 2013 ; kendu-ka, *Sans.*, D. 615 ; kerásya, *Arab.*, P. 1297 ; kerroa, *Arab.*, R. 369; kerrua, *Arab.*, R. 369 ; kesaraja (Wedelia), *Sans.*, W. 25 ; kesa023ja (Eclipta), *Sans.*, E. 7 ; késaramu, *Sans.*, M. 489; keshiní, *Sans.*, C. 1053; ketaka, *Sans.*, P. 26 ; ketaki, *Sans.*, P.26.

Kh Khabsul-hadíd, *Arab.*, I. 472; khadga, *Sans.*, R. 237; khadira, *Sans.*, A. 135 ; khadira khadirasára, *Sans.*, A. 181; khaggara, *Sans.*, S. 49; khaira-ka-jhor, *Pers.*, A. 880 ; khair-bava, *Arab.*, E. 151; khákshí, *Pers.*, S. 2219 ; khalanjan, *Arab.*, A. 853; khálávan, *Arab.*, Z. 50 ; khall, *Arab.*, A. 356 ; khallul-himmas, *Arab.*, C. 1061 ; khamar, *Arab.*, H. 414; khamr, *Arab.*, V. 243; khanda, *Sans.*, S. 65, S. 376 ; khandarús, *Arab.*, Z. 50; khanek-ul-kella, *Arab.*, S. 2943 ; kharbaqe-hindi, *Arab.* & *Pers.*, P. 700; khar-buzahe rubáh, *Pers.*, C. 1211 ; khar-buzahe-talkh, *Pers.*, C. 1211 ; kharbúzeh, *Pers.*, C. 2263 ; khardál, *Arab.*, B. 841 ; khardale-abyaz, *Arab.*, B. 800 ; khardale-í asvad, *Arab.*, B. 812 ; kháre-khasak, *Pers.*, T. 548 ; kháré-mughlán, *Pers.*, A. 101 ; kháre-vázhún, *Pers.*, A. 382 ; khár-í-shutr, *Pers.*, A. 745 ; kharjal, *Arab.*, S. 717; kharjjúraha, *Sans.*, P. 555 ; khar-jura, *Sans.*, P. 588 ; khark, *Pers.*, C. 170, C. 225 ; kharnúb shámi, *Arab.*, C. 933 ; kharpara, *Sans.*, Z. 190; kharvujá, *Sans.*, C. 2263 ; khar-zahrah, *Pers.*, C. 225, N. 80 ; khas (Andropogon), *Pers.*, A. 1097; khas (Lactuca), *Arab.*, L. 21; khasabul-hgyah, *Arab.*, A. 1567; khasak, *Arab.*, T. 548 ; khasake-kabír, *Arab.*, P. 363 ; khasake-kalán, *Pers.*, P. 363 ; khashkhásh, *Pers.*, P. 87 ; khash khásh-i-mansúr, *Arab.*, *Pers.*, P. 82 ; khasife-hindí, *Arab.*, A. 737 ; khaskhasa, *Sans.*, P. 87 ; khassuch, *Pers.*, F. 95 ; kháyahe-i-iblís, *Pers.*, C. 6 ; kher-beck-seeah, *Pers.*, H. 111 ; kherbekas-wed, *Arab.*, H. 111 ; kherefeh, *Pers.*, P. 1179 ; khertic, *Arab.*, H. 111 ; khiláf, *Arab.*, S. 550 ; khinjak, *Pb.*, P. 847 ; khirdal, *Arab.*, B. 841 ; khirnúb nubti, *Arab.*, C. 933 ; khirvá, *Arab.*, R. 369 ; khitmi, *Arab.*, M. 115 ; khitmi-i-kuchak, *Pers.*, M. 115 ; khitmi-ká-jhár, *Pers.*, A. 880 ; khiyár, *Pers.*, L. 556 ; khíyár-chanbar, *Pers.*, C. 756 ; khíyát-shanbur, *Arab.*, C. 756 ; kho-shahe-makkí, *Pers.*, Z. 50 ; khubázi, *Pers.*, M. 115 ; khubazí, *Arab.*, M. 115 ;

khujar, *Pers.*, L. 569 ; khúlanján, *Arab.*, A. 862 ; khúlanján-e-qasbi, *Arab.*, A. 853 ; khúlanján-e-kabír, *Arab.*, A. 853 ; khúne-síyávusháne-hindí, *Pers.*, P. 1370 ; khun(i)-siávesham, *Pers.*, C. 68 ; khun-jad, *N.-E. Persia*, P. 847 ; khunjada, *N.-E. Persia*, P. 847 ; khurdaní, *Pers.*, S. 602 ; khurfa, *Arab.*, P. 1179 ; khur-fáh, *Pers.*, P. 1179 ; khurjjúri, *Pers.*, P. 588 ; khurmae-khushk, *Pers.*, P. 555 ; khurmáe-yábis, *Arab.*, P. 555 ; khurne, *Sans.*, C. 1403 ; khusrave-dúrúé-kalán, *Pers.*, A. 853 ; khusro-dáru, *Pers.*, A. 862 ; khussuck, *Pers.*, T. 548 ; khusya-tus-saalab, *Arab.*, S. 521 ; khusyus saa-lab, *Arab.*, S. 521.

Ki Kibabeh, *Pers.*, P. 801 ; kíchaka, *Sans.*, B. 118 ; kils, *Arab.*, C. 489 ; kilz, *Pers.*, L. 483 ; kinbíl, *Arab.*, M. 71 ; kinjad, *N.-E. Persia*, P. 847 ; kinjada, *N.-E. Persia*, P. 847 ; kinjak, *N.-E. Persia*, P. 847 ; kinjalkamu, *Sans.*, M. 489 ; kin-nab, *Arab.*, *Pers.*, C. 331 ; kinneh, *Pers.*, P. 841 ; kinnoli, *Pers.*, P. 841 ; kinsuka, *Sans.*, B. 944 ; kirás, *Arab.*, A. 775 ; kirata, *Sans.*, A. 1064 ; kirata-tikta, *Sans.*, S. 3018 ; kiráth, *Arab.*, A. 775 ; kisht-burkísht, *Pers.*, H. 92 ; kitmakit, *Arab.*, C. 6 ; kizaz, *Arab.*, G. 229.

Kk Kkaangi, *Sans.*, A. 1242.

Ko Kodrava, *Sans.*, P. 332 ; kohal, *Arab.*, A. 1224 ; kohr-a-gaz, *Pers.*, T. 51 ; koki-láksha, *Sans.*, H. 508 ; kóknáresurkh, *Pers.*, P. 82 ; kola, *Sans.*, P. 805 ; kola, *Sans.*, Z. 231 ; kola-ballí, *Sans.*, S. 970 ; kolutha, *Sans.*, D. 758 ; kon-shad, *Pers.*, G. 167 ; konsso, *Arab.*, B. 856 ; korá-dusha, *Sans.*, P. 332 ; kosámra, *Sans.*, M. 209 ; koshátaki, *Sans.*, L. 563 ; kourfaka-ra-or, *Arab.*, P. 1179 ; koushikaha (Bal-samodendron), *Sans.*, B. 43 ; koushi-kaha (Shorea), *Sans.*, S. 1656 ; kouz-kunáe-siyah, *Pers.*, D. 151 ; kouz-másale-saféd, *Pers.*, D. 160.

Kr Krishná, *Sans.*, P. 805 ; krishnábhra, *Sans.*, M. 509 ; krishna chandanam, *Sans.*, S. 790 ; krishnachúrá, *Sans.*, C. 32 ; krishna dhattúra, *Sans.*, D. 151 ; krishna-jiraka, *Sans.*, N. 158 ; krishna-kámbojí, *Sans.*, P. 663 ; krishnala, *Sans.*; A. 51 ; krishna mirtika, *Sans.*, C. 1317 ; krishna-pak-phula, *Sans.*, C. 596 ; krishnasirish, *Sans.*, A. 686 ; kríshtna-surasa, *Sans.*, J. 116 ; krisnakeli, *Sans.*, M. 606 ; krist-na-nimba, *Sans.*, M. 800.

Ks Kshira, *Sans.*, P. 760 ; kshirini, *Sans.*, M. 583.

Ku Kuberaka, *Sans.*, C. 838 ; kuberákshí, *Sans.*, C. 6 ; kuehwa, *Arab.*, C. 1641 ; kuineh, *Arab.*, F. 95 ; kúka váivinta, *Sans.*, C. 1367 ; kúka-vumitie, *Sans.*, C. 1367 ; kukubha, *Sans.*, T. 282 ; kuku-radru, *Sans.*, B. 546 ; kuláhala, *Sans.*, C. 878 ; kulaka, *Sans.*, S. 2943 ; kulattha, *Sans.*, D. 758 ; kulbahebarrí, *Arab.*, S. 1714 ; kulf, *Arab.*, C. 1003 ; kúlianiní, *Sans.*, P. 625 ; kúllu-jana, *Sans.*, A. 053 ; kulja, *E. Turkistan*, S. 1276 ; kullalíc-div, *Pers.*, A. 590 ; kullalie-dio, *Pers.*, F. 725 ; kúmari, *Sans.*, A. 836 ; kumbhi, *Sans.*, C. 563 ; kumbhiká, *Sans.*, P. 874 ; kúmbh samarogh, *Pers.*, A. 590 ; kumkuma, *Sans.*, C. 2083 ; kumuda,

Sans., N. 200 ; kumudwutí, Sans., L. 355 ; kuna, Sans., P. 805 ; kunár, Pers., Z. 231, Z. 280 ; kúnda, Sans., J. 32 ; kundali (Azima), Sans., A. 1665 ; kundalí (Clerodendron), Sans., C. 1377 ; kundar, Pers., P. 847 ; kundar-i-rumí, Pers., P. 841 ; kundarud, Pers., P. 847 ; kundel, Sans., S. 493 ; kunderu, Pers., P. 847 ; kundur, Arab., Pers., B. 751, B. 771, B. 777 ; kunduru, Sans., B. 751 ; kunduru guggulu, Sans., B. 771 ; kunjad (Pistacia), Pers., P. 847 ; kunjad (Sesamum), Pers., S. 1078 ; kunjada, Pers., P. 847 ; kunjed, Pers., S. 1078 ; kunjidah, Pers., A. 1591, A. 1592 ; kun-kham, Arab., G. 116 ; kunkuma, Sans., C. 2083 ; kuntali, ᴐans., Z. 1 ; kunuk-champa, Sans., O. 1 ; kupilu, Sans., S. 2943 ; kur, Arab., C. 1732 ; kúrák, Pers., C. 431 ; kuravaka, Sans., L. 126 ; kurbuzah, Arab., C. 1954 ; kurfáh, Pers., P. 1179 ; kurkarú, Sans., C. 2331 ; kurkum, Arab., C. 2433 ; kurtamussul, Arab., C. 404 ; kusha, Sans., E. 252 ; kúshmánda, Sans., B. 430 ; kushníz, Pers., C. 1954 ; kúsh-pandaha, Sans., B. 430 ; kushtha, Sans., S. 910 ; kúshumbha, Sans., C. 637 ; kúst, Arab., Pers., S. 910 ; kusumbha, Sans., C. 637 ; kutaja, Sans., H. 294 ; kutan, Pers., L. 385 ; kutha, Sans., E. 252 ; kuzbarah, Arab., C. 1954.

La Laghu drákshá, Sans., V. 233 ; lajjálu, Sans., M. 557 ; lákshá, Sans., C. 1416, L. 1 ; lakucha, Sans., A. 1511 ; lále-moáb bári, Pers., S. 1656 ; lámajjaka, Sans., A. 1093 ; langali, Sans., H. 504 ; lángalíká, Sans., G. 243 ; langalin, Sans., C. 1520 ; lanjasavaram, Sans., I. 342 ; lardak lahori, Pers., I. 348 ; lasánul-aasafír, Arab., W. 122 ; lasánul-aasáfíᴜulhaló, Arab., W. 122 ; lasánul-aasáfíᴜul-murr, Arab., H. 294 ; lasuna, Sans., A. 779 ; latákaranja, Sans., C. 6 ; latakasturíkam, Sans., H. 168 ; latá-palása, Sans., B. 978 ; lauha, Sans., I. 440 ; lávana, Sans., S. 602 ; lavanga, Sans., C. 706 ; lavani, Sans., P. 627.

Li. Limeh, Pers., C. 1258 ; limpáka, Sans., C. 1296 ; límú, Arab., Pers., C. 1258, C. 1296 ; límúe-hámiz, Arab., C. 1258, C. 1296 ; límúe-tursh, Pers., C. 1258, C. 1296 ; límun, Arab., C. 1296 ; lisan-el-hamal, Arab., P. 930.

Lo Loabate-barbarí, Arab., C. 1703 ; loban, Arab., S. 2968 ; loban jáwí, Arab., S. 2968 ; lobiya (Vigna), Pers., V. 116 ; lobíyá (Dolichos), Pers., D. 789 ; lodhra, Sans., S. 3062 ; lóham, Sans., I. 440 ; loni, Sans., P. 1179 ; lonika, Sans., P. 1179 ; louz, Arab., P. 1274 ; louz-el-murr, Arab., P. 1280.

Lu Lubán, Arab., Pers., B. 751 ; lubán, Arab., B. 771 ; lufa, Pers., C. 1834 ; lufah, Arab., M. 128 ; luff, Arab., L. 569 ; lúlú, Arab., P. 355 ; lúnak, Pb., P. 1187 ; lúnia, Sans., P. 1179.

Ma Mabotpala, Sans., N. 39 ; mádalá, Sans., A. 658 ; madana, Sans., R. 1 ; madana-ghanti, Sans., S. 2515 ; maddhika, Sans., V. 243 ; madha-dút, Sans., M. 147 ; mádhavi, Sans., H. 285 ; madhu, Sans., H.

342 ; madhujam, Sans., C. 931, H. 342 ; madhuka (Bassia), Sans., B. 220, B. 265 ; madhuka (Glycyrrhiza), Sans., G. 278 ; madhukarkatiká, Sans, C. 1301 ; madhukshir, Sans., P. 588 ; madhuli, Sans., T. 634 ; madhu malati, Sans., D. 823 ; madhúriká, Sans., F. 659 ; madhu-yashtikam, Sans., G. 278 ; madhválu, Sans., D. 481 ; madhvi, Sans., J. 13 ; má-el-khílaf, Arab., S. 550 ; magháse-hindí, Arab., L. 483 ; maghath, Arab., L. 486 ; maghz-pipal, Pers., P. 805 ; magudhí, Sans., P. 805 ; mahágodhuma, Sans., T. 634 ; mahájambíra-karuna, Sans., C. 1286; mahákála, Sans., T. 600 ; mahalib, Arab., P. 1313 ; mahá-nimba, Sans., M. 393 ; maha-ushadha (Zingiber), Sans, Z. 201 ; maha-ushadha (Allium), Sans., A. 779 ; mahferfin, Arab., D. 253 ; mahí, Pers., R. 287 ; majdabh, Arab., Pers., B. 430 ; majuphul, Sans., Q. 43 ; mákhál, Sans., C. 1211 ; makhánna, Sans., E. 569 ; mákshika, Sans., H. 342 ; makushtaka, Sans., P. 468 ; málati, Sans., A. 584 ; malayaja, Sans., S. 790 ; malh-i-barut, Arab., S. 682 ; malika, Sans., L. 385 ; mallíká, Sans., J. 35 ; malira, Sans., V. 243 ; malura, Sans, A. 534 ; manahsila, Sans., R. 61 ; mánaka, Sans., A. 803 ; mandára, Sans., C. 170 ; mandu-kaparní, Sans., H. 486 ; manduki, Sans., H. 149 ; mandúram, Sans., I. 472 ; mandwah, Pers., E. 170 ; máni, Sans., H. 525 ; manikya, Sans., R. 619 ; manjarika, Sans., O. 31 ; manjishthá, Sans., R. 564 ; mann, Arab., F. 696 ; mansúr, Arab., Pers., P. 82 ; maragh, Pers., P. 350 ; marakata, Sans., P. 355 ; marchobah, Pers., A. 1567 ; mardakusch, Arab., O. 214 ; márgíyah, Pers., A. 1567 ; marg-mósh, Pers., A. 1425 ; maricha, Sans., P. 811 ; marichi-phalam, Sans., C. 448, C. 455 ; márisha, Sans., A. 938 ; márjana, Sans., S. 3062 ; marura, Sans., S. 785 ; marwarid, Pers., P. 355 ; marzangósh, Arab., Pers., A. 1469 ; mas, Pers., D. 364 ; más, Arab., S. 1290 ; masang, Pers., L. 100 ; másh, Arab., P. 513 ; másha, Sans., P. 513 ; masht-ul-ghoul, Arab., A. 89 ; masína, Sans., L. 385 ; masrind, Sans., L. 385 ; massel, Arab., D. 164 ; mastaká-i-rumí, Pers., P. 841 ; masúna, Sans., L. 385 ; masura, Sans., L. 252 ; mathil, Arab., D. 164 ; matsyandiká, Sans., S. 371 ; matulunga, Sans., C. 1270 ; maurid, Pers., M. 921 ; maya shutr, Arah., R. 73 ; may-ínekhurd, Pers., T. 51 ; mayúra, Sans., P. 350 ; máyúra-shikhá (Adiantum), Sans., A. 501 ; mayur asikha (Celosia), Sans., C. 873 ; mazariyun, Arab., Pers., D. 124 ; mázariyune-hindí, Arab., C. 1403 ; mazirium, Pers., D. 124 ; mazú, Pers., Q. 43.

Me Méaahe-sáyelah, Arab., A. 892 ; meah, Arab., L. 455 ; meati-lubani, Arab., L. 455 ; meetiya, Arab., B. 48 ; mei, Pers., V. 243 ; meih-sila, Pers., L. 455 ; mekanada, Sans., A. 950 ; mekhak, Pers., C. 706 ; mendhi, Sans., L. 126 ; mendiká, Sans., L. 126 ; merdum-geeah, Pers., M. 128 ; merdum seeah, Pers., A. 1614 ; merján, Pers., C. 1808 ; mesharingi, Sans., G. 748 ; méthí, Sans., T. 612 ; methiká,

Sans., 1. 612; mewiz, *Arab.*, V. 233;
mezereon, *Arab.*, D. 124.
Mh Mhaisabol, *Sans.*, B. 41.
Mi Miah-sáyelah, *Arab.*, L. 455; mighnátís,
Arab., I. 472; milh, *Arab.*, S. 602; mil-
hul-aajín, *Arab.*, S. 602; mílhul-qilí,
Arab., C. 541; milhunnár *Arab.*, À.962;
milhus-sághah, *Arab.*, B. 731; miqnatís,
Arab., I. 472; mirga, *Sans.*, S. 1226;
mirgal, *Sans.*, F. 4.0; mirjumak, *Pers.*,
L. 252; miromati, *Sans.*, P. 1078; mis,
Pers., C. 2361; mish, *Pers.*, S. 1278;
mishk, *Arab.*, D. 228; mishmish, *Pers.*,
P. 1285; misk, *Arab.*, D. 228; misreyá,
Sans., P. 460; mizangosh, *Pers.*, O.
220; mizunjúsh, *Arab.*, O. 214.
Mo Mochá, *Sans.*, B. 632; mokah, *Arab.*, M.
623; mokhátah, *Arab.*, C. 1940; mom,
Pers., C. 931; móm, *Pers.*, H. 342;
moql, *Arab.*, B. 43; moqlearzaqi aflátan,
Arab., B. 43; mouz, *Arab.*, *Pers.*, M.
811.
Mr Mriaviká, *Sans.*, V. 233; mrigala, *Sans.*,
F. 410; mriganábhi, *Sans.*, D. 228;
mriga-shinga, *Sans.*, H. 92; mrigasringa,
Sans., H. 50.
Mu Muasfir, *Pers.*, C. 637; muchukunda, *Sans.*,
P. 1397; mudga, *Sans.*, P. 496; mudga-
parni, *Sans.*, P. 523; mukhitah, *Arab.*,
C. 1940; muktá, *Sans.*, P. 355; múlaka,
Sans., R. 31; múndí, *Sans.*, S. 2518;
mun-ditiká, *Sans.*, S. 2518; munduka-
purna, *Sans.*, O. 233; mung, *Pers.*, P.
496; munjariki, *Sans.* O. 18; mun shír-
khist, *Arab.*, M. 235; mur, *Arab.*, B.
48; murján, *Pers.*, C. 1808; murr, *Arab.*,
B. 48; muruvá, *Sans.*, S. 785; mus-alí,
Sans., R. 110; mushali, *Sans.*, C. 2375;
mushk, *Arab.*, *Pers.*, D. 228; mushkaka,
Sans., S. 959; mushk-dana, *Pers.*, H.
168; mushke, zamín, *Pers.*, C. 2617;
muslí-e-siyah, *Pers.*, A. 1122; musta,
Sans., C. 2617; mustá, *Sans.*, C. 2612;
mustaka, *Sans.*, C. 2612; mustoka,
Arab., P. 841; mutra, *Sans.*, U. 45.

Na Naanáæ-hindí, *Arab.*, M. 447; naanaaul-
hind, *Arab.*, M. 447; nabat, *Arab.*, S.
376; nabatúl-khash-khashul-ahmar,
Arab., P. 82; nabátul-qunnab, *Arab.*,
Pers., C. 331; nabá-tussibr, *Arab.*, A.
838; nabiq, *Arab.*, Z. 231; nádika,
Sans., C. 1861; nága-balá, *Sans.*, S.
1714; nágadamani, *Sans.*, A. 1469; nága-
késaram, *Sans.*, M. 489; nága-késaram-
pushpam, *Sans.*, O. 6; nágaa, *Sans.*, Z.
201; nágaranga, *Sans.*, C. 1233; nágar-
mustaka, *Sans.*, C. 2617; nágavalli,
Sans., P. 775; nagbó, *Pers.*, M. 465;
naghyak, *Pers.*, M. 147; na, *Pers.*, B.
118; nainehavandi, *Pers.*, A. 1064; nai-
shakar, *Pers.*, S. 30; nakhl, *Pers.*, P.
555; nakhud, *Pers.*, C. 1061; naktamála,
Sans., P. 1121; nákuli, *Sans.*, V. 17;
nala, *Sans.*, P. 618; náli, *Sans.*, H. 177;
nalina, *Sans.*, N. 39; nálitá, *Sans.*, C.
1801; nalla ativasa, *Sans.*, C. 2301;
namak, *Pers.*, S. 602; namake, *Pers.*, S.
602; nandí-vriksha, *Sans.*, C. 838; nán-
i-kulágh, *Pers.*, M. 115; nánkhwah, *Pers.*,
C. 691; nar, *Pers.*, P. 1426; nárang,
Pers., C. 1233; náranj, *Arab.*, C. 1233;
nárendj, *Pers.*, C. 1233; nárgil, *Pers.*,

C. 1520; nárgíle bahrí, *Pers.*, L. 511;
nargilli, *Arab.*, C.1520; nári-kela, *Sans.*,
C. 1520; nárikela-tailam, *Sans.*, C.1520;
nári-keli, *Sans.*, C. 1520; nári-kera,
Sans., C. 1520; narif-ka-krute, *Arab.*, C.
1520; narjible, *Pers.*, C. 1520; nárjíl,
Arab., C. 1520; nárjíle-bahrí, *Arab.*, L.
511; náshpáti, *Pers.*, P. 1452; nashtar,
Pers., C. 846; navamáliká, *Sans.*, J. 35;
navamallika, *Sans.*, J. 13; navanía,
Sans., B. 983; nazbu, *Pers.*, O. 18.
Ne Nelumbium, *Sans.*, E. 569; nepala, *Sans.*,
J. 41; nespava, *Sans.*, D. 789.
Ni Níb, *Pers.*, M. 363; nichula, *Sans.*, B.
180; nidigdhika, *Sans.*, S. 2345; nift,
Arab., P. 436; nihsuki, *Sans.*, T.634;
nikumba, *Sans.*, J. 52; níl, *Pers.*, I. 145;
níla-ghiria, *Sans.*, C. 1403; nílaghírie
kurní, *Sans.*, C. 1403; nílah, *Pers.*, I.
145; nílaj, *Arab.*, I. 145; níla-nirundí,
Sans., J. 116; níli, *Sans.*, I. 145; nílika,
Sans., I. 145; nílini, *Sans.*, I. 145; níl-
kantha, *Sans.*, P. 350; nilotpala, *Sans.*,
N. 209; nilufar, *Arab.*, *Pers.*, N. 200;
nilufer, *Arab.*, *Pers.*, N. 39; nilufu, *Pers.*,
N. 39; nimba, *Sans.*, M. 363; nimba-
patram, *Sans.*, M. 800; nimba-vrikshaha,
Sans., M. 363; nimbuka, *Sans.*, C.1296;
nimu, *Arab.*, C. 1296; nípa, *Sans.*, A.
1192; nirgundi, *Sans.*, V. 164; nir-
notsjil, *Sans.*, C. 1377; nirvishá (Kyllinga), *Sans.*, K. 47; nir-visha (Delphinium),
Sans., D. 254; nisá, *Sans.*, C. 2433; nis-
hashta, *Pers.*, S. 2682; nishkooti, *Sans.*,
E. 151; nishpáva (Dolichos), *Sans.*, D.
789; nishpáva (Vigna), *Sans.*, V. 116;
nisomali, *Sans.*, P. 1078; nivára, *Sans.*,
O. 258.
No Nohás, *Arab.*, C. 2361; nokra, *Pers.*, S.
2205; nóshádar, *Pers.*, A. 962.
Nu Nubátussibi, *Arab.*, A. 829; numaketaám,
Pers., S. 602; nuqrah, *Pers.*, A. 1359;
núrah, *Pers.*, C. 489.

Oo Ood, *Arab.*, A. 1252; ood-hindi, *Arab.*, A.
1252.
Os Oschor, *Arab.*, C. 170; oshmor, *Arab.*, C.
170.
Ou Oushneh, *Pers.*, O. 220.

Pa Padma (Euryale), *Sans.*, E. 569; padma
(Nelumbium), *Sans.*, N. 39; padmachári,
Sans., N. 39; padmachárini, *Sans.*, H.
224; padmaka, *Sans.*, P. 1333; pad-
maksh, *Sans.*, P. 1333; paiwand-i-mir-
yam, *Pers.*, P. 1313; palah, *Pers.*, B.
944; palándu, *Sans.*, A. 769; palang,
Pers., T. 434; palanggini, *Sans.*, M.
232; palangmishk, *Pers.*, O. 28; pálanki,
Sans., B. 480; palása, *Sans.*, B. 944;
palitmandár, *Sans.*, E. 342; pambah,
Pers., G. 404; panasa, *Sans.*, A. 1489;
panceruba, *Sans.*, N. 39; panír bad,
Pers., C. 225; panír-bad, *Pers.*, W. 93;
panírmaya, *Pers.*, R. 73; panj-angusht,
Pers., V. 164; panj-angusha-ábí, *Pers.*,
V. 181; panna, *Pers.*, L. 126; páppána,
Sans., P. 338; párada, *Sans.*, M. 473;
páradaha, *Sans.*, M. 473; párasikaya,
Sans., H. 525; paravata-yeranda, *Sans.*,
J. 41; par-e-siyá-washán, *Pers.*, A. 510;
párijatáka, *Sans.*, N. 179; párisa, *Sans.*,
T. 392; parkati, *Sans.*, F. 216; parnása,

Sans., O. 31; parpata, *Sans.*, O. 132;
parvata, *Sans.*, C. 1263; parvatanimba-
vrikshaha, *Sans.*, M. 393; pásang, *Pers.*,
S. 1233; páshána bhedi, *Sans.*, C. 1715;
pata, *Sans.*, S. 1688; pátalá, *Sans.*, S.
2876; páthá, *Sans.*, S. 2794; pathmapu-
todami, *Sans.*, 1. 368; pathyá, *Sans.*, T.
325; patí, *Sans.*, P. 625; patmapu,
Sans., I. 368; patola, *Sans.*, T. 576, T.
586; patránga, *Sans.*, S. 2501; patta,
Sans., C. 1861, C. 1879; pattakarie,
Sans., E. 527; pattánga, *Sans.*, C. 35;
pattra bunga, *Sans.*, A. 1395; páusu,
Sans., M. 237.

Pe Pechak, *Pers.*, H. 92.

Ph Phalapúrá, *Sans.*, C. 1270; phala-traya,
Sans., G. 664; phánita, *Sans.*, S. 65;
phenila, *Sans.*, S. 808, S. 818; phushpa
kásis, *Sans.*, 1. 481.

Pi Pilpil, *Pers.*, P. 805, P. 811; pilpile-surkh,
Pers.. C. 448; pilu, *Arab.*, *Sans.*, S. 717;
píl-zahrah, *Pers.*, B. 443; pinda hari-tala,
Sans., O. 242; pindakharjura, *Sans.*, P.
555; pindálu, *Sans.*, D. 513; pindára,
Sans., T. 525; pindi, *Sans.*, R. 656; pin-
dítuka, *Sans.*, V. 25; pipal, *Pers.*, P.
805; pippali, *Sans.*, P. 805; pippalu,
Sans., P. 805; pippulí, *Sans.*, P. 805;
pista, *Pers.*, P. 858; pistá, *Pers.*, P. 858;
pistalik, *Pers.*, P. 858; pitabhringi, *Sans.*,
W. 25; pitachandana, *Sans.*, S. 790;
pithari, *Sans.*, G. 247; píti, *Sans.*, H.
414; pitshirish, *Sans.*, A. 695; pitta,
Sans., F. 48; pittavriksh, *Sans.*, S. 2651;
piyála, *Sans.*, B. 913; payáz, *Pers.*, A.
769; piyáze-dashtié-hindí, *Pers.*, U. 39;
piyáze-móshe-hindí, *Pers.*, U. 39.

Pl Plaksha, *Sans.*, F. 216.

Po Pópal, *Pers.*, A. 1294; porusha, *Sans.*, G.
663; póstékóknár, *Pers.*, P. 87; póstu-
béjam, *Sans.*, P. 87.

Pr Prabála, *Sans.*, C. 1808; prabúnátha,
Sans., C. 797; práchínama-laka, *Sans.*,
F. 603; prasáran, *Sans.*, 1. 342; prasá-
rani, *Sans.*, P. 4; pratápasa, *Sans.*, C.
170; pratihása, *Sans.*, N. 80; pravála,
Sans., C. 1808; prisniparni, *Sans.*, U.
23; prithweeka, *Sans.*, E. 151; priyangu
(Aglaia), *Sans.*, A. 644; priyangu
(Setaria), *Sans.*, S. 1212; priyunger,
Sans., P. 1313; pruthvítailam, *Sans.*,
P. 436.

Pu Púdinah, *Pers.*, M. 447; púdneh, *Pers.*, M.
465; púga-phalam, *Sans.*, A. 1294;
punaraví, *Sans.*, T. 537; punarnavá,
Sans., B. 619; punar-navi, *Sans.*, T. 530;
pundarika, *Sans.*, S. 30; púndra, *Sans.*,
S. 30; punnaga (Mallotus), *Sans.*, M.
71; punnága (Calophyllum), *Sans.*, C.
146; pushcara, *Sans.*, N. 39; pushkara
mulaka, *Sans.*, C. 2013; pushparakta,
Sans., P. 393; putiká, *Sans.*, B. 207;
pútikaranja, *Sans.*, C. 6; putranjiva,
Sans., P. 1433; putulika, *Sans.*, T. 586;
puvitrung, *Sans.*, E. 252.

Qa Qafral-yahúd, *Arab.*, P. 436; qahvá, *Arab.*,
Pers., C. 1641; qákilahekalán, *Pers.*, A.
976; qákilahe-kibár, *Arab.*, A. 976;
qalambak, *Arab.*, C. 1286; qalqand,
Arab., C. 2367; qand, *Arab.*, S. 370;
qande-suféd, *Pers.*, S. 376; qaqilah,
Arab., E. 151; qaqi-lahe-sighár, *Arab.*,

E. 151; qáqilahe zakar, *Pers.*, A. 975;
qarnulbahr, *Arab.*, A. 955; qasab,
Arab., B. 118; qasabhuvá, *Arab.*, A.
1064; qasabus-sakar, *Arab.*, S. 30; qasa-
buzzarírah (Andrographis), *Arab.*, A.
1064; qasabuzzarirah (Swertia), *Arab.*,
Pers., S. 3018.

Qi Qilí, *Arab.*, C. 541; qirfahe-sailáníyah,
Arab., C. 1196; qirtum, *Arab.*, C. 637;
qisául-barri, *Arab.*, M. 626; qishrulkhash-
khásh, *Arab.*, P. 87.

Qu Quandól, *Arab.*, M. 869; que-ábi, *Arab.*,
V. 181; qulqás, *Arab.*, C. 1732; qurtum,
Arab., C. 637.

Ra Racta-vinda-chada, *Sans.*, E. 549; rad,
Sans., P. 63; rági, *Sans.*, E. 170;
raiata, *Sans.*, S. 2205; rájádani, *Sans.*,
M. 583; rájakoshátaki, *Sans.*, L. 569;
rájamásha, *Sans.*, V. 116; rajanaku-
chandana, *Sans.*, P. 1381; rajanígandha,
Sans., P. 1044; rájanikasa, *Sans.*, N.
179; rajata, *Sans.*, A. 1359; ráiataru,
Sans., C. 756; rájika (Eleusine), *Sans.*,
E. 170; rájiká (Brassica), *Sans.*, B. 817,
B. 833, B. 841; rajiva, *Sans.*, N. 39;
raktachandana, *Sans.*, P. 1381; rakta-
chitraka, *Sans.*, P. 979; rakta éránda,
Sans., R. 369; raktaka, *Sans.*, 1. 513;
rakta khurnah, *Sans.*, S. 2424; rakta-
pósta-vrikshaha, *Sans.*, P. 82; rakta-
shikha, *Sans.*, P. 979; rakta-valli,
Sans., V. 55; rákuchi, *Sans.*, V. 73;
rála, *Sans.*, S. 1656; rambhá, *Sans.*, M.
811; raml, *Arab.*, S. 771; ráná, *Arab.*,
P. 1426; ranga, *Sans.*, T. 460; ranjani,
Sans., 1. 145; rasa, *Sans.*, M. 473;
rasagandhaha, *Sans.*, IB. 48; rasa-karpura,
Sans., M. 473; rasam, *Sans.*, M. 473;
rásan, *Arab.*, I. 333; rásná (Vanda), *Sans.*,
V. 17; rásná (Acampe), *Sans.*, A. 317;
rassás, *Arab.*, L. 143; ratah, *Pers.*, S.
818; rá-vande hindí, *Arab.*, R. 215; raza-
neh-rúmí, *Pers.*, P. 727; razani, *Pers.*,
P. 727; rázíyanah, *Pers.*, P. 727; razi-
yanahe- khatáí, *Pers.*, 1. 30; rázíyánaj,
Arab., P. 727.

Re Rechanaka, *Sans.*, M. 71; réghan-i-kanjak,
Pers., P. 847; resás, *Arab.*, T. 460;
reschad, *Arab.*, L. 283; résha-i-khitmí,
Pers., A. 880; révande-hindí, *Pers.*, R.
215.

Ri Rihan, *Arab.*, O. 18; ríme áhan, *Pers.*, I.
472; rita, *Arab.*, S. 818; riziyanaje,
Arab., I. 30.

Ro Rob-a-sus, *Pers.*, G. 278; rodan, *Pers.*,
R. 580; rodang, *Pers.*, R. 580; róghane-
balasán, *Pers.*, B. 54; róghane-bándinj,
Pers., C. 1520; roghane kunjad, *Pers.*,
S. 1078; róghane-nárgíl, *Pers.*, C. 1520;
róghane-shírín, *Pers.*, S. 1078; roghen,
Pers., S. 1078; rohituka, *Sans.*, A.1988;
rohuna, *Sans.*, S. 2501; roupya, *Sans.*,
A. 1359.

Ru Rubbi-revánd, *Arab.*, *Pers.*, G. 66; rub-
bussús, *Arab.*, G. 278; rudráksha, *Sans.*,
E. 57; rúmí, *Pers.*, P. 727; rumman,
Arab., P. 1426; rúnás, *Pers.*, R. 564;
rupya, *Sans.*, S. 2205; rusala, *Sans.*, S.
30; ruvuka, *Sans.*, R. 369; ruvya, *Sans.*,
D. 428.

Sa Saalabmisri, *Arab.*, *Pers.*, S. 521 ; sabaat azlaa, *Arab.*, P. 930 ; sabafrapatra, *Sans.*, N. 39 ; sabárá (Tamarindus), *Arab.*, T. 28 ; sabbárá (Aloe), *Arab.*, A. 829; sabir, *Arab.*, A. 824 ; sadápaha, *Sans.*, R. 663 ; safarjal, *Arab.*, C. 2546; safarjale-hindí, *Arab.*, *Pers.*, A. 534 ; saféda, *Pers.*, L. 143 ; safral, *Arab.*, F. 48 ; sagafiún, *Pers.*, S. 493 ; sagpistán, *Pers.*, C. 1940; sahadevi, *Sans.*, V. 79 ; sailaja, *Sans.*, L. 332; sailáníyah, *Pers.*, C. 1196 ; saileya, *Sans.*, L. 332; sáj, *Arab.*, T. 232 ; sáj, *Pers.*, T. 232 ; sáka, *Sans.*, T. 232 ; sakachara, *Sans.*, L. 126 ; sakbínaj, *Arab.*, S. 493; sákhotaka, *Sans.*, S. 2912 ; sakkarul-abyaz, *Arab.*, S. 375 ; sakkarul-hind, *Arab.*, S. 373; sákmúnia, *Arab.*, *Pers.*, C. 1783 ; sál, *Pers.*, T. 232 ; sala, *Sans.*, S.1656; salab, *Pers.*, O. 205 ; sala jet, *Arab.*, L. 455 ; salap, *Pers.*, O. 205 ; sála parni, *Sans.*, D. 339 ; salasiniryása, *Sans.*, B. 771 ; sallaki, *Sans.*, B. 771 ; salmali (Eriodendron), *Sans.*, E. 296; sálmalí (Bombax), *Sans.*, B. 632 ; saman, *Arab.*, J. 35 ; saman, *Sans.*, T. 634; samaratul-asl, *Arab.*, T. 51, T. 70 ; samaratul-tarfá, *Arab.*, T. 70 ; samárugh, *Pers.*, A. 590, F. 725 ; sámbala kshára, *Sans.*, A. 1425 ; sambuka-bhasma, *Sans.*, C. 489 ; samgh-i-arabi, *Arab.*, *Pers.*, A. 102 ; samgh-i-arabi, *Pers.*, A. 282 ; samghul, *Arob.* A., 102 ; sammulfár, *Arab.*, A. 1425 ; samratul-arraar, *Arab.*, J. 78 ; samudra-palaka, *Sans.*, A. 1362 ; san (Crotalaria), *Pers.*, C. 2105 ; san (Linum), *Sans.*, L. 385 ; sansa, *Sans.*, C. 2105 ; saná-e-hindí, *Arab.* & *Pers.*, C. 737 ; sandal-abiyaz, *Arab.*, S. 790 ; sandaleah-mar, *Arab.*, P. 1381 ; sandalesurkh, *Pers.*, P. 1381 ; sandal supéd, *Pers.*, S. 790 ; sandarús, *Pers.*, C. 142 ; sandhyaka, *Sans.*, N. 200 ; sangeáhanrubá, *Pers.*, I. 472 ; sange surma, *Pers.*, A. 1224 ; sangíchamak, *Pers.*, I. 472 ; sang-i-yashm, *Pers.*, J. 2 ; sankhaphuli, *Sans.*, V. 136 ; sangkhi, *Sans.*, V. 136 ; sang sabóyah, *Pers.*, C. 797; sanipát, *Sans.*, S. 964 ; sanjsabóyah, *Arab.*, C. 797 ; sankhabhasm, *Sans.*, C. 489 ; sankhálu, *Sans.*, P. 1 ; sankhapushpi, *Sans.*, C. 382 ; sankhavisha, *Sans.*, A. 1425 ; sanóbare-hindí, *Pers.*, C. 846 ; sanóbarul-hind, *Arab.*, C. 846 ; sanviránjana, *Sans.*, L. 143 ; sapistán, *Pers.*, C. 1940 ; sapta chhada, *Sans.*, A. 871 ; saptala (Jasminum), *Sans.*, J. 13 ; saptalá (Acacia), *Sans.*, A. 200 ; saptaparna, *Sans.*, A. 871 ; sarala, *Sans.*, P. 760 ; sarala drava, *Sans.*, P. 760 ; sarapun khá, *Pers.*, T. 270 ; sarasa, *Sans.*, N. 39 ; sarasiruba, *Sans.*, N. 39 ; sarata, *Sans.*, R. 110 ; saríqún, *Arab.*, *Pers.*, A. 1452, A. 1460 ; sárivá (Hemidesmus), *Sans.*, H. 119 ; sárivá (Ichnocarpus), *Sans.*, I. 1 ; sarivádvaya, *Sans.*, H.121 ; sarjikákshára, *Sans.*, C. 541 ; sarkara, *Sans.*, S. 65, S. 375 ; sarpagandhá, *Sans.*, R. 57 ; sarpákshi, *Sans.*, O. 180 ; sárshaf, *Pers.*, B. 841 ; sarshap, *Sans.*, B. 841 ; sarshapa, *Sans.*, B. 812 ; sarúreahmar, *Pers.*, M. 473 ; sarvajayá, *Sans.*, C. 321 ; sásam, *Arab.*, D. 64; sásim, *Arab.*, D. 64 ; satamúli, *Sans.*, A. 1575 ; satapatra, *Sans.*, N. 39 ; sataprása,

Sans., N. 80 ; satapushpi, *Sans.*, P. 460 ; sati, *Sans.*, C. 2422, C. 2499 ; satila, *Sans.*, P. 885 ; satte-giló, *Arab.*, T. 470, T. 483 ; saurab, *Sans.*, C. 2083 ; savaram, *Sans.*, I. 342.

Se Séb, *Pers.*, P. 1463 ; seba, *Sans.*, P. 1463 ; seehoondee, *Sans.*, E. 496 ; séf, *Pers.*, P. 1463 ; sehunda, *Sans.*, E. 520 ; semen, *Arab.*, S. 1078 ; semsem, *Arab.*, S.1078 ; sephalica, *Sans.*, N. 179 ; serag-al-coshrob, *Arab.*, M. 128 ; setapushpa, *Sans.*, P. 727 ; seubbára, *Arab.*, A. 603 ; seunti, *Sans.*, C. 1043 ; seva, *Sans.*, P. 1463.

Sh Shaaír, *Arab.*, H. 382 ; shad- grandhika, *Sans.*, C. 2381 ; shadgranthá, *Sans.*, A. 430 ; shahad, *Pers.*, H. 342 ; sháhasfaram, *Arab.*, O. 18 ; shah-sufiam, *Pers.*, M. 465 ; sháhtarah, *Pers.*, F. 721, F. 723 ; shair-ul-jin, *Arab.*, A. 498 ; shajratud-dévdár, *Arab.*, C. 846 ; shajratul-jouze-hindi, *Arab.*, C. 1520 ; shajratul-mouz, *Arab.*, M. 811 ; shajratun-nárjíl, *Arab.*, C. 1520 ; shajraturrummán, *Arab.*, P. 1426 ; shajzátultalh, *Arab.*, M. 811 ; shakákul, *Arab.*, *Pers.*, A. 1577 ; shakakul-misri, *Arab.*, E. 335 ; shakare-hindí, *Pers.*, S. 373 ; shakare-supéd, *Pers.*, S. 375 ; shakkar, *Pers.*, S. 375 ; shakull, *Pers.*, C. 49 ; shalgham, *Pers.*, B. 811 ; shaljam, *Arab.*, B. 811 ; shál-mali-vrikshaha, *Sans.*, M. 756 ; shama, *Arab.*, C. 931, H. 342 ; shamár, *Arab.*, P. 727 ; shambalíd, *Pers.*, C. 1704 ; shamlíd, *Pers.*, T. 612 ; shamlít, *Pers.*, T. 612 ; sham líthe-dash-tí, *Pers.*, S. 1714 ; shamlíz, *Pers.*, T. 612 ; shanah, *Pers.*, A. 89 ; shanbalíd, *Pers.*, T. 612 ; shanbalídebarri, *Pers.*, S. 1714 ; shangam-kuppi, *Sans.*, C. 1377 ; shangarf, *Pers.*, M. 473 ; shanjarf, *Arab.*, M. 473 ; shaqáqul, *Arab.*, *Pers.*, A. 1577 ; sháqáqul, *Pers.*, *Arab.*, A. 1575 ; shaqáqule-hindi, *Arab.*, *Pers.*, A. 1562 ; shará, *Sans.*, S. 6 ; sharáb, *Pers.*, V. 243 ; sharíiah, *Arab.*, *Pers.*, A. 1166 ; shateraj, *Arab.*, F. 721 ; shatra, *Pers.*, F. 723 ; shávaka, *Sans.*, T. 70 ; shayakah, *Pers.*, A. 1592 ; shaz, *Arab.*, C. 49 ; shazzir, *Arab.*, P. 355 ; sheá-bhra, *Sans.*, M. 509 ; shévániarba, *Sans.*, I. 111 ; shévantiká, *Sans.*, C. 1043 ; shib, *Arab.*, A. 897 ; shíb, *Arab.*, A. 1454 ; shíh, *Arab.*, *Pers.*, A. 1452, A. 1460 ; shihatuta, *Pers.*, M. 756 ; shikam-daridah, *Pers.*, P. 932 ; shikhamúlam, *Sans.*, D. 173 ; shikhár, *Pers.*, C. 541 ; shilm, *Pers.*, P. 847 ; shimbí (Canavalia), *Sans.*, C. 289 ; shimbí (Dolichos), *Sans.*, D. 789 ; shingshupa, *Sans.*, D. 64 ; shir, *Arab.*, A. 498 ; shírraj, *Arab.*, S. 1078 ; shírkhisht (Fraxinus), *Arab.*, *Pers.*, F. 696 ; shír-khist (manna), *Pers.*, M. 235 ; shirtothar, *Pers.*, E. 553 ; shirul-jibal, *Arab.*, A. 510 ; shirul-jir n, *Arab.*, A. 510 ; shíshah, *Pers.*, G. 229 ; shitaraj, *Arab.*, P. 986 ; shitaraje-ahmar, *Arab.*, P. 979 ; shitarak, *Pers.*, P. 986 ; shitarake-surkh, *Pers.*, P. 979 ; shitirah, *Pers.*, P. 986 ; shitturridge, *Arab.*, P. 979 ; shivás neem, *Sans.*, I. 111 ; shóbhánjana-vrikshaha, *Sans.*, M. 721 ; shód, *Sans.*, I. 107 ; shora, *Pers.*, S. 682 ; shórah, *Pers.*, S. 682 ; shór-gaz, *Pers.*, T. 70 ; shoshmir, *Arab.*, E. 151 ; shoukul-jamal, *Arab.*, A

745; shoūndi, *Sans.,* P. 805 ; sh-ouníz, *Arab.,* N. 158; showkrán, *Arab.,* C. 1765 ; shubit, *Arab.,* P. 460 ; shúftálú, *Pers.,* P. 1322 ; shúk, *Arab.,* A. 1425 ; shúklika, *Sans.,* O. 547 ; shukta, *Sans.,* A. 356 ; shul, *Arab., Pers.,* A. 534 ; shunaka-barbara, *Sans.,* C. 1367 ; shunkhiní, *Sans.,* C. 1053 ; shúrákasthíka, *Sans.,* P. 625 ; shutar-khár, *Pers.,* A. 745, C. 224; shutavedhí, *Sans.,* R. 650; shvéta-barbúra vrikshaha, *Sans.,* A. 249 ; shvétapanjasáram, *Sans.,* S. 375 ; shvetasarshapa, *Sans.,* B. 800 ; shvétas-harkará, *Sans.,* S. 375 ; shvéta-surasa, *Sans.,* V. 164; shyámaka, *Sans.,* P. 53.

Si Síb, *Pers.,* P. 1463 ; sibr, *Pers.,* A. 827, A. 829 ; sibr-sagótari, *Arab.,* A. 824 ; sibre-surkho, *Pers.,* A. 838 ; siddartha (Eruca), *Sans.,* E. 327 ; siddhártha (Brassica), *Sans.,* B. 800 ; sidhu, *Sans.,* S. 65 ; sidr, *Arab.,* Z. 231 ; sigru, *Sans.,* M. 721 ; siktha, *Sans.,* C. 931 ; silarumba, *Sans.,* C. 321 ; silhaka, *Sans.,* L. 455 ; sillarus, *Arab.,* L. 455 ; sim, *Pers.,* S. 2205 ; simáb, *Pers.,* M. 473 ; símáhang, *Pers.,* M. 626; simbi, *Sans.,* D. 758, D. 789; sim-sim, *Arab.,* S. 1078; sin, *Pers.,* A. 1359; sindhuvára, *Sans.,* V. 164; sindika (Boerhavia), *Sans.,* B. 619; sindika (Diospyros), *Sans.,* D. 582; sindura, *Sans.,* L. 143 ; sing giká, *Sans.,* C. 1861; sinjid-ijiláni, *Pers.,* Z. 280; sínsapá, *Sans.,* D. 64; sipandáne-siyah, *Pers.,* B. 812 ; sipandáne-supíd, *Pers.,* B. 800; sir, *Pers.,* A. 779 ; sireshame-mahi, *Pers.,* A. 393; sirkah, *Pers.,* A. 356 ; sirkahé-nakhúd, *Pers.,* C. 1061 ; sír-siápesháne, *Pers.,* A. 498; sísaka, *Sans.,* L. 143 ; sisbán, *Pers.,* V. 164 ; sisúmar, *Sans.,* W. 58; sitáfal, *Arab.,* A. 1166 ; sitalapatriká, *Sans.,* P. 625 ; sitali, *Sans.,* L. 355 ; sitopalá, *Sans.,* S. 65, S. 377; situnda, *Sans.,* E. 496; situshúka, *Sans.,* H. 382; siyahbiranj, *Pers.,* N. 158; siyáhdánah, *Pers.,* N. 158 ; siyáhgush. *Pers.,* T. 426 ; siyah músií, *Pers.,* A. 1122 ; síyáhzírah, *Pers.,* C. 697.

Sj Sjetun, *Arab.,* O. 145.

Sn Snehaphala, *Sans.,* S. 1078 ; snuhi, *Sans.,* E. 520.

So Soad, *Arab.,* C. 2617 ; soade-kúíí, *Arab.,* C. 2617 ; sóbhánjana, *Sans.,* M. 721 ; sodáda, *Arab.,* C. 402 ; soma, *Sans.,* S. 882 ; somalata, *Sans.,* R. 663 ; sómarája, *Sans.,* P. 1352 ; somaráji, *Sans.,* V. 73 ; sóma vallí, *Sans.,* T. 470, T. 483 ; sophore, *Sans.,* F. 351 ; sorasaw, *Sans.,* O. 31 ; sothaghni, *Sans.,* B. 619.

Sp Sphatikari, *Sans.,* A. 897.

Sr S'reyasí, *Sans.,* S. 970 ; srigála kantá, *Sans.,* A. 1351 ; srigála koli, *Sans.,* Z. 263; sríhastiní, *Sans.,* H. 102 ; srikhanda, *Sans.,* S. 790 ; sringátaka, *Sans.,* T. 516 ; sringavéra, *Sans.,* Z. 201 ; sringí, *Sans.,* A. 398 ; sriphal, *Sans.,* A. 534; sríphali, *Sans.,* I. 145 ; sripnari, *Sans.,* G. 287; sritálam, *Sans.,* C. 1995 ; srivása, *Sans.,* P. 760.

St Stategiló, *Pers.,* T. 470, T. 483 ; sthalapadma, *Sans.,* H. 224; sthala sringataka, *Sans.,* T. 548; sthulagranthi, *Sans.,* Z. 225.

Su Sudáb, *Pers.,* R. 663 ; sudhá, *Sans.,* C. 489 ; súdúmstra, *Sans.,* T. 548 ; sufir, *Arab.,* S. 855 ; sugandha kantak, *Sans.,* P. 1352 ; sugandha-muricha, *Sans.,* P. 801 ; sugandhi, *Sans.,* H. 119 ; sugbínuj, *Arab.,* S. 493 ; sugmonia, *Arab., Pers.,* C. 1783 ; súgpistan, *Pers.,* C. 1931 ; sujjádo, *Pers.,* H. 177 ; sukasa, *Sans.,* C. 2287 ; sukkar, *Arab.,* S. 375 ; sukkarul-ghushar, *Arab.,* M. 235 ; suktíbhasma, *Sans.,* C. 489 ; sum, *Arab.,* A. 779 ; sumak, *Arab., Pers.,* R. 287 ; suman, *Arab.,* J. 35; sumáque-amriquah, *Arab., Pers.,* C. 19 ; sumbulul-aasáffir, *Arab.,* N. 17 ; sumbulúl-hind, *Arab.,* N. 71 ; sum-el-himar, *Arab.,* N. 80; sun, *Pers.,* P. 1381 ; sunandá hari, *Sans.,* A. 1398 ; sunbuluttib, *Pers.,* N. 17 ; sunbuluttibehindi, *Arab.,* N. 17 ; sung-i-marmar, *Pers.,* M. 277 ; sunti, *Sans.,* Z. 201 ; súparsha-vaka, *Sans.,* T. 392 ; surá, *Sans.,* C. 960 ; surabhí, *Sans.,* M. 800 ; surabhí-nimbu, *Sans.,* M. 800 ; surasa-vrikshaha, *Sans.,* V. 181 ; surb, *Pers.,* L. 143; suria-mukhi, *Sans.,* H. 74; súringán, *Arab.,* C. 1703; surjávarta, *Sans.,* G. 753; surmah, *Pers.,* A. 1224 ; surmah-i-isfahani, *Pers.,* A. 1224 ; suryavarta, *Sans.,* C. 2211 ; súsán-i-ásmán-júní, *Pers.,* I. 429 ; sushave, *Sans.,* N. 158 ; sushavi (Carum), *Sans.,* C. 681 ; sushavi (Momordica), *Sans.,* M. 626; sutur, *Arab.,* O. 220 ; suvarna, *Sans.,* G. 317; suvarnaka, *Sans.,* C. 756; suvarnam, *Sans.,* A. 1622 ; suvarna-patram, *Sans.,* A. 1622.

Sv Svádu khanda, *Sans.,* S. 372; svana-burbárá *Sans.,* C. 1367 ; svarnajuthiká, *Sans.,* J. 24 ; svarnnamákshika, *Sans.,* I. 478 ; svaytaurkum, *Sans.,* C. 170 ; sveta, *Sans.,* S. 2205 ; svetakanchan, *Sans.,* B. 318.

Sw Swarna, *Sans.,* G. 317 ; swenti, *Sans.,* C. 1043.

Sy Syámák, *Sans.,* P. 53 ; syonáka, *Sans.,* O. 233.

Ta Taam, *Arab.,* S. 2424 ; tabashír, *Arab., Pers.,* B. 84; taftaf, *Arab.,* C. 551 ; tagara, *Sans.,* T. 3 ; táj-e-khurús, *Pers.,* A. 939; tála, *Sans.,* B. 663 ; tálamulika, *Sans.,* C. 2375 ; tali, *Sans.,* C. 1995 ; tálib-el-khubz, *Arab.,* C. 691 ; talikhahe, *Pers.,* C. 1196 ; tálisapatra, *Sans.,* A. 25 ; talisha, *Sans.,* F. 603 ; talis-patar, *Pers.,* F. 603 ; talk, *Pers.,* M. 509 ; tamál, *Sans.,* C. 1183 ; tamála, *Sans.,* G. 99 ; tamála-vrikshaha, *Sans.,* P. 1121; támarasa, *Sans.,* N. 39; tamar-i-hindí, *Arab., Pers.,* T. 28 ; tamból, *Pers.,* P. 775 ; támbula, *Sans.,* P. 775 ; támra, *Sans.,* C. 2361 ; tambák, *Arab.,* N. 101 ; tanbáku, *Pers.,* N. 101; tanból, *Arab.,* P. 775 ; tanduliya, *Sans.,* A. 941, A. 943 ; tankana, *Sans.,* B. 731 ; tankári, *Sans.,* P. 682; taon, *Arab.,* P. 350 ; táous, *Arab.,* P. 350 ; tar, *Pers.,* A. 1622 ; táramákshika, *Sans.,* I. 478 ; tarambuja, *Sans.,* C. 1221 ; taranjabín, *Arab.,* M. 235 ; tarfá, *Arab.,* T. 70; tarfál-ahmar, *Arab.,* T. 51 ; tári, *Pers.,* B. 663 ; táríye-nárgíl, *Pers.,* C. 1520 ; tarúlatá, *Sans.,* I. 405 ; tashmízaj, *Arab.,* C. 728 ; tátulah, *Pers.,*

D. 150; tátúrahesaféd, *Pers.*, D. 160;
tátúrahe-siyáh, *Pers.*, D. 151.
Te Tejanaka, *Sans.*, S. 6; ten, *Arab.*, F. 149;
téntráni, *Sans.*, T. 28; terenjabín, *Arab.*,
M. 235; terenjabiri, *Arab.*, M. 235; tes-
patra, *Sans.*, C. 1183; teys, *Arab.*, S.
1290.
Th Thil, *Pers.*, G. 317; thus, *Arab.*, *Pers.*,
B. 751.
Ti Tibr, *Arab.*, G. 317; tikta, *Sans.*, C. 1789;
tiktadugdha, *Sans.*, O. 600; tiktaka,
Sans., Z. 1; tikta-shaka, *Sans.*, C. 2039;
tila, *Sans.*, S. 1078; tilá, *Pers.*, A. 1622;
tilaha, *Sans.*, S. 1078; tilapari, *Sans.*,
P. 1381; tilataila, *Sans.*, S. 1078; tilla,
Pers., G. 317; tillaka, *Sans.*, S. 3062;
tindisa, *Sans.*, H. 196; tinduka, *Sans.*,
D. 582; tine-barri, *Arab.*, F. 202; tine-
gázur, *Pers.*, C. 541; tinisa sejanduna,
Sans., O. 537; tinkár tankár, *Pers.*, B.
731; tintidi, *Sans.*, T. 28; tintili, *Sans.*,
T. 28; tintiri, *Sans.*, T. 28.
To Tochém-keweh, *Pers.*, C. 1641; toffah,
Arab., E. 432; towdri, *Pers.*, L. 283;
towdrie, *Pers.*, M. 115; toyapippali,
Sans., S. 842.
Tr Trapu, *Sans.*, T. 460; trapusha, *Sans.*,
C. 2287; trayaman, *Pers.*, D. 271;
tráyamáná, *Sans.*, F. 194; trikantaka,
Sans., T. 548; tripakshí, *Sans.*, C. 1707;
triputá, *Sans.*, I. 415; triputi, *Sans.*, L.
100; trivrit, *Sans.*, I. 415.
Tu Tuda, *Sans.*, M. 756; tufah-ul-shi-tan,
Arab., M. 128; tuffah, *Arab.*, P. 1463;
tuffa-urmena, *Arab.*, P. 1285; tukhme-
ábi, *Pers.*, C. 2546; tukhmeahare-talkh,
Pers., H. 294; tukhmebalasán, *Pers.*, B.
54; tukhme-bikhehayát, *Pers.*, C. 1403;
tukhme-karafs, *Pers.*, C. 701; tukhme-
katán, *Pers.*, L. 385; tukhme-khiyár,
Pers., C. 2278; tukhme-khiyare-daraz,
Pers., C. 2278; tukhme-khiyarzah, *Pers.*,
C. 2278; tukhme-khurfah, *Pers.*, P. 1179;
tukhme-koknar, *Pers.*, P. 87; tukhme-
turrahtézak, *Pers.*, L. 283; tukhm-i-
búlangú, *Pers.*, L. 67; túkhmí khiyúrain,
Pers., C. 2290; tukin-i-nil, *Pers.*, I. 393;
tukm-i-khitmi, *Pers.*, A. 880; tukm-i-níl,
Pers., I. 384; tukm-i-panjangusht, *Pers.*,
V. 179; tukm-i-sarmak, *Pers.*, C. 1005;
tula, *Sans.*, M. 747, M. 756; tulashi,
Sans., O. 31; tulhtula, *Arab.*, *Pers.*, M.
811; tulí, *Sans.*, I. 145; tulk, *Pers.*, S.
2712; tumala, *Sans.*, D. 628; tumburu,
Sans., Z. 9; tumtum, *Arab.*, R. 287;
tunna, *Sans.*, C. 838; tunna-kuberaka,
Sans., C. 838; turábul-hálik, *Arab.*, A.
1425; turanj, *Pers.*, C. 1270; turb, *Pers.*,
R. 31; turbund, *Arab.*, I. 415; tureh-
korasani, *Pers.*, O. 18; túrg, *Sans.*, H.
414; túrk, *Pers.*, P. 1179; turmus, *Pers.*,
L. 578; turrah-tizkah, *Pers.*, L. 283;
tursak, *Pers.*, R. 650; turshah, *Pers.*, R.
650; túrshumuk, *Pers.*, R. 650; turuk,
Pers., P. 1179; turunjebín, *Arab.*, M.
933; tusha, *Sans.*, T. 293; tút, *Arab.*,
Pers., M. 747, M. 756; túth, *Arab.* &
Pers., M. 747; tuth-thanjanam, *Sans.*,
C. 2367; tuttha (Indigofera), *Sans.*, I.
145; túttha (copper sulphate), *Sans.*, C.
2367; tuverika, *Sans.*, B. 817, B. 822;
tuyus, *Arab.*, S. 1290.

Ub Ubdie narikaylum, *Sans.*, L. 511; ubkii,
Arab., S. 682.
Ud Úd, *Arab.*, A. 1251; údal-qarha, *Arab.*, A.
1026; úde-bálasán, *Pers.*, B. 54; úde-
hindí, *Arab.*, A. 1251; udumbara, *Sans.*,
F. 179; udumber, *Sans.*, F. 175.
Uf Uffes, *Arab.*, Q. 43.
Uj Ujáyá, *Sans.*, C. 331.
Uk Ukleel-ul-jilbul, *Arab.*, R. 547; ukti
khaman, *Arab.*, S. 769.
Ul Uluk baghdaní, *Arab.*, P. 841.
Um Uma, *Sans.*, L. 385; umblí, *Arab.*, T. 28;
ummatta-vrikshaha, *Sans.*, D. 160; um-
miatto-dumbara, *Sans.*, F. 202; ummughí-
lán, *Arab.*, A. 101; umrita, *Sans.*, P. 632.
Un Undum, *Arab.*, *Pers*, P. 1381; ungúsht-
kuni-zuckan, *Pers.*, O. 18; unmatta,
Sans., D. 151; unnáh, *Arab.*, Z. 280;
ununda, *Sans.*, C. 331.
Up Upadyki, *Sans.*, P. 1187; upakunchika,
Sans., E. 151; upana, *Sans.*, A. 1545;
upukúlya, *Sans.*, P. 805.
Ur Urista, *Sans.*, S. 808; urúk-el-káfúr, *Pers.*,
C. 2499; urúk-es-sabá-ghín, *Arab.*, C.
2433; urúk-es-subr, *Arab.*, C. 2433;
urzíz, *Pers.*, T. 460.
Us Usfar, *Arab.*, C. 637; ushak, *Arab.*, *Pers.*,
D. 810; ushana, *Sans.*, P. 811; ushar,
Arab., C. 170, C. 225; ushirah, *Arab.*, L.
332; ushtar-kbár, *Pers.*, A. 745; ushuna,
Sans., P. 805; usír, *Arab.*, A. 1097; usira,
Sans., A. 1097; 'ussulneelu-fir, *Arab.*, N.
39; ussul-ul-lufah, *Arab.*, M. 128;
usteruck, *Arab.*, L. 455; ustrung (Man-
dragora), *Arab.*, M. 128; ustrung
(Atropa), *Arab.*, A. 1614; lustúkhúdús,
Arab., L. 117; usturak, *Arab.*, S. 2978.
Ut Utpala, *Sans.*, N. 209; utraj, *Arab.*, C. 1270;
utrej, *Arab.*, C. 1270; utroj, *Arab.*, C.
1270; úturlnji, *Arab.*, C. 1270.

Va Vabbúla, *Sans.*, A. 101; vachá-ugra-gan-
dhaha, *Sans.*, A. 430; vag, *Sans.*, H.
414; vahisí, *Sans.*, M. 639; vaj, *Arab.*,
A. 430, C. 225; vaji, *Sans.*, H. 414;
vajidantakaha-atarusha, *Sans.*, A. 484;
vajjeturkí, *Pers.*, A. 401; vajrábhra,
Sans., M. 509; vajradantí, *Sans.*, B.
171; vajradru-vrikshaha, *Sans.*, C. 331;
vajrakantaka, *Sans.*, E. 496; vajravallí,
Sans., V. 219; vajri, *Sans.*, E. 496;
vakúchi, *Sans.*, P. 1352; vakula, *Sans.*,
M. 570; váláne-buzarg, *Pers.*, P. 727;
válika, *Sans.*, S. 25; vanaharidrá, *Sans.*,
C. 2406; vanakárpása, *Sans.*, H. 263;
vana malli, *Sans.*, J. 10; vanamalliká,
Sans., J. 35; vana methiká, *Sans.*, M.
422; vanapalán dam, *Sans.*, U. 39; van
árdraka, *Sans.*, Z. 199; vánari, *Sans.*,
M. 786; vanasrangátá, *Sans.*, T. 548;
vanatíktika (Cocculus), *Sans.*, C. 1452;
vanatiktika (Stephania), *Sans.*, S. 2794;
vanayamáni, *Sans.*, S. 1201; vánda,
Sans., L. 549; vanga, *Sans.*, T. 460;
vangana, *Sans.*, S. 2284; vanjula, *Sans.*,
S. 861; vansa, *Sans.*, B. 118l; vansa-
loohana, *Sans.*, B. 84; vansapatri haritala,
Sans., O. 242; vanua, *Sans.*, L. 549;
váráhak ránta, *Sans.*, M. 557; vard,
Arab., R. 504; vardáku, *Sans.*, S. 2284;
vardea-byaz, *Arab.*, J. 35; varhi, *Sans.*,
P. 350; varqe-nuqrah, *Pers.*, A. 1359;
varqesim, *Pers.*, A. 1359; varqul-fizah,

Arab., A. 1359 ; várshiki, *Sans.*, J. 35 ; vártáku, *Sans.*, S. 2284 ; varuna, *Sans.*, C. 2039 ; váruni, *Sans.*, C. 962 ; varvara, *Sans.*, O. 18 ; vásaka, *Sans.*, A. 484 ; vashira, *Sans.*, L. 451 ; vasira, *Sans.*, S. 970 ; vastuk, *Sans.*, C. 1003 ; vasuka, *Sans.*, I. 125 ; vasunta, *Sans.*, P. 468 ; vata, *Sans.*, F. 129 ; vata-dalla, *Sans.*, Z. 229 ; váta-ghní, *Sans.*, C. 1386 ; vátári, *Sans.*, R. 369 ; vatsanábha, *Sans.*, A. 397 ; vatsa priam, *Sans.*, C. 1752.

Ve Veli-man, *Sans.*, S. 1226 ; venú-lavanam, *Sans.*, B. 84 ; vetasa, *Sans.*, C. 104 ; vetra, *Sans.*, C. 77.

Vi Vibhitaka, *Sans.*, T. 293 ; vibhitaki, *Sans.*, T. 293 ; vidanga, *Sans.*, E. 199 ; vidári, *Sans.*, I. 379 ; videhí, *Sans.*, P. 805 ; vidruma, *Sans.*, C. 1808 ; vijapura, *Sans.*, C. 1270, C. 1296 ; vijáyá, *Sans.*, C. 331 ; vimba, *Sans.*, C. 919 ; vipitakaha, *Sans.*, T. 293 ; virana, *Sans.*, A. 1097 ; víra-vriksha, *Sans.*, D. 402 ; visalyakrit, *Sans.*, V. 12 ; visaprasuna, *Sans.*, N. 39 ; visha, *Sans.*, A. 397 ; visha kharpara, *Sans.*, B. 619 ; vishala, *Sans.*, C. 1211 ; visha-mandala, *Sans.*, C. 2062 ; víshamúshti, *Sans.*, S. 2943 ; vishashódhaní, *Sans.*, I. 107 ; vishnugandhi, *Sans.*, E. 581 ; vishnu kránta, *Sans.*, C. 1403 ; vishva-bhísha-gam, *Sans.*, Z. 201.

Vr Vriddhadáraka, *Sans.*, A. 11362 ; vrihati, *Sans.*, S. 2280 ; vrihi, *Sans.*, O. 258 ; vrihib-heda, *Sans.*, P. 63 ; vrijpatta, *Sans.*, C. 331 ; vrikshabhaksha, *Sans.*, L. 549 ; vriksha-daní, *Sans.*, L. 549 ; vrikshaha, *Sans.*, V. 164 ; vriksharuha, *Sans.*, L. 549 ; vrischikáli, *Sans.*, T. 509.

Vu Vujri, *Sans.*, E. 520 ; vulahwa, *Sans.*, P. 625.

Wa Wain, *Arab.*, V. 243 ; wanizad, *N.-E. Persia*, P. 847 ; wanjad, *Pers.*, P. 847 ; warahi, *Sans.*, C. 2375.

Wf Wfeheh, *Arab.*, R. 73.

Wo Worga, *Arab.*, P. 63 ; worglo, *Arab.*, P. 63.

Wu Wúlika, *Sans.*, S. 25.

Ya Yabrooz, *Pers.*, M. 128 ; yamani, *Sans.*, C. 691 ; yaqut, *Arab.*, R. 619 ; yasada, *Sans.*, Z. 190 ; yásaman, *Arab.*, J. 35 ; yása míne-barrí, *Pers.*, H. 119 ; yashm, *Pers.*, J. 2 ; yashti madhu, *Sans.*, G. 278 ; yava (Hordeum), *Sans.*, H. 382 ; yavá (Triticum), *Sans.*, T. 634 ; yavaka, *Sans.*, H. 382 ; yavakshára, *Sans.*, C. 527 ; yavakshra, (?) *Sans.*, S. 682 ; yava-nala (Zea), *Sans.*, Z. 50 ; yavanála (Sorghum), *Sans.*, S. 2424 ; yavása, *Sans.*, C. 224.

Yo Yoranná, *Arab.*, L. 126.

Yu Yugaphala, *Sans.*, D. 9 ; yuthika-purní, *Sans.*, R. 231 ; yuz, *Pers.*, T. 424 ; **yuz-palang**, *Pers.*, T. 424.

Za Zaafarán, *Arab.*, *Pers.*, C. 2083 ; zabáne-kunjashk, *Pers.*, W. 122 ; zabáne-kunjashke-shírín, *Pers.*, W. 122 ; zabáne-kunjashke-talkh, *Pers.*, H. 294 ; zabib, *Arab.*, V. 233 ; záfaránul-hadíd, *Arab.*, I. 472 ; zaghír, *Pers.*, L. 385 ; zaghú, *Pers.*, L. 385 ; zahab, *Arab.*, A. 1622, G. 317 ; zahrahe, *Pers.*, F. 48 ; zahr-ul-ajl, *Arab.*, M. 606 ; zaitún, *Pers.*, O. 145 ; zaiyán, *Arab.*, H. 119 zaj, *Arab.*, A. 897 ; záje-akhzar, *Arab.*, C. 2367 ; záje-asfar, *Arab.*, I. 481 ; zájul-akhzar, *Arab.*, C. 2367 ; zák, *Pers.*, A. 897 ; záke-sabz, *Pers.*, C. 2367 ; záke-safed, *Pers.*, A. 897 ; záke-zard, *Pers.*, I. 481 ; zakhm-haiyát, *Pers.*, B. 909 ; zaló, *Arab.*, *Pers.*, L. 245 ; zalók, *Pers.*, L. 245 ; zambak, *Pers.*, J. 35 ; zamin-kand, *Pers.*, A. 996 ; zangáre-ahan, *Pers.*, I. 472 ; zange-áhan, *Pers.*, I. 472 ; zanjabíl, *Arab.*, Z. 201 ; zanjabíle-khushk, *Pers.*, Z. 201 ; zanjabíle-ratab, *Arab.*, Z. 201 ; zanjabíle-tar, *Pers.*, Z. 201 ; zanjabíle-yábis, *Arab.*, Z. 201 ; zanjabíl-i-shámi, *Pers.*, I. 333 ; zanjafr, *Arab.*, M. 473 ; zanjarf, *Pers.*, M. 473 ; zanjárul-hadíd, *Arab.*, I. 472 ; zaqqume-hindí, *Arab.*, *Pers.*, E. 496, E. 553 ; zaquniyœ, *Pers.*, E. 553 ; zaqunniyœ-hindí, *Pers.*, E. 495 ; zarávand, *Arab.*, A. 1400 ; zarávand-daráz, *Pers.*, A. 1400 ; zarávande-gírd, *Pers.*, A. 1403 ; zarávande-hindi, *Arab.*, *Pers.*, A. 1398 ; zarávand-emudahraj, *Arab.*, A. 1403 ; zarávande-tavíl, *Arab.*, A. 1400 ; zardak, *Pers.*, D. 173 ; zard-chóbah, *Pers.*, C. 2433 ; zarír, *Arab.*, D. 271 ; zarishk, *Arab.*, *Pers.*, B. 443 ; zarishke-trush, *Pers.*, B. 469 ; zarnab (Cassia), *Arab.*, C. 1183 ; zarnab (Flacourtia), *Arab.*, F. 603 ; zarni-khe-asfar, *Arab.*, O. 242 ; zarnikhezard, *Pers.*, O. 242 ; zarsúd, *Arab.*, C. 2433 ; zaták-asturiká, *Sans.*, H. 168.

Ze Zeitun, *Arab.*, O. 145 ; zerumbád, *Pers.*, C. 2425.

Zi Zib, *Arab.*, R. 116 ; zibaq, *Arab.*, M. 473 ; zibl, *Arab.*, M. 237 ; zinián, *Pers.*, C. 691 ; zin-zeid, *Pers.*, E. 40 ; zir, *Pers.*, G. 317 ; zírá, *Pers.*, C. 2339 ; zírahe-kirmání, *Pers.*, C. 697 ; zírahe-siyáh, *Pers.*, C. 697 ; zirír, *Pers.*, B. 469 ; zirish-tursh, *Pers.*, B. 469 ; ziriskh, *Pers.*, B. 458.

Zo Zouz-el-mathil, *Arab.*, D. 150 ; zowr, *Pers.*, C. 225.

Zu Zubbí, *Arab.*, D. 212 ; zúfahyabis, *Arab.*, *Pers.*, H. 550 ; zughál, *Pers.*, C. 487, C. 1414 ; zúkhamsate-asábea, *Arab.*, V. 164 ; zukhamsatilouráq, *Arab.*, V. 164 ; zúra, *Arab.*, S. 2424 ; zurambád, *Arab.*, C. 2499 ; zúrna, *Sans.*, S. 2424 ; zurratul-makkah, *Arab.*, Z. 50.

Aa Aal, *Mar.*, M. 656; aamári, *Hind.*, A. 1215; aanbar, *Hind.*, A. 956; aaraar, *Hind.*, J. 78.

Ab Ababil-ka-ghoslah, *Hind.*, C. 1722; abai, *Mar.*, V. 233; abásí, *Pb.*, M. 606; abhal, *Dec.*, J. 78; abhasie, *Sind.*, M. 606; abhúl, *Pb.*, J. 78; abhúlas, *Sind.*, M. 921; abí, *Hind.*, C. 2546; abír, *Hind.*, A. 31, C. 2500; abnús, *Hind.*, *Mar.*, D. 569, *Hind.*, D. 615; abrak (mica), *Hind.*, M. 509; abrak (steatite), *Hind.*, S. 2712; abri, *Mar.*, M. 656; abú, *Mar.*, V. 22.

Ac Ach, *Beng.*, *Hind.*, M. 704, *Hind.*, M. 656; aich, *Beng.*, M.656; achár, *Hind.*, B. 913; achhu, *Beng.*, M. 656, *Uriya*, M. 704; achmehudi, *Hind.*, L. 589.

Ad Ada, *Hind.*, *Pb.*, Z. 201; adá, ada, *Beng.*, *Uriya*, Z. 201; adad, *Gus.*, P. 513; adah, *Afg.*, L. 252; adalsá, *Hind.*, *Mar.*, A. 484; adant, *Beng.*, *Hind.*, O. 574; adári, *Beng.*, *Hind.*, O. 574; adársa, *Hind.*, *Mar.*, A. 484; adah, *Afg.*, L. 252; addi, *Uriya*, F. 486; adei, (Abelia), *Pushtu*, A. 2; adei (Lonicera), *Pb.*, L. 535; aderay-ja-denay, *Sind*, S. 2345; adhabirni, *Beng.*, H. 149; adhruka, *Dec.*, *Hind.*, Z. 201; adhsarita-ka-jari, *Pb.*, A. 501; adhú, *Gus.*, Z. 201; adnára, *Hind.*, T. 434; adona, *Hind.*, C. 900; adrak, *Dec.*, *Hind.*, *Mar.*, *Pb.*, Z. 201; ádrok, *Beng.*, Z. 201; ádú, *Mar.*, *Gus.*, Z. 201; adulasá, *Hind.*, *Mar.*, A. 484; adulaso, *Hind.*, *Mar.*, A. 484; adulsa, *Mar.*, A. 484; aduso, *Gus.*, A. 484.

Af Afim, *Dec.*, *Hind.*, *Pb.*, P. 87; afiyun, *Hind.*, P. 87; afkar, *Sind*, C. 913; afsantín, *Mar.*, A. 1467; afsuntín, *Pb.*, A. 1469; áftímún, *Hind.*, *Pb.*, C. 2508; afú-ke-thar, *Mar.*, P. 87.

Ag Ág, *Hind.*, C. 170, 191; agal-górú, *Duk.*, A. 1026; agáni, *Hind.*, J. 104; agar (Dillenia), *Hind.*, D. 438; agar (Aquilaria), *Guj.*, *Hind.*, A. 1251; agaru, *Beng.*, A. 1251; agas, *Hind.*, M. 363; agástá, *Mar.*, S. 1186; agasti, *Beng.*, *Hind.*, S. 1186, *Mar.*, S. 1186; agathic, *Gus.*, S. 1186; agetha, *Hind.*, P. 1233; aggai, *Hind.*, D. 438; aggar, *Pb.*, C. 607; agháda, *Mar.*, A. 382; aghedo, *Guj.*, A. 382; aghzái, *Pushtu*, F. 6; aghzakai, *Pushtu*, P. 1259; aghzakár, *Pushtu*, P. 1259; agi-mali-gadi, (*Chanda*), *Hind.*, M. 232; aginbútí, *Duk.*, *Mar.*, A. 958; agiya, *Duk.*, *Mar.*, A. 958; ágiyu, *Mar.*, C. 617; agla (Mimosa), *Hind.*, M. 562; agla (Acacia), *Hind.*, A. 267; aglaia, *Kumaon*, S. 502; ágnád nemuka, *Beng.*, S. 2794; agniú, *Hind.*, P. 1242; agniun (Euonymus), *Hind.*, E. 479; ágniún (Premna), *Hind.*, P. 1242; agnu, *Kumaon*, E. 479; agrai-khaki, *Hind.*, A. 106; agsend, *Pb.*, W. 98; agumáki, *Hind.*, M. 791; agust, *Hind.*, S. 1186; agusta, *Beng.*, *Hind.*, S. 1186; aguyábát, *Uriya*, P. 1233; agyághás, *Beng.*, A. 1117; agzhan, *Pb.*, B. 33.

Ah Áhak, *Pb*, C. 489; ahalíva, *Mar.*, L. 283; ahalú, *Pb.*, V. 149; ahera, *Dec.*, T. 293; ah-hí, *Sind*, F. 546; ahír, *Mar.*, F. 506; ahirávana-mahiravana, *Guj.*, *Mar.*, B. 909; ahliva, *Mar.*, L. 283; ahreo, *Sind*, L. 283; ahsa, *Hind.*, P. 885; ahu, *Baluch.*, S. 1250.

Ai Áich, *Hind.*, M. 704; aila (Acacia), *Hind.*, A. 200; aila (Cæsalpinia), *Hind.*, C. 42; aimu, *Pb.*, S. 1264; ain, *Hind.*, *Mar.*, *Gus.*, T. 361; ainshe, *Mar.*, M. 656; ainskati, *Uriya*, V. 138; aintha, *Hind.*, S. 1024; ainthia dhamin, *Hind.*, H. 92; aira, *Pb.*, P. 702; airaj, *Uriya*, R. 127; airan, *Mar.*, C. 1386; airanamúla, *Mar.*, C. 1386; aisan, *Pb.*, T. 361; aisar, *Mar.*, C. 129; aita-lugala, *Hind.*, H. 285.

Aj Ajamá, *Guj.*, A. 1130; ajamo, *Guj.*, C. 691; ajánta bairula, *Hind.*, C. 1950; ajensak, *Beng.*, X. 8; ajmánu-pátru, *Guj.*, A. 1130; ajmod, *Guj.*, *Hind.*, C. 701; ajmódá-vóvá, *Mar.*, C. 701; ajmot, *Hind.*, C. 701; ajmúd, *Mar.*, A. 1227; ájmúd (Apium), *Hind.*, A. 1227; ájmúd (Carum), *Beng.*, *Hind.*, C. 701; ajmúdá, *Dec.*, *Hind.*, C. 701; ajmúdah-ajván, *Dec.*, C. 701; ajowan, *Hind.*, C. 691; ajván-ka-patta, *Duk.*, A. 1130; ajwain, *Hind.*, C. 691; ajwán, *Mar.*, C. 691; ajwanka-puta, *Guj.*, A. 1227.

Ak Ak (Withania), *Pb.*, W. 98; ak (Juglans), *Hind.*, J. 61; ak (Morinda), *Hind.*, M. 656; ák (Saccharum), *Beng.*, S. 30; ák (Calotropis), *Duk.*, *Hind.*, C. 170; *Hind.*, *Pb.*, *Sind*, C. 191, 225; ákadá, *Hind.*, C. 191; akado, *Guj.*, C. 170; akaka, *Hind.*, A. 115; akakia, *Hind.*, A. 101; akalbai, *Hind.*, D. 144; akalber, *Hind.*, D. 144; akalbír, *Hind.*, D. 144; ákan, *Hind.*, C. 170; akanádi (Cissampelos), *Beng.*, *Hind.*, C. 1205; ákanádi (Stephania), *Beng.*, S. 2794; akanda, *Beng.*, D. 387; ákanda, *Beng.*, *Mar.*, C. 170; akandá, *Mar.*, C. 170; akar-kanta, *Beng.*, A. 681; akarkará (Anacyclus), *Beng.*, *Hind.*, A. 1026, *Mar.*, A. 1026; ákarkára (Spilanthes), *Mar.*, *Ph.*, S. 2571; akásbel (Cassytha), *Beng.*, C. 805; akasbel (Cuscuta), *Hind.*, *Pb.*, C. 2508; ákás-gaddah, *Hind.*, C. 1834; akas-nim, *Hind.*, *Mar.*, M. 550; akaspawan, *Dec.*, C. 2508; akaswel (Cuscuta),

Guj., C. 2508 ; akáswel (Cassytha),`Mar.`,
C. 805; akata, *Hind.*, C. 896; akda
cha jhada, *Mar.*, C. 170 ; ákdámu jháda,
Guj., C. 170 ; akhákhíyá, *Hind.*, A.
115 ; akhí, *Pb.*, R. 590, 593 ; akhiári,
Pb , R. 533 ; akhil-ul malik, *Pb.*, A.
1585 ; akhod, *Mar.*, A. 737 ; akhoda,
Guj., A. 737 ; akhor, *Hind.*, *Kashmir*,
J. 61 ; ákhor, *Hind* , J. 61 ; akhórí,
Pb., J. 61 ; akhrót, *Beng.*, *Guj.*, *Hind.*,
Pb., J. 61 ; akhrerí, *Pb.*, R. 588 ; akik,
Guj., C. 618 ; akilbír, *Pb.*, D. 144 ;
akkalkádhá, *Mar.*, A. 1026 ; akkikaruká,
Pb., A. 1026 ; akkilákáram, *Pb.*, A. 1026 ;
aklel-ul-mulk, *Pb.*, C. 117 ; akola, *Hind.*,
A. 681, *Beng.*, *Hind.*, A. 737 ; akolshi,
Beng., M. 786 ; ákond, *Hind.*, C.170 ;
akoria, *Hind.*, R. 335 ; akorkaro, *Guj.*,
A. 1026 ; akra, *Hind.*, V. 114 ; ákra,
Duk., *Mar.*, C. 170 ; akri, *Hind.*, *Pb.*,
W. 93 ; akri, *Pb.*, C. 225 ; akróda,
Mar., J. 61 ; ákrót (Aleurites), *Beng.*,
Hind., *Mar.*, A. 737 ; ákrót (Juglans),
Hind , J. 61 ; akrota, *Guj.*, A. 737 ;
ákrút, *Beng.*, J. 61 ; aksan, *Pb.*, W. 98 ;
aksíswérai, *Afg.*, R. 362 ; aku, *Uriya*,
S. 30 ; akurkura, *Mar.*, C. 1047 ; akur
kurra, *Hind.*, C. 1043.

Al Al, *Hind.*, M. 656 ; al, *Mar.*, M. 711 ; ál,
Hind., F. 671 ; ál, *Dec.*, *Guj.*, *Hind.*,
Mar., M. 656, *Hind.*, M. 704 ; alai,
Mar., D. 94; alalun, *Hind.*, S. 1078 ;
álan, *Mar.*, M. 656 ; alash, *Pb.*, C.
756 ; alásí, *Mar.*, L. 385 ; alá thanda,
Uriya, P. 632 ; albinda, *Pb.*, C. 1227 ;
alch, *Mar.*, Z. 201 ; álé, *Mar.*, Z. 201 ;
alecha, *Hind.*, P. 1304 ; alei, *Mar.*,
D. 94; alem, *Mar.*, Z. 201 ; alen, *Mar.*,
Z. 201 ; alettié, *Pb.*, T. 528 ; aleverie,
Beng., L. 283 ; algusi, *Beng.*, C. 2508 ;
ali, *Pb.*, C. 756 ; aliár, *Hind.*, D. 725 ;
alise, *Uriya*, F. 528 ; alísh (Linum),
Kashmir, *Pb.*, L. 385 ; alish (Rubus),
Pb., R. 593 ; al-kaddu, *Hind.*, L. 30 ;
alkushi, *Beng.*, M. 784 ; alkutí, *Mar.*,
F. 400 ; alla (Girardinia), *Hind.*, G.
213 ; allá (Mimosa), *Raj.*, *Pb.*, M. 562 ;
alli, *Uriya*, F. 486 ; allian, *Hind.*, C.
1969 ; alli palli, *Pb.*, A. 1565 ; alli-phul,
Dec., N. 200 ; allocha, *Pushtu*, P.1304 ;
aloál, *Hind.*, C. 722 ; alombe, *Mar.*, A.
590 ; alombe, *Mar.*, T. 725 ; alphajan,
Mar., L. 117 ; alshi, *Dec.*, *Guj.*, L.
385 ; alsí, *Hind.*, *Pb.*, L. 385 ; alu
(Vangueria), *Mar.*, V. 22, 25 ; alu (Alo-
casia), *Mar.*, A. 809 ; alú (Prunus), *Pb.*,
P. 1285, *Hind.*, *Beng.*, P. 1304 ; álu
(Solanum), *Beng.*, *Hind.*, S. 2320 ; álú
(Colocasia), *Mar.*, *Pb.*, C. 1732 ; alu-
bálu, *Hind.*, P. 1297 ; álúbo, *Hind.*,
Beng., P. 1304 ; alu-bokhára, *Guj.*,
Pb., P. 1304 ; alu-bukhara,' *Kumaon*,
P. 1304 ; alúcha, *Kumaon*, *Pb.*, P. 1304 ;
alvinda, *Sind*, C. 1227.

Am Ám (Mangifera), *Beng.*, *Dec.*, *Hind.*,
Mar., *Pb.*, *Uriya*, M. 147 ; ám (Psi-
dium), *Hind.*, P. 1343 ; amabára, *Hind.*,
S. 2649; amádá, *Beng.*, C. 2381 ; ámalací,
Hind., P. 632 ; ámal-bel, *Hind.*, *Pb.*, V.
195 ; amalgúch, *Pb.*, P. 1333 ; amallata,
Beng., V. 195 ; amaltás, *Hind.*, *Duk.*,

C. 756 ; amaltás, *Hind.*, F. 671 ; amalu,
Pb., V. 164; amara, *Hind.*, S. 2644, S.
2649 ; amara, *Hind.*, *Mar.*, S. 2649 ;
amarbeli, *Hind.*, C. 805 ; amarkarh,
Hind., I. 503 ; amarvéla, *Mar.*, C. 805 ;
amarwel, *Dec.*, C. 2508 ; amaut-múl,
Hind., Z. 182 ; ámb (Mangifera), *Hind.*,
Mar., *Pb.*, *Sind*, M. 147 ; amb (Hippophae), *Pb.*, H. 277 ; amb (Spondias),
Mar., S. 2649 ; amba (Mangifera), *Dec.*,
Mar., M. 147 ; amba (Cicer), *Mar.*, C.
1061 ; ambada (Hibiscus), *Mar.* H. 177 ;
ambada (Spondias), *Mar.*, S. 2642,
2649 ; amba-haladar, *Mar.*, C. 2381 ;
amba-haldi, *Mar.*, C. 2503 ; ambal, *Pb.*,
P. 632 ; ámbala, *Gus.*, P. 632 ; ambali,
Mar., T. 28 ; ambára, *Mar.*, *Pb.*, S.
2649 ; ambara-bárisa, *Hind.*, B. 445 ;
ambárí (Hibiscus), *Hind.*, *Mar.*, H.
177 ; ambari (Rumex), *Dec.*, *Hind.*, R.
650 ; ambarvel, *Pb.*, P. 397 ; ambat (Embelia), *Mar.*, E. 202 ; ambat (Spondias),
Mar., S. 2642 ; ambat-bel, *Mar.*, V. 195 ;
ambecha jhar, *Dec.*, M. 147 ; ambé-haldí,
Mar., C. 2406 ; ambgul, *Mar.*, E. 48 ;
ambhota, *Uriya*, B. 318 ; ambia, *Hind.*,
D. 628 ; ambiabahar, *Hind.*, C. 1234 ;
ámblí (Phyllanthus), *Pb.*, P. 632 ; ámblí
(Tamarindus), *Beng.*, *Dec.*, *Gus.*, *Mar.*,
T. 28 ; amblú, *Pb.*, G. 240 ; ambo, *Gus.*,
Mar., M. 147 ; ambodha, *Hind.*, S. 2649 ;
ambolati, *Beng.*, P. 632 ; ambor, *Mar.*,
M. 756 ; ambra (Mangifera), *Beng.*, M.
147 ; ambra (Spondias), *Beng.*, S. 2649 ;
ámbrá, *Hind.*, S. 2649 ; ámbre, *Pb.*, V.
104 ; ambu, *Guj.*, C. 1061 ; ambula,
Uriya, S. 2649 ; ambut chúka, *Hind.*, R.
650 ; ambuti, *Dec.*, *Hind.*, *Mar.*, O. 547 ;
ambyapát, *Beng.*, H. 177 ; amchur,
Hind., M. 147 ; amdhauka, *Beng.*, V.
205 ; am-haldi, *Hind.*, C. 2381 ; ámí (Rumex), *Pb.*, R. 642 ; ámi (Rhus), *Hind.*,
R. 293 ; ámil, *Pb.*, C. 2508 ; amjour,
Beng., D. 834 ; am-kí-adrak, *Dec.*, C.
2381 ; am-ki-gúthlí, *Hind.*, M. 147 ; amlá
(Phyllanthus), *Beng.*, *Dec.*, *Guj.*, *Hind.*,
Mar., *Pb.*, P. 632 ; amla (Rumex), *Pb.*,
R. 642 ; amlái, *Pb.*, Z. 270, Z. 280 ;
ámlaki, *Beng.*, *Hind.*, *Uriya*, P. 632 ;
amlánch, *Pb.*, R. 355 ; amldandi, *Pb.*, P.
1106 ; amli (Tamarindus), *Hind.*, *Gus.*,
Mar., T. 28 ; amlí (Bauhinia), *Dec.*,
Hind., B. 304 ; amlí (Aspidoparia), *Dec.*,
F. 326 ; amliácha, *Kashmir*, V. 106 ;
amlicá, *Hind.*, T. 28 ; amlika, *Hind.*,
Pb., O. 547 ; amlí-ka-bót, *Dec.*, T. 28 ;
amlín, *Pb.*, D. 251 ; amlok, *Hind.*, *Pb.*,
D. 611 ; amlora, *Kumaon*, *Pb.*, R. 642 ;
amlósa, *Hind.*, B. 304 ; amlu, *Pb.*, O.
597 ; amlúk, *Pb.*, D. 611 ; amluki, *Beng.*,
A. 722 ; amna, *Beng.*, S. 2649 ; amni,
Pb., Z. 280 ; amnia, *Pb.*, Z. 270, Z. 280 ;
amoluka, *Beng.*, V. 205 ; ampath, *Beng.*,
S. 1934 ; ampatia tasar, *Beng.*, S. 1934 ;
ámrá (Spondias), *Beng* , *Hind.*, *Mar.*, S.
2649 ; amrá, *Pb.*, Z. 280 ; amrái, *Kashmir*, *Pb.*, U. 4 ; amrer, *Pb.*, D. 196 ;
amri, *Sind*, T. 28 ; amrique-ka-sumáq,
Duk., C. 19 ; amrit-phal, *Hind.*, C.
1233 ; amrola, *Pb.*, V. 104 ; amrool,
Hind., E. 439 ; amrú, *Pb.*, P. 1463 ;
amrucha, *Afg.*, P. 1452 ; amrúd, *Hind.*,

Pb., P. 1343; amrúd, *Afg.*, *Kashmir*, P. 1452; amrúl, *Beng.*, *Hind.*, *Pb.*, O. 547; ámrulsák, *Beng.*, *Hind.*, O. 547; amrút, *Hind.*, *Pb.*, *Raj.*, P. 1343; amrutphal, *Hind.*, C. 1300; amsánia, *Pb.*, E. 234; amsúl, *Mar.*, G. 36; amti, *Mar.*, E. 202; amú, *Sind*, M. 147; amúdanda, *Pb.*, B. 463; amulati, *Beng.*, P. 632; ámultás, *Beng.*, C. 756; amúr (Amoora), *Beng.*, A. 983; amur (Spondias), *Hind.*, S. 2649; amút, *Pb.*, L. 549.

An Án, *Pb.*, M. 775; anab, *Sind*, Z. 280; anánas, *Guj.*, *Hind.*, *Mar.*, A. 1045; anander, *Pb.*, P. 760; anandi, *Hind.*, O. 574; anannas, *Hind.*, A. 1045; anantamul, *Beng.*, *Mar.*, H. 119; anár, *Beng.*, *Dec.*, *Hind.*, *Pb.*, *Pushtu*, *Sind*, P. 1426; anara, *Mar.*, P. 1426; anáras (Ananas), *Beng.*, A. 1045; ánarás (Agave), *Beng.*, A. 603; ánárdána (Punica), *Pb.*, P. 1420; anardána (Amarantus), *Hind.*, A. 916, 925; anárka-chilka, *Beng.*, P. 1426; anár-ká-jar, *Dec.*, P. 1426; anár-ká-pér, *Hind.*, P. 1420; anásphal, *Dec.*, *Guj.*, *Hind.*, I. 30; anatomúl, *Beng.*, H. 119; anber, *Hind.*, A. 956; anbli, *Hind.*, T. 28; anbóti, *Hind.*, O. 547; anbóti-ki-bhají, *Dec.*, O. 547; anchu, *Hind.*, R. 613; anda, *Hind.*, R. 369; andho, *Raj.*, P. 382; ándí, *Pb.*, C. 42; andúga, *Hind.*, B. 771; anduku, *Kumaon*, B. 771; andúsi, *Pb.*, T. 562; anerú, *Pb.*, R.369; angan, *Hind.*, F. 693; angír, *Afg.*, F. 230; angjani, *Beng.*, F. 558; angnera, *Rajputana*, B. 868; angnera,*Hind.*, B. 868; angú, *Hind.*, *Pb.*, F. 693; angúr, *Afg.*, *Hind.*, *Pb.*, V. 233; angúri, *Hind.*, *Sind*, V. 243; angúrphal, *Beng.*, V. 233; angúr-shéfa, *Hind.*, A. 1614; anguza-kema, *Afg.*, F. 84; anhuri, *Hind.*, V. 108; anisa, *Gus.*, P. 727; anísun, *Hind.*, P. 727; anjabár, *Pb.*, P. 1112; anjan (antimony), *Dec.*, A. 1224; ánján (Girardinia), *Pb*, G. 213; anjan (Hardwickia), *Hind.*, *Mar.*, H. 16; anjan (Memecylon), *Mar.*, M. 439; anjan (Pennisetum), *Pb*, P 382; anjan (Terminalia), *Hind.*, *Mar.*, T. 282; anjana, *Mar.*, M. 439; anjani, *Hind.*, T. 282; anjír, *Beng.*, *Guj.*, *Hind.*, F. 149; anjíra (Ficus), *Mar.*, F. 149; anjíra (Astragalus), *Hind.*, A. 1591, 1592; anjiri, *Hind.*, F. 230; anjír zard, *Pb.*, P. 1343; anjudán, *Kashmir*, F. 76; anka koli, *Uriya*, C. 606; ankari, *Beng.*, V. 114; ankol (Alangium), *Mar.*, A. 681; ankol (Sageretia), *Pb.*, S. 505; ankola, *Mar.*, A. 681; ankra, *Hind.*, V. 114; ankránti, *Uriya*, S. 2345; ankren, *Pb.*, R. 588; ánkri, *Pb.*, R. 590; ankula, *Uriya*, A. 681; ánla, *Hind.*, P. 632; anne, *Pb.*, P. 63; anni-nar, *Mar.*, S. 2861; ánolá, *Hind.*, P. 632; anor, *Pushtu*, P. 1426; anrar, *Hind.*, R. 369; áns, *Pb.*, R. 365; ansale, *Mar.*, G. 682; ansarishá, *Beng.*, G. 753; ansun, *Uriya*, T. 361; antamúl, *Hind.*, T. 855; anter-ghunga, *Dec.*, P. 874; anteri, *Hind.*, H. 92; anthamul, *Mar.*, T. 855; anti, *Mar.*, C. 416; antomul, *Beng.*, T. 855; ánúli, *Hind.*, P. 632; ánv, *Hind.*, M. 147; ánvulá, *Hind.*, P. 632; ánvurah, *Hind.*, P. 632; ánwerd, *Hind.*, P. 632; ánwlásár, *Pb.*, S. 2999.

Ao Aojan, *Hind.*, H. 16; aonla, *Hind.*, P. 632; aonli, *Mar.*, *Pb.*, P. 632; aor (Macrones), *Beng.*, F. 485; aor (Prunus), *Pb.*, P. 1304, 1322; aos, *Mar.*, S. 30; áoula, *Hind.*, P. 632; aout láni, *Sind*, C. 224.

Ap Apáng, *Beng.*, A. 382; apata).ta, *Beng.*, *Hind.*, C. 1403; apatá, *Mar.*, B. 318; aphím, *Mar.*, P. 87; aphina, *Gus.*, P. 87; aphu, *Mar.*, P. 87; appo, *Mar.*, P. 87; áprájit, *Hind.*, *Pushtu*, C. 1403; aprang, *Hind.*, C. 68; apta, *Mar.*, B. 304; apta, *Mar.*, B. 318; aptú, *Mar.*, B. 334; apúrs, *Baluch.*, J. 92.

Ar Araba, *Hind.*, A. 956; arad, *Gus.*, P. 513; arak, *Hind.*, V. 243; arak, *Pb.*, S. 717; arambu, *Pushtu*, P. 1408; aran, *Mar.*, E. 73; arand, *Hind.*, *Pb.*, R. 369; arandkharbúza, *Pb.*, C. 581; arang, *Mar.*, E. 314; arar, *Hind.*, R. 1; arara, *Pb.*, R. 1; arar dal, *Hind.*, C. 49; ararut, *Gus.*, M. 267; ararut-ke-gadde, *Dec.*, C. 2385; arátórá, *Duk.*, A. 484; arbambal, *Pb.*, H. 52; arbi, *Pb.*, C. 1732; archaká, *Beng.*, S. 2362; archálwá, *Pb.*, C. 1958; archar-ru, *Pb.*, C. 1958; archu, *Hind.*, R. 215; ardanda, *Hind.*, *Sind*, *Duk.*, C. 416; ardawal, *Pb.*, R. 253; ardhi supárí, *Mar.*, C. 711: areká-jhár, *Duk.*, B. 318; arend, *Hind.*, *Raj.*, R. 369; arendi, *Hind.*, S. 1939; argana, *Hind.*, C. 909; arghawán, *Afg.*, S. 2376; arhai-ka-bél, *Pb.*, A. 233; arhand, *Pushtu*, R. 369; arhar, *Beng.*, *Hind.*, *Pb.*, C. 49; arhar, *Beng.*, *Hind.*, F. 671; ári, *Pb.*, V. 104; arind, *Pb.*, R. 369; arindi, *Beng.*, S. 2077; arinj, *Raj.*, A. 249; aritha, *Hind.*, S. 808; aritha, *Pb.*, S. 808; aritha, *Gus.*, S. 818; arithan, *Gus.*, S. 818; ariza, *Beng.*, F. 464; arjan (Terminalia), *Hind.*, *Pb.*, T. 282, T. 361; arján (Holaptelea), *Pb.*, H. 324; arjha san, *Hind.*, C. 2105; arjun (Ehretia), *Hind.*, E. 20; arjun (Terminalia), *Beng.*, *Hind.*, *Mar.*, *Uriya*, T. 282; arjuna (Terminalia), *Beng.*, *Hind.*, *Mar.*, T. 282; arjuna (Lagerstrœmia), *Hind.*, L. 42; arjuna-sadra, *Mar.*, T. 282; arjun ladada, *Mar.*, T. 282; arjun sádada, *Gus.*, T. 282; arjunna (Croton), *Hind.*, C. 2180; árk, *Hind.*, C. 170; arkahuli, *Beng.*, G. 753; arkar, *E. Turkistan*, S. 1276; arkhar, *Pb.*, R. 316, 318, 323, 335; arkmula, *Guj.*, A. 1398; arkmut, *Mar.*, P. 523; arkol, *Pb.*, R. 318, 335; arkhol, *Pb.*, R. 323; ark pushpí, *Pb.*, P. 397; arleí, *Pb*, C. 42; arlu (Cæsalpinia), *Hind.*, C. 42; arlu (Mimosa), *Pb.*, M. 562; arlu (Oroxylum), *Hind.*, O. 233; armúra, *Pb.*, C. 1958; arna, *Hind.*, O. 555; arnah, *Pb.*, C. 1394; arni (Clerodendron), *Guj.*, C. 1386, *Pb.*, C. 1394; arni (Bubalus), *Hind.*, O. 555; arni (Premna), *Guj.*, *Hind.*, *Mar.*, P. 1233; arrak, *Pb.*, R. 243; arrak, *Pb.*, H. 382; arriahalli, *Uriya*, F. 486; arsúl, *Mar.*, C. 397; artal, *Guj.*, O. 242; arti, *Pb.*, P. 1285; artichoke, *Mar.*, C. 2556; arts, *Pb.*, R. 215; artso, *Pb.*, R. 215; aru (Prunus), *Hind.*, *Pb.*, *Raj.*, P. 1322; aru (Rhododendron), *Pb.*, R. 253; arúa, *Hind.*, A. 658; aru bukhára, *Pb.*, P. 1304; arúí,

Pb., P. 1322; arund, *Hind.*, S. 1939; arund, *Pb.*, P. 1252; arur, *Pb.*, P. 702; arúsá, *Hind., Mar.*, A. 484; arushá (Adhatoda), *Hind., Mar.*, A. 484; arusha (Callicarpa), *Beng.*, C. 126; arvi, *Dec., Hind.*, C.1732; arwán, *Pb.*, P. 702; arwi, *Hind.*, C. 1732.

As　Asain, *Hind.*, T. 361; asainda, *Hind.*, O. 537; asálio, *Gus.*, L. 283; asáliya, *Mar.*, L. 283; asán (Populus), *Pb.*, P. 1148; ásan (Pterocarpus), *Mar.*, P. 1370; ásan (Terminalia), *Beng., Hind., Pb.*, T. 361; asan (Withania), *Gus.*, W. 98; asáná (Briedelia), *Gus., Mar.*, B. 863, 868; asana (Pterocarpus), *Mar.*, P. 1370; asanjiya, *Hind.*, V. 210; asano, *Guj., Mar.*, B. 863; asárun, *Hind., Pb.*, V. 5, V. 10; asata, *Hind.*, E. 248; asauna, *Mar.*, B. 868; asaunra, *Hind.*, E. 259; asbarg, *Hind., Pb.*, D. 271; aseka, *Uriya*, S. 861; asgand, *Hind.*, W. 98; asgand nagauri, *Pb.*, W. 98; asgand nágorí, *Pb.*, W. 98; asgund, *Mar.*, W. 98; ashádi-tal, *Mar.*, S. 1078; áshán, *Beng.*, T. 361; ashathwa, *Beng.*, F. 236; asheta, *Hind.*, S. 2341; ashok, *Mar.*, S. 861; ashoka, *Mar.*, S. 861; ashopálava, *Gus.*, S. 861; ashopalo, *Gus.*, P. 1052; ashphal, *Beng.*, N. 72; ashshoura, *Beng.*, G. 271; asht, *Mar.*, F. 265; ashta (Bauhinia), *Hind.*, B. 318; ashta (Ficus), *Mar.*, F. 265; ashvagandá, *Beng.*, W. 93; ashú-kachú, *Hind.*, C. 1732; ashú-kuchú, *Beng.*, C. 1732; ashwal, *Beng.*, V. 159; asid, *Hind.*, L. 55; ásin, *Hind.*, T. 361; asindrí, *Mar.*, B. 318; asindro, *Mar.*, B. 318; ask, *Baluch.*, S. 1250; askandha, *Mar.*, W. 98; askúta, (*Lahoul*), *Pb.*, R. 359; aslasús, *Pb.*, G. 278; asl-rai, *Hind.*, B. 841; asmáni, *Pb.*, S. 855; ásna, *Hind., Mar.*, T. 361; asnea, *Mar.*, T. 434; asoda, *Gus.*, W. 98; asoja, *Hind.*, V. 210; asok (Polyalthia), *Hind., Mar.*, P. 1052; asok (Saraca), *Beng., Hind., Mar., Pb., Uriya*, S. 861; asok (Terminalia), *Uriya*, T. 28, *Mar.*, S. 861; asoka, *Mar.*, P. 1052; asoka, *Beng., Mar.*, S. 861; ásopálav, *Mar.*, P. 1052; asotri, *Mar.*, B. 318; aspurk, *Hind.*, M. 419; asrelei, *Sind.*, T. 51; asri, *Sind*, T. 51; ássain, *Hind.*, T. 361; assana, *Mar.*, B. 868; assar sauna, *Hind.*, O. 233; asshar, *Hind.*, C. 617; assu, *Pb.*, E. 327; assumar, *Mar.*, S. 950; ast, *Baluch.*, S. 1250; asta patu, *Beng.*, B. 320; asuá-gandhá, *Uriya*, W. 98; asúd, *Beng.*, F. 236; as-ufír, *Hind.*, H. 294; ásun, *Pb.*, T. 361; asundro, *Guj.*, B. 334; ásupál, *Mar.*, P. 1052; asúpála, *Mar.*, P. 1052; asúr, *Kashmir*, B. 833; asvagandha, *Mar.*, W. 98; asvagandhá, *Beng.*, W. 98; asvattha, *Beng.*, F. 236; aswat, *Beng.*, F. 236.

At　Át, *Dec., Gus., Hind., Mar.*, A. 1166; ata, *Beng.*, A. 1166; atár, *Bihar*, D. 534; átá sítáphal, *Duk.*, A. 1166; átá sítáphal, *Gus., Hind., Mar.*, A. 1166; atavish, *Mar.*, A. 401; atavishni-kalí, *Guj.*, A. 401; atcapalí, *Beng.*, S. 2865; ati, *Uriya*, S. 1861; atipich, *Hind., Pb.*, H. 88; atis, *Hind., Mar.*, A. 401; atívakh, *Guj.*, A. 401; atívish, *Guj.*, A. 401; ativista, *Guj.*, A.401; atki, *Mar.*, M. 40;

atmatti, *Mar.*, B. 308; atmorá, *Beng.*, H. 92; atnún, *Pb.*, H. 100; atpar, *Beng.*, F. 530; atrílal, *Hind.*, P. 425; atrúna, *Mar.*, F. 624; átsú, *Pb.*, R. 215; attahbar, *Beng.*, I. 62; attahbar, *Beng.*, F. 165½; atta-jam, *Beng.*, O. 153; attak, *Mar.*, F. 613; atti, *Mar.*, F. 179; atu, *Mar.*, V. 25; atúlgán, *Pb.*, M. 910; aturni, *Mar.*, F. 615; atvíká, *Hind., Mar.*, A. 401; atwín, *Hind.*, H. 100.

Au　Auch, *Beng.*, M. 704; audar, *Beng., Hind.*, O. 574; auga, *Hind.*, F. 700; augusta, *Mar.*, S. 1186; áúngra, *Hind.*, P. 632; áunlah, *Beng.*, P. 632; áunra, *Hind.*, P. 632; áura, *Hind.*, P. 632; auri, *Hind.*, C. 1732; ausarahe-revan, *Dec., Hind.*, G. 66; aúsneh, *Pb.*, L. 332; aústakhadús, *Pb.*, B. 902.

Av　Avachibá-vachi, *Gus.*, O. 28; ava chiretta, *Hind.*, E. 589; aval, *Guj.*, C. 741; ávalá, *Mar.*, P. 632; avalasara-gandhka, *Mar.*, S. 2999; avalkatí, *Mar.*, P. 632; ávla, *Mar.*, P.632; avois, *Hind.*, C. 1732.

Aw　Awa, *Hind.*, G. 213; awa-bichhu, *Hind.*, G. 213; awal (Acacia), *Hind.*, A. 267; awal (Cassia), *Guj.*, C. 741; awala, *Guj.*, C. 741; awáni-búti,1*Pb.*, B. 33.

Ay　Ayár, *Hind.*, P. 702; ayatta, *Pb.*, P. 702; ayrun-kukrí, *Sind*, R. 369.

Az　Azkhun, *Pb.*, A. 1186; azmei, *Pushtu*, A. 554; azun, *Mar.*, T. 282.

Ba　Baag-aarí, *Beng.*, F. 329; baagat, *Pushtu*, F. 129; baal, *Uriya*, F. 584; bab, *Hind.*, E. 323; bábachi, *Hind.*, P. 1352; babal ják, *Hind.*, M. 299; bábar, *Hind.*, E. 323; babar-sher, *Hind.*, T. 428; babassa, *Tel.*, H. 486; babbar (Acacia), *Sind*, A. 101; babbar (Ischœmum), *Pb.*, I. 494; babbar-sher, *Hind.*, F. 764; bab-basant, *Pb.*, L. 381; babbil, *Pb.*, A. 238; babchí (Psoralea), *Guj., Hind., Pb.*, P. 1352; babchí (Vernonia), *Beng.*, V. 73; baberáng, *Hind.*, E. 199; bábhula, *Mar., Sind*, A. 101; bábilá, *Hind.*, E. 323; báblá, *Hind., Beng., Pb.*, A. 101; báblá-áts, *Beng.*, A. 102; bablá-gónd, *Beng.*, A. 102; baboi, *Beng.*, I. 494; babra, *Dec.*, T. 293; bábrang, *Pushtu*, E. 199; babri (Eclipta), *Hind.*, E. 7; babri (Ocimum), *Pb.*, O. 18; bábri (Phaseolus), *Pb.*, P. 530; babrung, *Pb.*, E. 199; babui, *Hind.*, A. 101; babui, *Beng.*, I. 494; babui tulsi, *Beng., Hind.*, O. 18; bábul (Ocimum), *Hind.*, O. 18; babúl (Acacia), *Beng., Hind., Pb.*, A. 101, *Pb.*, A. 238; babuna, babunah (Anthemis), *Dec., Hind., A.* 1184; babuna (Corchorus), *Pb.*, C. 1842; babuna (Cotula), *Hind., Pb.*, C. 2025; babuna (Matricaria), *Guj., Pb.*, M. 314; babur, *Hind., Beng., Pb.*, A. 101; babúri (Mentha), *Pb.*, M. 461; baburi (Ocimum), *Pb.*, O. 18; bacaila, *Beng.*, F. 392; bach (Acorus), *Beng., Hind.*, A. 430; bach (Salix), *Mar.*, S. 579; bacha, *Mar.*, S. 579; bachhi, *Beng., Hind.*, O. 574; bachhwa, *Beng., Hind.*, O. 574; bachita, *Hind.*, U. 29; bachnáb, *Pb.*, A. 399; bachnág, *Hind., Mar.*, A. 397; bachnak, *Hind.*, A. 397; bacho, *Pb.*, R. 580; bachra, *Pb.*, O. 574; bacuchi, *Hind.*, P. 4; bad, *Hind.*, A. 1093; bad, *Hind.*, H. 289;

bada, *Pb.*, S. 531, S. 544, S. 1906 ; badadí-shingácha-jháda, *Mar.*, M. 721 ; bádam (Prunus), *Dec., Hind., Mar., Pb.*,P. 1274; badâm, *Guj.*, P. 1274; bádám (Terminalia), *Mar., Uriya*, T. 312 ; bádáme-hindí, *Dec.*, T. 312 ; bádámí, *Hind., Mar.*, T. 312 ; badanjan, *Hind.*, S. 2284 ; bádanján bostáni, *Kashmir*, S. 2284 ; badapuncha, *Hind.*, H. 164 ; badar (Taxus), *Pushtu*, T. 93 ; badár (Pueraria), *Hind., Pb.*, P. 1401 ; bádar (Abies), *Kashmir*, A. 22 ; bádáward (Carduus), *Pb.*, C. 559 ; bádáward (Volutarella), *Mar.*, V. 279 ; bad-bó kí-yírangí, *Dec.*, M. 864 ; badha (Ficus), *Pb.*, F. 265 ; badha (Salix), *Pb., Sind*, S. 579 ; badhára, *Hind., Pb.*, G. 298 ; bádián (Illicium), *Mar.*, I. 30; badián (Pimpinella), *Afg.*, P. 727 ; bádiánkóhi, *Afg.*, P. 1221 ; badishep (Fœniculum), *Mar.*, F. 659 ; badishep (Pimpinella), *Mar.*, P. 727 ; badiyá, *Hind.*, O. 574 ; badlo, *Pb.*, C. 862 ; badol, *Beng.*, S. 375 ; badra-kéma, *Afg.*, F. 95 ; badrang, *Hind.*, Z. 23 ; badranj boya, *Pb.*, N. 64 ; badror (Machilus), *Pb.*, M. 22 ; badror (Phœbe), *Pb.*, P. 546; bádsháhirái, *Hind.*, B. 833 ; badu manu, *Hind.*, T. 522 ; badúri, *Beng.*, B. 978 ; baelo, *Uriya*, P. 1397 ; baer, *Hind.*, Z. 231 ; baga banósa, *Mar.*, V. 143 ; bága mushada, *Hind.*, T. 456 ; baga nella, *Hind.*, T. 504 ; bagaur, *Pb.*, C. 1043, 1047 ; bagbherenda, *Beng., Hind.*, J. 41 ; bágdos, *Beng.*, T. 441 ; bageyra, *Hind.*, O. 251 ; bagfal, *Beng.*, S. 1186; baggar, *Pb.*, I. 494 ; bag-giárí, *Pushtu*, X. 1 ; baggí bútí, *Pb.*, S. 2674 ; baggí lána, *Pb.*, C. 224 ; baggí lána, *Pb.*, S. 2985 ; bagh, *Hind.*, F. 768 ; bágh, *Hind.*, T. 437 ; baghachura, *Beng.*, P. 824 ; bagh-ankurá, *Beng.*, A. 681 ; bagha-nulla, *Hind.*, C. 2522 ; bágh-dásha, *Hind.*, T. 440 ; bághni, *Hind.*, T. 437 ; bagh nokí, *Beng.*, M. 308 ; baghra-bahri, *Uriya*, F. 361 ; bagh-runga, *Beng.*, U. 69 ; bagnai, *Hind.*, C. 416 ; bag-narri, *Pb.*, P. 618 ; bagnú, *Pb.*, P. 1148 ; bagrí, *Pb.*, P. 555 ; bagriwala darim, *Hind.*, C. 862 ; bahadha, *Uriya*, T. 293 ; bahamb, *Hind., Pb.*, S. 2649; bahan, *Pb., Pushtu, Sind*, P. 1153; báhárá, *Uriya*, T. 293 ; báhavá, *Mar.*, C. 756 ; bahera, *Pb.*, F. 671; bahera, *Beng., Hind., Mar., Pb.*, T. 293; baheri, *Beng.*, T. 293 ; bahikat, *Hind.*, A. 484 ; bahila, *Beng., Hind.*, O. 574 ; bahira, *Beng., Pb.*, T. 293 ; bahm, *Beng., Uriya, Pb., Sind*, F. 493 ; bahman safaid, *Hind., Mar.*, C. 912 ; bahra, *Pb.*, O. 574 ; bahri, *Uriya*, F. 360; bahru, *Uriya*, F. 494 ; bahrum, *Uriya*, F. 469 ; bahtahna, *Hind.*, P. 885 ; ba hubara, *Beng.*, C. 1931 ; bahud-da, *Mar.*, T. 293; báhúl, *Hind., Pb.*, G. 688; bahúphallí, *Pb.*, C. 1842 ; bahura, *Beng.*, T. 293; bai, *Beluch.*, B. 59 ; baib, *Hind.*, I. 494; baiba, *Hind.*, P. 382 ; baibarang, *Hind.*, E. 199 ; báibidanga, *Uriya*, E. 199 ; baibrang, *Hind.*, E. 199 ; baichua, *Beng.*, D. 842 ; baid, *Pb.*, P. 1138; baigab, *Orissa*, J. 41; baigan, *Kumaon, Hind.*, S. 2284; baigana, *Mar.*, S. 2284 ; baigun, *Hind., Uriya*, S. 2284; bai hira, *Pb.*, T. 434 ; baikal, *Hind.*, C. 860 ; baikar, *Pb.*, P. 1252; baikunti,

Kumaon, R. 355; bail (oxen), *Hind.*, *Pb.*, O. 574 ; bail (Salix), *Pb.*, S. 570; bailewa, *Hind.*, S. 2943; baili, *Uriya*, F. 615 ; bainch, *Hind.*, F. 624 ; bainchi, *Beng.*, C. 596 ; bain-cho, *Uriya*, F. 615 ; baingan, *Hind.*, S. 2284; bains, *Hind.*, S. 579; bainsa, *Pb.*, O. 558 ; bairala, *Kumaon*, C. 1931 ; bairda, *Dec.*, T. 293 ; bairi, *Hind.*, C. 725 ; baishi, *Hind., Mar.*, S. 579 ; bajan, *Pb.*, S. 497 ; bajar-bang, *Pb.*, P. 684 ; bajarbattú, *Mar.*, C. 1995 ; bajar-battuler, *Beng.*, C. 1995 ; bajar banj, *Pb.*, C.|1003; bajaro, *Sind*, P. 384 ; bajau-ri nímbú, *Pb.*, C. 1270; bájera, *Hind., Mar.*, P. 384; bajra (Pennisetum), *Hind., Kumaon, Mar.*, P. 384; bájra (Sorghum), *Hind.*, S. 2394; bajhol, *Pb.*, M. 22; bají, *Hind.*, S. 2574 ; bajír, *Pb.*, P. 1112 ; bajr, *Hind.*, D. 144 ; bajra jhopanwa, *Hind.*, S. 2424 ; bájra-jku panwa, *Pb.*, S. 2424 ; bajra-mula, *Uriya*, P. 384 ; bájra tangunanwa, *Hind.*, P. 384 ; bajri (carbonate of lime), *Pb.*, C. 489 ; bájri (Pennisetum), *Hind., Mar.*, P. 384 ; bájrí (Typha), *Mar.*, T. 875 ; bajur (Abies), *Afg.*, A. 17; bajúr (Corypha), *Beng.*, C. 2006 ; bajvaran, *Beng.*, E. 496 ; bajza, *Pb.*, P. 384, *Beng., Hind.*, S. 1186 ; bakáin, *Hind., Pb.*, M. 393 ; bakalpatta, *Kumaon*, K. 17 ; bakalpattia, *Kumaon*, D. 31 ; bakalwa, *Hind.*, G. 240; bakam (Cæsalpinia), *Beng., Guj., Hind.*, C. 35 ; bakam (Melia), *Pb.*, M. 363; bakan, *Pb.*, L. 451; bakanlimbodo, *Gus.*, M. 393; bakar (Premna), *Hind.*, P. 1242 ; bakár (Cornus), *Hind.*, C. 1973; bakarcha (Premna), *Hind.*, P. 1233, P. 1242; bakarcha (Securinega), *Hind.*, S. 1029; bakardharra, *Kumaon*, O. 525 ; bakarja (Melia), *Hind.*, M. 393 ; bakarja (Osyris), *Kumaon*, O. 525 ; bákas, *Beng.*, A. 484 ; bakayan, *Hind., Mar., Sind*, M. 393; baká-yan, *Mar.*, M. 363; bákchí, *Hind.*, V. 73; bakhra, *Pb.*, T. 544, T. 548; bakhrú, *Pb.*, L. 535; bakhtmal, *Pb.*, P. 1356; bakil, *Hind.*, S. 1939; bakla (Anogeissus), *Hind.*, A. 1149; bakla (Phaseolus), *Hind.*, P. 530; bakla (Vicia), *Hind., Pb.*, V. 108; bákla (Faba), *Hind.*, F. 1; bákla (Vicia), *Kumaon*, V. 108; bákli (Lagerstrœmia), *Hind., Pb.*, L. 55; bakli (Anogeissus), *Hind.*, A. 1149; bako, *Hind.*, S. 1186; bakra (Elæodendron), *Hind., Pb.*, E. 78; bakra (goat), *Hind.*, F. 758, S. 1290; bakra chimyaka, *Hind.*, P. 1009; bakri, *Hind.*, S. 1290; bakshel, *Pb.*, S. 579; bákuchi, *Uriya*, P. 1352; bákula (Phaseolus), *Hind.*, P. 530; bakúla (Mimusops), *Mar.*, M. 570; bakul-bakal, *Beng.*, M. 570; bakul mulsari, *Hind.*, M. 570; baky-ána, *Pushtu*, M. 393; bala (Andropogon), *Hind.*, A. 1097; bálá (Pavonia), *Beng., Hind., Mar.*, P. 344; bálá (Sida), *Beng., Mar.*, S. 1688, S. 1694; bálá (Valeriana), *Pb.*, V. 5, V. 10; balá-bahulá, *Beng.*, F. 194 ; balacharea, *Mar.*, N. 17; baladh, *Pb.*, O. 574; balah, *Hind.*, A. 1097 ; bála hírade, *Mar.*, T. 325; balah-kera, *Uriya*, F. 436 ; balá-látá, *Beng.*, F. 194 ; bala mushk, *Pb.*,

Hind., *Beng.,* M. 422 ; banmudga, *Beng.,* P. 468 ; banmussureya, *Hind.,* A. 1215 ; banna, *Pb.,* V. 164 ; ban narin-ga, *Hind.,* G. 158 ; ban-natí, *Beng.,* A. 923 ; bannatia, *Hind.,* P. 1078 ; banni, *Pb.,* Q. 13, 23 ; ban-níl-gáchh, *Beng.,* T. 270 ; ban nimbu, *Hind.,* G. 271 ; banogal, *Pb.,* F. 8 ; ban-okra, *Beng., Hind.,* X. 1 ; banosa, *Beng.,* V. 143, ban-pála, *Pb.,* P. 1460 ; banpálang, *Beng.,* S. 2354 ; banpalás, *Hind.,* S. 959 ; banpalti, *Hind., Kumaon,* P. 1460 ; ban pálu, *Pb.,* C. 1988 ; banpat, *Beng.,* C. 1858, *Beng., Sind,* C. 1861 ; banpa-tol, *Beng., Hind.,* T. 576 ; banpatrak, *Pb.,* S 924 ; banphal, *Hind., Pb.,* C. 1861 ; bán phúnt, *Pb.,* S. 3079 ; banpiaáj, *Beng.,* U. 39 ; banpindalu (Randia), *Hind.,* R. 16 ; banpindálú (Gardenia), *Hind.,* G. 124 ; banraj, *Beng.,* B. 318 ; banraji, *Beng.,* B. 318 ; ban-rithá, *Beng.,* A. 200 ; báns, *Beng., Hind.,* B. 118, *Hind.,* D. 292 ; bánsa (Adhatoda), *Hind.,* A. 484, 711 ; bánsa (Eragrostis), *Hind.,* E. 267 ; bánsa-bánsu, *Pb.,* T. 270 ; ban-san, *Beng., Hind.,* C. 2163 ; ban-sanjlí, *Pb.,* C. 2035 ; bánsá siyáh, *Pb.,* B. 165 ; bansh-pata-lál-natí, *Beng.,* A. 919 ; bansi, (*Bundel.*), *Hind.,* P. 63 ; banskapur, *Beng., Hind.,* B. 84 ; ban shágalí, *Hind., Pb.,* S. 2678 ; bans kaban, *Hind.,* D. 292 ; banskeora, *Beng., Hind.,* A. 603 ; bans khúrd, *Hind.,* D. 292 ; bans-lóchan, *Hind.,* B. 84 ; báns patá-natíya, *Beng.,* A. 928 ; bansu, *Pb.,* J. 29 ; bansua batana, *Beng.,* Q. 85 ; ban-sulpha, *Beng.,* F. 723 ; ban tamáкú, *Pb.,* V. 64 ; ban-tepariya, *Beng.,* P. 678 ; banti, *Guz.,* P. 50 ; bantí, *Pb.,* I. 39 ; bántphút, *Pb.,* H. 277 ; ban-tulsi, *Kumaon,* O. 214, *Pb.,* O. 31 ; banyarts, *Pb.,* C. 1101 ; baonli, *Raj.,* A. 239 ; baoti, *Hind.,* S. 1226 ; bá-pattra,(*Jhelam*), *Pb.,* C. 133 ; bapchie, *Beng.,* V. 73 ; báphalí (Peucedanum), *Mar.,* P. 456 ; baphalli (Convolvulus), *Pb.,* C. 1781 ; baphuli, *Hind., Pb.,* C. 1842 ; bar, *Beng., Hind. Pushtu,* F. 129 ; bara (Rhynchobdella), *Uriya,* F. 561 ; bara (Sorghum), *Hind.,* S. 2394 ; bara bet, *Beng.,* C. 77 ; bara bhurbhura, *Hind.,* E. 263 ; bara chakma, *Beng.,* Q. 81 ; bara chali, *Beng.,* P. 1058 ; barachar, *Pb.,* Q. 13 ; bará-charáyatah, *Hind.,* E. 585 ; barad, *Beng., Hind.,* O. 754 ; baradodak, *Pb.,* E. 549 ; bara-eláchi, *Beng.,* A. 976 ; barag, *Mar.,* P. 63 ; bara gach, *Beng.,* C. 2180 ; bara garri, *Kuma-on,* R. 14 ; bara gókhru, *Beng., Hind.,* P. 363 ; bara hal kasá, *Beng.,* L. 312 ; barahmi, *Hind.,* C. 1183 ; barain (Vitis), *Kumaon,* V. 215 ; baráin (Quercus), *Pb.,* Q. 13 ; bara jal-ganti, *Beng.,* P. 57 ; bara-juár, *Hind., Pb., Sind,* Z. 50 ; bará-kaliján, *Hind., Beng.,* A. 853 ; bará-kandá, *Dec.,* T. 15 ; bara-kanur, *Beng.,* C. 2062 ; bara kanwar, *Hind.,* A. 603 ; bará-khúlan-jan, *Duk,* A. 853 ; bara kúkúr chita, *Beng.,* L. 474 ; bará-kúlanján, *Hind., Beng.,* A. 853 ; bára-lai, *Hind.,* B. 817 ; bara-lóniya, *Beng.,* P. 1179 ; baraluniá, *Hind.,* P. 1179 ; bara-manda, *Beng.,* L. 549 ; bárá masiya, *Beng.,* M. 626 ; bara-

mattar, *Hind., Pb.,* P. 885 ; barambhi, *Hind.,* H. 149 ; baran, *Pb.,* T. 87 ; baran-dá, *Hind.,* O. 31 ; bara-nebu, *Beng.,* C. 1286 ; baranga, *Hind.,* K. 42 ; barangi, *Hind.,* C. 1388, 1394 ; bara nimbu, *Dec., Hind.,* C. 1286, *Beng., Hind.,* C. 1270 ; baráni múlí, *Pb.,* D. 674 ; baranipiaz, *Hind.,* A. 777 ? bara-pílú, *Hind.,* S. 705 ; barar, *Pb.,* C. 431 ; bararí (Zizyphus), *Pb.,* Z. 280 ; barárí (Capparis), *Pb.,* C. 431 ; bara-ritha, *Beng.,* S. 818 ; bara-salpan, *Beng., Hind.,* F. 633 ; bara sán-wak, *Pb.,* P. 48 ; bai a sarpot, *Hind.,* D. 445 ; bara singha, *Hind.,* D. 225, 238 ; barát, *Pb.,* D. 758 ; bara takria, *Pb.,* P. 77 ; bara-toriya-gadi, (*C. P.*), *Hind.,* I. 497 ; barau, *Hind.,* A. 692 ; baraulia, *Hind.,* A. 692 ; báráun, *Hind.,* T. 440 ; barbaru, *Pb.,* F. 260 ; barbat, *Hind.,* D. 32 ; barbati (Vigna), *Beng.,* V. 116 ; bar-batti (Embelia) *Mar.,* E. 202 ; barbed, *Mar.,* I. 131 ; barbutítí, *Hind.,* V. 116 ; barca, *Beng.,* F. 513 ; barchitta, *Pb.,* S. 1223 ; bardáel, *Beng., Hind.,* O. 574 ; bardanni, *Hind.,* S. 1223 ; bareli, *Beng.,* F. 360 ; bareskatú, *Pb.,* F. 10 ; barf-ka-rich, *Hind.,* F. 819 ; barg-a-bana, *W. Afghan,* P. 847 ; bargad, *Pb.,* F. 129 ; bargat, *Hind.,* F. 129 ; barghat, *Mar.,* F. 129 ; barghauna, *Pushtu,* L. 587 ; bar-gund, *Mar.,* C. 1931 ; barhal, *Hind.,* A. 1511 ; bárhang, *Mar.,* P. 926 ; barhantá (Solanum), *Hind.,* S. 2280 ; barhantá (Tragia), *Hind.,* T. 509 ; barí, *Pb.,* V. 164 ; bári, *Kumaon,* V. 164 ; bariál, *Hind.,* B. 356 ; bariala, *Beng.,* B. 118 ; bariar, *Hind.,* S. 1694 ; bariára, *Hind.,* S. 1688 ; bari boj, *Pb.,* A. 430, C. 225 ; bari bhodore, *Hind.,* P. 48 ; bari-chobchini, *Hind.,* S. 2245 ; bari gagli, *Raj.,* P. 42 ; barí-gumchí, *Duk., Guj.,* A. 471 ; barí-iláchí, *Hind.,* A. 976 ; bari-iláyechi, *Duk.,* A 976 ; bárika-ghola, *Mar.,* P. 1187 ; bari kander, *Pushtu,* E. 35 ; bari kári, *Hind.,* M. 545 ; barikasóndi, *Dec., Hind.,* C. 780 ; barik bhauri, (*Konkan*), *Mar.,* I. 393 ; barik-dágadaphúl, *Mar.,* L. 332 ; barik-moth, *Mar.,* C. 2612 ; barik tel, *Hind., Dec.,* S. 1078 ; bariktíl, *Mar.,* S. 1078 ; barili, *Beng.,* F. 359 ; barí-mahín, *Pb.,* T. 70 ; bari-mái, *Dec.,* T. 70 ; bari-máin, *Hind., Beng.,* T. 70 ; barí mauharí, *Pb.,* S. 2264 ; barín, *Pb.,* Q. 23 ; baringú, *Pb.,* C. 198 ; barini, *Kashmir, Pb.,* I. 93 ; bari-pán-ki-jar, *Dec., Mar.,* A. 853 ; bari-piplí, *Hind.,* S. 970 ; barirái, *Hind.,* B. 833 ; barisaunf, *Hind.,* F. 659 ; bari-shopha, *Mar.,* F. 659 ; bárj, *Kashmir, Z.* 280 ; barjajantis, *Uriya,* S. 1174 ; bar-kanghi, *Beng., Hind.,* A. 84 ; bár-ki-send, *Deccan,* E. 553 ; bar koli, *Uriya, Z.* 231 ; barkur, *Uriya,* F. 384 ; barlái, *Hind.,* B. 833 ; barma, *Kashmir, Pb.,* T. 93 ; bar-mat, *Kumaon,* T. 376 ; barmé-ké-patte, *Dec.,* C. 2040 ; barmera, *Pb.,* C. 1711 ; barmi (Cinnamomum), *Dec.,* C. 1183 ; barmí (Hydrocotyle), *Guj.,* H. 486 ; barmi (Taxus), *Mar.,* T. 93 ; barmí, *Pb.,* T. 93 ; barna (Barilius), *Beng.,* F. 360 ; barna (Cratæva), *Hind., Pb., Raj.,* C. 2039 ; barnáhí, *Pb., Raj.,* C. 2039

baro, *Hind.*, A. 717; barokolí, *Uriya,*
Z. 263; barola, *Beng.*, H. 317; baroli,
Mar., I. 141; baror, *Pb.*, P. 1449;
baro-shíálkánta, *Beng.*, A. 1351; barotri,
Pb., A. 1061; barra-al, *Hind.*, M.
656; barrarra, *Pb.*, *Pushtu*, P. 419;
barre, *Hind.*, C. 637; barrí, *Pb.*, P. 419;
barro-gaddi, *Uriya*, F. 457; barru, *Hind.*,
S. 2394; barrú, *Pb.*, S. 2223; barsaj,
Hind., T. 361; barsoli, (*Merwar*), *Hind.*,
M. 570; barsúngá, *Beng.*, M. 800; bart,
Pb., P. 1316; bartakú, *Beng.*, S. 2284;
bartang, *Pushtu*, P. 923, *Pb.*, P. 932;
bártang, *Mar.*, P. 926; barthoa, *Pb.*, H.
517; barti (Panicum), *Mar.*, P. 50; barti
(Setaria), *Hind.*, S. 1223; bartondi,
Mar., M. 656; bartu, *Pb.*, H. 517;
bártundi, *Mar.*, M. 656; bartung, *Beng.*,
P. 923; baru (Apluda), *Hind.*, A. 1232;
baru (Coix), *Hind.*, C. 1686; baru (Pani-
cum), *Pb.*, P. 42; barú (Sorghum), *Hind.*,
Pb., S. 2394; barua (Daphne), *Kumaon*,
D. 115; barua (Mastacembelus), *Hind.*,
F. 493; barúa (Sorghum), *Pushtu*, S.
2394; barun, *Beng.*, *Hind.*, C. 2039;
barwári, *Raj.*, P. 42; barwa, *Pb.*, S.
2394; barweza, *Pushtu*, H. 164; bas
(Dendrocalamus), *Mar.*, D. 292; bás
(Albizzia), *Hind.*, A. 711; basaha,
Beng., *Hind.*, O. 574; basak, *Hind.*,
D. 397; básang, *Uriya*, M. 800; basan-
guti, *Hind.*, F. 307; basant (Linum), *Pb.*,
L. 381; basant (Reinwardtia), *Pb.*, R.
71; basanti, *Hind.*, C. 840; basaunta,
Hind., P. 57; bashal, *Pb.*, S. 560;
bashangarús, *Kumaon*, A. 484; básingh,
Kumaon, E. 591; basinibans, *Beng.*, B.
149; basini bansh, *Beng.*, B. 147; bás-kí-
kasóndí, *Hind.*, C. 787; basla, *Pb.*, C.
909; basma, *Pb.*, I. 145; basna, *Hind.*,
Mar., S. 1186; basóta, *Hind.*, P. 1242;
bassa, *Mar.*, D. 292; bassant, *Hind.*, *Pb.*,
H. 544; bassar, *Pb.*, C. 431; bassari,
Mar., F. 216; bassein, *Hind.*, A. 711;
bastra, *Hind.*, C. 129; basung, *Uriya*,
A. 484; basúti (Adhatoda), *Pb.*, A. 484;
basutí (Colebrookia), *Pb.*, C. 1711; bata
(Securinega), *Pb.*, S. 1024; báta (Periploca), *Pb.*, P. 419; batan, *Beng.*, S.
833; batana, *Hind.*, P. 885; batang
(Pyrus), *Kashmir*, *Pb.*, P. 1452; bátang
(Rubus), *Pb.*, R. 588; batangí, *Pb.*, P.
1466; batank, *Kashmir*, *Pb.*, P. 1452;
batata, *Guz.*, *Mar.*, S. 2320; batate,
Mar., S. 2320; bátausí, *Hind.*, F. 307;
batávi nebu, *Beng.*, *Hind.*, C. 1263;
bat-bakri, *Pb.*, F. 725; batbar, *Pb.*, F.
179, 216; bat bel, *Pb.*, C. 1205; bathar,
Pb., G. 677; bather, *Pb.*, G. 705; bathor,
Pb., V. 91; báthú, *Pb.*, C. 1003; bathu
(Amarantus), *Pb.*, A. 925; bathu (Cheno-
podium), *Pb.*, C. 1003; bathúa, *Pb.*,
C. 1003; bathur, *Sind*, T. 87; bathú-sag,
Beng., *Hind.*, C. 1003; batia-rang, *Beng.*,
P. 427; batindu, *Pb.*, T. 470, T.
483; batindú páth, *Pb.*, C. 1205;
batkar, *Pb.*, C. 881, 886, 896, 898;
batkateya, *Pb.*, S. 2345; bator-nebú,
Beng., C. 1263; bat-phagár, *Pb.*, F.
168; bat piá, *Pb.*, S. 924; batra, *Dec.*,
T. 293; batraj, *Beng.*, C. 1377; batrén,
Pb., O. 38; bat sínjal, *Pb.*, R. 159;
batta, *Beng.*, F. 411; battal, *Pb.*, P.

1447; battamanku, *Pb.*, C. 896; battaní-
chola, *Hind.*, P. 885; batthal, *Pb.*,
L. 112; battijamb, *Beng.*, E. 412;
battí-sai, *Beng.*, D. 676; battuli, *Uriya*,
F. 546; batu (Amarantus), *Pb.*, A. 925;
batu (Chenopodium), *Pb.*, C. 1018;
batúl, *Beng.*, S. 833; batula, *Pb.*, S. 904;
batya gingaru, *Kumaon*, R. 14; bau-
jhánjhe, *Uriya*, P. 874; baul (Mimusops),
Uriya, M. 570; bául (Felis chaus), *Mar.*,
T. 427; baulo, *Uriya*, M. 570; baulu,
Pb., C. 1958; baum, *Beng.*, F. 791;
bauna, *Pb.*, C. 133; baunra (Eurya),
Hind., *Mar.*, E. 565; baunra (Ipomœa),
Hind., I. 384; baunri, *Hind.*, P. 50;
bau-piring, *Beng.*, M. 419; baurala,
Kumaon, C. 1931; bauri, *Pb.*, C. 431;
bauru, *Pb.*, R. 293; bauti, *Mar.*, P.
50; bávachá, *Guz.*, P. 1352; bávaché,
Sind, F. 615; bavachí, *Beng.*, *Mar.*,
P. 1352; bávachya, *Mar.*, P. 1352;
bával, *Guj.*, A. 101; bávan-chí, *Hind.*,
P. 1352; bávanchíyán, *Hind.*, P. 1352;
bávchiyán, *Hind.*, P. 1352; bávto, *Mar.*,
P. 53; bávto nágli, *Guz.*, E. 170; bawa,
Mar., C. 756; bawachi, *Guz.*, *Mar.*,
P. 1352; baya, *Mar.*, C. 756; baya-
birang, *Hind.*, E. 202; bayi (Aspidoparia), *Uriya*, F. 326; bayi (Balsamodendron), *Baluch.*, B. 59; bayisa-
gugul, *Mar.*, B. 59; bayrah, *Pb.*,
Sind, T. 293; baz-anjir, *Afg.*, R. 369;
bazrbang, *Pb.*, H. 525; bazrúl, *Beng.*,
Hind., H. 525; bazrulbanj, *Pb.*, H. 525.

Be Bebana, *Mar.*, M. 855; bebrang (Embelia),
Beng., *Hind.*, *Pb.*, E. 199, E. 202;
bebrang (Myrsine), *Pb.*, M. 910;
bebrang khatai, *Pb.*, N. 64; bed,
Pb., S. 544; bed, *Pb.*, S. 570; bed,
Hind., *Mar.*, *Pb.*, S. 579; bedá,
Mar., T. 293; bedanjir, *Pb.*, R. 369;
bedari kand, *Hind.*, P. 1401; bedh,
Afg., S. 531; bedi, *Pb.*, S. 560;
bedina, *Hind.*, M. 855; bed-i-siah,
Afg., S. 537; bed-khist, *Pb.*, M. 235;
bed leila, *Pb.*, S. 579; bed majú, *Pb.*,
S. 544; bed-mushk, *Hind.*, *Pb.*, S.
550; bedu, *Hind.*, F. 230; begana,
Hind., H. 324; beg-púra, *Beng.*, C.
1270; begti, *Beng.*, F. 476; begún,
Beng., S. 2284; beguniyá, *Uriya*,
V. 164; beh, *Sind*, N. 39; behada,
Mar., T. 293; behara, *Hind.*, T. 293;
behara, *Mar.*, T. 293; beharbáns,
Hind., B. 139; beharda, *Mar.*, T. 293;
beharia, *Hind.*, T. 293; behári nimbu,
Hind., C. 1286; behasá, *Guz.*, *Mar.*,
T. 293; behda, *Mar.*, T. 293; beh-
dánah, *Hind.*, *Duk.*, C. 2546; beheda,
Guz., *Mar.*, T. 293; behedan, *Guz.*,
T. 293; behedo, *Mar.*, T. 293; behéra,
Reng., *Hind.*, *Pb.*, T. 293; behhúl,
Pb., P. 1252; behikar, *Pb.*, A. 484;
behkul, *Pb.*, P. 1252; behli, *Hind.*,
A. 1166; behoor báns, *Beng.*, B. 148;
behor, *Beng.*, B. 139; behra, *Hind.*, C.
1031; behra (Terminalia), *Hind.*, T. 293;
behrah (Cirrhina), *Pb.*, F. 409; behru,
Hind., *Uriya*, C. 1031; behúrbáns,
Beng., B. 118; beis, *Pb.*, S. 579;
bekeda, *Mar.*, T. 293; bekh-akwar,
Hind., M. 650; bekhar, *Pb.*, P. 1252;
békh-gilló, *Kashmir*, T. 470, T. 483;

bekh-i-marján, *Pb.*, C. 1808 ; bekh-kurphus, *Hind.*, N. 17 ; bekh sosan, *Kashmir*, I. 425 ; bekh-unjubaz, *Beng.*, P. 1084 ; bekhwa, *Pb.*, P. 1252 ; bekkar, (*Salt Range*), *Pb.*, A. 484 ; bekklí, *Pb.*, P. 1252 ; bekkra, *Hind.*, P. 1252 ; bek-kut, *Uriya*, F. 476 ; bekling, *Pb.*, P. 1252 ; bekra baikur, *Mar.*, D. 219 ; bek-rúl, *Pb.*, P. 1252 ; bel, bela (Ægle), *Beng.*, *Hind.*, *Mar.*, A. 534, (Cra-tæva), *Hind.*, C. 2039, (Jasminum), *Beng.*, *Hind.*, J. 35 ; belambu, *Dec.*, *Hind.*, A. 1644 ; belanne, *Pb.*, M. 461 ; belatak, *Dec.*, S. 1041 ; belatak, *Hind.*, S. 1041 ; beli (Limonia), *Hind.*, L. 362 ; beli (Ribes), *Pb.*, R. 359 ; beli (Salix), *Pb.*, S. 560, S. 570 ; bellari, *Sind*, M. 791 ; belli-nandi, *Mar.*, L. 55 ; belli-pata, *Mar.*, H. 255 ; bélpalás, *Duk.*, B. 978 ; belpath, *Sind*, C. 1205 ; bemal, *Hind.*, G. 688 ; bem beimi, *Pb.*, P. 1322 ; bembi, *Pb.*, P. 1322 ; bena, *Pb.*, D. 227 ; bená, *Hind.*, A. 1097 ; bena-joni, *Beng.*, S. 2668 ; bender-wel, *Mar.*, V. 189 ; bendi, *Gus.*, *Mar.*, T. 392 ; bendrí, *Mar.*, V. 189 ; bengalí bádám, *Mar.*, T. 312 ; bengan, *Mar.*, *Pb.*, S. 2284 ; bengí, *Pb.*, C. 331 ; benkar, *Pb.*, H. 285 ; benkra, *Mar.*, S. 1287 ; bent, *Hind.*, S. 579, S. 2424 ; bentha, *Kashmir*, J. 78 ; bentí, *Pb.*, P. 1153 ; bér (Capparis), *Hind.*, *Pb.*, C. 431 ; bér (Ficus), *Hind.*, F. 129, F. 260 ; bér (Zizyphus), *Beng.*, *Dec.*, *Hind.*, *Kumaon*, *Pb.*, *Raj.*, Z. 231, Z. 254, Z. 280 ; bera (Ficus), *Pb.*, F. 129 ; bera (Glochidion), *Pb.*, G. 240 ; bera (Picrasma), *Pb.*, P. 693 ; bera (Zizyphus), *Hind.*, *Mar.*, Z. 231 ; berain, *Hind.*, U. 61 ; beral, *Uriya*, C. 1847 ; berda, *Mar.*, T. 293 ; bergandu tongur, *Kumaon*, *Pb.*, S. 919 ; ber-gherie, *Hind.*, O. 251 ; beri, *Hind.*, *Pb.*, Z. 231 ; beri, *Pb.*, Z. 270, Z. 280 ; béri, *Hind.*, Z. 290 ; bering, *Pb.*, P. 693 ; ber jangrí, *Sind*, Z. 231 ; berli, *Mar.*, C. 711 ; berli mád, *Mar.*, C. 711 ; berli mhár, *Mar.*, C. 711 ; berrú, *Pushtu*, Z. 231 ; bersa, *Hind.*, A. 711 ; bersinge, *Mar.*, S. 2884 ; bersu, *Pb.*, A. 1061 ; bertia, *Beng.*, *Dec.*, *Hind.*, S. 1212 ; beru, *Hind.*, F. 230 ; berula, *Hind.*, C. 1950 ; berwaja, *Pushtu*, C. 137 ; besenda, *Hind.*, N. 39 ; besu, *Pb.*, S. 544 ; bet (Calamus), *Beng.*, *Hind.*, *Mar.*, C. 104 ; bet (Peristrophe), *Beng.*, P. 427 ; bet (Salix), *Hind.*, S. 579 ; beta, *Mar.*, C. 104 ; betain, *Kumaon*, M. 363, *Hind.*, M. 393 ; betar, *Kashmir*, *Pb.*, J. 78 ; bethuá sák, *Beng.* & *Hind.*, C. 1003 ; bethuwa, *Hind.*, C. 1003 ; bet kukri, *Kumaon*, L. 535 ; betsa, *Pb.*, S. 560 ; bettar, *Pb.*, J. 104 ; betthal, *Pb.*, J. 78 ; bettir, *Hind.*, J. 104 ; bettú shak, *Beng.*, C. 1003 ; betu, *Kumaon*, A. 925 ; bewal, *Hind.*, G. 688 ; bewba, *Dec.*, P. 1370 ; bey-a-rah, *Mar.*, T. 325 ; beyguna, *Uriya*, V. 164.

Bh Bhábar (Girardinia), *Pb.*, G. 213 ; bhábar (Ischœmum), *Hind.*, *Pb.*, I. 494 ; bhabhur, *Hind.*, E. 323 ; bhabhuri, *Hind.*, E. 323 ; bhádali, *Mar.*, S. 1207 ; bhadi, *Beng.*, O. 38 ; bhadli, *Dec.*,

Mar., S. 1207 ; bhadrak, *Mar.*, S. 927 ; bhadu til, *Beng.*, S. 1078 ; bhai-birrung, *Beng.*, E. 199 ; bhái-koi, *Mar.*, S. 2819 ; bhail, *Pb.*, S. 560 ; bhains (oxen), *Beng.*, *Hind.*, O. 558 ; bhains (Salix), *Pb.*, S. 570 ; bhains, *Hind.*, S. 589 ; bhairá, *Beng.*, *Hind.*, *Mar.*, T. 293 ; bhaji (Trigonella), *Gus.*, T. 612 ; bhájí (Amarantus), *Mar.*, A. 938 ; bhakhra, *Pb.*, T. 548 ; bha-khúmba, *Beng.*, T. 573 ; bhála, *Hind.*, S. 1041 ; bhalian, *Hind.*, S. 1041 ; bhaliún, *Hind.*, R. 335 ; bhallia, *Hind.*, *Uriya*, S. 1041 ; bhalena, *Hind.*, H. 517 ; bhalia, *Beng.*, *Hind.*, F. 633 ; bhalu, *Hind.*, F. 819, 820, 822 ; bhalua, *Kumaon*, D. 115 ; bhamaruda, *Mar.*, B. 540 ; bhamína, *Hind.*, H. 517 ; bhamiri, *Hind.*, E. 267 ; bhamji, *Pb.*, U. 4 ; bhamni, *Pb.*, U. 4 ; bhan (Populus), *Pb.*, *Sind*, P. 1153 ; bhán (Rhus), *Pb.*, R. 293 ; bhanber, *Hind.*, F. 615 ; bhánd, *Pb.*, G. 177, *Hind.*, *Pb.*, G. 180 ; bhánda, *Hind.*, *Pb.*, G. 177, 180 ; bhándí, *Pb.*, S. 497 ; bhandír, *Hind.*, A. 711 ; bhan-dira, *Mar.*, C. 1380 ; bháng, *Beng.*, *Hind.*, *Pb.*, C. 331 ; bhángácha-jháda, *Mar.*, C. 331 ; bhangara, *Kumaon*, P. 413 ; bhángaro, *Gus.*, W. 25 ; bhangí, *Pb.*, C. 331 ; bhang-jala, *Hind.*, *Pb.*, D. 144 ; bhangli, *Hind.*, S. 589 ; bhangra (Indigofera), *Beng.*, *Mar.*, I. 134 ; bhangra (Eclipta), *Gus.*, *Hind.*, E. 7 ; bhangra (Viscum), *Pb.*, V. 149 ; bhán-gra (Wedelia), *Hind.*, W. 25 ; bhangria, *Kumaon*, E. 196 ; bhani, *Pb.*, P. 1153 ; bhanjira, *Hind.*, P. 413 ; bhanjra, *Hind.*, A. 1232 ; bhanjuri (Elionurus), *Hind.*, E. 192 ; bhan-juri (Apluda), *Hind.*, A. 1232 ; bhánrá, *Hind.*, W. 25 ; bhans (Oxybaphus), *Pb.*, O. 595 ; bháns (Bambusa), *Duk.*, B. 118 ; bhánt, *Beng.*, *Hind.*, C. 1380 ; bhanta, *Hind.*, S. 2284 ; bhánwar, *Hind.*, *Pb.*, I. 382 ; bháoga, *Mar.*, T. 427 ; bhara (Rhizophora), *Beng.*, R. 242 ; bhára (Terminalia), *Uriya*, T. 293 ; bharal, *Hind.*, S. 1273 ; bharalhe, *Pb.*, T. 439 ; bharam-dandi, *Duk.*, A. 1351 ; bharang, *Mar.*, C. 1388 ; bháranga-mula, *Mar.*, C. 1388 ; bharangi (Clerodendron), *Mar.*, C. 1388, C. 1394 ; bharangi (Picrasma), *Hind.*, P. 693 ; bharani, *Hind.*, R. 16 ; bharar, *Hind.*, S. 1273 ; bharatti, *Mar.*, C. 860 ; bharbhand, *Hind.*, A. 1351 ; bharbhunt, *Raj.*, P. 382 ; bhar-bhurwa, *Hind.*, A. 1351 ; bharburi, *Hind.*, E. 265 ; bharbusi, *Hind.*, E. 263 ; bhárgí-guda, *Beng.*, C. 1398 ; bhari, *Hind.*, C. 725 ; bhár jambol, *Duk.*, *Mar.*, IA. 958 ; bharlá, *Hind.*, T. 293 ; bharout, *Hind.*, C. 909 ; bharta, *Hind.*, P. 48 ; bharti, *Pb.*, P. 48, 50 ; bháru, *Pb.*, E. 184 ; bharua, *Pb.*, A. 1186 ; bharungi, *Guj.*, C. 1388 ; bharut, *Hind.*, S. 1273 ; bhar-waí, *Hind.*, I. 51 ; bhasamkand, *Hind.*, S. 902 ; bhát, *Hind.*, O. 258 ; bhát (Oryza), *Mar.*, O. 258 ; bhát (Polynemus), *Mar.*, F. 538 ; bhat (Clerodendron), *Hind.*, C. 1380 ; bhat (Glycine), *Hind.*, G. 263 ; bhata, *Pb.*, C. 2101 ; bhátavarná, *Mar.*, C. 2039 ; bhate,

Hind., C. 298 ; bháthi (Securinega), *Pb.*, S. 1024 ; bhati (Buddleia), *Kumaon*, B. 929 ; bhatia, *Hind.*, *Kumaon*, D. 94 ; bhat kateya, *Pb.*, A. 1351 ; bhat kúkra, *Kumaon*, L. 535 ; bhatmil, *Pb.*, A. 1351 ; bhat niggi, *Pb.*, W. 86 ; bhatreri, *Hind.*, R. 369 ; bhatta, *Mar.*, O. 258 ; bhatte, *Pb.*, L. 596 ; bhatuaghás, *Hind.*, O. 230 ; bhatwan, *Hind.*, G. 263 ; bhaulan, *Hind.*, H. 517 ; bhauphalí, *Mar.*, C. 1858 ; bhauri (Ipomæa), *Mar.*, I. 393 ; bhauri (Porana), *Dec.*, P. 1165 ; bhauri (Symplocos), *Beng.*, S. 3076 ; bhavan-bakra, *Hind.*, P. 1009 ; bhávanj, *Hind.*, P. 1352 ; bháwá, *Mar.*, C. 756 ; bhayrú, *Uriya*, C. 1031 ; bhe, *Pb.*, N. 39 ; bheckhol, *Mar.*, S. 2819 ; bhed, *Pb.*, S. 1320 ; bhedaira (Gmelina), *Hind.*, G. 298 ; bhedára (Juniperus), *Hind.*, J. 104 ; bhedas, *Mar.*, E. 464 ; bheji begun, *Uriya*, S. 2345 ; bhekal (Flacourtia), *Mar.*, F. 615 ; bhekal (Prinsepia), *Hind.*, *Mar.*, *Pb.*, P. 1252 ; bhekala, *Kumaon*, P. 1252 ; bhekar (Adhatoda), *Pb.*, A. 484 ; bhekar (Prinsepia), *Pb.*, P. 1252 ; bhekara, *Kumaon*, P. 1252 ; bhékhr, *Pb.*, P. 1252 ; bhekkar, *Pb.*, A. 484 ; bhekla, *Kumaon*, P. 1252 ; bhektí, *Beng.*, F. 476 ; bhel, *Pb.*, P. 702 ; bhela, *Hind.*, *Pb.*, S. 1041 ; bhelá, *Beng.*, *Hind.*, S. 1041 ; bhelatukı, *Beng.*, S. 1041 ; bheli, *Hind.*, *Pb.*, S. 370 ; bhend (Æschynomene), *Mar.*, A. 560 ; bhenda (Hibiscus), *Hind.*, *Mar.*, H. 196 ; bhendi, *Hind.*, H. 196 ; bhendi (Thespesia), *Hind.*, *Mar.*, T. 392 ; bhendu, H. 92 ; bheng, *Beng.*, F. 703 ; bhengal, *Hind.*, G. 688 ; bhengúl, *Hind.*, G. 688 ; bhenta, *Uriya*, L. 362 ; bhenwal, *Pb.*, G. 688 ; bheı, *Pb.*, S. 1320 ; bher, *Hind.*, Z. 231 ; bhera, *Hind.*, F. 757 ; bhéra, *Hind.*, S. 1320 ; bherda, *Dec.*, *Mar.*, T. 293 ; bherda, *Mar.*, T. 293 ; bherdha, *Mar.*, T. 293 ; bheri, *Hind.*, Z. 280 ; bheria, *Mar.*, C. 1031 ; bherawa, *Mar.*, C. 711 ; bherband, *Pb.*, A. 1351 ; bherendá, *Beng.*, R. 369 ; bheria, *Mar.*, C. 1031 ; bherki,ᴅ*Hind.*, D. 219 ; bherlá máda, *Mar.*, C. 711 ; bhernda, *Hind.*, J. 41 ; bhes, *Hind.*, A. 244 ; bhéul, *Pb.*, S. 560 ; bhewal, *Pb.*, G. 688 ; bheyla, *Hind.*, S. 1041 ; bheyrí, *Uriya*, C. 1031 ; bhiba, *Mar.*, S. 1041 ; bhika-purni, *Beng.*, H. 486 ; bhíl (Dolichandrone), *Hind.*, D. 748 ; bhil (Juniperus), *Hind.*, J. 101 ; bhiládar, *Pb.*, S. 1041 ; bhilama, *Mar.*, S. 1041 ; bhilámu, *Gus.*, S. 1041 ; bhilavan, *Dec.*, S. 1041 ; bhiláwa, *Hind.*, *Pb.*, S. 1041 ; bhillaur, *Hind.*, T. 525 ; bhillaura, *Hind.*, T. 525 ; bhıllauri, *Mar.*, T. 525 ; bhimal, *Hind.*, *Kumaon*, *Pb.*, G. 688 ; bhimb, *Hind.*, *Mar.*, C. 919 ; bhimráj, *Beng.*, W. 25 ; bhimúl, *Hind.*, G. 679 ; bhin datorí, *Pb.*, H. 196 ; bhindi (Hibiscus), *Hind.*, *Pb.*, H. 196 ; bhindi (Thespesia), *Gus.*, *Mar.*, T. 392 ; bhíndu, *Gus.*, H. 196 ; bhingule, *Mar.*, I. 125 ; bhin-sarpati, *Mar.*, O. 547 ; bhirand, *Mar.*, G. 36 ; bhirandel, *Mar.*, G. 36 ; bhıɾki, *Hind.*, S. 1287 ; bhirli mahad, *Mar.*, C. 711 ; bhirmi, *Hind.*, T. 93 ;

bhirmie, *Beng.*, T. 93 ; bhirra, *Hind.*, C. 1031 ; bhís khupra, *Dec.*, T. 530 ; bhobra, *Pb.*, E. 166, 188 ; bhœmag, *Guj.*, A. 1261 ; bhohár, *Hind.*, H. 517 ; bhoii kúmra, *Beng.*, T. 573 ; bhoj, (*Simla*), *Pb.*, P. 539 ; bhojapatra, *Guj.*, B. 501 ; bhojpatra, *Mar.*, B. 501 ; bhokar, *Dec.*, C. 1942 ; bhokar, *Hind.*, *Mar.*, C. 1931 ; bhokaɾa, *Mar.*, C. 1931 ; bhokɾa, *Hind.*, A. 692 ; bhokra, *Gus.*, S. 1287 ; bhokur, *Mar.*, C. 1931 ; bholia, *Hind.*, S. 3073 ; bholoni, *Hind.*, E. 263 ; bhólsarí, *Hind.*, M. 570 ; bhoma, *Mar.*, G. 238 ; bhoni-mug, *Sind*, A. 1261 ; bhoplabija. *Mar.*, L. 30 ; bhor (Ficus), *Pb.*, F. 236 ; bhor (Zizyphus), *Mar.*, Z. 231 ; bhor, *Raj.*, Z. 254 ; bhora, *Beng.*, R. 242 ; bhora álu, *Hind.*, *Beng.*, D. 515 ; bhorgoti, *Mar.*, Z. 290 ; bhorí, *Hind.*, I. 1 ; bhort, *Pb.*, C. 909 ; bhotheula, *Mar.*, D. 41 ; bhoti (Kydia), *Hind.*, K. 42 ; bhoti (Cordia), *Mar.*, C. 1927 ; bhotia badám (Corylus), *Kumaon*, C. 1988 ; bhotiya badám (Prunus), *Kumaon*, P. 1304 ; bhoursál, *Mar.*, H. 517 ; bhovarlit, *Mar.*, D. 292 ; bhóyachená, *Guj.*, A. 1261 ; bhoyaringani, *Gus.*, S. 2345 ; bhoza ámali, *Gus.*, P. 632 ; bhringga, *Hind.*, P. 468 ; bhringurája, *Mar.*, E. 7 ; bhudoi, *Hind.*, F. 202 ; bhuí, *Raj*, A. 554 ; bhuí-ámlá, *Beng.*, P. 657 ; bhuí-áola, *Uriya*, P. 657 ; bhuí-ávalá, *Mar.*, P. 657 ; bhuiavali (Phyllanthus), *Mar.*, P. 668 ; bhuiavali (Sesbania), *Mar.*, S. 1166 ; bhui áwali, *Mar.*, P. 632 ; bhui champa, *Beng.*, *Hind.*, *Guj.*, K. 8 ; bhuíchane, *Mar.*, A. 1261 ; bhui dalim, *Beng.*, C. 580 ; bhui-daɾí.*Mar.*, T. 860 ; bhui-dúmúr, *Beng.*, F. 194 ; bhuigholi, *Mar.*, P. 1179 ; bhuijám, *Beng.*, P. 1231 ; bhuíkándá, *Mar.*, S. 968 ; bhuikohala, *Mar.*, I. 379 ; bhúi-kumrá, *Beng.*, I. 379 ; bhui-múga, *Mar.*, A. 1261 ; bhui-okra, *Hind.*,ｊL. 451 ; bhuiringani, *Mar.*, S. 2345 ; bhuí-sarpatí, *Mar.*, O. 547 ; bhui-sheng, *Mar.*, A. 1261 ; bhúí tarwar, *Mar.*, C. 778 ; bhuinánvaláh, *Dec.*, *Hind.*, P. 657 ; bhúítarvada, *Mar.*, C. 737 ; bhuiterada, *Mar.*, L. 273 ; bhúj, *Pb.*, B.501 ; bhuja, *Hind.*, C. 2324 ; bhujapatra, *Guj.*, B. 501 ; bhuj patar, *Hind.*, B. 501 ; bhújpattra, *Hind.*, B. 496, 501 ; bhúk, *Pb.*, A. 786 ; bhukas, *Mar.*, E. 73 ; bhúkrí, *Pb.*, T. 548 ; bhulan, *Sind*, W. 58 ; bhulel, *Kumaon*, B. 212 ; bhulga, *Mar.*, W. 54 ; bhúmi, *Beng.*, I. 379 ; bhúmi kúmara, *Beng.*, T. 573 ; bhúmphor, *Pb.*, T. 845 ; bhunga, *Hind.*, C. 2331 ; bhungi, *Beng.*, C. 1861 ; bhunguru, *Kumaon*, S. 1078 ; bhunja (Benincasa), *Kumaon*, B. 430 ; bhunja (Cucurbita), *Kumaon*, C. 2331 ; bhunjí-pát, *Beng.*, C. 1861 ; bhún ke dúm, *Pb.*, V. 64 ; bhúnr-bhundá, *Hind.*, O. 574 ; bhúra, *Hind.*, S. 373 ; bhur-bhur, *Hind.*, C. 1412 ; bhú-ringni, *Mar.*, S. 2345 ; bhúrjapatra, *Mar.*, B. 501 ; bhúrkúr, *Hind.*, H. 517 ; bhúrkúrí, *Mar.*, W. 122 ; bhurmi, *Mar.*, Z. 231 ; bhursunga, *Uriya*, M. 800 ; bhurt, *Hind.*, F. 672 ; bhurt, *Hind.*, C. 909 ; bhurt-kasi, *Mar.*,

M. 855; bhúru kolu, *Guj.*, B. 430; bhurundi, *Mar.*, H. 102; bhurungi, *Beng.*, P. 693; bhut, *Kumaon, Pb.*, G. 263; bhúta-kesa, *Mar.*, M. 855; bhutankas, *Sind*, F. 615; bhutá-pálá, *Mar.*, E. 73; bhút-bhiravi, *Beng.*, P. 1233; bhút-jatt, *Kashmir*, N. 17; bhút jhata, *Pb.*, A. 1227; bhutkesi, *Hind.*, *Beng.*, C. 1980; bhútkis, *Hind.*, *Beng.*, C. 1980; bhutrak, *Hyderabad*, E. 73; bhutta (Solanum), *Kumaon*, S. 2284; bhutta (Zea), *Beng.*, *Hind.*, *Kumaon*, Z. 50; bhú-tulsí, *Beng.*, S. 746; bhutwa, *Hind.*, C. 1003.

Bi Bia (Pterocarpus), *Dec.*, *Gus.*, *Hind.*, P. 1370; biár (Pinus), *Pb.*, P. 737; biba, *Mar.*, S. 1041; bibala, *Mar.*, P. 1370; bibha, *Mar.*, S. 1041; bibia-bágh, *Mar.*, T. 434; bibla, *Dec.*, *Gus.*, *Mar.*, P. 1370; bibsar, *Hind.*, M. 797; bibu (Holigarna), *Mar.*, H. 306; bibu (Semecarpus), *Mar.*, S. 1041; bíbúcha, *Pb.*, D. 674; bibwa, *Mar.*, S. 1041; bichati, *Beng.*, T. 509; bichhoti, *Beng.*, M. 786; bichhra, *Hind.*, P. 618; bichhu, *Hind.*, U. 61; bichtárak, *Beng.*, A. 1362; bichu (Martynia), *Hind.*, *Pb.*, M. 308; bichu (Urtica), *Pb.*, U. 49; bichua (Girardinia), *Hind.*, G. 213; bichúa (Urtica), *Pb.*, U. 49; bida, *Pb.*, S. 570; bidá, *Pb.*, S. 579; bidah, *Sind*, F. 530; bidái, *Pb.*, S. 544, S. 560; bidárí-kand, *Pb.*, P. 1401; bidelganj, *Hind.*, J. 104; bidu, *Pb.*, S. 579; bigbund, *Hind.*, P. 1078; bigni, *Pb.*, C. 881, 886; bihara, *Hind.*, T. 293; bihi, *Hind.*, S. 1908; bihi, *Hind.*, C. 2546; bihra, *Hind.*, C. 1031; bihri, *Hind.*, C. 1031; bihul, *Sind*, G. 705; bija, *Hind.*, *Mar.*, P. 1370; bijapúra, *Mar.*, C. 1270; bij-asa, *Uriya*, P. 1370; bíja sah, *Hind.*, P. 1370; bíjasal, *Hind.*, P. 1370; bija-sál, *Hind.*, P. 1370; bijasár, *Hind.*, P. 1370; bijaura, *Beng.*, *Hind.*, C. 1270; bijavwi, *Pb.*, C. 629; bíjband (Sida), *Pb.*, S. 1694; bíjband (Rumex), *Pb.*, R. 645; bíj-elosha, *Sind*, C. 170; bijgái, *Pb.*, L. 535; bijindak, *Afg.*, L. 276; bijorí, *Mar.*, C. 1270; bíjoro, *Sind*, C. 1263; bijorú, *Guj.*, C. 1270; bij-palak, *Pb.*, S. 2574; bijri, *Hind.*, L. 385; bikh, *Hind.*, A. 397; bikhonda, *Kumaon*, S. 2394; biksa, *Hind.*, H. 113; bikuntia, *Uriya*, F. 492; bil (Juniperus), *Hind.*, J. 104; bil (Ægle), *Guj.*, A. 534; bila (Ægle), *Mar.*, *Sind*, A. 534; bila (Cratæva), *Hind.*, C. 2039; biláeti tamáku, *Beng.*, N. 98; bilái kand (Ipomæa), *Beng.*, *Hind.*, I. 379; bilái kand (Pueraria), *Hind.*, *Kumaon, Pb.*, P. 1401; bilambi (Averrhoa), *Mar.*, A. 1644; bilambi (Semecarpus), *Mar.*, S. 1041; bilangra, *Hind.*, F. 615; bilaran, *Hind.*, S. 1041; bilari, *Hind.*, *Kumaon*, M. 791; bilási, *Hind.*, C. 2039; biláti, *Hind.*, F. 615; bilatí-ánanúoh, *Beng.*, A. 603; biláti-badám, *Beng.*, P. 1274; biláti-mung, *Beng.*, A. 1261; bilatipát, *Beng.*, A. 603; bilaur, *Hind.*, C. 616; bílaurí, *Pb.*, P. 1112; bildí, *Pb.*, I. 384; bilga, *Pb.*, R. 495; bilgagora, *Beng.*, F. 477; bili, *Kumaon*, P. 1401; biliana, *Hind.*, C. 2039;

bilimbi, *Beng.*, *Hind.*, A. 1644; bilin, *Hind.*, *Pb.*, F. 53; biliri, *Hind.*, C. 1359; billa, *Beng.*, S. 830; billi (Pueraria), *Hind.*, P. 1401; billi (Setaria), *Hind.*, S. 1207; billi lotan (Nardostachys), *Dec.*, N. 17, *Pb.*, N. 64; billi-lotan (Valeriana), *Hind.*, V. 8; billú, *Mar.*, C. 1031; bil-nalita, *Beng.*, C. 1858; bilodar, *Pb.*, S. 838; biloja, *Pb.*, S. 838; bilphari, *Hind.*, P. 546; bilsa, *Hind.*, S. 579; biluga, *Uriya*, C. 1031; bilva, *Beng.*, A. 534; bímal, *Kumaon*, G. 688; bimbal, *Mar.*, C. 2612; bimbi, *Mar.*, C. 919; bimbu, *Beng.*, C. 919; bimla, *Hind.*, G. 673; bín, *Pb.*, S. 579; bina, *Beng.*, A. 1655, *Beng.*, *Hind.*, A. 1660; bincha, *Beng.*, F. 615; bind, *Hind.*, S. 6; bindak, *Hind.*, C. 1985; bindra, *Hind.*, S. 1207; bingharbij, *Pb.*, A. 1579; binna, *Pb.*, V. 164; bínsín, *Pb.*, M. 910; bipuwa-kánta, *Kumaon*, C. 416; birálipanwa, *Kumaon*, P. 1401; biráli-púna, *Kumaon*, P. 1401; biranga, *Beng.*, E. 199; biranjasif, *Guj.*, A. 367; birár, *Hind.*, C. 224; birár, *Hind.*, *Pb.*, Z. 254; bireja, *Hind.*, F. 95; bírgo, *Pb.*, P. 693; bírha, *Pb.*, T. 293; birhatta, *Hind.*, S. 2280; bi-rijeh, *Afg.*, F. 95; birli mád, *Mar.*, C. 711; birli-mhad, *Mar.*, C. 711; birl-mhar, *Mar.*, C. 711; birmi, *Hind.*, *Kashmir*, *Pb.*, T. 93; birm-kanwal, *Pb.*, S. 919; birota, *Pb.*, Z. 254; birotá, *Pb.*, C. 224; biroza, *Pb.*, P. 760; birralli, *Uriya*, I. 525; bírri, *Pb.*, C. 1361; bis (Aconitum), *Hind.*, A. 397; bís (Salix), *Pb.*, S. 537, S. 579; bís (Myricaria), *Pb.*, M. 881; bisa, *Pb.*, S. 1906; bisa, *Kashmir*, S. 544; bisa, *Pb.*, S. 544; biscobra, *Hind.*, R. 116; bisfáij, *Pushtu*, A. 498; bish (Melocanna), *Beng.*, M. 425; bish (Aconitum), *Beng.*, *Hind.*, A. 397; bisha, *Beng.*, G. 243; bishalanguli, *Beng.*, G. 243; bísh báns, *Beng.*, B. 417; bish kachú, *Beng.*, C. 1738; bish kapra, *Sind*, T. 541; bish-káprá, *Pb.*, T. 537; bishkáprá, *Mar.*, T. 537; bish-káprá, *Pb.*, T. 541; bish-kopra, *Kumaon*, P. 1250; bisír, *Pb.*, P. 1460; bislombhi, *Hind.*, C. 2310; bislúmbhí, *Hind.*, C. 2310; bísrú, *Pb.*, C. 838; bissarhí pála, *Pb.*, D. 611; bisténd, *Hind.*, D. 628; bísu, *Pb.*, S. 531; biswúl, *Hind.*, A. 267; bitchu-ka-mutchí, *Hind.*, F. 568; bíthú, *Pb.*, C. 1003; bithúa (Dalbergia), *Hind.*, D. 32; bithúa (Heliotropium), *Hind.*, *Pb.*, H. 100; bit pálang, *Beng.*, B. 480; bitsa, *Pb.*, *Sind*, S. 579; bitsu, *Pb.*, S. 570; bitsu, *Pb.*, S. 586; bitsú bes, *Pb.*, S. 544; biúgli, *Pb.*, C. 881; biúgu, *Pb.*, C. 886; biúl, *Hind.*, *Pb.*, G. 688; biuna, *Pb.*, V. 164; biúng, *Hind.*, G. 688; bíúns, *Pb.*, P. 1159; biur, *Pb.*, A. 1464.

Bl Blail, *Pb.*, S. 570; blair, *Mar.*, S. 1939; blimbi, *Beng.*, A. 1644; blimbu, *Guj.*, A. 1644.

Bo Boalli, *Uriya*, F. 601; boassa, *Hind.*, F. 384; bobri, *Mar.*, F. 469; bobriya, *Pb.*, E. 188; boch, *Mar.*, S. 579; boda, *Hind.*, L. 53; bodaga, *Mar.*, L. 120; boda-kanod, *Beng.*, C. 2062; bodál, *Kumaon*, *Hind.*, S. 2819; bodála, *Kumaon*, *Hind.*, S. 2819; bodal-mowa, *Kumaon*,

E. 208 ; bodará, *Pb.*, M. 289 ; bodha, *Mar.*, C. 551 ; bodiajamo, *Guj.*, C. 701 ; bodo-bodo-ria, *Uriya*, O. 127 ; bodoka, *Uriya*, H. 517 ; bodosi, *Uriya*, F. 508 ; bodula, *Hind.*, S. 2819 ; boe, *Mar.*, S. 2978 ; boga, *Beng.*, F. 465 ; bogalí, *Hind.*, G. 385 ; boggut, *Mar.*, F. 411 ; bohar (Ficus), *Pb.*, F. 129 ; bohar (Odina), *Beng.*, O. 38 ; bohari, *Beng.*, C. 1931 ; boharpótúr, *Hind.*, H. 517 ; bohera, *Beng.*, T. 293 ; bohir, *Pb.*, F. 129 ; bohl, *Beng.*, M. 570 ; boho-dari, *Beng.*, C. 1931 ; bohorá, *Beng.*, T. 293 ; boichand, *Mar.*, P. 588 ; boichind, *Mar.*, P. 551 ; boilam, *Beng.*, S. 3040 ; boilshura, *Beng.*, V. 45 ; boilsur, *Beng.*, S. 3040 ; bóín, *Kashmir*, P. 935 ; boj, *Pb.*, T. 864 ; boj, *Pb.*, T. 875 ; boja, *Uriya*, X. 16 ; bokat, *Pb.*, A. 1579 ; boke, *Mar.*, B. 520 ; bokenal, *Mar.*, L. 503 ; bó-kí-adrak, *Dec.*, C. 2381 ; bokkan, *Beng.*, H. 1 ; bokmo, *Uriya*, C. 35 ; ból, *Beng.*, *Duk.*, *Guj.*, *Hind.*, B. 48 ; bola (Barbus), *Beng.*, F. 362 ; bola (Hibiscus), *Beng.*, H. 255 ; bolas, *Beng.*, E. 210 ; bole, *Beng.*, F. 578 ; bol-kadam, *Beng.*, A. 1192 ; bolsari, *Gus.*, M. 570 ; bombaksing, *Kumaon*, P. 1316 ; bombali, *Kumaon*, P. 1316 ; bonbhérandá, *Beng.*, J. 41 ; bondaga, *Hind.*, L. 53 ; bondar, *Beng.*, F. 791 ; bondára, *Mar.*, L. 42, L. 55 ; bong, *Beng.*, S. 2284 ; bonga, *Pb.*, A. 401 ; bonicha, *Uriya*, F. 615 ; bonín, *Kashmir*, P. 935 ; bon-joí, *Beng.*, C. 1377 ; bunkuaso, *Uriya*, F. 424, 530 ; bonmethí, *Beng.*, S. 1688, S. 1714 ; bonpói, *Hind.*, B. 203 ; bonurlauri, *Hind.*, C. 756 ; bophallí, *Pb.*, C. 1842 ; bopotassi, *Uriya*, F. 546 ; bor (Ficus), *Beng.*, *Hind.*, *Pb.*, F. 129, F. 165, I. 62 ; bór (Zizyphus), *Beng.*, *Gus.*, *Mar.*, Z. 231 ; bora (Vigna), *Hind.*, V. 116 ; bora (Typha), *Kumaon*, T. 864 ; bóra (Zizyphus), *Mar.*, Z. 231 ; borailli, *Beng.*, G. 385 ; borara, *Uriya*, B. 356 ; bora-salli, *Gus.*, M. 570 ; borbacha, *Dec.*, T. 434 ; borboti, *Beng.*, *Hind.*, D. 789 ; bordi, *Mar.*, Z. 231 ; bordi, *Gus.*, Z. 231 ; borí (Typha), *Pb.*, T. 864 ; borí (Zizyphus), *Hind.*, Z. 231 ; boriajamoda, *Mar.*, A. 1227 ; borí-ajmúd karaís, *Hind.*, A. 1227 ; borla, *Kumaon*, C. 1931 ; bormala, *Beng.*, C. 123 ; boro, *Beng.*, F. 512 ; boro-khotiya, *Beng.*, D. 212 ; boromali, *Uriya*, H. 285 ; boronda, *Mar.*, C. 596 ; boropatri, *Uriya*, C. 123 ; borrur, *Mar.*, P. 45 ; borsali, *Mar.*, M. 570 ; boru, *Uriya*, F. 129 ; boruna, *Beng.*, V. 174 ; boruti, *Beng.*, P. 41 ; borz, *Pushtu*, S. 1233 ; borz-kuhi, *Baluch.*, S. 1233 ; bosanti, *Beng.*, H. 285 ; bosontogundi, *Uriya*, M. 71 ; bostán afraz, *Pb.*, C. 873 ; botahl, *Uriya*, F. 571 ; botat, *Mar.*, D. 481 ; botee-jam, *Beng.*, E. 453 ; bother, *Mar.*, E. 314 ; bothngt, *Pb.*, D. 758 ; botku, *Mar.*, E. 314 ; botsaka, *Mar.*, I. 120 ; bottuka, *Hind.*, H. 92 ; botya jharo, *Berar*, C. 1026 ; boul-la, *Pb.*, F. 455 ; bouna, *Pb.*, D. 409 ; bounce-puttri, *Uriya*, F. 307 ; bounchi, *Pb.*, F. 394 ; boura, *Beng.*, M. 12 ; bouro, *Beng.*, *Hind.*, S. 1939 ; bouro, *Uriya*, B. 632 ; bouro simuri, *Uriya*, B. 632 ; bowála, *Hind.*, *Pb.*, M. 800 ; bowári, *Hind.*, S. 2394 ; bowchí,

Dec., *Hind.*, F. 615 ; bowmaj, *Mar.*, C. 898 ; boyari, *Beng.*, F. 601 ; boyedi, *Gus.*, Z. 231 ; boyra, *Beng.*, T. 293 ; bozandán, *Pb.*, A. 1575 ; bóz-ghánj, *W. Afghan.*, P. 858 ; bozídán, *Pb.*, A. 1575.

Br Bráa, *Pushtu*, C. 1740 ; brahám, *Pb.*, S. 2394 ; brahám, *Kashmir*, S. 2394 ; brahmadandi, *Mar.*, L. 72 ; bráhma-manduki, *Beng.*, *Hind.*, H. 486 ; brahman-patta, *Beng.*, C. 1394 ; brahmáni, *Kashmir*, E. 479 ; brahmi (Herpestis), *Hind.*, H. 149, *Mar.*, H. 486 ; brahmi (Ulmus), *Pb.*, U. 4 ; bráhmi (Taxus), *Kumaon*, T. 93 ; brahmoka, *Beng.*, H. 88, *Mar.*, H. 74 ; bramhadandi, *Mar.*, T. 567 ; bramji, *Pb.*, C. 881 ; bran (Ulmus), *Pb.*, *Kashmir*, U. 4 ; brán (Quercus), *Pb.*, Q. 23 ; branchu, *Pb.*, M. 910 ; brankúl, *Kashmir*, *Pb.*, U. 4 ; brannú, *Pb.*, U. 4 ; bránti, *Pb.*, M. 910 ; brapú, *Pb.*, F. 19 ; brárí, *Kashmir*, U. 4 ; brás, *Kumaon*, *Pb.*, R. 253 ; bratta, *Pb.*, E. 232 ; bre (Desmodium), *Pb.*, D. 348 ; bre (Eremurus), *Pb.*, E. 273 ; bré (Quercus), *Pb.*, Q. 29 ; brekche, *Pb.*, Q. 29 ; brelá, *Beng.*, S. 1694 ; bren (Ulmus), *Kashmir*, U. 4 ; brén (Quercus), *Pb.*, Q. 23 ; brera, *Pb.*, U. 4 ; breri (Rosa), *Pb.*, R. 533 ; brerí (Ulmus), *Kashmir*, U. 4 ; brés, *Pb.*, F. 10 ; briálí, *Pb.*, C. 1711 ; brihat-chakramed, *Hind.*, S. 1166 ; brihmisak, *Beng.*, H. 149 ; brímdú, *Pb.*, C. 886 ; brimla, *Pb.*, C. 886 ; brimlu, *Pb.*, C. 881 ; brimposh, *Kashmir*, N. 192 ; brind, *Pb.*, A. 692 ; brinjal, S. 2284 ; brinkol, *Pb.*, S. 505 ; brisarí, *Pb.*, S. 2376 ; broá, *Pb.*, R. 253 ; brúmaj, *Pb.*, C. 898 ; brúmbrúm dakári, *Pb.*, H. 52 ; brúmij, *Pb.*, C. 886 ; brúndu, *Pb.*, C. 898 ; brus, *Kumaon*, R. 253.

Bu Buchachi, *Pb.*, G. 186 ; buchnága, *Guj.*, A. 397 ; buchua, *Hind.*, F. 547 ; buckra, *Hind.*, F. 311 ; buckshí, *Hind.*, V. 73 ; budá-nár, *Pb.*, M. 289 ; búdar, *Kashmir*, A. 22 ; budbhola, *Beng.*, B. 929 ; buddha narikella, *Beng.*, S. 2806 ; budha, *Sind*, S. 531 ; budhal, *Hind.*, M. 289 ; budhan, *Pb.*, E. 263 ; budiajiwan, *Guj.*, A. 1227 ; budjari-dha-mun, *Mar.*, E. 317 ; budshur, *Pb.*, E. 234 ; buggarah, *Hind.*, F. 362 ; buggush, *Uriya*, F. 362 ; bughy, *Beng.*, S. 1934 ; bugra, *Pb.*, C. 1367 ; bugut rori, *Hind.*, S. 2501 ; buhal, *Beng.*, C. 1931 ; buhéra, *Hind.*, T. 293 ; buhura, *Hind.*, T. 293 ; buhuru, *Beng.*, T. 293 ; búi (Ærua,) *Sind*, A. 554 ; búi (Plectranthus), *Pb.*, P. 959 ; búi (Pulicaria), *Pb.*, P. 1406 ; búi (Anabasis), *Pb.*, A. 1005 ; búi (Ballota), *Pb.*, B. 33 ; búi (Kochia), *Pb.*, K. 30 ; búi (Swertia), *Kashmir*, S. 3014 ; búi chhotí, *Pb.*, K. 30 ; búi-kallan, *Pb.*, A. 554 ; búi láthia, *Pb.*, C. 2101 ; búi mádarán, *Pb.*, A. 1469 ; búi máderán, *Afg.*, A. 367 ; búin, *Pb.*, *Kashmir*, P. 935 ; buin-owla, *Dec.*, *Hind.*, P. 663 ; buin-phal, *Pb.*, F. 725 ; bujina, *Hind.*, A. 1117 ; bujjerbhang, *Hind.*, N. 101 ; bújlo, *Pb.*, G. 186 ; buka, *Beng.*, S. 1186 ; bukaín, *Hind.*, M. 393 ; búkan, *Pb.*, L. 451 ; bukchí (Psoralea), *Hind.*, P. 1352 ; bukchie (Vernonia), *Beng.*, V. 73 ; bukha (Adansonia), *Guj.*, A. 455 ; bukhain, *Pb.*, M. 363 ; bukhar (Dendrocalamus), *Hind.*, D. 292 ; búki, *Pb.*, E. 241 ; bukká (Abir),

Deccan, A. 37 ; buko, *Beng.*, S. 1186 ;
bukoki, *Pb.*, V. 73 ; buksha, *Beng.*, H.
113 ; búl, *Beng.*, F. 579 ; bulat, *Beng.*,
P. 496 ; búldu, *Pb.*, A. 22 ; bulgar jangli,
Kashmir, A. 596 ; búlkkora, *Beng.*, C.
987 ; bulla, *Dec.*, *Hind.*, T. 293 ; bulmuj,
Kashmir, D. 173 ; bumalo, *Beng.*, F.
458 ; bumbal, *Pb.*, R. 588 ; bummi, *Beng.*,
Uriya, F. 493 ; bún, *Hind.*, *Mar.*, C.
1641 ; búna (Albizzia), *Pb.*, A. 692 ; búna
(Platanus), *Kashmir*, *Pb.*, P. 935 ; búna
(Sophora), *Pb.*, S. 2376 ; bunbur-butti,
Beng., P. 489 ; búnch, *Hind.*, F. 329 ;
bun-chichinga, *Beng.*, T. 596 ; búnd,
Dec., *Guj.*, *Mar.*, C. 1641 ; bundal,
Beng., V. 195 ; búndei, *Uriya*, F. 328 ;
bundi (Carthamus), *Raj.*, C. 637 ; bundi
(Eleotris), *Uriya*, F. 436 ; búndí (lead),
Beng., L. 143 ; bunga-ghundeena, *Beng.*,
A. 786 ; bungka-chael, *Pb.*, F. 396 ; bunj,
Dec., *Hind.*, F. 615 ; bun-join, *Beng.*,
C. 1377 ; bun-jumat, *Beng.*, C. 1377 ;
bunkuai, *Uriya*, F. 330 ; bun-kuchú,
Beng., C. 1732 ; bun-lung, *Beng.*, J. 114 ;
bun-ochra (Urena), *Beng.*, U. 29 ; bun-
okra (Triumfetta), *Beng.*, T. 839 ; bun-
pala, *Pb.*, P. 1458 ; bun-palung, *Beng.*,
R. 645 ; bun-shim, *Beng.*, D. 789 ; bun-
sputta, *Uriya*, F. 307 ; búnt-ká-sirkah,
Hind., C. 1061 ; búnt-nu-zirko, *Guj.*,
C. 1061 ; bunun, *Pb.*, F. 682 ; búnún
musrini, *Pb.*, F. 678 ; bur (Bambusa),
Hind., B. 139 ; bur (Ficus), *Sind*, F.
129 ; búr (Andropogon), *Hind.*, *Pb.*, A.
1093 ; búr (Gerbera), *Pb.*, G. 186 ; búr,
Pb., T. 51 ; búr (Typha), *Sind*, T. 864 ;
búr (Zizyphus), *Mar.*, Z. 231 ; bura,
Hind., *Pb.*, S. 374 ; bura chucha, *Beng.*,
C. 2603 ; búrade abnús, *Pb.*, D. 656 ;
búrajanwar, *Hind.*, H. 289 ;;búrak, *Dec.*,
C. 1117 ; bura keru, *Beng.*, E. 531 ; bura-
kúnda, *Beng.*, J. 13 ; bura-manda, *Beng.*,
L. 549 ; bura-shakkar, *Dec.*, S. 374 ; bura
shama, *Beng.*, P. 48 ; burasúksúng,
Beng., B. 546 ; buráwa, *Pushtu*, C. 2558 ;
buraye, *Sind*, P. 419 ; burburra, *Mar.*,
I. 134 ; búrdiya, *Hind.*, C. 1026 ; bureta,
Hind., S. 1226 ; burg-sadab, *Beng.*, E.
514 ; búrgú, *Pb.*, P. 688 ; burh, *Pb.*, F.
260 ; burhna, *Hind.*, P. 1406 ; buri
(Symplocos), *Beng.*, S. 3073 ; burí (Vitis),
Pb., V. 233 ; búrí (Typha), *Sind*, T.
864 ; buriya, *Beng.*, D. 212 ; búrj, *Pb.*,
B. 501 ; burkas, *Mar.*, E. 73 ; burla, *Pb.*,
A. 22 ; burmie, *Beng.*, T. 93 ; búrnak,
Pb., A. 1462 ; buro, *Kumaon*, D. 328 ;
buro-keruee, *Beng.*, E. 531 ; buro-mussúr,
Beng., L. 252 ; búrra, *Pb.*, A. 22 ; burra-
mattar, *Beng.*, P. 885 ; burra shím, *Duk.*,
C. 289 ; burrayra, *Sind*, S. 1694 ; bur-
reah, *Pb.*, F. 361 ; burrel-hay, *Pb.*, F.
770 ; burrí-shep, *Dec.*, P. 727 ; bursha,
Sind, C. 1707 ; bursunga, *Hind.*, M.
800 ; buru, *Kumaon*, S. 2394 ; burua,
Beng., M. 9 ; burúl, (*Bhajji*), *Pb.*, A.
22 ; búrúndi, *Mar.*, H. 102 ; burzal, *Pb.*,
B. 501 ; bushan, *Pb.*, S. 537, S. 560 ;
búsk, *Baluch.*, L. 276 ; bustán-afróz,
Kashmir, A. 939 ; but (Notopterus), *Pb.*,
F. 511 ; but (Ficus), *Sind*, F. 129 ; bút
(Cicer), *Beng.*, C. 1061 ; bút (Populus),
Pb., P. 1142 ; buta (Zea), *Mar.*, Z. 50 ;

butá, *Uriya*, Z. 50 ; búta (Crotalaria),
Pb., C. 2101 ; butana, *Hind.*, P. 885 ;
butchua, *Hind.*, F. 549, *Uriya*, F. 450 ;
bute, *Mar.*, E. 314 ; bute, *Mar.*, E. 312 ;
bútí-ka-mochka, *Pb.*, A. 596 ; but kalái,
Beng., C. 1061 ; bút pesh, *Pb.*, S. 906 ;
butshur, *Pb.*, E. 234 ; buttani, *Hind.*, P.
885 ; butwal, *Hind.*, C. 1234 ; búz, *Pb.*,
S. 1242 ; búza, *Pb.*, H. 382 ; buz-anjir,
Afg., R. 369 ; buzghanj, *Hind.*, *Mar.*,
P. 847 ; buzkuhi, *Baluch.*, S. 1239 ; buzlí,
Pb., G. 186 ; búzrú, (*Sutlej*), *Pb.*, P. 540.

Bw Bwir, *Pb.*, S. 589.
By Byakur, *Beng.*, S. 2280 ; byákura, *Beng.*, S.
2280 ; byans, *Kumaon*, P. 737 ; byasa,
Uriya, P. 1370 ; bysabol, *Mar.*, B. 41.

Ca Caffi, *Mar.*, C. 1641 ; cahwa, *Duk.*, C.
1641 ; cairava, *Hind.*, L. 355 ; cajupúta,
Mar., M. 342 ; cajuputte, *Beng.*, M.
342 ; camral-nebú, *Beng.*, C. 1296 ;
canchra, *Hind.*, P. 338 ; caphi, *Mar.*, C.
1641 ; cappi, *Guj.*, C. 1641 ; carnicára,
Hind., P. 338 ; cart-kuntea, *Uriya*, F. 329 ;
cart-kana, *Uriya*, F. 311 ; cartua-gorai,
Uriya, F. 516 ; cat'hachampa, *Hind.*,
P. 338 ; catla, *Hind.*, *Beng.*, *Pb.*, F. 384.

Ce Cenia, *Sind.*, F. 451.
Ch Chab, *Hind.*, P. 797 ; chaba, *Hind.*, P. 797 ;
chábuk, *Pb.*, H. 285 ; chachar, *Pb.*, B.
465 ; cháchenda, *Hind.*, T. 569 ; chache-
on, *Pb.*, R. 253 ; chachi bet, *Beng.*, *Hind.*,
C. 104 ; chachinda, *Hind.*, T. 569 ; chachinda, *Kumaon*, T. 569 ; chachinga,
Hind., T. 569 ; chachri, *Pb.*, M. 910 ;
chachundar (musk-rat), *Hind.*, R. 54 ;
chacunda (Chatoessus), *Beng.*, F. 387 ;
chadua, *Hind.*, R. 152 ; chaedri, *Beng.*,
F. 359 ; chagal-batí, *Beng.*, N. 8 ; chá-
gulbánti, *Beng.*, D. 9 ; chagul-nadí,
Beng., S. 2518 ; chagulpatí, *Beng.*, C.
2554 ; chaguni, *Beng.*, *Hind.*, F. 334 ;
chahna, *Sind*, C. 1061 ; chai, *Beng.*, P.
797 ; chaikath, *Beng.*, P. 797 ; chaina,
Mar., D. 513 ; chainchar, *Pb.*, D. 196 ;
chairyili, *Pb.*, D. 196 ; chaitra, *Kashmir*,
T. 376 ; chaiúra, *Kumaon*, B. 212 ; chak
(Hordeum), *Pb.*, H. 382 ; chak (Rumex),
Beng., R. 650 ; chák (chalk), *Guj.*, C.
489 ; chakchakóti ka'jhâr, *Duk.*, A. 376 ;
chakemdia, *Beng.*, D. 32 ; chakkarnitta-
gadi, *Hind.*, S. 1223 ; chakkí (Artocarpus),
Hind., A. 1489 ; chakki (sugar-cane),
Hind., S. 370 ; chakmak (agate), *Hind.*,
C. 617 ; chakmak (flint), *Hind.*, F. 652 ;
chakotra, *Beng.*, *Hind.*, *Pb.*, C. 1263 ;
chakra-bhenda, *Duk.*, A. 89 ; chakra-bhenda, *Mar.*, A. 80, A. 89 ; cháksie, *Mar.*,
C. 728 ; cháksu, *Hind.*, *Dec.*, C. 728 ; cha-
kua, *Beng.*, A. 722 ; chákuliá, *Beng.*, U.
23 ; chakunda, *Beng.*, *Hind.*, *Pb.*, C.
797 ; chákút, *Dec.*, *Hind.*, C. 728 ; chak-
wa, *Beng.*, A. 1146 ; chakwit, *Mar.*, C.
1003 ; chal (Rosa), *Kashmir*, R. 538 ;
chál (Oryza), *Beng.*, O. 258 ; chalai, *Pb.*,
J. 92 ; chál-anár, *Pb.*, P. 1426 ; chalcha,
Hind., B. 944 ; chal-chalíra, *Pb.*, L.
332 ; chaldua, *Uriya*, E. 342 ; chálé-
michhri, *Beng.*, S. 521 ; chalhatti, *Hind.*,
P. 760 ; chal-kumra, *Beng.*, *Pb.*, B. 430 ;
challa manta, *Hind.*, S. 1024 ; chalmeri,
Hind., P. 627 ; chalmori, *Hind.*, *Kuma-*

on, O. 547 ; chálmúgra, *Hind.*, G. 761 ;
chalniya, *Kumaon*, P. 1148 ; chalrúndar,
Hind., Pb., I. 434 ; chalodra, *Pb.*, E.
170 ; chalon, *Pb.*, P. 1148 ; chalonwa,
Pb., P. 1148 ; cháltá, *Beng., Hind.*, D.
428 ; chamaggai, *Hind.*, D. 424 ; cha-
mak, *Hind.*, I. 472 ; chamakhrí, *Pb.*,
M. 517 ; chámal, *Mar.*, B. 334 ; chamar,
Pushtu, E. 35 ; chamar gular, *Hind.*, M.
71 ; chámári, *Mar.*, P. 1233 ; chamari-ki-
vel, *Hind.*, V. 12 ; chámb, *Pb.*, A. 801 ;
chamba (Jasminum), *Pb.*, J. 18, 24,
Hind., Kashmir, J. 29, *Pb.*, J. 35 ;
chamba (Michelia), *Pb.*, M. 517 ; chamba
(Prinsepia), *Pb.*, P. 1252 ; chambadi,
Mar., P. 1239 ; chambal, *Pushtu*, E.
35 ; chambara, *Mar.*, P. 1248 ; chambari,
Mar., P. 1239 ; chambar máya, *Hind.*,
U. 4 ; chambelí, *Pb.*, J. 18, 35, *Kumaon*,
J. 29 ; chambelí, *Mar.*, J. 18 ; chamboi,
Kumaon, D. 115 ; chambolli, *Duk.*, B.
342 ; chambra, *Pb.*, A. 1469 ; chambul
(Ranunculus), *Pb.*, R. 26 ; chambuli
(Bahinia), *Mar.*, B. 342 ; chambúra,
Mar., B. 342 ; chameli (Plumeria), *Hind.,
Mar.*, P. 989 ; chameli (Tabernœmon-
tana), *Beng.*, T. 3 ; chámeli (Jasminum),
Hind., J. 18 ; chamgidari, *Hind.*, F.
655 ; cham-guddri, *Beng.*, F. 655 ; cham-
hún, *Pb.*, H. 440 ; chamiári, *Pb.*, P. 1333 ;
chamkát, *Hind.*, D. 348 ; chamkharak,
Hind., C. 631 ; chamkúl, *Hind.*, D.
348 ; chamkúré-ka-gaddah, *Duk.*, C. 1732 ;
chamlia, *Kumaon*, W. 86 ; chámmá, *Pb.*,
S. 537 ; chamoti (Michelia), *Pb*, M. 517 ;
chamotí (Tulipa), *Pb.*, T. 845 ; champ,
Pb., I. 487 ; champa, *Beng., Dec.*, *Pb.*,
Hind., Mar., Uriya, M. 517 ; champac,
Hind., M. 517 ; champaka, *Beng.*, M.
517 ; chámpánatíya, *Beng.*, A. 941 ; cham-
per, *Mar.*, F. 613 ; champo, *Gus.*, M.
517 ; chamrá (Desmodium), *Hind., Pb.*,
D. 348 ; chamrá (hides), *Hind.*, H. 265 ;
chamrá (leather), *Hind.*, L. 156 ; chamrár,
Hind., E. 25 ; chamror, *Pushtu*, E. 35 ;
chamrui, *Hind.*, E. 25 ; chamúní, *Pb.*,
T. 845 ; chamútí, *Pb.*, M. 517 ; chamyár,
Hind., Pb., D. 348 ; chan (Bauhinia),
Mar., B. 334 ; chan (Phœbe), *Pb.*, P.
546 ; chan (Populus), *Kumaon*, P. 1148 ;
chana, *Guj., Hind., Mar., Raj.*, C.
1061 ; chana-amba, *Mar.*, C. 1061 ;
chanangi, *Hyderabad*, M. 800 ; chanár,
Pb., P. 935 ; chánch, *Kashmir*, R. 588 ;
chanching, *Pb.*, L. 252 ; chánda, *Mar.*,
M. 12 ; chandal, *Hind.*, S. 790 ; chandan
(Santalum), *Beng., Hind., Mar., Pb.*,
S. 790 ; chandan (Symplocos), *Beng.,
Hind.*, S. 3059 ; chandana (Hibiscus),
Beng., H. 177 ; chandana (adjutant
bird), *Hind.*, S. 2907 ; chandan betu,
Beng., Hind., C. 1003 ; chandan lái,
Pb., P. 1381 ; chandápushpi, *Beng.,
Hind., Mar.*, A. 862 ; chándcha, *Beluch.*,
F. 577 ; chandi (Ambassis), *Uriya*, F.
310, *Hind.*, F. 312 ; chándi (silver),
Beng., Duk., Guj., Hind., Mar., A.
1359, S. 2205, *Beng.*, F. 312 ; chandi-
ari, *Hind.*, S. 2907 ; chándí-ká-varaq,
Hind., A. 1359 ; chandirí, *Sind*, M. 109 ;
chándkudá, *Mar.*, A. 1200 ; chándla,
Mar., A. 1200 ; chandna (Litsæa), *Pb.*,

L. 483 ; chandna (Machilus), *Pb.*, M.
22 ; chandoie, *Hind.*, S. 790 ; chandora,
Mar., M. 15 ; chandra (Panicum), *Pb.*,
P. 53 ; chandra (Phœbe), *Pb.*, P. 546 ;
chandra (Pœonia), *Hind.*, P. 11 ; chan-
dra, *Mar.*, R. 57 ; chandrá (Rauwolfia),
Beng., R. 57 ; chandramúla, *Hind.*,
K. 3 ; chandrasa, *Mar.*, C. 142 ; chánd-
uí, *Hind.*, T. 3 ; chandú múlá, *Beng.*,
K. 3 ; chándwar, *Mar.*, M. 15 ; chané-
ká-sirkah, *Hind.*, C. 1061 ; chang, *Pb.*,
H. 382 ; changma (Populus), *Pb.*,
P. 1142 ; changma (Salix), *Pb.*, S.
537 ; changtaw, *Pb.*, J. 2 ; changua,
Beng., F. 319 ; chánhel, *Hind.*, L. 300 ;
chania, *Guj.*, C. 1061 ; chaniari-dhauk,
Beng., S. 2906 ; chaniát, *Hind.*, R. 293 ;
chankan buti, *Pb.*, E. 267 ; chankar, *Pb.*,
S. 560 ; channa, *Pb.*, C. 1061 ; channi
niggi, *Pb.*, D. 130 ; chano, *Sind*, C. 1061 ;
chanoti, *Guj.*, A. 51 ; chánsá, *Duk.*, B.
118 ; chansaur, *Hind.*, L. 283 ; chanú
(Apium), *Beng.*, A. 1227 ; chanu (Carum),
Beng., C. 701 ; chanún, *Pb.*, P. 1148 ;
chanúní, *Pb.*, P. 1138 ; chánval, *Dec.*, O.
258 ; chánvol, *Beng.*, O. 258 ; chánwal,
Hind., O. 258 ; chánwal kangni, *Pb.*, S.
1212 ; chánwar, *Sind*, O. 258 ; chaoli,
Mar., V. 116 ; chaori gao, *Hind.*, O. 571 ;
cháp, *Pb.*, A. 801 ; chapala, *Kumaon*, R.
541 ; chapar (slate), *Pb.*, S. 2235 ; chápar
(Panicum), *Hind.*, P. 41 ; chápha, *Mar.*,
M. 517 ; chapkali, *Dec.*, R. 110 ; chapkia,
Pb., O. 247 ; chapkiye, *Kumaon*, O. 247 ;
chaplash, *Beng.*, A. 1479, I. 57 ; chaplis,
Beng., A. 1479 ; chappal, *Dec.*, O. 193 ;
chappal-send, *Dec.*, O. 193 ; chappar
tang, *Kumaon*, V. 202 ; chappu tattu,
Hind., H. 102 ; chapralákh, *Pb.*, L. 1 ;
chaprur, *Hind.*, P. 41 ; chaprura, *Hind.*,
P. 41 ; chaptis, *Beng.*, F. 573 ; chápu,
Pb., A. 801 ; chár (Buchanania), *Hind.*,
B. 913 ; char (Valeriana), *Hind., Pb.*, V.
5, V. 10 ; chara (Eruca), *Kumaon*, E.
327 ; chára (Bassia), *Kumaon*, B. 212 .
charái (Chenopodium), *Hind.*, C. 1003 ;
charái (Juniperus), *Pb.*, J. 92 ; charangi,
Hind., P. 693 ; charas, *Hind., Pb., C.*,
331 ; charátté, *Mar.*, T. 548 ; charaya-
tah (Swertia), *Hind., Dec.*, S. 3018 ;
charáyatah (Erythræa), *Hind.*, E. 338 ;
charáyetah (Andrographis), *Dec., Hind.*,
A. 1064 ; chárbor, *Mar.*, B. 342 ; char-
charíla, *Pb.*, L. 332 ; char godar, *Pb.*, V.
10 ; chari, *Beng., Pb.*, S. 2424 ; charká,
Pb., L. 490 ; charkeint, *Pb.*, P. 1452 ;
charkhri, *Pb.*, C. 631 ; chár-kí-cháróli,
Duk., B. 913 ; char-kuchú, *Beng.*, C.
1732 ; charkuli, *Uriya*, M. 583 ; charl,
Pb., F. 558 ; charlay, *Mar.*, F. 415 ; chár-
maghz, *Pb.*, J. 61 ; charmara, *Hind.*, P.
77 ; charmo, *Hind.*, L. 156 ; chároli,
Guj., Mar., B. 913 ; charr, *Raj.*, P.
1121 ; charrei, *Afg.*, C. 224, Q. 29 ;
charríaddí, *Hind.*, F. 413 ; char tukhm,
Pb., O. 23 ; charu, *Uriya*, B. 913 ; cha-
rúngli, *Pb.*, B. 782 ; charvár mádá, *Mar.*,
A. 1200 ; charwa, *Hind.*, P. 382 ; char-
wari, *Hyderabad*, B. 913 ; chástang, *Pb.*,
F. 1 ; chástang, *Pb.*, V. 108 ; chastang
raiún, *Pb.*, V. 108 ; chatáí, *Pb.*, M. 319 ;
chataveli, *Mar.*, D. 522 ; chaterni, *Pb.*,

R. 159 ; chatida, *Beng.*, D. 212 ; chatiun, *Beng.*, *Hind.*, A. 871 ; chato, *Hind.*, R. 152 ; chatr, *Pb.*, R. 152 ; chatra, *Pb.*, L. 312 ; chatriwal, *Pb.*, E. 509 ; chatta, *Pb.*, P. 45, 57 ; chattai, *Guj.*, *Hind.*, M. 319 ; chattri, *Afg.*, A. 590, *Pb.*, F. 725 ; chatúng, *Kashmir*, *Pb.*, T. 93 ; chatwan, *Beng.*, A. 871 ; cháu, *Pb.*, M. 22 ; chauho, *Sind*, C. 1221 ; chául, *Uriya*, O. 258 ; chaulai (Amarantus), *Hind.*, A. 934, *Pb.*, A. 916 ; chaulai (Wendlandia), *Hind.*, W. 33 ; chauli, *Hind.*, E. 73 ; chaulmugra, *Mar.*, G. 761 ; chaulmúgrí, *Beng.*, G. 761 ; chauniya, *Kumaon*, P. 1148 ; chaunkra, *Hind.*, P. 1259 ; chaun-ro, *Sind*, V. 116 ; chaunsh, *Pb.*, S. 505 ; chauri (Ceriops), *Sind*, C. 964 ; chauri (Elæodendron), *Hind.*, E. 73 ; chaurila, *Hind.*, P. 72 ; chavai, *Mar.*, M. 844 ; chavaicha kanda, *Mar.*, M. 844 ; chával, *Hind.*, O. 258 ; chaval-ke-bhaji, *Mar.*, P. 1187 ; chavara, *Mar.*, C. 2428 ; cha-vaya, *Mar.*, M. 809 ; chavel, *Mar.*, C. 2057 ; chavi, *Hind.*, P. 797 ; chavlya kand, *Guj.*, M. 844 ; chavulmungri, *Hind.*, G. 761 ; chawír, *Sind*, A. 531 ; chaya, *Beng.*, A. 554 ; cháyena, *Mar.*, D. 537 ; chay-la-rí, *Sind*, F. 424 ; chayruka, *Hind.*, C. 413 ; chayung, *Uriya*, F. 514 ; chechar, *Pb.*, R. 318 ; chechra, *Beng.*, F. 368 ; ched-du-ah, *Pb.*, F. 311 ; ched-wala, *Hind.*, R. 152 ; chehna, *Hind.*, P. 63 ; chehur, *Beng.*, B. 342 ; chein, *Pb.*, M. 393 ; chel, *Pb.*, C. 331 ; chel-hul, *Hind.*, F. 394 ; chel-lí, *Pb.*, F. 546 ; *Sind*, F. 450 ; chelliah, *Hind.*, F. 392 ; chelluah, *Hind.*, F. 326 ; chelunatípa, *Beng.*, A. 942 ; chelwa, *Beng.*, H. 255 ; chemri, *Pb.*, E. 184 ; chena (Ophiocepha-lus), *Beng.*, F. 517 ; chena (Panicum), *Hind.*, *Pb.*, P. 63 ; chena (Setaria), *Beng.*, *Dec.*, *Hind.*, S. 1212 ; chenah (Panicum), *Mar.*, P. 63 ; chench, *Pb.*, R. 593 ; chen-da-la, *Pb.*, F. 494 ; chen-da-lah, *Hind.*, F. 566 ; chendi, *Mar.*, H. 196 ; chenga, *Uriya*, F. 514 ; chenjúl, *Pb.*, D. 196 ; chenna (Cicer), *Duk.*, C. 1061 ; chenna (Setaria), *Mar.*, S. 1212 ; chenyel, *Mar.*, D. 537 ; chera (Erinocar-pus), *Mar.*, E. 278 ; chera (Thalictrum), *Pb.*, T. 376 ; cherai, *Pushtu*, Q. 29 ; cherailu, *Hind.*, R. 261 ; cherara, *Hind.*, P. 1252 ; cheraya, *Kumaon*, L. 535 ; cherayta, *Hind.*, S. 3018 ; cheretta, *Hind.*, S. 3027 ; cheriá, *Sind*, A. 1660 ; cheri chara, *Hind.*, C. 898 ; cherkúsh, *Kashmir*, *Pb.*, P. 1285 ; cherolí, *Pb.*, P. 1285 ; cherú-natía, *Beng.*, A. 942 ; cheru-pinai, *Mar.*, C. 162 ; cheshamdár, *Mar.*, C. 617 ; cheshamdár, *Hind.*, C. 618 ; chetabúta, *Pb.*, A. 2 ; chetain, *Pb.*, R. 152 ; chetchua-porah, *Uriya*, F. 411 ; cheulí, *Hind.*, B. 212 ; chevari, *Mar.*, A. 1536 ; chevati, *Mar.*, C. 1047 ; chewa, *Pb.*, E. 234 ; chhá, *Hind.*, C. 244 ; chha-chhindará, *Uriya*, T. 569 ; chhágulkúrí, *Beng.*, I. 302 ; chhagupuputi, *Beng.*, E. 505 ; chhale, *Pb.*, Z. 50 ; chhalmúgra, *Hind.*, G. 761 ; chharilá, *Hind.*, L. 332 ; chhat, *Pb.*, I. 487 ; chhatin, *Beng.*, A. 871 ; chhatnia, *Uriya*, A. 871 ; chhatra, *Pb.*, S. 1320 ; chhela, *Pb.*, S. 1290 ; chheli, *Pb.*, S. 1290 ; chhembar, *Pb.*, E. 184 ;

chhimbar, *Hind.*, E. 184 ; chhimbar, *Hind.*, F. 672 ; chhími, *Hind.*, P. 496 ; chhinchra, *Hind.*, S. 1207 ; chhinke (Eleusine), *Raj.*, E. 190 ; chhinke (Pani-cum), *Raj.*, P. 79 ; chhinkri, *Hind.*, C. 1026 ; chhoa, *Hind.*, S. 318 ; chhoel, *Hind.*, B. 958 ; chhota álu, *Kumaon*, P. 1304 ; chhóta-dúdhí-lata, *Beng.*, G. 748 ; chhota-gokhrú, *Hind.*, X. 1 ; chhótá-janglí-ánanash, *Beng.*, A. 836 ; chhótá-kanvár, *Hind.*, *Duk.*, A. 836 ; chhota-kulpha, *Hind.*, T. 562 ; chhótá-laslasá, *Hind.*, C. 1940 ; chhótá-lasórá, *Hind.*, C. 1940 ; chhota-pan-ki-jar, *Beng.*, *Dec.*, *Hind.*, A. 862 ; chhota-pílú, *Beng.*, *Hind.*, S. 717 ; chhota-pindálu, *Dec.*, D. 481 ; chhota ván, *Pb.*, S. 717 ; chhótí-góndni, *Dec.*, C. 1940 ; chhoti iláyechi, *Hind.*, E. 151 ; chhoti láni, *Pb.*, C. 224 ; chhótí-máyi, *Dec.*, T. 51 ; chhótí-máyín, *Hind.*, T. 51 ; chhoti van, *Pb.*, C. 224 ; chhotká váchoyá, *Hind.*, F. 549 ; chhoto-boh-naári, *Beng.*, C. 1940 ; chian, *Hind.*, E. 219 ; chibbur, *Sind*, C. 2558 ; chiber, *Sind*, C. 2306 ; chibúda, *Mar.*, C. 2263 ; chibunda, *Mar.*, C. 2263 ; chicha, *Mar.*, T. 28 ; chicháda, *Duk.*, *Mar.*, A. 711 ; chichia, *Kumaon*, J. 78 ; chichinda, *Pb.*, I. 569 ; chichingá, *Beng.*, T. 569 ; chichna, *Duk.*, *Mar.*, A. 711 ; chichola, *Mar.*, A. 695 ; chichora. *Hind.*, S. 2341 ; chicht (Girardinia), *Hind.*, G. 213 ; chichra (Achyranthes), *Hind.*, A. 382 ; chichra (Butea), *Hind.*, B. 944 ; chíchrí, *Pb.*, P. 959 ; chichrú, *Pb.*, U. 49 ; chichwi, *Hind.*, P. 50 ; chickana, *Mar.*, L. 483 ; chida sáwan, *Raj.*, P. 50 ; chidgu, *Beng.*, F. 655 ; chiggak, *Afg.*, P. 1272 ; chihúnt, *Hind.*, B. 978 ; chihut lar, *Hind.*, S. 2508 ; chijá-kri, *Pb.*, P. 1009 ; chika, *Hind.*, P. 67 ; chikakai, *Guj.*, A. 200 ; chikalí, *Mar.*, A. 376 ; chikan (Phaseolus), *Hind.*, P. 496 ; chíkan (Euonymus), *Pb.*, E. 475 ; chikana, *Mar.*, S. 1688, S. 1694 ; chikára (Eleusine), *Hind.*, E. 166 ; chikára (Gazella), *Hind.*, S. 1250 ; chikna, *Hind.*, L. 385 ; chikna bara, *Hind.*, S. 1223 ; chikni matí, *Hind.*, C. 1317 ; chikrassi, *Beng.*, C. 1021 ; chikri, *Kash-mir*, B. 985 ; chikti (Eragrostis), *Hind.*, E. 263 ; chikti (Triumfetta), *Hind.*, T. 835, T. 839 ; chikun, *Beng.*, T. 522 ; chil (Arthrocnemum), *Guj.*, A. 1475 ; chíl (Pinus), *Hind.*, *Pb.*, P. 760, *Hind.*, *Kashmir*, *Pb.*, P. 737 ; chila (Oryza), *Hind.*, O. 258 ; chila (Wendlandia), *Hind.*, W. 33 ; chíla (Pinus), *Hind.*, P. 737 ; chilana, *Pb.*, P. 1460 ; chilanghati kármar, *Pb.*, S. 3079 ; chilara, *Hind.*, C. 725 ; chilári, *Mar.*, A. 233 ; chilassi-tamáku, *Pb.*, N. 98 ; chilauni, *Hind.*, S. 940 ; chilaya, *Hind.*, S. 1223 ; chilbil, *Hind.*, H. 324 ; chilbing, *Mar.*, S. 2960 ; chilbinge, *Hind.*, S. 2943 ; chil-binj, *Dec.*, S. 2960 ; chilchil, *Pb.*, C. 868 ; chilgoza, *Afg.*, *Hind.*, P. 746 ; chílirágha, *Hind.*, A. 22 ; chilkiya, *Hind.*, W. 33 ; chil-kiyá, *Pb.*, W. 33 ; chilla (Casearia), *Hind.*, C. 722, 725 ; chilla (Holoptelea), *Hind.*, H. 324 ; chillára, *Mar.*, C. 42 ; chillur, *Mar.*, C. 42 ; chilmil, *Hind.*, H. 324 ; chilotú raulí, *Pb.*, L. 490 ; chílrao, *Hind.*, A. 22 ; chilrow, *Pb.*, A. 22 ; chilta,

Hind., D. 428; chílu, *Hind.*, P. 737; chilúchí, *Hind.*, *Pb.*, I. 434; chilwa, *Pb.*, F. 326; chim, *Beng.*, V. 116; chimakal gadi, *Hind.*, O. 185; chimar, *Guj.*, C. 728; chimbari, *Pb.*, E. 166, 184; chími, *Hind.*, D. 789; chim-kani, *Sind.*, C. 756; chimnánú, *Pb.*, P. 1322; chimr, *Guj.*, C. 728; chimu, *Pb.*, M. 775; chimul, *Hind.*, R. 261; chimyáka, *Pb.*, P. 1009; chin (Dioscorea), *Mar.*, D. 485; chín (Fagopyrum), *Pb.*, F. 19; chin (Panicum), *Hind.*, P. 63; china (Panicum), *Beng.*, *Hind.*, *Pb.*, P. 63; chína (Setaria), *Kumaon*, S. 1212; chíná (Dioscorea), *Mar.*, D. 534; china ghauza, *Beng.*, X. 23; chinai-ghás, *Mar.*, G. 653; chinai katha, *Mar.*, U. 11; chinai sálit, *Mar.*, P. 829; chinaka (Diplotaxis), *Pb.*, D. 674; chináka (Malcolmia), *Pb.*, M. 63; chinár, *Pb.*, *Pushtu*, P. 935; chinch (Pithecolobium), *Mar.*, P. 900; chinch (Tamarindus), *Mar.*, T. 28; chincha, *Mar.*, T. 28; chínchá, *Pb.*, T. 28; chindar, *Hind.*, C. 2062; chindi, *Pb.*, L. 490; chin-do-lah, *Pb.*, F. 558; chine alú, *Beng.*, I. 348; chine ghás, *Beng.*, X. 23; chiner-bádáme, *Beng.*, A. 1261; chingan butai, *Pushtu*, N. 62; chinghar, *Hind.*, F. 791; chingrá, *Beng.*, C. 2232; chinh, *Hind.*, P. 63; chiní (Panicum), *Pb.*, P. 63; chíní (Saccharum), *Hind.*, *Pb.*, S. 375, 376; chini naranghi, *Hind.*, T. 631; chíníshakkar, *Hind.*, S. 375; chinjara, *Hind.*, S. 2907; chinkara, *Hind.*, S. 1250; chínki-tút, *Hind.*, M. 756; chinna, *Hind.*, P. 63; chinni, *Hind.*, *Mar.*, M. 747; chinní-ká-jhar, *Duk.*, A. 304; chinnkabale, *Guj.*, *Mar.*, P. 801; chino, *Gus.*, *Mar.*, P. 63; chínól, *Guj.*, C. 728; chinta, *Hind.*, C. 1950; chintar, *Pushtu*, P. 935; chintz, *Mar.*, T. 28; chinu (Panicum), *Sind*, P. 63; chínú (Syringa), *Pb.*, S. 3079; chinwa, *Hind.*, *Kashmir*, P. 63; chinwári, *Hind.*, P. 41; chínyágónd, *Hind.*, B. 945; chipa-chi-magadi, *Hind.*, I. 300; chipal (Ulmus), *Pb.*, U. 4; chipkulí, *Hind.*, R. 110; chippal (Eragrostis), *Hind.*, E. 263; chippuah, *Hind.*, F. 326; chir (Eragrostis), *Hind.*, E. 248, 252; chir (Pinus), *Hind.*, *Pb.*, P. 737, 760; chira (Erinocarpus), *Mar.*, E. 278; chira (Litsæa), L. 490; chiráita, *Mar.*, S. 3018; chira kura, *Hind.*, A. 942; chíran, *Pb.*, P. 1285; chirand, *Mar.*, G. 36; chiraputi, *Mar.*, Z. 1; chirara (Prinsepia), *Kumaon*, P. 1252; chírara (Litsæa), *Hind.*, L. 490; chiráti, *Sind*, M. 791; chirauli, *Pb.*, B. 913; chiraunda, *Pb.*, X. 21; chirayata, *Gus.*, S. 3018; chirayitá, *Mar.*, S. 3018; chirboti, *Mar.*, P. 678; chírchíl, *Afg.*, P. 746; chirchira (Achyranthes), *Hind.*, A. 382; chirchira (Litsæa), *Hind.*, *Kumaon*, L. 490; chirchira (Setaria), *Pb.*, S. 1223; chirchitta (Achyranthes), *Hind.*, A. 382; chirchitta (Lycium), *Pb.*, L. 589; chirchitta (Setaria), *Hind.*, S. 1223; chireta (Thalictrum), *Pb.*, T. 376; chirétá (Swertia), *Beng.*, S. 3018; chiretta, *Pb.*, S. 3014; chiri (Badis), *Pb.*, F. 328; chírí (Pinus), *Pb.*, P. 746; chirichog, *Kashmir*, J. 29; chirika-chanwalia,

Raj., E. 267; chiri-ka-khet, *Raj.*, E. 263, 267; chiri-ko-bajro, *Raj.*, E. 263; chiri piazi, *Hind.*, A. 777; chiriya-chaina, *Hind.*, S. 1210; chiriya-ka-dána, *Hind.*, S. 2668; chiriya-ke-chaolai, *Hind.*, E. 257; chirmiti, *Hind.*, A. 51; chírmútti, *Pb.*, A. 1061; chírndi, *Pb.*, X. 21; chirodheli, *Pushtu*, T. 61; chironji, *Beng.*, *Hind.*, *Pb.*, B. 913; chirónji, *Hind.*, B. 913; chiror, *Pb.*, B. 463; chirphal, *Mar.*, Z. 35; chirpoti, *Hind.*, Z. 1; chirput, *Mar.*, P. 682; chírrú, *Pb.*, X. 1; chirua, *Hind.*, P. 998; chírudi, *Pb.*, L. 490; chirúnda, *Pb.*, X. 21; chirval, *Hind.*, O. 137; chirwa, *Hind.*, P. 63; chírwí, *Pb.*, P. 555; chita (hunting leopard), *Hind.*, T. 424, T. 434; chitá (molasses), *Beng.*, S. 372, 767; chítá (plumbago), *Beng.*, *Hind.*, *Uriya*, P. 986; chita-bágh, *Hind.*, T. 434; chita bagnú, *Pb.*, P. 1138; chitabánsa, *Pb.*, I. 415; chita billa, *Hind.*, T. 425; chita-billi, *Hind.*, F. 760; chital, *Hind.*, D. 212; chitala, *Beng.*, F. 510; chitámúl-níl, *Uriya*, P. 986; chitana, *Pb.*, P. 1458, 1460; chítarak, *Hind.*, P. 986; chitarmul, *Dec.*, P. 986; chitar-mulam, *Dec.*, P. 986; chitaro, *Gus.*, P. 986; chitawála, *Pb.*, S. 1059; chitawár, *Hind.*, P. 986; chit-batto, *Kashmir*, T. 605; chitbatto, *Pb.*, T. 607; chiti (Marsdenia), *Beng.*, M. 299; chíti (plumbago), *Hind.*, P. 986; chitijari, *Pb.*, A. 401; chitiphúl, *Hind.*, H. 106; chíti-phul, *Pb.*, H. 106; chiti-sirin, *Pb.*, C. 838; chitol, *Beng.*, F. 510; chit pattra, *Pb.*, M. 209; chitra, *Mar.*, P. 986; chitra (Berberis), *Hind.*, *Pb.*, B. 443, *Hind.*, B. 458; chitra (Drosera), *Pb.*, D. 837; chitra (plumbago), *Hind.*, P. 979, 986, *Beng.*, P. 979; chitra (spotted deer), *Hind.*, D. 212; chitra (Staphylea), *Kashmir*, S. 2678; chitra (Thalictrum), *Pb.*, T. 376; chitrack, *Mar.*, P. 986; chitrah, *Beng.*, S. 372; chitrak, *Pb.*, P. 986; chitrak, *Hind.*, P. 979; chitraka, *Mar.*, P. 986; chitramúl (Thalictrum), *Pb.*, T. 376; chitra-múla (plumbago), *Mar.*, P. 986; chitruk sufaid, *Beng.*, P. 986; chittania, *Hind.*, Z. 290; chitul, *Uriya*, F. 510; chitz, *Mar.*, T. 28; chiú, *Pb.*, R. 253; chiúla, *Hind.*, B. 944; chiúndi, *Pb.*, X. 21; chiúra, *Kumaon*, B. 212; chiúrr, *Pb.*, S. 1212; cho, *Pb.*, P. 1463; choak-si, *Beng.*, F. 429; chobchíni, *Hind.*, S. 2252, *Beng.*, *Hind.*, *Mar.*, *Pb.*, S. 2240; choca, *Dec.*, P. 811; choca mirch, *Hind.*, P. 811; chochar, *Pb.*, B. 465; choda, *Pb.*, P. 1449; chodhará, *Mar.*, A. 1132; chódhári, *Gus.*, V. 219; chogak, *Afg.*, P. 1272; chogu, *Pb.*, *Kashmir*, T. 93; chohara, *Guj.*, C. 691; choi, *Beng.*, P. 797; chók, *Pushtu*, R. 650; chóka, *Pushtu*, R. 650; chokha, *Gus.*, O. 258; chokhóta-téla, *Mar.*, S. 1078; choklu, *Pb.*, R. 316, 323; choku, *Pb.*, C. 881; chola (Barbus), *Beng.*, F. 335; chola (Prunus), *Kumaon*, P. 1285; chola (Pyrus), *Pb.*, P. 1460; chola (Vigna), *Mar.*, *Gus.*, V. 116; cholá (Cicer), *Beng.*, *Pb.*, *Raj.*, C. 1061; chólái, *Hind.*, A. 943; chonrh, *Beng.*, *Hind.*, O. 574; chonrhi, *Beng.*, *Hind.*, O. 574; chopar (Hiptage), *Pb.*, H. 285; chopar (Panicum), *Baluch.*, P. 79; chopchini, *Mar.*, S.

1186; chopra (Euonymus), *Hind.*, E. 482,
Pb., E. 485; chopra (Xylosma), *Pb.*, X.
21; chopri alu, *Mar.*, D. 513; chora
(Angelica), *Pb.*, A. 1127; chora (Quercus),
Pb., Q. 13; chora, *Gu.*, V. 116; chora-
onva, *Mar.*, A. 1130; chorgu, *Dec.*, V.
55; chorie-ajowan, *Duk.*, C. 1367; chor-
kántá, *Beng.*, C. 1053; chorpatta, *Beng.*,
L. 79; chota (Pyrus), *Pb.*, P. 1460; chota
(molasses), *Hind.*, S. 372; chota bekkut,
Uriya, F. 479; chota-bhánkta, *Raj.*, E.
246; chota chand, *Mar.*, R. 57; chota-
chánd, *Hind.*, R. 57; chota-chárd, *Beng.*,
R. 57; chota chikiya, *Hind.*, S. 1210;
chota chiretta, *Hind.*, E. 218; chhotá-ghi-
kanvar, *Dec.*, *Hind.*, A. 836; chota gokhrú, *Hind.*, T. 548; chota-hal-kúsa,
Hind.,*Beng.*, L. 309; chota-jamb, *Beng.*,
E. 428; chota kalia, *Raj.*, P. 964; chota
kanval, *Hind.*, N. 200; chota-kiráyata,
Hind., E. 217; chota-kulijan, *Beng.*,
Hind., *Mar.*, A. 862; chhota lunia, *Hind.*,
Beng., P. 1179, *Beng.*, P. 1187; chota
mandiya, *Hind.*, E. 166; chota mattar,
Beng., *Hind.*, P. 882; chota sarsata,
Hind., S. 1210; chota sundi, *Beng.*, N.
200; choti, *Pb.*, A. 1005; chotiál, *Pb.*, R.
215; chotie láchi, *Hind.*, E. 151; choti-
juar, *Hind.*,*Pb.*, S. 2424; choti junri, *Pb.*,
S. 2424; choti-kulijan, *Beng.*, *Hind.*,
Mar., A. 862; choti láni, *Pb.*, S. 2985;
choti mauhari, *Pb.*, S. 2345; choti semai,
Hind., P. 72; chotka dudhi, *Hind.*, E.
549; choto-jhunjhun, *Beng.*, C. 2154;
choto-keruee, *Beng.*, E. 518; choto-kulpa,
Beng., T. 562; choto-kut, *Beng.*, S. 510;
chotra, *Hind.*, B. 443; chótia, *Pb.*, P.
1460; chouka, *Hind.*, S. 1287; chounláyí,
Hind., P. 1187; chounláyí-kí-bhájí, *Dec.*,
P. 1187; chour-ká-namak, *Duk.*, C. 541;
chour-kí-mattí, *Duk.*, C. 541; chousingha,
Hind., S. 1287; chowar, *Mar.*, C. 2428;
chowlaí-ka-bhají, *Hind.*, A. 941; chowli
(Portulaca), *Dec.*, P. 1187; chowli (Vigna), *Dec.*, *Hind.*, *Mar.*, V. 116; chowun,
Sind, C. 728; choyanda, *Hind.*, S. 373;
choyari chinch, *Mar.*, A. 455; chritri-
jhánk, *Hind.*, D. 212; chu, *Pb.*, *Him.*,
E. 538; chúa (Amarantus), *Hind.*, A. 916,
A. 925; chúa (Rosa), *Pb.*, R. 543; chual,
Pb., E. 479; chúal, *Hind.*, S. 2678;
chúa-mársa, *Hind.*, A. 925; chuara, *Mar.*,
P. 555; chúari, *Hind.*, P. 1285; chúáru,
Kumaon, P. 1285; chubrei, *Pb.*, E. 166,
184; chúch, *Pb.*, J. 104; chúchí (Polygo-
num), *Pb.*, P. 1106; chúchi (Rheum), *Pb.*,
R. 215; chucka, *Hind.*, C. 2180; chúgú,
Pb., P. 959; chúhára, *Hind.*, P. 555; chui
(Juniperus), *Pb.*, J. 78; chúí (Pyrus),
Pb., P. 1463; chuj, *Pb.*, F. 700; chúk
(Hippophae), *Hind.*, H. 277; chúk
(Rumex), *Beng.*, R. 650; chúka, *Beng.*,
Dec., *Hind.*, *Sind*, R. 650, chúka
(Rumex), *Hind.*, R. 650; *Mar.*, R.
650; chúka (Sida), *Pb.*, S. 1694; chúka-
pálak, *Hind.*, R. 650; chuka-pálang,
Beng., R. 650; chuka-tripati, *Beng.*,
O. 547; chukeka sák, *Hind.*, R. 650;
chukha, *Pb.*, O. 547; chuko, *Guj.*, A.
916; chúkrí, *Afg.*, *Pb.*, R. 215; chula,
Pb., E. 538; chulajuti, *Mar.*, M. 797;
chulcherila, *Hind.*, L. 332; chúle,

Pb., P. 1316; chúli (Limnanthenum),
Beng., L. 355; chúli (Prunus), *Pi.*, P.
1285; chuma, *Hind.*, H. 277; chuma,
Pb., H. 281; chumati patí, *Bing.*,
C. 2609; chúmbar, *Pb.*, A. 1162;
chumbul, *Sind*, E. 25; chúmlság,
Hind., A. 941; chun (Euphorbia),
Pb., E. 538; chún (lime), *Beng.*, C.
489; chun (Morus), *Hind.*, M. 717;
Pb., M. 775; chuná, *Beng.*, *Hind.*,
Pb., C. 489; chúnak, *Duk.*, C. 48;
chúncha, *Beng.*, C. 2584; chunder ka-
mal, *Kumaon*, l. 434; chundrus, *Beng.*,
V. 31; chundun, *Kumaon*, J. 92; chuug,
Pb., S. 537; chung (Hordeum), *Pb.*, H.
382; chung (Pyrus), *Pb.*, P. 1463; chung
(Caralluma), *Pb.*, C. 478; chunga, *Pi.*,
E. 538; chunga pippá, *Pb.*, C. 47;
chungi (Boucerosia), *Pb.*, B. 782; chungi
(Lagerstrœmia), *Dec.*, L. 55; chun-haí,
Beng., A. 51; chúni, *Pb.*, J. 78; chúni-
gónd, *Hind.*, B. 945; chun-na (Cicer,
Hind., C. 1061; chunná (lime), *Dec,
Hind., *Mar.*, C. 489; chuno (lime,
Guj., C. 489; chuno (Oryza), *Sind*, C.
258; chunt, *Pb.*, P. 1463; chupein, *Pb.*,
P. 1197; chupra, *Kumaon*, M. 919.
chúpra, *Hind.*, M. 910; chúpri álu,
Hind., *Beng.*, D. 513; chúr, *Pb.*, C.
224, Q. 29; chura (Commelina), *Pb.*, C.
1748, 1752; churá (Angelica), *Pb.*, A.
1127; chúrai-ajwani, *Duk.*, C. 1367;
churál, *Pb.*, L. 100; churál chapa, *Hind.*,
L. 100; churi, *Pb.*, H. 285; churial, *Pb.*,
A. 1273; churí-ki-bhájí, *Duk.*, A. 950;
chúri saroj, *Pb.*, A. 1464; churna,
Hind., Z. 273; churota, *Hind.*, G. 753;
churt kasi, *Mar.*, M. 855; chusa, *Hind.*,
O. 185; chush-maidar, *Hind.*, C. 622;
chuti, *Pb.*, A. 1573; chuti-ál, *Pb.*, R.
215; chúti-sirin, *Pb.*, C. 838; chútsalé,
Pb., B. 731; chye, *Pb.*, B. 131; chyúra,
Kumaon, B. 212.

Ci   Cindra, *Mar.*, R. 16; cir-re-oh, *Sind*, F. 470.

Ck   Ckintá, *Hind.*, B. 944.

Cl   Clanuko, *Sind*, P. 555.

Co   Coi, *Beng.*, *Uriya*, F. 317; coilia, *Uriya*,
F. 596; coir, *Hind.*, C. 1520; coitor,
Beng., F. 571; collúse, *Hind.*, F. 470;
conác, *Kashmir*, Z. 50; corsula, *Beng.*,
F. 496; cowa, *Hind.*, G. 22; cown,
Uriya, F. 317.

Cu   Cuchia, *Beng.*, F. 316; cuchia, *Uriya*,
F. 316; cuggera, *Beng.*, F. 466; cuja,
Beng., F. 572; cumuda, *Hind.*, L. 355;
cún-che-li-e, *Hind.*, F. 531; cundahla,
Uriya, F. 448, 449; cungúr, *Sind*, F.
367; cunta-gagah, *Uriya*, F. 552; cuntea,
Uriya, F. 487; curru, *Sind*, F. 409;
cursa, *Hind.*, F. 407; cursua, *Uriya*, F.
470; cusso, *Hind.*, B. 856; cusunt,
Hind., F. 633; cutturpoh, *Uriya*, F. 341;
cut-wálursi, *Uriya*, F. 418.

Da   Da, *Kumaon*, S. 2394; dáán, *Pb.*, P. 1426;
dab (Eragrostis), *Hind.*, E. 252; dab
(Cornus), *Pb.*, C. 1973; dab, *Pb.*, T. 864;
dáb (Cocos), *Beng.*, C. 1520; dáb (Eragrostis), *Hind.*, *Pb.*, E. 252; dáb (Imperata), *Pb.*, I. 51; dáb (Vibrunum), *Pb.*,
V. 104; daba, *Beng.*, S. 1934; da-bah,
Hind., F. 424; dábali, *Guj.*, A. 89;

daberi, *Hind.*, E. 73 ; dábh, *Hind.*, I. 51 ; dabhat, *Hind.*, E. 252 ; dabhir, *Raj.*, C. 1686 ; dábi dúba, *Hind.*, X. 23 ; dabí-dubí, *Beng.*, X. 23 ; dabmo, *Pb.*, S. 1242 ; daboi, *Hind.*, E. 252 ; dábrá, *Hind.*, U. 26 ; dábriá, *Guj.*, A. 1149 ; dabúr, *Beng.*, C. 943 ; dabvi, *Hind.*, E. 252 ; dacca, *Mar.*, P. 1442 ; dadá, *Pb.*, C. 846 ; dádam, *Gus.*, P. 1426 ; dádamardana, *Mar.*, C. 732 ; dádam-nu-jháda, *Gus.*, P. 1426 ; dadár, *Hind.*, *Kumaon*, *Kashmir*, *Pb.*, C. 846 ; dadhuri, *Pb.*, F. 179 ; dadia, *Hind.*, L. 375 ; dádmardan, *Beng.*, C. 732 ; dádmári (Ammannia), *Beng.*, *Hind.*, A. 958, *Pb.*, A. 960 ; dádmari (Cassia), *Beng.*, C. 732 ; dád-murdan, *Hind.*, C. 732 ; dadúr (Punica), *Guj.*, P. 1426 ; dádúr (Rhamnus), *Pb.*, R. 152 ; dadúri, *Pb.*, F. 202, 260 ; da-ghauri, *Pb.*, C. 42 ; daghún-bán, *Pb.*, O. 35 ; daháí, *Pb.*, W. 106 ; dahan, *Hind.*, *Raj.*, T. 489 ; dahar karanja, *Beng.*, P. 1121 ; daheo, *Pb.*, A. 1511 ; daheyá, *Hind.*, S. 2912 ; dahi, *Hind.*, D. 15 ; dahi, *Hind.*, C. 1927 ; dahipalás, *Hind.*, C. 1927 ; daholia, *Pb.*, F. 230 ; dah-rah, *Pb.*, F. 361 ; dahu, *Beng.*, *Hind.*, A. 1511 ; dahya, *Pb.*, S. 2912 ; dai, *Guj.*, F. 466 ; dain (Brassica), *Hind.*, B. 822 ; dain (Heptapleurum), *Hind.*, H. 131 ; dáin-lai, *Hind.*, B. 822 ; daira, *Hind.*, *Mar.*, W. 131 ; daiwas, *Mar.*, C. 1927 ; daiyeti, *Pb.*, *Sind*, S. 1939 ; dajkar (Celastrus), *Pb.*, C. 860 ; dajkar (Flacourtia), *Pb.*, F. 624 ; dák, *Pb.*, R. 365 ; dakachrú, *Pb.*, S. 924 ; dakh, *Hind.*, D. 16 ; dákh, *Hind.*, *Pb.*, V. 233 ; dak-hangú, *Pb.*, D. 249 ; dakhani babúl, *Hind.*, P. 900 ; dakhini babúl, *Hind.*, P. 900 ; dakhmila, *Hind.*, R. 318 ; dakh nirbisi, *Hind.*, C. 1205 ; dakkí, *Pb.*, V. 233 ; dakkúri, *Hind.*, F. 693 ; da-kóm, *Beng.*, A. 514 ; dal, *Hind.*, D. 18 ; dal (Cajanus), *Hind.*, C. 49 ; dál (Cedrela), *Hind.*, C. 835 ; dálá, *Pb.*, V. 10 ; dalchini (Cinnamomum), *Hind.*, C. 1183 ; *Mar.*, C. 1196, *Beng.*, *Dec.*, *Guj.*, *Hind.*, *Mar.*, C. 1195 ; dalchini (Machilus), *Pb.*, M. 22 ; dalchini tiki, *Mar.*, C. 1183 ; dalím, *Beng.*, *Uriya*, P. 1426 ; dalímba, *Mar.*, *Uriya*, P. 1426 ; dalímba-jháda, *Mar.*, P. 1426 ; dálím-gáchh, *Beng.*, P. 1426 ; dalkaramcha, *Beng.*, P. 1121 ; dalli, *Hind.*, C. 835 ; dalmara, *Beng.*, C. 1021 ; dalme, *Hind.*, S. 1029 ; dalunchi, *Pb.*, P. 539 ; dalúng, *Pb.*, A. 2 ; damá, *Pb.*, *Sind*, F. 6 ; damáhán, *Hind.*, F. 6 ; dam-áhár, *Hind.*, F. 6 ; dáman, *Mar.*, G. 714 ; dámana, *Mar.*, G. 714 ; daman-papar, *Hind.*, O. 132 ; dámar, *Hind.*, S. 1656 ; damba, *Sind*, F. 577 ; damiyá, *Pb. & Sind*, F. 6 ; damlakwaypi, *Hind.*, C. 68 ; dampel, *Hind.*, *Mar.*, G. 99 ; damtura, *Pb.*, *Sind*, H. 525 ; damú, *Hind.*, P. 1426 ; dam-ul-akhwain, *Hind.*, C. 68 ; damún, *Hind.*, G. 714 ; dana, *Hind.*, D. 113 ; dána (Anabasis), *Pb.*, A. 1005 ; dána, *Pb.*, C. 224 ; dána (Haloxylon), *Pb.*, H. 6 ; dána, *Pb.*, S. 2985 ; danadar, *Beng.*, S. 371 ; dandál, *Hind.*, X. 21 ; dandarwa, *Pb.*, P. 684 ; dándashi, *Mar.*, D. 32 ; dandelo, *Mar.*, H. 517 ; dandoria, *Hind.*,

P. 546 ; dandous, *Mar.*, *Sind*, D. 32 ; dandú, *Kashmir*, R. 564 ; dandúra, *Pb.*, H. 525 ; dandúsa, *Mar.*, D. 32 ; danga gurgur, *Beng.*, C. 1683 ; dángar, *Mar.*, O. 258 ; dangara, *Sind*, F. 476 ; dángrí, *Pb.*, C. 49 ; danikoní, *Beng.*, F. 558 ; dankena, *Beng.*, F. 531 ; dánkuni, *Beng.*, C. 382 ; danmo, *Pb.*, S. 1242 ; dannahrah, *Hind.*, F. 531 ; dant, *Guj.*, *Mar.*, A. 938 ; dantí (Artemisia), *Beng.*, A. 1464 ; dántí (Baliospermum), *Beng.*, *Hind.*, *Mar.*, B. 28 ; dántimul, *Beng.*, *Guj.*, *Mar.*, B. 28 ; dant játhi, *Hind.*, *Pb.*, C. 1744 ; dant-ké-bhájí, *Duk.*, A. 938 ; danú, *Pb.*, P. 1426 ; dánu-jhada, *Gus.*, M. 363 ; dar, *Kashmir*, C. 846 ; dara (Polynemus), *Mar.*, F. 538 ; dara (Lagerstrœmia), *Mar.* L. 42 ; darab, *Pb.*, C. 838 ; darachini, *Mar.*, C. 1196 ; dára-halada, *Hind.*, B. 445 ; darákh, *Gus.*, V. 233 ; daram, *Guj.*, *Hind.*, P. 1426 ; dáran, *Pb.*, P. 1426 ; dararhi, *Mar.*, C. 2514 ; daráu, *Pb.*, F. 10 ; darbh, *Mar.*, E. 252 ; darbha, *Mar.*, E. 252 ; darchini, *Mar.*, C. 1183 ; darchíní, *Hind.*, C. 1158, *Pb.*, C. 1196 ; dárchír, *Pb.*, P. 737 ; dár-chób, *Hind.*, B. 443 ; darengri, *Kashmir*, D. 142 ; darengri, *Kashmir*, R. 293 ; darfilfil, *Pb.*, P. 805 ; darga (Juglans), *Pb.*, J. 61 ; dargá (Saccharum), *Pb.*, S. 6 ; dargola, *Pb.*, S. 505 ; darhalad, *Mar.*, B. 458 ; dár-hald, *Hind.*, B. 443 ; darhú, *Sind* P. 1426 ; dari ¹(carpet), *Hind.*, C. 626 ; dári (Pueraria), *Mar.*, P. 1401 ; darim (Securinega), *Hind.*, S. 1029 ; dárim (Punica), *Beng.*, *Hind.*, P. 1426 ; darímpashk, *Pb.*, P. 1426 ; dárím pushp, *Hind.*, P. 1426 ; dariún, *Pb.*, P. 1426 ; darkaranja, *Beng.*, P. 1121 ; darkonah, *Pb.*, M. 363 ; dárma, *Kumaon*, A. 6 ; darmar, *Hind.*, Z. 6, Z. 9 ; darmí, *Beng.*, P. 1426 ; dárnharidrá, *Beng.*, M. 704 ; dárú, *Pb.*, P. 1426 ; darú-bij, *Sind*, P. 1426 ; dárudi, *Guj.*, A. 1351 ; dáru-garm, *Afg.*, P. 811 ; dáruharidrá, *Beng.*, M. 652 ; darú-jo-kul, *Sind*, P. 1426 ; darúnaj-akrabí, *Pb.*, D. 813 ; darúní, *Pb.*, P. 1426 ; darúr, *Hind.*, E. 25 ; dárúri, *Dec.*, *Mar.*, A. 1351 ; daryá-ká-nárél, *Dec.*, L. 511 ; daryá-ká-náríyal, *Hind.*, L. 511 ; dárya-kí-páchi, *Dec.*, G. 653 ; daryá-nu-naríyal, *Gus.*, L. 511 ; das, *Hind.*, D. 420 ; dasakarantod, *Uriya*, B. 171 ; dasan, *Hind.*, R. 305 ; dasarni, *Hind.*, R. 305 ; dasáundu, *Hind.*, D. 628 ; dasí, *Beng.*, B. 175 ; dasinda-cha-phula, *Mar.*, H. 227 ; dasmúli, *Mar.*, D. 6 ; dasni, *Hind.*, R. 305 ; dassi, *Pb.*, J. 29 ; dastá, *Hind.*, Z. 190 ; dáswila, *Hind.*, R. 318 ; dat, *Duk.*, A. 938 ; dather, *Kashmir*, D. 160 ; datia, *Hind.*, P. 48 ; datír, *Mar.*, F. 175 ; dát-ká-pát, *Hind.*, C. 732 ; dát-ká-pattá, *Duk.*, C. 732 ; datki, *Pushtu*, W. 106 ; datrang, *Mar.*, E. 25 ; datranga, *Hind.*, E. 25 ; datte-phal, *Mar.*, B. 180 ; datúra (Datura), *Afg.*, *Pb.*, D. 166 ; datúra (Hyoscyamus), *Pb.*, H. 525 ; datwan, *Gus.*, P. 663 ; dau, *Raj.*, A. 1149 ; dáula (Phyllanthus), *Hind.*, P. 632 ; daulah (Ophiocephalus), *Pb.*, F. 515 ; dauldhák, *Hind.*, E. 356 ;

dauna, *Mar.*, A. 1467; daurgarm, *Afg.*, P. 811; dauri, *Hind.*, C. 835; daurva, *Pb.*, C. 2558; davala, *Mar.*, U. 23; davan, *Hind.*, R. 305; davoli, *Hind.*, E. 252; dawa, *Beng.*, W. 106; dawai, *Beng.*, W. 106; daw-ái-mubarík, *Pb.*, C. 1394; dáwaka-jhar, *Mar.*, D. 725; dawal, *Mar.*, L. 503; dawáná, *Mar.*, A. 1460; dáwi, *Hind.*, *Pb.*, W. 106; daya, *Kumaon*, C. 133; dáye, *Mar.*, D. 339; daza, *Beng.*, S. 1934.

De Debari, *Beng.*, F. 424; debdari, *Hind.*, P. 1052; debdáru (Cedrus), *Beng.*, C. 846; deb-dáru (Polyalthia), *Uriya*, P. 1052; debkotí, *Hind.*, C. 868; debunsha, *Beng.*, O. 18; deccani babul, *Mar.*, P. 900; ded ún, *Beng.*, S. 2390; degar, *Pb.*, F. 202; dehla, *Pb.*, V. 233; dehua, *Beng.*, A. 1511; deikna, *Hind.*, M. 393; dein, *Kashmir*, O. 258; dekámáli, *Hind.*, G. 116; delha, *Pb.*, C. 402; demúr, *Beng.*, F. 260; dendlu, *Hind.*, *Pb.*, H. 545; dendrú, *Pb.*, L. 535; dengua, *Beng.*, A. 927; dengullar, *Beng.*, C. 75; denthar, *Pb.*, C. 133; dentúrá, *Pb.*, H. 525; dentúrú, *Pb.*, P. 688; deo, *Hind.*, G. 385; deodágri, *Mar.*, L. 574; deodar (Cupressus), *Pb.*, C. 2358; deodár (Cedrus), *Hind.*, *Kashmir*, *Kumaon*, *Pb.*, C. 846; deodari, *Mar.*, C. 838; deodhán (Oryza), *Hind.*, O. 258; deo-dhan (Sorghum), *Hind.*, S. 2397; deo-dhan, *Hind.*, *Mar.*, S. 2405; deo kapas, *Hind.*, *Mar.*, G. 385; deo khádir, *Pb.*, M. 562; deonal, *Mar.*, L. 503; deowar, *Hind*, *Kashmir*, *Kumaon*, *Pb.*, C. 846; dephal, *Beng.*, A. 1511; der, *Pb.*, C. 838; dera, *Kumaon*, C. 123; deri, *Pb.*, C. 838; desi-badám, *Hind.*, T. 312; desi-kulfah, *Hind.*, P. 1179; desi-mattar, *Beng.*, *Hind.*, P. 882; deterdane, *Pb.*, U. 26; deura, *Mar.*, E. 565; dévadár, *Beng.*, *Hind.*, P. 1052; dévadáru (Cedrus), *Beng.*, C. 846; dévadáru (Pinus), *Beng.*, *Hind.*, *Mar.*, P. 1052; dévadárúcha-jháda, *Mar.*, C. 846; deva kanchan, *Beng.*, B. 308; devakanchana, *Mar.*, B. 308; deva kápúsa, *Mar.*, C. 385; deva-keli, *Mar.*, C. 321; devbábhul, *Mar.*, A. 244; dévdár, *Beng.*, *Guj.*, C. 846; dévdáru, *Duk.*, C. 846; devi bábul, *Mar.*, P. 322; devi-diár, *Pb.*, C. 2358; dewadar, *Mar.*, C. 846; dewdar, *Hind.*, *Kashmir*, *Kumaon*, *Pb.*, C. 846; dewan, *Hind.*, C. 1927; dewdár, *Pb.*, C. 846.

Dh Dha, *Hind.*, W. 106; dhá, *Pb.*, W. 106; dhab, *Hind.*, *Pb.*, E. 252; dha-balakain, *Orissa*, N. 200; dhadki, *Beng.*, W. 106; dhadonjra,'*Pb.*, A. 344; dhái, *Beng.*, *Hind.*, *Pb.*, *Sind*, W. 106; dhaim, *Mar.*, C. 1927; dháian, *Hind.*, C. 1927; dhainti, *Beng.*, W. 106; dhaiti, *Hind.*, W. 106; dhaiwan, *Mar.*, C. 1927; dhák (Butea), *Hind.*, B. 944; dhák (stork), *Hind.*, S. 2903; dhákar, *Beng.*, *Hind.*, O. 574; dhakki, *Hind.*, A. 1215; dhákti dudhi, *Mar.*, E. 512, dhakur, *Beng.*, C. 943; dhákutá kunvára, *Mar.*, A. 836; dhalá-kura, *Beng.*, A. 681; dhala múg, *Uriya*, P. 496; dhála tulasi, *Uriya*, O. 18; dhalím, *Hind.*, *Sind*, P. 1426; dham,

Hind., A. 1149; dhamá, *Pb.* & *Sind*, F. 6; dhaman (Cenchrus), *Hind.*, C. 909; dhaman (Cordia), *Hind.*, *Mar.*, C. 1927; dhaman (Ehretia), *Pushtu*, E. 35; dháman (Grewia), *Hind.*, *Pb.*, G. 663, 673, 688, *Mar.*, *Uriya*, G. 714; dhaman (Pennisetum), *Pb.*, P. 382; dhamana, *Guj.*, G. 714; dhamánh, *Pb.*, *Sind*, F. 6; dhamani, *Beng.*, *Hind.*, G. 673, *Hind.*, G. 714; dhamásá, *Mar.*, F. 2; dhamaso, *Guj.*, F. 2; 6; dhamin, *Hind.*, G. 714; dhamna, *Hind.*, H. 324; dhamni (Helicteres), *Dec.*, H. 92; dhamni (Portulaca), *Pb.*, P. 1179, *Sind*, P. 1191; dhamni (Grewia), *Hind.*, G. 663; dhamni, *Hind*, G. 714; dhámnú, *Kumaon*, G. 688; dhamono, *Uriya*, G. 714; dhámorá, *Mar.*, A. 1149; dhamru, *Hind.*, G. 663; dhamun, *Hind.*, *Pb.*, G. 673, 688; dhamun, *Hind.*, G. 663; dhan (Woodfordia), *Beng.*, W. 106; dhán (Oryza), *Beng.*, *Hind.*, *Pb.*, *Uriya*, O. 258; dhana, *Mar.*, C. 1954; dhanap, *Pb.*, O. 574; dhanattar, *Pushtu*, C. 1403; dhand, *Hind.*, *Pb.*, P. 48; dhanda, *Pb.*, O. 574; dhandain, *Hind.*, S. 1166; dhandal, *Kumaon*, L. 569; dhandiáin, *Hind.*, S. 1174; dhane, *Beng.*, C. 1954; dhanera, *Hind.*, P. 50; dhani (Danio), *Beng.*, F. 423; dhani (Securinega), *Hind.*, S. 1029; dhaniá, *Hind.*, C. 1954; dhanicha, *Beng.*, S. 1166; dhanie, *Hind.*, S. 579; dhan-lungka-murich, *Beng.*, C. 466; dháno, *Sind*, C. 1954; dhanphari, *Pb.*, F. 10; dhanttar, *Pushtu*, C. 1403; dhanua-maror, *Hind.*, I. 503; dhanya, *Hind.*, *Mar.*, C. 1954; dhao (Woodfordia), *Beng.*, W. 106; dháo (Artocarpus), *Kumaon*, A. 1511; dhápa, *Mar.*, S. 1203; dharauli, *Hind.*, W. 131; dharimb, *Hind.*, *Sind*, P. 1426; dhar karela, *Pb.*, M. 639; dharki-karer, *Pb.*, C. 42; dhárla, *Kumaon*, W. 106; dharmar, *Beng.*, S. 2865; dharposh, *Pb.*, S. 924; dhárú, *Hind.*, L. 117; dharúr, *Pb.*, D. 503; dhas, *Pb.*, W. 106; dhatela, *Hind.*, *Kumaon*, P. 1252; dhattiki, *Uriya*, G. 705; dhaturo, *Sind*, D. 151; dhaturo ghas, *Raj.*, M. 232; dhau (Anogeissus), *Hind.*, A. 1149, 1153; dhau (Artocarpus), *Hind.*, A. 1511; dhau (Lagerstrœmia), *Pb.*, L. 55; dhauk, *Hind.*, A. 1149, 1153; dhaukra, *Raj.*, A. 1149; dháukra, *Hind.*, A. 1153; dhaula, *Kumaon*, B. 929; dhaula (Chrysopogon), *Hind.*, C. 1055; dhaula, *Kumaon*, W. 106; dhaula (Woodfordia), *Hind.*, W. 106; dhaula khejra, *Hind.*, A. 291; dhaula phindawri, *Raj.*, N. 172; dhaulí, *Hind.*, H. 517; dhaulian, *Pb.*, C. 1055; dhauntika, *Pb.*, O. 38; dhaur, *Pb.*, W. 106; dháura (Anogeissus), *Hind.*, A. 1149; dhaura (Lagerstrœmia), *Hind.*, *Pb.*, L. 55; dhaura (Woodfordia), *Beng.*, *Hind.*, *Kumaon*, W. 106; dhaura (Zizyphus), *Hind.*, Z. 273; dhauri (Anogeissus), *Hind.*, A. 1149; dhauri (Woodfordia), *Mar.*, W. 106; dhauri (Zizyphus), *Hind.*, Z. 273; dhauta, *Hind.*, A. 1149; dhává (Anogeissus), *Dec.*, *Hind.*, A. 1149; dháává (Artocarpus), *Hind.*, A. 1511; dhavadina, *Guz.*, W. 106; dhaval, *Mar.*,

R

L. 503 ; dhávdá, *Mar.*, A. 1149 ; dhávdo, *Guj.*, A. 1149 ; dhávri, *Mar.*, W. 106 ; dhavshi, *Mar.*, W. 106 ; dháwa, *Hind.*, S. 1939 ; dhawayi, *Beng.*, W. 106 ; dháwí, *Pb.*, W. 106 ; dhayatí (Woodfordia), *Mar.*, W. 106 ; dháyti (Lagerstroemia), *Mar.*, L. 52 ; dhazatichi, *Mar.*, W. 106 ; dhe daumaro, *Gus.*, F. 202 ; dhedu, *Mar.*, F. 202 ; dhedumbara, *Mar.*, F. 216 ; dhedumera, *Mar.*, F. 202 ; dhék, *Pb.*, M. 393 ; dhén, *Hind.*, O. 574 ; dhengan, *Hind.*, C. 1927 ; dheniani, *Hind.*, O. 127 ; dhenras, *Beng.*, H. 196 ; dher, *Pb.*, R. 588 ; dhera, *Hind.*, A. 681 ; dheras, *Beng.*, H. 196, 215 ; dheri dhok, *Hind.*, F. 514 ; dherimurl, *Hind.*, F. 517 ; dhewti, *Hind.*, W. 106 ; dhíngra, *Pb.*, C. 49 ; dhoalí, *Sind*, F. 516 ; dhobein, *Hind.*, D. 53 ; dhobela, *Hind.*, C. 1742 ; dhobel kirat, *Uriya*, G. 136 ; dhobin, *Hind.*, D. 53 ; dhobu, *Uriya*, A. 1149 ; dho-guru, *Pb.*, F. 429 ; dhohan, *Hind.*, G. 723 ; dhohein, *Hind.*, D. 53 ; dhokri, *Raj.*, A. 1149 ; dhokridan, *Hind.*, S. 1939 ; dhol, *Mar.*, L. 371 ; dhola akdo, *Guj.*, C. 170 ; dhola sindur, *Beng.*, M. 71 ; dholi musali, *Guj.*, A. 1562 ; dholi-musali, *Mar.*, A. 1562 ; dholisá-turdi, *Guj.*, B. 619 ; dholo dhatúro, *Guj.*, D. 160 ; dholsamudra, *Beng.*, *Hind.*, L. 232 ; dholtu, *Pb.*, B. 933 ; dhon-ga-nu, *Sind*, F. 547 ; dhop-chamni, *Beng.*, H. 149 ; dhóp-chíní, *Beng.*, S. 375 ; dhóp-raí, *Beng.*, B. 800 ; dhorára, *Hind.*, B. 318 ; dhorbenla, *Hind.*, P. 1370 ; dhorbeulá, *Mar.*, P. 1370 ; dhor dangar, *Hind.*, O. 574 ; dhotará, *Mar.*, D. 160 ; dhoti, *Mar.*, S. 2515 ; dhoula, *Raj.*, P. 382 ; dhoura (Chloroxylon), *Hind.*, C. 1031 ; dhoura (Ficus), *Hind.*, F. 230 ; dhowa, *Beng.*, W. 106 ; dhowai, *Hind.*, W. 106 ; dhowda, *Gus.*, H. 294 ; dhowra, *Hind.*, W. 106 ; dhudhali, *Hind.*, E. 335 ; dhúd (Ficus), *Pb.*, F. 230 ; dhúdi (Holarrhena), *Hind.*, H. 294 ; dhúdiá pathar, *Hind.*, C. 621 ; dhud kalmí, *Beng.*, I. 415 ; dhuluá, *Beng.*, S. 374, 375 ; dhúna, *Pb.*, S. 1656 ; dhúná, *Hind.*, S. 1656 ; dhúná, *Beng.*, *Mar.*, S. 1656 ; dhunchí, *Beng.*, S. 1166 ; dhunda, *Hind.*, A. 1076 ; dhundul, *Beng.*, L. 569 ; dhun-sha, *Beng.*, S. 1166 ; dhúnu (Abies), *Pb.*, A. 22 ; dhúnú (Taxus), *Pb.*, T. 93 ; dhúp (Boswellia), *Guj.*, B. 771 ; dhúp (Cana-rium), *Mar.*, C. 285 ; dhúp (Juniperus), *Pb.*, J. 78, *Hind.*, J. 92 ; dhúp (Juri-nea), *Kashmir*, *Pb.*, J. 111 ; dhúp (Oli-banum), *Hind.*, B. 768 ; dhúp (Pinus), *Hind.*, P. 760 ; dhúpa, *Pb.*, J. 111 ; dhúp ral-dhup, *Mar.*, C. 285 ; dhupsa, *Hind.*, C. 2558 ; dhúr, *Beng.*, *Hind.*, O. 574 ; dhúra, *Kumaon*, F. 202 ; dhúrá-drú, *Pb.*, C. 873 ; dhúr chúk, *Hind.*, H. 277 ; dhúr-chuk, *Pb.*, H. 281 ; dhúr dánger, *Beng.*, *Hind.*, O. 574 ; dhurpisag, *Hind.*, L. 312, 321 ; dhúru, *Pb.*, F. 230 ; dhusa, *Hind.*, S. 1207 ; dhúturá, *Beng.*, D. 160 ; dhuvi, *Hind.*, W. 106.

Di Di, *Pb.*, F. 466 ; diár (Cedrus), *Hind.*, *Kumaon*, *Kashmir*, *Pb.*, C. 846 ; diár (Salvadora), *Pb.*, *Sind*, S. 705 ; diár jhal, *Hind.*, S. 705 ; dib (Eragrostis),

Pb., E. 252 ; dib (Typha), *Pb.*, T. 864 ; diddani, *Afg.*, A. 1588 ; didrian, *Pb.*, C. 42 ; didriár, *Pb.*, M. 562 ; dier, *Hind.*, C. 1452 ; dihgan, *Hind.*, C. 1927 ; di-hí, *Sind*, F. 466 ; diho, *Baluch.*, T. 434 ; dikamáli, *Guj.*, *Mar.*, G. 116, *Guj.*, *Hind.*, *Mar.*, G. 128 ; dikámlí, *Hind.*, G. 116 ; dikmalí, *Hind.*, G. 116 ; dila (Cyperus), *Pb.*, C. 2592 ; dila (Odina), *Pb.*, O. 38 ; dila, *Pb.*, F. 671 ; dila, *Pb.*, S. 977 ; díla (Scirpus), *Pb.*, S. 981 ; dile, *Pb.*, T. 875 ; dilpa-sand (Citrullus), *Pb.*, *Sind*, C. 1227 ; dilpasand (Luffa), *Hind.*, L. 569 ; dimeri, *Hind.*, *Uriya*, F. 179 ; dimmon, *Sind*, F. 368 ; dinda, *Mar.*, L. 232 ; dindása, *Pb.*, J. 61 ; dingala, *Mar.*, C. 2150 ; dinger, *Pb.*, C. 49 ; dingoda, *Mar.*, C. 2150 ; dinsa, *Pb.*, I. 14 ; dintili, *Kumaon*, P. 1252 ; dio, *Pb.*, P. 1359 ; dípmal, *Mar.*, L. 266 ; dissi, *Pb.*, S. 1207 ; diúsa, *Pb.*, I. 14 ; diva, *Mar.*, T. 15 ; diva kanda, *Mar.*, T. 15 ; dívárí-mullí, *Duk.*, B. 546 ; diveli, *Gus.*, R. 369 ; diveligo, *Gus.*, R. 369.

Do Do (Populus), *Pb.*, P. 1159 ; do (Triticum), *Pb.*, T. 634 ; doarrah, *Pb.*, F. 514 ; dob, *Raj.*, C. 2558 ; dobein, *Hind.*, D. 53 ; dobra, *Pb.*, P. 79 ; doda (Papaver), *Pb.*, P. 87 ; doda (Pyrus), *Pb.*, P. 1458, 1460 ; doda (Tetraceros), *Hind.*, S. 1287 ; dodak, *Pb.*, C. 1781 ; dodak, *Pb.*, S. 2357 ; dodan (oxen), *Beng.*, *Hind.*, O. 574 ; dodan (Sapindus), *Hind.*, *Pb.*, S. 808 ; do-dant, *Beng.*, O. 574 ; do-dant, *Hind.*, O. 574 ; dodhak, *Guj.*, E. 7 ; dodhi (Dregea), *Mar.*, D. 823 ; dodhi (Leptadenia) *Mar.*, L. 292 ; dodhmará-gha, *Kumaon*, *Hind.*, A. 22 ; dodi, *Mar.*, H. 442 ; dodru, *Pb.*, I. 14 ; dodúr, *Pb.*, C. 42 ; dohu, *Hind.*, *Uriya*, A. 1149 ; doká-násiju, *Uriya*, E. 496 ; dola, *Hind.*, C. 618 ; dola, *Mar.*, C. 617 ; dola álu, *Hind.*, D. 542 ; dolakúra, *Dec.*, *Mar.*, H. 294 ; dolanku, *Uriya*, A. 681 ; dol-chilla, *Kumaon*, P. 737 ; doli, *Mar.*, A. 695 ; dolochápá, *Mar.*, P. 989 ; dolo shamlo, *Guj.*, E. 304 ; dolu, *Hind.*, R. 215 ; dom-sál, *Hind.*, F. 671 ; domsál, *Hind.*, M. 545 ; dom-sál, *Hind.*, M. 545 ; dona (Artemisia), *Pb.*, A. 1464, *Hind.*, A. 1469 ; dona (Daphne), *Pb.*, D. 130 ; donda, *Hind.*, A. 1076 ; dondru, *Mar.*, H. 517 ; dongole, *Hind.*, R. 351 ; doomoor, *Beng.*, F. 260 ; dopahariya, *Hind.*, P. 393 ; dopati-latá, *Hind.*, I. 362 ; dopohoriá, *Hind.*, P. 393 ; dor (Arisæma), *Pb.*, A. 1369 ; dor (Rhus), *Pb.*, R. 316 ; dorabyara, *Beng.*, S. 1223 ; dorádár, *Mar.*, C. 617 ; dorá kiram, *Sind*, C. 402 ; dorbenla, *Mar.*, P. 1370 ; dord, *Hind.*, P. 513 ; dorí (Cedrela), *Pb.*, C. 838 ; dorí (Polygonum), *Pb.*, P. 1112 ; dorka, *Hind.*, L. 556 ; dorlí, *Mar.*, S. 2280 ; doro, *Sind*, C. 402 ; douná, *Hind.*, *Duk.*, A. 1469 ; dowdowlá, *Mar.*, F. 633 ; dowla, *Mar.*, U. 23.

Dr Drab (Cedrela), *Pb.*, C. 838 ; drab (Eragrostis), *Afg.*, *Pb.*, E. 252 ; dráb chír, *Pb.*, P. 760 ; drakasha, *Gus.*, V. 233 ; drakh, *Hind.*, V. 233 ; drákh, *Mar.*, V. 233 ; drákh, *Sind*, V. 233 ;

drakhyaluta, *Beng.*, V. 233 ; draksha, *Mar.*, V. 233 ; dral, *Guj.*, B. 663 ; dramah, *Pb.*, *Sind*, F. 6 ; drammaho, *Sind.*, F. 2 ; drange, *Pb.*, S. 502 ; drangu, *Pb.*, S. 505 ; drawi, *Hind.*, C. 835 ; drawí, *Pb.*, C. 838 ; drawo, *Pb.*, F. 19 ; drek (Melia), *Pb.*, M. 393, *Hind.*, *Mar.*, *Pb.*, *Sind*, M. 393 ; drek (Pistacia), *Kashmir*, *Pb.*, P. 833 ; drendú, *Pb.*, X. 21 ; drik, *Kashmir*, L. 245 ; drín, *Kashmir*, F. 754 ; drinkhari, *Pb.*, D. 144 ; dro, *Pb.*, T. 634 ; droga-puttiah, *Uriya*, F. 595 ; drúkrí, *Pb.*, V. 195 ; drúmbí, *Pushtu*, P. 618 ; drum-mahú, *Sind*, F. 2 ; drúnda, *Pb.*, I. 14 ; drunnu, *Sind*, C. 2101 ; drúss, *Pb.*, C. 133.

Du Dua, *Kumaon*, E. 327 ; duan, *Hind.*, E. 327 ; dub (Cynodon), *Beng.*, *Pb.*, C. 2558 ; dúb, *Pb.*, F. 672 ; duba, *Hind.*, C. 2558 ; dubein, *Hind.*, D. 53 ; dúbla, *Beng.*, C. 2558 ; dubra (Cynodon), *Pb.*, C. 2558 ; dubra (Eleusine), *Pb.*, E. 184 ; dubra (Panicum), *Pb.*, P. 77 ; dúdagrú, *Pb.*, F. 168 ; dúdal (Euphorbia), *Pb.*, E. 509 ; dúdal (Taraxacum), *Pb.*, T. 87 ; du-dant, *Beng.*, *Hind.*, O. 574 ; dudeli, *Gus.*, E. 531 ; dúdfras, *Pb.*, P. 1148 ; dudha álu, *Hind.*, D. 542 ; dudháli, *Mar.*, S. 2378 ; dudháni, *Mar.*, O. 600 ; dudha-pár, *Pb.*, E. 475 ; dudhar, *Beng.*, *Hind.*, O. 574 ; dudhari, *Beng.*, *Hind.*, O. 574 ; dúdh batthal, *Pb.*, I. 87 ; dudhi (Euphorbia), *Hind.*, *Mar.*, E. 531 ; dudhi (Ichnocarpus), *Beng.*, *Hind.*, I. 1 ; dudhi (Oxystelma), *Beng.*, *Hind.*, *Sind*, O. 600 ; dudhi (Wrightia), *Mar.*, W. 131 ; dudhi, *Hind.*, *Pb.*, W. 122, 131 ; dúdhi (Vallaris), *Kumaon*, V. 12 ; dudhia bish, *Kashmir*, *Pb.*, A. 413 ; dudhia latá, *Beng.*, *Hind.*, *Uriya*, O. 600 ; dúdhiká, *Mar.*, O. 600 ; dudhio vachhanág, *Guj.*, A. 413 ; dudh-koraiya, *Beng.*, W. 131 ; dúdhlak, *Pb.*, L. 112 ; dudhlutta, *Beng.*, O. 600 ; dudhya, *Mar.*, L. 30 ; dudia maddi, *Dec.*, T. 361 ; dudi maddi, *Dec.*, T. 361 ; dudiya, *Beng.*, E. 549 ; dudiya-kalmí, *Beng.*, I. 415 ; dudiya-kulmi, *Beng.*, I. 368 ; dúdla (Prunus), *Pb.*, P. 1316 ; dúdla (Sapium), *Mar.*, *Pb.*, S. 838 ; dúdla (Syringa), *Pb.*, S. 3079 ; dúdla kakkarí, *Pb.*, R. 318 ; dúdlí, *Pb.*, T. 87 ; dudlutta, *Hind.*, O. 600 ; dúd shambar, *Pb.*, D. 348 ; dudurli, *Mar.*, H. 328 ; dugdhicá, *Hind.*, O. 600 ; dugdugia, *Hind.*, E. 453 ; duggerful, *Sind*, C. 146 ; dugurphort, *Sind*, C. 146 ; dúgúrú, *Pb.*, F. 168 ; duh-rie, *Sind*, F. 566 ; duiya-khuiya, *Beng.*, A. 914 ; dúkar, *Mar.*, H. 289 ; dukri-e, *Hind.*, I'. 424 ; dúkú, *Hind.*, P. 456 ; dul, *Beng.*, P. 48 ; dula, *Pb.*, H. 215 ; dulal labhá, *Beng.*, A. 745, C. 224 ; dulí champa, *Beng.*, M. 51 ; dullúnga, *Pb.*, F. 516 ; dulshat, *Kumaon*, C. 1711 ; dumar, *Beng.*, F. 202 ; duma-so, *Mar.*, F. 2 ; dúmba, *Pb.*, S. 1320 ; dumbar, *Mar.*, F. 202 ; dumbur, *Beng.*, F. 156 ; dumeli, *Hind.*, F. 179 ; dumki mirchi, *Dec.*, P. 801 ; dumni, *Pb.*, J. 29 ; dumra, *Pb.*, F. 414 ; dumrez, *Hind.*, C. 1234 ; dúmsal, *Kumaon*, B. 771 ; dúmtúli, *Kashmir*, A. 498 ; dún, *Kashmir*, *Pb.*, J. 61 ; dun-de-a, *Sind*, F.

405 ; dundikerri, *Uriya* ; F. 558 ; dún-dúl, *Beng.*, L. 569 ; dúngari, *Guj.*, *Sind*, A. 769 ; dúngla, *Hind.*, R. 311 ; dungra, *Hind.*, C. 2263 ; dunnahrí, *Hind.*, F. 396 ; dúpati, *Beng.*, I. 39 ; dupri, *Kumaon*, J. 92 ; durána, *Afg.*, C. 2035 ; durang, *Hind.*, S. 2508 ; dúrbá, *Beng.*, C. 2558 ; durbui, *Hind.*, R. 351 ; dureshta, *Afg.*, M. 334 ; durhie, *Hind.*, F. 350 ; durruah, *Uriya*, F. 476 ; durva (Cynodon), *Mar.*, C. 2558 ; durva (Eragrostis), *Hind.*, E. 252 ; dúsera-sag, *Beng.*, M. 615 ; dúss (Colebrookia), *Pb.*, C. 1711 ; dúss (Elscholtzia), *Pb.*, E. 196 ; dusta (adjutant bird), *Hind.*, S. 2906 ; dusta (zinc), *Hind.*, Z. 190 ; dutundi, *Mar.*, X. 1.

Dw Dwárena, *Pushtu*, P. 618.

Dy Dyowa báns, *Beng.*, B. 142.

Dz Dzaral, *Pushtu*, C. 862.

Eb Ebans, *Hind.*, D. 569.

Ec Echalat, *Beng.*, M. 435 ; echra, *Mar.*, P. 727.

Ed Edia, *Raj.*, R. 369.

Eg Egorea, *Guj.*, B. 13.

Eh Ehandrajot, *Dec.*, J. 41.

Ei Eilan, *Pb.*, P. 702 ; eilaur, *Pb.*, P. 702 ; ein, *Pb.*, G. 213 ; eisúr, *Mar.*, M. 412.

Ek Ekharo, *Gus.*, H. 508 ; eklbír, *Pb.*, V. 64.

El Elachi (Amomum), *Guj.*, A. 976 ; eláchi (Elettaria), *Beng.*, E. 151 ; elaich, *Beng.*, E. 151 ; elchi, *Mar.*, E. 151 ; elcho, *Guj.*, A. 976 ; eliyá, *Dec.*, *Guj.*, A. 824 ; ellal, *Pb.*, P. 702, 829 ; élvá, *Duk.*, A. 824.

Em Em, *Pb.*, C. 1003.

En Endra, *Pb.*, H. 285 ; endraní, *Hind.*, P. 1078.

Eq Eqilbír, *Pb.*, D. 144.

Er Er, *Pb.*, P. 1304 ; eran, *Pb.*, P. 702 ; erana. *Pb.*, P. 702 ; erand, *Hind.*, R. 369 ; erandá-gáchh, *Beng.*, J. 41 ; eranda kakdi, *Guj.*, C. 581 ; erand (Rhus), *Mar.*, R. 293 ; erandi (Ricinus), *Mar.*, R. 369 ; erend, *Hind.*, R. 369 ; erendi, *Mar.*, R. 369 ; eri, *Pb.*, V. 106 ; ermul, *Beng.*, R. 663 ; ersá, *Kashmir*, I. 425 ; eru, *Hind.*, M. 237 ; erund, *Deccan*, R. 369 ; erundi, *Dec.*, J. 41 ; ervados, *Mar.*, P. 727.

Es Esar, *Hind.*, R. 590 ; esar, *Kumaon*, R. 590 ; esbi, *Pb.*, S. 1260 ; esbu, *Pb.*, S. 1260 ; esesh, *Mar.*, B. 751 ; eshopgól, *Beng.*, P. 932.

Et Etári, *Duk.*, *Mar.*, A. 89.

Fa Fagári, *Pb.*, E. 149 ; fagora, *Pb.*, R. 162 ; fágu, *Pb.*, F. 149, 230 ; fagúrí, *Pb.*, F. 149 ; fagwara, *Hind.*, F. 230 ; faharghás, *Kashmir*, R. 564 ; falís, *Pb.*, P. 1148 ; falsh, *Kashmir*, *Pb.*, P. 1148 ; farangí-aúd, *Duk.*, B. 777 ; farangi dhatúra, *Dec.*, *Hind.*, A. 1351 ; faras (Butea), *Hind.*, B. 944 ; farás (Tamarix), *Pb.*, T. 51, T. 61 ; farash (Cupressus), *Sind*, C. 2354 ; farásh (Tamarix), *Pb.*, I. 51, T. 70 ; faríd-búti, *Hind.*, C. 1452 ; farídmúlí, *Pb.*, F. 35 ; farid-búti (Farsetia), *Pb.*, F. 35 ; farid-búti (Pedalium), *Hind.*, P. 363 ; farrí, *Pb.*, G. 673 ; farsh, *Pb.*, P. 1159 ; farw, *Pb.*, P. 77 ; farwá, *Pb.*, T. 51 ; farwan, *Pb.*, T. 61 ; fauglimehndí, *Pb.*, A. 960.

Fe Fease, *Baluch.*, *Sind*, N. 1; fessah, *Beng.*, F. 443.

Fi Filfil, *Pb.*, *Kumaon*, V. 164; filfil daráz, *Pb.*, P. 805; filfildray, *Sind*, P. 805; filfilglrd, *Hind.*, P. 811; findak, *Hind.*, C. 1985; findora, *Hind.*, L. 353; firangi dhotra, *Mar.*, A. 1351; fisáúni, *Pb.*, H. 13; fitrásálúm, *Pb.*, *Sind*, P. 1221; fitrasúliun, *Pb.*, *Sind*, P. 1221; fituras-aliyún, *Mar.*, P. 1221.

Fl Flassu, *Pb.*, P. 1148.

Fr Fraga, *Pb.*, F. 682; fras, *Kashmir*, *Pb.*, P. 1138; frast, *Kashmir*, *Pb.*, P. 1159.

Fu Fulungí, *Mar.*, F. 408; furrunj-mushk, *Pb.*, O. 18.

Fy Fysur láni, *Sind*, C. 224.

Ga Gab, *Mar.*, D. 582; gab (Ricinus), *Uriya*, R. 369; gáb (Diospyros), *Beng.*, *Hind.*, D. 582; gabdi, *Hind.*, C. 1512; gabhin, *Beng.*, *Hind.*, O. 574; gabna, *Beng.*, N. 163; gach, *Pb.*, G. 769; gach-karan, *Mar.*, R. 231; gách-marich, *Beng.*, C. 448, 455, *Hind.*, C. 466; gáchmirich, *Hind.*, C. 448, 455; gada-bani, *Hind.*, *Beng.*, T. 530; gadal, *Hind.*, F. 655; gadámbal, *Pb.*, R. 335; gadambikanda, *Mar.*, C. 2075; gadancha, *Beng.*, T. 470, T. 483; gadar puchha, *Hind.*, S. 1223; gádar punch, *Pb.*, E. 261; gadda (donkey), *Hind.*, H. 414; gadda (Phœnix), *Pb.*, P. 555; gaddí siúngar, (*Rav.*), *Pb.*, C. 1003; gadgundi, *Gus.*, C. 1940; gadha, *Hind.*, E. 186; gadha-charwa, *Hind.*, E. 186; gádha púrna, *Beng.*, B. 619; gádkúji, *Pb.*, P. 1466; gadi, *Hind.*, C. 1686; gado, *Gus.*, T. 470, T. 483; gadoreji, *Sind*, S. 1166; gadru, *Hind.*, C. 1927; gaduwa, *Hind.*, C. 2316; gae, *Hind.*, *Pb.*, O. 574; gaer-honar-patta, *Beng.*, C. 2062; gagandhul, *Hind.*, P. 26; gágar, *Beng.*, *Hind.*, D. 173; gaggar yurmí, *Kashmir*, R. 261; gagjaira, *Hind.*, F. 265; gáglí (Colocasia), *Pb.*, C. 1732; gáglí (Dolichos), *Pb.*, *Sind*, D. 758; gagora, *Beng.*, F. 321; gaha, *Hind.*, F. 700; gahat, *Hind.*, *Kumaon*, D. 758; gahu, *Mar.*, T. 634; gahula, *Mar.*, P. 1313; gahuma, *Hind.*, P. 384; gahung, *Mar.*, T. 634; gai, *Hind.*, O. 574; gaiaswát, *Beng.*, F. 265; gainda, *Hind.*, R. 235, 237; gainti, *Hind.*, B. 615; gair, *Kumaon*, O. 160; gaira, *Beng.*, Z. 9; gaivara, *Mar.*, C. 289, C. 290; gaiya, *Hind.*, O. 574; gaj, *Beng.*, E. 83; gaja, *Mar.*, C. 6; gajá, *Duk.*, C. 6; gajachinní, *Hind.*, C. 860; gajagá, *Mar.*, C. 6; gajahanda, *Hind.*, T. 392; gajakarní, *Mar.*, R. 231; gajapipal, *Beng.*, *Hind.*, S. 970; gájar, *Guj.*, *Hind.*, *Pb.*, D. 173; gajarlahorí, *Sind*, I. 348; gajel, *Mar.*, O. 38; gájga, *Guj.*, C. 6; gajir, *Hind.*, D. 542; gajiún, *Hind.*, F. 265; gajna, *Hind.*, F. 265; gajpipali (Plantago), *Pb.*, P. 919; gajpípali (Scindapsus), *Hind.*, S. 970; gajpiplí, *Hind.*, S. 970; gajpipul, *Beng.*, S. 970; gajra (Cochlospermum), *Hind.*, C. 1512; gajrah (Strychnos), *Mar.*, S. 2960; gakhurá, *Uriya*, T. 548; gal, *Pb.*, *Pushtu*, S. 1212; gálá, *Beng.*, C. 1491, L. 1; gál-anar, *Gus.*, P. 1426; gálar

torí, *Pb.*, T. 569; galay, *Mar.*, R. 1; galli, *Hind.*, C. 1686; galboja, *Pb.*, P. 746; galdam, *Kumaon*, R. 355; galdu, *Kumaon*, O. 160; galédu, *Guj.*, C. 919; galgal (Cochlospermum), *Hind.*, C. 1512; galgala (Eragrostis), *Hind.*, E. 263; galgoja, *Pb.*, P. 746; gali, *Gus.*, I. 145; gálícha, *Hind.*, C. 626; galion, *Hind.*, P. 1460, *Kumaon*, P. 1460; galka, *Pushtu*, R. 598; galla jári, *Hind.*, S. 2394; gallí, *Pb.*, P. 555; gallu, *Hind.*, *Kumaon*, T. 93; galmora, *Beng.*, P. 1440; galmorre, *Beng.*, P.1440; galo, *Gus.*, T. 470; galo, *Gus.*, T. 483; galphula, *Hind.*, S. 2668; galu, *Hind.*, C. 1686; gam, *Beng.*, T. 634; gamari, *Beng.*, *Hind.*, G. 287; gambari, *Hind.*, *Uriya*, G. 287; gámbhár, *Hind.*, G. 287; gametta, *Mar.*, Z. 182; gamhar (Gmelina), *Hind.*, G. 287; gamhár (Trewia), *Hind.*, T. 525; gámni chíní-búro, *Gus.*, S. 319; gámnikhánd, *Gus.*, S. 373; gamni-révanchini, *Gus.*, R. 215; gámniskkar, *Gus.*, S. 373; gamur, *Hind.*, P. 42; gán, *Pb.*, C. 606; ganam, *Afg.*, T. 634; ganára, *Kumaon*, P. 45; ganari, *Hind.*, L. 219; ganasur, *Mar.*, C. 2180; ganasura, *Mar.*, C. 2180; gánaura, *Hind.*, M. 237; ganchi-shím, *Beng.*, D. 789; gándá, *Deccan*, S. 30; ganda-biroza (Ferula), *Hind.*, F. 95; ganda biroza (Pinus), *Hind.*, P. 760; gandá búte, *Pb.*, E. 509; gandah-birozah, *Mar.*, P. 760; gandah ferozah, (Boswellia), *Hind.*, B. 772; gandak, *Pb.*, S. 2999; gandal, *Hind.*, A. 1632; gandali, *Uriya*, P. 4; gándalú (Murraya), *Pb.*, M. 800; gandalún (Daphne), *Pb.*, D. 130; gandam, *Afg.*, T. 634; gandamdar, *Afg.*, S. 1015; gandamgúndú, *Kashmir*, L. 602; gandán, *Hind.*, *Duk.*, A. 1395; gandana, *Afg.*, *Hind.*, *Pb.*, A. 767; ganda-ním, *Pb.*, M. 800; gandati, *Hind.*, *Duk.*, A. 1395; gand-bábúl, *Hind.*, *Duk.*, A. 217; ganddula, *Hind.*, C. 1686; gandel, *Hind.*, I. 489; gandelí, *Pb.*, V. 233; gandera, *Pb.*, R. 166; ganderái, *Pushtu*, N. 80; gandha bena, *Beng.*, A. 1079, 1117; gandhabhádulia, *Beng.*, P. 4; gandhácha-koda, *Mar.*, S. 790; gandha gokal, *Beng.*, T. 444; ganadhak-ka-phula, *Mar.*, S. 2999; gandháli, *Hind.*, P. 4; gandhan, *Pb.*, A. 767; gandhana, *Gus.*, P. 4; gandha-nákuli, *Beng.*, O. 180; gandha-rasaha, *Beng.*, B. 48; gandhátí, *Mar.*, A. 1395; gandha-trina, *Hind.*, A. 1079; gándhaumbara, *Mar.*, F. 216; gandhílovaj, *Gus.*, A. 430; gandhomálati, *Beng.*, A. 584; gandi (Chloris), *Hind.*, C. 1026; gandi (Murraya), *Pb.*, M. 800; gandi, *Beng.*, X. 8; gándi (Iseilema), *Pb.*, I. 487; gandial, *Pb.*, A. 1332; gandiali, *Pb.*, C. 1732; gandí bútí, *Pb.*, M. 615; gandkí-lakrí, *Duk.*, A. 430; gandla, *Hind.*, *Kumaon*, *Pb.*, M. 800; gando-gaula, *Beng.*, T. 444; gandra, *Kumaon*, S. 1212; gandrichri, *Mar.*, F. 312; gandrok, *Beng.*, S. 2999; gandru-papura, *Mar.*, G. 124; ganegan, *Raj.*, G. 702; ganer, *Hind.*, A. 1632; ganeri, *Mar.*, C. 1512; ganerjeí, *Pb.*, A. 1632; gangal,

Hind., C. 1512 ; gangan, *Raj.*, G. 702 ; ganger (Grewia), *Pb.*, G. 702 ; ganger (Ehretia), *Pushtu*, E. 35 ; ganger (Lycium), *Pb.*, *Sind*, L. 589 ; ganger (Sageretia), *Pb.*, S. 497 ; gangerun, *Raj.*, G. 702 ; gangí, *Pb.*, *Sind*, G. 702 ; gángichu, *Pb.*, E. 520 ; gango, *Sind*, G. 702 ; gangr, *Sind*, Z. 254 ; gangr, *Sind*, C. 224 ; gangro, *Sind*, L. 589 ; gang-siuli, *Uriya*, N. 179 ; ganguli, *Hind.*, A. 1093 ; gangwa, *Beng.*, E. 593 ; ganhar, *Hind.*, A. 934 ; ganhár, *Pb.*, A. 916 ; ganhíla, *Pb.*, P. 1242 ; ganhin, *Pb.*, P. 1242 ; ganhíra, *Hind* , *Pb.*, C. 225 ; gan hira, *Pb.*, N. 80 ; ganhúla gándal, *Pb.*, S. 763 ; gani (Murraya), *Hind.*, *Kumaon*, M. 800 ; ganí (Oxystelma), *Pb.*, O. 600 ; ganiár (Cochlospermum), *Hind.*, C. 1512 ; ganiári (Premna), *Beng.*, P. 1233 ; ganira, *Mar.*, N. 80 ; ganj (Millettia), *Pb.*, M. 549 ; gánja (Cannabis), *Beng.*, *Dec.*, *Guj.*, *Hind.*, C. 331, (Cutch), 331 ; ganjar, *Hind.*, F. 265 ; ganji, *Hind.*, S. 2682 ; ganjira, *Hind.*, D. 542 ; ganjní, *Dec.*, *Hind.*, A. 1107 ; ganna, *Beng.*, *Hind.*, *Pb.*, S. 30 ; ganni (Chloris), *Pb.*, C. 1026 ; ganni (Iseilema), *Pb.*, I. 489 ; ganniari, *Hind.*, P. 1233 ; ganrar, *Hind.*, A. 1097 ; gant, *Hind.*, M. 800 ; gant-baháraní, *Hind.*, C. 1388 ; gantha, *Hind.*, S. 959 ; gánth dob, *Raj.*, E. 184 ; ganthia, *Raj.*, E. 184 ; ganthian, *Pb.*, I. 343 ; ganthya, *Raj.*, E. 188 ; ganti ghás, *Raj.*, E. 188 ; ganyeri, *Pb.*, Z. 280 ; gao, *Pb.*, O. 574 ; gaoiya, *Mar.*, O. 567 ; gaoj, *Beng.*, D. 240 ; gaoni, *Hind.*, D. 238 ; gaorishiora, *Beng.*, F. 194 ; gaozabán (Macrotomia), *Hind.*, M. 32 ; gaozabán (Onosma), *Beng.*, O. 168 ; gaozaban (Trichodesma), *Sind*, T. 562 ; gao-zubán, *Hind.*, A. 1132 ; garaj-phal, *Dec.*, C. 1834 ; garambi, *Mar.*, E. 219 ; garán, *Beng.*, C. 965, C. 972 ; garand, *Baluch. & Sind*, S. 1278 ; garandu, *Pb.*, P. 1252 ; garani, *Guj.*, C. 1403 ; garar, *Afg.*, *Pb.*, R. 132 ; garári, *Hind.*, *Mar.*, L. 219 ; garaykhíri, *Sind*, O. 600 ; gar badero, *Hind.*, C. 1038, I. 60 ; garbi, *Mar.*, E. 219 ; garbijaur, *Hind.*, L. 483 ; gardal, *Mar.*, E. 219 ; gardalu, *Pb.*, P. 1304 ; gardhan, *Pb.*, R. 162 ; gardul, *Mar.*, E. 219 ; garelú, *Pb.*, F. 168 ; gargadan, *Hind.*, R. 237 ; gargas (Grewia), *Pb.*, G. 705 ; gargas (Securinega), *Pb.*, S. 1024 ; gargela, *Hind.*, B. 606 ; garges, *Pb.*, G. 677 ; gar gira, *Pushtu*, P. 1340 ; gargú-narú, *Hind.*, B. 905 ; garham, *Pb.*, T. 470, T. 483 ; garigond, *Hind.*, A. 597 ; gariha, *Hind.*, O. 574 ; gari kulay, *Beng.*, G. 263 ; garinda, *Pb.*, C. 606 ; garinga, *Hind.*, C. 596 ; gariya (Ipomæa), *Mar.*, I. 393 ; gariya (Porana), *Dec.*, P.1165 ; gariya (molasses), *Hind.*, S. 372 ; garian, *Beng.*, D. 676, 701 ; garjun, *Beng.*, D. 683 ; garkath, *Kumaon*, I. 510 ; garkum, *Hind.*, M. 289 ; garm, *Pb.*, P. 42 ; garmal (Cassia), *Guj.*, C. 756 ; garmal (Coleus), *Mar.*, C. 1719 ; garmálá, *Guj.*, C. 756 ; garmehal, *Hind.*, S. 2910 ; garna, *Pb.*, C. 606 ; gar-pípal, *Kumaon*, P. 1148 ; garrar, *Hind.*,

L. 219 ; garri (Arundinaria), *Hind.*, A. 1523 ; garri (Oryza), *Raj.*, O. 258 ; garri (Cotoneaster), *Kumaon*, C. 2023 ; garso, *Hind.*, A. 695, 717 ; gar tashiára, *Kumaon*, V. 131 ; garua, *Beng.*, F. 547 ; garudar, *Pb.*, E. 196 ; garúm, *Pb.*, T. 470, T. 483 ; garun, *Hind.*, C. 1686 ; garúr (Euonymus), *Hind.*, E. 482 ; garúr (Litsæa), *Beng.*, L. 483 ; garúr (Olea), *Kumaon*, O. 160 ; garur (adjutant bird), *Hind.*, S. 2906 ; gathil, *Pb.*, E. 184 ; gatte, *Pb.*, N. 39 ; gatval, *Mar.*, V. 116 ; gau, *Beng.*, *Hind.*, *Pb.*, O. 574 ; gauli, *Hind.*, B. 868 ; gaungchi, *Hind.*, A. 51 ; gauni, *Hind.*, G. 186 ; gaúnrí, *Kashmir*, *Pb.*, T. 516 ; gaunta, *Hind.*, M. 919 ; gaur, *Hind.*, O. 567 ; gaur, *Beng.*, O. 574 ; gaur, *Hind.*, O. 574 ; gauri, *Mar.*, C. 2514 ; gauri-gai, *Hind.*, O. 567 ; gauri-phul, *Kashmir*, R. 590 ; gausam, *Hind.*, *Pb.*, S. 950 ; gavacha malmandi, *Mar.*, I. 131 ; gávala, *Mar.*, P. 1313 ; gavari, *Mar.*, C. 289 ; gavria, *Guj.*, C. 289 ; gavung, *Hind.*, C. 1026 ; gawán, *Raj.*, E. 192 ; gawar, *Mar.*, C. 2514 ; gawn, *Guz.*, T. 634 ; gay, *Beng.*, *Hind.*, O. 574 ; gayal, *Hind.*, O. 565 ; gáy goru, *Beng.*, *Hind.*, O. 574 ; gaywah, *Hind.*, F. 467 ; gaz, *Sind*, T. 51, T. 61 ; gazanj-bin (manna), *Mar.* M. 235 ; gazanj-bín (Tamarix), *Pb.*, T. 51. gázara, *Mar.*, D. 173 ; gaz khera; *Baluch.*, T.70 ; gazlan, *Sind*, T. 51, gazlei, *Sind*, T. 79 ; gaz pipal (Piper); *Hind.*, P. 805 ; gaz pipal (Plantago), *Pb.*, P. 926 ; gaz-surkh, *Pushtu* ; T. 70 ; gaz-surkh, *Baluch.*, T. 79.

Ge Géang, *Pb.*, L. 528 ; gedúri, *Mar.*, C. 1931 ; geh, *Pb.*, C. 1988 ; gehela, *Dec.*, *Mar.*, R. 1 ; gehma, *Afg.*, E. 228 ; gehún, *Dec.*, *Hind.* ; *Pb.*, T.634 ; gehún, *Hind.*, T. 634 ; geia, *Hind* , B. 863 ; gél, *Hind.*, V. 25 ; gelaphal, *Mar.*, R. 1 ; gelaphala, *Mar.*, R. 1 ; gélarpatr, *Hind.*, L. 69 ; gelha, *Hind.*, E. 219 ; geli, *Hind.*, *Kumaon*, T. 93 ; genda, *Hind.*, *Beng.*, T. 17 ; gendasad bargí, *Pb.*, T. 17 ; gendi, *Pb.*, C. 1047 ; gendu, *Uriya*, T. 17 ; gengáru, *Pb.*, C. 2033 ; gengri, *Mar.*, D. 32 ; genthi, *Hind.*, *Kumaon*, D. 542 ; genti, *Hind.*, D. 542 ; genti gajír, *Kumaon*, D. 542 ; geor. *Beng.*, E. 593 ; gera, *Mar.*, R. 1 ; geredi, *Uriya*, E. 219 ; geria (Excæcaria), *Beng.*, E. 593 ; geriá (Suæda), *Uriya*, S. 2994 ; gero, *Hind.*, A. 1523 ; géru, *Hind.*, I. 472 ; getan, *Mar.*, D. 420 ; geti, *Hind.*, B. 615 ; geva, *Mar.*, E. 593 ; geyápal, *Mar.*, C. 2192 ; geyár, *Pb.*, C. 846.

Gh Ghabajarin, *Guz.*, T. 864 ; ghach-haldi, *Beng.*, C. 2008 ; ghadi, *Beng.*, O. 38 ; ghafiz, *Pb.*, D. 271 ; ghagari, *Mar.*, C. 2101 ; ghagharu tág, *Mar.*, C. 2105 ; ghain, *Pb.*, E. 51 ; ghain-chú, *Hind.*, L. 355 ; ghále, *Pb.*, P. 882 ; ghalme, *Pb.*, A. 1005 ; ghalme, *Pushtu*, C. 224 ; ghambeli, *Guz.*, J. 18 ; ghamor, *Hind.*, P. 42 ; ghamrur, *Pb.*, P. 42 ; ghamur, *Pb.*, F. 672 ; ghamur, *Pb.*, P. 42 ; ghanjín, *Kumaon*, D. 542 ; ghannasaphan, *Mar.*, S. 785 ; ghant (Rhamnus), *Hind.*, R. 162 ; ghant (Schrebera),

Hind., S. 959; ghánta-chi-baji, *Mar.*, S. 2515; ghantá párul, *Beng.*, S. 959; ghanta patali, *Hind.*, S. 959; ghantiáli, *Hind.*, C. 1359; 1360; ghantil, *Pb.*, E. 184; ghár, *Guj.*, C. 618; ghara, *Mar.*, *Hind.*, L. 219; gharam, *Pb.*, P. 42; gharayt, *Sind.*, C. 1386; ghárbi, *Mar.*, E. 219; gharei kashmálú, *Hind.*, *Pb.*, L. 67; ghargashtáí, *Pushtu*, P. 1322; ghar-guru, *Pushtu*, Z. 270; ghari, *Hind.*, S. 1029; ghárikún, *Pb.*, A. 596; gharnangoi, *Pushtu*, P. 1426; gharol. *Mar.*, T. 470; T. 483; gharo-te, *Pb.*, O. 600; gharsa, *Pb.*, S. 1290; gharúr gundolí, *Pb.*, L. 556; ghátí, *Mar.*, S. 2299; gháti-pitpáprá, *Mar.*, J. 120; ghati-pitta papada, *Mar.*, P. 425; ghato, *Hind.*, Z. 290; ghat-yárí, *Hind. & Pb.*, A. 1093; ghausar, *Hind.*, C. 257; gha-vum, *Gus.*, T. 634; ghawn, *Mar.*, T. 634; ghawut-ghum, *Mar.*, T. 634; ghayamári, *Guj.*, *Mar.*, B. 909; ghazlei, *Pb.*, T. 61; ghechu, *Hind.*, A. 1242; gheea-sim, *Beng.*, D. 789; ghela, *Mar.*, R. 1; ghénti-natí, *Beng.*, A. 949; ghentol *Pushtu*, T. 845; ghentú, *Beng.*, C. 1380; ghesuwa, *Kumaon*, U. 39; ghet-kochu, *Beng.*, T. 883; ghétu, *Hind.*, R. 1; ghetulí, *Mar.*, B. 619; ghí, *Hind.*, G. 189; ghia, *Kumaon*, S. 3079; ghiá kaddu, *Pb.*, C. 2321; ghiá-tarui, *Hind.*, L. 569; ghía torí, *Pb.*, L. 569; ghí gandolí, *Pb.*, L. 569; ghigvár, *Hind.*, A. 829; ghi-kanvar, *Dec.*, *Hind.*, A. 829, *Dec.*, A. 838; ghi-kavár, *Hind.*, A. 829; ghi-kawár, *Hind.*, A. 838; ghila, *Hind.*, B. 318; ghimá sák, *Beng.*, M. 613; ghi-nalitá-pat, *Beng.*, C. 1846; ghirri, *Pb.*, P. 42; ghirta kamári, *Beng.*, A. 838; ghirta-kanvár, *Beng.*, A. 838; ghirta-kumári, *Beng.*, A. 829; ghisoda, *Gus.*, L. 556; ghí turái, *Pb.*, L. 569; ghivala, *Guj.*, C. 123; ghiwáin, *Kumaon*, E. 48, *Pb.*, E. 51; ghiwala, *Kumaon*, C. 123; ghíya taroi, *Hind.*, *Kumaon*, L. 569; ghíya tori, *Hind.*, L. 569; ghlanchá-ki-jar, *Hind.*, T. 470; T. 483; ghobe, *Guj.*, C. 919; ghodá, *Gus.*, W. 98; ghodila, *Hind.*, E. 259; ghod-khúrí, *Mar.*, A. 506; ghogar, *Hind.*, G. 143, *Mar.*, G. 124; ghogiya, *Hind.*, C. 868; ghókaru, *Mar.*, T. 548; ghókrú, *Dec.*, T. 548; ghol (Cephalandra), *Pb.*, C. 919; ghol (Portulaca), *Hind.*, P. 1179; ghól-kí-bhájí, *Dec.*, P. 1187; ghólsarí, *Hind.*, M. 570; gholú, *Hind.*, P. 1179; gholyel, *Mar.*, S. 2252; ghona-saphan, *Mar.*, S. 785; ghor, *Hind.*, A. 430; ghora, *Hind.*, H. 414; ghora chela, *Beng.*, F. 394; ghora muga, *Beng.*, P. 496; ghorán, *Beng.*, C. 972; ghoránim, *Beng.*, M. 393; ghoráskai, *Pushtu*, D. 725; ghorayal, *Hind.*, I. 487; ghor-chubba, (*C. P.*), *Hind.*, O. 185; ghor-hé, *Pb.*, T. 434; ghori, *Pb.*, C. 618; ghorila, *Hind.*, E. 259; ghork, *Beng.*, S. 2518; ghorkaram, *Hind.*, A. 658; ghorla, *Hind.*, S. 2670; ghor-masán, (*C. P.*), *Hind.*, I. 849; ghorrái, *Hind.*, B. 841; ghorúmba, *Pb.*, C. 1211; ghorvel, *Mar.*, V. 189; ghorwel, *Mar.*, V. 189; ghosálí, *Mar.*, L. 569; ghosal-phal, *Hind.*, M. 639; gho-samp, *Beng.*, *Dec.*, *Hind.*, *Pb.*, R. 116; ghoshá-latá, *Beng.*,

L. 563; ghot, *Mar.*, Z. 290; ghótághaubá, *Hind.*, C. 66; ghoti-suara, *Beng.*, F. 194; ghous, *Eastern Bengal*, D. 240; ghóut, *Hind.*, Z. 290; ghoutí, *Hind.*, Z. 290; ghoya, *Hind.*, C. 1732; ghua, *Hind.*, S. 6; ghúbot, *Hind.*, T. 634; ghugharo, *Gus.*, C. 2101; ghui (Ficus), *Hind.*, F. 156; ghui (Eragrostis), *Hind.*, E. 259; ghuiya, *Hind.*, C. 1732; ghúndí, *Pb.*, S. 2518; ghungchi, *Mar.*, A. 51; ghunja, *Hind.*, G. 143; ghunt, *Hind.*, Z. 290; ghunta, *Beng.*, S. 2876; ghúr, *Hind.*, M. 237; ghuraskai, *Pushtu*, D. 725; ghuráske, *Pushtu*, D. 725; ghurg, *Pushtu*, T. 609; ghúrga, *Hind.*, G. 136; ghuri, *Mar.*, A. 1475; ghur-khur, *Hind.*, H. 414; ghúrúmba, *Pb.*, C. 1211; ghurúsh, *Hind.*, *Pb.*, P. 486; ghus, *Hind.*, R. 51; ghuter, *Hind.*, Z. 280; ghúttia, *Pb.*, B. 933; ghutti-ki-jar, *Dec.*, C. 1403; ghuya, *Hind.*, C. 1732; ghuyan, *Pb.*, C. 1732; ghuz, *Pb.*, T. 51; ghuzbe (Plantago), *Pb.*, *Pushtu*, P. 926; ghuzhbe (Alnus), *Pushtu*, A. 801; ghwá, *Pb.*, *Pushtu*, T. 51; ghwanza, *Pushtu*, C. 2035; ghwardza, *Pushtu*, C. 2035; ghwareshtaí, *Pushtu*, P. 1322; ghwarga, *Pushtu*, P. 618; ghweia, *Kumaon*, C. 1055; ghwhaz, *Pushtu*, T. 51; ghwi, *Hind.*, F. 156; ghwiṛa, *Hind.*, C. 1732.

Gi Gían, *Pb.*, P. 1242; gíánru, *Hind.*, C. 2033; giáshúk, *Pb.*, J. 78; gid, *Pb.*, F. 467; gidamrí, *Sind.*, T. 28; gidar, *Hind.*, F. 755; gídardák (Sageretia), *Kashmir*, S. 502; gídardák (Vitis), *Pb.*, V. 195; gidar-dák (Prunus), *Pb.*, P. 1316; gidar drak, *Hind.*, V. 195; gidar-rúkh, *Hind.*, M. 545; gidartamákú (Heliotro-pium), *Hind.*, *Pb.*, H. 100; gidar-tamákú (Verbascum), *Hind.*, *Pb.*, V. 64; giddah, *Pb.*, F. 467; giddi-kaoli, *Hind.*, F. 333, 350; gidhro, *Sind.*, C. 2263; gídí, *Pb.*, P. 1406; gidúri, *Sind.*, C. 1931, 1940; giggar, *Hind.*, Z. 270; gih, *Sind.*, T. 634; gil, *Hind.*, C. 1317; gila-gach, *Beng.*, E. 219; gilás, *Kumaon*, P. 1295, *Pb.*, P. 1297; gilheri, *Hind.*, F. 807; gili, *Beng.*, F. 341; gilla, *Beng.*, E. 219; gilland, *Beng.*, F. 359; gilo, *Hind.*, *Pb.*, T. 470; T. 483; giloe, *Beng.*, *Hind.*, T. 470; giloe, *Beng.*, T. 483; gilo-gularich, *Pb.*, T. 470, T. 483; giloi, *Kumaon*, T. 470, T. 483; gima, *Beng.*, E. 338; gin, *Pushtu*, E. 35; gingárú, *Hind.*, C. 2033; gingelin, *Hind.*, S. 1078; gingelli, *Hind.*, S. 1078; gingili, *Hind.*, S. 1078; gingli, *Hind.*, S. 1078; gini gawat, *Mar.*, P. 59; gini ghas, *Gus.*, P. 59; ginvan, *Hind.*, *Mar.*, O. 38; gíra, *Afg.*, A. 801; girari, *Uriya*, C. 725; girbútí, *Mar.*, A. 1614; girchhatra, *Pb.*, F. 725, M. 647; girgitti, *Hind.*, G. 271; giri, *Hind.*, I. 440; giringa, *Uriya*, P. 1397; girk, *Pb.*, S. 1024; girmálah, *Hind.*, *Duk.*, C. 756; girmi, *Beng.*, E. 338; giroli, *Mar.*, T. 470, T. 483; girta-kun-vár, *Beng.*, A. 829; girthan (Securinega), *Pb.*, S. 1024; gírthan (Sageretia), *Pb.*, S. 502; girúí, *Pb.*, P. 42; girya, *Hind.*, C. 1031; girya, *Hind.*, C. 1031; gish, *Hind.*, C. 629; gíthi, *Hind.*, D. 542; gitoran, *Hind.*, C. 416; gium, *Mar.*, T.

634; giúm, *Beng.*, *Hind.*, T. 634; giur, *Kashmir*, S. 1906; giúr, *Kashmir*, S. 544; gívían, *Pb.*, C. 1732.

Gl Gluru, *Guj.*, C. 919.

Gn Gniu, *Pb.*, C. 1003; gnua, *Uriya*, E. 593.

Go Goa, *Hind.*, M. 237; goa, *Pb.*, S. 1264; góaáchhi-phal, *Beng.*, P. 1343; goacháppi, *Uriya*, F. 311; goagari-lakei, *Mar.*, S. 2936; goalilata, *Beng.*, V. 217; goá-sál, *Hind.*, M. 545; gob, *Raj.*, O. 38; gobar-mowa, *Kumaon*, E. 208; gobay goru, *Beng.*, O. 565; gobhi, *Hind.*, E. 80; gobla, *Hind.*, *Kumaon*, F. 202; góbra, *Hind.*, S. 1939; gobura-natí, *Beng.*, A. 933; goburchamp, *Hind.*, P. 989; gocinda, *Beng.*, C. 2061; goda, *Beng.*, V. 159, V. 174; godadi, *Mar.*, I. 120; godávaj, *Guj.*, A. 430; godchabba, *C. P.*, *Hind.*, E. 186; godela, *Hind.*, C. 1927; goden, *Mar.*, C. 1931; gód-gadala, *Pb.*, S. 2841; gód-gúdála, *Pb.*, S. 2861; godhai, *Uriya*, F. 418; godhi, *Pb.*, M. 306; goen, *Hind.*, D. 238; goenjak, *Hind.*, D. 238; gogána, *Kumaon*, S. 896; goganda, *Hind.*, S. 896; gogar, *Hind.*, G. 124; gogarli, *Mar.*, G. 124; gogatti, *Kumaon*, A. 2; gogera, *Hind.*, D. 31; gogil, *Hind.*, B. 43; gogina, *Hind.*, *Kumaon*, S. 896; gogird, *Pb.*, S. 2999; gogí ság, *Pb.*, M. 105; gog·ah, *Mar.*, F. 564; gogsa (Mæsa), *Hind.*, M. 38; gogsa (Myrsine), *Hind.*, M. 919; gogsa (Rhamnus), *Hind.*, *Pb.*, R. 152; goha, *Beng.*, F. 362; goher, *Pb.*, S. 497; gohinla, *Pb.*, H. 13; gohnasarsón, *Hind.*, B. 833; gohum, *Mar.*, T. 634; goí, *Hind.*, O. 574; goín, *Hind.*, O. 574; goinda, *Mar.*, D. 628; goindú, *Mar.*, D. 628, goira, *Uriya*, A. 249; gojiálat·, *Beng.*, E. 80; gokaran, *Mar.*, C. 1403; gokarnamul, *Mar.*, C. 1403; gokarni, *Mar.*, C. 1403; gokharú, *Guj.*, T. 548; gokhru (Hygrophila), *Guj.*, H. 508; gokhrú (Tribulus), *Beng.*, *Guj.*, *Hind.*, *Mar.*, *Sind*, T. 548; gokhrudesi, *Pb.*, T. 544; gokhrú desi, *Pb.*, T. 548; gokhrú kalán, *Pb.*, X. 1; gokhrú kallán, *Sind*, X. 1; gokhulakanta, *Hind.*, H. 508; gokhuríkalen, *Hind.*, T. 544; gokhuru, *Beng.*, *Hind.*, T. 548; góki, *Hind.*, S. 959; gokru, *Pb.*, X. 1; gokshra, *Uriya*, T. 548; gókshura (Hygrophila), *Hind.*, H. 508; gok-shura (Tribulus), *Beng.*, T. 548; gokurna-bija, *Mar.*, C. 1403; gokurnamula, *Mar.*, C. 1403; gokurrah, *Hind.*, R. 124; gol (Portulaca), *Mar.*, P. 1179; gol (Saccharum), *Mar.*, S. 30; gol (Trema), *Mar.*, T. 522; gola bet, *Beng.*, C. 92; goladára, *Mar.*, S. 2837; golainchi, *Hind.*, P. 989; golak, *Beng.*, C. 115; gola methí, *Beng.*, C. 2585; gola-mohaní, *Beng.*, A. 914; golancha, *Hind.*, *Beng.*, F. 671; goláp, *Beng.*, R. 504; golápjám, *Uriya*, E. 432; golarú, *Sind*, C. 919; goldár, *Mar.*, S. 2837; goldarú, *Már.*, S. 2824; goldin, *Raj.*, A. 1149; gulka, *Kumaon*, B. 570; gol-kaddu (Benincasa), *Hind.*, *Pb.*, B. 430; gol-kaddu (Lagenaria), *Hind.*, *Pb.*, L. 30; gólkadú (Benincasa), *Mar.*, B. 430; gol kamíla, *Pb.*, G. 240; gol-kándra, *Hind.*,

M. 639; gol-kankra, *Hind.*, M. 639; gol-kuddú, *Sind*, B. 430; gollund, *Uriya*, F. 566; gol-mattar, *Hind.*, P. 885; gol-mirich, *Pb.*, P. 811; gól-morich, *Beng.*, P. 811; golpatta, *Beng.*, P. 582; gol-phal, *Beng.*, N. 163; gólra, *Raj.*, A. 1149; gol-shingra, *Beng.*, C. 812; gom, *Beng.*, T. 634; goma, *Dec.*, L. 323; gomed sannibh, *Hind.*, C. 621; gometti, *Mar.*, Z. 182; gomuk, *Beng.*, C. 2306; goncha, *Hind.*, M. 786; gond (Phœnix), *Pb.*, P. 555; gond (Typha), *Pb.*, T. 864, T. 875; gonda, *Beng.*, R. 237; gondála, *Mar.*, P. 874; gondan, *Mar.*, C. 1931; gondani, *Mar.*, C. 1944; gon-daúra, *Hind.*, M. 237; gondhona (Phyllanthus), *Uriya*, P. 632; gondhona (Premna), *Uriya*, P. 1239; gondi, *Hind.*, *Uriya*, C. 1931, *Hind.*, *Pb.*, C. 1944; gondní, *Hind.*, C. 1944; gondu (Cordia), *Raj.*, C. 1927; gondu (rhinoceros), *Beng.*, R. 235; gondui, *Mar.*, C. 1944; gondula, *Beng.*, P. 67; gongo, *Uriya*, N. 179; goni, *Beng.*, F. 470; gonjha, *Pb.*, M. 549; gonsali, *Mar.*, S. 556, L. 569; gonta, *Hind.*, *Mar.*, E. 565; gooloo, *Hind.*, S. 2850l; goolum, *Beng.*, E. 450; goorloo, *Hind.*, S. 2850; gophla, *Kumaon*, H. 304; gorabel, *Raj.*, P. 1401; goráchakra, *Beng.*, S. 785; gorádu, *Mar.*, *Dec.*, D. 534; gorádu, *Mar.*, D. 485; górakamalí, *Hind.*, A. 455; gorak-cháuliá, *Beng.*, S. 1714; gorakhaamli, *Guj.*, *Mar.*, A. 455; gorakh-amli, *Hind.*, A. 455; gorakh chinch, *Guj.*, A. 455; gorakhchincha, *Mar.*, A. 455; gorakh chintz, *Mar.*, A. 455; gorakhmundi (Sphœranthus), *Mar.*, S. 2518; gorakh múndi (Lippia), *Pb.*, L. 451; gorakhpámo, *Pb.*, H. 106; gorakhpánu, *Pb.*, H. 106; gorakh pánw, *Pb.*, C. 1781; gorak mundi, *Hind.*, S. 2518; goral, *Kumaon*, S. 1247; gora lana (Haloxylon), *Pb.*, H. 6; gora lane (Anabasis), *Pb.*, A. 1005; gorá láne (Salsola), *Pb.*, S. 596; goral-hé, *Pb.*, T. 434; gor-amli chora, *Hind.*, A. 455; gorán, *Beng.*, C. 964; goran álu, *Beng.*, D. 530; gora nebu, *Beng.*, C. 1286; goranimb, *Mar.*, M. 800; gora-tél, *Hind.*, A. 1263; górá-tíl, *Hind.*, A. 1263; gor aunsa, *Kumaon*, W. 6; gorbach (Acorus), *Hind.*, A. 430; gorbacha (leopard), *Dec.*, T. 434; gordi, *Mar.*, D. 534; gordon, *Hind.*, E. 531; goria (Carpinus), *Hind.*, C. 631; goria (Excæcaria), *Beng.*, E. 593; goria (antelope), *Hind.*, S. 1226; gorikachú, *Hind.*, C. 1732; gorissa, *Uriya*, F. 516; gorkan, *Mar.*, *Dec.*, D. 534; gorkatri, *Kashmir*, I. 119; gormi-kawat, *Uriya*, A. 658; goroma, *Beng.*, A. 1232; gorpadvel, *Mar.*, V. 217; gorpjiba, *Hind.*, B. 165; goru, *Beng.*, *Hind.*, O. 574; gorur-champa, *Beng.*, P. 989; gosam, *Hind.*, *Mar.*, S. 950; gosh-wára, *W. Afghan.*, P. 847; gossiporah, *Uriya*, F. 508; got, *Uriya*, Z. 290; gota begún, *Beng.*, S. 2313; gótá ganbá, *Hind.*, G. 66; gotáha, *Hind.*, Z. 290; gotho, *Hind.*, C. 596; goti, *Hind.*, *Mar.*, Z. 290; goting, *Mar.*, T. 293; gotoboro, *Uriya*, Z. 290; gouí-ním, *Dec.*, M. 393; goundhan, *Mar.*, D. 628; gourí-ním

Dec., M. 393 ; gourkassi, *Uriya*, B. 873 ; gourubati, *Uriya*, B. 304 ; gouti, *Beng.*, *Uriya*, F. 493 ; go-vágh, *Beng.*, T. 437 ; govarnellu, *Pb.*, B. 520 ; govila, *Beng.*, V. 213 ; gowaí mirchi, *Mar.*, C. 449 ; gowali, *Mar.*, G. 696 ; gowara, *Mar.*, C. 290 ; gowindi, *Mar.*, C. 416 ; gowli, *Mar.*, G. 696; gozang, *Pb.*, A. 1632.

Gr Grau, *Kashmir*, F. 788 ; gray, *Pb.*, B. 308; grelu, *Pb.*, F. 168 ; grim, *Pb.*, H. 375 ; gro-age, *Pb.*, *Sind*, F. 493, *Pb.*, F. 494 ; grui, *Pb.*, E. 475 ; grundi, *Hind.*, R. 369.

Gu Gua (Areca), *Beng.*, A. 1294 ; gua koli, *Uriya*, M. 3 ; guar (Dolichos), *Pb.*, D. 758 ; guár (Cyamopsis), *Hind.*, C. 2514 ; guara, *Beng.*, E. 48 ; gua supari, *Beng.*, A. 1330 ; guava, *Dec.*, P. 1343 ; guázárá-kha, *Pushtu*, S. 2374 ; gubábhul, *Mar.*, A. 217 ; gúch (Coriaria), *Pb.*, C. 1958 ; gúch (Viburnum), *Pb.*, V. 91, V. 96 ; gúdal, *Pb.*, X. 1 ; gúdá-pandú, *Dec.*, T. 600 ; gud batal, *Pb.*, R. 71 ; guddi kúm, *Pb.*, M. 324 ; gudél, *Dec.*, H. 227 ; gudgega, *Duk.*, C. 6 ; gudha-púrna, *Beng.*, B. 619 ; gúdi, *Pb.*, M. 324 ; gudji-curama, *Uriya*, F. 580 ; gudlei, *Pb.*, R. 162 ; gudtha, *Beng.*, F. 508 ; gudúa, *Hind.*, C. 2324; gudúríchá-kánda, *Mar.*, C. 2013; gugah-alli, *Uriya*, F. 486 ; gugal, *Guj.*, B. 43 ; gugal (Balsamoden-dron), *Beng.*, *Dec.*, *Guj.*, *Hind.*, *Sind*, B. 43, B. 62 ; gugal (Jurinea), *Pb.*, J. 111 ; gugala, *Beng.*, B. 62 ; gugali, *Guj.*, B. 771 ; gugani, *Beng.*, F. 342 ; gugar, *Sind*, B. 62 ; gúggal, *Hind.*, J. 104 ; guggala, *Mar.*, B. 43 ; gúggar, *Kumaon*, B. 771 ; guggilu, *Mar.*, S. 1656 ; guggul, *Beng.*, *Dec.*, *Guj.*, *Hind.*, *Sind*, B. 43 ; guggula, *Mar.*, B. 771 ; gugguladhup, *Mar.*, A. 670 ; gúgli, *Mar.*, F. 368, 548 ; gugu, *Hind.*, A. 567 ; gúgul (Canarium), *Mar.*, C. 285 ; gúgul (Callichrous), *Mar.*, F. 368 ; gúgul (Myrsine), *Pb.*, M. 910 ; guguli, *Beng.*, A. 1362 ; gúhbabúl, *Hind.*, *Duk.*, A. 217 ; guhera, *Hind.*, I. 497 ; guhria, *Hind.*, P. 41 ; gúhú, *Hind.*, S. 2850; guj, *Hind.*, E. 83 ; gujar, *Mar.*, A. 1591, 1592 ; gujráti eláchi, *Beng.*, E. 151 ; gu-kíkar, *Hind.*, *Duk.*, A. 217 ; gúl (Cichorium), *Pb.*, C. 1108 ; gúl (mo-lasses), *Guj.*, S. 372 ; gúl (Plantago), *Kashmir*, *Pb.*, P. 926 ; gúl (Rosa), *Mar.*, R. 501, *Afg.*, R. 508 ; gula, *Pb.*, P. 760 ; guláb, *Hind.*, *Mar.*, *Pb.*, *Afg.*, R. 501, 504, 508, 533 ; gúlá-bás, *Beng.*, M. 606 ; gul-abbas, *Mar.*, M. 606 ; gulabbás, *Hind.*, *Pb.*, M. 606 ; gulabi, *Pb.*, C. 2151 ; gulábí, *Pb.*, C. 2152 ; guláb-jáman, *Hind.*, E. 432 ; guláb-jamb, *Beng.*, E. 432 ; gulab-zurdi, *Hind.*, R. 504 ; gúl-achin, *Hind.*, P. 989 ; gulan, *Uriya*, F. 455 ; gulakri, *Pb.*, Z. 182 ; gulál, *Afg.*, R. 508 ; gulál-akali, *Guz.*, *Mar.*, R. 508 ; gulale-kukri, *Pb.*, Z. 182 ; gúla-methi, *Beng.*, C. 2622 ; gulanár, *Beng.*, *Pb.*, P. 1426 ; gulancha, *Beng.*, *Hind.*, *Kumaon*, *Uriya*, T. 470, T. 483 ; gular (Sterculia), *Hind.*, *Mar.*, S. 2850 ; gular (Ficus), *Hind.*, F. 179, 230 ; gulashruf, *Pb.*, R. 71 ; gulattí, *Pb.*, D. 758 ; gula-vela, *Mar.*, T. 470, T. 483 ; gula-vélí, *Mar.*, T. 470, T. 483 ; gul bahar, *Pb.*,

W. 106 ; gúlbánsa, *Hind.*, M. 606 ; gúlbas, *Uriya*, B. 523 ; gul-bél, *Dec.*, *Hind.*, T. 470 ; T. 483 ; gulbhaji, *Mar.*, M. 606 ; gulbodla, *Pb.*, S. 2861 ; gul-chandni, *Mar.*, I. 368 ; gulchéri, *Hind.*, *Mar.*, P. 1044; gúl-chíní, *Hind.*, *Dec.*, C. 1043 ; guldár (Cedrela), *Pb.*, C. 838 ; gúldar (Staphylea), *Hind.*, *Pb.*, S. 2678 ; gúl-daudí, *Beng.*, *Guj.*, *Hind.*, C. 1043, *Hind.*, C. 1047 ; gul daur, *Pb.*, W. 106 ; guldháwi, *Pb.*, W. 106 ; guldoda, *Pb.*, L. 312 ; gul-dupaharia, *Pb.*, P. 393 ; gule, *Pb.*, E. 475 ; gule-aabbás, *Hind.*, M. 606 ; gule anár, *Dec.*, P. 1426 ; gulesewati, *Mar.*, C. 1043 ; gúlgá, *Beng.*, N. 163 ; gul gollop, *Mar.*, G. 679 ; gulgoto, *Mar.*, T. 17 ; gulgul, *Pb.*, C. 1286 ; gúl-gúlli, *Pb.*, F. 521 ; guli *Mar.*, I. 145 ; gul-i-ajaib, *Pb.*, H. 224 ; gulibádáwurd, *Kash-mir*, C. 559 ; gul-i-gao-zabán, *Kashmir*, O. 168 ; gul-i-khadmi, *Afg.*, M. 115 ; gúlíli, *Pb.*, O. 160 ; gul-i-pista, *Hind.*, *Mar.*, P. 847 ; gul-i-surkh, *Pb.*, R. 504 ; gulja, *Turkistan*, S. 1276 ; gul-jáfari, *Mar.*, T. 17 ; guljafaripurnka, *Pb.*, L. 358 ; gul-jatil, *Mar.*, D. 271 ; guljháro, *Guz.*, T. 17 ; gul-kakra (Momordica), *Hind. & Beng.*, M. 634; gúl-kákrú (Po-dophyllum), *Pb.*, P. 1009 ; gul-kandar, *Pb.*, S. 2861 ; gulkhair, *Guj.*, *Mar.*, A. 880 ; gulkhairo, *Duk.*, *Hind.*, *Mar.*, A. 880 ; gul-khand, *Pb.*, R. 504 ; gulkhandar, *Mar.*, S. 2861 ; gúlkheir, *Hind.*, M. 115 ; gulkuk-ru, *Pb.*, Z. 182 ; gulla, *Pb.*, C. 2358 ; gulli (Sterculia), *Hind.*, S. 2850 ; gúlli (coral), *Dec.*, C. 1808 ; gúl-mendí, *Hind.*, I. 39 ; gúlmirch, *Hind.*, P. 811 ; gúlmirien, *Sind*, P. 811 ; gulmohr, *Mar.*, P. 1036 ; gulnár, *Hind.*, P. 1426 ; gulnari, *Hind.*, C. 840 ; gúlnashtar, *Pb.*, E. 356 ; gul-nasta-ran, *Afgh.*, R. 538 ; gul-nastran, *Afgh.*, R. 538 ; gulo, *Dec.*, T. 470, T. 483 ; guloe, *Mar.*, T. 470, T. 483 ; gulrai, *Pb.*, C. 2358 ; gulsakarí, *Hind.*, S. 1714 ; gulseotí, *Pb.*, R. 501 ; gulshabbá, *Hind.*, P. 1044 ; gulshabbo, *Hind.*, *Pb.*, P. 1044 ; gul-sham, *Hind.*, D. 5 ; gúl-tun, *Pb.*, C. 838 ; gulu (Sterculia), *Hind.*, *Mar.*, S. 2850 ; gúlú (Cannabis), *Pb.*, C. 331 ; gúlú (Gobius), *Hind.*, *Sind*, F. 455 ; gúlul, *Beng.*, D. 608, 648, 664 ; gú-lú-wah, *Pb.*, F. 455 ; gulvél, *Gus.*, T. 470, T. 483 ; gulwail, *Dec.*, *Mar.*, T. 470 ; T. 483 ; gúlwél, *Hind.*, T. 470 ; T. 483 ; gúma, *Hind.*, L. 269, L. 323 ; gúmár, *Beng.*, G. 287 ; gúmbar, *Beng.*, G. 287 ; gúmbhar, (*C.P.*), *Hind.*, G. 287 ; gumchí, *Duk.*, A. 51 ; gúmhár, *Pb.*, G. 287 ; gumi, *Beng.*, A. 640 ; gún (Æsculus), *Pb.*, A. 567 ; gún (Cressa), *Sind*, C. 2057 ; gún (Dioscorea), *Kumaon*, D. 503 ; gun-ácha, *Pb.*, R. 598 ; gunara, *Hind.*, P. 42 ; gunch (Abrus), *Beng.*, A. 51 ; gúnch (Bagarius), *Hind.*, F. 329 ; gúnchgaji, *Pb.*, M. 786 ; gunda, *Uriya*, F. 566 ; gunda-gilla, *Beng.*, B. 301 ; gundhak, *Hind.*, S. 2999 ; gundhun, *Beng.*, A. 767 ; gundi, *Guj.*, *Hind.*, *Mar.*, C. 1944 ; gundo, *Gus.*, C. 1931 ; gundo moto, *Gus.*, C. 1931 ; gundun, *Sind*, F. 510 ; gúngárí, *Pushtu*, R. 538 ; gungatiya, *Beng.*, D. 420 ; gungituri, *Uriya*, F. 366,

461 ; gunglay, *Mar.*, C. 1512 ; gungrú, *Pb.*, D. 503 ; gúngú, *Pushtu*, C. 225 ; gúngwah, *Pb.*, F. 368 ; gung-wah-ri, *Hind.*, F. 368 ; gúngwari, *Mar.*, F. 548 ; gunj, *Pb.*, D. 330 ; gunjá (Abrus), *Mar.*, A. 51 ; gúnja (Odina), *Hind.*, O. 38 ; gúnjan (Briedelia), *Mar.*, B. 868 ; gunkírí, *Pb.*, A. 501 ; gunober, *Hind.*, P. 746 ; gunpalos, *Pb.*, P. 1458, *Pushtu*, P. 1460 ; gunta, *Beng.*, F. 566 ; guntea, *Uriya*, F. 325, 487 ; gúnteah, *Beng.*, F. 477 ; gúnyún, *Kashmir*, *Pb.*, C. 681 ; gupkarí, *Uriya*, F. 477 ; gúr (Cannabis), *Hind.*, C. 331 ; gúr (Saccharum), *Beng.*, *Dec.*, *Hind.*, *Pb.*, S. 316, 318 ; gurach (Tinospera), *Beng.*, *Hind.*, T. 470, 483 ; guracha (Rubus), *Pb.*, R. 590, 598 ; gura-niya álu, *Beng.*, D. 530 ; gurar, *Hind.*, A. 717 ; gurázáh, *Afg.*, L. 533, *Pushtu*, L. 535 ; gúr-balchor-ák, *Afg.*, V. 10 ; gur-bárí, *Hind.*, A. 717 ; gur-begun, *Hind.*, *Beng.*, L. 596 ; gúr-bhanga, *Hind.*, C. 331 ; gúrbheli, *Kumaon*, G. 708 ; gúrbíání, *Pb.*, T. 376 ; gurcha, *Hind.*, *Kumaon*, T. 470, 483 ; gurcháwa, *Hind.*, E. 186 ; gúrchí, *Pb.*, F. 494 ; gúrdah, *Hind.*, F. 566 ; gurdal-shim, *Beng.*, D. 789 ; gur-dalu, *Kashmir*, *Pb.*, P. 1285 ; gurdub, *Hind.*, E. 184 ; guren, *Duk. Mar.*, A. 958 ; gúrganna, *Pb.*, V. 64 ; gurgú, *Kashmir*, *Pb.*, P. 833 ; gúrguli, *Pb.*, A. 1061 ; gur-gumna, *Pb.*, S. 738 ; gurgunna, *Pb.*, E. 270 ; gurgur (Coix), *Beng.*, C. 1686 ; gurgura (Reptonia), *Pb.*, *Pushtu*, R. 132 ; gurh, *Hind.*, S. 316 ; gurhul, *Hind.*, H. 261 ; guri (Stephegyne), *Hind.*, S. 2799 ; gúri (Colocasia), *Beng.*, C. 1732 ; guria, *Beng.*, K. 21 ; guriál, *Hind.*, B. 318, 356 ; gurikaram, *Hind.*, S. 2799 ; gúrin, *Pb.*, A. 1369 ; gurinda, *Pb.*, P. 1252 ; gurja, *Hind.*, G. 143 ; gurjun, *Gus.*, D. 701 ; gurkámái, *Beng.*, S. 2280 ; gurkámái, *Beng.*, S. 2299 ; gurkats, *Pb.*, D. 348 ; gurkur, *Hind.*, A. 717 ; gurlpata, *Kumaon*, S. 2223 ; gurlu, *Hind.*, C. 1686 ; gúr-mussureya, *Hind.*, A. 1215 ; gúr-ounsh, *Kumaon*, P. 486 ; gurragadi, *Hind.*, E. 186 ; gurrie, *Uriya*, F. 516 ; gurshagal, *Hind.*, D. 348 ; gurtákand, *Beng.*, C. 170 ; guruk mundi, *Pb.*, S. 2518 ; gúrúsh, *Kumaon*, P. 486 ; gurwa, *Pb.*, A. 1523 ; gussir, *Sind*, F. 578 ; gusvakendhu, *Uriya*, D. 582 ; gutáchin, *Mar.*, P. 989 ; gutchka, *Duk.*, C. 6 ; gúteah, *Beng.*, F. 477 ; gutea-shuk-china, *Beng.*, S. 2248 ; guter, *Kumaon*, Z. 231 ; gutí (Zizyphus), *Mar.*, Z. 290 ; gútí (Smilax), *Mar.*, S. 2252 ; guti tasar, *Beng.*, S. 1934 ; gutti, *Uriya*, F. 561 ; gútwel, *Mar.*, S. 2252 ; guvaini, *Hind.*, M. 910 ; guwár, *Guj.*, C. 2514 ; guwch, *Mar.*, F. 329 ; guya, *Kumaon*, V. 91 ; guya, *Kumaon*, V. 96 ; gúyabábúla, *Beng.*, A. 217.

Gw **Gwa** (Litsæa), *Pb.*, L. 474, 483 ; gwa (Pistacia), *Baluch.*, P. 847 ; gwala, *Hind.*, S. 1029 ; gwála darim, *Hind.*, C. 862 ; gwála-kakri, *Hind.*, *Kumaon*, M. 791 ; gwálam, *Hind.*, P. 1449 ; gwala-mehal, *Kumaon*, P. 1449 ; gwál-dakh, *Hind.*, R. 362 ; gwalidar, *Pb.*, D. 611 ; gwal kakri, *Pb.*, T. 576 ; gwana, *Baluch.*, P. 847 ; gwándish, *Pb.*, S. 763 ; gwayral,

Hind., B. 330 ; gwia, *Kumaon*, V. 91 ; gwiar, *Hind*, B. 356 ; gwira, *Mar.*, S. 2850 ; gwíya, *Hind.*, M. 545.

Ha **Haba**, *Mar.*, M. 299 ; habat ul khizra, *Pb.*, R. 323 ; habbíkáknaj, *Pb.*, P. 678 ; hab-bul-balasán, *Mar.*, B. 54 ; habbulbán, *Pb.*, M. 393 ; habhúl, *Pb.*, M. 921 ; hab-úl-ás, *Beng.*, *Pb.*, M. 921 ; hábul bálasán, *Mar.*, B. 54 ; hab-ul-kalkal, *Pb.*, C. 551 ; habul-kilkils, *Beng.*, P. 1426 ; habush, *Hind.*, P. 811 ; hádar, *Pb.*, R. 359 ; háda-varná, *Mar.*, C. 2039 ; haddú, *Hind.*, C. 1969 ; hadga, *Mar.*, S. 1186 ; hadjora-*Hind.*, V. 219 ; hadri, *Gus.*, *Hind.*, T. 361 ; hádru, *Mar.*, D. 628 ; hadu-karanda, *Mar.*, D. 494 ; hænu-greeb, *Beng.*, T. 612 ; hág, *Guj.*, *Hind.*, T. 361 ; hagai, *Pushtu*, F. 700, R. 251 ; haiza-ka-patta, *Pb.*, *Hind.*, K. 17 ; haji-lag-lag, *Hind.*, S. 2903 ; hajr-ul-ya húdi, *Pb.*, C. 489 ; hakik, *Pb.*, C. 321 ; hakna, *Hind.*, I. 141 ; haksha, *Pb.*, P. 1187 ; hakúch (Psoralea), *Beng.*, P. 1352 ; hákuch (Vernonia), *Beng.*, V. 73 ; hakúm, *Beng.*, *Hind.*, B. 28 ; hakún, *Beng.*, *Hind.*, B. 28 ; hala (Picrasma), *Pb.*, P. 693 ; hala (Rhus), *Pb.*, R. 323 ; halá (Cornus), *Hind.*, C. 1973 ; halada, *Gus.*, C. 2433 ; haladhwán, *Guj.*, A. 514 ; haladwail, *Mar.*, H. 285 ; halai, *Pb.*, R. 323 ; halang, *Hind.*, L. 283 ; hal-bambar, *Pb.*, H. 52 ; halda (Chloroxylon), *Mar.*, C. 1031 ; haldá (Terminalia), *Dec.*, T. 325 ; haldar (Curcuma), *Pb.*, C. 2433 ; haldi (Adina), *Hind.*, A. 514 ; haldí (Curcuma), *Hind.*, C. 2433 ; haldi-algusi-lutta, *Beng.*, C. 2508 ; haldí-gach, *Beng.*, C, 2007 ; haldimáti, *Hind.*, I. 472 ; haldu, *Hind.*, A. 514 ; halede, *Mar.*, C. 2433 ; halela, *Pb.*, T. 325 ; haleo, *Hind.*, *Pb.*, C. 1969 ; halepan, *Hind.*, S. 1024 ; hal-hal, *Dec.*, G. 753 ; halim, *Beng.*, *Dec.*, *Guj.*, *Hind.*, *Pb.*, L. 283 ; halinda, *Hind.*, C. 1221 ; halj, *Kashmir*, S. 1264 ; halja, *Pb.*, C. 2433 ; hal-ká-bij, *Pb.*, Z. 231 ; halkasá, *Beng.*, L. 323 ; hal-khusa, *Hind.*, L. 323 ; halrá, *Dec.*, T. 325 ; halsi, *Beng.*, A. 531 ; halu (Salvia), *Pb.*, S. 738 ; hálú (Impatiens), *Pb.*, I. 39 ; halud, *Beng.*, C. 2433 ; halwa, *Hind.*, A. 839 ; halwa, *Hind.*, C. 232 ; halyún, *Hind.*, A. 1567 ; hama, *Kashmir*, O. 258 ; hamáz, *Pb.*, S. 1694 ; hambar máya, *Pb.*, B. 496 ; hambúkh, *Pb.*, M. 881 ; hamer, *Pb.*, F. 693 ; hamra, *Gus.*, P. 1259 ; hanchu, *Pb.*, E. 475 ; hand, *Pb.*, C. 1108 ; handi, *Hind.*, F. 615 ; háne, *Kashmir*, A. 567 ; hangra, *Beng.*, S. 37 ; hangul, *Kashmir*, D. 225 ; hanjal, *Uriya*, T. 282 ; hanóch, *Pb.*, F. 700 ; hanspadi, *Guj.*, A. 498 ; hansráj, *Hind.*, A. 498, A. 510, *Guj.*, *Mar.*, A. 506 ; hanudún, *Kashmir*, A. 567 ; hanuman, *Uriya*, C. 1847 ; hanuz, *Pb.*, F. 700 ; hanzal, *Pb.*, C. 1211 ; haoul, *Kumaon*, B. 496 ; hápar máli, *Beng.*, V. 12 ; har (Terminalia), *Hind.*, *Pb.*, *Sind*, T. 325 ; har (Vitis), *Beng.*, V. 219 ; hár (Nyctanthes), *Hind.*, N. 179 ; hara, *Hind.*, I. 382 ; haragaura, *Uriya*, I. 39 ; haraira, *Hind.*, T. 325 ; haran, *Hind.*, S. 1226 ; harancha, *Beng.*, F. 512 ; haran-dorí, *Mar.*, D. 823 ; haran-khúri, *Hind.*,

l. 382; harar, *Pb.*, F. 671; harara, *Hind.*,
T. 325; harásingara, *Mar.*, N. 179;
harathi, *Beng.*, *Hind.*, O. 574; harbara,
Mar., C. 1061; har-bare-ká-sirká, *Duk.*,
C. 1061; hár-bhángá, *Beng.*, *Uriya*, V.
219; harcuch kanta, *Beng.*, A. 324;
hardá, *Mar.*, T. 325; hardi (Chloroxy-
lon), *Mar.*, C. 1031; hardi (Terminalia),
Gus., T. 325; hardu, *Hind.*, A. 514;
hardua, *Hind*, A. 514; harduli, *Mar.*,
O. 127; harein, *Pb.*, L. 474; harfarauri,
Hind., P. 627; hargesa, *Beng*, D. 428;
hargila, *Hind.*, S. 2906; harhuch, *Hind.*,
E. 213; harhuria, *Mar.*, C. 1367; hárí,
Pb., P. 1285; hariáli, *Hind.*, C. 2558;
harían, *Pb.*, P. 1285; haribával, *Guj.*,
A. 249; harido, *Uriya*, W. 131; haridra,
Uriya, T. 325; hariha, *Hind.*, T. 349;
harik, *Mar.*, P. 332; hari-múng, *Beng.*,
P. 496; harin, *Hind.*, S. 1226; hariná-
shúk-china, *Beng.*, S. 2245; harin hara
(Amoora), *Hind*, A. 988; harin hara
(Briedelia), *Beng.*, B. 873; harin khana,
Hind., A. 988; harin-pádí, *Pb.*, *Hind.*,
C. 1777; hariphul, *Beng.*, P. 627; harira,
Hind., T. 349; haríra, *Uriya*, T. 325;
háritáki, *Beng.*, T. 325, T. 349; haritála,
Mar., O. 242; haritha, *Pb.*, S. 808; har-
jeuri, *Hind.*, C. 1205; hárjorá, *Hind.*,
Mar., V. 219; harkai, *Mar.*, R. 57;
harkaya, *Mar.*, R. 57; harké, *Beng.*, I.
60; harki, *Beng.*, C. 1038; harkú ríkhali,
Pb., R. 335; harlá, *Dec.*, T. 325; harle,
Gus., T. 325; harlephúl, *Gus.*, T. 325;
harmal (Peganum), *Hind.*, P. 372; harmal (Vitis), *Hind.*, V. 228; harmala (Peganum), *Mar.*, P. 372; harna, *Hind.*, S.
1226; harnauli (Solanum), *Pb.*, S. 2345;
harnaulí (Ricinus), *Pb.*, R. 369; harni,
Hind., S. 1226; hárpîlé, *Hind.*, T. 325;
harra, *Beng.*, T. 349; harrá, *Hind.*, T.
325; harrana, *Hind.*, C. 1847; harráni,
Mar., D. 32; harrar, *Pb.*, T. 325; harreri,
Mar., A. 711; harri, *Hind*, M. 800;
harri múng, *Hind.*, P. 496; harru, *Hind.*,
C. 1969; harsankar, *Guj.*, *Hind.*, *Mar.*,
V. 219; harsinghar, *Hind.*, F. 671;
harsinghar, *Beng.*, *Hind.*, *Pb.*, N. 179;
hartál, *Hind.*, O. 242; hartal warki, *Pb.*,
O. 242; hartál wilayití, *Pb.*, O. 242; har-
tho, *Hind.*, S. 1024; harúntutia, *Pb.*, S.
3014; harwan, *Pb.*, T. 61; harwánh chota,
Pb., V. 116; harwari, *Uriya*, W. 106;
haryá, *Hind.*, O. 574; haryeli, *Mar.*, C.
2558; hasak, *Pb.*, T. 544; hásha, *Pb.*, T.
61; hasjora, *Beng.*, V. 219; hassan dhúp,
Pb., S. 2999; hastipata, *Mar.*, E. 80; hat,
Hind, H. 294; hatbaha, *Gus.*, H. 294;
hath (Holarrhena), *Gus.*, H. 294; háth
(Scindapsus), *Hind.*, S. 970; hatha jori,
Hind., *Pb.*, M. 308; háthi, *Hind.*, E.
83; háthi chingár, *Hind.*, A. 636; háthí
khatyán, *Duk.*, A. 455; hathi-sengar,
Hind., A. 603; hati, *Hind.*, E. 83; háti-
ánkusá, *Uriya*, P. 824; hatichinch, *Mar.*,
P. 900; hati-choke, *Beng.*, *Hind.*, C.
2556; hátisurá, *Beng.*, *Uriya*, H. 102;
hátmúl, *Pb.*, S. 3014; hatta-júrie, *Hind.*,
H. 102; hatta-súra, *Hind.*, H. 102; hat-
tian, *Hind.*, E. 289; hattí-charátté, *Mar*,
P. 363; hattichók, *Beng.*, *Hind.*, *Mar.*,
H. 88; hatti ghókrú, *Dec.*, P. 363; hatti
gumchí, *Duk.*, *Guj.*, A. 471; hattipipli,

Dec., S. 970; haulber, *Pb.*, J. 78; haulia,
Hind., P. 546; háur, *Hind.*, B. 496;
hauza, *Afg.*, A. 238; báwar, *Hind.*, D.
748; hazárdána, *Pb.*, E. 512. 549; hazar
maní, *Beng.*, *Hind.*, P. 673; hazár-ma-
sálah, *Duk.*, A. 1079.

He He-chi, *Afg.*, P. 1273; hedu, *Mar.*, A. 514;
hegu, *Raj.*, M. 717; hejurchei, *Beng.*, L.
266; helá, *Mar.*, T. 293; hellounda, *Mar.*,
C. 2494; helu, *Mar.*, V. 22; hembra, *Pb.*,
U. 4; hemságar, *Hind.*, *Beng.*, K. 14;
hen, *Raj.*, P. 77; héna, *Hind*, L. 126;
henzil, *Duk.*, C. 1211; hérán, *Sind*, R.
369; herar, *Hind.*, *Beng.*, F. 725; heriss,
Hind., G. 705; heru, *Pb.*, Q. 29; heta,
Mar., S. 1186; hetenuria, *Hind.*, T. 565;
hewar, *Hind.*, *Mar.*, A. 249.

Hh Hharaoli, *Mar.*, S. 2912

Hi Hier, *Hind.*, C. 1452; hijál, *Beng.*, B. 180;
hijala, *Beng.*, F. 512; hij-dáona, *Beng.*,
E. 520; hijlí-bádá m, *Beng.*, A. 1014;
hika gadi, *Hind.*, C. 1028; hikal, *Mar.*,
C. 780; hillúa, *Beng.*, A. 1567; hilsa,
Beng., F. 414; hilwa, *Hind.*, C. 293;
himagihira, *Gus.*, T. 325; himsí-míre,
Mar., P. 801; himu, *Hind.*, M. 775;
híndiagara, *Mar.*, A. 1251; hindí-bádam,
Dec., *Hind.*, T. 312; hindí-chobchini,
Hind., S. 2248; hindí-kaknaj, *Dec.*, 93;
W. 98; hindí-katérá, *Hind.*, C. 1512;
híndí-révand chíní, *Hind.*, R. 215; hindí-
sálsá, *Hind.*, H. 119; hindí-sana, *Hind.*,
C. 737; hindí-saná-ká-pát, *Hind.*, C. 737;
hindwana, *Hind.*, *Pb.*, C.1221; hing,
Mar., F. 70; hing, *Hind.*, *Guj.*, F. 76,
Afg., *Hind.*, F. 84; hinga, *Beng.*, A.
580; hingan, *Dec.*, *Mar.*, B. 13; hingana,
Mar., B. 13; hinganbet, *Dec.*, *Guj.*,
Mar., B. 13; hingat, *Hind.*, B. 13;
hingchá, *Beng.*, E. 213; hingen, *Hind.*,
B. 13; hinger, *Guj.*, B. 13; hingol
(Balanites), *Hind.*, B. 13; hingol
(mercury), *Hind.*, M. 473; hingon,
Beng., B. 13; hingot, *Hind.*, B. 13;
hingota, *Hind.*, B. 13; hingra, *Hind.*, F.
76, F. 84; hingra, *Mar.*, F. 84; hingu,
Hind., B. 13; hinjara, *Uriya*, B. 180;
hinna, *Pb.*, L. 126; hintál, *Beng.*, P. 582;
hírá, *Hind.*, D. 364; hírábol, *Beng.*,
Guj., B. 48; hiradá, *Mar.*, T. 325; hírá
dakhan, *Guj.*, *Mar.*, C. 68; hiradá-phúla,
Mar., T. 325; hírádókhí (Pterocarpus),
Hind, P. 1370; hıradukhí (Calamus),
Guj., *Hind.*, *Mar.*, C. 68; hira-hing,
Hind., F. 79; hírákashísh, *Mar.*, I. 481;
híra-kasis, *Guj.*, *Hind.*, I. 481; hírá-kos,
Beng., I. 481; hírákosís, *Beng.*, I. 481;
hirandodi, *Mar.*, D.823; hirandori, *Mar.*,
H. 442; hirankhorí, *Mar.*, C. 1858; hiran
paddí, *Pb.*, *Hind.*, C. 1777; hiranvel,
Mar., P. 4; hírda, *Mar.*, T. 325; hirdí,
Hind., T. 325; hirek, *Pb.*, D. 628;
hirmji, *Hind.*, I. 472; hirn, *Raj.*, I. 499;
hirn-pug, *Sind*, C. 1777; hirojah, *Hind.*,
C. 756; hirruseeah, *Hind*, E. 509; hirtiz,
Kashmir, E. 543; hiru, *Mar.*, S. 1226;
hirun, *Hind.*, S. 1226; hirvacha, *Mar.*,
A. 1079; hís, *Pb.*, C. 416; hisalu, *Hind.*,
Kumaon, R. 590; hísálu, *Kumaon*, R.
596; hísára, *Kashmir*, R. 590; hish alu,
Hind., *Kumaon*, R. 590; hital, *Beng.*,
P. 582; hitterlu, *Mar.*, S. 984; hiun-
garna, *Pb.*, C. 416, 427; hiúnsew, *Pb.*,

L. 332 ; hius, *Pb.*, C. 427 ; hívar, *Dec.*, *Mar.*, A. 249.

Hk Hkarshu, *Pb.*, Q. 35.

Ho Hodthai, *Duk.*, B. 57 ; hoglá, *Beng.*, T. 864 ; hokmchil, *Pb.*, P. 555 ; hol, *Afg.*, *Pb.*, M. 331 ; holashi, *Pb.*, R. 323 ; holda, *Dec.*, T. 361 ; holgeri, *Mar.*, H. 306 ; holma, *Pb.*, L. 224 ; holonda, *Uriya*, A. 514 ; homa, *Raj.*, P. 50 ; hong-i-saféd, *Kashmir*, A. 401 ; honglu, *Kashmir*, D. 225 ; honi, *Mar.*, P. 1370 ; honne, *Mar.*, P. 1370 ; honsá nebu, *Beng.*, C. 1270 ; hopári, *Guj.*, A. 1294 ; hórá, *Beng.*, T. 325 ; horina, *Beng.*, V. 159 ; horitál, *Beng.*, O. 242 ; hórjórá, *Beng.*, V. 219 ; horma, *Rajputana*, P. 48 ; hotai, *Duk.*, B. 57.

Hu Hub-ul-jaráb, *Pb.*, S. 2943 ; hub-ul-níl, *Kashmir*, *Sind*, I. 384 ; húdúm, *Beng.*, C. 102 ; huh, (*Sylhet*) *Beng.*, W. 58 ; hujírú, *Sind*, M. 562 ; hukaragadi, *Hind.*, H. 164 ; hukmandáz (Carpesium), *Pb.*, C. 624 ; hukmandáz (Rhynchospermum), *Pb.*, R. 348 ; húlagiri, *Mar.*, H. 317 ; húlá obúl, *Pb.*, R. 645 ; hulása, *Pb.*, R. 335 ; hulashing, *Pb.*, R. 318, 323 ; hulda, *Mar.*, C. 1031 ; huldi-kunj, *Beng.*, M. 656 ; húldi-múrga, *Beng.*, C. 873 ; hulga, *Mar.*, D. 758 ; hulgeri, *Mar.*, H. 306 ; hulhul, *Hind.*, *Pb.*, C. 1367 ; hulia, *Pb.*, *Kumaon*, P. 1456 ; hullowla, *Mar.*, P. 484 ; hulúg, *Pb.*, R. 318 ; húlugiri, *Mar.*, H. 317 ; húlúl, *Hind.*, G. 753 ; hum (Ephedra), *Afg.*, E. 228 ; hum (Periploca), *Afg.*, P. 419 ; húm (Fraxinus), *Pb.*, F. 693 ; húm (Polyalthia), *Mar.*, P. 1048 ; húm, *Mar.*, S. 487 ; húm (Saccopetalum), *Mar.*, S. 487 ; huma (Ephedra), *Afg.*, E. 228 ; huma (Periploca), *Afg.*, P. 419 ; húmb, *Mar.*, S. 487 ; humbu, *Pb.*, M. 878, 881 ; humcatchari, *Uriya*, F. 394 ; humúla, *Beng.*, K. 3 ; hun, *Dec.*, T. 270 ; húní, *Mar.*, P. 1370 ; hunráj, *Hind.*, P. 1078 ; hura, *Mar.*, S. 3062 ; hurdi, *Beng.*, M. 656 ; hurd-wahre, *Pb.*, F. 496 ; hurf, *Hind.*, L. 283 ; hurh, *Pb.*, T. 325 ; húrhúr (Cleome), *Hind.*, C. 1367 ; húrhúr (Gynandropsis), *Hind.*, G. 753 ; húrhuria, *Beng.*, C. 1367 ; G. 753 ; húrhúriya, *Mar.*, C. 1367 ; hurmal (Ficus), *Pb.*, F. 260 ; hurmal (Peganum), *Mar.*, P. 372 ; hurmaro, *Dec.*, *Mar.*, P. 372 ; hurmul, *Hind.*, *Pb.*, *Sind*, P. 372 ; hurná, *Mar.*, S. 833 ; hurna nús, *Hind.*, R. 51 ; hurnurgullar, *Beng.*, C. 103 ; hurrea, *Mar.*, V. 116 ; hurrea-kadu, *Sind*, L. 30 ; húruá, *Beng.*, S. 833 ; husket kangni, *Pb.*, S. 1212 ; hussí, *Hind.*, S. 2968 ; bussuk, *Hind.*, T. 548.

Ib Ibharankusha, *Beng.*, *Hind.*, *Pb.*, A. 1093.

Ig Igoreá, *Guj.*, B. 13.

Ij Ijal, *Mar.*, B. 180 ; ijál, *Hind.*, *Duk.*, B. 180 ; ijar, *Hind.*, B. 180 ; ijjul, *Hind.*, B. 193.

Ik Ik, *Beng.*, S. 30 ; ikar (Saccharum), *Hind.*, S. 6 ; ikara (bandicoot), *Beng.* R. 51 ; ikh, *Hind.*, *Pb.*, S. 30 ; ikhari (Saccharum), *Hind.*, S. 30 ; ikhari (Barleria), *Mar.*, B. 170 ; ikria, *Beng.*, R. 51.

Il Iláchi (Elettaria), *Beng.*, E. 151 ; iláchi bari (Amomum), *Pb.*, A. 968 ; ilachie (Mela-

leuca), *Beng.*, M. 342 ; ilál kalmí, *Beng.*, I. 368 ; iláyechi, *Hind.*, E. 151 ; iláyechidáné, *Hind.*, *Duk.*, A. 980 ; ilisha, *Beng.*, F. 414 ; iliya, *Mar.*, K. 42 ; illáchi, *Mar.*, E. 151 ; illar billar, *Pb.*, C. 1448 ; ilvá, *Hind.*, A. 824.

Im Ímar, *Pb.*, C. 629 ; imbri, *Pb.*, Q. 23 ; imlá, *Pb.*, Z. 280 ; imli, *Hind.*, *Pb.*, T. 28.

In Inai, *Hind.*, Q. 23 ; ináraun-maraghúne, *Pb.*, C. 1211 ; ind, *Dec.*, *Hind.*, R. 369 ; indai, *Mar.*, G. 243 ; indák, *Hind.*, C. 1950 ; indarjau (Holarrhena), *Pb.*, H. 294 ; indarjau (Hordeum), *Hind.*, H. 382 ; indarjou (Wrightia), *Beng.*, *Dec.*, *Hind.*, *Guj.*, *Mar.*, W. 122 ; inderjan tulkh, *Hind.*, H. 294 ; indrajab, *Beng.*, *Hind.*, H. 294 ; indrajau, *Beng.*, W. 122 ; indrak, *Guj.*, C. 1211 ; in-drámai, *Uriya*, O. 38 ; indra-maris, *Uriya*, A. 306 ; indravana, *Guj.*, *Mar.*, C. 1211 ; indrávena, *Guj.*, C. 1211 ; indrawan, *Duk.*, C. 1211 ; indráyan (Citrullus), *Beng.*, *Hind.*, *Mar.*, *Pb.*, C. 1211 ; indráyan (Cucumis), *Hind.*, C. 2130 ; indráyan (Trichosanthes), *Hind.*, *Kumaon*, T. 600 ; indráyan makal, *Hind.*, T. 600 ; indrendi, *Kumaon*, R. 369 ; indri, *Pb.*, Q. 23 ; indzar, *Pushtu*, G. 702 ; inganí, *Hind.*, *Pb.*, M. 131 ; ingar, *Dec.*, *Hind.*, *Mar.*, B. 180 ; in-ge-lí, *Beng.*, F. 496 ; inghra, *Hind.*, C. 2237 ; ingrach, *Pb.*, F. 678. 682 ; ingua, *Hind.*, B. 13 ; ingur, *Hind.*, *Pb.*, L. 143 ; injani, *Pb.*, M. 131 ; insra, *Pb.*, R. 588 ; inzar, *Afg.*, *Pb.*, F. 230 ; inzarra, *Pushtu*, G. 723 ; ínzarre, *Pb.*, G. 702.

Ip Ippícha-jháda, *Mar.*, B. 265.

Ir Ira, *Pb.*, V. 106 ; irai, *Mar.*, C. 162 ; iráo, *Sind*, L. 30 ; irisa (Iris), *Hind.*, I. 423, *Pb.*, I. 425 ; irisa (Narcissus), *Pb.*, N. 10 ; irkor, *Hind.*, I. 497 ; irr, *Pb.*, C. 1003 ; írrí, *Pb.*, C. 224, Q. 29 ; irsa, *Hind.*, I. 425 ; irum, *Hind.*, B. 520 ; irun, (Clerodendron), *Guj.*, C. 1386 ; irún (Zizyphus), *Hind.*, Z. 263.

Is Isabagóla, *Mar.*, P. 932 ; isabghol, *Pb.*, P. 932 ; isabgul, *Beng.*, *Hind.*, *Uriya*, P. 932 ; isafghol, *Pb.*, P. 919, *Kashmir*, *Pb.*, P. 926 ; *Guj.*, *Pb.*, *Pushtu*, P. 932 ; isamdhárí, *Duk.*, C. 1377 ; isapghol, *Dec.*, *Guj.*, *Mar.*, P. 932 ; isarbadí, *Deccan*, S. 1688 ; isarmul, *Beng.*, A. 1398 ; isbadí, *Deccan*, S. 1688 ; isbaghól, *Hind.*, P. 932 ; isband. *Hind.*, C. 225, P. 372 ; isbandlahour, *Hind.* P. 372 ; isbund, *Sind*, C. 1875 ; isbund-lahouri, *Pb.*, *Sind*, P. 372 ; iser, *Kashmir*, P. 1285 ; isfanáj, *Hind.*, *Mar.*, S. 2574 ; infanák, *Pb.*, S. 2574 ; isgand, *Pb.*, W. 98 ; isha-langulya, *Beng.*, H. 504 ; isharmúl, *Dec.*, *Hind.*, A. 1398 ; ishpagul, *Pushtu*, P. 932 ; ishpecha, *Pb.*, I. 384 ; ishwarg, *Sind*, R. 166 ; iskíl, *Hind.*, U. 39 ; iskin, *Pb.*, S. 1242 ; is-mógul, *Kashmir*, P. 932 ; ispaghól, *Pb.*, P. 932 ; ispaghúl. *Beng.*, *Hind.*, P. 932 ; ispand, *Mar.*, P. 372 ; ispanthan, *Baluch.*, P. 372 ; ispoghul, *Dec.*, P. 932 ; ispun, *Gus.*, P. 372 ; ispund. *Beng.*, R. 663 ; ispungur, *Sind*, P. 932 ; issharmúl, *Duk.*, A. 1398 ; issufgúl, *Hind.*, P. 932 ; isus, *Mar.*, S. 493.

It Itá, *Uriya*, S. 808 ; itola. *Hind.*, C. 756 ; itsaka, *Mar.*, C. 2522 ; itsit (Plectranthus), *Pb.*, P. 959 ; itsit (Trianthema), *Pb.*, T. 541.

lv Ivak-chhár, *Hind.*, C. 527.
lz Izkhir, *Mar.*, A. 1093.

Ja Jab, *Beng.*, H. 382; jablota, *Pb.*, J. 41; jadiya, *Hind.*, B. 812; jadu-pálan, *Beng.*, A. 1475; jaephal, *Beng.*, *Hind.*, M. 885; jáfrán, *Beng.*, C. 2083; jagat-madan, *Beng.*, J. 116; jaggam, *Mar.*, F. 603; F. 611; jaggarwah, *Hind.*, C. 756; jagla, *Kashmir*, S. 1260; jagrái, *Hind.*, B. 841; jagyadumar, *Beng.*, F. 179; jáhari-naral, *Mar.*, L. 511; jaharí sontakká, *Mar.*, A. 762; jáhi, *Hind.*, J. 18; jai (Avena), *Hind.*, *Pb.*, A. 1639; jái (Jasminum), *Pb.*, J. 24, 29; jaida-rumi, *Hind.*, C. 68; jaimangal, *Hind.*, S. 2884; jainghani, *Hind.*, A. 1035; jaint, *Hind.*, *Pb.*, S. 1174; jaintar, *Pb.*, S. 1166; jaiphal, *Mar.*, *Pb.*, M. 885; jáiphala, *Mar.*, M. 885; jaiphal-jari, *Hind.*, P. 1214; jaishbomodhu, *Beng.*, G. 278; jait, *Hind.*, *Pb.*, F. 671; jait, *Hind.*, *Mar.*, S. 1174; jajinga, *Hind.*, L. 556; jajyadomur, *Beng.*, F. 156; jál, *Beng.*, *Hind.*, *Pb.*, S. 705; 717; jala (leech), *Guj.*, L. 245; jála (Hydrilla), *Pb.*, H. 484; jalalakan, *Hind.*, V. 8; jal bágu, *Pb.*, V. 106; jaldáru, *Kumaon*, *Pb.*, P. 1285; jaldárú chúlí, *Pb.*, P. 1285; jalganti, *Beng.*, P. 57; jalghóza, *Afg.*, P. 746; jalgundya, *Raj.*, I. 503; jalidar (Cotoneaster), *Pb.*, C. 2021; jalidar (Grewia), *Pb.*, G. 723; jalidar (Rhamnus), *Pb.*, R. 157; jaliddhar (Celastrus), *Hind.*, C. 862; jaljatang-jhara, *Mar.*, S. 1223; jalkhumbi, *Hind.*, P. 874; jalkúkar, *Pb.*, T. 845; jal-kunbhí, *Hind.*, P. 874; jal-kutra, *Kumaon*, P. 1250; jallaur, *Hind.*, B. 342; jallur, *Hind.*, B. 342; jal manjar, *Mar.*, O. 534; jal-manus, *Mar.*, O. 534; jal ním (Herpestis), *Hind.*, H. 149; jal ním (Lippia), *Pb.*, L. 451; jal ním (Lycopus), *Kashmir*, L. 602; jalpai, *Beng.*, E. 65, E. 67; jalpipari, *Hind*, C. 1759; jalsawank, *Hind.*, P. 48; jám (Eugenia), *Beng.*, *Hind.*, E. 419; jám (Psidium), *Dec.*, P. 1343; jamalagota, *Mar.*, C. 2192; jamál-gota (Baliospermum), *Guj.*, *Mar.*, B. 28; jamál-gota (Croton), *Hind.*, C. 2192; jamál-goh (Jatropha), *Guj.*, *Pb.*, J. 41; jaman (Eugenia), *Hind.*, E. 404, 419; jaman (Prunus), *Kashmir*, P. 1316; jaman (Punica), *Pb* P. 1426; jamana, *Hind.*, *Kumaon*, P. 1316; jamára, *Pb.*, V. 96; jamawa, *Hind.*, E. 453; jámb, *Deccan*, E. 432; jámba (Xylia), *Mar.*, X. 16; jámba (Psidium), *Mar.*, P. 1343; jámbha, *Mar.*, X. 16; jambho, *Sind*, E. 327; jámbhul, *Mar.*, E. 419; jambíra, *Hind.*, *Duk.*, C. 1286; jambo, *Beng.*, E. 396; jambu (Eugenia), *Gus.*, E. 419; jambu (Xylia), *Hind.*, X. 16; jámbu, *Mar.*, E. 419; jámbudi, *Gus.*, E. 419; jambudo, *Mar.*, E. 419; jambul, *Mar.* E. 419; jambura, *Gus.*, E. 419; jamin, *Hind.*, E. 419; jamír, *Pb.*, F. 230; jamíra, *Kumaon*, C. 1286; jam-johara, *Hind.*, O. 251; jámkuli, *Uriya*, E. 419; jamla, *Hind.*, *Mar.*, T. 282; jammú, *Pb.*, P. 1316; jamna, *Pb.*, P. 1316; jamnai, *Pb.*, E. 327; jamni, phaláni, *Hind.*, E. 419; jamo (Eugenia), *Uriya*, E. 419;

jámo (agate), *Mar.*, C. 617; jamoa (Schleichera), *Pb.*, S. 950; jamoá (Elæodendron), *Pb.*, E. 73; jamrasi, *Hind.*, E. 73; jamroo, *Hind.*, E. 439; jamrúd, *Gus.*, P. 1343; jamrukh, *Gus.*, P. 1343; jamtí-kí-bel, *Hind.*, C. 1452; jamu (Eugenia), *Sind*, E. 432; jamu (Prunus), *Pb.*, P. 1316; jamul, *Hind.*, E. 419; jamun (Eugenia), *Hind.*, *Raj.*, E. 416, 419; jamun (Prunus), *Pb.*, P. 1316; jamuwa, *Hind.*, E. 73; janab, *Dec.*, C. 2105; janar, *Beng.*, Z. 50; jánbu, *Hind.*, E. 416; jand (Prosopis), *Hind.*, *Pb.*, P. 1259; jand (Zizyphus), *Pb.*, Z. 254; jánd, *Hind.*, Z. 254; jandar lamba, *Pb.*, A. 1383; jandí (Ballota), *Pb.*, B. 33; jandí (Prosopis), *Pb.*, P. 1259; janera, *Hind.*, S. 2424; janewar, *Hind.*, A. 1076; jangali-badam (Sterculia), *Mar.*, S. 2824; jangalí bádám (Canarium), *Hind.*, C. 279; jangalí bédáná, *Guj.*, C. 279; jangal ka parúngi, *Pb.*, Q. 70; jan-gama, *Mar.*, F. 603; jangi, *Pb.*, C. 1988; jangli (Desmodium), *Mar.*, D. 354; janglí (Corchorus), *Beng.*, C. 1858; jangliá krót, *Hind.*, *Beng.*, A. 737; janglí ákhrota, *Mar.*, A. 737; jangliám, *Mar.*, S. 2649; janglí-anár-ká-jhar, *Dec.*, H. 517; jangli angir, *Gus.*, F. 202; jangli-angur, *Hind.*, *Dec.*, V. 205; jangli-arandi, *Hind.*, J. 41; janglí-araudi, *Gus.*, J. 41; jangli-aushbah, *Hind.*, S. 2252; jangli-badám (Hydnocarpus), *Dec.*, *Hind.*, H. 468, 472; janglí-badám (Sterculia), *Hind.*, S. 2824; jangli-badám (Terminalia), *Dec.*, *Hind.*, *Mar.*, T. 312; janglí-bádáma, *Mar.*, T. 312; jangli-billi, *Hind.*, F. 761, T. 427; jangli-bukra, *Hind.*, D. 219; jangli-chachinda, *Kumaon*, T. 576; janglí-chanbélli, *Hind.*, H. 119; jangli chichinda, *Hind.*, T. 576, T. 596; janglí-chichóndá, *Hind.*, T. 576; janglídal, *Hind.*, H. 513; janglí dál-chíní, *Dec.*, C. 1158; janglí-dárchíní, *Hind.*, C. 1158; jangli eranda (Aleurites), *Mar*, A. 737; janglí-erandí (Jatropha), *Mar.*, J. 52; janglí-frast, *Pb.*, P. 1138; janglí-haldí, *Hind.*, C. 2406; jangli harhar, *Hind.*, C. 1367; jangli hulvul, *Duk.*, C. 1367; jangli-ilichi, *Duk.*, A. 976; jangli jaiphal, *Mar.*, M. 904; jangli jamalgota, *Hind.*, B. 28; janglí jháú, *Dec.*, C. 826; janglí-kali-mirch, *Dec.*, *Hind.*, T. 489; janglí-kálí-mirchí, *Hind.*, T. 489; janglí-kálí-mirchí-ki-jar-kí-chhál, *Dec.*, *Hind.*, T. 489; janglí-kánda, *Gus.*, *Mar.*, U. 39; janglí-karanj, *Dec.*, T. 361; jangli kariátu, *Gus.*, E. 338; jangli-kásní, *Duk.*, B. 546; jangli khajúr, *Hind.*, P. 551; jangli khulga, *Hind.*, O. 567; jangli kulthi, *Dec.*, *Mar.*, T. 270; janglí-kunvára, *Guj.*, A. 603; jangli-kuta, *Hind.*, F. 759; jangli malicha, *Raj.*, E. 188; janglí mandira, *Kumaon*, P. 45; jangli-matar, *Beng.*, L. 96; jangli-methí, *Mar.*, S. 1688; janglí-méthí, *Dec.*, *Hind.*, S. 1714; jangli-mohá, *Dec.*, *Hind.*, B. 220; jangli-mohvá, *Hind.*, B. 220; jangli-mudrika, *Mar.*, P. 82; jangli-múlí, *Hind.*, B. 546; janglí-nimbu, *Duk.*, A. 1601; janglí-padavala, *Mar.*, T. 576; jangl

Beng., T. 70; jháv-nu-jháda, *Guz.*, T.
70; jháw, *Hind.*, S. 959; jhenku indúr,
Beng., R. 51; jhera, *Dec.*, T. 293; jhijan,
Pb., S. 1166; jhijan, *Hind.*, S. 1174;
jhikráí, *Beng.*, V. 129; jhil (Chenopodium), *Sind*, C. 1003; jhila (Tephrosia),
Guz., T. 270; jhind, *Hind.*, A. 249;
jhinga, *Beng.*, *Pb.*, L. 556; jhingan,
Hind., O. 38; jhirang, *Mar.*, M. 800;
jhingri, *Hind.*, E. 186; jhinja, *Hind.*,
B. 318; jhinjhor, *Hind.*, E. 186; jhinkar,
Pb., D. 238; jhira, *Hind.*, L 300;
jhit, *Pb.*, S. 717; jhojhrú, *Pb.*, T. 270,
jhora, *Hind.*, J. 104; jhotá, *Pb.*, O.
558; jhotak, *Pb.*, H. 382; jhóto, *Uriya*,
C. 1879; jhula, *Pb.*, S. 1260; jhúngara,
Hind., P. 53; jhunjhuni-ankari, *Hind.*,
V. 112; jhusa, *Hind.*, E. 263; jhut,
Uriya, C. 1879; jhutela, *Kumaon*, P.
413; jhútó, *Uriya*, C. 1879.

Ji Jial, *Beng.*, O. 38; jíapota, *Pb.*, P. 1433;
jia púta, *Beng.*, *Hind.*, P. 1433; jibán,
Hind., O. 38; jidkar, *Pb.*, F. 624; jíjan,
Pushtu, C. 754; jikjik, *Pb.*, R. 533;
jikri, *Pb.*, D. 130; jil, *Sind*, I. 145;
jil-lung, *Beng.*, *Uriya*, F. 581; jilo,
Uriya, F. 558; jindí, *Pb.*, S. 2912;
jinga, *Beng.*, *Hind.*, *Mar.*, L. 556;
jingan, *Hind.*, O. 38; jinjrú, *Pb.*, L.
528; jíntí, *Pb.*, P. 1252; jintiána (Saxifraga), *Pb*, S. 924; jintiyána (Gentiana),
Mar., G. 167; jiol, *Beng.*, O 38;
jír, *Beng.*, O. 38; jira (Carum), *Beng.*,
C. 681; ira (Cuminum), *Beng.*, C 2339;
jirani, *Mar.*, M. 800; jíra-utmi, *Guz.*,
C. 2339; jíre gire, *Mar.*, C. 2339; jirka,
Pb., P. 688; jirrag, *Kumaon*, P. 688;
jírú, *Guz.*, C. 2339; jis, *Kashmir*, F.
521; jit, *Pb.*, C. 224; jít, *Pb.*, S. 717; jítí,
Beng., M. 299; jiunti, *Pb.*, C. 1113; jiva,
Hind., P. 1433; jival, *Beng.*, O. 38;
jivputrak, *Hind.*, *Mar.*, P. 1433; jiwa,
Beng., H. 227; jiyal, *Beng.*, O. 38;
jiyaputra, *Pb.*, P. 1433.

Jo Joár, *Hind.*, *Mar.*, *Pb.*, S. 2424; joar-
aktse, *Pb.*, M. 881; joba, *Beng.*, H. 227;
jogiya-hísálu, *Kumaon*, R. 590; jogiya-
hísálu, *Hind.*, R. 590; jogmodon, *Beng.*,
J. 116; johra, *Mar.*, F. 361; jojre, *Pb.*,
X. 1; jók, *Hind.*, L 245; jo-kul, *Sind*,
M. 756; jonár, *Beng.*, Z. 50; jondhala
(Sorghum), *Mar.*, S. 2424; jondha
(Coix), *Mar.*, C. 1686; jondhariya, *Hind.*,
P. 384; jondla, *Dec.*, *Hind.*, *Mar.*, S.
2424; jondri, *Mar.*, A. 1218; jongli piaáj,
Beng., U. 39; jonir, *Guj.*, F. 558;
jonk, *Beng.*, *Hind.*, L. 245; jonkapha,
Hind., H. 92; jon-khár, *Hind.*, C. 527;
jóo, *Beng.*, H. 382; joon-gah, *Sind*,
F. 357; jora, *Beng.*, F. 357; jorah,
Mar., F. 395; joti, *Hind.*, P. 1433; jótri,
Beng., M. 885; joufra, *Beng.*, C. 2175;
joutrí, *Dec.*, M. 885; jouz-ul-maindal,
Sind, R. 1; jowa, *Beng.*, B. 142;
jowádi manjúr, *Mar.*, T. 444; jowa
khar, *Hind.*, H. 382; jowan, *Beng.*,
C. 691; jowár, *Afg.*, *Guj.*, *Hind.*,
Kumaon, S. 2424; jowári, *Dec.*, *Hind.*,
Mar., S. 2424; jowasa, *Raj.*, F. 2; jow-
thak-thak, *Afg.*, S. 1015.

Ju Juár (cattle), *Hind.*, O. 574; juár (Sorghum), *Beng.*, *Hind.*, S. 2424; júar
(Zea), *Pb.*, Z. 50; juára, *Hind.*, O.

574; juari (Jasminum), *Pb.*, J. 24; juárí
(Sorghum), *Mar.*, S. 2424; jubbi-cowri,
Uriya, F. 477; judwar, *Hind.*, D. 253;
júfa, *Sind*, N. 56; ju-gar, *Hind.*, F.
494; jugni, *Pb.*, M. 131; juhí, *Pb.*, D.
269; júi, *Beng.*, P. 338; júi-pana, *Beng.*,
R. 231; jui-pani, *Hind.*, R. 231; júk,
Pb., I. 39; julnár, *Hind.*, P. 1426; júl-
palum, *Hind.*, R. 645; jum (Elæodendron), *Hind.*, E. 73; júm (Garuga),
Beng., G. 143; jumla, *Pb.*, T. 282;
jumnapari, *Hind.*, F. 758; junala,
Kumaon, Z. 50; júnala, *Hind.*, Z. 50;
júnali, *Kumaon*, S. 2424; jundri, *Hind.*,
S. 2424; jundúrí, *Pb.*, F. 350; jungl,
Mar., T. 875; jungla, *Beng.*, F. 451;
jungli-am, *Dec.*, S. 2649; jungli-badam,
Mar., S. 2824; jungli-bendi, *Mar.*, T.
372; jungli-methí, *Guz.*, S. 1688; junglí-
mullí, *Duk.*, B. 546; jungli-palak, *Pb.*,
R. 645; junka, *Beng.*, S. 1699; jún-lí-kálí-
mirchí, *Mar.*, T. 489; junri (Sorghum),
Hind., *Pb*, S. 2424; junri (Zea), *Hind.*,
Z. 50; juntiyánah, *Dec.*, G. 167; junvásá,
Hind., C. 224; junvásá, *Hind.*, *Mar.*,
A. 745; junvásá, *Pb.*, C. 224; junvásá,
Mar., C 224; júríjur, *Sind*, C. 826;
jurkunkundalu, *Hind.*, G. 213; juti
(Murraya), *Hind.*, M. 797; juti (Pu-
tranjiva), *Hind.*, P. 1433; jutru, *Pb.*,
M. 910; jutuk, *Dec*, *Hind.*, D. 9; jutup,
Dec., D. 9; juvaní, *Beng.*, C. 691;
juvari jondhla, *Hind.*, S. 2424; juvashur,
Hind., O. 189; juwa, *Beng.*, H. 227;
juwásá, *Hind.*, *Mar.*, A. 745; juwásá,
Hind., *Mar.*, *Pb.*, C. 224; juwashur,
Mar., O. 189.

Jw Jwarán-tika, *Hind.*, Z. 9.
Jy Jy-chee, *Beng.*, E. 505.

Ka Ka (Sorghum), *Pb.*, S. 2424; ká (Juglans),
Pb., J. 61; kabába, *Pb.*, Z. 9; kábáb-
chíní, *Beng.*, *Dec.*, *Guj.*, *Hind.*,
Mar., P. 801; kabaipipal, *Hind.*, *Kumaon*, F 265; kaban, *Mar.*, D. 292;
kabar, *Mar.*, C. 431; kabar (Ficus),
Hind., F., 265; kabar (Salvadora), *Sind*,
C. 224; kabarra, *Afg.*, *Pb.*, C. 431;
kabawa, *Afg.*, C. 431; kabbar, *Hind.*,
Pb., C. 224; kabbar (Cynodon), *Pb.*, C.
2558; kabbar (Salvadora), *Sind*, S. 705,
S. 717; kabbu, *Mar.*, S. 30; kabdai,
Hind., P. 79; kabit, *Mar.*, F. 53; kábli
bakla, *Pb.*, V. 108; kabonan, *Raj.*, P.
663; ka-botang, *Pb.*, J. 61; kabra,
Hind., *Pb.*, C. 431; kabri, *Guj.*, C. 643;
kabsún, *Hind.*, B. 856; kábúd, *Pb.*, S.
855; kabuda, *Afg.*, P. 1159; kábulí
kaddú, *Pb.*, L. 30; kábulí mastaki,
Hind., *Mar.*, P. 847; kabútar-ka-jhár,
Dec., R. 231; kach (catechu), *Pb.*, A.
181; kach (Daucus), *Kashmir*, D. 173;
kacha chíní, *Hind.*; S. 375; kachakra,
Hind., *Guz.*, R. 97; kachal, *Kashmir*
A. 17, *Pb.*, A. 17; kachálú (Colocasia)
Pb., C. 1732; kachálu (Saxifraga), *Pb.*,
S. 924; kachám, *Pb.*, H. 324; kachan.
Kashmir, *Pb.*, A. 17; kachen, *Pb.*,
M. 393; kachera, *Mar.*, S. 977; kachhi,
Beng., F. 530; kachhur, *Hind.*, C. 2514;
kachia udal, *Beng.*, H. 221; kachír
(Cornus), *Hind.*, C. 1969; kachír (Pinus),
Pb., P. 737; kachki, *Guj.*, C. 6; kach-

kula, *Beng., Hind.,* M. 811 ; kách-kúri, *Dec.,* M. 786 ; kachlei, *Hind., Pb.,* T. 61 ; kachlora, *Hind., Mar.,* P. 806 ; káchmách, *Pb.,* S. 2299 ; kachnál, *Dec., Hind.,* B. 295, *Hind.,* B. 318 ; kachnár, *Dec., Hind.,* B. 295, *Hind.,* B. 334, 356 ; kachola, *Afg.,* D. 166 ; kachra, *Hind.,* C. 2274 ; káchrádám, *Beng.,* C. 1748 ; kachras, *Hind.,* S. 368 ; kachri (Holarrhenna), *Hind., Kumaon,* H., 294 ; kachrí (Cucumis), *Pb.,* C. 2306 ; kachu, *Beng.,* C. 1739 ; kachú, *Beng., Hind., Mar.,* C. 1732 ; kachúr (Hedychium), *Pb.,* H. 59 ; kachúr (Zingiber), *Pb., Z.* 225 ; kachura, *Beng., Hind., Mar.,* C. 2499 ; kachúria cháll, *Uriya,* W. 22 ; kachúr-kacha, *Hind.,* H. 59 ; kachur-kachu, *Pb.,* H. 59 ; kachwassal, *Pb.,* U. 39 ; kada, *Gus.,* S. 2850 ; kadai, *Mar.,* S. 2850 ; kadal, *Hind.,* D. 628 ; kadali, *Mar.,* M. 811 ; kadam, *Beng., Hind., Mar.,* A. 1192 ; kadamb (Anthocephalus), *Guj., Hind., Mar.,* A.1192 ; kadamb (Stephegyne), *Mar.,* S. 2799 ; kadamba, *Mar.,* A. 1192 ; kadambo, *Uriya,* A. 1192 ; kadámí, *Hind.,* A. 514 ; kadanda, *Pb.,* V. 64 ; kadash niangna mandrí, *Pb.,* R. 359 ; kada-todali, *Beng.,* T. 489 ; kadavanchi, *Mar.,* M. 637 ; kadavi (sámbar), *Mar.,* D. 240 ; kadavi (Swertia), *Mar.,* S. 3025 ; kadavi-nai, *Mar.,* E. 338 ; kadavinayi, *Mar.,* E. 217 ; kaddam, *Hind., Mar.,* F. 671 ; kaddam, *Hind., Mar.,* S. 2799 ; kaddú (Lagenaria), *Hind., Sind,* L. 30 ; kaddú (Cucurbita), *Hind.,* C. 2325 ; kaddukankri, *Mar.,* C. 1211 ; kadenrú, *Pb.,* T. 93 ; kadera, *Pb.,* I. 14 ; kadewar, *Pb.,* C. 862 ; kadhi, *Gus.,* M. 800 ; kadímah, *Beng., Hind.,* C. 2331 ; kadivi, *Mar.,* D. 240 ; kadloli, *Pb.,* H. 52 ; kadotrí, *Sind,* T. 569 ; kadu (Picrorhiza), *Guj.,* P. 700 ; kadú (Cucurbita), *Hind.,* C. 2316 ; kadú (Swertia), *Mar.,* S. 3023 ; kadú bhopalá, *Mar.,* L. 30 ; kadú-dodaká, *Mar.,* L. 563 ; kadu-dorka, *Mar.,* L., 563 ; kadu-inderjao (Wrightia), *Mar.,* W. 131 ; kadú-indrajou (Holarrhena) *Mar.,* H. 294 ; kadukavata, *Mar.,* H. 472 ; kadu khajur, *Gus.,* M. 412 ; kadu khajur, *Mar.,* M. 363 ; kadú padavala, *Mar.,* T. 576 ; kadú-sirola, *Mar.,* L. 563 ; kadu vrindavana, *Mar.,* C. 1211 ; kadval, *Mar.,* S. 2424 ; kadvoindarjou, *Gus.,* H. 294 ; kadvo-jiri, *Gus.,* V. 73 ; káephal, *Dec., Hind., Sind.,* M. 869 ; kaff, *Pb.,* G. 186 ; kaffí, *Kumaon,* kafi, *Pb.,* G. 186 ; kafrí-murich, *Beng., Hind ,* C. 464 ; kafsún, *Hind.,* B. 856 ; káfúr, *Hind.,* C. 257 ; káfúr-ká-pát, *Hind.,* M. 485 ; kafúr-ká-pattá, *Dec., Mar.,* M. 485 ; kag, *Pb.,* B. 131 ; kagara, *Beng., Hind., Mar.,* S. 49 ; kagdana chhatra, *Guj.,* A. 590, F. 725 ; kaghák, *Hind.,* R. 362 ; kaghania, *Hind., Pb.,* S. 2678 ; kagh dák, *Pb.,* R. 365 ; kaghzi nímbú, *Beng.,* C. 1296 ; kagoha, *Kumaon,* F. 202 ; kágphala, *Pb.,* S. 2943 ; kágsari, *Pb.,* D. 130 ; kágsha (Ficus), *Hind., Kumaon,* F. 202 ; kágsha (Cornus), *Hind.,* C. 1969 ; kagshi (Cornus), *Pb.,* C. 1973 ; kagshi, *Kumaon,* C. 1969 ; kagshi (Villebrunia), *Kumaon,* V. 131 ; kagúji-nebu,

Beng., C. 1296 ; kagya, *Hind.,* C. 1028 ; kahbang, *Pb.,* V. 149 ; káhi, *Pb.,* S. 49 ; kahí guláb ghure, *Pushtu,* R. 538 ; kahi kahela, *Pb.,* M. 869 ; kahimal, *Hind.,* F. 216 ; káhi-máti, *Hind.,* I. 481 ; káhi-sabz, *Hind.,* I. 481 ; káhi-saféd, *Hind.,* I. 481 ; kahi-siyá, *Hind.,* I. 481 ; kahi-zard, *Hind.,* I. 481 ; kahli-bundahni, *Uriya,* F. 328 ; kahlí-poi, *Uriya,* F. 328 ; kahm, *Pb., Sind,* F. 493 ; kahruba (Vateria), *Hind.,* V. 31 ; kahrubá (amber), *Dec., Hind.,* A. 955 ; káh taroí, *Hind.,* L. 556 ; káhú (Lactuca), *Beng., Hind., Pb., Sind,* L. 21 ; káhu (Saccharum), *Sind,* S. 49 ; kahú (Terminalia), *Beng., Hind.,* T. 282 ; kahúa, *Hind.,* T. 282 ; kahwa, *Hind., Mar.,* C. 1641 ; kái (Carpinus), *Pb.,* C. 631 ; kái (Celtis), *Pb.,* C. 881 ; kái (Ulmus), *Pb.,* U. 4 ; kaiar, *Kashmir, Pb.,* P. 737 ; kaiger, *Mar.,* A. 229 ; kaikar, *Hind.,* G. 143 ; kaikra, *Hind.,* G. 143 ; kaikun, *Hind., Mar.,* F. 615 ; kail (Abies), *Kumaon,* A. 17 ; kail (Pinus), *Hind.,* P. 737 ; kail, (Garhwal), *Hind.,* A. 17 ; kaim (Ficus), *Hind., Mar.,* F. 216 ; kaim (Stephegyne), *Hind.,* S. 2799 ; kaim-bil, *Kashmir,* M. 71 ; kaimil, *Hind.,* O. 38 ; kaimu, *Hind.,* B. 330 ; káin, *Pb.,* U. 4 ; kaincho, *Uriya,* M. 786 ; kanri, *Hind.,* D. 542 ; kari, *Hind.,* H. 523 ; kainshing, *Kumaon,* A. 328 ; kainth, *Pb.,* P. 1466 ; kaiphal, *Mar.,* M. 869, M. 904 ; káiphal, *Beng., Dec., Hind., Pb., Sind,* M. 869 ; kaira, *Gus.,* M. 583 ; kait, *Beng., Hind., Pb.,* F. 53 ; kaith, *Hind.,* F. 53 ; káiun (Faba), *Kashmir,* F. 1 ; káiún (Vicia), *Kashmir,* V. 108 ; kaiwal, *Pb.,* C. 846 ; kai-zabán, *Pb.,* R. 251 ; kájali, *Mar.,* C. 1403 ; kajar, *Beng.,* P. 588 ; kájar-wel, *Mar.,* S. 2936 ; kajei, *Pushtu,* R. 538 ; kajirah, *Beng.,* C. 637 ; kajoli, *Beng.,* F. 307 ; kajra (Prosopis), *Raj.,* P. 1259 ; kajra (Strychnos), *Hind., Mar., Pb.,* S. 2943 ; kajrauta, *Hind.,* M. 545 ; kájú, *Beng., Dec., Guj., Hind., Mar.,* A. 1014 ; kájúcha-bi, *Mar.,* A. 1014 ; káju-kalíyá, *Mar.,* A. 1014 ; kajúli, *Beng.,* S. 30 ; kajura, *Pushtu,* P. 555 ; kak (Daphne), *Pb.,* D. 130 ; kák (Ficus), *Ph.,* F. 230 ; kákáchiá, *Guj.,* C. 6 ; kakachia-kerundi, *Uriya,* F. 354 ; kakad (Flacourtia), *Mar.,* F. 615 ; kákad (Garuga), *Mar.,* G. 143 ; kakadashingi, *Mar.,* R. 323 ; kakada-shingi, *Mar.,* R. 323 ; kákadi, *Mar.,* C. 2278 ; kákamári, *Beng.,* A. 1038 ; kakammal, *Pb.,* F. 179 ; kákaphala, *Mar.,* A. 1038 ; kakar (Cervulus), *Hind.,* D. 219 ; kakar (Pistacia), *Pb.,* P. 833 ; kákara-singi, *Dec.,* R. 323 ; kakari, *Gus.,* C. 2287 ; kakariya, *Hind.,* E. 186 ; kakaróndá, *Hind.,* B. 540 ; kakarsing, *Hind.,* R. 323 ; kákatundí, *Mar.,* A. 1558 ; kakdi, *Dec.,* C. 2278, *Mar.,* C. 2287 ; kak-dumar, *Beng.,* F. 202 ; kakei (Flacourtia), *Mar.,* F. 615 ; kakei (Pteris), *Pb.,* P. 1359 ; kaker, *Mar.,* F. 615 ; kakhan, *Mar.,* S. 705 ; kákhan, *Mar.,* S. 717 ; kakhash, *Pb.,* P. 1359 ; kakhum, *Pb.,* M. 910 ; kaki (Grewia), *Hind.,* G. 679 ; káki (Citrus), *Beng.,* C. 1234 ; kákjanghá, *Hind.,*

Beng., L. 229; kakkar, *Kashmir*, *Pb.*, P. 833; kakkar-tamáku, *Pb.*, N. 98; kakkeran (Pistacia), *Pb.*, P. 833; kakkeran (Rhus), *Pb.*, R. 318; kakkrangche, *Pb.*, P. 833; kakkrei, *Pb.*, P. 833; kakkrein, *Pb.*, R. 316; kakkrín, *Pb.*, R. 323; kakkru, *Pb.*, A. 344; kakla, *Pb.*, P. 833; kákmáchi, *Beng.*, S. 2299; kákmári, *Hind.*, *Duk.*, A. 1038; kaknai, *Uriya*, C. 2287; káknaj (Physalis), *Pb.*, P. 678; káknaj (Withania), *Mar.*, W. 93; kakni (Cucumis), *Hind.*, C. 2278; kakni (Setaria), *Beng.*, *Dec.*, *Hind.*, S. 1212; kak-nuj, *Beng.*, W. 98; kakoa, *Pb.*, F. 615; kakodumar, *Beng.*, F. 202; kákódúmbári, *Pb.*, G. 287; kakor, *Hind.*, Z. 290; kákphal, *Mar.*, A. 1038; kákra (Bruguiera), *Beng.*, B. 898; kakra (Gelonium), *Uriya*, G. 156; kákrá (Pistacia), *Guj.*, *Hind.*, P. 833; kákrá (Podophyllum), *Pb.*, P. 1009, *Hind.*, *Mar.*, P. 833; kakrain, *Pb.*, P. 833; kákrá singí (Pistacia), *Guj.*, *Hind.*, *Mar.*, *Pb.*, P. 833; kákrá singí (Rhus), *Hind.*, R. 323; kakra sringi (Pistacia), *Beng.*, P. 833; kákrá sringi (Rhus), *Beng.*, R. 323; kakri (Capparis), *Pb.*, C. 431; kakri (Cucumis), *Beng.*, *Hind.*, *Mar.*, C. 2278, 2287; 2306; *Pb.*, C. 2306; kakria (Butea), *Hind.*, B. 944; kakria (Lagerstrœmia), *Guj.*, L. 55; kákrol, *Hind.*, *Beng.*, M. 634; kakró ndá, *Hind.*, B. 546; kakshama, *Pb.*, V. 73; kaksh kachúr, *Hind.*, C. 1969; káktundí, *Pb.*, A. 1558; kakua, *Pb.*, A. 1332; kakún (Securinega), *Pb.*, S. 1024; kákun (Setaria), *Beng.*, *Dec.*, kákun, *Hind.*, S. 1212; kakunda, *Uriya*, F. 496; kákundañrangul, *Hind.*, C. 854; kákúr, *Beng.*, C. 2278; kakúri, *Hind.*, L. 474; kákuriya, *Uriya*, E. 166; kakursinghi, *Pb.*, R. 323; kal, *Pb.*, G. 213; kala (Indigofera), *Hind.*, I. 119; kala (Musa), *Beng.*, M. 811; kala (antelope), *Hind.*, S. 1226; kala (Shorea), *Pb.*, S. 1656; kala-adulsa, *Mar.*, J. 116; kála-akhi, *Pb.*, R. 613; kalá akolá, *Mar.*, A. 681; kala aunsa, *Kumaon*, W. 6; kála-bánsa, *Hind.*, B. 165; kala-batta, *Beng.*, F. 409; kala-battali, *Uriya*, F. 465; kalabeinse, *Hind.*, *Uriya*, F. 466; kálábís, *Pb.*, H. 281; kálabísa, *Pb.*, H. 277; kala chakma, *Beng.*, Q. 18; kálá-dámar, *Hind.*, *Beng.*, *Dec.*, S. 1682; kálá dammar, *Beng.*, *Guj.*, *Hind.*, C. 285; kála dáná (Ipomæa), *Guj.*, I. 384; kálá-dánah, *Hind.*, I. 384; kálá-dánah, *Beng.*, I. 384; kálá dánah (Nigella), *Hind.*, N. 158; káládánah (Ipomæa), *Mar.*, I. 384; kala dhatúrá, *Dec.*, *Guj.*, *Hind.*, *Mar.*, D. 151; kala dhaukra, *Hind.*, A. 1153; kala dhuturá, *Beng.*, D. 151; kaládri, *Pb.*, H. 324; kalafnáth, *Duk.*, A. 1064; kalagori, *Mar.*, S. 2876; kalahád, *Guj.*, B. 540; kálá haldí, *Beng.*, C. 2422; kála hísálu, *Kumaon*, R. 598, 613; kalai, *Pb.*, C. 489; kalai-ka-pattar' *Hind.*, *Pb.*, C. 489; kalain, *Pb.*, C. 846; kala inderjan, *Mar.*, W. 131; kálajám, *Beng.*, E. 419; kala jira (Nigella), *Beng.*, *Hind.*, N. 158; kala-jira (Vernonia), *Beng.*, V. 79; kala-jíra, *Dec.*, V. 73; kalak, *Mar.*, B. 118; kálá-kado, *Mar.*, W. 122; kalá-kadu, *Mar.* H.,

517; kala káliya, *Hind.*, R. 351; kala kangni, *Hind.*, S. 1212; kálakát, *Pb.*, P. 1316; kala katwa, *Mar.*, S. 1078; kálá khajur, *Mar.*, M. 412; kála-kuda, *Mar.*, W. 122; kála kúdú, *Mar.*, W.122; kálakúra, *Mar.*, W. 122; kala kútkí, *Beng.*, *Dec.*, H. 111; kalalag (Berchemia), *Kumaon*, B. 471; kala lag (Ventilago), *Kumaon*, V. 48; kála lobia, *Pb.*, D. 789; kalam (Stephegyne), *Hind.*, *Mar*, *Pb.*, S. 2799; kalam (Anthocephalus), *Mar.*, A. 1192; kala madwan, *Hind.*, F. 405; kálámari, *Gus.*, P. 811; kálambar, *Gus.*, F. 124; kalambe, *Mar.*, A. 590; kalambe, *Mar.*, F. 725; kalmewa, *Pb.*, S. 2341; kala-miri, *Mar.*, P. 811; kála-mohare, *Mar.*, B. 812; kálámorich, *Beng.*, P. 811; kalá múche, *Beng.*, S. 2390, S. 2394; kalá múg, *Uriya*, P. 496; kala mukha, *Hind.*, E. 73; kála muttar, *Pb.*, P. 882; kalán, *Pb.*, A. 968; kalanchí, *Pb.*, D. 348; kalandar zatar, *Pb.*, T. 416; kalao, *Pb.*, P. 882; **kalaon**, *Pb.*, G. 240; kalap, *Beng.*, S. 2682; kala palas, *Mar.*, O. 537; **kalaph**, *Beng.*, S. 2682; kala phalas, (*C. P.*), *Hind.*, O. 537; kala-pú-ti-ah, *Pb.*, F. 328; kálaráí, *Guj.*, B. 812; kalaruk, *Mar.*, D. 41; kalarukh, *Mar.*, D. 41; kala sahájú, *Uriya*, T. 361; kala samp, *Hind.*, R. 124; kálásar, *Pb.*, C. 489; kala-shim, *Beng.*, C. 290; kalath, *Kumaon*, D. 758; kalathi, *Gus.*, D. 758; kala til (Sesamum), *Beng.*, S. 1078; kálá-tíl (Guizotia), *Hind.*, G. 735; kala titmaliya, *Kumaon*, V. 88; kála trúmba, *Pb.*, F. 19; kalatt, *Pb.*, D. 758; kala tulsi, *Beng.*, *Hind.*, O. 31; kalaúrd, *Dec.*, P. 513; kalauri, *Mar.*, S. 2850; kálávála (Aristolochia), *Mar.*, A. 1409; kálávála (Pavonia), *Mar.*, P. 344; kálávála (Valeriana), *Mar.*, V. 8; kálawar, *Kumaon*, R. 598; kálá zíra (Vernonia), *Pb.*, V. 73; kálá-zírá (Nigella), *Hind.*, N. 158; kálban, *Pb.*, M. 22; kalbandá, *Duk.*, A. 838; kalbasu, *Beng.*, F. 466; kalbir, *Hind*, D. 144; kalbriskh, *Hind.*, A. 455; kalchan, *Pb.*, S. 537; kále-hár, *Hind.*, T. 325; kálémadh-ká-jhar, *Dec.*, kale-madh-ka-per, *Hind.*, P. 663; kálenjire (Nigella), *Mar.*, N. 158; kalenjiri, *Mar.*, V. 73; kalen-jírí (Vernonia), *Mar.*, V. 73; kalé-pán-ki-jor, *Duk.*, A. 862; kálé-ráí, *Hind.*, B. 812; kálé-ráyan, *Duk.*, B. 812; kaleri-hisálu, *Hind.*, R. 598; kalfah, *Mar.*, C. 1195; kalga, *Pushtu*, R. 598, *Pb.*, R. 609; kalghoza, *Hind*, P. 760; kali, *Hind.*, G. 213; káliakará, *Beng.*, C. 427; kali andi jahria, *Hind.*, P. 425; kaliár, *Hind.*, B. 308; kálí basúti, *Pb.*, C. 1380; kálícha, *Hind.*, C. 626; kalichhad, *Guj.*, N. 17; kálí chúna, *Mar.*, C. 489; kalií dudhi, *Mar.*, C. 2325; kálí ghas, *Hind.*, C. 2558; kálí halada, *Mar.*, C. 2422; kálí haldí, *Hind.*, C. 2422; kali harr, *Hind.*, V. 79; kálijhánp, *Hind.*, A. 510; kálí-jhánt, *Beng.*, *Hind.*, A. 506; kalijiei, *Gus.*, V. 73; kálijíri, *Dec.*, *Guj.*, *Mar.*, *Kumaon*, V. 73; kálikachúná, *Hind.*, C. 489; káli-kikar, *Dec.*, *Mar.*, *Sind*, A. 101; kali-korafi, *Hind.*, F. 314; kálkutki, *Dec.*, *Pb.*, P. 700; kálímirch, *Hind.*, *Mar.*, P. 811; káli-mirchi, *Dec.*, P. 811; kali mirchingay, *Dec.*, P. 811; kalimitti, *Pb.*, P. 974; kálímort, *Pb.*, D. 348; kálímúsli, *Hind.*, *Mar.*, C. 2375; kálín, *Hind.*, C. 626; kalindra, *Pb.*, A. 328; kalinga, *Mar.*, C. 1221; kalingad, *Mar.*, C. 1221; kalingada, *Mar.*, C. 1221; kálírái, *Guj.*, *Hind.*, B. 841; káli ríng, *Pb.*, Q. 13; kali-saras, *Mar.*, A. 711; káli sarsón, *Beng.*, *Hind.*, B. 812; kálíshanbalí, *Dec.*, J. 116; káli-taroi, (*Bundel.*), *Hind.*, L. 556; kali-til, *Mar.*, G. 735; kálí torí, *Pb.*, L. 556; kali-tulsi, *Hind.*, O. 18; kaliun, *Pb.*, N. 1; kálizer, *Dec.*, *Hind.*, C. 1403; kalizewar, *Pb.*, B. 936; kálí zíri (Nigella), *Hind.*, N. 158; kalí zíri (Saussurea), *Pb.*, S. 904; kálí-zíri (Vernonia), *Beng.*, *Hind.*, *Dec.*, *Pb.*, V. 73; káli-zirkí, *Dec.*, I. 384; kaljendru, *Pb.*, S. 947; kalkalín, *Pb.*, R. 588; kal-kam, *Pb.*, S. 2799; kál-káshundá, *Beng.*, C. 780, 787; kalkatiya tamáku, *Beng.*, N. 98; kalkattia-tamáku, *Hind.*, N. 98; kalkora, *Beng.*, A. 692; kallai (Dillenia), *Hind.*, D. 438; kallai (tin), *Hind.*, T. 460; kallain, *Pb.*, C. 2358; kalla-jati, *Beng.*, D. 5; kallak, *Mar.*, B. 118, B. 149; kallam, *Hind.*, S. 2799; kallar, *Pb.*, R. 67; kallar, *Pb.*, M. 237; kalliachi, *Pushtu*, R. 598; kalli jarri, *Pb.*, S. 738; kallríbútí, *Pb.*, T. 562; kálmegh, *Beng.*, A. 1064; kalmi, *Beng.*, F. 671; kalmi-ság, *Beng.*, I. 343; kalmi-sák, *Beng.*, I. 343; kálo-dhatúro, *Guj.*, D. 151; kalo-kera, *Beng.* C. 441; kalo-kuchú, *Beng.*, C. 1732; kálo-mirich, *Gus.*, P. 811; kalon (Pisum), *Hind.*, P. 882, P. 885; kálon (Cedrus), *Pb.*, C. 846; kalonji, *Hind.*, *Mar.*, N. 158; kalor, *Beng.*, *Hind.*, O. 574; kalo-sarasio, *Guj.*, A. 711; kalp-brishk, *Hind.*, A. 455; kalpunch, *Hind.*, S. 1250, kalrei, *Pb.*, A. 22; kalru, *Hind.*, S. 2850; kalsar, *Hind.*, S. 1226; kalsipi *Mar.*, S. 1250; kalsis (Albizzia), *Hind.*, A. 695; kalsís (Mæsa), *Kumaon*, M. 40; kal sunda, *Mar.*, B. 171; kalthaun (Ehretia), *Pb.*, E. 20; kalthaun (Stereospermum), *Pb.*, S. 2876; kalu, *Hind.*, O. 616; kalúchia, *Uriya*, D. 655; kálu-chilu, *Hind.*, *Kumaon*, A. 17; kalúcho, *Pb.*, I. 14; kaluganthi, *Guj.*, E. 7; kaluí, *Beng.*, V. 129; kaluku, *Mar.*, S. 1207; kalunji, *Hind.*, E. 267; kálusura, *Hind.*, S. 2672; kalvári, *Sind*, C. 431; kalwit, *Hind.*, S. 1226; kálzar, *Pushtu*, C. 1403; kálzira, *Beng.*, N. 158; kam (Dioscorea), *Hind.*, D. 547; kám (Stephegyne), *Pb.*, S. 2799; kámách, *Beng.*, M. 786; kamá-khér, *Beng.*, A. 1107; kamal, *Pb.*, M. 71; kamala (Mallotus), *Hind.*, *Mar.*, M. 71; kamala (Nelumbium), *Mar.*, N. 39, M. 71; kamalágundi, *Beng.*, M. 71; kamalágurí, *Beng.*, M. 71; kamalá moglai, *Beng.*, C. 1234; kamalphul, *Pb.*, G. 165; kam álu, *Mar.*, D. 485; kamand; *Sind*, S. 30; kamáud, *Pb.*, S. 30; kamarak, *Beng.*, A. 1646; kamarakha, *Guj.*, *Mar.*, A. 1646; kámaráli nibu, *Beng.*, C. 1300; kamaranga, *Hind.*, A. 1646; kamarghwal, *Pushtu*, S. 924; kamarkas, *Hind.*, B. 945; kamarri, *Gus.*, G. 116; kamba, *Hind.*, C. 563; kámbaila,

Pushtu, M. 71; kambal (Mallotus), *Pb.*, M. 71; kambal (Odina), *Pb.*, O. 38; kambal (Rhus), *Pb.*, R. 335; kámbei, *Pb.*, S. 2299; kambhal, *Hind.*, M. 71; kambhar, *Hind.*, G. 287; kambilá, *Hind.*, M. 71; kambul, *Afg.*, S. 993; kamela, *Hind.*, *Pb.*, M. 71; kamela, *Hind.*, F. 671; kamilá, *Beng.*, *Hind.*, M. 71; kamini, *Beng.*, M. 797; kamkúi, *Beng.*, B. 868; kamla, *Pb.*, C. 862; kamlai, *Hind.*, *Pb.*, O. 38; kamlá nembu, *Beng.*, C. 1233; kámlatá, *Beng.*, *Hind.*, l. 405; kammar-kas, *Mar.*, S. 746; kammarri, *Hind.*, G. 116, kamo, *Beng.*, *Sind*; R. 242; kámohí, *Sind*, P. 663; kampti, *Hind.*, H. 285; kámrángá, *Beng.*, A. 1646; kamrup, *Beng.*, S. 1939; kamrup, *Beng.* F. 253; kamúd (Mallotus), *Hind.*, M. 71; kamúd (Nymphæa), *Kashmir*, N. 192; kámuni, *Mar.*, S. 2299; kan (Coffea), *Mar.*, C. 1641; kan (Olea), *Hind.*, *Pb.*, O. 145; kána (Commelina), *Hind.*, C. 1756; kána (Saccharum), *Pb.*, S. 6; kanachi, *Pb.*, R. 593; kanagalu, *Mar.*, D. 438; kanaji, *Gus.*, P. 1121; kanak (Sageretia), *Kashmir*, S. 502; kanak (Triticum), *Pb.*, T. 634; kánakach, *Pb.*, F. 725, M. 647; kanak-champa (Ochna), *Mar.*, O. 1; kanak-champa (Pterospermum), *Beng.*, *Hind.*, *Mar.*, P. 1389; kana-kuri, *Uriya*, F 457; kánálá, *Beng.*, G. 753; kanalla, *Hind.*, B. 330; kanalu, *Duk.*, A. 771; kana-pachethi, *Mar.*, B. 180; kan-aucha, *Pb.*, M. 786; kanaunji, *Hind.*, S 2850; kán-bher, *Mar.*, A. 317; kanch (glass), *Hind.*, G. 229; kancha (Mucuna), *Gus.*, M. 786; kanchan (Bauhinia), *Beng.*, B. 295, *Mar.*, B. 356; kanchan (Diospyros), *Hind.*, D. 628; kanchanamu, *Uriya*, M. 517; káncharái, *Beng.*, C., 1748; kanchá-ri, *Pb.*, C. 559; kánchkúre, *Dec.*, T. 509; kánchkúri (Tragia), *Mar.*, T. 509; kánchkúri (Mucuna), *Dec.*, M. 786; kanchli, *Hind.*, A. 344; kanchora-, *Hind.*, S. 368; kanchura, *Beng.*, C. 1748; kándá (Allium), *Guj.*, *Mar.*, A. 769; kandá (Amorphophallus), *Dec.*, *Hind.*, A. 996; kanda (Meconopsis), *Pb.*, M. 324; kanda (Prosopis), *Pb.*, P. 1259; kánda (Saccharum), *Pb.*, S. 6; kánda (Sageretia), *Pb.*, S. 505; kánda (Urginea), *Hind.*, U. 39; kandágar, *Kumaon*, H. 339; kandahári-tamáku, *Pb.*, N. 98; kandahár-kakkar, *Pb.*, N. 98; kandal, *Bokhara*, D. 810; kandalái, *Pushtu*, l. 548; kandan, *Hind.*, B. 308, 356; kán-dár (Shorea), *Hind.*, S. 1656; kandara (Cornus), *Hind.*, C. 1969; kandara (Ilex), *Pb.*, I. 14; kandaru (Acer), *Pb.*, A. 334; kanda-tella, (*Garhwal*), *Hind.* T. 875; kandául, *Pb.*, R. 215; kánda-vela, *Mar.*, V. 219; kándawel, *Hind.*, *Mar.*, V. 219; kande (Coriaria), *Pb.*, C. 1958; kánde (Urginea), *Hind.*, U. 39; kandeb, *Beng.*, C. 152; kandei (Astragalus), *Ph.*, A. 1588; kandei (Flacourtia), *Pb.*, F. 615; kander (Capparis), *Pb.*, C. 431; kander (Celastrus), *Pb.*, C. 862; kander (Rhamnus), *Pb.*, R. 157; kandero, *Sind*, C. 224; kandhára, *Hind.*, X. 21; kandi, *Hind.*, *Pb.*, *Pushtu*, *Sind*, P. 1259; kandiára (Astragalus), *Pb.* A.

F

1588; kandiára (Carthamus), *Pb.*, C. 633; kandiára (Rubus), *Pb.*, R. 598; kandiári (Argemone), *Pb.*, A. 1351; kandiári (Celastrus), *Pb.*, C. 862; kandiári (Cousinia), *Pb.*, C. 2027; kandiári (Rubus), *Kashmir*, R. 598; kandiári (Solanum), *Pb.*, S. 2345; kandiári (Zizyphus), *Hind.*, *Pb.*, Z. 280; kandika, *Hind.*, *Pb.*, Z. 280; kandla, *Hind.*, B. 330; kandlar, *Pb.*, I. 14; kandlu, *Pb.*, P. 1460; kando (Cæsalpinia), *Hind.*, C. 42; kando (Hippophae), *Pb.*, H. 277; kándo (Sterculia), *Mar.*, S. 2850; kandol, *Mar.*, S. 2850; kandori (Securinega), *Mar.*, S. 1029; kankra (Trichosanthes), *Hind* , T. 576; kandrá, *Dec.*, U. 39; kandre, *Hind.*, *Kumaon*, A. 17; kandru, *Hind.*, H. 324; kandu (Celastrus), *Pb.*, C. 862; kandú (Eryngium), *Pb.*, E. 335; kándúla, *Mar.*, S. 2850; kandur, *Afg.*, P. 847; kandurí, *Hind.*, *Pb.*, *Sind*, C. 919; kandurí-kí-bél, *Hind.*, C. 919; kandwa, *Hind.*, C. 1313; kandyárí, *Pb.*, S. 2280; kanél, *Hind.*, N. 80; kaner, *Hind.*, *Pb.*, C. 225, N. 80; kanéra (Hamiltonia), *Pb.*, H. 13; kanera (Nerium), *Guj.*, N.80; kaneri, *Mar*, N. 80; kaner zard, *Pb.*, T. 9; kanfodi-kánphodi, *Mar.*, C. 1367; káng, *Guz.*, *Mar.*, S. 1212; kangach, *Pb.*, F. 725; kangahi, *Hind.*, A. 80; kangai, *Hind.*, S. 2799; kangani, *Mar.*, C. 854; kangar (Anthistiria), *Hind.*, A. 1186; k a n g a r (Dioscorea), *Mar.*, D. 507; kangar (Pistacia), *Pb.*, P. 833; kangar (Rhus), *Pb.*, R. 316; kangei, *Mar.*, S. 2799; kanger (Grewia), *Pb.*, G. 702; kanger (Sageretia), *Pb.*, S. 497; kanghai, *Pb.*, A. 501; kanghani, *Hind.*, A. 89; kanghi (Abutilon), *Hind.*, A. 80, 89; kanghi (Pyrus), *Pb.*, P. 1460; kanghol mirch, *Pb.*, C. 886; kanghuní, *Hind.*, S. 1212; kanghur, *Dec.*, *Mar.*, kangi (E u p h o r b i a), *Pb.*, E. 505; kangi (Lycium), *Pb.*, L. 589; kangi (Stephegyne), *Hind.*, S. 2799; kangkila, *Beng.*, F. 366; kangna, *Pb.*, P. 50; kangni (Manisurus), *Hind.*, M. 232; kángni (Setaria), *Beng.*, *Dec.*, *Hind.*, *Mar.*, *Pb.*, S. 1212; kangoi, *Duk.*, *Mar.*, A. 89; kangori, *Duk.*, *Mar.*, A. 89; kangori, *Mar.*, A. 80; kangra (Sorghum), *Mar.*, S. 2424; kangra (Corchorus), *Uriya*, C. 1847; kangra (Panicum), *Dec.*, P. 53; kangri, *Raj.*, E. 190; kangsmúki, *Beng.*, I. 478; kangu (Setaria), *Mar.*, S. 1212; kángu, *Beng.*, *Dec.*, *Hind.*, S. 1212; kangú (Flacourtia), *Pb.*, F. 915; kangú (Lycium), *Pb.*, L. 589; kanguni, *Mar.*, C. 854; kánh, *Pb.*, S. 49; kanha bíchu, *Pb.*, F. 725; kanhera, *Mar.*, C. 225; kanhera, *Mar.*, N. 80; kaniár (Bauhinia), *Hind.*, B. 308, 356; kaniár (Cassia), *Pb.*, C. 756; kaniár (Pterospermum), *Hind.*, P. 1389; kani-magur, *Beng.*, F. 536; kani-pabda, *Beng.*, F. 368; kanir, *Mar.*, N. 80; kanír, *Mar.*, C. 225; kaníra, *Hind.*, *Pb.*, C. 225, *Pb.*, N. 80; kanírkejur, *Dec.*, N. 80; kaniúri, *Pb.*, H. 52; kanj, *Hind.*, *Kumaon*, T. 489; kanjak, *Afg.*, P. 847; kanján-búra, *Hind.*, *Beng.*, K. 1; kánjar, *Pb.*, A. 344; kanjara, *Pb.*, A. 334; kanji (Malva), *Hind.*, M. 115; kanji (Rhamnus), *Pb.*,

R. 152; kánjlá, *Mar.*, A. 531; kanjru, *Pb.*, I. 14; kanju (Holoptelea), *Kumaon*, H. 324; kanjú (Celosia), *Pb.*, C. 873; kanjú (Flacourtia), *Hind.*, F. 615; kanjura, *Hind.*, *Kumaon*, C. 1756; kank (Triticum), *Sind*, T. 634; kánk (Flacourtia), *Hind.*, F. 615; kankadi, *Mar.*, N. 39; kankala, *Mar.*, P. 797; kankar, (lime), *Pb.*, C. 489; kankar (Grewia), *Hind.*, G. 143; kankhina, *Mar.*, S. 705; kan khúra, *Beng.*, B. 576; kánki, *Hind.*, F. 615; kankola (Piper), *Mar.*, P. 801; kankoli (Elæagnus), *Pb.*, E. 51; kankol mirch, *Pb.*, E. 51; kankor, *Hind.*, Z. 290; kankra (Bruguiera), *Beng.*, B. 898; kankra (Pavetta), *Hind.*, P. 338; kankra (Webera), *Beng.*, W. 22; k a n k r e i, *Hind.*, B. 944; kankri, *Mar.*, C. 2287; kánkur, *Beng.*, C., 2278; kán-kuti, *Mar.*, C. 728; kanlai, *Hind.*, D. 402; kanlao, *Hind.*, B. 330; kanmar, *Hind.*, *Mar.*, S. 808; kanmu, *Hind.*, C. 2062; kanna, *Pb.*, C. 1748, 1752; kanne-ki-gond, *Hind.*, O. 39; kanóch (Fraxinus), *Pb.*, F. 700; kanocha (Phyllanthus), *Hind.*, P. 654; kanor, *Hind.*, *Pb.*, A. 567; kan perún, *Sind*, S. 2299; kanphúl, *Pb.*, T. 87; kánphútí (Cardiospermum), *Mar.* C. 551; kánphútí (Cleome), *Hind.*, *Mar.*, C. 1367; kanrája, *Mar.*, B. 318; kanrak, *Pb.*, C. 886; káns (Dioscorea), *Pb.*, D. 503; káns (Saccharum), *Hind.*, *Pb.*, S. 49; kánsa, *Hind.*, S. 49; kansaráj, *Mar.*, A. 506; kansárinata, *Orissa*, I. 362; kanseri, *Mar.*, D. 748; kanséri, *Hind.*, D. 748; kán-sevari, *Mar.*, S. 1166; kanshira, *Beng.*, C. 1748; kanshura, *Hind.*, C. 1748; kánsi (Saccharum), *Hind.*, S. 49; kánsí (Ribes), *Pb.*, R. 355; kánsian, *Pb.*, D. 130; kanta (Corchorus), *Uriya*, C. 1847; kanta (Meconopsis), *Pb.*, M. 324; kanta (Rosa), *Beng.*, R. 531; kanta (Zizyphus), *Hind.*, Z. 254; kánta, *Mar.*, D. 481; kanta-álu, *Hind.*, *Mar.*, D. 522; kánta-ber, *Hind.*, Z. 254; kanta bohul, *Uriya*, S. 1718; kanta bohul, *Uriya*, Z. 290; kanta gola batana, *Beng.*, O. 6; kantá-gúíkamai (Azima), *Hind.*, A. 1665; kantá-gúr-kamai (Capparis), *Beng.*, C. 427; kántájáti, *Beng.*, B. 171; kantaka, *Mar.*, F. 615; kantakalika, *Beng.*, H. 508; kantakari, *Beng.*, S. 2345; kanta-katchú, *Beng.*, L. 84, kanta kúlika, *Hind.*, H. 508; kantá, kusham, *Uriya*, A. 1351; kantala, *Hind.*, A. 603; kanta lal batana, *Beng.*, C. 818; kantálú, *Pb.*, H. 13; kánta-maris, *Beng.*, A. 943; kántánaté, *Beng.*, A. 943; kantanch, *Kashmir*, R. 588; kántá-nu-dánt, *Guj.*, A. 943; kánta nutía, *Beng.*, A. 943; kántá shelio, *Guj.*, B. 171; kántedhotrá, *Mar.*, A. 1351; kánte gotti, *Mar.*, Z. 290; kántékángi, *Mar.*, D. 481; kantela, *Hind.*, A. 1351; kántemát, *Mar.*, A. 943; kantena, *Hind.*, A. 1288; kanteri samar, *Mar.*, B. 632; kante savar, *Mar.*, B. 632; kánthál, *Beng.* *Hind.*, I. 58; kánthál, *Beng.*, *Hind.*, A. 1489; kantħan, *Pb* D. 130; kanthár, *Guj.*, C. 427; kanthi-rikhu, *Hind.*, S. 30; kantiári, *Pb.*, C. 633; kanti kapali, *Uriya*, C. 427; kántí-sénbal, *Hind.*, B. 632; kantolán, *Guz.*, M. 639;

kántón-ká-khatyán, *Duk.*, B. 632 ; kántón-ká-sémul, *Duk.*, B. 632 ; kántosariyo, *Guj.*, A. 261 ; kanujera, *Hind.*, A. 722 ; kanura, *Hind.*, H. 164 ; kanuraka, *Beng.*, C. 1748 ; kanuriya, *Orissa*, H. 177 ; kanval (Nelumbium), *Hind.*, N. 39 ; kanval (Nymphæa), *Guj.*, *Hind.*, N. 200 ; kanvár, *Hind.*, A. 838, *Sind*, A. 829 ; kauvar-patha, *Duk.*, A. 829 ; kanvár phod, *Duk.*, A. 838 ; kanwail, *Mar.*, V. 55 ; kanwal (Crinum), *Hind.*, C. 2062 ; kanwal (Nelumbium), *Hind.*, N. 39 ; kanwal (Saussurea), *Pb.*, S. 919 ; kanwal(a) (Litsæa), *Hind.*, L. 490 ; kánwal kakri, *Pb.*, N. 39 ; kanyúr, *Kumaon*, N. 80 ; kanyúrts, *Pb.*, A. 1458 ; kanzal, *Pb.*, A. 328, 344 ; kanzars, *Pb.*, F. 678, 682 ; kanzla, *Pb.*, A. 334 ; kao, *Pb.*, *Pushtu*, O. 145 ; kaogrum, *Beng.*, D. 62 ; kaoli, *Hind.*, F. 356 ; kapale, *Mar.*, T. 634 ; kapas, *Mar.*, G. 404 ; kapás, *Pb.*, G. 385, *Beng.*, *Dec.*, *Guj.*, G. 404 ; kapási (Corylus), *Kumaon*, C. 1988 ; kapasi (Gerbera), *Kumaon*, *Pb.*, G. 186 ; kapasiya (Gerbera), *Kumaon*, G. 186 ; kapasiya (Hibiscus), *Hind.*, H. 215 ; kapela, *Mar.*, M. 71 ; kapfí, *Pb.*, G. 186 ; kaphal, *Hind.*, *Pb.*, M. 869 ; kaphi, *Mar.*, C. 1641 ; kápi, *Beng.*, C. 1641 ; kapilo, *Guz.*, M. 71 ; kap-o-chist, *Afg.*, C. 2221 ; kappárphodi, *Mar.*, P. 678 ; káppúr, *Beng.*, C. 257 ; kapra, *Hind.*, C. 121 ; kapsí, *Mar.*, D. 200 ; kapua kanwál, *Kumaon*, P. 548 ; kapur, *Dec.*, *Guj.*, *Hind.*, *Mar.*, C. 257 ; kapur kachali, *Guz.*, C. 2406 ; kapur kachri, *Hind.*, *Pb.*, H. 59 ; kapur krachari, *Guj.*, *Mar.*, H. 59 ; kápurlí, *Mar.*, A. 1130 ; kapur madhura, *Mar.*, A. 554 ; kar (Carthamus), *Hind.*, C. 637 ; kar (Celtis), *Pb.*, C. 881, 885 ; kara (Limonia), *Hind.*, L. 362 ; kara (Strobilanthes), *Mar.*, S. 2925 ; kara (Strychnos), *Mar.*, S. 2943 ; karaal, *Pb.*, F. 492 ; karabi, *Beng.*, C. 225, N. 80 ; karada, *Uriya*, L. 219 ; karah, *Beng.*, F. 575 ; karai, *Guz.*, *Mar.*, S. 2850 ; karail (Dendrocalamus), *Beng.*, D. 292 ; karaila (Gynandropsis), *Hind.*, G. 753 ; karain, *Guj.*, *Kashmir*, *Pb.*, P. 882 ; karâita, *Pb.*, V. 70 ; karakana, *Mar.*, G. 714 ; karakaya, *Dec.*, T. 361 ; karala (Cynodon), *Mar.*, C. 2558 ; karalá (Momordica), *Beng.*, M. 626 ; karale, *Mar.*, M. 626 ; karálla (Sapium), *Pb.*, S. 838 ; karalli, *Pb.*, B. 308 ; karallu, *Mar.*, A. 717 ; káralye, *Mar.*, V. 73 ; karam, *Hind.*, A. 514 ; karamal, *Mar.*, D. 438 ; karamala, *Mar.*, D. 438 ; karamara, *Mar.*, A. 1646 ; karamarda, *Guj.*, C. 596 ; karamb, *Mar.*, S. 2799 ; karambel, *Mar.*, D. 428 ; karambu, *Mar.*, O. 153 ; karamchá, *Beng.*, C. 596 ; karam-kallá, *Hind.*, B. 851 ; karam-ká-ság, *Duk.*, B. 851 ; karam-kí-bhájí, *Duk.*, B. 851 ; karamm, *N. Pb.*, D. 815 ; karan, *Pb.*, M. 756 ; karanda (Dioscorea), *Mar.*, D. 494 ; karándá (Carissa), *Mar.*, C. 596 ; karanfal, *Pb.*, C. 706 ; karáng, *Guz.*, S. 1212 ; karanga, *Hind.*, P. 1252 ; karangal, *Pb.*, C. 756 ; karanj, *Dec.*, *Guj.*, *Hind.*, *Mar.*, P. 1121 ; karanja, *Beng.*, *Mar.*, P. 1121 ; karanjaca, *Hind.*, P. 1121 ; karanjavá, *Hind.*, C. 6 ; karanj-gáchh, *Beng.*, P.

1121 ; karanjh, *Dec.*, *Hind.*, P. 1121 ; karanji (Albizzia), *Hind.*, A. 717 ; karanji (Holoptelea), *Hind.*, H. 324 ; karanjíchajháda, *Mar.*, P. 1121 ; karanj-nu, *Guz.*, P. 1121 ; karanjú, *Hind.*, C. 6 ; kárán-kusa, *Beng.*, A. 1093 ; karankusha, *Hind.*, *Pb.*, A. 1093 ; karansá, *Uriya*, P. 1121 ; karanta, *Hind.*, C. 2247 ; karantolí, *Mar.*, M. 639 ; karaoli, *Mar.*, S. 2912 ; karaptir, *Guj.*, G. 143 : karár (Bauhinia), *Pb.*, B. 308 ; karar (Carthamus), *Pb.*, C. 633, 637 ; karar-gandhel-dungarko, *Hind.*, I. 487 ; kararhi, *Mar.*, C. 637 ; karas, *Pb.*, L. 100 ; karáthrí, *Pb.*, V. 64 ; karatola, *Mar.*, M. 639 ; karaunda, *Hind.*, S. 1939 ; karaundá, *Hind.*, C. 596 ; karaunj, *Kumaon*, C. 6 ; karaunji, *Hind.*, S. 2850 ; karavanda, *Mar.*, C. 596 ; karavi, *Mar.*, R. 57 ; karbaru, *Pb.*, H. 52 ; karber, *Hind.*, N. 80 ; karbi, *Mar.*, S. 2424 ; karbuj, *Hind.*, C. 1221 ; karchanna, *Mar.*, S. 2912 ; karchanua, *Hind.*, S. 2912 ; karchi, *Hind.*, H. 294 ; karchilla, *Hind.*, P. 737 ; karda, *Mar.*, T. 434 ; kardahi, *Hind.*, A. 1153 ; kardai, *Mar.*, *Sind*, C. 637 ; karedha, *Uriya*, T. 325 ; karei, *Mar.*, C. 725 ; kárék, *Guz.*, P. 555 ; karél, *Hind.*, C. 402 ; karela, *Hind.*, M. 626 ; karélá, *Dec.*, *Guj.*, *Hind.*, *Pb.*, M. 626 ; kareli, *Hind.*, M. 626 ; karelo, *Guj.*, *Sind*, M. 626 ; karelu (Ilex), *Pb.*, I. 14 ; karelu (Momordica), *Guj.*, M. 626 ; karena, *Uriya*, M. 626 ; karendera, *Pb.*, A. 353 ; karenja, *Beng.*, C. 596 ; karé-pák, *Dec.*, M. 800 ; karé-pákácha, *Mar.*, M. 800 ; karér (Capparis), *Hind.*, C. 402 ; karer (Rosa), *Hind.*, R. 538, *Pb.*, R. 588 ; karera, *Mar.*, S. 2912 ; karet (Plantago), *Pb.*, P. 926 ; kareta (Sida), *Hind.*, S. 1688 ; kargam, *Pb.*, C. 886 ; karghauna, *Pushtu*, L. 587 ; kargnalia, *Hind.*, B. 863 ; karguha, *Hind.*, I. 440 ; karhar, *Hind.*, *Kumaon*, R. 1 ; karhár (Gardenia), *Hind.*, G. 136 ; karhár (Randia), *Hind.*, R. 1 ; karhi-nimb, *Mar.*, M. 800 ; kari (Capparis), *Mar.*, C. 402 ; kari (Clerodendron), *Mar.*, C. 1380 ; kari (Glochidion), *Hind.*, G. 240 ; kari (Holarrhena), *Hind.*, H. 294 ; kari (Litsæa), *Hind.*, L. 474 ; kári (Miliusa), *Hind.*, M. 545 ; kári (Rhamnus), *Pb.*, R. 159 ; karí (Saccopetalum), *Hind.*, S. 487 ; kari (Konkan), *Mar.*, M. 6 ; karia, *Pb.*, C. 402 ; karíal, *Pb.*, D. 9 ; karianag, *Mar.*, G. 243 ; kariaphulli, *Beng.*, M. 800 ; kariaput, *Mar.*, M. 412 ; kariári, *Hind.*, *Pb.*, G. 243 ; kariga, *Mar.*, G. 124 ; karigo, *Sind*, C. 1221 ; karihári, *Hind.*, C. 243 ; karik (Celtis), *Pb.*, C. 886 ; kárik (Vitis), *Pb.*, V. 195 ; káriki-send, *Dec.*, E. 553 ; karil (Lathyrus), *Pb.*, L. 100 ; karíl (Capparis), *Mar.*, *Pb.*, C. 402 ; karílá, *Pb.*, M. 626 ; kárí lání, *Sind*, C. 224 ; kári-lání, *Sind*, H. 10 ; kari-mattal, *Hind.*, O. 537 ; karin, *Kashmir*, P. 53 ; karinda (Carissa), *Mar.*, C. 596, karinda (Dioscorea), *Mar.*, D. 494 ; karinga, *Guj.*, C. 1221 ; karinga, *Mar.*, H. 486 ; karipat, *Mar.*, M. 800 ; kariphal, *Guz.*, M. 869 ; karír (Acacia) *Hind.*, A. 249 ; karír (Rosa), *Pushtu*, R. 538 ; ka-rír-re, *Sind*, F. 326 ; karis, *Pb.*, C. 402 ; karit, *Mar.*, C. 2310 ;

karivana, *Mar.*, H. 486; karí-wágetí, *Mar.*, P. 316; kariyátu, *Guj.*, A. 1064; kark, *Beng.*, D. 292; karka, *Kumaon*, P. 618; karkacha, *Hind.*, C. 756; karkan, *Pushtu*, Z. 254; karkana, *Pushtu*, C. 224; katkana, *Pushtu*, Z. 254; karkan ber, *Pushtu*, Z. 280; karkani, *Mar.*, L. 241; karkanna, *Afg.*, C. 224; karkannie, *Mar.*, E. 199; karkanrá, *Pushtu*, Z. 254; karkar, *Pb.*, I. 430; karkath, *Mar.*, O. 233; karkaya, *Dec.*, 'l. 361; karkí, *Kumaon*, G. 186; karkotta, *Beng.*, D. 438; karkú, *Pb.*, A. 677; karkún, *Pb.*, R. 71; kárla (Girardinia), *Pb.*, G. 213; kárlá (Momordica), *Mar.*, M. 626; karlí, *Hind.*, *Mar.*, M. 626; karmabres, *Pb.*, F. 19; kar-madhana, *Pb.*, E. 166; karmai, *Beng.*, B. 304; karmał (Averrhoa), *Hind.*, A. 1646; karmal (Dillenia), *Mar.*, D. 428, 438; karmal (Peganum), *Pb.*, C. 225; karmarri, *Hind.*, G. 116; karmbal, *Pb.*, F. 168; karmbel, *Mar.*, D. 428; karmbru, *Pb.*, A. 711; karmora, *Kashmir*, H. 52; karmuj, *Beng.*, P. 1121; kárn, *Pb.*, R. 159; karna, *Pb.*, C. 1234; karna nebu, *Beng.*, C. 1286; karndol, *Pb.*, F. 156; karngúra, *Pb.*, P. 1252; karni kara, *Mar.*, P. 1389; karo, *Hind.*, A. 717; karólá (Luffa), *Hind.*, L. 567; karólá (Momordica), *Hind.*, M. 626; karolio, *Guj.*, C. 551; karolu (Albizzia), *Hind.*, A. 717; karolu (Garuga), *Hind.*, G. 143; karonda, *Hind.*, C. 596; kározgi, *Pushtu*, S. 2299; karpa, *Mar.*, M. 439; karphal, *Hind.*, M. 869; karpúr, *Beng*, *Guj.*, C. 257; kárpúra, *Mar.*, C. 257; karra (Albizzia), *Hind.*, A. 717; karra (Holarrhena), *Hind.*, H. 294; karrai, *Hind.*, S. 2850; karralura, *Hind.*, C. 416; karrat, *Hind.*, F. 806; karrauth, *Hind.*, T. 456; karre, *Hind.*, S. 6; karri, *Hind.*, *Mar.*, S. 487; karria-pat, *Mar.*, M. 800; karri-ním, *Mar.*, M. 800; karroná, *Hind.*, C. 596; karrú (Gentiana), *Pb.*, G. 165; karrú (Picrorhiza), *Pb.*, P. 700; karundi, *Beng.*, F. 335; karsh, *Pb.*, Q. 13; karshu, *Kumaon*, *Pb.*, Q. 70; karsi, *Pb.*, G. 769; karsi, *Hind.*, M. 237; karsúi, *Pb.*, Q. 70; kart, *Pb.*, S. 1260; kartál, *Dec.*, S. 2394; kartoii, *Mar.*, M. 639; kárt-tút, *Pb.*, M. 775; karu (Capparis), *Hind.*, C. 402; karú (Gentiana), *Beng.*, *Hind.*, G. 165; karuk (Myrsine), *Pb.*, M. 910; karúk (Cordia), *Pb.*, C. 1950; karu-karida, *Dec.*, D. 494; karumba, *Raj.*, *Hind.*, G. 136; karún (Euonymus), *Pb.*, E. 479; karún (Morus), *Pb.*, M. 775; karúnda, *Hind.*, C. 596; karunjá, *Hind.*, C. 6; karur (Hedera), *Pb.*, H. 52; karúr (Sageretia), *Pb.*, S. 505; karusura-ghás, *Hind.*, S. 2672; karvá-indarjou, *Dec.*, *Hind.*, *Mar.*, H. 294; karvan, *Mar.*, C. 2039; karva, *Mar.*, A. 1200; karvati, *Mar.*, S. 2912; karvi, (Hymenodictyon), *Hind.*, H. 523; karvi (Strobilanthes), *Mar.*, S. 2925; karvíla, *Pb.*, C. 416; karvíturái, *Dec.*, L. 563; karvisturí, *Hind.*, L. 563; karvíziri, *Dec.*, V. 73; karwah, *Hind.*, A. 1351; karwai, *Mar.*, H. 523; karwanr *Mar.*, C. 2039; karwand *Mar.*, C. 596;

karwanth, *Hind.*, T. 456; karware-*Pushtu*, R. 593; karwat (Antiaris), *Mar.*, A. 1200; karwat (Ficus), *Mar.*, F. 124; karwinai (Corallocarpus), *Mar.*, C. 1834; karwí-naí (Pueraria), *Gus.*, P. 1401; karwitarui, *Hind.*, L. 563; karyal, *Duk.*, C. 402; karya matti, *Pb.*, C. 489; karyani, *Hind.*, M. 786; karyá-pák, *Dec.*, M. 800; karyápát, *Dec.*, M. 800; karzu, *Pb.*, Q. 70; kas (Acacia), *Hind.*, A. 103; kas (Cannabis), *Pb.*, C. 331; kás (Saccharum), *Beng.*, *Hind.*, S. 49; kása, *Hind.*, L. 100; kasa-jonar (Pennisetum), *Hind.*, P. 384; kasa jona (Sorghum), *Beng.*, S. 2424; kásamm, *Pb.*, A. 1632; kásani, *Guj.* C. 1108; kasári, *Hind.*, L. 100; kasaurí, *Pb.*, C. 1732; kasayapalla, *Beng.*, H. 221; kasbál, *Pb.*, S. 906; kasb ul zaríra, *Pb.*, S. 3014; kasdí, *Mar.*, C. 637; kasei, *Hind.*, C. 1686; kaseru (Scirpus), *Pb.*, S. 977; kaserú (Cyperus), *Pb.*, C. 2592; kash, *Beng.*, S. 49; kash, *Hind.*, S. 49; kásh, *Kumaon*, *Hind.*, S. 49; kashí, *Hind.*, S. 49; káshín, *Pb.*, R. 318; kashiphal, *Hind.*, L. 30; kashísh, *Mar.*, I. 481; kashiya, *Beng.*, S. 49; kashkásh, *Hind.*, P. 87; kashmal (Berberis), *Hind.*, *Pb.*, B. 443, 456, 458, 465; kashmala (Odína), *Hind.*, O. 38; kashmírí, *Pb.*, P. 1285; kashshing, *Hind.*, P. 693; kashti, *Afg.*, *Pb.*, P. 746; kashtúrí, *Beng.*, D. 228; kashú, *Pb.*, P. 1463; kasi, *Uriya*, B. 868; kasílpathá, *Pb.*, H. 382; kasini, *Beng.*, *Hind.*, *Mar.*, C. 1104; kasni, *Hind.*, C. 1104; kasír (Albizzia), *Pb.*, A. 722; kasir (Corcus), *Hind.*, C. 1969; kasis, *Hind.*, I. 481; kaskai, *Afg.*, I. 128; kaskeri, *Pb.*, I. 123; kaskúsri, *Pb.*, G 723; kasmal, *Pb.*, B. 443; kasmal-rasout, *Guj.*, B. 458; kasmol, *Hind.*, C. 1973; kásní, *Pb.*, C. 1108; kasóndí, *Dec.*, *Hind.*, C. 780; kaspat (Dioscorea), *Pb.*, D. 503; kaspat (Fagopyrum), *Hind.*, F. 19; kassa, *Hind.*, F. 156; kassai-bija, *Mar.*, C. 1686; kassar, *Hind.*, V. 195; kassar tiuri, *Hind.*, L. 100; kasschra, *Beng.*, H. 504; kassi, *Hind.*, B. 868; kassod, *Mar.*, C. 785; kassuma, *Mar.*, S. 950; kassúr, *Beng.*, *Hind.*, L. 100; kastel, *Mar.*, H. 463; kastori-kaman, *Pb.*, E. 219; kastura, *Hind.*, D. 227; kasturé, *Kashmir*, D. 227; kasturi (civet), *Hind.*, T. 444; kasturí (musk deer), *Hind.*, *Mar.*, D. 228; kastúrí, *Hind.*, D. 245; kásumba, *Hind.*, C. 637; kásundá, *Dec.*, *Hind.*, C. 780, *Hind.*, C. 787; kasuru, *Hind.*, *Beng.*, S. 977; kasús, *Hind.*, *Pb.*, C. 2508; kat (Acacia), *Hind.*, *Mar.*, A. 137; kat (Lagerstrœmia), *Hind.*, L. 55; katai (Solanum), *Hind.*, S. 2345; katái (Flacourtia), *Beng.*, F. 615; katáia (Celtis), *Hind.*, C. 806; katail, *Beng.*, F. 615; kataila, *Hind.*, Z. 273; katakaránja, *Hind.*, C. 6; katambal, *Pb*, R. 642; katambi, *Mar.*, C. 36; katan, *Hind.*, E. 289; katang, *Hind.*, G. 705; katangkári, *Hind.*, S. 2280; katári (Prunus), *Pb.*, P. 1466; katári (Xylosma), *Hind.*, X. 21; katárí (Saccharum) *Hind.*, S. 30; kátar-kanda, *Pb.*, A'

1588; katás, *Hind.*, T. 444; katavandai, *Hind.*, H. 196; kat-bel, *Beng., Hind.*, F. 53; katber, *Mar.*, F. 671; katber, *Hind., Mar.*, Z. 290; kat bhewal, *Hind.*, G. 679; katbhilawa, *Hind.*, B. 913; kat-bish, *Beng.*, A. 397, *Kashmir, Pb.*, A. 413; katcha karawa, *Hind.*, F. 335, 352; kátchámpá, *Uriya*, P. 989; katcí, *Pb.*, A. 1351; katela, *Pb.*, S. 2345; katelí, *Hind.*, S. 2345; katéré-ké-jhárkí-rúí, *Dec.*, C. 1512; kateru, *Pb.*, I. 14; kat-guláb, *Beng.*, R. 531; kat-gularia, *Hind.*, F. 202; katgúli, *Hind.*, S. 589; kath (catechu), *Hind.*, A. 137, *Beng., Hind., Mar., Pb., Sind*, A. 181; kath (Cæsalpinia), *Hind.*, C. 6; katha, *Hind., Duk.*, A. 135; kathachampa, *Hind.*, P. 1389; katha chibhado, *Sind*, C. 581; katháglí, *Pb.*, X. 21; káthál, *Beng., Hind.*, A. 1489; kathal bat, *Beng.*, F. 118; kathalparhar, *Hind.*, G. 753; kathalyá gonda, *Mar.*, C. 1512; kathár, *Bihar*, D. 534; kathaul, *Hind.*, C. 756; kath-bel, *Beng.*, F. 53; káthbol, *Mar.*, A. 188; kath chibda, *Guj.*, C. 581; kathel, *Hind.*, T. 460; kathera, *Hind.*, C. 722; katherti, *Pb.*, P. 1322; kathewat, *Pb.*, I. 123; kathgúlar, *Pb.*, F. 179; kathi (Alpinia), *Sind*, A. 853; kathi (Desmodium), *Pb.*, D. 348; kathi (Indigofera), *Kashmir*, I. 119; *Pb.*, I. 128; kathi (Myricaria), *Pb.*, M. 881; kathí (Panicum), *Dec.*, P. 53; kathi (Sophora), *Pb.*, S. 2376; kathí, *Pb.*, S. 2376; kathir, *Hind.*, T. 460; kathjular, *Pb.*, F. 156; kath kutha, *Hind.*, U. 11; kathli, *Dec.*, P. 53; kátho, *Guj.*, A. 137, 181; kathshim, *Beng.*, C. 290; kathu (Indigofera), *Pb.*, I. 128; káthu (Fagopyrum), *Pb.*, F. 10, F. 19; kathun bán, *Pb.*, C. 224; Q. 29; kathú niár, *Hind.*, C. 898; kati, *Pb.*, I. 123, 128; katiain, *Mar.*, B. 868; katila (Garuga), *Hind.*, G. 143; katila (Sterculia), *Hind.*, S. 2850; katilá (Astragalus), *Hind.*, A. 1585; katilla katerá, *Hind.*, A. 1597; katíra (Bassora), *Hind.*, B. 283; katíra (Salix), *Pb.*, S. 544; katíra-í-hindí, *Hind.*, C. 1512; katjang, *Pb.*, D. 789; katkaleja, *Hind.*, C. 6; katkalijá, *Hind.*, C. 6; katkaranga, *Hind.*, C. 6; katkaranj, *Hind.*, C. 6; katkaraunj, *Hind.*, C. 6; katki (Coptis), *Beng.*, C. 1791; katki (Panicum), *Hind.*, P. 41; katki (Picrorhiza), *Beng., Hind.*, P. 700; katkulijí-ságar-ghôta, *Hind.*, C. 6; kátlálá, *Beng.*, P. 393; katmal, *Pb.*, R. 663; katmanli, *Hind.*, B. 318; kat marra, *Hind.*, L. 474; katmédh, *Hind.*, L. 474; katmoría, *Hind.*, L. 474; katnim, *Hind.*, M. 800; katol (Artocarpus), *Hind.*, A. 1489, *Beng., Hind.*, I. 58; kátonda, *Pb.*, V. 91; katonj, *Kumaon*, C. 818; katori (Cissampelos), *Pb., Sind*, C. 1205; katorí (Ægle), *Sind*, A. 534; katorí (Feronia), *Sind*, F. 53; katráin, *Pb.*, S. 303; katrar, *Kumaon*, A. 233; kátri, *Mar.*, V. 164; katsareyá, *Hind.*, B. 171; katsávar, *Mar.*, E. 289; katsirsa, *Hind.*, D. 53; katsol, *Kumaon*, R. 605; kétsú, *Pb.*, I. 128; katta mítha, *Pb.*, R. 650; kattang, *Hind.*, B. 118; kattár, *Hind.*, F. 615; kattáwa, *Hind.*, X. 21; katthá, *Hind.*, A. 137, *Beng., Hind.*, A. 181;

katti, *Hind.*, F. 615; kát til, *Beng.*, S. 1078; kattorí, *Sind*, C. 1367; kátu, *Pb.*, F. 19; katúl, *Hind., Mar.*, R. 16; katula, *Kumaon, Pb.*, G. 143; katumbri, *Hind.*, F. 202; katwal, *Hind.*, M. 639; kau, *Hind., Pb.*, O. 145; kauka, *Kumaon*, L. 30; kauki, *Mar.*, M. 590; kaula, *Hind.*, P. 546; kaula (Cucurbita), *Mar.*, C. 2331; kaula (Pygeum), *Mar.*, P. 1442; kauli (Sageretia), *Pb.*, S. 505; káulí (Abies), *Pb.*, A. 17; kaundal, *Mar*, T. 600; kauni. *Beng., Dec., Hind.*, S. 1212; kaunki, *Hind*; R. 335; kaunphal, *Mar.*, D. 534; kauntel, *Mar.*, C. 410; kaunuí, *Hind.*, R. 335; kaur (Capparis), *Pb.*, C. 431; kaur (Picrorhiza), *Pb.*, P. 700; kaur (Reinwardtia), *Pb.*, R. 71; kaur (Roylea), *Pb.*, R. 561; kaura (Holarrhena), *Hind.*, H. 294; káura (Morus), *Pb.*, M. 775; kauraro, *Pb.*, K. 30; kaurchi, *Mar.*, D. 32; kaureyá, *Hind.*, H. 294; kauri (Acer), *Pb.*, A. 328; kauri (Roylea), *Kumaon, Pb.*, R. 561; kauri (Cyamopsis), *Hind.*, C. 2514; kauri bóti, *(Jhelam), Pb.*, A. 677; kaurí jál, *Pb.*, S. 717; kaurí-jál, *Pb.*, C. 224; kaurio, *Mar.*, R. 16; kauri ván, *Pb.*, C. 224; kaurí ván, *Pb.*, S. 717; kauti, *Mar.*, H. 472; kautí, *Mar.*, H. 468; kautú, *Pb.*, T. 93; kava (Coffea), *Beng.*, C. 1641; kava (Hydnocarpus), *Mar.*, H. 472; kavacha, *Mar.*, M. 786; kavalee, *Mar.*, S. 2850; kavalí (Gymnema), *Mar.*, G. 748; kávali (Sterculia), *Mar.*, S. 2850; kavandala, *Mar.*, T. 600; kavasitengara, *Beng.*, F. 487; kavatela, *Mar.*, H. 472; kavatha, *Mar., Sind*, F. 53; kavá-thénthi, *Hind.*, C. 1403; kavdi, *Beng.*, F. 315; kavili, *Uriya*, S. 2850; kavít, *Gus., Mar.*, F. 53; kavith, *Mar.*, F. 53; kavitha, *Hind.*, F. 53; kavsi, *Mar.*, H. 371; kawa, *Hind., Mar.*, C. 1641; káwal, *Hind.*, P. 546; kawala, *Hind., S.* 1911; kawala, *Hind.*, S. 2103; kawala, *Hind.*, M. 22; kawalechedole. *Mar.*, B. 905; kawanch, *Pb.*, M. 786; kawar, *Pb.*, H. 294; kawára, *Hind.*, C. 2515; kawári, *Guj.*, C. 797; kawat (Feronia), *Mar.*, F. 53; kawat (Limonia), *Mar.*, L. 362; kawátúntí, *Pushtu*, C. 1403; kawíd. *Pb.*, H. 382; káyákutí, *Mar.*, M. 342; káyaphala, *Mar.*, M. 869; káyaphul, *Beng.*, M. 869; kayaputí, *Hind.*, M. 342.

Ke Keautiah, *Hind.*, R. 124; kecara, *Mar.*, C. 2083; kedári, *Guj.*, L. 312; kedarí chúa, *Pb.*, A. 921; keddú, *Pb.*, L. 30; kedgi, *Dec.*, P. 26; keharsu, *Pb.*, Q. 29; kehimu, *Pb.*, S. 3079; kehú, *Pb.*, C. 49; keiao, *Beng.*, P. 882; keim, *Hind., Pb.*, S. 2799; kein, *Hind.*, B. 520; keindu, *Pb.*, D. 628; keint, *Pb.*, P. 1466; keiri (Feronia), *Raj.*, F. 53; keiri (Limonia), *Hind.*, L. 362; keitha, *Pb.*, P. 1466; kejur, *Beng., Hind.*, P. 588; kekar *Mar.*, D. 219; kekti, *Mar.*, A. 603; kél, *Dec., Mar.*, M. 811; kéla, *Guj.*; *Hind., Mar., Pb.*, M. 811; kelai, *Pb.*, C. 846; kelí, *Pb.*, C. 846; kelikadam, *Beng.*, A. 514; kelmang, *Pb.*, C. 846; kelu, *Pb.*, C. 846; kemal, *Pb.*, O. 38; kemball, *Pb.*, O. 38; kembri, *Raj.*, F. 230; kemdo, *Sind*, P. 1259; kemuka, *Mar.*, C. 2013; kena, *Mar.*, C. 1752; kend, *Beng.*, D. 615,

656 ; kenda, *Mar.*, P. 26 ; kendakeri, *Uriya*, C. 596 ; kendhu, *Uriya*, D. 569, 615, 656 ; kendu, *Hind.*, D. 615, *Pb.*, D. 628, 656 ; kenjal, *Mar.*, T. 361 ; kent, *Pb.*, P. 1466 ; kenwal, *Pb.*, C. 846 ; keol, *Hind.*, F. 216 ; keoli (Cedrus), *Himálayas*, C. 846 ; keoli (Costus), *Hind.*, C. 2013 ; keoli-kel mang, *Pb.*, C. 846 ; keonge, *Beng.*, S. 2850 ; keonla, *Hind.*, C. 1235 ; keor, *Pb.*, W. 131 ; keora, *Hind.*, *Mar.*, P. 26 ; keore-ká-múl, *Hind. Mar.*, I. 428 ; keori (Cedrus), *Pb.*, C. 846; keori (Pandanus), *Beng.*, P. 26 ; keot, (*C. P.*), *Hind.*, V. 55 ; keowra, *Beng.*, S. 2369 ; ker (Capparis), *Guj.*, C. 402 ; ker (Ipomæa), *Pb.*, I. 384 ; kerá, *Mar.*, C. 402 ; ke-raad, *Pb.*, F. 549 ; keiaita, *Pb*, T. 376 ; kerak jayat, *Beng.*, C. 115 ; kerani, *Mar.*, G. 735 ; kerauli, *Hind.*, L. 474 ; kerendo kuli, *Uriya*, C. 596 ; kerí (Capparis), *Pb.*, C. 431; kerí(Girardinia), *Pb.*, G. 213 ; kerni, *Mar.*, M. 583 ; kerra, *Uriya*, S. 2943 ; kerula, *Hind.*, L. 563 ; kerze, *Pb.*, L. 252 ; késar (Crocus), *Hind.*, C. 2083 ; késar (Mallotus), *Beng.*, M. 71 ; kesarája, *Beng.*, W. 25 ; kesaraya, *Beng.*, E. 7 ; kesardá, *Uriya*, E. 7 ; kesare, *Mar.*, B. 523 ; kesar-kirote, *Hind.*, C. 2083 ; keshar, *Guz.*, C. 2083 ; keshwri, *Beng.*, E. 7 ; kesraj, *Beng.*, W. 25 ; kesrú, *Pb.*, P. 1078 ; kessar, *Mar.*, C. 2083 ; késú, *Hind.*, B. 950 ; kesuda, *Hind.*, B. 950 ; kesui (Vitis), *Beng.*, V. 195 ; kesúr (Scirpus), *Beng.*, *Hind.*, S. 977 ; kesuri (Bixa), *Mar.*, B. 523 ; kesuri (Fimbristylis), *Beng.*, F. 281 ; kesuriá, *Beng.*, W. 25 ; kesuti, *Beng.*, E. 7 ; ketári, *Hind.*, S. 30 ; ketgi, *Hind.*; P. 26 ; keti, *Beng.*, D. 92 ; ketki, keya, *Beng.*, P. 26 ; ketuki, *Beng.*, P. 26 ; keú, *Beng.*, *Hind.*, C. 2013 ; keun, *Kashmir*, L. 385 ; keur, *Mar.*, P. 26 ; keute sap, *Beng.*, R. 124 ; kevan, *Dec.*, H. 92 ; kevana, *Mar.*, H. 92 ; kewai, *Hind.*, P. 77, 79 ; kewan, *Mar.*, H. 92 ; kewanné, *Dec.*, H. 92 ; kewar, *Pb.*, H. 294 ; kewiro, *Sind*, M. 811 ; kewoda, *Guz.*, P. 26 ; keyá, *Beng.*, P. 26 ; keysuria, *Beng.*, E. 7 ; keyu, *Hind.*, C. 2013.

Kh Khabájhi, *Sind*, M. 115 ; khabar (Ficus), *Hind.*, F. 216, 265 ; khabára, *Hind.*, F. 230 ; khabáre, *Pb.*, F. 230 ; khabarra (Ehretia), *Pushtu*, E. 35 ; khabazi, *Sind*, M. 109 ; khabbal, *Pb.*, C. 2558 ; khabiún, *Pb.*, R. 215 ; khad, *Hind.*, M. 237 ; khádar, *Hind.*, M. 237 ; khadaur, *Hind.*, M. 237 ; khadchampo, *Mar.*, P. 989 ; khaddhi, *Hind.*, M. 237 ; khaderí, *Mar.*, A. 135 ; khádrí, *Pb.*, R. 365 ; khagal, *Pb.*, T. 51 ; khagar, *Hind.*, S. 49 ; khagarwal, *Pb.*, X. 1 ; khágin, *Hind.*, C. 1403 ; khágrá, *Beng.*, S. 49 ; khaija, *Kumaon*, P. 618 ; khailuwa, *Kumaon*, P. 618 ; khair, *Dec.*, *Hind.*, A. 135 ; khaira, *Mar.*, A. 135 ; khaira-ka-jhor, *Dec.*, *Hind.*, *Mar.*, A. 880 ; khair-babúl, *Dec.*, *Hind.*, A. 135 ; khair-champa, *Mar.*, P. 989 ; khair gónd, *Hind.*, A. 136 ; khairposh, *Pb.*, L. 358 ; khairuwa, *Hind.*, A. 1562 ; khairwál, *Hind.*, B. 308, 356 ; khája, *Hind.*, B. 863, 868 ; kháj-goli-cha-vel, *Mar.*, V. 228 ; khají, *Pb.*, *Sind*, P.

555 ; *Hind.*, *Pb.*, P. 588 ; kháj-kolli, *Mar.*, T. 509; khájoti, *Mar.*, A. 306 ; khajú, *Pb.*, P. 1463 ; khajúr, *Beng.*, *Guj.*, *Mar.*, P. 555, *Hind.*, *Pb.*, P. 555, 588 ; khajura, *Mar.*, P. 588 ; khajuri, *Hind.*, P. 551, *Mar.*, P. 588 ; khajuria, *Hind.*, E. 259 ; khajuwa, *Beng.*, G. 263 ; khaka, *Hind.*, T. 232 ; khakananutela, *Guz.*, S. 705 ; khakar, *Guj.*, B. 944 ; khákara, *Guj.*, *Mar.*, B. 944 ; khákar-gónd, *Mar.*, B. 945 ; khakhado, *Guj.*, B. 944; khakhan, *Mar.*, S. 705 ; khákharnu-jhada, *Guj.*, B. 944 ; khakharo, *Mar.*, B. 944; khakkar, *Pb.*, P. 833 ; khákshí, *Mar.*, *Pb.*, S. 2219 ; khalátra, *Pb.*, E. 270 ; khalis, *Kumaon*, Z. 231 ; khalis, *Hind.*, G. 280 ; khalk, *Pb.*, C. 881, 886 ; khalshi, *Beng.*, A. 531 ; khám, *Beng.*, *Hind.*, D. 485 ; khama, *Pb.*, E. 234 ; khamach, *Beng.*, M. 784 ; khamádrús, *Pb.*, S. 2518 ; khám-alu, *Hind.*, *Beng.*, D. 485 ; khamara, *Hind.*, *Kumaon*, T. 525 ; khamaraka, *Mar.*, A. 1646 ; khamazora, *Pushtu*, W. 93 ; khambhári, *Hind.*, G. 287 ; khámbúr, *Pb.*, F. 725, *Pushtu*, A. 590 ; khambarna, *Hind.*, D. 534 ; khamjaria, *Pb.*, W. 93 ; khamjíra-*Pb.*, W. 93 ; khámlákh, *Pb.*, L. 1 ; kham, mara, *Hind.*, G. 287 ; khám mittí, *Pb.*, A. 962 ; khamrak, *Dec.*, *Hind.*, A. 1646 ; khan (Állium), *Pb.*, A. 786 ; khan (Olea), *Sind*, O. 145 ; khán (Saccharum), *Hind.*, *Sind*, S. 49; khanam, *Pb.*, C. 838; khand (Saccharum), *Pb.*, S. 30 ; khanda (Ephedra), *Pb.*, E. 234 ; khandodi, *Mar.*, D. 823 ; khandú, *Pb.*, P. 885 ; khandura, *Hind.*, A. 1186 ; khand-vel, *Mar.*, V. 55 ; khangar, *Pb.*, P. 833 ; khan-ge, *Pb.*, F. 424 ; khanna, *Sind*, C. 1748 ; khanphutia, *Hind.*, C. 1367 ; khánr, *Hind.*, S. 378 ; khántián, *Pb.*, R. 543 ; khápará, *Mar.*, B. 619 ; khápato, *Sind*, A. 89; khappar kadu *Hind.*, C. 978, *Mar.*, C. 982 ; khápra, *Mar.*, T. 537 ; khár (Barilla), *Guj.*, B., 156 ; khár (Carthamus), *Pb.*, C. 637 ; khár (Chrysopogon), *Pb.*, C. 1055 ; khár (Haloxylon), *Pb.*, *Pushtu*, H. 10 ; khár (Heteropogon), *Hind.*, H. 164 ; khár (Prosopis), *Pb.*, P. 1259 ; khará, *Hind.*, *Beng.*, P. 1304 ; kharabúja, *Mar.*, C. 2263 ; kharai (Celastrus), *Pb.*, C. 860 ; kharai (Heliotropium), *Pb.*, H. 106 ; kharak (Celtis), *Hind.*, C. 881, *Pb.*, C. 886, 898 ; kharak (Phœnix), *Guj.*, P. 588 ; kharakírásna, *Mar.*, T. 855 ; kharanja, *Pb.*, C. 224, Q. 29 ; kharat, *Hind.*, J. 61 ; khara-tua, *Pb.*, C. 1018 ; kharawat, *Mar.*, F. 202 ; kharawíne, *Pb.*, S. 2341 ; kharawune, *Pushtu*, E. 35 ; kharbuj, *Mar.*, C. 2263 ; kharbuja (Cucumis), *Mar.*, C. 2263 ; kharbuza (Carica), *Pb.*, C. 581 ; kharbuza (Cucumis), *Hind.*, C. 2263 ; khardag, *Pb.*, P. 684 ; khardháwa, *Hind.*, A. 1149 ; khardi, *Mar.*, C. 2057 ; kharebútí, *Pb.*, G. 186 ; kharen, *Kumaon*, Q. 70 ; kharent, *Pb.*, S. 1694 ; khareo, *Pb.*, Q. 29, 70 ; khareti, *Hind.*, S. 1694 ; kha-reu, *Pb.*, Q. 29, 70 ; kháreza, *Pb.*, C. 633 ; kharg, *Pb.*, C. 886 ; khar-gaingí, *Pushtu*, O. 230 ; khargosh (hare), *Hind.*, F. 777, 779, H. 31, 35 ; khargosh (Verbascum), *Pb.*, V. 64 ; khargul, *Mar.*, T. 522 ; khari (Eragrostis),

Hind., E. 248 ; khári (Cicer), *Guj.*, C.
1061 ; kharial, *Sind*, D. 9 ; kharián maraghúne, *Pb.*, S. 2345 ; khariára, *Kashmir,*
R. 588 ; kharídjar, *Sind*, S. 717, C. 224 ;
khar-i-jinghak, *Afg.*, P. 1272 ; kharíka,
Hind., C. 881 ; kha-rí-láni (Suæda), *Sind,*
S. 2994 ; khárí-láni (Haloxylon), *Sind,*
H. 10 ; khari-máti, *Beng.*, C. 489 ; kharí-
mittí, *Hind.*, *Pb.*, C. 489 ; kharimbar,
Pb., E. 184 ; kharipírú, *Sind*, S. 717 ;
kháriyu, *Mar.*, C. 617 ; kháriz, *Pb. Hills,*
C. 2021 ; khariz lúni, *Kashmir*, C. 2023 ;
kharjúr, *Mar.*, P. 555 ; khark, *Pb.*, C.
881, 886 ; kharkána (Saccharum), *Pb.*, S.
6 ; kharkath, *Hind.*, O. 233 ; khar kharnár
(Verbascum), *Pb.*, V. 64 ; kharkhura,
Hind., S. 1207 ; kharkhusa, *Pb.*, S. 2985 ;
kharlei, *Pb.*, T. 51 ; kharmach, *Pb.*, R.
598 ; kharmo, *Pb.*, L. 533 ; kharmuch,
Kashmir, R. 598 ; kharmuj, *Beng.*, C.
2263 ; kharnúb, *Pb.*, C. 933 ; kharnúb
hindí, *Pb.*, P. 1259 ; kharnúb núbti, *Pb.*,
C. 933 ; kharoa, *Pushtu*, T. 93 ; kharo-
makro, *Raj.*, E. 188 ; kharot, *Kumaon*, J.
61 ; kharoti, *Mar.*, F. 124 ; kharpat, *Hind.*,
Pb., G. 143 ; khárpata, *Pushtu*, Q. 29 ;
kharpata serei, *Pb.*, Q. 35 ; kharra, *Beng.*,
Hind., H. 35 ; kharrei, *Hind.*, M. 545 ;
kharri lani, *Sind*, S. 2990 ; kharsa, *Hind.*,
F. 463 ; khár-sajji, *Hind.*, B. 158 ; kharsan
kauriála, *Pb.*, C. 2101 ; kharshú, *Pb.*, Q.
70 ; kharsing, *Mar.*, S. 2884 ; khartua sag,
Hind., C. 1003 ; khártuma, *Pb.*, C. 1211 ;
kharvat, *Mar.*, A. 1200 ; kharwala (Debregeasia), *Pushtu*, D. 196 ; kharwala
(Salix), *Pb.*, S. 537 ; kharwat, *Mar.*, F.
124 ; kharwé, *Pushtu*, C. 2021 ; kha.
(Andropogon), *Hind.*, A. 1097 ; khas
(Lactuca), *Hind.*, L. 21 ; khasa (Sesamum), *Beng.*, *Uriya*, S. 1078 ; khasa-
khasa, *Mar.*, A. 1097 ; khashbar, *Pb.*, P.
693 ; khashlala-chodí, *Sind*, P. 1426 ;
khash-khash, *Dec.*, *Pb.*, P. 87 ; khas-khas
(Andropogon), *Beng.*, *Dec.*, *Hind.*, A.
1097 ; khas-khas (Papaver), *Mar.*, P. 87 ;
khaskhasa (Suæda), *Pb.*, S. 2985 ; khaslá
til, *Beng.*, S. 1078 ; khas-rai, *Hind.*, B.
833 ; khát, *Hind.*, M. 237 ; khát, *Guz.*,
V. 195 ; khatái, *Pb.*, F. 624 ; khátajamir,
Beng., C. 1234 ; khata limbu, *Guj.*, C.
1296 ; khatara, *Mar.*, L. 355 ; khatás,
Beng., *Hind.*, T. 427, T. 441 ; khatbírí,
Pb., R. 650 ; khatechawal, *Hind.*, P. 1187 ;
khatip, *Pb.*, V. 91 ; khat-khataya, *Hind.*,
S. 2345 ; khatkhati, *Hind.*, C. 2101 ; khát
kúra, *Pb.*, M. 237 ; khatmi, *Beng.*, M.
115 ; khatta mithá, *Pb.*, O. 547 ; khatta
níbu, *Hind.*, C. 1300 ; khattíkan, *Pb.*, R.
645 ; khattímal, *Pb.*, R. 642 ; khattítan,
Pb., R. 650 ; khatumdre, *Guz.*, V. 195 ;
khatyán, *Duk.*, E. 289 ; khatyan-kakalli,
Dec., E. 295 ; khau (manure), *Hind.*, M.
237 ; khau (Olea), *Sind*, O. 145 ; khau
(Saccharum), *Sind*, S. 49 ; khauní, *Pb.*,
S. 1212 ; khaun phal, *Mar.*, D. 513 ; khávi,
Hind., *Pb.*, A. 1093 ; khawa, *Hind.*, T.
282 ; khawàri, *Beng.*, P. 1121 ; kháwi,
Hind., *Pb.*, A. 1093 ; khayer, *Beng.*, A.
135 ; khejra, *Raj.*, P. 1259 ; khejuri, *Uriya,*
P. 588 ; khéla, *Pb.*, M. 811 ; kheli, *Pb.*, D.
503 ; khem, *Hind.*, S. 2799 ; khemra,
Hind., G. 136 ; khen (Dichrostachys),
Raj., D. 402 ; khen (Tecoma), *Sind*, T.

227 ; khenan, *Hind.*, F. 156 ; khenda,
Uriya, D. 569 ; khenti, *Pb.*, I. 123, 128 ;
kheo, *Pb.*, S. 2672 ; khep, *Pb.*, C. 2101 ;
kher (Acacia), *Guj.*, A. 135 ; kher (Heteropogon), *Beng.*, H. 164 ; kher (Setaria),
Pb., S. 1212 ; khera (Acacia), *Mar.*, A.
135 ; kherbaba, *Pb.*, C. 2021 ; kheri (Dichrostachys), *Hind.*, D. 402 ; kheri (Phaseolus), *Beng.*, P. 468 ; kheri (steel),
Hind., I. 440 ; kheroa, *Pb.*, C. 2021 ;
kherúya, *Beng.*, P. 496 ; khesári, *Beng.*,
Hind., L. 100 ; khetiya, *Hind.*, B. 822 ;
khetki, *Hind.*, A. 636 ; khetpapra, *Beng.*,
O. 132 ; khet sunsuni, *Hind.*, D. 343 ;
khewnau, *Hind.*, F. 156 ; khhaji, *Hind.*,
P. 555 ; khif, *Pb.*, C. 2101 ; khijado, *Guz.*,
P. 1259 ; khijro, *Guz.*, P. 1259 ; khí-
khowa, *Sind*, L. 114 ; khímor, *Pb.*, V.
91 ; khin, *Kumaon*, A. 247 ; khína lienda,
Hind., S. 838 ; khinam, *Hind.*, C. 835 ;
khinna, *Hind.*, S. 838, 1889 ; khíp (Crotalaria), *Pb.*, C. 2101 ; khíp (Orthanthera), *Pb.*, *Sind*, O. 247 ; khíp (Pæderia), *Hind.*, P. 4 ; khippí, *Pb.*, C. 2101 ;
khir (Mimusops), *Hind.*, M. 583 ; khir
(Sporobolus), *Pb.*, S. 2670 ; khira, *Beng.*,
Hind., *Pb.*, C. 2287 ; khirai, *Beng.*, O.
600 ; khiran, *Mar.*, H. 92 ; khircha, *Pb.*,
Pushtu, G. 702 ; khírd, *Mar.*, F. 329 ;
khirg, *Pb.*, C. 881 ; khirk, *Hind.*, *Pb.*, C.
881, *Pb.*, 886 ; khirkhejur, *Beng.*, M. 583 ;
khirni, *Hind.*, *Mar.*, W. 122 ; (Wrightia)
khirni (Mimusops), *Hind.*, F. 671, *Hind.*,
Raj., *Guj.*, M. 583, *Hind.*, *Mar.*, *Pb.*,
M. 590 ; khírurh, *Mar.*, F. 564 ; khírwá,
Uriya, S. 2369 ; khishing, *Hind.*, C. 835 ;
khish-khash, *Pb.*, P. 87 ; khista, *Pushtu,*
P. 1285 ; khit, *Hind.*, I. 440 ; khitmi-ká-
jhár, *Dec.*, *Hind.*, *Mar.*, A. 880 ; khiyar,
Pb., C. 2287 ; khíyár-shánbur, *Hind.*, C.
756 ; kho, *Pb.*, G. 186 ; khoi, *Hind.* &
Pb., A. 1093 ; khoinbo, *Sind*, C. 637 ;
khoiru, *Uriya*, A. 135 ; khokalí, *Hind.*,
Mar., A. 306 ; khokar, *Pb.*, S. 705 ;
khokli, *Hind.*, *Mar.*, A. 306 ; khoksa,
Beng., F. 361 ; khólar-mandá, *Hind.*, P.
1370 ; khóprá, *Dec.*, *Hind.*, C. 1520 ;
khópru, *Guj.*, C. 1520 ; khor (Acacia),
Sind, A. 273 ; khor (Hedychium), *Pb.*,
H. 59 ; khor (Juglans), *Pb.*, J. 61 ; khora-
kema, *Afg.*, F. 84 ; khórásáni-ajmo, *Guz.*,
H. 525 ; khcrasaniajowan, *Beng.*, H. 525 ;
khorásáni ajván, *Guz.*, H. 525 ; khorasáni
ajwáin, *Hind.*, H. 525 ; khorasaní kútkí,
Hind., H. 111 ; khorasáni owa, *Mar.*,
H. 525 ; khórásáni-vóvá, *Mar.*, H. 525 ;
khoréti, *Mar.*, F. 124 ; khorí djhar, *Sind,*
C. 224 ; khotbir, *Mar.*, C. 1954 ; khotmir,
Mar., C. 1954 ; khowsey, *Mar.*, S. 2819 ;
khubani, *Hind.*, *Pushtu*, P. 1285 ; khu-
bazi, *Hind.*, M. 109, *Mar.*, M. 115 ;
khúb kalam, *Beng.*, O. 18 ; khúb-kalán,
Hind., *Pb.*, S. 2219 ; khúdijamb, *Beng.*,
A. 1218 ; khúdiokra, *Beng.*, C. 2211 ;
khudrí, *Mar.*, F. 346 ; khueri, *Beng.*, C.
997 ; khujiyan, *Hind.*, *Pb.*, P. 555 ; khu-
kan, *Pb.*, M. 910 ; khul, *Duk.*, A. 554 ;
khulakhudi, *Hind.*, H. 486 ; khulen, *Pb.*,
H. 324 ; khulfé-ké-bínj, *Dec.*, P. 1179 ;
khulfé-ki-bhájí, *Dec.*, P. 1179 ; khuljeh ke
baji, *Duk.*, C. 1003 ; khúlti (Cyamopsis),
Hind., C. 2514 ; khúlti (Dolichos), *Hind.*,
D. 758 ; khúm, *Pb.*, L. 535 ; khúmaza·e,

Pb., W. 93 ; khúmb (Hiptage), *Pb.,* H. 285 ; khúmb (Morche'la), *Pb.,* M. 647 ; khúmbah, *Afg.,* A. 590, *Pb., Sind,* F. 725 ; khumbi, *Hind.,* C. 563 ; khumha, *Sind,* A. 590 ; khunda (Ischœmum), *Hind.,* I. 501 ; khunda (Prosopis), *Pb.,* P. 1259 ; khuni, *Pb.,* C. 576 ; khunjad, *Afg.,* P. 847 ; khunjada, *Afg.,* P. 847 ; khunseráia, *Pb.,* M. 63 ; khunuk, *Hind.,* D. 151 ; khupyabágh, *Hind.,* T. 440 ; khura, *Mar.,* I. 515 ; khurásáni-ajvan, *Dec.,* H. 525 ; khurá-sáni-ajváyan, *Hind.,* H. 525 ; khurá-sánijamani, *Hind.,* H. 525 ; khurásh, *Pb.,* P. 77 ; khurasli, *Mar.,* N. 179 ; khurbanri, *Pb* , A. 677 ; khurbúj, *Hind.,* C. 2263 ; khurbuza, *Pb.,* C. 2263 ; khúrd, *Pb.,* W. 106 ; khurfah, *Hind.,* P. 1179 ; khurfé-ká-ság, *Hind.,* P. 1179 ; khurfé-ké-bínj, *Hind.,* P. 1179 ; khurhur, *Hind.,* F. 156 ; khuri (Ischœmum), *Hind.,* I. 501 ; khuri (Rubia), *Pb.,* R. 564 ; khuri (Saccharum), *Beng.,* S. 25, 49 ; khuriari, *Hind.,* C 136 ; khuríu, *Pb.,* Q. 70 ; khurma, *Beng., Hind.,* P. 555 ; khurphendra, *Mur.,* G. 136 ; khúrrúr, *Hind.,* G. 136 ; khursa, *Hind.,* P. 1179 ; khurti (Cyamopsis), *Hind.,* C. 2514 ; khurti (Dolichos), *Hind.,* D. 758 ; khusam, *Hind.,* E. 208 ; khúshin (Myrsine), *Pb.,* M. 910 ; khúshíng (Cedrela), *Pb.,* C. 838 ; khusiyár, *Hind.,* S. 30 ; khus-khus, *Guj., Mar.,* P. 87 ; khutrau, *Hind.,* A. 17 ; khutta, *Pb.,* C. 1286 ; khwa, *Pushtu,* T. 61 ; khwagawala, *Pushtu,* S. 550 ; khwairaal, *Hind.,* B. 356 ; khwan, *Baluch.,* O. 145.

Ki Kiain, *Pb.,* A. 596 ; kiámil, *Hind., Pb.,* O. 38 ; kiar, *Pb.,* C. 756 ; kiárí, *Pb.,* C. 431 ; kiat, *Pb.,* P. 1466 ; kibrit, *Pb.,* S. 2999 ; kich-uk-lonar, *Sind,* F. 413 ; kidámárí, *Mar.,* A. 1395 ; kierpa, *Beng.,* C. 474 ; k:kalnshing, *Kumaon,* A. 353 ; kikar, *Beng., Hind., Pb.,* A. 101, F. 671 ; kikkri, *Pb.,* M. 562 ; kikra, *Hind.* C. 1183 ; kíkúr, *Sind.,* P. 322 ; kílan, *Hind.,* C. 846 ; kilar, *Himálaya,* C. 846 ; kilaunta, *Himálaya,* A. 22 ; kiláwa, *Pb.,* W. 131 ; k:lei,!*Himálaya,* C. 846 ; kilgach, *Hind.,* C. 6 ; kilik, *Hind.,* S. 25 ; killah (Ficus), *Mar.,* F. 216 ; killar (Parrottia), *Pb.,* P. 327 ; killo-debdháor, *Beng.,* S. 2390 ; kílmích, *Pb* , V. 91, V. 96 ; kilmira, *Pb.,* G. 143 ; kilmora, *Kumaon,* B. 453 ; kilonj, *Hind.,* Q. 13 ; kilu (Abies), *Hind., Kumaon,* A. 17 ; kilu (Acer), *Kumaon,* A. 328 ; kilu (Nannorhops), *Pb.,* N. 1 ; kilut, *Hind.,* S. 25 ; kimlú, *Pb.,* O. 38 ; kimri, *Pb.,* F. 149 ; kimti, *Pb.,* C. 1286 ; kimu, *Hind., Pb.,* M. 775 ; kimúl, *Hind., Mar.,* O. 38 ; kin, *Pb.,* S. 1242 ; kinai, *Mar.,* A. 717 ; kinai tihiri, *Mar.,* A. 717 ; kindal, *Mar.,* T. 355 ; kingaro, *Pb.,* C. 860 ; kíngí, *Pb.,* G. 213 ; kingin, *Mar.,* C. 2556 ; king khak durunga, *Pb.,* A. 1464 ; kíngli, *Hind.,* M. 562 ; kíngrei, *Hind.,* M. 562 ; kingro, *Pb.,* F. 624 ; kínhai, *Mar.,* A. 717 ; kinjad, *Afg.,* P. 847 ; kinjada, *Afg.,* P. 847 ; kinjak, *Afg.,* P. 847 ; kínjal, *Mar.,* T. 355 ; kinjolo, *Uriya,* B. 180 ; kinkanela, *Mar.,* S. 705 ; kinkar, *Pb* , A. 238 ; kinnab, *Hind.,* C. 331 ; kinnú, *Pb.,* D. 656 ;

kinro (Gynandropsis), *Sind,* G. 753 ; kínro (Salvia), *Sind,* S. 746 ; kinton, *Beng.,* C. 1165 ; kinyá-gond, *Hind.,* B. 945 ; kioch, *Pb.,* E. 475 ; kios-a-gi, *Afg.,* S. 2424 ; kip (Leptadenia), *Sind,* L. 296 ; kip (Orthanthera), *Sind,* O. 247 ; kirab, *Sind,* C. 402 ; kiráíta, *Mar.,* S. 3018 ; kirakal, *Pb.,* A. 1369 ; kirakang, *Beng., Dec., Hind.,* S. 1212 ; kiral, *Sind,* C. 402 ; kiralu, *Pb.,* A. 1379 ; kiramál, *Hind., Mar.,* P. 1121 ; kiramáni owa, *Mar.,* A. 1452 ; kírámár, *Dec , Hind.,* A. 1395 ; kiran, *Sind,* S. 1024 ; kiranda, *Hind.,* C. 1458 ; kirang, *Sind,* S. 1212 ; kiranj, *Beng., Dec., Hind.,* S. 1212 ; kirara, *Pb.,* M. 639 ; kirbut, *Sind,* C. 6 ; kírch, *Pb.,* D. 409 ; kírdamána, *Mar.,* C. 1765 ; kirfa, *Pb.,* C. 1196 ; kiri (Ceriops), *Sind,* C 964 ; kiri (Jasminum), *Hind.,* J. 29 ; kiri-ki-kukri (Arisæma), *Pb.,* A. 1379, 1381 ; kírímar, *Pb.,* S. 2674 ; kirindur, *Beng.,* G. 53 ; kirki, *Pb.,* C. 886 ; kirkichálú, *Pb.,* A. 1369 ; kirkiria, *Hind.,* C. 1183 ; kirkundí, *Mar ,* J. 59 ; kirm, *Pb.,* C. 1458 ; kirmai, *Pb.,* L. 1 ; kirmaz, *Mar.,* C. 1458 ; kírmdana, *Beng.,* C. 1458 ; kirmi, *Pb.,* F. 230 ; kírminji-ajván, *Mar ,* S. 1201 ; kirmira, *Mar.,* G. 271 ; kirmoli, *Hind.,* A. 341 ; kirna, *Hind., Mar.,* S. 487 ; kírni, *Mar.,* C. 393 ; kirpa, *Beng.,* L. 576 ; kirpáwa, *Pb.,* I. 384 ; kirra kerin, *Pb.,* C. 402 ; kirrari, *Sind,* C. 964 ; kirri, *Baluch.,* T. 51, 79 ; kírrú, *Pb.,* P. 327 ; kirrur, *Sind,* C. 402 ; kirsel, *Mar.,* S. 2865 ; kir-thag, *Afg.,* E. 252 ; kirui, *Beng.,* O. 600 ; kirwatzei, *Pushtu,* A. 498 ; kiryát, *Hind.,* A. 1064 ; kiryáta, *Guj.,* A. 1064 ; kiryáto, *Guj.,* A. 1064 ; kisára, *Hind ,* L. 100 ; kisárí, *Pb.,* L. 100 ; kishing, *Kumaon,* A. 567 ; kishmish, *Mar.,* V. 233 ; kishmishi. *Hind.,* V. 243 ; kishta, *Pb.,* P. 1285 ; kishta bahíra, *Kashmir,* P. 1452 ; kishur, *Beng.,* M. 71 ; kismis, *Beng., Hind.,* V. 233 ; kisri, *Mar.,* B. 523 ; kithi, *Pb.,* D. 503 ; kithú, *Pb* , P. 1466 ; kitmira, *Kumaon,* G. 143 ; kitola, *Kumaon,* C. 756 ; kitoli, *Hind.,* C. 756 ; kitwáli, *Hind.,* C. 756 ; kiu (Diospyros), *Beng.,* D. 615 ; kiú (Rosa). *Pushtu,* R. 538 ; kiur, *Pb.,* A. 786 ; kiváchh, *Hind.,* M. 786 ; kivánch, *Guj., Hind.,* M. 786 ; kiwára, *Hind.,* C. 321 ; kiyo, *Pushtu,* R. 538.

Kl Klandru, *Pb.,* S. 947 ; klenchu, *Pushtu,* R. 598 ; klíúntí, *Pb.,* L. 535.

Km Kmárkábij, *Pb.,* Z. 231.

Kn Knebáwal, *Sind,* A. 217 ; kníss, *Pb.,* D. 503 ; knítrí, *Pb.,* R. 318.

Ko Ko, *Pb.,* O. 145 ; koá, *Pb.,* T. 70 ; koame, *Pb.,* O. 170 ; koámil, *Pb.,* G. 240 ; koamla, *Pb.,* O. 38 ; koán, *Pb.,* T. 61 ; kóbí, *Hind.,* B. 851 ; kóbiá, *Guj.,* B. 851 ; kóbra cha-téla, *Mar.,* C. 1520 ; kocham, *Mar.,* S. 950 ; kochan, *Hind.,* C. 1969 ; koch gad, *Baluch., Sind,* S. 1278 ; kochi, *Hind.,* A. 200 ; kochinda, *Mar.,* U. 39 ; kochora, *Mar.,* C. 2503 ; koda (Cordia), *Hind.,* C. 1931 ; koda (Ehretia), *Hind.,* E. 20, 25 ; koda (Eleusine), *Pb.,* E. 170, F. 672 ; koda (Paspalum), *Hind.,* P. 332 ; kodai, *Hind.,* P. 332 ; kodaka, *Hind.,* P. 332 ; kodalia, *Beng.,* D. 354 ; kodarsi,

Mar., S. 1029; kodí, *Pb.*, L. 533; kodie, *Hind.*, P. 332; kodo, *Hind.*, *Pb.*, P. 332; kodoá dhán, *Beng.*, P. 332; kodon (Eleusine), *Pb.*, E. 170; kodon (Paspalum), *Hind.*, *Pb.*, P. 332; kodra (Eleusine), *Pb.*, E. 170; kodra (Paspalum), *Guj.*, *Mar.*, *Pb.*, P. 332; kodrám, *Hind.*, *Kumaon*, P. 332; kodri, *Mar.*, P. 332; kodro, *Mar.*, P. 332; kodroakora, *Mar.*, P. 332; kodru, *Mar.*, P. 332; kodu, *Beng.*, L. 30; koe, *Pb.*, A. 797, 801; koelo, *Guj.*, C. 487, 1414; koeta, *Uriya*, F. 53; koeva, *Pb.*, H. 294; kogar, *Pb.*, H. 294; koha, *Hind.*, T. 282; koha!á (Benincasa), *Mar.*, B. 430; kohala (Cucurbita), *Mar.*, C. 2331; koham, *Mar.*, S. 950; kohen, *Pb.*, S. 2376; kohér, *Pb.*, S. 497; kohi, *Pb.*, A. 797; koh-i-dumba, *Afg.*, S. 1278; kohlú, *Pb.*, C. 49; koholú, *Guj.*, B. 430; kohoranj, *Hind.*, C. 200; koh-tor, *Baluch.*, L. 587; kohú, *Pb.*, O. 145; kohula, *Guj.*, B. 430; kohumba, *Gus.*, M. 363; koi, *Beng.*, N. 200; koilari, *Hind.*, B. 308; koileka, *Uriya*, B. 165; koinar, *Hind.*, B. 308; koiral, *Beng.*, *Pb.*, B. 308; kojba, *Hind.*, S. 950; kok (Ficus), *Pb.*, F. 230; koka (Semecarpus), *Hind.*, S. 1041; koka buradí, *Beng.*, S. 746; kokam, *Hind*, *Mar.*, G. 36; kokam chatel, *Mar.*, G. 36; kokamb, *Dec.*, G. 36; kokan (Celosia), *Hind.*, C. 873; kokan (Garcinia), *Guj.*, G. 36; kokanber, *Pb.*, Z. 231, 254, 27c, 280; koketi, *Mar.*, S. 2837; kokhur (Embelia), *Pb.*, E. 199; kokhúr (Myrsine), *Pb.*, M. 910; kokkita, *Hind.*, G. 128; kokli, *Mar.*, Z. 35; kokni-ber, *Pb.*, Z. 254; kokní bér, *Pb.*, C. 224; kokoaru, *Beng.*, O. 127; kokoranj, *Hind.*, C. 122; kokra, *Beng.*, A. 1245; koksimá, *Beng.*, C. 878; koku, *Hind.*, *Pb.*, C. 224; kokwa, *Beng.*, D. 281; kol, *Pb.*, D. 758; kola, *Hind.*, F. 755; kolain, *Hind.*, *Kumaon*, P. 760; kolan, *Hind.*, *Kumaon*, P. 760; kolaréchi kal, *Mar.*, L. 355; kólasé, *Mar.*, C. 487; kole-ján, *Mar.*, V. 205; kole-zán, *Mar.*, V. 184; koliar, *Hind.*, B. 308, 356; koli-che-chútar, *Mar.*, L. 273; kolijána, *Beng.*, *Hind.*, *Mar.*, A. 862; kolinjan, *Guj.*, A 853; kólís, *Mar.*, F. 411; kolisha, *Beng.*, F. 596; kolissura, *Dec.*, S. 1934; kolistá, *Mar.*, B. 170; kolkánda, *Mar.*, U. 39; kolkaphul, *Beng.*, T. 410; kolon, *Hind.*, *Kumaon*, P. 760; kolsa (carbon), *Dec.*, C. 487; kolsa (coal), *Dec.*, C. 1414; kolsa-ka-pathar (manganese), *Hind.*, M. 131; kolso, *Guj.*, C. 487, 1414; kolsunda, *Mar.*, H. 508; kolth, *Pb.*, D. 758; kolukung, *Beng.*, P. 811; komál, *Afg.*, P. 1221; komba, *Mar.*, S. 1939; kómbi wakumba, *Mar.*, C. 563; kome, *Uriya*, F. 390; kon, *Mar.*, D. 485; konch kari, *Pb.*, M. 786; konda (Cucurbita), *Beng.*, *Hind.*, C. 2331; kondai (Flacourtia), *Hind.*, F. 624; kondákuri, *Mar.*, D. 219; konda-mangà, *Hind.*, G. 128; konda-pulla, *Hind.*, C. 1026; kondaturi, *Uriya*, F. 477; kóndhá, *Hind.*, B. 430; konea-dumbar, *Hind.*, F. 202; kong, *Kashmir*, C. 2083; kongora, *Uriya*, X. 16; konháiah, *Hind.*, B. 426; koni (Panicum), *Kumaon*, P. 53; koni (Setaria), *Hind.*, *Kumaon*, S. 1212; koniari, *Uriya*, O. 1; konkan ber, *Kashmir*, Z. 280; konoo, *Uriya*, H. 517; konpalsehnd, *Mar.*, E. 553; konsso, *Hind.*, B. 856; kontopaláš, *Uriya*, C. 1512; koochuri, *Beng.*, E. 589; kopar, *Hind.*, D. 292; kopásia, *Uriya*, K. 42; kópi, *Beng.*, B. 851; kóp-pátá, *Beng.*, B. 909; kora (Holarrhena), *Hind.*, H. 294; kora (Setaria), *Beng.*, *Dec.*, *Hind.*, S. 1212; korafi-kaoli, *Hind.*, F. 429; korakanda, *Sind*, A. 829; kora-kang, *Mar.*, S. 1212; korak bet, *Beng.*, C. 89; korake, *Pb.*, A. 1606; korala, *Mar.*, B. 304; koral baor, *Beng.*, F. 476; korani, *Kashmir*, P. 882; koranjú, *Uriya*, P. 1121; koranza, *Mar.*, C. 701; kora-phad, *Mar.*, *Sind*, A. 829; kóré-kí-jhár, *Dec.*, C. 2612; koreta, *Mar.*, B. 171; korgani, *Beng.*, A. 60; korhántí, *Mar.*, B. 171; kori, *Beng.*, A. 717; koria, *Hind.*, G. 240; kor-kand, *Duk.*, A. 829; korkol-jodi, *Hind.*, P. 77; korna, *Hind.*, C. 1756; koroh, *Hind.*, S. 1656, 1939; koroi, *Beng.*, A. 717; koron, *Hind.*, S. 1656; korphad, *Mar.*, A. 838; korra, *Uriya*, S. 2943; korsa, *Hind.*, B. 520; kosa, *Hind.*, S. 49; kosam, *Mar.*, S. 950; kosamb, *Mar.*, S. 950; kosh, *Pb.*, A. 801; kosham, *Beng.*, *Hind.*, M. 209; koshimb, *Mar.*, S. 950; koshta, *Beng.*, *Hind.*, C. 1861; kosht-kulinjan, *Mar.*, A. 853; koshú, *Pb.*, M. 461; kosi, *Uriya*, B. 868; kosla, *Mar.*, F. 759; kosto, *Mar.*, H. 472; kostra, *Uriya*, C. 1847; kosum, *Hind.*, S. 950; kosúndra, *Pb.*, B. 318; koswati, *Beng.*, F. 339; kot (Plectranthus), *Pb.*, P. 959; kot (Saussurea), *Hind.*, *Pb.*, S. 910; kota, *Mar.*, P. 1187; kotagandhal, *Hind.*, I. 515; kotaj, *Pb.*, C. 854; kotaku, *Uriya*, S. 2960; kotan, *Duk.*, C. 805; kota-ranga, *Uriya*, G. 124; kotari, *Hind.*, S. 1287; kotha, *Gus.*, F. 53; kothamira, *Mar.*, C. 1954; kotiar, *Pb.*, C. 489; kothíla, *Beng.*, S. 2813; koth-mir, *Mar.*, C. 1954; kotra, *Beng.*, S. 372; kotrí, *Hind.*, F. 356; kotti kang, *Duk.*, A. 1242; kottruk, *Sind*, M. 615; kotu (Fagopyrum), *Hind.*, F. 10; kotu (Setaria), *Pb.*, S. 1207; koundel, *Dec.*, T. 600; kovariya, *Guj.*, *Mar.*, C. 797; kovidara, *Mar.*, B. 356; kovit, *Mar.*, F. 53; kow, *Hind.*, T. 282; kowa (Clitoria), *Hind.*, C. 1403; kowa (Terminalia), *Hind.*, *Mar.*, T. 282; kowah, *Hind.*, T. 282; kowár, *Pb.*, S. 1694; kowaria, *Mar.*, C. 797; kowa-theti, *Hind.*, C. 1403; kowha, *Hind.*, T. 282; kowin, *Pb.*, P. 57; kowit, *Hind.*, *Mar.*, F. 53; kowria, *Uriya*, C. 1847; kowti, *Mar.*, H. 472; koyalá, *Beng.*, C. 487, 1414; koyam, *Uriya*, T. 28; koyan (Agave), *Beng.*, A. 603; koyan (Tamarindus), *Uriya*, T. 28; koyelah, *Hind.*, C. 487, 414.

Kr Kramal, *Pb.*, P. 1142; kramali, *Pb.*, P. 1148, 1159; krambal, *Pb.*, P. 1148; krammal, *Pb.*, P. 1148; krás, *Kashmir*, S. 1160; krau, *Kashmir*, F. 788; kraunti, *Pb.*, L. 535; kre, *Kashmir*, O. 258; kreu, *Pb.*, Q. 70; kripa, *Beng.*, L. 576; krish, *Pb.*, D. 503; krishnachúrá, *Beng.*, C. 32; krishna múga, *Beng.*, P. 496; krishna surmá, *Hind.*, L. 143; krishnatil, *Beng.*, *Hind.*, *Mar.*, S. 1078; krishno-kéli,

Beng., M. 606; kríshún, *Kashmir*, I. 423; kríss, *Pb.*, D. 503; krok, *Pb.*, A. 17, A. 22; krot, *Pb.*, *Kashmir*, J. 61; krowee, *Beng.*, D. 328; krúcho, *Pb.*, I. 14; krúi, *Pb.*, Q. 70; krúm, *Pb.*, M. 775; krumbal, *Pb.*, F. 179; krún, *Pb.*, P. 1316; krúnda, *Afg.*, T. 548; krushál, *Kumaon*, P. 1295.

Ks Ksharisiju, *Uriya*, E. 553; kshiri, *Hind.*, M. 583.

Ku Ku (Celtis), *Pb.*, C. 881, 886; ku (Pyrus), *Pb.*, P. 1466; kúaka neshasteh, *Dec.*, M. 267; kuar, *Hind.*, H. 294; kuayral, *Hind.*, B. 330; kubbur, *Hind.*, *Pb.*, S. 705; kuber, *Sind*, C. 224; kubi, *Gus.*, L. 312; kubjak, *Pb.*, R. 504; kubra, *Hind.*, G. 213; kubrah sál, *Pb.*, F. 515; kuchan (Asparagus), *Pb.*, A. 1573; kuchan (Ephedra), *Pb.*, E. 232; kúchi-kanta, *Beng.*, M. 562; kuchila, *Beng.*, *Pb.*, S. 2943; kuchila lata, *Beng.*, *Hind.*, S. 2936; kuchlá, *Guj.*, *Hind.*, *Uriya*, S. 2943; kuchle-ka-malang, *Hind.*, V. 154; kuchlé-kí-sookan, *Dec.*, V. 154; kúchni, *Pb.*, R. 157; kuchra, *Pb.*, Z. 231; kuchú, *Beng.*, C. 1732; kuchugundubí, *Hind.*, H. 337; kúchunduna, *Beng.*, P. 1381; kuda (Holarrhena), *Guj.*, *Mar.*, H. 294; kuda (Pisum), *Beng.*, P. 885; kuda (Wrightia), *Mar.*, W. 122; kudak, *Mar.*, G. 143; kudaka, *Mar.*, C. 838; kudaliya, *Hind.*, D. 354; kudari, *Beng.*, Z. 182; kud chámpa, *Mar.*, M. 517; kuddia-khár, *Guj.*, B. 731; kudhal, *Dec.*, H. 227; kudhu, *Mar.*, C. 868; kudímah, *Hind.*, B. 430; kúdi mankúni, *Beng.*, G. 175; kudji-kerundi, *Uriya*, F. 347, 356; kúdkí, *Hind.*, D. 758; kudna, *Uriya*, A. 1288; Kudran, *Hind.*, *Kumaon*, A. 17; kudrum, *Hindi*, H. 177; kudsumber abye, *Duk.*, C. 289; kudumí, *Hind.* P. 1048; kuela, *Hind.*, C. 1414; kuer, *Hind.*, H. 294; kúfra, *Pb.*, G. 186; kug-ga, *Pb.*, F. 412; kugína, *Pb.*, R. 543; kuhili, *Mar.*, M. 786; kuh-nah-ní, *Sina*, F. 350; kúj, *Hind.*, B. 868; kuja, *Pb.*, J. 24; kujai, *Hind.*, R. 538; kúji, *Hind.*, R. 538; kujran, *Hind.*, P. 555; kujya, *Beng.*, C. 838; kukai (Flacourtia), *Pb.*, F. 615; kukai (Rhamnus), *Pb.*, R. 157; kukar (Cenchrus), *Hind.*, C. 909; kukar (Sterculia), *Mar.*, S. 2837; kúkar phalí, *Pb.*, R. 564; kukarwel, *Mar.*, L. 574; kúkerai, *Pushtu*, M. 109; kukhiah, *Pb.*, F. 357; kukil-i-pot, *Kashmir*, N. 17; kukkurbandá, *Hind.*, B. 546; kukoa, *Pb.*, F. 615; kukri (Cucumis), *Pb.*, C. 2278; kúkrí (Zea), *Hind.*, *Pb.*, Z. 50; kukronda (Crozophora), *Pb.*, C. 2221; kukronda (Pluchea), *Beng.*, P. 961; kúkshím (Vernonia), *Beng.*, V. 79; kúkshima (Celsia), *Beng.*, C. 878; kúksím, *Beng.*, V. 79; kukuliya, *Himálaya*, R. 351; kúkúrachúra, *Beng.*, P. 338; kúkúrálu, *Beng.*, D. 488; kúkúrát, *Hind.*, H. 517; kukurbicha, *Hind.*, G. 696; kúkúr chita, *Beng.*, L. 483; kúkúrkat, *Hind.*, H. 517; kukursungá, *Beng.*, B. 546; kukyán, *Hind.*, *Pb.*, P. 555; kúl, *Hind.* S. 1939; kúl, *Beng.*, Z. 231; kúl, *Hind.*, Z. 280; kula-aja, *Beng.*, E. 20; kulah, *Pb.*, P. 882; kulai, *Hind.*, P. 885; kulaibatana, *Hind.*, P. 882; kúla jhád,

Mar., M. 6; kúlanján, *Hind.*, *Beng.*, A. 853; kulára, *Pb.*, V. 96; kula sunda, *Mar.*, B. 171; kúlat, *Pb.*, D. 758; kuláwan, *Pb.*, P. 882; kulfa, *Hind.*, *Pb.*, P. 1179; kulfadodak, *Pb.*, E. 509; kulfi, *Beng.*, P. 1179; kulhárí, *Hind.*, G. 243; kuli (Barbus), *Hind.*, F. 358; kuli (Sterculia), *Hind.*, S. 2850; kulisija, *Hind.*, L. 55; kuliákhárá, *Beng.*, H. 508; kulíbegún, *Beng.*, S. 2284; kulinján, *Beng.*, *Hind.*, A. 853, 862, *Mar.*, A. 862; kulith, *Dec.*, *Mar.*, D. 758; kulitha, *Sind*, D. 758; kuljud, *Hind.*, A. 1632; kul-ka-batta, *Beng.*, F. 467; kul-ke-jar, *Duk.*, A. 554; kullin, *Mar.*, S. 2850; kullounda, *Mar.*, P. 484; kullua, *Beng.*, S. 30; kulmi luta, *Beng.*, R. 487; kulnar, *Hind.*, G. 769; kulo, *Uriya*, G. 677; kulon, *Hind.*, *Kumaon*, P. 882; kult, *Pb.*, D. 758; kulte, *Mar.*, D. 758; kúlth *Pb.*, D. 758; kulthi, *Dec.*, *Hind.*, *Kumaon*, *Mar.*, D. 758; kultolia, *Kumaon*, H.322; kúltú, *Hind.*, F. 10; kulu, *Hind.*, *Mar.*, S. 2850; kulumb, *Hind.*, O. 153; kulus-nar, *Beng.*, P. 41; kúm (Adina), *Beng.*, A. 519; kúm (Fraxinus), *Pb.*, F. 685; kúm (Sageretia), *Pb.*, S. 505; kumad, *Hind.* S. 30; kumala (Mallotus), *Uriya*, M. 71; kumála (Leea), *Hind.*, L. 224; kumali, *Kumaon*, L. 224; kumanta, (*Lahoul*), *Pb.*, S. 586; kumár (Aloe), *Guj.*, A. 838; kumár (Gmelina), *Hind.*, G. 287; kúmara, *Beng.*, *Hind.*, C. 2331; kumári, *Dec.*, *Hind.*, A. 829; kúmarika, *Beng.*, S. 2252; kúmba, *Hind.*, H. 324; kúmbal, *Mar.*, G. 311; kúmbh, *Beng.*, *Hind.*, F. 725; kúmbha, *Mar.*, C. 563; kumbhár, *Hind.*, G. 287; kumbhásála, *Mar.*, C. 563; kúmbhí, *Hind.*, *Pb.*, C. 563; kúmbi (Careya), |*Hind.*, *Mar.*, C. 563; kúmbi (Cochlospermum), *Hind.*, *Pb.*, C. 1512; kumbía, *Mar.*, C. 563; kumbya, *Mar.*, C. 563; kúmhár, *Pb.*, G. 287; kumhir, *Hind.*, C. 2077; kumhrá (Benincasa), *Kumaon*, B. 430; kumhrá (Cucurbita), *Kumaon*, C. 2331; kúmila, *Pb.*, M. 71; kum-jameva, *Beng.*, S. 2981; kumki, *Kumaon*, A. 2; kumkuma, *Hind.*, D. 417; kúmla, *Mar.*, C. 2039; kumla nebu, *Hind.*, C. 1233; kummar, *Pb.*, U. 4; kúmpaimán, *Hind.*, C. 1950; kumra (Stephegyne), *Hind.*, S. 2799; kumrá (Beninçasa), *Beng.*, *Hind.*, B. 430; kumrá (Cucurbita), *Beng.*, *Hind.*, C. 2331, *Hind.*, C. 2325; kumri, *Hind.*, C. 563; kúmta, *Raj.*, A. 273; kumtia, *Hind.*, A. 291, kumuda, *Mar.*, L. 355; kún, *Pb.*, S. 2376; kunáchí, *Pb.*, R. 590; kunak, *Hind.*, T. 634; kunba, *Pb.*, F. 725; kunch (Abrus), *Beng.*, A. 51; kunch (Coix), *Beng.*, C. 1686; kúnch (Mucuna), *Pb.*, M. 786; kúnch (Vibernum), *Pb.*, V. 96; kunchon pungti, *Beng.*, F. 338; kunda (Chirocentrus), *Uriya*, F. 402; kunda (Ischæmum), *Hind.*, I. 501; kunda (Jasminum), *Beng.*, *Hind.*, J. 32; kunda (Urginea), *Hind.*, U. 39; kunda buttam godi, *Hind.*, P. 48; kundah, *Uriya*, F. 402; kundalu, *Hind.*, G. 213; kundar (Pistacia), *Afg.*, P. 847; kundar (Portulaca), *Pb.*, P. 1179; kúndar

(Typha), *Pb.*, T. 864, T. 875 ; kundarud, *Afg.*, P. 847 ; kúndash, *Pb.*, A. 801 ; kundaye, kundayi, *Dec.*, *Hind.*, F. 615; kundel, *Hind.*, S. 493 ; kunderu, *Afg.*, P. 847 ; kundi (Cajanus), *Pb.*, C. 49 ; kúndí (Prosopis), *Sind*, P. 1259 ; kundna, *Beng.*, F. 466 ; kúndo, *Hind.*, J. 32 ; kúnd-phul, *Beng.*, *Hind.*, J. 32 ; kúndri, *Hind.*, U. 39 ; kundro, *Beng.*, B. 771 ; kúndrú, *Pb.*, C. 919 ; kundu, *Mar.*, D. 628 ; kundul, *Pb.*, F. 394 ; kundur, *Hind.*, B. 751, *Dec.*, *Hind.*, B. 771, 772, 777 ; kúndurrúmí, *Beng.*, *Hind.*, P. 841 ; kun-gí, *Pb.*, F. 596 ; kung-gi, *Pb.*, F. 308 ; kungku, *Hind.*, E. 485; kúngkúng, *Hind.*, M. 919 ; kungni (Panicum), *Kumaon*, P. 53 ; kungni (Setaria), *Hind.*, S. 1212 ; kungu (Panicum), *Hind.*, P. 67; kúngú (Lycium), *Pb.*, L. 589 ; kungwelka-gudda, *Dec.*, N. 39 ; kungyí, *Hind.*, S. 1694 ; kuni, *Sind*, N. 200 ; kunia, *Kumaon*, F. 156 ; kúnich, *Pb.*, A. 801 ; kúnj, *Hind.*, H. 324 ; kunjad (Pistacia), *Afg.*, P. 847 ; kunjad (Sesamum), *Pb.*, S. 1078 ; kunjada, *Afg.*, P. 847; kúnja náli, *Hind.*, H. 324 ; kunja (Alpinia), *Sind*, A. 853 ; kunjar (Sageretia), *Pb.*, S. 497 ; kunji, *Pb.*, R. 159 ; kunjia, *Beng.*, U. 35 ; kunjit, *Afg.*, S. 1078 ; kunjuya, *Hind.*, U. 35 ; kunkanatí, *Beng.*, A. 919 ; kúnku, *Hind.*, E. 482 ; kunnul, *Pb.*, F. 361 ; kunrat, *Hind.*, D. 402 ; kunro, *Sind*, R. 242 ; kúnsa, *Pb.*, A. 801 ; kúnsh, *Pb.*, A. 801; kunta pudena, *Uriya*, E. 569 ; kunti (Cæsalpinia), *Uriya*, C. 27 ; kunti (Murraya), *Mar.*, M. 797 ; kuntola, *Gus.*, M. 639 ; kunvar, *Guj.*, A. 829 ; kúpald, *Pb.*, C. 1012 ; kupás, *Hind.*, G. 404 ; kupásí, *Pb.*, H. 92 ; kuppi, *Hind.*, *Mar.*, A. 306 ; kur, *Beng.*, *Hind.*, S. 910 ; kura (Celastrus), *Hind.*, C. 862 ; kura (Holarrhena), *Hind.*, *Pb.*, H. 294; kura(Ixora), *Mar.*, I. 515 ; kurada, *Mar.*, C. 868; kúrak, *Dec.*, *Mar.*, G. 143 ; kuraki, *Mar.*, A. 1558 ; kural (Ocimum), *Beng.*, O. 31 ; kurál (Bauhinia), *Pb.*, B. 330, *Hind.*, B. 356 ; kúrand, *Pb.*, C. 1842 ; kurang, *Pb.*, M. 295 ; kúrasanna, *Sind*, P. 964 ; kúrat, *Mar.*, I. 515 ; kura-tuka, *Hind.*, P. 50 ; kurbí, *Beng.*, S. 2424 ; kurchi (Holarrhena), *Beng.*, H. 294 ; kurchi (Labeo), *Beng.*, F. 470 ; kurdi, *Mar.*, C. 637 ; kurdu (Celosia), *Mar.*, C. 868 ; kurdu (Sterculia), *Dec.*, S. 2850 ; kurela, *Kumaon*, M. 626 ; kurelí (Aneilema), *Beng.*, A. 1122 ; kurelí (Hydrilla), *Hind.*, H. 484 ; kurelo-jangro, *Sind* , M. 623 ; kures, *Beng.*, D. 59 ; kureyá, *Hind.*, H. 294 ; kurfá, *Hind.*, P. 1179 ; kurfah, *Mar.*, P. 1179 ; kúrg, *Pb.*, C. 886 ; kurgnulia, *Kumaon*, B. 863 ; kúrgotar, *Pb.*, S. 924 ; kuri (Ficus), *Pb.*, F. 156 ; kuri (Hedera), *Pb.*, H. 52 ; kuri (Nyctanthes), *Pb.*, N. 179 ; kuri (Panicum), *Hind.*, *Mur.*, P. 37, *Hind.*, P. 63 ; kuri (Sterculia), *Pb.*, S. 2861 ; kúrí (Xanthium), *Pb.*, X. 1 ; kuri chinke, *Raj.*, E. 190 ; kúrihári, *Hind.*, G. 243 ; kuriya, *Hind.*, P. 57 ; kúrka, *Pb.*, F. 429 ; kurkán, *Pb.*, P. 382 ; kurkní (Andrachne), *Pb.*, A. 1061 ; kurkní (Marlea), *Pb.*, M. 289 ; kurkní (Staphylea), *Kashmir*, S. 2678 ;

kurku, *Hind.*, F. 129 ; kurkuli, *Pb.*, A. 1061 ; kurkuna, *Hind.*, E. 20 ; kúrkun bér, *Afg.*, Z. 270 ; kurkunda, *Pb.*, Z. 231; kurkundái, *Pushtu*, T. 544; kurkurjihwa, *Hind.*, *Beng.*, L. 241 ; kurkuti, *Beng.*, L. 343 ; kurma, *Mar.*, *Sind*, P. 555 ; kurmáli (Leea), *Kumaon*, L. 224 ; kurmáli (Lonicera), *Kumaon*, L. 535; kurmru, *Pb.*, A. 692 ; kurne-kema, *Afg.*, F. 84 ; kurol, *Pb.*, H. 52 ; kurpa, *Mar.*, H. 124 ; kurpah, *Mar.*, F. 455 ; kurpúra-silasit, *Hind.*, G. 769 ; kurreah, *Sind.*, F. 357 ; kurrél, *Hind.*, C. 402 ; kurrí khár, *Pb.*, M. 237 ; kurru khajur, *Mar.*, M. 412 ; kursan, *Sind*, C. 1452 ; kursáni, *Himálaya*, C. 455 ; kursí, *Beng.*, F. 470 ; kúrsi, *Hind.*, W. 33 ; kursingh, *Mar.*, S. 2884 ; kursumbulle-pullie, *Hind.* P. 489 ; kúrtam, *Pb.*, C. 637 ; kurtamma, *Pb.*, C. 1211 ; kurti-kalai, *Beng.*, D. 758 ; kurtoli, *Dec.*, *Mar.*, M. 639 ; kurtum, *Sind*, C. 637 ; kurú (Coptis), *Beng.*, C. 1792 ; kuru (Picrorhiza), *Beng.*, *Hind.*, P. 700; kúrú (Gardenia), *Hind.*, G. 128 ; kúrú (Limnanthemum), *Hind.*, L. 358 ; kurú chuntz, *Mar.*, C. 1875 ; kurud, *Beng.*, L. 343 ; kúrúk (Cedrela), *Mar.*, C. 838 ; kúrúk (Garuga), *Mar.*, G. 143 ; kuru kanda, *Hind.*, D. 494 ; kurukarlu, *Mar.*, J. 41 ; kurumia, *Beng.*, S. 1939 ; kurúnai, *Mar.*, E. 338 ; kurund (Chenopodium), *Pb.*, C. 1018 ; kurund (corundum), *Hind.*, C. 1978 ; kurunda, *Hind.*, S. 1939 ; kurunji, *Hind.*, P. 1121 ; kuruvira, *Hind.*, N. 80 ; kuruwa, *Kumaon*, P. 700 ; kurwan, *Uriya*, I. 515 ; kurwi wágetí, *Mar.*, P. 316 ; kus, *Hind.*, S. 49 ; kusa (Eragrostis), *Pb.*, E. 252 ; kusa (Pennisetum), *Hind.*, P. 382 ; kusal, *Hind.*, H. 164 ; kusáli, *Hind.*, H. 164 ; kúsum, *Beng.*, *Pb.*, C. 637 ; kusara, *Mar.*, J. 13 ; kush (Eragrostis), *Hind.*, E. 252 ; kush (Prunus), *Pb.*, P. 1285 ; kusha, *Beng.*, *Hind.*, E. 252 ; kushi-ála, *Hind.*, P. 1426 ; kúshiar, *Beng.*, S. 30 ; kúshmand, *Guj.*, B. 430 ; kushm-áru, *Kumaon*, P. 1285 ; kushmul, *Hind.*, B. 458 ; kushi, *Pb.*, S. 1212 ; kushumbha, *Pb.*, C. 637 ; kusi (Briedelia), *Hind.*, B. 863 ; kusí (Diospyros), *Hind.*, D. 582 ; kúsimb, *Gus.*, G. 143 ; kussar, *Mar.*, J. 13 ; kússeer, *Sind*, S. 717 ; kussuah, *Uriya*, F. 596 ; kússum, *Hind.*, S. 950 ; kússúmb, *Pb.*, S. 950 ; kúst (Costus), *Beng.*, *Hind.*, C. 2013 ; kust (Saussurea), *Hind.*, S. 910 ; kust talkh, *Pb.*, S. 910 ; kust talkputchuk, *Hind.*, S. 910 ; kustura, *Hind.*, O. 616 ; kusum (Carthamus), *Beng.*, *Hind.*, C. 637 ; kusum (Schleichera), *Hind.*, S. 950 ; kusumb (Schleichera), *Mar.*, S. 950 ; kusumba (Carthamus), *Guj.*, *Mar.*, C. 637 ; kusúmbo, *Guj.*, C. 637 ; kút, *Guj.*, *Hind.*, *Pb.*, S. 910 ; kutaki (Eragrostis), *Hind.*, E. 261 ; kutaki (Picrorhiza), *Mar.*, P. 700; kutchú, *Beng.*, C. 2426 ; kuter, *Pb.*, C. 854 ; kuterni, *Mar.*, F. 503; kuth (Acacia), *Beng.*, A. 135; kúth (Saussurea), *Pb.*, S. 910 ; kuthí, *Pb.*, Z. 50 ; kuti bubhá, *Sind*, F. 725 ; kútílál (Daphne), *Pb.*, D. 130; kútílál (Withania), *Pushtu*, W. 98 ; kútílána, *Pb.*, W. 93 ; kuti leúbhá, *Sind*, A. 590 ; kutkí (Celosia), *Mar.*,

C. 878 ; kutki (Gentiana), *Beng., Hind.,* G. 165 ; kutki (Panicum), *Hind., Pb.,* P. 67 ; kutki (Picrorhiza), *Beng., Hind.,* P. 700 ; kutla, *Hind.,* C. 1270 ; kutra, *Pb., Hind.,* E. 170 ; kutri, *Pb.,* A. 382 ; kútsái, *Pb.,* A. 2 ; kútshírín, *Hind.,* C. 2013 ; kutta, *Pb.,* S. 1223 ; kutta bari, *Hind.,* S. 1223 ; kutta choti, *Hind.,* S. 1207 ; kuttahnimbú, *Pb.,* C. 1296 ; kuttahrah, *Hind.,* F. 492 ; kutte-ki-jibh-ka-patta, *Deccan,* E. 520 ; kutto-ki-jibh-kis-end, *Deccan,* E. 520 ; kuttipushli, *Pb.,* E. 259 ; kutz, *Pb.,* I. 128 ; kuwádice, *Guj.,* C. 787 ; kuwára, *Hind.,* C. 2514 ; kuwí, *Hind.,* S. 2960.

Kw Kwangere, *Pb.,* P. 959 ; kwan saf safei, *Pb.,* S. 2299 ; kwar, *Pushtu,* V. 233 ; kwer, *Pb.,* J. 29 ; kwia, *Hind.,* R. 538 ; kwiala, *Hind.,* R. 538 ; kwillar, *Hind.,* B. 308 ; kwíllim, *Pb.,* V. 96 ; kwíspre, *Pb.,* V. 64.

Ky Kyirin, *Hind.,* O. 600 ; kyl, *Kashmir,* S. 1242 ; kyonti, *Hind.,* V. 48 ; kyou, *Beng.,* D. 615, 656 ; kysur, *Beng.,* F.281 ; kyth, *Hind.,* F. 53.

La Labang, *Dec.,* C. 706 ; laber, *Hind., Pb.,* D. 348 ; labhán, *Pb.,* P. 1153 ; lablab (Dolichos), *Beng..* D. 789 ; lablab (Hedera), *Hind.,* H. 52 ; labri (Abrus), *Pb.,* A. 51 ; lábri (ruby), *Pb.,* R. 619 ; lachú, *Pb.,* R. 215 ; lad, *Mar.,* O. 534 ; ladákhí badam, *Kumaon,* P. 1304 ; ladakí-révanda-chíní, *Mar.,* R. 215 ; lád-an, *Hind.,* O. 574 ; laddoi, *Hind.,* S. 377 ; laduri, *Pb.,* N. 179 ; laggar, *Hind.,* T. 424 ; laghme, *Pushtu,* C. 224 ; laghme, *Pushtu,* H. 10 ; laghúne, *Afg.,* D. 130 ; lag-lag, *Hind.,* S. 2903 ; láhá, *Beng.,* I. 440 ; lahan, *Raj.,* T. 489 ; lahana, *Mar.,* S. 968 ; lahana gokharu, *Mar.,* T. 548 ; lahana gokrú, *Mar.,* T. 548 ; lahána kalpa, *Mar.,* T. 562 ; laháni-kumári, *Mar.,* A. 836 ; lahan popti, *Mar.,* P. 678 ; láhán shivan, *Mar.,* Q. 298 ; lahasaniá, *Hind.,* C. 622 ; lahéro, *Sind,* T. 227 ; lahi, *Hind.,* B. 822, 841 ; lahota, *Hind.,* B. 812 ; lahouri-hurmul, *Hind., Pb., Sind,* P. 372 ; lahra, *Hind.,* P. 384 ; lahsan, *Hind.,* A. 779 ; lahúra, *Pb.,* T. 227 ; lahúrd, *Pb.,* F. 568 ; lai (Brassica), *Hind.,* B. 822 ; lái (Tamarix), *Pb., Sind,* T. 61, 70 ; laila, *Pb.,* S. 544 ; laila, *Hind.,* S. 579 ; lain, *Hind.,* O. 574 ; lainja, *Mar.,* C. 725 ; laínyá, *Pb.,* T. 70 ; laita, *Hind.,* B. 812 ; lájak, *Beng.,* M. 557 ; lájálu (Mimosa), *Guj., Hind., Mar.,* M. 557 ; laj-alú (Neptunia), *Hind.,* N. 76 ; laj burud, *Hind.,* L. 75 ; lajjá vati, *Hind.,* M. 557 ; láiri, *Mar.,* M. 557 ; lájwanti, *Kumaon, Pb.,* M. 557 ; lajwárd, *Hind.,* L. 75 ; lák, *Gus.,* L. 1 ; lakati, *Hind.,* F. 792 ; lákh (lac), *Hind.,* C. 1491, L. 1 ; lákh (Lathyrus), *Mar.,* L. 100 ; lakhandi, *Mar.,* H. 124 ; lakhar, *Pb.,* R. 323 ; lakhar bagha, *Pb.,* T. 434 ; lákh dáná, *Pb.,* L. 1 ; lakhtei, *Pb.,* C. 2027 ; laki, *Pb.,* E. 267 ; lakshmana, *Pb.,* D. 409 ; lakshmí-am, *Beng.,* M. 209 ; lakúch, *Hind.,* A. 1511 ; lakúcha, *Beng.,* A. 1511 ; lál (Ipomæa), *beng.,* I. 405 ; lál (ruby), *Hind.,* R. 619 ; lálá, *Guj., Hind.,*

P. 82 ; lála ambádí, *Sind,* H. 233 ; lála-chandana, *Mar.,* P. 1381 ; lal-alú, *Beng.,* I. 348 ; lál ambári, *Dec., Hind., Mar.,* H. 233 ; lál-bachlé-kí-bháji, *Duk.,* B. 207 ; lálbachlú, *Hind.,* B. 207 ; lál-báli, *Hind.,* E. 259 ; lál bethí, *Hind.,* C. 1003 ; lálbherenda, *Beng.,* J. 52 ; lal-bhopali, *Mar.,* C. 2316 ; lal-bhuín-ánvalha, *Hind.,* P. 673 ; lal-bunlunga, *Beng.,* J. 114 ; lál-chámpá-natíya, *Beng.,* A. 941 ; lál-chámpá-nutí, *Beng.,* A. 941 ; lal-chandan (Symplocos), *Hind., Beng.,* S. 3059 ; lál chandan (Pterocarpus), *Beng., Dec., Hind.,* P. 1381 ; lál-chandí, *Uriya,* F. 312 ; lál chítá, *Beng., Hind., Uriya,* P. 979 ; lál-chítarak, *Hind.,* P. 479 ; lál chitarmul, *Dec.,* P. 979 ; lál-chitarmulam, *Dec.,* P. 979 ; lál-chitra, *Hind., Mar.,* P. 979 ; lál-chitrak, *Gus.,* P. 979 ; lal chuniin, *Hind.,* C. 154 ; laldanah, *Hind.,* I. 413 ; láldanah, *Hind.,* I. 376 ; lal-dudiya, *Mar.,* C. 2316 ; lal-ghígavár, *Hind.,* A. 838 ; lal-ghi-kanvar, *Duk.,* A. 838 ; lal-gul-makmal, *Beng.,* A. 914 ; lal-gurania-álu, *Hind., Beng.,* D. 526 ; láli, *Mar.,* A. 686 ; lalia, *Hind.,* V. 79 ; lál-indrávan, *Dec.,* T. 600 ; lál-indráyan, *Hind.,* T. 600 ; lalita pát, *Beng.,* C. 1861 ; lál-jari, *Pb.,* O. 170 ; lal jhau, *Beng.,* T. 61 ; lál-jháv, *Dec., Hind ,* T. 51 ; láljháv-nu-jháda, *Gus.,* T. 51 ; lálkhair, *Mar.,* A. 295 ; lál-khash-khash-ká-jhár, *Dec.,* P. 82 ; lál khas-khas-nu-jháda, *Gus.,* P. 82 ; lál-kkatyán, *Duk.,* B. 632 ; lál kudsum-bal, *Hind.,* C. 289 ; lal-kumárí, *Hind.,* A. 838 ; lál-kúmra, *Hind.,* C. 2323 ; lál lanká murich, *Beng.,* C. 455 ; lál marich, *Beng.,* C. 448, 455 ; lál mirch, *Dec.,* C. 466, *Hind., Pb.,* C. 448 ; lál-mircha, *Hind.,* C. 455 ; lál mirchí, *Mar.,* C. 455 ; lál mirich, *Guj.,* C. 448, *Hind.,* C. 455 ; lál mirich marchá, *Guj.,* C. 466 ; lál-mista-bij, *Beng.,* H. 233 ; lál-morich, *Beng.,* C. 466 ; lál múrgá, *Beng.,* C. 873 ; lál-murghka, *Hind.,* C. 873 ; lál-natí, *Beng.,* A. 919 ; lalphul-jamb, *Beng.,* E. 409 ; lál-phúl-ké-kólsé-ká-pattá, *Duk.,* B. 171 ; lál-póshta, *Beng.,* P. 82 ; lál-poshtér-gáchh, *Beng.,* P. 82 ; lál-póst, *Hind.,* P. 82 ; lal-sabuni, *Hind.,* T. 537 ; lál-ság, *Hind ,* A. 927 ; lál sarbo jaya, *Beng.,* C. 321 ; lal-shák, *Beng.,* A. 927 lal-shakar, *Hind.,* S. 373 ; lal-shakar-kand-alú, *Beng.,* I. 348 ; lal siris, *Hind.,* A. 692 ; lál tit-maliya, *Kumaon,* V. 106 ; lalu, *Hind.,* E. 327 ; lama, *Mar.,* S. 882 ; lamb, *Pb.,* H. 164 ; lamcha, *Hind.,* E. 259 ; lámjak, *Hind., Pb.,* A. 1093 ; lam-kana, *Hind., Raj.,* B. 868 ; lamma, *Hind.,* H. 35 ; lamp (Aristida), *Pb.,* A. 1383 ; lamp (Heteropogon), *Hind.,* H. 164 ; lampa (Chrysopogon), *Hind.,* C. 1053 ; lampa (Heteropogon), *Hind.,* H. 164 ; lampar, *Hind.,* H. 164 ; lamshing, *Kumaon,* P. 737 ; lána (Anabasis), *Pb.,* A. 1005 ; lana (Ballota), *Pb.,* B. 33 ; lana (Barilla), *Pb.,* B. 160 ; lana (Haloxylon), *Hind., Pb.,* H. 6 ; lana (Suæda), *Mar.,* S. 2990 ; láná góra, *Pb.,* S. 596 ; lánan, *Sind,* S. 596 ; lándachúta, *Mar.,* M. 855 ; lánebar, *Pb.,* O. 247 ; lánetsúrú, *Kashmir,* X. 1 ; láng, *Gus.,* L. 100 ; langa, *Beng.,* C. 706 ; langshúr, *Pb.,* J.

78 ; langshúr thélu, *Pb.*, J. 78 ; lang, thang, *Pb.*, P. 684 ; lánguli, *Hind.*, G. 243 ; languli-latá, *Beng.*, I. 399 ; langúr, *Hind.*, F. 809 ; lani, *Pb.*, S. 2990 ; láning, *Pb.*, V. 233 ; lanísah, *Sind*, R. 65 ; lánjai, *Hind.*, C. 1377 ; lanka sij, *Beng.*, D. 387 ; lanka (Capsicum), *Beng.*, C. 455 ; lanka (Cucurbita), *Beng.*, *Hind.*, C. 2331 ; lanka (Cuphorbia), *Uriya*, E. 553 ; lanká-ám, *Uriya*, A. 1023 ; lanká-marich, *Beng.*, C. 448 ; lanká-mirchi, *Hind.*, C. 455 ; lanká-morich, *Beng.*, C. 466 ; lanka sij, *Beng.*, D. 387, E. 553 ; lano, *Sind*, C. 224 ; láo, *Sind*, T. 61 ; lap (Andropogon), *Hind.*, A. 1090 ; lap (Heteropogon), *Hind.*, H. 164 ; láp (Celastrus), *Pb.*, C. 862 ; lápadi, *Guj.*, C. 868 ; laphra, *Pb.*, S. 738 ; lapta (Cenchrus), *Pb.*, C. 909 ; lapta (molasses), *Hind.*, S. 318 ; lapti, *Hind.*, S. 1223 ; largá, *Pb.*, R. 293 ; lariya-dáona, *Beng.*, E. 496 ; lár-kána, *Sind*, P. 885 ; las, *Sind*, P. 1176 ; lasaj, *Pb.*, A. 1464 ; lasan, *Guj.*, *Hind.*, A. 779 ; lasania, *Hind.*, C. 618 ; lashan, *Beng.*, A. 779 ; laskar, *Pb.*, D. 245 ; laskara, *Pb.*, D. 245 ; lasniyán, *Hind.*, C. 622 ; lasora, *Hind.*, Raj., C. 1931 ; lasrín, *Hind.*, A. 695, *Pb.*, A. 711 ; lassar, *Pb.*, J. 78 ; lastúk, *Pb.*, E. 232 ; lasun, *Beng.*, A. 779 ; lasunas, *Mar.*, A. 779 ; lasúrá, *Hind.*, C. 1931 ; laswara, *Pb.*, C. 1931 ; láta-dáona, *Beng.*, E. 553 ; latak, *Sind*, T. 544 ; lata kasturi, *Beng.*, P. 1352 ; latá-palásh, *Beng.*, B. 978 ; latáphatkarí, *Beng.*, C. 551 ; latar, *Bihar*, D. 534 ; latha, *Beng.*, F. 508 ; láthia, *Pb.*, F. 35 ; latí-am, *Beng.*, I. 78 ; latjirá, *Hind.*, A. 382 ; latká, *Beng.*, B. 4 ; latkhan, *Beng.*, *Hind.*, B. 523 ; latman, *Hind.*, A. 914 ; latmí, *Beng.*, A. 983 ; latrí, *Hind.*, L. 100 ; láu, *Beng.*, *Hind.*, L. 30 ; laudar, *Pb.*, S. 3046 ; laudgah, *Hind.*, F. 757 ; lauká, *Hind.*, L. 30 ; lauki, *Hind.*, *Pb.*, L. 30 ; laung, *Hind.*, *Pb.*, C. 706 ; laur, *Pb.*, A. 344 ; lavang, *Mar.*, C. 706 ; lavanga, *Beng.*, C. 706 ; lavinga, *Guj.*, *Mar.*, C. 706 ; lawála, *Mar.*, C. 2617 ; lawanga, *Mar.*, *Guj.*, C. 706 ; lawange. *Pb.*, A. 1464 ; láyí, *Hind.*, A. 1591 ; layubuka, *Beng.*, F. 531.

Le Leá, *Pb.*, C. 909 ; leauri, *Hind.*, C. 2358 ; lebu, *Beng.*, C. 1270, *Beng.*, *Guj.*, *Hind.*, C. 1296 ; leddil, *Pb.*, M. 22 ; ledra, *Pb.*, S. 838 ; leh, *Pb.*, T. 61 ; lehan, *Pb. Hills*, C. 2021 ; lehí, *Hind.*, O. 258 ; lei (Celastrus), *Pb.*, C. 862 ; lei (Tamarix), *Pb.*, T. 61 ; lei, *Sind*, T. 70 ; leila, *Pb.*, S. 3079 ; lelka, *Hind.*, F. 179 ; lendi (Solenanthus), *Pb.*, S. 2352 ; lendi (Lagerstrœmia), *Hind.*, *Mar.*, L. 55 ; lendwa, *Hind.*, S. 838 ; lendya, *Hind.*, S. 1939 ; lendya, *Hind.*, L. 55 ; lenr, *Hind.*, R. 369 ; lenwa, *Pb.*, S. 2352 ; lepcha (Ribes), *Kumaon*, R. 355 ; lepcha phal (Machilus), *Hind.*, M. 21 ; lephee, *Sind*, A. 829 ; lepistan, *Guz.*, C. 1931 ; lersima, *Hind.*, S. 2819 ; leru, *Beng.*, *Hind.*, O. 574 ; lesúri, *Sind*, C. 1931 ; lesúri gedúri, *Mar.*, C. 1931 ; leswa, *Pb.*, D. 420 ; lete, *Hind.*, C. 402 ; lévu, *Guz.*, I. 440 ; lewar, *Pb.*, J. 78, *Kumaon*, *Pb.*, J. 92.

Lh Lhíjo, *Pb.*, P. 1449.

Li Li (Celastrus), *Pb.*, C. 862 ; li (Pyrus), *Pb.*, P. 1452 ; liár, *Sind*, C. 1944 ; liári, *Sind*, C. 1944 ; liasada, *Sind*, L. 569 ; libado, *Guz.*, M. 363 ; libi-dibi, *Mar.*, C. 19 ; lichakhro, *Pb.*, C. 1958 ; lichi, *Mar.*, N. 68 ; lidra, *Pb.*, O. 38 ; lí-gur, *Baluch.*, F. 415 ; lijhar, *Hind.*, E. 186 ; lilacha, *Guj.*, A. 1079 ; lilichá, *Guj.*, A. 1079 ; liljahri, *Hind.*, G. 184 ; lím (Cedrela), *Hind.*, C. 838 ; lím (Pinus), *Hind.*, *Pb.*, P. 737 ; limanza, *Pushtu*, P. 737 ; limb (Cedrela), *Mar.*, C. 838 ; limb (Melia), *Guz.*, M. 363 ; limba, *Guj.*, *Mar.*, M. 363, *Mar.*, M. 439 ; limbácha-jháda, *Mar.*, M. 363 ; limbado (Ailanthus), *Hind.*, A. 658 ; limbado (Melia), *Guj.*, M. 363 ; limbara bakánanimb, *Mar.*, M. 393 ; limbarra, *Mar.*, M. 412 ; limbdo, *Guz.*, M. 800 ; limbo, *Hind.*, M. 363 ; limbu, *Guj.*, *Mar.*, C. 1258 ; *Hind.*, *Mar.*, C. 1270 ; *Beng.*, *Hind.*, *Mar.*, C. 1296 ; limbu-nimbu, *Guj.*, C. 1296 ; limri, *Mar.*, T. 489 ; limtoá, *Beng.*, A. 1218 ; límú, *Hind.*, *Mar.*, C. 1270 ; límú, *Dec.*, *Hind.*, *Sind*, C. 1258, *Dec.*, *Hind.*, C. 1296 ; límún, *Dec.*, *Hind.*, C. 1258, 1296 ; lín. *Pb. Hills*, C. 2021 ; lingúr, *Mar.*, V. 164 ; linú, *Pb.*, C. 2021 ; liokpa, *Pb.*, D. 245 ; lisan-ul, *Hind.*, H. 294 ; litchi, *Hind.*, N. 68 ; litsí, *Pb.*, P. 1449, 1460 ; liú, *Pb.*, P. 1449 ; líwár, *Pb.*, P. 1449.

Lm Lmanza, *Afg.*, C. 846.

Lo Loabate-barbarí, *Mar.*, C. 1703 ; loabate-barbarí, *Beng.*, C. 1703 ; loabate-barbarí, *Hind.*, C. 1703 ; loanní, *Beng.*, F. 473 ; loari, *Beng.*, A. 1076 ; loba, *Hind.*, P. 530 ; loban, *Guz.*, S. 2968 ; lobaní úd, *Hind.*, S. 2968 ; lobeh, *Dec.*, *Mar.*, V. 116 ; lobia (Dolichos), *Hind.*, D. 789 ; lobiá (Vigna), *Hind.*, V. 116 ; lobiya (Dolichos), *Hind.*, D. 789 ; lobiyá (Phaseolus), *Pb.*, P. 489 ; lobiyá (Vigna), *Pb.*, V. 116 ; lobiya riánsh, *Kumaon*, V. 116 ; loda, *Beng.*, P. 627 ; lodar, *Pb.*, S. 3046 ; lodar, *Guz.*, S. 3062 ; lodh, *Kumaon*, S. 1889 ; lodh, *Beng.*, *Hind.*, *Kumaon*, *Mar.*, S. 3046, 3062, 3073 ; lodhuka-sijhu, *Uriya*, E. 553 ; lodh pathání, *Pb.*, S. 3046 ; lodhra, *Mar.*, S. 3062 ; lodri síyáh, *Pb.*, C. 997 ; lodur, *Sind*, S. 3046 ; lohá, *Beng.*, *Hind.*, I. 440 ; lohá chúr, *Hind.*, I. 440 ; loha jangia, *Hind.*, I. 515 ; lohánuzang, *Guz.*, I. 472 ; lohar, *Beng.*, O. 38 ; lohar, *Sind*, F. 568 ; lohár-gú, *Beng.*, I. 472 ; lohari, *Hind.*, D. 628 ; lohárjhangár, *Beng.*, I. 472 ; lohása, *Pb.*, R. 335 ; lóhé-ka-gú, *Hind.*, I. 472 ; lóhé-kazang, *Hind.*, I. 472 ; lohero, *Mar.*, T. 227 ; loherú, *Sind*, T. 232 ; lohíra, *Sind*, T. 227 ; lohuri, *Mar.*, *Sind*, T. 227 ; loidan-siput, *Hind.*, P. 41 ; loj, *Pb.*, S. 1889 ; loj, *Pb.*, S. 3046 ; loja, *Pb.*, S. 1889 ; loja, *Pb.*, S. 3046 ; lokandi (Ventilago), *Mar.*, V. 55 ; lokandi (Ixora), *Mar.*, I. 515 ; lokaneli, *Mar.*, H. 124 ; lokhanda, *Mar.*, I. 440 ; lokhan-dhácha-katai, *Mar.*, I. 472 ; lokhandi, *Mar.*, V. 33, lokri, *Hind.*, F. 811 ; lolti, *Pb.*, S. 3079 ; lón, *Hind.*, S. 602 ; lonah, *Mar.*, F. 454 ; lonak, *Pb.*, P. 1179 ; lonar, *Sind*, F. 415 ; lóng, *Hind.*, C. 706 ; longarbis thiras, *Mar.*, V. 159 ; longarbi thiras, *Mar.*, V. 159 ;

loní (butter), *Mar.*, B. 983 ; loni (Portu-
laca), *Guj.*, P. 1179 ; lonia, *Hind.*, P.
1179 ; loniyá, *Hind.*, P. 1187 ; lónk,
Sind, P. 1179 ; lor, *Pushtu*, E. 35 ; lorhi,
Sind, N. 200 ; losh, *Pb.*, S. 3046 ; lot,
Mar., S. 902 ; lotak, *Pb.*, T. 544, T.
548 ; lotlotí, *Hind.*, U. 35 ; louná, *Hind.*,
A. 1158 ; lovi, *Duk.*, *Mar.*, A. 1511.

Lu Lú, *Pb.*, S. 3046 ; lúár, *Pb.*, T. 227 ; lubán,
Hind., B. 751 ; lubán (Boswellia), *Beng.*,
Hind., B. 771 ; lubán (Styrax), *Beng.*,
Hind., *Mar.*, *Pb.*, S. 2968 ; lúbani-úd,
Dec., S. 2968 ; lúbar, *Pb.*, P. 688 ; lubis
firmun, *Hind.*, L. 467 ; luch, *Sind*, P.
1176 ; luckmuna (Atropa), *Hind.*, A.
1614 ; luckmuna (Mandragora), *Hind.*,
M. 128 ; lud, *Beng.*, *Hind.*, C. 838 ;
ludhu, *Uriya*, S. 3062 ; lúdra, *Hind.*, L.
451 ; lúdút, *Pb.*, C. 1638 ; lúet, *Kumaon*,
T. 93 ; lufah, *Hind.*, M. 128 ; lúgar, *Pb.*,
H. 382 ; luhuriya, *Hind.*, *Kumaon*, P.,
926 ; luinji, *Pb.*, I. 487 ; luir, *Pb.*, J. 92 ;
luit-marz, *Kashmir*, P. 801 ; lukh, *Pb.*,
T. 864, T. 875 ; lulai, *Mar.*, A. 686 ;
lumba-nuli-jamb, *Beng.*, E. 407 ; lumpeil,
Hind., H. 164 ; lumra, *Pb.*, E. 259 ;
lumri, *Hind*, F. 811 ; lúna, *Beng.*, A.
1166 ; lúnak (Chenopodium), *Pb.*, C.
1003 ; lúnak (Portulaca), *Hind, Kumaon*,
P. 1179 ; lúnak (Suæda), *Pb.*, S. 2985 ;
lundi, *Mar.*, S. 1210 ; lungar, *Pb.*, P.
1359 ; lúní (Cotoneaster), *Pb.*, C. 2021 ;
luni (Portulaca), *Guj.*, P. 1187 ; luni
(Prunus), *Pb.*, P. 1304 ; lúnia, *Hind.*,
Pb., P. 1179 ; lúniya, *Hind.*, P. 1179 ;
lúniya-kúlfah, *Hind.*, *Kumaon*, P. 1179 ;
lún-ki-búti, *Pb.*, P. 1187 ; luntak. *Mar.*,
E. 338 ; lúnuk, *Hind.*, P. 1179, *Sind*, P.
1191 ; lúrjúr, *Beng.*, C. 720 ; lussunia,
Hind., C. 622 ; lúst, *Hind.*, *Kumaon*, T.
93 ; luta mahawria, *Beng.*, D. 420 ; lutch-
mi, *Mar.*, D. 725 ; lutco, *Hind.*, B. 4 ;
lutiam, *Beng.*, W. 88 ; lút-putiah, *Dec.*,
N. 28.

Ly Lyi, *Sind*, T. 61.

Ma Ma, *Pb.*, C. 637 ; máal, *Pb.*, P. 1142 ;
maapul, *Dec.*, Q. 43 ; maar, *Mar.*, C. 1520 ;
máblí, *Mar.*, G. 753 ; mabura, *Beng.*, A.
534 ; macao, *Hind.*, P. 682 ; mach,
Baluch., P. 555 ; machauri, *Hind.*, I. 487 ;
mach-bágral, *Beng.*, T. 440 ; mach-bag-
rul, *Hind.*, F. 771 ; mach-bhondar, *Beng.*,
T. 441 ; machchhi-ka-sirish, *Dec.*, *Hind.*,
A. 393 ; machmach, *Uriya*, M. 550 ;
machni, *Hind.*, L. 523 ; machola, *Mar.*,
A. 1475 ; machoti, *Hind.*, P. 1078 ;
máchul, *Mar.*, A. 1475 ; machútie, *Beng.*,
P. 1078 ; mackundi, *Uriya*, F. 388 ; mád,
Mar., C. 1520 ; máda (Cocos), *Mar.*, C.
1520 ; mada (wax), *Mar.*, H. 342 ; madan,
Beng., R. 1 ; madana, *Pb.*, E. 166 ; ma-
danaghantí, *Hind.*, S. 2515 ; madanchur,
Beng., S. 2907 ; madanmast (Amorpho-
phallus), *Mar.*, A. 997 ; madan-mast
(Artabotrys), *Duk.*, A. 1431 ; madánú,
Pb., S. 537 ; madanya, *Hind.*, E. 186 ;
madar (Rubia), *Mar.*, R. 564 ; madár,
Hind., D. 387 ; mádár (Artocarpus),
Beng., A. 1511 ; madár (Calotropis),
Hind., C. 170, 191, *Hind.*, *Pb.*, *Sind*,
C. 225 ; madára, *Hind.*, E. 356 ; madare,
Pb., A. 1061 ; madar-patí, *Beng.*, P. 625 ;

madat, *Mar.*, T. 361 ; madh (wine),*Beng.*,
V. 243 ; madh (honey), *Guj.*, *Hind.*, H.
342 ; madhána, *Pb.*, E. 166 ; mádhavilatá,
Beng., *Hind.*, H. 285 ; madhu, *Beng.*,
H. 342 ; madlatáh, *Hind.*, B. 873 ; mad-
malti, *Hind.*, H. 285 ; madmánti, *Duk.*,
A. 1431 ; madúbhi, *Beng.*, H. 285 ; ma-
dúbhlúta, *Beng.*, H. 285 ; madyá, *Beng.*,
V. 243 ; mag, *Guz.*, P. 496 ; maga, *Mar.*,
Sind, P. 513 ; magar, *Pb.*, B. 118 ; ma-
garbáns, *Hind.*, B. 118 ; mageer, *Hind.*,
O. 38 ; maggru-gadi, (*C. P.*), *Hind.*, I.
503 ; mághal, *Pb.*, P. 1142 ; maghz-
pipal, *Pb.*, P. 805 ; magi, *Uriya*, F. 503 ;
magina muniya, *Kumaon*, D. 522 ;
magíya, *Kumaon*, D. 528 ; magiya-máin,
Mar., T. 51, T. 70 ; magr, *Hind.*, C.
2077 ; magrabu, *Hind.*, H. 119 ; magsher,
Pb., S. 579 ; magúr, *Beng.*, F. 412 ; ma-
gurah, *Uriya*, F. 412 ; mah, *Pb.*, *Sind*,
P. 513, *Sind*, P. 496 ; maha (sámbar),
Hind., D. 240 ; maha (Bassia), *Mar.*, B.
220 ; mahabal, *Hind.*, H. 13 ; maha bari
bach, *Hind. & Beng.*, Z. 225 ; mahabi,
Hind., E. 509 ; mahad, *Mar.*, C. 1520 ;
mahadeoka-phúl, *Pb.*, D. 115 ; máhála,
Uriya, A. 658 ; mahalan, *Hind.*, B. 342 ;
mahalib, *Sind*, P. 1313 ; maha limbo
(Melia), *Hind.*, *Mar.*, M. 393 ; mahá-
limbo (Cedrela), *Hind.*, C. 838 ; maha-
limbu, *Uriya*, C. 838 ; mahá-lunga, *Mar.*,
C. 1270 ; mahanim (Ailanthus), *Uriya*,
A. 658 ; maháním (Cedrela), *Hind.*, *Mar.*,
C. 838 ; mahá-ním (Melia), *Pb.*, M. 363,
Beng., M. 393 ; mahá níbu, *Hind.*, C.
1263 ; mahánimb, *Mar.*, A. 658 ; mahá-
nimbu, *Beng.*, C. 1263 ; maha-nínb,
Hind., M. 393 ; maha-rang'a, *Pb.*, O. 170 ;
mahárúkh, *Duk.*, A. 658 ; mahárukha,
Hind., A. 658 ; mahárut, *Hind.*, G. 143 ;
mahasaula, *Beng.*, F. 357 ; mahasir,
Beng., F. 357 ; mahátíta, *Beng.*, *Hind.*,
A. 1064 ; mahaul, *Pb.*, P. 1458, 1460 ;
mahauli, *Hind.*, B. 318 ; mah-gur, *Beng.*,
F. 412 ; mahín, *Pb.*, T. 70 ; mahiya,
Hind., S. 377 ; mah-kák, *Afg.*, V. 10 ;
mahkoa, *Beng.*, Z. 263 ; máhlun, *Hind.*,
C. 838 ; mahmira, *Sind*, C. 1789 ; mah-
múdah, *Pb.*, C. 1783 ; mahori (Pimpi-
nella), *Hind.*, P. 727 ; mahorí (Sola-
num), *Pb.*, S. 2264, 2345 ; mahotí
hinpoli, *Beng.*, S. 2284 ; mahturi, *Uriya*,
F. 456 ; mahua, *Mar.*, B. 220 ; mahúa,
Hind., B. 220 ; mahuda, *Guj.*, B. 265 ;
mahúda, *Guj.*, B. 220, 265 ; mahulá,
Beng., *Hind.*, B. 220 ; mahur, *Hind.*,
O. 247 ; máhúr, *Hind.*, O. 247 ; mahura,
Guj., B. 220 ; mahwa, *Mar.*, B. 265 ;
mahwá, *Beng.*, *Hind.*, B. 220 ; mai,
Pushtu, P. 496 ; máí chhótí, *Pb.*, T. 51 ;
maidah, *Hind.*, S. 970 ; maidálakadí,
Mar., L. 483 ; maidá-lakrí, *Beng.*, *Hind.*,
Mar., L. 483 ; maidasak, *Pb.*, L. 483 ;
mail, (Behari), *Hind.*, S. 377 ; máil, *Pb.*,
P. 1460 ; maila (Saccharum), *Hind.*, S.
377 ; maila (Pyrus), *Pb.*, P. 1460 ; máil
tang, *Pb.*, P. 1460 ; maimúna, *Afg.*, S.
497 ; main (Randia), *Hind.*, R. 1 ; maín
(Chickrassia), *Hyderabad*, C. 1021 ; máín
(Tamarix), *Pb.*, T. 61 ; mainá, *Pb.*, M.
329 ; mainh (buffalo), *Pb.*, O. 558 ; main-
húri, *Hind.*, R. 1 ; maini, *Hind.*, R. 1;
mainphal (Randia), *Hind.*, *Kumaon*, R.

1 ; mainphal (Vangueria), *Beng., Hind.,* V. 25 ; maiphal, *Mar., Q.* 43 ; mái varí, *Pb., T.* 51 ; maizurrye *Pushtu, N.* 1 ; mája, *Mar., Q.* 43 ; majethi, *Kumaon, R.* 564 ; majh, *Pb., O.* 558 ; majít, *Pb., R.* 564 ; majíth, *Beng., Hind., R.* 564 ; majiti, *Hind.,* I. 40 ; majnúm, *Pb., S.* 544 ; maj-nún, *Pb., S.* 544 ; majori, *Pb., S.* 6 ; máj-tari, *Hind., A.* 1469 ; májún, *Hind., A.* 1623 ; májuphal, *Beng., Hind., Q.* 43 ; mak, *Pb., Z.* 50 ; maka (Zea), *Mar., Z.* 50 ; maká, *Hind., Uriya, Z.* 50 ; máká (Eclipta), *Mar.,* E. 7 ; mákadlimbu, *Mar., A.* 1601 ; makai (Zea), *Hind., Mar., Z.* 50 ; makai (Zizyphus), *Hind., Z.* 263 ; mákál (Trichosanthes), *Beng., T.* 600 ; mákál (Citrullus), *Hind., C.* 1211 ; maka-makna, *Hind., E.* 166 ; makar-sing, *Mar., C.* 480 ; makar-tendi, (*Banda*), *Hind., D.* 628 ; makh, *Afg., G.* 278 ; makhál (Trichosanthes), *Hind., T.* 600 ; makhal (Citrullus), *Beng., C.* 1211 ; makham shim, *Beng., C.* 289 ; mákhan, *Beng., Guj., Hind., B.* 983 ; makhan-sim, *Beng., Hind., D.* 789 ; makhana, *Beng., Hind., E.* 569 ; makhazura, *Pb., W.* 93 ; mákhi, *Hind., P.* 663 ; makhmal, *Mar., T.* 17 ; makhtúmí-shakkar, *Dec., S.* 373 ; makhzan, *Hind., A.* 510 ; maki, *Sind, T.* 61 ; makká, *Hind., Z.* 50 ; makkai, *Guz., Sind, Z.* 50 ; makka-járí, *Dec., Z.* 50 ; makka-jowárí, *Dec., Z.* 50 ; makkal, *Pb., P.* 1142, 1159 ; makkei, *Pb., Z.* 50 ; makkí, *Pb., Z.* 50 ; makkúna, *Hind., B.* 318 ; mako (Zizyphus), *Hind., Z.* 263 ; mako (Solanum), *Beng., Mar., S.* 2299 ; makoi, *Hind., S.* 2299 ; makola, *Hind., C.* 1958 ; makra, *Hind., E.* 166, 170 ; makra, *Hind., F.* 672 ; makraila, *Hind., E.* 186 ; makra-rái, *Hind., B.* 841 ; mak-reru, *Pb., F.* 168 ; makri, *Hind., E.* 166 ; makriya-chilauni, *Hind., S.* 940 ; makshári (Ulmus), *Pb., U.* 4 ; makshéri (Betula), *Pb., B.* 496 ; makurjali, *Beng., P.* 79 ; makur-kendi, *Hind., Beng., D.* 582 ; makusal, *Hind., S.* 940 ; mál, *Beng., Hind., O.* 574 ; mál, *Pb., P.* 1138 ; mala (Bryonia), *Beng., B.* 907 ; mala (Raphanus), *Guj., R.* 31 ; malabari-elachi, *Mar., E.* 151 ; malabaripánki-jar, *Mar., A.* 853 ; maláka jamrul, *Beng., E.* 444 ; málan, *Pb., S.* 2376 ; mal-ankuri, *Hind., E.* 186 ; málati, *Beng., A.* 584 ; málati, *Hind., A.* 584 ; málchang, *Pb., S.* 537 ; maldoda, *Pb., L.* 312 ; maldung, *Pb., U.* 4 ; male, *Pushtu,* P. 42 ; malghán, *Hind., B.* 342 ; malghi, *Hind., E.* 186 ; máli, *Mar., P.* 1024 ; malicha, *Raj., E.* 166 ; malighah, *Raj., E.* 166 ; mal jal, *Beng., Hind., O.* 574 ; maljan, *Hind., B.* 342 ; maljhanji, *Hind., D.* 445 ; málkákni, *Hind., Kumaon, C.* 854 ; malkangana, *Guj., C.* 854 ; mál kánganítela, *Mar., C.* 854 ; mál kangní, *Beng., Hind., Mar., C.* 854 ; málkang, ní-ka-jantar, *Duk., C.* 854 ; mál kangoni, *Mar., C.* 860 ; mál-kángóní, *Mar., C.* 854 ; malkonia, *Beng., V.* 129 ; málkundai, *Pushtu, T.* 548 ; málkungi, *Hind., C.* 854 ; malla, *Hind., Z.* 254 ; mallá, *Pb., Z.* 254 ; malla-bér, *Pb., Z.* 254 ; mallán, *Pb., Z.* 254 ; malla ním, *Hind., M.* 393 ; mál-le, *Pb., F.* 424 ; mallik, *Beng.,* J. 35 ; malmúriya, *Beng.,*

M. 40 ; malok, *Hind., D.* 611 ; malorígáh, *Pb., R.* 642 ; malto, *Hind.,* J. 24 ; malú, *Hind., B.* 342 ; malúk, *Pb., D.* 611 ; malwa bakchí, *Pb., V.* 73 ; malwajari, *Hind.,* I. 487 ; mambre, *Pb., F.* 168 ; ma-mech, *Pb., P.* 1112 ; mamekh, *Pb., P.* 11 ; mámíjwá, *Guz., E.* 217 ; mamíra (Thalictrum), *Kumaon, Pb., T.* 376 ; mamírá (Coptis), *Hind., C.* 1789 ; mamíran, *Mar., T.* 376 ; mámírán (Coptis), *Hind., C.* 1789 ; mam-í-rán (Geranium), *Afg., G.* 184 ; mamiri, *Pb., C.* 198 ; mamoli, *Pb., S.* 2345 ; mamral, *Pb., R.* 152 ; mámre, *Pb., V.* 233 ; mamri, *Hind., E.* 73 ; mána, *Pushtu, P.* 1463 ; manabina, *Pb., H.* 517 ; manak, *Pb., R.* 619 ; manakká, *Hind., V.* 233 ; manakká, *Beng., V.* 233 ; man-álu, *Hind., D.* 481 ; manchi-malwa, *Hind.,* I. 487 ; manchingi, *Mar., D.* 748 ; mándá, *Mar., D.* 539 ; mandadúpa, *Pb., A.* 670 ; mandal (Acer), *Pb., A.* 344 ; mandal (Eleusine), *Pb., E.* 170 ; mandál (Rhododendron), *Pb., R.* 253 ; mandála, *Pushtu, P.* 1322 ; mandao, *Afg., E.* 327 ; mandar, *Pb., A.* 328 ; mándárá (Calotropis), *Mar., C.* 170, C. 191 ; mandára (Erythrina), *Hind., E.* 342 ; mandata, *Pb., Pushtu, P.* 1285, *Pushtu, P.* 1322 ; mandavi, *Kumaon, E.* 186 ; manderung, *Pb., U.* 4 ; mandgay, *Mar., B.* 118 ; mandh, *Hind., M.* 549 ; mandia, *Kashmir, H.* 52 ; man-diál, *Hind., E.* 186 ; mandira (Setaria), *Kumaon, S.* 1212 ; mandira (Panicum), *Kumaon, P.* 53 ; mandiya, *Hind., P.* 79 ; mandjiro, *Sind, E.* 188 ; mand-kolla, *Pb., R.* 1 ; mandór, *Hind.,* I. 472 ; mandrá, *Pb., M.* 289 ; mandú, *Pb., U.* 4 ; mandua, *Hind., F.* 672 ; manduá, *Hind., E.* 170 ; mandwa, *Raj., E.* 186 ; manehingi, *Hind., D.* 748 ; maner, *Pb., A.* 344 ; mangai, *Mar., M.* 721 ; mangarwal, *Pb., E.* 232 ; mangastín, *Mar., G.* 55 ; manghati, *Uriya, L.* 126 ; mángle, *Hind., R.* 351 ; man-glí-ah-ní, *Sind, F.* 307 ; mangostín, *Mar., G.* 55 ; mangrí, *Hind., F.* 412 ; mángri, *Pushtu, S.* 497 ; mangrúr, *Pb., P.* 42 ; mangús, *Hind., F.* 774 ; mangustán, *Beng., Hind., Mar., G.* 55 ; manhrí, *Pb., L.* 252 ; manjishta, *Mar., M.* 704 ; manjít, *Mar., R.* 564 ; manjéshta, *Mar., R.* 564 ; manjistá, *Beng., Uriya, R.* 564 ; manjít, *Beng., Dec., Hind., Kumaon, Pb., R.* 564 ; manka, *Hind., C.* 617 ; mánkachú, *Beng., A.* 809 ; mánkanda, *Hind., A.* 809 ; manneal, *Hind., R.* 1 ; mannú, *Pb., U.* 4 ; mannua, *Hind., G.* 385 ; manra, *Pushtu, P.* 1463 ; mansa, *Raj., E.* 166 ; mán saru, *Uriya, C.* 1736 ; mansa-sij, *Beng., D.* 387 ; mansa-sij, *Beng., E.* 520 ; mansil, *Hind. & Pb., R.* 61 ; mánskhel, *Kashmir, A.* 590, *Kashmir, Pb., F.* 725 ; mánú (Ulmus), *Pb., U.* 4 ; manú (Rhus), *Pb., R.* 293 ; manua, *Hind., G.* 385 ; manucha, *Sind, E.* 217 ; manyar, *Pb., R.* 543 ; mányí, *Pb., U.* 4 ; mányúl, *Hind., Kumaon, R.* 1 ; manyunth, *Sind, R.* 580 ; manzakhta, *Pushtu, R.* 598 ; mao, *Kumaon, E.* 208 ; máoz, *Dec., M.* 811 ; maoz-kula, *Hind., M.* 811 ; mapúri bet, *Beng., C.* 81 ; mápursika, *Mar., A.* 448 ; már, *Mar., C.* 1520 ; mara, *Pb., E.* 475 ; mara ber, *Pb., Z.* 254 ; mará bér, *Pb., C.* 224 ; maradsing, *Dec., H.* 92 ; maraghúne (Sola-

mum), *Pb.*, S. 2264; maraghune (Ehretia), *Pushtu*, E. 35; maraharalu, *Mar.*, J. 41; marál, *Pb.*, U. 4; márapasapolí, *Mar.*, D. 517; maraphali, *Hind.*, H. 92; marára, *Hind.*, *Pb.*, D. 348; marári, *Pb.*, U. 4; marayadavel, *Mar.*, I. 362; marazh, *Pb.*, U. 4; march, *Afg.*, P. 811; marcha, *Hind.*, *Pb.*, C. 448; márchob, *Pushtu, Afg.*, S. 2678; márchol, *Hind.*, *Pb.*, S. 2678; marchu, *Guj.*, C. 448; marchula, *Hind.*, M. 797; marchula juti, *Mar.*, M. 797; marda, *Hind.*, L. 474; maredi, *Mar.*, E. 276; mareila, *Pb.*, C. 860; marghang, *Pb.*, Q. 13; mar-ghum, *Mar.*, T. 634; marghwalawa, *Pb.*, V. 91; marghwalwa, *Pushtu*, V. 91; mari, *Hind.*, C. 711; marí-ka-gúr, *Hind.*, S. 370; marikájhár, *Dec.*, C. 711; marízha, *Pb.*, T. 416; marjádvel, *Mar.*, I. '362; marjal, *Kashmir*, I. 423; marjavel, *Mar.* I. 362; ma
k (Acer), *Pb.*, A. 341; mark (Briedelia), *Pb.*, B. 868; markhor, *Kashmir*, *Afg.*, S. 1239; marleya, *Beng.*, M. 66; marmandai, *Pb.*, P. 964; marmati, *Mar.*, A. 215; maror, *Dec.*, *Hind.*, H. 92; marorphal, *Hind.*, H. 92; maror-phalli, *Hind.*, *Pb.*, H. 92; marorí-ke-phalli, *Dec.*, H. 92; már-páspoli, *Mar.*, D. 517; marpol, *Pb.*, P. 1460; marrún, *Pb.*, U. 4; martan, *Hind.*, D. 348; marthi, *Mar.*, T. 361; mártí, *Pb.*, J. 24, 29; martili, *Kumaon*, P. 1456; martz, *Kashmir*, P. 811; márú, *Pb.*, Q. 13, 35; marúa (Artemisia), *Pb.*, A. 1464; maruá, *Hind.*, E. 170; maruá (Eleusine), *Beng.*, E. 170; marúl, *Hind.*, S. 785; marúya, *Uriya*, R. 663; marvil, *Hind.*, B. 318; marwa, *Pb.*, V. 164; marwan, *Pb.*, V. 164; marwandaí, *Pushtu*, V. 164; marwande, *Pushtu*, P. 964; masán, *Hind.*, I. 487; masar, *Pb.*, L. 252; másh, *Pb.*, P. 513; mashk-billa, *Hind.*, T. 444; máshkulái, *Beng.*, P. 513; masho, *Pb.*, T. 416; mashúr, *Pb.*, D. 130; mási, *Hind.*, N. 17; masiná, *Beng.*, L. 385; masjot, *Beng.*, B. 893; maska, *Mar.*, B. 983; maslún (Saxifraga), *Pb.*, S. 924; maslún (Polygonum), *Kashmir*, *Pb.*, P. 1112; masna, *Pb.*, *Pushtu*, P. 833; más-patrí, *Duk.*, A. 1469; massandari, *Beng.*, C. 129; massei, *Mar.*, C. 725; massú, *Pb.*, S. 2861; mastáki, *Hind.*, P. 847; mastaru, *Hind.*, A. 1469; mastiara, *Pb.*, S. 998; masúrchana, *Kumaon*, V. 112; masúri (Coriaria), *Hind.*, C. 1958; masúri (Lens), *Beng.*, *Hind.*, *Pb.* L. 252; masúri (Vicia), *Kumaon*, V. 112; masuridál, *Guz.*, L. 252; mat, *Hind.*, P. 468; mataki, *Mar.*, P. 528; mátangnár, *Mar.*, A. 1601; matar, *Beng.*, *Hind.*, P. 885; matar rewari, *Pb.*, P. 882; matazor, *Pb.*, P. 688; matela, *Beng.*, B. 142; math (Phaseolus), *Guj.*, *Mar.*, P 468; máth (molasses), *Beng.*, S. 372; matha, *Mar.*, P. 468; mathágar, *Pb.*, F. 168; mathaniya, *Hind.*, C. 1028; mathara, *Beng.*, C. 133; mathi, *Sind*, T. 612; mathi-dudhi, *Mar.*, E. 549; mathira, *Pb.*, C. 1221; mathirshi, *Hind.*, A. 695, *Pb.*, A. 692; mathna, *Hind.*, E. 166; mát hú, *Pb.*, P. 693; máti-jer, *Mar.*, L. 266; mátisúl, *Mar.*, L. 266; matitsa-wangru, *Kumaon*,

C. 448; mátiyá-tail, *Beng.*, P. 436; mátkalái, *Beng.*, A. 1261; matki, *Mar.*, P. 468; matra, *Hind.*, P. 885; mattar (Lathyrus), *Hind.*, *Sind*, L.100; mattar (Pisum), *Hind.*, P. 882, *Hind.*, *Pb.*, P. 885; mattar rewari. *Hind.*, P. 882; matti (Equisetum), *Pb.*, E. 241; mattí (Orthanthera), *Pb.*, O. 247; mattí-cha-téla, *Mar.*, P. 436; mattí-ká-tailam, *Dec.*, P. 436; mattí-ká-tél, *Dec.*, P. 436; mattí-nu-tél, *Guz.*, P. 436; mattisa, *Hind.*, *Pb.*, C. 448; mattranja, *Beng.*, C. 133; mattu, *Pb.*, I. 123, 128; mattz-rewari, *Pb.*, V. 108; matura, *Beng.*, A. 534; matzbang, *Pb.*, A. 2; maualu, *Hind.*, D. 481; maul (Bassia), *Beng.*, *Hind.*, B. 220; maul (Pyrus), *Pb.*, P. 1460; mául (Bauhinia), *Hind.*, B. 342; maula (Bauhinia), *Hind.*, B. 318; maula (Spatholobus), *Hind.*, S. 2508; máuli, *Hind.*, P. 1460; maulsarau, *Hind.*, M. 570; maulsári, *Hind.*, M. 570; maulsarí, *Pb.*, M. 570; maulser, *Hind.*, M. 570; maulsirí, *Pb.*, M. 570; maun, *Beng.*, C. 725; maur, *Pb.*, Q. 13; máura, *Pb.*, V. 164; maurain, *Hind.*, B. 342; mauri (Fœniculum), *Beng.*, F. 659; maúri (Lens), *Pb.*, L. 252; máúrn mamjí, *Pb.*, U. 4; mavalung, *Mar.*, C. 1270; mawa (Quercus), *Sind*, Q. 43; mawá (Vitex), *Pb.*, V. 164; máwal, *Kashmir*, *Pb.*, C. 873; mawashi, *Pb.*, M. 147; maweshi, *Beng.*, *Hind.*, O. 574; maya (Cervulus), *Beng.*, D. 219; máyá (Quercus), *Sind*, Q. 43; mayarawa, *Hind.*, A. 1288; mayurshika, *Guj.*, A. 501; mazar, *Baluch.*, T. 437; mazri, *Hind.*, N. 1; mázú, *Dec.*, *Hind.*, Q. 43.

Me Mechitta, *Beng.*, C. 913; meda, *Hind.*, L. 474; medachob, *Pb.*, L. 483; médalakri, *Pb.*, L. 474, 483; medasak, *Pb.*, L. 483; medh, *Hind.*, L. 483; medhola, *Guz.*, R. 1; médí, *Guz.*, L. 126; meenaharma, *Mar.*, B. 57; mehal, *Hind*; *Kumaon*, *Pb.*, P.1466; mehali, *Kumaon*, P. 1460; mehat, *Pb.*, I. 503; méhédí, *Beng.*, L. 126; mehendi, *Raj.*, L. 126; méhndá, *Hind.*, S. 1320; mehndi (Elscholtzia), *Pb.*, E. 196; mehndi (Lawsonia), *Pb.*, L. 126; meho, *Sind*, C. 1221, 1227; mehrwán, *Pushtu*, V. 164; meih-síla, *Guj.*, *Hind.*, L. 455; meiní, *Pb.*, C. 2101; mekhun, *Beng.*, C. 289; memoká, *Pb.*, M. 289; ména (wax), *Mar.*, C. 931; ména (honey), *Mar.*, H. 342; menda, *Hind.*, S. 1273; méndar, *Hind.*, L. 483; mendal, *Hind.*, *Mar.*, D. 748; mendar, *Pb.*, D. 725; mendi (Tylophora), *Uriya*, T. 855; méndí (Lawsonia), *Beng.*, *Guj.*, *Mar.*, *Sind*, L. 126; menorú, *Pb.*, D. 725; ménhdí, *Hind.*, L. 126; mennie, *Hind.*, F. 792; menphal, *Beng.*, R. 1; mentog, *Pb.*, S. 1065; mentok, *Pb.*, T. 17; menya, *Guz.*, P. 332; merádú, *Hind.*, P. 1062; méra-singí, *Beng.*, *Hind.*, G. 748; merino, *Pb.*, P. 1201; mermahaul, *Pb.*, E. 485; mersingh, *Hind.*, D. 748; mersingi, *Mar.*, D. 748; meru (sámbar), *Mar.*, D. 240; merú (Ulmus), *Pb.*, U. 4; mesaki, *Pb.*, V. 131; mesh, *Wakhan*, S. 1276; messinge, *Mar.*, I). 748; mesta, *Beng.*, H. 233; mestapát, *Beng.*, H. 177; meta limbu, *Guj.*, C. 1286; méthi, *Beng.*,

Guz., *Hind.*, *Pb.*, T. 612 ; methika, *Beng.*, T. 612 ; methini, *Guz.*, T. 612 ; methishak, *Beng.*, T. 612 ; meth-kalai, *Hind.*, P. 468 ; methri, *Pb.*, T. 612 ; methun (Trigonella), *Pb.*, T. 612 ; methun (gayal), *Hind.*, O. 565 ; methúri, *Mar.*, M. 431 ; metkur, *Hind.*, E. 73 ; metralana (Haloxylon), *Pb.*, H. 6 ; metra láne (Anabasis), *Pb.*, A. 1005 ; metunga, *Beng.*, M. 425 ; mewri, *Hind.*, V. 164.

Mh Mhains, *Hind.*, O. 558 ; mhaishabola, *Mar.*, B. 62 ; mhaner, *Hind.*, G. 136 ; mhar, *Mar.*, E. 166 ; mharengala, *Hind.*, A. 341 ; mhenda, *Hind.*, S. 1320 ; mhéndí, *Hind.*, L. 126 ; mhindí, *Hind.*, L. 126 ; mhór-tuttah, *Dec.*, C. 2367 ; mhowa, *Hind.*, B. 220.

Mi Michren, *Kanáwar*, P. 371 ; middiáwal, *Guj.*, C. 737 ; mighri, *Hind.*, P. 67 ; mijhaula, *Kumaon*, E. 48 ; mijhri, *Hind.*, F. 672 ; mijhri, *Hind.*, P. 67 ; mılech, *Pb.*, H. 277 ; mile-lo-ah, *Hind.*, F. 558 ; miles, *Pb.*, H. 277 ; mimarari, *Pb.*, R. 159 ; mín, *Guj.*, C. 931, H. 342 ; minaguta, *Sind*, E. 520 ; minak tanah, *Mar.*, P. 436 ; mindhal, *Guj.*, *Pb.*, R. 1 ; mindhala, *Guz.*, R. 1 ; mindla, *Pb.*, R. 1 ; mindri, *Beng.*, B. 875 ; mingut, *Mar.*, E. 520 ; minguta, *Mar.*, E. 520 ; mini-chambeli, *Hind.*, M. 550 ; mira, *Mar.*, F. 202 ; miragbahar, *Hind.*, C. 1234 ; mirandú (Elæodendron), *Pb.*, E. 73 ; mírandú (Dodonæa), *Pb.*, D. 725 ; mirch (Capsicum), *Hind.*, *Pb.*, C. 448 ; mirch (Piper), *Hind.*, P. 811 ; mirchai, *Beng.*, I. 384 ; mirchai, *Hind.*, I. 384 ; mirchai, *Mar.*, I. 384 ; mirchi, *Guj.*, *Mar.*, C. 455, *Dec.*, C. 466 ; mirchia gand, *Hind.*, A. 1117 ; mirchía-gard, *Hind.*, A. 1117 ; mirchu, *Guj.*, C. 448 ; mirch-wángum, *Kashmir*, C. 448 ; miré, *Mar.*, P. 811 ; mir-goo, *Pb.*, E. 73 ; miri, *Guz.*, *Mar.*, P. 811 ; miringa, *Pb.*, O. 233 ; miriyaban, *Hind.*, A. 1093 ; mirrgah, *Uriya*, F. 410 ; mírri, *Pb.*, P. 746 ; mirsingá, *Mar.*, C. 448 ; mirtz-a-vangun, *Kashmir*, C. 448 ; mirzanjosh, *Pb.*, O. 220 ; mishkdána, *Mar.*, H. 168 ; misrí, *Beng.*, *Hind.*, *Pb.*, S. 367 ; mísrí leí, *Pb.*, T. 51 ; missi siyá, *Hind.*, M. 131 ; miswak, *Pushtu*, S. 705 ; míta-alú, *Hind.*, I. 348 ; míta-nimbú, *Pb.*, C. 1301 ; mitenga, *Beng.*, B. 142 ; mítha (Trigonella), *Sind*, T. 612 ; mítha (salt), *Mar.*, S. 602 ; míthá (raisins), *Pb.*, V. 233 ; mitha amritphal, *Hind.*, C. 1301 ; mítha dúdia, *Hind.*, P. 1074 ; mítha gokhru, *Guz.*, T. 548 ; míthái, *Hind.*, S. 370 ; míthá indarjou, *Dec.*, *Hind.*, W. 122l; míthájtrá, *Beng.*, P. 727 ; mitha-kaddu, *Hind.*, C. 2325 ; míthá-kaddú, *Dec.*, *Hind.*, C. 2316 ; mitha limbu, *Guj.*, *Mar.*, C. 1301 ; mitha nebu, *Beng.*, *Hind.*, C. 1301 ; mitha nimbu, *Hind.*, C. 1300 ; míthá-tél, *Dec.*, *Hind.*, S. 1078 ; mitha tendu, *Hind.*, D. 656 ; mítha-zahar, *Hind.*, A. 397, *Kashmir*, *Pb.*, A. 413 ; mítházahar, *Hind.*, A. 399 ; mithi diár, *Sind*, S. 705 ; mithi diár, *Sind*, C. 224 ; mithiga, *Pb.*, L. 528 ; mithitumbi, *Hind.*, L. 30 ; míthí van, *Pb.*, C. 224 ; míthí van, *Hind.*, C. 224 ; míthí ván, *Pb.*, S. 705 ; mitholaní, *Sind*, C. 224 ; mithpatta, *Pb.*, M. 22 ; míthú, *Guz.*, S. 602 ; mithu tél, *Guz.*, S. 1078 ; mitthilakri,

Dec., G. 278 ; mittí-ka-tél, *Hind.*, P. 436 ; mittíkhám, *Pb.*, A. 962 ; mittúa, *Pb.*, E. 335 ; mítú, *Pb.*, R. 564.

Mm Mmánraí, *Pushtu*, S. 502.

Mo Mo, *Hind.*, P. 380 ; moakurra, *Beng.*, C. 989 ; moal, *Beng.*, V. 40 ; mócharas, *Beng.*, B. 633 ; mochkand, *Hind.*, E. 7 ; mochna, *Hind.*, F. 463 ; moda, *Mar.*, C. 722 ; modhú, *Beng.*, H. 342 ; modun-tiki, *Beng.*, S. 2907 ; mogalíeranda, *Mar.*, J. 41; mogali eranda, *Mar.*, J. 41 ; mogbíre-kapattá, *Duk.*, A. 1132 ; mogaa, *Mar.*, J. 32, *Beng.*, *Hind.*, *Mar.*, J. 35 ; mogri (Jasminum), *Mar.*, J. 35; mogri (Raphanus), *Mar.*, R. 43 ; mogro, *Guz.*, J. 35 ; mogul, *Hind.*, S. 2850; moh, *Hind.*, *Pb.*, F. 511 ; moha, *Dec.*, *Uriya*, B. 220 ; mohá, *Mar.*, B. 220; mohá, *Dec.*, *Hind.*, B. 265 ; móháchajháda, *Mar.*, B. 265 ; mohakri, *Pb.*, T. 576 ; mohand-i-gúj saféd, *Kashmir*, A. 401 ; moháni, *Kumaon*, C. 710 ; mohar, *Sind*, P. 468 ; mohari, *Mar.*, B. 833 ; mohaylí, *Hind.*, F. 467; mohí (Bassia), *Mar.*, B. 265 ; mohi (Garuga), *Uriya*, G. 143 ; mohin, *Hind.*, O. 38 ; moho, *Mar.*, B. 220 ; mohr (Lens), *Pb.*, L. 252 ; mohr (peacock), *Hind.*, P. 350 ; mohri (Aconitum), *Kashmir*, *Pb.*, A. 413 ; mohrí, *Pb.*, L. 252 ; mohuá, *Hind.*, B. 265 ; mohuvá, *Beng.*, B. 265; moi (Eragrostis), *Raj.*, E. 246; moi (Odina), *Mar.*, O. 38; moina (Vangueria), *Beng.*, *Hind.*, V. 25 ; moina (Odina), *Mar.*, O. 38 ; moinsia-ballia, *Uriya*, F. 601 ; moiyar, *Raj.*, P. 380 ; moja, *Mar.*, O. 38 ; moka, *Hind.*, S. 959 ; moka gantha, *Mar.*, S. 959 ; mokha (Schrebera), *Hind.*, S. 959 ; mokha (Momordica), *Hind.*, M. 623 ; mokna, *Pb.*, L. 451 ; moksha, *Pb.*, A. 590 ; moksha, *Pb.*, F. 725 ; mol, *Hind.*, *Kumaon*, P. 1466 ; molarda, *Mar.*, O. 38; mole, *Pushtu*, D. 674 ; móm, *Beng.*, *Dec.*, *Hind.*, C. 931 ; móm, *Beng.*, *Dec.*, *Hind.*, H. 342 ; momádrú chopándiga, *Kashmir*, A. 367 ; momanna, *Afg.*, S. 497 ; momchina (Croton), *Beng.*, C. 2191 ; mom-china (Sapium), *Beng.*, S. 842 ; monda, *Uriya*, T. 525 ; mondia-jori, *Hind.*, E. 265 ; moni, *Kumaon*, C. 710 ; mont, *Pb.*, P. 693 ; montá bhokar, *Mar.*, C. 1931 ; mooi, *Uriya*, O. 38 ; mopsha, *Pb.*, F. 725 ; mór, *Mar.*, P. 350 ; mora (Vitex), *Pb.*, V. 164 ; mora (Bassia), *Mar.*, B. 220 ; mor-ah-ki, *Sind*, F. 410 ; morala, *Beng.*, F. 463 ; mórang-iláchí, *Beng.* & *Hind.*, A. 965 ; morann, *Pb.*, V. 164 ; morar, *Beng.*, F. 326 ; morara, *Uriya*, F. 315 ; morari, *Beng.*, F. 326 ; morasa, *Mar.*, S. 2985 ; morasa, *Mar.*, S. 2994 ; moráun, *Pb.*, V. 164 ; moravela, *Mar.*, C. 1363 ; morbhaga, *Hind.*, C. 1028 ; mored, *Hind.*, U. 4 ; moriel, *Mar.*, C. 1363 ; morinda, *Hind.*, *Kumaon*, A. 17, A. 22 ; moriya, *Hind.*, H. 294 ; mor múj, *Kashmir*, D. 173 ; mor-pach, *Hind.*, A. 448 ; mor-pankhi, *Hind.*, A. 448 ; morphal, *Pb.*, P. 1460 ; mor-ri-ah, *Pb.*, F. 530 ; morrul, *Hind.*, F. 517 ; morta, *Pb.*, S. 1063 ; morthan, *Hind.*, P. 380 ; mór-tútá, *Guz.*, C. 2367 ; mórtuttá, *Dec.*, C. 2367 ; moru, *Hind.*, *Pb.*, Q. 13 ; morúa, *Pb.*, R. 251 ; morún, *Pb.*, U. 4 ; morunda, *Hind.*, A. 22 ; morvel, *Mar.*, C. 1363 ;

G

morwa, *Mar.*, S. 785 ; móshabbar, *Beng.*,
A. 824 ; moshiki, *Mar.*, T. 862 ; moso-
nea, *Uriya*, E. 25 ; mot, *Hind.*, P. 468 ;
motá bhokar, *Mar.*, C. 1931 ; motabon,
Mar., L. 42 ; mota-bondara, *Mar.*, L.
42 ; motä dágada phúl, *Mar.*, L. 332 ;
mota karmal, *Mar.*, D. 428 ; mota kar-
mel, *Mar.*, D. 428 ; mote, *Dec.*, P. 468 ;
moté-veldode, *Mar.*, A. 976 ; moth, *Hind.*,
P. 468 ; moth, *Hind.*, F. 671 ; moth,
Hind., *Pb.*, P. 468 ; motha (Cyperus),
Beng., *Guj.*, C. 2612 ; motha (Desmodium), *Hind.*, D. 348 ; motha (mats),
Kumaon, M. 319 ; mothá, *Hind.*, F. 671 ;
mothá siras, *Mar.*, A. 695 ; mothan gokharu, *Guz.*, P. 363 ; mothe gokharu, *Mar.*,
P. 363 ; mothe karamala, *Mar.*, D. 428 ;
mothi, *Hind.*, P. 468 ; mothidudhí, *Mar.*,
E. 531 ; mothi-kuhili, *Mar.*, M. 781 ;
moti, *Hind.*, P. 355 ; motia (Panicum),
Hind., P. 41 ; motía (Jasminum), *Hind.*,
J. 35 ; motighol, *Mar.*, P. 1179 ; moti
khabbal, *Pb.*, P. 77 ; motí khajati, *Mar.*,
G. 213 ; motí láne, *Pb.*, S. 596 ; motí
lání, *Pb.*, C. 224 ; motíringí, *Mar.*, S.
2280 ; motki, *Beng.*, S. 371 ; motoaduso,
Guj., A. 658 ; moto-iláchi, *Guj.*, A. 976 ;
moto sarsio, *Guj.*, A. 686 ; moto satodo,
Guj., B. 619 ; mottoghókru, *Guz.*, P.
363 ; motto-piper, *Guz.*, S. 970 ; mótu-
límbu, *Guj.*, C. 1286 ; mótu-nimbu, *Guj.*,
C. 1286 ; motusi, *Beng.*, F. 549 ; mou
álu, *Beng.*, D. 481 ; mouni, *Hind.*, O.
38 ; mouz, *Dec.*, M. 811 ; mová, *Mar.*,
B. 220 ; mová-nujháda, *Guj.*, B. 265 ;
mowa (Bassia), *Hind.*, B. 220 ; mowa (En-
gelhardtia), *Kumaon*, E. 208 ; mowa (Glochidion), *Hind.*, G. 240 ; mowá (Orthanthera), *Pb.*, O. 247 ; mowa (Pennisetum),
Hind., P. 380 ; mowda, *Mar.*, B. 220 ;
mowen, *Hind.*, O. 38 ; moya (Ischœmum), *Hind.*, I. 494 ; moya (Odina),
Mar., *Sind*, O. 38 ; moye, *Mar.*, O. 38 ;
moyeen, *Hind.*, O. 38 ; moyen, *Hind.*,
O. 38.

Mr Mrál, *Pb.*, L. 589 ; mrig, *Hind.*, S. 1226 ;
mrigala, *Hind.*, F. 410 ; mrigale, *Uriya*,
F. 410.

Mu Múah, *Sind*, M. 717 ; mubárak, *Mar.*,
A. 506, A. 510 ; mubáraka, *Hind.*,
Kumaon, A. 498 ; muchkand, *Beng.*,
Hind., P. 1397 ; muchukunda, *Mar.*,
P. 1397 ; muchunda, *Mar.*, P. 1397 ;
muckní (Ambassis), *Pb.*, F. 311 ; muckní
(Rita), *Hind.*, F. 566 ; mudanu, *Pb.*,
S. 560 ; mudar-ktai, *Beng.*, C. 2624 ;
mudhár, *Hind.*, C. 170 ; múdhírí, *Sind*,
C. 1842 ; mudnu, *Pb.*, P. 540 ; mudo-
pun, *Sind*, T. 875 ; muduna, *Beng.*,
Hind., V. 25 ; mudú-nírbisha, *Hind.*,
Beng., K. 1 ; múg (Phaseolus), *Beng.*,
Hind., *Mar.*, P. 496 ; múg (Stereo-
spermum), *Beng.*, S. 2876 ; mugah,
Beng., F. 509 ; mugáni, *Beng.*, *Hind.*,
P. 523 ; muglai bedáná, *Guj.*, C. 2551 ;
múgra (Jasminum), *Pb.*, J. 35 ; mugra
(Raphanus), *Hind.*, R. 43 ; mugrela,
Beng., N. 158 ; muhishabole, *Guj.*, B.
41 ; muhli, *Hind.*, M. 393 ; muhri,
Sind, P. 468 ; muhúrí, *Beng.*, P. 727 ;
mujethi, *Hind.*, I. 39 ; mújí, *Pb.*, P.
496 ; mujjun, *Sind*, C. 826 ; múkani,
Mar., P. 528 ; mukhajali, *Hind.*, D.

836 ; mukká (Zea), *Raj.*, Z. 50 ; muk-
ka (chela), *Sind*, F. 396 ; mukni (Zea),
Kumaon, Z. 50 ; mukni (Amblypharyn-
godon), *Pb.*, F. 315 ; muktajuri, *Beng.*,
A. 306 ; mukta maya (Sapindus), *Uriya*,
S. 818 ; muktamoya (Erioglossum),
Uriya, E. 310 ; muktá-páta, *Beng.*, P.
625 ; mukul, *Beng.*, *Dec.*, *Guj.*, *Hind.*,
Sind, B. 43 ; mukul salai, *Guj.*, B.
771 ; mukund babri, *Pb.*, A. 677 ;
mukuya, *Mar.*, P. 523 ; mula (Ra-
phanus), *Beng.*, *Mar.*, R. 31 ; mula
(Spatholobus), *Hind.*, S. 2508 ; múla
(Sterculia), *Beng.*, S. 2819 ; múlandah,
Mar., F. 329 ; mulatthi-kásras, *Hind.*,
G. 278 ; mulei, *Pb.*, F. 35 ; mulethi,
Hind., C. 278 ; muleti, *Pb.*, G. 278 ;
mulhatti, *Hind.*, C. 278 ; múli, *Hind.*,
R. 31 ; múli (Melocanna), *Beng.*, M.
425 ; múli (Raphanus), *Hind.*, *Pb.*, R.
31 ; mulím, *Pb.*, G. 243 ; mulín, *Pb.*,
O. 233 ; mulín, *Pb.*, S. 2352 ; muljuyáti,
Uriya, L. 126 ; mulkácha, *Mar.*, C. 737 ;
mulká-cha-révalchinní, *Mar.*, К. 215 ;
mulkácha-sakhar, *Mar.*, S. 373 ; mul-la,
Sind, F. 601 ; mullí, *Dec.*, R. 31 ; mulsári,
Hind., M. 570 ; mumkára, *Hind.*, E.
259 ; muna, *Hind.*, C. 2148 ; munagácha-
jháda, *Mar.*, M. 721 ; múncha, *Hind.*,
P. 1179 ; múndhrí, *Dec.*, S. 2518 ; mundi,
Dec., *Hind.*, *Mar.*, S. 2518 ; múndi,
Beng., *Pb.*, S. 2518 ; múndla, *Pb.*, P.
1322 ; mundúa, *Kumaon*, S. 1212 ; mún-
dwál, *Pb.*, D. 245 ; mung (buffalo),
Hind., O. 555 ; múng (Phaseolus), *Beng.*,
Hind., *Kumaon*, *Pb.*, *Raj.*, *Sind*, P.
496 ; múnga, *Hind.*, C. 1808 ; mungari,
Hind., Z. 50 ; mungas kajur, *Hind.*,
S. 2310 ; mungé-ká-jhár, *Dec.*, M. 721 ;
múngu, *Pb.*, P. 496 ; múngphali, *Hind.*,
A. 1261 ; mungra, *Pb.*, R. 31, 43 ; múngu,
Hind., P. 496 ; munídi, *Mar.*, O. 38 ;
munigha, *Uriya*, M. 721 ; múníla, *Pb.*, D.
253 ; muniya, *Kumaon*, D. 528 ; munj
(Saccharum), *Pb.*, S. 6 ; munj (Phaseo-
lus), *Hind.*, P. 496 ; múnja, *Hind.*, S.
6 ; munjhú rukha, *Beng.*, P. 961 ; munji
(Ischœmum), *Pb.*, l. 494 ; munjí (Oryza),
Hind., *Pb.*, O. 258 ; munmuna, *Pb.*, I.
503 ; munri, *Kumaon*, A. 2 ; múnya,
Beng., *Hind.*, P. 1179 ; munyú, *Pb.*, A.
1462 ; múnzat, *Pb.*, R. 564 ; múphal,
Hind., Q. 43 ; múr, *Hind.*, P. 350 ; mura,
Guz., R. 31 ; múrad (Myrtus), *Hind.*, *Pb.*,
M. 921 ; murádh (Ribes), *Pb.*, R. 359,
365 ; murahara, *Beng.*, S. 785 ; muraiti-
ka-jur, *Hind.*, G. 278 ; múrak, *Pb.*, S.
981 ; murba, *Beng.*, S. 785 ; murchob,
Kumaon, M. 797 ; murdasang, *Hind.*,
Pb., L. 143 ; murdásing, *Guz.*, H. 92 ;
murdi, *Hind.*, I. 503 ; murgá, *Beng.*,
S. 785 ; murgábi, *Beng.*, S. 785 ; mur-
gali, *Dec.*, *Mar.*, S. 785 ; murga murji,
Beng., A. 603 ; murgli, *Beng.*, S. 785 ;
murhoa, *Kumaon*, S. 1212 ; muri, *Mar.*,
Sind, R. 31 ; muria (Vitex), *Uriya*, V.
177 ; muriá (Buchanania), *Hind.*, B.
913 ; muricha, *Beng.*, P. 811 ; muri-
chung, *Beng.*, P. 811 ; muri-muri, *Uriya*,
H. 92 ; murivacha, *Beng.*, *Uriya*, F.
549 ; murján, *Hind.*, C. 1808 ; murkanta,
Beng., A. 306 ; murkíla, *Kumaon*, M.
295 ; múkrú, *Pb.*, M. 22 ; murkúla,

Hind., M. 295 ; murl, *Hind.*, F. 517 ; murlai, *Pushtu*, P. 1179 ; murmuriá, *Beng.*, S. 2518 ; muro, *Hind.*, *Mar.*, R. 31 ; murra, *Pb.*, C. 2520 ; murrú, *Beng.*, O. 214 ; murrul, *Mar.*, F. 515 ; múrt, *Hind.*, D. 348 ; murudásenga, *Mar.*, H. 92 ; múrvá, *Hind.*, S. 785 ; murwa, *Dec.*, O. 214 ; murwo, *Sind*, O. 214 ; mús, *Beng.*, P. 1389 ; musabbar, *Hind.*, A. 824 ; musal, *Hind.*, V. 213 ; musambarból, *Mar.*, A. 824 ; musán, *Hind.*, I. 489 ; musanbar, *Duk.*, A. 824 ; muscat-ka-halwa, *Hind.*, C. 232 ; musel (Andropogon), *Hind.*, A. 1090, 1117 ; musel (Anthistiria), *Hind.*, A. 1188 ; musel (Heteropogon), *Hind.*, H. 164 ; musel (Iseilema), *Hind.*, I. 487 ; mushak-dana, *Beng.*, *Guj.*, H. 168 ; múshalí, *Hind.*, *Mar.*, C. 2375 ; mushk, *Hind.*, D. 228 ; mushk-bhéndí-kébij, *Dec.*, *Mar.*, H. 168 ; mushk-dana, *Hind.*, H. 168 ; mushkiára, *Pb.*, S. 763 ; mushk náfá, *Pb.*, D. 228 ; mushk tara, *Pb.*, M. 461 ; mushkwálí, *Pb.*, V. 10 ; musiál, *Hind.*, I. 487 ; muskei, *Pb.*, H. 13 ; múslí-e-siyah, *Hind.*, A. 1122 ; múslí-kand, *Hind.*, *Mar.*, C. 2375 ; musna, *Hind.*, S. 850 ; múss, *Hind.*, D. 348 ; mussabar, *Deccan*, A. 826 ; muss-ayahri, *Uriya*, F. 562 ; mussoassah, *Pb.*, F. 508 ; mussulkund, *Hind.*, C. 2375 ; mustá, *Mar.*, C. 2612 ; mustakh, *Kashmir*, C. 1003 ; mustarú, *Hind.*, G. 660 ; musubbar, *Hind.*, A. 839 ; musúr chuna (Vicia), *Beng.*, V. 112 ; musúr-chúna (Lathyrus), *Beng.*, L. 96 ; mút, *Dec.*, *Guj.*, P. 468 ; muta-bela, *Hind.*, J. 13 ; mutchlí, *Hind.*, R. 287 ; muthá, *Beng.*, C. 2612 ; múthí, *Hind.*, T. 612 ; mút-kar, *Pb.*, A. 1061 ; mutki, *Mar.*, C. 2514 ; mútní, *Pb.*, R. 152 ; mutruk, *Hind.*, S. 2819 ; mutta, *Beng.*, A. 1215 ; mutti, *Guz.*, P. 355 ; muyá muyá, *Beng.*, S. 510 ; muyna, *Beng.*, *Hind.*, V. 25 ; muyuana, *Beng.*, *Hind.*, V. 25 ; múz, *Pb.*, I. 811.

Mw Mwari, *Hind.*, J. 10.

My Myal-ki-bhaji, *Hind.*, B. 203 ; myoukphal, *Mar.*, D. 513.

Mz Mzarái, *Pushtu*, N. 1.

Na Nabamallíká, *Beng.*, J. 35 ; nábar, *Pb.*, R. 359, 365 ; nabhi-ánkuri, *Uriya*, S. 2315 ; nábre, *Pb.*, R. 365 ; nachani, *Mar.*, E. 166 ; nachiri, *Mar.*, E. 170 ; nachni, *Sind*, E. 170 ; nadiyá, *Hind.*, O. 574 ; nag, *Hind.*, R. 124 ; naga, *Pb.*, E. 479 ; naga-champa, *Mar.*, M. 489 ; nagadavana, *Mar.*, C. 2062 ; nágakuda, *Mar.*, M. 656 ; nága kundá, *Mar.*, M. 656 ; nagala dudhí, *Guj.*, D. 9 ; nágalkuda, *Mar.*, S. 2824 ; nágani, *Mar.*, C. 156 ; nágárjamán, *Pb.*, F. 168 ; nágar-móthá, *Dec.*, *Hind.*, C. 2617 ; nágar-mútha, *Beng.*, C. 2617 ; nág bal, *Mar.*, A. 911 ; nág bala, *Beng.*, A. 911 ; nág champa (Mesua), *Mar.*, M. 489 ; nág chámpa (Calophyllum), *Mar.*, C. 146 ; nág-chápha, *Mar.*, M. 489 ; nagdamani, *Guz.*, C. 2062 ; nagdaun (Crinum), *Beng.*, C. 2062 ; nágdaun (Staphylea), *Hind.*, *Pb.*, S. 2678 ; nágdoná, *Beng.*, A. 1469 ; nág-

doun, *Hind.*, A. 1567 ; nágdouná, *Hind.*, A. 1469 ; nag-dowana (Centipeda), *Beng.*, *Hind.*, *Mar.*, C. 913 ; nagdown (Crinum), *Mar.*, C. 2062, C. 2068 ; nágésar (Mesua), *Beng.*, M. 489 ; nágésar (Ochrocarpus), *Beng.*, O. 6 ; nagesh-voro, *Uriya*, M. 489 ; nágeswar, *Uriya*, M. 489 ; nagfan, *Mar.*, S. 785 ; naghas, *Hind.*, M. 489 ; nagin-kapatta, *Duk.*, C. 2062 ; nágkaria, *Mar.*, G. 243 ; nag-kesar (Mesua), *Beng.*, *Hind.*, *Pb.*, M. 489 ; nag kesar (Ochrocarpus), *Hind.*, O. 6 ; nagkeshur, *Hind.*, M. 489 ; nágli, *Mar.*, E. 166, 170 ; nagoda, *Gus.*, V. 164 ; nágorigond, *Hind.*, L. 455 ; nág-phaná, *Beng.*, *Hind.*, O. 193 ; nag-phansi, *Dec.*, O. 193 ; nag-pút, *Beng.*, B. 297 ; nagre, *Hind.*, A. 1523 ; nagri, *Mar.*, E. 48 ; nag samp, *Hind.*, R. 124 ; nágur-vel, *Gus.*, P. 775 ; nahání, *Pb.*, V. 5 ; nahání-kunvar, *Guj.*, A. 836 ; nahar, *Hind.*, F. 768 ; náhar, *Hind.*, T. 437 ; naharm, *Hind.*, F. 357 ; naharn, *Uriya*, F. 495 ; nahni-máyí, *Dec.*, T. 51 ; nahnuí-máyín, *Hind.*, T. 51 ; nahotara, *Gus.*, I. 415 ; nái, *Hind.*, V. 17 ; nái (barley), *Pb.*, H. 382 ; nái (Phragmites), *Pb.*, P. 618 ; nái (Vanda), *Beng.*, V. 17 ; nai-bél, *Mar.*, L. 362 ; naim, *Hind.*, F. 410 ; nai-sakar, *Gus.*, *Hind.*, S. 30 ; nak, *Hind.*, P. 1452 ; nák, *Afg.*, *Pb.*, P. 1452 ; nákacinkaní, *Mar.*, C. 913 ; nakbel, *Sind.*, B. 619 ; nakchhikní, *Hind.*, D. 823 ; nakel, *Mar.*, P. 555 ; nakhtar (Cedrus), *Afg.*, C. 846 ; nakhtar (Pinus), *Pb.*, *Pushtu*, P. 760 ; nakhter, *Pushtu*, P. 1285 ; nakk-chikní, *Beng.*, *Hind.*, *Mar.*, C. 913 ; nakríze, *Pb.*, L. 126 ; naktrúsa, *Pb.*, S. 2219 ; nal (Nymphæa), *Beng.*, N. 200 ; nal (Phragmites), *Beng.*, *Kumaon*, *Pb.*, P. 618 ; nál (Bambusa), *Pb.*, B. 118 ; nal báns, *Beng.*, B. 134, *Hind.*, B. 118 ; nálí (Tephrosia), *Dec.*, T. 270 ; nálí (Ipomæa), *Pb.*, I. 343 ; nalichilbaji, *Mar.*, I. 343 ; nálitá, *Uriya*, C. 1847 ; nálitá (Corchorus), *Beng.*, C. 1879 ; nálitá (Hibiscus), *Hind.*, H. 177 ; náliyer, *Guj.*, C. 1520 ; nallar, *Hind.*, *Mar.*, V. 219 ; nalla tapeta, *Pb.*, S. 2354 ; nallér, *Dec.*, V. 219 ; nalu, *Pb.*, P. 618 ; namak, *Hind.*, *Mar.*, S. 602 ; namuti, *Beng.*, G. 660 ; nana, *Mar.*, L. 53, 55 ; nanbhantúr, *Beng.*, C. 2171 ; nandi, *Mar.*, L. 55 ; nandi-butchua, *Uriya*, F. 450 ; nandin, *Beng.*, F. 472 ; nando-jangro, *Sind*, Z. 254 ; nandrú, *Pb.*, P. 684 ; nandruk, *Mar.*, F. 253 ; nang, *Hind.*, C. 1969 ; nangke, *Pb.*, R. 362 ; nangke hádar, *Pb.*, R. 365 ; nangli, *Mar.*, *Sind*, E. 170 ; náni-janglí-kándo, *Mar.*, S. 968 ; nanna, *Pb.*, M. 105 ; nannári, *Dec.*, H. 119 ; nanugúndi, *Gus.*, C. 1940 ; nanu-witi, *Beng.*, G. 311 ; nanvachi-wel, *Mar.*, P. 678 ; naolli, *Uriya*, F. 456 ; napo, *Sind*, N. 200 ; nar (iron), *Hind.*, I. 440 ; nar (Phragmites), *Pb.*, P. 618 ; nara, *Pb.*, P. 618 ; nárakel, *Beng.*, C. 1520 ; nárakya-úd, *Mar.*, C. 894 ; naral, *Mar.*, C. 1520 ; naral-cha-jháda, *Mar.*, C. 1520 ; nárali-cha-jháda, *Mar.*, C. 1520 ; náralichatéla, *Mar.*, C. 1520 ; náralmád, *Mar.*, C. 1520 ; naral-tela,

Mar., C. 1520; náranghi cantra, *Mar.*, C. 1233; nárangi, *Guj.*, *Hind.*, nárangi, *Mar.*, *Pb.*, C. 1233; náranj, *Pb.*, C. 1233; naraseja, *Mar.*, E. 496; narasij, *Beng.*, E. 496; narchá, *Beng.*, C. 1846; narel, *Mar.*, C. 1520; nárél, *Dec.*, *Hind.*, *Mar.*, C. 1520; narela, *Mar.*, C. 1520; nárélí, *Hind.*, C. 1520; nárengá, *Beng.*, C. 1233; nárenj, *Hind.*, C. 1233; nargis, *Pb.*, N. 10; nargosh, *Pushtu*, P. 1426; nargunda, *Mar.*, V. 164; narguni, *Uriya*, A. 1601; nari (Equisetum), *Pb.*, E. 241; nári, *Hind.*, I. 343; nárí (Ipomæa), *Pb.*, I. 343; naria, *Pb.*, P. 618; narich sag, *Beng.*, C. 1861; náriel, *Guj.*, *Hind.*, C. 1520; náriera, *Guj.*, C. 1520; nárikel, *Beng.*, C. 1520; naril, *Mar.*, C. 1520; naril-ká-tél, *Hind.*, *Duk.*, C. 1520; na-rillie, *Duk.*, C. 1520; náringa, *Mar.*, C. 1233; náringhi, *Beng.*, C. 1234; nárin-ghie, *Duk.*, C. 1233; naringi (Limonia), *Mar.*, L. 362; náringi (Citrus), *Hind.*, *Mar.*, *Pb.*, C. 1233; náringsála, *Mar.*, C. 1233; náriyal, *Beng.*, *Hind.*, C. 1520; nariyal-ka-gúr, *Hind.*, S. 370; náríyel, *Hind.*, C. 1520; nariyéla, *Guj.*, C. 1520; nar-kachúr (Zingiber), *Beng.*, *Hind.*, *Pb.*, Z. 225; nar-kachúra (Curcuma), *Hind.*, *Mar.*, C. 2422; narkat, *Hind.*, A. 1523; narkul, *Hind.*, P. 618; narkuli, *Uriya*, P. 627; narkurat, *Mar.*, I. 515; narlei, *Pb.*, T. 51; narma, *Pb.*, T. 541; náro (Casearia), *Hind.*, C. 722; naro (Ipo-mæa), *Sind*, I. 343; narqual, *Hind.*, A. 1523; nárr, *Pb.*, M. 105; narra, *Hind.*, E. 20; narri (Arundinaria), *Hind.*, A. 1523; narrí (Polygonum), *Pb.*, P. 1084; narruk, *Baluch.*, S. 593; nárula, *Mar.*, C. 1520; nárungasála, *Mar.*, C. 1233; nárúnge, *Hind.*, C 1233; nárungi, *Beng.*, *Guj.*, C. 1233; nárvel, *Mar.*, P. 1233; narwa, *Sind*, T. 541; nárwel, *Mar.*, V. 99; náryal, *Guj.*, C. 1520; nasa bhaga, *Beng.*, P. 425; nashotar, *Guz.*, I. 415; náshpáti, *Hind.*, *Pb.*, P. 1452; nashtar, *Pb.*, *Pushtu*, P. 760; naskarkáni, *Uriya*, C. 1847; násoná, *Beng.*, O. 233; náspál, *Hind.*, *Pb.*, P. 1426; naspáti (Pyrus), *Hind.*, *Kashmir*, P. 1452; *Pb.*, S. 1906; nasurjinghi ke jurr, *Dec.*, T. 537; nasút, *Hind.*, E. 356; náta (bullock), *Hind.*, O. 574; nátá (Cæsalpinia), *Beng.*, C. 6; nátá karanja, *Beng.*, C. 6; natar, *Guj.*, C. 104; nát-bá-dám, *Mar.*, T. 312; natch-ni, *Mar.*, E. 166; nathur, *Guz.*, C. 298; natíyá-ság, *Beng.*, A. 938; nát-ká-aushbah, *Dec.*, H. 119; nát-ká-bachhnág, *Dec.*, G. 243; nát-ká-dam-mul-akhvain, *Dec.*, P. 1370; nát-ká-katérá, *Dec.*, C. 1512; nát-kí-asgand, *Dec.*, W. 93, 98; nát-kí-révan-chíní, *Deccan*, R. 215; nátkí-sana, *Duk.*, C. 737; natmi, *Beng.*, A. 983; nátú-koranza, *Beng.*, C. 6; naugei, *Pb.*, J. 24; naugri, *Pb.*, J. 29; naural, *Mar.*, C. 1520; naushádar, *Pb.*, A. 962; naushádar kání, *Hind.*, *Pb.*, R. 61; nava-mallika, *Hind.*, J. 13; navananji-chapála, *Mar.*, D. 413; navaságar, *Guj.*, *Mar.*, A. 962; navsár, *Guj.*, A. 962; nawal, *Mar.*, I. 421; nayá-phatki, *Beng.*, C. 551; nayata, *Mar.*, E. 549; nayeti, *Mar.*, E. 512, E. 531, E. 549; nazbo, *Sind*, O. 18; nazpat, *Sind*, P. 425.

Ne Ne, *Pb.*, H. 382; neal boti, *Tank*, C. 2211; nebu, *Beng.*, C. 1258, 1270, *Beng.*, *Hind.*, C. 1296; négli, *Mar.*, P. 1062; nehar, *Kumaon*, S. 2223; nehare, *Beng.*, F. 458; neimal, *Hind.*, S. 2960; neja, *Hind.*, S. 25; nelkar, *Pb.*, D. 64; nelmal, *Hind.*, S. 2960; nembú, *Hind.*, C. 1301; nemuká, *Beng.*, C. 1205; nengar, *Hind.*, V. 164; neori, *Hind.*, S. 1207; neoza, *Hind.*, *Kumaon*, *Pb.*, P. 746; nepál, *Guz.*, C. 2192; nepáli dhaniá, *Beng.*, Z. 9; nepárí, *Kumaon*, D. 245; ner, *Pb.*, S. 2223; nera, *Pb.*, R. 251; neva-dunga, *Mar.*, E. 520; newala, *Pb.*, V. 233; newar (Barringtonia), *Mar.*, B. 180; newar (Juniperus), *Kumaon*, J. 92; newarang, *Mar.*, E. 520; newr, *Hind.*, *Pb.*, P. 746; newra, *Hind.*, F. 774; new-rang, *Mar.*, E. 527; newul, *Hind.*, F. 774.

Nh Nhana gokharu, *Guz.*, T. 548; nhare, *Ku-maon*, T. 93; nhio, *Mar.*, A. 1192; nhiu, *Mar.*, A. 1192; nhiv, *Mar.*, A. 1192; nhyú, *Mar.*, A. 1192.

Ni Niála, *Pb.*, P. 1109; niálo, *Pb.*, P. 1109; nibari, *Hind.*, N. 179; níbú, *Dec.*, C. 1258, *Hind.*, C. 1296; níchní, *Pb.*, R. 251; nidhu, *Uriya*, S. 3062; nigál, *Hind.*, A. 1523; nigand, *Pb.*, O. 18; niggi (Daphne), *Pb.*, D. 115; niggi (Hamil-tonia), *Pb.*, H. 13; nijní, *Pb.*, M. 131; nikari, *Beng.*, C. 812; nikasanwak, *Pb.*, E. 261; nikki, *Pb.*, R. 157; nikkí-bekkar, *Pb.*, G. 677, 705; nikki-kurkan, *Pb.*, E. 232; nil (nilgai), *Hind.*, S. 1229; nil (Indigofera), *Hind.*, I. 109; níl, *Pb.*, I. 111, *Beng.*, *Hind.*, *Guj.*, *Pb.*, *Sind*, I. 145; nila, *Mar.*, I. 145; níla ghiria, *Pushtu*, C. 1403; níla kantha, *Beng.*, C. 2422; nilakil, *Pb.*, G. 165; nílak rái (Trichodesma), *Kashmir*, *Pb.*, T. 562; nílakrái (Crozophora), *Pb.*, C. 2211; nílakrái (Cynoglossum), *Pb.*, C. 2569; nilam, *Hind.*, *Pb.*, S. 855; nilan, *Pb.*, C. 2221; níl aparájitá, *Beng.*, C. 1403; níla thárí, *Pb.*, C. 2508; nílá-thútha, *Hind.*, C. 2367; nila-túsya, *Pb.*, C. 2361; nílá-tútá, *Hind.*, C. 2367; nil bhadi, *Beng.*, G. 143; nilgai, *Hind.*, S. 1229; nilgao, *Hind.*, S. 1229; níli, *Mar.*, I. 145; níli-nargandí, *Hind.*, J. 116; nil-isband, *Pushtu*, C. 1403; níl-kalmi, *Beng.*, I. 384; nil kamál, *Kumaon*, I. 434; nilkant (Gentiana), *Pb.*, G. 165; níl-kanth (Cli-toria), *Pushtu*, C. 1403; níl-kantha (Cur-cuma), *Beng.*, C. 2422; nilkantihi (Ajuga), *Pb.*, A. 1677; nílkantti, *Pb.*, A. 677; nil kanwal, *Hind.*, I. 434; níl kattei, *Hind.*, *Pb.*, H. 100; nílkbantí, *Pb.*, C. 2211; nill-dub, *Raj.*, C. 2558; nílwal, *Hind.*, *Beng.*, I. 145; nilofár, *Kashmir*, N. 192; nilofir, *Sind*, N. 39; nilon, *Hind.*, A. 1076; nilophal, *Guj.*, N. 200; nil-padma, *Hind.*, N. 209; nílpatie, *Beng.*, *Hind.*, I. 145; nilsáphala, *Beng.*, N. 209; nilsápla, *Beng.*, N. 209; níltá-tutiya, *Hind.*, C. 2367; ním, *Beng.*, *Dec.*, *Hind.*, *Mar.*, *Pb.*, M. 363; nimak, *Mar.*, S. 602; nimb, *Mar.*, M. 393; nímb, *Hind.*, M. 363; nímbar (Senecio), *Pb.*, S. 1065; nimbar (Acacia), *Hind.*, A. 249; nímbara (Me-lia), *Mar.*, M. 412; nimbarra, *Mar.*, M. 412; nimbay, *Mar.*, M. 363; nimbu,

Guj., C. 1258, *Hind.*, C. 1270, *Beng.*,
Dec., *Hind.*, *Pb.*, C. 1296 ; nimda,
Beng., B. 929 ; nimgachh, *Beng.*, M.
363 ; ním-gilo, *Beng.*, T. 470, T. 483 ;
nimi-chambeli, *Mar.*, M. 550 ; nimok,
Beng., S. 602 ; ním púteli, *Beng.*, M.
87 ; ní-much, *Hind.*, F. 450 ; nimúrdi,
Mar., B. 546 ; nimuri, *Sind*, M. 363 ;
ninai, *Mar.*, D. 560 ; nínb, *Hind.*, M.
363 ; nindo-trikund, *Sind*, T. 544 ; nin-
gur, *Hind.*, M. 71 ; nior, *Pb.*, R. 152 ;
nipálidhanya, *Hind.*, Z. 6, Z. 9 ; nir,
Sind, l. 145 ; nírádhar, *Pb.*, C. 2508 ;
nirbishi (Kyllinga), *Beng.*, *Hind.*, K.
47 ; nirbisí (Aconitum), *Pb.*, A. 428 ;
nirbisi (Cissampelos), *Dec.*, C. 1205 ;
nirbisi (Delphinium), *Beng.*, *Hind.*, D.
253 ; nirgal, *Hind.*, A. 1523 ; nirgandí,
Hind., V. 164 ; nírgari, *Gus.*, V. 164 ;
nirguda, *Mar.*, V. 164 ; nirgunda, *Mar.*,
V. 164 ; nirgúndi, *Beng.*, *Mar.*, V. 164 ;
nirgúr, *Mar.*, V. 164 ; nirmal, *Hind.*,
S. 2943 ; nirmali, *Beng.*, *Hind.*, *Mar.*,
Pb., S. 2960 ; nirmulí, *Mar.*, C. 2508 ;
niruri, *Sind*, P. 657 ; nís, *Pb.*, T. 634 ;
nísan, *Mar.*, Z. 199 ; nisana, *Mar.*, Z.
199 ; nishedal, *Beng.*, A. 962 ; nishindá,
Beng., V. 164 ; nishotar, *Mar.*, I. 415 ;
nishottara, *Mar.*, I. 415 ; nisinda, *Beng.*,
Mar., V. 164 ; nisindá, *Hind.*, V. 164 ;
nisomali, *Hind.*, P. 1078 ; nisot, *Pb.*, I.
415 ; nisoth, *Hind.*, I. 415 ; niú (Alnus),
Pb., A. 801 ; niú (Rubus), *Pushtu*, R.
598 ; niurtsi, *Pb.*, A. 1462 ; nivadunga,
Sind, E. 520 ; nivar, *Mar.*, B. 180, B.
193 ; niwal, *Mar.*, E. 553 ; niwalí, *Mar.*,
S. 2960 ; niwar, *Mar.*, B. 180 ; niyazbo,
Pb., O. 18.

No Noalatá, *Beng.*, D. 330 ; noaphutki, *Beng.*,
C. 551 ; noárí, *Beng.*, P. 627 ; noktowa,
Hind., S. 1210 ; nóna, *Beng.*, A. 1158 ;
nonak, *Pb.*, S. 2668 ; nonkha, *Hind.*, P.
1179 ; nonkhalunuk, *Hind.*, P. 1179 ;
nóshágar, *Beng.*, A. 962 ; nousádar,
Hind., A. 962.

Nu Nubari, *Uriya*, D. 94 ; nuch (Fraxinus), *Pb.*,
F. 700 ; núch (Juniperus), *Kashmir*, *Pb.*,
J. 78 ; nukachúni, *Beng.*, S. 2792 ; nuk-
patar, *Hind.*, I. 415 ; nukta, *Mar.*, F.
374 ; núl, *Mar.*, C. 1021 ; nuli, *Beng.*,
Hind., T. 28 ; nulla kashina, *Beng.*, S.
2259 ; nullie, *Pb.*, F. 568 ; númáni, *Afg.*,
S. 497 ; nun, *Beng.*, S. 602l; núnbora,
Beng., I. 335 ; nuniári, *Uriya*, A. 1215 ;
núniya, *Hind.*, P. 1179 ; núniya, *Beng.*,
P. 1187 ; núrálam, *Pb.*, E. 335 ; nuriya,
Beng., A. 558 ; nutma, *Hind.*, G. 385 ;
nuthrini haran, *Beng.*, D. 216 ; nutia,
Beng., C. 1861 ; nuva-mullika, *Beng.*,
J. 13.

Ny Nyái phulánch, *Pb.*, R. 362 ; nyamdal, *Pb.*,
T. 93 ; nyangha, *Pb.*, R. 362 ; nyul,
Hind., F. 774.

Oa Oao, *Uriya*, D. 428.
Ob Uba kotru, *Guj.*, C. 1263 ; obál, *Pb.*, F. 10.
Oc Oche, *Kashmir*, R. 598.
Od Uda-bilui, *Hind.*, S. 2259 ; odi, *Mar.*, V.
195.
Oe Oe, *Pb.*, A. 722.
Og Ogái, *Pb.*, A. 1595 ; ogal, *Pb.*, F. 10 ; ogul,
Kumaon, F. 10.
Oh Ohalu, *Uriya*, P. 632.

Oi Oi, *Pb.*, A. 722.
Ok Okharada, *Gus.*, C. 2211.
Ol Ol, *Beng.*, A. 996 ; olá, *Pb.*, S. 2341 ;
olancha, *Mar.*, A. 1079 ; olchi, *Pb.*, P.
1297, 1304; olenkiráyat, *Mar.*, A. 1064 ;
olikiryát, *Guj.*, A. 1064.
Om Oman, *Pushtu*, E. 228 ; omlóti, *Beng.*,
O. 547.
On Onei, *Hind.*, A. 1097 ; ongwá, *Pb.*, C. 42 ;
onkla, *Guj.*, A. 681 ; onth, *Mar.*, G. 99.
Oo Oos, *Guj.*, B. 156 ; ootrum, *Dec.*, D. 10.
Or Ora, *Beng.*, B. 130 ; orangen, *Duk.*, C.
1233 ; orasmaro, *Uriya*, C. 2531 ; orcha,
Beng., S. 2362 ; ormul, *Pushtu*, F. 179 ;
orol, *Beng.*, C. 49 ; oror, *Beng.*, C. 49 ;
oru, *Beng.*, H. 227.
Os Osar, *Beng.*, *Hind.*, O. 574; oserwa, *Uriya*,
C. 416 ; osha, *Pb.*, S. 2861 ; osth, *Mar.*,
G. 99.
Ou Oudhuphulé, *Guj.*, C. 2569 ; ouk-chhár,
Hind., C. 527 ; ouplate, *Mar.*, S. 910.
Ov Ova, *Mar.*, C. 691 ; ovallí, *Mar.*, M. 570.
Ow Owa (Carum), *Mar.*, C. 691 ; owa (Coleus),
Mar., C. 1715 ; owla, *Dec.*, P. 632 ;
ownla, *Dec.*, P. 632.

Pa Páán, *Pb.*, R. 293 ; paba (Ischœmum),
Hind., I. 497 ; pabba (Chickrassia),
Beng., *Mar.*, C. 1021 ; pabban, *Sind*, N.
39 ; pabbin, *Hind.*, N. 39 ; pabe, *Pb.*,
P. 1148 ; pabha, *Mar.*, C. 1021 ; pabuna,
Hind., U. 4 ; paburpaní, *Sind*, T. 560 ;
pachak, *Beng.*, *Hind.*, S. 910 ; páchan,
Pb., M. 63 ; pachápát, *Beng.*, P. 1024 ;
pachar, *Beng.*, *Hind.*, O. 574 ; pachin,
Baluch., S. 1239 ; pachittie, *Beng.*,
Hind., C. 913, *Mar.*, C. 913 ; pachólí,
Hind., P. 1024 ; pachpanadi, *Gus.*, P.
1024 ; packur-múl, *Beng.*, P. 1095 ; pád
(Betula), *Pb.*, B. 501 ; pád (Stereosper-
mum), *Hind.*, S. 2876 ; pada (Mollugo),
Mar., M. 613 ; pada (Populus), *Afg.*, P.
1153 ; pada (Trichosanthes), *Mar.*, T.
569 ; padai, *Mar.*, B. 118 ; padak, *W.
Afg.*, P. 1153 ; pádal, *Hind.*, *Mar.*, *Pb.*,
S. 2865, 2876 ; padálí, *Pb.*, H. 132 ; padam,
Hind., J. 92 ; padam, *Uriya*, N. 39 ;
padama, *Beng.*, N. 39; padar (Anemone),
Pb., A. 1125 ; padar (Saccharum), *Hind.*,
S. 49 ; padaria, *Hind.*, S. 2876 ; padar-
suh, *Kashmir*, T. 437 ; padda jalla gudi,
Hind., S. 2394 ; paddal, *Mar.*, S. 2876 ;
paddam, *Hind.*, *Pb.*, P. 1333 ; padda-
tunga, *Hind.*, P. 50 ; pader, *Hind.*, S.
2865 ; padera (Hamiltonia), *Kumaon*, H.
13 ; padera (Pavetta), *Hind.*, P. 338 ;
pádhri, *Mar.*, S. 2865 ; padiá, *Mar.*, A.
256 ; padiála, *Hind.*, S. 2876 ; padialú,
Mar., S. 2876 ; padína, *Hind.*, M. 461 ;
padlú, *Pb.*, M. 289 ; padma, *Beng.*, N.
39 ; padmak, *Gus.*, P. 1333 ; padmaka,
Mar., P. 1333 ; padma káshtha, *Hind.*,
Mar., P. 1333 ; padmakasta, *Mar.*, P.
1333 ; padma kathi, *Gus.*, P. 1333 ;
padmak surqi, *Hind.*, J. 92 ; padri (Dalbergia), *Mar.*, D. 53 ; padri (Stereospermum), *Hind.*, S. 2865 ; padrián (Bauhinia), *Hind.*, B. 356 ; padriún (Elædendron), *Pb.*, E. 73 ; padual, *Mar.*, T.
569 ; pádul, *Mar.*, S. 2865 ; padúna, *Pb.*,
T. 845 ; paduro, *Sind*, N. 39 ; padval
(Trichosanthes), *Mar.*, T. 569 ; padvale
(Stereospermum), *Mar.*, S. 2865 ; pañlú,

Kashmir, M. 306 ; pagua, *Pb.*, F. 10 ; pagun, *Hind.*, B. 632 ; pagúnai, *Pb.*, R. 598 ; pagu-tulla, *Beng.*, B. 417 ; pahad, *Mar.*, S. 2876 ; pahádí kiraita, *Mar.*, S. 3016 ; pahadipudina, *Mar.*, M. 465 ; pahar, *Mar.*, F. 615 ; pahari cha, *Hind.*, M. 910 ; pahári erand, *Beng.*, *Hind.*, J. 41 ; pahari gájar, *Pb.*, E. 335 ; pahári-indrayan, *Hind.*, C. 2310 ; pahári kaghazi, *Hind.*, C. 1286 ; pahári-kaghzi, *Hind.*, *Duk.*, C. 1286 ; pahárikánda, *Mar.*, S. 968 ; paharí kiretta, *Hind.*, S. 3016 ; pahári-nímbu, *Hind.*, *Duk.*, C. 1286 ; pahári pípal (Piper), *Beng.*, P. 821 ; pahári pípal (Populus), *Pb.*, P. 1148 ; pahari pípal (Thespesia), *Pb.*, T. 392 ; pahári pudína, *Hind.*, *Pb.*, M. 465 ; pahári pudína, *Hind.*, M. 465 ; pahár-phuta, *Kumaon*, V. 210 ; pahilaunth gay, *Beng.*, *Hind.* O. 574 ; pahiloth, *Beng.*, *Hind.*, O. 574 ; páhú, *Pb.*, P. 327 ; paiman (Eugenia), *Hind.*, E. 453 ; paiman (seir fish), *Hind.*, E. 419 ; pair, *Mar.*, F. 265 ; páiya, *Kumaon*, P. 1333 ; pájá, *Pb.*, P. 1333 ; pajerra, *Pb.*, C. 1958 ; pajia, *Pb.*, P. 1333 ; pajja (Prunus), *Pb.*, P. 1333 ; pajja (Rhamnus), *Pb.*, R. 152 ; pakána, *Pb.*, R. 598 ; pakánbed, *Mar.*, G. 165 ; pakání, *Pb.*, R. 598 ; pakar, *Beng.*, F. 216, *Hind.*, F. 265 ; pakh, *Pb.*, P. 1142 ; pakhána, *Pb.*, R. 593 ; pakhán béd (Gentiana), *Guj.*, *Hind.*, G. 167 ; pakhan bed (Saxifraga), *Hind.*, *Pb.*, S. 924 ; pakhar, *Hind.*, *Pb.*, F. 216 ; pakhshu, *Pb.*, P. 1142 ; pakhshu-bút, *Pb.*, P. 1159 ; pakhur (Ficus), *Hind.*, F. 194 ; pákhur (Lonicera), *Kashmir*, L. 535 ; pakod, *Mar.*, P. 332 ; pakodi, *Mar.*, P. 332 ; pakri, *Mar.*, F. 216 ; pákri, *Hind.*, F. 216 ; pak-tah, *Pb.*, F. 361 ; paktawar, *Pushtu*, A. 2 ; pakur (Ficus), *Beng.*, *Hind.*, F. 216 ; pakura (Nyctanthes), *Pb.*, N. 179 ; pakya, *Mar.*, F. 798 ; pála (Ehretia), *Hind.*, *Mar.*, E. 23 ; pálá (Zizyphus), *Pb.*, Z. 254 ; palách, *Kashmir*, *Pb.*, P. 1148 ; palai, *Pb.*, R. 3161; palak (Beta), *Beng.*, *Hind.*, B. 480 ; palák (Ficus), *Pb.*, F. 179, 216, 265 ; pálak (Spinacia), *Hind.*, *Mar.*, *Pb.*, *Sind*, S. 2574 ; pálá-khari, *Hind.*, A. 1117 ; pálak-juhi, *Hind.*, R. 231 ; pala-kuda, *Mar.*, L. 292 ; palandu, *Beng.*, A. 769 ; pálang, *Beng.*, S. 2574 ; pálang ság, *Beng.*, B. 480 ; palaó, *Hind.*, K. 42 ; palara, *Mar.*, C. 1021 ; palás, *Beng.*, *Guj.*, *Hind.*, *Mar.*, B. 944 ; palása, *Mar.*, B. 944 ; palásavéla, *Guj.*, *Mar.*, B. 978 ; palási, *Mar.*, B. 978 ; pálas pipal, *Beng.*, T. 3921; pálas piplo, *Mar.*, T. 392 ; palawat, *Hind.*, P. 576 ; paldua, *Uriya*, E. 342 ; pale, *Dec.*, E. 23 ; palengi, *Hind.*, F. 672 ; pali, *Hind.*, Z. 254 ; palichhi, *Raj.*, E. 261 ; palik-juhia, *Hind.*, R. 231 ; palinji, *Pb.*, E. 263 ; palitá-mádár, *Beng.*, E. 342 ; palit-mandar, *Beng.*, E. 342 ; paljor, *Pb.*, F. 678 ; palkhí (Ficus), *Pb.*, F. 216 ; palki (Spinacia), *Hind.*, S. 2574 ; pal kurwan, *Uriya*, W. 131 ; pálla paggar gadi, *Hind.*, C. 1055 ; pallu (butter fish), *Pb.*, F. 368 ; pallú (Impatiens), *Pb.*, I. 39 ; palo, *Beng.*, *Dec.*, *Hind.*, *Pb.*, *Sind*, T. 470, 483 ; pálon, *Pb.*, P 50 ; palosa, *Afg.*, A. 261 ; paltu (Prunus), *Pb.*, P. 1331 ; paltu (Pyrus), *Hind.*, *Pb.*, P. 1460 ;

palu, *Pb.*, P. 1460, 1463 ; paluah, *Hind.*, C. 1026 ; paluch, *Pb.*, P. 1148 ; palúdar (Abies), *Pb.*, A. 22 ; palúdar (Cedrus), *Pb.*, C. 846 ; palurr, *Pb.*, C. 846 ; palval, *Hind.*, T. 586, 598 ; palwa, *Hind.*, F. 672 ; palwa, *Hind.*, S. 6 ; palwal, *Hind.*, *Pb.*, T. 586, 600 ; palwal, *Pb.*, T. 586 ; palwal jarga, *Hind.*, A. 1092 ; palwa-minyár, *Pb.*, A. 1114 ; palwán, *Pb.*, A. 1114 ; páma, *Kashmir*, *Pb.*, J. 78 ; pam-anke, *Pb.*, B. 782 ; pambash, *Pb.*, R.215 ; pamphunia, *Uriya*, S. 2865 ; pamposh, *Pb.*, N. 39 ; pámúkh, *Pb.*, V. 70 ; pan (Typha), *Pb.*, T. 864 ; pan (Viscum), *Hind.*, V. 152 ; pán (Piper), *Beng.*, *Dec.*, *Guj.*, *Hind.*, *Mar.*, P. 775 ; panalavanga, *Mar.*, J. 114 ; pánan, *Hind.*, O. 537 ; panár, *Hind.*, *Mar.*, R. 16 ; panaraweo, *Guz.*, E. 342 ; panarvo, *Guz.*, E. 342 ; panas, *Beng.*, *Hind.*, I. 58 ; panas, *Hind.*, A. 1489 ; panasa, *Hind.*, *Uriya*, A. 1489 ; pan-babiyo (Eriophorum), *Kumaon*, E. 323 ; panbabiyo (Ischœmum), *Hind.*, *Kumaon*, I. 494 ; panchi, *Uriya*, A. 1146 ; panchoti palu, *Mar.*, D. 379 ; panchsim, *Beng.*, D. 789 ; pand, *Pb.*, L. 549 ; panda, *Pb.*, L. 549 ; pandair, *Beng.*, S. 2865 ; panden, *Hind.*, E. 20 ; pandhara-sakhar, *Mar.*, S. 375 ; pandhari (Eriodendron), *Mar.*, E. 289 ; pándhari (Croton), *Mar.*, C. 2189 ; pándhari dháman, *Mar.*, G. 708 ; pándhári-miri, *Mar.*, P. 811 ; pandharipale, *Mar.*, I. 134 ; pándharisálá, *Mar.*, C. 2189 ; pandhí, *Mar.*, I. 134 ; pandhora-mohare, *Mar.*, B. 800 ; pándhrá khair, *Mar.*, A. 229 ; pandhrakúra, *Dec.*, *Mar.*, H. 294 ; pándhra saur, *Mar.*, E. 304 ; pand-kanda, *Mar.*, N. 39 ; pando, *Pb.*, *Sind*, T. 569 ; pandolu, *Mar.*, T. 569 ; pandri, *Hind.*, S. 2865, S. 2876 ; pánd-rúk, *Mar.*, S. 2850 ; pándrúka, *Mar.*, S. 2850 ; panel, *Sind*, P. 1024 ; panelra, *Mar.*, R. 16 ; panevár, *Beng.*, *Hind.*, C. 797 ; pangára, *Dec.*, E. 356, *Hind.*, *Mar.*, E. 342 ; pangaru, *Mar.*, E. 342 ; pang-chak, *Beng.*, F. 457 ; pángla, *Mar.*, P. 1020 ; pangra, *Hind.*, *Mar.*, E. 342, 356 ; pangut, *Mar.*, F. 339 ; pánharyá bábhuliche jháda, *Mar.*, A. 249 ; panhawa, *Hind.*, S. 1207 ; paniah, *Hind.*, R. 16 ; paniálá, *Beng.*, F. 603 ; panialajamb, *Beng.*, E. 460 ; pániámalak, *Hind.*, F. 603 ; pani-aonvola, *Hind.*, F. 603 ; pani-gara, *Mar.*, D. 670 ; pání jamá, *Beng.*, S. 579 ; páni-kakharu, *Uriya*, C. 2331 ; pani-kánchirá, *Beng.*, C. 1759 ; pání-ki-sambhálú, *Hind.*, V. 164 ; pání kí-san-bhálú, *Hind.*, V. 181 ; pání-kí-shanbálír *Dec.*, V. 181 ; pani kuta, *Hind.*, F. 782 ; pani kutta, *Hind.*, O. 534 ; páni-lájak, *Beng.*, N. 76 ; páni-lájak, *Mar.*, N. 76 ; páni-najak, *Beng.*, N. 76 ; paníphal, *Beng.*, T. 516 ; panír, *Pb.*, W. 93 ; panír, *Pb.*, *Sind*, C. 225 ; panírak, *Pb.*, M. 105 ; paniri, *Uriya*, A. 1395 ; panír maya, *Guz.*, *Hind.*, *Pb.*, R. 73 ; panis, *Mar.*, C. 1263 ; pánisamálú, *Beng.*, V. 181 ; paniya, *Hind.*, R. 16 ; panj, *Pb.*, T. 61 ; panjam-bul, *Mar.*, E. 416 : pan-jira, *Hind.*, E. 342 ; panjírí-ká-pát, *Hind.*, A. 1130 ; panjírí-ká-pattá, *Duk.*, A. 1130 ; panjolí, *Hind.*, P. 663 ; panj-pilchi, *Pb.*, T. 61 ; panjulí, *Beng.*, *Pb.*, P. 663 ; pán-kar, *Hind.*, A. 567; pán-ki-jer, *Beng.*, *Hind.*, *Mar.*, A. 862; pankul, *Mar.*, I. 513; pánmahúri, *Uriya*, P. 727; pan-manjar, *Mar.*, O.534; pan-mohuri, *Mar.*, F. 659; pan-muhori, *Beng.*, F. 659; panni (Andropogon), *Hind.*, *Pb.*, A. 1093, 1097; panni (Delphinium), *Pb.*, D. 245; pan-niári, *Dec.*, *Hind.*, B. 180; pannir, *Dec.*, G. 733; panr, *Pb.*, O. 18; pansaura, *Beng.*, *Hind.*, G. 685; pansherú, *Beng.*, H. 113; pansi, *Mar.*, C. 474; pansíra, *Pb.*, W. 33; pansra, *Hind.*, C. 1711; panwa, *Hind.*, F. 179; panwar, *Hind.*, C. 797; panwár (Cassia), *Pb.*, C. 797; panwár (Lawsonia), *Pb.*, L. 126; pa-o-char, *Pb.* ; F. 326; páote, *Mar.*, D. 789; pápadi, *Mar.*, P. 338; papai, *Mar.*, C. 581; papaiya, *Hind.*, C. 581; papaiya amba, *Hind.*, C. 581; papanass, *Mar.*, C. 1263; papar (Buxus), *Pb.*, B. 985; papar (Euonymus), *Pb.*, E. 475, 479; papar (Holoptelea), *Kumaon*, H. 324; papar (Pongamia), *Hind.*, *Kumaon*, P. 1121; papar (Ribes), *Kumaon*, R. 359; pápari, *Hind.*, P. 338; papát, *Mar.*, P. 338; pápat kalam, *Pb.*, V. 91; papaya, *Guj.*, *Mar.*, C. 581; papáyi, *Guj.*, C. 581; páphar, *Hind.*, G. 124; paphri, *Pb.*, P. 1121; papia, *Guj.*, C. 581; papnasa, *Mar.*, C. 1263; papnass, *Mar.*, C.1263; pappa nasa, *Mar.*, C. 1263; pap-prí, *Sind*, F. 350; papra (Gardenia), *Hind.*, G. 116; papra (Salvia), *Pb.*, S. 738; pápra (Fumaria), *Pushtu*, F. 723; pápra (Gardenia), *Hind.*, G. 124; pápra (Podophyllum), *Hind.*, P. 1009; paprang, *Pb.*, B. 985; papri (Holoptelea), *Hind.*, *Pb.*, H. 324; papri (Myrsine), *Pb.*, M. 910; papri (Podophyllum), *Hind.*, *Pb.*, P. 1009; papri (Ventilago), *Hind.*, V. 48; pápri (Ammonium chloride), *Pb.*, A. 962; pápri (Buxus), *Pb.*, B. 985; papría, *Hind.*, L. 474; papur, *Pb.*, B. 985; paputa, *Sind*, C. 581; pár, *Hind.*, S. 2876; párá (hog deer), *Hind.*, D. 216; párá (mercury), *Beng.*, *Hind.*, *Mar.*, M. 473; parah, *Hind.*, F. 403; paral, *Mar.*, S. 2865, S. 2876; páral, *Hind.*, S. 2876; parand, *Pb.*, L. 549; parang, *Hind.*, A. 1585; paras (Butea), *Hind.*, *Mar.*, B. 944; páras (Thespesia), *Pb.*, P. 1316; párasa-píplo, *Gus.*, T. 392; para-siyávashán, (*Salt-range*), *Pb.*, A. 498; paráspipal (Ficus), *Raj.*, F. 265; páras-pípal (Thespesia), *Hind.*, *Pb.*, T. 392; páras-pippal, *Dec.*, T. 392; paraura, *Hind.*, H. 164; parba, *Hind.*, H. 164; parba parbi, *Hind.*, A. 1090; parbar pangí, *Pushtu*, P. 923; parbáti, *Hind.*, D. 32; parbi, *Hind.*, H. 164; parbik, *Pb.*, C. 1205; pardesi da wano, *Guj.*, A. 1460; páre, *Pb.*, P. 327; par-e-siyá-washán, *Hind.*, A. 510; pares pipal, *Beng.*, T. 392; pargái, *Pb.*, Q. 29; pargái, *Pushtu*, C. 224; parhar pangi, *Pushtu*, P. 921, 923; pári, *Hind.*, C. 1205; pariára, *Pb.*, E. 356; parijátaka, *Mar.*, N. 179; paris, *Dec.*, T. 392; par-jamb, *Mar.*, O.153; parjan (Sisymbrium), *Hind.*, S. 2219; parjan (Diplotaxis), *Hind.*, D. 674; párkánd, *Mar.*, A. 603; parlú, *Beng.*, S. 2876; parna-bij, *Mar.*, K. 14; páro, *Gus.*, M. 473; paroa,

Hind., F. 179, S. 2865; parosi, *Mar.*, L. 569; parpatrah, *Dec.*, G. 748; parral, *Hind.*, S. 2865; parrí, *Mar.*, F. 581; parsacha-jháda, *Mar.*, T. 392; parsál, *Hind.*, H. 513; parsha-warsha, *Pb.*, A. 498; parsid, *Mar.*, H. 16; parsipu, *Hind.*, *Mar.*, T. 392; parú, *Beng.*, A. 775; parúl, *Mar.*, S. 2876; párul, *Beng.*, S. 2876; parula (Trichosanthes), *Mar.*, T. 569; parula (Luffa), *Mar.*, L. 569; parúnjí, *Pb.*, Q. 13; parur, *Hind.*, S. 2876; parva, *Hind.*, A. 1090; parvar, *Hind.*, T. 586, 598; parwana, *Hind.*, M. 919; parwar, *Hind.*, T. 600; parwar, *Mar.*, T. 569; parwata, *Pushtu*, H. 52; parwatti (Cocculus), *Pb.*, C. 1448; par-watti (Dioscorea), *Pb.*, D. 503; parwel, *Mar.*, C. 1452; pasáhí, *Hind.*, O. 258; pasend, *Hind.*, D. 628; pasendu, *Pb.*, D. 628; pásh, *Pb.*, E. 475; pashan-abheddie, *Beng.*, O. 18; pashanbheda, *Mar.*, S. 924; pashkand, *Pb.*, C. 191; pashmaran, *Pb.*, T. 376; pasi (Anogeis-sus), *Uriya*, A. 1146; pásí (Dalbergia), *Mar.*, D. 53; passáhi, *Hind.*, H. 513; passai, *Hind.*, H. 513; passarí (Hygro-ryza), *Hind.*, H. 513; passarí (Oryza), *Hind.*, O. 258; pássi, *Hind.*, D. 32, 53; pastal (Taxus), *Pb.*, T. 93; pastal (Hy-groryza), *Pb.*, H. 513; pastaoni, *Pushtu*, G. 663; pastawana, *Pushtu*, G. 688; pastawanai, *Pushtu*, G. 688; pastuwanne, *Pushtu*, G. 688, 723; pát, *Beng.*, C. 1861, 1879; páta, *Mar.*, S. 1688; patájan, *Pb.*, P. 1433; patákhan, *Pb.*, C. 2035; patáki (Celastrus), *Pb.*, C. 862; patáki (Cissampelos), *Pb.*, C. 1205; patak-ri, *Mar.*, A. 897; patal, *Uriya*, T. 586; patalatum-bari, *Mar.*, C. 978; patali, *Hind.*, S. 959; patal kohnda, *Hind.*, P. 1401; pátal tumbdi, *Mar.*, C. 982; patana, *Gus.*, P. 885; patang, *Beng.*, *Dec.*, *Guj.*, *Hind.*, *Mar.*, C. 35; patangalia, *Hind.*, A. 341; patar phor, *Hind.*, T.528; páta-shij, *Beng.*, E. 520; patáwar, *Hind.*, S. 6; patayatbágh, *Mar.*, T. 437; patcha, *Mar.*, P. 1024; patchpan, *Mar.*, P. 1024; pater, *Hind.*, T. 864; patera, *Hind.*, T. 875; páter chúr, *Beng.*, C. 1715; patha, *Hind.*, V. 152; patha (Loranthus),*Hind.*, L. 549; patha (Nannorhops), *Pb.*, N. 1; pathaní lodh, *Sind*, S. 3046; pathanni, *Hind.*, C. 617; páthara-suva, *Mar.*, G. 247; pathar-ka-phúl, *Hind.*, L. 332; patharola, *Hind.*, R. 613; pathor (Brie-delia), *Pb.*, B. 868; pathor (Marsdenia), *Pb.*, M. 295; páthor chur, *Hind.*, *Mar.*, C. 1715; pathri (Launæa), *Mar.*, L. 114; pathrí (Taraxacum), *Dec.*, T. 87; pathúr chúr, *Mar.*, C. 1715; patí, *Beng.*, *Hind.*, C. 2601; patiakerundi, *Uriya*, T. 315, 352; patigia, *Hind.*, P. 1433; patí-khori, *Beng.*, S. 25; pátí-nebu, *Beng* C. 1296; patí-patá, *Beng.*, P. 625; patíra, *Pb.*, T. 864, T. 875; patís, *Pb.*, A. 401; pativa, *Uriya*, R. 1; patji, *Hind.*, P. 1433; patkarru, *Hind.*, R. 561; patki, *Baluch.*, P. 1153; pát kuári, *Kumaon*, K. 17; pat-mossu, *Uriya*, S. 487; pátola, *Hind.*, L. 474; patol, *Beng.*, T. 573, *Hind.*, T. 576; patola (Gerbera), *Pb.*, G. 186; patola (Trichosanthes), *Guj.*, T. 576, *Hind.*, L. 474; patpatula, *Pb.*, G. 186; pát-phanas,

Mar., A. 1482 ; patrang, *Pb.*, A. 526 ;
patrís, *Pb.*, A. 401 ; patsalan, *Kashmir*,
T. 429 ; patsan, *Hind.*, *Pb.*, H. 177 ;
pattali, *Pb.*, E. 475 ; patteoon, *Dec.*, E.
527 ; pattewar, *Hind.*, D. 628 ; patthar-
man, *Pb.*, C. 133 ; patthra, *Pb.*, M. 63 ;
pattia, *Hind.*, A. 722 ; patton-ki-send,
Hind., E. 520 ; pattrá, *Hind.*, K. 42 ;
patúl, *Uriya*, S. 2876 ; patwa, *Beng.*,
Dec., *Hind.*, *Mar.*, H. 233 ; patwan,
Hind., D. 628 ; paunda, *Pb.*, S. 30 ;
paungsi, *Beng.*, F. 463 ; páus, *Hind.*, M.
237 ; páusá, *Hind.*, M. 237 ; pauti, *Mar.*,
D. 789 ; pavana, *Mar.*, P.663 ; pawanne
(Boucerosia), *Pb.*, B. 782 ; pawanne (Clematis), *Pb.*, C. 1361 ; pawár, *Pb.*, C. 797 ;
pawás, *Pb.*, C. 797 ; paya (Alnus), *Kumaon*, A. 801 ; páya (Prunus), *Hind.*, P.
1333 ; payála, *Hind.*, B. 913 ; páyar,
Mar., F. 265.

Pe Pe, *Pb.*, A. 22 ; pease, *Baluch.*, *Sind*, N. 1 ;
pech, *Sind*, D. 130 ; peduman, *Mar.*, S.
950 ; peholi, *Hind.*, P. 1024 ; peira,
Hind., B. 913 ; pek (Plectranthus), *Pb.*,
P. 959 ; peka (Bambusa), *Hind.*, B. 142 ;
pencha, *Beng.*, F. 443 ; pendari, *Mar.*,
R. 16 ; pendra, *Uriya*, R. 16 ; pendri,
Mar., G. 136 ; peng, *Sylhet*, C. 2574 ;
pengla, *Kumaon*, I. 376 ; pengusiya,
Beng., F. 473 ; peni, *Pb.*, A. 2 ; penle-
kosht, *Mar.*, Z. 199 ; pentgul, *Mar.*, D.
89 ; penva, *Mar.*, C. 2013 ; peori-wilayti,
Hind., L. 143 ; pepa, *Hind.*, C. 104 ;
pepar, *Mar.*, F. 216 ; pepero, *Hind.*, G.
124 ; pepeiya, *Beng.*, *Hind.*, C. 581 ;
pepri, *Guj.*, F. 216, 230 ; perala, *Mar.*,
P. 1343 ; perapast-awane, *Afg.*, S. 1024 ;
perasárá, *Hind.*, V. 17 ; percí, *Hind.*, F.
359 ; pérrah, *Hind.*, B. 913 ; persárá,
Hind., V. 17 ; peru, *Guj.*, *Mar.*, P. 1343 ;
perula, *Mar.*, T. 576 ; pesh, *Baluch.*,
Sind, N. 1 ; pesháb, *Hind.*, U. 45 ;
pesho, *Pb.*, P. 693 ; pessara, *Hind.*, P.
496 ; pesu, *Uriya*, L. 385 ; petagulí, *Mar.*,
D. 89 ; pétaígágar, *Sind*, D. 173 ; peta-
kara, *Beng.*, C. 1050 ; petari (Abutilon),
Beng., *Mar.*, A. 80 ; petári (Trewia),
Mar., T. 525 ; petarkura, *Beng.*, G. 761 ;
petha (Cucurbita), *Pb.*, C. 2331 ; péthá
(Bánincase), *Hind.*, *Pb.*, B. 430 ; pethra,
Kashmir, J. 78 ; petiah, *Sind*, F. 357 ;
petinar, *Beng.*, P. 41 ; petparia, *Beng.*,
A. 514 ; petthar, *Pb.*, J. 78 ; petthrí, *Pb.*,
J. 78 ; pet-toh-í, *Sind*, F. 355 ; peyara,
Beng., P. 1343.

Pf Pfarra, *Baluch.*, *Sind*, N. 1 ; pfis, *Baluch.*,
Sind, N. 1 ; pfudnah, *Sind*, M. 447.

Ph Phag (Ficus), *Pb.*, F. 230 ; phag (Sorghum),
Hind., S. 2424 ; phagorú, *Pb.*, F. 230 ;
phagwara, *Pb.*, *Pushtu*, F. 230 ; phag-
wari, *Pb.*, F. 129 ; phák, *Beng.*, A. 31 ;
phalanda, *Hind.*, E. 419 ; phalári, *Mar.*,
A. 722 ; phalásá-cha-jhádá-kakrá-cha-
jháda, *Mar.*, B. 944 ; phálasi, *Mar.*, G.
663 ; phaláwa, *Kumaon*, P. 1252 ; phaldu
(Hymenodictyon), *Hind.*, H. 517 ; phaldu
(Stephegyne), *Hind.*, S. 2799 ; phalel,
Hind., B. 212 ; phálí, *Pb.*, A. 962 ; pha-
liant, *Hind.*, Q. 23 ; phaligawar, *Hind.*,
C. 2514 ; phalijarí, *Pb.*, T. 376 ; phalinda,
Hind., E. 419 ; phálja, *Pb.*, P. 1148 ;
phalphura, *Mar.*, O. 233 ; phálsa, *Beng.*,

Guj., *Hind.*, *Pb.*, G. 663 ; phalsh, *Pb.*,
O. 160 ; phálsi, *Mar.*, G. 663 ; phalwá,
Pb., G. 673 ; phalwara, *Hind.*, B. 212 ;
phán, *Pb.*, R. 293 ; phanas, *Mar.*, A.
1489, I. 58 ; phanasa-alambé, *Guj.*, *Mar.*,
A. 598 ; phanát, *Hind.*, Q. 23 ; phanda,
Mar., G. 136 ; phandayat, *Mar.*, S. 1226 ;
phandra, *Mar.*, E. 342 ; pháng, *Mar.*, P.
1020 ; phángla, *Mar.*, P. 1020 ; phanja,
Mar., R. 487 ; phánk, *Hind.*, S. 372 ;
phansámba, *Guj.*, *Mar.*, A. 598 ; phansi,
Mar., C. 474 ; pháphar, *Pb.*, F. 10 ;
phapharchor, *Pb.*, C. 1958 ; pha-phor,
Pb., U. 39 ; phaphra, *Hind.*, *Pb.*, F. 10,
F. 19 ; phaprúsag, *Pb.*, I. 384 ; phar,
Bihar, D. 534 ; pharad, *Hind.*, E. 342 ;
pháraho, *Sind*, G. 663 ; pharai, *Mar.*, C.
1221 ; pharenda, *Hind.*, E. 419 ; phargur,
Hind., H. 517 ; pharkath, *Hind.*, O.
233 ; pharli, *Beng.*, *Hind.*, O. 574 ; pha-
roah, *Hind.*, G. 663 ; pharonj, *Hind.*, Q.
23 ; pharra, *Baluch.*, *Sind*, N. 1 ; pharsa,
Hind., G. 663, 714 ; pharsia (Combretum),
Hind., *Pb.*, C. 1744 ; pharsia, pharsiya
(Grewia), *Hind.*, G. 663, 673, *Pb.*, G.
708 ; pharwa, *Pb.*, G. 688 ; pharwán, *Pb.*,
T. 51 ; phás (Anogeissus), *Mar.*, A. 1146 ;
phas (Potamogeton), *Sind*, P. 1197 ; phasa
phasah, *Beng.*, F. 442, 443, 556 ; phasai,
Hind., O. 258 ; pháshánveda, *Mar.*, G.
165 ; phashin pachin, *Baluch.*, S. 1233 ;
phásí, *Mar.*, D. 53 ; phaskela, *Hind.*, S.
1250 ; phasrúk, *Pb.*, V. 64 ; phassi, *Mar.*,
A. 1146 ; phatak (Betula), *Pb.*, B. 501 ;
phatak (rock crystal), *Guj.*, C. 616 ;
phatarphod, *Mar.*, B. 868 ; phatki, *Mar.*,
A. 897 ; phatkirí, *Beng.*, A. 897 ; phatmer,
Pb., P. 1406 ; phattars-álam, *Mar.*, P.
1221 ; phatusuva, *Mar.*, G. 247 ; phaunda,
Hind., E. 419 ; phe, *Pb.*, N. 39 ; phedú,
Pb., F. 230 ; phegra, *Pb.*, F. 230 ; pheni-
mama, *Beng.*, O. 193 ; phenk, *Mar.*, F.
566 ; phesak láne, *Pb.*, S. 2985 ; phesak
láné, *Pb.*, C. 224 ; phetain, *Beng.*, *Hind.*,
O. 574 ; phetra (Gardenia), *Mar.*, G. 136 ;
phetra (Randia), *Mar.*, R. 16 ; phikai,
Hind., P. 63 ; phíkí-ki-jar, philkí-ka-jhár,
Dec., C. 1403 ; philkú, *Pb.*, L. 528 ; phillu,
Pb., H. 13 ; phindák, *Pb.*, C. 2035 ;
phipai, *Pb.*, R. 152 ; phiphar, *Mar.*, G.
124 ; phísbekkar, *Pb.*, C. 1711 ; phitkarí,
Hind., A. 897 ; phitní, *Kashmir*, *Pb.*, Z.
280 ; phlankur, *Pb.*, V. 202 ; phog (Calli-
gonum), *Pb.*, *Sind*, C. 137 ; phog (Ficus),
Pb., F. 230 ; phogallí, *Pb.*, *Sind*, C. 137 ;
phogrí, *Pb.*, F. 168 ; phok (Calligonum),
Pb., *Sind*, C. 137 ; phok (Ephedra), *Pb.*,
E. 234 ; phokra, *Guz.*, S. 1287 ; pholoe,
Beng., F. 511 ; phomphli, *Pb.*, S. 505 ;
phopéti, *Mar.*, P. 678 ; phophal, *Guj.*,
A. 1294 ; phopti, *Mar.*, P. 682 ; phudino,
Guj., *Sind*, M. 465 ; phúl (Hamiltonia),
Pb., H. 13 ; phúl (sal-ammoniac), *Pb.*, A.
962 ; phúl (Verbasum), *Pb.*, V. 64 ; phu-
lahi, *Pb.*, A. 261 ; phúlan, *Pb.*, F. 10 ;
phulánch, *Pb.*, R. 365 ; phularwa, *Hind.*,
E. 263 ; phúl-bhanga, *Hind.*, C. 331 ;
phulchela, *Beng.*, F. 396 ; phul dhok,
Hind., F. 516 ; phulel, *Hind.*, B. 212 ;
phulgánjá, *Hind.*, C. 331 ; phul-jamb,
Beng., E. 409 ; phulkia (Chloris), *Hind.*,
C. 1026 ; phulkia (Leptochloa), *Hind.*, L.

300; phulla, *Pb.*, R. 162; phulsá,*Hind.*,
G. 663; phulsan, *Hind.*, C. 2105; phul-
satti, *Mar.*,W. 106; phulsel, *Kashmir*,
V. 106; phulsha, *Dec.*, G. 663; phul-
shola, *Beng.*, A. 560; phulu, *Hind.*, J.
104; phulvárá, *Hind.*, B. 212; phulwa,
Hind., B. 212; phulwáí, *Pb.*, C. 42; phúl-
wára, *Pb.*, P. 1252; phulwari, *Kashmir*,
R. 538; phúman, *Pb.*, L. 312; phundi,
Hind., C. 1026; phúndú, *Pb.*, A. 786;
phunphuna, *Uriya*, O. 233; phun, *Hird.*,
C. 2274; phúntar, *Pb.*, V. 64; phúpárí,
Pb., C. 862; phurush, *Hind.*, *Beng.*, L.
52; phurz, *Pb.*, B. 501; phús, *Pb.*, P.
1197; phúsar-patta, *Kumaon*, V. 131;
phusera, *Hind.*, M. 38, *Pb.*, M. 44;
phut (Cucumis), *Hind.*, C. 2274; phut
(Lonicera), *Pb.*, L. 535; phúta (Saxifraga),
Pb., S. 924; phúthíá, *Hind.*, B. 430;
phuti, *Beng.*, C. 2274; phutini pungti,
Beng., F. 347; phútkanda, *Pb.*, B. 33;
phútkaɪi, *Mar.*, l. 415.

Pi Piá bansh, *Beng.*, B. 146; piák, *Pb.*, A.801;
piál, *Hind.*, B. 913; piáɪ,(*Hind.*, B. 913;
píar-alú, *Dec.*, R. 1; piasal (Pterocarpus), *Hind.*, *Mar.*, *Uriya*, P. 1370;
piásál (Terminalia), *Beng.*, T. 361; piáshál, *Beng.*, T. 361; piáz, *Pb.*, I. 430;
piazi, *Pb.*, A. 1579; pichka, *Pb.*, O. 38;
pichle, *Beng.*, B. 134; pich-ru, *Sind*, F.
596; pí-dah, *Sind*, F. 312; pigaví,
Mar., C. 854; pi-i-kí, *Sind*, F. 601; pij,
Kashmir, S. 1247; pijur, *Kashmir*, S.
1247; píkapírú, *Sind*, P. 663; pil, *Pb.*,
S. 705; píl (elephant), *Hind.*, *Pushtu*,
E. 83; píl (Salvadora), *Hind.*, *Pb.*, C.
224; píla-baréla, *Beng.*, S. 1714; pila-
barélá-shikar, *Beng.*, S. 1688; pilabung-
ra, *Dec.*, W. 25; pila champa, *Hind.*,
M. 537; píla-dhatúra, *Dec.*, *Hind.*, A.
1351; píla-gokundu, *Hind.*, F. 564;
pílá-halra, *Dec.*, T. 325; píla hisálu,
Hind., R. 609; pila-jari, *Kumaon*, T.
376; pílajau, *Pb.*, A. 1464; pílajur, *Pb.*,
C. 1205; pilak, *Pb.*, S. 2345; pílá kanér,
Dec., *Hind.*, *Mar.*, T. 410; píla-murgh-
ka, *Hind.*, C. 873; pílásarsón, *Hind.*,
B. 817; pilavan, *Guz.*, U. 26; pilchí,
Hind.,*Pb.*, T. 61, 70; pílé-har, *Hind.*,
T. 325; píle-phúlka-kanér, *Dec.*, T. 410;
pílé-ráyán, *Duk.*, B. 817; pílíjari (Cissampelos), *Pb.*, C. 1205; pílíjari (Thalictrum),
Hind.,*Pb.*, T. 376; pílí karbír, *Pb.*, pílí-
kaner, *Mar.*, A. 762; pílí-kapás-kí-rúí,
Dec., C. 1512; pílí-ráí, *Hind.*, B. 817;
pilkhan, *Hind.*, F. 216, *Pb.*, F. 265; pil-
kin, *Pb.*, F. 216; pílo bhangro, *Guz.*, W.
25; pílo-harle, *Guz.*, T. 325; pilokanera,
Guz., T. 410; pilosarshio, *Guj.*, A. 695;
pilpápra, *Mar.*, E. 219; pílrú, *Pb.*, L.
528; pilrupotala, *Hind.*, G. 271; pílsa,
Pb., R. 355; pílu, *Beng.*, *Hind.*, *Mar.*,
Pb., S. 705, 717; pilu (Careya), *Hind.*,
C. 563; pílú (Salvadora), *Hind.*, *Mar.*,
Pb., C. 224; piludi, *Guz.*, S. 717; pilúgu,
Hyderabad, L. 55; pilvu, *Mar.*, S.717;
pimpal, *Mar.*, F. 236; pimpala, *Mar.*,
F. 236; pimpli, *Mar.*, P. 805; pimpri
(Casearia), *Hind.*, C. 722; pimpri (Ficus),
Mar., F. 145, 276; pín (Andrachne), *Pb.*,
A. 1061; pín (Cordia), *Hind.*, C. 1950;
píncho (Debregeasia), *Pb* , D. 196; pincho
(Euonymus), *Hind.*, E. 482; pind, *Pb.*,

P. 555; pindakhejúr, *Beng.*, *Hind.*, P.
555; pind-álu (Ipomæa), *Hind.*, I. 355;
pindálu (Randia), *Hind.*, R. 16; pindar
(Crinum), *Hind.*, C. 2062; pindar (Randia), *Hind.*, R. 16; pindára, *Hind.*, T.
525; pindaru, *Kumaon*, R. 16; pind-
chirdi, *Sind*, P. 555; pind khajúr, *Hind.*,
Pb., P. 551; pindra, *Mar.*, R. 16; pin-
drai, *Pb.*, A. 22; pindrau, *Pb.*, A. 22;
píngí, *Kashmir*, S. 1212; pingi natchi,
Beng., S. 1207; píngyát, *Pb.*, C. 2035;
pinj, *Mar.*, S. 2819; pinjari, *Hind.*, T.
376; pinju (Capparis), *Pb.*, C. 402; pinju
(Salvadora), *Pb.*, S. 705; pinnga, *Mar.*,
C. 2013l; pinnis, *Hind.*, S. 2574; pinnis,
Beng., S. 2574; pinvalá-dhotrá, *Mar.*, A.
1351; oinyát, *Pb.*, C. 2035; pípa, *Pb.*, C.
478; pipal (Ficus), *Hind.*, *Pb.*, F. 236,
Hind., F. 265; pipal (Piper), *Hind.* *Pb.*,
P. 805; pípal (1hespesia), *Hind.*, T.
392; pipalu, *Pb.*, D. 725; pipal, pípat-
butí, *Pb.*, H. 109; pipér, *Guz.*, P. 805;
piperi, *Pb.*, T. 845; pipita, *Beng.*, *Hind.*,
Mar., S. 2940; pipla-mor, *Beng.*, P. 805;
pipla-múl, *Beng.*, *Hind.*, *Mar.*, *Pb.*, P.
805; piplí (Ficus), *Mar.*, F. 216; piplí
(Piper), *Beng.*, *Guj.*, *Hind.*, *Mar.*, P.
805; piplo, *Mar.*, F. 236; pippal, *Pushtu*,
F. 236; pippal-j'hanca, *Hind.*, S. 970;
pippal-yang, *Hind.*, S. 842; pippira-sarí,
Hind., P. 1214; pip-plie, *Dec.*, P. 805;
pippú, *Pb.*, C. 478; pipul (Ficus), *Guj.*,
F. 236, *Hind.*, F. 265; pipul (Piper),
Beng., P. 805; pipulmul, *Dec.*, *Hind.*,
P. 805; pípúli-jhunjhun, *Beng.*, C. 2158;
pipur, *Sind*, F. 236; piralo, *Beng.*, R.
16; pírhí, *Pb.*, D. 348; píriya, *Hind.*,
A. 1093; piriya-hálim, *Hind.*, N. 28;
piru, *Pb.*, P. 702; pírun, *Hind.*, C. 1386;
pisa, *Mar.*, A. 452; písar, *Pb.*, A. 1588;
pishor, *Pb.*, P. 327; pisina, *Uriya*, M. 3;
pismarum, *Dec.*, *Hind.*, R. 663; pissi-
babul, *Dec.*, *Hind.*, A. 217; pista (Pistacia), *Afg.*, *Beng.*, *Hind.*, *Mar.*, P.
858; pista (Sapium), *Trans-Indus*, S.
842; pistalik, *Afg.*, P. 858; pistan, *Guj.*,
Mar., C. 1931; pita gohum, *Uriya*, M.
617; pitakári, *Mar.*, T. 855; pitáli,
Beng.. T. 525; pitallu, *Uriya*, F. 319;
pitári, *Mar.*, T. 525; pitavan, *Guz.*, U.
26; pitbalá, *Beng.*, S. 1703; pitchandan,
Beng., S.790; pithapra, *Mar.*, G. 247;
pithi, *Hind.*, E. 263; pit-kárí, *Dec.*, T.
855; pitmalti, *Hind.*, J. 24; pitmári,
Mar., T. 855; pitni, *Pb.*, Z. 270, Z. 280;
pitni-bér, *Hind.*, Z. 280; pítochampo,
Guz., M. 517; pitohri, *Hind.*, I. 415;
pitpápada, *Mar.*, J. 120; pitpápara, *Hind.*,
F. 721, *Dec. Hind.*, F. 723; pitpapda,
Guz., F. 723; pitpápra (Fumaria), *Hind.*,
Mar., *Pushtu*, F. 723; pit-pápra (Peristrophe), *Mar.*,P.425; pit-pápra(Rungia),
Mar., R. 656; pitraj, *Beng.*, A. 988;
pit-ras, *Beng.*, C. 2433; pít-sál, *Beng.*,
P. 1370; pít shál, *Beng.*, P. 1370; pit-
shola, *Hind.*, P. 1370; pitso, *Pb.*, A. 1523;
pitta-pápada (Glossocardia), *Mar.*, G.247;
pitta-pápada (Rungia), *Mar.*, R. 656;
pittha-kerrundi, *Uriya*, F. 335; pitti,
Hind., V. 55; pittmári, *Mar.*, N. 23;
pittpápra, *Mar.*, N. 23; pittvel, *Mar.*,
N. 23; pitul-kas, *Hind.*, F. 449; pitvan,

Hind., U. 23; pitwa, *Hind.,* H. 177; pitz, *Pb., Kashmir,* T. 864; piumár, *Pb.,* P. 959; píuni, *Pushtu,* P. 737; pivala bhangra, *Mar.,* W. 25; pivalá-cháph, *Mar.,* M. 517; pivala kaner, *Mar.,* T. 410; pivala-kanhera, *Mar.,* T. 410; piválákún, *Mar.,* B. 334; pivala-tilá-vana, *Mar.,* C. 1367; pivli kanher, *Mar.,* A. 762; piwala koranta, *Mar.,* B. 171; piwalá máká, *Mar.,* W. 25; piwar, *Mar.,* B. 180; piyáj, *Beng. Mar.,* A. 769; piyál, *Beng., Hind., Mar.,* B. 913; piyála, *Hind.,* B. 913; piyár (Buchanania), *Hind.,* B. 913; piyára (Psidium), *Beng., Guj., Hind.,* P. 1343; piyáz, *Hind.,* A. 769; piz, *Kashmir,* T. 875.

Pk Pkalsh, *Pb.,* P. 1142.

Pl Plewan, *Pushtu,* S. 705; plewan, *Pushtu,* S. 717; plewane, *Pb.,* C. 224, S. 705.

Pm Pmadni, *Dec., Mar.,* A. 89.

Po Po, *Pb.,* P. 327; pobtah, *Uriya,* F. 368; podina, *Beng.,* M. 447, *Hind.,* M. 461; pohwa, *Hind.,* S. 1207; poi, (Streblus), *Mar.,* S. 2912; pói (Basella), *Guj., Hind., Sind,* B. 203, *Hind.,* B. 207; pói (Maoutia), *Hind.,* M. 260; poidhaula, *Kumaon,* V. 131; pokarmúl (Spilanthes), *Pb.,* S. 2571; pokharmul (Jurinea), *Pb.,* J. 111; pola (Crotalaria), *Pb.,* C. 2101; pola (Kydia), *Hind., Pb.,* K. 42; polach, *Pb,* A. 711; polang, *Uriya,* C. 146; poli (Carthamus), *Pb.,* C.633; poli (Eryngium), *Pb.,* E. 335; polian, *Pb.,* C. 637; polie, *Hind.,* H. 342; polí kandieri, *Pb.,* C. 2027; poliyán, *Pb.,* C. 633; pom-pi-ah, *Hind.,* F. 311; pomponia, *Uriya,* O. 233; pona (Ischœmum), *Hind.,* I. 499, 505; pona (Pueraria), *Hind.,* P. 1401; poncha geraldi, *Uriya,* F. 427; pondarra, *Mar.,* D. 53; popái, *Duk.,* C. 581; popaiyá, *Hind.,* C. 581; popal, *Pb.,* S. 924; popat bútí, *Hind., Pb.,* H. 100; popli, *Mar.,* O. 525; porál, *Pb.,* H. 132; pórash, *Beng.,* T. 392; porásu, *Uriya,* B. 944; porprang (Convolvulus), *Pb.,* C. 1781; porprang (Mollugo), *Pb.,* M. 615; porush, *Hind.,* T. 392; poshkar, *Kashmir,* S. 1061; post, *Kumaon,* P. 87; poshúr, *Beng.,* C. 482; poshwa, *Pb.,* S. 2861; post, *Beng., Hind., Mar., Pb.,* P. 87; posta, *Guj., Hind., Mar.,* P. 87; postal, *Kashmir,* T. 93; post-anár, *Hind.,* P. 1426; postíl, *Kashmir,* T. 93; post-khai, *Kashmir,* S. 910; potah, *Hind.,* F. 350; potala, *Guz.,* T. 586; potali, *Hind.,* G. 271; potari (Abutilon), *Beng.,* A. 89; potari (Kydia), *Hind.,* K. 42; potassah-fessah, *Beng.,* F. 556; pothí, *Pb.,* E. 196; poti-kakar, *Beng.,* M. 626; potiya haran, *Hind.,* D. 238; potodhamun, *Hind.,* G. 673; potól, *Beng.,* T. 586; pottiah, *Hind.,* F. 352; potúr, *Hind.,* H. 517; porraya, *Sind,* C. 146; powárijo-bij, *Sind,* C. 637; powasi, *Beng.,* B. 523; pownia, *Hind.,* I. 505.

Pr Prabba, *Hind.,* C. 104; prál, *Pb.,* S. 1212; prasáram, *Mar.,* P. 4; prash, *Beng.,* T. 392; prashní, *Mar.,* P. 874; prásni, *Mar.,* P. 874; prastí, *Pb.,* P. 1138; prátshú, *Pb.,* M., 910; prau, *Pb.,* E. 273; prenji, *Hind.,* C. 1026; prí, *Hind.,* D. 348; prín, *Pb.,* D. 196; prisniparni, *Mar.,* U. 26; prist, *Pb.,* P. 1138; prita, *Pb.,* P. 746;

priyangu, *Beng., Hind.,* A. 644; prong, *Hind.,* A. 1523; prora, *Pb.,* M. 22; prost, *Pb.,* P. 1159; prot, *Kashmir,* M. 289.

Ps Psher, *Pb.,* P. 327.

Pu Pú, *Pb.,* A. 573; púa, *Hind.,* M. 260; puchownda, *Mar.,* C. 410; puck-wah-rí, *Pb.,* F. 361; pudárí, *Pb.,* H. 13; puddu, *Sind,* F. 311; pudhárá, *Pb.,* W. 33; pudína. *Guj., Hind., Mar.,* M. 447, *Hind.,* M. 461, *Mar., Pb.,* M. 465; púdína, kúhí, *Pb.,* M. 465; púdna-kúshma, *Pb.,* M. 461; pudo ganla, *Beng.,* T. 441; pudola, *Hind.,* T. 569; pudoli, *Mar.,* T. 576; púdú, *Hind.,* V. 152; pufta, *Hind., Pb.,* F. 368; púisák, *Beng.,* B. 207; pukana, *Pb.,* R. 598; pukána, *Pb.,* R. 590; pukko, *Sind,* C. 402; púlá, *Hind., Pb.,* K. 42; pulákh, *Pb.,* F. 265; pulipatha, *Hind.,* K. 42; pulla (Clupea), *Sind,* F. 414; pulla (Rubus), *Pb.,* R. 613; pulli (Notopterus), *Uriya,* F. 511; pulli (Kydia), *Pb.,* K. 42; pullung, *Mar.,* H. 159; púlmú, *Pb.,* V. 96; pulsha, *Dec.,* G. 663; pulsu-malwa-gadi, *Hind.,* I. 489; pultosi, *Beng.,* F. 547; pulu, *Hind.,* H. 177; puluah, *Hind.,* A. 1114; pulwal, *Hind.,* A. 1114; pumag, *Mar.,* C. 146; pumila, *Hind.,* A. 844; pumne, *Hind.,* C. 631; pu murl, *Hind.,* F. 515; pun (Abies), *Pb.,* A. 22; pún (Calophyllum), *Mar.,* C. 156; pún (Jatropha), *Pb.,* J. 41; pún (Sterculia), *Mar.,* S. 2824; pun (Typha), *Sind,* T. 864; puna, *Hind.,* E. 20; punac. *Pb.,* O. 595; punag champa, *Beng.,* A. 860; punang, *Uriya,* C. 146; puna-rikhu, *Hind.,* S. 30; punarnabá, *Beng.,* B. 619; punarnavá, *Mar.,* B. 619; punárnuwa, *Guj.,* B. 619; púndia, *Beng.,* M. 465; púndna, *Pb.,* G. 240; pungti, *Beng.,* F. 341, 349, 351; puni, *Sind,* N. 200; punia-buchua, *Uriya,* F. 547; punír, *Hind.,* W. 93, W. 98; punír-band, *Sind,* W. 93; punír-ja-fota, *Sind,* W. 93; punjah-gaggah, *Uriya,* F. 488; punjlawái, *Hind.,* E. 20; punk, *Hind.,* C. 1742; punna, *Pb.,* E. 20; punnág (Calophyllum), *Beng.,* C. 146; punnág (Ochrocarpus), *Mar.,* O. 6; punnikaú, *Uriya,* F. 495; punnur, *Sind.,* F. 430; punra, *Pushtu,* E. 20; punschi, *Mar.,* C. 474; púnyan, *Hind.,* E. 20; punya-safed, *Raj.,* E. 246; puráhiya, *Beng., Hind.,* O. 574; purain, *Kumaon,* V. 210; purbia, *Mar.,* S. 1078; purebha, *Hind.,* O. 574; púri, *Beng.,* S. 30; púr-ing, *Pb.,* J. 24, 29; púrjlú, *Pb.,* G. 186; puroahung, *Hind.,* M. 71; purohapalás, *Hind.,* B. 944; purram, *Mar.,* F. 601; pur-ran-dah, *Pb.,* F. 424; purreya, *Sind,* C. 146; purri, *Pb.,* F. 511; purruwa, *Uriya,* F. 484; pursan, *Pb.,* E. 20; pursha, *Hind.,* A. 498; púrtúk, *Hind.,* A. 1585; purula (Luffa), *Hind.,* L. 569; purula (Stereospermum), *Hind.,* S. 2876; puruni, *Hind.,* D. 196; puruni-ság, *Uriya,* P. 1179; purwa, *Mar.,* F. 368; purwul, *Hind.,* T. 569; pusai, *Hind.,* O. 258; pushí, *Baluch.,* F. 401; pussai, *Uriya,* F. 442; pussí, *Sind,* C. 402; pussur, *Beng.,* C. 482; pústbœni, *Hind.,* D. 339; pust-burn, *Hind.,* A. 1107; puta, *Hind.,*

K. 42 ; puta-jan, *Mar.*, P. 1433 ; puth-orín, *Pb.*, P. 693 ; pút kanda, *Pb.*, C. 2211 ; pútra, *Hind.*, P. 1433 ; putra-jiva, *Mar.*, P. 1433 ; pútrajívak, *Pb.*, P. 1433 ; putranjiva, *Beng.*, P. 1433 ; putri lát, *Hind.*, S. 372 ; pútr-jiva, *Hind.*, P. 1433 ; put-tah-re, *Hind.*, F. 546 ; putter-chettah, *Hind.*, F. 429 ; putteriki, *Uriya*, F. 571 ; putti, *Uriya*, F. 358 ; puttiah, *Uriya*, F. 593 ; puttiya, *Hind.*, K. 42 ; put-to lak, *Sind*, F. 311 ; puttosi, *Beng.*, F. 547 ; puttu, *Sind*, F. 566 ; puttuh-chettah, *Uriya*, F. 451 ; put-tul, *Pb.*, F. 546 ; puttuli, *Uriya*, F. 307 ; puttur chattah, *Mar.*, F. 564 ; pútúr, *Hind.*, H. 517 ; púnjan, *Pb.*, A. 1469 ; púya (Maoutia), *Kumaon*, M. 260 ; púya (Prunus), *Kumaon*, P. 1333 ; púya udish, *Pb.*, B. 496.

Py Pyál-chár, *Mar.*, B. 913.

Qu Qualami-dár-chíní, *Hind.*, *Beng.*, *Duk.*, C. 1196.

Ra Raan-kaal-le, *Pb.*, F. 558 ; ráb (Colocasia), *Pb.*, C. 1732; ráb (Saccharum) *Hind.*, *Pb.*, S. 317 ; raban, *Pb.*, O. 160 ; rábana, *Uriya*, O. 258 ; rabésus, *Pb.*, C. 278 ; rab-ki-shakkar, *Dec.*, S. 371 ; rábshakkar, *Dec.*, S. 371 ; race gundo, *Guz.*, C. 1931 ; radam, *Pb.*, T. 87 ; ráde, *Pb.*, R. 365 ; radhia, *Mar.*, G. 385 ; radhuni, *Beng.*, C. 701 ; ráe champo, *Guz.*, M. 517 ; rág, *Pb.*, A. 17, 22 ; ragatarohado, *Guz.*, V. 55 ; ragat chandan, *Hind.*, P. 1381 ; ragat sémal, *Hind.*, B. 632 ; ragat-sénbal, *Hind.*, B. 632 ; rágha, *Hind.*, A. 17 ; rágha, *Kumaon*, A. 22 ; rágha, *Kumaon*, A. 17 ; ragota, *Mar.*, C. 410 ; ragtá kan-chan, *Mar.*, B. 308, 356 ; ráha, *Hind.*, P. 79 ; rahan, *Hind.*, C. 49 ; raho, *Pb.*, A. 22 ; rái (Abies), *Hind.*, *Kumaon*, *Pb.*, A. 17 ; rái (Brassica), *Guj.*, *Hind.*, *Mar.*, B. 833, 841 ; rai (Dillenia), *Uriya*, D. 428, 438 ; rái (Rhizophora), *Uriya*, R. 242 ; rai (goral), *Kashmir*, S. 1247 ; ráiála, *Hind.*, *Kumaon*, A. 17 ; raiang, (*Sutlej*), rai banj, *Kumaon*, Q. 55 ; raidawan, *Pushtu*, T. 227 ; rai dhani, *Hind.*, V. 48 ; rai-jáman, *Hind.* E., 453 ; raikura, *Mar.*, I. 515 ; rail, *Pb.*, A. 22 ; raila, *Hind.*, C. 756 ; rain, *Raj.*, M. 583 ; raira, *Guj.*, B. 817 ; raisalla (Abies), *Kumaon*, A. 22 ; raisalla (Cupressus), *Kumaon*, C. 2358 ; raisalla (Pinus), *Kumaon*, P. 737 ; ráisalla. *Beng.*, A. 22 ; rái sarishá, *Beng.*, B. 833, 841 ; ráish, *Kumaon*, V. 116 ; rai-túng, *Hind.*, *Pb.*, R. 311 ; raiyang, *Pb.*, A. 17 ; raj (Eleusine), *Mar.*, E. 166 ; raj (Euphorbia), *Raj.*, E. 527 ; rájab, *Pb.*, V. 91 ; rajagaro, *Guj.*, A. 939 ; rájahans, *Mar.*, A. 506 ; rájahrar, *Hind.*, G. 243 ; rajáin (Alnus), *Pb.*, A. 801 ; rajáin (Holoptelea), *Pb.*, H. 324 ; rája-jíra, *Mar.*, C. 1875 ; rájan, *Mar.* M. 583 ; rajana (Ixora), *Beng.*, I. 513 ; rájana (Mimusops), *Mar.*, M. 583 ; rajanígandha, *Beng.*, P. 1044 ; rájávaral, *Guz.*, L. 75 ; raj-briksk, *Kumaon*, C. 756 ; rájgirá, *Duk.*, A. 939 ; rájib, *Mar.* P. 555 ; rájiká, *Mar.*, B. 833 ; rak, *Pb.*, V. 243 ; rákas-gaddah, *Hind.*, C. 1834 ; rakas pattah, *Hind.*, A. 603 ; rakat rohan, *Hind.*, S.

2501 ; rakhal, *Pb.*, T. 93 ; rakht-reora, *Mar.*, T. 227 ; rakkas-gaddah, *Dec.*, C. 1834 ; rakkas-patta, *Duk.*, A. 603 ; rakta-chandan (Adenanthera), *Beng.*, A. 471 ; rakta-chandan (Bauhinia), *Mar.*, B. 308 ; rakta-chandan (Pterocarpus), *Beng.*, *Mar.*, *Uriya*, P. 1381 ; raktachandana, *Beng.*, P. 1381 ; rakta-chitrá, *Hind.*, P. 979 ; rakta-chondon, rakta-jháv, *Beng.*, T. 51 ; rakta kambal (Adenanthera), *Beng.*, A. 471 ; rakta kambal (Nymphæa), *Beng.*, N. 200 ; rakta kanchan (Adenan-thera), *Beng.*, A. 471 ; rakta kanchan (Bauhinia), *Beng.*, B. 308, 356 ; raktálu, *Hind.*, I. 355 ; rakt angliya, *Beng.*, D. 141 ; raktapita, *Kumaon*, V. 48 ; rakta-pita, *Beng.*, V. 55 ; rakta pitta, *Uriya*, V. 55 ; rakta rohida (Polygonum), *Mar.*, P. 1091 ; rakta-rohida (Rhamnus), *Mar.*, R. 164 ; rakta til, *Beng.*, S. 1078 ; rakt-chandan, *Beng.*, D. 141 ; rakto chíta, *Beng.*, P. 979 ; rakto-chitra, *Beng.*, P. 979 ; raktrúra, *Mar.*, M.6 ; rál (Mimosa), *Pb.*, M. 562 ; ral (Shorea), *Beng.*, *Dec.*, *Hind.*, *Guj.*, *Mar.*, S. 1656; rál (Vateria), *Mar.*, V. 31 ; rala (Setaria), *Beng.*, *Dec.*, *Hind.*, S. 1212 ; rala (Shorea), *Mar.*, S. 1656 ; rále, *Mar.*, S. 1212 ; rali, *Hind.*, P. 63 ; ral kala, *Pb.*, S. 1656 ; ral sufed, *Pb.*, S. 1656 ; ral zard, *Pb.*, S. 1656 ; rámabána, *Mar.*, T. 864 ; ráma-káti, *Mar.*, A. 101 ; ramanigi-kula, *Beng.*, M. 809 ; ramanjir, *Hind.*, F. 216 ; ráma-tíla, *Mar.*, G. 735 ; ramatta, *Mar.*, L. 89 ; ramatulasa, *Mar.*, O. 28 ; rambal, *Mar.*, F. 202 ; ramban, *Mar.*, T. 864 ; rámbána, *Mar.*, T. 864 ; rambegun, *Beng.*, S. 2273 ; rámdána, *Hind.*, A. 925 ; rametha, *Mar.*, L. 89 ; rámetta, *Mar.*, L. 89 ; rám ghás, *Hind.*, C. 2558 ; ramgua, *Beng.*, A. 1330 ; ramhikolai, *Beng.*, V. 116 ; rámita, *Mar.*, L. 89 ; ramjani, *Beng.*, M. 40 ; rám kantá, *Hind.*, A. 603, *Pb.*, A. 17 ; ram kurthi, *Hind.*, G. 263 ; ram-kuta, *Hind.*, F. 759 ; ramnia, *Pb.*, C. 2035 ; rámpatri, *Mar.*, M. 904 ; rámphal, *Dec.*, *Guj.*, *Hind.*, *Mar.*, A. 1158 ; ramraj, *Hind.*, I. 472 ; rámsar (Saccharum), *Hind.*, S. 6 ; rámsar (Vallaris), *Beng.*, *Hind.*, V. 12 ; ramshing, *Kumaon*, Q. 13 ; ramtezpat, *Beng.*, C. 1165 ; rám-tíl, *Beng.*, G. 735 ; rámtorái, *Beng.*, H. 196 ; rámtulsi, *Beng.*, *Dec.*, *Hind.*, O. 28 ; rámturái, *Hind.*, *Pb.*, H. 196 ; rámu, *Kashmir*, S. 1264 ; ran, *Mar.*, M. 60 ; ranácha-dál-chini, *Mar.*, C. 1158 ; ránácha-ippécha-jháda, *Mar.*, B. 220 ; ránácha-jiré, *Mar.*, V. 73 ; ránácha-kándé, *Mar.*, U. 39 ; ránácha móhách-a jháda, *Mar.*, B. 220 ; ráná-cha-padavali, *Beng.*, *Mar.*, T. 576 ; ranái, *Pb.*, E. 479 ; ránakándá (Urginea), *Mar.*, U. 39 ; ránakándá (Scilla), *Mar.*, S. 968 ; ran-amb, *Dec.*, S. 2649; ranamba (Lit-sæa), *Mar.*, L. 474 ; ránamba (Spondias), *Mar.*, S. 2649 ; ranatulasi, *Mar.*, O. 28 ; rána-yerandi, *Mar.*, J. 41 ; ránbhenda (Hibiscus), *Mar.*, H. 253 ; rán-bhendi (Thespesia), *Hind.*, *Mar.*, T. 387, 392 ; ránbor, *Mar.*, Z. 280 ; rand, *Hind.*, S. 1939; rand, *Hind.*, R. 369; ran-dal-chini, *Mar.*, C. 1158; rándhoni, *Beng.*, C. 701 ; randhuni (Carum), *Hind.*, C. 701 ; rán-dhuni (Apium), *Beng.*, A. 1227 ; rand-

kari, *Hind.*, L. 474; randkarri, *Hind.*,
L. 474; randrak-sha, *Mar.*, V. 205;
randu, *Baluch.*, S. 593; randuk, *Baluch.*,
S. 593; ráng (tin), *Hind.*, T. 460; ráng
(Peristrophe), *Beng.*, P. 427; ranga,
Hind., T. 460; ranga-alú, *Beng.*, l. 348;
rangan, *Beng.*, l. 513, 515; rangashák,
Beng., A. 927; rangbarat (Pterocarpus),
Hind., P. 1370; rangbharat (Calamus),
Hind., C. 68; rangchari, *Pb.*, E. 196;
rang chul (Syringa), *Pb.*, S. 3079; rang
chúl (Euonymus), *Pb.*, E. 475; rang-
dúni, *Pb.*, H. 47; ranghan túrb, *Beng.*,
Dec., *Hind.*, *Guj.*, *Sind*, B. 43; rangkain,
Orissa, N. 200; rangoe, *Hind.*, T. 456;
rangrek, *Pb.*, P. 1447; rangrúnt, *Pb.*,
H. 47; rángsbúr, *Pb.*, T. 416; rangún-
ki-bel, rangún-ki-bíl, *Dec.*, *Hind.*, Q.
88; rán hald, *Mar.*, C. 2406; raníkrún,
Pb., S. 3079; ranj, *Kumaon*, Q. 55;
ranja, *Mar.*, F. 560; ranjáe, *Mar.*, C.
1363; ránjai (Clematis), *Mar.*, C. 1363;
ranjaiphal (Myristica), *Mar.*, M. 904;
ranjana (Adenanthera), *Beng.*, A. 471;
ranjana (Pterocarpus), *Beng.*, P. 1381;
ránjondhala, *Mar.*, C. 1686; ran-kando,
Guz., U. 39; rankela, *Mar.*, M. 809;
rankokari, *Guz*, M. 583; rán-limbú,
Mar., L. 362; rán-makkai, *Mar.*, C. 1686;
ranmethi, *Mar.*, R. 354; rán parul, *Mar.*,
T. 576; rán-phanas, *Mar.*, A. 1482; ran
sher, *Mar.*, S. 882; ransherú, *Beng.*, H.
113; rán shevari, *Mar.*, S. 1166; rán-
shewrá, *Mar.*, S. 1166; ransla, *Kumaon*,
A. 22; ran-tánkala, *Mar.*, C. 787; rán-
thúl, *Pb.*, P. 1447; rántikhí, *Mar.*, S.
2219; rántondla, *Mar.*, C. 919; rán-
tulsi, *Mar.*, O. 28; ranturi, *Hind.*, H.
196; ran undi, *Mar.*, O. 6; ranzuru, *Pb.*,
P. 760; ráo, *Hind.*, *Kumaon*, *Pb.*, A. 17;
raongi, *Pb.*, V. 116; ráorágha, *Kumaon*,
A. 22; rapesho, *Pb.*, L. 533; rara,
Hind., S. 49; rárápod-bara karanj, *Pb.*,
P. 1121; rárarada, *Hind.*, B. 817; rára-
sarsón, *Hind.*, B. 817; rare, *Pb.*, M.
22; rár (Prinsepia), *Pb.*, P. 1252; rárí
(Rhamnus), *Pb.*, R. 159; ras, *Hind.*, S.
368; rásan-na, *Pb.*, P. 964; rasaurah,
Hind. E. 259; rasavanti, *Hind.*, B. 445;
rasbora, *Beng.*, F. 557; raselwa, *Pb.*, C.
1958; rashtu, *Pb.*, R. 316, 318; rasi,
Beng., *Uriya.*, S. 1078; rásin, *Hind.*, S.
1174; ras-kapúr, *Hind.*, M. 473; rásná
(Acampe), *Beng.*, *Hind.*, *Mar.*, A. 317;
rásná (Vanda), *Beng.*, *Hind.*, *Mar.*, V.
17; rasno, *Guz.*, V. 17; rasota, *Hind.*,
B. 445; rasout, *Hind.*, B. 458; rass,
Pamir, S. 1276; rassaul, *Hind.*, A. 200;
rassi, *Hind.*, E. 73; rasún, *Beng.*, A.
779; rasunia, *Uriya*, K. 21; rasuria,
Uriya, K. 21; rasvat, *Hind.*, B. 443;
raswanti, *Hind.*, B. 458; rata, *Hind.*, D.
884; rátadin, *Mar.*, C. 617; ratak, *Pb.*,
A. 51; ratáli, *Mar.*, I. 348; ratálu, *Bihar*,
D. 534; ratalú (Ipomæa), *Mar.*, I. 348;
rátálú (Dioscorea), *Guj.*, *Hind.*, *Pb.*, D.
485, 534; rátambí, *Dec.*, *Mar.*, G. 36;
ratambusála, *Mar.*, G. 36; ratánjli, *Guz.*,
Mar., P. 1381; ratanjot (Onosma),
Hind., *Pb.*, O. 170; ratanjota (Jatropha),
Guz., J. 41; ratan-purus, *Dec.*, *Hind.*, I.
335; ratar, *Bihar*, D. 534; rati, *Guj.*,
Hind., A. 51; rátinágkesar, *Guz.*, O. 6;

ratisurkh, *Kashmir*, T. 562; ratmandi,
Kumaon, *Pb.*, T. 562; ratmandú, *Pb.*,
T. 562; rátobával, *Guj.*, A. 238; ratoilyá,
Guj., *Mar.*, L. 451; rato-shemalo, *Guj.*,
B. 632; ratpathá, *Kumaon*, A. 677; rat-
suhara, *Pb.*, C. 1958; ratta, *Pb.*, C. 1003;
rattanjog, *Pb.*, A. 1125; rattanjot (Ja-
tropha), *Pb.*, J. 41; rattanjot (Macro-
tomia), *Pb.*, M. 35; rattanjot (Poten-
tilla), *Pb.*, P. 1203; rattanjot (Vinca),
Pb., V. 138; rattanjote (Clausena),
Hind., C. 1311; rattankát (Pieris), *Pb.*,
P. 702; rattankát (Rhododendron), *Pb.*,
R. 251; rattiasan, *Hind.*, H. 177; ratua,
Hind., S. 2670; ratún, *Beng.*, L. 483;
rau (Abies,) *Pb.*, A. 17; rau (Coriaria),
Pb., C. 1958; ráu (Cotoneaster), *Pb.*, C.
2021; raundra, *Hind.*, A. 249; raunel,
Hind., R. 311; raung, *Kashmir*, C. 706;
rauni, *Hind.*, *Kumaon*, M. 71; raunj,
Hind., A. 249; raúns (Andropogon), *Pb.*,
A. 1117; raúns (Cotoneaster), *Hind.*, C.
2019, 2021; raunsla, *Beng.*, A. 22; rausa,
Hind., V. 116; ravand chíní, *Pb.*, V. 64;
ráwa, *Hind.*, S. 371; rawan (Dolichos),
Pb., D. 758; rawan (Lathyrus), *Pb.*, L.
96; rawán (Vigna), *Pb.*, V. 116; rawangan, *Pb.*, V. 116; rawángi, *Pb.*, V. 116;
rawari, *Pb.*, L. 96; rawás, *Hind.*, V. 116;
rawásh, *Afg.*, R. 215; rawla, *Beng.*,
Dec., *Hind.*, S. 1212; rayadodi, *Mar.*, L.
292; ráya gundo, *Guz.*, C. 1931; ráyán
(Brassica), *Mar.*, B. 833; rayan (Mimusops), *Guj.*, M. 583; rayani, *Mar.*, M.
583; raysinganí, *Guz.*, S. 1174; razbam,
Pb., L. 535; razli júarí, *Pb.*, S. 3079.

Re   Re, *Hind.*, *Kumaon*, *Pb.*, A. 17, 22; rebdan, *Pushtu*, T. 227; rebdún, *Pushtu*, T.
227; rebhri, *Sind*, T. 569; regdáwan,
Pushtu, T. 227; réghan-i-kanjak, *W.
Afg.*, P. 847; reg-mahi, *Pb.*, R. 114;
rehan, *Pb.*, O. 18; reini, *Pb.*, M. 71;
rek, *Hind.*, P. 1322; relme, *Pb.*, C. 42;
relnú, *Pb.*, Z. 280; relú, *Hind.*, C. 42;
renak, *Hind.*, F. 703; rendi, *Hind.*, R.
369; reng, *Pb.*, V. 149; rengnie, *Hind.*,
S. 2345; renr, *Hind.*, R. 369; renrwári,
Hind., R. 369; renu kabij, *Mar.*, V.
179; reodán, *Pushtu*, T. 227; reoni,
Hind., M. 71; reori, *Pb.*, V. 149; reri,
Hind., R. 369; rerú (Acacia), *Hind.*, A.
249; rerú (Mallotus), *Kumaon*, M. 71;
résha-i-khitmí, *Mar.*, A. 880; resham-
búti, *Pb.*, P. 964; reshamí, *Pb.*, P. 964;
reshœ, *Pb.*, P. 964; reteon, *Pb.*, R. 152;
reti, *Hind.*, S. 771; rettia, *Uriya*, S. 818;
reun, *Pb.*, M. 71; reús, *Pb. Hills*, C.
2021; reúsh, *Pb. Hills*, C. 2021; reva-
chinnisírá, *Mar.*, G. 66; rewah, *Beng.*,
Hind., F. 410, 411; rewand chíní (Ere-
mostachys), *Pb.*, E. 270; rewand chíní
(Rheum), *Pb.*, R. 215; rewari, *Pb.*, A. 17,
A. 22; rewund chini, *Hind.*, R. 649.

Rh   Rhadachampo, *Guz.*, P. 989; rhái, *Hind.*,
Kumaon, A. 17; rhakhan, *Mar.*, S. 717;
rhanamb, *Mar.*, S. 2649; rhanbhendy,
Mar., H. 250; rhi, *Hind.*, *Pb.*, P. 746;
rhita, *Mar.*, S. 818; rhurucháphá, *Mar.*,
P. 989.

Ri   Rí, *Pb.*, P. 746; rí (Cotoneaster), *Pb.*, C.
2021; rí (Jasminum), *Pb.*, J. 24, 29; riálla,
Hind., *Kumaon*, A. 17; rian, *Pb.*, L.
474, 483; ríanish, *Hind.*, V. 116; rianj,

Kumaon, Q. 55 ; riaul, *Pb.*, M. 562 ; riaungi, *Pb.*, S. 2299 ; ribas, *Pb.*, R. 215 ; rich, *Hind.*, F. 820 ; richang, *Pb.*, S. 560 ; richh kas, *Pb.*, S. 763 ; ríchhábí, *Pb.*, V. 91 ; richh úklú, *Pb.*, V. 91 ; richni, *Pb.*, E. 505 ; riensh, *Kumaon*, V. 116 ; rigana, *Guz.*, S. 2284 ; rikaling, *Pb.*, T. 93 ; rikhái, *Pb.*, T. 93 ; rikhon, *Kumaon*, S. 2394 ; rikhu, *Hind.*, S. 30 ; rikhú, *Kumaon*, *Hind.*, S. 30 ; ríkhú, *Hind.*, S. 30 ; ríkhúl, *Pb.*, R. 323, 335 ; ríkkan, *Pb.*, P. 1138, 1148 ; rikúnra, *Pb.*, A. 801 ; rín, *Pb.*, Q. 35 ; ríng, *Pb.*, C. 2035 ; ringa, *Hind.*, A. 249 ; ringál, *Hind.*, A. 1523, A. 1526, A. 1535 ; ringall, *Hind.*, T. 379 ; ringani, *Mar.*, S. 2280 ; ríngí, *Pb.*, V. 149 ; ringin, *Mar.*, S. 818 ; ringli, *Hind.*, S. 2280 ; ringni, *Guz.*, S. 2284 ; ringni, *Dec.*, *Hind.*, *Mar.*, S. 2345 ; ringo, *Pb.*, C. 2035 ; rini, *Pb.*, V. 149 ; rinj (Acacia), *Hind.*, A. 249 ; rinj (Quercus), *Pb.*, Q. 35 ; rinjal, *Hind.*, S. 1656 ; rínság, *Pb.*, P. 688 ; rís, *Pb.*, V. 104 ; risámani, *Guz.*, M. 557 ; rish, *Pb.*, C. 2021 ; rísh bargad, *Pb.*, F. 129 ; rishka, *Afg.*, *Pb.*, M. 331 ; rishka, *Afg.*, M. 334 ; rishku, *Afg.*, M. 336 ; riskawa, *Hind.*, H. 164 ; rita, *Mar.*, S. 818 ; rita (cobalt), *Pb.*, C. 1444 ; rita (Rita), *Beng.*, F. 562 ; rithá (Acacia), *Hind.*, A. 200 ; ritha (Erioglossum), *Hind.*, E. 210 ; ritha, rithá (Sapindus), *Beng.*, *Dec.*, *Hind.*, *Mar.*, *Pb.*, S. 808, 818 ; rithe, *Mar.*, S. 818 ; rithei, *Pb.*, S. 1024 ; rithia, *Hind.*, S. 818 ; rithu, *Pb.*, E. 479 ; riú, *Hind.*, *Pb.*, C. 2019, 2021 ; riúna, *Kumaon*, M. 71 ; riús, *Hind.*, C. 2019 ; riwás-an, *Pb.*, S. 1174 ; riyong, *Beng.*, M. 302.

Ro Ro, *Pb.*, A. 17 ; roang ching, *Pb.*, S. 560 ; rodan, *Afg.*, R. 580 ; rodang, *Afg.*, R. 580 ; rohan, *Beng.*, *Hind.*, *Mar.*, S. 2501; rohani, *Hind.*, A. 249 ; rohedo, *Mar.*, C. 6 ; rohi, *Hind.*, E. 73 ; rohina, *Beng.*, *Guz.*, S. 2501 ; rohing, *Mar.*, S. 2501 ; rohíra, *Pb.*, T. 227 ; rohiaha, *Mar.*, A. 1117 ; rohita, *Beng.*, F. 474 ; rohni (Mallotus), *Hind.*, M. 71 ; rohní (Symida), *Hind.*, S. 2501 ; rohra, *Beng.*, S. 2501 ; rohtí, *Mar.*, T. 567 ; rohun, *Dec.*, *Hind.*, S. 2501 ; rohunna, *Dec.*, *Hind.*, S. 2501 ; roi, *Pb.*, A. 17 ; ró-ín, *Pb.*, V. 116 ; roíong, *Pb.*, D. 758 ; roír, *Pb.*, T. 227 ; roira, *Hind.*, *Mar.*, T. 227 ; rojh, *Hind.*, S. 1229 ; rojia chaphúl, *Mar.*, T. 17 ; roimarí, *Mar.*, A. 367 ; rojra, *Hind.*, S. 1229 ; rokto-simul, *Beng.*, B. 632 ; roku, *Pb.*, C. 881 ; rolecha, *Kumaon*, P. 746 ; roli, *Hind.*, *Kumaon*, M. 71 ; rom, *Kashmír*, S. 1247 ; romúsk, *Pb.*, R. 152 ; ronecha, *Kumaon*, P. 746 ; roníunchi, *Pb.*, P. 746 ; rori (carnelian), *Hind.*, C. 618, 622 ; rori (Mallotus), *Beng.*, *Hind.*, M. 71 ; ror (jaggery), *Pb.*, S. 370 ; roan, *Guj.*, S. 2501 ; rosbang, *Pb.*, P. 959 ; roseigavat, *Mar.*, A. 1117 ; roshel, *Mar.*, A. 1117 ; roti, *Beng.*, F. 566 ; rotí-alú, *Hind.*, M. 217 ; rotka, *Hind.*, E. 170; rouch, *Beng.*, M. 656 ; rouen, *Dec.*, S. 2501 ; rous, *Kashmír*, D. 227 ; rousá-ghás, *Hind.*, A. 1117 ; rovi, *Hind.*, S. 1229 ; row, *Pb.*, A. 22 ; rowanra, *Hind.*, E. 356 ; rowi, *Beng.*, F. 474 ; rowíl, *Pb.*, G. 177 ; royadar, *Beng.*, S. 371 ; roz, *Hind.*, S. 1229 ; rozatt, *Pb.*, T. 634.

Ru Ru, *Guz.*, G. 404 ; rúba barík, *Pb.*, S 2268 ; rúchia, *Hind.*, C. 1969 ; ruchiya, *Kumaon*, C. 1969 ; ruconi-chanda, *Beng.*, F. 447 ; rudrák, *Hind.*, E. 57, 69 ; rudraksh, *Mar.*, E. 57 ; rudrákya, *Beng.*, E. 57 ; rúen, *Kumaon*, M. 71 ; rugtrora, *Hind.*, T. 227 ; rugtrora,; *Mar.*, T. 227 ; rugtrorar, *Mar.*, R. 164 ; ruhimula, *Guj.*, A. 1398 ; ruhin, *Dec.*, S. 2501 ; ruhu, *Uriya*, F. 474; ruí (Calotropis), *Mar.*, C. 170 ; ruí (Gosaypium), *Hind.*, *Mar.*, *Pb.*, G. 404; ruí (Labeo), *Beng.*, F. 474 ; rúin, *Hind.*, M. 71 ; rúinia kamela, *Hind.*, M. 71 ; ruinish, *Hind.*, C. 2019 ; ruinsh, *Hind.*, C. 2021 ; rukar, *Hind.*, A. 1114 ; rúkh, *Pb.*, T. 51, T. 61, T. 70 ; rukhtochandan, *Hind.*, P. 1381 ; rukt chundun, *Beng.*, P. 1381 ; rukto-gurániya álu, *Hind.*, *Beng.*, D. 526 ; rukto-púi, *Beng.*, B. 207 ; rúktupita, *Beng.*, V. 48 ; rúl, *Pb.*, H. 277 ; rúlú, *Hind.*, M. 71 ; rúlyá, *Pb.*, M. 71; rumaid, *Mar.*, F. 179 ; rúmbal, *Pb.*, F. 179, 202, 265 ; rúmí-mastiki, *Hind.*, P. 841 ; rúmí mastungi, *Beng.*, P. 841 ; ruminche, *Pb.*, P. 746 ; rúna, *Pb.*, R. 564 ; rúnang, *Pb.*, R. 564 ; rund, *Deccan*, R. 369 ; rungara, *Kumaon*, E. 340 ; rúngra, *Hind.*, E. 356 ; rungrú, *Pb.*, *Hind.*, K. 17 ; rungtra, *Hind.*, C. 1234 ; runjra, *Hind.*, A. 249 ; runjuni, *Beng.*, P. 1044 ; rupá (silver), *Beng.*, *Dec.*, *Guj.*, *Hind.*, *Mar.*, A. 1359 ; rupa (silver), *Dec.*, *Hind.*, S. 2205 ; rúpali, *Beng.*, A. 1359 ; rúpamakhi, *Hind.*, I. 478 ; rupérítagat, *Duk.*, A. 1359 ; ruperivarakh, *Guj.*, A. 1359 ; rúpi-chanda, *Beng.*, F. 430; rupo, *Sind*, A. 1359 ; ruppa, *Beng.*, *Dec.*, *Guj.*, *Hind.*, *Mar.*, A. 1359 ; rupu, *Guj.*, A. 1359 ; rupyáchá-varkh, *Mar.*, A. 1359 ; rúrí, *Hind.*, F. 414 ; rús (Adhatoda), *Hind.*, A. 484 ; rús (musk deer), *Kashmír*, D. 227 ; rúsa, *Hind.*, *Mar.*, S. 2912 ; rúsá ghás, *Hind.*, A. 1117 ; rusaka-tel, *Mar.*, A. 1117 ; rusam, *Uriya*, S. 950 ; rush, *Pamir*, S. 1276 ; rusi-gugar, *Kash.*, F. 796 ; ruskar, *Pb.*, D. 245 ; rusmarí, *Hind.*, R. 547 ; rusot, *Hindi*, B. 445 ; ruswul, *Hind.*, B. 445; rutripul, *Kumaon*, P. 1449 ; rutta, *Uriya*, F. 318.

Sa Saba, *Beng.*, I. 494 ; sabai, *Hind.*, I. 494 ; sabajhi, *Sind*, O. 18 ; sabarsinghadú, *Guj.*, H. 50 ; sabba jaya, *Hind.*, C. 321 ; sabji, *Hind.*, C. 331 ; sábudáná, *Hind.*, C. 2531 ; sabuni (Trianthema), *Hind.*, T. 537 ; sabuni (Saponaria), *Beng.*, S. 850 ; sabzá sabzah, *Hind.*, *Mar.*, O. 18 ; sabzí, *Hind.*, *Pb.*, C. 331 ; sach, Kashmír, G. 769 ; sádada, *Mar.*, T. 361 ; sada dhútúrá, *Beng.*, D. 66 ; sádado, *Guj.*, T. 282 ; sádá guláb, *Pb.*, R. 531 ; sada-hazur-maní, *Beng.*, *Hind.*, P. 657 ; sada-jati, *Beng.*, B. 165; sada-natía, *Beng.*, A. 938 ; sadáphal (Cítrus), *Hind.*, C. 1263 ; sadaphúl (Vinca), *Mar.*, V. 138 ; sáda-raí, *Beng.*, B. 812 ; sadáb, *Hind.*, R. 663 ; sadáf, *Deccan*, R. 663 ; sádah-dhatúrá, *Hind.*, D. 160 ; sádáhurhuriá, *Beng.*, G. 753 ; saddr, *Guj.*, *Hind.*, T. 361 ; saddra, *Mar.*, T. 361 ; sáder, *Hind.*, *Guj.*, T. 361 ; sadha-gandhaka, *Mar.*, S. 2999 ; sadhi, *Mar.*, C. 637; sadnidjar, *Sind*, C. 224 ; sadora, *Dec.*, T. 361 ; sádri, *Guj.*, *Hind.*, *Mar.*, T. 361 ; sadul kou, *Beng.*, M. 435 ; sadur, *Hind.*, T. 361 ; sadura, *Mar.*, T. 282 ; saer, *Mar.*, B. 632 ;

saféd, *Pb.,* Z. 190 ; saféd ák, *Hind.,* C.
170, 191 ; saféd bábúl, *Beng.,* A. 249 ;
saféd bachlá, *Hind.,* B. 203 ; saféd bahman,
Hind., Mar., C. 912; saféd bhangra,
Hind., Pb., H. 106 ; saféd bhurki, *Hind.,*
E. 263 ; saféd chamni, *Hind.,* H. 149 ;
saféd chandan, *Beng., Hind., Mar.,* S.
790 ; saféd dámar, *Dec., Hind.,* V. 31 ;
saféd dhátúra, *Hind.,* D. 160 ; saféd gul-
makmal, *Beng.,* A. 914 ; saféd ind, *Beng.,*
J. 41 ; saféd kaddu, *Beng., Hind.,* C. 2331 ;
saféd kanphál, *Mar.,* D. 513 ; saféd
khair, *Hind.,* F. 671 ; saféd khatyán,
Dec., E. 289 ; saféd kikar, *Hind., Pb.,*
A. 249 ; saféd kudsumbal, *Hind.,* C. 289;
saféd lobeh, *Mar.,* V. 116 ; saféd mirch,
Hind., P. 811 ; saféd miri, *Mar.,* P. 811 ;
saféd múrgha, *Hind.,* C. 868 ; safédmuslí,
Dec., A. 1577 ; saféd pai, *Beng.,* E. 59 ;
saféd pán, *Dec.,* A. 853 ; saféd pattar, *Pb.,*
C. 489, M. 277 ; saféd rai, *Dec., Hind.,*
B. 800 ; saféd rayán, *Dec., Hind.,* B. 800 ;
saféd sanbhálú, *Hind.,* V. 181 ; saféd
sanbul, *Dec., Hind.,* A. 1425 ; saféd semal,
Hind., E. 289 ; saféd shakar, *Hind.,* S.
375 ; saféd shorshi, *Beng.,* E. 327 ; saféd
siris, *Hind.,* A. 717 ; saféd soná, *Hind.,*
P. 940 ; saféd túta, *Dec.,* Z. 190 ; saféda
(lead), *Hind., Pb.,* L. 143 ; saféda
(Populus), *Afg.,* P. 1159, *Pb.,* P. 1138,
1148, 1153, 1159, *Sind,* P. 1153 ; saféda
(Salix), *Pb.,* S. 579 ; saféda maddawa,
Hind., F. 406 ; saféda musalí, *Mar.,* A.
1562; safodínd, *Hind.,* J. 41 ; safra, *Hind.,*
F. 48 ; safran, *Mar.,* C. 2083 ; sá (Amaran-
tus), *Beng.,* A. 934 ; ság (Spinacia), *Hind.,*
S 2574 ; ság (Tectona), *Guz., Hind., Mar.,*
T. 232 ; ság (Terminalia), *Guj.,* T. 293,
Hind., Mar., T. 361 ; sága, *Guj., Mar.,*
T. 232 ; sagach, *Guj.,* T. 232 ; sag-angúr,
Hind., A. 1614 ; sagar, *Guj., Mar.,* I. 3 ;
sagarabatna, *Uriya,* E. 464 ; sagaragota,
Guj., Mar., C. 6 ; saggar, *Pushtu,* E. 35 ;
sag-i-al, *Pb.,* O. 534 ; sagon, *Hind.,* T. 232 ;
sagoná, *Hind ,* T. 293 ; sagor, *Beng ,*
F. 324 ; sagowani, *Hind.,* D. 9 ; ság-
pálak, *Hind.,* S. 2574 ; sagu-chawul,
Hind., S. 514 ; sagu-dana, *Beng., Hind.,*
S. 514 ; ságún, *Hind., Pb.,* T. 232 ; ságur
ghota, *Mar.,* C. 6 ; ságván, *Dec., Mar.,*
T. 232 ; ságwán (Tectona), *Mar., Pb.,*
T. 32 ; sagwan (Terminalia), *Mar.,* T. 293 ;
sáh, *Pb.,* S. 1247 ; sahadebi (Sida), *Hind.,*
S. 1703 ; sahadevi (Vernonia), *Pb.,* V. 79 ;
sahadevi bari, *Hind.,* S. 2354 ; sahajna,
Hind., M. 721 ; saháju, *Uriya,* T. 361 ;
saherwa, *Hind.,* N. 179 ; sahine, *Hind.,*
L. 55 ; sah-lun, *Uriya,* F. 329 ; saht, *Pb.,*
H. 342 ; sahuda, *Uriya,* S. 2912 ; sahut,
Hind., H. 342 ; sahwan, *Hind.,* E. 327 ;
sái, *Pb.,* L. 535 ; sai-kanta, *Beng.,* A. 291 ;
sain, *Hind., Pb.,* T. 361 ; sainad, *Hind.,*
I. 499 ; sainjna, *Mar., Raj.,* M. 717 ; sair,
Baluch., S. 1233 ; sairan, *Raj.,* F. 672, I.
499 ; sáita, *Hind.,* C. 1444 ; sáj, *Hind.,*
T. 232, *Hind., Mar., Uriya,* T. 361 ; séja,
Hind., T. 361 ; sajina, *Beng., Uriya,* M.
721 ; sajjekhára, *Mar.,* C. 541 ; sajji,
Beng., Hind., C. 541 ; sajjíkhár, *Hind.,*
B. 158, C. 541 ; sajjí-mátí, *Beng.,* B. 151 ;
sajjí-mittí, *Hind.,* C. 541 ; sajru, *Beng.,*
P. 1171 ; sák, *Hind.,* A. 103 ; sakal yel,
Mar., V. 48 ; sakaria, *Guz.,* I. 348 ; sakem,

Sind, T. 51 ; sakena, *Hind.,* I. 119, 141 ;
sakher, *Hind.,* S. 1656 ; sákhu, *Hind.,*
S. 1656 ; sákhú, *Hind.,* T. 232 ; sakhua,
Hind., S. 1656 ; saki (Populus), *Pb.,* P.
1148 ; sákí (Hedychium), *Pb.,* H. 59 ;
sakin, *Pb.,* S. 1242 ; sakkar, *Hind.,* S.
373 ; sakkur, *Pushtu,* E. 35 ; sák múnia,
Hind., Pb., Sind, C. 1783 ; sakna, *Hind.,*
I. 119 ; sakoh. *Hind.,* S. 1656 ; saku-
limba, *Mar.,* C. 1233 ; sakwa, *Hind.,* S.
1656 ; sal, *Mar.,* T. 223 ; sál (Ophiocephalus), *Uriya,* F. 515 ; sál (Oryza), *Raj.,*
O. 258 ; sál (Salvadora), *Pb.,* S. 705 ; sál
(Shorea), *Beng., Hind., Mar., Pb.,* S.
1656 ; sála, *Hind.,* S. 1656 ; sála, *Hind.,*
S. 1939 ; salab, *Hind.,* O. 205, *Afg.,* O.
207, S. 521 ; sálá-bhir, *Kashmir,* S. 1264 ;
salab-misri, *Hind.,* S. 521 ; salád, *Beng.,*
Hind., L. 21 ; salai, *Beng., Hind., Mar.,*
B. 771 ; salajet, *Mar.,* L. 455 ; salájit,
Kumaon, P. 436; sala-kodam gadi, *Hind.,*
C. 1028 ; sálama-misri, *Mar.,* S. 521 ;
sálam misri, *Guj., Mar.,* S. 521 ; sálan
(Panicum), *Pb.,* P. 63 ; sálan (Setaria),
Pb., S. 1212 ; salanker, *Pb.,* A. 2 ; salap,
Afg., Hind., O. 205, *Afg.,* O. 207, S. 521 ;
salaphali, *Mar.,* B. 771 ; sálar (Boswel-
lia), *Hind.,* B. 771 ; sálar (Panicum),
Pb., P. 63 ; sálaya-dhup, *Mar.,* B. 771 ;
salbia sefakuss, *Hind.,* S. 743 ; sáldan,
Hind., O. 537 ; saldbargh, *Pb ,* C. 117 ;
saldhol, *Mar.,* S. 2861 ; salei, *Hind.,* B.
771 ; sále manta, *Hind.,* S. 1024 ; salen-
dra, *Mar.,* P. 1171 ; salepan, *Hind.,* S.
1024 ; salep misrie, *Beng.,* O. 209 ; saleri,
Hind., A. 1227 ; saleya, *Hind.,* B. 771 ;
salga, *Hind., Mar.,* B. 771 ; salgára, *Pb.,*
C. 868 ; salhe, *Hind.,* B. 771 ; salhi, *Pb.,*
B. 777 ; sálib misri, *Pb.,* E. 467, S. 521 ;
salie, *Pb.,* A. 17 ; salíma, *Pb.,* A. 328 ;
saliya gugul, *Guj.,* B. 771 ; salla (Abies),
Pb., A. 17 ; salla (Pinus), *Hind.,*
Kumaon, Kashmir, P. 760 ; salle, *Pb.,*
A. 22 ; salma, *Hind.,* P. 588 ; sálmalí,
Mar., E. 289 ; saloha, *Pb.,* P. 1401; salopa,
Uriya, C. 711 ; salor (Pueraria), *Pb.,* P.
1401 ; salora (Lagerstrœmia), *Uriya,* L.
155 ; salpan, *Hind.,* D. 339 ; salpáni,
Beng., salparni, *Mar.,* D. 339 ; salpe,
Hind., B. 771 ; salú, *Mar.,* S. 2405 ;
sáluk, *Beng.,* N. 200 ; salum (Orchis),
Guj., O. 209 ; sálum (Eulophia), *Guj.,*
E. 467 ; salún, *Hind.,* D. 339 ; salúní,
Pb., R. 650 ; salwa, *Hind., Uriya,* S.
1656, *Hind.,* S. 1939 ; sálwan, *Mar.,* D.
339 ; salzat, *Dec.,* O. 18 ; sama (Glochidion), *Pb.,* G. 240 ; sama (Oryza),
Kashmir, O. 258 ; sama (Panicum), *Raj.,*
P. 48, *Hind.,* P. 50, 53, *Pb.,* P. 53,
Mar., P. 63 ; samada, *Guj.,* C. 1978 ;
samagk, *Afg.,* R. 287 ; samáhk, *Afg.,* R.
287 ; samak (Panicum), *Hind.,* P. 53 ;
samágk (Rhus), *Kashmir,* R. 311; sámálú,
Beng., V. 164 ; sáman, *Hind.,* P. 53 ;
samandar-ká-pattá, *Dec.,* A. 1362 ;
samandar-ká-pát, *Hind.,* A. 1362 ;
samandar phaind, *Hind.,* A. 1362 ;
samandar phal, *Mar.,* B. 180; samandar-
sóf, *Hind.,* A. 1362 ; samandar-sokh,
Hind., A. 1362 ; samanka, *Hind.,* C.
1221 ; samar (Bombax), *Mar.,* B. 632 ;
samar (Opuntia), *Mar.,* O. 193 ; samarogh,
Afg., Beng., Hind., F. 725 ; samarri,

Hind., Mar., S. 2819 ; sámba lakshára, *Beng.,* A. 1425 ; sambar, *Hind.,* D. 240, 348 ; sambhal, *Hind.,* V. 164 ; sambhálu, *Hind.,* V. 164 ; sambúng, *Mar.,* S. 3040 ; samdulun, *Hind.,* E. 80 ; samei, *Hind.,* P. 53 ; samghul, *Hindi,* A. 102 ; sámli, *Guz.,* P. 63 ; samma, *Pb.,* S. 950 ; sammal-khár, *Beng.,* A. 1425 ; samp-ki-khumb, *Pb.,* A. 1379 ; samprú, *Pb.,* C. 1711 ; samra-shama, *Beng.,* P. 53 ; samsem, *Pb.,* J. 29 ; samsundra, *Hind.,* A. 722 ; samu, *Uriya,* P. 53 ; samudraca, *Hind.,* L. 232 ; samudra phal, *Beng.,* B. 193 ; samudra phala, *Mar.,* B. 180 ; samudra soka, *Mar.,* A. 1362 ; samuka, *Pb.,* P. 53 ; samunaar láni, *Sind.,* C. 224 ; samundar, *Beng., Dec., Hind., Guj., Mar.,* B. 180 ; samúndar-sok, *Pb.,* S. 746 ; samundra phal, *Dec., Hind., Guj.,* B. 180 ; san (Andropogon), *Hind., Pb.,* A. 1093 ; san (Crotalaria), *Beng. Dec., Guj., Hind., Mar.,* C. 2105, *Hind.,* C. 2160 ; sanai, *Beng., Hind.,* C. 2105 ; sanáke, *Pb.,* V. 164 ; sanatha, *Pb.,* D. 725 ; sanatta, *Pb.,* D. 725 ; sanbhálu, *Hind.,* V. 164 ; sanbul-khár, *Hind., Duk.,* A. 1425 ; sandal, *Hind., Mar.,* S. 790 ; sándan, *Hind., Mar., Pb.,* O. 537 ; san-darach, *Sind,* C. 142 ; sandaras, *Sind,* C. 142 ; sandari (Cassia), *Uriya,* C. 756 ; sandári (Debregeasia), *Pb.,* D. 196 ; sandhur, *Hind., Pb.,* L. 143 ; sandóle-ka-gúr, *Hind.,* S. 370 ; sandole-ka-nar, *Dec.,* P. 588 ; sandras, *Hind.,* V. 31 ; sandugaza, *Beng.,* T. 372 ; sanga, *Mar.,* M. 721 ; sangal, *Pb., Himálaya,* A. 17 ; sangar, *Pb.,* P. 1259 ; sangcha, *Pb.,* C. 489 ; sang-goah, *Pb.,* F. 486 ; sanggye, *Pb.,* S. 1065 ; sang-i-ákík, *Hind. & Pb.,* C. 618 ; sang-i-dalam, *Hind.,* C. 1317 ; sang-i-irmali, *Pb.,* C. 489 ; sang-i-jeráhat, *Pushtu,* G. 769 ; sang-i-kas, *Pb.,* J. 2 ; sangi-khurús, *Pb.,* C. 489 ; sang-i-marján, *Pb.,* C. 1808 ; sang-i-marmar, *Pb.,* C. 489 ; sanginphrú, *Mar.,* S. 3040 ; sang-i-palaun, *Hind.,* S. 2712 ; sang-i-sar-mai, *Pb.,* C. 489 ; sang-i-yashab, *Pb.,* J. 2 ; sang-kupi, *Dec., Hind.,* C. 1377 ; sangran, *Hind.,* L. 474 ; sangri, *Raj.,* P. 1259 ; sangtara, *Hind.,* C. 1233 ; sang-yahúdi, *Pb.,* C. 489 ; sanh, *Pb.,* O. 574 ; sani, *Beng., Hind.,* C. 2105 ; sanipát, *Hind., Mar.,* S. 964 ; san-i-shadauj, *Pb.,* C. 489 ; sanj, *Pb.,* O. 70 ; sanjad, *Pb.,* P. 1466 ; sanjata, *Afg.,* E. 40 ; sanjít, *Afg.,* E. 40 ; sánjna, *Pb.,* M. 721 ; sanka (arsenic), *Beng.,* A. 1425 ; sánka (Panicum), *Hind.,* P. 50 ; sánkálu, *Beng.,* P. 1 ; sankar, *Hind.,* S. 373 ; sankar-jata, *Beng.,* U. 26 ; san-karunda, *Uriya,* C. 596 ; sankásúra, *Mar.,* P. 1032 ; sankhá huli, *Hind.,* C. 382 ; sankhii, *Pb.,* C. 854 ; sankhpushpi, *Pb.,* E. 581 ; sankhri, *Pb.,* P. 1259 ; sankhú, *Pb.,* C. 854 ; sankhyá, *Hind., Duk.,* A. 1425 ; sankhyá-sanbul, *Dec., Hind.,* A. 1425 ; sánkí til, *Beng.,* S. 1070, sánklu, *Pb.,* C. 1606 , sankokla, *Pb.,* H. 177 ; sankru, *Hind.,* C. 1686 ; sankush, *Uriya,* F. 598 ; sán madat, *Mar.,* T. 282 ; sanmali, *Pb.,* A. 1573 ; sanna, *Pb.,* C. 754 ; sanna makki, *Beng.,* C. 737 ; sannan (Ougeinia), *Pb.,* O. 537 ; sannan

(Populus), *Pb.,* P. 1138 ; sanní, *Pb.,* O. 157 ; sanolí, *Pb.,* G. 213 ; sanpatti, *Beng.,* O. 233 ; sanrh, *Beng., Hind.,* O. 574 ; sansárú, *Pb.,* D. 196 ; sanshi, *Beng.,* B. 812 ; sansri, *Hind.,* I. 440 ; sánt, *Hind.,* B. 619 ; san tág, *Mar.,* C. 2105 ; san-tara, *Hind.,* C. 1234 ; santara, *Pb.,* C. 1243 ; santha (Woodfordia), *Hind.,* W. 106 ; sánthá (Dodonæa), *Pb.,* D. 725 ; santij, *Afg.,* E. 40 ; sanwa, *Beng., Dec., Hind.,* P. 53 ; sanwak, *Hind., Pb.,* F. 672, *Hind.,* P. 48, *Pb.,* P. 45, 53 ; sánwán, *Hind.,* F. 672, P. 53 ; saochála, *Kumaon,* B. 572 ; saon, *Beng., Dec., Sind,* P. 53 ; saónf, *Hind.,* P. 727 ; saora, *Hind.,* S. 1174 ; saori (porcupine), *Guj.,* P. 1171 ; saori (Sesbania), *Mar.,* S. 1174 ; saouké cha-wal, *Duk.,* C. 2531 ; sápasan, *Mar.,* A. 1398 ; sapfalú, *Pb.,* D. 245 ; saphari kumhra, *Hind.,* C. 2325 ; saphéd-musli, *Guj.,* A. 1562 ; saphéd-sakkar, *Guz.,* S. 375 ; sápheta musali, *Mar.,* A. 1562 ; saphetasávara, *Mar.,* E. 289 ; saphúri-komra, *Beng.,* C. 2316 ; sapin, *Hind., Kumaon,* P. 760 ; sapistán, *Mar.,* C. 1931 ; sapotá, *Hind., Beng.,* A. 376 ; saprotri, *Pb.,* S. 924 ; saptala, *Hind.,* J. 13 ; sar (Saccharum), *Sind,* S. 6 ; sar (goral), *Pb.,* S. 1247 ; sara (Abies), *Pb.,* A. 22 ; sara (Cupressus), *Hind.,* C. 2354 ; sara (Litsæa), *Hind.,* L. 490 ; sara (Sac-charum), *Beng., Hind.,* S. 6 ; saráh (Sind ibex), *Baluch., Sind,* S. 1233, *Baluch.,* S. 1239 ; saragavo, *Guj.,* M. 721 ; saragvo, *Mar.,* M. 721 ; sara-hati, *Hind.,* O. 180 ; sarai, *Kumaon,* C. 2358 ; saral, *Hind., Pb.,* P. 760 ; sarana, *Beng., Uriya,* F. 350 ; sarana pungti, *Beng.,* F. 350 ; sárang, *Mar.,* D. 212 ; saráo, *Pb.,* S. 1264 ; saraoji, *Guz.,* M. 656 ; saráp, *Pushtu,* T. 93 ; sarapatri, *Uriya,* A. 717 ; sarapuna, *Mar.,* C. 162 ; sarár, *Hind.,* P. 1401 ; sarári, *Pb.,* A. 1090 ; sararhi, *Hind.,* S. 6 ; saras (Albizzia), *Mar.,* A. 695 ; saras (Dalbergia), *Hind.,* D. 31 ; saras (Grewia), *Hind.,* G. 705 ; sarás (Cupressus), *Hind.,* C. 2354 ; sarashire, *Guj.,* B. 817 ; saraté, *Mar.,* T. 548 ; sarawán, *Pb., Pushtu,* P. 833 ; sarba-jaya, *Beng.,* C. 321 ; sarda (Cucu-mis), *Afg.,* C. 2263 ; sarda (Sterculia), *Mar.,* S. 2861 ; sarei, *Hind.,* S. 1656 ; sarei, *Pb.,* A. 17 ; sarhar, *Hind.,* S. 6 ; sarí (Oryza), *Sind,* O. 258 ; sárí (Prunus), *Pb.,* P. 1285 ; sariála, *Pb.,* A. 1090 ; sarí-kasóndí, *Duk.,* C. 787 ; sarivan, *Hind.,* D. 339 ; sariyu chibhars, *Sind* C. 2263 ; sarjbar, *Pb.,* S. 6 ; sarjika, *Beng.,* C. 527 ; sarkanda, *Pb.,* S. 4, *Hind.,* S. 6 ; sarkara, *Hind., Pb.,* S. 6, *Pb.,* S. 49 ; sarl, *Kashmir,* P. 760 ; sarlakhtei, *Pb.,* P. 702 ; sarmal, *Pushtu,* H. 164 ; sarma safaid, *Pb.,* C. 489 ; sármei, *Pb.,* P. 964 ; sarmúl, *Pb.,* A. 1588 ; sarngar, *Pb.,* R. 261 ; saroka jhar, *Mar.,* C. 826 ; saroli, *Pb.,* A. 801 ; saron, *Sind,* P. 53 ; sarop, *Pushtu,* T. 93 ; sarora, *Pb.,* G. 143 ; sarova, *Mar.,* C. 826 ; sarpankh (Teph-rosia), *Pb.,* T. 270 ; sarpankha (Celosia), *Pb.,* C. 868 ; sarpat, *Hind.,* S. 4, 6 ; sar-patta, *Hind.,* S. 6 ; sarphonka, *Pb.,* T. 270 ; sarphónká, *Beng., Hind.,* T. 270 ;

sarphúnkha, *Mar.*, T. 270 ; sarpot,
Hind., P. 41 ; sarpúhala, *Mar.*, C. 826;
sarpur, *Hind.*, P. 72 ; sarput, *Hind.*, P.
41 ; sarra, *Pb.*, M. 237 ; sárrí, *Pb.*, C. 1101 ;
sarsan, *Mar.*, B. 841 ; sarsán banda,
Hind., O. 230 ; sarsata, *Hind.*, S. 1223 ;
sarsawa, *Guj.*, B. 812 ; sarshoti, *Hind.*,
A. 1215 ; sarson, *Hind.*, B. 817, *Hind.*,
Mar., B. 833, *Hind.*, F. 671 ; sarsón-
lahi, *Hind.*, B. 833 ; sarsón-zard, *Hind.*,
B. 817 ; sarsu, *Raj.*, B. 812 ; saru, *Uriya*,
C. 1732 ; saruála, *Pb.*, H. 164 ; sarúboke,
Mar., C. 2354 ; sarum eutur, *Duk.*, C.
1394 ; sarun, *Hind.*, I. 499 ; sarunga, *Pb.*,
P. 688 ; sarwadh, *Mar.*, M. 855 ; sarwak,
Hind., P. 45 ; sarwála (Andropogon),
Pb., A. 1090 ; sarwála (Heteropogon),
Hind., *Raj.*, H. 164 ; sarwála (Pueraria),
Hind., P. 1401 ; sarwáli, *Hind.*, *Pb.*, C.
868 ; sarwar (Andropogon), *Hind.*, A.
1090 ; sarwar (Heteropogon), *Hind.*, H.
164 ; sarwaír, *Hind.*, C. 868 ; sasa,
Beng., C. 2287 ; sásníálu, *Beng.*, D. 507 ;
sasrú, *Beng.*, H. 35 ; sassa, *Mar.*, H.
31 ; sassái, *Pushtu*, A. 554 ; satamúli,
Beng., A. 1575 ; satáp, *Mar.*, R. 663 ;
satápa, *Guj.*, *Mar.*, R. 663 ; sata-
phuspha, *Mar.*, P. 727 ; satávar, *Hind.*,
A. 1562 ; satavari, *Guj.*, A. 1575 ; satari,
Dec., *Hind.*, R. 663 ; satávarí-múl, *Mar.*,
A. 1575, 1577 ; satáwar, *Hind.*, *Pb.*, A.
1575 ; satawri, *Hind.*, A. 1577 ; satbalon,
Pb., P. 1076 ; sat-biroza, *Pb.*, P. 760 ;
sat-gilo, *Dec.*, *Pb.*, T. 470, T. 483 ;
sathete, *Beng.*, F. 505 ; sáthi (Oryza),
Hind., O. 258 ; sathí (Salvia), *Pb.*, S.
746 ; sathiya, *Hind.*, P. 50 ; sáthra,
Hind., O. 220 ; sathui, *Beng.*, D. 507 ;
sati, *Beng.*, C. 2499 ; satiún, *Hind.*, A.
871 ; satmulí, *Beng.*, A. 1577 ; satní,
Hind., A. 871 ; satodiputchu, *Guj.*, B.
619 ; satpatiya, *Hind.*, L. 556 ; satpúra,
Hind., D. 115 ; satpuria, *Hind.*, D. 53 ;
satranjí, *Hind.*, C. 626 ; sat-sarila,
Beng., M. 869 ; satte-gilo, *Hind.*, T.
470, T. 483 ; satu, *Dec.*, *Mar.*, H. 382 ;
satur, *Mar.*, M. 747 ; satvin, *Mar.*, A.
871 ; satwín, *Hind.*, A. 871 ; satyanasa,
Pb., A. 1351 ; satzarra, *Pb.*, A. 1565 ;
sa-uké-chawal, *Dec.*, S. 514 ; saul, *Pb.*,
F. 521 ; saumúgli, *Uriya*, P. 496 ; sauna,
Hind., O. 233 ; sauna assar, *Mar.*, O.
233 ; saundad, *Mar.*, P. 1259 ; saunder,
Mar., P. 1259 ; saunf, *Hind.*, F. 659 ;
saunspaur, *Pb.*, A. 1565 ; saur, *Mar.*,
B. 632 ; sauraf, *Hind.*, P. 727 ; saurif,
Hind., P. 727 ; sáva, *Dec.*, *Mar.*, P. 63 ;
savandal, *Mar.*, *Uriya*, P. 1259 ; savar,
Mar., E. 289 ; sávara, *Mar.*, B. 632 ;
savari, *Mar.*, B. 632 ; sawa (Panicum),
Dec., *Hind.*, P. 53, *Mar.*, P. 63 ; sawa
(Peucedanum), *Hind.*, P. 460 ; sawa
(Setaria), *Hind*, S. 1210 ; sáwach, *Mar.*,
T. 428 ; sawál, *Pb.*, P. 1194 ; sáwala,
Hind., V. 14 ; sawálí, *Pb.*, A. 801 ;
sáwan, *Hind.*, P. 45, 53 ; sáwan-bha-
deha, *Hind.*, P. 53 ; sáwan-chaitwa,
Hind., P. 63 ; sáwan-jethwa, *Hind.*,
P. 63 ; sawánk. *Hind.*, F. 672, P. 45,
Pb., P. 53 ; sáwank saun, *Kumaon*, P.
45 ; sa wi, *Dec.*, P. 63 ; sawuk, *Sind*,
P. 45.

Sb Sbama, *Pb.*, J. 78.

Sc Sco, *Hind.*, P. 1463.

Se Sé, *Raj.*, P. 1259 ; sea, *Pb.*, R. 543 ; seb,
Beng., *Hind.*, *Afg.*, P. 1463 ; sebist,
Afg., M. 334 ; sedwa, *Hind.*, I. 499 ;
segat, *Mar.*, M. 721 ; segora, *Raj.*, M.
717 ; segu, *Raj.*, M. 717 ; segum kati,
Mar., D. 402 ; segun, *Hind.*, T. 232 ;
según, *Beng.*, T. 232 ; ségvá, *Hind.*, M.
721 ; seh (Óxalis), *Hind.*, O. 547 ; seh
(Prosopis), *Pb.*, P. 1259 ; sehnd, *Mar.*,
E. 553 ; sehnr, *Hind.*, E. 553 ; sehora,
Hind., S. 2912 ; sehta, *Hind.*, C. 1444 ;
sehud, *Hind.*, D. 387, E. 553 ; sehund
Hind., E. 520, *Mar.*, E. 553 ; sein
(Ischæmum), *Hind.*, I. 499 ; sein (Ter-
minalia), *Hind.*, T. 361 ; sein (Vicia),
Pb., V. 108 ; seina, *Hind.*, L. 55 ; seind,
Hind., I. 499 ; seindí, *Hind.*, *Mar.*, P.
588 ; seju, *Uriya*, E. 553 ; sekto, *Mar.*,
M. 721 ; seláras, *Guz.*, L. 455 ; selavágh,
Hind., T. 437 ; sela vagh, *Hind.*, F.
768 ; selé, *Beng.*, F. 538 ; sél-gónd, *Hind.*,
B. 771 ; selliah, *Beng.*, F. 538 ; seloo,
Garhwal, B. 345 ; sem (Bauhinia), *Hind.*,
B. 357 ; sem (Canavalia), *Hind.*, *Pb.*, C.
289 ; sem (Dolichos), *Hind.*, D. 789 ;
sema, *Hind.*, P. 53 ; sémal, *Hind.*, B.
632 ; semálu, (Behar), *Mar.*, V. 164 ;
semar (Kombax), *Hind.*, B. 632 ; semar
(Cordia), *Mar.*, C. 1931 ; sem-ki gónd,
Hind., B. 308, 357 ; sémbi, *Hind.*, D.
789 ; semla, *Hind.*, B. 330 ; semla
gónd, *Hind.*, B. 357 ; semru, *Guz.*, P.
1259 ; semul, *Hind.*, *Mar.*, B. 632,
Beng., *Hind.*, S. 1939 ; semul-tula,
Beng., B. 629 ; semur, *Hind.*, B. 632,
S. 1939 ; sén, *Pb.*, P. 885 ; sená makhi,
Guj., C. 737 ; send (Apluda), *Hind.*, A.
1232 ; send (Euphorbia), *Dec.*, *Mar.*, E.
553 ; sendhí, *Hind.*, *Mar.*, *Pb.*, P. 588 ;
sendri (Bixa), *Mar.*, B. 523 ; sendri
(Phœnix), *Mar.*, P. 588 ; séndúr, *Dec.*,
L. 143 ; seni, *Hind.*, S. 1939 ; senjna,
Hind., *Pb.*, M. 721 ; sensar pál, *Pb.*,
A. 1565, 1573 ; senth, *Hind.*, A. 1097 ;
sentha, *Pb.*, S. 6 ; senu, *Pb.*, P. 1463 ;
seo, *Hind.*, P. 1463 ; seoli, *Hind.*,
Uriya, N. 179 ; seonikar, *Hind.*, T.
634 ; seotí, *Mar.*, C. 1043 ; sephalika,
Beng., N. 179 ; sepistar, *Mar.*, C. 1931 ;
seral, *Pb.*, S. 1656 ; serdi, *Guj.*, *Mar.*,
S. 30 ; serei, *Afg.*, C. 224, Q. 29 ; seri,
Hind., G. 247 ; seriss, *Hind.*, G. 705 ;
serissa, *Hind.*, G. 705 ; ser kuji, *Pb.*,
P. 1285 ; serow, *Hind.*, S. 1264 ; serowa,
Hind., S. 1264 ; serr, *Beng.*, P. 1170 ;
serrí, *Pb.*, C. 1101 ; sessal, *Mar.*, Z.
35 ; sesti, *Mar.*, M. 485 ; seta, *Pb.*, R.
152 ; set barúwa, *Hind.*, D. 115 ; setur,
Mar., M. 756 ; sev, *Hind.*, P. 1463 ;
sévalá, *Mar.*, D. 821 ; sevari, *Mar.*, S.
1174 ; seveta punarnabá, *Beng.*, B. 619 ;
sewa, *Guz.*, P. 727 ; sewan (Elionurus),
Pb., E. 192 ; sewan (Gmelina), *Hind.*, *Pb.*,
G. 287 ; sewar, *Sindhi*, R. 166 ; sewri,
Mar., S. 1174 ; seyára, *Mar.*, B. 318 ;
seyr, *Mar.*, E. 553 ; seyr-teg, *Mar.*, E.
553.

Sh Shá, *Pb.*, P. 327 ; shabbi, *Pushtu*, P. 419 ;
shada-búri, *Beng.*, S. 1018 ; shadevi,
Hind., *Sind*, C. 2211, 2221 ; sháfrí, *Pb.*,
S. 3079 ; sháítal, *Pb.*, T. 609 ; shaft-álu,
Pb., P. 1304, 1322, *Pushtu*, P. 1322 ;

shag, *Pb.*, B. 496 ; 501 ; shágali, *Pb.*, I. 123, 128 ; shahad *Dec.*, *Hind.*, H. 342 ; sháh-balút, *Afg.*, C. 224, Q. 29 ; shahd, *Pb.*, H. 342 ; shah-maksadi, *Hind.*, *Pb.*, C. 489, M. 277 ; shá htara, *Pushtu*, F. 723 ; sháhtút, *Mar.*, M. 747, *Hind.*, M. 756, *Pb.*, M. 772, 775 ; sháhzáda-rái, *Hind.*, B. 833 ; sháh-zírah, *Dec.*, *Hind.*, C. 697 ; sháili, *Pb.*, N 179; shajná, *Hind.*, M. 721 ; shak (Setaria), *Pb.*, S. 1212 ; shák (Betula), *Pb.*, B. 501 ; shakai, *Pushtu*, D. 196 ; shakákal, *Afg.*, P. 1074 ; shakákalmisrí, *Hind.*, P. 1074 ; shakákul, *Hind.*, A. 1575, 1577 ; shakar, *Hind.*, S. 373, *Pb.*, S. 374 ; shakar-al-lighal, *Pb.*, C. 191 ; shakardána, *Pb.*, C. 1711 ; shakar-kand, *Hind.*, *Pb.*, I. 348 ; shakar-kandu, *Mar.*, I. 348 ; shakar-pitan, *Hind.*, *Pb.*, E. 538; shakar surkh, *Pb.*, S. 30 ; shakar-tagnár, *Pb.*, M. 235 ; shakar-ul-úshar, *Pb.*, C. 191 ; shakei, *Pb.*, T. 416 ; sháktekas, *Pb.*, R. 359 ; shal (Nauclea), *Beng.*, N. 32 ; shal (Shorea), *Beng.*, S. 1656 ; shalá, (Ravi), *Pb* (C. 1958 ; shálakát, *Pb.*, M. 881 ; sha-langhi, *Pb.*, P. 546 ; shalanglí, *Pb.*, S. 2223; shalanglú (Litsæa, *Pb.*, L. 490); shalanglú (Machilus), *Pb.*, M. 22 shalangri, *Pb.*, D. 130 shalapara *Hind.*, H. 224 ; shalgham, *Beng.*, *Hind.*, B. 811 ; shalí (Oryza), *Pb.*, O. 258 ; shálí (Setaria), *Kashmir*, *Pb.*, S. 1212 ; shálian, *Pb.*, O. 258 ; shálídag gánch, *Pb.*, R. 593 ; shál-ké-pandú-ká jhár, *Dec.*, B. 523 ; shállú, *Mar.*, S. 2397 ; shalme, *Pushtu*, H. 6 ; shálpurní, *Pb.*, D. 339 ; shálu (Coriaria), *Pb.*, C. 1958 ; shálu (Sorghum), *Dec.*, S. 2405, *Mar.*, S. 2424 ; sháluk, *Beng.*, N. 200 ; shalwi, *Pb.*, H. 59 ; shama, *Beng.*, P. 45, *Dec.*, *Hind.*, P. 53 ; shamakh, *Dec.*, P. 63, *Hind.*, P. 45 ; shamálú, *Dec.*, V. 164 ; shambálí, *Dec.*, V. 164 ; sham-baloo kabij, *Hind.*, V. 179 ; shamdulun, *Beng.*, E. 80 ; shami, *Beng.*, *Mar.*, *Uriya*, P. 1259; shamieula, *Mar.*, E. 289; shámí-ka-bij, *Pushtu*, C. 1403 ; shamli, *Afg.*, T. 612 ; ahamor, *Pb.*, Z. 270, 280, shamri, *Mar.*, P. 1259 ; shamru, *Hind.*, D. 348 ; shamshabái, *Pushtu*, M. 461 ; shamshád (Buxus), *Pb.*, B. 985 ; shamshád (Myrsine), *Pb.*, M. 910 ; shamúke, *Pb.*, T. 87 ; shamukei, *Pb.*, T. 87 ; shamukha, *Pushtu*, P. 42 ; shamuki, *Pushtu*, V. 70 ; shamula, *Beng.*, *Dec.*, *Hind.* P., 53 ; shan, *Pb.*, H. 177 ; shanadér-jhar, *Beng.*, *Dec.*, *Hind.*, A. 1097 ; shanbálí, *Dec.*, V. 164 ; shanchi, *Beng.*, A. 914 ; shandái gúl, *Pushtu*, T. 845 ; shanda laghúne, *Afg.*, B. 985 ; shang, *Pushtu*, F. 700 ; shangal, *Pb.*, F. 700 ; shangala, *Pb.*, I. 14 ; shángar, *Pb.*, P. 1259 ; shangarf, *Hind.* ; M. 473 ; shang-cha, *Pb.*, P. 1202 ; shangraf, *Beng.*, *Hind.*, M. 473 ; shangti, *Pb.*, P. 746 ; shánjan, *Hind.*, O. 537 ; shan keshvara, *Mar.*, X. 1 ; shankha palita, *Mar.*, A. 1551 ; shankhávalli, *Mar.*, F. 581 ; shanku, *N. Pb.*, D. 815 ; shanshobai. *Pushtu*, P. 429; shaoul, *Kumaon*, B. 514 ; shápiang, *Pb.*, W. 93 ; shápránga, *Pushtu*, W. 93 ; shapri duden, *Pb.*, S. 3079 ; shaprochí, *Pb.*, S. 924 ; sha- pussundo, *Hind.*, I. 413 ; shaqáqul, *Hind.*

A. 1577 ; shaqáqule hindi, *Duk.*, A. 1562 ; shaquáqul-e-misri, *Duk.*, A. 1575 ; shar, *Beng.*, S. 6; sharáb, *Hin.*, *Pb.*, V. 243 ; sharáb-ki-kkari, *Duk.*, A. 249 ; sharapunkha, *Mar.*, T. 270 ; shargar, *Pb.*, R. 261 ; shargundei, *Pb.*, L. 283 ; shári, *Pb.*, P. 1285 ; sharífa, *Dec.*, *Guj.*, *Hind.*, *Mar.*, A. 1166 ; |sharmindi billi, *Hind.*, F. 789 ; sharoli, *Pb.*, C. 1988 ; shatávarí, *Guj.*, *Mar.*, A. 1577 ; shátrá, *Dec* , *Sind*, F. 721, 723; shaukarjata, *Guj.*, C. 711 ; shául, *Hind.*, B. 496 ; shauriya, *Kumaon*, E. 73 ; shaursi, *Kumaon*, E. 20 ; shavír, *Hind.*, M. 473 ; shavíram, *Hind.*, M. 473 ; shawa (Dalbergia), *Pushtu*, D. 64 ; sháwa (Populus), *Pb.*, P. 1148 ; shawáli, *Hind.*, V-164 ; sháwalí ríngyál, *Pb.*, R. 543 ; shawan, *Pushtu*, O. 145 ; shea, *Pb.*, L. 531 ; shedúri, *Pb.*, H. 59 ; shegal, *Pb.*, P. 1466 ; shegúl, *Pb.*, P. 1466 ; shegva, *Mar.*, M. 721 ; shej, *Hind*, L. 55 ; shekra, *Mahr.*, F. 802 ; shelim, *Dec.*, F. 237 ; shembal, *Hind.*, *Mar.*, B. 632 ; shembar, *Mar.*, A. 722 ; shembat, *Mar.*, O. 38 ; shemi, *Mar.*, P. 1259 ; shemolo, *Guj.*, B. 632 ; shemri, *Mar.*, P. 1259 ; shemú, *Mar.*, P. 1259; shendorwel, *Mar* , D. 522 ; shendri (Bixa), *Mar.*, B. 523 ; shendri (Mallotus), *Mar.*, M. 71 ; shení, *Pb.*, R. 564 ; sheo (wine), *Pb.*, V. 243 ; sheo (Pyrus), *Hind.*, P. 1463 ; sheodur, *Mar.*, D. 53 ; sheora, *Beng.*, S. 2912 ; heoris, *Pb.*, B. 496 ; sher (lion), *Hind.*, F. 764 , T. 428 ; sher (Pyrus), *Pb.*, P. 1463; sher (tiger), *Hind.*, F. 768, T. 437; shera, *Mar.*, E. 553 ; sheradi, *Gus.*, S. 30 ; sheras, *Mar.*, V. 159; sherasa (Vitex), *Mar.*, V. 159 ; sherasa (Brassica), *Mar.*, B. 812 ; shera-wane (Celastrus), *Pushtu*, C. 860 ; shera-wane (Flacourtia), *Pb.*, F. 624 ; shera-wane (Rhamnus), *Afg.*, R. 157 ; sher-betí, sherbetee, *Hind.*, C. 1300; sherdi, *Gus.*, S. 30, shere, *Pb.*, C. 1958 ; sher-i-darakht-i-khurma, *Pb.*, P. 555 ; sherni, *Hind.*, T. 437 ; sher núi, *Hind.*, M. 308 ; shetashirsa, *Hind.*, B. 817 ; shetur, *Gus.*, M. 756 ; sheuti gu'ab, *Beng.*, R. 501 ; shevaga, *Mar.*, M. 721 ; shevari (Bombax), *Mar.*, B. 632 ; shevar, (Sesbania), *Mar.*, S. 1174 ; 1186 ;'shevati, *Mar.*, C. 1047 ; sheveri, *Dec.*, S. 1174 ; shevgi, *Mar.*, M. 721 ; shewa (Dalbergia), *Pb.*, *Pushtu*, D. 64 ; shewa (Lonicera). *Pb.*, L. 531 ; shewa (Pyrus), *Pushtu*, P, 1463 ; shewa dáru, *Afg.*, N. 158 ; shewan. *Hind.*, *Mar.*, G. 287 ; shewar, *Pb* , E, 192 ; shewari, *Dec.*, *Mar.*, S. 1174 ; shewun, *Mar.*, G. 287 ; sh1, *Hyderabad*, E. 73 ; shia-*Pb.*, D. 64 ; shiah-kanta, *Beng.*, *Hind.*, M. 562 ; shiá-jirá, *Hind.*, C. 681 ; shiál . kántá, *Hind.*, A. 1351 ; shib-jal, *Mar.*, C. 551 ; shiblách makhán bed, *Pb.*, S. 924 ; shigaroti, *Gus.*, P. 397 ; shika. *Mar.*, A. 200 ; shikákái, *Mar.*, A. 200 ; shikari-mewa, *Pb.*, G. 702 ; shikarimai- wah, *Pushtu*, G. 663 ; shiláras, *Mar.*, A. 892 ; shili, *Pb.*, E. 273 ; shili1, *Pb.*, F. 700 ; shilling, *Kumaon*, O. 518 ; shilm, *Afg.*, P. 847 ; shim (Vigna), *Beng.*, V', 116 ; shím (Dolichos), *Beng.*, D. 789 ; shim ar, *Guj.*, B. 632 ; shimarra, *Hind.*, C. 756 ; shimat, *Mar.*, O. 38 ; shími,

Hind., D. 789 ; shimía batrají, *Beng.*, P. 1401 ; shimlo, *Guj.*, B. 632 ; shimti, *Mar.*, O. 38 ; shimul, *Guj.*, B. 632 ; shín, *Pb.*, D. 64 ; shindar (Pyrus), *Pb.*, P. 1466 ; shindar (Quercus), *Pb. ; Q.* 35 ; shindi, *Mar.*, P. 588 ; shindur, *Mar.*, M. 71 ; shing (Daphne), *Pb.*, D. 130 ; shíng (Jasminum), *Pb.*, J. 24 ; 29 ; shingádá, *Mar.*, T. 516 ; shingal, *Beng.*, T. 428 ; shingári (Rosa), *Pb. ; R.* 533 ; 543 ; shingári (Trapa), *Dec.*, T. 516 ; shingodá, *Guø.*, T. 516 ; shingr, *Beng.*, C. 2577 ; shíngtík, *Pb.*, L. 531 ; shinguti, *Mar.*, L. 292 ; shinh, *Sind*, T. 437 ; shinwala, *Pb.*, R. 261 ; shinwan, *Raj*, E. 192 ; shioli, *Uriya*, B. 342 ; shir (Ficus), *Pb.*, F. 129 ; shír (Machilus), *Pb.*, M. 22 ; shíra, *Hind.*, S. 372 ; shiral, *Mar.*, G. 682 ; shiralli, *Hind.*, N. 179 ; shiran *Pb.*, P. 1285 ; shiras, *Dec.*, *Mar.*, A. 711 ; shirásh, *Pb.*, C. 629 ; shirchin, *Pb.*, J. 101 ; shirian, *Pb.*, F. 725 ; shírín, *Pb.*, C. 1703 ; shir-khist (Fraxinus), *Dec.*, *Hind.*, F. 696 ; shirkhist (manna), *Beng.*, *Hind.*, *Pb.*,M. 235 ; sbir-khushk, *Hind.*, M. 235 ; shírlan, *Pb. ;* B. 632 ; shiro, *Hind.*, I. 51 ; shiroka, *Fb.*, H. 382 ; shirol, *Hind.*, C. 1988 ; shirola, *Mar.*, L. 556 ; shirrus, *Sind*, A. 695 ; shirsh, *Pb.*, A. 692 ; shirsha, *Pb.*, A. 722 ; shir thohar, *Mar.*, E. 553 ; shirúlí, *Pb.*, F. 168 ; shisao, *Mar.*, D. 41 ; shisar, *Mar.*, D. 41 ; shísh, *Dec.*, *Hind.*, L. 143 ; shíshái, *Pb.*, D. 64 ; shísham, *Hind.*, *Mar.*, *Pb.*, D. 41, *Hind.*, *Raj.*, *Pb.*, D. 64 ; shishi, *Pb.*, A. 692 ; shishona *Hind.*, U. 61 ; shishuna, *Kumaon*, G. 213 ; shissam, *Mar.*, D. 41 ; shísu, *Beng.*, D. 64 ; shitranj, *Kashmir*, P. 979 ; shitray, *Kashmir*, P. 979 ; shiuchana, *Hind.*, P. 530 ; shiu-lichhop, *Beng.*, L. 355 ; shiúlik, *Hind.*, E. 40 ; shiúntra, *Kumaon*, B. 929 ; shivan, *Mar.*, G. 287 ; shivardole, *Mar.*, M. 855 ; hiwajatá, *Guj.*, C. 711 ; shiwáli (Vitex), *Kumaon*, V. 164 ; shiwali (Callicarpa), *Kumaon*, C. 133 ; shiwari, *Mar.*, V. 164 ; shiwari, *Hind.*, V. 164 ; shívun, *Hind.*, G. 287 ; shko, *Pb.*, U. 4 ; shmalak, *Sind*, U. 471 ; shné, *Pb.*, *Pushtu*, P. 833 ; *Pb.*, P. 847 ; shobánjŷn, *Hind.*, C. 1403 ; shobri, *Pb.*, S. 738 ; shodar, *Pushtu*, P. 1148 ; shogul, *Pb.*, P. 1466 ; shokh, *Mar.*, A. 1362 ; shol, *Pb.*, O. 258 ; shola, *Beng.*, A. 560 ; holars, *Pb.*, P. 684 ; sholongákuspi, *Mar.*, C. 1403 ; shomfol, *Pb.*, R. 152 ; shon, *Beng.*, *Hind.*, C. 2105 ; shónámakhí, *Mar.*, C. 737 ; shón-pát, *Beng.*, C. 737 ; shoondul, *Beng.*, A. 580 ; shorá, *Pushtu*, S. 596 ; *Guj.*, *Hind.*, S. 682 ; shorag (Salsola), *Pushtu*, S. 596 ; shorag (Suæda), *Pushtu*, S. 2985 ; shorah, *Guj.*, *Hind.*, S. 682 ; shora kalmi, *Guj.*, *Hind.*, S. 682 ; shóra-mitha, *Mar.*, S. 682 ; shorí, *Beng.*, C. 2499 ; shoɪɪ lána, *Pb.*, H. 6 ; shorlana, *Pb.*, A. 1005 ; shotí, *Hind.*, *Pb.*, I. 434 ; shotul, *Pb.*, T. 609 ; shriphula, *Beng.*, A. 534 ; shrol, *Pb.*, A. 801 ; shrúl, *Pb.*, I. 425 ; shta, *Pb.*, M. 775 ; shudí, *Beng.*, L. 126 ; shuha, *Pb.*, J. 78 ; shuk, *Baluch.*, F. 430 ; shúkchína, *Beng.*, S. 2240 ; shúkpá, *Pb.*, J. 92 ; shukrí, *Benø.*, *Hind.*, G. 663 ; shulúpa,

Beng., P. 460 ; shumaj, *Pb.*, B. 985 ; shumanjra, *Pb.*, A. 328 ; shumeo, *Hind.*, *Kumaon*, V. 5 ; shumí, *Hind.*, P. 1259 ; shumshad, *Pushtu*, D 725 ; shun (Fraxinus), *Pb.*, F. 693 ; shún (Salix), *Pb.*, S. 560 ; shunam, *Duk.*, A. 779 ; shundri, *Beng.*, H. 134 ; shúngura, *Kumaon*, S. 1212 ; shunkhapushappi, *Guj.*, C. 382 ; shúpa, *Pb.*, J. 78 ; shuprak, *Hind.*, T. 376 ; shúr, *Pb.*, J. 92 ; shúrgu, *Pb.*, J. 92 ; shúria múkti, *Beng.*, H. 74 ; shurli, *Pb.*, C. 1988 ; shurmá, *Beng.*, A. 1224 ; shurshí, *Beng.*, B. 812 ; shwan, *Bluch.*,; O. 145 ; shwárí, *Pb.*, V. 164 ; shwet busunta, *Beng.*, A. 306 ; shwet chamni, *Hind.*, H. 149 ; shwet-gothúbi, *Beng.*, *Hind.*, K. 47 ; shwet-hulí, *Beng.*, Z. 188 ; shwet-keruee, *Beng.*, E. 549 ; shwet-púrna, *Beng.*, B. 619 ; shwet-rái, *Beng.*, B. 817 ; shwet-simul, *Beng.*, E. 289 ; shwet-sursha, *Beng.*, E. 327 ; shweta-garjan, *Beng.*, D. 676 ; shyakúl, *Beng.*, Z. 263 ; shyámá, *Beng.*, I. 1 ; shyin, *Pb* , S. 1242 ; shyona, *Hind.*, O. 233.

Si Sia, *Pb.*, R. 543 ; siagosh, *Hind.*, F. 762 ; siákul, *Beng.*, Z. 263 ; siáli (Dæmia), *Pb.*, D. 9 ; siálí (Pueraria), *Hind.*, *Pb.*, P. 1401 ; síal-kántá, *Beng.*, *Pb.*, A. 1351 ; siálú, *Kashmir*, *Pb.*, M. 289 ; siama-latá, *Hind.*, I. 1 ; síárú, *Fb.*, D. 196 ; sibjhúl, *Beng.*, C. 551 ; síchú, *Pb.*, C. 2021 ; sicka, *Hind.*, H. 440 ; sida, *Beng.*, *Hind.*, L. 55 ; sída atsú, *Pb.*, D. 144 ; siddhí, *Hind.*, C. 331 ; sidha, *Hind.*, L. 55 ; sidhera, *Pb.*, E. 479 ; sidhí, *Beng.*, *Dec.*, C. 331 ; sidodi, *Mar.*, H. 328 ; sidori, *Mar.*, H. 328 ; sígwan, *Hind.*, T. 232 ; sihar (Bauhinia), *Hind.*, B. 342 ; sihar (Rhazya), *Sind*, R. 166 ; siháru (Debregeasia), *Pb.*, D. 196 ; siháru (Nyctanthes), *Hind.*, N. 179 ; sihora, *Hind.*, S. 2912 ; sij, *Beng.*, D. 387, *Hind.*, E. 520, 527 ; sijra *Hind.*, T. 361 ; sika, *Mar.*, A. 200 ; sikanda, *Pb.*, R 543 ; sike-kái, *Dec.*, *Mar.*, A. 200 ; siki, *Pb* , E. 475 ; 479 ; sil (slate), *Hind.*, S. 2235 ; síl (Amarantus), *Pb.*, A. 916 ; síl (Celosia), *Pb.*, C. 868 ; síl (Imperata), *Hind.*, I. 51 ; siláí, *Hind.*, B. 771 ; silájít, *Beng.*, S. 2978 ; *Dec.*, S. 3023 ; silang, *Kumaon*, O. 518 ; silapoma, *Hind.*, E. 210 ; siláras (Altingia), *Dec.*, *Guj.*, *Hind.*, A. 892 ; silá ras (liquidambar), *Beng.*, *Hind.*, *Mar.*, L. 455 ; silarasa, *Mar*, L. 455 ; sila supári, *Kashmir*, Q. 35 ; sílein, *Pb.*, A. 801 ; silgai, *Hind.*, S. 1229 ; silgao, *Hind.*, S. 1229 ; silha, *Beng.*, L. 455 ; silkanti, *Hind.*, C. 1183 ; silkhari, *Pb.*, S. 2712 ; sil koroi, *Beng.*, A. 709 ; sillangti, *Kumaon*, S. 947 ; sillum, *Mar.*, F. 581 ; silond, *Beng.*, *Pb.*, *Uriya*, F. 581 ; silphora, *Hind.*, S. 924 ; sím (Cassia), *Hind.*, C. 756 ; sím (Dolichos), *Hind.*, D. 789 ; sím (Jasminum), *Pb.*, J. 24, 29 ; símák, *Pb.*, S. 1694 ; simáli, *Kumaon*, V. 164 ; simati, *Mar.*, *Sind*, O. 38 ; simbal, *Pb.*, B. 632 ; siming, *Kashmir*, T. 428 ; simli, *Hind.*, Z. 280 ; simlo, *Mar.*, B. 632 ; simlu, *Pb.*, B. 443 ; simrung, *Pb.*, R. 261 ; simul, *Beng.*, B. 632 ; sin (Elionurus), *Pb.*, E. 192 ; sin (Withania), *Pushtu*, W. 98 ; sína, *Mar.*, L. 55 ; sind, *Pb.*, D. 130 ; sindan harallú, *Hind.*, O. 38 ; sindelwan, *Mar.*, C. 2494 ;

sinderbur, *Mar.,* C. 2494; sinderwani, *Mar.,* C. 2494; sindhí, *Hind.,* Z. 201; sin dhuca, *Hind.,* V. 164; sindrol, *Pb.,* R. 152; sinduari, *Hind.,* V. 164; sinduri, *Beng.,* M. 71; sindúria, *Hind.,* M. 71; sindwar, *Hind.,* V. 164; sing, *Guj., Hind.,* H. 408 ;singala, *Pb.,* F. 486; singarota, *Mar.,* P. 397; singgi, *Beng.,* F. 568; singgin-janascha, *Hind.,* C. 1862; singh, *Hind.,* T. 428; singha, *Hind.,* F. 764; singhal, *Mar.,* T. 434; singhala, *Mar.,* F. 486; singhár (Nyctanthes), *Beng.,* N. 179; singhara (Castanopsis), *Beng.,* C. 818; singhará (Trapa), *Hind., Pb.,* T. 516; singhari, *Sind,* F. 486; singhí, *Beng.,* F. 568; singi (Euonymus), *Pb.,* E. 479; singi (Saccobranchus), *Hind., Uriya,* F. 568; singin janascha, *Hind.,* C. 1861; singli, *Hind.,* Z. 280; singodi, *Guz.,* T. 516; singraf, *Hind.,* L. 474; singrauf, *Hind.,* L. 483; singti, *Mar.,* F. 487; singuru, *Uriya,* T. 232; singyá, *Hind.,* A. 397; singyá-bis, *Hind.,* A. 397; sini, *Sind,* C. 2105; sinji, *Pb.,* M. 422; sínjili, *Hind.,* Z. 280; sinji (Cratægus), *Pb.,* C. 2035; sínjlí (Zizyphus), *Pb., Kashmir,* Z. 280; sinjubárá, *Pb.,* H. 177; sinkami, *Hind.,* C. 1183; sinri. *Hind.,* P.79; sinth, *Kashmir,* C. 931; sintra, *Hind.,* C. 1234; siotá, *Hind.,* S. 2912; sipar, *Hind.,* E. 263; siphal, *Hind.,* A. 534; sipi, *Hind.,* O. 616; sip'l, *Pb.,* B. 936; sír (Hedychium), *Mar.,* H. 59; sír (Imperata) *Hind.,* I. 51; sira (Albizzia), *Hind.,* A. 711; sira (Ischœmum). *Hind.,* I. 499; sira (Olea), *Pb.,* O. 160; sirai, *Hind.,* A. 695; siráli, *Hind.,* C. 868; siralu, *Hind.,* N. 179; siran, *Hind.,* A. 722; sirár, *Hind.,* A. 695; siras (Albizzia), *Hind.,* A. 695; *Dec., Mar.,* A. 711; siras (Dalbergia), *Hind.,* D. 41; sirasa, *Sind,* A. 695; sirda palíz, *Afg.,* C. 2263; sirgochi, *Hind., Pb.,* R. 355; sirgúja, *Beng.,* G. 735; sirgúza, *Beng.,* G. 735; sirai, *Beng.,* B. 875; siriári(Chenopodium), *Pb.,* C. 1003; siriári (Heliotropium), *Hind.,* H. 102; sirid, *Hind.,* H. 523; siri lasht, *Baluch.,* L. 301; sirín, *Hind.,* A. 695; *Pb.,* A. 692; 722; stíngrí, *Pb.,* P. 959; si iphal, *Hind.,* A. 534; siris *Beng., Dec , Hind., Mar.,* A. 695; 711; sirisha, *Beng.,* A. 695; sirka, *Hind.,* A. 356; sirki, *Pb.,* S. 6; sirkuchi, *Kumaon,* R. 355; sirma, *Pb.,* H. 277; sirmakar, *Pb.,* P. 45; sírola, *Mar.,* L. 556; sirpon, *Mar.,* C. 156; sirsa, *Hind.,* D. 41; sirú, *Pb.,* l. 51; sís, *Pb.,* C. 2101; sisa, *Beng., Hind.,* L. 143; sísa liús, *Pb.,* L. 312; sísam, *Guj., Hind.,* D. 64; sisan, *Sind,* C. 2077; sisha, *Pb.,* L. 143; sishuk, *Beng.,* W. 58; siske, *Pb.,* S. 763; sismuliá, *Guj.,* A. 1122; sissa, *Uriya,* D. 41; sissai (Crotalaria), *Pb.,* C. 2101; síssai (Dalbergia), *Hind.,* D. 64; sissu, *Guj., Hind., Mar.,* D. 41, *Hind., Mar.,* D. 64; sissua, *Uriya,* D. 41; sissúi, *Mar.,* D. 41; sisu (Dalbergia), *Mar.,* D. 41, *Hind., Sind, Uriya,* D. 64; sisu (lead), *Hind.,* L. 143; sisva, *Mar.,* D. 41; siswa, *Mar.,* D. 41; sitabér, *Hind.,* Z. 290; sitáfal, *Dec., Guj., Hind., Mar.,* A. 1166; sítáki-panjír, *Hind.,* A. 1130; sital-patí, *Beng.,* P. 625; sitalpátir gách, *Beng.,* P. 625; sitaphal, *Hind.,* C. 2325; sitiya, *Hind.,* P. 50; sit-rutí, *Hind.,* H. 59; sitsál, *Beng., Hind.,* D. 41; sittú, *Pb.,* C. 478; sitún, *Pb.,* C. 478; sivaen, *Hind.,* P. 45; siwan, *Raj.,* E. 192; siyah-bhura, *Hind.,* A. 106, siyah-dáru, *Afg.,* N. 158; siyah-gush, *Hind.,* T. 426; siyáh-muslí (Aneilema), *Hind.,* A. 1122; siyáh-músli (Curculigo), *Hind., Mar.,* C. 2375; siyáh-tút, *Kumaon,* M. 772; siyákhúl, *Hind.,* Z. 263; sizgai, *Pushtu,* D. 674.

Sk Skecho, *Pb.,* I. 425; skin, *Pb.,* S. 1242; skinung, *Pb.,* E. 241; skioch, *Pb.,* E. 479.

Sn Snrari, *Pb.,* H. 164.

So Soá (Hordeum), *Pb.,* H. 382; soá (Morus), *Pb.,* M. 775; soak, *Kashmir, Pb.,* P. 53; soanjna, *Hind., Pb.,* M. 721; sodhera, *Hind.,* B. 933; sohágá (Amoora), *Hind.,* A. 988; sohágá (borax), *Beng., Dec., Hind., Pb.,* B. 731; soh-li, *Beng.,* F. 570; sohund, *Kumaon,* E. 538; soi, *Kashmir,* P. 460; sojná, *Beng.,* M. 721; sokhta, *Pb.,* G. 186; sokutia, *Guj., Mar.,* L. 53; sol, *Beng.,* F. 517; sola (Æschynomene), *Beng.,* A. 560; sola (Ophiocephalus), *Uriya,* F. 517; sola (Plectranthus), *Pb.,* P. 959; solára, *Hind., Pb.,* A. 1093; solei, *Kashmir, Pb.,* P. 959; soltraj, *Hind.,* C. 2522; som (Ischæmum), *Hind.,* I. 494; soma (Sarcostemma), *Beng., Mar.,* S. 882; *Hind.,* S. 1207; somal, *Guj., Mar.,* A. 1425; somal-khár, *Guj., Mar.,* A. 1425; somaráji, *Hind.,* P. 4; somi, *Beng.,* P. 1259; somlatá, *Beng., Hind.,* S. 882; somní, *Pb.,* D. 409; somp, *Hind., Mar.,* P. 727; somr, *Hind., Mar.,* B. 632; sómráj (Pæderia), *Hind.,* P. 4; sómráj (Vernonia), *Beng., Hind., Uriya,* V. 73; som shing, *Pb.,* P. 737; somun, *Pb.,* J. 29; son, *Hind.,* C. 2160; sona (Bauhinia), *Hind.,* B. 308; sona (Oroxylum), *Beng.,* O. 233; soná (gold), *Beng., Dec., Guj., Hind., Mar.,* A. 1622; *Hind., Mar.,* G. 317; sonachá-pha, *Mar.,* M 517; sonachita, *Hind.,* T. 434; sonagáravi, *Mar.,* M. 781; sonái, *Pb.,* D. 130; sonajahi, *Kumaon,* J. 24; sona-kuta, *Hind.,* F. 759; sonali (Cassia), *Beng.,* C. 756; sonali (gold), *Beng.,* A. 1622; sona múga, *Beng.,* P. 496; sona-nu-varaq, *Guj.,* A. 1622; sonar-pát, *Beng.,* A. 1622; sonballi, *Hind., Sind,* C. 2211; sonchal, *Pb.,* M. 105; sonchala, *Hind.,* M. 109; sondarpadal, *Hind.,* S. 2884; sone, *Beng.,* V. 195; soné-cha-varuq, *Guj.,* A. 1622; soné-ká-tagat, *Duk.,* A. 1622; sóné-ká-varaq, *Hind.,* A. 1622; sonehrívaraq, *Hind.,* A. 1622; sonf (Ilicum), *Hind.,* I. 30; sonf (Pimpinella), *Dec., Mar.,* P. 727; songarbi, *Mar.,* V. 159; sonkhair, *Mar.,* A. 229; sonp, *Hind.,* F. 659; son-pát (Cassia), *Beng.,* C. 737; son-pát (Schweinfurthia), *Sind,* S. 964; sonsali, *Mar.,* E. 276; sont (Fœniculum), *Hind.,* F. 659; són (Zingiber), *Beng., Dec., Mar.,* Z. 201; sónth, *Hind., Pb.,* Z. 201; sopári, *Guj.,* A. 1294; sórá, *Beng.,* S. 682; sori, *Pb.,* O. 233; soringhi, *Uriya,* S. 1656; sóro-khar, *Guz.,* S. 682; sosan, *Hind., Pb.,* I. 434; sosun, *Hind.,* I. 423, 425; souf-ka-jur, *Beng.,* P. 727; soundar, *Dec.,* P. 1259; sounder, *Mar.,* P. 1259;

sounf, *Pb.*, P. 727 ; sous súsá, *Hind.*, W. 58 ; souta, *Hind.*, *Kumaon*, *Pb.*, V. 116 ; sowa, *Beng.*, *Hind.*, P. 460 ; soya, *Hind.*, *Pb.*, P. 460.

Sp Spail anai, *Pushtu*, P. 372 ; spalaghzai, *Pushtu*, F. 6 ; spalmak, *Pushtu*, C. 191, 225 ; spalmei, *Afg.*, C. 225 ; spalwakka, *Afg.*, C. 191 ; spand, *Afg.*, P. 372 ; spangjhá, *Pb.*, P. 1201 ; spanj, *Baluch.*, P. 372 ; spastu, *Pushtu*, M. 334 ; speda, *Afg.*, P. 1138 ; spelane, *Pb.*, C. 225, P. 372 ; spelda, *Afg.*, P. 1138 ; speraghunai, *Pushtu*, S. 2674 ; sperái, *Afg.*, L. 533 ; sperawan, *Pb.*, P. 1153 ; spera wuna, *Afg.*, B. 933 ; spercherei, *Pushtu*, C. 224 ; spercherei, *Pb.*, Q. 29 ; sperdor, *Afg.*, P. 1138 ; spet, *Pb.*, D. 245 ; spiág, *Pb.*, A. 1523 ; spíghwol, *Pb.*, P. 919 ; spikso, *Pb.*, A. 1523 ; spilecha, *Pb.*, *Pushtu*, P. 327 ; spinaj, *Afg.*, S. 2574 ; spín bajja, *Pb.*, *Pushtu*, W. 93 ; spinkhalak, *Pb.*, A. 1383 ; spín kharnár, *Pb.*, V. 64 ; spinwege, *Pb.*, A. 1383 ; spírke, *Pushtu*, A. 554 ; spiti, *Kumaon*, R. 152 ; spúdukeí, *Pb.*, L. 112 ; spulmei, *Pushtu*, C. 191 ; spun, *Pb.*, A. 22.

Sr Srighas, *Hind.*, T. 434 ; srikhanda, *Beng.*, S. 790.

St Starbú, *Pb.*, H. 277 ; starga, *Pb.*, J. 61 ; sthalkamal, *Hind.*, H. 224 ; sthalpadma, *Beng.*, H. 224.

Su Sua bháji, *Mar.*, C. 617 ; súah, *Guz.*, P. 460 ; suáli (Colebrookia), *Pb.*, C. 1711 ; suáli (Populus), *Pb.*, P. 1148 ; subali, *Hind.*, *Sind*, C. 2211, 2221 ; subdikaim, *Uriya*, N. 209 ; subjah, *Dec.*, O. 18 ; subze, *Dec.*, O. 18 ; suchal, *Pb.*, C. 1108 ; súchi, *Pb.*, A. 1614 ; sudáb (Euphorbia), *Pb.*, E. 505, 514 ; sudáb (Ruta), *Pb.*, R. 663 ; sudburg, *Hind.*, *Mar.*, R. 508 ; sudhari, *Hind.*, B. 933 ; sudrabibo, *Mar.*, H. 317 ; súf, *Sind*, P. 1463 ; sufir, *Pb.*, C. 637 ; sufurí-kumra, *Hind.*, C. 2323 ; sugan, *Hind.*, T. 93 ; sugandh, *Beng.*, T. 93 ; súgandh-bala, *Beng.*, P. 344 ; sugandha-bálá, *Hind.*, P. 344 ; sugandí-pálá. *Dec.*, H. 119 ; sugdásí, *Sind*, O. 258 ; sugmonia, *Hind.*, *Sind*, C. 1783 ; súh, *Kashmir*, T. 428, 434 ; suhágá, *Beng.*, B. 731 ; suha-rúk, *Hind.*, D. 438 ; súhigandhal, *Pb.*, C. 478 ; súj, *Hind.*, H. 382 ; sujjádo, *Sind*, H. 177 ; sujna, *Hind.*, *Mar.*, M. 721 ; sujuna, *Beng.*, M. 721 ; sújuniya, *Hind.*, P. 11 ; sukat sing, *Kumaon*, S. 874 ; sukdarshan, *Beng.*, C. 2068 ; súkét, *Guz.*, S. 790 ; sukhad, *Sind*, S. 790 ; súkh chein, *Kumaon*, *Pb.*, P. 1121 ; sukh-darsan (Amaryllis), *Pb.*, A. 954 ; sukh-darsan (Crinum), *Beng.*, C. 2075 ; súkhihari, *Pb.*, A. 401 ; sukhud, *Guz.*, S. 790 ; sukká-pát, *Duk.*, A. 1665 ; sukna, *Pb.*, A. 1538 ; súkurtothí, *Beng.*, R. 556 ; sulálí, *Pb.*, P. 1148 ; sulambra, *Pb.*, O. 38 ; sulea, *Beng.*, F. 538 ; suli, *Pb.*, *Him.*, E. 538 ; súlia (Pyrus), *Pb.*, P. 1456 ; suliah (Polynemus), *Beng.*, F. 538 ; súliya, *Kumaon*, P. 1456 ; súlpa, *Beng.*, P. 460 ; súlpha, *Beng.*, P. 460 ; sultána champa, *Beng.*, *Hind.*, C. 146 ; sum (Ehretia), *Pb.*, E. 20 ; súm (Fraxinus), *Pb.*, F. 685, 693 ; sumagh, *Afg.*, R. 287 ;

sumahk, *Afg.*, R. 287 ; sumak (Pistacia), *Pb.*, P. 833 ; sumák (Rhus), *Mar.*, R. 287 ; súmáli, *Pb.*, C. 133 ; sumbul, *Mar.*, N. 17 ; súmlú, *Pb.*, B. 443 ; summe, *Pb.*, S. 2876 ; summun, *Pb.*, J. 24 ; sumok, *Beng.*, R. 287 ; sumrí, *Mar*, P. 1259 ; sumsum, *Beng.*, S. 1678 ; sun (Briedelia), *Dec.*, B. 868 ; sun (Crotalaria), *Beng.*, *Hind.*, C. 2105 ; sún (Pyrus), *Pb.*, P. 1463 ; suna, *Guz.*, C. 2105 ; sunari, *Uriya*, C. 743, 756 ; súnd, *Kumaon*, S. 2910 ; sundálí, *Beng.*, C. 756 ; súndar, *Pb.*, C. 1012 ; sundaras, *Guj.*, C. 142 ; súndari, *Hind.*, H. 134 ; sundarí gnuá, *Uriya*, S. 2362 ; súnder (Connarus), *Mar.*, C. 1768 ; súnder (Heritiera), *Beng.*, H. 137 ; súndia, *Mar.*, S. 2390, 2397 ; *Guj.*, S. 2424 ; súndra, *Beng.*, H. 134 ; súndragundi, *Uriya*, M. 71 ; súndri, *Hind.*, H. 134 ; *Beng.*, *Mar.*, H. 137 ; sung, *Pb.*, G. 186 ; sungal, *Kashmir*, *Pb.*, T. 93 ; sung busri, *Dec.*, Z. 190 ; sung-misrie, *Beng.*, E. 467 ; sungotta, *Beng.*, C. 73 ; sungtara, *Pb.*, C. 1234 ; súngtú, *Pb.*, X. 1 ; suni, *Pb.*, J. 29 ; súnjna, *Raj.*, M. 717 ; sun kanwál, *Kumaon*, P. 546 ; sún káwal, *Hind.*, P. 546 ; sun-keint, *Pb.*, P. 1452 ; sunkeiut, *Pb.*, S. 1906 ; sunkerchor, *Beng.*, R. 127 ; sunní, *Pb.*, *Sind*, F. 411 ; sunnu (Fraxinus), *Pb.*, F. 693 ; súnnú (Prunus), *Pb.*, P. 1322 ; súnsar, *Sind*, W. 58 ; sunt, *Guz.*, *Mar.*, Z. 201 ; sunta, *Mar.*, Z. 201 ; suntara, *Mar.*, C. 1234 ; sunthura, *Hind.*, C. 1233 ; sunwar, *Hind.*, R. 166 ; supalú, *Pb.*, D. 245 ; supári (Areca), *Beng.*, *Dec.*, *Hind.*, *Mar.*, A. 1294 ; supári-ká-phul (Bombax), *Hind.*, B. 633 ; suphadie-khus, *Beng.*, S. 968 ; supida, *Pb.*, P. 1148 ; supra, *Pb.*, M. 105 ; supat, *Hind.*, F. 633 ; supyárí, *Dec.*, *Hind.*, A. 1294 ; sur (Eragrostis), *Pb.*, E. 259 ; súr (pig), *Hind.*, H. 289 ; súra, *Pb.*, H. 525 ; sura-gaz, *Pushtu*, T. 70 ; surah (Brassica), *Guj.*, B. 812 ; surah (Macrones), *Mar.*, F. 487 ; surain, *Kumaon*, A. 6 ; surajmaka, *Mar.*, H. 74 ; surajmaki, *Mar.*, H. 74 ; surajmúkhí, *Hind.*, H. 74 ; suraka, *Pb.*, A. 1606 ; súral, *Hind.*, *Pb.*, P. 1401 ; surált, *Hind.*, P. 1401 ; suran (Zizyphus), *Hind.*, Z. 273 ; súran (Amorphophallus), *Mar.*, A. 996 ; surangi (Morinda), *Mar.*, M. 656 ; surangru, *Pb.*, A. 692 ; suranjí (Calo-phyllum), *Mar.*, *Sind*, C. 146 ; surat (*Beng.*, L. 79 ; surata, *Hind.*, R. 631 ; suráti khára, *Pushtu*, C. 542 ; surb, *Pb.*, L. 143 ; surband, *Mar.*, A. 1469 ; surbuli, *Beng.*, *Uriya*, O. 137 ; súrch, *Pb.*, H. 281 ; surchi, *Pb.*, O. 547 ; surfan, *Dec.*, C. 146 ; surganch, *Kashmir*, *Pushtu*, R. 598 ; súrgi, *Kumaon*, J. 92 ; surgúja, *Hind.*, G. 735 ; suri, *Sind*, A. 695 ; suria, *Mar.*, X. 16 ; suriakhar, *Guj.*, *Hind.*, S. 682 ; suriala, *Pb.*, H. 164 ; súrí-chakka *Dec.*, V. 55 ; súrij-ká-jhar, *Dec.*, H. 74 ; suringán, *Beng.*, *Hind.*, *Mar.*, *Pb.*, C. 1703 ; suringi, *Mar.*, O. 6 ; surjamúki, *Beng.*, H. 74 ; surjmukha, *Hind.*, C. 1311 ; súrka-chúp, *Pb.*, R. 355 ; surkhi, *Hind.*, R. 619 ; súr-ki-charbi, *Hind.*, H. 289 ; surmá, surmah (antimony), *Beng.*, *Hind.*, A. 1224 ; surmá (lead), *Hind.*, L. 143 ; surmai, *Pb.*, S. 855 ; surmainlí,

Hind., I. 109; surmn-áo-pahro, *Guj.,*
A. 1224; surma safed, *Pb.,* G. 769;
surmo, *Guj.,* A. 1224; súrmoyí, *Hind.,*
C. 2528; suro, *Pb.,* E. 538; surpan,
Hind., C. 146; surpanka, *Dec.,* C. 146;
surpunka (Tephrosia), *Sind,* T. 270;
surpunka (Calophyllum), *Hind.,* C.
146; surs, *Pb.,* E. 538; suran, *Mar.,* Z.
273; sursha, *Beng.,* B. 812; sursi, *Beng.,*
Hind., B. 812; sursínjlí, *Pb.,* C. 2035;
surtári, *Hind.,* W. 106; suru, *Mar.,* C.
826; surva, *Guz.,* P. 460; súrwala,
Hind., C. 1053; surwalí, *Sind,* C. 868;
súryakánta, *Mar.,* H. 74; sús (dolphin),
*Hind.,*W. 58; sús (Glycyrrhiza), *Afg.,*
G. 278; súsni-shak, *Beng.,* M.306; súss
(Debregeasia), *Pb.,* D. 196; sússú, *Pb.,*
V. 91; sústa, *Uriya,* F. 48 2; súsú,
Hind., W. 58; súsúk, *Beng.,* W. 58;
sutei, *Pb.,* P. 1406; suteigul, *Pb.,* M.
314; sut-gilo, *Hind.,* *Sind,* T. 470, 483;
suthni, *Hind.,* D. 507; sútí, *Mar.,* Z.
290; sutopsha, *Hind.,* P. 460; sutri,
Hind., P. 486; sutrsowa, *Beng.,* M.
921; súts, *Pb.,* H. 277, 281; sutti,
Mar., H. 59; súwar, *Hind.,* H. 289.

Sv **Svarnajui,** *Beng.,* J. 24; svet-berela koreta,
Beng., S. 1688; svet berelá, *Beng.,* S.
1703.

Sw **Swadu,** *Mar.,* F. 615; swána, *Pb.,* D. 130;
swanján, *Pb.,* V. 164; swanjera, *Sind,*
M. 721; swarna, *Mar.,* G. 317; swar-
namukhi, *Mar.,* I. 478; swarsingh, *Mar.,*
H. 50; swet (Rosa), *Beng.,* R. 501; swét
(Trianthema), *Hind.,* T. 537; swétá
(Ipomæa), *Beng.,* I. 405; swet-ákond,
Beng., C. 170; swetaparájitá, *Beng.,*
C. 1403; swetapunarnava, *Mar.,* T. 537;
swétaturúlatá, *Beng.,* I. 405; swet-berela,
Hind., S. 1703; swet-kerua, *Beng.,* E.
549; swet-múrga, *Beng.,* C. 868.

Sy **Syámá dhán,** *Beng.,* P. 53 ; syamalatá,
Beng., I. 1; syansundari, *Hind.,* C.
2514; syáorá, *Beng.,* S. 2912; syala,
Hind., V. 14.

Ta **Taha,** *Beng.,* C. 1296; tabá-khir, *Pb.,*
C. 489; tabsi, *Hind.,* S. 2850; tachla,
Hind., A. 1235; tád, *Guj.,* B. 663; táda,
Mar., B. 663; tada-mirí, *Guj.,* P. 801;
tád-nu-jháda, *Guj.,* B. 663; tadrelú,
Pb., B. 165; tadrelú balel, *Kashmir,*
C. 1958; tadru, *Pb.,* R. 152, 159; tág,
Dec., C. 2105; tagar (Asarum) *Hind.,* A.
1545; tagar (Tabernæmontana), *Beng.,*
Guj., *Mar.,* T. 3; tagara, *Guj.,* *Mar.,*
A. 1545; tagarmul, *Beng.,* *Hind.,* L.
355; taggai, *Hind.,* T. 3; taggar
(Asarum), *Hind.,* A. 1545; taggar
(Tabernæmontana), *Hind.,* T. 3;
taggar (Valeriana), *Pb.,* V. 5, V. 10;
tágger, *Beng.,* *Hind.,* V. 5, tagger-
ganthoda, *Mar.,* V. 5; tagha, *Sind.,* C.
881, *Afg.,* C. 896; tagho, *Afg.,* C. 881,
Pb., C. 886; *Sind.,* C. 900; taghum,
Pb., C. 886; tág-san, *Sind.,* C. 2105;
taguna, *Kumaon,* D. 522; tagur (Dioscorea), *Kumaon,* D. 532; tagur (l'abernæ-
montana), *Beng.,* T. 3; tahir-hé, *Pb.,*
T. 434; táhlí safeda, *Pb.,* D. 64; tai
(Diospyros), *Mar.,* D. 569; táí (Oryza),
Pb., O. 258; tair (Carissa), *Beng.,* C.
596.; tair (Dioscorea), *Kumaon,* D. 532;

tairi, *Hind.,* *Beng.,* C. 35; taitimúl,
Uriya, S. 1174; taitu, *Mar.,* S. 2865;
taj, *Mar.,* C. 1196; taj, *Guj.,* C. 1183,
1196; táj-e-bádshah, *Hind.,* A. 1585;
táj-e-khurús, *Pushtu,* A. 939; taj kalam,
Hind., C. 1183; taj kalmi, *Hind.,* C.
1183; tajpat, *Hind.,* C. 1183; tají
khoros, *Pb.,* C. 873; tak, (Vitis), *Afg.,*
V. 233; ták (Salvadora), *Pb.,* S. 705;
takálá, *Mar.,* C. 797; tákápáná, *Beng.,*
Hind., P. 874; taker, *Pb.,* C. 431;
takhum, *Afg.;* C. 881; *Pb.,* C. 886;
takkri, *Pb.,* P. 77; takkum, *Afg.,* C. 900;
tákla (Cassia), *Mar.,* C. 797; takla
(Lonicera), *Pb.,* L. 535; takoli, *Hind.,*
D. 32; takmaki, *Mar.,* C. 2306; takoli,
Hind., *Mar.,* D. 32; takpa, *Pb.,* B.
501; takpun, *Pb.,* C. 886; takri, *Hind.,*
F. 672; tákrí, *Mar.,* F. 550; takri takria,
Hind., P. 77.; táksha, *Pb.,* O. 613;
táku, *Pb.,* C. 1003; takuli, *Kumaon,* D.
522; tál (Borassus), *Beng.,* *Hind.,* B.
663; tal (Sesamum), *Guz.,* *Mar.,* S.
1078; tála, *Hind.,* B. 663; tála mulí,
Beng., C. 2375; talasi, *Guj.,* O. 31;
talat-mád, *Mar.,* B. 663; talbaval, *Guj.,*
A. 217; talhang, *Pb.,* V. 96; tali
(Corypha), *Beng.,* *Mar.,* C. 1995; tali
(Dalbergia), *Sind,* D. 41; *Pb.,* *Sind,*
D. 64; tali (Dichopsis), *Beng.,* D. 387,
892; tálimakhána, *Mar.,* H. 508; tálim-
khana, *Mar.,* H. 508; taliori, *Pb.,* R.
278,; talísa, *Pb.,* R. 251; talísfar, *Pb.,*
R. 251; talisfur, *Hind.,* R. 278; tálispatr
(Taxus), *Mar.,* T. 93; tálíspatra (Abies),
Beng., *Hind.,* A. 25; tálispátra (Flacourtia), *Guz.,* F. 603; talispatri (Cinnamomum), *Hind.,* C. 1183; talíspatri
(Flacourtia), *Hind.,* F. 603; talisputar,
Hind., C. 1183; talísrí, *Pb.,* R. 251;
talkar, *Pb.,* C. 860; talkh, *Pb.,* C. 1703;
talla, *Pb.,* C. 2558; talle, *Pb.,* P. 1331;
tallier, *Beng.,* C. 1995; tallon, *Pb.,* P.
1148; tálmakhána, *Hind.,* H. 508;
tálmúlí, *Uriya,* C. 2375; talor dach,
Pb., V. 233; talsiari, *Pb.,* D. 196;
tálúng, *Pb.,* P. 1148; talúni, *Pushtu,*
R. 650; tama, *Beng.,* *Dec.,* *Hind.,*
C. 2361; tamák, *Beng.,* N. 101; tamáku, *Dec.,* *Guj.,* *Hind.,* *Mar.,* *Sind,*
N. 101; tamál, *Beng.,* *Hind.,* *Mar.,*
G. 66, *Beng.,* *Hind.,* G. 99; tamálá,
Mar., C. 1183; tamálpatra, *Guj.,* C.
1183; taman, *Mar.,* L. 42; tamana,
Mar., L. 42; tamanya, *Guz.,* V. 195;
tamáqu, *Dec.,* N. 101; tamara, *Mar.,*
P. 555; tamarak, *Guj.,* A. 1646; tam-
arhindí, *Pb.,* T. 28; tamari, *Mar.,* B.
632; tamáte, *Mar.,* L. 596; tamati,
Beng., *Hind.,* L. 596; tamba, *Mar.,*
L. 309; támbada chandana, *Mar.,* P.
1381; tambada-chitramúla, *Mar.,* P.
979; támbada-gand-háchachekká, *Mar.,*
P. 1381; támbada-khasa-khasá-cha-jháda,
Mar., P. 82; támbadámáth, *Mar.,* A.
938; támbáka, *Hind.,* N. 101; tam-
bakhu, *Mar.,* N. 101; támbáku, *Hind.,*
N. 101; tambat, *Mar.,* F. 611, F. 615;
tambat, *Mar.,* F. 603, *Beng.,* F. 615;
támbat, *Mar.,* F. 624; támbath, *Mar.,*
F. 603; tambhuda mírchingay, *Mar.,*
C. 455; tambol, *Hind.,* E. 25; tam-
boli, *Mar.,* E. 25; tambolli, *Beng.,*

E. 15 ; tambra (Catla), *Mar.,* F. 384 ;
tambra (garnet), *Pb.,* R. 619 ; támbra
(copper), *Mar.,* C. 2361 ; támbra-
nágkesar, *Mar.,* O. 6 ; tambridupári,
Mar., P. 393 ; tambul, *Beng.,* Z. 6 ;
támbuli, *Hind.,* P. 775 ; tamiya, *Pb.,*
C. 224 ; tampara, *Uriya,* F. 442,
443 ; támrug, *Guj.,* D. 615 ; tamruj,
Mar., E. 73 ; tamrulhindi, *Hind.,* T.
28 ; tan, *Dec.,* L. 451 ; tanach (Dalbergia), *Guj.,* D. 64 ; tanach (Ougéinia),
Mar., O. 57 ; tanaur, *Pb.,* V. 233 ;
tánbá, *Dec. ; Hind.,* C. 2361 ; tandái,
Pb., A. 692 ; tandala (Digera), *Pb.,*
D. 420 ; tandala (Ephedra), *Pb.,* E.
232 ; tandei, *Pb.,* V. 96 ; tandhári-send,
Guj., E. 496 ; tandra, *Pb.,* R. 159 ;
tandua, *Pb.,* S. 2672 ; tandula, *Mar.,*
O. 258 ; tandulai, *Pushtu,* P. 1078 ;
tandús, *Hind.,* C. 1227 ; tang, *Pb. ; S.*
1906 ; tang, *Kashmir, Pb.,* P. 1452 ; *Pb.,*
P. 1466 ; tángan, *Beng., Hind., Dec.,*
S. 1212 ; tanghing, *Kumaon,* A. 6 ; tángi,
Pb., P. 1452, 1466 ; tangla, *Pb.,* T. 17 ;
tangori, *Beng.,* G. 385 ; tangrol, *Pb.,* S.
1242 ; tangun, *Hind., Uriya,* S. 1212 ;
tánhári, *Pb.,* P. 833 ; táni, *Kashmir,* O.
258 ; tánkalá, *Mar.,* C. 797 ; tankan-
khár, *Guj.,* B. 731 ; tankír, *Hind.,* C.
2399 ; tánkli, *Mar.,* C. 797 ; tansala,
Pb., C. 616 ; tantarik (Pistacia), *Pb.,* P.
841 ; tantarik (Rhus), *Pb.,* R. 311 ; tantia
(Albizzia), *Hind.,* A. 695 ; tantia
(Dalbergia), *Hind., Kumaon,* D. 31 ;
tantun, *Mar.,* O. 233 ; tanuku, *Hind.,*
S. 2850 ; tapkoté, *Mar.,* U. 35 ; tappad-
dar, *Pb.,* E. 196 ; tappal búti, *Pb.,* C.
2221 ; tar (Borassus), *Dec., Hind.,* B.
663 ; tar (Dioscorea), *Pb.,* D. 503 ; tar
(Marsdenia), *Pb.,* M. 295 ; tar (Phœnix),
Sind, P. 555 ; tara (Corypha), *Beng.,* C.
1995 ; tara (Eruca), *Hind., Pb.,* E. 327 ;
tára (Prunus), *Pb.,* P. 1331 ; tarali,
Hind., Z. 182 ; taramira, *Hind.,* E.
327 ; tárá míra (Brassica), *Hind.* B.
841 ; taramira (Eruca), *Hind.,* E. 327 ;
tara mira (Raphanus), *Pb.,* R. 31 ; taran-
jabin, *Hind.,* A. 745 ; tarar, *Pb.,* D.
503 ; tarar puttr, *Pb.,* D. 496 ; taravada,
Mar., C. 741 ; tarbuch, *Guj.,* C. 1221 ;
tarbucha, *Gus.,* C. 2263 ; tarbud, *Hind.,*
I. 415 ; tarbuj, *Mar.,* C. 1221 ; tarbuja,
Mar., C. 2263 ; tarbuz, *Hind., Pb.,*
C. 1221 ; tarbuza, *Beng., Hind.,* C.
1221 ; tardí, *Pb.,* D. 503 ; tari, *Hind.,*
C. 962, C. 963 ; tári (Borassus), *Hind.,*
B. 663 ; tari (Phœnix), *Pb.,* P. 588 ;
traít, *Beng.,* C. 1995 ; tarízha, *Pb.,* L.
112 ; tar-ka-gur, *Hind.,* S. 370 ; tarkak-
di, *Mar.,* C. 2278 ; tarkhá, *Pb.,* A.
1469 ; tarkhana, *Pb.,* A. 328, 344 ;
tarmuj, *Beng.,* C. 1221 ; tarní, *Pb.,* S.
2376 ; taro, *Beng.,* A. 849 ; tarod,
Kumaon, L. 569 ; tarota, *Dec., Mar.,*
C. 797 ; taroi, *Hind.,* L. 556 ; tarota,
Mar., C. 741 ; tarrú, *Pb.,* H. 277 ;
tarsi, *Mar.,* C. 1050 ; tarsiphala, *Mar.,*
C. 1050 ; tartara, *Pb.,* D. 420 ;
tarui, *Beng.,* F. 500 ; taruko, *Beng.,*
Λ. 849 ; tarulatá, *Beng.,* I. 405 ; tarur,
Hind., Kumaon, D. 532 ; tarvar,
Hind., Duk., C. 741 ; tárwá, *Hind.,*
H. 277 ; tarwa-chuk, *Pb.,* H. 281 ; tar-

war, *Hind., Duk.,* C. 741 ; tasar,
Beng., S. 1934 ; tásar, *Pb.,* S. 763 ;
tasb, *Pb.,* F. 682 ; tashari siar,
Kumaon, D. 196 ; tashiári, *Kumaon,*
D. 196 ; tat, *Hind.,* S. 25 ; tatára, *Pb.,*
Hind., K. 17 ; tataur, *Pb.,* A. 1469 ;
tátichá-jháda, *Mar.,* B. 663 ; tatmo-
rang, *Pb.,* O. 233 ; tátpalang, *Pb.,* O.
233 ; tatrak, *Hind.,* R. 287 ; tatri, *Pb.,*
R. 318, 3231; tattunúa, *Hind.,* O. 233 ;
tattur, *Pb.,* D. 166 ; tatúa, *Pb.,* P.
1252 ; tatúra, *Pb.,* I. 39 ; tatwen, *Pb.,*
A. 1462 ; tau, *Pb.,* W. 106 ; taur, *Pb.,*
B. 318, 342 ; táura (Machilus), *Pb.,*
M. 22 ; taura (Pennisetum), *Pushtu,* P.
382 ; tavakhíra, *Mar.,* C. 2385 ; tavkil,
Mar., M. 267 ; tawai, *Pb.,* F. 678, F.
682 ; tawal, *Pb.,* A. 916 ; táwi, *Pb.,*
W. 106 ; tay-lí, *Sind,* F. 384 ; tazak-
tsun, *Kashmir,* R. 251.

Te Teak, (*C. P.*), *Hind.,* T. 232 ; tégu,
Mar., T. 232 ; tehr, *Pb.,* S. 1260 ; teila,
Pb., R. 355 ; tej, *Mar.,* E. 553 ; tejbal,
Hind., Pb., Z. 9 ; tejpát, *Beng.,* C.
1198 ; tejpát, *Beng., Dec.,* C. 1183 ;
tej-phal, *Hind.,* Z. 9 ; teka, *Hind.,* T.
232 ; tekári, *Beng.,* P. 682 ; tekáta sij
Beng., E. 496 ; tél, *Beng.,* S. 1078 ; telá,
kúchá, *Beng.,* C. 919 ; telara, *Beng.,* F.
443 ; telas, *Mar.,* O. 537 ; tel doaka,
Hind., L. 569 ; telinga-china, *Hind.,* S.
1939 ; telinga-chína, *Hind., Beng.,* L.
52 ; telingchi, *Pb.,* H. 525 ; teliya, *Garhwal), Hind.,* A. 22 ; télíyá-bis, *Hind.,*
A. 397 ; tekku, *Mar.,* T. 232 ; tellarrí,
Pb., F. 409 ; tellu, *Uriya,* I. 515 ; télni,
Hind., M. 864 ; télní-makkhi, *Hind.,*
M. 864 ; telphetru, *Mar.,* R. 16 ; telsur,
Beng., D. 834 ; telus, *Pb.,* O. 537 ; tem-
bhuran, *Mar.,* D. 582; tembhurni, *Mar.,*
D. 615 ; temburni, *Mar.,* D. 582, 628 ;
temru, *Hind., Mar.,* D. 569, 615,
628l; tendiya, *Hind.,* L. 55 ; tendli, *Mar.,*
C. 919 ; tendu (Citrullus), *Hind.,* C.
1227 ; tendu (Cucurbita), *Hind.,* C. 2329 ;
tendu (Diospyros), *Beng., Hind.,*
Mar., D. 569, 582, 615, 628 ; tendua,
Hind., F. 767 ; tenduli, *Mar.,* C. 919 ;
tendús, *Hind.,* C. 2329 ; tendwa, *Hind.,*
T. 434 ; teng, *Beng.,* S. 4 ; tengara,
Beng., F. 492 ; tenginmar, *Mar.,* C.
1520 ; tengrah, *Beng.,* F. 492 ; tensi,
Hind., C. 1227 ; tenti, *Pb.,* C. 402 ; tentí,
delpha, *Pb.,* C. 402 ; téntúl, *Beng.,* T.
28 ; tentúli, *Uriya,* T. 28 ; teo, *Mar.,* J.
116 ; teora, *Beng.,* L. 100l; teori, *Beng.,*
I. 415 ; tepáriyo, *Hind.,* P. 682 ; tephal,
Mar., Z. 35 ; tepíriya, *Beng.,* P. 682 ;
tepuriá, *Beng.,* P. 682 ; ter, *Sind.,* S.
1233 ; ter, *Pb.,* P. 1331 ; teradá, *Mar.,* I.
39 ; terem, *Mar.,* C. 1732 ; teringole, *Pb.,*
S. 1242 ; teriya-bhanggan, *Beng.,* F. 540 ;
tessul, *Mar.,* Z. 35 ; tesú-ká-jhar, *Duk.,*
B. 944 ; tésú-ká-pér, *Hind.,* B. 944 ; tétai,
Beng., T. 28 ; tetar, *Pb.,* R. 316, 318 ;
teter, *Hind.,* T. 28 ; tetta manga, *Hind.,*
G. 128 ; tetu, *Guj., Mar.,* O. 233 ; tewas,
Mar., O. 537 ; teyrúr, *Hind.,* C. 1311 ;
tezak, *Pb.,* L. 283 ; tezbal, *Pb.,* Z. 9 ;
tezma, *Pb.,* I. 430 ; tezmal, *Hind.,* Z. 9 ;
tezpat, *Beng.,* C. 1165, 1183.

Th Thab (Erythrina), *Pb.,* E. 356 ; thab (Hymen
odictyon), *Pb.,* H. 517 ; tháí, *Pb., Kash*

mir, W. 106; tháil, *Pb.*, S. 560; thaila-
ankúl, *Hind.*, A. 681; thakil, *Hind.*, P.
588; thakola, *Pb.*, X. 21; thakra, *Beng.*,
Hind., O. 574; thal-kesur, *Beng.*, S.
2943; thalma, *Hind.*, P. 588; thalot, *Pb.*,
R. 152; thalun, *Pb.*, V. 104; tháman,
Pb., G. 688; thammal, *Hind.*, C. 1966;
tham-ther, *Pb.*, C. 723; thana (Debregeasia), *Pb.*, D. 196; thana (Feronia),
Mar., F. 53; thanárí, *Hind.*, *Pb.*, S.
2678; thanella, *Hind*, G. 136; thaner,
Kumaon, T. 93; thangi, *Kashmir*, *Pb.*,
C. 1988; thanka, *Pb.*, J. 61; thán ki, *Pb.*,
P. 1460; thankoli, *Kashmir*, *Pb.*, C. 1988;
thánmori, *Mar.*, P. 678; thansa, *Pb.*, P.
760; thanthán, *Pb.*, J. 61; thánzatt, *Pb.*,
H. 382; thapur, *Hind.*, *Pb.*, F. 230; thar
(Sind Ibex), *Pb.*, S. 1233; thar (Euphorbia), *Pb.*, E. 538; thara, *Uriya*, T. 293;
tharbal, *Hind.*, C. 1966; tharrí, *Pb.*, D.
503; thartahrni, *Pb.*, S. 1260; tharwar,
Hind., *Pb.*, C. 1966; thaur (Bauhinia),
Hind., B. 318; thaur (Ficus), *Pb.*, F.
168; thawa, *Kumaon*, W. 106; thawi,
Hind., W. 106; tháwí, *Kashmir*, W. 106;
thegi, *Mar.*, C. 2605; thelain, *Pb.*, V. 90;
thelu, *Hind.*, J. 104; thélu, *Pb.*, J. 78;
theluphulu, *Pb.*, J. 104; themanna bed
khist, *Pb.*, S. 577; thembi, *Mar.*, A. 267;
theot, *Pb.*, I. 123, I. 128; therna, *Pb.*, E.
190; thesi, *Hind.*, *Pb.*, C. 1966; thikirí,
Hind., P. 513; thikrí-ká-jhár, *Duk.*, B.
619; thilák, *Pb.*, W. 86; thili, *Pb.*, S. 6;
thilkain, *Pb.*, V. 96; 104; thirr, *Sind*,
S. 1078; thissa, *Pb.*, R. 318; thohar
(Cajanus), *Hind.*, C. 49; thohar (Euphorbia), *Hind.*, E. 520; thohur, *Sind*, E.
520; thol-kuri, *Beng.*, H. 486; thona, *Pb.*,
T. 93; thont wa, *Hind.*, S. 1207; thor
(Cajanus), *Hind.*, C. 49; thor, *Guj.*,
Mar., *Pb.*, E. 520, 538, 553; thora, *Mar.*,
Sind, E. 553; thóralimbu, *Mar.*, C. 1286;
thora-pimpli, *Mar.*, S. 970; thordandalio,
Guz., E. 553; thorinjal, *Sind*, S. 882;
thorla-champa, *Mar.*, M. 489; thorligunj,
Mar., A. 471; thorli, indráyan, *Mar.*, C.
1211; thulpadma, *Beng.*, H. 224; thum
(Fraxinus), *Pb.*, F. 700; thúm (Sageretia),
Pb., S. 505; thun, *Pb.*, P. 57; thúner,
Hind., *Kumaon*, T. 93; thungtu, *Hind.*,
Kumaon, V. 147; thúno, *Hind.*, T. 93;
thúnu, *Kashmir*, *Pb.*, T. 93; thur, *Hind.*,
C. 49; thur-dúribaji, *Dec.*, L. 309; thurí
(Dioscorea), *Pb.*, D. 503; thuri (Rhyn-
chobdella), *Uriya*, F. 561; thurwágh,
Kashmir, T. 439; thus, *Hind.*, B. 751;
thut, *Pb.*, S. 738; thuthera, (*C. P.*),
Hind., S. 2424.

Ti Tiamb, *Pb.*, F. 260; tian, *Pb.*, A. 344;
tianlandhá, *Pb.*, V. 96; tiári, *Pb.*, S.2341;
tickar, *Mar.*, C. 2385; tickhur, *Hind.*,
M. 267; tidhára-sehnd, *Hind.*, E. 496;
tidhári-send, *Deccan*, E. 496; tidj-pat,
Hind., C. 1189; tidua, *Hind.*, T. 434;
tíhya gurjun, *Beng.*, D. 701; tigri, *Hind.*,
C. 1055; tij, *Hind.*, C. 1189; tijabal,
Mar., Z. 35; tik, *Sind*, E. 7; tílıaní,
Dec., F. 393; tík-chana, *Beng.*, L. 110;
tikhi, *Mar.*, C. 1158, C. 1196; tikhor,
Hind., M. 267; tikhur, *Hind.*, C. 2385;
tiki-okra, *Beng.*, M. 429; tíkkar, *Hind.*,
B. 858; tikkoe, *Hind.*, A. 514; tikor,

Beng., C. 2431; tikri, *Pb.*, C. 1205;
tiktaláu, *Beng.*, L. 30; tikta-raj, *Beng.*,
A. 988; tikto-shak, *Beng.*, C. 2039;
tikúl *Beng.*, G. 82; tikúr, *Ben.?*, G.
82; tikuri, *Dec.*, I. 415; til (Saccharum),
Pb., S. 6; til (Saxifraga), *Pb.*, S. 924;
tíl (Sesamum), *Afg.*, *Beng.*, *Dec.*, *Guz.*,
Mar., *Pb.*, *Sind*, S. 1078; til, *Hind.*, S.
1078; tíla, *Mar.*, S. 1078; tíla, birsa,
Hind., W. 33; tilak, *Pb.*, S. 6; tilanga,
Hind., Q. 13; tilaparni, *Beng.*, P. 1381;
tiláts, *Pb.*, V. 96; tilávana, *Mar.*, G. 753;
tilbora, *Hind.*, L. 490; tilchang, *Pb.*, S.
574; tilchuni, *Hind.*, M. 115; tíle, *Beng.*,
F. 506; tilhanj, *Pb.*, V. 96; tili, *Hind.*,
Pb., S. 1078; tilia cachang, *Kashmir*, *Pb.*,
A. 413; tiliakora, *Beng.*, T. 456; tiliakoru,
Beng., T. 456; tiliya, *Hind.*, P. 41; til-
ká-tél, *Hind.*, S. 1078; tilki, tilai, *Hind.*,
W. 33; tilla, *Pb.*, C. 2558; tilli, *Mar.*,
W. 98; tilli, *Uriya*, W. 38; tilluk, *Hind.*,
S. 25; tilmí, *Beng.*, S. 1078; tilon, *Pb.*,
S. 6; tilonj, *Hind*, *Kumaon*, Q. 13;
tilora, *Sind*, A. 1577; tilpattar (Acer),
Pb., A. 328, 344; tilpattra (Marlea), *Pb.*,
M. 289; tílphár, *Pb.*, I. 39; til-pungti,
Beng. F. 355; tilún, *Pb.*, S. 2376; tilwan,
Guj., C. 1367; tilya, *Pb.*, S. 2397; timal,
Hind., F. 260; tímra rákh, *Pb.*, E. 208;
timbal, *Pb.*, F. 260; imbaroa, *Guz.*, D.
628; timbarran, *Guj.*, C. 596; timberni,
Mar., D. 615; timbírí, *Mar.*, D. 582;
timl orí, *Mar.*, D. 582; timburni, *Guj.*,
Mar., D. 615; *Mar.*, D. 628; timbúr
timúr, *Hind.*, Z. 9; timbwini, *Mar.*, D.
582; timla, *Hind.*, F. 260; timmal, *Pb.*,
Z. 9; timmer, *Sind*, A. 1660; tímoti,
Beng., *Hind.*, L. 596; timra, *Guz.*, D.
628; timrú (Zanthoxylum), *Pb.*, Z. 9;
timur (Diospyros), *Guj.*, *Mar.*, D. 628;
timsa (Ougeinia), *Hind.*, *Mar.*, O. 537;
timsha (Quercus), *Hind.*, Q. 13; timukhia
Hind., C. 596; timukhia, *Hind.*, C. 596;
timúr, *Hind.*, Z. 9; tin, *Hind.*, A. 1097;
tinani, *Afg.*, A. 1588; tinda, *Pb.*, *Sind*,
C. 1227; tind albinda, *Hind.*, C. 1227;
tindhára sehund, *Hind.*, E. 496; tin-
dhári-send, *Deccan*, E. 496; tindú, *Pb.*,
C. 2329; tindú (Diospyros), *Pb.*, D. 656;
tindu Heliotropium), *Pb.*, H. 106; ting-
ga-rah, *Pb.*, F. 492; tingí, *Pb.*, S. 2264;
tinia, *Uriya*, A. 695, 717; tinis, *Beng.*, O.
537; tinkál, *Hind.*, *Pb.*, B. 731; tinkár,
Pb., B. 731; tinmani, *Guj.*, C. 1367;
tinnas, *Hind.*, O. 537; tinni (Hygroryza),
Hind., H. 513; tinní (Oryza), *Hind.*,
O. 258; tinpáni, *Mar.*, N. 23; tinsa
(Ougeinia), *Hind.*, O. 537; tinsah (Symplocos), *Hind.*, S. 3062; tintil, *Beng.*, T.
28; tin-til, *Baluch.*, F. 406; tintírí, *Beng.*,
T. 28; tintúri, *Beng.*, T. 28; tipakia,
Hind., E. 166; tipári, *Hind.*, P. 682; ti-
páriya, *Hind.*, P. 682; tir, *Hind.*, S.
1078; tira (Eruca), *Hind.*, E. 327; tírá
(Brassica), *Hind.*, B. 841; tiradá, *Mar.*,
I. 39; tircorai-kalai, *Beng.*, P. 513; tírkú,
Ph., H. 277; tirmí, *Pb.*, F. 260; tírnal,
Pb., F. 260; tirní, *Pb.*, *Sind*, C. 137;
tirphal, *Mar.*, Z. 35; tirra, *Hind.*, P.
1401; tirunitru, *Dec.*, O. 18; tisal, *Mar.*,
Z. 35; tísi, *Beng.*, *Hind.*, *Kumaon*, *Pb.*,
L. 385; tíso, *Pb.*, C. 559; tít, *Pb.*, *Sin i*,

C. 402 ; títa (Corchorus), *Uriya*, C. 1847;
tita (Gen-tiana), *Pb.*, G. 172; títa baterí,
Kashmir, L. 535 ; títá-indarjou, *Beng.*,
H. 294 ; titakhana, *Hind.*, E. 589 ; tita-
kunga, *Beng.*, D. 823 ; tita láu, *Hind.*,
L. 30 ; titaliya, *Hind.*, S. 2357; títápát,
Beng., C. 1840 ; títar, *Pb.*, R. 323 ; títari,
Pb., R. 316 ; titávalí, *Mar.*, D. 89; tithu,
Pb., P. 693 ; tit-kunga, *Beng.*, D. 823 ;
titni, *Hind.*, Z. 280 ; títo-dhundul, *Beng.*,
L. 563 ; títo-jhingá, *Beng.*, L. 563 ; títo-
torai, *Beng.*, L. 563 ; tit-patti, *Kumáon*,
R. 561 ; títrí, *Pb.*, R. 318, 323 ; tittri, *Pb.*,
R. 293 ; tiún, *Pb*, A. 1511 ; tiun dheu,
Pb., A. 1511 ; tiura, *Hind* ., L. 100;
tiúrú, *Pb.*, R. 564 ; tivar, *Mar.*, *Sind*, A.
1655 ; A. 1660 ; tivas, *Mar.*, D. 41 ; tiwar,
Mar., B. 180 ; tiwas, *Mar.*, O. 537; tízhú,
Pb., C. 1101.

To To, *Pb.*, T. 634 ; toandí, *Hind.*, T. 293 ;
tochém-keweh, *Duk.*, C. 1641 ; todri líla,
Pb., M. 320 ; todri nafarmání, *Pb.*, C. 997;
todri safed, *Pb.*, *Sind*, M. 320 ; todri
surkh, *Hind.*, C. 997 ; todri surukh, *Pb.*,
C. 997 ; tohar, *Pb.*, C. 49; toli, *Raj.*, I.
503 ; tomila, *Kumaon*, F. 202 ; tondali,
Mar., C. 919; tongus, *Hind.*, M. 299 ;
tookm kudú, *Pb.*, C. 2316 ; tophli, *Beng.*,
P. 682 ; topia, *Mar.*, D. 53 ; toposwi,
Hind., F. 539; tor (Cajanus), *Hind.*, C.
49 ; tor (Euphorbia), *Raj.*, E. 527 ; tora
bajja, *Pushtu*, A. 484 ; toradana, *Pushtu*,
D. 151 ; torai, *Hind.*, L. 556 ; toran, *Mar.*,
Z. 273 ; torban, *Pb.*, V. 164 ; tórbanna,
Pb., V. 164 ; tordanda, *Pb.*, E. 538 ; tori,
Hind., B. 822 ; torié, *Kumaon*, L. 556;
tóriya, *Hind.*, B. 822 ; torjaga, *Pushtu*,
A. 567 ; torki, *Hind.*, *Mar.*, *Pb.*, I. 134 ;
torne, *Mar.*, Z. 273 ; tos, *Pb.*, A. 17, A.
22 ; totmila, *Hind.*, F. 202.

Tr Trambashirin, *Kashmir*, F. 10 ; trambú,
Guz., C. 2361 ; tras, *Guj.*, C. 1211 ;
trawuke, *Pb.*, O. 547 ; trayamán, *Mar.*,
D. 271 ; trekhan, *Pb.*, A. 328, 344; tremal,
Pb., F. 260 ; trepatra, *Pb.*, T. 607 ;
triangúli, *Hind.*, P. 523 ; trikanna, *Pb.*, A.
344 ; trikanta ga tí, *Beng.*, A. 1665 ;
tríkhgandere, *Pu htu*, R. 253 ; tríkh-
shawan, *Pushtu*, O. 145 ; trikunda, *Pb.*,
A. 344 ; trikundrí, *Sind*, T. 544, T. 548 ;
trímal, *Pb.*, F. 260 ; trimbal, *Pb.*, F. 216,
260 ; trimmal, *Hind.*, F. 260 ; trind, *Pb.*,
R. 533 ; trindus, *Sind*, C. 1227 ; trinpali,
Hind., M. 232 ; tri-pakshí, *Mar.*, C. 1707 ;
tripattra, *Pb.*, M. 306 ; triphal, *Mar.*, Z.
35 ; trip ungkí, *Hind.*, C. 1707 ; tripungti,
Beng., F. 354 ; tripunkhí, *Hind.*, C. 1707;
triwa kka, *Pb.*, R. 650 ; tro, *Pb.*, H. 382 ;
tror, *Pb.*, P. 1106 ; trotak, *Pb.*, E. 241 ;
t₁ otu, *Pb.*, D. 9 ; trúal, *Pb.*, I. 39 ; tru-
j o-gosht, *Sind*, C. 1211 ; tru-jo-par, *Sind*,
C. 1211 ; trumbal, *Pb.*, F. 156; trúná deda,
Guj., C. 1211.

Ts Tsábrí, *Pb.*, F. 19 ; tsalé, *Pb.*, B. 731 ;
tsalémentog, *Pb.*, B. 731 ; tsápú, *Pb.*, A.
801 ; tsar, *Pb.*, G. 186 ; tsarap, *Pb.*, H.
277 ; tsarbs, *Pb.*, P. 959 ; tsarmang, *Pb.*,
H. 277 ; tsarma nfechak, *Pb.*, H. 277 ;
tsátín, *Pb.*, A. 1061 ; tselain, *Pb.*, V. 96 ;
tser, *Kashmir*, *Pb.*, P. 737 ; tserdkar, *Pb.*,
H. 281 ; tserkar, *Pb.*, H. 277 ; tsjovanna
amelpodi, *Mar.*, R. 57 ; tsonu, *Pb.*, J. 24 ;

tsui, *Pb.*, *Him.*, E. 538 ; tsúing, *Kashmir*,
C. 487 ; tsuk, *Pb.*, H. 277 ; tsuna,
Kashmir, C 487 ; tsúnt, *Pb.*, P. 1463 ;
tsúnú, *Pb.*, P. 1322 ; tsúr, *Kashmir*, X. 1 ;
tswak, *Pb.*, H. 277.

Tu Tuatuka, *Mar.*, S. 2865 ; tudje, *Hind.*, C.
1189; tue, *Hind* , F. 179 ; túghar, *Pushtu*,
C. 886 ; tugur, *Beng.*, T. 3 ; túin, *Pb.*, V.
96 ; túk, *Pb.*, F. 396 ; tukati, *Mar.*, S.
1688 ; tukhm-bálangú, *Pb.*, L. 67 ; túkhm-
ferunjmishk, *Hind.*, D. 817 ; tukhm-i-
bálangú, *Mar.*, L. 67 ; tukhm-i-bhang,
Hind., H. 177 ; tukhm-kúlpha, *Beng.*, P.
1179; tukhm-lealanga, *Hind.*, L. 67 ;
tukhm-malanga, *Pb.*, L. 67 ; tukhm
malanga, *Pb.*, S. 733 ; tukkhm túmma, *Pb.*,
C. 1211 ; túkm-i-gandna, *Kashmir*, N. 158;
tukmi-i-bádanjánrumi, *Kashmir*, S. 2284 ;
tukm-i-khitmi, *Mar.*, A. 8S0 ; túl, *Hind.*,
M. 747, *Pb.*, M. 756 ; tula, *Beng.*, G. 404 ;
tula ambor, *Mar.*, M. 756 ; túla-lodh,
Beng., W. 38 ; túlanch, *Kashmir*, R. 598;
tulas, *Mar.*, O. 31 ; tulasa, *Mar.*, O. 31 ;
tulas-icha-jháda, *Mar.*, O. 31 ; túlati-pati,
Hind., P. 678 ; tulatule, *Mar.*, H. 328 ;
tulda, *Beng.*, B. 142 ; tulenni, *Pb.*, H. 13 ;
tulidun, *Beng.*, S. 2299; tulklu, *Hind.*, M.
747, *Pb.*, M. 775 ; tulshi, *Beng.*, O. 31 ;
tulsi, *Pb.*, O. 18, *Mar.*, O. 28; tulsi, *Beng.*,
Dec., *Hind.*, *Pb.*, O. 31 ; tultuli, *Mar.*, H.
328 ; túlúkúl, *Pb.*, M. 775 ; tum, *Hind.*,
G. 143 ; tumadá, *Guz.*, L. 30 ; tumák,
Hind., N. 101 ; tumal, *Hind.*, D. 656;
túmari (Castanopsis), *Kumaon*, C. 818 ;
tumari (Premna), *Hind.*, P. 1242 ; tumba, *Pb.*, L. 30 ; túmbhúl, *Hind.*, H.
543 ; túmbi (Citrullus), *Pb.*, C. 1211 ;
túmbi (Rosa), *Pb.*, R. 533 ; túinbri,
Hind., M. 289 ; tumburní, *Mar.*, D.
615 ; túm kharpat, *Beng.*, G. 143 ;
túmma, *Pb.*, V. 91 ; túmra (Rhus), *Hind.*,
Pb., R. 311 ; tumra (Zanthoxylum),
Hind., Z. 6; tumri (Diospyros), *Mar*.,
D. 615; tumri (Lagenaria), *Hind.*, *Ku-
maon*, L. 30; tumri (Marlea), *Kumaon*,
M. 289 ; tumri (Trewia), *Hind.*, *Kumaon*,
Mar., T. 525 ; tumru, *Hind.*, Z. 9;
tún (Cedrela), *Beng.*, *Hind.*, *Mar.*, *Pb.*,
C. 838 ; tun (Zanthoxylum), *Beng.*,
Hind., Z. 9 ; túna, *Hind.*, C. 838 ; tunani
zanani, *Pb.*, R. 159 ; tunánizenáni, *Pb.*,
V. 96; tundapará, *Uriya*, I. 489 ; túndhe,
Pb., V. 96 ; tundhi, *Pb.*, R. 159 ; túndu,
Mar., C. 838 ; tundupara, *Uriya*, T. 489 ;
túng (Abies), *Kashmir*, A. 22 ; túng
(Mallotus), *Beng.*, M. 71 ; túng (Rhus),
Hind., *Pb.*, R. 293, 311 ; túng (Taxus),
Kashmir, *Pb.*, T. 93 ; túnga. *Hind.*, R.
293, 311 ; túngla, *Hind.*, *Pb.*, R. 311 ;
tungú, *Pb.*, P. 833, *Beng.*, *Hind.*, *Mar.*,
C. 838 ; tunka-chandí, *Uriya*, F. 447 ;
túnká-jhár, *Hind.*, C. 838 ; tunna, *Mar.*,
C. 838 ; túnna, *Beng.*, C. 838 ; tunnia,
(*Banswara*), *Hind.*, *Mar.*, O. 537 ; túnnú,
Pb., F. 693 ; tún-túní-nati, *Beng.*, A. 923 ;
tunus, *Mar.*, O. 537 ; túpa, *Mar.*, C. 838;
tupkaria, *Mar.*, S. 1688 ; túp-kél, *Mar.*,
P. 1343 ; tupsi, *Beng.*, F. 539 ; túr, *Dec.*,
Hind., *Mar.*, C. 49 ; túra, *Mar.*, C. 49;
turai, *Dec.*, *Mar.*, *Pb.*, L. 556 ; turai,
Kumaon, L. 569 ; turan, *Mar.*, Z. 273;
turanj, *Beng.*, *Dec.*, *Guj.*, *Hind.*, C.

1270; túrapanli, *Afg.*, V. 149; turbuch, *Guj.*, C. 1221; turbud, *Pb.*, l. 415; turbuj, *Mar.*, C. 1221; turbúza, *Hind.*, C. 2324; túrdal, *Guj.*, C. 49; turi (Cajanus), *Mar.*, C. 49; turi (Mastacembelus), *Uriya*, F. 494; turi (Luffa), *Hind.*, *Sind*, L. 556; *Mar.*, *Sind*, L. 569; turia, *Guz.*, L. 569; turin, *Guz.*, L. 556; turmás, *Hind.*, L. 578; turmuz (Citrullus), *Hind.*, C. 1221; turmuz (Lupinus), *Beng.*, *Pb.*, L. 578; tursbuk, *Pushtu*, P. 1179; tursiphal, *Mar.*, C. 1043; turti, *Mar.*, A. 897; turun-jabín, *Sind*, T. 61; túsi, *Pb.*, F. 260; tústús, *Pb.*, V. 91; tusuru, *Hind.*, S. 1934; tút, *Beng.*, *Dec.*, *Hind.*, *Kumaon*, *Mar.*, *Pb.*, *Sind*, M. 747, 756, 772, 775; tuthai, *Pb.*, P. 693; tuti, *Hind.*, C. 2274; tutia, *Beng.*, C. 2367; tutiri, *Pushtu*, E. 35; tútíyá, *Beng.*, C. 2367; tutla, *Mar.*, M. 747; tútri, *Hind.*, *Mar.*, *Pb.*, M. 756; tuturala, *Hind.*, F. 784; tuvar, *Hind.*, C. 49; tuver, *Mar.*, C. 49; tuvéro, *Guj.*, C. 49; tuvvar, *Duk.*, C. 49; tuwar, *Mar.*, B. 180.

Tw Twara, *Mar.*, B. 180; tworsing, *Kumaon*, S. 3079.

Ua Uá, *Pb.*, H. 375; úájaó, *Pb.*, H. 375.
Ub Úbúsha, *Pb.*, A. 1469.
Uc Ucha-kukur, *Sind*, C. 868; uehchhc, *Beng.*, M. 626.

Ud Ud (Ailanthus), *Mar.*, A. 670; ud (Lutra), *Hind.*, O. 534; úd (styrax), *Beng.*, *Dec.*, *Hind.*, *Mar.*, S. 2968; údahdhatúrá, *Dec.*, D. 151; udal, *Hind.*, *Mar.*, S. 2861; udala, *Mar.*, A. 722; udant, *Beng.*, *Hind.*, O. 574; udar, *Hind.*, *Mar.*, S. 2861; ud biláo, *Hind.* O. 534; úde-bálasán, *Mar.*, B. 54; udesh, *Kumaon*, A. 801; udha, *Mar.*, D. 292; úd-hindí, *Pb.*, C. 607; udi, *Guj.*, *Mar.*, C. 146; udid, *Dec.*, *Mar.*, P. 513; udis, *Hind.*, A. 797; udí-san-bhálú, *Hind.*, J. 116; udísh, *Hind.*, *Kumaon*, A. 797; udrack, *Mar.*, S. 818; ud-sálap, *Hind.*, P. 11.

Ug Ugáda, *Mar.*, T. 372; ugal, *Pb.*, F. 19; ugar, *Beng.*, A. 1251; ughai, *Hind.*, *Pb.*, C. 224; ughz, *Afg.*, *Pb.*, J. 61; úgla, *Pb.*, F. 10; uguru, *Beng.*, E. 593.

Uj Újan, *Pb.*, H. 375; ujar-kantá, *Hind.*, A. 1351; ujlá-dhatúrah, *Dec.*, D. 160; ujli, *Hind.*, S. 2903; ujli-musli, *Guj.*, A. 1562; ujlo-búro-khand, *Guz.*, S. 375; ujlo-chíní, *Guz.*, S. 375; ujlorái, *Guj.*, B. 800.

Uk Úk, *Beng.*, *Hind.*, S. 30; úkh, *Hind.*, S. 30; úkhán, *Pb.*, T. 51; úkhari, *Hind.*, S. 30; ukhi, *Hind.*, S. 30; ukilbar-ki-munker, *Duk.*, C. 321; úklú, *Pb.*, V. 96; ukshi, *Mar.*, C. 200.

Ul Úlasí, *Mar.*, D. 522; ulat-chandal, *Beng.*, G. 243; ulatkambal, *Beng.*, *Guj.*, *Hind.*, A. 41; ulgo, *Pb.*, F. 19; uljíshanbáli, *Dec.*, V. 181; ullah (Bœhmeria), *Hind.*, B. 571; úllak (Anthistiria), *Hind.*, A. 1186; ullar-hillar, *Sind*, C. 1448; ullu, *Hind.*, O. 233; ulsi, *Mar.*, D. 522; ulta kánta, *Kumaon*, C. 416, 431; ulu (Oroxylum), *Mar.*, O. 233; ulú (Anthistiria), *Hind.*, A. 1186; úlú (Imperata), *Beng.*, *Pb.*, l. 51; ulu chakma, *Beng.*, Q. 60.

Um Um, *Baluch.*, P. 419; uma, *Baluch.*, P. 419; umar, *Hind.*, F. 179; umbar, *Mar.*, F. 179; umbar, *Guj.*, F. 175; 179; umbara, *Mar.*, F. 179; umbargular, *Hind.*, F. 179; umbi, *Hind.*, S. 487; umbia, *Hind.*, S. 487; umble, *Mar.*, G. 311; úmblí, *Mar.*, G. 311; úmbú, *Pb.*, M. 878, 881; umla bela, *Beng.*, A. 995; umra, *Hind.*, S. 2644; umrái, *Hind.*, F. 179; umtoá, *Hind.*, A. 1218; umulbet, *Beng.*, O. 547; umul-kúchi, *Beng.*; C. 26.

Un Unab, *Pb.*, Z. 231; únarjal, *Kashmir*, I. 423; undag, *Mar.*, C. 146; undala, *Hind.*, F. 496; undarbibi, *Mar.*, J. 52; undar-punchha (Eragrostis), *Hind.*, E. 250; undar punchha (Setaria), *Hind.*, S. 1210; ungliya, *Hind.*, S. 970; undela, *Mar.*, C. 146; undergúpa, *Hind.*, B. 873; undi, *Dec.*, *Hind.*, *Mar.*, *Sind*, C. 146; undira-cha-kan, *Mar.*, L. 19; undirkáni, *Mar.*, I. 407; undum, *Dec.*, *Hind.*, P. 1381; unhali, *Mar.*, T. 270; unnáb, *Mar.*, Z. 280; unnum, *Beng.*, P. 1426; úns, *Guz.*, S. 30; únt, *Hind.*, C. 205; untia-bagh, *Guz.*, T. 428; untrahi, *Uriya*, F. 399; ununta-múl, *Hind.*, H. 119.

Up Upalasərí, *Mar.*, H. 119; upaleta, *Guz.*, S. 910; uparájitá, *Beng.*, C. 1403; uparsára, *Mar.*, H. 119; upila-kamal, *Mar.*, N. 209; upphing, *Beng.*, W. 19; úpwa, *Pb.*, A. 1632.

Ur Urad, *Hind.*, P. 468; *Pb.*, *Sind*, P. 513; úran, *Pb.*, C. 42; urbúl, *Pb.*, F. 260; urchirri, *Mar.*, O. 127; urd, *Hind.*, F. 671; urd, *Kumaon*, P. 513; urdiya, *Hind.*, P. 41; úrí, *Hind.*, C. 42; urial, *Pb.*, S. 1278; urialli, *Uriya*, F. 417; urid, *Hind.*, P. 513; uri-gáb, *Dec.*, D. 648; urishnapárni, *Orissa*, H. 149; urkúr, *Pb.*, R. 335; urmúl, *Pb.*, F. 260; urn, *Hind.*, C. 42; urni (Clerodendron), *Hind.*, C. 1386; urni (Corylus), *Pb.*, C. 1988; ursi, *Uriya*, F. 528; urthamujirú, *Guz.*, P. 932; urud, *Dec.*, *Hind.*, P. 513; úrún, *Hind.*, J. 104.

Us Ús, *Dec.*, *Mar.*, S. 30; usa, *Mar.*, S. 30; usadhan, *Mar.*, A. 1107; usak laní, *Pb.*, S. 2985; usak láni, *Sind*, C. 224; usan (Eruca), *Pb.*, E. 327; usán (Terminalia), *Beng.*, T. 361; usar, *Hind.*, R. 67; usar-ki-ghás, *Hind.*, S. 2672; ushuklaní, *Sind*, S. 2985; úshur, *Beng.*, V. 5; usirh, *Hind.*, I. 51; ustabunda, *Hind.*, P.1233; ústar-khár, *Hind.*, F. 2; ustúkhúdús (Brunella), *Sind*, B. 902; ustúkhúdús (Lavandula), *Mar.*, L. 117; usturak, *Mar.*, S. 2978; usturgar, *Hind.*, F. 2; usturuk, *Mar.*, L. 455.

Ut Út, *Hind.*, C. 205; utaniya, *Hind.*, O. 185; utarana, *Mar.*, *Sind*, D. 9; utarni, *Mar.*, D. 9; ute-sirkum, *Hind.*, E. 166; ute-sirla, *Hind.*, E. 166; utigun ka bíj, *Hind.*, L. 79; utran, *Dec.*, *Hind.*, D. 9; uttrui, *Uriya*, D. 9.

Va Vach, *Duk.*, A. 430; váchá (Eutropiichthys), *Beng.*, F. 450; vachhá (ox), *Pb.*, O. 574; vachha-nág, *Guj.*, A. 397; vachhí, *Pb.*, O. 574; vachnág, *Guj.*, A. 397; vad, *Mar.*, F. 129; vada, *Mar.*, F. 129; vadgunda, *Guz.*, C. 1931; vadhál, *Pushtu*,

T. 61 ; vadhul, *Hind.*, F. 508 ; vaghayani,
Sind, F. 84 ; vahi, *Kashmir*, A. 430 ;
vaivarang, *Mar.*, E. 199 ; vajra dant,
Guj., B. 171 ; vajra dantí, *Hind.*, *Mar.*, B.
171 ; vajraduhu, *Mar.*, E. 553 ; vákamba,
Hind., C. 563 ; vákerichebháte, *Mar.*, C.
26 ; vákerí-múl, *Hind.*, C. 26 ; vákerí-
múla, *Mar.*, C. 26 ; vakha khaparo, *Guj.*,
B. 619 ; vakumba (Orobanche), *Guj.*, O.
230 ; vákumbhá (Careya), *Mar.*, *Pb.*, C.
563 ; vál (Adenanthera), *Mar.*, A. 471 ;
vál (Dolichos), *Guj.*, *Hind.*, *Mar.*, D.
789 ; vala (Trichosanthes), *Mar.*, T. 569 ;
válá (Andropogon), *Guj.*, *Mar.*, A. 1097 ;
válá bálá, *Beng.*, *Dec.*, *Hind.*, A. 1097 ;
válapápadi, *Mar.*, D. 789 ; vallárí, *Dec.*,
H. 486 ; vallúr (Vitis), *Pb.*, V. 195 ; vallúr
(Cocculus), *Pb.*, C. 1448 ; val-milaku,
Hind., P. 801 ; válo, *Guj.*, A. 1097 ;
valpapri, *Mar.*, D. 789 ; vaman, *Pb.*, W.
98 ; ván, *Pb.*, S. 705 ; vana-bhenda, *Mar.*,
U. 29 ; vana-gao, *Beng.*, O. 567 ; vana-
jai, *Mar.*, C. 1377 ; vanchhi kánto, *Guj.*,
A. 306 ; vandá (Vanda), *Hind.*, V. 17 ;
vánda (Loranthus), *Mar.*, L. 549 ; vando,
Gus., L. 549 ; vánge, *Mar.*, S. 2284 ;
vángi, *Mar.*, S. 2284 ; váni, *Pb.*, S. 705 ;
vání jhál, *Hind.*, *Pb.*, C. 224 ; vansa-
kalaka, *Mar.*, B. 149 ; vánseo-deodár,
Guj., C. 846 ; vánskapúr, *Guj.*, B. 84 ;
vanták, *Gus.*, S. 2284 ; vantha-karatola,
Mar., M. 639 ; vanúthi, *Pb.*, S. 1024 ;
vanveri, *Pb.*, P. 397 ; vapchí, *Hind.*, V.
73 ; varaha, *Beng.*, H. 289 ; várala,
Mar., C. 1520 ; varandá, *Hind.*, O. 31 ;
varanga, *Mar.*, K. 42 ; várangada, *Mar.*,
K. 42 ; vardárá, *Mar.*, R. 558 ; vargalum,
Pushtu, R. 166 ; vargund, *Mar.*, C.
1931 ; vargúnd, *Mar.*, C. 1931 ; *Guj.*, C.
1940 ; vari (Panicum), *Guj.*, *Mar.*, P.
63 ; vari (Quercus), *Pb.*, Q. 35 ; vari
(Panicum), *Mar.*, P. 63 ; variali, *Gus.*, P.
727 ; variari, *Gus.*, F. 659 ; varika-anu,
Mar., P. 63 ; varísáva, *Mar.*, P. 63 ;
variyali, *Gus.*, F. 659 ; varkhihartála,
Mar., O. 242 ; vartalau, *Mar.*, M. 461 ;
varul, *Duk.*, A. 658 ; vas, *Mar.*, B. 118 ;
vásá, *Hind.*, A. 484 ; vásaka, *Beng.*, A.
484 ; vasana-vela, *Mar.*, C. 1452 ; vás-nu-
mítha, *Guj.*, B. 84 ; vassa, *Mar.*, D. 292.
vasu, *Mar.*, B. 619 ; vatána, *Gus.*, P.
885 ; vátána, *Mar.*, P. 885 ; vátáne,
Mar., P. 885 ; vátoli, *Mar.*, A. 1038 ;
vauhr, *Pb.*, O. 574 ; vaum, *Sind*, G. 404 ;
vavading, *Guj.*, E. 199 ; vavadinga,
Mar., E. 199 ; vavani, *Mar.*, S. 1212 ;
vávarang, *Pb.*, M. 910 ; vavoli, *Mar.*, M.
570 ; vavut, *Kashmir*, B. 731 ; vay,
Beng., B. 417 ; váya-varná, *Mar.*, C.
2039 ; váyni, *Mar.*, P. 1032 ; vaysha,
Beng., B. 417.

Ve Vehkali, *Mar.*, P. 912 ; vehrí, *Pb.*, C. 1448 ;
vehyenti, *Mar.*, P. 912 ; vekhand, *Guj.*,
Mar., A. 430 ; vekhanda, *Guj.*, A. 430 ;
vekhanda, *Mar.*, C. 225 ; vékhariyo,
Mar., I. 131 ; vekriavas, *Raj.*, I. 120 ;
veldode (Amomum), *Mar.*, A. 965 ;
veldode (Elettaria), *Mar.*, E. 151 ; vél-
khákar, *Guj.*, B. 978 ; vellajung, *Beng.*,
P. 811 ; velloda, *Mar.*, E. 151 ; vena,
Pb., R. 166 ; vengni, *Gus.*, S. 2284 ;
venivel, *Mar.*, C. 1205 ; veravena, *Pushtu*,
D. 725 ; verí (Convolvulus), *Hind.*, *Pb.*,

C. 1777 ; verí (Marsdenia), *Pb.*, M. 295 ;
veta, *Mar.*, C. 104 ; vet-vángi, *Mar.*, L.
596.

Vi Vídecha-pána, *Mar.*, P. 775 ; vien, *Pb.*, M.
461 ; vikhári, *Mar.*, P. 912 ; vilaiti ámli
(Pithecolobium), *Mar.*, P. 900 ; vilaiti-
amli (Garcinia), *Mar.*, G. 14 ; vilaití imlí,
Hind., P. 900 ; vilaiti-níl, *Hind.*, I. 107 ;
vilarjuti vakundi, *Mar.*, C. 2253, I. 61 ;
viláti-chunó   *Guj.*, C. 489 ; viláyatí-
afsantín, *Hind.*, *Duk.*, A. 1441 ; viláyatí-
agati, *Duk.*, C. 732 ; vilayatibábúl, *Hind.*,
Duk., A. 217 ; vilayati chambeli, *Mar.*,
Q. 88 ; vilayatí-chámpa, *Mar.*, A. 1431 ;
viláyati chuna, *Dec.*, *Mar.*, C. 489 ;
viláyati isband, *Dec.*, P. 372 ; viláyatí-
kangai, *Hind.*, M. 115 ; vilayatí-kangói,
Dec., M. 115 ; vilayatikikar, *Hind.*,
Duk., A. 217 ; viláyati mehndi (Myrtus),
Hind., *Pb.*, M. 921 ; viláyatí-mhéndí
(Peganum), *Dec.*, P. 372 ; viláyatisaro,
Mar., C. 826 ; viláyetí múng, *Hind.*, A.
1261 ; vilayti ambi, *Mar.*, P. 900 ; vilayti,
babúl, *Sind.*, P. 322 ; vilayti-ghas, *Gus.*,
M. 334 ; vilaytíimlí, *Hind.*, A. 455 ;
vilayti kikar, *Mar.*, P. 322 ; viláyti
vengan, *Gus.*, L. 596 ; villiah, *Mar.*, U.
29 ; vilva, *Beng.*, A. 534 ; vilyadele,
Mar., P. 775 ; vinchú, *Mar.*, M. 308 ;
vind, *Hind.*, S. 6 ; visesh, *Mar.*, B. 751 ;
vish-lúmba, *Pb.*,-C. 1211 ; vishnú-kanté,
Pushtu, C. 1403 ; vishnukrant (Ipomæa),
Mar., I. 405 ; visnukranti (Clitoria),
Hind., C. 1403.

Vo Vomb, *Mar.*, N. 72 ; votamba, *Mar.*, A.
1511.

Vr Vrásh, *Pb.*, R. 318.

Vu Vubbina, *Mar.*, P. 1048 ; vúlr, *Pb.*, V. 64 ;
vurtuli, *Hind.*, D. 402.

Vw Vwír, *Kashmir*, S. 537.

Wa Wa (Saxifraga), *Pb.*, S. 924 ; wa (bharal),
Pb., S. 1273 ; wad, *Mar.*, F. 129 ; wadi,
Mar., P. 63 ; waftangel, *Kashmir*, D. 144 ;
wag, *Mar.*, C. 416 ; wagati (Wagatea),
Mar., W. 1 ; wagati (leopard-cat), *Mar.*,
T. 425 ; wagatti, *Mar.*, C. 416 ; waghz,
Afg., *Pb.*, J. 61 ; wahág, *Mar.*, T. 437 ;
wahal, *Pb.*, V. 149 ; wáhití, *Mar.*, B. 175 ;
wahran gur, *Hind.*, C. 854 ; waingi, *Mar.*,
S. 2284 ; wairká, *Pb.*, O. 574 ; wairki, *Pb.*,
O. 574 ; wai varang, *Pb.*, E. 199 ; wakan,
Sind, L. 451 ; wakandi, *Mar.*, G. 748 ;
wákerí, *Mar.*, W. 1 ; wakiry, *Mar.*, W. 1 ;
wala (Salix), *Pb* , S. 544 ; wálá (Valeriana),
Pb., V. 10 ; walasura, *Mar.*, W. 13 ;
walena, *Hind.*, *Mar.*, S. 2819 ; wall,
Hind., *Sind*, D. 789 ; walli múng, *Hind.*,
P. 496 ; wallunj, *Mar.*, S. 579 ; wallursi,
Mar., W. 13 ; wal-wangi, *Mar.*, L. 596 ;
wámpú-lítsí, *Pb.*, P. 1447 ; wan (Pistacia),
Baluch., P. 847 ; wan (Salvadora), *Pb.*, S.
705 ; wana, *Pb.*, V. 164 ; wana, *Baluch.*,
P. 847 ; wandak, *Pb.*, C. 1361 ; wánder-
roti, *Mar.*, N. 174 ; wangan, *Sind*, S.
2284 ; wangan tsúrú, *Pb.*, X. 1 ; wangi,
Dec., S. 2284 ; wángrú, *Hind.*, *Pb.*, C.
448 ; wanizad, *Afg.*, P. 847 ; wanjad, *Afg.*,
P. 847 ; wans, *Guj.*, B. 118 ; wánsb, *Pb.*,
R. 318 ; wan-wángan. *Pb.*, P. 1009 ; war
(Ficus), *Mar.*, *Pb.* ; F. 129 ; 216 ; war
(bharal), *Pb.*, S. 1273 ; warai, *Mar.*, P. 67 ;
wara wane, *Pushtu*, R. 365 ; wara wi,

Kashmir, Pb., C. 1988 ; warch, *Pb.,* A. 430 ; warga, *Hind.,* C. 756 ; wargalion, *Pushtu,* R. 166 ; wári, *Dec. ;* P. 63 ; wariaree, *Gus.,* F. 659 ; warkhárai, *Pushtu,* P. 1179 ; warmande, *Pushtu,* V. 164 ; warras, *Mar.,* H. 159; warúmba, *Pb.,* S. 2345 ; waruna, *Mar.,* C. 2039 ; warung, *Mar.,* K. 42 ; wásho, *Pb.,* R. 318 ; wassanwel, *Mar.,* C. 1452 ; watal, *Pb.,* E. 479 ; watana, *Dec., Mar.,* P. 885 ; watani-ya, *Hind.,* O. 185 ; watkana, *Beng., Hind.,* B. 523 ; wátpan, *Pb.,* T. 852 ; wattahna, *Hind.,* P. 885 ; wattamman, *Pb.,* C. 886 ; wattunah, *Mar.,* F. 409; wa-upla-bij, *Gus.,* L. 574 ; wawali, *Mar.,* H. 324 ; wawrung, *Hind.,* E. 199.

We Wesha, *Afg.,* A. 17 ; wetyar, *Pb.,* J. 78, 104.

Wi Wiláyati baigan, *Hind.,* L. 596 ; wilayati jan, *Hind., Pb.,* A. 1639 ; wilayáti kikar, *Pb.,* P. 322 ; wilayatí-ním, *Mar.,* M. 393 ; wilayati-nimb, *Mar.,* M. 393 ; wiláyatí-sarú, *Mar.,* C. 826 ; wiláyatí-zírah, *Mar.,* C. 681 ; wilayti-gawuth, *Hind.,* M. 334 ; wilyati-kaitalu, *Pb.,* A. 603 ; wínri, *Kashmir,* C. 1988 ; *Pb.,* C. 1988 ; wiri, *Pb.,* C. 1988 ; wiri, *Kashmir,* C. 1988.

Wo Wonding, *Hind.,* E. 199 ; wora, *Hind.,* 129 ; worshúrah, *Mar.,* F. 601 ; wotiangil, *Kashmir,* C. 624 ; wowli, *Mar.,* M. 570.

Wu Wúman, *Hind.,* A. 22 ; wumb, *Mar.* N. 72 ; wumb-ashphal, *Mar.,* N. 72 ; wumbusing, *Hind.,* A. 22 ; wúndí, *Mar.,* C. 146 ; wur, *Sind,* F. 129 ; wurak, *Pushtu,* R. 152 ; *Afg., Pb.,* R. 157 ; wuraskai, *Pushtu,* D. 725 ; wúriya, *Kashmir, Pb.,* C. 1988 ; wurmah, *Dec.,* T. 537 ; wussanthau-ganda, *Hind.,* M. 71 ; wusta (Croton), *Uriya,* C. 2171 ; wústa (Alnus), *Hind.,* A. 797.

Ya Yajnadumbar, *Beng.,* F. 179 ; yak, *Hind.,* F. 793 ; yakayela, *Mar.,* D. 89 ; yal, *Pb.,* R. 533 ; yalachí, *Kan.,* Z. 231 ; yálki, *Mar.,* M. 439 ; yalvá, *Hind.,* A. 824 ; yamni, *Hind.,* C. 617 ; yamskollung, *Gus.,* D. 534 ; yamu, *Pb.,* S. 1264 ; yange, *Ph.,* R. 362 ; yangtarsh, *Pb.,* F.678 ; yáqút, *Hind.,* R. 619 ; yara, *Kashmir,* P. 737 ; yarand, *Deccan ;* R. 369 ; yarandícha, *Mar.,* R. 369 ; yári, *Kashmir, Pb.,* P. 737 ; yárpa, *Pb.,* P. 1142 ; yarta, *Pb.,* P. 702 ; yashm, *Hind.,* J. 2 ; ya vásá, *Hind., Mar.,* A. 745 ; yavásá, *Hind., Mar., Pb.,* C. 224.

Ye Yebruj (Atropa), *Beng.,* A. 1614 ; yebruj (Mandragora), *Beng.,* M. 128 ; yehala, *Dec.,* T. 293 ; yehela béhadá, *Mar.,* T. 293 ; yekaddi, *Mar.* P. 912 ; yekal, *Mar.,* C. 860 ; yekdi, *Mar.,* P. 912 ; yek-gisamká-bachlá, *Dec.,* V. 228 ; yel, *Mar.,* T. 293 ; yelá, *Mar.,* T. 293 ; yelíyo, *Gus.,* A. 824 ; yella, *Mar.,* T. 293 ; yelnir-kapani, *Duk.,* C. 1520 ; yél-parás, *Mar.,* B. 978 ; yén, *Mar.,* T. 361 ; yenki, *Beng.,* M. 260 ; yeonlah, *Beng.,* P. 632 ; yerindi, *Mar.,* D. 884 ; yero, *Pb.,* P. 737 ; yerrul *Mar.,* X. 16 ; yerwa, *Hind.,* O. 185.

Yi Yir, *Pb.,* S. 570 ; yír, *Kashmir,* S. 579 ; yira, *Kashmir, Pb.,* T. 864 ; 875 ; yiro, *Kash-*

mir, P. 737 ; yírú, *Pb.,* Q. 29 ; yírú khareo, *Pb.,* C. 224.

Ym Ymwah, *Gus. ;* H. 382.

Yu Yúpo, *Pb.,* lA. 1632 ; yúr (Pinus), *Kashmir,* P. 737 ; yúr (Salix), *Kashmir,* S. 560 ; yür, *Pb.* S. 537 ; yúra, *Pb.,* M. 461 ; yurk, *Hind.,* H. 382 ; yuru, *Pb.,* Q. 29.

Za Zaffre, *Pb.,* C. 1444 ; zafran, *Hind.,* C. 2083 ; zagar, *Pushtu,* D. 64 ; zagúkeí, *Pb. ;* R. 645 ; zahr mohra (magnesia), *Hind., Pers., Pb.,* M. 53; zahr muhra (serpentine), *Hind. ;* S. 2712 ; záih, *Kafiristan,* Q. 13 ; zaisí, *Afg.,* G. 278 ; zaitun (Gyrocarpus), *Hind.,* G. 780 ; zaitun (Psidium), *Sind,* P. 1343 ; zaitún (Olea), *Pushtu,* O. 145 ; zakhm-haiyát, *Hind.,* B. 909 ; zakhm-haiyát-ka-pattá, *Duk.,* B. 909 ; zakhmhyát, *Mar.,* K. 14 ; zakhmi, *Mar.,* D. 725 ; zakhmíhai yát, *Pb.,* S. 2518 ; zakhmí haiya (Tinospora), *Pb.,* T. 470, 483 ; zakhmí hai yát (Cissampe los), *Pb.,* C. 1205 ; zakhuní haiyat (Mollugo), *Pb.,* M. 615 ; zam, *Pb.,* P. 1316 ; zamái, *Pb.,* S. 2985 ; zamái, (*Trans-Indus*), *Pushtu,* C. 224 ; zambchule, *Kashmir, Pb.* 1316 ; zambu, *Pb.,* P. 1316 ; zaminkand (Amorphophallus), *Hind.,* A. 996 ; zamín kand (Dioscorea), *Hind., Pb.,* D. 494 ; zamir, *Sind,* C. 1452 ; zanda, *N. Pb.,* D. 815 ; zanghíhar, *Hind.,* T. 325 ; zan-ghóz, *Afg.,* P. 746 ; zan-ghóza, *Afg.,* P. 746 ; zangi-halré, *Dec.,* T. 325 ; zangzabíl, *Pb.,* Z. 201 ; zanumu, *Hind.,* C. 331 ; zarbútí, *Pb.,* C. 2508 ; zárdák *Afg.,* D. 173 ; zard-alu, *Afg., Hind., Pb.,* P. 1285 ; *Pb.,* P. 1304 ; zard-áru, *Kumaon,* P. 1285 ; zard kunel, *Mar.,* T. 410 ; zard kunél, *Hind.,* T. 410 ; zargal, *Pb.,* F. 624 ; zarí, *Pb.,* Z. 254 ; zarish, *Dec.,* B. 443 ; zarishk, *Hind.,* B. 443 ; zatar, *Mar.,* A. 1577 ; zaz, *Pushtu,* C. 224.

Zb Zbang (Abelia), *Pb.,* A. 2 ; zbang (Lonicera), *Pb.,* L. 535 ; zbiir, *Pb.,* A. 1462.

Ze Zeeberwo, *Gus.,* D. 582 ; zemardac an, *Pb.,* V. 202 ; zemaro, *Pb.,* V. 202 ; zergul, *Pb.,* C. 117 ; zéro, *Sind,* C. 2339.

Zh Zhand, *Pushtu,* M. 557 ; zhangar, *Pb.,* J. 111 ; zharas, *Mar.,* M. 617 ; zhbang, *Pb.,* P. 540 ; zhi, *Pb.,* D. 130 ; zhiko, *Pb.,* L. 533 ; zhorhatheylo, *Gus.,* O. 193.

Zi Zidadi, *Mar.,* C. 919 ; zilechá til, *Mar.,* S. 1078 ; zímbíl, *Pb.,* P. 1197 ; zimeh, *Pushtu,* S. 2985 ; zir (Ficus), *Beng.,* F. 253 ; zir (Melilotus), *Sind,* M. 422 ; zíra (Carum), *Hind.,* C. 68 1 ; zíra (Cuminum), *Hind.,* C. 2339 ; zirangi, *Dec.,* M. 864 ; zira síyah, *Pb.,* C. 681 ; zirishk (raisins), *Pb.,* V. 233 ; zirishk (Berberis), *Pb.,* B. 465 ; zirki, *Dec.,* I. 384 ; zirnub birmí, *Hind.,* T. 93.

Zo Zœnil, *Pb.,* C. 1043 ; zosho, *Pb.,* D. 130 ; zoz, *Pushtu,* C. 224 ; zozan, *Pushtu,* C. 224.

Zu Zucum-yeat, *Pb.,* C. 1205 ; zud, *Pb.* T., 634 ; zúfahyabis, *Hind.,* H. 550 ; zúfa, yabis, *Pb.,* N. 56 ; zúm, *Pb.,* P. 1316 ; zunjbel, *Pb.,* Z. 201 ; zurmish, *Pb.,* L. 578 ; zuwasha, *Gus.,* A. 745.

BHIL, GOND, KOL, KURKU, MUNDA, SANTAL, URAON.

Ah Ahauna, *Santal*, C. 170.
Ai Ain, *Kol*, F. 156 ; aita, *Gond*, H. 92.
Ak Akh, *Santal*, S. 30.
Al Alag jari (Cassytha), *Santal*, C. 805 ; alag jarí (Cuscuta), *Santal*, C. 2508 ; alali, *Baori*, S. 1226 ; alawa, *Kurku*, F. 179 ; ali, *Gond*, F. 236 ; alkusi, *Kol*, M. 786.
Am Ama, *Baigas*, M. 147 ; amba bhósa, *Bhil*, B. 318 ; ambe, *Kurku*, M. 147 ; ambera, *Kurku*, S. 2649 ; ambotha, *Kurku*, B. 304 ; amburri, *Kol*, S. 2649.
An Andiadhurup arak, *Santal*, L. 312 ; anduli, *Gond*, D. 560 ; anjan, *Kol*, H. 16 ; ankol, *Kol*, A. 681 ; ankola, *Gond*, A. 681.
Ap Apúng, *Kol*, *Santal*, H. 328.
Ar Arak (Amarantus), *Santal*, A. 927 ; arak (Leucas), *Santal*, L. 321 ; arak jháwár, *Santal*, U. 67 ; arak kudúm, *Santal*, H. 233 ; arengebaung, *Kol*, O. 233 ; arengi banu, *Kol*, O. 233 ; arjun, *Kurku*, T. 282 ; arma, *Gond*, A. 1149 ; arma-suri, *Gond*, F. 615.
As Asána, *Bhil*, B. 868 ; astra, *Gond*, B. 318 ; asunda, *Kol*, C. 563.
At At, *Santal*, Z. 182 ; atená, *Santal*, C. 1743 ; athna, *Kurku*, T. 361 ; atkir, *Santal*, S. 2252 ; atkura, *Santal*, W. 131 ; atnak, *Santal*, T. 361 ; ato sang, *Santal*, D. 510, 534.
Au Aunre, *Kurku*, P. 632 ; áunri, *Gond*, P. 632 ; aura, *Kol*, P. 632.

Ba Babur, *Gond*, C. 860 ; bachkron, *Santal*, I. 494 ; badar zhapni, *Santal*, P. 673 ; bádú, *Kol*, S. 1226 ; bæphol, *Santal*, D. 345 ; bag lucha, *Santal*, M. 308 ; bagni, *Santal*, C. 6 ; bahu tuturi, *Santal*, V. 79 ; bai, *Kol*, F. 129 ; baichua, *Santal*, D. 842 ; bakrelara, *Paharia*, H. 68 ; bana etka, *Santal*, I. 415 ; banag, *Kurku*, C. 756 ; bana hatak, *Santal*, O. 233 ; banda (Loranthus), *Santal*, L. 549 ; banda (Pterocarpus), *Santal*, P. 1370 ; banda (Viscum), *Santal*, *Kol*, V. 156 ; bando, *Santal*, S. 2508 ; bangru, *Kurku*, C. 756 ; banjir, *Gond*, T. 293 ; bankatari, *Santal*, M. 704 ; bara churcheri, *Paharia*, V. 202 ; barangom (Vernonia), *Santal*, V. 79 ; barangom bir barangom (Glossogyne), *Santal*, G. 250 ; bare, *Santal*, F. 129 ; bare baha, *Santal*, P. 393 ; barelli, *Gond*, F. 129 ; barge khode baha, *Santal*, P. 425 ; bariar, *Santal*, S. 1699 ; baringa, *Gond*, S. 2861 ; barni, *Santal*, C. 1380 ; barsa pakor, *Santal*, C. 1711 ; baru, *Santal*, *Kurku*, S. 950 ; barúi, *Kol*, G. 116 ; barurí, *Kol*, G. 116 ; baswesa, *Kol*, F. 216 ; bata karas, *Bhil*, E. 73 ; batha, bijir, *Mundari*, C. 2612.
Be Bela (Ægle), *Kurku*, A. 534 ; bela (Bauhinia), *Gond*, B. 342 ; belaunja, *Santal*, S. 2518 ; bengo nári, *Santal*, D. 513 ; berenjo, *Santal*, T. 410 ; beriju, *Santal*, B. 318.
Bh Bhadu, *Santal*, V. 174 ; bhainsa, *Santal*, C. 1711 ; bhandra, *Gond*, G. 124 ; bharangeli, *Kurku*, E. 202 ; bharbari, *Santal*, O. 18, 26 ; bhatua arak', *Santal*, C. 1003 ; bhedi janetet, *Santal*, A. 1398 ; bhernda, *Santal*, J. 41 ; bhidi janetet, *Santal*, U. 29 ; bhir, *Gond*, S. 1287 ; bhira, *Gond*, C. 1031 ; bhir kura, *Gond*, S. 1287 ; bhiru, *Baigas*, D. 292 ; bhirul, *Bheel*, S. 1287 ; bhirwa, *Baigas*, C. 1031 ; bhita, *Kurku*, R. 1 ; bhoga kuskom, *Santal*, G. 385 ; bhonder, *Gond*, E. 318 ; bhor khond, *Santal*, H. 517 ; bhotuk, *Bhil*, D. 41 ; bhringeli, *Kurku*, E. 199.
Bi Biba, *Gond*, S. 1041 ; bibla, *Bauris*, T. 434 ; bijo, *Gond*, P. 1370 ; bir, *Santal*, S. 1699 ; birbarangon, *Santal*, S. 2354 ; bir but, *Santal*, F. 633 ; bir jhawar, *Santal*, D. 332 ; bir kana arak, *Santal*, C. 1759 ; birkapi, *Santal*, D. 347 ; bir-kauni, *Santal*, S. 1223 ; birkod, *Santal*, E. 428 ; bir lopong arak' (Rungia), *Santal*, R. 656 ; bir lopong arak' (Vernonia), *Santal*, V. 79 ; birmalla, *Santal*, L. 110 ; bir móch (Phaseolus), *Santal*, P. 468 ; bir móch (Zornia), *Santal*, Z. 300 ; bir mung (Phaseolus), *Santal*, P. 468 ; bir munga (Dalbergia), *Santal*, D. 94 ; bir-mungara, *Santal*, A. 471 ; birmutha, *Santal*, C. 2610 ; bir sang, *Santal*, P. 513 ; bir sangi, *Santal*, D. 481 ; birsing, *Kol*, E. 342 ; bir surajmukhi, *Santal*, I. 335 ; bittia, *Kol*, K. 42.
Bo Bod-lar-nari, *Santal*, V. 184 ; boi bindi, *Santal*, R. 1 ; bomair, *Gond*, F. 202 ; bomudu, *Kol*, A. 871 ; bone, *Kurku*, H. 16 ; bonga-sarjom, *Santal*, V. 48 ; bonga-sarjun, *Kol*, V. 48 ; bonga tiain, *Santal*, E. 467 ; bon kapsi, *Santal*, T. 387 ; bonur-lati, *Uraon*, C. 756 ; bor, *Baigas*, Z. 231 ; bor-salei, *Gond*, B. 771 ; bosha (Bauhinia), *Gond*, B. 318 ; bosha (Kydia), *Gond*, K. 42 ; bossai, *Kurku*, B. 318 ; bot, *Gond*, C. 1927.
Br Bruru, *Bhumij*, G. 116.
Bu Buch, *Santal*, C. 1931 ; budhi ghasit, *Santal*, P. 338 ; budhi tilai, *Santal*, P. 338 ; budi kaskom, *Santal*, G. 385 ; búndún (Callicarpa), *Kol*, C. 123 ; bún-dún (Eriolæna), *Kol*, E. 314 ; burkál, *Gond*, T. 434 ; burkunda, *Bhumij*, H. 517 ; bursa, *Kol*, G. 702 ; bursu, *Kol*, G. 705 ; buru asaria, *Santal*, C. 416 ; buru ekasira nari, *Santal*, F. 633 ; buruju, *Kol*, B. 308 ; buruk, *Gond*, K. 42 ; buru-katkom-charec', *Santal*, E. 241 ; buru machkunda, *Santal*, W. 131 ; buru mat, *Santal*, D. 292 ; buru raher, *Santal*, C. 2514 ; but, *Santal*, C. 1061 ; buti, *Kurku*, S. 2861 ; butisa, *Kol*, F. 253.
By Byong, *Kol*, D. 549 ; byong-kullú, *Kol*, D 548.

Ch Chailí, *Santal*, M. 704 ; chakaoda arak', *Santal*, C. 797 ; champa baha, *Santal*, O. 1 ; champa pungár, *Gond*, P. 989 ; chapa, *Kurku*, B. 304 ; chapot siris (Albizzia), *Santal*, A. 695 ; chapot siris (Dalbergia), *Santal*, D. 32 ; chapún, *Kol*, A. 722 ; chari, *Baori*, S. 1250 ;

chatin, *Kol*, A. 871 ; chatom aruk, M. 306 ; chattisghari tasar, *Santal*, S. 1934 ; chaulia, *Santal*, R. 635 ; chauric arak, *Santal*, L. 509 ; chekerey, *Kurku*, L. 55 ; cheli, *Kol*, G. 705 ; chhatnia, *Santal*, A. 871 ; chhota dundhera, *Gond*, H. 16 ; chicha, *Kurku*, T. 28 ; chichola, *Gond*, A. 711 ; chichora, *Kurku*, A. 711 ; chichwa, *Gond*, A. 711 ; chihut lar, *Santal*, S. 2508 ; chikhari, *Kurku*, P. 77 ; chimman, *Bhil*, G. 287 ; chindi, *Gond*, P. 551 ; chíta, *Gond*, T. 28 ; chitra, *Gond*, T. 424 ; chorcho, *Santal*, C. 725 ; chota jám, *Santal*, E. 404 ; choti khidi, *Kurku*, E. 248 ; chuduk'-bad, *Santal*, E. 419 ; churni, *Kurku*, Z. 273 ; chutia chandbol, *Santal*, I. 340.

Co Corvalum, *Kurku*, A. 534.

Da Dadki, *Bhumij*, W. 106 ; dahu, *Santal*, *Kol*, A. 1511 ; dakichak, *Santal*, J. 114 ; dan, *Bhil*, S. 2876 ; dandukit, *Santal*, G. 136 ; dangraseya, *Santal*, P. 1239 ; danta, *Gond*, C. 860 ; dare banki, *Santal*, V. 17 ; dare dhompo, *Santal*, L. 266 ; dare-huter, *Santal*, I. 141 ; dare jhapak', *Santal*, S. 970 ; dare kudrum, *Santal*, H. 177 ; dare orsa, *Santal*, C. 1761 ; datti, *Bhil*, W. 106.

De Debrelara, *Paharia*, P. 1401 ; dedhaori janum, *Santal*, Z. 231 ; dél, *Kol*, B. 632 ; dela, *Santal*, A. 681.

Dh Dhai, *Santal*, T. 15 ; dhakha (Schrebera), *Gond*, S. 959 ; dhakka (Elædendron), *Gond*, E. 73 ; dhamin, *Kol*, G. 714 ; dhamnak, *Bhil*, G. 714 ; dhamni, *Kurku*, G. 714 ; dhatte, *Gond*, O. 233 ; dhatti, *Bhil*, C. 860 ; dhatura, *Santal*, D. 151 ; dhaundak, *Bhil*, A. 1149 ; dhaura, *Kurku*, A. 1149 ; dhawa, *Baigas*, A. 1149 ; dhawe, *Kol*, W. 106 ; dhi, *Kurku*, W. 106 ; dhobi-ghás, *Santal*, C. 2558 ; dhoka, *Kol*, O. 38 ; dhondel, *Gond*, B. 304 ; dhondri, *Gond*, B. 318 ; dhótamara, *Gond*, S. 2884 ; dhotte, *Gond*, S. 2884 ; dhundera, *Gond*, B. 318.

Di Dimbu baha, *Santal*, O. 18 ; disti (Ehretia), *Gond*, E. 25 ; disti (Ixora), *Gond*, I. 515.

Do Dodhari, *Santal*, C. 996 ; doe-saraj, *Kol*, V. 48 ; doka, *Santal*, O. 38 ; dotti, *Gond*, E. 25 ; doudoukit, *Santal*, G. 136 ; dowka, *Kol*, O. 38.

Du Dudha-malida-kand, *Khand*, N. 39 ; dudhi, *Kol*, I. 1 ; dudhia-phul, *Santal*, E. 518 ; dudhilota, *Santal*, I. 1 ; duduri, *Kol*, G. 136 ; dum kotokoi, *Santal*, C. 123 ; durya arak', *Santal*, V. 79.

Ed Edel, *Santal*, B. 632.
Eh Ehúri, *Kol*, V. 164.
Em Embrúm, *Kol*, C. 1931
En Enga dhurup, *Santal*, L. 321.
Er Eradom, *Santal*, R. 369 ; erba, *Santal*, S. 1212.
Et Etka, *Santal*, M. 786 ; etkec, *Santal*, E. 496.

Fe Feley, *Kol*, S. 2850.
Fu Fulsa, *Paharia*, G. 663 ; fursu, *Paharia*, G. 663.

Ga Gabur, *Santal*, A. 101 ; 217 ; gada, *Santal*, S. 579 ; gadahundbaha, *Santal*, J. 13 ;

gada-kalha, *Santal*, S. 2923 ; gada-phassa, *Kurku*, E. 356 ; gadasigrik (Erythræa), *Santal*, E. 338 ; gada sigrik (Salix), *Santal*, S. 579 ; gada terel, *Santal*, D. 628 ; gaighura, *Santal*, P. 1062 ; gambu, *Kurku*, E. 416 ; gandhari, *Santal*, A. 927 ; gandoli, *Baori*, S. 1226 ; ganerí, *Bhil*, C. 1512 ; ganga, *Gond*,,B. 771 ; gangam, *Gond*, C. 1512 ; gangáru, *Kurku*, R. 16 ; gangru, *Kurku*, R. 16, gan jher, *Santal*, S. 2861 ; gaphni, *Kol*, G. 723 ; gara hatana, *Kol*, T. 282 ; gara-hesel, *Kol*, A. 1146 ; gara-kud, *Santal*, E. 416 ; gara-kuda, *Kol*, E. 416 ; gara lohadaru, *Kol*, T. 525 ; gara nim, *Kol*, M. 393 ; ga-ranji, *Gond*, P. 1121 ; gara-saikre, *Kol*, L. 42 ; garaterél, *Santal*, D. 582 ; garbha gojha, *Santal*, C. 390 ; gárrah, *Gond*, B. 13 ; garúm kúla, *Kol*, T. 437 ; gatonli, *Uraon*, B. 318.

Ge Geggar, *Gond*, G. 124 ; gera erumi, *Gond*, O. 555.

Gh Ghangra, *Santal*, V. 116 ; gharri (Garuga), *Gond*, G. 143 ; gharri (Odina), *Gond*, O. 38 ; ghato, *Uraon*, S. 959 ; ghattól ghotia, *Gond*, Z. 290 ; ghenra, *Paharia*, S. 30 ; ghora lidi *Santal*, V. 231 ; ghota, *Kurku*, Z. 290.

Gi Gilchi, *Gond*, E. 25 ; girchi (Casearia), *Gond*, C. 722 ; girchi (Holarrhena), *Gond*, H. 294 ; gitilarak, *Santal*, L. 327.

Go Gogar, *Bhil*, G. 124 ; goit, *Bhumij*, Z. 290 ; gokhula-janumi(Argemone), *Santal*, A. 1351 ; gokhula-janum (Hygrophila), *Santal*, H. 508 ; golari, *Gond*, D. 330 ; góm, *Santal*, A. 1158 ; gonyer, *Kol*, K. 42 ; gora tiril, *Kol*, D. 615 ; gordág, *Gond*, T. 434 ; góré, *Santal*, S. 2799 ; gote, *Santal*, C. 2180.

Gu Gua, *Kol*, V. 158 ; gua-goli, *Santal*, E.314 ; gui, *Kol*, S. 2799 ; gulanj baha, *Santal*, P. 989 ; gulgal, *Kol*, C. 1512 ; gumar, *Gond*, C. 563 ; gumher, *Kol*, G. 287 ; gumi, *Gond*, L. 323 ; gummar, *Gond*, C. 563 ; gumpri, *Gond*, O. 38 ; gumudu masúrbauri, *Gond*, A. 1215 ; gundli, *Santal*, P. 67 ; gúngat, *Gond*, G 124 ; gúnjan, *Bhil*, B. 868 ; gúpni, *Gond*, G. 143 ; guraya, *Gond*, S. 1229 ; gurgoti, *Kurku*, F. 615 ; gúrúr, *Gond*, M. 549 ; gutra, *Gond*, D. 219 ; gutri, *Gond*, D. 219.

Ha Hadra, *Kol*, T. 325 ; hake húmú, *Kol*, S. 487 ; halpa, *Gond*, D. 292 ; hamara, *Gond*, S. 2649 ; hana, *Paharia*, T. 325 ; handi khandi, *Santal*, P. 678 ; hara-mada, *Santal*, L. 237 ; hara saíjung, *Kol*, S. 2912 ; hardu, *Gond*, A. 514 ; hari, *Kol*, C. 756 ; harna pakor, *Santal*, S., 2923 ; harro, *Gond*, T. 325 ; hat, *Santal*, *Kol*, H. 294 ; hatana, *Kol*, T. 361 ; hat-kan, *Santal*, L. 232, 237.

He Hed, *Kol*, P. 1370 ; hehel, *Santal*, M. 549 ; hél, *Kol*, M. 549 ; hend, *Gond*, E. 416 ; hendedisomhorec, *Santal*, G. 263 ; herka, *Gond*, B. 913 ; hesak, *Santal*, F. 236 ; hesar, *Kol*, F. 236 ; hesel, *Kol*, *Santal*, A. 1149 ; hetmudia *Santal*, T. 562.

Hi Hid, *Kol*, P. 1370 ; hilum, *Kol*, P. 1370 ; hindi, *Gond*, P. 551 ; hinjol, *Santal*, B. 180 ; hir, *Gond*, T. 325 ; hitta,

kesari, *Santal*, E. 7 ; lallal, *Gond*, P. 632 ; lama, *Kol*, B. 342 ; larchuter, *Kol*, L. 219 ; larka baha, *Santal*, A. 925 ; lasséri, *Baigas*, C. 1931 ; lauri kassamár, *Kurku*, C. 1927 ; layo-gundli, *Santal*, P. 42.

Le   Leinja, *Kurku*, L. 474 ; leja, *Gond*, L. 474 ; lendha, *Santal*, P. 384 ; lóré, *Santal*, F. 237 ; leria, *Gond*, L. 55.

Li   Lihúng, *Kol*, T. 293 ; lilibichi, *Santal*, I. 141 ; lil kathi, *Santal*, P. 1064 ; lisi, *Kol*, S. 2861.

Lo   Loa, *Kol*, *Santal*, F. 179 ; lodam, *Santal*, S. 3062 ; lodh, *Kol*, S. 3062 ; lohagasi, *Kol*, A. 534 ; lopong, *Santal*, T. 293 ; loso, *Kol*, S. 1041 ; loto, *Santal*, R. 1 ; lowa, *Kol*, *Santal*, F. 179.

Lu   Ludam, *Kol*, S. 3062 ; lumam, *Santal*, S. 1934 ; lumang, *Santal*, S. 1934 ; lupi, *Gond*, D. 212 ; lupúng, *Kol*, T. 293.

Ma   Machkunda, *Santal*, P. 1389 ; madge, *Bhil*, T. 361 ; mahaka, *Gond*, A. 534 ; mahoka, *Gond*, T. 325 ; mahu, *Baigas*, B. 220 ; mahurá, *Bhil*, B. 220 ; maika, *Gond*, A. 534 ; mák, *Santal*, B. 142 ; makarkenda, *Santal*, D. 582 ; malhan, *Santal*, D. 789 ; mali buha, *Santal*, O. 18 ; mandar góm, *Santal*, A. 1166 ; mandukum, *Kol*, B. 220 ; mangi, *Gond*, T. 282 ; manjurjuti, *Santal*, E. 80 ; maoo, *Gond*, D. 240 ; marak, *Santal*, V. 174 ; marang kongat, *Santal*, D. 823 ; marar-baha, *Santal*, E. 342 ; marka, *Gond*, M. 147 ; marria, *Gond*, W. 33 ; maru, *Gond*, T. 361 ; marudi, *Santal*, I. 503 ; mat, *Santal*, B. 118 ; matasure, *Kol*, A. 1218 ; matha, *Santal*, A. 1215 ; mathan, *Kol*, D. 292 ; mathomarak, *Santal*, H. 506 ; matkom, *Santal*, B. 220 ; matnak, *Kol*, T. 361.

Me   Mendah, *Gond*, L. 474 ; meral, *Santal*, P. 632 ; merlec, *Kol*, F. 615 ; merlee, *Santal*, F. 615 ; merom chunchi, *Santal*, C. 2525 ; merom met (Ixora), *Santal*, I. 515 ; merom met (Olax), *Santal*, O. 125.

Mi   Mihndi, *Santal*, L. 126 ; milli, *Gond*, P. 632 ; mirchi, *Baigas*, D. 438 ; miri, *Kol*, E. 73 ; mirubaha, *Santal*, A. 89.

Mo   Moch, *Santal*, P. 468 ; mohu, *Kurku*, B. 220 ; mohul, *Bhumij*, B. 220 ; mokha, *Gond*, S. 959 ; mokkak, *Bhil*, S. 959 ; molol, *Gond*, H. 35 ; morouraak, *Santal*, H. 328 ; moru, *Kol*, S. 2508 ; mota bhedi janetet', *Santal*, U. 35 ; mota uric' alang, *Santal*, P. 1179.

Mu   Mulgia, *Kol*, M. 721 ; mundi, *Gond*, S. 2799 ; múng, *Santal*, P. 496 ; munga arak, *Santal*, M. 721 ; murari, *Kurku*, M. 549 ; murga, *Santal*, P. 1370 ; murim, *Kol*, S. 1229 ; murr, *Gond*, *Kurku*, B. 944 ; múrrd, *Kol*, S. 2508 ; murup, *Santal*, B. 944 ; murut, *Kol*, B. 944 ; musna, *Santal*, S. 850 ; mussuck-naba, *Paharia*, D. 227.

Na   Nachal, *Kol*, S. 579 ; naindi, *Gond*, E. 419 ; nalavail, *Gond*, D. 330 ; nalli, *Gond*, P. 632 ; nangtháda, *Kurku*, E. 356 ; nanha, *Santal*, C. 996 ; nanha bindi mutha, *Santal*, F. 283 ; nanha jhunka, *Santal*, C. 2154 ; nanha-pusi-toa, *Santal*, E. 549 ;

nari murup, *Santal*, B. 978 ; nari siris, *Santal*, D. 94.

Ne   Nelli, *Gond*, L. 55 ; neouri, *Santal*, E. 73 ; nerinda, *Gond*, R. 369 ; neuri, *Santal*, E. 73 ; newri, *Santal*, C. 722.

Ni   Niajowa, *Kol*, G. 143 ; nilli, *Gond*, P. 632 ; nim, *Kol*, *Santal*, M. 363 ; nirgiri, *Gond*, V. 164 ; nirgudi, *Kurku*, V. 164 ; niru, *Kurku*, E. 73 ; nisur, *Gond*, E. 73.

No   Noduúr, *Kol*, V. 48.

Nu   Nuruic, *Santal*, C. 756.

Oi   Oit bulung, *Kol*, E. 314.

Ol   Olat, *Santal*, G. 673, 714.

Om   Ome (Miliusa), *Santal*, M. 545 ; omé (Saccopetalum), *Kol*, S. 487.

Or   Orop, *Santal*, C. 2013.

Ot   Ot, *Santal*, A. 590, F. 725 ; ot dhompo, *Santal*, L. 273.

Pa   Padar, *Gond*, *Kurku*, S. 2876 ; paddi, *Gond*, H. 289 ; pader, *Santal*, S. 2876 ; padri, *Gond*, *Bhil*, D. 53 ; padurni, *Bhil*, S. 2865 ; paedebiri, *Paharia*, P. 4 ; paisar, *Kol*, P. 1370 ; pakhar, *Kurku*, F. 216 ; pan, *Bhil*, S. 2876 ; pandrai, *Kol*, A. 717 ; pandri, *Kol*, A. 1146 ; pango nari, *Santal*, C. 2211 ; pani-lara, *Paharia*, V. 184 ; panjon, *Santal*, P. 1048 ; panjot, *Santal*, C. 1386 ; panjra, *Gond*, G. 136 ; panniabhil, *Gond*, G. 124 ; pansi, *Kol*, A. 1146 ; papar, *Kol*, G. 124 ; papasar, *Kol*, G. 124 ; papra, *Kol*, G. 124 ; parasu, *Kol*, L. 219 ; paror jhinga, *Santal*, L. 556 ; parwa, *Santal*, E. 505 ; pas, *Kol*, L. 219 ; paspu, *Gond*, A. 514 ; passerginni, *Gond*, A. 717 ; passi, *Kurku*, *Baigas*, D. 53 ; pasu, *Kol*, L. 219 ; patadhamin, *Kol*, K. 42 ; pata kohnda, *Santal*, P. 1401 ; pato, *Kol*, R. 1 ; patpatta, *Paharia*, T. 24 ; patwa-ghás, *Santal*, C. 775 ; paur, *Gond*, B. 342.

Pe   Peddei, *Gond*, P. 1370 ; pella, *Gond*, A. 1215 ; pender, *Gond*, R. 16 ; pendra, *Gond*, G. 136 ; pepere, *Kurku*, F. 216 ; pepesiman, *Kol*, B. 142 ; perú-maú, *Gond*, O. 567 ; petchamra, *Santal*, H. 92 ; pet chamra banda, *Santal*, V. 154 ; pété, *Kol*, l. 515 ; petisurali, *Gond*, W. 106 ; petrada, *Santal*, J. 114.

Ph   Phalgataitu, *Kurku*, S. 2876 ; phangera, *Gond*, E. 356 ; pharsa, *Baigas*, B. 944 ; phassi, *Kurku*, D. 53 ; phetrak, *Bhil*, G. 136 ; phurpata, *Kurku*, G. 136.

Pi   Piaj, *Santal*, A. 769 ; pial, *Bhumij*, B. 913 ; pichari baha, *Santal*, W. 33 ; pichka, *Kol*, D. 551 ; pichkú, *Kol*, D. 548 ; pindar, *Kol*, R. 16 ; pinde, *Santal*, R. 16 ; pinoh, *Gond*, S. 2850 ; pipar, *Kol*, F. 236 ; pipri, *Kurku*, F. 236 ; píri, *Kol*, D. 32 ; piro, *Kol*, S. 2861 ; pironja, *Mundai*, S. 2861 ; píronja, *Kol*, S. 2861 ; pisi, *Kol*, S. 2319 ; piska, *Santal*, D. 494, 517 ; pitoj, *Santal*, P. 1433 ; pitta, *Gond*, W. 106 ; pitua arak', *Santal*, S. 2515.

Po   Pochati, *Kurku*, H. 164 ; pojo, *Santal*, L. 474 ; pondair, *Kol*, S. 2865 ; pond disom, *Santal*, G. 263 ; pond-gandhari, *Santal*, A. 950 ; pondri, *Kol*, D. 53 ; ponra, *Uraon*, E. 314 ; popro, *Santal*, G. 124 ; porok podha, *Uraon*, F. 156 ;

poros, *Kol,* A. 1489, I. 58 ; porponda, *Kol,* C. 1927 ; porsya, *Baeri,* S. 1250 ; portoho, *Kol,* R. 1 ; poskaolat, *Santal,* K. 42 ; poter, *Kol,* C. 2180 ; potra, *Gond,* D. 330.

Pu Púli, *Gond,* T. 437 ; punk, *Gond,* C. 1742 ; pusi-toa, *Santal,* E. 531 ; púskú, *Gond,* S. 950 ; putri, *Kol,* C. 2180.

Ra Radatbera, *Bhil,* P. 1370 ; raella, *Baigas,* C. 756 ; rai, *Kol,* D. 438 ; raila baha, *Santal,* B. 175 ; raini, *Gond,* M. 583 ; rajbaka, *Santal,* N. 80 ; rakat rohen, *Kol,* S. 2501 ; raketberár, *Gond,* M. 797 ; ralli, *Santal,* P. 805 ; ramra, *Santal,* P. 513 ; rangaini janum, *Santal,* S. 2345 ; raníphul, *Santal,* P. 1104 ; rari, *Kol,* C. 722 ; rasun, *Santal,* A. 779 ; ratop, *Kurku,* M. 232.

Re Rend, *Kurku,* E. 20 ; renga, *Bhil,* Z. 231 ; enta, *Kol,* H. 92 ; rera, *Gond,* C. 756 ; reúnja, *Gond,* A. 249 ; reuta, *Kol,* C. 1927 ; rewat, *Kurku,* C. 722.

Ri Ridi, *Baigas,* E. 20 ; rimmel, *Kol,* O. 127 ; ringa, *Gond,* Z. 231 ; rinja, *Gond,* A. 249 ; riu, *Kol,* F. 156.

Ro Rol, *Santal,* T. 325 ; rola, *Kol,* T. 325 ; rora, *Santal,* M. 71 ; roré, *Kol,* C. 725 ; rot, *Santal,* O. 537 ; royta, *Bhil,* S. 2501.

Ru Ruhen, *Santal,* S. 2501 ; rukni, *Baigas,* T. 522 ; rung, *Kol,* B. 342 ; ruta, *Kol,* O. 537 ; ruti, *Kol,* D. 41 ; rutok, *Kurku,* O. 537.

Sa Sag, *Bhil,* G. 287 ; sag, *Bhil,* T. 232 ; saga, *Kurku,* H. 164 ; saguna, *Santal,* T. 232 ; sahra, *Santal,* S. 2912 ; saikre, *Kol,* L. 55 ; sakar-kenda, *Santal,* I. 348 ; sakomsang, *Kol,* H. 92 ; salga, *Santal,* B. 771 ; sali, *Kol,* H. 517 ; salla, *Gond,* B. 771 ; sama, *Santal,* P. 45 ; samoka, *Gond,* H. 294 ; samur, *Gond,* B. 978 ; sandapsing, *Kol,* S. 959 ; sandi kuya, *Kol,* W. 131 ; sandiomé, *Kol,* P. 1058 ; sanko, *Kol,* A. 1192 ; saparom, *Santal,* N. 179 ; saparung, *Kol,* N. 179 ; saprung, *Kol,* D. 180 ; sar, *Santal,* S. 6 ; sáraka, *Gond,* B. 913 ; saram lutur, *Santal,* C. 1388 ; sargi, *Bhumij,* S. 1656 ; saring, *Kol,* D. 292 ; sarjom, *Santal, Kol,* S. 1656 ; sarlarkha, *Kol,* F. 615 ; satsaiyar, *Santal,* D. 41 ; sauri arak', *Santal,* P. 1091 ; saurighás, *Santal,* H. 164 ; saweli, *Kurku,* P. 45.

Se Sega janum, *Santal,* M. 562 ; segum kati, *Gond,* D. 402 ; sekra (Lagerstrœmia), *Santal,* L. 42 ; sekra (Zizyphus), *Santal,* Z. 273 ; sekrec, *Santal,* L. 55 ; sekura, *Kol,* S. 1656 ; sekwa, *Uraon,* S. 1656 ; selte, *Gond,* C. 1931 ; sengel sali, *Kol,* C. 1031 ; sengel sing, *Santal,* T. 509 ; sér, *Gond,* O. 537 ; serali, *Kol.,* F. 615 ; serilli, *Gond,* F. 216 ; seris, *Gond,* D. 41 ; serisso, *Kurku,* D. 41 ; seta, *Paharia,* G. 263 ; seta andir, *Santal,* G. 696 ; setakata, *Santal,* G. 696 ; seta kata arak, *Santal,* G. 753.

Sh Shermana, *Gond,* O. 537.

Si Sigrik, *Santal,* S. 579 ; siju, *Santal,* E. 553 ; sikriba sikérúp, *Kol,* P. 338 ; sikru, *Kol,* A. 988 ; sikuar, *Santal,* U. 32 ; sikyom baha, *Santal,* C. 2072 ; silu,

Kurku, C. 1931 ; sim busak, *Santal,* F. 644 ; sim matha sura, *Santal,* D. 341 ; simyanga, *Kol,* V. 158 ; sina, *Gond,* L. 55 ; sindi, *Gond,* P. 588 ; sinduari, *Santal, Kol,* V. 164 ; sindwor, *Kol,* V. 164 ; singhindamin *Kol,* G. 663 ; síngya, *Kol,* B. 356 ; singyara, *Santal,* B. 308 ; sipna, *Kurku,* T. 232 ; sír, *Bhil,* B. 913 ; sirgit arak, *Santal,* C. 868 ; siric-samano, *Santal,* G. 243 ; sirom, *Santal,* A. 1097 ; sisi, *Kol,* S. 2819, 2861 ; sísír, *Uraon,* S. 2861 ; siso, *Kol,* D. 41 ; sitanga, *Santal,* G. 705 ; sitapelu, *Kol,* G. 705 ; sitapordóh, *Santal,* F. 202 ; sitta, *Gond,* T. 28.

So Sogot, *Ho. Kol,* T. 444 ; soimi, *Gond,* S. 2501 ; so-kod, *Santal,* E. 419 ; somepatta, *Kol,* O. 233 ; son jhunka, *Santal,* C. 2160 ; sonora, *Kurku,* T. 434 ; soso, *Santal,* S. 1041 ; sosokera, *Kol,* F. 202.

Su Suaruk, *Kurku,* D. 438 ; sukriruin, *Kol,* H. 131 ; suman, *Kol,* F. 265 ; suman-pipar, *Kol,* F. 265 ; sunamjor, *Santal,* F. 265 ; sundeh, *Gond,* H. 339 ; sunonijar, *Santal,* F. 145 ; suntu-bukrui, *Santal,* E. 166 ; sunumjon, *Santal,* F. 253 ; sura, *Santal,* C. 2622 ; surgúja, *Kol,* *Santal,* G. 735 ; surteli, *Baigas,* D. 53 ; surteyli, *Gond,* W. 106 ; susam, *Kol,* S. 1229 ; sutri, *Santal,* P. 486.

Ta Tahaka, *Gond,* T. 293 ; taka (silver), *Santal,* A. 1359 ; taka (Terminalia), *Gond,* T. 293 ; takli, *Kurku,* S. 2850 ; tale, *Santal,* B. 663 ; tamák, *Bhil,* S. 2799 ; tandi, *Santal,* S. 1699 ; tandi bhedi janetet, *Santal,* D. 339 ; tandi chatom arak (Desmodium), *Santal,* D. 343 ; tandi chatom arak (Oxalis), *Santal,* O. 547 ; tandi, jhapni, *Santal,* Z. 300 ; tandi kode baha (Evolvula), *Santal,* E. 581 ; tandi khode baha (Indigofera), *Santal,* I. 134 ; tandi meral, *Santal,* P. 668 ; tandisol, *Santal,* I. 335 ; tandi sunsuni, *Santal,* D. 343 ; tandi sura, *Santal,* C. 2612 ; tapkél, *Bhil,* R. 16 ; tarbuj, *Santal,* C. 2263 ; taro, *Kurku,* B. 913 ; taróp, *Santal,* B. 913 ; tarsekotap, *Santal,* G. 723 ; tarum, *Kol,* B. 913.

Te Tejo malla, *Santal,* C. 1205 ; teka, *Gond,* T. 232 ; teley, *Kol,* S. 2850 ; telhec, *Santal,* S. 2850 ; tendu, *Baigas,* D. 615 ; tenduwa, *Bauris,* T. 434 ; teon-kula, *Kol,* T. 434 ; teóri khair, *Bhil,* A. 229 ; terel, *Santal,* D. 615 ; teri, *Santal,* C. 35 ; teto, *Kurku,* S. 2884 ; tewar, *Uraon,* B. 330 ; tewsa, *Bhil,* O. 537.

Th Thanki, *Kol,* E. 73 ; thaur, *Gond,* B. 330 ; thoja, *Gond,* F. 179 ; thoska, *Gond,* S. 487 ; thuiak' arak', *Santal,* M. 429 ; thundri, *Gond,* C. 725.

Ti Tihon, *Santal,* C. 289 ; tikkoe, *Gond,* A. 514 ; til, *Kol,* S. 1078 ; tilai, *Kol, Santal,* W. 38 ; tilliah, *Baigas,* W. 33 ; tilmin, *Santal,* S. 1078 ; tinsai, *Gond,* O. 537 ; tiril, *Kol,* D. 656 ; tiririte, *Santal,* V. 112 ; tirra, *Santal,* P. 1401 ; tirup-sing, *Mandari,* T. 565.

To Togot arak', *Santal,* H. 233 ; togri, *Bhil,* I. 141 ; topa, *Kol,* E. 453 ; tor-chandbol, *Santal,* E. 250 ; toska, *Gond,* L. 474 ;

totka bendi, *Kol,* J. 41 ; totonopak ',
    Santal, E. 453.

Ts Tshirka, *Kol,* Z. 273.

Tu Tudi, *Kurku,* I. 503 ; tumki, *Gond,* D.
    615 ; tumma, *Gond,* A. 249 ; tummer,
    Gond, D. 615 ; tumri, *Gond,* D. 615 ;
    tunang, *Kurku,* B. 978 ; túndri, *Gond,*
    C. 722 ; turi sim, *Santal,* O. 230.

Tw Twar, *Uraon,* B. 330.

Ul Ul, *Santal,* M. 147 ; uli, *Kol,* M. 147.

Un Undaru, *Santal,* A. 267 ; unt katar, *Gond,*
    S. 2876.

Ur Urí, *Santal,* O. 258 ; úrí horo, *Santal,*
    O. 258 ; uru, *Gond,* A. 681 ; urumin,
    Kol, E. 333.

Us Usir, *Gond,* P. 632 ; ustumri, *Gond,* S.
    2960.

Ut Uterr, *Kol,* I. 141 ; utri dudhi, *Santal,*
    C. 2274 ; utru-banda, *Uraon,* C. 2612.

Va Vadur, *Gond,* D. 292 ; varni, *Santal,* C.
    1380.

Ve Veddar, *Gond,* D. 292 ; verendi, *Kol,* J.
    52.

Wa Walkóm, *Kol,* S. 2861 ; wallaiki, *Gond,*
    B. 632.

Ye Yehera, *Bhil,* T. 293 ; yenu, *Gond,* E.
    83 ; yerjoohetta, *Gond,* A. 711 ; yerma,
    Gond, A. 1149.

KANARESE, MALAYALAM, TAMIL, TELUGU.

Aa Aadai-otti, *Tam.*, T. 835 ; aadai-otti, *Tam.*, T. 839 ; aatlaria, *Tam.*, P. 1091.

Ab Abalu, *Kan.*, C. 397 ; ab-bro-ny, *Kan.*, F. 455 ; abhini, *Tel.*, P. 87 ; abini, *Tam.*, P. 87 ; ábúba, *Tel.*, T. 600 ; ábuvva, *Tel.*, T. 600.

Ac Acatsjabulli, *Mal.*, C. 805 ; acha (Diospyros), *Tam.*, D. 569 ; achá (Hardwickia), *Tam.*, H. 16 ; achchellu, *Kan.*, S. 1078 ; acherunáranna, *Mal.*, C. 1286 ; achı, *Tam.*, O. 233.

Ad Adaka, *Mal.*, A. 1294 ; adakodien, *Mal.*, H. 328 ; adala-vitala, *Tel.*, L. 283 ; adamarathu, *Tam.*, B. 868 ; adamboe, *Mal.*, L. 42 ; ada-modien, *Mal.*, H. 328 ; adasara, *Tel.*, A. 484 ; ádasyamáli, *Tel.*, H. 92 ; adavalantiriki, *Tam.*, F. 597 ; adavi, *Kan.*, E. 278 ; adavi-ámudam, *Tel.*, B. 28 ; adavi-atti, *Kan.*, F. 202 ; adavi bende, *Kan.*, T. 392 ; adavi-bíra, *Tel.*, L. 563 ; adavicheruku, *Tel.*, S. 4 ; adavi-chikkudu, *Tel.*, D. 789 ; adaviy-élakáya, *Tel.*, A. 976 ; adavigoránta, *Tel.*, E. 370 ; adavi-ippechettu, *Tel.*, B. 220 ; adavi-irulli, *Kan.*, U. 39 ; adavi-jílakara, *Tel.*, V. 73 ; adavi kákara, *Tel.*, M. 634 ; adavilavangapatta, *Tel.*, C. 1158 ; adavi-lavanga-patte, *Kan.*, C. 1158 ; adavimalle, *Tel.*, J. 13 ; adavi-munaga, *Tel.*, M. 721 ; adavi nábhi, *Tel.*, G. 243 ; adavinimbe, *Kan.*, A. 1601 ; adavinimma, *Tel.*, A. 1601 ; adavi-pálatíge, *Tel.*, C. 2247 ; adavi ponna, *Tel.*, R. 242 ; adavipotla, *Tel.*, T. 576 ; adavipratti, *Tel.*, T. 387 ; adavi-puch-cha, *Tel.*, C. 2310 ; adavitéku, *Tel.*, T. 232 ; adaviyéla-káya, *Tel.*, A. 976 ; adda, *Tel.*, B. 318 ; B. 342 ; áddabukkudu, *Tel.*, E. 25 ; addalai, *Tam.*, J. 52 ; addannú, *Tel.*, G. 229 ; addasaram, *Tel.*, A. 484 ; ádeli, *Tel.*, L. 283 ; adevie-mallie, *Tel.*, J. 10 ; adhatodai, *Tam.*, A. 484 ; adike, *Kan.*, A. 1294 ; aditya bhakti-chettu, *Tel.*, H. 74 ; adit-yalu, *Tel.*, L. 283 ; adiviea-mida, *Tel.*, J. 41 ; adivie kunda gudda, *Tel.*, D. 821 ; adivio-mamadie, *Tel.*, S. 2649 ; ádíyalu, *Tel.*, L. 283 ; adonda, *Tel.*, C. 416 ; adugu-pal-sorrah, *Tam.*, F. 377 ; adul, *Mal.*, S. 888 ; adulay-kai, *Tam.*, C. 2313 ; adutina-pálai, *Tam.*, A. 1395 ; advikuri, *Kan.*, D. 219 ; advimulangi, *Tel.*, B. 546.

Af Afím, *Kan.*, P. 87 ; afiun, *Mal.*, P. 87.

Ag Agákara, *Tel.*, M. 639 ; agal, *Tam.*, C. 1021 ; agar, *Tam.*, A. 1251 ; agaru, *Kan.*, P. 1381 ; agasatamaré, *Tam.*, P. 874 ; agase, *Kan.*, S. 1186 ; agath-thinár, *Tam.*, S. 1186 ; agáti, *Tam.*, S. 1186 ; aggalichandana, *Tam.*, A. 1251 ; aglay, *Tam.*, C. 1021 ; agle-marum, eleutharay, *Tam.*, C. 1021 ; agnimáta, *Tel.*, P. 986 ; agnisikha (Carthamus),

Tel., C. 637 ; agni-skikha (Gloriosa), *Tel.*, G. 243 ; agnivenda-páku, *Tel.*, A. 958 ; agru, *Tel.*, A. 1251 ; ágúba, *Tel.*, T. 600 ; agúre, *Tam.*, H. 444 ; aguskitti, *Tam.*, F. 361.

Ah Ah-ku-lah, *Tam.*, C. 2528 ; ah-ku-lah, *Tam.*, F. 419 ; ahlada, *Kan.*, F. 129 ; ahsunkar, *Kan.*, S. 861 ; ahuau, *Kan.*, A. 514.

Ai Aila-cheddi, *Tam.*, E. 151 ; ailum chedy, *Mal.*, E. 151 ; aima, *Tam.*, B. 913 ; aini (Terminalia), *Kan.*, T. 361 ; aini (Artocarpus), *Mal.*, A. 1482 ; aivanam, *Tam.*, L. 126 ; aiyanepela, *Tam.*, A. 1482.

Aj Ajjanapatte, *Kan.*, A. 1200 ; ájmodá vómá, *Kan.*, C. 701 ; ajumóda-vomam, *Tel.*, C. 701 ; ajumoda vomaru, *Tel.*, C. 701.

Ak Akása támara, *Tel.*, P. 874 ; ákáshagaruda-gadde, *Kan.*, C. 1834 ; ákáshagarudan, *Tam.*, C. 1834 ; aken-parah, *Tel.*, F. 403 ; akkala-karé, *Kan.*, A. 1026 ; akkalkará. *Tel.*, *Tam.*, A. 1026 ; akkara carum, *Tam.*, C. 1047 ; akki, *Kan.*, O. 258 ; akkirakáram, *Tel.*, *Tam.*, A. 1026 ; akku-jella, *Tel.*, F. 546 ; ak-kurati, *Tel.*, F. 311 ; akota, *Kan.*, S. 950 ; akiódu, *Kan.*, J. 61 ; akrottu, *Tel.*, J. 61 ; aku jemudu, *Tel.*, E. 520, E. 527.

Al A'l, *Tam.*, F. 129 ; ala, *Tam.*, *Kan.*, F. 129 ; alá-buvu, *Tel.*, L. 30 ; alada, *Kan.*, F. 129 ; alale-huvvu, *Kan.*, T. 325 ; alale-kayi, *Kan.*, T. 325 ; alalepínda, *Kan.*, T. 325 ; ala-mottah, *Tel.*, F. 432 ; alangi, *Tam.*, A. 681 ; alaranji, *Tel.*, C. 1780 ; alari, *Tam.*, *Mal.*, N. 80 ; alasále, *Tel.*, P. 950 ; alasandi, *Kan.*, V. 116 ; alashí, *Kan.*, L. 385 ; alava, *Kan.*, F. 129 ; alie, *Tam.*, O. 616 ; ali-verai, *Tam.*, L. 283 ; allabatsalla, *Tel.*, B. 207 ; allagiligich-cha, *Tel.*, C. 2163 ; allam, *Tel.*, Z. 201 ; alli (Antiaris), *Tam.*, A. 1200 ; alli (Memecylon), *Tel.*, M. 439 ; alliaku, *Tel.*, M. 439 ; allibija, *Kan.*, L. 283 ; al-liki, *Tel.*, S. 974 ; alli payaru, *Tel.*, G. 679 ; alli-támara, *Tel.*, N. 200 ; talli-támarai, *Tam.*, N. 200 ; alli topalu, *Tel.*, M. 439 ; állu, *Tel.*, P. 332 ; alpam, *Mal.*, B. 791 ; alpogádá-pandlu, *Tel.*, P. 1304 ; alpogádá-pazham, *Tam.*, P. 1304 ; alsanda, *Tel.*, D. 789 ; alshi, *Kan.*, L. 385 ; alshi-virai, *Tam.*, L. 385 ; alti, *Tam.*, H. 16 ; alu (Paspalum), *Tel.*, P. 332 ; álú (Solanum), *Kan.*, S. 2320 ; alu-ba:hehali, *Tel.*, B. 203 ; álú guddalu, *T.l.*, S. 2320 ; alúngú-thadú, *Mal.*, R. 97 ; alusundí, *Tel.*, V. 116.

Am Amadam, *Tel.*, R. 369 ; aman, *Tam.*, C. 691 ; amanakkam, *Tam.*, R. 369 ; amate. *Kan.*, S. 2649 ; amatum, *Tel.*, S. 2649 ; amba, *Kan.*, M. 147 ; ambal, *Tam.*, N,

I z

39, N. 200; ambála chettu, *Tel.*, S. 2649; ambati mádu, *Tel.*, T. 537; ambatte mara, *Kan.*, S. 2649; amberi, *Kan.*, N. 170; ambru, *Mal.*, B. 318; ambutan-wahlah, *Tam.*, F. 511; amdi, *Tel.*, R. 369; amkolam-chettu, *Tel.*, A. 681; amkudu, *Tel.*, W. 122; amkudu-vittulu, *Tel.*, H. 294; amkudu-vittum, *Tel.*, H. 294; am-kúlang, *Tam.*, W. 98; amla, *Tel.*, C. 77; ammah, *Mal.*, R. 97; ammera verai, *Tam.*, C. 741; amrita-balli, *Kan.*, T. 470, T. 483; amrúta balli, *Kan.*, T. 470, T. 483; amruta valli, *Mal.*, T. 470, 483; amte, *Kan.*, S. 2649; amti, *Tel.*, M. 811; amtuasag, *Mal.*, A. 1215; ámuda-pu, *Tel.*, R. 369; amukkura, *Tam.*, W. 93; amukkurá, *Tam.*, W. 98; amúlang kalung, *Tam.*, W. 98; amumpatchay-arissí, *Tam.*, E. 531.

An Anai-gundumani, *Tam.* A. 471; anaikkat rázhai, *Tam.*, A. 603; ánai-nerunji, *Tam.*, P. 363; anai-puli, *Tam.*, A. 455; anai-puliya-marram, *Tam.*, A. 455; anakúva, *Mal.*, C. 2406; anánasu-hannu, *Kan.*, A. 1045; ana-neringie, *Tam.*, P. 363; ána-nerinnil, *Mal.*, P. 363; ánapa chettu, *Tel.*, L. 30; anapa chikkudu, *Tel.*, D. 789; ánapa káya, *Tel.*, L. 30; ana-sandra, *Tel.*, A. 229; anása-pandu, *Tel.*, A. 1045; anásapuvu, *Tel.*, I. 30; aná-shap-pazham, *Tam.*, A. 1045; anasho-vadi, *Tam.*, E. 80; anáshuppu, *Tam.*, I. 30; anasí-pú, *Tam.*, I. 30; anati, *Tel.*, M. 811; anay, *Tam.*, *Tel.*, *Kan.*, *Mal.*, E. 83; anbóti-kura, *Tel.*, O. 547; ancor-uthai, *Tam.*, T. 600; andipunar, *Kan.*, C. 474; ándu, *Tel.*, B. 771; anduga-pisunu, *Tel.*, B. 771, B. 777; anduku, *Tel.*, B. 771; andúvan, *Nilgiris*, V. 1; ánem, *Tel.*, B. 863; anemui, *Tam.*, T. 361; angelí, *Tam.*, A. 1482; ani, *Tam.*, *Tel.*, *Kan.*, *Mal.*, E. 83; ani-nar, *Tam.*, S. 2861; anisaruli, *Kan.*, A. 681; anjalli, *Tam.*, A. 1482; anjanakkallu, *Tam.*, A. 1224; anjana-ráyi, *Tel.*, A. 1224; anjura, *Kan.*, F. 149; anjúrí, *Kan.*, F. 149; ankadósa, *Tel.*, L. 241; ankola, *Kan.*, A. 681; ankudu kurra, *Tel.*, U. 11; anna-bédi, *Tam.*, I. 481; anna-bhédi, *Kan.*, I. 481; anna-bhédi, *Tel.*, I. 481; anna bhédi, *Mal.*, I. 481; annanakkalla, *Mul.*, A. 1224; annanas, *Mal.*, A. 1045; annapa, *Tel.*, D. 789; annegalu-gidá, *Kan.*, P. 363; ansaroli, *Kan.*, A. 681; ansjeli, *Mal.*, A. 1482; ansjeni, *Mal.*, A. 1482; antala, *Kan.*, S. 818; antara-tamara, *Tel.*, L. 355; antawala, *Kan.*, S. 818; antika-dúndiawah, *Tel.*, F. 483; anti-ma·du·am, *Tam.*, G. 278; anti-malari, *Mal.*, M. 606; anti-mantáram, *Mal.*, M. 606; antintalu, *Tam.*, D. 336; antisha, *Tel.*, A. 382; anugakaya, *Tel.*, L. 30; anumulu, *Tel.*, D. 789.

Ap Apa márgamu, *Tel.*, A. 382; appakovay kalung, *Tam.*, R. 341; appracam, *Tam.*, *Tel.*, M. 509; appracum, *Tel.*, *Tam.*, S. 2712.

Ar Ara-bevu, *Kan.*, M. 412; aradal, *Kan.*, G. 14, G. 66; arakku, *Mal.*, L. 1; aral, *Tam.*, F. 561; arali, *Kan.*, F. 236; aralivayr, *Tam.*, N. 80; aralu, *Tam.*, F. 532; arana-maram, *Mal.*, C. 838; ararút-gaddalu, *Tel.*, C. 2385; ararút-kishangu,

Tam., C. 2385; arasa, *Tam.*, F. 236; araya-angely, *Mal.*, A. 1200; archi, *Tam.*, B. 318; áre, *Tel.*, B. 318 : areka, *Tam.*, B. 318; are-maram, *Tam.*, B. 318; are tíge, *Tel.*, B. 318; ari (Bauhinia), *Tel.*, B. 318; arí (Chrysophrys), *Mal.*, F. 405; ari (Oryza), *Mal.*, O. 256; aria-víla, *Mal.*, C. 1367; arikalu, *Tel.*, P. 332; árike, *Tel.*, P. 332; arikota, *Tel.*, C. 1742; arishi, *Tam.*, O. 252; arishina, *Kan.*, C. 2433; arisina burga, *Kan.*, C. 1512; ari-táram, *Tam.*, O. 242; ariti, *Tel.*, M. 811; ariya-véppa, *Mal.*, M. 363; ariza, *Tel.*, F. 465; arkagida, *Kan.*, C. 170; arku-konissi, *Tel.*, F. 530; arle, *Kan.*, F. 236; arni, *Tam.*, S. 2861; arrakíah, *Mal.*, F. 420; arriti-ki, *Tel.*, F. 580; arsanatega (Anthocephalus), *Kan.*, A. 1192; arsanatéga (Adina), *Kan.*, A. 514; arsinagurgimara, *Kan.*, C. 66; arsintega, *Kan.*, A. 514; artala, *Kan.*, S. 818; ártcandí, *Tam.*, F. 363; aru, *Mal.*, C. 826; arudonda, *Tel.*, C. 416; arudu, *Tel.*, R. 663; aruga, *Tel.*, P. 332; arugam-pilla, *Tam.*, C. 2558; aru kanla kachóram, *Tel.*, C. 2381; árukanupula-kránuga, *Tel.*, S. 30; aru-nelli, *Tam.*, P. 627; aruzu, *Tel.*, F. 407; arvada, *Tam.*, R. 663.

As Asam, *Kan.*, T. 28; asana-gurgi, *Kan.*, H. 16; asek (Tamarindus), *Tel.*, T. 28; asek (Saraca), *Tel.*, S. 861; asha, *Kan.*, T. 392; asham tágam, *Tam.*, C. 701; ashamtá-óman, *Tam.*, C. 701; ashoka, *Kan.*, S. 861; ashumadágavóman, *Tel.*, C. 701; asoge, *Kan.*, S. 861; asoka (Polyalthia), *Tel.*, *Kan.*, P. 1052; asoka (Saraca), *Kan.*, S. 861; assothi, *Tam.*, P. 1052; asuna, *Kan.*, B. 868; asva-gandhi, *Kan.*, W. 93; asvalta, *Kan.*, F. 236; aswagandhí, *Tam.*, W. 98; áswa-láyana, *Tel.*, E. 252; aswartham, *Tam.*, F. 236.

At Atakka, *Mal.*, A. 1294; atakká maniyan, *Mal.*, S. 2518; atalari, *Tam.*, P. 1084; atalótakam, *Mal.*, A. 484; atanday, *Tam.*, C. 416; atasi, *Tel.*, L. 385; athi-balla chettu, *Tam.*, S. 1703; atí, *Tam.*, B. 318; atika-mámidi, *Tel.*, B. 619; ati-madhurá, *Kan.*, G. 278; ati madhu-ram, *Mal.*, G. 278; ati-madhuramu, *Tel.*, G. 278; ati-maduram, *Tam.*, G. 278; atiparich-cham, *Tam.*, C. 854; ativa-dayam, *Tam.*, A. 401; ati-vasa, *Tel.*, A. 401; atrupálai, *Tam.*, S. 579; átru-sha-vukku, *Tam.*, T. 70; átta (Anona), *Mal.*, A. 1166; atta (leech), *Mal.*, L. 245; attai, *Tam.*, L. 245; attalu, *Tel.*, L. 245; attandax, *Tam.*, C. 416; atta patti, *Tel.*, M. 557; atti, *Tam.*, *Tel.*, *Kan.*, F. 179; attitippili, *Tam.*, *Mal.*, S. 970; atúkoia, *Tel.*, F. 585; atukula baddu, *Tel.*, V. 231; atunete, *Tam.*, A. 560; atutintap-pála, *Mal.*, A. 1395.

Au Aulanche, *Tel.*, D. 560; aupta, *Kan.*, B. 318; autara támara, *Tel.*, P. 874; authoondy kai, *Tam.*, C. 441.

Av Avaduta, *Tel.*, T. 600; avagude-hannu, *Kan.*, T. 600; avak, *Kan.*, E. 25; avalo, *Tel.*, B. 841; avanakku, *Mal.*, R. 369; avára, *Mal.*, C. 741; ávara-gidá, *Kan.*, C. 741; avarai, *Tam.*, D. 789; avare, *Kan.*, D. 789; avareke, *Kan.*,

C. 741 ; avári, *Tam.,* C. 741 ; avasinana, *Tel.,* S. 1186 ; avatenga tige, *Tel.,* D. 517 ; avesi, *Tel.,* S. 1186 ; ávirai, *Tam.,* C. 741 ; avre, *Kan.,* D. 789 ; avvagúda, *Tel.,* T. 600 ; avvagúda-pandu, *Tel.,* T. 600 ; avvuru-gaddi-véru, *Tel.,* A. 1097.

Ay Aya, *Tam.,* H. 324 ; ayach-chenduram, *Tam.,* I. 472 ; aya-shindúramu, *Tel.,* I. 472 ; ayendal, *Tel.,* R. 369 ; ayma, *Tam.,* S 1939 ; ayni, *Tam.,* A. 1482.

Az Azhinji, *Tam.,* A. 681 ; azinghi-maram, *Tam.,* A. 681.

Ba Babassa, *Tam.,* H. 486 ; babola, *Mal.,* A. 101 ; bachange, *Kan.,* D. 670 ; badadam, *Tel.,* E. 356 ; bádam, *Mal.,* P. 1274 ; bádámi, *Kan.,* P. 1274 ; badamiyanne, *Kan.,* C. 279 ; badamvittulu (Terminali-), *Tel.,* T. 312 ; bádam-vittulu (Prunus), *Tel.,* P. 1274 ; badane kayi, *Kan.,* S. 2284 ; badanike, *Tel ,* L. 549 ; bádapu, *Tel.,* E. 342 ; badidapu, *Tel.,* E. 342 ; badimottah, *Tel.,* F. 569 ; bádise, *Tel.* E. 342 ; badishaya, *Tel.,* F. 392 ; badisopu, *Kan.,* F. 659 ; badnekái, *Kan.,* S. 2284 ; bága, *Tam.,* A. 135,A. 295 ; baga-dhúp, *Kan.,* A. 670 ; bagana, *Kan.,* A. 722 ; bagni, *Kan.,* C. 711 ; bahadha, *Tel.,* T. 293 bahadrha, *Tel.,* T. 293 ; bahel-chulli *Mal.,* H. 508 ; baini, *Kan.,* C. 711 ; bajé, *Kan.,* A. 430 ; bakamu, *Tel ;* C. 35 ; bakánu-chekka, *Tel.,* C. 35 ; bakapu, *Tel.,* C.35 ; bálabándi tíge, *Tel.,* I. 362 ; bálage, *Kan ,* G. 143 ; balai, *Kan.,* D. 615 ; bála-menasu, *Kan.,* P. 801 ; bálarakkasi-gida, *Kan.,* P. 344 ; bálé, *Kan.,* M. 811 ; bál-énaru ; *Kan.,* M. 811 ; balgay, *Kan.,* V. 158 ; bálimtra-pólam, *Tel.,* B. 48 ; balk-uniki, *Kan.,* D. 628 ; ballangi, *Kan.,* P. 1016 ; balli-mut-taga, *Kan.,* B. 978 ; balpalé, *Kan.,* L. 546 ; balsú, *Tam.,* Tel., C. 393, C. 397 ; bálusu, *Tel.,* C. 393 ; bamari, *Tel.,* F. 202 ; banapu, *Kan.,* T. 361 ; bandára (Hymenodictyon), *Tel.,* H. 517 ; bandára (Lagerstrœmia), *Kan.,* L. 53, L. 55 ; bandarayanni, *Kan.,* H. 517 ; bandaru (Adina), *Tel.,* A. 514 ; bandaru (Dodonæa), *Tel.,* D. 725 ; bandi, *Tel.,* C. 122, C. 200 ; bandigarjana, *Tel.,* D. 94 ; bandigúlivinda, *Tel.,* F. 600 ; bandi guriginja, *Tel.,* D. 94 ; bandi gurivenda, *Tel.,* A. 471 ; bandi káttu tige, *Tel.,* C. 1746 ; bandi kóta, *Tel.,* C. 1746 ; bandrike, *Kan.,* D. 725 ; bandu, *Kan.,* D. 725 ; bandurgi, *Kan.,* D. 725 ; bangáru, *Tel. & Kan.,* A. 1622 ; bangáru-réku, *Tel. & Kan.,* A. 1622 ; bangi-aku, *Tel.,* C. 331 ; bangi-ilai, *Tam.,* C. 331 ; banka-jella, *Tel.,* F. 527 ; banka munnú, *Tel.,* C. 1317 ; banki yeddu, *Tel.,* F. 564 ; banni, *Kan.,* A. 229, A. 295 ; bannimara, *Kan.,* A. 291 ; banti, *Tel. ;* T. 17 ; bápana-búri, *Tel.,* E. 23 ; bápanamushti, *Tel.,* O. 127 ; bappayi, *Tel.,* C. 581 ; barabutsali, *Tel.* V. 228 ; baragadam, *Tel.,* I. 131 ; baragu, *Kan.,* P. 63 ; bara kuram, *Mal.,* A. 514 ; bárankichettu, *Tel.,* B. 978 ; barapatálu, *Tel.,* I. 131 ; barapatam, *Tel.,* I. 131 ; bargund, *Tam.,* C. 1940 ; barigalú, *Tel.,* P. 530 ; barijamu, *Tel.,* E. 342 ; bariniki, *Tel.,* S. 2912 ; bari venka, *Tel.,* S. 2912 barjapu, *Tel.,* E. 342 ; barli-arishi, *Tam.,* H. 382 ; barli-biyam, *Tel.,* H. 382 ; baro-kala-goru,

Tam., H. 159 ; barranki, *Tel.,* S. 2912 ; barre bach-chali, *Tel.,* V. 228 ; barumbiss, *Tel.,* I. 51 ; basari-mara, *Kan.,* S. 30 ; basavanapáda, *Kan.,* B. 304 ; basella-kíra, *Mal.,* B. 203 ; basri, *Kan.,* F. 216, F. 236 ; bassari, *Kan., Tel.,* F. 216 ; bata ganapu, *Tel.,* S. 2799 ; batái náringa pandu, *Tel.,* C. 1234 ; bata-jania, *Tel.,* E. 428 ; batala, *Kan.,* G. 714 ; batáte, *Kan.,* S. 2320 ; batchaliaku, *Tam.,* P. 1179 ; batgadle, *Kan.,* P. 885 ; batta, *Tel.,* B. 180 ; batta-parra, *Mal.,* F. 371 ; bavungie, *Tel.,* C. 854 ; bawang, *Mal.,* A. 769.

Be Beami, *Tam.,* H. 149 ; beati, *Tam.,* C. 785), hédisa tiva, *Tel.,* V. 213 ; beina, *Kan.,* C. 711 ; bejalu, *Kan.,* A. 1149 ; bekaro, *Kan.,* S. 2806 ; bél, *Kan.,* F. 53 ; bélada, *Kan.,* F. 53 ; belandi, *Coorg,* C. 838 ; belanji havulige, *Kan. ;* A. 440 ; belerica, *Mal.,* C. 170 ; beli-korava, *Kan.,* F. 516 ; bella, *Kan. ;* S. 370 ; bellaka, *Kan.,* K. 42 ; bellamu, *Tel.,* S. 370 ; bella-schora; *Mal.,* L. 130 ; belli, *Kan.,* A. 1359 ; belli ? rekhu, *Kan.,* A. 1359 ; belluli, *Kan.,* A. 779 ; bellum *Tel.,* S. 370 ; belori, *Nilgiris,* L. 493 ; belpatri, *Kan.,* A. 534 ; benáde, *Nilgiris,* E 370 ; bend, *Tel.,* A. 560 ; bendakainaru, *Kan.,* H. 196 ; benda-káya, *Tel.,* H. 196 ; bendé-káyi, *Kan.,* H. 196 ; bendekdí, *Kan.,* H. 196 ; bende-naru, *Kan.,* K. 42 ; bendí, *Kan.,* K. 42 ; benga (Pterocarpus), *Kan.,* P. 1370 ; bengha (Albizzia), *Kan.,* A. 695 ; bentaek, *Tam.,* L. 53 ; beppale, *Kan ,* H. 294 ; beppalli, *Kan.,* W. 122 ; beratu mara, *Kan.,* M. 550 ; béri, *Tel.,* L. 266 ; beta mu, *Tel.,* C. 104 ; betha, *Kan.,* C. 115 ; bettada-bevina, *Kan.,* M. 393 ; bettadaharalu, *Kan.,* J 41 ; bettada-pada-vala, *Kan.,* T. 576 ; betta-ganapa, *Tel.,* A. 514 ; bettakanagala, *Kan.,* D. 428 ; bettam, *Tel.,* C. 104 ; bettapu chettu, *Tel.,* C. 104 ; betta tovare, *Kan.,* C. 1512 ; bévina mara, *Kan.,* M. 363 ; bévu, *Kan.,* M. 393 ; bévu bettabévu, *Kan.,* M. 412.

Bh Bhadrákshi, *Tel.,* M. 606 ; bhadramuste, *Tel.,* C. 2612 ; bhallú-chettu, *Tel.,* C. 1031 ; bhangi, *Kan.,* C. 331 ; bhangigida, *Kan.,* C. 331 ; bhangra, *Tel.,* S. 2354 ; bhárangi, *Tel.,* C. 1388 ; bhatala penari, *Kan.,* S. 2824 ; bhatta, *Kan.,* O. 258 ; bhávanji, *Tel.,* P. 1352 ; bhendi, *Tam.,* H. 196 ; bherda, *Kan.,* T. 293 ; bhudoi, *Mal.,* F. 202 ; bhujapatri chettu, *Tel.,* B. 501 ; bhumichekri-gadde, *Kan.,* I. 379 ; bhúmi-tayilam, *Tel.,* P. 436 ; bhutan kusam, *Tel.,* C. 2180 ; bhuttále, *Kan.,* A. 603 ; bhú-tulasi, *Tel.,* O. 18 ; bhyni, *Kan.,* C. 711.

Bi Bibla, *Kan.,* P. 1370 ; bída, *Tel.,* M. 562 ; bidarie, *Tel.,* E. 531 ; bidaruppu, *Kan.,* B. 84 ; bidungulu, *Kan.,* B. 118 ; biduru nána biyyam, *Tel.,* E. 549 ; bíja púra, *Tel.,* C. 1270 ; bija-puramu, *Tel.,* C. 1270 ; bikka gida, *Kan.,* G. 116 ; bikki (Elæocarpus), *Nilgiris,* E. 63 ; bikki (Gardenia), *Tel.,* C. 1941 ; bikiro, *Kan.,* S. 2837 ; bilapatri, *Kan.,* A. 534 ; bile padri, *Kan.,* S. 2865 ; bile sasive, *Kan.,* B. 841 ; bilgu, *Tel.,* C. 1031 ; bili, *Kan.,* R. 31 ; bili baragu, *Kan.,* P. 63 ; bili-barlu, *Kan.,* E. 289 ; bili-bili-káyalu, *Tel.,* A. 1644 ; bili burga, *Kan.*, E. 289 ; biligárá, *Kan.*, B. 731 ; bili-jáli, *Kan.*, A. 249 ; bilijali topáli, *Kan.*, A. 249 ; bili manavare, *Kan.*, D. 789 ; bilimbikáy, *Tam.*, A. 1644 ; bili-sakkare, *Kan.*, S. 375 ; bili-sásave, *Kan.*, B. 800 ; bilitigadu, *Kan.*, I. 415 ; bil-kam-bi, *Kan.*, A. 686 ; billa, *Kan.*, S. 2876 ; billa gan-néru, *Tel.*, V. 138 ; billa ilei, *Kan.*, R. 51 ; billa kora, *Tel.*, C. 1031 ; billawar, *Kan.*, A. 711 ; billi, *Nilgiris*, R. 253 ; billi matti, *Kan.*, T. 282 ; billu, *Tel.*, C. 1031 ; billu-chettu, *Tel.*, C. 1031 ; billuda, *Tel.*, C. 1031 ; billugaddi, *Tel.*, S. 49 ; billukura, *Tel.*, C. 1031 ; bilu, *Tel.*, C. 1031 ; biluga, *Tel.*, C. 1031 ; bilvapandu, *Tel.*, A. 534 ; bilvara, *Kan.*, A. 711 ; bilwara, *Tam.*, A. 711 ; bimbiká, *Tel.*, C. 919 ; biné, *Kan.*, C. 1995 ; bintangor, *Mal.*, C. 146 ; bíra, *Tam.*, E. 73 ; bira-káya, *Tel.*, L. 556 ; birama-dandu, *Tam.*, A. 1351 ; bíranerijamanu, *Tel.*, E. 73 ; biridi, *Kan.*, D. 64 ; bísha, *Mal.*, B. 417 ; bisi, *Mal.*, A. 1288 ; biti, *Kan.*, D. 41 ; biyam, *Tel.*, O. 258 ; biyawak, *Mal.*, R. 116.

Bl Blerong, *Mal.*, S. 2999.

Bo Bobbi, *Kan.*, C. 162 ; boberlu (Dolichos), *Tel.*, D. 789 ; boberlú (Vigna), *Tel.*, V. 116 ; bobra, *Tel.*, V. 116 ; bockada, *Tel.*, C. 1380 ; boda-mamadi, *Tel.*, F. 202 ; bodanta-chettu, *Tel* , B. 308 ; bóda-sarum, *Tel.*, S. 2792 ; bóda-tarapu, *Tel.*, S. 2518 ; bodda, *Tel.*, F. 179 ; boddakúra, *Tel.*, P. 1191 ; boddamámidi, *Tel.*, H. 124!; boddi, *Tel.*, R. 487 ; boddi chettu, *Tel.*, M. 15 ; bod dikúra, *Tel.*, R. 487 ; boddu malle, *Tel.*, J. 35 ; boddu-pávilikúra, *Tam.*, *Tel.*, P. 1179 ; boddu-pávili-kurá-vittulu, *Tam.*, P. 1179 ; bodina gidda, *Kan.*, A. 1288!; bodu, *Tel.*, O. 230 ; boggu, *Tel.*, C. 487, C. 1414 ; bója, *Tel* , X. 16 ; bójeh, *Tel.*, X. 16 ; bokal-boklu, *Kan.*, M. 570 ; boka sorrah, *Tam.*, F. 381 ; boke-nakú, *Tel.*, L. 451 ; boki-sorrah, *Tel.*, F. 401 ; bokkudu, *Tel.*, H. 486 ; bokkudu-chettu, *Tel.*, H. 486 ; bólá, *Kan.*, B. 48 ; boli, *Kan.*, C. 122 ; bolla gadimi, *Tel.*, W. 65 ; bolundúr, *Kan.*, L. 53 ; bolur, *Kan.*, T. 372 ; bombalinas, *Tam.*, C. 1263 ; bombelimarunga, *Mal.*, C. 1263 ; bomma jemudu, *Tel.*, E. 496 ; bomma kachika, *Tel.*, C. 2013 ; bomma médi, *Tel.*, F. 202 ; bomma sári, *Tel.*, P. 1060 ; bommiday, *Tel.*, F. 561 ; bomri, *Tel.*, F. 561 ; bonda-bíja, *Kan.*, C. 1641 ; bonda-janu, *Tel.*, S. 2424 ; bondgu, *Tel.*, H. 159 ; bondh-vála, *Kan.*, S. 2865 ; bónga, *Tel.*, B. 118 ; bonga, veduru, *Tel.*, B. 118 ; bongeri, *Kan.*, G. 136 ; bonta-ariti, *Tel* , M. 811 ; bonta chámalu, *Tel.*, P. 53 ; bonta-chemudu, *Tel.*, E. 496 ; bontah, *Tel.*, F. 386 ; bonta-kalli, *Kan.*, E. 553 ; bonta-shama, *Tel.*, P. 53 ; bonta vempali, *Tel.*, T. 270 ; bonthshama, *Tel.*, P. 53 ; bontú, *Tel.*, F. 579 ; boomidapu, *Tel.*, I. 131 ; bootigi, *Tel.*, E. 73 ; boppayi, *Tel.*, C. 581 ; bor, *Mal.*, F. 129 ; botanskam, *Tel.*, E. 73 ; botchí, *Tel.*, F. 384.

Br Brahma-dandi-chettu, *Tel.*, A. 1351 ; abrahmdanti, *Mal.*, A. 1351 ; brahma-médi, *Tel.*, F. 202 ; brah-mari mari, *Tel.*, C. 1388.

Bu Buchakarum, *Tel.*, A. 1146 ; bucha-pattai, *Kan.*, C. 726 ; buda darini, *Tel.*, C. 563 ;

búdadegummadi, *Tel.*, C. 2331 ; budadurmi, *Tel.*, C. 563 ; budamara, *Tel.*, G. 705 ; budári, *Kan.*, S. 1250 ; buddabúsara, *Tel.*, P. 682 ; búdekum-bala-káyi, *Kan.*, B. 430 ; búdide gum-madi, *Tel.*, B. 430 ; búdide-vuppu, *Tel.*, C. 527 ; búdthí kíray, *Tam.*, R. 487 ; budukattalenaru, *Kan.*, A. 603 ; bukkapu-chekka, *Tel.*, C. 35 ; bulali, *Tam.*, G. 226 ; bullie, *Tel.*, R. 110; bulpum, *Tam.*, S. 2712 ; buma-chumadoo, *Kan.*, E. 496 ; bummarri, *Tel.*, F. 202 ; bundédu, *Tel.*, D. 725 ; bungárú, *Tel.*, G. 317 ; bungárum, *Tel.*, G. 317 ; buraga, *Tel.*, E. 295 ; buraga, *Tel.*, S. 1939 ; buraga-pintha, *Tel.*, E. 295 ; buragasánna, *Tel.*, E. 289 ; búrakas, *Tam.*, F. 448 ; búrá-shakkara, *Tel.*, S. 374 ; búrásharukkarai, *Tam.*, S. 374 ; búrda-gúmúdú, *Tel.*, B. 430 ; búrga, *Tel.*, S. 1939 ; búrja, *Tel.*, H. 517 ; burkai, *Tel.*, L. 556 ; burla, *Kan.*, B. 632 ; buróni, *Tel.*, F. 194 ; burranuge, *Kan.*, O. 153 ; bursh, *Toda.*' T. 437; burudu jinka, *Tel.*, S. 1250, búruga, *Tel.*, E. 289 ; burús, *Tam.*, C, 1031 ; búsara káya, *Tel.*, P. 682 ; búsi, *Tel.*, V. 177 ; buta-kadambe, *Tam.*, S.2799; bútále, *Kan.* G. 714 ; butalli ; *Tam.*, G. 226.

By Byala, *Kan.*, F. 53 ; byaladahannu, *Kan.*, F. 53.

Ca Caar-noochie, *Tam.*, J. 41 ; caat karnaykaloung, *Tam.*, D. 821 ; caat-mallica, *Tam.*, J. 10 ; calamoiapota, *Tel.*, F. 453 ; cambegida, *Kan.*, C. 289 ; cambú-kelletí, *Tam.*, F. 486 ; canari, *Mal.*, C. 279 ; caugu, *Tam.*, S. 1682 ; capi, *Tam.*, *Tel.*, C. 1641 ; capie-cottay, *Tam.*, C.'1641 ; capúr ; *Mal.*, C. 489; carambu, *Mal.*, J. 114-caramunnypyre, *Tam.*, V. 116 ; carie, kíray, *Tam.*, P. 1179; caril-kíray; *Tam.*, P. 1179 ; caripe, *Tel.*, F. 543 caroua ; *Tam.*, F. 542 ; carramanní, *Tam.*, F. 471 ; carúkúva, *Tam.*, Z. 229 ; carú-múlí-candí, *Tam.*, F. 471 ; casacasa, *Tam.*, *Tel.*, P. 87 ; casanelai, *Tam.*, M. 439 ; casara kaia, *Tel.*, C. 2313 ; cashamaram, *Tam.*, M. 439; cashi-mara, *Tel.*, F. 449 ; catappa, *Mal.*, S. 1939 ; cate, *Tam.*, A. 137 ; catimandu, *Tam.*, D. 387 ; caval sorrah, *Tam.*, F. 381 ; cavalum, *Mal.* S. 2809 ; cavara-pullu, *Mal.*, E. 166 ; cay-vang- dang, *Mal.*, F. 113.

Ch Chabai, *Mal.*, C. 455 ; chabe-lombok, *Mal.*, C. 455 ; chadda-vakku, *Tam.*, W. 13 ; chadu-perigi, *Tel.*, F. 351, 352 ; chága, *Tel.*, S. 785 ; chakimuki, *Tam.*, *Tel.*, F. 652 ; chakota, *Kan.*, S. 950; chakrakéliariti, *Tel.*, M. 811 ; chalapachchi, *Tel.*, I. 125 ; chalavamiri-yálu, *Tel.*, P. 801 ; chalie, *Mal.*, C. 466 ; challa (Asparagus), *Tel.*, A. 1575 ; challá (Lagerstroemia), *Kan.*, L. 42 ; challagaddalu, *Tel.*, A. 1577 ; challagaddu, *Tel.*, A. 1575 ; challa-gumudu, *Tel.*, C. 298 ; challani, *Tam.*, D. 701 ; challe, *Kan.*, Z. 290 ; challemara, *Kan.*, C. 1931 ; chalmarí, *Kanara*, A. 306 ; chama, *Tel.*, P. 53 ; chama dumpa, *Tel.*, C. 1732 ; cháma-gadda, *Tel.*, C. 1732 ; chámai, *Tam.*, P. 67 ; chámakúra, *Tel.*, C. 1732 ; chama-kuru, *Tel.*, C. 1732 ; chámalu, *Tel.*, P. 53 ; chámanti, *Tel.*, C. 1403 ; chami, *Tel.*, P. 1259 ; chamma,

Tel., C. 289; champai, *Tam.,* S. 1174; champakam, *Mal.,* M. 517; champakamu, *Tel.,* M. 517; chámpéyamu, *Tel.,* M. 517; chamundi-huvvu, *Kan.,* A. 1184; chámunti, *Tel.,* C. 1047; chanam, *Mal.,* C. 2105; chanangi, *Kan.,* L. 252; chanchali kúra, *Tel.,* D. 420; chanda, *Tel.,* T. 15; chandam, *Tel.,* P. 1381; chandanam, *Tel.,* S. 790; chandana mutti, *Mal.,* S. 790; chanda napu chettu, *Tel.,* S. 790; chandarasa, *Kan.,* F. 179; chandkal, *Kan.,* M. 15; chanjra, *Tel.,* M. 562; chandra-kánta, *Tel.,* M. 606; chandramalli, *Tel.,* M. 606; chandra-mallige, *Kan.,* M. 606; chandra-poda, *Tel.,* A. 1362; changala, *Tel.,* S. 910; changalamparanda, *Mal.,* V. 219; changan-chedi, *Tam.,* A. 1665; chani, *Tel.,* P. 1259; chanki, *Mal.,* C. 706; channangi *Kan.,* L. 55; chanupála-vittulu, *Tel.,* C. 728; chappanum, *Mal.,* C. 35; chara, *Tam.,* D. 569; *Tel.,* B. 913; chára-chettu, *Tel.,* B. 913; charachi, *Tel.,* G. 714; cháramámidi, *Tel.,* B. 913; chára-pandu, *Tel.,* B. 913; chára-puppu, *Tel.,* B. 913; charki, *Tel.,* S. 30; charu mamudí, *Tel.,* B. 913; cháta-kattu-tiva, *Tel.,* I. 376; chatarashi, *Tel.,* F. 721; cháta-ráshi, *Tel.,* F. 723; chatirak-kalli, *Mal.,* E. 496; chattuelupa, *Tam.,* T. 293; chattu mallika, *Tam.,* J. 10; chavaka-maram, *Mal.,* C. 826; chavela, *Mal.,* S. 2424; chavukumánú, *Tel.,* C. 826; chavuku-patta, *Tel.,* C. 826; chaw-a-manu, *Tel.,* A. 988; chaya-vellachí, *Tam.,* F. 391; cháyilam, *Mal.,* M. 473; chayud pottah, *Tel.,* T. 576; chébíra, *Tel.,* P. 425; chechu, *Tam.,* C. 123 ;3 chédu-bíra, *Tel.,* L. 563; chédukodisha-vittulu, *Tel.,* H. 294; chedu paddu dumpa, *Tel.,* D. 494; chédu-potla, *Tel.,* T. 576; chéga gadda, *Tel.,* V. 25; chékurti tivva, *Tel.,* P. 396; chelagada, *Tel.,* I. 348; chelahwahlah, *Tam.,* F. 368; chella, *Kan.,* C. 19 31; chella kassu, *Tam.,* F. 448; chelutímara, *Kan.,* C. 1931; chema, *Tel.,* C. 1732; chemelpaniché, *Tam.,* D. 564; chemlú-pamú, *Tam.,* F. 500; chémpa-kizhanna, *Mal.,* C. 1732; chempa-rattip-púva, *Mal.,* H. 227; chempara valli, *Mal.,* V. 205; chemudu, *Tel.,* E. 527; chéna, *Mal.,* T. 883; chénakizhangu, *Mal.,* T. 15; chenangi, *Tel.,* L. 55; chenchali kúra, *Tel.,* D. 420; chendala, *Coorg.,* S. 950; chendra-sinduri, *Tel.,* M. 71; cheninge, *Kan.,* L. 55; chennanáyakam, *Mal.,* A. 824; chennangi, *Tel.,* L. 42; chenthakanni, *Kan.,* M. 15; chepavaj-ramú, *Tel.,* A. 393; cheppuru, *Kan.,* B. 304; cheppu tatta, *Tel.,* D. 336; cheputdtaku, *Tel.,* A. 1545; chera-gaddam, *Tel.,* I. 125; cheratali badu, *Tel.,* D. 330; cherivelu, *Tel.,* O.137; cheriya-ela-vanna-toli, *Mal.,* C. 1196; cherru puli, *Mal.,* T. 427; cheru-chána-vittinte-vitta, *Mal.,* L. 385; cheru-chunta, *Mal.,* S. 2280; cheru-katruvazha, *Mal.,* A. 836; cheruku, *Tel.,* S. 30; cherukulo-bhedam, *Tel.,* S. 30; cherunaranna, *Mal.,* C. 1296, cheru-náranná, *Mal.,* C. 1258; cherun kuru, *Mal.,* S. 1041; cheru pinnay, *Tam.,* C. 162; cherupoiaar, *Mal.,* P. 513; cheru-pullate, *Mal.,* I. 125; cheru tékka, *Mal.,* C. 1388; cheta kum karra, *Tel.,* C. 1021; chetenta, *Tel.,*

D. 332; chéti-potla, *Tel.,* T. 576; chétippa, *Tel.,* H. 517; chetni maragu, *Kan.,* M. 447; chettu (Strychnos), *Tel.,* S. 2960; chettu (Ophiorrhiza), *Tel.,* O. 180; chettú (Canthium), *Tel.,* C. 393; chevalle, *Tam.,* F. 340; chéva manu, *Tel.,* S. 2501; chevulapilli tige, *Tel.,* I. 362; chhaparbadne, *Kan.,* L. 596; chhatnia, *Mal.,* A. 871; chick-lenta, *Tel.,* S. 1223; chick-linta-kura, *Tel.,* M. 306; chigri, *Kan.,* S. 1226; chikaya, *Tel.,* A. 200; chik bévu, *Kan.,* M. 393; chikul, *Kan.,* A. 717; chilaka dúdúga, *Tel.,* P. 1058; chilaka-kúra, *Tel.,* A. 923; chilaka-tóta-kúra, *Tel.,* A. 923; chilikeswarapu, *Tel.,* P. 1032; chilkadúdú, *Tel.,* S. 487; chilkadúdúgú, *Tel.,* P. 1048; chilla, *Tel.,* E. 593; chilla chettu, *Tel.,* S. 2960; chilta-eita, *Tam.,* P. 576; chilta tumiki, *Tel.,* D. 656; chilu, *Kan.,* S. 2960; chima-púnji, *Mal.,* C. 1512; china-dúla gondi, *Tel.,* T. 509; china karinguva, *Tel.,* G. 128; chinaká-ringuva, *Tel.,* G. 116; chin-amam, *Tam.,* E. 549; china muttama, *Tel.,* S. 1714; chíná-shakkara, *Tel.,* S. 375; chindaga, *Tel.,* A. 722; chindu. *Tel.,* A. 711; chinna, *Tel.,* T. 434; chinna botuka, *Tel.,* C. 1944; chinna-botuku, *Tel.,* C. 1940; chínna-kalabanda, *Tel.,* A. 836; chinnakalinga, *Tel.,* D.438; chinnamekkera-chettu, *Tel.,* C. 1940; chinna mora, *Tel.,* B. 913; chinna muddapulagam, *Tam.,* P. 347; chinni, *Tel.,* A. 304; chinniáka, *Tel.,* A. 304; chinta, *Tel.,* T. 28; chinta-pandu, *Tel.,* T. 28; chipo, *Kan.,* A. 101; chippa-gaddi, *Tel.,* A. 1079; chípudi, *Tel.,* C. 705; chípuru-tíge, *Tel.,* C. 1452; chiranji, *Tel.,* R. 564; chiratala bódi, *Tel.,* D. 330; chircha, *Kan.,* T. 424; chiri-ánem, *Tel.,* B. 873; chiribenda, *Tel.,* S. 1694; chiri-bikka, *Tel.,* G. 116; chiri giligichcha, *Tel.,* C. 2148; chirimalle, *Tel.,* J. 10; chirímánu, *Tel.,* A. 1149; chiri pála, *Tel.,* O. 600; chiri-palléru, *Tel.,* T. 548; chirisanagalu, *Tel.,* L. 252; chiri vanga, *Tel.,* S. 2284; chiri-veru, *Tel.,* O. 137; chiron, *Mal.,* C. 854; chirtah, *Tel.,* F. 482; chiru dekku, *Tam.,* C. 1 388; chiru-kizhukánelli, *Mal.,* P. 673; chirupála, *Tel.,* O. 600; chirupalléru, *Tel.,* T. 548; chiruthai, *Tam.,* T. 434; chitacha, *Tam.,* H. 16; chitakamraku, *Tel.,* G. 99; chiankalú, *Tel.,* W. 122; chita puli, *Tel.,* T. 424; chitchilli, *Tel.,* F. 406; chitt, *Tam.,* M. 299; chitigina soppu, *Kan.,* M. 306; chiti-késwarum, *Tel.,* P. 1032; chiti-mutti, *Tam.,* S. 1688; chitlinta kúra, *Tel.,* M. 306; chitra-múlá, *Kan.,* P. 986; chitra-múlam, *Tel.,* P. 986; chitrika, *Tel.,* B. 941; chitta burkani, *Tel.,* R. 51; chitta-dudaga, *Tel.,* P. 1048; chitta ganda, *Tel.,* R. 51; chittagong chettu, *Tel.,* C. 1021; chittagong karru, *Tel.,* C. 1021; chittahri, *Tel.,* F. 411; chittamatta, *Tel.,* C. 116; chitta, mruta, *Mal.,* T. 470, 483; chittá. mudapu, *Tel.,* R. 369; chittamutti, *Tam.,* P. 347, chitte-duru, *Tel.,* V. 17; chitti-pápara, *Tel.,* C. 1211; chittira, *Tam.,* P. 986; chittira-múlam, *Tam.,* P. 986; chittita, *Tel.,* P. 576; chittira, *Tam.,* B. 771; chittur-mol, *Tam.,* P. 979; chityelka, *Tel.,* R. 51; chivanna-pulachi-chira-vitta,

Mal., H. 233 ; choarí, *Tam.*, F. 363 ; cholam, *Mal.*, Z. 50 ; cholam, *Tam.*, S. 2424 ; chómara, *Mal.*, A. 1130; chórapulla, *Mal.*, A. 1107 ; chotah wahlah, *Tam.*, F. 368, F. 370, F. 511 ; chotte, *Kan.*, C. 1931 ; chouk, *Tam.*, C. 826 ; chovanna-kasha-kashach-cheti, *Mal.*, P. 82 ; chovanna-katru-vazha, *Mal.*, A. 838 ; chovvauna-basella-kíra, *Mal.*, B. 207 ; chowra, *Kan.*, E. 278 ; chuddupaddaka, *Tel.*, F. 335 ; chukanna-kízhánelli, *Mal.*, P. 673 ; chukanna-kotuvéli, *Mal.*, P. 979 ; chukka kúra, *Tel.*, R. 650 ; chukke, *Kan.*, A. 101 ; chukku, *Mal.*, Z. 201 ; chumbum, *Mal.*, C. 2528 ; chúnámbú. *Tam.*, C. 489 ; chúndai kai, *Tel.*, S. 2313 ; chundawah, *Tel.*, F. 475 ; chu-vanna, avilpori, *Mal.*, R. 57 ; chyaapotta, *Tel.*, T. 576.

Ci Ciga sorrah, *Tel.*, F. 382 ; cittra-molum, *Tam.*, P. 979.

Cl Clanta, *Mal.*, Z. 231.

Co Coal, *Tam.*, F. 464 ; coal-arinza-candí, *Tam.*, F. 465 ; coat comul, *Tam.*, C. 129 ; coatí candí, *Tam.*, F. 345 ; cobri, *Kan.*, C. 1520 ; coco-mín, *Tam.*, F. 366, F. 367 ; cocomottah, *Tel.*, F. 458 ; coddapani, *Tam.*, C. 1995 ; collúngíe-pullum, *Tam.*, C. 1233 ; commárí, *Tam.*, F. 578 ; commúmadu, *Tel.*, C. 982; cconda-mayúru, káki-nérédu, *Tel.*, A. 1288 ; ondapanna, *Tam.*, C. 711 ; conda patti, *Tel.*, T. 387 ; congilium-marum, *Tam.*, C. 285 ; congo, *Tam.*, S. 1682 ; congú, *Kan.*, S. 2799 ; corake, *Tam.*, F. 542 ; corangansorrah, *Tam.*, F. 401 ; corunga-manje, *Kan.*, M. 71 ; cottan, *Tam.*, C. 805 ; cowlie, *Tam.*, F. 426 ; coya, *Tel.*, P. 1343.

Cr Crishmabage, *Kan.*, D. 64.

Cu Cucahsawahri, *Tel.*, F. 458 ; cucuma-dunda, *Tel.*, R. 341 ; cuddapah, *Tam.*, *Mal.*, B. 193 ; cuddú-lavanga, *Kan.*, C. 1158 ; cuddul-verarl, *Tam.*, F. 433 ; cul, *Tam.*, F. 561 ; culakelletti, *Tam.*, F. 527 ; culcúndallum, *Tam.*, F. 436 ; culim-poun, *Tel.*, F. 507 ; cullí mulayan, *Tam.*, C. 476 ; cul-nahmacunda, *Tam.*, F. 569 ; cummi, *Tam.*, G. 287 ; cúndal-lum, *Tam.*, *Tel.*, F. 456 ; cundung katric, *Tam.*, S. 2345 ; cunnesí, *Tel.*, F. 497 ; cunnyila, *Tam.*, F. 575 ; cúringí-kírai, *Tam.*, D. 823 ; currapu-mattawa, *Tam.*, F. 405 ; currie, *Tam.*, F. 405 ; currumay, *Tam.*, F. 483 ; currupu verarl, *Tam.*, F. 517 ; currutche, *Tam.*, F. 543 ; currúway tiriki, *Tam.*, F. 306 ; cúrúwa, *Tam.*, F. 574 ; cutta, *Tam.*, F. 487 ; cut timín *Tam.*, F. 490.

Da Dabba, *Tel.*, C. 1270 ; dabha, *Tel.*, E. 252 ; dacer-karah, *Tel.*, F. 445 ; dadima, *Tel.*, P. 1426 ; dádima-pandu, *Tel.*, P. 1426 ; dadsal, *Kan.*, C. 714 ; daduga, *Tel.*, A. 514 ; dála-chini, *Kan.*, C. 1196 ; dalchini, *Kan.*, C. 1158, C. 1183 ; dálchini yanne, *Kan.*, C. 1158 ; dálimba, *Tel.*, P. 1426 ; dalimbapandu, *Tel.*, P. 1426 ; dálimbegidá, *Kan.*, P. 1426 ; dalimbe káyi, *Kan.*, P. 1426 ; dalmara, *Kan.*, C. 1021 ; damádi, *Tel.*, D. 615 ; dampara, *Tam.*, O. 38 ; danimma, *Tel.*, P. 1426 ; danimmapandu, *Tel.*, P. 1426 ; danki-bura, *Tel.*, B. 868, B. 873 ; dantáusi, *Tel.*,C. 860 ;

danti, *Tel.*, C. 860 ; danyalu, *Tel.*, C. 1954 ; darbha, *Tel.*, E. 252 ; dargu, *Tel.*, O. 537 ; dári, *Tel.*, P. 1401 ; dárigummadi, *Tel.*, P. 1401 ; darisanchai, *Tel.*, A. 295 ; darshana, *Tel.*, A. 695 ; darsuk, *Kan.*, C. 708 ; dásána, *Tel.*, H. 227 ; dásavala, *Kan.*, H. 227 ; datturi, *Kan.*, A. 1351 ; datturi gida (Datura), *Kan.*, D. 166 ; datturi gidda (Argemone), *Kan.*, A. 1351 ; davananu, *Tel.*, *Kan.*, A. 1469.

De Deddi-jella, *Tel.*, F. 325 ; deokurpas *Kan.*, C. 385 ; désaválipendalam, *Tel.*, D. 526 ; dévadáram, *Tam.*, E. 370 ; dévadári, *Tam.*, E. 370 ; dévadári-chedi, *Tam.*, C. 846 ; dévadári-chettu, *Tel.*, C. 846 ; dévadári-mará, *Kan.*, C. 846 ; dévadáru (Erythroxylon), *Tel.*, E. 370 ; dévadáru (Polyalthia), *Tel.*, P. 1052 ; dévatá dhányamu, *Tel.*, S. 2397, S. 2405 ; devátá-malle, *Tel.*, R. 16 ; dévatáram, *Mal.*, C. 846 ; devdari, *Kan.* ; C. 838, C. 1021 ; dewantsi-pilli, *Tel.*, F. 780.

Dh Dhánya bhédam, *Tel.*, H. 382 ; dhataki, *Tel.*, W. 106 ; dhau, *Mal.*, A. 1149 ; dhunirapatramu, *Tel.*, N. 101 ; dhúp, *Kan.*, A. 670 ; dhupa, *Kan.*, V. 31 ; dhupadamara, *Kan.*, V. 31 ; dhúpadíenné, *Kan.*, G. 36 ; dhup maram, *Tam.*, V. 31.

Di Diár, *Tam.*, C. 224 ; diká-málli, *Tam.*, C. 116 ; dikke-malli, *Kan.*, C. 116 ; dindal, *Kan.*, A. 1149 ; dindiga, *Kan.*, A. 1149 ; dindlu, *Kan.*, A. 1149 ; dinduga, *Kan.*, A. 1149 ; dintana, *Tel.*, C. 1403 ; dirasan, *Tel.*, A. 695 ; dirisana, *Kan.*, A. 695.

Do Dodal-gatti-gadu, *Tel.*, S. 2907 ; dodalkonga, *Tel.*, S. 2907 ; dodda badane, *Kan.*, S. 2284 ; dodda-hipalli, *Kan.*, S. 970 ; dodda-jirage, *Kan.*, P. 727 ; doddánimbe-hannu, *Kan.*, C. 1286 ; dodda-patri, *Kan.*, A. 1130 ; doddá-yalakki, *Kan.*, A. 976 ; doggali-kúra, *Tel.*, A. 941 ; donda (Trichosanthes), *Tel.*, T. 600 ; donda (Cephalandra), *Tel.*, C. 919 ; dondlup, *Tel.*, O. 233 ; dondu-paum, *Tam.*, F. 316 ; donn-mullina-jali, *Kan.*, A. 244 ; dosa, *Tel.*, C. 2278 ; dosray, *Tam.*, C. 2278 ; dovana, *Kan.*, G. 660 ; dovedah, *Mal.*, C. 1021 ; doza-kaia, *Tel.*,C. 2287.

Dr Dracha, *Tel.*, V. 233 ; draksha-pondu, *Tel.*, V. 233 ; drakshi, *Kan.*, V. 233.

Du Dúdagorai, *Kan.*, P. 1179 ; dúdagú, *Tel.*, A. 514 ; dudduka, *Tel.*, P. 1048 ; dudi máddi, *Tel.*, B. 868 ; dudi-palla (Diegea), *Tel.*, D. 823 ; dudi-palla (Oxystemma), *Tel.*, O. 600 ; dudippa, *Tel.*, H. 517 ; dudippi, *Tel.*, C. 563 ; dudiyetta, *Tel.*, H. 517 ; dúka-dúmú, *Tel.*, F. 368 ; dúla-gondi, *Tel.*, M. 786 ; dúla-góvela, *Tel.*, A. 1398 ; dumer, *Mal.*, F. 179 ; dumpa, *Tel.*, C. 1732 ; dumpa-bachchali, *Tel.*, S. 2574 ; dumpa-rásmi, *Kan.*, A. 853 ; dumper, *Tel.*, O. 38 ; dum-pini, *Tel.*, O. 38 ; dumpri, *Tel.*, O. 38 ; dúndiawah, *Tel.*, F. 485 ; dundigapu, *Tel.*, J. 52 ; dundillam, *Tel.*, O. 233 ; duntú-pesalú, *Tel.*, V. 116 ; dupada, *Tel.*, V. 31 ; dúpa-dámaru, *Tel.*, V. 31 ; dupa maram, *Kan.*, V. 31 ; dupt, *Tel.*, D. 212 ; durada-kanda-gadda (Typhonium), *Tel.*, T. 883 ; durada-kanda-godda (Amorphophallus), *Tel.*,A. 996 ; duriyamaddi, *Tel.*, B. 868 ; durpa, *Tel.*, E. 252 ; duruda-gunti, *Tel.*, T. 509 ; dusa, *Tel.*,

P. 41 ; dúsari-tíge, *Tel.,* C. 1452 ; dushta-puchattu, *Kan.,* P. 833 ; dushtupu, *Tel.,* D. 9 ; duttúramu, *Tel.,* D. 160.

Ed Edakula, *Tel.,* V. 217; édákula·ariti, *Tel.,* A. 871 ; édákula-pála, *Tel.,* A. 871 ; édákula-ponna, *Tel.,* A. 871 ; edapandú, *Tel.,* C. 1263 ; eddi (Andropogon), *Tel.,* A. 1090 ; éddi (Hetero-pogon), *Tel.,* H. 164 ; eddi-gaddi, *Tel.,* A. 1090 ; eddu-málike chettu, *Tel.,* E. 80 ; eddu mukka-dumpa, *Tel.,* P. 1216 ; eddu-mutte dumpa, *Tel.,* P. 1216 ; eddu tóka dumpa, *Tel.,* D. 510 ; edikkol, *Tam.,* C. 728 ; edudi, *Tel.,* G. 404.

Ee Eerál, *Tam.,* C. 2237 ; eeti, *Tam.,* D. 41.
Eg Egisa, *Tel.,* P. 1370.
Ei Eíam, *Tam.,* L. 143.
Ej Ejú, *Mal.,* A. 1335.
Ek Ekkemále, *Kan.,* C. 170.
El Elaka (Feronia), *Tel.,* F. 53 ; ela-ká (Elettaria), *Tam.,* E. 151 ; ela-kalli, *Mal.,* E. 527 ; ela-kay, *Tam.,* E. 151 ; ela-káya, *Tel.,* E. 151 ; elakayvirai, *Tam.,* E. 151 ; élaki chettu, *Tel.,* E. 151 ; elakulu, *Tel.,* A. 980 ; elam, *Tam.,* A. 980 ; elamarunga, *Mal.,* B. 909 ; ela-marunna, *Mal.,* B. 909 ; elamávi, *Tel.,* M. 147 ; elanda, *Tam.,* Z. 231 ; elandap, *Tam.,* Z. 231 ; elavum, *Tam ,* E. 289 ; elegáram, *Tel.,* B. 731 ; elemitchum, *Tam.,* C. 1296, C. 1301 ; elengi, *Mal.,* M. 570 ; elentha, *Mal.,* Z. 231 ; elettari, *Mal.,* E. 151 ; elevam, *Tam.,* E. 295 ; ellaay, *Tam., Tel.,* E. 151 ; elládu, *Tam.,* Z. 231 ; ellakay, *Tam., Tel.,* E. 151 ; ella·kura ; *Tel.,* S. 596 ; ellepa, *Tam.,* B. 266 ; ellu, *Tam., Kan., Mal.,* S.1078 ; ellupi, *Mal.,* B. 265 ; elumich-cham-pazham, *Tam.,* C. 1258, 1270, 1296; elumich-cham-tolashi, *Tam.,* O. 28 ; elupa, *Tam.,* B. 220, B. 265.

Em Embúdi chettu, *Tel.,* P. 824; embure cheddi, *Tam.,* O. 137 ; emiga-junum *Tel.,* T. 864.
En Ena-kalabanda, *Tel.,* A. 838 ; enuga bira, *Tel.,* E. 80 ; enu-ga-palléru mullu, *Tel.,* P. 363 ; enuga-pippalu, *Tel.,* S. 970.
Ep Epe, *Tel.,* H. 16.
Er Eramudapu, *Tel.,* R. 369 ; erica, *Mal.,* C. 170 ; erim-panna, *Tam.,* C. 711 eriwadi, *Tam.,* D. 41 ; erra-allu-bach-chali, *Tel.,* B. 207 ; erra-bondala, *Tel.,* C. 1520 ; erra chan-danum, *Tel.,* P. 1381 ; erra chiratali, *Tel.,* V. 55 ; erra-chitramúlam, *Tel.,* P. 979 ; erra-érusaru, *Tel.,* T. 51 ; erra gali-jéru, *Tel.,* T. 537 ; erragandhapu-chekka, *Tel.,* P. 1381 ; erra-gasa-gasala-chethe, *Tel.,* P. 82 ; erragoda, *Tel.,* D. 628 ; erra-gomgura, *Tel.,* H. 233; erra-gom-kaya, *Tel.,* H. 233 ; erra-góng-áka, *Tel.,* H. 233 ; erra-góng-kúra, *Tel.,* H. 233 ; erra-góngúru, *Tel.,* H. 233 erra gummadi, *Tel. ;* C. 2316; errajilma vadlu, *Tel.,* O. 258 ; erra-jilgua, *Tel.,* S. 1166 ; erra jíluga, *Tel.,* S. 1166 ; erra kaluva, *Tel.,* N. 200 ; erra-kodí-utta-totakuru, *Tel.,* C. 873 ; erra kúti, *Tel.,* P. 344 ; erra, lodduga, *Tel.,* S. 3062; erra maildi, *Tel.,* T. 282; erra-mulu-góranta, *Tel.* A. 943 ; erra pachchári, *Tel.,* D. 32 ; erra-posta-káya-chethe, *Tel.,* P. 82 ; erra-pulike, *Tel.,* S. 2861 ; erra-puniki chettu, *Tel.,* S. 2850 ; erra-shiri-saru, *Tel.,* T. 51 ; erra-támara-veru, *Tel.,* N. 39; erra-tóta-kúra, *Tel.,* A. 938; erra-usirika,

Tel., P. 673 ; erra-vĕgisa, *Tel.,* P. 1363 ; erukkam, *Tam.,* C. 170, erukku, *Tam., Mal.,* C. 170; erumitchi narracum, *Mal.,* C. 1296, 1301; eru-pĭtchecha, *Tel.,* C. 1377; erru puchcha, *Tel.,* C. 1377 ; éru-saru, *Tel.,* T. 70; eruvalli, *Tam.,* M. 3; eruvalu, *Tel.,* X. 16.

Et Etipála, *Tel.,* S. 579; eti-pisinka, *Tel.,* C. 1377; eti-puch-cha, *Tel.,* C. 1211 ⸴ ettik-kottai, *Tam.,* S. 2943.

Ez Ezhilaippálai, *Tam.,* A. 871.

Ga Gába, *Tel.,* D. 341 ; gabbday, *Tel.,* F. 655 ; gabbu nelli, *Tel.,* P. 1246; gachcha, *Tel.,* C. 6 ; gach-chakáya, *Tel.,* C. 6 ; gadamu, *Tel.,* W. 67; gada-nelli, *Tel.,* T. 522; gaddaisinka, *Tel.,* W. 106; gaddí janú, *Tel.,* S. 2394; gaddi janumu, *Tel.,* S. 239 ; gadi chikkudu káya, *Tel.,* R. 346; gádide-gada-para-áku, *Tel.,* A. 1395 ; ga-disugandhi, *Tel.,* H. 119 ; gaggera-chettu, *Tel.,* O. 31 ; gaiger, *Tel.,* G. 124; gaija soppu, *Kan.,* T.530; gajagakayi, *Kan., C.* 6; gájangi, *Tel.,* P. 26 ; gajanimma, *Tel.,* C.1301 ; gaja-pippallu, *Tel.,* S. 970 ; gajega ballí, *Kan.,* C. 6 ; gáji, *Tel.,* W. 106 ; gajjara gedda, *Tel.,* D. 173; gájjara kelangu, *Tam.,* D. 173; gajjari, *Kan ,* D. 173; gajkai, *Kan.,* C. 6 ; gáju chettu, *Tel.,* S. 2299 ; galagara, *Tel.,* E. 7; gáli chekka, *Tel.,* S. 2240; galijéru, *Tel.,* T. 530, T. 537; galkaranda, *Kan.,* C. 390 ; gamgudu, *Tel.,* C. 725 ; gams, *Mal.,* S. 2682; gamudu, *Tel.,* G. 208; ganagalu, *Kan.* P. 989; gana-patináranna, *Mal.,* C. 1270; gandada, *Kan.,* S. 790 ; ganda-hanchi-khaddi, *Kan.,* A. 1107; gadakam, *Tam.,* S. 2999; gandala, *Tel.,* C. 2612; gándámrugam-netturu, *Tel.,* P. 1370; gandha, *Kan.,* S. 790; gandhaka, chekke, *Kan.,* S. 790; gandhakam, *Tel.* S. 2999; gandha-phallí, *Tel.,* M. 517 ; gandhapu-chekka, *Tel.,* S. 790; gandu-bhárangí, *Tel.,* C. 1388; gandumenu, *Tel.,* F.469; gangah, *Tel.,* C. 331 ; ganga-pávilí-kúra, *Tel.,* P. 1179; gangarávi, *Tel.,* T. 392 ; gangaraya, *Tel.,* T. 392 ; ganga régu, *Tel.,* Z. 231 ; gangarénu, *Tel.,* T. 392 ; gangirana, *Tel.,* T. 392 ; gánja, *Tam.,* C. 331 ; gánja-chedi, *Tem.* C. 331 ; ganja-chettu, *Tel.,* E. 331 ; ganjai, *Tel.,* C. 331; ganja-ilai, *Tam.,* C. 331 ; ganjanimma, *Tel.,* C. 1233 ; ganja-pávili-kúra, *Tam.,* P. 1179; ganja-phal, *Tam.,* C. 331 ; ganja-rasham, *Tam.,* C. 331; gánjari-chettu, *Tel.,* C. 331; gannéru, *Tel.,* N. 80 ; gan-ribija, *Kan.,* I. 384; gantelú, *Tel.* P. 384; ganti malle, *Tam.,* C. 1021 ; gantu-bhárangí, *Tel.,* C. 1388 ; ganuga, *Tel.,* P. 1121; gao-zaban, *Tam.,* O. 168 ; gára-chettu, *Tel.,* B. 13 ; garaga, *Tel.,* G. 116 ; garagada-sappu, *Kan.,* E. 7; garaka nattu, *Tel.,* I. 501 ; gára-pandu, *Tel.,* B. 13 ; gardhi, *Tel.,* H. 414; gardundi, *Kan.,* O. 6 ; gárgá, *Tel.,* G. 143; gárí, *Tel.,* B. 13 ; gariti kamma, *Tel.,* V. 79 ; garpa-shola, *Anaimalais,* S. 838 ; garuda mukku, *Tel.,* M. 308 ; garugu, *Tel.,* G. 143 ; gasagasala-tólu, *Tel.,* P. 87 ; gasagasálú, *Tel.,* P. 87 ; gasagase, *Kan.,* P. 87; gashagasha, *Tam.,* P. 87 ; gashagasha-tól, *Tam.,* P. 87 ; gathiri, *Tel.,* E. 370 ; gatrinta, *Tel.,* H. 444 ; gavuldu, *Kan.,* C. 563 ; gáyapu áku, *Tel.,* S. 1699 ; gaynara, *Tel ,* H. 177.

Ge Gechchakkáy, *Tam.*, C. 6; gédangi, *Tel.*, P. 26; geja pushpam, *Tel.*, M. 489; genasu (Ipomæa), *Kan.*, I. 348; genasu (Dioscorea), *Kan.*, D. 481; geneopullu, *Tam.*, P. 59; ger, *Kan.*, S. 1041; gerabija, *Kan.*, A. 1014; gerapoppu geruváte, *Kan.*, A. 1014; gerra chandan, *Tel.*, P. 1381; gerú, *Kan.*, S. 1041.

Gh Ghansing, *Kan.*, S. 2884; ghattavare, *Kan.*, R. 346; ghebu-nelli, *Tel.*, P. 1233; ghelegherinta, *Tel.*, C. 2163; gheneru, *Tel.*, N. 80; ghensikanda, *Tel.*, A. 996; ghericha, *Tel.*, C. 2558; gheru, *Kan.*, S. 1041; ghunia, *Tel.*, S. 717; ghurie-ghénzá, *Tel.*, A. 51.

Gi Gid hágalu, *Kan.*, M. 639; gidpakke, *Kan.*, F. 333, 350; gila góranta, *Tel.*, C. 2163; gingelin, *Mal.*, S. 1078; gingelli, *Mal.*, S. 1078; gingi-lacki-lacki, *Mal.*, C. 331; gini hullu, *Kan.*, P. 59; giniya-chettu, *Tel.*, A. 745; ginjil-achi-lachi, *Mal.*, C. 331; girikarmika, *Tel.*, A. 745; giruka táti, *Tel.*, P. 582; gíta naram, *Tel.*, D. 339; gittigadda, *Tel*, S. 974.

Go Gobli, *Kan.*, A. 101; gobur rí-koya, *Tel.*, C. 1520; godambe mas, *Tam.*, S. 2682; godári, *Tel.*, W. 106; godda hunshe, *Kan.*, A. 695; goddatipalusu, *Tel.*, S. 2876; goddu pavili, *Tel.*, P. 1187; goddutunga kodu, *Tel.*, C. 2585; gódhi, *Kan.*, T. 634; godla, *Kan.*, F. 368; gódumai, *Tam.*, T. 634; godumbay-arisi, *Tam.*, T. 634; gódumulu, *Tel.*, T. 634; godú tunga kúda, *Tel.*, C. 2585; gogu, *Tel.*, A. 200; goguldhúp, *Kan.*, A. 670; goindú, *Kan.*, D. 628; goin jol, *Kan.*, Z. 50; gojal, *Kan.*, O. 38; gojé, *Kan.*, B. 868; gokarna mul, *Kan.*, C. 1403; gola gandí, *Tel.*, T. 504; golakonda, *Tel.*, C. 455; golattu koku, *Tel.*, R. 51; gollan-kóvaik-kizhangu, *Tam.*, C. 1834; golla pulleda, *Tel.*, D. 725; golugu, *Tel.*, G. 271; gondu-gogu, *Tel.*, C. 1512; góngúra, *Tel.*, H. 177; goni, *Kan.*, F. 223; gonji pandu, *Tel.*, G. 271; gonkura, *Tel.*, H. 177; goragú, *Tel.*, C. 711; góranta, *Tel.*, L. 126; górante, *Kan.*, L. 126; gorantlu, *Kan.*, L. 126; goriti, *Tel.*, P. 950; goriti donka, *Tel.*, P. 950; gorivi, *Kan.*, I. 515; gorklu, *Kan.*, T. 522; gorregu, *Tel.*, C. 711; goshtam, *Tam.*, S. 910; gós-kíre, *Tam.*, B. 851; gós-kúra, *Tel.*, B. 851; góstaní dráksha, *Tel.*, V. 233; goti, *Tel.*, Z. 290; gotte, *Tel.*, Z. 290; govila, *Tel.*, A. 1398; goyyápandu, *Tel.*, P. 1343; goyyá-pazham, *Tam.*, P. 1343.

Gr Grandi tagarapu, *Tel.*, T. 3.

Gu Gua-kasi, *Mal.*, E. 314; gúáku, *Tel.*, R. 16; gubadarra, *Tel.*, H. 92; gubba dára, *Tel.*, S. 3042; gubba kaya, *Tel.*, L. 30; gudama tíge, *Tel.*, V. 184; gudha, *Tel.*, C. 121; gudimi donda pendalam, *Tel.*, D. 485; gudí muralú, *Tel.*, C. 2257; gudla-jella, *Tel.*, F. 562; gudúchi, *Tel.*, T. 470; gudúchi, *Tel.*, T. 483; guga, *Kan*, D. 685; gúgal, *Tel.*, S. 1656, S. 1939; guggala (Balsamodendron), *Kan.*, B. 43; guggala (Shoiea), *Kan.*, S. 1656; guggilamu, *Tel.*, S. 1656; guggilapu-chettu, *Tel.*, B. 777; gugg-mámidi, *Tel.*, M. 147; gugil, *Tel.*, B. 777; gugul, *Tel.*, B. 43; gúgúlu, *Tam.*, B. 771; gujju-narek-adam, *Tel.*, C. 1520; gukkal, *Tam.*, B.

43; gukkulu, *Tam.*, B. 43; gúkul, *Tam.*, B. 62; gulábi, *Kan.*, R. 504; gulamaji, *Kan.*, M. 606; guláppá, *Tam.*, R. 508; guli, *Tel.*, C. 410; gullem chettu, *Tel.*, C. 410; gulni, *Tel.*, P. 382; gulúchi, *Tel.*, T. 470, T. 483; gumadi, *Tam.*, G. 287; gúmar-tek, *Tel.*, G. 287; gummadi, *Tel.*, C. 2316; gúmodi, *Tel.*, P. 1401; gumpena-chettu, *Tel.*, O. 38; gumpini, *Tel.*, O. 38; gumpna, *Tel.*, O. 38; gumudu, *Tam.*, *Tel.*, G. 287, G. 298; gumudu téku, *Tam.*, *Tel.*, G. 287; gúnapendálam, *Tel.*, D. 513; gunda, *Tel.*, S. 977; gundebingula, *Tel.*, B. 863; gundra, *Tel.*, S. 6; gundumaní, *Tam.*, A. 51; gundu-meda, *Tel.*, C. 854; gúndúsani-ghelú, *Tel.*, P. 885; gúngú, *Tel.*, C. 1512; guntagalijeru, *Tel.*, E. 7; guntaka-lagara, *Tel.*, E. 7; gunta kaminam, *Tel.*, S. 2792; gunti paringaie, *Tel.*, C. 1388; guoroka, *Tel.*, F. 541; guramu, *Tel.*, H. 414; gurapu-badam, *Tel.*, S. 2824; guricha, *Tel.*, T. 470; guricha, *Tel.*, I. 483; guri genza chettu, *Tel.*, C. 321; gúroda, *Kan.*, G. 271; gurrapu sakatunga, *Tel.*, S. 981; guruga-pála-tíge, *Tel.*, C. 2247; gurugu, *Tel.*, C. 868; gurugu chettu, *Tel.*, C. 2211; guruti-chettu, *Tel.*, D. 9; gurwa, *Mal.*, A. 267; gutti bíra, *Tel.*, L. 569; guvva-gutti, *Tel.*, T. 562.

Gw Gwel, *Mal.*, C. 919.

Ha Hadang, *Kan.*, E. 314; haiga, *Kan.*, H. 371; hala, *Kan.*, G. 143; haladí, *Kan.*, E. 278; hála-kóratíge, *Kan.*, D. 9; halasu, *Kan.*, A. 1482; hale (Wrightia), *Kan.*, W. 122; hale (Chry santhemum), *Kan.*, C. 1043; hali (Chry-sophyllum), *Kan.*, C. 1050; halibachcheli, *Kan.*, P. 1187; hali-maru, *Kan.*, C. 1050; halivana, *Kan.*, E. 342; háli-wára, *Kan.*, E. 342; halla naddi, *Tel.*, T. 361; halsina, *Kan.*, A. 1489; halsu, *Kan.*, A. 1489; hama-padi, *Tel.*, C. 1707; hamsa-padu, *Tel.*, C. 1707; handi, *Kan.*, H. 289; hannusampige, *Kan.*, F. 613; hantige, *Kan.*, A. 440; hanumantabira, *Tel.*, L. 266; happusa, vaga, *Kan.*, S. 2837; haralu, *Kan.*, R. 369; hariáli, *Tam.*, C. 2558; hari chandanam, *Tel.*, S. 790; haridalam, *Tel.*, O. 242; harik, *Kan.*, P. 332; harimandha-kam, *Tel.*, C. 1061; haritamanjiri, *Tel.*, A. 306; harlu, *Kan.*, R. 369; haro, *Kan.*, E. 593; harsing, *Kan.*, N. 179; haryali-*Tel.*, C. 2558; hasaru, *Kan.*, P. 513; hasísunthi, *Kan.*, Z. 201; haspath, *Kan.*, F. 236; hasti-kasaka, *Tel.*, E. 80; hasu-guniri, *Kan.*, D. 53; hattut-tumatti, *Tam.*, C. 2310; havija, *Kan.*, C. 1954.

He Heb, *Kan.*, A. 1482; hebalsu, *Kan.*, A. 1482; heb-havu, *Kan.*, M. 363; heb-bu-ti, *Tel.*, F. 537; heb-helsu, *Kan.*, A. 1489; hedde, *Kan.*, A. 514; hedu, *Kan.*, S. 2799; heggarjige, *Kan.*, C. 596; hegge-nasu, *Kan.*, D. 534; heggin, *Kan.*, R. 51; heltega, *Kan.*, A. 1192; hémángamu, *Tel.*, M. 517; hennugorvi, *Kan.*, I. 515; hesaru hesarn-bele, *Kan.*, P. 496; hessain, *Kan.*, A. 1482; hessare, *Kan.*, S. 487; hesswa, *Kan.*, A. 1482.

Hi Hintalamu, *Tel.*, P. 582; hippal-neralí, *Kan.*, M. 756; hippe, *Kan.*, B. 220, 265; hírda, *Kan.*, T. 325; hírékáyi, *Kan.*, L. 556.

Ho Hogesappu, *Kan.*, N. 101 ; hogni, *Kan.*, B. 220 ; holada, *Kan.*, H. 177 ; holedásál, *Kan.*, L. 42 ; holeger, *Kan.*, H. 317 ; hole kauva, *Kan.*, B. 180 ; hole matti, *Kan.*, T. 282 ; holletupra, *Coorg*, D. 582 ; honal, *Kan.*, T. 355 ; honge ; *Kan.*, P. 1121 ; honiga, *Kan.*, T. 434 ; honné, *Kan.*, P. 1370, 1381 ; honnemaradabanke, *Kan.*, P. 1870 ; hoolooni, *Nilgiris*, E. 565 ; hotebaghi, *Kan.*, A. 722 ; hote-jali, *Kan.*, A. 244 ; hotsigé, *Kan.*, C. 42.

Hu Huchchellu, *Kan.*, G. 735 ; húday, *Kan.*, S. 2876 ; húdigolla, *Kan.*, S. 1718 ; hulékara, *Kan.*, S. 1226 ; huli, *Kan.*, E. 370 ; hulichel-lu, *Kan.*, M. 71 ; huluvá, *Kan.*, T. 355 ; hulvé, *Kan.*, T. 355 ; humbilli, *Tam.*, M. 3 ; hu-mín, *Kan.*, F. 386 ; hunáb, *Kan.*, T. 355 ; hunase, *Kan.*, T. 28 ; hunashé-hannu, *Kan.*, T. 28 ; húngé, *Kan.*, P. 1121 ; huragalu, *Kan.*, C. 1031 ; hurali, *Kan.*, D. 758 ; hurlí, *Kan.*, D. 758 ; hurvashi, *Kan.*, T. 392 ; hushi-dálimbe, *Kan.*, P. 1426 ; huvarase, *Kan.*, T. 392 ; húvina murl, *Kan.*, F. 515.

Ic Ichal, *Kan.*, P. 576, 588 ; ichalu mara, *Kan.*, P. 588 ; ich-cha-vellam, *Tam.*, S. 370 ; ichchi, *Tam.*, F.|276 ; ichchura-múli, *Tam.*, A. 1398 ; ich-churamúli-vér, *Tam.*, A. 1398.

Id Iddali kalu, *Tel.*, M. 131 ; iddallu, *Kan*, C. 487, 1414.

Ie I-ech-chak-kirai, *Tam.*, M. 447.

Ig Íga-engili-kúra, *Tel.*, M. 447.

Ij Ije, *Kan.*, P. 1248.

Ik Ikam, *Tam.*, L. 143.

Il Ila kura, *Tel.*, S. 2990 ; ilámich-chamuér, *Tam.*, A. 1097 ; ilang-ilang, *Mal.*, C. 271 ; ilanji, *Kan.*, Z. 231 ; ilavam, *Tam.*, E. 289, E. 295 ; ilavangappú, *Tam.*, C. 706 ; ilemose, *Tam.*, F. 411 ; illaik-kalli, *Tam.*, E. 520 ; illaku, *Tam.*, E. 289 ; illavam, *Tam.*, S. 1939 ; illinda, *Tel.*, D. 560 ; illukatte, *Tel.*, I. 1 ; illupi, *Tam.*, B. 220, 265 ; il-tenkí, *Tel.*, F. 306 ; iluppai, *Tam.*, B. 265.

Im Imbural, *Tam.*, O. 137.

In Inapa chittam, *Tel.*, I. 472 ; inchl, *Mal.*, Z. 201 ; induga, *Tel.*, S. 2960 ; indupa, *Tel.*, S. 2960 ; indupu, *Tel.*, S. 2943, S. 2960 ; ingaliká, *Kan.*, M. 473 ; ingelf, *Tel.*, F. 535 ; ingiligamu, *Tel.*, M. 473 ; inji, *Tam.*, Z. 201 ; inumu, *Tel.*, I. 440.

Ip Ippa, *Tel.*, B. 220, 265 ; ippa wajna, *Tel.*, L. 549 ; ippe chettu, *Tel.*, B. 265 ; ippi. *Tel.*, B. 220, 265 ; ippigidá, *Kan.*, B., 265.

Ir Iramballi, *Tam.*, M. 3 ; irangún-malli, *Tam.*, Q. 88 ; irapú, *Tam.*, C. 2577 ; irasham, *Tam.*, M. 473 ; irattai-péy, *Tam.*, A. 1132 ; iratti-madhuram, *Mal.*, G. 278 ; iravengáy-am, *Tam.*, A. 769 ; iréval-chinip-pál, *Tam.*, G. 66 ; iriki, *Tel.*, C. 1931 ; iripa, *Mal.*, C. 2572 ; irippa, *Mal.*, B. 265 ; irki, *Tel.*, C. 1931 ; irkuli, *Tam.*, E. 73 ; irojáppú, *Tam.*, R. 508 ; irri, *Tel.*, S. 1226 ; irrwa, *Tel.*, F. 534 ; irubogam, *Mal.*, H. 368 ; irugudu, *Tel.*, D. 41 ; irúl (Xylia), *Tam.*, X. 16 ; irúl (Beesha), *Mal.*, B. 419 ; irulli, *Tam.*, A. 769 ; irumba, *Mal.*, I. 440 ; irumbu, *Tam.*, I. 440 ; irum-buchittam, *Tam.*, I. 472 ; irumbukkítam, *Mal.*, I. 472 ; irumpalai, *Tam.*, M. 3 ; irung-kell ettí, *Tam.*, F. 356 ; iruppai,

Tam., B. 265 ; iruppúttu, *Tam.*, D. 41 ; iruvudu, *Tel.*, D. 41.

Is Isabakólu, *Kan.*, P. 932 ; isabgólu, *Kan.*, P. 932 ; isakadásari kúra, *Tel.*, G. 220 ; isakarási, *Tel.*, E. 310 ; isapagála-vittulu, *Tel.*, P. 932 ; ishan-chedi, *Tel.*, P. 588 ; ishappu-kol-virai, *Tam.*, P. 932 ; ishvará-múii, *Mal.*, A. 1398 ; ishvara-véru, *Tel.*, A. 1398 ; ishverí-vérú, *Kan.*, A. 1398 ; isikedunti kúra, *Tel.*, G. 220 ; iskól-virai, *Tam.*, P. 932 ; isoppitay, *Tel.*, F. 508 ; isphagula, *Tel.*, P. 932 ; ispoghol, *Tam.*, P. 932 ; ispoghol verai, *Tam.*, P. 932 ; issakí-dúndú, *Tel.*, F.455 ; istarakula, *Tel.*, H. 328 ; istarákupála, *Tel.*, H. 294.

It Ita (Phœnix), *Tel.*, P. 588 ; itah (Helicteres), *Tel.*,H. 92 ; ita-koyya, *Tel.*, P. 576 ; ítanara, *Tel.*, P. 588 ; itcham-nar, *Tam.*, P. 588 ; itcham thattu, *Tam.*, P. 576 ; itchumpannay, *Tam.*, P. 588 ; itham pannay, *Tam.*, P. 588 ; itf, *Tam.*, D. 41.

Iu Ium síndúrum, *Tam.*, L. 143.

Iv Iveni, *Tel.*, L. 126 ; ivuru mámidi, *Tel.*, S. 2649.

Iw Iwara memadí tamalamu, *Tel.*, G. 99.

Ja Jabmé, *Kan.*, X. 16 ; jádí, *Kan.*, T. 232 ; jádikkáy, *Tam.*, M. 885 ; jádi-lingam, *Tam.*, M. 473 ; jádi-pattiri, *Tam.*, M. 885 ; jafra-virai-maram, *Tam.*, B. 523 ; jafra-vittulu-chettu, *Tel.*, B. 523 ; jag, *Tel.*, W. 106 ; jagalagante, *Kan.*, D. 628 ; jaggari, *Tel.*, F. 452 ; jagúri, *Kan.*, A. 1200 ; jaji (Jasminum), *Tel.*, J. 18 ; jaji kaya (Myristica), *Tel.*, M. 885 ; jájikáyi, *Kan.*, M. 885 ; jál, *Tam.*, C. 224 ; jalada, *Kan.*, S. 1679 ; jalaranda, *Kan.*, S. 1679 ; jalari, *Kan.*, S. 1679 ; jalari, *Tel.*, S. 1679 ; jáli, *Kan.*, A. 101, 217 ; jallugu, *Tel.*, F. 438 ; jam, *Mal.*, E. 419 ; jama, *Tel.*, P. 1343 ; jambu, *Tam.*, P. 1259 ; jamcana, *Tel.*, C. 626 ; jammi,*Tel.*, P. 1259 ; jammu gaddi, *Tel.*, T. 864 ; jám-pandu, *Tel.*, P. 1343 ; janapa, *Tam.*, *Tel.*, *Mal.*, C. 2105 ; jána-palaseru, *Tel.*, A. 1218 ; jangli-bhendi, *Kan.*, E. 278 ; jángli tamáku, *Tel.*, S. 2354 ; janíkalam, *Tam.*, C. 626 ; jan-thalla, *Kan.*, A. 871 ; janu, *Tel.*, S. 2424 ; janumú, *Tam. & Tel.*, C. 2105 ; japápush-pamu, *Tel.*, H. 227 ; japatri, *Tel.*, *Kan.*, M. 885 ; jáphara chettu, *Tel.*, B. 523 ; japhra-maram, *Tam.*, B. 523 ; jára, *Tel.*, G. 705 ; jargi, *Tel.*, W. 106 ; járumámidi, *Tel.*, B. 913 ; jatamámshi, *Tel.*, N. 17 ; jatamáshi, *Tam.*, N. 17 ; jati, *Mal.*, T. 232 ; jati, *Kan.*, D. 212 ; játikká, *Mal.*, M. 885 ; játi-lingam, *Mal.*, M. 473 ; játi pattri, *Mal.*, M. 885 ; játí-phalamu, *Tel.*, M. 885 ;java pushpamu, *Tel.*, H. 227 ; javegodhi, *Kan.*, H. 382 ; jaynkatala, *Kan.*, S. 2806 ; jazúgri, *Kan.*, A. 1200.

Je Jelagalu, *Tel.*, L. 245 ; jellow, *Tel.*, F. 582 ; jemudu, *Tel.*, E. 553 ; jemudu-kádalu, *Tel.*, E. 553 ; jenappa, *Tam. & Tel.*, C. 2105 ; jénu, *Kan.*, H. 342 ; jeriku, *Tel.*, L. 245 ; jermála, *Kan.*, T. 07a ; jéroc-nanis, *Mal.*, C. 1233 ; jerrí-potú, *Tel.*, F. 532 ; jerúc, *Kan.*, C. 1233 ; jérúk, *Mal.*, C. 1296 ; jeru-tika, *Mal.*, C. 1388 ; jetamánchi, *Mal.*, N. 17 ; jeta-mávashi, *Kan.*, N. 17 ; jewadi, *Kan.*, H. 289 ; jewi, *Tel.*, F. 216.

Jh Jhár, *Tam.*, C. 224.

Ji Jiburai, *Tel.*, F. 655 ; jiddu us te , *Tel.*, S.
2345 ; jidi. *Tel.*, S. 1041 ; jídi-ánti, *Tel.*,
A. 1014 ; jídi chettu, *Tel.*, S. 1041 ; jídi-
mámidi-vittu, *Tel.*, A. 1014 ; jídivate,
Kan., A. 1014 ; jidi-vittulu, *Tel.*, S. 1041;
jíedí pundú, *Tel.*, A. 1014 ; jigani, *Kan.*,
L. 245 ; jílakarra. *Tel.*, C. 2339 ; jili,
Mal., F. 253 ; jilledu, *Tel.*, C. 170 ; jille-
duchettu, *Tel.*, C. 170 ; jilleru, *Tel.*, C.
170 ; jimmimara, *Kan.*, Z. 35 ; jinka,
Tel., S. 1226 ; jinnagow, *Tel.*, F. 438 ;
jírage, *Kan.*, C. 2339 ; jíraka, *Tel.*, C.
2339 ; jírakam, *Mal.*, C. 2339 ; jiri, *Tel.*,
S. 1041 ; jiringe, *Kan.*, C. 2339 ; jirki-
virai, *Tam.*, I. 384 ; jirúgú, *Tel.*, C. 711 ;
jitangi, *Tel.*, D. 41 ; jitegí, *Tel.*, D. 41 ;
jiti, *Tam.*, M. 299 ; jitregí, *Tel.*, D. 41 ;
jittagé, *Tam.*, D. 41 ; jittupáku, *Tel.*, D.
9.

Jo Jolah, *Kan.*, S. 2424 ; jonakam-náranná,
Mal., C. 1258, C. 1296; jonna, *Tel.*, S.
2424 ; jonnalu, *Tel.*, S. 2424 ; jooi, *Tam.*,
F. 216 ; jovi, *Tel.*, F. 276.

Jr Jradap-patta, *Tel.*, C. 1117.

Ju Jubbo, *Kan.*, F. 558 ; judapa, *Tel.*, S. 2960;
jumika, *Tel.*, D. 582 ; jummina, *Kan.*, Z.
35 ; jurka pilli, *Tel.*, T. 427 ; juvvi,
Tam., *Tel.*, F. 276.

Ka Kaada vailu, *Kan.*, A. 1192 ; kaatamunck,
Tam., J. 41; kaat-juti, *Tam.*, M. 128 ;
kabbar, *Tam.*, C. 224 ; kabbu, *Kan.*, S.
1656; kabina, *Kan.*, I. 440 ; kach-chólam,
Mal., C. 2499 ; káchchúri-kizhanna,
Mal., C. 2499 ; kachí, *Tel.*, S. 2299 ;
kachipadél, *Tel.*, I., 515 ; kachórá (Curcuma), *Kan.*, C. 2499 ; kachóram, *Tel.*,
C. 2499 ; kachóram (Kæmpferia), *Tel.*,
K. 3; kachu, *Kan.*, A. 137, 181; kadaga,
Kan., A. 1192; kádágaruganie, *Tam.*,
H. 111 ; kadagho, *Tam.*, B. 841 ; káda,
jemudu, *Tel.*, E. 553 ; kadakái, *Tam.*, T.
325 ; kadalai, *Tam.*, C. 1061 ; kadalai-
kádi, *Tam.*, C. 1061 ; kadalaipulippu,
Tam., C. 1061 ; kadale-kádi, *Kan.*,
Mal., C. 1061 ; kadale-kayí, *Kan.*, A.
1261 ; kadali (Lagerstrœmia), *Tam.*, L.
42; kadali (Musa), *Tel.*, M. 811 ; kadal-
pách-chi, *Tam.*, G. 653 ; kada má, *Tam.*,
C. 943 ; kadamba, *Tel.*, A. 1192 ; kada-
mic, *Tel.*, B. 180; kadani, *Kan.*, S. 2799 ;
kadapara, *Tel.*, A. 1395; kadapilva, *Mal.*,
M. 656; kadaralai, *Tam.*, C. 943 ; kada-
tathie, *Tam.*, D. 748 ; kadat-réngáy,
Tam., L. 511 ; kadbevina mara, *Kan.*,
M. 363, 393 ; kád-bévu, *Kan.*, M. 412 ;
kádepatíge, *Tel.*, V. 217 ; kádépa tíge,
Tel., V. 195 ; kádi, *Tam.*, A. 356 ;
kadigga-garaga, *Kan.*, E. 7 ; kadishen,
Tel., I. 219 ; kad-jemudu, *Tel.*, E.
553 ; kadkanagola, *Kan.*, D. 438 ; kad-
kanagula, *Kan.*, D. 428 ; kadli, *Kan.*,
C. 1061 ; kadrajuvi, *Tel.*, P. 1433 ; kad-
sige, *Coorg*, A. 686 ; kadu, *Tam.*, C. 943;
kadughú, *Tam.*, G. 753 ; kádu-ippe-gida,
Kan., B. 220 ; kádujirage, *Kan.*, V. 73 ;
kadukar, *Tel.*, T. 325 ; kaduk-kay, *Tam.*,
T. 325 ; kaduk-káy-pinji, *Tam.*, T. 325 ;
kadumenthyá, *Kan.*, S. 1714; kadwal,
Kan., A. 1192; kadwar, *Kan.*, S. 2799 ;
kafur, *Mal.*, C. 257 ; kág-ala-káyi, *Kan.*,
M. 626; kagali, *Kan.*, A. 135; kaggalibija,
Kan., C. 279; kaggera, *Tel.*, P. 1121 ;

kagira, *Kan.*, H. 306 ; kagli (Acacia),
Kan., A. 135 ; kagli mara (Canarium),
Kan., C. 279 ; kai, *Tam.*,. M. 711 ;
kaida, *Mal.*, P. 26 ; kaidaryamu, *Tel.*,
M. 869 ; kai-donda, *Tel.*, C. 919 ;
kaikeshi, *Tam.*, E. 7 ; kaipa-kotakap-
pála-vitta, *Mal.*, H. 294 ; kaippa-pate-
lam, *Mal.*, T. 576 ; kaippa-valli, *Mul.*,
M. 626 ; kairt, *Tam.*, F. 53 ; kaita-
chakka, *Mal.*, A. 1045 ; kaivartakamuste,
Tel., C. 2612 ; kaivishi-ilai, *Tam.*, E. 7 ;
kaka, *Tel.*, F. 615 ; káká-koliviari, *Tam.*,
A. 1038 ; kakamachi, *Tel.*, S. 2299 ;
káká-mári, *Tel.*, A. 1038 ; kákamári-bija,
Kan., A. 1038 ; kákamuchi, *Tel.*, S.
2280; káka pála, *Tel.*, T. 855 ; kákara,
Tel., M. 626; kákara-shingí, *Tel.*, P. 833 ;
kakari-kai, *Tam.*, C. 2274; kaka-tati,
Tam., D. 569 ; kaka ulimera, *Tel.*, D.
628 ; káka-valli, *Mal.*, C. 1403; káka-
villa, *Mal.*, C. 1403 ; kakee, *Kan.*, C.
756 ; kákíbíra, *Tel.*, H. 444; kákichampa,
Tel., A. 1038 ; káki-donda, *Tel.*, C. 919;
káki donda, *Tel.*, I. 600 ; kakí veduru,
Tel., S. 49 ; kákka múllu, *Mal.*, P. 363 ;
kákkanan-kodi, *Tam.*, C. 1403 ; kákka-
nam-koti, *Mal.*, C. 1403 ; kakkarik,
Tam., C. 2278 ; kákkatán-virai, *Tam.*,
I. 384 ; kakkatashingi (Pistacia), *Tam.*,
P. 833; kakkatashingi (Rhus), *Tam.*, R.
323 ; kákka-tutari-véttin-thól, *Mal.*, T.
489 ; kakkáy-kolli-virai, *Tam.*, A. 1038 ;
kaku-múllu, *Mal.*, C. 30 ; kakú-pala,
Tel., Z. 229 ; kalabanda, *Tel.*, A. 829 ;
kala bhangra, *Tel.*, S. 2354 ; kala i,
Mal., C. 1732 ; kala ginja, *Tel.*, P. 1352;
kala goindú, *Kan.*, D. 628 ; kala gorú,
Tel., S. 2865, S. 2876; kalaippaik-kishan-
gú, *Tam.*, G. 243 ; kalaka, *Tam.*, S.
1939 ; kalaka, *Tam.*, C. 596 ; kalamara,
Tel., F. 405 ; kálamaram, *Mal.*, B. 913 ;
kalambir, *Mal.*, C. 1520 ; kalapa (Cocos),
Mal., C. 1520 ; kalapa (Carissa), *Tam.*,
C. 596 ; kalapa minak, *Mal.*, C. 1520 ;
kalappa-gadda, *Tel.*, G. 243 ; kalarva,
Tam., S. 717 ; kalasa, *Tel.*, L. 292 ; kala-
vankabija, *Kan.*, H. 508 ; kalay, *Tel.*,
F. 330 ; kalayána murukku, *Tam.*, E.
342 ; kála-yatte, *Kan.*, H. 124 ; kal-
baghi, *Kan.*, A. 695, 722 ; kalbon, *Kan.*,
H. 371 ; kalda, *Tam*, H. 414 ; kalgante,
Coorg, S. 959 ; kalhá ramu, *Tel.*, N. 200 ;
kalichi, *Tam.*, C. 6 ; kalíhútrú, *Kan.*, S.
2865 ; kali maruthai, *Tam.*, T. 361 ;
káli-munnu, *Tam.*, C. 1317 ; kalinga
(Wrightia), *Tel.*, W. 122 ; kalinga (Dillenia), *Tel.*, D. 428 ; kalinji, *Nilgiris*, A.
440 ; kal'shikkáy, *Tam.*, C. 6 ; kalivi
kaya, *Tel.*, C. 596 ; kaliyána-púshinik-
káy, *Tam.*, B. 430 ; kalkanda, *Kan.*,
S. 323 ; kal kilingi, *Nilgiris*, C. 838 ;
kall-alun, *Tam.*, F. 216 ; kalli, *Tam.*,
E. 553 ; kalli chemudu, *Tel.*, E. 553 ;
kallu, *Tam.*, R. 619 ; kall-udi, *Kan.*,
S. 2865 ; kallurivi, *Tam.*, A. 958 ; kal-
lúrvanchi, *Mal.*, A. 958 ; kalp alagi,
Tam., S. 2235 ; kalpam, *Tam.*, C. 331 ;
kalpam-chettu, *Tel.*, C. 331 ; kalpún,
Kan., C. 162 ; kal-shivani, *Kan.*, G. 298;
kal-támara, *Mul.*, S. 2252 ; kaltega,
Kan., D. 438 ; kalthuringi, *Tam.*, A. 711;
kálu-gechcha, *Tel.*, P. 1352 ; kalugudu
Tel., G. 143 ; kalung, *Tam.*, D. 485;

kalyanámurukku, *Tam.,* E. 342 ; kámák-
shikasuvu, *Tel.,* A. 1107 ; kámákshí-pulla,
Mal., A. 1107 ; kamákshí-pullu, *Tam.,*
A. 1107 ; kamal, *Kan.,* T. 28 ; kamalá
pandu, *Tel.,* C. 1234 ; kámanchi, *Tel.,*
S. 2299 ; kamanchi chettu, *Tel.,* S. 2299 ;
kámanchi-gaddi, *Tel.,* A. 1107 ; kam-
anji, *Tam.,* B. 868 ; kamarak, *Kan.,*
A. 1646 ; kambali, *Tel.,* M. 756 ; kam-
bali-búchi, *Tel.,* M. 756 ; kambilipúch,
Tam., M. 756 ; kambu, *Tam.,* P. 384 ;
kamela-mávu, *Tam.,* M. 71 ; kam kasturi, *Kan.,* O. 18 ; kamma régu, *Tel.,*
A. 1511 ; kam-pira, *Mal.,* S. 1041 ;
kampu-túmma, *Tel.,* A. 217 ; kamrá,
Kan., H. 16 ; kamugu, *Tam.,* A. 1294 ;
kam-wepila, *Tam.,* M. 800 ; kanaga,
Kan., P. 1121 ; kanagala (Dillenia),
Kan., D. 438 ; kanagala (Plumeria),
Kan., P. 989 ; kanagale, *Kan.,* N. 80 ;
kanagi, *Kan.,* M. 904 ; kanagole, *Kan.,*
D. 438 ; kanagurta, *Tel.,* F. 575 ; kana-
kaia búdhakakara, *Tel.,* C. 551 ; kanang
kirai, *Tam.,* C. 1752 ; kanapa, *Tel.,* B.
180 ; kanapa-chettu, *Tel.,* B. 180 ; kanapa
chettu badanike, *Tel.,* V. 17 ; kanapa-
tíge, *Tel.,* V. 195 ; kánarégu, *Tel.,* F.
624 ; kánchan, *Tel.,* B. 308 ; kánchana,
Tel., B. 295 ; kánchanamu, *Tel.,* M.
517 ; kancháva-chetti, *Mal.,* C. 331 ;
kánchini, *Tam., Tel.,* B. 334 ; kanchi-
pundu, *Tel.,* S. 2299 ; kan chivála,
Kan., B. 308 ; kanchivala-do, *Kan.,* B.
356 ; kánchu, *Tel.,* A. 137, 181 ; kan-
chupranthi, *Kan.,* M. 15 ; kan churi-
vayr, *Tam.,* T. 509 ; kanda (Amorphophallus), *Tel.,* A. 996 ; kanda (Tacca),
Tel., T. 15 ; kanda-gadda, *Tel.,* T.
883 ; kanda gariga mara, *Kan.,* C. 838 ;
kanda-godda, *Tel.,* A. 996 ; kandalanga,
Tam., C. 482 ; kandalu, *Tel.,* C. 49 ;
kándámiruga-mirat-tam, *Tam.,* P. 1370 ;
kandan-kattiri, *Tam.,* S. 2345 ; kandar-
ola, *Kan.,* T. 392 ; kandu-rellu gaddi,
Tel., S. 25 ; kanemi, *Tel.,* E. 73 ; kanga,
Tel., P. 1121 ; káni ápa tíge, *Tel.,* V. 217 ;
kaniga, *Tel.,* P. 1121 ; kanja, *Kan.,* M.
570 ; kanji, *Mal.,* S. 2682 ; kanka (Bambusa), *Tel.,* B. 118 ; kanka (Dendrocalamus), *Tel.,* D. 292 ; kanka-pati, *Kan.,*
F. 655 ; kankéli, *Tel.,* S. 861 ; kanki
putri, *Tel.,* P. 824 ; kannadi, *Telegu,* D.
240 ; kanná-kampu, *Tam.,* W. 13 ; kan-
naku, *Tel.,* F. 350 ; kannem, *Tel.,* V.
217 ; kánnimbe, *Kan.,* A. 1601 ; kanni-
rak-kuru, *Mal.,* S. 2943 ; kannu palle,
Tam., M. 583 ; kanregu, *Tel.,* F. 615 ;
kanru, *Tel.,* F. 624 ; kánsana, *Tel.,* B.
295 ; kanta-bháranní, *Mal.,* C. 1388 ; kanta kachó-ramu, *Tel.,* L. 84 ; kantúkelangú,
Tam., D. 481 ; kánuga, *Tel.,* P. 1121 ;
kanugamanu, *Tel.,* P. 1121 ; kanupula-
cheruku, *Tel.,* S. 30 ; kaori, *Kan.,* G.
679 ; kaphi, *Kan.,* C. 1641 ; kápi-bija,
Kan., C. 1641 ; kapidh, *Tel.,* F. 53 ; kapi-
kottai, *Tam.,* C. 1641 ; kapila, *Tam.,*
Tel., M. 71 ; kapi-vittulu, *Tel.,* C. 1641 ;
kapli, *Tam.,* M. 71 ; kapor-barus, *Mal.,*
C. 257 ; kappakka, *Mal.,* M. 626 ; kap-
pal-chérun-kuru, *Mal.,* A. 1014 ; kappal-
melaka, *Mal.,* C. 466 ; kappamávakuru,
Mal., A. 1014 ; kappa tige, *Tel.,* T. 456 ;
kappa-tiva, *Tel.,* I. 376 ; kappal-melaka,

Mal., C. 448, 455 ; kappí-kirri, *Tam.,* A.
950 ; kappikura, *Mal.,* C. 1641 ; kappura,
Kan., E. 25 ; kappúram, *Mal.,* C. 257 ;
kappu-sasoe, *Kan.,* B. 812 ; kapsí, *Kan.,*
D. 200 ; kapú-mologú, *Mal.,* C. 448 ;
kapur, *Mal.,* C. 257 ; karachi, *Kan.,* H.
16 ; karacho, *Kan.,* H. 16 ; karachunai,
Tam., T. 15 ; karadi-pongn, *Tam.,* H.
124 ; karai-cheddi, *Tam.,* C. 393 ; karaingi,
Tel., G. 128 ; karaka (Terminalia), *Tel.,*
T. 325 ; karaka (Sterculia), *Tel.,* S. 2319 ;
karakarbúda, *Tel.,* F. 124 ; kara-kartan,
Tam., C. 1403 ; karak-káya, *Tel.,* T.
325 ; karak-káya-puvvulu, *Tel.,* T. 325 ;
karaku, *Tel.,* T. 325 ; kara-kuduruk-kan,
Mal., S. 1682 ; káralasana, *Tel.,* P. 484 ;
karalekam, *Mal.,* A. 1398 ; karalli, *Tel.,*
C. 474 ; káralsona, *Tam.,* P. 484 ; karal-
vekam, *Mal.,* A. 1398 ; káram, *Mal.,* C.
527 ; karamútí sorrah, *Tel.,* F. 382 ; karan-
dai, *Tam.,* S. 2518 ; karangalli, *Tam.,* A.
135, 295 ; karangi, *Kan.,* T. 28 ; karan-
kuttai, *Tam.,* I. 515 ; karanta-kattin-káya,
Mal., A. 1038 ; karappu-pillánji, *Tam.,*
P. 663 ; karapu dammar, *Tam.,* C. 285 ;
karapu kongiliam, *Tam.,* C. 285 ; kara-
puti, *Tam.,* S. 370 ; kara-vélá, *Mal.,* G.
753 ; karbujá dósa, *Tel.,* C. 2263 ; kar-
chia, *Tel.,* N. 179 ; kard-ardu, *Kan.,* S.
1257 ; kard-kadrai, *Kan.,* S. 1229 ; kare
(Diospyros), *Kan.,* D. 569 ; karé (Randia), *Kan.,* R. 1, 16 ; karea-pela, *Mal.,*
M. 800 ; kare-jirage, *Kan.,* N. 158 ; kare-
kai, *Kan.,* C. 596 ; karelakkí-gidá, *Kan.,*
J. 116 ; karellu chitrallu, *Mal.,* S. 1078 ;
kare menasu, *Kan.,* P. 811 ; karepak,
Tel., M. 800 ; kari (carbon), *Tam., Mal.,*
C. 487 ; kari (coal), *Tam., Mal.,* C.
1414 ; kari (Ficus), *Kan.,* F. 216 ; kari-
baragu, *Kan.,* P. 63 ; kari-bévana, *Kan.,*
M. 800 ; karibevin, *Kan.,* M. 412 ; kari-
bevu, *Kan.,* M. 800 ; karichakka, *Mal.,*
A. 1644 ; karie-chíra, *Mal.,* P. 1179 ;
karig anne, *Kan.,* C. 854 ; kari-gheru,
Kan., S. 1041 ; karijirigi, *Kan.,* N. 158 ;
kari kadale, *Kan.,* C. 1061 ; kari-maruta,
Mal., T. 361 ; kari-matti, *Tam.,* T. 361 ;
karimsiragam, *Kan.,* N. 158 ; kari mutal,
Kan., O. 537 ; karinga, *Tel.,* G. 128 ;
karinghota, *Mal.,* S. 749 ; karinkolla,
Mul., C. 728 ; karinpa, *Mal.,* S. 30 ;
karinta-kali, *Mal.,* G. 175 ; karinthagara,
Mal., P. 1370 ; karintúmba, *Mal.* A.
1132 ; karipal, *Tel.,* I. 515 ; kariram,
Mal., S. 2943 ; karisasive, *Kan.,* B. 841 ;
karisha-langanni, *Tam.,* E. 7 ; kari-vépa,
Tel., M. 800 ; kariya pólam, *Tam.,* A.
824 ; karjúrukáya, *Tel.,* P. 555 ; karka-
kartum, *Tam.,* C. 1403 ; karkan-dhavu,
Tel., Z. 231 ; karkandu, *Tam.,* S. 376 ;
karkapilli, *Tam.,* P. 900 ; kár-karunaik-
kizhangu, *Tam.,* T. 883 ; karkattam,
Tam., Z. 229 ; karkava, *Tam.,* E. 73 ; kar
kisellimara, *Kan.,* C. 679 ; kárkol, *Tam.,*
S. 717 ; kar-kona, *Kan.,* O. 567 ; karlay,
Mal., F. 415 ; karmi, *Tel.,* S. 2799 ;
karna-kita, *Tam.,* F. 575 ; karnang-
luullutan, *Tam.,* T. 575 ; karomonga,
Tel., A. 1646 ; karona, *Tam.,* F. 476 ;
karpa, *Kan.,* H. 124 ; kárpó-karishi,
Tam., P. 1352 ; karpúra, *Kan.,* C. 257 ;
karppúram, *Tam., Tel.,* C. 257 ; karp-
púra-valli, *Tam., Tel.,* A. 1130 ; kar-

pushpú, *Tel.*, Z. 199 ; kárpuvá-arishi, *Tam.*, P. 1352; karra, *Tel.*, D. 64; karra-antinta, *Tel.*, D. 347 ; karra marda, *Tam.*, T. 361 ; karra sirli, *Tel.*, D. 63 ; karre vembú, *Tam.*, C. 143 ; karrijáli, *Kan.*, A. 101 ; karril, *Kan.*, V. 159 ; karruwa, *Tam.*, C. 1196 ; karsaar, *Tam.*, F. 449 ; kar-shunnambu, *Tam.*, C. 489 ; kárttikaik-kishangu, *Tam.*, C. 243 ; kárttu-kiz-hangu, *Tam.*, D. 173 ; káru alachanda, *Tel.*, P. 484 ; káru allamu, *Tel.*, Z. 199 ; karu-bach-chali, *Tel.*, B. 203 ; karúbógi, *Tel.*, P. 1352 ; karu boppayi, *Tel.*, S. 2819 ; káru-chiya, *Tel.*, N. 179 ; káru-chodi, *Tel.*, E. 186 ; káruguggillam, *Tel.*, G. 156 ; karukak-pulla, *Mal.*, A. 1398 ; karukatá, *Tam.*, Z. 229 ; karukiti karin-guva, *Tel.*, C. 124 ; karu-maradu, *Tam.*, T. 361 ; karú marúthú, *Tam.*, T. 361 ; karúmbú, *Tam.*, S. 30 ; karu-minumulu, *Tel.*, P. 513 ; karumsembai, *Tam.*, S. 1174 ; káru munaga, *Tel.*, M. 721 ; karu-naik-kizhangu (Typhonium), *Tam.*, T. 883 ; karu-naik-kizhangu (Amorphophallus), *Tam.*, A. 996; karuna-kalang, *Tam.*, A. 996 ; karuna-karang, *Mal.*, A. 996 ; karuna-kizhanna, *Mal.*, A. 996 ; karun-chirakam, *Mal.*, N. 158 ; karung, *Tel.*, G. 128 ; karu-nili, *Tel.*, *Kan.*, I. 109 ; karunkáli, *Tam.*, D. 569; karunká-nam, *Tam.*, C. 728 ; karunnochi, *Mal.*, J. 116 ; karu-noch-chi, *Tam.*, J. 116 ; karun-shíra-gam, *Tam.*, N. 158 ; karunthumb, *Tam.*, D. 615 ; karupale, *Tam.*, P. 1433 ; karu-patti, *Tel.*, H. 263 ; káru pendalam, *Tel.*, D. 515 ; káru-pógáku, *Tel.*, B. 546; karup-pa-katuka, *Mal.*, B. 812; karuppu dámar (Canarium), *Tam.*, C. 285 ; karuppu-dámar (Shorea), *Tam.*, S. 1682; karuppu-kadugu, *Tam.*, B. 812 ; karuppu-noch-chi, *Tam.*, J. 116 ; karup-púram, *Tam.*, C. 257 ; karúpú-voval, *Tam.*, F. 584 ; karu-umaté, *Tam.*, D. 151 ; karu-um-matta, *Mal.*, D. 151 ; karuvaga, *Tam.*, A. 711 ; karu vage, *Tel.*, A. 711 ; karuváp-pattai, *Tam.*, C. 1196 ; karuváp-puicrambú, *Tam.*, C. 706 ; karúvelum, *Tam.*, A. 101 ; karu-vémbu, *Tam.*, M. 800 ; karu-vengé, *Tam.*, A. 711 ; karu-véppilai, *Tam.*, M. 800 ; karú veylam, *Tam.*, A. 101 ; karu-vúmattai, *Tam.*, D. 151 ; kar vaghe, *Tam.*, A. 711 ; kar-vaila, *Tam.*, A. 101 ; karvambú, *Tam.*, G. 143 ; kásá, *Tam.*, M. 439 ; kasaka, *Tel.*, S. 2299; kásámardhakamu, *Tel.*, C. 787 ; kasangu, *Tam.*, P. 576 ; kasaragadde, *Kan.*, S. 2943 ; kasaraka, *Kan.*, S. 2943 ; kasha-kasha-karuppa, *Mal.*, P. 87 ; kasha-kashak-kuru, *Mal.*, P. 87 ; kashappu-vetpá-larishi, *Tam.*, H. 294 ; káshi-katti, *Mal.*, A. 137, 181 ; kashinda, *Tel.*, T. 489; kashíni-virai, *Tam.*, C. 1104, 1108 ; kashkashat-tól, *Mal.*, P. 87 ; kashlikire, *Tam.*, H. 250 ; káshu, *Tam.*, A. 137, 181 ; káshukatta, *Tam.*, A. 137 ; káshu-katti, *Tam.*, A. 181 ; kasindhá, *Tel.*, C. 780 ; kasini-vittulu, *Tel.*, C. 1108 ; kashkukatti, *Tam.*, A. 135 ; kasmaryamu, *Tam.*, G. 287; kasmiri, *Kan.*, G. 287; kasrike, *Kan.*, C. 826 ; kasturí, *Mal.*, *Tam.*, *Tel.*, D. 228 ;

kastúrí-arishiná, *Kan.*, C. 2406 ; kastúri-benda-vittulu, *Tel.*, H. 168 ; kasturí-manjal, *Tam.*, C. 2406 ; kastúrí pasupa, *Tel.*, C. 2406 ; kas-turi-patte, *Tel.*, N. 80; kasturi-vendaik-kay-virai, *Tam.*, H. 168 ; kastúri-venta-vitta, *Mal.*, H. 168 ; ka-sundara, *Tel.*, F. 328 ; katakamí, *Tel.*, S. 2960 ; kataka-nai, *Tam.*, P. 63 ; kata-kelenga, *Tel.*, D. 481 ; katak-kalli, *Mal.*, E. 496 ; kataláti, *Mal.*, A. 382 ; katalli kai, *Tam.*, C. 416 ; kat-alluri, *Mal.*, H. 339 ; kat ambo-lam, *Mal.*, S. 2649; kat-arali (Cerbera), *Tam.*, C. 943 ; kát-aralie (Tabernæmontana), *Tam.*, T. 9 ; kat-avéri, *Tam.*, I. 109 ; kát-carva, *Mal.*, C. 1158 ; katcha-catta-marum, *Tam.*, L. 55 ; kat-chelí, *Tel.*, F. 573 ; katch-ka-ıawa, *Kan.*, F. 352 ; kate chettu, *Tel.*, C. 1053 ; kathalai, *Tam.*, A. 636 ; katharum cheddy, *Tam.*, G. 244 ; káthe-nerinnil, *Mal.*, P. 363 ; káthi-iluppai, *Tam.*, B. 220 ; kati, *Tam.*, A. 137; káti ámudapu, *Tel.*, J. 52 ; kat illipi, *Tam.*, B. 220 ; kátimandu, *Tam.*, D. 387 ; katimango, *Tam.*, B. 913 ; kat kuddaghu, *Mal.*, C. 1367 ; kat-kumbla, *Kan.*, T. 525 ; katle gaddi, *Tel*, C. 1053 ; katle-tige, *Tel.*, C. 1452 ; katmaá (Buchanania), *Tam.*, B. 913 ; kat-maá (Spondias), *Tam.*, S. 2649 ; kat-malli, *Tam.*, M. 550 ; kátma-maram, *Tam.*, B. 913 ; kat-maparam, *Tam.*, B. 913 ; kátma-parpu, *Tam.*, B. 913 ; kátma-payam, *Tam.*, B. 913 ; kat-mara, *Tam.*, S. 2649 ; katrazh-ai, *Tam.*, A. 829 ; kat santhanam, *Tam.*, E. 370 ; kát-siragam, *Tam.*, V. 73 ; katsjal-kelungu, *Tam.*, D. 547 ; katsjolum, *Tam.*, K. 3 ; katsjulum, *Mal.*, K. 3 ; kats-jút-kelangu, *Tam.*, D. 485 ; kattakambu, *Tam.*, A. 181 ; kattála, *Mal.*, A. 829 ; kattalai, *Tam.*, A. 829 ; kattámanakku, *Tam.*, J. 41 ; kátta-vanakka, *Mal.*, J. 41 ; katthu-olupœ, *Tel.*, T. 293 ; katti, *Mal.*, A. 137, 181; katti-mandu, *Tel.*, E. 561 ; kattiríppa bonam, *Mal.*, B. 220 ; kat-truvázha, *Mal.*, A. 829 ; kattuy-elakkáy, *Tam.*, A. 976 ; káttu-elumichcham-param, *Tam.*, A. 1601 ; kattu-elupœ, *Tam.*. T. 293 ; kattu-elupay, *Tam.*, T. 293 : káttuiluppai, *Tam.*, B. 220 ; kattu-inchi kúva, *Mal.*, Z. 225 ; káttu-irrupai, *Tam.*, B. 220 ; káttu-jírakam, *Mal.*, V. 73 ; káttu-karuváp pattai, *Tam.*, C. 1158 ; kattu-karuvátoli, *Mal.*, C. 1158 ; káttu-kastúri, *Mal.*, H. 168 ; kattuk-kastúri, *Tam.*, H. 168 ; káttuk-kodi, *Tam.*, C. 1452 ; káttukkol, *Tam.*, C. 728 ; káttu-kúrkká, *Mal.*, A. 1130 ; káttulli, *Mal.*, U. 39; kattu-mannal, *Tel.*, C. 2406 ; kattu-mannar, *Mal.*, C. 2406 ; kattu-mullángi, *Tam.*, B. 546 ; káttup-pépu-dal, *Tam.*, T. 576 ; kat turanji, *Tam.*, A. 722 ; kattu-shíragam, *Tam.*, V. 73 ; káttu-tultuva, *Mal.*, O. 28 ; kattu-vullie-kelangu, *Tam.*, D. 522 ; káttu-yelak-káy, *Tam.*, A. 976 ; katú-bala, *Mal.*, C. 321 ; kat-udugu, *Tam.*, H. 16 ; katúkaró-gani, *Tel.*, H. 111 ; katuka-rogani, *Tel.*, P. 700 ; kattukká, *Mal.*, T. 325 ; katuku-rogani, *Tam.*, P. 700 ; katuku-roni, *Tel.*, P. 700 ; katu-pitsje-

gam, *Mal.*, J. 10; katu-punai, *Tam.*,
T. 427; katupuveras, *Tam.*, F. 279;
katu-tsjiregam-mulla, *Mal.*, J. 32; katu-
yeni, *Tam.*, O. 567; kat vaghe, *Tel.*,
A. 695; katyalu, *Tam.*, A. 1601;
kaulay, *Kan.*, D. 656; kaulla-kuri,
Kan., S. 1287; kaval, *Kan.*, C. 563;
kavanchi, *Tel.*, H. 92; kavargi, *Kan.*,
H. 92; kávartam-pullu, *Tam.*, A. 1107;
kavile, *Tel.*, S. 2850; kavirisandra, *Tel.*,
A. 135; kavit, *Tam.*, F. 53; kávu, *Tel.*,
L. 21; kavugu, *Mal.*, A. 1294; kavva-
gumudu, *Tel.*, C. 298; kawa, *Mal.*, C.
1641; kawili, *Tam.*, S. 2837; káya
pendalam, *Tel.*, D. 499; kayap-pankot-
tai, *Tam.*, S. 2940; kayápute, *Tam.*,
M. 342; kayi (Acacia), *Kan.*, A. 200;
kayi (Citrullus), *Kan.*, C. 1211; kayi
(Stereospermum), *Kan.*, S. 2876; kayo-
gadis, *Mal.*, C. 1170; kayur, *Tam.*, E.
170; kazhanchik-kuru, *Mal.*, C. 6;
kazhar-shikkáy, *Tam.*, C. 6; kazuthai,
tumbai, *Tam.*, T. 562.

Ke Kékku-virai, *Tam.*, C. 681; kélahú-húdin-
gana, *Kan.*, C. 321; kelvaragú, *Tam.*,
E. 170; kembu, *Tam.*, R. 619; kempu-
chitramúlá, *Kan.*, P. 979; kempú gan-
dagheri, *Kan.*, C. 838; kempu gandha-
cheke, *Kan.*, P. 1381; kempu géru-bíja,
Kan., A. 1014; kempu kadale, *Kan.*,
C. 1061; kempu khairada, *Kan.*, A. 295;
kempu-khasa-khasi-gída, *Kan.*, P. 82;
kempu-kiranelli, *Kan.*, P. 673; kempu-
lóla-sará, *Kan.*, A. 838; kempu rai,
Tel., R. 619; kend, *Mal.*, D. 615;
kenjol, *Kan.*, S. 2424; kenneggilu,
Kan., I. 125; kerkal, *Kan.*, T. 434;
kesar chettu, *Tel.*, C. 2062, 2068;
keshavaná-gadde, *Kan.*, C. 1732; ketáki,
Kan., *Tel.*, P. 26; kethi-kanni, *Tam.*,
D. 628; kevalee, *Tel.*, S. 2850.

Kh Khabbanada-kittá, *Kan.*, I. 472; khabbu,
Kan., S. 30; khargas, *Kan.*, F. 124;
kharjúra, *Kan.*, P. 555; kharjúrapu, *Tel.*,
P. 555; kharkar, *Mal.*, G. 136; khasa-
khasi, *Kan.*, P. 87; khasca, *Kan.*, S.
2943; khomanfg, *Nilgiri Hills*, C. 218;
khorkhorendna, *Mal.*, C. 705; khurásáni-
vádakki, *Kan.*, H. 525; khurásáni-vómá,
Kan., H. 525.

Ki Kícha viri chettu, *Tel.*, C. 1940; kichchilick-
kizhanghu, *Tam.*, C. 2499; kich-chili-
gaddala, *Tel.*, C. 2499; kich-chili-pandu,
Tel., C. 1233; kich-chilip-pazham, *Tam.*,
C. 1233; kichidi, *Tel.*, C. 1233; kichidi-
pandu, *Tel.*, C. 1233; kichige, *Kan.*, E.
354; kichili, *Tam.*, C. 1233; kichlie-pul-
lum, *Tam.*, C. 1233; kigingan, *Tam.*, F.
580; kíjápúté, *Tam.*, M. 342; kilávari,
Tam., A. 1577; kilay, *Tel.*, F. 565;
kilinjan, *Tam.*, F. 580; kimri, *Mal.*, A.
681; kimsukamu, *Tel.*, B. 944; kínan-
chikkuru, *Mal.*, C. 6; kinda-min, *Tam.*,
F. 465; kin-ghenna, *Kan.*, C. 1520;
kírai, *Tam.*, T. 537; kirai-tand, *Tam.*,
A. 938; kiralbogi, *Kan.*, H. 368, 371;
kiramber, *Tam.*, C. 706; kirámbu, *Tam.*,
C. 706; kiranelli-gídá, *Kan.*, P. 657;
kiraruga, *Tel.*, P. 332; kiray, *Tam.*, S.
2994; kirballi, *Kan.*, P. 1016; kirgunna,
Kan., C. 1403; kiridpodla káyi, *Kan.*,
T. 576; kiri-purandán, *Tam.*, O. 180;
kiriyattu, *Mal.*, A. 1064; kirnelli, *Kan.*,

P. 627; kiru-wahlah, *Tam.*, F. 402;
kistapa tamara, *Tel.*, S. 2252; kisumí-
suchettu, *Tel.*, V. 233; kitchli, *Tam.*,
Tel., C. 1233; kithlí, *Tam.*, C. 1234;
kittaboippe, *Kan.*, C. 1233; kit tale-
pannu, *Kan.*, C. 1233; kittalesippe, *Kan.*,
C. 1233; kittali, *Tel.*, C. 1233; kittali-
pandu, *Tel.*, C. 1233; kittan, *Tam.*, *Tel.*,
C. 399; kízhá-nelli, *Tam.*, P. 657; kízhá-
nelli, *Mal.*, P. 657; kizhkáy-nelli, *Tam.*,
P. 657.

Ko Koatamunak, *Tam.*, J. 41; koaya, *Tam.*, P.
1343; ko-bari, *Kan.*, C. 1520; kcbbarait-
téngáy, *Tam.*, C. 1520; kobbari, *Kan.*,
Tel., C. 1520; kobbari chettu, *Tel.*, C.
1520; kobbera, *Tel.*, C. 1520; kobbera-
ténkáya, *Tel.*, C. 1520; kobri chullú, *Tel.*,
C. 1520; koch-chit-tamarttai, *Tam.*, A.
1644; kodaga saleh, *Tam.*, R. 660;
kodan, *Toda*, F. 808; kodawá, *Tam.*, C.
1031; kodawá-porasham, *Tam.*, C. 1031;
kodi, *Tam.*, I. 384; kódi-budama, *Tel.*,
C. 2306; kodíbu-dinga, *Tel.*, C. 2306;
kodic-palay, *Tam.*, D. 823; kodi-juttu-
tota-kura, *Tel.*, C. 873; kodimúli, *Tam.*,
P. 979; kodi-mun-dirrip-pazham, *Tam.*,
V. 233; kodi-murukkam, *Tam.*, B. 978;
kodipalásham, *Tam.*, B. 978; kodipungi,
Tel., F. 554; kodisha, *Tel.*, W. 122;
koditáni, *Tam.*, G. 218; kodi-vér, *Tam.*,
T. 470, T. 483; kod, murki, *Kan.*, W.
122; kodrimúndrie, *Tam.*, V. 233;
koduwa, *Tam.*, F. 476; koila-mukri, *Tel.*,
W. 131; koilangu, |*Tam.*, I. 343; kok,
Kan., R. 51; kokkíta, *Tel.*, A. 1362;
kokkitayárálu, *Tel.*, V. 184; koku, *Tam.*,
C. 224; koku, *Tam.*, S. 705; kolák-
ponna, *Tel.*, D. 339; kólánji-narakam,
Mal., C. 1233; kóla ponna, *Tel.*, U. 23;
kola sampige, *Kan.*, M. 517; kólatunga-
muste, *Tel.*, C. 2617; kolávu, *Tam.*, H.
22; koli, *Tam.*, O. 153; koli kuki, *Kan.*,
B. 4; kolinji-marum, *Tam.*, C. 1233; kol-
kalli, *Mal.*, E. 553; kollan-kóvakizhauna,
Mal., C. 1834; kolli-vittulu, *Tel.*, I. 384;
kollú, *Tam.*, D. 758; kolluk-káy-vélai,
Tam., T. 270; kolupu, *Tel.*, G. 677;
komara-parah, *Tam.*, F. 373; koma-
sorra, *Tel.*, F. 602; kombú, *Tam.*, H.
408; kombu-kalli, *Tam.*, E. 553; kombu-
pudalai, *Tam.*, T. 586, T. 598; kombur-
ruki, *Tam.*, L. 1; kome, *Tel.*, F. 390;
komi, *Tel.*, W. 22; kommolaka, *Tel.*, L.
1; kommú, *Tel.*, H. 408; kommu-ariti,
Tel., M. 811; kommu-potla, *Tel.*, T.
586; kommu-potta, *Tel.*, T. 598; konam,
Tam., *Tel.*, C. 2528, F. 419; konda
(Toddalia), *Tel.*, T. 489; konda (Ipomæa), *Tel.*, I. 342; konda (Sorghum),
Tel., S. 2424; konda-ámudam, *Tel.*, B.
28; konda benda, *Tel.*, H. 194; konda-
cahínda, *Tel.*, T. 489; konda chiragu,
Tel., A. 722; konda dantena, *Tel.*, S.
2252; konda-gang, *Tel.*, H. 194; konda-
gógu-banka, *Tel.*, C. 1512; konda-gógu-
patti, *Tel.*, C. 1512; konda-gogu-pisunu,
Tel., C. 1512; konda golugu, *Tel.*, G.
271; konda góngña, *Tel.*, II. 219;
konda-gori, *Tel.*, S. 1287; konda-gum-
mudu, *Tel.*, D. 522; konda gurava tíge,
Tel., S. 2252; konda kahínda, *Tel.*, T,
489; konda kalava, *Tel.*, K. 8; konda.
kanamoo, *Tel.*, S. 4; kondakasin-da-

véru-patt, *Tel.*, T. 489 ; konda mulle, *Tel.*, P. 1084 ; konda manga, *Tel.*, G. 124 ; konda-murga-rattam, *Tam.*, C. 68 ; konda nakkera, *Tel.*, X. 12 ; konda pála, *Tel.*, S. 882 ; konda támara, *Tel.*, S. 2252 ; konda tangédu, *Tel.*, X. 16 ; konda tantepu chettu, *Tel.*, C. 769 ; konda tekkali, *Tel.*, S. 3042 ; kondavaghe, *Tam.*, A. 717 ; konda-vepa, *Tel.*, M. 393 ; konda-yeutawa, *Tel.*, F. 775 ; kondguri, *Kan.*, S. 1287 ; kone, *Tam.*, C. 756 ; kong (Cochlospermum), *Tel.*, C. 1512 ; kong (Hopea). *Tam.*, H. 360 ; konga, *Tel.*, S. 2903 ; kongillam, *Tam.*, C. 1512; kongu, *Tam.*, H. 360 ; konki, *Tel.*, P. 824 ; konkúdú, *Tel.*, S. 818 ; konnak-káya, *Mal.*, C. 756 ; konnári-gadde, *Kan.*, C. 2617 ; konraih-káy, *Tam.*, C. 756 ; kopi, *Mal.*, C. 1641 ; kóppara, *Mal.*, C. 1520 ; kora (Ophiocephalus), *Tel.*, F. 517 ; kora (Setaria), *Tel.*, S. 1212 ; korah-mottah, *Tel.*, F. 514 ; kórai, *Tam.*, C. 2612 ; kóraik-kizhangu, *Tam.*, C. 2617 ; kórakizhanna, *Mal.*, C. 2617 ; koralú, *Tel.*, S. 1212 ; kora-madi, *Tel.*, B. 868 ; kora-mánu, *Tel.*, B. 868 ; korangu, *Kan.*, F. 808 ; korasana, *Tel.*, F. 202 ; korattai, *Tam.*, T. 600 ; korava, *Tam.*, F. 516 ; koray kalung, *Tam.*, C. 2617 ; kore-gadi, *Tel.*, S. 49 ; kórendam, *Tel.*, A. 233 ; korgi, *Kan.*, I. 515 ; koricha, *Tel.*, L. 219 ; korimipála, *Tel.*, I. 515 ; korinta, *Tel.*, A. 233 ; koriti, *Tel.*, P. 950 ; korivi, *Tel.*, H. 124 ; korivipála, *Tel.*, I. 515 ; korjashtam, *Tel.*, P, 1352 ; korkkar-muli, *Tam.*, C. 331 ; korkotta, *Mal.*, D. 438 ; korsi, *Tel.*, L. 219 ; korukápuli, *Tam.*, P. 900 ; korún, *Tam.*, F. 335 ; kosramba, *Mal.*, G. 143 ; kostum, *Tam.*, S. 910 ; kotah, *Tam.*, F. 507 ; kotakappála, *Mal.*, W. 122 ; kotamalli, *Tam.*, C. 1954 ; kotambári, *Kan.*, C. 1954 ; kotanpam, *Mal.*, T. 634 ; kotap-pana, *Mal.*, C. 1995 ; kot-aralu, *Tam.*, F. 422 ; kóta-shavukku, *Tam.*, T. 70 ; kote, *Mal.*, C. 2180 ; koteka, *Tel.*, N. 200 ; kotiki, *Tel.*, O. 127 ; kotimiri *Tel.*, C. 1954 ; kot-karva, *Mal.*, C. 1158 ; kottai (Ana-cardium), *Tam.*, A. 1014 ; kottai (Ter-minalia), *Tam.*, T. 312 ; kottai-mundiri, *Tam.*, A. 1014 ; kottai pákku, *Tam.*, A. 1294 ; kottaip-panai, *Tam.*, C. 1995 ; kóttak, *Tam.*, S. 2518 ; kotta-kuru, *Mal.*, T. 312 ; kottei, *Tam.*, R. 369 ; kotti-gaddanama, *Tel.*, A. 1242 ; kotti-katang, *Tam.*, A. 1242 ; kotti-kizhangu, *Tam.*, A. 1242 ; kotu-véli, *Mal.*, P. 986 ; kot vaghe, *Tam.*, A. 695 ; kouda-júvee, *Tel.*, F. 175 ; kousu kandíra, *Tel.*, S. 2936 ; kóva, *Mal.*, C. 919 ; kóvai, *Tam.*, C. 919 ; koyalú, *Tel.*, S. 527 ; koyilamokiri, *Tel.*, W. 131 ; koyya, *Tam.*, P. 1343 ; koyya-jemudu, *Tel.*, E. 553 ; koyyapippili, *Tel.*, A. 1475 ; koyya-tota-kúra, *Tel.*, A. 950 ; kozhinhil, *Mal.*, T. 270 ; koz-hunjip-pazham, *Tam.*, C. 1233.

Kr Kránuga, *Tel.*, P. 1121 ; kreata. *Kan.*, A. 1064 ; krishna-tamarah, *Tel.*, C. 321 ; krishna-tulsi, *Mal.*, O. 31 ; krushna-tulasi, *Tel.*, O. 31.

Ku Kua (Curcuma), *Tam.*, C. 2385 ; kúa (Maranta), *Mal.*, M. 267 ; kuamau, *Tam.*, M. 267 ; kubera-kashi, *Tel.*, S 2876 ; kubyakam, *Tel.*, T. 516 ; kuchan-danam, *Tel.*, P. 1381 ; kúchi-namurl, *Kan.*, F. 517 ; kudagu, *Tel.*, A. 681 ; kuddera, *Tel.*, F. 367 ; kuddurai-pud-duki, *Tam.*, S. 2824 ; kudira-gullu, *Mal.*, C. 1053 ; kudoly, *Kan.*, C. 1061 ; kudrajinie, *Tel.*, P. 1433 ; kudra-plukku, *Tam.*, S. 2824 ; kudri, *Tam.*, H. 414 ; kúdua, *Mal.*, T. 437 ; kuduru juvir, *Tel.*, P. 1433 ; kuduwa, *Tam.*, F. 476 ; kugate, *Kan.*, S. 818 ; kujarra, *Kan.*, S. 2943 ; kuka-gori, *Tel.*, D. 219 ; kuka-omintaw, *Tel.*, C. 1367 ; kúkate-káyi, *Kan.*, S. 818 ; kukha-avalu, *Tel.*, C. 1367 ; kukka-pála, *Tel.*, T. 855 ; kukka pálakúra, *Tel.*, T. 528 ; kukka tulasi, *Tel.*, O. 26 ; kúkudu, *Tel.*, S. 818 ; kukudu-kayalu, *Tel.*, S. 818 ; kúlá, *Tam.*, S. 950 ; kulap-pálai-virai, *Tam.*, H. 294 ; kul-aral, *Tam.*, F. 493 ; kuli (tiger), *Kan.*, T. 437 ; kuli (Gmelina), *Kan.*, G. 287 ; kulingah, *Tam.*, F. 580 ; kulivi-pambú, *Tam.*, F. 507 ; kulkantu, *Mal.*, S. 376 ; kul-korava, *Tam.*, F. 429 ; kulla-kith, *Kan.*, F. 179 ; kullarávi, *Tel.*, F. 236 ; kull-ponné, *Kan.*, C. 162 ; kullu, *Kan.*, P. 588 ; kullvalei-mani, *Tam.*, C. 321 ; kulpasi, *Tam.*, L. 332 ; kumara, *Nilgiris*, P. 548 ; kumári, *Tam.*, A. 838 ; kumarpulki, *Tel.*, G. 780 ; kumbai, *Tam.*, G. 116 ; kumbala, *Kan.*, C. 2316 ; kumbala kagi, *Kan.*, C. 2331 ; kumbay, *Tam.*, G. 124 ; kumbi, *Tam.*, G. 128 ; kummara baddu (Dioscorea), *Tel.*, D. 481 ; kummara baddu (Smilax), *Tel.*, S. 2252 ; kummara ponuku, *Tel.*, G. 780 ; kumpalam, *Mal.*, B. 430 ; kumpalanná, *Mal.*, B. 430 ; kumpole, *Kan.*, A. 376 ; kunbali, *Kan.*, A. 769 ; kúncúma-pesalú, *Tel.*, P. 468 ; kunda-jungura, *Tel.*, C. 717 ; kundala, *Tel.*, F. 328 ; kúndal-panai, *Tam.*, C. 711 ; kundamani cheddi, *Tam.*, C. 321 ; kundá-nuga, *Tel.*, L. 30 ; kún-dar-panai-vellam, *Tam.*, S. 370 ; kúndéli, *Tel.*, H. 31 ; kundinga, *Tel.*, F. 495 ; kúndricum, *Tam.*, V. 31 ; kúndrikam morada, *Tam.*, B. 771 ; kundrow, *Mal.*, B. 308 ; kua-duru-kam-pishin, *Tam.*, B. 771, 777 ; kun-duru nattu, *Tel.*, I. 501 ; kungiliyam (Vateria), *Tam.*, V. 31 ; kungiliyam (Shorea), *Tam.*, S. 1656 ; kungli, *Tam.*, B. 771 ; kungumapu, *Tam.*, C. 2083 ; kunkudu chettu, *Tel.*, S. 818 ; kunkudu-káyalu, *Tel.*, S. 818 ; kunkúma, *Kan.*, *Tel.*, M. 71 ; kun-kuma-donda, *Tel.*, B. 907 ; kunkum apave, *Tel.*, C. 2083 ; kunnadi, *Tam.*, G. 229 ; kunnaku, *Tam.*, F. 350 ; kunnay, *Mal.*, F. 495 ; kunninga, *Tel.*, F. 567 ; kunnu katti pillu, *Tam.*, C. 1752 ; kunurakkam-pishin, *Tam.*, B. 751 ; kupanti, *Tel.*, P. 678 ; kuppai-chettu, *Tel.*, A. 306 ; kuppaimeni, *Tam.*, A. 306 ; kuppa-mankala, *Kan.*, B. 523 ; kuppi (Acalypha), *Kan.*, A. 306 ; kuppi (Pimpinella), *Tel.*, P. 727 ; kura (Portu-laca), *Tel.*, P. 1187 ; kura (Colocasia), *Tel.*, C. 1732 ; kúra palléru, *Tel.*, V. 191 ; kúra pasúpu, *Tel.*, Z. 199 ; kúrásání,

yomam, *Tam.,* H. 525 ; kúrásháni-
vámam, *Tel.,* H. 525 ; kurinji vámam,
Tel., H. 525 ; kurka, *Tel.,* T. 325 ;
kúrkká, *Mal.,* A. 1130 ; kurku (Ficus),
Tam., F. 216 ; kurku (Mallotus), *Kan.,*
M. 71 ; kuror, *Tel.,* E. 186 ; kurpá, *Tel.,*
B. 180 ; kurpah, *Kan.,* H. 124 ; kurpodur,
Tel., O. 127 ; kurpúra maram, *Tam.,* E.
382 ; kurpura-pulla, *Tam.,* A. 1079 ;
kurriminu, *Kan.,* F. 466 ; kurruppu-
maruta, *Tam.,* T. 361 ; kursan-pyro,
Kan., V. 116 ; kuru dinne, *Tel.,* V. 195 ;
kurukkum-ckedi, *Tam.,* A. 1351 ; kuru
mulaka, *Mal.,* P. 811 ; kurungu-munjil-
varai-maram, *Tam.,* B. 523 ; kurungu-
múnji-vittulu-chettu, *Tel.,* B. 523 ; kuru-
vakumara, *Kan.,* E. 370 ; kuruvingi,
Tam., E. 23 ; kurví-kírai, *Tam.,* A.
950 ; kusa, *Tel.,* P. 382 ; kusa-darbha,
Tel., E. 252 ; kusambi, *Kan.,* C. 637 ;
kusharta, *Kan.,* D. 582 ; kushumbá,
Tam., C. 637 ; kushumba-virai, *Tam.,*
C. 637 ; kúshumbá-vittulu, *Tel.,* C. 637 ;
kustam, *Tel.,* S. 910 ; kusturi, *Tel.,*
A. 217 ; kusumamu, *Tel.,* W. 106 ;
kusumba, *Kan.,* C. 637 ; kutaru-bu-
dama, *Tel.,* M. 791 ; kuthirekai, *Tam.,*
S. 2284 ; kuti, *Tam.,* A. 137 ; kútili,
Tam., F. 592 ; kuttukkárchammatti,
Tam., I. 137 ; kutugeri, *Kan.,* H.
306 ; kúvalap-pazham, *Mal.,* A. 534 ;
kuva mavú, *Tam.,* M. 267 ; kuve-
gadde, *N. Kanara,* C. 2385.
Kw Kwai, *Tam.,* C. 919 ; kwe, *Tam.,* C.
919 ; kwel, *Mal.,* C. 919.
Ky Kyad-age-gida, *Kan.,* P. 26 ; kyam, *Tam.,*
F. 76.

La Ladamera, *Mal ,* C. 455 ; ladamera china,
Mal., C. 455 ; ladu mira, *Mal.,* C. 455 ;
láhán shivan, *Kan.,* G. 298 ; lakki, *Kan.,*
V. 164 ; lakki gidá, *Kan.,* V. 164 ; lakkle,
Kan., V. 164 ; lakote, *Kan.,* E. 285 ;
lakshminarayaná chettu, *Tel.,* C. 2062 ;
lakuchamu nakka-rénu, *Tel.,* A. 1511 ;
lal, *Kan.,* C. 1021 ; lal chandan, *Tel.,* P.
1381 ; lamaj-jakamuvéru, *Tel.,* A. 1097 ;
laringi, *Kan.,* O. 6 ; latte-terla, *Tel.,* F.
430 ; lavanam, *Mal.,* S. 602 ; lavanam,
Tel., S. 602 ; lávanchá, *Kan.,* A. 1097 ;
lavangada patte, *Kan.,* C. 1183 ; lavan-
gadz yale, *Kan.,* C. 1158 ; lavangalu, *Tel.,*
C. 706 ; lavanga-patta, *Mal.,* C. 1196 ;
lavanga-patta, *Tel.,* C. 1196 ; lavangap-
pattai, *Tam.,* C. 1196 ; lavanga patte,
Kan., C. 1183, 1196 ; lavanga yale, *Kan.,*
C. 1158 ; lavucheruku, *Tel.,* S. 30.
Le Ledi, *Tel.,* S. 1226 ; lelin, *Mal.,* C. 931 ;
levagani-galu, *Kan.,* N. 80.
Li Limb-toli, *Kan.,* M. 439 ; límbu, *Kan.,* C.
1270 ; limowe, *Mal.,* C. 1296 ; linga donda, *Tel ,* B. 905 ; lingam, *Tam.,* M. 473 ;
linga miriyam, *Tel.,* C. 2211 ; linga potla,
Tel., T. 569.
Lo Loabate-barbarí, *Tam.,* C. 1703 ; loda-china,
Mal., C. 466 ; lolagu, *Tel.,* P. 1397 ;
lola-sorá, *Kan.,* A. 829 ; lonnáhadakana-
·gidá, *Kan.,* B. 909 ; lotá-sach-chi, *Tel.,*
C. 541 ; loti-pitta, *Tel.,* C. 205.
Lu Ludduga, *Tel.,* S. 3062 ; luki, *Tel.,* V. 159 ;
lung amú, *Tel.,* C. 1270.

Ma Maá, *Tam.,* M. 147 ; mabheri-chinarana-
bheri, *Tel.,* A. 1132 ; machakai, *Tam.,*
Q. 43 ; machil, *Kan.,* D. 438 ; machi-
pattarí, *Mal., Kan., Tam., Tel.,* A.
1469 ; mada, *Tel.,* A. 1655 ; madachettu,
Tel., A. 1660 ; madagari vembu, *Tel.,* C.
1021 ; máda-lada-hammu, *Kan.,* C. 1270 ;
mádalai, *Tam.,* P. 1426 ; mádalaip-
pazham, *Tam.,* P. 1426 ; madalam, *Tam.,*
P. 1426 ; madal tútum, *Tam.,* Z. 190 ;
madana, *Tel.,* S. 2515 ; madana-anapa-
káya, *Tel.,* C. 581 ; mudana budata káda,
Tel., S. 2515 ; madana grandhi, *Tel.,* S.
2515 ; madana séku, *Tel.,* C. 2247 ;
madanginjalu, *Tel.,* L. 385 ; maddi (Terminalia), *Kan., Tel.,* T. 282, 361 ; madd
(Morinda), *Kan., Tel.,* M. 656 ; maddi
chekhe, *Kan.,* M. 711 ; maddi chettu,
Tel., M. 704 ; maddi pál, *Tam.,* A. 670 ;
maddi-pálu, *Tel.,* A. 670 ; mádhavi tíge,
Tel., H. 285 ; mádhípala-pandu, *Tel.,* C.
1270 ; madinawah bontú, *Tel.,* F. 579 ;
madki, *Kan.,* P. 468 ; maduga, *Tam.,* E.
356 ; madu-karray, *Tam.,* R. 1 ; madura-
káméshvari, *Mal.,* A. 1431 ; magabíra,
Tel., A. 1132 ; maga-boshi, *Tel.,* F. 538 ;
maga-jellí, *Tel.,* F. 540 ; magila-maram,
Tam., M. 570 ; magilan, *Tam.,* P. 1426 ;
magunigadde, *Kan.,* R. 31 ; mahameda,
Tel., E. 342 ; maháputra jíví putrajíví,
Tel., P. 1433 ; mahisáksh, *Tel.,* B. 43 ;
mah korava, *Kan.,* F. 514 ; máhura-nár-
anná, *Mal.,* C. 1233 ; mah-wu-laachi,
Tam., F. 419 ; mah-wu-luachi, *Tam.,* C.
2528 ; mai, *Tam.,* O. 574 ; maida-laktí,
Tam., L. 483 ; mail, *Tam.,* P. 350 ;
maila, *Tam.,* V. 158 ; mailai, *Tel.,* S.
785 ; mail-tutyá, *Kan.,* C. 2367 ; main-
am, *Tel.,* H. 342 ; mai-nam, *Tel.,* C.
931 ; mairu, *Kan.,* S. 1229 ; maisáchi,
Tel., B. 43 ; máisákshi, *Tel.,* B. 43 ;
mai sháchi, *Tam.,* B. 43 ; maishákshi,
Tam., B. 43 ; maiyala erikut, *Tel.,* C.
854 ; majjige-gadde, *Kan.,* A. 1575, 1577;
makandamu, *Tel.,* M. 147 ; makánin,
Tel., M. 393 ; makarung-kai, *Tam.,* R.
1 ; makkam mokob, *Tel.,* S. 959 ; makká-
shôlam, *Tam.,* Z. 50 ; mákká zonnalu,
Tel., Z. 50 ; makki, *Tam.,* G. 66 ; makor,
Mal., B. 142 ; mala, *Hind.,* S. 2508 ;
malahcota, *Kan.,* N. 72 ; malai, *Tam.,*
M. 393 ; malai-kalli, *Tam.,* B. 909 ;
malai-kone, *Tam.,* A. 440 ; malaitangi,
Tam., S. 1688 ; malait-támara, *Tam.,* S.
2252 ; malai-veppam, *Tam.,* M. 393 ;
malákákáya-pendalam, *Tel.,* D. 494 ;
malakanda, *Tel.,* S. 376 ; malakul,
Tam., K. 14 ; malampongu, *Tam.,* C.
95 ; malan-kua, *Mal.,* K. 8 ; malé geru,
Coorg, D. 438 ; malí, *Tam.,* C. 838 ;
máli, *Tam.,* C. 838 ; málkanginitailamu,
Tel., C. 854 ; málkanguni-vittulu, *Tel.,*
C. 854 ; malla, *Kan.,* H. 31 ; mallaita-
nák, *Tam.,* H. 523 ; mallali (Dicspyros),
Kan., D. 569 ; mallali (Stereospermum),
Coorg., S. 2865 ; mallani padman, *Tel.,*
E. 569 ; mallay nangal, *Tam.,* M. 489 ;
mallay vembu, *Tam.,* M. 412 ; malle
(Jasminum), *Tel.,* J. 35 ; malle (Olax),
Tel., O. 127 ; malle-nerale, *Coorg,* E.
432 ; malligaip-pú, *Tam.,* J. 35 ; mallige,
Kan., J. 35 ; mallippu, *Tam.,* J. 35 ;

K

mal-náraugá, *Mal.*, A. 1601 ; maluramu, *Tel.*, A. 534 ; mámadi, *Tel.*, M. 147 ; mamadichitú, *Tel.*, M. 147 ; mámíd, *Tel.*, M. 147 ; mámidi, *Tel.*, M. 147 ; mamidi allam, *Tel.*, C. 2381 ; mammarum, *Tam.*, M. 147 ; mam-palam, *Mal.*, M. 147 ; manalie kirai, *Tam.*, G. 220; manapála, *Tel.*, T. 470, T. 483; manat-takkali, *Tam.*, S. 2299 ; manavare, *Kan.*, D. 789 ; manawak, *Mal.*, R. 116 ; manche, *Tel.*, E. 553 ; manchetti, *Mal.*, R. 564 ; manchikanda, *Tel.*, A. 996; manchi-núne, *Tel.*, S. 1078 ; mandá, *Tel.*, R. 1 ; mandadhup, *Kan.*, C. 285 ; manda motuku, *Tel.*, O. 537 ; mándap-pullu, *Tam.*, A. 1107 ; mandaramu, *Tel.*, C. 170 ; mandareh, *Tam.*, B. 308 ; mandúka-bramha-kúráku, *Tel.*, H. 486 ; mandula, *Tel.*, V. 217 ; mandulamárí tíge, *Tel.*, V. 195 ; maneru, *Tel.*, C. 854; manga, *Tel.*, R. 1 ; mangal, *Tam.*, B. 118 ; mangarúli, *Kan.*, V. 219 ; mangas, *Tam.*, M. 147 ; mangi, *Tel.*, S. 785 ; mánik-kam, *Tam.*, R 619 ; manipangam, *Tam.*, E. 310 ; manjadí, *Kan.*, A. 471 ; manjakadambe, *Tam.*, A. 514 ; manjal, *Tam.*, C. 2433 ; manjalmullángi, *Tam.*, D. 173 ; manja-pu, *Tam.*, N. 179 ; manjati, *Mal.*, A. 471 ; manje konne, *Tam.*, C. 785 ; manjishta, *Tel.*, R. 564; manjishtige, *Tel.*, R. 564; manjítti, *Tam.*, R. 564 ; manjúnda, *Tel.*, G. 136 ; manjushta, *Kan.*, R. 564 ; mánna, *Mal.*, M. 147 ; mannakungilíyam, *Mal.*, B. 777 ; mannal, *Mal.*, C. 2433 ; mannuyanne, *Kan.*, P. 436 ; manoranjatam, *Kan.*, A. 1431 ; manoranjitam, *Tam.*, *Tel.*, A. 1431 ; man-tailam, *Mal.*, P. 436; man-tayilam, *Tam.*, P. 436 ; mant-bek, *Kan.*, T. 427 ; manthulli, *Kan.*, G. 14; manti-núne, *Tel.*, P. 436; manti-tayilam, *Tel.*, P. 436 ; mánu pasupu, *Tel.*, C. 2007, 2422 ; manupendalam, *Tel.*, M. 220 ; manú-pota, *Tam.*, S. 1229 ; manupotu, *Tel.*, S. 1229; manuranjitam, *Mal.*, A. 1431 ; manu ulava, *Tel.*, F. 559 ; manu-ulavi, *Tam.*, F. 559 ; mánu-vuppu, *Tel.*, C. 527 ; man-yenney, *Tam.*, P. 436 ; mara, *Kan.*, A. 101 ; marachini, *Mal.*, M. 216 ; marada-arishiná, *Kan.*, C. 2007; maradakarji, *Kan.*, S. 2865 ; maradauppu, *Kan.*, C. 527 ; maraharalu, *Kan.*, J. 41 ; mara-lingam, *Tam.*, C. 2039 ; maram, *Tam.*, C. 410 ; mara-manjal, *Tam.*, C. 2007 ; mara-narulle, *Kan.*, J. 41 ; maratatti, *Nilgiris*, H. 463 ; maráti mogga, *Tel.*, S. 2571 ; marati moggu, *Tam.*, E. 295, 304; maráti tíge, *Tel.*, S. 2571 ; mara-uppa, *Mal.*, C. 527 ; maravetti, *Tam.*, H. 472 ; maravetti, *Mal.*, H. 472; maravi, *Kan.*, S. 1229 ; maravulí, *Tam.*, M. 216 ; mara-vuppu, *Tam.*, C. 527 ; mardáru, *Tel.*, V. 17 ; marédu, *Tel.*, A. 534 ; margina-hulimara, *Kan.*, C. 36 ; mari (Caryota), *Tel.*, C. 711 ; mari (Ficus), *Tel.*, F. 129, 179 ; mar-i-kurondu, *Tam.*, A. 1469; mariman-chedi, *Tam.*, S. 2649; marinalu, *Mal.*, C. 2433 ; marithondi, *Tam.*, L. 126 ; marlumutta, *Tam.*, X. 1 ; marpu, *Tel.*, F. 568 ; marpú, *Tel.*, F. 412 ; marravittai, *Tam.*, H. 468; marri, *Tel.*, F. 129 ; marrú, *Tam.*, O. 214 ; marsada, *Kan.*, C. 122; marsada

boli, *Kan.*, C. 200 ; marudampattai, *Tam.*, M. 869 ; marudarsinghie, *Tam.*, L. 143 ; marukkallán-kay,*Tam.*, R. 1 ; marúl, *Tam.*, S. 785 ; marula mátangi, *Tel.*, X. 1 ; marúl-kalung, *Tam.*, S. 785 ; marulu jada, *Tel.*, X. 1 ; marulu tíge, *Tel.*, X. 1 ; marum, *Tam.*, B. 913 ; marupindi, *Tel.*, A. 306 ; marutai, *Tam.*, T. 361 ; marutamtoli, *Mal.*, M. 869 ; marutónri, *Tam.*, L. 126 ; marutti, *Tam.*, A. 1132 ; maruva, *Kan.*, L. 42 ; marvilingá, *Tam.*, C. 2039 ; marwa, *S. Kanara*, T. 355 ; mashik-kai, *Tam.*, Q. 43 ; máshipatri, *Tam.*, G. 660 ; mashudla, *Kan.*, C. 1031 ; massur, *Kan.*, L. 252 ; mátalam, *Mal.*, P. 1426 ; mátalam-pa-zham, *Mal.*, P. 1426; matáyen sampráni, *Mal.*, H. 22 ; matsakanda,*Tel.*, P. 1389 ; mattanga, *Mal.*, C. 2316 ; matta-pal-tiga, *Tel.*, I. 379 ; matti, *Kan.*, T. 361 ; mattipál, *Tam.*, A. 670 ; mattip-pál, *Mal.*, A. 670; mattur bachchali, *Tel.*, S. 2574 ; mava, *Mal.*, M. 147 ; mávalingamáku, *Tel.*, C. 2040 ; mávalingam-ilai, *Tam.*, C. 2040; rnávi, *Tel.*, M. 147 ; mavína, *Kan.*, M. 147 ; mávu, *Kan.*, M. 147 ; may, *Tel.*, S. 950 ; mayi, *Tel.*, S. 950; mayilánchi, *Mal.*, L. 126 ; mayil-kondai-pullu, *Tam.*, C. 1029; mayil-tutta, *Mal.*, C. 2367 ; mayil-tuttam, *Tam.*, C. 2367 ; mayilumánikyam, *Tel.*, S. 1714; mayilu-tuttam, *Tel.*, C. 2367 ; mayir-rámkkam, *Tam.*, S. 1714 ; mayir-mankham, *Tam.*, S. 1688j; mayir-sikki, *Tam.*, A. 510 ; maylú, *Tam.*, P. 350.

Me Meda, *Tel.*, L. 483 ; medi, *Tel.*, F. 179 ; mékamettani chettu, *Tel.*, V. 195 ; meka muaduga, *Tel.*, I. 399 ; mekkejola, *Kan.*, Z. 50 ; meladi-kurundu, *Tam.*, F. 53 ; mella, *Tel.*, C. 1403 ; méllugú, *Tam.*, C. 931 ; men, *Tel.*, C. 725 ; ména (wax), *Kan.*, C. 931 ; ména (honey), *Kan.*, H. 342, *Tam.*, *Tel.*, M. 235 ; méná (Fraxinus), *Tam.*, *Tel.*, F. 696 ; menashiná káyi, *Kan.*, C. 455 ; ménasiná-káyi, *Kan.*, C. 448 ; menasu, *Kan.*, P. 811 ; mente, *Kan.*, T. 612 ; mente-palle, *Kan.*, T. 612 ; mente soffu, *Kan.*, T. 612 ; ménthyá, *Kan.*, T. 612; menti kúra, *Tel.*, T. 612; mentulu, *Tel.*, T. 612 ; merapukai, *Tel.*, C. 448 ; metta kákara, *Tel.*, M. 626 ; metta túti, *Tel.*, I. 410 ; metta vanke, *Tel.*, S. 2284 ; mezhuka, *Mal.*, C. 931 ; mezhuka, *Mal.*, H. 342.

Mh Mhár mardi, *Tel.*, C. 711.
Mi Micha tummurra, *Tel.*, D. 628 ; mikka, *Kan.*, D. 212, H. 289 ; milagáy, *Tam.*, C. 448 ; milágu, *Tam.*, P. 811 ; mila-karanai, *Tam.*, T. 489 ; milakarnai-vér-pattai, *Tam.*, T 489 ; milkaranai, *Tam.*, T. 489 ; minak, *Mal.*, C. 1520 ; minak nur, *Mal.*, C. 1520 ; mína-maram, *Tam.*, M. 656 ; mínángáni, *Mal.*, S. 2518 ; mini, *Tam.*, T. 522 ; minum, *Tel.*, C. 931 ; minumulu, *Tel.*, P. 468, P. 513 ; mínvajjaram, *Tam.*, A. 393 ; mirapah, *Tel.*, C. 455 ; mirapagandra, *Tel.*, T. 489 ; mirápa káia, *Tel.*, C. 455 ; mirapakándra, *Tel.*, T. 489 ; mirapa kándravéru-patta, *Tel.*, T. 489 ; mirapa káya, *Tel.*, C. 448 ; mirialu, *Kan.*, P. 811 ; miriyálu, *Tel.*, P. 811; mirkadambe, *Tel.*, S. 2799 ; miryála tige, *Tel.*, P. 811 ; misur-pappu, *Tel.*, L. 252 ; misurpurpur, *Tam.*, L. 252 ; míthí van, *Tam.*, C. 224 ; mitli, *Kan.*, S. 2912.

Mo Mochai, *Tam.*, D. 789 ; modaga mardulu, *Tel.*, B. 944 ; modagerri vembu, *Tam.*, B. 520 ; modavagaddi, *Tel.*, I. 51 ; modina, *Tel.*, S. 2515 ; módira-kani-ram, *Mal.*, S. 2936 ; móduga, *Tel.*, B. 944 ; móduga chettú, *Tel.*, B. 944 ; modugu, *Tel.*, E. 342 ; moga-bíra, *Tel.*, A. 1132 ; moga-bírakú, *Tel.*, A. 1132 ; mogadam, *Tam.*, M. 570 ; moga-linga, *Tam.*, S. 959 ; mogali-sandlu, *Tel.*, P. 26 ; mogang voval, *Tam.*, F. 585 ; mogbíra, *Tel.*, A. 1132 ; mohtu, *Tel.*, B. 944 ; mohul, *Mal.*, B. 220 ; moieri-kata, *Tel.*, C. 854 ; moka-yapa, *Tel.*, S. 2865 ; mokkajonna, *Tel.*, Z. 50; molagan, *Tam.*, W. 54; molagha, *Tel.*, M. 656 ; molam pullum, *Tam.*, C. 2263 ; moleuppa, *Mal.*, B. 84 ; mollaghai, *Tam.*, C. 448 ; mollagu, *Tam.*, C. 448 ; molúvukodi, *Kan.*, P. 811 ; monah-aral, *Tam.*, F. 561 ; mootoo, *Tam.*, N. 37 ; moran cundai, *Tam.*, F. 495 ; mori, *Mal.*, H. 92 ; morli, *Tel.*, B. 913 ; morunga, *Tam.*, M. 721 ; mosúl, *Tel.*, F. 470 ; motta sirli, *Tel.*, D. 330 ; motuku, *Tel.*, B. 944 ; mouricon, *Kan.*, E. 354 ; mowda, *Tam.*, B. 913 ; mowe, *Tam.*, F. 585 ; moydi, *Tel.*, F. 179 ; mozhukku, *Tam.*, C. 931, H. 342.

Mr Mridu-maruvamu, *Tel.*, O. 220.

Mu Muchchu góni, *Kan.*, T. 537 ; muchi tanki, *Tel.*, D. 628 ; múda-cottan, *Tam.*, C. 551 ; mudah, *Kan.*, O. 1 ; mudari, *Kan.*, S. 1250 ; mudda kharjúrapu, *Tel.*, P. 555 ; mudda pulagam, *Tam.*, P. 344 ; muddírú, *Tel.*, F. 387 ; muddu-candai, *Tam.*, F. 390 ; mudi-bom-mi-day, *Tel.*, F. 493 ; mudla, *Kan.*, O. 153 ; mudu, *Kan.*, E. 496 ; múdúda, *Tam.*, C. 1031 ; múdúdad, *Tam.*, C. 1031 ; múdúga, *Tel.*, B. 959 ; mudu-gudavare, *Kan.*, M. 557 ; mudugu támara, *Tel.*, M. 306 ; mugali (Mimusops), *Kan.*, M. 570 ; mugalí (Pandanus), *Tel.*, P. 26 ; mugali-soppu, *Kan.*, A. 291 ; mugli, *Kan.*, A. 291 ; muhevehri, *Tam.*, C. 2287 ; múkalí-parah, *Tel.*, F. 592 ; mukampala, *Mal.*, A. 871 ; mukarattekire, *Tam.*, B. 619 ; mukhan sorrah, *Tam.*, F. 381 ; mukúkrattai, *Tam.*, B. 619 ; mukul-jellah, *Tel.*, F. 486 ; múlacarnai, *Tam.*, T. 489 ; múlaga chettu, *Tel.*, M. 704 ; mulagáy, *Tam.*, C. 448 ; mulagu, *Mal.*, P. 805 ; mulaip-pál-virai, *Tam.*, C. 728 ; mula-jemudu, *Kan.*, E. 496 ; mulaku-tháni, *Mal.*, T. 489 ; mulampandu, *Tel.*, C. 2263 ; múlasari, *Tel.*, L. 84 ; mulharave-soppu, *Kan.*, A. 943 ; mulilavu, *Tam.*, B. 632 ; múlíli, *Tam.*, P. 1048 ; mulinghie, *Tam.*, R. 31 ; mulkas, *Tel.*, B. 118 ; mulláátú, *Mal.*, S. 1257 ; mulla-dantu, *Kan.*, A. 943 ; mullá-ghái, *Tam.*, C. 455 ; mulla-muste, *Tel.*, S. 2315 ; mullanchíra, *Mal.*, A. 943 ; mullangi, *Kan.*, *Tel.*, R. 31 ; mullangiyanne, *Kan.*, R. 31 ; mullappú, *Mal.*, J. 35 ; mulli, *Tam.*, S. 2280 ; mullila-púla, *Mal.*, B. 632 ; mul-lilava, *Mal.*, B. 632 ; mullu-búraga-mará, *Kan.*, B. 632 ; mullu káre, *Kan.*, Z. 290 ; mulluk-kírai, *Tam.*, A. 943 ; mullu pengdalam, *Tel.*, D. 522 ; mullu-vengay, *Tam.*, B. 868 ; mulú-alley, *Tam.*, F. 402 ; múlúghúdú, *Tel.*, M. 711 ; mulugogu, *Tel.*, H. 250 ; mulugolimidi, *Tel.*, L. 266 ; mulugoranta, *Tel.*, B. 171 ;

mulu modugu, *Tel.*, E. 356 ; mulu vempali, *Tel.*, T. 270 ; munaga, *Tel.*, M. 721 ; munda dhup, *Kan.*, V. 31 ; munda-valli, *Mal.*, I. 368 ; mundíri, *Tam.*, A. 1014 ; mundla-búraga-chettú, *Tel.*, B. 632 ; mundla-tota-kura, *Tel.*, A. 943 ; munduttí, *Mal.*, F. 337 ; múney kíray, *Tam.*, P. 1233 ; munga-luppu, *Tam.*, B. 84 ; muni, *Tam.*, E. 356 ; muni-ganga rávi, *Tel.*, T. 392 ; munja-pavattay, *Tam.*, M. 656 ; munkuro-kuri, *Mal.*, C. 725 ; munnatakali-pullum, *Tam.*, S. 2299 ; munnay, *Tam.*, P. 1233 ; munnivayr, *Tam.*, P. 1233 ; munrú, *Tam.*, F. 514 ; munta gajjanamu, *Tel.*, I. 1 ; muntamámidi-vittu, *Tel.*, A. 1014 ; munta mandu, *Tel.*, D. 354 ; munugu támara, *Tel.*, M. 306 ; múraga, *Tel.*, M. 721 ; muráka, *Tam.*, E. 342 ; múrgala, *Kan.*, G. 36 ; múrgal mara, *Tam.*, G. 36 ; muri-kúti, *Mal.*, B. 909 ; murkalu, *Kan.*, B. 913 ; murkandachettu, *Tel.*, A. 306 ; muiki, *Tel.*, O. 127 ; murkitumma, *Tel.*, A. 217 ; muru-donda, *Tel.*, C. 1834 ; murududu, *Tel.*, C. 122, 200 ; muruká, *Tam.*, E. 342 ; murukkamaram, *Mal.*, B. 944 ; murukkan, *Tam.*, B. 944 ; múrúkonda, *Tel.*, A. 306 ; muru mámidi, *Tel.*, H. 124 ; murun-gai, *Tam.*, M. 721 ; músal, *Tam.*, H. 31 ; musalli, *Tam.*, F. 479 ; mushaippé-yetti, *Tel.*, L. 483 ; múshámbaram, *Tel.*, A. 824 ; mushti, *Kan.*, S. 2943 ; mushti, *Tel.*, S. 2943 ; mushu kattai, *Tam.*, M. 756 ; musidi, *Tel.*, S. 2943 ; mustakamu, *Tel.*, C. 2612 ; musuku jola, *Kan.*, Z. 50; musu-musuk-kai, *Tam.*, M. 791 ; mutcheh, *Tam.*, D. 789 ; múthera, *Mal.*, D. 758 ; muthu, *Tam.*, P. 355 ; muthu-chippi, *Tam.*, P. 355 ; múthú vullay, *Tam.*, L. 143 ; mutiamu, *Tel.*, P. 355 ; muti-jella, *Tel.*, F. 486, F. 487 ; muti-pál, *Tam.*, A. 672 ; mutirai, *Tam.*, C. 1031 ; mutriknujayvie, *Tam.*, A. 1545 ; muttaga-mará, *Kan.*, B. 944 ; muttageddasa, *Tel.*, F. 517 ; muttah, *Tel.*, F. 516, F. 517 ; muttah-kách, *Tam.*, C. 2617 ; muttapulgam, *Tel.*, H. 119 ; muttava, *Tel.*, S. 1694 ; muttava pulagum, *Tel.*, S. 1714 ; mutten gapilloo, *Tel.*, E. 166 ; mutthi, *Kan.*, *Mal.*, F. 415 ; muttu, *Tam.*, R. 369 ; muttu-chengan-chedi, *Tam.*, A. 1665 ; muttuga, *Kan.*, B. 944 ; muttuga-gidá, *Kan.*, B. 944 ; mutu-va-pulogum, *Tam.*, S. 1688.

My Myladí, *Tam.*, V. 177 ; mylekondai, *Tam.*, A. 501 ; myrole, *Kan.*, V. 158.

Na Nábhi, *Tel.*, A. 397 ; nach-churuppán, *Tam.*, T. 855 ; náchu, *Tel.*, H. 484 ; naga, *Tam.*, E. 419 ; naga betha, *Kan.*, C. 115 ; nága-dali, *Tam.*, O. 193 ; nágadalisappu, *Kan.*, R. 663 ; nága-donda, *Tel.*, C. 1834 ; naga-golugu, *Tel.*, M. 797 ; naga golunga, *Tel.*, M. 797 ; nága késara, *Tel.*, M. 489 ; nága késaramu, *Tel.*, M. 489 ; nagal, *Tel.*, P. 1248 ; naga-malli, *Tam.*, *Tel.*, R. 231 ; nagamallich-cheti, *Mal.*, R. 231 ; nága mallige, *Kan.*, R. 231 ; nagamúghatei, *Tam.*, I. 368 ; naga-mulla, *Mal.*, O. 193 ; nága-musadi, *Tel.*, S. 2936 ; naga mushini, *Tel.*, T. 456 ; nágap-pu (Ochrocarpus), *Tam.*, O. 6 ; naga-pu (Pentapetes), *Tam.*, P. 393 ; nagara-

múkuttykai, *Tel.*, l. 368 ; nagaru, *Tel.*,
P. 1248; naga sampigi, *Kan.*, M. 489 ;
naga-sara maitantos, *Tel.*, P. 618 ; naga-
sháp-pú (Mesua), *Tam.*, M. 489 ; naga-
sháp-pú (Ochrocarpus), *Tam.*, O. 6 ;
nágatumma, *Tel.*, A. 217 ; nágavalli, *Tel.*,
P. 775 ; nágésar-pu, *Tam.*, O. 6 ; nagetta,
Nilgiris, G. 379 ; nágin-kada, *Kan.*, T.
3 ; nagsampige, *Kan.*, M. 489 ; nahi-
kuddaghú, *Tam.*, C. 1367 ; nahra-jella,
Tel., F. 487 ; nai-bél, *Kan.*, L. 362 ;
nairla, *Kan.*, E. 428 ; nairuri, *Tel.*, E.
419 ; nai-ték, *Tam.*, D. 438 ; nai-udi,
Coorg, S. 2865 ; naka dosa, *Tel.*, C.
2278 ; nakakora, *Tel.*, S. 1207 ; naka-
náru, *Tel.*, É. 259 ; nakkanaregu,
Tel., F. 615 ; nakka vulli-gadda, *Tel.*, U.
39 ; nakkera, *Tel.*, C. 1931 ; nakulsi,
Tam., P. 1048 ; nakurmaral, *Tel.*, E. 259 ;
naladindi, *Tel.*, F. 559 ; nalaika, *Tel.*, R.
16 ; nala kanugida, *Kan.*, N. 23 ; nala-
sandawah, *Tel.*, F. 584 ; nala tige, *Tel.*,
D. 330 ; nala usereki, *Tel.*, P. 654 ; nali,
Tel., H. 324 ; nalla, *Tam.*, *Tel.*, C. 397 ;
nalla-ariti, *Tel.*, M. 811 ; nalla ativasa,
Tel., C. 2422 ; nalla-áválu, *Tel.*, B. 812 ;
nalla-budama, *Tel.*, C. 2304 ; nalla chámalu, *Tel.*, P. 67 ; nalla dadúgá, *Tel.*, M.
545 ; nalla-dámar, *Tel.*, S. 1682 ; nalla-
doggali, *Tel.*, A. 943 ; nalla gúlisienda,
Tel., C. 551 ; nalla-jidi chettu, *Tel.*, S.
1041 ; nalla-jilakra, *Tel.*, N. 158 ; nalla
kákasi, *Tel.*, R. 16 ; nallaka-lava, *Tel.*, N.
209 ; nalla katchelí, *Tel.*, F. 573 ; nalla
kokkita, *Tel.*, I. 397 ; nallak-tattah, *Kan.*,
F. 511 ; nalla-kupi, *Tel.*, C. 1377 ; nalla-
mada, *Tel.*, A. 1655, 1660 ; nallamaddi,
Tel., T. 361 ; nalla nílámbari, *Tel.*, D. 4 ;
nalla-nochchili, *Tel.*, J. 116 ; nalla-puru-
gudu, *Tel.*, P. 663 ; nallaputiki, *Tel.*, A.
1383 ; nalla-renga, *Tel.*, A. 686 ; nalla-
rojan (Shorea), *Tel.*, S. 1682 ; nalla-rójan
(Canarium), *Tel.*, C. 285 ; nalla-sandra,
Tel., A. 135, 295; nalla tady gudda, *Tel.*,
C. 2375 ; nallatiga, *Tel.*, I. 1 ; nallatíge,
Tel., I. 1 ; nalla torriti, *Tam.*, F. 444 ;
nallatummakara, *Tel.*, A. 101 ; nalla-uli-
mera, *Tel.*, D. 628 ; nalla-ummetta, *Tel.*,
D. 151 ; nalla vávili, *Tel.*, V. 164 ; nalla-
vávili, *Tel.*, J. 116 ; nallekaruna-karang,
Tam., A. 996 ; nallenna, *Mal.*, S. 1078 ;
nal-lenny, *Tam.*, S. 1078; nalléru, *Tel.*, V.
219 ; nallo-chengan-chedi, *Tam.*, A. 1665 ;
nal valanga, *Tam.*, D. 32 ; namali pilli,
Tam., T. 435 ; namilí adogú, *Tel.*, V.
158 ; namli, *T l.*, H. 324 ; namme, *Tam.*,
A. 1149 ; nánabála, *Tel.*, E. 531 ; nana-
beeam, *Tel.*, E. 531 ; nanchuntá, *Mal.*, B.
13 ; nandi (Cedrela), *Tel.*, C. 838 ; nandi
(Lagerstrœmia), *Kan.*, L. 53 ; nandi-
chettu, *Tel.*, C. 838 ; nandiréka, *Tel.*, F.
253 ; nandivardhana, *Tel.*, T. 3 ; nang,
Tam., M. 489 ; nangal, *Tam.*, M. 489 ;
nanja-murich-chán, *Tam.*, T. 855 ; nan-
jundá, *Tam.*, B. 13 ; nannári, *Tam.*, H.
119 ; nannári-kizhána, *Mal.*, H. 119 ;
naoni, *Kan.*, S. 1212 ; naoru, *Tel.*, P.
1248 ; nápálu, *Mal.*, C. 1388 ; nara, *Tel.*,
L. 474 ; nára dabba, *Tel.*, C. 1270 ; nára
épe, *Tel.*, H. 16 ; nárak-karandaí, *Tam.*,
B. 546 ; narala, *Kan.*, E. 419 ; nara mamidi, *Tel.*, L. 483 ; nara mamúdí, *Tel.*, L.
474 ; náranga-pandu, *Tel.*, C. 1233 ; nára
tega, *Tel.*, D. 510 ; narbotku, *Tel.*, E.

314 ; naredu, *Tel.*, E. 419 ; nareyr, *Tel.*,
E. 419 ; nargarmollay, *Tel.*, R. 231 ; nari
Coorg, T. 437 ; nari kadam, *Tel.*, C. 1520-
nárínja-pandu, *Tel.*, C. 1233 ; naripayir
Tam., P. 523 ; nari-vengáyam, *Tam.*, U
39 ; narlingi, *Tel.*, A. 686 ; narole, *Kan.*
O. 1 ; narra alagi, *Tel.*, L. 483 ; nárttam.
pazham, *Tam.*, C. 1270 ; narum-pánál,
Mal., U. 71 ; naru-nínti, *Mal.*, H. 119 ;
narvala, *Tam.*, C. 2039 ; narvel, *Tam.*, E.
419 ; narvilli, *Tam.*, C. 1944 ; naryepi,
Tel., H. 16 ; nasag uni-gidá, *Kan.*, M.
786 ; nasedu, *Tel.*, E. 419 ; nasodu, *Tel.*,
E. 419 ; nát-akródu, *Kan.*, A. 737 ; nát-
bádámi, *Kan.*, T. 312 ; nathe-badam-
vittulu, *Tel.*, T. 312 ; nathe-vadam-kottai,
Tam., T. 312 ; nati-shambu, *Mal.*, E.
444 ; nát-ká-deodár, *Tam.*, E. 370 ; nat-
ram-takara, *Mal.*, C. 780 ; nat-révá-chinni,
Kan., R. 215 ; nattai-chúri, *Tam.*, S.
2515 ; nattam-takarai, *Tam.*, C. 780 ;
nattu, *Tel.*, F. 440, 501 ; náttu-akrótu-
kottai, *Tam.*, A. 737 ; náttu-bádam, *Mal.*,
T. 312 ; náttu-iréval-chinni, *Tam.*, R. 215 ;
náttu-manjat-chínak-kishangu, *Tam.*, R.
215 ; náttunilá-virai, *Tam.*, C. 737 ; nattu-
panjasára, *Mal.*, S. 373 ; náttu-pasupu-
chína-gadda, *Tel.*, R. 215 ; náttu-réval
chinni, *Tel.*, R. 215 ; náttu-shakkara, *Tel.*,
S. 373 ; nattu-sharkkara, *Mal.*, S. 373 ;
náttu-sharukkarai, *Tam.*, S 373 ; nattuvadam-kottai, *Tam.*, T. 312 ; nattu-vadom,
Tam., T. 312 ; nátuakrótu-vittu, *Tel.*, A.
737 ; natu sengote, *Tel.*, S. 1057 ; natva.
dom, *Tam.*, T. 312 ; nauladi, *Kan.*, V.
158 ; naura, *Tel.*, P. 1248 ; nava-charum,
Tam., A. 962 ; navách-chárum, *Tam.*, A.
962 ; nával, *Tam.*, E. 419 ; navaládi,
Kan., V. 174 ; navamálika, *Tel.*, J. 35 ;
navani, *Kan.*, S. 1212 ; navásagara, *Kan.*,
A. 962 ; navá-sagáram, *Tel.*, A. 962 ;
navasáram, *Mal.*, A. 962 ; navelu, *Kan.*,
P. 350 ; navili, *Tel.*, H. 324 ; navuru, *Tel.*,
P. 1248 ; nawal, *Tam.*, E. 398, E. 419 ;
nawar, *Tam.*, E. 419 ; nayavaylie, *Tam.*,
C. 1367 ; nayibela (Cleome), *Kan.*, C.
1367 ; nayi béla (Acacia), *Kan.*, A. 249 ;
nayka-dughu, *Tam.*, C. 1367 ; náy-pálai
Tam., T. 855 ; naypawlum, *Tel.*, B. 28 ;
ná-yurivi, *Tam.*, A. 382 ; nazel-nagai,
Tam., G. 161.

Ne Neckanie, *Tam.*, *Tel.*, C. 397 ; neddean,
Kan., F. 558 ; negalu, *Kan.*, T. 548 ;
nehoemaka, *Mal.*, B. 905 ; neivaylla,
Tam., G. 753 ; nekota, *Tam.*, H. 124 ;
néla alumu, *Tel.*, R. 346 ; néla ámudamu,
Tel., J. 52 ; néla benda, *Tel.*, H. 168 ;
nela-bevinágidá, *Kan.*, A. 1064 ; nelabevu,
Kan., S. 3018 ; nela-guli, *Tel.*, E. 217 ;
nela-gulimidi, *Tel.*, E. 217 ; néla jammi,
Tel., D. 402 ; nelak-katalá, *Mal.*, A.
1261 ; nelampala, *Mal.*, G. 660 ; nela
mulaka, *Tel.*, S. 2345 ; nela-naregam,
Mal., N. 23 ; nela-naringu, *Kan.*, N. 23 ;
néla nírédu, *Tel.*, P. 1231 ; nélá-nuga,
Tel., L. 30 ; néla pála, *Tel.*, O. 600 ;
nelapoka, *Tel.*, C. 75 ; nélasampenga,
Tel., P. 1044 ; néla tádi, *Tel.*, C. 2375 ;
néla-tangédu, *Tel.*, C. 737 ; néla-táti-gad-
dalu, *Tel.*, C. 2375 ; nela-táti-gadde, *Kan.*,
C. 2375 ; néla-usirika, *Tel.*, P. 657 ; nelá-
varike, *Kan.*, C. 737 ; néla-vávili, *Tel.*, J.
116 ; nélavelaga, *Tel.*, F. 53 ; nela vem-

pali, *Tel.*, T. 270 ; nela-vému, *Tel.*, A. 1064 ; nélavusari, *Tel.*, P. 657 ; nelepanny kalung, *Tam.*, C. 2375 ; nelgale-káyi, *Kan.*, A. 1261 ; nella, *Tam.*, P. 1239 ; nellabenda, *Tam.*, H. 215 ; nella-ghentána, *Tel.*, C. 1403 ; nella-ghentana vayrú, *Tel.*, C. 1403 ; nella-jedi, *Tel.*, S. 1041 ; nella-jilledu, *Tel.*, C. 170 ; nella-madi, *Tel.*, M. 3 ; nella-madu, *Tel.*, T. 361 ; nella molunga, *Tel.*, S. 2345 ; nella-purúdúdú, *Tel.*, P. 663 ; nella-shamalu, *Tel.*, P. 67; nella-túma, *Tel.*, A. 101 ; nella-uppi, *Tel.*, C. 427 ; nella vekal sorrah, *Tam.*, F. 381 ; nellay piku, *Tam.*, C. 2313 ; nelle, *Mal.*, P. 632 ; nelli (Phyllanthus), *Tam.*, *Tel.*, *Kan.*, *Mal.*, P. 627, P. 632 ; nelli (Premna), *Tel.*, P. 1239 ; nellie (Clerodendron), *Tel.*, C. 1386 ; nelli-kái, *Tam.*, P. 632 ; nelli kúra, *Tel.*, P. 1239 ; nellu, *Kan.*, O. 258 ; nellú, *Tam.*, O. 258 ; némilie, *Tel.*, P. 350 ; nemma pandu, *Tel.*, C. 1301 ; némmapúndú, *Tel.*, C. 1296 ; nemmi chettu, *Tel.*, O. 537 ; nempali, *Tel.*, T. 270 ; nepála (Croton), *Kan.*, C. 2192 ; népálam (Jatropha), *Tel.*, J. 41 ; nepála-vitua, *Tel.*, C. 2192 ; neradi, *Tel.*, E. 73 ; nerale, *Kan.*, F. 419 ; nerasi,*Tel.*, E. 73 ; neri-arishippál (Liquidambar), *Tam.*, L. 455; nerinjil, *Mal.*, T. 548 ; neriuri-shippál (Altingia), *Tam.*, A. 892 ; nerkal, *Kan.*, E. 464 ; nerlu, *Kan.*, E. 419 ; nerrenjí kíray, *Tam.*, T. 548 ; nerunjí, *Tam.*, T. 548 ; nerunji-mullu, *Tam.*, T. 548 ; nerválam, *Tam.*, C. 2192 ; netavil, *Mal.*, A. 1200 ; néti bíra, *Tel.*, L. 569 ; nettávil, *Mal.*, A. 1200 ; nettávil maram, *Tam.*, A. 1200 ; nettelí, *Tam.*, F. 393, 440 ; neval-adugu mánu, *Tel.*, V. 177 ; neva-lédi, *Tel.*, V. 159; nevali adugu, *Tel.*, V. 177 ; nevari dhanyamu, *Tel.*, O. 258 ; newli, *Kan.*, E. 553 ; neyi, *Tam.*, *Tel.*, G. 189.

N₁ Nidra-yung, *Tel.*, N. 76 ; niella tirtua, *Mal.*, O. 31 ; niepa, *Tam.*, S. 749 ; nila, *Nil-ghiris*, T. 847 ; níla-dintana, *Tel.*, C. 1403 ; nilai sedachi, *Tam.*, P. 1060 ; nilak-kadalai, *Tam.*, A. 1261 ; nilak-kumazh, *Mal.*, G. 298 ; nilak-kumizh, *Tam.*, G. 298 ; nílakobari, *Tel.*, I. 335 ; nílam (Indigofera), *Tam.*, *Mal.*, I. 145 ; nilam (sapphire), *Mal.*, *Tam.*, S. 855 ; nilap-panaik-kizhangu, *Tam.*, C. 2375 ; níla váká, *Mal.*, C. 737 ; nila-vákai, *Tam.*, C. 737 ; níla vém, *Tel.*, S. 3018 ; nilavémbu (Swertia), *Tam.*, S. 3018 ; nila-vémbu (Andrographis), *Tam.*, A. 1064 ; nila-véppa, *Mal.*, S. 3018 ; nila-veppu, *Mal.*, A. 1064 ; nilá virai, *Tam.*, C. 737 ; níli, *Kan.*, I. 145 ; nilika, *Kan.*, P. 632 ; níli-mandu, *Tel.*, I. 145 ; nilunu, *Tel.*, D. 547 ; nilusu, *Tel.*, F. 344 ; niluvu-pendalum, *Tel.*, D. 485 ; nilwa pendalam, *Tel.*, D. 545 ; ním-bamu, *Tel.*, M. 363 ; nimbe-hanna, *Kan.*, C. 1258 ; nimbehannu, *Kan.*, C. 1296 ; nimbe hanu, *Kan.*, C. 1270 ; nímíri, *Tel.*, T. 355 ; nimma-gaddi, *Tel.*, A. 1079 ; nimma-pandu, *Tel.*, C. 1258, 1270, 1296 ; nimma-tulási, *Tel.*, O. 28 ; nípís, *Mal.*, C. 1296 ; niradi, *Tel.*, H. 468 ; niradi-mattu, *Tam.*, H. 468; niradimuttu, *Tam.*, H. 472 ; níradi-muttu-enney, *Tam.*, H. 472 ; níradi-vittulú, *Tel.*, H. 472 ; níradi-vittulú-núne, *Tel.*, H. 472 ; níra-lakki-gidá, *Kan.*, V. 181 ; nirambali, *Tam.*, P. 1004; nirangi (Poinciana), *Kan.*,

P. 1032 ; niranji ((Salix), *Kan.*, S. 579 ; nirasi, *Tel.*, E. 73 ; nirbrami, *Tam.*, H. 149 ; nir-chappay, *Nilgiri Hills*, D. 138 ; nirgol, *Kan.*, S. 2424 ; nirguvíveru, *Tel.*, H. 508 ; nirija, *Tel.*, E. 73 ; nirjílúza, *Tel.*, A. 560 ; nirmalli, *Tam.*, H. 508 ; nir-nai, *Kan.*, O. 534 ; nirnochchi, *Tam.*, V. 164 ; nir-noch-chi, *Tam.*, *Mal.*, V. 181 ; nír-núchie, *Tam.*, V. 73 ; nirpa, *Tel.*, B. 330 ; nírpirimie, *Tam.*, H. 149 ; nirpulli, *Tam.*, C. 2522 ; niruag-nivéndrapáku, *Tam.*, J. 14 ; níru boddi, *Tel.*, R. 487 ; níru budiki, *Tel.*, P. 874 ; niru gannéru, *Tel.*, P. 1084 ; nirujani, *Coorg*, C. 2039 ; niru kassuvu, *Tel.*, C. 1752 ; niru-kuka, *Tel.*, O. 534 ; nírulli, *Kan.*, *Tel.*, A. 769 ; nírumélner-uppu, *Tam.*, A. 958 ; nirunai, *Tam.*, O. 534 ; niru prabba, *Tel.*, C. 104 ; niru-tal-vapu, *Tel.*, N. 76 ; níru vanga, *Tel.*, S. 2284 ; níru-vávili, *Tel.*, V. 181 ; nirválá, *Kan.*, *Mal.*, C. 2039 ; nirválam, *Mal.*, C. 2192 ; nir-vallipullu, *Mal.*, H. 513.

No Noch-chi, *Mal.*, V. 164 ; noch-chi, *Tam.*, V. 164; nogé, *Coorg*, C. 838; nolaitali, *Tam.*, A. 1212; nowli eragu, *Tel.*, V. 177.

Nu Nucka kura, *Tel.*, R. 553 ; nugge-gidá, *Kan.*, M. 721 ; nugu-benda, *Tel.*, A. 89 ; nugu-benda chettu, *Tel.*, A. 80 ; núkku-kattái, *Tam.*, D. 64 ; nulitali, *Mal.*, A. 1212 ; nuliti, *Tel.*, H. 92 ; nulla-budinga, *Tel.*, C. 2304 ; nulla-gandu-menu, *Tel.*, F. 466 ; nullah-ramah, *Tel.*, F. 319 ; nulla vellam, *Tam.*, S. 370 ; nullerútigeh, *Tel.*, V. 219 ; nullútí, *Tam.*, D. 569 ; numma, *Tam.*, A. 1146; nummnelli kírai, *Tam.*, G. 220 ; núna, *Tam.*, M. 711, *Tel.*, M. 656 ; núne bíra, *Tel.*, L. 569 ; núne-pápata, *Tel.*, P. 338 ; núní gatcha, *Tel.*, C. 26 ; nur, *Mal.*, C. 1520 ; nur, *Kan.*, C. 1520 ; núra, *Mal.*, C. 489 ; nuraikal, *Tam.*, C. 1808 ; nur-minak, *Mal.*, C. 1520 ; nurrey pitten keeray, *Tam.*, R. 553 ; nuskul, *Kan.*, B. 913 ; núti kashindha, *Tel.*, C. 787 ; nuvvu, *Tcl.*, S. 1078; nuvvulu, *Tam.*, *Tel.*, S. 1078.

Ny Nyadale-huvu, *Kan.*, N. 200 ; nyor, *Mal.*, C. 1520.

Od Odai-mánu, *Tel.*, O. 38 ; odallam, *Mal.*, C. 943 ; odasale, *Tel.*, A. 217 ; odiyamaram, *Tam.*, O. 38 ; olam-min, *Tam.*, F. 414.

Oe Oepata, *Mal.*, A. 1655.

Ok Okánu-katta, *Tel.*, C. 35.

Om Oma, *Kan.*, C. 691 ; omamí, *Tel.*, C. 691 ; omamu, *Tel.*, C. 691; ómamuáku, *Tel.*, A 1130 ; oman, *Tam.*, C. 691 ; omu, *Kan.*, C. 691.

Oo Oomarie keeray, *Tam.*, S. 527 ; oosulay, *Mal.*, A. 686.

Op Opa, *Tam.*, C. 224 ; opa, *Tam.*, S. 717.

Or Orilaihámarai, *Tam.*, I. 335.

Os Osirka, *Tel.*, P. 632.

Ot Otiyam, *Tam.*, O. 38 ; ottagam, *Tam.*, C. 205 ; ottàh, *Tam.*, F. 588.

Ou Ourupalay, *Tel.*, O. 600 ; ouru-véru, *Tel.*, A. 1097.

Ov Ovaricandí, *Tam.*, F. 558.

Pa Pachaku, *Tel.*, C. 1158 ; pachári, *Kan.*, D. 53 ; pachcha adavimolla, *Tel.*, J. 24 ; pach-cha-gannéru, *Tel.*, T. 410 ; pach-ch-ai-alari, *Tam.*, T. 410 ; pach-cha-mullangi, *Tel.*, D. 173 ; pach cha-pedda góranta,

Tel., L. 126; pachchári, *Tel.*, D. 53; pachchayava, *Tel.*, H. 382; páchi, *Tel.*, H. 484; páchimánu, *Tel.*, A. 1146; páda-rasá, *Kan.*, M. 473; páda-rasam, *Tel.*, M. 473; padari, *Tel.*, S. 2876; padavala káyi, *Kan.*, T. 569; padenarayan, *Tam.*, P. 1032; padiri, *Tel.*, S. 2876; padma káshtam, *Tel.*, S. 882; padri (Dalbergia), *Kan.*, D. 53; pádri (Stereospermum), *Tam.*, S. 2865, S. 2876; paducan, *Tam.* F., 377; padúmenú, *Tel.*, F. 476; págá-dam, *Tel.*, C. 1808; paghada, *Tel.*, N. 179; pahmum kolah, *Tam.*, F. 365; paidi, *Tel.*, F. 179; paidi-tangédu, *Tel.*, C. 787; pailœpúta tammi, *Tam.*, C. 563; paillie, *Tam.*, R. 110; painamara, *Mal.*, V. 31; paini, *Kan.*, V. 31; painipasha, *Mal.*, V. 31; painí-pishin, *Tam.*, V. 31; painní-mín, *Tam.*, F. 476; paipli-chakka, *Kan.*, V. 55; pakituma, *Tel.*, A. 244; pakké, *Tel.*, T. 70; pakki (Tamarix), *Tel.*, T. 70; pakki (Streblus), *Tel.*, S. 2912; pákku, *Tam.*, A. 1294; pála (Alstonia), *Mal.*, A. 871; pala (Mimusops), *Tel.*, M. 583; pálá (Sideroxylon), *Tam.*, S. 1718; pálá (Wrightia), *Tam.*, W. 122; pála-chuk-kani-déru, *Tel.*, H. 119; pála dantam, *Tel.*, E. 25; pala-garuda, *Tel.*, A. 871; palah, *Tel.*, F. 386; palak, *Tam.*, W. 122; palakodsa, *Tel.*, H. 294; palakúdé, *Tam.*, L. 292; palakuna, *Tam.*, C. 987; pála kúra, *Tel.*, O. 600; palakura-pála-gurugu, *Tel.*, H. 328; pála malle tivva, *Tel.*, V. 12; palam-pási, *Tam.*, S. 1699; pálas, *Tel.*, B. 944; palasah, *Tel.*, F. 414; pála-samudra, *Tel.*, A. 1362; palá-sham, *Tam.*, B. 944; paláshamu, *Tel.*, B. 944; palasorrah, *Tel.*, F. 380; palasu-gandhi, *Tel.*, H. 119; pálatíge, *Tel.*, L. 292; palavay-raynú, *Tam.*, W. 122; palay, *Mal.*, C. 2253, I. 61; palaykirai, *Tam.*, H. 328; pa-liakiri, *Tam.*, O. 547; paliwára, *Kan.*, E. 342; palla, *Tam.*, M. 583; palla-chinta, *Tel.*, O. 547; palla-pandu, *Tel.*, M. 583; pallepanlo, *Tel.*, M. 583; palléru, *Tel.*, T. 548; palléru-mullu, *Tel.*, T. 548; palpirai, *Tam.*, S. 2912; pal-sorrah, *Tel.*, F. 377; palungu, *Tam.*, H. 177; palúpaghel-kalung, *Tam.*, M. 639; pamania, *Tel.*, O. 233; pám-budda, *Tel.*, P. 682; pam-budinga, *Tel.*, C. 2301; pampaí, *Tam.*, B. 663; pam-pana, *Tel.*, O. 233; pampi, *Tel.*, C. 2433; pana (Oroxylum), *Tam.*, O. 233; paná (Borassus), *Mal.*, B. 663; panai-maram, *Tam.*, B. 663; panai-vellam, *Tam.*, S. 370; panam, *Tam.*, B. 663; panam-kat-rázha, *Mal.*, A. 603; panasa-pandu, *Tel.*, A. 1489; páncha-dárá, *Tam.*, S. 375; panchadub, *Tel.*, V. 14; panche chettu, *Tel.*, C. 868; panchman, *Tel.*, A. 1146; panchonta, *Kan.*, D. 379; panchotí, *Tam.*, D. 387; panchotí pala, *Tam.*, D. 379; pandi, *Tel.*, H. 289; pandiki, *Tel.*, K. 42; pandili dosa, *Tel.*, C. 2278; pandi mukku dumpa, *Tel.*, D. 522; pandi-pakke, *Kan.*, F. 429; pandkoku, *Tel.*, R. 51; pandri, *Kan.*, R. 16; pandu kopah, *Tel.*, F. 476; pandu-menu, *Tel.*, F. 476; panémara, *Kan.*, B. 663; paniá, *Mal.*, E. 289; paniala, *Mal.*, E. 289; pani bíra, *Tel.*, L. 574; panichchi, *Mal.*, D. 582; panichekai, *Tam.*, D. 582; pani-

chika, *Tam.*, D. 582; paninir, *Mal.*, R. 504; pani-pyre, *Tam.*, P. 523; panír, *Tam.*, G. 733; panjiri, *Tam.*, F. 350; panna maram, *Tam.*, B. 663; pan-nerale, *Coorg*, E. 432; pannerali, *Kan.*, E. 432; pannie, *Tam.*, B. 663; panniru puvvu, *Tel.*, G. 733; panny-pyre, *Tam.*, P. 513; panri-pullu, *Tam.*, A. 1090; pansa, *Tel.*, A. 1489; pansadarry, *Tam.*, S. 375; pan-tangi, *Tel.*, B. 863; papara búdama, *Tel.*, C. 1211; paparapuli, *Tam.*, A. 455; pápata, *Tel.*, P. 338; papli, *Kan.*, *Tam.*, V. 55; pappadi, *Kan.*, P. 338; pappálí, *Tam.*, C. 581; pappanti, *Tel.*, A. 306; pappara-mulli, *Tam.*, S. 2280; pappatak-káram, *Mal.*, C. 527; pappáya, *Mal.*, C. 581; pappáyi, *Tam.*, C. 581; pappili-chakka, *Tam.*, V. 55; pappu-kura (Cheno-podium), *Tel.*, C. 1003; pappu-kura (Portulaca), *Tel.*, P. 1179; pappu-kúra-vittulu, *Tam.*, P. 1179; páputtavayrú, *Tel.*, P. 338; paraga, *Kan.*, F. 314; paragi, *Tel.*, Z. 263; para korava, *Tam.*, F. 514; parakorawa, *Tam.*, F. 516; para-mutty, *Tam.*, P. 344; paranda, *Tam.*, F. 500; parangi-sámbráui, *Tel.*, B. 751, 771, 777; parangi-shámbiráni, *Tam.*, B. 751, 771, 777; paranki-mávakuru, *Mal.*, A. 1014; paranu, *Tel.*, Z. 263; parapalanam, *Tel.*, G. 247; parasa, *Tam.*, B. 944; paratti, *Tel.*, G. 404; paravala-damara, *Kan.*, E. 342; parawa-mará, *Kan.*, C. 390; pári-játamu, *Tel.*, N. 179; parimi. *Tel.*, Z. 263; paringay, *Tam.*, S. 2240; paringi, *Tel.*, Z. 263; parinta, *Tel.*, C. 1861; paritt, *Tel.*, G. 404; parjamb, *Kan.*, O. 153; parpadagum, *Tam.*, M. 613; par-pa-raal, *Tel.*, F. 494; parpalli gidda, *Kan.*, Z. 254; parparam, *Tam.*, P. 396; parpataka, *Tel.*, M. 613; parpu-kíre, *Tam.*, P. 1179; parpu-kíre-virai, *Tam.*, P. 1179; parrúwai sorrah, *Tam.*, F. 376; párswapu, *Tel.*, X. 1; partanga, *Kan.*, H. 1; párujátamu, *Tel.*, N. 179; pa-runta, *Mal.*, V. 219; parúpú, *Tam.*, P. 1179; parupu-benda, *Tam.*, H. 215; pa-rupu kire, *Tam.*, C. 1003; parutti, *Tam.*, G. 404; pasarai-kírai, *Tam.*, P. 1187; pasarganni, *Tel.*, D. 32; pasha, *Mal.*, *Tel.*, A. 102; pasikende, *Tam.*, F. 500; paspukadimi, *Tel.*, A. 514; paspu kandi, *Tel.*, A. 514; passeliekíray, *Tam.*, P. 1179; passeli-kírai, *Tam.*, P. 1187; pasu-pu, *Tel.*, C. 2433; pata, *Tel.*, C. 1205; páta arige, *Tel.*, P. 332; pátala gandhi, *Tel.*, R. 57; pátála garuda, *Tel.*, R. 57; patali, *Tel.*, S. 2876; patanga, *Tam.*, C. 3; patanga-chekke, *Kan.*, C. 35; pa-tanga-katta, *Tel.*, C. 35; patanlu, *Tel.*, P. 885; páta veru, *Tel.*, T. 456; patcha arise, *Tam.*, E. 549; patchalai, *Tam.*, D. 53; patcha-pessalú, *Tel.*, P. 496; patcha-pessara, *Tel.*, P. 496; patchay-pyre, *Tam.*, P. 513; patche paira, *Tam.*, P. 496; pátéru, *Tel.*, T. 456; pathiri, *Tam.*, S. 2865; pati-káram, *Tam.*, *Tel.*, A. 897; patik-káram, *Mal.*, A. 897; patir, *Tel.*, A. 534; patólam, *Mal.*, T. 586, 598; patólamu, *Tel.*, T. 576; patólas, *Tel.*, T. 576; patsa-pesalu, *Tel.*, P. 513; patsuru, *Tel.*, D. 53; pattanie, *Tam.*, P. 885; pattaráshu, *Tam.*, M. 606; patti, *Tel.*, G.

385 ; patu-kúrkka, *Mal.*, A. 1130 ; pau, *Tam.* & *Mal.*, A. 1622 ; paunch figa, *Tel.*, C. 805 ; paunchinan,*Tel.*, A. 1146 ; pauni-eyri, *Tam.*, F. 317 ; pauttie, *Tel.*, G. 404 ; pává, *Tam.*, S. 950 ; pava-kai, *Tam.*, M. 626 ; pávakká-chedi, *Tam.*, M. 626 ; pávakká-cheti, *Mal.*, M. 626 ; pávalam, *Tam.*, C. 1808 ; páva-mekke, *Kan.*, C. 1211 ; pavati, *Kan.*, P. 338 ; pávili, *Tel.*, P. 1187 ; pavut-tay-vayr, *Tam.*, P. 338 ; payalaku, *Tel.*, P. 1187 ; payana, *Mal.*, V. 31 ; paycúmuti, *Tam.*, C. 1211.

Pe Pe-attiss, *Tam.*, F. 202 ; pedda, *Tel.*, A. 658 ; pedda-ánem, *Tel.*, B. 868 ; pedda-áre, *Tel.*, B. 308 ; pedda-bach-chali, *Tel.*, B. 203 ; pedda-baketu, *Tel.*, C. 1931 ; pedda boku, *Tel.*, C. 1931 ; pedda chilaka-daduga, *Tel.*, S. 487 ; pedda chilka dúdúga, *Tel.*, M. 545 ; pedda chintú, *Tel.*, C. 860 ; pedda-dosrai, *Tel.*, C. 2274 ; pedda duchirram, *Tel.*, A. 695 ; pedda-dúlagondi, *Tel.*, M. 786 ; pedda-dumpa-rásh-trakam, *Tel.*, A. 853 ; pedda-galli-gísta, *Tel.*, C. 2148 ; pedda-ganti, *Tel.*, P. 384 ; peddagi, *Tel.*, P. 1370 ; pedda gomru, *Tel.*, G. 287 ; pedda gumudu téku, *Tel.*, G. 287 ; pedda-guriginga, *Tel.*, A. 471 ; pedda illinda, *Tel.*, D. 560 ; pedda íta, *Tel.*, P. 588 ; pedda-jilakara, *Tel.*, P. 727 ; pedda-jila-kurra, *Tel.*, F. 659 ; pedda-kai, *Tel.*, C. 2274 ; pedda kalinga, *Tel.*, D. 428 ; pedda-kanda-gadda, *Tel.*, T. 15 ; pedda karinga, *Tel.*, G. 124 ; pedda-kome, *Tel.*, F. 390 ; pedda kunji, *Tel.*, K. 42 ; peddamánu-patta, *Tel.*, A. 658, 670 ; pedda-mattu-neatku-batsala, *Tel.*, B. 207 ; pedda-mottah, *Tel.*, F. 433 ; pedda-nella-kúra, *Tel.*, P. 1239 ; pedda nellikúra, *Tel.*, P. 1239 ; pedda-neredu, *Tel.*, E. 419 ; pedda nidra kanti, *Tel.*, M. 557 ; pedda-nimma-pandu, *Tel.*, C. 1286 ; pedda-nowlieragu, *Tel.*, H. 324 ; pedda-pail-kuru, *Tam.*, P. 1179 ; pedda pála, *Tel.*, W. 131 ; pedda-palléru, *Tel.*, P. 363 ; pedda pattseru, *Tel.*, A. 717 ; pedda-pávila-kurá-vittulu, *Tan.*, P. 1179 ; pedda pávili, *Tel.*, P. 1187 ; pedda-pávili-kura, *Tel.*, P. 1179 ; peddapotri, *Tel.*, K. 42 ; pedda-púli, *Tel.*, T. 437 ; pedda-pulimera, *Tel.*, E. 25 ; pedda-púrawah, *Tel.*, F. 442 ; pedda sadapa, *Tel.*, P. 727 ; pedda shaka, *Tel.*, C. 2597 ; pedda sópara, *Tel.*, D. 32 ; pedda teku, *Tel.*, T. 232 ; pedda-tóla-kúra, *Tel.*, A. 938 ; peddatumni, *Tel.*, L. 312 ; pedda vara góki, *Tel.*, S. 717 ; pedda-warago-wenki, *Tel.*, C. 224 ; pedda wúndú, *Tel.*, P. 48 ; pedda-yélai-káyalu, *Tel.*, A. 976 ; peddimari, *Tel.*, F. 129 ; pedéga, *Tel.*, P. 1370 ; pedéi, *Tel.*, P. 1370 ; pedu, *Tel.*, A. 658 ; pedumpa, *Tel.*, D. 500 ; pedyegí, *Tel.*, P. 1370 ; pee, *Tam.*, A. 658 ; pe-karakai, *Tam.*, T. 355 ; pella-gumudu, *Tel.*, A. 1215 ; pem-pri, *Mal.*, G. 124 ; pemu, *Tel.*, C. 104 ; penári, *Kan.*, S. 2850 ; penar-valli, *Mal.*, Z. 1 ; pendalam, *Tel.*, D. 485 ; pondli gummadi-káya, *Tel.*, B. 430 ; pendri, *Kan.*, R. 16 ; pennéru, *Tel.*, W. 98 ; pennérugadda, *Tel.*, W. 93, W. 98 ; pennika, *Tel.*, C. 1377 ; pente-veduru, *Tel.*, B. 118 ; penti táti *Tel.*, B. 663 ;

pépálam, *Tel.*, J. 41 ; pépatolam, *Mal.*, T. 576 ; pépírk-kam, *Tam.*, L. 563 ; pepre, *Tam.*, F. 216 ; pepu, *Tel.*, C. 104 ; péra, *Mal.*, P. 1343 ; peragú, *Mal.*, C. 1380 ; perainpúli, *Mal.*, T. 437 ; pérakka, *Mal.*, P. 1343 ; perambu, *Tam.*, C. 104 ; perambu, *Tam.*, *Mal.*, C. 77 ; peram-pilli, *Tam.*, T. 437 ; perámútiver, *Tam.*, P. 344 ; pera mutti, *Tel.*, P. 347 ; perangi, *Kan.*, C. 581 ; peraratta, *Mal.*, A. 853 ; péra rattai, *Tam.*, A. 853 ; peratti-kíraí, *Tam.*, C. 1861 ; pérá-virai, *Tam.*, C. 787 ; pérélam, *Mal.*, A. 976 ; peria-ítcham, *Tam.*, P. 588 ; périch-chankay, *Tam.*, P. 555 ; perina, *Mal.*, F. 156 ; perin-chírakam, *Mal.*, P. 727 ; perinji, *Kan.*, C. 581 ; perinkara, *Kan.*, E. 67 ; périnta-kúra, *Tel.*, C. 1861 ; períta, *Tel.*, P. 555 ; periya-elattari, *Mal.*, A. 976 ; periya-elumich-c ham-pazham, *Tam.*, C. 1286 ; periya karunaik-kizhangu, *Tam.*, T. 15 ; periya-takarai, *Tam.*, C. 787 ; periya-yélak-káy, *Tam.*, A. 976 ; perretay kíray, *Tam.*, I. 407 ; pershk, *Toda*, F. 808 ; peru, *Tam.*, A. 658, 670 ; peru-marat-toli, *Mal.*, A. 670 ; peru-marattup-pattai, *Tam.*, A. 670 ; peru-marindu, *Tam.*, A. 1398 ; perúmarum, *Tel.*, *Mal.*, A. 658, 670 ; perumbe, *Tam.*, *Kan.*, P. 1259 ; perumaruttú, *Tam.*, A. 658 ; perum-bai, *Kan.*, P. 1259 ; perum-kizhangu, *Tam.*, A. 1398 ; perúmtiriki, *Tam.*, F. 555 ; perum vullie kalangu, *Tam.*, D. 485 ; perundei codie, *Tam.*, V. 219 ; peru-nerunji, *Tam.*, P. 363 ; perungayam, *Tam.*, F. 76 ; perunshí-ragam, *Tam.*, P. 727 ; peruntumba, *Mal.*, A. 1132 ; perun-tutti, *Tam.*, A. 80, 89 ; petha-kalabantha erikatali, *Tel.*, A. 636 ; petluppu, *Kan.*, S. 682 ; petluppu, *Tel.*, S. 682 ; peuya áre, *Tam.*, B. 308 ; pey, *Tel.*, A. 658 ; pe-yalti-paraka, *Mal.*, F. 202 ; peyá-veri, *Tam.*, C. 780 ; péya-verutti, *Tam.*, A. 1132 ; peycom muttie, *Mal.*, C. 1211 ; péyimeratti, *Mal.*, A. 1132 ; pey-ko-mattitu-matti, *Tam.*, C. 1211 ; peymarutti, *Tam.*, A. 1132 ; péyp pálai, *Tam.*, T. 855 ; péy-pudal, *Tam.*, T. 576 ; peyt-tumatti, *Tam.*, C. 1211.

Ph Phala-sampenga, *Tel.*, A.1431 ; phal-modek, *Mal.*, I. 379 ; phansi, *Kan.*, C. 474 ; pháshána, *Kan.*, A. 1425 ; phulsr, *Tel.*, P. 663 ; phutiki, *Tel.*, G. 663.

Pi Piktúmi, *Tel.*, A. 217 ; píkunkai, *Tam.*, L. 556 ; píl, *Tam.*, C. 224 ; pilá (Artocarpus), *Tam.*, A. 1489, I. 58 ; pilá (Wrightia), *Tam.*, W. 122 ; pilála, *Kan.*, F. 253, S. 1939 ; píla-marda, *Tam.*, T. 325 ; pilápa-zham, *Tam.*, A. 1489 ; pilappu-shíragam, *Tam.*, C. 697 ; pil-aringan, *Tam.*, F. 411 ; piliya mankena, *Tel.*, U. 35 ; pillai mardá, *Tam.*, T. 355 ; pillánji, *Tam.*, P. 663 ; pilli-adugu, *Tel.*, M. 786 ; pilli-dumpa, *Tam.*, P. 1216 ; pillinchan, *Tam.*, F. 449 ; pilli persara, *Tel.*, P. 523 ; pilli-píchara, *Tel.*, A. 1575, 1577 ; pillivéndram, *Tel.*, W. 98 ; pillu-samei, *Tam.*, P. 68 ; pílú, *Tam.*, C. 224 ; pinaii, *Tam.*, C. 1377 ; pinári, *Tam.*, S. 2824 ; pinari-marum, *Tam.*, S. 2824 ; píná-shengam-kuppi, *Tam.*, C. 1377 ; pinda karak-káya, *Tel.*, T. 325 ; pindie conda, *Tel.*, A. 554 ; pindi-kai, *Kan.*, M. 904 ; pindi kúnda, *Tel.*, R. 656 ; pinekai, *Kan.*, C. 146 ; piney maram, *Tam.*, V. 31 ; pini-gala-konga, *Tel.*, S. 2906 ; pinna bija, *Kan.*, C. 146 ; pinna-élaki-chettu, *Tel.*, H. 486 ; pinnai, *Tam.*, D. 438 ; pinna-ippa, *Tel.*, B. 265 ; pinna mulaka, *Tel.*, S. 2345 ; pinna-nelli, *Tel.*, P. 1233 ; pinna-vara-gógu, *Tel.*, S. 717 ; pinnay, *Tam.*, C. 146 ; pinne, *Kan.*, C. 146 ; pinsttariní, *Tam.*, M. 864 ; pinval, *Kan.*, F. 253 ; pipili, *Tam.*, *Tel.*, P. 805 ; pippali katte, *Tel.*, P. 805 ; píppallu, *Tam.*, P. 805 ; pirandai, *Tam.*, V. 219 ; pirangi chekka, *Tel.*, S. 2240 ; pirri, *Toda*, T. 437 ; písangi, *Tel.*, C. 1377 ; pishin, *Tam.*, A. 102 ; pishina, *Tel.*, M. 3 ; pishinika, *Tel.*, C. 1377 ; pishin-pattai, *Tam.*, L. 483 ; pisíngha, *Tel.*, C. 1377 ; písinika, *Tel.*, M. 3 ; pisúl, *Tel.*, S. 2865 ; pitachan-danam, *Tel.*, S. 790 ; píta-kanda, *Tel.*, D. 173 ; píta vrikshamu, *Tel.*, S. 2649 ; pitcha, *Tam.*, C. 1221 ; pitha kalabun-tha, *Tam.*, A. 603 ; pitta-pisiniki, *Tel.*, E. 23 ; piuna-pála, *Tel.*, O. 600.

Pl Plách-cha, *Mal.*, B. 944 ; plewane, *Tam.*, C. 224.

Po Poallkáya, *Tel.*, T. 569 ; poarah-cunjú-candí, *Tam.*, F. 425 ; poarí candí, *Tam.*, F. 333 ; pod, *Tam.*, M. 71 ; podala-manu, *Tel.*, A. 135 ; podapatra, *Tel.*, G. 748 ; podda-kanru, *Tel.*, F. 615 ; podda-trin-gudda-chettu, *Tel.*, H. 74 ; podla káyi, *Kan.*, T. 598 ; podun-gali, *Tam.*, P. 1018 ; podútalei, *Tam.*, L. 451 ; pogada, *Kan.*, *Tel.*, M. 570 ; pogada-mánu, *Tel.*, M. 570 ; pogáku, *Tel.*, N. 101 ; poga-yellei, *Tam.*, N. 101 ; pogháko, *Tam.*, N. 101 ; poghei, *Tam.*, N. 101 ; pokala, *Mal.*, N. 101 ; póka-vakka, *Tel.*, A. 1294 ; polam-bachchali, *Tel.*, B. 203 ; polári, *Tel.*, A. 1218 ; pólechí, *Mal.*, H. 233 ; póle-kala, *Tam.*, F. 538 ; pollai, *Tel.*, A. 1218 ; polla nuvvulu, *Tei.*, S. 1078 ; pollí maun, *Tam.*, D. 212 ; pollium, *Tam.*, T. 28; polluk-kaya, *Mal.*, A. 1038 ; pollv káya, *Tam.*, T. 569 ; polun-kala, *Tam.*, F. 540 ; pomígra, *Tam.*, W. 44 ; pomúshti, *Tam.*, C. 1205 ; ponán-kottai, *Tam.*, S. 818 ; poné kelítí, *Tam.*, F. 524 ; pong, *Kan.*, P. 1121 ; pongá, *Tam.*, P. 1121 ; pongú, *Tam.*, C. 156 ; ponika, *Tel.*, S. 6 ; ponkáram, *Mal.*, B. 731 ; ponna, *Tel.*, *Mal.*, C. 146 ; ponna-chettu, *Tel.*, C. 146 ; ponnagam, *Mal.*, M. 71 ; ponnám-takara, *Mal.*, C. 787 ; ponnari-tárakam, *Tam.*, O. 242 ; ponná-viraí, *Tam.*, C. 787 ; ponnáviram, *Mal.*, C. 741 ; ponna vittulu, *Tel.*, C. 146 ; ponnú, *Tam.*, G. 317 ; pon-padira, *Tam.*, S. 2865 ; ponuku, *Tel.*, G. 780 ; ponundi, *Tel.*, F. 596 ; poonam, *Mal.*, B. 220 ; popli-chukai, *Kan.*, V. 55 ; porasan, *Tam.*, B. 944 ; porilla sápara, *Tel.*, D. 53 ; porilla sopara, *Tel.*, D. 53 ; poris, *Tam.*, T. 392 ; porki, *Tel.*, Z. 263 ; poroh, *Mal.*, F. 156 ; portia, *Tam.*, T. 392 ; postaka-tol, *Tam.*, P. 87 ; postaley-kaantagerai, *Tam.*, W. 25 ; posuku, *Tel.*, S. 950 ; potari, *Tel.*, K. 42 ; poti-kanda, *Tel.*, A. 996 ; potla, *Tel.*, T. 569 ; potla káya, *Tel.*, T. 569 ; pot-luppu, *Tam.*, S. 682 ; potrum, *Tel.*, D. 53 ; potti dumpa, *Tel.*, G. 243 ; pottigummadi,

Tel., C. 2331 ; pottikanapu, *Tel.*, S. 30 ; potti-luppu, *Tam.*, S. 682 ; pótuágákara, *Tel.*, M. 639 ; pótu kandulu, *Tel.*, M. 639; potu táti, *Tel.*, B. 663 ; pótuvadla, *Tel.*, H. 285 ; pou, *Tam.*, A. 1622 ; pou, *Mal.*, A. 1622 ; pounanga, *Tam.*, S. 818.

Pr Prabba chettu, *Tel.*, C. 104 ; prabhali, *Tel.*, C. 104 ; prakke, *Tel.*, T. 70 ; prayám, *Tam.*, S. 2912 ; priyadarsini, *Tam.*, P. 1259 ; proddu-tiru-gudú-chettu, *Tel.*, H. 74.

Ps Psetta-kelleté, *Tam.*, F. 491.

Pu Pú, *Tam.*, S. 950 ; puágákara, *Tel.*, M. 639 ; púarasú, *Tam.*, T. 392 ; pucha-payarú, *Tam.*, P. 496 ; pudacarpan, *Tam.*, C. 894 ; pudel, *Tam.*, T. 576 ; pudiná, *Tam.*, M. 447 ; pudíná, *Tel.*, M. 447 ; 465 ; puerri-danimma, *Tel.*, P. 1426 ; pugai-ilai, *Tam.*, N. 101 ; puíchenggah, *Mal.*, L. 556 ; puj-velam, *Tam.*, A. 217 ; pú-karunkáli, *Tam.*, D. 646 ; puka-yila, *Mal.*, N. 101 ; pulá, *Tam.*, B. 632 ; pulachakiri, *Kan.*, H. 233 ; pula chapa, *Tel.*, F. 515 ; púlachi, *Tam.*, S. 950 ; púlá-kizhanna, *Mal.*, C. 2499 ; pulalani, *Mal.*, M. 277 ; pulama-ram, *Mal.*, B. 632 ; púlán-kizhanga, *Tam.*, C. 2499 ; púla pála, *Tel.*, P. 396 ; pulasa, *Tel.*, F. 414 ; palasu, *Tel.*, F. 414 ; púlatummi, *Tel.*, L. 323 ; púlavayr-puttay, *Tam.*, P. 663 ; pulcha, *Tel.*, C. 2301 ; pul-colli, *Mal.*, R. 231 ; pulbari, *Tel.*, B. 304 ; púli (Tamarindus), *Tam.*, *Mal.*, T. 28 ; puli (Felis), *Mal.*, *Tel.*, T. 434, T. 437 ; púlia, *Tam.*, T. 28 ; pulichar-kírai, *Tam.*, H. 177 ; pulich-chakkáy, *Tam.*, A. 1644 ; pulichi, *Tam.*, H. 177 ; puli-chintá, *Tel.*, O. 547 ; puli dumpa, *Tel.*, D. 500 ; puli máda, *Tel.*, V. 217 ; pulimanji, *Tam.*, H. 177 ; puli-naravi, *Tam.*, V. 228 ; puli-pérandai, *Tam.*, V. 228 ; púli redda-púli, *Tam.*, T. 437 ; puli shinta, *Tel.*, B. 304 ; puli-tenke, *Tel.*, F. 598 ; pulivanji, *Mal.*, C. 2105 ; puliyam-pazham, *Mal.*, T. 28 ; puliyam-pazham, *Tam.*, T. 28 ; puliyárai, *Tam.*, O. 547 ; puliyárai-kírai, *Tam.*, O. 547 ; pulla bach-chali, *Tel.*, V. 228 ; pulla-chanchali, *Tel.*, O. 547 ; pulla-dabba, *Tel.*, C. 1270 ; pulla dondur, *Tel.*, B. 304 ; pullam-purachí-sappu, *Kan.*, O. 547 ; pullarazu, *Tel.*, F. 411 ; pulla tíge, *Tel.*, S. 882 ; pulli, *Tam.*, F. 430 ; pullibaghi, *Kan.*, A. 711 ; pullum, *Tam.*, C. 1221 ; pulluri, *Tam.*, V. 154 ; pul-lurivi, *Tel.*, V. 154 ; pulsur, *Tel.*, A. 1218 ; pulusu-káya-lu, *Tel.*, A. 1644 ; pú-madalai, *Tam.*, P. 1426 ; pú-máda-lam, *Mal.*, P. 1426 ; pumágamu, *Tel.*, C. 146 ; pumaram, *Nilgiris*, R. 253 ; pú mardá, *Anamalais*, T. 355 ; puma-rum, *Tam.*, S. 950 ; pú-mín-candí, *Tam.*, F. 357 ; púmplemús, *Mal.*, *Kan.*, C. 1263 ; pún, *Mal.*, C. 156 ; puna (Nephelium), *Tam.*, N. 72 ; púna (Calophyllum), *Tel.*, C. 146 ; púnachu, *Tel.*, H. 484 ; punagam, *Tam.*, C. 146 ; púna-gin bek, *Kan.*, T. 444 ; púnagú pilli, *Tel.*, T. 444 ; punaik-káli, *Tam.*, M. 786 ; púnakapúndú, *Tam.*, R. 656 ; punaku cheddy, *Tam.*, C. 1861 ; punarpuli, *Kan.*, G. 66 ; púnás, *Tel.*, C. 146 ; punatsu, *Tel.*, V. 14 ; púnattí, *Tam.*, F. 581 ; pundi (Crotalaria), *Kan.*, C. 2105 ;

pundi (Spondias), *Kan.*, S. 2649 ; pundi-
bija, *Kan.*, H. 233 ; pundrikegida, *Kan.*,
H. 177 ; púnduouringa, *Tam.*, F. 431 ;
púnduringa, *Tam.*, F. 413 ; púne (Calophyllum), *Mal.*, C. 156 ; púne (Ochrocarpus), *Kan.*, O. 6 ; pungam-maram,
Tam., P. 1121 ; pungella, *Tam.*, F. 350 ;
pungikírai, *Tam.*, A. 925 ; pungu, *Tel.*,
P. 1121 ; púngú, *Mal.*, C. 156 ; púnil,
Kan., O. 38 ; punje, *Kan.*, S. 2912 ;
punkirai, *Tam.*, I. 515 ; pun-mulla, *Mal.*,
J. 35 ; pún múshtie, *Tam.*, C. 1205 ;
punna, *Mal.*, C. 146 ; punnágam, *Tam.*,
C. 146 ; punnah, *Tam.*, F. 537 ; punnai,
Tam., C. 146 ; punnai-virai, *Tam.*, C.
146 ; punnam, *Mal.*, P. 1121 ; punni-
calawah, *Tam.*, F. 579 ; púr, *Tel.*, E. 289 ;
puraishu, *Tam.*, B. 944 ; pura-gadí, *Tel.*,
C. 2605 ; púram, *Tam.*, M. 873 ; púramu,
Tel., M. 473 ; purasa, *Tam.*, T. 392 ;
púrasha, *Tam.*, T. 392 ; purashu, *Tam.*,
B. 944 ; púrawah, *Tel.*, F. 439 ; pureea,
Tel., G. 124 ; púr-halihulla, *Kan.*, A. 1079 ;
purititige, *Tel.*, I. 410 ; puropú-kíray, *Tam.*,
P. 1179 ; púr-relan, *Tam.*, F. 441 ; purrh,
Tam., C. 1031 ; purruvú-kende, *Tam.*, F.
558 ; pursa, *Tam.*, T. 392 ; pursung, *Tam.*,
T. 392 ; purugudu, *Tel.*, P. 663 ; purus-
burus, *Tam.*, C. 1031 ; purusharatnam,
Tel., I. 335 ; purúsh-múdudad-marum,
Tam., C. 1031 ; pushini, *Tam.*, C. 2316 ;
pushinik-kay, *Tam.*, C. 2316 ; pushpa-
kedal, *Mal.*, R. 231 ; púskú, *Tel.*, S. 950 ;
putchuk, *Tam.*, S. 910 ; pútengí, *Tam.*,
C. 162 ; putiki, *Tel.*, G. 663, 677, 705 ;
putla-podara, *Tel.*, G. 748 ; putol, *Mal.*,
C. 2180 ; pútrajauví, *Tel.*, F. 145 ; putra-
jiva, *Kan.*, P. 1052 ; putribudinga, *Tel.*,
M. 791 ; putsakayachoythú-putsa, *Tel.*,
C. 1211 ; putstrangali, *Tel.*, A. 1232 ; pútat
jillédu, *Tel.*, W. 131 ; puttapála, *Tel.*,
I. 515 ; putta podara yárala, *Tel.*, V. 12 ;
puvandi, *Tam.*, S. 818 ; púvanti, *Tam.*,
S. 818 ; puvarasam, *Tam.*, T. 392 ;
púvarashu, *Mal.*, 392 ; púvarasu, *Tam.*,
T. 392 ; puvati, *Tam.*, N. 72 ; puverari,
Tam., F. 515 ; puvu, *Tam.*, S. 950 ;
puzhuk-kolli, *Mal.*, R. 231.

Pw Pwitíge, *Tel.*, I. 382 ; pwon, *Tam.*, G. 317.

Qu Qualar katcheli, *Tam.*, F. 599 ; qucilú,
Tel., S. 527.

Ra Racha mamidi, *Tel.*, M. 147 ; ráchaneredu,
Tel., E. 419 ; rácha usirike, *Tel.*, P. 627 ;
ragi, *Kan.*, F. 672 ; rági (copper), *Tel.*,
Mal., C. 2361 ; rági (Eleusine), *Kan.*,
Tam., E. 170 ; rági (Ficus), *Tel.*, *Kan.*,
F. 236 ; ragota, *Tel.*, C. 410 ; rágulu, *Tel.*,
E. 170 ; rahtijellah, *Tel.*, F. 329 ; rai (Dillenia), *Mal.*, *Tam.*, D. 438 ; rái (Ficus),
Tel., F. 236 ; raiga, *Tel.*, F. 236 ; rajuma,
Tel., P. 1060 ; rákáshimatalu, *Tel.*, A. 603 ;
rakta-candan, *Kan.*, *Tam.*, *Tel.*, P. 1381 ;
rakta-channanam, *Mal.*, P. 1381 ; rakta
gandham, *Tel.*, P. 1381 ; rakta krishna
chandanam, *Tel.*, S. 790 ; raldhupada,
Kan., C. 285 ; rálla sunnamu, *Tel.*, C.
489 ; rámá-chandar-pandu, *Tel.*, A. 1158 ;
ramachcham-vér, *Mal.*, A. 1097 ; rama
karé, *Tam.*, F. 445 ; raman sorrah, *Tam.*,
F. 381 ; rámá-pandu, *Tel.*, A. 1158 ; ramá-
phalam, *Tel.*, A. 1158 ; ráma tulási, *Tel.*,

O. 28 ; rámi, *Kan.*, L. 89 ; ramphal, *Kan.*,
A. 1158 ; rámsitá, *Tam.*, A. 1158 ; rámsítu-
plam, *Tam.*, A. 1158 ; ram-tíl, *Kan.*, G.
735 ; ranabhéri, *Tel.*, L. 266 ; ranagu,
Tel., P. 1121 ; rangamále, *Kan.*, M.
71 ; rangamali, *Kan.*, B. 523 ; rangamali-
hannu, *Kan.*, B. 523 ; rangi, *Kan.*, F.
236 ; rangú, *Tel.*, F. 480 ; ran sorrah,
Tam., F. 81 ; ráp-shakkara, *Tel.*, S. 371 ;
ráp-sharukkarai, *Tam.*, S. 371 ; rasagadi
manu, *Tel.*, S. 2341 ; rasa-karpúramu,
Tel., M. 473 ; rasam, *Tel.*, M. 473 ; rasa-
mála, *Mal.*, A. 892 ; rasbija, *Kan.*, H.
324 ; rasha-karup-púram, *Tam.*, M. 473 ;
ra-sorrah, *Tel.*, F. 401 ; rathapu, *Tel.*, L.
332 ; rátipáchi, *Tel.*, L. 332 ; ratipalaka,
Tel., S. 2235 ; ratnagandi, *Kan.*, C. 32 ;
ratrinta, *Tel.*, S. 2357 ; ráva kada, *Tel.*,
S. 2994 ; rávaná suruni mísálu, *Tel.*, S.
2580 ; ravi, *Tel.*, F. 236 ; rawadan, *Tel.*, D.
438.

Re Reddi vári mánu bála, *Tel.*, E. 549 ; regi,
Tel., Z. 231 ; régu, *Tel.*, Z. 231 ; regutti,
Tel., C. 410 ; réku, *Mal.*, *Tam.*, A. 1622 ;
réla-káyalu, *Tel.*, C. 756 ; rélarálá, *Tel.*,
C. 756 ; rellu-gaddi, *Tel.*, S. 49 ; rengha,
Tel., Z. 231 ; révalchini-pál, *Tel.*, G. 66 ;
révati-dula gondi, *Tel.*, T. 509 ; reyi káda,
Tel., S. 2994 ; reylu, *Tel.*, C. 756 ; rayyi
pappu, *Tel.*, W. 106 ; reza-kutta, *Tel.*, F.
759.

Rh Rhetsa, *Tel.*, Z. 35 ; rhetsa-maram, *Tel.*, Z.
35 ; rhetsamaum, *Tel.*, Z. 35.

Ri Ringri, *Tel.*, B. 13.

Ro Roatanga, *Tel.*, S. 950 ; róga-chettu, *Tel.*,
A. 1130 ; rohitakah, *Tel.*, A. 988 ;
roiclu, *Tel.*, C. 2237 ; roja, *Tel.*, R.
504 ; rondapatti, *Tel.*, T. 387 ; rongdi,
Mal., D. 94 ; rotan, *Mal.*, C. 104 ; rotanga,
Tel., S. 950 ; rotan-jarang, *Mal.*, C. 68.

Ru Ruchu, *Tel.*, F. 469 ; rudra-challu, *Tel.*,
E. 57 ; rudra jada, *Tel.*, O. 18 ; rudrakai,
Tam., E. 57 ; rudrák, *Kan.*, E. 69 ; rudrak-
sha, *Tel.*, G. 726 ; rudrakshakamba, *Tel.*,
A. 1192 ; rudraksha pendalam, *Tel.*, D.
499 ; rudrakshe, *Kan.*, G. 726 ; rudrasum,
Tam., G. 726 ; rúna-kalli, *Tam.*, B. 909.

Sa Saal candí, *Tam.*, F. 333 ; sadápa, *Tel.*, R.
663 ; sada-parauda, *Tam.*, F. 497 ; sada-
vaku, *Tam.*, C. 987 ; sádhanapu venduru.
Tel., D. 292 ; sága, *Tel.*, S. 785 ; sagade,
Kan., S. 950 ; sagapu, *Tam.*, H. 517 ;
sagdi, *Kan.*, S. 950 ; sagu, *Mal.*, C. 2531 ;
sagwani, *Kan.*, T. 232 ; sajjalú, *Tel.*, P.
384 ; sajje, *Kan.*, P. 384 ; sajjebiya, *Kan.*,
O. 18 ; sakala-phala-sampenga, *Tel.*, A.
1431 ; sakkarei-vellei-kelangu, *Tam.*, I.
348 ; sakotra hannu, *Kan.*, C. 1263 ; sak-
rela, *Mal. (S. P.)*, C. 123 ; saku-jella, *Tel.*,
F. 492 ; sal, *Tel.*, A. 269 ; sálá-mishri, *Mal.*,
S. 521 ; sálá-misiri, *Tel.*, S. 521 ; salava-
jella, *Tel.*, F. 550 ; salé, *Tel.*, A. 269 ; sali,
Tel., P. 950 ; salitunga, *Tel.*, C. 2584 ;
salla-wúdú, *Tel.*, P. 57 ; samai, *Tam.*, P.
67 ; sambara, *Tel.*, V. 205 ; sambrán
chettu, *Tel.*, H. 149 ; sáme, *Kan.*, P. 53 ;
sammanathi, *Tam.*, E. 370 ; sampaga-
pala, *Kan.*, V. 158 ; sampage-huvvu, *Kan.*,
M. 517 ; sampige, *Kan.*, M. 517 ; samudra,
Tam., *Mal.*, B. 193 ; samudra-pach-chai,
Mal., A. 1362 ; samudra-pala, *Mal.*, *Tel.*,
A. 1362 ; samudra-yógam, *Mal.*, A. 1362 ;

samudrapu-páchi, *Tel.,* G. 653 ; samudra-
pu-tenkáya, *Tel.,* L. 511 ; sanabu, *Kan.,*
C. 2105 ; sana-linga, *Tel.,* C. 1196 ; san-
dani vembu, *Tam.,* C. 838 ; sandra, *Tel.,*
A. 295 ; sangana mara, *Kan.,* S. 2861 ;
sanheesare, *Kan.,* P. 1048 ; sanja-mallige,
Kan., M. 606 ; sank-húlé, *Kan.,* S. 1250 ;
sannaelumparásh trakam, *Tel.,* A. 862 ;
sannagalu, *Tel.,* C. 1061 ; sanna-lavanga-
patta, *Tel.,* C. 1196 ; sanna-pappu, *Tel.,*
P. 1187 ; sannapávili, *Tel.,* P. 1187 ; sanna-
payala, *Tel.,* P. 1187 ; sannasolti, *Kan.,*
H. 463 ; sante kayi, *Kan.,* C. 2287 ; santi,
Kan., T. 293 ; sanua élaki, *Tel.,* E. 151 ;
sappanga, *Kan.,* C. 35 ; sára-chettu, *Tel.,*
B. 913 ; sarala-dévadaru, *Tel.,* B. 474 ;
sarayam, *Tam.,* V. 243 ; sarayi, *Tel.,* V.
243 ; sarga, *Kan.,* D. 212 ; sari-kullah,
Tel., F. 574 ; sarkarei valfi, *Tam.,* I. 343 ;
sarnakasari, *Kan.,* M. 71 ; sarong-burong,
Mal., C. 1722 ; sarsive, *Kan.,* B. 812 ;
sarugáta, *Tel.,* C. 158 ; sarúl, *Kan.,* B.
308 ; sasive, *Kan.,* B. 841 ; sata kuppi,
Tam., P. 460 ; saumai, *Tel.,* P. 67 ; sauola-
mara, *Kan.,* C. 838 ; savaga, *Kan.,* S.
2861 ; sa-vale, *Tam.,* F. 595 ; sáve, *Kan.,*
P. 53, 63 ; savé, *Tel.,* G. 660 ; savirela,
Tel., P. 4 ; sawa (Panicum), *Tel.,* P. 53 ;
sawa (python), *Tel.,* R. 129 ; sawaal-candi,
Tam., F. 340 ; sawala, *Tel.,* F. 593 ; saya,
Tam., O. 137 ; sayawer, *Tam.,* O. 137 ;
sazza, *Tel.,* P. 384.

Se Sebe, *Kan.,* P. 1343 ; sedi, *Tel.,* S. 1226 ;
segapú, *Tam.,* L. 143 ; segapu (Cana-
valia), *Tam.,* C. 289 ; segapu (Psidium)
Tam., P. 1343 ; segapu-munthari, *Tam.,*
B. 356 ; segapu thumbatin, *Tam.,* C. 289 ;
sela vanjai, *Tam.,* A. 711 ; selintan, *Tam.,*
F. 309 ; selupa, *Tam.,* E. 73 ; sembela puli
pilla, *Tam.,* C. 1158 ; sembú, *Tam.,* C.
2361 ; sempagum, *Tam.,* M. 537 ; sempan-
gam, *Tam.,* M. 517 ; semparuthi, *Tam.,*
G. 385 ; sem sorrah, *Tel.,* F. 376 ; senai-
kilangu, *Tam.,* C. 1732 ; senan karawa,
Tam., F. 480 ; sendu-bir-kai, *Tel.,* L. 563 ;
senduérum, *Tam.,* L. 143 ; sendurgam,
Tam., C. 637 ; sendurkun, *Tam.,* C. 637 ;
senel-kas, *Tel.,* F. 449 ; sengeni, *Kan.,*
V. 159 ; senkane, *Kan.,* V. 159 ; sennal,
Tam., F. 317 ; sépachettu, *Tel.,* O. 600 ;
sepa-kilangu, *Tam.,* C. 1732 ; septi, *Tel.,*
F. 508 ; serappadi, *Tam.,* C. 1707 ; séra-
shengalanír, *Tam.,* V. 79 ; seregad, *Tel.,*
E. 25 ; seri-gally-gista, *Tel.,* C. 2154 ;
serinji, *Tel.,* W. 106 ; seru-padi, *Tam.,*
C. 1707 ; serva, *Tel.,* C. 826 ; sévu, *Kan.,*
P. 1463 ; seyapu chandanum, *Tam.,* *Tel.,*
P. 1381.

Sh Shaal, *Tam.,* F. 469 ; shach-chi-káram,
Tam., C. 541 ; shadhurak-kalli, *Tam.,*
E. 496 ; shadray kullie, *Tam.,* E. 496 ;
shaing, *Tam.,* S. 1041 ; shákha tungavéru,
Tel., C. 2612 ; shakkán-kírai, *Tam.,*
R. 650 ; shálá mishirí, *Tam.,* S. 521 ;
shalláttu, *Tam.,* L. 21 ; shálu, *Kan.,*
S. 2424 ; shamai, *Tam.,* P. 67 ; shámak-
kizhangu, *Tam.,* C. 1732 ; shámantip-pú,
Tam., C. 1043 ; shámáthúmpa, *Tel.,*
C. 1732 ; shámbal-vuppu, *Tam.,* C. 527 ;
shambhárapulla, *Mal.,* A. 1079 ; sham-
biráni, *Tam.,* S. 2968 ; sháme-gadde,

Kan., C. 1732 ; shampang, *Tam.,* M. 517 ;
shampangi-puvvu, *Tel.,* M. 517 ; sha-
muddirap-pach-chai, *Tam.,* A. 1362 ;
shanabiná, *Kan.,* C. 2105 ; shanagakádi,
Tel., C. 1061 ; shan-aga-pulusu, *Tel.,*
C. 1061 ; shanapam, *Tam. & Tel.,*
C. 2105 ; shanárisí, *Tam.,* S. 514 ; shand-
anak-kattai, *Tam.,* S. 790 ; shangali
gaddi, *Tel.,* P. 79 ; shangamkupi, *Tam.,*
C. 1377 ; shanga-pushpam, *Mal.,* C. 1403 ;
shannámbu, *Tam.,* C. 489 ; shappat-tup-
pu, *Tam.,* H. 227 ; sharkkara, *Mal.,*
S. 370 ; sharunnai, *Tam.,* T. 537 ; sha-
ta-rah, *Tam.,* F. 493 ; shatávali, *Mal.,*
A. 1575 ; shatávali, *Mal.,* A. 1577 ;
shatávari-kiz-hanna, *Mal.,* A. 1577 ;
shavalai, *Tam.,* T. 537 ; shavari-pazham,
Tam., T. 600 ; shaviramu, *Tel.,* M. 473 ;
shavu-ku-maram, *Tam.,* C. 826 ; shavu-
ku-pattay, *Tam.,* C. 826 ; shav-víram,
Tam., M. 473 ; shayrang, *Tam.,* S. 1041 ;
shekram, *Tam.,* A. 686 ; shekrani, *Tel.,*
A. 686 ; shellel, *Tam.,* F. 448 ; shellí,
Tam., F. 333 ; shem, *Tam.,* S. 2501 ;
shémai-tutti, *Tam.,* A. 880 ; shema-
kalenga, *Tam.,* C. 1732 ; shembara-valli,
Tel., V. 205 ; shembugha, *Tam.,* M. 517,
537 ; shemhara, *Tam.,* F. 483 ; shemi,
Kan., A. 295 ; shemi-velvel, *Tam.,* A. 229 ;
shemmarum, *Tam.,* S. 2501 ; shemmuli,
Tam., B. 171 ; shem maram, *Tam., Mal.,*
A. 988 ; shempangan, *Tam.,* M. 537 ;
shem paratie, *Tam.,* G. 385 ; shenba,
Mal., Tel., C. 2361 ; shenbú, *Tam.,*
C. 2361 ; shen curungi, *Tam.,* G. 257 ;
shengan-kuppi, *Tam.,* C. 1377 ; shengú-
tan, *Tam.,* D. 569 ; shén-kottai, *Tam.,*
S. 1041 ; shenkupashanam, *Tel.,* A. 1425 ;
shen kurani, *Tam.,* G. 257 ; shen-shand-
anam, *Tam.,* P. 1381 ; shépe, *Kan.,*
P. 1343 ; shep-punerunji, *Tam.,* I. 125 ;
sherán-kottai, *Tam.,* S. 1041 ; sherimán,
Tel., A. 1149 ; shervu, *Tel.,* H. 113, 164 ;
sheshumu, *Tel.,* L. 143 ; shevelli, *Tam.,*
R. 564 ; shevenar-vaymbú, *Tam.,* I. 111 ;
shewney, *Kan.,* G. 287 ; shi-anvige, *Kan.,*
S. 2861 ; shibé hannu, *Kan,* P. 1343 ;
shigapp-upós-taká-chedi, *Tam.,* P. 82 ;
shíka, *Tam.,* A. 200 ; shíkáya, *Tel.,* A. 200 ;
shikha-mulamu, *Tel.,* D. 173 ; shilandi,
Tam., C. 2605 ; shilá-rasam (Altingia),
Tel., A. 892 ; shilá-rasam (liquidambar),
Tel., L. 455 ; shilve, *Kan.,* X. 16 ; shimaa,
Tam., C. 489 ; shíma-akatti. *Mal.,*
C. 732 ; shíma-gómgura, *Tel.,* H. 233 ;
shímai-agati, *Tam.,* C. 732 ; shímai-
aravandi-virati, *Tam.,* P. 372 ; shímai-
azha-vanai virai, *Tam.,* P. 372 ; shímai-
chamantipú, *Tam.,* A. 1184 ; shímai-
eluppai, *Tam.,* A. 376 ; shímai-kashuruk-
virai, *Tam.,* H. 233 ; shimai-kich-chilik-
kishangu, *Tam.,* H. 59 ; shímai-madalai-
virai, *Tam.,* C. 2546 ; shímai-sapu, *Tel.,*
C. 681 ; shímai-shíragam, *Tam.,* C. 697 ;
shimai-shombu, *Tam.,* C. 681 ; shimai-
velve-lam, *Tam.,* A. 229 ; shimaiya-viri,
Tam., I. 107 ; shíma-jevanti-pushpam,
Mal., A. 1184 ; shíma-karpúram-aku,
Tam., M. 485 ; shíma-pangi-parutti, *Mal.,*
C. 1512 ; shíma-pangi-pashá, *Mal.,*
C. 1512 ; shíma-sunná, *Kan.,* C. 489 ;

shíma-sunnum, *Tel.,* C. 489; shimbu, *Tam.,* M. 517; shíme-agase, *Kan.,* C. 732; shíme-dálimba-vittulu, *Tel.,* C. 2546; shíme-jerage, *Kan.,* C. 697; shímekattáli, *Kan.,* A. 836; shímensli, *Kan.,* l. 107; shíme-shyá-mantige, *Kan.,* A. 1184; shimli, *Kan.,* O. 38; shimtek, *Tam.,* C. 1388; shimti, *Kan.,* O. 38; shíndil, *Tam.,* T. 483; shíndil, *Tam.,* T. 470; shíndil-kodi, *Tam.,* T. 470, T. 483; shíndil-shakkarai, *Tam.,* T. 470, T. 483; shinduga, *Tel.,* A. 711; shipur-gaddi, *Tel.,* A. 1385; shíragam, *Tam.,* C. 2339; shirat-kuchchi, *Tam.,* A. 1064; shirat-kuch-chi, *Tam.,* S. 3018; shiri-saru, *Tel.,* T. 70; shiru-katrázh-ai, *Tam.,* A. 836; shiru-kattálai, *Tam.,* A. 836; shiru-kírai, *Tam.,* A. 950; shiru-kurunjá; *Tam.,* G. 748; shiru-nágap-pú, *Tam.,* M. 489; shiru-nari-vengayam, *Tam.,* S. 968; shiru-naruvili, *Tam.,* C. 1940; shiru-noch-chi, *Tam.,* V. 181; shirushavukku, *Tam.,* T. 70; shiruvavíli, *Tel.,* V. 181; shíshum, *Tel.,* L. 143; shitta-rattaí, *Tam.,* A. 862; shivadai, *Tam.,* l. 415; shiva-malli, *Kan.,* I. 111; shivánarvembu, *Tam.,* I. 111; shivani, *Kan.,* C. 287; shivapp-kuttalai, *Tam.,* A. 838; shivappu-átru-shavukku, *Tam.,* T. 51; shivappu-chittira, *Tam.,* P. 979; shivappu-gasha-gasháchedi, *Tam.,* P. 82; shivappu-káshuruk-kai, *Tam.,* H. 233; shivappu-káshuruk-virai, *Tam.,* H. 233; shivappu-katrázhai, *Tam.,* A. 838; shivappu-kóta-sha-vukku, *Tam.,* T. 51; shivappu-neli, *Tam.,* P. 673; shivappu-shhóttu-katrazh-ai, *Tam.,* A. 838; shivappu-shiru-sha-vukku, *Tam.,* T. 51; shivappu-támara-ver, *Tam.,* N. 39; shivappu-vasla-kíre, *Tam.,* B. 207; shlonga-kuspi, *Mal.,* C. 1403; sholang kelleté, *Tam.,* F. 491; sholongo-kusbi, *Mal.,* C. 1403; sholongo-kuspi, *Mal.,* C. 1403; shómbu, *Tam.,* P. 727; shora-kai, *Tam.,* L. 30; shóttu-katrazhai, *Tam.,* A. 829; show árisí, *Tam.,* C. 2531; shrimudrigida, *Kan.,* A. 89; shritalam, *Tel.,* C. 1995; shritale, *Kan.,* C. 1995; shudan, *Tam.,* C. 257; shukku, *Tam.,* Z. 201; shulundu-kora, *Tam.,* I. 515; shúmak, *Tam.,* C. 19; shumi, *Tel.,* P. 1259; shúnansli, *Tel.,* I. 107; shund apana, *Mal.,* C. 711; shunkú-pushpa, *Mal.,* C. 1403; shun-kur-puspa, *Mal.,* C. 1403; shúnnámbu, *Tam.,* C. 489; shunnárip-pullu, *Tam.,* A. 1107; shúrákáram, *Tel.,* S. 682; shurappattai, *Tam.,* C. 1117.

Si Sibaydú, *Tel.,* L. 143; sigé, *Kan.,* A. 200; sikkai, *Tel.,* A. 686; silandi-arisi, *Tam.,* C. 2605; sillaú, *Tel.,* F. 484; sima, *Tel.,* C. 455; sima avisl, *Tel.,* C. 732; sima boggu, *Tel.,* C. 1414; síma-chámanti-purvu, *Tel.,* A. 1184; síma-chámanti-pushpam, *Tel.,* A. 1184; síma chinduga, *Tel.,* P. 900; síma chinta, *Tel.,* P. 900; síma-gorontí-vittulu, *Tel.,* P. 372; simai karri, *Tam.,* C. 1414; síma-ippa, *Tel.,* A. 376; símajamudu, *Tel.,* B. 909; síma-jilakara, *Tel.,* C. 697; síma jíluga, *Tel.,* P. 322; simao-manis, *Kan.,* C. 1233; sima-shatávari, *Tel.,* A. 1575; sime hunase, *Kan.,* P. 900; sime-mara, *Kan.,* S. 2501; sime tangadi, *Kan.,* C. 785; simie-takalje-

palam, *Tam.,* L. 596; simpuliccai, *Tam.,* E. 370; simpu-licham, *Tam.,* E. 370; simur, *Mal.,* B. 632; sinban-karawa, *Tel.,* P. 801; singara (Lutjanus), *Tam.,* F. 480; singhara (Trapa), *Tam.,* T. 516; sinní-marum, *Tam.,* A. 304; siragadam, *Tel.,* E. 25; sira njikadi, *Kan.,* M. 656; siri, *Tam.,* E. 73; siriánem, *Tel.,* B. 873; siri jana, *Tel.,* C. 677; síri-mánu, *Tel.,* A. 1149; siri púne kuve, *Kan.,* C. 156; siris, *Kan.,* F. 671; sirpha, *Mal.,* A. 1166; sirrúghá kuttalay, *Tam.,* A. 838; sirrú-kánchorivayr, *Tam.,* T. 509; sirru-kiraí, *Tam.,* A. 950; sirru-kúra, *Tel.,* A. 950; sirrú-púlay-vayr, *Tam.,* A. 554; sirru-vullie-kelangu, *Tam.,* D. 481; siru-padi, *Tam.,* C. 1707; siru-pasarai-kírai, *Tam.,* P. 1187; sirú-payarú, *Tam.,* P. 496; sirutalí, *Tam.,* I. 397; sisik-kurakura, *Mal.,* R. 97; sisik-panu, *Mal.,* R. 97; síssú, *Tel.,* D. 64; síssukarra, *Tel.,* D. 64; sítama purgonalu, *Tel.,* C. 2508; sítamma pógu núlu, *Tel.,* C. 2508; sita-palam, *Tam.,* A. 1166; sítapandu, *Tel.,* A. 1166; sítápázham, *Tam.,* A. 1166; sitapu chettu, *Tel.,* S. 2252; sítasavaram, *Tel.,* l. 342; sittamindi, *Tel.,* R. 369; sittamunuk, *Tam.,* R. 369; sittrapaladi, *Tam.,* E. 549; sivagrandi, *Tam.,* E. 584; sivungi, *Kan.,* T. 424.

Sk Skarak-konrask-káy, *Tam.,* C. 756; skimai-shadavari, *Tam.,* A. 1575.

So Soda, *Tel.,* P. 41; sodee, *Tam.,* E. 166; sogadaheru, *Kan.,* H. 119; sohan, *Mal.,* S. 2501; sohikire, *Tam.,* F. 659; sosn-parpu-kírai, *Tam.,* P. 1187; solomanim, *Tam.,* A. 711; sóma latá, *Tel.,* S. 882; sombú, *Tam.,* P. 727; some, *Kan.,* S. 2501; sómida manu, *Tel.,* S. 2501; sómpú, *Kan.,* P. 727; sona-kah-tiriki, *Tam.,* F. 598; sona-ka-wahlah, *Tam.,* F. 593; sonawir, *Mal.,* C. 756; sondeli, *Kan.,* R. 54; sonmakki, *Tel.,* I. 478; sonthi, *Tel.,* Z. 201; sonti, *Tel.,* Z. 201; soogúndaraju gida, *Kan.,* C. 321; sópu, *Tel.,* P. 727; sorakáya, *Tel.,* L. 30; sora-parah, *Tel.,* F. 373; soriai-kai, *Tam.,* L. 30; soring, *Tel.,* F. 580; soundalay únna, *Tam.,* H. 141; sowarah, *Tel.,* F. 515; sowarah, *Tel.,* F. 517.

Su Suam, *Mal.,* S. 2501; suámi, *Kan.,* S. 2501; subba dumpa, *Tel.,* D. 537; sub-basije, *Kan.,* P. 727; sudi-mirapakaia, *Tel.,* C. 455; sudi-sandawa, *Tel.,* F. 583; sudmirapa kaia, *Tel.,* C. 466; sudumu, *Tel.,* F. 475; sugandha-pálada-gidá, *Kan.,* H. 119; sugandhi-pála, *Tel.,* H. 119; suggaalhtú-bontú, *Tel.,* F. 578; sug-gipatte, *Kan.,* O. 38; suiminta sominta, *Tel.,* S. 1174; súli, *Tam.,* C. 838; súmí, *Tel.,* S. 2501; sumpaghy, *Kan.,* M. 17; sunari, *Tel.,* O. 1; súndai kai, *Tam.,* S. 2313; sundara badinika, *Tel.,* V. 156; sunday-kíray, *Tam.,* N. 76; sundel, *Tam.,* S. 790; sung-ilai, *Tam.,* A. 1665; sunkanthemara, *Kan.,* P. 1032; sunkés-waram, *Tel.,* P. 1032; sunna, *Kan.,* Tel., C. 489; sunnam, *Tel.,* C. 489; sunpail-kura, *Tel.,* P. 1187; sunúru, *Tel.,* O. 1; sura, *Kan.,* C. 826; surabunne, *Kan.,* C. 146; suragada, *Tel.,* G. 156; sura-gonne, *Kan.,* C. 146; surala chaki, *Tam.,*

V. 55; surala-tége-patta, *Tel.*, V. 55; surala tíge, *Tel.*, V. 55; suralpattai, *Tam.* V. 55; suranji, *Tam.*, A. 686; surapádi, *Tel.*, L. 229; sura-ponna, *Tel.*, O. 6; súrate eheka, *Tam.*, V. 55; surati pette tige, *Tel.*, V. 55; súringán, *Tam.*, C. 1703; suringi, *Kan.*, O. 6; suriti-pette-chakka, *Tel.*, V. 55; surla tíge, *Tel.*, V. 55; surna-múki, *Tel.*, I. 476; surponne bobbi, *Kan.*, C. 156; surúdu, *Tel.*, S. 3042; súrúghúndu-putta, *Tel.*, V. 55; surugudu yerra chicatli, *Tel.*, V. 55; surúl, *Kan.*, B. 308; suryákánti, *Tel.*, I. 335; suvarnam, *Tel.*, C. 756.

Sw Swarm, *Nilgiris*, Z. 273.

Sy Syamali, *Tel.*, H. 92.

Ta Tabasi, *Tel.*, S. 2850; tabsu, *Tel.*, S. 2850; ta-but-ti, *Kan.*, F. 537; tada, *Tel.*, G. 714; tadáchi, *Tam.*, G. 663; tadagunny, *Kan.*, V. 116; taddapallu, *Tel.*, I. 515; taddo, *Tam.*, P. 1397; tádi, *Tel.*, T. 293; tagada, *Tel.*, S. 2865; tagara-chettu, *Tel.*, C. 787; tagarai-virai, *Tam.*, L. 306; taga-ram, *Tam.*, T. 460; tagarisha-chettu, *Tel.*, C. 797; tagumúda, *Tel.*, G. 287; tahlun-kala, *Tam.*, F. 538; tai, *Tam.*, D. 569; takke, *Tam.*, A. 560; takkólapu-chettu, *Tel.*, C. 1377; tala-bon, *Tel.*, F. 507; tálakam, *Tam.*, O. 242; tálakamu, *Tel.*, O. 242; tala-neli, *Tam.*, I. 342; tala noppi, *Tel.*, X. 1; talari, *Tam.*, S. 1679; talbsu, *Tel.*, S. 2850; tale mara, *Kan.*, P. 26; tále (Borassus), *Kan.*, B. 663; tále (Corypha), *Kan.*, C. 1995; talip-panai, *Tam.*, C. 1995; tálísapatri (Flacourtia), *Tam.*, F. 603; tálisha-pattiri (Cassia), *Tam.*, C. 1183; táll, *Kan.*, B. 663; talla, *Tel.*, S. 2424; talla-maya, *Tel.*, F. 315; talnopi, *Tel.*, X. 1; taludala, *Tam.*, C. 1386; talúdalel, *Tam.*, C. 1386; talum, *Tam.*, P. 26; talura, *Tam.*, S. 1679; tamál, *Kan.*, G. 66; tamalapáku, *Tel.*, P. 775; tamara, *Mal.*, N. 39; tamarattúka, *Mal.*, A. 1646; tamarta, *Tam.*, *Tel.*, A. 1646; tambachi-marum, *Tam.*, H. 324; támbra, *Kan.*, C. 2361; tamida, *Tam.*, E. 166; tamidelu, *Tam.*, E. 170; tamkai, *Tam.*, T. 293; tamma, *Tel.*, C. 289; tammichetta, *Tel.*, O. 1; tamramu, *Mal.*, *Tel.*, C. 2361; támra-valli, *Tel.*, R. 564; tanaku, *Tam.*, C. 1512; tanaku-parutti, *Tam.*, C. 1512; tanaku-pishin, *Tam.*, C. 1512; tandassir, *Kan.*, C. 1875; tandi, *Tel.*, T. 293; tandi tonda, *Tam.*, T. 293; tand-kírai, *Tam.*, A. 938; tándra-káya, *Tel.*, T. 293; tangádi-gida, *Kan.*, C. 741; tanga-réku, *Tam.*, A. 1622; tanga-réku, *Mal.*, A. 1622; tangédu, *Tel.*, C. 741; tani, *Tam.*, *Tel.*, T. 293; táni-kái, *Tam.*, T. 293; tanikaia, *Tam.*, T. 293; tanikayi, *Kan.*, T. 293; taniki, *Tel.*, H. 339; tanikoi, *Tam.*, T. 293; tanjédu, *Tel.*, X. 16; tanjedu mánu, *Tel.*, X. 16; tannab, *Mal.*, C. 1317; tanni-muttán-kalan-gu, *Tam.*, A. 1577; tannir-muttan-kizhan-gu, *Tam.*, A. 1575; tannír-vittán-kizhan-gu, *Tam.*, A. 1577; tánrik-káy, *Tam.*, T. 293; tansi, *Tel.*, S. 2850; tansu-paum, *Tel.*, S. 2936; tanuku mánu (Sterculia), *Tel.*, S. 2850; tanuku mánu (Gyrocarpus), *Tel.*, G. 780; taosee, *Tel.*, S. 2850; tappu cúti,

Tel., F. 555; taraka vépa, *Tel.*, M. 393; táraní, *Tam.*, O. 242; tarate, *Kan.*, C. 410; taravadagida, *Kan.*, C. 741; tare (Acacia), *Kan.*, A. 135; táre (Terminalia), *Kan.*, T. 293; tari, *Kan.*, S. 1939; tarí, *Kan.*, T. 293; tarí, *Kan.*, T. 312; tári-káyi, *Kan.*, T. 293; tarotah, *Tam.*, C. 797; tartoore, *Tel.*, F. 518; taru, *Kan.*, T. 312; taruka, *Tel.*, M. 363; táti-chettu, *Tel.*, B. 663; tavaksha, *Kan.*, M. 267; tavakshírá, *Kan.*, B. 84; tavaribija, *Kan.*, N. 39; tavarigadde, *Kan.*, N. 39; tavashú múrunghie, *Tam.*, R. 656; tawúlú pinnel, *Tel.*, F. 428; taynga, *Tam.*, C. 1520; taynga-nunay, *Tam.*, C. 1520; tazhai, *Tam.*, P. 26.

Te Tedlapál, *Tel.*, W. 122; tedlapála, *Tel.*, W. 122; téga (Tectona), *Kan.*, T. 232; téga (Dioscorea), *Tel.*, D. 510; tégada, *Tel.*, I. 415; tégálu, *Tel.*, D. 517; teg-gummadu, *Tam.*, G. 287; tegina, *Kan.*, T. 232; ték, *Tam.*, *Tel.*, T. 232; tekkali, *Tel.*, C. 1386; tekku, *Tam.*, S. 1939; tékku, *Tam.*, T. 232; tékkumaram, *Tam.*, T. 232; téku, *Tel.*, T. 232; téku-mánu, *Tel.*, T. 232; telaki, *Tel.*, C. 1386; telanga chína, *Tam.*, L. 52; tele, *Tam.*, F. 601; teleki, *Tel.*, C. 1386; telél, *Tel.*, G. 136; tel-kodduki, *Tam.*, H. 102; tella (Alhagi), *Tel.*, A. 745; tella (Clitoria), *Tel.*, C. 1403; tella (Sorghum), *Tel.*, S. 2424; tella antisa, *Tel.*, S. 1694; tella-avalu, *Tel.*, B. 800; tellabarinka, *Tel.*, F. 175; tella chandanam, *Tel.*, S. 790; tella-chettu, *Tel.*, E. 593; tella-chikkudu, *Tel.*, D. 789; tella-chikurkai, *Tel.*, D. 789; tella chin-dagu, *Tel.*, A. 717; tella-chitra-múlam, *Tel.*, P. 986; tella dámaru, *Tel.*, V. 31; telladintana, *Tel.*, C. 1403; tella dura dagondi, *Tel.*, T. 509; tella galijéru, *Tel.*, T. 530; tella goda, *Tel.*, D. 655; tella-janular, *Tel.*, S. 2424; tella jonna, *Tel.*, S. 2397, S. 2405; tellakáká-mushti, *Tel.*, C. 902; tella kákara, *Tel.*, M. 626; tella kakkisa, *Tel.*, G. 136; tella-kalava, *Tel.*, N. 200; tellamaddi, *Tel.*, T. 282; tellamadu, *Tel.*, T. 282; tella-manga, *Tel.*, G. 116, 128; tellamódugu, *Tel.*, B. 944; tellamotuku, *Tel.*, O. 537; tella mulaka, *Tel.*, S. 2280; tellanéla mulaka, *Tel.*, S. 2280; tella pal, *Tel.*, W. 122, W. 131; tella-páshánam, *Tel.*, A. 1425; tella pat-saru, *Tel.*, D. 53; tellaponuku, *Tel.*, G. 226; tella púnkí, *Tel.*, G. 226; tella sandra, *Tel.*, A. 291; tella-shakkara, *Tel.*, S. 375; tella sopara, *Tel.*, A. 717; tella-soring, *Tel.*, F. 580; tellasugan-dhipála, *Tel.*, H. 119; tella tégada, *Tel.*, I. 415; tella tíge, *Tel.*, tella-tóta-kúra, *Tel.*, A. 938; tella-tuma, *Tel.*, A. 249; tella uli-midi, *Tel.*, C. 2039; tella-uppí, *Tel.*, A. 1665; tella uste, *Tel.*, S. 2315; tella-vávili, *Tel.*, V. 164, V. 181; tella vem-pali, *Tel.*, T. 270; tella vúle, *Tel.*, C. 2039; tella yampali, *Tel.*, T. 270; tellay tumbetten kaza, *Tel.*, C. 289; telli-sandawa, *Tel.*, F. 583; telmunnie, *Tam.*, H. 102; telsu, *Tel.*, A. 711; télu kond, chettu, *Tel.*, M. 308; ten, *Tam.*, *Mal.*, H. 342; tenai, *Tam.*, S. 1212; téne, *Tel.*, H. 342; tenga, *Tam.*, *Mal.*, C. 1520; tengái-yenne, *Tam.*, C. 1520; téngá kallu, *Tam.*, C. 1520; téngáy,

Tam., C. 1520; téngá-yenney, *Tam.*, C. 1520; tengedui, *Kan.*, C. 741; tengina, *Kan.*, C. 1520; tengina chippu, *Kan.*, C. 1520; tenginá-gidá, *Kan.*, C. 1520; tenginákáyi, *Kan.*, C. 1520; tenginararu, *Kan.*, C. 1520; tenginay amne, *Kan.*, C. 1520; tanginá-yanne, *Kan.*, C. 1520; tenkáia, *Tel.*, C. 1520; tenkáia-chettu, *Tel.*, C. 1520; tenkáia gurtu, *Tel.*, C. 1520; tenkáia-kallu, *Tel.*, C. 1520; tenkaia nar, *Tel.*, C. 1520; tenkáia núnay, *Tel.*, C. 1520; tenkala, *Tel.*, C. 1520; tenkeia-chettu-puthie, *Tel.*, C., 1520; tenkí-kunsul, *Tel.*, F. 555; tenna, *Mal.*, *Tam.*, C. 1520; ténna-enna, *Mal.*, C. 1520; tennai, *Tam.*, S. 1212; tenna maruttú pungie, *Tam.*, C. 1520; tennam kurtu, *Tam.*, C. 1520; tennam nar, *Tam.*, C. 1520; tennam-púppa, *Mal.*, C. 1520; tennanchedi, *Tam.*, C. 1520; tennan-kallu, *Tam.*, C. 1520; tennang-kallú, *Tam.*, C. 1520; tenna-vellam, *Tam.*, S. 370; ténn-maram, *Mal.*, C. 1520; terangúni, *Tam.*, F. 440; teregam (Callicarpa), *Mal.*, C. 129; teregam (Ficus), *Mal.*, F. 156; ternalla benda, *Tel.*, S. 1714; teta, *Kan.*, O. 233; tetan-kottai, *Tam.*, S. 2960; téttá, *Tam.*, S. 2960; tetta gorra chettu, *Tel.*, S. 1694; tettamparel, *Mal.*, S. 2960; tettamperel marum, *Mal.*, S. 2943; tettian, *Tam.*, S. 2960.

Th Thada, *Tam.*, G. 714; thadsal, *Kan.*, G. 714; thágedu tangar, *Tel.*, C. 741; thágu, *Tel.*, S. 2865; thain-púchie, *Tam.*, G. 726; thalay, *Tam.*, P. 26; thalay marathu, *Kan.*, S. 818; thalí kirai, *Tam.*, I. 410; thá-maroja, *Kan.*, E. 73; thambá, *Tel.*, S. 1682; thambatin, *Tam.*, C. 289; thana, *Tel.*, T. 293; thanddi, *Tel.*, T. 293; thandra, *Tel.*, T. 293; thangan, *Tam. & Mal.*, A. 1622; thani, *Tam.*, T. 293; thani, *Tel.*, T. 293; thánni, *Mal.*, T. 293; thaontay, *Nilgiris*, R. 282; thapsi, *Coorg*, H. 324; thapsi, *Kan.*, H. 324; tharli, *Tam.*, F. 568; tharpai-pullu, *Tam.*, I. 51; tharra, *Tam.*, *Tel.*, G. 714; thasadaram, *Tam.*, E. 370; thay-li, *Tam.*, F. 568; thekku-maram, *Mal.*, T. 232; thelli, *Mal.*, C. 285; thenpinna, *Kan.*, C. 1520; thetti, *Tel.*, F. 430; thidsal, *Kan.*, G. 688; thodágatti, *Kan.*, D. 41; thodapga-pullu, *Tam.*, A. 1385; thombay-keeray, *Tam.*, L. 309; thondi, *Tam.*, B. 520; thorás, *Kan.*, B. 944; thortay, *Kan.*, H. 469; thovaray, *Tam.*, C. 49; thubisee, *Tel.*, S. 2850; thúnamaram, *Tam.*, C. 838; thuringi, *Tam.*, A. 686; thuthuvelai, *Tam.*, S. 2315; thuvarai, *Tam.*, M. 3.

Ti Tid-danda, *Tel.*, Z. 182; tiga jílúga, *Tel.*, A. 565; tiga mushadí, *Tel.*, T. 456; tiga-tshumudú, *Tel.*, S. 882; tige (Dioscorea), *Tel.*, D. 510; tíge (Rubia), *Tel.*, R. 564; tige (Tiliacora), *Tel.*, T. 456; tíge jemudu, *Tel.*, S. 882; tíge-móduga, *Tel.*, B. 978; tíge motku, *Tel.*, B. 978; tige mushidi, *Tel.*, T. 456; tige mushini, *Tel.*, T. 456; tígepaláshama, *Tel.*, B. 978; tilaka, *Tel.*, C. 1386; tillac-heruku, *Tel.*, S. 30; tilla káda, *Tel.*, E. 505; timmuri, *Tel.*, D. 615; tina, *Mal.*, T. 460; tinduki, *Tel.*, D. 582; tipi ulavi, *Tel.*, F. 560; tippa-tége-véru, *Tel.*, T. 470;

T. 483; típpa-tíge, *Tel.*, T. 470, T. 483; tippa-tíge-sattu, *Tel.*, T. 470, T. 483; tippatingai, *Tel.*, T. 470, T. 483; tippili, *Tam.*, *Mal.*, P. 805; tippili, *Mal.*, P. 805; tippilimúlam, *Tam.*, P. 805; tirawa, *Kan.*, X. 16; tirikalli, *Tam.*, E. 496; tirim, *Mal.*, O. 616; tirman, *Tel.*, A. 1149; tirnutpatchie, *Tam.*, O. 18; tirpu, *Kan.*, H. 368; tirukalli, *Mal.*, *Tam.*, D. 387, E. 553; tirunitru, *Tam.*, O. 18; tirutali, *Mal.*, I. 410; tiruvách chip-pú, *Tam.*, T. 410; tiska, *Kan.*, S. 1250; títránparala, *Mal.*, S. 2960; tivoapotike, *Tel.*, H. 444; tivva-máduga, *Tel.*, B. 978; tivva mushidi, *Tel.*, T. 456; tiya mámidi, *Tel.*, M. 147.

To Toal-parah, *Tam.*, F. 403; toandi, *Tel.*, T. 293; todda-maram, *Mal.*, C. 2538; togari, *Kan.*, C. 49; togarike, *Tel.*, D. 560; togral naibela, *Kan.*, A. 249; toinuatalí, *Tel.*, I. 407; tóka-miriyálu, *Tel.*, P. 801; tolparah, *Tel.*, F. 404; tolu, *Tel.*, H. 265; tomagarika, *Tel.*, S. 2670; tomi-tomi, *Mal.*, F. 611; tondeballi, *Kan.*, C. 919; tondi (Terminalia), *Tel.*, T. 293; tondi (Callicarpa), *Mal.*, C. 129; toper, *Tel.*, D. 53; tóppi, *Tam.*, P. 632; torathi, *Kan.*, H. 463; tór-elaga, *Tel.*, L. 362; tore matti madi, *Kan.*, T. 361; tormatti, *Kan.*, T. 282; torriti, *Tam.*, F. 430; torugu, *Tel.*, M. 656; tótakatti, *Tam.*, D. 41; tótakúra, *Tel.*, A. 927, 938; total-vadi, *Tam.*, M. 557; toura-mamidi, *Tel.*, S. 2649; tovaray, *Kan.*, C. 49.

Tr Trinadula gondi, *Tel.*, T. 509; trikála-malle, *Tel.*, M. 109; troja, *Tam.*, R. 504.

Ts Tshama-oada, *Tel.*, S. 785; tshil-ankalú, *Tel.*, W. 122; tsiag ri-nuren, *Mal.*, D. 539; tsikideondoa, *Tel.*, F. 455; tsinnataliaku, *Tel.*, I. 397; tsiroupanna, *Mal.*, C. 162; tsjana-kua, *Mal.*, C. 2013; tsjandana marum, *Mal.*, S. 790; tsjéron kárá, *Mal.*, C. 393; tsjerú cans-java, *Mal.*, C. 331; tsjeru-teka, *Mal.*, C. 1388; tsjuriacranti, *Mal.*, I. 405.

Tu Tubiki, *Tel.*, D. 582; tudavullay, *Tam.*, S. 2315; tuki, *Tel.*, D. 569; tulashi, *Tam.*, O. 31; tulashi-gidá, *Kan.*, O. 31; tulasi, *Tam.*, *Tel.*, O. 31; tulkapyre, *Tam.*, P. 468; tulu-candal, *Tam.*, F. 386; túma, *Tel.*, A. 101; tumba, *Mal.*, L. 312; tumbai-chedi, *Tam.*, L. 309; tumbali, *Tam.*, D. 615; tumbay kayi, *Kan.*, C. 289; tumbi, *Tam.*, *Tel.*, D. 569, D. 615; tumbika, *Tam.*, D. 582; tumbilik-kay, *Tam.*, D. 582; tumbugai, *Tam.*, S. 1682; tumbugai-pishin, *Tam.*, S. 1682; tumi, *Tel.*, D. 615; tumida, *Tel.*, D. 615; tumika, *Tel.*, D. 582; tumiki, *Tel.*, D. 582; tumil, *Tel.*, D. 582; tumki, *Tel.*, D. 615; tummachettu, *Tel.*, L. 309; tummeda mámidi, *Tel.*, S. 1041; tummi kura, *Tel.*, L. 309; tumni, *Tel.*, L. 312; tumpa-kotu-véli, *Mal.*, P. 986; tumti kayi, *Kan.*, C. 1211; túm, *Tam.*, C. 838; tunda, *Kan.*, C. 838; tundu, *Kan.*, C. 838; tunga gaddala-veru, *Tel.*, C. 2617; tunga gaddi, *Tel.*, S. 977; tunga-musle, *Tel.*, C. 2612; tuni, *Tam.*, C. 121; tunkí, *Tel.*, D. 615; tura, *Tam.*, F. 723; turachi-gidá, *Kan.*, M. 786; turichu, *Tam.*, C. 2367; turisha, *Mal.*, C. 2367; turka-vepa, *Tel.*, O. 127; turu, *Tam.*, F. 721; turuni, *Toda*, F. 808; tútánagam,

Tam., Z. 190 ; tutari-véru, *Mal.,* T. 489;
tuthí nar, *Tam.,* A. 80 ; tútikúra, *Tel.,* I.
343 ; tutiri-chettu, *Tel.,* A. 89 ; tuttam-
turichi, *Tam.,* C. 2367 ; tutti, *Tam.,* A.
80, 89 ; tuttura-benda, *Tel.,* A. 80, 89 ;
tútum, *Tel.,* Z. 190 ; tuvara, *Mal.,* C. 49 ;
tuvarai, *Tam.,* C. 49.

Ty Tyágada mara, *Kan.,* T. 232.

Ub Ubbay, *Tam.,* F. 432 ; ubbolu, *Kan.,* F.
611.

Uc Uchchinta, *Tel.,* S. 2315 ; uchchi usirika,
Tel., P. 668.

Ud Uda-gaddí, *Tel.,* P. 50 ; udagina-gida, *Kan.,*
A. 681 ; udalai, *Tam.,* J. 52 ; udan, *Tam.,*
F. 452 ; udda, *Tel.,* D. 748 ; uddu, *Kan.,*
P. 513 ; udi, *Kan.,* O. 38 ; údrik-patta,
Tel., G. 726 ; údugachettu, *Tel.,* A. 681 ;
udumbu, *Tel.,* R. 116 ; udumu, *Tam.,* R.
116 ; údúpai, *Tel.,* G. 714.

Ug Ugá, *Tam.,* S. 717 ; ughai, *Tam.,* C. 224 ;
ughai, *Tam.,* S. 705, S. 717.

Ul Ular, *Mal.,* R. 129 ; ularí, *Tam.,* F. 314;
ulava, *Tel.,* D. 758 ; ulavi, *Tel.,* F. 560 ;
ulimera, *Tel.,* D. 560 ; ulimidi, *Tel.,* C.
2039 ; ullahti, *Tam.,* F. 438 ; ullinda, *Tel.,*
D. 560 ; ullú súlú, *Tel.,* L. 385 ; ulpathy,
Tam., F. 534 ; ulsa, *Tel.,* F. 326 ; ulum,
Tam., F. 414 ; uluva, *Mal.,* T. 612 ;
ulúway, *Tam.,* F. 455, 569, *Tel.,* F. 456.

Um Umari, *Tam.,* A. 1475 ; umatai, *Tam.,* D.
160, 166 ; umbrúth ballé, *Kan.,* C. 311 ;
ummam, *Mal.,* D. 160 ; ummatta, *Mal.,*
D. 160 ; ummatte gidá, *Kan.,* D. 160 ;
ummetta, *Tel.,* D. 160, D. 166.

Un Undra, *Tel.,* M. 562 ; undurugu, *Tel.,* E.310 ;
unna-mam, *Mal.,* P. 1121.

Up Upalkai, *Kan.,* M. 15 ; uparanthi, *Kan.,* H.
144 ; upligi, *Kan.,* M. 15 ; uppa, *Mal.,*
S. 602 ; uppi-áku, *Tel.,* A. 1665 ; uppu,
Kan., Tam., Tel., S. 602 ; uppu-dali,
Mal., R. 633 ; uppu nute, *Kan.,* M. 747 ;
uppusanaga, *Tel.,* C. 2057 ; upputti, *Mal.,*
A. 1660 ; upupoma, *Tel.,* R. 242.

Ur Úra kakara, *Tel.,* M. 626 ; úranechra, *Tel.,*
X. 12 ; uranke, *Tel.,* E. 259 ; uranka, *Tel.,*
E. 259 ; urgu, *Tel.,* A. 681 ; urimidi (Cas-
sia), *Tel.,* C. 773 ; urimidi (Cratæva), *Tel.,*
C. 2039 ; urlú, *Tel.,* O. 258 ; urumitti,
Tel., C. 2039 ; urvanjik kya, *Mal.,* S.
818.

Us Usamaduga, *Tam.,* B. 334 ; usereki, *Tel.,* P.
632 ; ushí-collarchi, *Tam.,* F. 367 ; ushit-
tagarai, *Tam.,* C. 797 ; usiki, *Tel.,* C.
2039 ; usiki mánu, *Tel.,* C. 2039 ; úsi-
mulaghai, *Tam.,* C. 466 ; usirika, *Tel.,* P.
632 ; uskia, *Tel.,* C. 2039 ; uskiamen, *Tel.,*
C. 773 ; usri, *Tel.,* P. 632 ; uste, *Tel.,*
S. 2315.

Ut Útalay gudda, *Tel.,* S. 2320 ; utalipanna,
Tam., C. 711 ; utarni, *Tam.,* D. 9 ; uti-
chettu (Maba), *Tel.,* M. 3 ; úti chettu
(Clerodendron), *Tel.,* C. 1377 ; utránigida,
Kan., A. 382 ; uttámaní, *Tam.,* D. 9 ;
uttaráne, *Kan.,* A. 382 ; utta-réni, *Tel.,*
A. 382.

Uv Uva, *Tam., Tel.,* D. 428.

Va Vádagannéru, *Tel.,* P. 989 ; vadaja, *Tel.,*
A. 430 ; vádámbram, *Tel.,* D. 4 ; vádam-
kottai, *Tam.,* P. 1274 ; vadampirí, *Tam.,*
H. 92 ; vadatalla, *Tam.,* D. 402 ; vadatara,
Tam., D. 402 ; vadencarni, *Tam.,* S.

2884 ; vadla-yárála, *Tel.,* H. 285 ; vágáti,
Kan., W. 1 ; vaghe, *Tam.,* A. 695 ; vake-
nar, *Tam.,* S. 2861 ; vakka, *Tel.,* A. 1294;
vakkavanji, *Mal.,* C. 2105 ; valaga, *Tel.,*
F. 601 ; valampúri, *Tam.,* H. 92 ; va-
lange, *Tam.,* D. 53 ; valanku-chámbráni,
Mal., B. 777 ; válarasi, *Tel.,* W. 13 ;
valei, *Tam.,* M. 811 ; valesulú, *Tel.,* G.
735 ; valiy, *Mal.,* C. 1286 ; vallai-murdú,
Tam., T. 293 ; vallai-púndu, *Tam.,* A.
779 ; vallai sharun-nai, *Tam.,* T. 530 ; val-
langa, *Tam.,* F. 53 ; vallárai, *Tam.,* H.
486 ; vallari, *Tam.,* E. 217; vellayanne,
Kan., S. 1078 ; valli kalangu, *Tam.,* I.
359 ; vallikilángu, *Tam.,* I. 348 ; valli-
murukka, *Mal.,* B. 978 ; vallipála, *Mal.,*
T. 855 ; valliplách-cha, *Mal.,* B. 978 ; val-
mellaghu, *Tam.,* P. 801 ; val-miláku,
Tam., P. 801 ; válmulaka, *Mal.,* P. 801 ;
valsanábhi, *Mal.,* A. 397 ; váluluvai-
taílam, *Tam.,* C. 854 ; válulvai, *Tam.,* C.
854 ; valumherí, *Tam.,* H. 92 ; valúthala,
Mal., S. 2284 ; valúthalay, *Tam.,* S.2284 ;
válu-zhuva, *Mal.,* C. 854; váminta, *Tel.,*
G. 753 ; vana-sunthi, *Kan.,* Z. 201 ; van-
dukolli, *Tam.,* C. 732 ; vanga, *Tam.,* O.
233 ; vanga chettu, *Tel.,* S. 2284 ; vanga-
chiri-vangu, *Tel.,* S. 2284 ; vangarreddi
kúra, *Tel.,* S. 1203 ; váníjhál, *Tam.,* C.
224 ; vankaya, *Tam.,* S. 2284 ; vankuda,
Tel., S. 2345 ; vanní, *Tam.,* P. 1259 ;
vanuturu, *Tel.,* D. 402 ; varagalu, *Tel.,* P.
63 ; varagu, *Tam.,* P. 63 ; vaiamullí,
Tam., B. 171 ; vari katchelí, *Tam.,* F.
574 ; varragoki, *Tel.,* T. 489 ; varra kasimi,
Tel., T. 489 ; vartangi, *Tam.,* C. 35 ;
vasa, *Tel.,* A. 430 ; vasanábhi, *Tam.,*
A. 397, *Tel.,* A. 413 ; vasanap-pulla,
Mal., A. 1079 ; vasha, *Mal.,* M. 812 ;
vashambu, *Tam.,* A. 430 ; vasha návi,
Tam., A. 397 ; vashanpa, *Mal.,* A. 430 ;
vashanup-pulla, *Tam.,* A. 1079 ; vasla-kíre,
Tam., B. 203 ; vassuntagunda, *Tel.,* M.
71 ; vátangolli, *Mal.,* J. 116 ; váta-tirupie,
Tam., C. 1205 ; vátham addakki, *Tam.,* C.
1388 ; vattángi, *Tam.,* C. 35 ; vittatirippi,
Tam., S. 1688 ; vatte kanni, *Tam.,* M.
15 ; vattéku, *Tam.,* C. 35 ; vattí-véru,
Tel., A. 1097 ; vavali-padú, *Tel.,* V. 164 ;
vavani, *Kan.,* S. 1212 ; vávili, *Tel.,* V.
164 ; vayivalanga, *Kan.,* E. 199 ; vayilettu,
Tam., V. 143 ; vaykkavalai, *Tam.,* T.
280 ; váyu-vilamgam, *Tam., Tel.,* E. 199 ;
vazhaip pazham, *Tam.,* M. 811 ; vazhap-
pagham, *Mal.,* M. 811.

Ve Veckali, *Tam.,* A. 1149 ; vedam, *Tel.,* T.
312 ; vedam, *Tam.,* S. 1939 ; veddavala,
Tam., A. 217 ; vedittalung kolildu, *Tam.,*
D. 402 ; vedú-kunari, *Tam.,* D. 639 ;
veduru, *Tel.,* B. 118 ; veduruppu, *Tel.,*
B. 84 ; végi, *Tel.,* P. 1370 ; végisa, *Tel.,*
P. 1370 ; vekhariyo, *Tel.,* I. 131 ; velá,
Tam., F. 53 ; vélá, *Mal.,* G. 753 ; velagá,
Tel., F. 53 ; velaga bada-nika, *Tel.,* L.
549 ; velai, *Tam.,* G. 753 ; velai-thuthi,
Tam., A. 98 ; velakura, *Tel.,* G. 753 ; vel-
amín, *Tam.,* F. 544 ; vélam-pishin, *Tam.,*
F. 54 ; vela-padri, *Tam.,* S. 2865 ; vela-
tambalin, *Tel.,* C. 289 ; vélip-paritti, *Mal.,*
D. 9 ; vélípparutti, *Tam.,* D. 9 ; vellache-
kende, *Tam.,* F. 393 ; vellachi-candí, *Tam.,*
F. 391 ; vella-ellay, *Tam.,* M. 855 ; vellaí-
cadamba, *Tam.,* A. 1192 ; vellaidámar,

Tam., V. 31 ; vellai-kadugu, *Tam.,* B. 800 ; vellai-karun káli, *Tam.,* D. 637 ; vellai-kungiliyam, *Tam.,* V. 31 ; vellai-kunrikam, *Tam.,* V. 31 ; vellai maruda, *Tam.,* T. 282 ; vellai, noch-chi, *Tam.,* V. 164 ; vellai-páshánam, *Tam.,* A. 1425 ; vellaip-pólam, *Tam.,* B. 48 ; vellai-sharu-karai, *Tam.,* S. 375 ; vellaítoaratt, *Tam.,* C. 410 ; vella kalada, *Tel.,* F. 307 ; vella-káram, *Mal.,* B. 731 ; vella-katuka, *Mal.,* B. 800 ; vella, kelleti, *Tam.,* F. 487 ; vella-ketchelí, *Mal.,* F. 571 ; vella-kudrikum, *Mal.,* V. 31 ; vellaí-kundirukkam, *Mal.,* B. 771 ; vella-kunturukkam, *Mal.,* V. 31 ; vellal, *Tam.,* E. 199 ; vellam, *Tam.,* F. 53 ; vellam, *Tam.,* S. 370 ; vella marda, *Tam.,* S. 1939 ; vella marda, *Tam.,* T. 282 ; vella-maruta, *Mal.,* T. 282 ; vella ma-rúthú, *Tam.,*T. 282; vellamattawal, *Tam.,* F. 405; vellamatti, *Tam.,* T. 282 ; vel-langú, *Tel.,* F. 506 ; vella-noch-chi, *Mal.,* V. 164 ; vella patani, *Tam.,* P. 885 ; vellap-páshánam, *Mal.,* A. 1425 ; vellarí-veri, *Tam.,* C. 2263 ; vellathorasay, *Tam.,* C. 902 ; vella-vengáyam, *Tam.,* A. 769 ; vella voval, *Tam.,* F. 583 ; vellay bootali, *Tam.,* S. 2850 ; vellay naga, *Tam.,* A. 1149 ; vellay pútali, *Tam.,* S. 2850 ; velláy thun betten, *Tam.,* C. 289 ; vellay toyeray, *Tam.,* D. 598 ; vellerku, *Tam.,* C. 191 ; velli, *Tam.,* S. 2205 ; velli, *Mal.,* *Tam.,* A. 1359 ; velligáram, *Tel.,* B. 371 ; Velli kúndricum, *Tam.,* V. 31 ; velli-rékkui, *Tam.,* A. 1359 ; velliri, *Tam.,* C. 2278 ; vellit-takita, *Mal.,* A. 1359 ; vellituru kon-alu, *Tel.,* D. 402 ; vellulli tellagadda, *Tel.,* A. 779 ; vel-noch-chi, *Mal.,* V. 164 ; vel-utta-chenpakam, *Mal.,* M. 489 ; veluturu, *Tel.,* D. 402 ; velvaila, *Kan.,* A. 249 ; vel-vel, *Tam.,* A. 249 ; vel-velam, *Tam.,* A. 249 ; vembádam, *Tam.,* V. 55 ; vembu, *Tam.,* M. 363, 393 ; venai, *Tel.,* S. 2205 ; vencandí, *Tam.,* F. 469 ; ven-chittiramú-lam, *Tam.,* P. 986 ; vendaik-kay, *Tam.,* H. 196 ; venda-kaya, *Tel.,* H. 196 ; ven-dále, *Tam.,* G. 226 ; vendapa, *Tel.,* H. 444 ; vendayam, *Tam.,* T. 612 ; vendi (Hihiscus), *Tam.,* H. 196 ; vendi (silver), *Tel.,* A. 1359 ; vendi-réku, *Tel.,* A. 1359 ; vengai, *Tam.,* P. 1370 ; vengáram, *Tam.,* B. 731 ; vegàyam, *Kan.,* A. 769 ; vengé, *Tam.,* B. 863 ; ven-kándi, *Tam.,* F. 407 ; venkáram, *Tam.,* B. 731 ; vénna, *Mal.,* P. 1370 ; venna-devi kura, *Tel.,* C. 1752 ; vénnamaram, *Mal.,* P. 1370 ; venna mu-dra, *Tel.,* C. 1752 ; vennap-pasha, *Mal.,* P. 1370 ; venna vedara, *Tel.,* C. 1752 ; ven-panjasára, *Mal.,* S. 375 ; ven shark-kra, *Mal.,* S. 375 ; ventak-káya, *Mal.,* H. 196 ; ventaku, *Kan.,* L. 55 ; venta-yam, *Tam.,* T. 612 ; venteak, *Tam.,* L. 53 ; ventra, *Tel.,* M. 562 ; venuturu, *Tel.,* D. 402 ; vépa, *Tel.,* M. 363 ; véppa, *Mal.,* M. 363 ; véppam, *Tam.,* M. 363 ; vépudu pachcha, *Tel.,* O. 18 ; verarlu, *Tam.,* F. 517 ; verasu, *Tam.,* C. 1931 ; véré, *Tam.,* H. 115 ; verewa puni, *Tam.,* T. 435 ; veriecúm-uttie, *Tam.,* C. 1211 ; vori púch-cha pútsakaia, *Tel.,* C. 1211 ; verit-tumatti, *Tam.,* C. 1211 ; vérk-kadalai, *Tam.,* A. 1261 ; vérk-kalá, *Mal.,* A. 1261 ; verri-bíra, *Tel.,* L. 563 ; verrichá-tarási, *Tel.,* M. 617 ; verri cheruku, *Tel.,* S. 49 ; verri

nélu vému, *Tel.,* O. 132 ; verri-pála, *Tel.,* T. 855 ; verri tala noppi, *Tel.,* X. 1 ; verrughung kalung, *Tam.,* C. 1737 ; véru-panasa, *Tel.,* A. 1489 ; vérusampenga, *Tel.,* P. 1044 ; verushanaga, *Tel.,* A. 1261 ; vetasawmu, *Tel.,* C. 77 ; veti-uppa, *Mal.,* S. 682 ; vetpála, *Tam.,* W. 122 ; vetpá-larishi, *Tam.,* W. 122 ; vettila, *Mal.,* P. 775 ; vettilai, *Tam.,* P. 775 ; vetti-vér, *Tam., Mal.,* A. 1097 ; veturu, *Tel.,* D. 402 ; vevay-lam, *Tam.,* A. 249 ; veyala, *Tel.,* V. 164 ; veypalé, *Tam.,* W. 122.

Vi Vibudi-patri, *Tel.,* O. 18 ; vidavali-véru, *Tel.,* A. 1097 ; vidi, *Tam.,* C. 1931 ; vid-puné, *Tam.,* C. 2168 ; vilám, *Tam.,* F. 53 ; vilám-pishin, *Tam.,* F. 54 ; vilári, *Tam.,* H. 517 ; vilayati-aldekayi, *Kan.,* C. 19 ; vilaytí-hullu, *Kan.,* M. 334 ; vile-dele, *Kan.,* P. 775 ; vilimbi, *Mal.,* A. 1644 ; vilunbikká, *Mal.,* A. 1644 ; vilva-pazham, *Tam.,* A. 534 ; vinivel-getta, *Tam.,* C. 2008 ; virali, *Tam.,* D. 725 ; víram, *Tam.,* M. 473 ; víramu, *Tel.,* M. 473 ; víranam, *Tam.,* A. 1097 ; virgi, *Tel.,* C. 1931 ; virugadu, *Tel.,* D. 41 ; víru malle, *Tel.,* I. 376 ; visha-boddi, *Tam.,* S. 1688 ; visha-kanta-kálu, *Tel.,* V. 73 ; visha mungali, *Tel.,* C. 2062 ; visha-mungil, *Tam.,* C. 2062 ; vishaul, *Mal.,* E. 199 ; vishnugran-die, *Tam.,* E. 584 ; vishnu-kantisoppu, *Kan.* C. 1403 ; vishnu-karandi, *Tam.,* E. 581 ; vishnu-kranta, *Tel.,* E. 581 ; vishnukranti, *Kan.,* E. 581 ; vistarákupála, *Tel.,* H. 294 ; vittu, *Tel.,* A. 1038 ; vittula, *Tel.,* E. 151 ; vitusi, *Kan., Mal.,* C. 2039 ; vizhalvér, *Tam.,* A. 1097.

Vo Vodalai, *Tam.,* A. 135 ; vódalam, *Tam.,* A. 135 ; vodamovettilla, *Tel.,* T. 312 ; volugiritenkí, *Tel.,* F. 597 ; vonangadi, *Kan.,* L. 304 ; von-de-lagá, *Kan.,* H. 486 ; vonte, *Kan.,* A. 1511.

Vu Vubbïna, *Kan.,* P. 1048 ; vúda gaddi, *Tel.,* P. 50 ; vudam vittulu, *Tel.,* T. 312 ; vudlu, *Tel.,* O. 258 ; vukkana, *Tam.,* D. 628 ; vulava, *Tam.,* D. 758 ; vulisi, *Tel.,* G. 735 ; vullay, *Tam.,* L. 143 ; vullay tútam, *Tam.,* Z. 190 ; vullie, *Tam.,* D. 522 ; vulli-gaddalu, *Tel.,* A. 769 ; vulunantri marada, *Kan.,* S. 2876 ; vuma, *Kan.,* C. 146 ; vummaai-pora-sham, *Tam.,* C. 1031 ; vummay, *Tam.,* C. 1031 ; vummray, *Tam.,* C. 1031 ; vungaravasí, *Tam.,* S. 1203 ; vunjurrum, *Tel.,* C. 2528 ; vunne, *Kan.,* *Tam.,* P. 1259 ; vun-paratie, *Tam.,* G. 404 ; vusayley-kiray, *Tam.,* S. 2574 ; vusirika, *Tel.,* P. 632 ; vutta thamary, *Tam.,* M. 16 ; vutti-khillokillupi, *Tam.,* C. 2163.

Wa Waaka, *Tel.,* C. 596 ; wada madichi, *Tam.,* C. 1386; wadata-toka gadí, *Tel.,* D. 445 ; waddan, *Tam.,* L. 219 ; wadinika, *Tel.,* L. 549 ; wadume, *Tel.,* S. 2903 ; wagata, *Tam.,* R. 16 ; waghutty, *Mal.,* C. 410 ; wahlah, *Tam.,* F. 601, *Tel.,* F. 402 ; wahlah-kuddera, *Tel.,* F. 365 ; wajn, *Tel.,* L. 549, wakuilu, *Tel.,* C. 606 ; walawah-tenkí, *Tel.,* F. 560 ; wale, *Tam.,* F. 593 ; walivara, *Kan.,* D. 685 ; wallagú, *Tel.,* F. 601 ; wallake-kellette, *Tam.,* F. 581 ; wallarai kilangu, *Tam.,* S. 2320 ; walsura, *Tam.,* W. 13 ; walsurai, *Tam.,* W. 13 ;

walurasi, *Tel.*, W. 13 ; wanai, *Kan.*, O.
6 ; wana-motta, *Tel.*, F. 458 ; wangara-
was, *Tam.*, F. 458 ; wang-kai, *Tel.*, S.
2284 ; wangon, *Tel.*, F. 581 ; wanjon, *Tel.*,
F. 581 ; wara-gudu, *Tel.*, C. 2538 ; waragu-
wenki, *Tel.*, S. 717 ; warri-ádú, *Tam.*, S.
1257 ; warri-ádú, *Tam.*, S. 1257 ; warri-
parah, *Tam.*, F. 371 ; warsi-pala, *Tel.*,
V. 228 ; wasa (Acorus), *Tel.*, A. 430 ;
wasa (Casearia), *Tel.*, C. 725 ; watte,
Anamalais, O. 612.

Wi Widjin, *Mal.*, S. 1078 ; wingeram, *Tel.*, F.
420.

Wo Wodahalle, *Tam.*, A. 135 ; wodalior, *Tam.*,
A. 135 ; wodayu, *Tam.*, L. 219 ; wodesha,
Tel., L. 219 ; wodi, *Tel.*, D. 748 ; wodier,
Tam., O. 38 ; wodlah-muku, *Tel.*, F. 367 ;
wodrase, *Tam.*, A. 871 ; wolga-tenki, *Tel.*,
F. 597 ; wol-lelu, *Kan.*, S. 1078 ; wond,
Tam., S. 2501 ; wongu, *Tel.*, W. 48 ;
wonte, *Tel.*, C. 205 ; wooda-tallum, *Tel.*,
E. 246 ; woodiya-chettoo, *Tel.*, A. 681 ;
woragú, *Tel.*, F. 373 ; worahwah, *Tel.*, F.
588 ; worga, *Tel.*, P. 63 ; wothalay, *Tam.*,
A. 135.

Wu Wuckú, *Mal.*, C. 2105 ; wúdaahwah, *Tel.*,
F. 452 ; wúdacha-marum, *Tam.*, L. 219 ;
wúdan, *Tel.*, F. 452 ; wude, *Tam.*, O.
38 ; wulawallí, *Tel.*, D. 758 ; wúlawúlú,
Tel., D. 758 ; wuma, *Kan.*, C. 146 ;
wundmarum, *Tam.*, S. 2501 ; wund, *Tel.*,
P. 45 ; wúni, *Tel.*, A. 229 ; wúnja, *Tam.*,
A. 686 ; wunjúli, *Tam.*, C. 838 ; wusel,
Tam., A. 686 ; wúsheriko, *Tel.*, P. 632 ;
wúthúlú, *Tel.*, P. 496 ; wutla-callawah,
Tam., F. 578.

Ya Yádala chettu, *Tel.*, C. 1746 ; yagaohi, *Kan.*,
Z. 231 ; yahla, *Tel.*, F. 544 ; yálakki,
Kan., E. 151 ; yalekalli, *Kan.*, E. 520 ;
yallu, *Kan.*, S. 1078 ; yampali, *Tel.*, T.
270 ; yamskalung, *Tam.*, D. 485 ; yapa
(Hardwickia), *Tel.*, H. 16 ; yapa (Melia),
Tel., M. 363 ; yarala, *Tel.*, P. 1433 ;
yashti madhuká, *Kan.*, G. 278 ; yashti-
madhukam, *Tel.*, G. 278 ; yashti-madhu-
kam, *Mal.*, G. 278 ; yashti-maduram-pálu,
Tel., G. 278 ; yava, *Tel.*, H. 382 ; yavaka,
Tel., H. 382 ; yavala, *Tel.*, H. 382.

Ye Yeanga, *Tam.*, *Tel.*, P. 1370 ; **yeanuga-**
pullerú, *Tel.*, P. 363 ; yed, *Kan.*, P. 1171 ;
yeddi, *Tel.*, H. 164 ; yeddupandi, *Tel.*, P.
1171 ; yeggi, *Tel.*, P. 1370 ; yegísa, *Tel.*,
P 1370 ; yehela, *Kan.*, T. 293 ; yekka,
Kan., *Tel.*, C. 170 ; yekkada beru, *Kan.*,
C. 170 ; yekkada-gidá, *Kan.*, C. 170 ;
yelaki, *Kan.*, E. 151 ; yelchi, *Kan.*, Z.
231 ; yelé, *Kan.*, M. 800 ; yelinga, *Tel.*,
L. 549 ; yellagada, *Kan.*, C. 1158 ; yella
kíray, *Tam.*, S. 2990 ; yellal, *Kan.*, C.
390 ; yellamaddi, *Tel.*, A. 1149 ; yellamal-
lakai, *Tam.*, H. 523 ; yellande, *Tam.*, Z.
231 ; yellanga, *Tel.*, F. 53 ; yella nir,
Tam., C. 1520 ; yella-niru, *Tal.*, C. 1520 ;
yellantha, *Tam.*, Z. 231 ; yellú-cheddie,
Tam., S. 1078 ; yeltu, *Tel.*, D. 402 ;
yengara, *Kan.*, S. 2424 ; yennai, *Tam.*,
N. 37 ; yenné, *Kan.*, H. 22 ; yenuga
palleru, *Tel.*, P. 363 ; yeppa (Bassia), *Tel.*,
B. 220, 265 ; yeppa (Melia), *Tel.*, M. 363 ;
yerakki, *Kan.*, E. 151 ; yercum, *Tam.*,
C. 170 ; yerica, *Mal.*, C. 170 ; yerkoli,
Tam., C. 390 ; yer muddí, *Tel.*, T. 282 ;
yerra-chakatli-chakka, *Tel.*, V. 55 ; yerra
chictalí, *Tel.*, V. 48 ; yerrachitra, *Tel.*, P.
979 ; yerra galijéru, *Tel.*, T. 537 ; yerra-
goda, *Tel.*, D. 628 ; yerra juri, *Tel.*, S.
1939 ; yerrajuvi, *Tel.*, F. 253 ; yerra-juvi,
Tel., O. 1 ; yerra-kala, *Tam.*, F. 540 ;
yerrakodivaylie, *Tam.*, P. 979 ; yerra-
palleru, *Tel.*, I. 125 ; yerra patsaru, *Tel.*,
D. 32 ; yérra sindúram, *Tel.*, L. 143 ;
yerra tambalin, *Tel.*, C. 289 ; yerrivale,
Tam., F. 412 ; yerruchinta, *Tel.*, A. 711 ;
yerugudu, *Tam.*, *Tel.*, D. 41 ; yetega,
Kan., S. 2799 ; yettada, *Kan.*, A. 514 ;
yettama, *Tel.*, A. 1149 ; yette, *Tam.*, D.
64 ; yettéga-pettega, *Kan.*, A. 514 ; yetti
(Strychnos), *Tam.*, S. 2943 ; yetti (Hydnocarpus), *Tam.*, H. 472 ; yetti-maram,
Tam., S. 2943.

Yi Yippali, *Kan.*, P. 805 ; **yiremaddinagadde**,
Kan., W. 98 ; yíta, *Tel.*, P. 551.

Yo Yokada, *Kan.*, C. 170.

Yu Yuri, *Tel.*, F. 216.

Za Zanumu, *Tel.*, C. 331.

Ze Zevangam, *Tel.*, M. 885.

Zo Zolim-buriki, *Tam.*, S. 950 ; zowbíum, *Tel.*,
C. 2531.

L

B. 615 ; dasta, *Nepal,* Z. 190 ; date, *Magh,* F. 228, I. 71 ; dauli, *Rajbanshi,* P. 1239 ; dawa, *Cachar,* A. 1511 ; dayshing, *Bhutia,* D. 115.

De Debrelara, *Nepal,* S. 2508 ; denga, *Assam,* D. 284 ; denyok, *Lepcha,* A. 1286, 1290 ; deo chágal, *Assam,* S. 1247 ; deo-muga, *Cachar,* S. 1739 ; deschú, *Lepcha,* J. 104; dewa, *Assam,* A. 1511.

Dh Dhagerako, *Nepal,* W. 106 ; dhaoli, *Michi,* P. 1246 ; dheugr, *Nepal,* D. 424 ; dhúna, *Assam,* C. 273 ; dhúp, *Nepal,* P. 760 ; dhúpi, *Nepal,* J. 92; dhúpri, *Nepal,* J. 92.

Di Digilati, *Michi,* L. 481 ; diglotti, *Assam,* L. 481 ; dimmuk, *Ladak,* S. 2352 ; dimok, *Bhutia,* A. 1411 ; dingan, *Khasia,* S. 940 ; dingdah, *Khasia,* B. 926 ; dingim, *Khasia,* Q. 26 ; dingjing, *Khasia,* Q. 18, Q. 81 ; dingkain, *Khasia,* R. 323; dingkurlong, *Assam,* M. 412 ; dinglaba, *Khasia,* E. 210 ; dinglatterdop, *Khasia,* C. 1173 ; dingleen, *Khasia,* B. 496 ; dingori, *Assam,* D. 889 ; dingpingwai (Lindera), *Khasia,* L. 375 ; dingpingwait (Machilus), *Khasia,* M. 22, S. 1911 ; dingri, *Rajbanshi,* A. 577 ; dingrit-tiang, *Assam,* Q. 77 ; dingsa, *Khasia,* P. 757 ; dingsableh (Podocarpus), *Khasia,* P. 1006 ; dingsableh (Taxus), *Khasia,* T. 93 ; dingsaot, *Khasia,* C. 818; dingsning, *Khasia,* Q. 52 ; dingsolir, *Khasia,* M. 869 ; dingsong, *Khasia,* E. 340.

Do Doba-kari, *Michi,* C. 1931 ; dogal, *Gáro,* S. 869 ; doh-ni-ko-nah, *Assam,* F. 558 ; doika, *Rajbanshi,* C. 123 ; domhyem, *Bhutia,* H. 304; dopatti, *Assam,* C. 1183 ; doron, *Assam,* F. 10 ; dosal, *Lepcha,* T. 431 ; dosúl, *Nepal,* C. 1711 ; dowari (Ehretia), *Nepal,* E. 38 ; dowari (Luculia), *Nepal,* L. 552 ; dowkipoma, *Assam,* A. 868 ; doyá, *Assam,* F. 546.

Dr Dron, *Assam,* L. 323.

Du Dudcory, *Assam,* H. 294 ; dúdela, *Nepal,* H. 52 ; dudhali, *Michi,* H. 294 ; dudhkuri, *Michi,* H. 294 ; dudíka, *Nepal,* F. 168 ; dudri, *Nepal,* M. 21 ; dullicham, *Cachar,* C. 244 ; dumri, *Nepal,* F. 179 ; dumshing, *Bhutia,* A. 22 ; dumsi, *Nepal,* P. 1171 ; dunkotah, *Nepal,* D. 115 ; dupatti (Cinnamomum), *Michi,* C. 1165 ; dupatti (Phœbe), *Michi,* P. 546 ; dúr, *Lepcha,* D. 842 ; dursul, *Nepal,* S. 2341.

Dz Dzatsutt, *Ladak,* U. 59 ; dzu, *Angam Naga,* C. 1732.

Ec Ech, *Ladak,* T. 426 ; echalat, *Khasia,* M. 435.

Ek Ekuhea, *Assam,* A. 577.

En Enhu, *Naga,* R. 564.

Er Eri, *Assam,* R. 369, S. 2077.

Es Escalu, *Nepal,* R. 590.

Et Etok, *Bhutia, Lepcha,* R. 253 ; etok-amal, *Lepcha,* R. 276.

Ez Ez, *Tibet,* F. 765.

Fa Falleto, *Nepal,* E. 354 ; famsikol, *Lepchu,* E. 409.

Fi Filing, *Nepal,* P. 1342.

Fl Flotungchoug, *Lepcha,* E. 563, 567.

Fu Fukum, *Nepal,* I. 348 ; fúlinazur, *Gáro,* S. 3054 ; fullidha, *Nepal,* E. 340, 354, 356 ; fussi, *Nepal,* G. 663.

Ga Gad-gud-di, *Assam,* F. 508 ; gaik, *Magh,* P. 1389 ; galáp machú, *Manipur,* C. 637 ; galdam, *Tibet,* P. 1285 ; galeni (Leea), *Nepal,* L. 237 ; galeni (Sambucus), *Nepal,* S. 767 ; galeni (Loepa), *Nepal,* S. 1917; gallah, *Cachar,* C. 88 ; gallah, *Cachar,* C. 115 ; gamari, *Nepal,* T. 525 ; gambari, *Nepal,* G. 287 ; gampé aselu, *Nepal,* R. 601 ; gande, *Nepal,* D. 115 ; gangai, *Assam,* M. 71 ; ganné, *Nepal,* V. 94 ; garjo, *Nepal,* T. 470 ; garjo, *Nepal,* T. 483 ; garodosal, *Michi,* D. 87 ; garum, *Nepal,* T. 525 ; gash-alú, *Assam,* M. 217 ; gaya, *Magh,* P. 1343.

Ge Gebokanak, *Lepcha,* D. 397 ; geio, *Nepal,* B. 863, 868 ; gendeli poma (Garuga), *Assam,* G. 143 ; gendelli poma (Dysoxylum), *Assam,* D. 886.

Gh Ghariám, *Assam,* M. 147 ; ghesi, *Nepal,* Q. 70 ; ghonás, *Nepal,* R. 253.

Gi Gia, *Michi,* G. 143 ; giam, *Tibet,* C. 846 ; giatsah, *Bhutia,* A. 962 ; gineri, *Nepal,* P. 1233, 1239.

Gn Gniet, *Lepcha,* P. 760 ; gniú, *Ladaki,* C. 1003 ; gnow, *Tibet,* S. 1269.

Go Goa, *Tibet,* S. 1252 ; gobia, *Nepal,* C. 925 ; gobria, *Nepal,* E. 3; gobr iasulah, *Nepal,* A. 22 ; goechassi, *Nepal,* S. 940 ; goehlo, *Nepal,* C. 123 ; gogen, *Neoal,* S. 896 ; gogul dhúp, *Nepal,* C. 273 ; gohora, *Assam,* P. 1246 ; gók, *Bhutia,* M. 535 ; gokpas, *Bhutia,* A. 779 ; gola bet, *Assam,* C. 88 ; gomari, *Assam,* G. 287 ; gonjo, *Nepal,* M. 549 ; gonyúch, *Ladaki,* L. 280 ; gopi, *Nepal,* C. 925 ; gor, *Assam,* R. 237 ; goreah, *Assam,* F. 329 ; goria alú, *Assam,* I. 348 ; gorideomunga, *Assam,* S. 1934 ; gorissa, *Assam,* F. 516 ; govorpongyota, *Assam,* D. 889 ; gouribet, *Nepal,* C. 98.

Gr Greem (Calamus), *Lepcha,* C. 94 ; grím (Hordeum), *Ladaki,* H. 375.

Gs Gser, *Tibet,* G. 317.

Gu Gugera, *Assam,* S. 940 ; gúhor, *Nepal,* P. 1469; gulsima, *Nepal,* I. 14 ; gumai, *Cachar,* G. 287 ; gumbengfong, *Michi,* P. 950; gumbong, *Magh,* D. 328 ; gumi, *Gáro,* A. 640 ; gundroi, *Cachar,* C. 1155 ; gundrow, *Michi,* R. 1 ; gung, *Magh,* A. 471 ; gunhi, *Magh,* A. 709 ; gunserai, *Michi, Assam,* C. 1155 ; gúnsi, *Nepal,* P. 1006 ; gurás, *Nepal,* R. 253, 259 ; gurjo, *Lepcha,* T. 470 ; gurjo, *Lepcha,* T. 470; gurm, *Bhutia,* M. 53 ; gurol, *Assam,* R. 1 ; gurrie, *Assam,* F. 5166.

Gw Gwyheli, *Nepal,* P. 1246.

Gy Gy, *Tibet,* T. 429; gya, *Nepal,* G. 263 ; gya, *Bhutia,* S. 1264; gyesa, *Lepcha,* E. 340 ; gyong, *Lepcha,* S. 3073.

Ha Ha-al, *Assam,* F. 515; hais, *Nepal,* C. 425 ; hakpo-chanko, *Tibet,* F. 756 ; haktapatia, *Assam,* C. 900; halloray, *Nepal,* O. 38; hardi, *Nepal, Lepcha,* M. 652 ; harrari, *Nepal,* A. 233 ; harray baer, *Nepal,* Z. 273 ; harré, *Nepal,* T. 24; haswa, *Nepal,* N. 17; hatchanda,

Assam, S. 2850; hatmuli, *Assam*, A. 1577; hattipaila (Eulophia), *Nepal*, E. 467; hattipaila (Pterospermum), *Nepal*, P. 1389; hazína, *Garo*, S. 2245.

He Heibúng, *Manipuri*, G. 82; hemalú-alúá, *Assam*, M. 217; hendol, *Assam*, B. 180; henduri poma, *Assam*, C. 838; her-bag-gi, *Assam*, F. 531; herpa, *Bhutia*, G. 213; herro, *Nepal*, T. 325.

Hi Hiho, *Assam*, W. 58; hikpi, *Lepcha*, I. 141; hila auwal, *Cachar*, V. 174; hilika, *Assam*, T. 349; hilikha, *Assam*, T. 325; hílkat, *Assam*, C. 244; himal cheri (Antidesma), *Nepal*, A. 1212; himal cheri (Embelia), *Nepal*, E. 199; hingori (Castanopsis), *Assam*, C. 815; hingori (Quercus), *Cachar*, Q. 52; hingua, *Nepal*, C. 239.

Hl Hlo sa, *Lepcha*, P. 1316; hlosiri, *Lepcha*, Q. 63; hlosungli, *Lepcha*, B. 496; hlotagbret, *Lepcha*, V. 202; hlot kúng, *Lepcha*, P. 1316; hlyamban, *Magh*, C. 711.

Hn Hneingpyoing, *Magh*, O. 38.

Ho Hodung, *Ladaki*, P. 1153; hol, *Ladaki*, M. 334; hollock, *Assam*, T. 353; hollong, *Assam*, D. 692; hona, *Cachar*, H. 154; hortaki, *Cachar*, T. 349; horú surat, *Assam*, G. 213; hotung, *Ladaki*, P. 1153; howka, *Assam*, C. 115.

Hr Hrupruk-ban, *Magh*, T. 525.

Hu Huara, *Kachar*, L. 474; hulluch, *Assam*, T. 293; húmwah, *Michi*, M. 800; hundi bet, *Assam*, C. 81; hunich, *Nepal*, F. 790; hurchu, *Nepal*, V. 149, V. 152; hurrea kon, *Naga*, T. 434.

Hy Hyan, *Tibet*, S. 1269.

Ig Iger, *Tibet*, F. 812.

Ik Ikar, *Tibet*, T. 439; i'ker, *Tibet*, F. 770; iktibi, *Lepcha*, S. 1029.

In Indrajao, *Nepal*, H. 294.

Is Iskin, *Tibet*, S. 1242; isos, *Tibet*, S. 1283.

Ja Jadwár-khatái, *Ladaki*, A. 428; jaggarú, *Assam*, M. 71; jagguchal, *Nepal*, P. 702; jagrikat, *Nepal*, M. 22; jála, *Ladaki*, P. 1197; jali, *Cachar*, C. 114, C. 115; jalla bet, *Assam*, C. 114; jam, *Cachar*, S. 940; jamnemunda, *Nepal*, B. 463; jamu, *Nepal*, S. 1939; jamu (Eugenia), *Assam*, E. 419; jamu (Ficus), *Nepal*, F. 253; jámun, *Nepal*, E. 442; jarat, *Assam*, B. 523; jarika, *Nepal*, P. 21; jarila, *Nepal*, E. 48; jarul, *Assam*, L. 42; jarúl-jhalna, *Cachar*, D. 842; jatámángsi, *Nepal*, N. 17; jatamansi, *Bhutia*, N. 17; jati bengani, *Assam*, S. 2284; jati-koroi, *Assam*, A. 711.

Je Jegachu, *Gáro*, M. 147.

Jh Jhan, *Rajbanshi*, T. 361; jharal, *Nepal*, S. 1260; jhenok, *Lepcha*, C. 425; jhingni, *Nepal*, E. 565; jhinli, *Assam*, B. 165; jhunok, *Lepcha*, P. 316; jhuri, *Nepal*, O. 525.

Ji Jía, *Assam*, C. 838; jibang, *Magh*, Z. 231; jigini, *Assam*, T. 522; jinari, *Cachar*, P. 1006; jiushing, *Bhutia*, B. 134.

Jo Jogi, *Nepal*, M. 9; jogi mallata, *Nepal*, M. 9, 66; joki, *Cachar*, B. 520, jolan-

dhar, *Assam*, B. 523; jonua, *Magh*, F. 156; jora tenga, *Assam*, C. 1234; jori, *Assam*, S. 1739.

Ju Jung-song, *Lepcha*, E. 458; jupong, *Assam*, T. 522; juripakri, *Assam*, F. 145; jutili, *Assam*, A. 892.

Ka Kabashi, *Nepal*, A. 331, 334, 351; kabra (Capparis), *Ladaki*, *Tibet*, C. 431; kabra (Ficus), *Nepal*, F. 145; kabsing, *Assam*, J. 61; kabúl, *Ladaki*, P. 1159; kachik, *Lepcha*, B. 308; kaching, *Lepcha*, Q. 81; kadam, *Nepal*, J. 41; kadmero, *Nepal*, L. 474; kadu, *Nepal*, G. 761; kadungbi, *Lepcha*, G. 1374; kaghuti (Daphne), *Nepal*, D. 115, *Bhutia*, D. 119; kaghuti (Edgeworthia), *Nepal*, E. 15; kagi, *Nepal*, D. 834; kagiri, *Khasia*, F. 165, I. 62; kago, *Nepal*, F. 790; kagphulai, *Nepal*, R. 301; kah-nípotiak, *Assam*, F. 356; kaibú, *Nepal*, L. 96; kaidai, *Lepcha*, S. 3062; kaiday, *Michi*, S. 3062; kainggo, *Magh*, M. 489; kainjal, *Nepal*, B. 520; kaisho, *Assam*, B. 863; kajengla, *Manipuri*, T. 434; kajo, *Magh*, T. 325; kajutalam, *Lepcha*, R. 598; kakiral, *Assam*, M. 626; kakral, *Assam*, M. 626; kala aselu, *Nepal*, R. 598; kala bogoti, *Nepal*, B. 4; kala champ, *Nepal*, M. 515; kala kharani, *Nepal*, S. 3070; kala kiamoni, *Nepal*, O. 153; kala salajit, *Nepal*, P. 436; kala siris, *Nepal*, A. 722; kalay, *Nepal*, N. 212; kalay bogoti, *Nepal*, E. 202; kalchang, *West Tibet*, S. 560; kaledzo, *Lepcha*, C. 2242; kálí kutki, *Bomb.*, P. 700; kalet, *Lepcha*, E. 38; kalhenyok (Hedyotis), *Lepcha*, H. 68; kalhenyok (Oldenlandia), *Lepcha*, O. 137; kali guras, *Nepal*, R. 276; kali-katha, *Nepal*, M. 919; kalilara (Combretum), *Nepal*, C. 1742; kali lara (Marsdenia), *Nepal*, M. 302; kalligangdin, *Lepcha*, F. 804; kalma, *Bhutia*, R. 271; kalman, *Assam*, I. 382; kalzang, *Ladaki*, C. 1043, C. 1047; kamaung, *Magh*, L. 42; kamboong, *Magh*, E. 48; kamhyem, *Lepcha*, D. 200; kamli, *Nepal*, B. 572, 606.; kamli mallata, *Nepal*, M. 87; kanai, *Michi*, F. 156; kanaizu, *Magh*, B. 4; kancha, *Lepcha*, V. 94; kandiapiu, *Tibet*, F. 753; kangi, *Nepal*, W. 33, W. 38; kangji, *Lepcha*, F. 129, 216; kangshior, *Lepcha*, P. 702; kangu, *Nepal*, F. 693; kanhil, *Lepcha*, L. 55; kanhya, *Nepal*, F. 156; kanhlyem, *Lepcha*, S. 2819, S. 2861; kaninchi, *Tibet*, P. 746; kannuchi, *Tibet*, P. 746; kanom, *Lepcha*, T. 293; kanshin, *Tibet*, A. 328; kanta singar, *Assam*, C. 818; kánthál, *Assam*, I. 58; kantjer, *Lepcha*, A. 1212, 1215; kanú, *Lepcha*, E. 575; kanukoa, *Nepal*, E. 575; kanyoung, *Magh*, D. 701; kanyu, *Lepcha*, B. 426; karam, *Nepal*, A. 514; karamkanda, *Nepal*, O. 233; karamm, *Ladaki*, D. 815; kaıaput, *Nepal*, I. 14; kardai, *Assam*, A. 1646; karin, *Ladaki*, H. 45; karingi (Holarrhena), *Nepal*, H. 294; karingi (Wrightia), *Nepal*, W. 131; karsiár, *Bhutia*, D. 219; kashi, *Gáro*, B. 868; kashi-endúng, *Lepcha*, Q. 18; kashiorón, *Lepcha*, C. 812; kashyem, *Lepcha*, R. 590;

kasí, *Naga*, C. 1686 ; kasmir, *Khasia*, F. 165, l. 62; kasrekan, *Nepal*, F. 260; kasru, *Nepal*, Q. 70; kasser, *Bhutia*, P. 555 ; kasúr, *Lepcha*, S. 896 ; kasuri, *Nepal*, E. 485; kat, *Assam*, A. 135; kata kuchi, *Michi*, B. 868; kátál, *Assam*, A. 1166; kat-bíj, *Nepal*, H. 294; katior, *Lepcha*, S. 2816; kat phula, *Assam*, A. 590, F. 725 ; kath, *Nepal*, A. 484; kathálu, *Assam*, D. 494; katheik, *Magh*, E. 342 ; kathia-nyal, *Nepal*, F. 786; katiang, *Lepcha*, E. 356; katkura, *Assam*, S. 1934; katol, *Assam*, l. 58; katong, *Lepcha*, W. 9; katongzu, *Lepcha*, D. 884; kattra, *Assam*, B. 304; katúr, *Lepcha*, M. 209; katús, *Nepal*, Q. 18; kau-dú-lí, *Assam*, F. 511 ; kaula (Ilex), *Nepal*, l. 14; kaula (Phœbe), *Lepcha*, P. 546; kaúll, *Ladaki*, P. 1159; kaviang, *Assam*, S. 3062; kawala, *Nepal*, M. 22; S. 1911; kay-oungwa, *Magh*, M. 425 ; kazu, *Lepcha*, C. 213.

Kc Kchaitun, *Assam*, M. 652.

Kd Kdung, *Lepcha*, C. 1380.

Ke Kechu, *Naga*, D. 789; kechung, *Lepcha*, R. 268; kedi, *Naga*, R. 235; kégu, *Bhutia*, R. 271; keh-va, *Limbu*, T. 437; kei, *Manipuri*, T. 437; kekhi, *Naga*, T. 434; keli-kadam, *Assam*, A. 514; kema, *Lepcha*, R. 268; kemma, *Nepal*, E. 462; kému, *Bhutia*, R. 259; kendu, *Assam*, D. 582; kenia, *Naga*, P. 413 ; kering, *Assam*, O. 233; kerkalú, *Nepal*, l. 348; kesa prúm, *Gáro*, V. 3; kesari, *Nepal*, L. 100; ke-sí, *Naga*, C. 1686 ; kesseru, *Assam*, H. 154.

Kh Khansing, *Nepal*, A. 334; kharani, *Nepal*, S. 3076; khar-laguna, *Nepal*, D. 216; kharsani, *Nepal*, P. 1456; kharwa, *Nepal*, F. 202; kháwe, *Ladaki*, L. 28; kheu, *Manipuri*, M. 352; khíchar, *Ladaki*, L. 594; khili, *Gáro*, A. 717; khítsar, *Ladaki*, L. 594; khlap, *Assam*, C. 244; khoidai, *Lepcha*, S. 3062; khoira, *Assam*, A. 135; khoja, *Assam*, C. 123; kholaruis, *Nepal*, H. 339; khoskadumar, *Assam*, F. 202; khour, *Nepal*, A. 229; khreik, *Magh*, H. 317; khudi, *Naga*, T. 437; khum, *Manipur*, S. 2930; khuma, *Manipur*, S. 2930; khur, *Lepcha*, A. 1330; khwairalo, *Nepal*, B. 308.

Ki Kiamoni, *Nepal*, E. 458; kienki, *Lepcha*, M. 260; kihur, *Assam*, B. 873; kili, *Gáro*, A. 717; kilok, *Bhutia*, A. 331 ; kimbu, *Nepal*, M. 756; kinkoit, *Cachar*, B. 139; kirma, *Nepal*, V. 131; kironli, *Nepal*, C. 152; kirra (Wrightia), *Nepal*, W. 131; kírra (Holarrhena), *Nepal*, H. 294; kirth, *Angami Naga*, C. 1732; kisi, *Nepal*, E. 567; kissi, *Nepal*, C. 239; kítserma, *Ladaki*, L. 594; kiyi thekera tenga, *Assam*, G. 82.

Kl Kla, *Khasia*, T. 437.

Kn Knáthál, *Assam*, A. 1489.

Ko Kobusi, *Nepal*, M. 869; kochan, *Assam*, D. 842; kodo, *Michi*, C. 123; kodum, *Michi*, A. 1192; kohi, *Assam*, B. 873 ; koh-lí-hona, *Assam*, F. 506; koir, *Assam*, A. 135; kokan, *Assam*, D. 842; kokullak, *Ladak*, T. 548; kongki (Prunus), *Lepcha*, P. 1333; kongki (Semecarpus), *Lepcha*, S. 1041; konikath, *Nepal*, C. 631; koni-únchi, *Tibet*, P. 746; konkra, *Nepal*, L.

30 ; korang, *Assam*, F. 362 ; kori, *Assam*, A. 717; kor-jam, *Michi*, E. 419; koroi, *Assam*, A. 717; koroi, *Cachar*, A. 711 ; kosse gokpa, *Ladaki*, A. 786; koto (Bambusa), *Assam*, B. 139; koto (Pinus), *Nepal*, P. 760; kotur, *Nepal*, C. 818; kouatch, *Nepal*, M. 786; kowal (Alnus), *Lepcha*, A. 797; kowal (Juglans), *Lepcha*, J. 61 ; kozo, *Michi*, C. 123.

Kr Kraipang, *Magh*, C. 81; krapchi, *Michi*, F. 228, l. 71 ; krawru, *Magh*, V. 174; krim, *Lepcha*, T. 3; kring, *Magh*, C. 114; krowai, *Magh*, C. 1165.

Ku Kúail, *Nepal*, T. 522; kubindé (Eriolæna), *Nepal*, E. 321 ; kubindé (Kydia), *Nepal*, K. 42 ; kudhati, *Naga*, C. 1686 ; kudhia thia, *Naga*, C. 1686 ; kuhi, *Assam*, *Manipur*, Q. 18; kuhilá, *Assam*, A. 560 ; kujitekra, *Assam*, C. 474; kulain, *Bhutia*, H. 482; kulumsag, *Nepal*, l. 343; kulyatzo, *Lepcha*, S. 877; kumby-úng, *Lepcha*, A. 1221 ; kúmkoi, *Chakma*, A. 519; kúmsúm, *Lepcha*, C. 900; kunda, *Naga*, R. 235; kundoung, *Lepcha*, F. 260; kung, *Bhutia*, T. 432, T. 441 ; kung-chen-rik, *Lepcha*, V. 184; kunhip, *Lepcha*, F. 145; kunkirkola, *Assam*, A. 577; kúnsúng, *Lepcha*, G. 673; kup, *Lepcha*, C. 6; kurkum, *Bhutia*, C. 2083; kurlinga, *Nepal*, R. 271 ; kurmang, *Michi*, B. 356; kurong, *Nepal*, T. 525 ; kuruchipat, *Assam*, L. 343; kúshú, *Ladaki*, P. 1463; kutki, *Nepal*, P. 700; kutkuri konkuri mung, *Assam*, S. 1934.

Kw Kweda, *Naga*, R. 235; kwei, *Tibet*, A. 1523.

Ky Kyingbi, *Lepcha*, P. 1218; kyinki, *Lepcha*, M. 260.

La Labshi, *Nepal*, C. 838; lachú, *Ladaki*, R. 215; ládara, *Ladaki*, D. 245; laffa, *Assam*, M. 125; lah-bo-e, *Assam*, F. 576; lahokúng, *Lepcha*, B. 944; laider, *Michi*, P. 1389; laigongron, *Michi*, T. 24; laikeza, *Michi*, M. 12 ; laiphauzeh, *Michi*, A. 453; lakoú, *Nepal*, W. 84; la-lawa, *Tibet*, D. 227; lalchamp, *Nepal*, M. 49; laldairo, *Nepal*, W. 106; lal gurás, *Nepal*, R. 253; lali (Machilus), *Nepal*, M. 22; lali (Phœbe), *Lepcha*, P. 546; lal kabashi, *Nepal*, A. 336; lal kainjal, *Nepal*, S. 830; lal koi-púra, *Assam*, S. 806; lal mallata, *Nepal*, M. 12; lal totilla, *Nepal*, H. 154; lamchitia, *Nepal*, T. 432; lampatia, *Nepal*, D. 842 ; lamshing, *Bhutia*, P. 737; langura, *Bhutia*, C. 1992; laokri, *Michi*, A. 717; lapaing, *Magh*, B. 632; lapi, *Nepal*, U. 2 ; lapongdong, *Khasia*, S. 3062; lapshi, *Nepal*, M. 412; larborna, *Assam*, P. 1091 ; larnong, *Assam*, M. 704; lassím, *Assam*, F. 411, 463; lasuni, *Nepal*, I. 19; lat, *Lepcha*, C. 91; latecku, *Assam*, B. 4; late, mahwa, *Nepal*, A. 640; latuwani, *Assam*, A. 51.

Le Leingang, *Manipur*, L. 249; lesu, *Nepal*, F. 165, l. 62; letbarwa, *Bhutia*, D. 119.

Lh Lhála, *Bhutia*, *Lepcha*, II. 281.

Li Li, *Lepcha*, P. 1466; lijai, *Assam*, C. 115; likh-aru, *Nepal*, P. 1316; likúng, *Lepcha*, D. 721; lilima, *Nepal*, B. 873; lipiah, *Nepal*, V. 134; lipic, *Nepal*, V. 134.

Lo Lobura, *Assam*, F. 343, 357; lokriah, *Nepal*, F. 803; longsoma, *Magh*, F. 27; lood,

Cachar, S. 1889; lota amari, *Assam*, A. 988.

Lu Lúdúma, *Bhutia*, D. 206; luk, *Tibet*, S. 1320; lukunah, *Michi*, C. 1380; lutki, *Cachar*, S. 1888; lut-ter, *Nepal*, A. 1479.

Ma Mab-ja, *Bhutia*, P. 350; machu-gan, *Gáro*, M. 71; madaewah, *Magh*, B. 142; madala, *Michi*, P. 1426; madar, *Cachar*, E. 342; madhu fulong, *Assam*, P. 1101; madhuriam, *Assam*, P. 1343; mad ling, *Lepcha*, A. 440; madmandi, *Gáro*, *Cachar*, H. 294; madu suleng, *Assam*, P. 1101; magkal, *Tibet*, P. 1142; maha, *Nepal Tarai*, D. 238; máhal, *Tibet*, P. 1142; mahlbans, *Nepal*, B. 134; mahlí, *Assam*, F. 466; mahlu, *Lepcha*, B. 134; mahow, *Michi*, K. 42; mahsír, *Assam*, F. 357; mahua, *Nepal*, E. 210; maidal, *Nepal*, R. 1, 16; maidal-lara, *Nepal*, P. 950; mail, *Tibet*, R. 152; maimati, *Assam*, V. 195; maina-kat, *Nepal*, T. 372; mainakat-lara, *Nepal*, P. 950; maiulok, *Magh*, F. 202; majathi, *Assam*, R. 564; majetti, *Assam*, R. 564; makai, *Assam*, S. 1647; mak-anchi, *Gáro*, C. 123; makriah, *Assam*, S. 940; maku, *Nepal*, C. 818; makusal, *Assam*, S. 940; malchang, *Tibet*, S. 560; maldit, *Lepcha*, G. 143; malet, *Lepcha*, C. 710; maling, *Nepal*, A. 1532; mallata, *Nepal*, M. 9; malligiri, *Nepal*, C. 1155; mal-sampra, *Nepal*, F. 784; mandal, *Gáro*, E. 356; mandania, *Nepal*, A. 440; man-nola, *Gáro*, G. 99; mani-muni, *Assam*, H. 486; maniphtyol, *Lepcha*, D. 332; mánkachú, *Assam*, A. 809; mantet, *Lepcha*, B. 875; márgut, *Lepcha*, T. 847; marisgiri, *Nepal*, C. 1155; marlea, *Assam*, M. 289; marliza, *Assam*, M. 289; mar-marati, *Assam*, V. 195; maspati, *Nepal*, A. 51; masur moha, *Assam*, L. 252; máte-kissi, *Nepal*, B. 453; matí álu, *Assam*, D. 494; matsá, *Gáro*, T. 437; matti-kalaie, *Assam*, P. 468; ina-ú, *Magh*, A. 1192; mayhell, *Nepal*, P. 1469.

Me Mealum-ma, *Lepcha*, L. 79; mehul, *Nepal*, D. 721; meinkara, *Nepal*, T. 489; mekrap, *Lepcha*, M. 756; mekuri, *Naga*, L. 30; messguch, *Assam*, A. 709; methuna, *Manipur*, O. 565; mezankuri, *Assam*, S. 1868.

Mi Michapgong, *Lepcha*, P. 1239; mihul, *Nepal*, E. 283; mikrum-rik, *Lepcha*, V. 210; milkisse, *Nepal*, B. 463; mimri, *Nepal*, W. 33; mipit-múk, *Lepcha*; F. 633; misa phlap, *Assam*, C. 244; mishmí títá, *Assam*, C. 1793; misi, *Cachar*, T. 437.

Mo Moah, *Assam*, F. 315; modula, *Assam*, M. 12; moh-do-ní-konah, *Assam*, F. 531; mongthel, *Lepcha*, H. 339; mong-yung, *Lepcha*, P. 350; mon kyourik, *Lepcha*, D. 243; moqchini, *Nepal*, B. 796; morhal, *Assam*, V. 40; moringi, *Nepal*, L. 79; morolia, *Assam*, S. 2789; morrh, *Naga*, T. 434; moshungon, *Michi*, K. 42; mou-ah, *Assam*, F. 326; mowa, *Nepal*, E. 210; mowhitta, *Assam*, D. 328; moyen, *Assam*, V. 22; moyúm, *Manipur*, N. 564, R. 578.

Mr Mroung-shisha, *Magh*, G. 143.

Mu Mú, *Nepal*, P. 496; múga, *Assam*, S. 2100; mugasong, *Michi*, L. 474; múhuriam, *Assam*, P. 1343; mukial, *Assam*, B. 134;

mula sinki, *Nepal*, R. 31; munasi, *Nepal*, I. 14; mung (Coix), *Manipur* C. 1686; múng (Phaseolus), *Nepal*, P. 496; múnga, *Assam*, S. 2100; murtenga, *Assam*, B. 941; musa, *Nepal*, G. 263; muslindi, *Nepal*, A. 4531; musré katús, *Nepal*, C. 818.

My Mya, *Nepal*, E. 283; myooma, *Bhutia*, A. 1532; myoosay, *Bhutia*, T. 379.

Na Na, *Ladak*, S. 1273; nabat, *Assam*, F. 328; nagagola bet, *Assam*, C. 79; naga tenga, *Naga*, C. 1234; nagdana, *Cachar*, S. 1888; nagpat, *Nepal*, T. 847; naharu, *Assam*, A. 779; nahor, *Assam*, M. 489; nahshor, *Michi*, M. 489; nai illi, *Nepal*, B. 297; nakhthan, *Ladak*, V. 108; nakouli, *Bhutia*, V. 94; nakshan (Faba), *Ladaki*, F. 1; nákshan (Vicia), *Ladak*, V. 108; nalshuna, *Nepal*, E. 20; namby-ong, *Lepcha*, M. 756; namlang, *Khasia*, A. 1531; nanda, *Rajbanshi*, B. 868; nao gnao, *Bhutia*, S. 1273; napay, *Magh*, A. 1146; napoo, *Manipur*, F. 115; narockpa, *Lepcha*, C. 273; nas, *Ladaki*, H. 375; *Bhutia*, H. 382; naskí, *Nepal*, C. 425; naswa, *Nepal*, N. 17; natkanta, *Nepal*, P. 316.

Ne Nepura, *Assam*, F. 467; nervati, *Nepal*, S. 1273; neverra, *Nepal*, F. 235; lnewar, *Nepal*, E. 485; newar maharangi, *Nepal*, O. 170; newarpati, *Nepal*, B. 929.

Ng Nga-ain, *Magh*, F. 516, nga-ain-di, *Magh*, F. 517; nga-congrí, *Magh*, F. 575; nga-gyi, *Magh*, F. 568; nga-khú, *Magh*, F. 412; nga-kyí, *Magh*; F. 468; nga-pay-ing-ki, *Magh*, F. 534, nga-pri, *Magh*, F. 317; nga-pus-súnd, *Magh*, F. 545; nga-rui, *Magh*, F. 580; nga-saki, *Magh*, F. 457; ngraem (Acacia), *Lepcha*, A. 233; ngraem (Albizzia), *Lepcha*, A. 709.

Nh Nhare, *Tibet*, T. 93.

Ní Niar, *Tibet*, S. 1269; niargal, *Ladaki*, O. 613; niasbo, *Nepal*, N. 64; nila-thokar, *Bhutia*, C. 2361; nilay, *Nepal*, G. 685; nilo bikh, *Nepal*, D. 253; nimat, *Lepcha*, C. 1931; nirmali, *Nepal*, S. 2943; nit biralu, *Nepal*, T. 441.

No Nomorchi, *Lepcha*, D. 206.

Nt Ntan, *Assam*, M. 652.

Nu Nuan, *Tibet*, S. 1269; number, *Lepcha*, O. 516; numbon, *Lepcha*, G. 287; numbong, *Lepcha*, M. 66; mumboongkor, *Lepcha*, M. 69; 71; numing rik, *Lepcha*, R. 613; numro, *Lepcha*, M. 9; nuin, *Assam*, M. 756; nuni ajhar, *Gáro*, P. 546; nupsor (Cinnamomum), *Lepcha*, C. 1165; nupsor (Lindera), *Lepcha*, L. 375.

Ny Nyan, *Ladak*, S. 1269; nyand, *Tibet*, S. 1269; nyang, *Tibet*, S. 1269; nyanmo, *Ladak*, S. 1269.

Od Odela, *Assam*, S. 2861; odla, *Assam*, S. 2850.

Oe Ó-é, *Limbu*, P. 1172.

Oh Oh, *Lepcha*, W. 6.

Ok Okhioungza, *Magh*, A. 988; okhyang, *Magh*, A. 988; okoi, *Assam*, D 438.

Om Omak, *Assam*, S. 2861; omlu, *Nepal*, R. 319, 325; omra, *Nepal*, S. 2819.

Op Opang, *Assam*, T. 522.

Or Orer, *Nepal*, R. 369; orjun, *Assam*, T. 282.

Os Osai, *Assam*, V. 174.

Ot Otengah, *Assam*, D. 428.
Ou Oulia champ, *Nepal*, M. 517.
Oz Oz-o-la, *Assam*, F. 360.

Pa Pád, *Ladaki*, B. 501; padam-chal, *Nepal*, R. 215; padebiri, *Nepal*, P. 4; pahar lampati, *Nepal*, S. 877; pah-boh, *Assam*, F. 368; paieli, *Nepal*, L. 470, paighambari, *Tibet*, H. 377; pakar, *Nepal*, F. 265; pakhori, *Assam*, C. 115; pakri, *Assam* F. 265; palamkat, *Nepal*, C. 474; palashu, *Michi*, B. 944; palási, *Nepal*, B. 944; palé, *Lepcha*, A. 336; palegnyok, *Lepcha*, A. 349; palet, *Lepcha*, M. 289; pal-kaitaik-rau, *Michi*, F. 156; palo, *Nepal*, L. 569; palok *Lepcha*, O. 522; pálti, *Bhutia*, F. 10; palu, *Bhutia*, R. 251; palyok, *Lepcha*, S. 3062; pama, *Tibet*, J. 104; pampe, *Bhutia*, N. 17; panas, *Assam*, I. 58; panchidung, *Gáro*, E. 409; panchu, *Gáro*, B. 632, pandúr, *Lepcha*, A. 1192; pangra, *Nepal*, E. 219; pani, *Magh*, A. 1479, I. 57; panisaj, *Nepal*, T. 353; panji, *Lepcha*, R. 1; pankakro, *Gáro*, T. 24; panpui, *Gáro*, D. 428; pantom, *Lepcha*, L. 237; pao, *Lepcha*, D. 281; paphok, *Lepcha*, P. 1342; param, *Michi*, T. 522; parari, *Nepal*, S. 2865, S. 2876; pareya-auwal, *Cachar*, S. 2865, pariam, *Lepcha*, R. 53; pa-shing, *Bhutia*, D. 281; pash-kouli, *Rajbanshi*, D. 438; passi (Pyrus), *Nepal*, P. 1466; passy (Docynia), *Nepal*, D. 721; patagari, *Bhutia*, M. 49; pátási, *Assam*, F. 546; patharua, *Assam*, P. 1091; pat-hioo, *Nepal*, A. 1532; patichanda, *Assam*, C. 1165; patimil, *Nepal*, A. 1215; pating, *Bhutia*, P. 1285; patlé, *Nepal*, Q. 18; patle katús, *Nepal*, Q. 52; patmaro, *Nepal*, C. 1969; 2242; patni-alu, *Assam*, D. 494; pattikashmiri, *Tibet*, R. 261; patú-swa, *Nepal*, P. 1102; paumaia, *Nepal*, P. 486; paumpe, *Bhutia*, N. 17; pawaing, *Magh*, M. 9; payir, *Lepcha*, A. 233; payomko, *Lepcha*, T. 372; payong (Betula), *Nepal*, B. 496; payong (Cephal-ostachyum), *Lepcha*, C. 925.
Pe Peluk, *Nepal*, F. 790; pendder, *Bhutia*, M. 49; pendre, *Lepcha*, M. 535; peng, *Cachvr*, C. 2574; pengji, *Lepcha*, B. 868; penma, *Ladaki*, P. 1201; pershuajelah, *Michi*, G. 673.
Ph Phaco singali, *Nepal*, Q. 81; phadupjoh, *Michi*, M. 40; phakram, *Assam*, T. 522; phaku, *Assam*, A. 31; phalat, *Nepal*, Q. 23; phamlet (Litsæa), *Lepcha*, L. 470; phamlet (Machilus), *Lepcha*, M. 22; S. 1911; phamsikol, *Lepcha*, D. 428; phani, *Lepcha*, M. 21; pharat-singhali, *Nepal*, Q. 47; phatak, *Ladaki*, B. 501; phegran, *Gáro*, F. 118; phekori bet, *Nepal*, C. 94; phirpbiri, *Nepal*, S. 2819; phise karin, *Ladaki*, H. 45; phlap, *Assam*, C. 244; phoberkúng, *Lepcha*, E. 419; phulamphi, *Nepal*, A. 1620; phulsappa, *Assam*, M. 541; phusi, *Nepal*, L. 30; phuspat, *Nepal*, B. 501; phuspat, *Bhutia*, *Nepal*, B. 501; phusri, *Nepal*, L. 470; phusri mallata, *Nepal*, M. 87.
Pi Piampiya, *Bhutia*, F. 794; piaz-ay, *Bhutia*, P. 702; pich kati, *Assam*, S. 2299; píndik, *Lepcha*, C. 1742; pínjúng, *Ladaki*, P. 1201; pipla mol, *Nepal*, P. 805; pipli (Bucklandia), *Nepal*, B. 926; pipli (Ficus), *Nepal*, F. 236; piriengo, *Nepal*, O. 132;

pithogarkah, *Assam*, C. 1043; pitho-garkh,; *Assam*, C. 1050; piyás, *Assam*, A. 769.
Po Poah, *Nepal*, B. 576; poguntig, *Lepcha*, B. 28; poh, *Tibet*, J. 101; pohor, *Lepcha*, A. 453; pokoh, *Magh*, A. 722; poksha, *Michi*, F. 202; pokuttia, *Nepal*, H. 482; poma, *Assam*, C. 838; pondám, *Lepcha*, B. 929; popal pipal, *Nepal*, P. 805; por-bong, *Lepcha*, L. 498.
Pr Prab, *Gáro*, F. 118, 216, 265; prangos, *Ladaki*, P. 1221; praong, *Lepcha*, A. 1528; pronchadik, *Lepcha*, H. 304; prong, *Lepcha*, A. 1528; prongnok, *Lepcha*, A. 1523; prongzam (Heptapleurum), *Lepcha*, H. 126; prongzam (Macropanax), *Lepcha*, M. 30; prongzam (Pittosporum), *Lepcha*, P. 912; prústi, *Lepcha*, L. 549.
Pu Púalay, *Nepal*, L. 470; puddum, *Assam*, M. 71; pudlikat, *Nepal*, S. 830; pugila, *Nepal*, A. 341; pummoon, *Lepcha*, A. 1532; pungcha, *Ladak*, I. 93; pungchu, *Ladak*, T. 93; pungmar, *Lepcha*, T. 432; puraj-blakut, *Nepal*, F. 797; purbo, *Lepcha*, A. 871; purbong, *Lepcha*, C. 2039; purisingbatti, *Nepal*, B. 931; púrkar, *Ladak*, T. 82; purmiok, *Lepcha*, T. 379; purmo, *Lepcha*, M. 40; puroa, *Lepcha*, M. 71; purphiok, *Lepcha*, P. 1342; purpuray timur, *Nepal*, Z. 26; purúni, *Nepal*, D. 203; púteli, *Nepal*, L. 490; púti-keintah, *Assam*, F. 334; put-linga, *Nepal*, R. 276; puya, *Nepal*, M. 260.
Py Pya-shing, *Bhutia*, B. 931.

Qu Quabi bet, *Assam*, C. 96.

Ra Rabáb tenga, *Assam*, C. 1234; rabi bet, *Nepal*, C. 79; rá-ché, *Ladak*, S. 1239; ragao, *Tibet*, S. 1252; ragdi, *Naga*, T. 437; ra-giyu, *Bhutia*, S. 1247; raidana, *Assam*, C. 115; rajahmas, *Assam*, F. 576; rajbirij, *Nepal*, C. 756; raj-briksha, *Nepal*, C. 756; raklop, *Lepcha*, R. 369; ramani, *Magh*, G. 287; rámpát, *Assam*, S. 2930; ramphal, *Nepal*, D. 428; ramtoroi, *Nepal*, L. 556; ran-doh-nt, *Assam*, F. 328; ranga (Areca), *Assam*, A. 1330; ranga (Ipomæa), *Assam*, I. 348; rangi, *Assam*, C. 115; rangirata, *Cachar*, D. 884; ranigovindhi, *Nepal*, D. 417; raniwalai, *Nepal*, R. 323; ranket, *Gáro*, F. 129; I. 62; rá-phoché, *Ladak*, S. 1239; rásha, *Ladak*, T. 548; rasúk, *Ladak*, T. 87; rasull, *Tibet*, H. 377; rata-gogen, *Lepcha*, S. 900; ratmanti, *Nepal*, L. 474; ratwa, *Nepal*, D. 219; rauket, *Gára*, F. 165; ráwa-che, *Ladak*, S. 1239.
Re Reem, *Lepcha*, C. 79; reipom, *Manipur*, B. 523; rek, *Tibet*, F. 778; re-mó, *Naga*, C. 2263.
Rg Rgeeta, *Tibet*, T. 70; rgelta, *Ladak*, T. 61.
Rh Rha, *Lepcha*, B. 356; rhea, *Assam*, B. 576.
Ri Rib-jo, *Ladaki*, D. 227; rigong, *Tibet*, F. 778; rik, *Lepcha*, A. 233; riphin, *Assam*, C. 115; risan, *Assam*, C. 115.
Rj Rjao, *Ladaki*, F. 19.
Ro Robhay, *Bhutia*, R. 351; rodinga, *Nepal*, E. 340; roghu (Adina), *Assam*, A. 514; roghu (Anthocephalus), *Assam*, A. 1192; rohu, *Lepcha*, C. 1155; ronchiling, *Lepcha*, S. 2649; rong, *Lepcha*, C. 111.

Ru Rue, *Lepcha,* C. 98 ; ruebee, *Lepcha,* C. 94 ; ruglim, *Lepcha,* C. 854 ; rui, *Assam,* F. 474 ; rukh baer, *Nepal,* Z. 273 ; rúm, *Assam,* S. 2930 ; rumgach, *Assam,* E. 210 ; rungbong, *Lepcha,* C. 711 ; rung-yeong rik, *Lepcha,* B. 471 ; runúl, *Lepcha,* P. 956 ; rupá, *Bhutia,* A. 1359 ; ruppa, *Bhutia,* A. 1359 ; rusa, *Naga,* T. 434 ; rusta, *Bhutia,* S. 910 ; ruzerap, *Michi,* D. 41.

Ry Rynoí, *Khasia,* R. 564 ; ryom, *Lepcha,* M. 302.

Sa Sa, *Naga,* T. 437 ; saar, *Lepcha,* L. 82 ; sachak, *Tibet,* T. 439 ; sacheng, *Magh,* T. 293 ; sadul kou, *Khasia,* M. 435 ; safed champ, *Nepal,* M. 535 ; safed kabra, *Nepal,* F. 216 ; safed mallata, *Nepal,* M. 69, 71 ; safhyi, *Lepcha,* P. 1444 ; sah (Larix), *Lepcha,* L. 82 ; sáh (snow leopard), *Tibet,* T. 439 ; sahu hingori, *Assam,* Q. 81 ; saikamehhia, *Magh,* A. 1146 ; sain, *Magh,* C. 89 ; saiphra, *Magh,* C. 1021 ; saitu, *Magh,* B. 629 ; saiyar, *Nepal Tarai,* T. 444 ; sakalang, *Assam,* E. 59 ; sakin, *Tibet,* S. 1242 ; sakwa, *Nepal,* S. 1656 ; sal, *Tibet,* B. 731 ; sala, *Nepal,* P. 760 ; sallah, *Bhutia,* C. 487 ; salua, *Assam,* S. 2861 ; sam, *Assam,* A. 1479 ; 1. 57 ; sambeíng, *Magh,* S. 2861 ; samkoh, *Assam,* A. 451 ; sampat, *Nepal,* L. 481 ; sana-kadan, *Lepcha,* C. 89 ; sandanpipli, *Nepal,* O. 537 ; sangaipru, *Magh,* D. 834 ; sangji, *Lepcha,* F. 156 ; sangryn, *Magh,* D. 834 ; sanpau, *Gáro,* A. 1245 ; sanu arkaula, *Nepal,* Q. 6 ; sanujhingni, *Nepal,* E. 563 ; sanyepang, *Magh,* A. 1192 ; sapai, *Magh,* F. 260 ; saphijirik, *Lepcha,* T. 489 ; saphiong, *Lepcha,* T. 441 ; sapong, *Assam,* T. 522 ; sappa (Michelia), *Assam,* M. 541 ; sappa (Talauma), *Assam,* T. 26 ; sapro, *Naga,* S. 2930 ; sarrú, *Tibet,* C. 2358 ; sási, *Assam,* A. 1251 ; sasin, *Nepal,* S. 1226 ; saslendi, *Nepa,* A. 338 ; sat-bur, *Cachar,* F. 265 ; satchuk, *Lepcha,* T. 432 ; sathong, *Lepcha ;* T. 437 ; sathung, *Lepcha,* P. 1172 ; satian, *Assam,* D. 387 ; satiana, *Assam,* A. 871 ; sattni, *Cachar,* A. 871 ; sau, *Assam,* A. 722 ; sauer, *Nepal,* B. 496 ; saul-kuri, *Assam,* E. 71 ; saver, *Nepal,* B. 496 ; sawáli, *Ladaki,* P. 1197.

Sc Schap, *Lepcha,* P. 551 ; schiap, *Lepcha,* P. 586.

Se Sea, *Ladaki,* R. 543 ; sealposra, *Nepal,* G. 673 ; sedangtaglar, *Lepcha,* K. 42 ; sedeng, *Lepcha,* B. 615 ; seho, *Assam,* W. 58 ; sehshing, *Bhutia,* A. 17 ; seing laing, *Singphu,* M. 704 ; sejjiak, *Lepcha,* T. 434 ; selcho, *Gáro,* A. 722 ; selemnyok, *Lepcha,* W. 131 ; sel-konah, *Assam,* F. 396 ; sema-dung, *Lepcha,* A. 6 ; sempak, *Assam,* T. 522 ; semul, *Assam,* S. 1939 ; semur, *Assam,* S. 1939 ; senén, *Lepcha,* D. 203 ; sen-ní, *Assam,* F. 350 ; sentórí, *Assam,* F. 576 ; serai-guti, *Assam,* B. 426 ; serang, *Assam,* C. 812 ; serh, *Lepcha,* R. 301 ; serhnyok, *Lepcha,* R. 323 ; serpa, *Bhutia,* G. 213 ; sey barasi, *Magh,* C. 1021.

Sh Sha, *Ladak,* S. 1278 ; shag, *Ladaki,* B. 501 ; shák, *Ladaki,* B. 501 ; shakab, *Gáro,* F. 202 ; shakshín, *Tibet,* B. 496 ; shal, *Tibet,* B. 731 ; shalshi, *Nepal,* Q. 47 ; shama baringi, *Nepal,* P. 693 ; shamo, *Ladak,*

S. 1278 ; shangdong, *Gáro,* A. 514 ; shangti, *Tibet,* P. 746 ; shanku, *Ladaki,* D. 815 ; shánmá, *Ladaki,* P. 885 ; shapo, *Ladak,* S. 1278 ; shapti, *Michi,* M. 359 ; S. 1888 ; shechin (Caryopteris), *Nepal,* C. 710 ; shechin (Dædalacanthus), *Nepal,* D. 4 ; shechin (Phlogacanthus), *Nepal,* P. 544 ; shedbarwa, *Bhutia,* D. 119 ; *Nepal,* D. 122 ; shelangri, *Gáro,* V. 174 ; shembal, *Assam,* S. 1939 ; shempati, *Nepal,* H. 285 ; sheto seru, *Nepal,* P. 530 ; sheu-shong, *Gáro Hills,* E. 48 ; shibíku, *Chakma,* S. 3040 ; shida, *Gáro,* L. 55 ; shidu, *Michi,* E. 496 ; shingali, *Nepal,* C. 818 ; shingra, *Gáro,* Q. 52 ; shiní, *Assam,* F. 568 ; shítzem, *Lepcha,* M. 797 ; shomá, *Ladak,* S. 2352 ; shon, *Tibet,* D. 223 ; shosho, *Bhutia,* D. 119 ; shour, *Ladaki,* P. 1208 ; shrolo, *Ladak,* S. 1035 ; shruk, *Michi,* T. 634 ; shukni, *Lepcha,* D. 438 ; shúkpa, *Nepal, Tibet,* J. 92 ; shúrbúta, *Tibet,* J. 92 ; shúrgú, *Tibet,* J. 92 ; shrúzbed, *Chakma,* C. 838 ; shústí, *Ladak,* S. 996 ; shyin, *Tibet,* S. 1242.

Si Sia, *Ladaki,* R. 543 ; sibri, *Nepal,* B. 375 ; sichi, *Lepcha,* S. 1264 ; sida, *Assam,* *Michi,* L. 55 ; siffú, *Lepcha,* T. 24 ; sigu-grip, *Lepcha,* M. 535 ; sigumgrip, *Lepcha,* M. 49 ; sik, *Tibet,* T. 434 ; sikmar, *Bhutia,* T. 431 ; siku, *Lepcha,* D. 219 ; silika, *Assam,* T. 349 ; likka, *Assam,* T. 349 ; silim, *Lepcha,* T. 325 ; silim-kung, *Lepcha,* T. 325 ; silingi, *Nepal,* S. 3070 ; silingi, *Nepal,* O. 520 ; silingia, *Nepal,* F. 790 ; sil-karai, *Assam,* A. 709 ; silli, *Khasia,* C. 925 ; sill-kurta, *Cachar,* D. 387 ; 392 ; sillkurta, *Cachar,* D. 387 ; símal, *Lepcha,* C. 838 ; simali, *Nepal,* M. 797 ; simbrangrip, *Lepcha,* L. 552 ; simong, *Lepcha,* C. 711 ; simrú, *Bhutia,* L. 126 ; sinakadang, *Lepcha,* A. 640 ; sinduria, *Nepal,* M. 71 ; sin-durpong, *Michi,* M. 71 ; singen, *Lepcha,* S. 3062 ; singhani, *Nepal,* A. 1528 ; singhata (Heptapleurum), *Nepal,* H. 131 ; singhata (Schizandra), *Lepcha,* S. 947 ; singi, *Assam,* F. 568 ; singka, *Bhutia,* P. 1469 ; singliang, *Lepcha,* B. 926 ; singna, *Lepcha,* A. 1620 ; sing-namúk, *Bhutia,* D. 397 ; singnok (Turpinia), *Lepcha,* T. 847 ; singnok (Wendlandia), *Lepcha,* W. 38 ; singríang, *Lepcha,* A. 722 ; singtok, *Bhutia,* M. 756 ; sing-yan, *Bhutia,* S. 3062 ; singyen, *Lepcha,* S. 2865, 2876 ; sinjang, *Bhutia,* R. 611 ; sinjang-lho, *Bhutia,* R. 611 ; sinna, *Nepal,* B. 933 ; sinong, *Lepcha,* B. 520 ; sipha-rúng, *Lepcha,* S. 900 ; sirgúllum, *Assam,* M. 87 ; sirhútúngchir, *Lepcha,* S. 806 ; siri, *Lepcha,* Q. 23, 52 ; sirikishu, *Lepcha,* C. 815 ; siriokhtem, *Lepcha,* H. 154 ; siris, *Assam,* A. 711 ; sir-nat, *Assam,* L. 79 ; sirong, *Singpho,* T. 437 ; sirpang, *Michi,* S. 2865 ; sirshing, *Tibet,* E. 40 ; sirsing, *Tibet,* E. 40 ; sisi, *Nepal,* L. 375 ; sisoo (Albizzia), *Gáro,* A. 711 ; sissu (Dalbergia), *Assam,* D. 64 ; síta más, *Nepal,* P. 486 ; sitnyok, *Lepcha,* F. 253 ; sitto siris, *Nepal,* A. 717 ; sitto udal, *Nepal,* S. 2816, S. 2819 ; sítul,, *Assam* F. 510 ; situyok, *Lepcha,* S. 1939 ; sivor, *Lepcha,* S. 2341.

Sk Skepkyew, *Lepcha,* E. 59 ; skin, *Tibet,* S. 1242 ; skodze, *Ladak,* A. 786.

Sn Sna, *Ladak,* S. 1273.

So Soah, *Tibet,* H. 382 ; sodoi, *Magh,* F. 156 ;

sohmyrlain, *Assam*, P. 632; sohriu, *Assam*, C. 1686; soi-zong, *Rajbanshi*, O. 233; somso, *Bhutia*, C. 838; sonalu, *Gáro*, C. 756; son dah, *Assam*, F. 311; soondi, *Cachar*, C. 115; soora-goy, *Tibet*, O. 571; sop, *Nepal*, P. 727; sopho, *Khasia*, D. 721; soplong, *Khasia*, P. 1004; sóth, *Bhutia*, R. 564; so-tsa, *Naga*, C. 1686; sow, *Assam*, A. 722; sowka bent, *Assam*, C. 115.

Sp Spót, *Bhutia*, P. 811.

St Star-bu, *Tibet*, H. 277; stokpo tsodma, *Ladak*, U. 59.

Su Sua, *Nepal*, P. 544, sualu, *Assam*, L. 474; sufokji, *Lepcha*, R. 605; sugoria, *Nepal*, D. 216; sugrúmúk, *Lepcha*, C. 135; sug-vat, *Lepcha*, C. 720; suhging, *Lepcha*, S. 1247; suilá, *Assam*, C. 2105; súkha sag, *Assam*, R. 650; suku, *Lepcha*, D. 219; sula, *Nepal*, P. 760; sullea, *Khasia*, C. 925; súm, *Assam*, S. 1911 (súm Litsæa), *Assam*, L. 481; súm (Machilus), *Assam*, M. 22; sumbling, *Lepcha*, B. 4; sumbrong, *Lepcha*, S. 940; sumcher, *Lepcha*, P. 544; sunakhari, *Nepal*, F. 27; sunaru, *Assam*, C. 756; sundari, *Assam*, F. 576; sundi-bet, *Assam*, C. 96; sundók, *Lepcha*, P. 338; súnga, *Lepcha*, C. 123; sung cha, *Ladak*, T. 93; súngen, *Lepcha*, S. 3062; sungli, *Lepcha*, B. 514; sungloch, *Lepcha*, T. 353; sunglú, *Lepcha*, B. 632; sunglyer, *Lepcha*, C. 152; sungna, *Lepcha*, P. 1246; sungnai, *Manipuri*, D. 239; sungrai, *Manipuri*, D. 239; sungribong, *Lepcha*, P. 1148; sungrú-kúng, *Lepcha*, Z. 9; sungsúm, *Lepcha*, C. 900; sungungrik, *Lepcha*, B. 342; sunkong-kúng, *Lepcha*, F. 223; sunkrong, *Lepcha*, L. 79; sunom, *Lepcha*, E. 442; E. 462; súnt, *Nepal*, Z. 201; suntala, *Nepal*, C. 1233; suntolah, *Nepal*, C. 1234; suntong (Brassaiopis), *Lepcha*, B. 796; suntong (Heptapleurum), *Lepcha*, H. 128; suntri, *Nepal*, C. 423; suom, *Lepcha*, P. 632; suphut, *Lepcha*, L. 474; suppatnyok, *Lepcha*, L. 483; súrah vyu, *Tibet*, C. 2358; súrú-phamsah, *Bhutia*, C. 448; suviak, *Lepcha*, E. 210.

Sw Swa, *Nepal*, A. 797.

Sy Syiak, *Lepcha*, T. 434; syik, *Lepcha*, T. 434.

Ta Tabongdeing, *Magh*, A. 451; tág, *Tibet*, T. 437; tagashing, *Bhutia*, J. 61; taggú, *Nepal*, R. 253; tailadu, *Michi*, C. 2039; tailainyom, *Lepcha*, L. 498; tailo, *Cachar*, C. 812; taisoh, *Michi*, B. 520; takali chaku, *Nepal*, S. 30; takbli-rik, *Lepcha*, V. 195; takbret (Bœhmeria), *Lepcha*, B. 574; takbret (Villebrunea), *Lepcha*, V. 131; takcha-brik, *Lepcha*, C. 2171; takh-ril, *Lepcha*, R. 318; taki, *Nepal*, B. 356; taki-bet, *Nepal*, P. 956; takmur, *Lepcha*, A. 717; takosu, *Nepal*, W. 6; takpa, *Bhutia, Ladaki*, B. 501; takpo, *Lepcha*, P. 1322; takpœdrik, *Lepcha*, P. 4; tak-sielrik, *Lepcha*, S. 947; taksor, *Lepcha*, T. 361; taksot, *Lepcha*, F.168; 202; taktokhyem, *Lepcha*, E 219; tama (Dendrocalamus), *Nepal*, D. 101; tama (Caragana), *Ladaki*, C. 471; tamákú hulas, *Tibet*, R. 261; tambul, *Assam*, A. 1294; tamomban, *Magh*, M. 40; tangarúk, *Lepcha*, A. 986, 988; tangshing, *Bhutia*, A. 6; tanprengjan, *Magh*, A. 1245; tapathyer, *Lepcha*, A. 1620; tapch, *Gáro*,

B. 426; tapria-siris, *Nepal*, A. 709; tarsing, *Nepal*, B. 426; tashiari, *Nepal*, D. 200; tatebiri, *Nepal*, D. 28, 87; tatri, *Nepal*, D. 438; tatwen, *Ladaki*, A. 1462.

Tc Tcheiray gulab, *Nepal*, T. 93; tcheiray sulah, *Nepal*, T. 93; tchenden, *Bhutia*, C. 2352; tchokpo, *Lepcha*, J. 101; tchongtay, *Lepcha*, F. 179; tchuka, *Lepcha*, R. 223.

Te Teadong (Pinus), *Bhutia*, P. 760; tedong (Albizzia), *Lepcha*, A. 711; teeshæ, *Naga*, A. 484; tekhu, *Naga*, T. 437; tekhu khuia, *Naga*, T. 434; tekreng, *Gáro*, C. 273; telta, *Tibet*, T. 70; teotosa, *Nepal*, R. 261; tepor, *Assam*, G. 99; teprong, *Gáro*, A. 1489; tesma, *Bhutia*, I. 423; teteli, *Assam*, T. 28; tetrobrik, *Lepcha*, S. 2508; teturl, *Lepcha*, S. 1656; teturldum, *Lepcha*, I. 509; tezpur, *Assam*, G. 99.

Th Thabyepauk, *Burm.*, E. 464; thainba, *Magh*, D. 393; thaing, *Magh*, A. 514, 519; thaka, *Michi*, S. 1029; thakal, *Nepal*, C. 2534; thali, *Nepal*, T. 847; thalikabashi, *Nepal*, A. 338; thamaga, *Assam*, M. 393; thapru, *Magh*, D. 428; thar, *Nepal*, S. 1264; thau, *Magh*, L. 500; thekera, *Assam*, I. 511; thekri, *Magh*, A. 1146; thingánisúla, *Nepal*, A. 6; thingia, *Nepal*, A. 6; thstoriyá, *Assam*, B. 84.

Ti Tibiliti, *Nepal*, P. 912; tid bhagnri, *Assam*, S. 2280; tigroht, *Michi*, S. 806; tíhur, *Assam*, G. 99; tíl, *Nepal*, S. 1078; tilki, *Nepal*, W. 33; timburnyok, *Lepcha*, S. 2223; timil, *Nepal*, M. 289; timmue, *Nepal*, P. 801; tiinur, *Nepal*, Z. 32; ting-ga-rah, *Assam*, F. 492; tingschi, *Bhutia*, T. 93; tingúrisalla, *Nepal*, A. 6; tissi, *Nepal*, S. 544, 1906; títá, *Assam*, C. 1789; titapat, *Nepal*, A. 1469, S. 1888; titaphapur, *Nepal*, *Lepcha*, F. 10; tita phul, *Assam*, P. 544; titasappa, *Assam*, M. 517; titinigala, *Nepal*, A. 1523; titri, *Nepal*, T. 28.

To Tokár, *Michi*, T. 634; toko pat, *Assam*, L. 498; tokra, *Magh* V. 159; toldúng, *Lepcha*, A. 200; tolrik, *Lepcha*, A. 267; tolu aselu, *Nepal*, R. 590; tomár, *Michi*, T. 634; tongrong, *Gáro*, S. 2649; tongschí, *Bhutia*, P. 737; tónnyok, *Lepcha*, D. 87; topatnyok, *Lepcha*, D. 4; toponi, *Magh*, A. 1479; tosa, *Nepal*, H. 382; totilla, *Nepal*, O. 233; totnye, *Nepal*, P. 1102; tourah, *Assam*, F. 561.

Tr Tráma, *Ladaki*, C. 471; trans, *Ladaki*, E. 234; tráo, *Ladaki*, F. 19; tro, *Michi*, T. 634.

Ts Tsaingtsa, *Magh*, S. 2865; tsallu, *Bhutia*, *Tibet*, R. 280; tsaluma, *Bhutia*, R. 278; tsapatt, *Ladaki*, E. 234; tsápo, *Tibet*, R. 152; tsaratpang, *Magh*, M. 147; tsashing, *Bhutia*, T. 93; tse, *Ladaki*, E. 234; tseikpoban, *Magh*, T. 372; tsema, *Bhutia*, B. 443; tsítkado, *Magh*, C. 838; tsogde, *Tibet*, T. 426; tsuma, *Bhutia*, R. 278; tsus, *Tibet*, S. 1283.

Tu Tú, *Nepal*, S. 30; tugla, *Lepcha*, T. 522; tugom-kúng, *Lepcha*, M. 44; tuhasi, *Nepal*, F. 693; tuk (*Bhot.*), *Bhutia*, T. 437; túk-kung, *Lepcha*, G. 761; tukla, *Lepcha*, M. 71; tuknu, *Nepal*, P. 1102; tuksur, *Lepcha*, B. 608; túkt, *Bhutia*, T. 437; túla, *Assam*, S. 2806; tulacmyom, *Lepcha*, L. 498; tuma,

Khasia, B. 133; tumberh, *Lepcha*, M. 855; tumbomri, *Lepcha*, H. 540; tumbrúng, *Lepcha*, N. 212; tumbúk, *Lepcha*, C. 1966; tún, *Assam*, C. 838; tungbram, *Lepcha*, S. 1888; tungbram, *Lepcha*, M. 359; tungcheong, *Lepcha*, M. 919; tungcher, *Lepcha*, A. 1221; tungchong, *Lepcha*, S. 3070; tungchung, *Lepcha*, E. 565; tungd, *Assam*, C. 838; tungnyok, *Lepcha*, A. 338; tungplam, *Lepcha*, T. 525; túngpung, *Magh*, G. 761; tungrung, *Lepcha*, O. 518; tungrútrikup, *Lepcha*, V. 217; tungsing, *Bhutia*, A. 6; túni, *Nepal*, C. 838; tunka, *Michi*, M. 359; tupi, *Nepal*, J. 104; tu-rah, *Assam*, F. 494; turgu-wah, *Magh*, B. 130; turmong, *Magh*, C. 123; tussar, *Assam*, S. 1934; tuttealy, *Assam*, E. 71.

Tz   Tzedze, *Ladaki*, P. 63; tzu-dza, *Naga*, G. 263.

Uc   Uchkai, *Gáro*, D. 438.
Ud   Udal, *Cachar*, S. 2861; udare, *Gáro*, S. 2861; udis, *Nepal*, A. 797.
Uh   Úh, *Lepcha*, W. 6.
Uk   Uk, *Nepal*, S. 30; ukieng, *Michi*, S. 1029; ukotang, *Assam*, D. 284.
Ul   Ullo, *Nepal*, G. 213.
Um   Úmbú, *Ladaki*, C. 681.
Un   Ungnai, *Magh*, S. 2912.
Ur   Uri, *Assam*, D. 789; uriam, *Assam*, B. 520; urohi, *Assam*, D. 789; urohi mahorpat, *Assam*, V. 116; urshi, *Assam*, D. 789; urúm, *Michi*, B. 520; urva, *Nepal*, F. 823.
Us   Usken, *Khasia*, B. 137; uso komphor, *Khasia*, C. 1234; uso mongar, *Khasia*, C. 1234; uso santra, *Khasia*, C. 1234; uso sim, *Khasia*, C. 1234; uso yanpriang, *Khasia*, C. 1234; ussey, *Assam*,

D. 284; ussú, *Bhutia*, C. 1954.
Ut   Utis (Alnus), *Nepal* A. 797; utis (Betula), *Nepal*, B. 496.

Va   Vakru, *Gáro*, E. 210.
Vh   Vhyem, *Lepcha*, R. 564.

Wa   Wah (Dendrocalamus), *Michi*, I D. 281; wah (red caf bear), *Tibet*, F. 751; wahghi, *Gáro*, B. 142; wah-kanteh, *Gáro*, B. 118; wahnok, *Gáro*, D. 281; wamer, *Nepal*, F. 813; wamoo, *Nepal*, F. 817; wampu litsi. *Lepcha*, P. 1456; wa-nah, *Magh*, B. 118; wasma, *Bhutia*, I. 145.

Ya   Yak, *Tibet*, O. 571; yalishin, *Bhutia*, A. 334; yamaghí khá, *Ladak*, T. 87; yang, *Lepcha*, C. 6; yarpa, *Tibet*, P. 1142; *Ladaki*, P. 1159; yatli, *Lepcha*, A. 331.
Ye   Yel, *Lepcha*, B. 212; yelnyo, *Lepcha*, E. 283; yel pote, *Lepcha*, B. 212; yet ghás, *Bhutia*, P. 11; yewcron, *Nepal*, F. 800.
Yi   Yi, *Lepcha*, C. 1388..
Yo   Yok, *Lepcha*, F. 165, I. 62; yokchounrik, *Lepcha*, C. 1038, l. 60; yokdúng, *Lepcha*, F. 118.
Yu   Yúlatt, *Ladaki*, P. 1159; yúmbok, *Ladak*, U. 4.

Za   Zabra, *Ladaki*, H. 45; zaghún, *Ladaki*, C. 2263; zambrún, *Magh*, D. 438; zanda, *Ladaki*, D. 815; zangs, *Bhutia*, C. 2361; zatúd, *Ladak*, U. 59.
Ze   Zebri, *Magh*, E. 412, 419.
Zi   Zig, *Tibet*, T. 439; zik, *Bhutia*, F. 766; zik, *Limbu*, T. 432.
Zu   Zú, *Naga*, C. 960.

BURMESE.

Aá Aálu-paká, P. 1304.
Ai Aipmwaynway, E. 202.
Ak Akyau, A. 1251.
Al Álu-pakárá, P. 1304.
Am Ambeng, A. 955; amé-mnnniyén-zi, B. 812.
An Anan, F. 23; ananbo, C. 2240; a-nan-pho, S. 940; anitia, R. 231.
Ap Aphiyu-munniyé-zi, B. 800.
At At-ká-let, C. 1768.
Au Aukchinsa, D. 580; auza, A. 1166.
Av Ava-tazávon-le-pa, A. 838.
Aw Awza, A. 1166.
Ay Ay-kayet, M. 550.

Ba Bádan, P. 1274; baibya, C. 2055; baikyo, C. 1388; ba-la, E. 151; ballowa, G. 66; ba-lu-wa, H. 168; bambway, C. 563; bamw-baynee, P. 916; ban-bwai, C. 563; banbwe, C. 563; ban-kha, T. 293; bava-net, J. 116; bawtamaka, M. 363.
Be Bé, *Karen*, C. 1686; bebya, C. 1388; bek-shá, G. 213; be-ma, *Karen*, C. 1686; ben (Amomum), A. 976; ben (Cannabis), C. 331; bet-mœshau, H. 221; bet-than, D. 756; betya, G. 213.
Bh Bhain, P. 87; bhain-bin-amí, P. 82; bhain-zi, P. 87; bhálá, E. 151; bhan-bhwai, C. 563; bhénbin, C. 331; bhin-bin-amí, P. 82; bhudína, M. 447.
Bi Biluletwa, H. 131; bin, C. 331; bingah, S. 2796.
Bn Bnuméza, A. 722.
Bo Bokemaiza, K. 42; bongmaiza, R. 57; bon-kho-e, A. 89; boo-dee-na, M. 461; botha-o, *Karen*, T. 437.
Br Bringilobán, B. 777.
Bu Budda-tha-rá-na, C. 321; bujiphyú, C. 1380; búkyu, C. 1403; búmaiza, A. 722; bú-sin-swai, L. 30; búlayet, A. 531.
Bw Bwaycheng, B. 356; bwaygyin, B. 304; bwéchin, B. 304, B. 356.
By Bya, C. 474; byaing-che-piu, G. 302; byaitsin, A. 1218; byu, R. 242; byú, D. 443; byúben, D. 424; byubo, B. 898; byuma, R. 242.

Ca Cabal-márá-gass, A. 722.
Ch Chán, O. 258; charatte, T. 548; chaung-lélu, O. 127; chayan-ka-you, A. 988; che, S. 1041; cheik, C. 1686; china cham-pac, P. 989; chinbaung, H. 233; chin-douk-nway-zouk, V. 213; chin-poung-ni, H. 233; chinyok, G. 143; chinyop, G. 143; chonoo, C. 2166; chop-pin, D. 628; cho-sa, X. 1; chuvondacodu-vallie, P. 979; chyai-beng, S. 1041.
Cl Clay ben, S. 1041.

Da Daintha, M. 721; dan (Lawsonia), L. 126; dán (iron), I. 440; danoung, C. 66; danbin, L. 126; dan-da-let, I. 39; dándalonbin, M. 721; da-ne, N. 163; dang-we, C. 797; dan-gywe, C. 797; dank, D. 87; danón, C.

66; danthalone, M. 721; danyin, P. 906; danyinnie, M. 549; daukyama, T. 847; daung-sat-pya, C. 123; daungsop, C. 32; dawéhmaing, Q. 88.
Di Didu, B. 632.
Do Douk-loung, D. 59; douk-ta-louk, D. 51; doung-sap-pya, C. 123; douthá, C. 2367.
Du Duyin, D. 876.
Dw Dwabók, *Shan*, K. 42; dwabote, K. 42; dwani, E. 312.
Dy Dyauthaukyeng, C. 115.

Ei Eing-hmyoung, R. 110.
Ek Ekarit, M. 322.
En Eng, D. 696; engkyeng, C. 115; eng-kyn, S. 1672; en-khyen, S. 1656.

Ga Gan-gau, M. 489; gán nyin, E. 219.
Ge Gee, D. 219.
Gi Gínsi-khiáv, Z. 201; gin sín, Z. 201; giyonsabá, T. 634.
Gn Gnan-pok, R. 127; gnathiet, C. 1722; gna yoke, C. 466; gna yoke-nopmyan, C. 466; gnoo-kyee, C. 756; gnooshway, C. 756; gnung-myit, C. 1053; gnu-theing, C. 777.
Go Gonnyin, E. 219; goun, B. 576.
Gu Gung, M. 570.
Gw Gway tankpin, D. 823; gwe, S. 2649; gwedoak, C. 1774.
Gy Gyengmaope, A. 1288; gyin-ywe, A. 51; gyo, S. 950; gyopho, W. 19; gyungsa-ba, T. 634; gyútbeng, D. 628; gyútnwé, G. 311.

He Heboo, C. 637; hen-ka-la, S. 2571; hen-tha-pada-yaing, M. 473.
Hi Hinkanoe-súba, A. 943; hinnoe-súbá, A. 943.
Hk Hkaw-kwa, C. 410.
Hl Hlyanpyoo, S. 2824.
Hm Hmaing, L. 576; hman, F. 53; hmanbyu, *Shan;* R. 16; hmanthin, C. 1177; hmyaseik, A. 1200.
Hn Hnabé, O. 38; hnan, S. 1078; hnan-bai; O. 38; hnan-lón-gyaing, A. 217; hnan-lón-kyaing, A. 101; hnaubeng, A. 514; hnaw, A. 514; hnen-ben, P. 1044; hnyet, H. 414.
Hp Hpalan, B. 318; hpan-kha, T. 361; hpa-young-ban, T. 410; hpekwoon, B. 477; hpet-woon, B. 474.
Hs Hsae-dan, O. 242; h'sang, L. 143; hsa-nwen, C. 2433; hsay-ma-kyí, V. 25; hsay-than-paya, R. 1; hsee-mee-touk, G. 243; hseik-ba-lu, N. 179; hsele, D. 115; hsen, E. 83; hseng neng thayet-M. 209; hsen-gno-myeet, E. 186; hsen-hmyau, S. 1000; hsen way, O. 1; hsen-youk, G. 143; hshú, C. 637.
Ht Htan, B. 663; htan-myouk-lu, L. 498; htat-ta-ya, C. 117; htein thay, S. 2799; hte-ka-yung, M. 557; htenrú, P. 771; htouk-sha, V. 177; htoukshar, V. 159; htso, *Shan*, T. 437.

Hu Huga-pyau, M. 811.
Hz Hzi-phyú, P. 632.

In In, D. 696; inbo, D. 689; ingyin, S. 1672.
Is Isa-pyit-ya, V. 243.

Jo Joe-boe, W. 13; joung-lai-lú, O. 127.

Ka Ka-aunggyl, C. 1380; ka away, C. 1179; kabaing, C. 972; kabeng, C. 2577; kabong, (*Mergui*), F. 772; ká-bunway, D. 499; kadapgnam, C. 271; ká-dat-ká-let, C. 1763; kadat-ngan, C. 271; kadet, C. 2039; kado, D. 228; kadon-kadet, C. 1774; kadot, F. 202; kadut, F. 202; kadwœ-oo, D. 507; kaing, P. 618; kaiyení, C. 2361; kajeng, R. 235; kakadit, F. 476; ka-ku-yan, F. 538; kala-khenboun, R. 650; kalamet, C. 1925; kalan, C. 780; kalapai, C. 1061; kala-saghia, S. 373; kala-tigiya, S. 373; ka la zaw, O. 193; kala-zoun-ya-si, A. 1644; kalein, C. 6; kalet, L. 241; kaletthein, L. 226; kaliendza, C. 6; kalithí; C. 1686; kal-lawhso, H. 466; ka-lwah, C. 943; kamáká, M. 363; ka-ma-ka, M. 393; kam-ba-la, S. 2369; kamon-yeki, S. 2960; kan, S. 2999; kanako, C. 2192; kanazo, B. 3, B. 4; kan-chop-ní, P. 979; kan-chop-phíjú, P. 986; kansan, B. 220; kánsó, B. 265; kanyinbyu, D. 676; kanyin-kok, D. 689; kanyin-ní, D. 701; kanyo-mi, A. 1577; kanzannu, B. 265; kaphet-theing, L. 226; káphi-si, C. 1641; kappalí, M. 603; ka-pwot, C. 1641; ka-pyaing, C. 972; ka-ra-mai, S. 790; katat, C. 2039; káteh, D. 510; katha, F. 538; katha-boung, F. 450; kathai, S. 749; ka-thay, S. 759; kathit, E. 342; kathitka, P. 390; ká-toopin, E. 80; kat-sae-nai, U. 29; kát-saynai, S. 1688; kaung-yan, H. 227; kaya, A. 324; ka-yam-my-pong, L. 596; kayan, C. 1763; kayau, E. 593; kayu, P. 961; kazwan, I. 348.

Ke Ke-hin-gá-bin, M. 626; ken-bung, C. 919; kenbwon, A. 200; ken-gau, M. 489; kengwa, C. 930; ken-khyoke-ní, P. 979; keo-khin, A. 897; kesú, R. 369; késu-gí, J. 41; kesúm-phiu, A. 779; kesun-ni, A. 769; keyán, S. 30; keyllowa, B. 130.

Kh Khaboung (Calamus), C. 115; kha-boung (Strychnos), S. 2943; kha-boung-yackyie, S. 2960; khai-ma, T. 460; khai-maphyu, T. 460; khai-ma-pok, L. 143; khai-pok, L. 143; kha-mung, K. 3; kha-oung gyí, C. 1380; khaya, M. 570; kha-yan, S. 2284; khejijk, L. 1; khi, *Karen*, T. 437; khiá-sí, C. 1211; khiáti, C. 1211; khi-sí, S. 1041; khoyari, H. 517; khuele, M. 786; khungiche, H. 92; khwele, M. 786; khyáa, A. 1351; khye, G. 252; khyee-paung, L. 549; khyen-seing, Z. 201.

Ki Kiahong, *Karen*, M. 352; kiché-phong, *Karen*, T. 434; kígisamyá-si, C. 1286; kimbo, C. 711; kimpalin, A. 1215; kinbun (Acacia), A. 200; kinbun (Desmodium), D. 341; kingalun, P. 17; kin-khen-ní, P. 979; kin-khen-phiú, P. 986; kin-pa-lin, A. 1221; kiva-lá-mon, A. 382; kiya-ni, N. 200; kiyásanoin, C. 2406; kiyási, C. 1211; kiyoubhán-bin, V. 181; kiyow-bhán-bin, V. 164; kiyubán-bin, V. 164.

Kl Kla-hla, *Karen*, T. 425.
Ko Koholen, B. 430; kokko, A. 695; konazota-lú, L. 541; kone-pyinma, L. 42; kongnyin-nway, E. 219; kouk-pin, X. 1; koung-ka-do, T. 444; koung-mhoo, S. 1676; koung-yan, H. 227; koyangali, A. 836; kóyángí, C. 2062.
Ks Kshauma, L. 385.
Ku Kujáv-pœn, G. 653; kujne, C. 797; kúkko, A. 695; kú-ku, S. 2252; kun, A. 1294; kune-lá-mon, A. 382; kung-nyen, E. 219; kun-lin-net, D. 819; kúnsi, A. 1294; kun-theí-bin, A. 1294; kúnyoe, P. 775; kusan, H. 517.
Kw Kwam-lin-nek, D. 819; kwam-thee-beng, A. 1294; kwán, P. 775; kwanynet, P. 775; kwaytanyeng, P. 900; kwey-yeng, (*Tavoy*), F. 538; kwon-rwet, P. 775.
Ky Kya, F. 752; kya, T. 437; kyabaing, C. 972; kya-bet-gyí, L. 232; kya-gyúk, T. 425; kya-hin-ka-lae-nway, I. 421; kyahphet-kyí, L. 232; kyah-phyú, N. 200; kyai-beng, B. 193; kyaigyee, B. 198; kyaitha, B. 180; kyakatwa, B. 118; kyalak, T. 434; kyán, S. 30; kyandoo, C. 154; kyan-hsen, R. 237; kyankchin, A. 897; kyantsa, C. 818; kya-nyu, N. 209; kyan-zah, T. 489; kyathaungwa, B. 138; kya-thit, T. 434; kya-thoungwa, B. 138; kyat-thou-bega, A. 779; kyat-thwe, C. 618; kyauk-pa-yon, B. 430; ky-a-vekhet, C. 1808; kyaw, R. 236; kyaw-shaw, R. 236; kyeik-phun, C. 1686; kyeingkha, C. 77; kyeing-ni, C. 68, C. 83; kyeit, C. 1686; kyeksu, R. 369; kyenbankyen, C. 66; kyengbot, B. 115; kyéni, B. 180; kyéni-kyíbeng, B. 180; kyenka, C. 77; kyétha, B. 192; kyethen-kha, M. 626; kye-thit, T. 425; kyetmauk, N. 68, N. 72; kyet-monk, C. 873; kyet-mouk, S. 950; kyet-poung-hpo, I. 75; ky-et-thwon-ni, A. 769; kyet-thwunbya, A. 779; kyetyo, V. 174; kyet-yob, V. 177; kyi, B. 198; kyjantsheng, R. 235; kyouk-nees ballamya, R. 619; kyouk-nya-lu, F. 465; kyouk-pan, C. 384; kyouk-seing, S. 1068; kyouktheyga-kakadit, F. 579; kyoung, C. 622; kyoung, (*Aracan*), T. 425; kyoungdouk, P. 17; kyoungmiku, B. 929; kyoungmyeng, T. 441; kyoung-myeng, T. 443; kyoung-sha, O. 233; kyoung tset-kun, (*Aracan*), T. 427; kyoungya-beng, O. 233; kyú, T. 349; kyum, S. 1939; kyún, T. 232; kyúnboc, G. 287; kyunnalin, P. 1248; kywae, S. 2649; kywai, O. 558, O. 574; kyway, D. 500; kyway-nway, D. 500; kyway-pin, D. 500, D. 537; kywégyi, B. 193; kywétanyin, P. 900; kywet-shov-byan, (*Aracan*), F. 799; kywon, T. 232; kywonpho, G. 287.

La Labri, R. 619; laieya, L. 62; lai-lu, O. 127; laipinkha, H. 506; lakhiya, B. 731; la-maing, W. 68; lambo, B. 913; lamboben, B. 913; lameb, A. 1655; la-móte, M. 145; lámyá-sí, C. 1258.
Le Leik-kyœ, R. 100; leik-pyen-won, R. 100; leinben, T. 310; lekkyouk temengnee, F. 555; lélu, O. 127; le-lun-pen, S. 830; leme, C. 115; len-kyan, C. 1158; lepán-bín, B. 632; let-khok, S. 2824; letkhya,

nga-thaing, F. 384; nga-tha-ley-doh, F. 477; nga-tha-louk, F. 414; nga-than-chyeik, F. 546; nga-thine-glay, F. 353; nga-towktú, (*Aracan*), F. 579; nga-towktú-shweydú, (*Aracan*), F. 578; ngaungyat, *Shan*, I. 71; nga-wet-sat, (*Aracan*), F. 452; nga-yan-dyne, F. 515; nga-yaw, F. 517; nga-yeh, F. 321, F. 322; nga-yin-pounsa, F. 472; nga-youn, F. 321, F. 322; nga-young, F. 320; nga-zen, F. 496; nga-zen-bya, F. 529; nga-zen-zap, F. 315; nga-zin-sap, F. 550; nga-zin-zat, F. 308; nga-zin-zine, F. 487, F. 492; ngetpyau, M. 811.

Nh Nhan-ben, A. 514; nhingpen, A. 514.
Ni Nie-pa-hsæ, M. 656.
No Noe-khiyu, G. 278; noe-khiyu-asui, G. 278; noye, A. 1359; noye-saku, A. 1359.
Nu Nunec, O. 565; nu-wa, G. 385.
Nw Nwa-mani-than-lyet, C. 416; nway-ka-zún-a-phyú, I. 368; nway-sat, S. 3042; nweka-zumbyí, I. 368.
Ny Nyaeh, S. 1166; nyah-gyi, M. 656; ny-an-gyee, R. 16; nyapaw, S. 1747; nyaung, F. 118; nyaungbawdi, F. 165, F. 236, I. 62; nyaungbyu, F. 265; nyaungchin, F. 216; nyaunggyat, *Shan*, F. 228; nyaung-gyat, F. 27; nyaungok, F. 253; nyaung-thabieh, F. 145; nya-yinboun-za, F. 398; nya-yoke-koung, P. 811; nyoungbaudi, F. 236; nyoungchin, F. 216; nyoungkyap, F. 27, F. 228, I. 71; nyoungop, S. 1939; nyoung-thabyeh, F. 253.

Oh Ohay, *Karen*, C. 255.
Ok Okshit, A. 534.
On Ŏn, C. 1520; ondi, C. 1520; ong, C. 1520; ong-ká-to, H. 253; ong-tong, L. 483; onsí, C. 1520; ontí, C. 1520.
Op Op-nai, S. 2912.
Ou Ouk-chingza, D. 580; ouk-chin-ya, D. 615; ouk thapha-ya, S. 855; oung mai phyú, C. 1403.

Pa Padá, M. 473; padagoji, A. 853; pa-daing-ame, D. 151; pa-daing-kyet-thwon, U. 39; padauk, P. 1363; padáyínkhatte, D. 151; padáyin-phiu, D. 160; pa-dung-ma, N. 39; pa-ga-gyis, A. 860; pa-gyay-theing, G. 86; pai (Dolichos), D. 789; pai (Phaseolus), P. 496; pai (Pisum), P. 885; pai ka lag, C. 289; paikhsan, C. 2105; paik piven, C. 2105; pai-len-mwae, T. 569; painouk, P. 496; pai noung ni, C. 1403; pai-pázoon, C. 2514; pai-si-gong, C. 49; pala (Amomum), A. 976; pala (Elettaria), E. 151; palan, B. 318; pálán toung-wœ, C. 2013; palawa, G. 86; pan, C. 2105; pangah, T. 325; pan-gan, T. 293; pánkhadé, T. 15; panlat, E. 151; pán-sá-yeik, I. 513; pánsáyeip, I. 515; panshít, I. 39; panta-ka, C. 154; panthitya, V. 40; panyaung, F. 129; parabaik, P. 313; parawa, G. 66; parouk, C. 257; pa-taing, C. 2070; patauk, P. 1367; paukbyu, A. 560; paukhya, S. 1186; paukpan (Æschy-nomene), A. 560; paukpan (Sesbania), S. 1186; páv, B. 944; payé, C. 1221; pay-ín-chong, C. 49; payo, C. 257; pa-yók, C. 257; paypa, *Shan*, S. 1264.

Pe Pebin, C. 1995; peik chin, P. 805; peik-khyen, P. 805; peingnai, A. 1489; peinné, I. 58; pembwaú, T. 15; pe-nán-ta-zi, T. 612; pen-bwa, M. 267; penglai-kanazo, H. 137; peng-lay-oang, C. 482; peng-le-ka-na-tso, H. 139; pengnyet, C. 146; pengtarow, G. 673; penlay-hsí, X. 12; penlaykathit, E. 342; penle-on-si, L. 511; pésigón, C. 49; petthan (Dolichandrone), D. 756; petthan (Heterophragma), H. 157; petwun, B. 474; pezinng-oun, P. 805.
Ph Phálá, E. 151; phangah, T. 293; phángá-si, T. 293; phánkhá-si, T. 293; phá-tan-phyu, F. 644; phá-yaíthi, C. 1221; phayonii, C. 931; H. 342; phetcwoon, C. 1879; phetyákyí, G. 213; pheytakyee, L. 79; pho-sáo, D. 522; phoungga, S. 4; phounniya, C. 146; phúm-masin, B. 542; phung-nyet, C. 146.
Pi Pido-sin, M. 800; pienné, A. 1489; pilaw, pinanwa, B. 133; pímbo-si, C. 581; pin, B. 944; pindo-sin, M. 800; pin-lai-ka-zum, I. 362; pin-lai-kú-yin, X. 12; pinlay-nga-ba-mah, F. 328; pinlayoung, C. 482; pinlaytsí, X. 12; pinlekanazo, H. 134, H. 137; pinletan, S. 927; pinle-thitkauk, G. 780; pinlézi, X. 12; pinlón, C. 482; pin pwa, P. 625; pintayan, G. 673; pintayo, G. 673; pirolai kyout, C. 1377; piyá-ye, H. 342; piyo, C. 257.
Po Poi noung ni, C. 289; pongnyet, C. 146; pong-pin, P. 682; ponmathein, B. 540; poonyet, P. 1118; poon-yet, D. 108; posa, M. 756; pouk (Butea), B. 944; pouk (Sesbania), S. 1166; poukgnwe, B. 978; pouknway, B. 978; pouk-nway, S. 2508.
Pu Puki, S. 1212; pulu pinan myouk, M. 216; pung-ben, P. 682; pung-ma-theing, B. 542; putchaw, D. 200; puve-kain-yoe, C. 737.
Pw Pwai-nget, P. 1118; pwá-sá-o, D. 522; pwenyet, P. 1118; pwenyet, D. 108; pwot-chaubeng, D. 200.
Py Pyaungbú, Z. 50; pyen-dan-gna-len, S. 1688; pyinkado (Xylia), X. 16; pyinma, L. 42; pyintagar-ne-thi, Q. 43; pyin-vaung, F. 129; pymmah, L. 50; pyn-kado (Afzelia), A. 580; pyoung, O. 567; pyoung, S. 2424; pyoung-lay-kouk, S. 1212; pyu, R. 242.

Qu Quindah, B. 220.

Ru Ruhíra, T. 293.

Sa Sa, S. 602; saba, O. 258; saba-gyee, R. 129; sabay, J. 35; sabi-si, V. 233; sa-byet, M. 639; sa-byit, V. 233; sac, N. 101; sacpín, N. 101; saga, M. 517; saghia-phiu, S. 375; sakauk, C. 30; sakyeik, C. 1686; sala-misri, S. 521; salé-bin, P. 1426; salé-sí, P. 1426; sa-lu, L. 343; sa-mong nway, M. 634; samon-né, N. 158; sa-mung-nee, L. 283; sa-mung-sa-ba, P. 727; samusaba, P. 727; sámyá-sí, C. 1296; samyeit, P. 460; san, O. 258; sanato-sí, G. 66; sandaku, P. 1381; san-da-ku, S. 790; sánkhí, I. 472; sanœ, C. 2433; sánpiyá, I. 472; san-ta-ku, S. 790; sapsha, B. 608; sap-sha-pen, T. 522; sa-pyih, V.

233 ; sapyit, M. 639 ; sa-tha-khiva, M. 791 ; sát jo yit, N. 8 ; sat sha yuet, M. 260 ; saunggya, A. 1646 ; sa yo mai, P. 811.

Sc Schap, D. 240.

Se Se, N. 101 ; sechaub, C. 331 ; seet, A. 717 ; sehoong, (*Aracan*), A. 1146 ; seikche, B. 868 ; seikgyí, B. 868 ; séjáv-bin, C. 331 ; sekhági, S. 3018 ; sen-bwon, D. 424 ; setahanbaya, G. 158 ; sethanbaya (Gardenia), G. 105 ; sethanbaya (Randia), R. 1.

Sh Sha, A. 135 ; shabju, P. 632 ; sha-ma-say-nway, S. 2794 ; shanh, E. 83 ; shank-tones, C. 1263 ; shán mai, I. 145 ; shap-sha, S. 869 ; shasaung, E. 520 ; sha-shoung-leknyo, E. 553 ; sha-soung, E. 520, E. 527 ; sha-soung-leknyo, D. 387 ; sha-soung-lekh-nyo, D. 387, E. 553 ; sha soung lit wa, O. 193 ; shasoung-pya-thal, E. 496 ; shatoo beng, B. 426 ; shauk ta kera, C. 1270 ; shawbyu, S. 2824 ; shawni, S. 2861 ; shazán-lese, V. 219 ; shazánv-ji, E. 496 ; shazaon-le-pá, A. 829 ; shazávn-le-pa, A. 838 ; shazávn-mina, E. 520 ; she-che, H. 317 ; shengna-roét, (*Aracan*), F. 430 ; sheu, (*Tenasserim*), F. 801 ; shíntan, P. 940 ; sh-on-si, C. 1233 ; shon-takhavá, C. 1270 ; shouh-ton-oh, C. 1263 ; shouk-ta-kwoh, C. 1270 ; shoungpang, C. 1233 ; sh-ousa khavá, C. 1270 ; showbju, S. 2824 ; shue, A. 1622 ; shue saka, A. 1622 ; shwae, G. 317 ; shway-nway-pin, C. 805 ; shwé-pay-on, C. 2316 ; shwéphyu, P. 940 ; shwoaygjo, A. 451.

Si Síhosayesi, A. 1014 ; sikiyabo, C. 1158 ; síma-don, G. 243 ; simbo-maizali, C. 732 ; simbo-si, C. 581 ; simbo-sikiyabo, C. 1196 ; simizu, P. 1121 ; símmi-dái, C. 243 ; sinban-karawa, P. 801 ; sin-gno-myet, E. 186 ; singo-moné, T. 470, T. 483 ; sing-ou-miá, A. 1107 ; sin-ka-de, S. 2273 ; sínkozí, C. 1181 ; sin-ma-no-pyin, B. 873 ; sinmir thayet, M. 209 ; sin-tha-hpan, F. 260 ; sinza-manne, T. 470 ; sinza-manne, T. 483 ; siskhyá-si, J. 61 ; sit, A. 717.

So Somblón-zi, P. 555 ; son, H. 414 ; soop-wotnway, A. 200.

Su Su, C. 637 ; subán, C. 637 ; subóknwé, A. 200 ; su-kwot-nwé, A. 200 ; sukyanbo, C. 42 ; sule-anén, T. 548 ; sule-gí, P. 363 ; sunguen, C. 1233 ; sunletthé, C. 26 ; supán, C. 637 ; súyit, A. 267.

Sw Swonpalwon, P. 555.

Ta Ta-bu, S. 2362 ; tachansa, H. 154 ; tagíyá-phíú, S. 375 ; tagúyi, C. 563 ; tagyet, P. 21 ; ta-hat, T. 268 ; taingthe, S. 2796 ; tainngiya, C. 35 ; tainniya, C. 35 ; tai-pen, D. 558 ; takayung, M. 557 ; takhva, C. 2278 ; takya-lai-wa, H. 131 ; talaguwa, G. 211 ; ta-lapí, O. 14 ; tala-ungnwi, D. 87 ; ta-le-té, C. 1768 ; talí-bin, P. 1426 ; taline-no-thee, P. 314 ; talí-sí, P. 1426 ; táma-bin, M. 363 ; ta-ma-ka, M. 393 ; tamakanwe, C. 1763 ; tamayoke, W. 38 ; támbiyá, I. 472 ; tambiyá-si, C. 1296 ; tám buyu-sí, C. 1258 ; ta-mu, S. 2362 ; tan, B. 663 ; tana, C. 1940 ; ta-nap, T. 268 ; tanato-así, G. 66 ; tanaung, A. 249 ; tankaet-

tva, U. 39 ; tánkhí, I. 472 ; tankma, D. 59 ; tan-kyet-thoon, A. 775 ; tansapai, I. 1 ; tanshouk, G. 271 ; tan-thie-den, M. 71 ; tanun, C. 2433 ; tanusi, C. 1940 ; tan-wet, H. 289 ; ta-nyen, P. 896 ; tanyeng, P. 906 ; tapoukben, D. 53 ; tapuya, M. 289 ; taroksaga, P. 989 ; tasha, P. 632 ; ta-tway-u, D. 499, D. 507 ; tau-ahnyeen, A. 639 ; tau-ahnyeen, A. 639 ; taubot, D. 628 ; tau-hzí, Z. 263 ; taukkyan, T. 361 ; taukshama, T. 847 ; taumagyee, E. 65 ; tau-myin, (*Pegu*), S. 1267 ; taungbeinné, I. 57 ; taung kathit, E. 354 ; taung-ong, A. 1335 ; taung-petwan, M. 9 ; taung-tha-bye, E. 412 ; taungthálé, G. 22 ; tau-tshiek, S. 1267 ; tawmaiyain, I. 141 ; tawmeyiang, I. 141 ; tawtan, L. 500 ; taw-tee-cteng, M. 71 ; tawthadin, M. 71 ; taw-yeng-ma, C. 1021 ; taw yimma, C. 1021 ; tawzinmé, Z. 263 ; tay, D. 600, D. 642 ; tayau, E. 593 ; tayopsagah, P. 989 ; tazávon-le-pá, A. 829.

Te Te, D. 558 ; té, D. 642 ; teakah, D. 603 ; teathaby-ay, E. 453 ; tee, D. 600 ; teh, D. 600 ; teinkala, A. 519 ; teinn-nyet, C. 35.

Th Thabaw, P. 21 ; thabí-ben, B. 771 ; tha-butgyí, M. 545 ; tha-bwot, I. 569 ; tha-bwot-kha, M. 791 ; tha-bwot-kha, T. 576 ; tha-bwot-kha-wai, L. 556 ; thabwot-nway, U. 69 ; thabyay, P. 1116 ; tha-byay-chin, E. 453 ; thabyay-nee, E. 450 ; thabyebyu, E. 419 ; thabyegyi, E. 412 ; thabyoo-thabyay, E. 444 ; thabyú, D. 428 ; thadi-ben, B. 941 ; thagwa, C. 2287 ; thagya, C. 1050 ; thaikwa, B. 117, B. 142 ; thaing, C. 73 ; tha khwahu-mway, C. 2274 ; tha-khwa-thee, C. 2287 ; tha-khwothpo, S. 2865 ; thakúppo, S. 2865 ; thakutma, D. 753 ; tha kya naí, F. 633 ; thakya-nway-than, L. 224 ; tha-lai, U. 2 ; thalé, P. 1426 ; tha-ma-chók, A. 89 ; tha-majamwai-soke, P. 1389 ; thamáká, M. 363 ; thamé, A. 1655 ; tha-men-gút, G. 66 ; thamin, D. 239 ; thaminsaní, G. 136 ; thán, I. 440 ; thanat, C. 1931 ; tha-nat-dau, G. 31 ; tha-nat-kha, M. 797 ; thanat-tau, G. 31 ; thanba-ya, C. 1270, C. 1296, C. 1301 ; than-day, S. 2874 ; thankya, C. 1050 ; than-kya-pen, C. 1050 ; thanthat, A. 709 ; than-that, S. 2872 ; thanu-wen, C. 2499 ; than-wai, C. 2083 ; thapwot, L. 556 ; tha-bwot-kha, L. 569 ; tharapi, O. 14 ; thau-ba-ya, C. 1233 ; thaukjot, A. 787 ; tha-wen, P. 1121 ; thawgabo, S. 861 ; thayet (Anacardium), A. 1014 ; thayet (Mangifera), M. 147 ; thayet san, S. 3040 ; thayet-thitsé, G. 252 ; G. 255 ; thee-hot, A. 1014 ; thee-noh, A. 1014 ; theeshe, B. 117 ; theet-khya, C. 812 ; theet-men, D. 111 ; theetya, S. 938 ; theet-ya, S. 933 ; 940 ; theing, C. 66, C. 73 ; theinkyeng, C. 115 ; theiwa, B. 142 ; thek-kay-gyee, S. 49 ; thek-kay-nyen, I. 51 ; them-ban-ma-hnyoban, V. 138 ; then, P. 625 ; théngan, H. 364 ; thengben, H. 255 ; thetkia kyn, S. 49 ; thetpyu, X. 10 ; the-yin, C. 2180 ; thi-deng, B. 523 ; thidin, B. 523 ; thihay-aza, L. 362 ; thimbaw, C. 581 ; thimbaw-thi, C. 581 ; thin, M. 363 ; thinban, H.

255 ; thinban-kyeksu, J. 41 ; thinbau-kyeksu, J. 41 ; thinbaw, C. 581 ; thin-baw-kyetsu, J. 41 ; thinbawle, E. 289 ; thínbawnibyu, P. 627 ; thin-boung, P. 551, P. 582 ; thinbozihpyú, P. 627 ; thinganshway, H. 367 ; thin pin, P. 625 ; thinwin, M. 549 ; thinwin, P. 1121 ; thit-cha, Q. 60, Q. 81 ; thit-cho, S. 1718 ; thit-hpaloo, F. 29 ; thit-hswé-lwé, S. 959 ; thitka, P. 390 ; thitkadoe, C. 838 ; thitkya (Diospyros), D. 603 ; thitkya (Juglans), J. 61 ; thitkya (Quercus), Q. 18, Q. 74 ; thit-kya-bo, C. 1183, C. 1196 ; thít-kyam-bo, C. 1158 ; thit kyouk nway, I. 78 ; thit kyouk nway, W. 88 ; thit-kyoung, T. 432 ; thit-kyúk, T. 425 ; thitlinda, H. 162 ; thitmagyí, A. 711 ; thitmin, P. 1004, P. 1006 ; thitmyoke, T. 525 ; thitni, A. 983, A. 988 ; thit-payoung, A. 519 ; thitpót, D. 57 ; thit-pouk, T. 372 ; thitpyú, X. 8 ; thít-sae-yaing, M. 350 ; thitsanweng, D. 49 ; thitsawnwin, D. 49 ; thitsein, T. 293 ; thitsi, M. 350, M. 352 ; thitsibin, M. 352 ; thit-than, M. 901 ; thitto, S. 776 ; thittú, D. 672 ; thit-ya, S. 1652 ; thityin, C. 2180 ; thityúben, P. 693 ; thœmbau-mali, J. 35 ; thombiyu, C. 489 ; thón-phiyu, C. 489 ; thoukwa, B. 142 ; thu-guay-khyoe, H. 92 ; thúmay, G. 255 ; thwot, Z. 190.

Ti Tie-thie, F. 149 ; tihotiya-si, A. 1014 ; tikyá-zi, J. 61'; tikyobo, C. 1158 ; timbo-likyobo, C. 1196 ; timbó-mezali, C. 732 ; timbo-si, C. 581 ; timizu, P. 1121 ; tin-wa, C. 930 ; tin-yu, C. 826 ; tinyu, P. 757, P. 771 ; tissein, T. 293.

To To-kesún, U. 39 ; tó-pelen-moye, T. 576 ; to-sikya-sí, A. 737 ; toukkyan, T. 282 ; touk-kyan, T. 361 ; touk-ta, T. 15 ; touktai, R. 110 ; toukyap, P. 1433 ; toukyat, P. 1433 ; toung-da-lai, G. 22 ; toungdama, C. 834 ; toungkalamet, C. 1925 ; toung-kha-yai, P. 1363 ; toung-kpek-wan, M. 9 ; toungkyeng, C. 115 ; toungletpet, E. 565 ; toung maizalee, C. 795 ; toung-mayobeng, A. 871 ; toung-ong, A. 1335 ; toungpeingnai, A. 1479 ; toungpeinné, A. 1479 ; toungpetwún, P. 1389 ; toung-ta-min, D. 347 ; toung thanat, C. 1931 ; toung-than-gyee, P. 1233 ; touta, T. 15.

Ts Tsam-belai, L. 55 ; tsa-tha-khwa, C. 919 ; tsat-tha-pu, P. 26 ; tsatya, S. 869 ; tsaybeeloo, N. 179 ; tsaythambyah, C. 108 ; tseichyee, B. 868 ; tseikchay, E. 310 ; tseikchyi, B. 868 ; tshaw, S. 2813 ; tsoing, O. 569 ; tsoukmayba, W. 19.

Tu Tupuli, T. 437.

Tw Twottapat, A. 376.

Un Undung, L. 474 ; ung, C. 1520 ; ung-bin, C. 1520 ; ung-dung, L. 483 ; ungdungnet, L. 483.

Us Ushitben, A. 534.

Ut Uta-long, C. 977.

Va Vá-chhá, B. 84 ; vasan, B. 84 ; vathegá-kiyo, B. 84 ; vathegasá, B. 84.

Ve Veng-daik, D. 21 ; veng-khat, G. 133.

Vo Vomon niu, C. 2617.

Wa Wa (Amorphophallus), A. 996 ; wa (Gossypium), G. 404 ; wabo, B. 130 ; wah, C. 404 ; waklí, D. 277 ; wanet, G. 212 ; wa-nway, D, 448 ; wa-pyoo-galay, O. 608 ; wapyugale, G. 209 ; waya, D. 277 ; wa-ya, D. 286 ; wa-yai, D. 288 ; wa-young-khyen, V. 191 ; wa-young-kyoung-byouk, (*Aracan*), T. 444.

We Wek chan, C. 2591 ; wek kyup, C. 1752 ; wek-tamyet, C. 2584 ; welmá chinpoung, H. 250 ; wet-khyae-pa-nai, U. 29 ; wetla, C. 2624 ; wet-shaw, S. 2819.

Ya Yae-chinya, S. 1029 ; yat-ea-gyiben, H. 339 ; yae-ta-kyee, H. 339 ; yae-tha-phan, F. 179 ; yagine, B. 520 ; yaiyœ, M. 656 ; yakatwa, B. 139 ; ya khaing, M. 811 ; yákút rumáni, R. 619 ; yamanai, G. 287 ; ya-ma-ta, C. 89 ; ya-mein, A. 1247 ; yanoung, C. 115 ; yán-zin, S. 682 ; ya-paw-nga, F. 530 ; yaseng-shaw, S. 2819 ; ya-thi-lan, M. 811.

Ye Yechinya, D. 84 ; ye-kha-ong, F. 156 ; ye-kín, H. 515 ; yé-kiyuban-bin, V. 181 ; yémené, G. 287 ; ye-myot, T. 525 ; yé-ná, P. 436 ; yenan, P. 436 ; yen-boung-za, F. 326 ; yendike, D. 21 ; yen-doung, V. 205 ; yeng-khat, G. 108 ; yengma, L. 42 ; yeng-ma, C. 1021 ; yengyé, L. 576 ; yethabyay, E. 453 ; ye-tha-pan, F. 179 ; yethugyi, S. 1174 ; yetpyai, S. 1888 ; yetwoon, H. 221.

Yi Yimmá, C. 1021 ; yindaik, D. 21 ; yingat, G. 133 ; yinhnaung peinne, V. 191 ; yinkat, G. 133 ; yin-yé, L. 576 ; yiyo, M. 652.

Yo Yodayah, O. 4 ; yoe-kiya-pin-ba, B. 909 ; yón, A. 1146 ; young-zalai, G. 55 ; youn-padi-sí, H. 196 ; young, H. 34.

Yu Yung, (*Aracan*), A. 1146 ; yungben, A. 1146.

Yw Ywaygyee, A. 471 ; yweguwe, A. 51 ; ywegyi, A. 471.

Za Zadetiho, M. 901 ; zádiphu, M. 885 ; zádiphu-apóén, M. 885 ; zalúben, L. 343 ; zami, S. 1212 ; zanón, C. 66 ; zanoung, W. 6 ; zarasa-dza-wet-tha, A. 962 ; záyap, L. 84.

Ze Zengbywoon, D. 438.

Zi Zí, Z. 231 ; zimbyun, D. 438 ; zimma, C. 1021 ; zinhlún, S. 830 ; zíphiyu-sí, P. 632 ; ziya, C. 2339.

Zo Zoun-si, A. 1646 ; zoun-ya-si, A. 1646.

ANDAMANESE.

Ah Ah-búd-dah, A. 1294; ah-pur-rud-dah, A. 1294.
Al Al-abada, M. 431.
Am Amdah, C. 115.
An An-na-kah-ro-dah, F. 482.
Ar A-ra-wud-dah, F. 534; ar-dah, F. 553; a-rig-dah, F. 437; arrodah, C. 1021.

Ba Badoh, P. 1116; baila da, P. 916; bairada, R. 242; bájada, S. 2861; baladah, O. 233.
Be Berdá, S. 2819; betina-da, M. 432; beymadá, A. 695.
Bo Boledah, C. 115; bondah, F. 536; boschi, F. 507.
Bu Búrdá, A. 717.

Ch Chad-dah, F. 514; chah-ti-ing-ud-dah, F. 600; chák-mud-dah, F. 453; cha-la-dah, F. 533; chalangada, P. 1363; cha-ra-wud-dah, F. 591; chardah, C. 115; chau-ur-dah, F. 534; chíb-ta-ta-dah, F. 554; chooglum, M. 899; cho-to-dah, F. 457; chowdah, C. 64; chow-lud-dah, F. 589.

Da Dakar táladá, C. 154.
Do Do-dah, F. 504; dod-da, B. 198; dogola, M. 603; do-gota, M. 583; domdomah, G. 733.
Ds Dsagundá, A. 580.

Ga Gachodá, A. 695.
Gu Gú-na-dah, F. 533; gun-na-to-dash, F. 430.

Jo Jobetah, C. 115; jobetahdah, C. 115.
Ju Jumuda, R. 242; ju-ru-cart-dah, F. 440; ju-win-dah, F. 481.

Ka Kaita-da, A. 1479, I. 57; kápadah, L. 343; kaukonda, P. 17.
Ko Ko-lig-dah, F. 584; kol-lid-dah, F. 444; ko-re-dah, F. 327; kore-paig-dah, F. 387; kow-lid-dah, F. 444.
Ku Ku-du-rock-o-dah, F. 459; kur-ku-to-dah, F. 542.

Ky Ky-tha-thong-dah, F. 551.
Li Lí-mí-dah, F. 536.
Lu Lúk-wa-dah, F. 575.

Ma Machalla, M. 797; madá, C. 964; mawtda, H. 137.
Mu Mú-rú-kí-dah, F. 405; mu-túk-dah, F. 437; mutwindá, M. 899.

Op O-pul-dah, F. 431, 441, 495.
Or O-ro-tam-dah, F. 579.

Pa Pábdá, L. 50; pah-nú-dah, F. 385; palug-dah, F. 506; pa-pa-dah, F. 593.
Pe Pecha-da, D. 603.
Pi Pilita, G. 311; pirijdá, A. 580; pi-tanig, M. 439.
Po Podák, G. 210; poothadah, N. 163; po-ra-chal-dah, F. 452; po-tang-dah, F. 478.
Pu Pú-dah, F. 455.

Ra Ráb-nadah, F. 579; ra-ta-charm-dah, F. 306.
Re Rechedá, A. 471.
Ri Rimda, H. 364.
Ro Ro-thul-dah, F. 372; row-je-dah, F. 579.

Si Sirivadi-babila, S. 1688.

Ta Tah-mír-dah, F. 590; talib-dá, C. 2166; tana-hál, S. 1212.
Th Thal-lib-dah, F. 582; thar-oar-dah, F. 587, 588; thol-o-dah, F. 580; thúk-o-dú-nú-dah, F. 367.
To To-bro-dah, F. 540; todah, F. 476; to-go-re-dah, F. 480; toung-hmayo, L. 541.

Uc U-chra-dah, F. 545.
Ur U-rug-nud-dah, F. 542; urungdah, F. 543.

We Welimáda, P. 1006.

Ye Yekin, E. 593.
Yo Yolba, A. 1032.

CINGHALESE.

Aa Aalu, T. 325.
Ab Abba, B. 833 ; abin, P. 87.
Ac Acháriyapalbe, M. 786.
Ae Æm-bærælla, S. 2649 ; aetrillapalla, G. 220.
Ah Ahalla, C. 756 ; ahatte, C. 563 ; ahilla, C. 756 ; ahu-gaha, M. 656.
Ai Aika-waireya, R. 57.
Ak Akmalla, S. 2571 ; akuru, S. 370.
Al Alanga, I. 368 ; allandoo-gass, A. 764 ; allia, E. 83 ; aludel, A. 1517 ; alúpilla, T. 278.
Am Amba, M. 147 ; amu-inguru, Z. 201 ; amúkkará, W. 98.
An Andara (Acacia), A. 249 ; andara (Dichrostachys), D. 402 ; anguru, C. 487, C. 1414 ; aníl, T. 278 ; ankenda, E. 577 ; annási, A. 1045 ; anoda-gaha, A. 89 ; anta-wála, S. 818 ; antenna, D. 151.
Ar Aralu, T. 325 ; aralu-mal, T. 325 ; arimaru, C. 563 ; arremene, C. 795 ; aruda, R. 663.
As Assamodum, C. 691.
At At-addeya, E. 80 ; at-demmata, G. 287 ; ati-maduram, G. 278 ; ati-muktamu, H. 285 ; ati-neranchi, P. 363 ; atta, A. 1166 ; attaireya, M. 797 ; atta-meeriya, E. 333 ; attana, D. 151 ; atteeka, F. 179 ; attora, C. 732.
Aw Awusada-nelli, P. 632.
Ay Ayma, C. 563.

Ba Baireya, L. 576 ; bakmi, S. 866 ; bála, N. 170 ; balát, P. 775 ; balu-dan, A. 1288 ; balu-nakuta, C. 989 ; bandaká, H. 196 ; batala, I. 348 ; batatdomba, E. 453 ; batta, B. 418 ; battú-karawilla, M. 626.
Be Beligobel, H. 255 ; bélli, A. 534 ; bellipattá, H. 255 ; be-lúlabba, D. 32.
Bh Bhutápálá, E. 73.
Bi Bimpól, C. 1029 ; bin-dada-kuriya, E. 549 ; binnúg, G. 748 ; bin-nuga, T. 855.
Bo Bo, F. 236 ; bokaara-gass, G. 373 ; bólam, B. 48 ; bowitteya, M. 359.
Bu Búambilla, A. 1218 ; búbálo, C. 1808 ; bú-dada-kiriya, E. 531 ; búgatteya, H. 444 ; bu-kattu, C. 2569 ; bukenda, M. 15 ; búlú, T. 293 ; bulu-gaha, T. 293 ; búnuga, F. 223 ; bú-pilla, T. 280 ; búrúla-gass, L. 241 ; búrúta, C. 1031 ; búrútch gala, C. 1031 ; burute, C. 1031 ; búsairu, P. 1248 ; bú-tóra, C. 728 ; bútsarana, C. 321.

Ca Calateya, P. 896 ; calukeale, B. 944 ; caunagona, A. 1511 ; cavati, O. 616.
Ch Champakamu, M. 517 ; chámpéyamu, M. 517 ; china-alla, S. 2240 ; chutayá, E. 73.
Co Copi cottá, C. 1641 ; corallia, F. 449 ; cos, A. 1489.

Da Daanga, D. 753 ; daawú, A. 1149 ; dada-hirilla, H. 324 ; dal-kattiya, O. 180 ; daluk, E. 496 ; dambala, P. 530 ; daminne, G. 714 ; dammala, S. 1656 ; dan, A. 1284 ; dan, E. 404 ; dáng, E. 404 ; davette, C. 474 ; dawal kúrúndú, L. 493 ; dawata, C. 474 ; dayúpalú, N. 8.
De Dehi, C. 1258, C. 1296, C. 1301 ; del, A. 1517 ; dellun, P. 1426 ; delun, P. 1426 ; delun-gahá, P. 1426 ; déva-duru, P. 727 ; deya-danga, D. 753 ; deya-kirilla, H. 504 ; deya-mainaireya, C. 1748 ; deya-midella, B. 193 ; deya-mitta, C. 1205 ; deya-ná, M. 489 ; deya-parandella, P. 874 ; deya-wawúl-atteya, C. 30.
Di Diatalia, M. 311 ; dimkola, N. 101 ; dirvi-kaduru, T. 9 ; divi-adiya, I. 399 ; divi-pahuru, I. 399 ; divul, F. 53 ; diwul, F. 53 ; diya-laba, L. 30 ; diya-menériya, C. 1748 ; diyapara, W. 118 ; diya-ratmal, S. 861 ; diye ratembelá, S. 861.
Do Dodan, C. 1233 ; dodang, C. 1233 ; dœdi-kaha, M. 439 ; domba, C. 146 ; domba-gahá, C. 146 ; domba gass, C. 146 ; dowaniya, G. 663.
Du Duda-kaha, C. 2406 ; dúl, A. 1141 ; dúmmaala, T. 576 ; dummele, E. 371 ; dungazha, N. 101 ; dungkola, N. 101 ; duru, C. 2339.

Ee Eepatta, A. 681.
Ek Eka, O. 180 ; ek-soetiya, H. 102.
El Ela-dada-kiriya, E. 512 ; ela-palol, S. 2865 ; ella-battu, S. 2345 ; ella-midella, B. 180 ; ella-nitul, P. 986.
En Enasal, E. 151 ; endaru, R. 369 ; ensal, E. 151.
Er Erandu, J. 41 ; ereya, M. 899 ; errabadu, E. 342 ; erraminya-wel, Z. 263.
Et Et-adi, E. 80 ; etaheralíya, A. 1511 ; etamba, M. 147 ; eta-werella, D. 725 ; etora, P. 75 ; etúna, H. 137.

Ga Gahala, C. 1732 ; gal ehi, C. 2585 ; gal. karanda, C. 390 ; gal mendóra, C. 2577 ; gal-mora-gass, P. 1440 ; galúchi, T. 470 ; galúchi, T. 483 ; gam-miris, P. 811 ; gam. mirris-wil, P. 811 ; ganaba, B. 841 ; gandarassa, B. 48 ; gand hap half, M. 517 ; gang-kolang-kola, P. 1024 ; ganjá-gahá, C. 331 ; ganmalu, P. 1370 ; gansuri-gahá, T. 392 ; gaskoela, B. 944 ; gas miris, C. 455 ; gas-penela, S. 818 ; gass-kahambilya, G. 213 ; gasskeala, B. 944 ; gatta-colla, H. 66 ; gatta-demmatta, G. 298 ; gatta-túmba, L. 329.
Ge Gedde-killala, S. 2362 ; gendakola, P. 1179 ; geta-netul, S. 2912.
Gi Gim-pól, N. 163.
Go Goda kadúra, S. 2943 ; goda-kadura-atta, S. 2943 ; goda mahanel, S. 910 ; goda-mánil, C. 2075 ; gojabbá, H. 513 ; golu-mora, C. 2245 ; gona-wel, A. 1059 ; gong-

Me Mediya-wel, V. 217; mendi, O. 180; menéri, P. 67.

Mi Mí, B. 265; middí, P. 1239; middí-gass, P. 1233; miettie, C. 931; mígong-karapichí-gass, C. 1309; milla, V. 158; millilla, V. 158; mínirum, M. 509; minirum, S. 2712; miris, C. 448, C. 466; miwana-kola, A. 1582.

Mo Mokú-nú-wanna, A. 877, A. 914; monera-kúdimbeya, V. 79; mora, N. 72.

Mu Mudang, E. 419; muda ura, W. 49; muddrap, V. 233; múdú-bíntambúrú, I. 362; mudu keyiya, P. 26; múdú-pol, L. 511; muduru-tulla, O. 31; múdu wará, C. 170; múnemal, M. 570; munmœ, P. 496; murale, N. 72; murungá, M. 721; múrúta-gass, L. 42; murute, L. 42; muruvá-dúl, M. 299; muthu-chippi, P. 355; muwa-kíriya, S. 882.

Na Ná. M. 489; naapiritta, H. 250; na-gaha, M. 489; naga walli, O. 182; naha, L. 89; nárang-ká, C. 1233; nara-wella, N. 8; nát-sakkare, S. 319; navá-cháram, A. 962; navahandi, E. 553; nawa-handi, E. 553.

Ne Nebede, E. 577; nelá-vari, C. 737; nelli, P. 632; nellika, P. 632; nelum, N. 39; nepora, C. 711; neranchi, T. 548; neranji, T. 548; nerrelu, E. 73; neya-dasse, E. 565; neyangalla, G. 243; neyang-natta-colú, L. 569.

Ni Nik-ka, V. 164; nil, S. 855; nil-andana-híruja, C. 2163; nilá-vari, C. 737; níl-katarolu, C. 1403; nil-katta-rodú, C-1403; nilpitcha, G. 733; nilpuruk, R. 633; nimbu-nimba-gahá, M. 363; nin-bin-kohomba, A. 1064; niyanda, S. 785.

Od Odials, B. 705; odú-talan, M. 447.

Ok Okaiyega, P. 21; okuru, C. 1348; okúrú-gass, C. 1348.

Ol Olu, N. 200.

Pa Pailae, C. 563; painaira-wel, C. 551; palati, H. 144; palé, M. 583; palmáni-kam, C. 2367; palú, M. 583; palutu, H. 144; pambúrú, A. 1599; pání, H. 342; panichai, C. 563; panneh-dodang, C. 1233; panúkondol, D. 494; papaw, C. 581; patangi, C. 35; pat-paadagau, M. 613; pattaappele, U. 29; patuk, E. 520; pawetta, P. 338.

Pe Peddi, S. 2205; pehimbia, F. 279; peni-tóra, C. 780; penquin, A. 1079; péra, P. 1343; péra-gadi, P. 1343; peti-tora, C. 797.

Pi Pich-chí-mal, J. 35; pieri, E. 73; pimatos, B. 702; píta-karosana, C. 1789; pitta-súdúpala, H. 71; pittawáka, P. 657.

Po Pœpol, C. 581; pol, C. 1520; pol-gahá, C. 1520; pol-gass, C. 1520; pol-nawasi, C. 1520; pol-tel, C. 1520; porawa-márá, C. 390; posta-tammi, C. 563; pot-lunu, S. 682.

Pu Punatú, B. 702; púndal, L. 245; punerai, S. 818; puskara, B. 731; pus-wel, E. 219; púta-tammi, C. 563; puta-tana, E. 166; púta-tanni, C. 563; puvakka, A. 1294; puvella, S. 818; puwak, A. 1294.

Ra Rabu, R. 31; rada-liya, C. 1768; rallia, F. 448; rambuk, S. 4; rameneidelle, W.

36; ran, A. 1622; rana-vará, C. 741; ran-tahadu, A. 1622; ran-taqadu, A. 1622; ranwan-kíkirindi, W. 25; rasadiyá, M. 473; rasa-karpúram, M. 473; rasa-tel-kola, I. 410; ras-ni, L. 503; rassakin-da, T. 470; rassakinda, T. 483; rassu kú-rúndu, C. 1196; rata-del, A. 1486; rata-dummula, B. 43; ratagoradiya, P. 885; ratainnala, S. 2320; rata-kœkana, C. 279; rata-kaju, A. 1261; rata-katambá, P. 1274; ratakekuna, H. 472; rata-miris, C. 448; rata-nelli, P. 627; rata-tana, P. 59; rata-tekola, C. 244; rata-tora, C. 49; ratau-hunu, C. 489; rat-handun, P. 1381; rat-hihiri, S. 790; ratkihiri, A. 135; rat-net-tol, P. 979; rat-nitúl, P. 979; ratoo-waa, C. 773; rat-pittawáka, P. 673; rawan-idala, W. 36.

Ri Ridi, A. 1359; ridi-tagadu, A. 1359; ridi-tahadu, A. 1359; riti, A. 1200.

Ro Rookerah, F. 805.

Ru Ruct-handún, P. 1381; rúkattana, A. 871; run, G. 317; ruta-komárika, A. 833.

Sa Sádika, M. 885; sakarí, S. 376; salada, L. 21; salunáyi, R. 660; samadara, S. 749; sana-kola, C. 737; sanni-náegam, V. 73; sanni-násang, V. 73; sanni-nayam, V. 73; sappu, M. 517; sapú, M. 537; sa-vandra-múl, A. 1097; sawandatala, O. 18; sayanmull, O. 137.

Se Sedaran, C. 1270; sekuhme, C. 2531.

Sh Shavíram, M. 473; sheti-putsa, C. 1211; shúne-dalimba-bíja, C. 2546.

Si Sindrika-gahá, M. 606; sinhala-asamoda-gan, P. 727; sinnah, A. 1251; siyambu-la, T. 28; siyembela, T. 28.

Su Suddu-abbe, B. 800; suduattana, D. 160; sudu-duru, C. 2339; sudu-lúnú, A. 779; súdú-nikka, V. 164; sudu-nitul, P. 986; sudu-pásánam, A. 1425; sureya, T. 392; súrí, R. 650; súriya-gaha, T. 392.

Ta Tadala, C. 1732; tal, B. 663; tala, C. 1995; tal-gass, B. 663; taliennoe, G. 761; talla, S. 1078; talla-atta, S. 1078; talla-goyá, R. 116; talla-maha, F. 773; talla mala, W. 49; támaruja, E. 73; tambili, C. 1520; tana-hál, S. 1212; tandala, D. 420; tarana, W. 22.

Te Tebu, C. 2013; tekka, T. 232; teldomba, C. 146; telemboo, S. 2824; tella kwiya, E. 593; tel-tala, S. 1078.

Th Thalay marutha, S. 818; theluja, F. 561; thovar, E. 553.

Ti Tibbatu, S. 2280; S. 2299; timbiri, D. 582; tippili, P. 805; tiringu, T. 634; titta-commodú, C. 1211; titta-hondala, T. 600; tittaval, A. 1038.

To Tokei, P. 350; tolabo, C. 2062; totilla, O. 233; tot punatú, B. 702.

Tr Trasta-wálu, I. 415.

Tu Túmba-karawilla, M. 639; tun-pattala, S. 1078; tuttiri, C. 1053.

Ug Ugurassa, F. 615.

Uk Uk, S. 30.

Ul Ulundu-mae, P. 513; uluva, T. 612.

Un Una, B. 149; úna, B. 118; una-kapura, B. 84; uná lunu, B. 84.

Ur Uralawa, T. 444; úru-tora, C. 787; úru-wi, O. 258.

Uw Úwus, V. 233.

G. I. C. P. O.—No. 134 R. & A.—4-3-96.—1380.—B. N. D.

Printed in the United States
By Bookmasters